DEVELOPMENTS IN QUATERNARY SCIENCE 1
SERIES EDITOR: JIM ROSE

THE QUATERNARY PERIOD IN THE UNITED STATES

Developments in Quaternary Science
Series editor: Jim Rose

Volumes in this series

1. The Quaternary Period in the United States
   Edited by A.R. Gillespie, S.C. Porter, B.F. Atwater
   0-444-51470-8 (hardbound); 0-444-51471-6 (paperback) – 2004

2. Quaternary Glaciations – Extent and Chronology
   Part I Europe; Part II North America; Part III South America, Asia, Africa, Australia, Antarctica
   Edited by J. Ehlers, P.L. Gibbard
   0-444-51462-7 (hardbound + CD-ROM) – 2004

3. Ice Age Southern Andes – A Chronicle of Paleoecological Events
   Authored by C.J. Heusser
   0-444-51478-3 (hardbound) – 2003

4. Spitsbergen Push Moraines – Including a translation of K. Gripp: Glaciologische und geologische Ergebnisse der Hamburgischen Spitzgbergen-Expedition 1927
   Authored by J. van der Meer
   Forthcoming – 2004
   0-444-51544-5

5. Tropical West Africa – Marine and Continental Changes during the Late Quaternary
   Authored by P. Giresse
   Forthcoming – 2004

For further information as well as other related products, please visit the Elsevier homepage (http://www.elsevier.com)

Developments in Quaternary Science, 1
Series editor: Jim Rose

THE QUATERNARY PERIOD IN THE UNITED STATES

by

A.R. Gillespie and S.C. Porter

University of Washington
Seattle

Washington
USA

and

B.F. Atwater

U.S. Geological Survey
Seattle

Washington
USA

2004

ELSEVIER

Amsterdam – Boston – Heidelberg – London – New York – Oxford – Paris
San Diego – San Francisco – Singapore – Sydney – Tokyo

ELSEVIER B.V.
Sara Burgerhartstraat 25,
P.O. Box 211, 1000 AE Amsterdam, The Netherlands

First edition 2004

Library of Congress Cataloging in Publication Data
A catalogue record from the British Library has been applied for.

ISBN: 0-444-51470-8 (Hardbound)
ISBN: 0-444-51471-6 (Paperback)
ISSN: 1571-0866 (Series)

⊗ The paper used in this publication meets the requirements of ANSI/NISO Z39.48-1992 (Permanence of Paper). Printed in The Netherlands.

Foreword

The "*Quaternary Period in the United States*" edited by Alan Gillespie, Steve Porter and Brian Atwater is the first volume in the *Elsevier Book Series* on "*Developments in Quaternary Science*." This book is to be published in 2004 to coincide with the very successful XVI INQUA Congress held in Reno, Nevada, in July 2003, and follows the well-known and well-used volume, published for the 1965 VII INQUA Congress, on "*The Quaternary of the United States*" edited by H. E. Wright and D. G. Frey.

Developments in Quaternary Science is a book series designed to bring together important texts within the continually expanding field of Quaternary science. This series is linked to the journals: *Quaternary Science Reviews*, *Quaternary Research* and *Quaternary International*, and provides an outlet for texts that require individual or special consideration, not achieved most effectively within an academic journal. In this respect the series will provide an outlet for topics that require substantial space, are linked to major scientific events, require special production facilities (i.e. use of interactive electronic methods), or are archival in character. The subjects covered by the series will consider Quaternary science across different parts of the Earth and with respect to the diverse range of Quaternary processes. The texts will cover the response to processes within the fields of geology, biology, geography, climatology archaeology and geochronology. Particular consideration will be given to issues such as the Quaternary development of specific regions, comprehensive treatments of specific topics such as global scale consideration of patterns of glaciation, and compendia on timely topics such as dating methodologies, environmental hazards and rapid climate changes. This series will provide an outlet for scientists who wish to achieve a substantial treatment of major scientific concerns and a venue for those seeking the authority provided by such an approach.

We are very proud indeed that The "*Quaternary Period in the United States*" edited by Alan Gillespie, Steve Porter and Brian Atwater will be the first issue of the series. This book has a distinguished lineage, and will be a benchmark publication for a part of the world that forms the basis of many models used to interpret Quaternary processes and patterns of environmental change throughout the globe.

The underlying aim of the book is to update our understanding of the Quaternary of the United States over the 38 years since the 1965 INQUA. With this in mind, it is interesting to consider the scope of this text. In addition to the traditional concerns with topics like glacial, aeolian, coastal, fluvial, lake, permafrost and soil processes and changes in patterns of vegetation, insects and mammals including humans, a number of topics are given special attention for the first time. On the physical side, there are chapters on outburst floods, groundwater and speleothem, palaeoseismology and volcanism, while on the biological side chapters consider vegetation dynamics, fire history, as well as a much-enlarged examination of Quaternary insects. Perhaps it is most marked by the attention given to modelling throughout, and there are notable chapters on coupling of ice sheet and climate models for the simulation of former ice sheets and modelling palaeoclimate.

The editors are to be congratulated on bringing together such a vast amount of up-to-date information together within such a short time, and to make available to the Quaternary community at large, a book that will act as a reference text for many years to come.

Jim Rose
Series Editor

Contents

Preface

This book is devoted to advances in understanding of the past ca. two million years of Earth history – the Quaternary Period – in the United States. It is being published in the year of INQUA XVI, the first Congress of the International Union for Quaternary Research to meet in the U.S. since INQUA VII, in 1965. The intervening 38 years has brought enormous growth in Quaternary research, as illustrated by journals dedicated to it. In 1965, there was one such journal, Japan's *Daiyonki Kenkyu* (*Quaternary Research*). Its U.S. namesake, *Quaternary Research*, appeared in 1970, followed by *Quaternary Science Reviews* in 1982, *The Journal of Quaternary Science* in 1985, *Quaternary International* in 1989, and *The Holocene* in 1990.

"The Quaternary Period in the United States" has ample precedent. For INQUA VII, Herbert E. Wright, Jr. and David G. Frey edited a review volume on the Quaternary of the United States that has served as a standard reference for nearly four decades (Wright & Frey, 1965). Nearly 20 years later, Steve Porter and Herb Wright edited updated reviews on Late Quaternary Environments of the United States (Porter, 1983; Wright, 1983). Vladimir Sibrava, David Q. Bowen and Gerald M. Richmond then surveyed Quaternary glaciation in the Northern Hemisphere (Sibrava *et al.*, 1986), and the Geological Society of America's Decade of North American Geology (DNAG) series delved into geologic aspects of Quaternary geologic and geomorphic history in detail. "The Quaternary Period in the United States" is the first general overview since the DNAG series appeared in the 1980s.

No review volume can provide comprehensive treatment for a topic as vast and rich as the Quaternary. We therefore had to limit this book to highlights from an admittedly incomplete selection of fields that are of widespread current interest today. The book begins with sections on ice and water – as glaciers, permafrost, oceans, rivers, lakes, and aquifers. Six chapters are devoted to the high-latitude Pleistocene ice sheets, to mountain glaciations of the western United States, and to permafrost studies. Other chapters discuss ice-age lakes, caves, sea-level fluctuations, and riverine landscapes.

With a chapter on landscape evolution models, the book turns to essays on geologic processes. Two chapters discuss soils and their responses to climate, and wind-blown sediments. Two more describe volcanoes and earthquakes, and the use of Quaternary geology to understand the hazards they pose.

The next part of the book addresses plants and animals. Five chapters, organized mainly by region, consider the Quaternary history of vegetation in the United States. Other chapters treat forcing functions and vegetation response at different spatial and temporal scales, the role of fire as a catalyst of vegetation change during rapid climate shifts, and the use of tree rings in inferring age and past hydroclimatic conditions. Three chapters address vertebrate paleontology and the extinctions of large mammals at the end of the last glaciation, beetle assemblages and the inferences they permit about past conditions, and the peopling of North America.

A final chapter addresses the numerical modeling of Quaternary climates, and the role paleoclimatic studies and climatic modeling has in predicting future response of the Earth's climate system to the changes we have wrought.

The resulting volume was made possible through the combined efforts of authors, reviewers, and a responsive production team. We are especially indebted to Karin Stewart-Perry. We hope the book stands alongside its predecessors as an enduring and useful resource for investigators of the Quaternary Period.

Alan R. Gillespie, Stephen C. Porter & Brian F. Atwater

Seattle, March, 2003

References

Porter, S.C. (ed.) (1983). *Late Quaternary environments of the United States, Vol. 1, The Late Pleistocene*. Minneapolis, University of Minnesota Press, 407 pp.

Sibrava, V., Bowen, D.Q. & Richmond, G.M. (eds) (1986). Quaternary glaciations in the northern hemisphere. *Quaternary Science Reviews*, **5**, 534 pp.

Wright, H.E., Jr. (ed.) (1983). *The Late Quaternary of the United States, Vol. 2, The Holocene*. Minneapolis, University of Minnesota Press, 277 pp.

Wright, H.E., Jr. & Frey, D.G. (1965). *The Quaternary of the United States*. New Jersey, Princeton University Press, 922 pp.

The southern Laurentide Ice Sheet

David M. Mickelson[1] and Patrick M. Colgan[2]

[1] *Department of Geology and Geophysics, University of Wisconsin Madison, Weeks Hall, 1215 West Dayton Street, Madison, WI 53706-1692, USA; mickelson@geology.wisc.edu*
[2] *Department of Geology, Grand Valley State University, 1 Campus Drive, Allendale, MI 49401, USA; colganp@gvsu.edu*

Changing Attitudes and Approaches

The publication of the *Quaternary of the United States* (Wright & Frey, 1965) for the 1965 INQUA Congress in Denver was a milestone that summarized our knowledge of the Quaternary of the U.S. in a single volume. Glacial geology was a major component of the volume, and it contained 125 pages on the Laurentide Ice Sheet (LIS) in the U.S. In the present volume, almost 40 years later, many fewer pages are devoted to the same topic, indicating the vast increase in other aspects of Quaternary studies. In the U.S., glacial geology has expanded greatly in knowledge and interest, and now glacial geologists have a much richer field and variety of techniques with which to study Quaternary history. The other chapters in this book are a clear indication of the diversity of fields that now make up what traditionally was classified as glacial geology or did not exist before 1965.

Radiocarbon dating remains the most important tool for determining the chronology of the last glaciation. Accelerator mass spectrometry (AMS) has allowed dating of smaller and somewhat older samples. Tree-ring calibration of the radiocarbon time scale has resulted in dating accuracy not possible 40 years ago. Newer dating methods such as thermoluminescence, amino-acid racemization, paleomagnetism, and cosmogenic-isotope methods have yielded mixed results, but have the potential to improve our interpretations of glacial chronology, especially those of the pre-late Wisconsin.

There have been revolutionary changes in the way we study glacial sediments and reconstruct their depositional environments. Genetic classifications have been replaced by descriptive lithofacies approaches, which focus on modern-process analogs for interpretations of depositional environments. Correlations of till units from one area to another are now approached with more caution, and the use of facies models facilitates the understanding of complex glacial sequences. Geophysical techniques have been used to explore lake basins and subsurface stratigraphy. Models of glacial landform genesis are also driven by modern analogs and interpretations made in modern glacier settings.

There have been major changes in our understanding of pre-late Wisconsin events since 1965. The terms "Nebraskan" and "Kansan" are no longer used, and now there is evidence of at least six pre-Illinoian glaciations in the continental record. Flint (1971), in a widely used textbook, hinted that there might be problems with correlations of what was called "Nebraskan Drift," but nevertheless used the terminology accepted at that time. Based on the oxygen-isotope record from the oceans, we now know that there were clearly more than four glacial and interglacial episodes during the Pleistocene. Much work remains to be done to unravel this continental record of early glaciations and correlate it to the ocean record. There has been debate about the extent of the early Wisconsin Glaciation as well. In many areas, particularly in northern Illinois and southern Wisconsin, deposits thought to be of this age now appear to be older.

Probably one reason for our rather poor understanding of pre-late Wisconsin glacial events is the decline of purely stratigraphic studies from the 1980s through the 1990s. State geologic surveys have in many cases reduced their staff and mapping has been displaced by topical studies and an emphasis on applied research related to groundwater and mineral extraction. In academia, traditional mapping and glacial stratigraphy have not been as common in the last 40 years as previously. Instead they have been replaced by, or combined with, studies of sediment genesis, glacial process, development of conceptual and quantitative models, and details of local chronology, and integrative studies of glacial deposits and other aspects of Quaternary history such as ice-marginal lakes, paleoclimatology, loess, soils, and the paleontologic record. In the 1990s the U.S. Geological Survey began to fund mapping projects in academia (EDMAP) and state geologic surveys (STATEMAP), and this has revitalized mapping of glacial deposits in the northern U.S. Since the late 1990s the Great Lakes Mapping Coalition, a joint effort of the U.S. Geological Survey and several state geologic surveys, has focused on detailed three-dimensional mapping that includes subsurface investigation. If this program continues to grow, it may re-stimulate interest in mapping glacial deposits.

Much of our improved understanding of the southern LIS has come from studies of the new field of paleoglaciology. Reconstruction of ice-sheet surfaces, interpretation of former bed conditions and discussions of sliding vs. subglacial deforming beds, estimates of sediment fluxes, interpretations of the nature and distribution of subglacial meltwater, new interpretations of landform genesis, and modeling have all been major areas of research in the last 40 years.

There have been several extensive compilations of the glacial record of the southern LIS since 1965 and we make no attempt to repeat these here. Instead, we highlight what we view as advances in our understanding of the southern LIS and its deposits since publication of the *Quaternary of the United States* in 1965. Events and processes along the southern margin of the LIS are closely tied to the behavior of the ice sheet in Canada, but much of this literature is not discussed here because of the scope of the book. Likewise, we refrain from discussing ancillary topics, like loess and

DEVELOPMENTS IN QUATERNARY SCIENCE
VOLUME 1 ISSN 1571-0866
DOI:10.1016/S1571-0866(03)01001-7

the history of ice-marginal lakes, because these are covered elsewhere in the volume.

Mapping and Compilations of Glacial Geology and Geomorphology Since 1965

Since 1965 there have been several comprehensive reviews of the geology and geomorphology of the southern LIS. The "state-of-knowledge" was summarized by Flint in 1971 in part of his classic text. At this time, the deep marine record was only beginning to be discovered. Our understanding of modern ice-sheet dynamics in Greenland and Antarctica was in its infancy, and an understanding of surging glaciers and modern glacial environments was just emerging. Records of global climate change such as ice cores, pollen databases, loess records, and lake cores were fragmentary, few, and far between. Much of what was known was based on the incomplete terrestrial record of continental and mountain glaciation.

Throughout the 1970s and 1980s numerous records of the last glaciation were collected, analyzed, and combined into a global database (e.g. CLIMAP, 1976, 1984; COHMAP, 1988). For the INQUA meeting in Moscow, the U.S. INQUA Committee produced two volumes of edited papers on the Quaternary of the U.S. (Wright, 1983). These include a comprehensive review of the glacial record and a chronology of glacial and periglacial events during the late Wisconsin glaciation (Mickelson *et al.*, 1983). Included were maps showing the nature of the glacier bed, moraines and other ice-margin positions, and a generalized map of landform regions. Andrews (1987) summarized major issues in understanding the whole LIS: the thickness of the ice sheet, the extent of ice during the mid-Wisconsin (marine oxygen isotope stage (OIS) 3), the timing of the late-glacial maximum, and the chronology of deglaciation. The publication *Quaternary Glaciations in the Northern Hemisphere* (Sibrava *et al.*, 1986), which contains several review papers on deposits of the LIS in the United States (Brown *et al.*, 2001; Eschman & Mickelson, 1986; Fullerton, 1986; Fullerton & Colton, 1986; Hallberg, 1986; Hallberg & Kemmis, 1986; Johnson, 1986; Lasemi & Berg, 2001; Matsch & Schneider, 1986; Stone & Borns, 1986), is the most recently published compilation covering all of the area of the LIS in the United States. Dyke

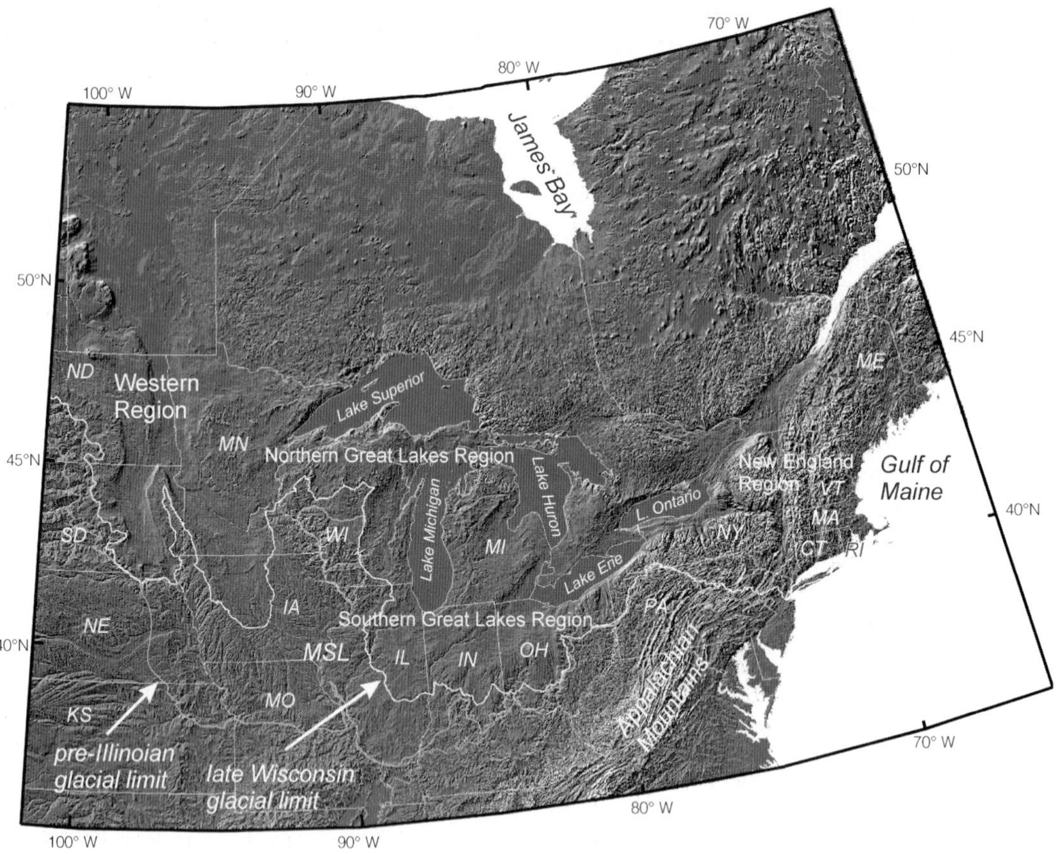

Fig. 1. A shaded relief image created from digital elevation data (USGS ETOPO5) showing southern limits of selected ice advances, location of Great Lakes, and names of states (CT, Connecticut; IA, Iowa; IL, Illinois; IN, Indiana; KS, Kansas; MA, Massachusetts; ME, Maine; MI, Michigan; MN, Minnesota; MO, Missouri; ND, North Dakota; NE, Nebraska; NH, New Hampshire; NY, New York; OH, Ohio; PA, Pennsylvania; RI, Rhode Island; SD, South Dakota; WI, Wisconsin). MSL is location of isotope record of Dorale et al. (1998). Scale in km is given in Fig. 2.

& Prest (1987) and Dyke *et al.* (2002) summarize what is currently known about the extent and timing of the entire LIS during the last glacial maximum (LGM).

A major contribution to our understanding of regional aspects of the glacial record is the compilation of Quaternary geologic maps at 1:1,000,000 scale for all of the area. Organized by G.M. Richmond and D.S. Fullerton, these maps, with many authors, are published as U.S. Geological Survey Miscellaneous Investigations Series I-1420. All of the southern LIS is covered by these maps, and although they were in most cases compiled from older mapping, they represent the most up-to-date regional maps available. Soller (1998) and Soller & Packard (1998) published a map of the thickness and character of surficial deposits across the area of the U.S. covered by the LIS. It portrays very thick deposits filling pre-late Wisconsin valleys and in interlobate areas.

Many states have mapping programs that have added to our knowledge of glacial deposits and several different map scales are being used. Much of this mapping has been driven by the need for geologic information to help solve environmental problems or to help locate groundwater and mineral resources. Many states (Fig. 1) also have developed specific derivative maps (e.g. contamination potential, aggregate resources) based on maps of Quaternary deposits. Almost all states have glacial or Quaternary geology maps at 1:500,000 or smaller. A summary of recently active mapping programs is shown in Table 1.

Massachusetts and Connecticut (Fig. 1) have been mapped mostly by the U.S. Geological Survey in cooperation with state agencies, and many of these maps are available at a scale of 1:24,000 as open file reports. Several states have had more limited recent mapping of glacial deposits (Vermont, New Hampshire, Michigan, Iowa, and South Dakota). The U.S. Geological Survey in cooperation with Ohio, Indiana, Illinois, and Michigan recently formed the Central Great Lakes Geologic Mapping Coalition. They are sharing resources to produce detailed (1:24,000) three-dimensional maps of Quaternary deposits in the southern Great Lakes area. This effort is reinvigorating interest in mapping of glacial deposits in the area, and in the future it may extend across all of the area covered by the LIS.

Table 1. Typical map scales used in states with recent published mapping and a representative reference to each.

Illinois	1:100,000 or 1:24,000	Curry *et al.* (1997), Grimley (2002)
Indiana	1:24,000	Brown & Jones (1999)
Maine	1:24,000	Thompson (1999)
Minnesota	1:100,000	Hobbs (1995)
New Jersey	1:24,000	Stanford *et al.* (1998)
New York	1:250,000	Muller & Cadwell (1986)
North Dakota	1:125,000 or 1:250,000	Harris & Luther (1991)
Ohio	1:62,500	Totten (1988)
Vermont	1:24,000	DeSimone & Dethier (1992)
Wisconsin	1:100,000	Clayton (2001)

New Data Sources

During the last four decades the number of new data sources with which to study Quaternary landforms and sediments has increased greatly. These include better topographic maps, digital raster graphics, digital elevation models, remotely sensed data acquired by satellite-based systems, and improved drilling and geophysical techniques. Most of the advances in this area have been driven by advances in satellite technology, low-cost computing power, the increased availability of geographic information systems (GIS), and the efforts of government agencies in creating large spatial databases.

A new series of 1:24,000 U.S.G.S topographic quadrangle maps has made it possible to compare landforms over all of the glaciated United States. These maps are now available as digital raster graphics (DRG) for the conterminous United States, Hawaii and portions of Alaska. By far the most important new data source is the digital elevation model (DEM). These arrays of elevations have revolutionized the way topography can be visualized, categorized, and analyzed. The first 3 arc-second DEMs (1×1 degree) were created from the processing of 1:250,000 scale topographic maps and, as a result, had relatively low spatial resolution (90×90 m), and contained numerous processing artifacts. The most recent series of 30- and 10-meter DEMs have improved on these problems. DEMs allow classification of slope, aspect, and many other parameters of interest to Quaternary scientists. Images derived from the raw DEMs allow the creation of visually stunning shaded-relief views, contour maps, and orthographic perspective views with overlying drapes of color-coded contour intervals or satellite images. Fairly subtle geomorphic features (such as small moraines and flutes) have been discovered from these DEMs. They have also allowed recognition of regional patterns by allowing areas much larger than a single 7.5-minute quadrangle to be viewed at once. An early example of a shaded relief image is the *Digital Shaded Relief Map of the Conterminous United States* (Thelien & Pike, 1989), and a smaller scale one is portrayed in Fig. 1. In the last 10 years numerous workers have used DEMs for mapping glacial features.

Satellite images such as those from the Landsat Multispectral Scanner and Thematic Mapper have been used in a few studies (e.g. Boulton & Clark, 1990), but their use has been limited because of their relatively low spatial resolution ($\sim$80 and 30 m respectively) compared to traditional low-altitude aerial photography ($<$1 m). In addition, there are now many other satellite image systems available, producing visible bands, infrared, and radar images. Traditional low-altitude aerial photographs, although still essential for mapping landforms, have always been limited by their lack of georeferencing. Image processing and GIS software now allow all aerial photos (and stereopairs) to be rectified and georeferenced. Traditional low-altitude aerial photographs are very useful because they are available back into the 1930s and can be used in determining land use changes as well as mapping subtle geomorphic features, which are only visible during certain times of the year. Computer mapping from digital sources is now routinely used as GIS software has advanced and because many data sources are now in digital format. This combination of digital data, GIS, and cheap computing power has led to regional scale mapping and landform genesis studies that focus on an entire continent (e.g. Aber *et al.*, 1995; Soller & Packard, 1998).

Geophysical methods have been used to study lake deposits and subsurface geology. Probably the best example is the work of Mullins & Hinchey (1989) and Mullins *et al.* (1996). They used seismic reflection techniques to describe the sediment filling the Finger Lakes of New York. The sediment fills the deep bedrock valleys to depths as great as 306 meters below present sea level. Coleman *et al.* (1989) also provided geophysical data from Lake Michigan to show the thickness of glacial sediment in the south end of the lake.

Advances in Glacial Sedimentology and Geomorphology

In the last 40 years, there have been major advances in the way glacial sediments are studied. In the 1960s classification of glacial sediments was commonly driven by a genetic terminology based on interpretations of depositional environment. Much of the descriptive information about boreholes and outcrops was never published leaving later workers little on which to judge these classifications. Research carried out in modern glacier environments (e.g. Anderson & Sollid, 1971; Boulton, 1970a, b, 1971; Clayton, 1964; Goldthwait, 1974; Hooke, 1970, 1973; Lawson, 1979, 1981) helped to focus glacial geologists' attention on the reconstruction of past depositional environments (e.g. Gustavson & Boothroyd, 1987). In the 1970s and 1980s glacial geologists began to use lithofacies analysis and modern-process approaches similar to those being used by sedimentologists in other fields. At this time the term "diamicton" was more widely used as a purely descriptive term for the several genetic types of till (e.g. lodgment till, meltout till). The papers included within special volume 23 published by the Society of Economic Paleontologists and Mineralogists exemplify the trend toward modern sedimentologic methods in glacial geology (Jopling & McDonald, 1975). Eyles *et al.* (1983) presented an important synthesis of the lithofacies approach and Johnson & Hansel (1990) have since applied this method to the glacial sequence in Illinois.

As advances in glacial sedimentology followed modern trends, so did interpretations of landform genesis. An important early paper by Clayton & Moran (1974) established the idea of the process-form model in glacial geology. They emphasized that glacial-landscape form is directly related to the glacial processes active in the past. They also stressed the fact that glacial environments are composites, including both palimpsest landforms (landforms partially hidden by later forms) and superimposed forms (most recent forms). Finally, they showed that postglacial processes must be understood in order to understand the current form of glacial landscapes.

Our expanding knowledge of the Antarctic and Greenland ice sheets has greatly influenced our current interpretations of the southern LIS margin (e.g. Hughes *et al.*, 1985). Papers included in a 1987 Journal of Geophysical Research Special Volume on fast glacier flow helped to focus attention on the physics of real ice sheets, especially the ice streams and outlet glaciers that drain modern ice sheets (e.g. Bentley, 1987; Clarke, 1987; Raymond *et al.*, 1987; Whillans *et al.*, 1987). Interpretations of the southern LIS have also been influenced by new discoveries on the mechanics of surging glaciers as well as the sediment and bed conditions beneath these enigmatic glaciers (e.g. Clarke *et al.*, 1984; Kamb *et al.*, 1985). The use of modern analogs as a guide to the interpretation of past ice sheets has led to the development of a new field in glacial geology called paleoglaciology.

Paleoglaciology Comes of Age

The study of the physical nature of past glaciers, what is now called paleoglaciology, has blossomed in the last 40 years. There has been a major focus on understanding the paleoclimate, flow dynamics, and physical processes responsible for the landforms and deposits that we see along the southern LIS margin. There are clearly different landform zones across the area covered by the southern LIS (Colgan *et al.*, 2003; Mickelson *et al.*, 1983), but confidently ascribing the formation of certain landforms to a certain set of glaciological conditions has been elusive.

Although there had been a few attempts to reconstruct physical characteristics of the southern LIS previous to 1965 (e.g. Harrison, 1958), Wright's (1973) suggestion that the Superior lobe had a low ice-surface profile and Mathews' (1974) reconstruction of ice-surface profiles in the southwest part of the ice sheet were the first modern attempts at reconstructing ice-surface slopes of lobes of the LIS based on glacial geomorphic data. They represented a major step forward in understanding ice dynamics of the southern LIS. Because ice-surface slope is controlled by, among other things, resistance at the bed, the shape of the ice surface is intimately tied to bed conditions. Was there water present? Was the bed soft and deforming or hard and rigid? Was the ice surging? Were there ice streams? Were there large areas of stagnant ice near the margin or did ice remain active during retreat? Did the nature of the bed control the formation of drumlins, tunnel channels and other landforms? How much were ice-margin fluctuations controlled by climate, and how much was climate controlled by the ice? All of the above questions have been asked and speculated upon from various points of view and we review these below.

Deforming Bed or Sliding of Basal Debris-Rich Ice?

Subglacial sediment transport mechanisms have been a focus of research all over the world, and the southern LIS is no exception. Since the importance of a soft deforming bed was suggested by, among others, Boulton & Jones (1979) and

Boulton & Hindmarsh (1987), for glaciers in general, and Alley *et al.* (1986, 1987) for Whillans Ice Stream (formerly Ice Stream B), the extent and thickness of a wet, deforming bed beneath the southern LIS have been debated. Boulton & Jones (1979) suggested that the south margin of the LIS was very thin because of low basal shear stress (ca. 5 kPa). Beget (1986) suggested a conceptual model for the Lake Michigan lobe that incorporated a combination of sliding and soft sediment deformation. He suggested that deforming till with a low yield strength (*ca.* 8 kPa) produced ice-surface profiles with an ice thickness of only 500 m about 200 km north of the terminus. Alley (1991) hypothesized that all of the southern LIS was underlain by a thick deforming layer and further argued that thick till in ice marginal areas could only be explained by the high sediment fluxes produced by a soft deforming bed. Clark (1992a), building on the approach of Mathews (1974), stressed the widespread distribution of low ice-surface slopes and related these to the existence of a soft deforming bed. This concept was expanded (Clark, 1994, 1997) when he pointed out the interrelationships between the distribution of sedimentary bedrock, and the continuity of soft, fine-grained sediments, low driving stresses, and relative lack of eskers that would have formed in "R" channels. Colgan & Mickelson (1997) and Colgan (1999a) showed that ice-surface profiles in the Green Bay lobe changed with time and profiles were probably steepest during the LGM and during readvances. Ice surface slopes were very low during retreat. Another approach was taken by Hooyer & Iverson (2001), who combined field observation and experiments of deformation in a ring shear device. Ice-surface profiles of the southeastern LIS have also been reconstructed by Ridky & Bindschadler (1990) in the Ontario lobe and by Shreve (1985) in Maine.

A major issue has been the interpretation of evidence of deformation from the sediments themselves. We cannot review all of the literature on this topic, but some is pertinent to the southern LIS. Clearly diamictons have been deformed. The issue of whether the deformation was as a thick wet sediment layer, a very thin wet sediment layer, or as debris-rich ice is still debated (Piotrowski *et al.*, 2001). Clayton *et al.* (1989) have argued that careful examination of the sediment record suggests that if a soft deforming bed were present, it must have been a thin layer based on the continuity of sedimentary layers and the presence of sand lenses and sand clasts in diamicton. They suggest that most sediment was transported as debris-rich ice and deposited as meltout till, based mostly on observations along the southwest edge of the ice sheet and in the Green Bay lobe (Fig. 2). In fact, there may be real differences between the sediments deposited in the far southern part of the ice sheet, in Ohio, Indiana, and Illinois, compared to just a short distance farther north. The landform record is certainly different (Colgan *et al.*, 2003;

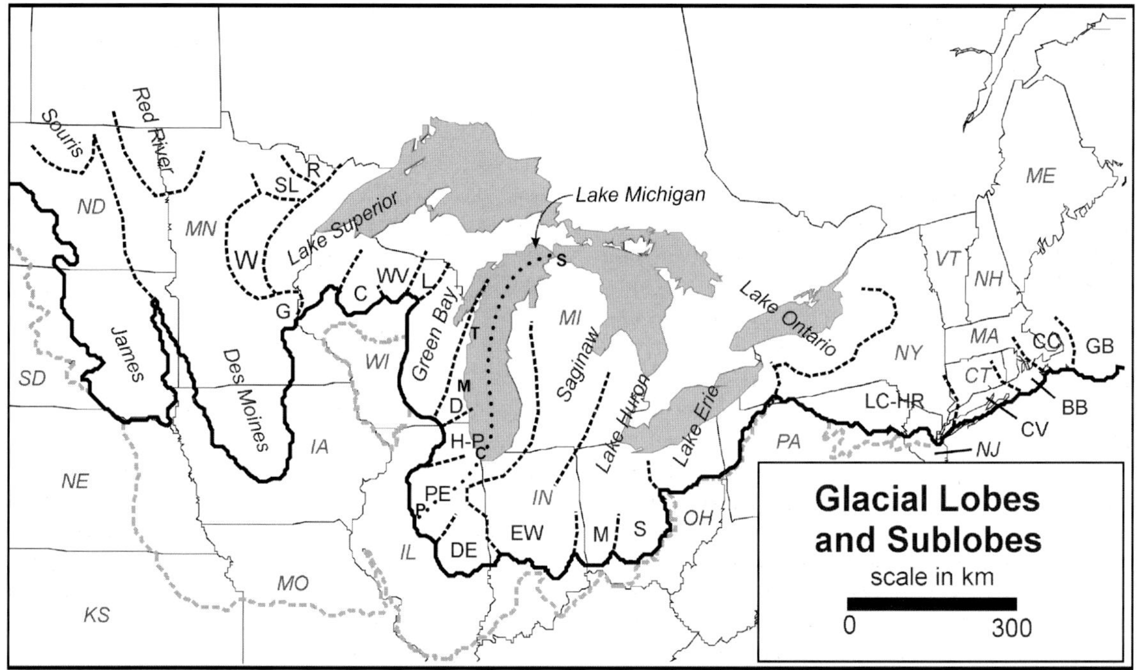

Fig. 2. Glacial lobes and sublobes of the southern Laurentide Ice Sheet during the late Wisconsin Glaciation. State abbreviations are explained in caption of Fig. 1. Major lobes are labeled and sublobes are as follows: G = Grantsburg, W = Wadena, SL = St. Louis, R = Rainey, C = Chippewa, WV = Wisconsin Valley, L = Langlade, D = Delevan, H-P = Harvard-Princeton, PE = Peoria, DE = Decatur, EW = East White, M = Miami, S = Scioto, LC = Lake Champlain, HR = Hudson River, CV = Connecticut Valley, BB = Buzzards Bay, CC = Cape Cod, GB = Georges Bank. Dotted line along axis of Lake Michigan lobe shows location of time-distance diagram shown in Fig. 3 (P, Peoria; C, Chicago; M, Milwaukee, T, Two Rivers; S, Straits of Makinac). Latitude and longitude are given in Fig. 1. Light dashed line shows the maximum limit of the ice-sheet.

Mickelson *et al.*, 1983), and modeling (discussed below) also suggests there were substantially different subglacial conditions from north to south in the southern LIS.

There have been relatively few detailed sediment descriptions, and even then, they describe deformation features without a definitive demonstration of conditions under which the sediments were deformed. Hicock & Dreimanis (1992a) examined three widespread diamicton units on the north sides of Lakes Ontario, Erie, and Superior where pre-advance lake sediment was overridden and incorporated into the sediment. They found evidence of viscous, ductile and brittle deformation based on fold types and orientation, fractures, the nature of striations on clasts, distribution of sand clasts, and other sedimentary features and concluded that there was evidence for soft-sediment deformation, but also deposition by lodgement of basal melt-out. They suggest that the evidence of the relative importance of deforming soft bed and sediment deposition from debris-rich glacial ice is indeterminate. Based on laboratory measurements, reconstructed ice-surface profiles, effective stress reconstructions, and clast fabrics, Hooyer & Iverson (2003) suggest that plowing was an important process at the base of the Des Moines Lobe (Fig. 2). In the area covered by the southern Lake Michigan lobe (Fig. 2), diamicton genesis has been studied extensively at Wedron, Illinois (Hansel & Johnson, 1987; Hansel *et al.*, 1987; Johnson & Hansel, 1990) and later at other sites (Hansel & Johnson, 1999). Although these authors point out that there is still room for doubt about the origin of much of the diamicton in moraines in Illinois, they favor accumulation of subglacially deformed wet sediment as a primary mechanism of deposition. They base this on the consistency of their observations with theoretical predictions of what the sedimentary record should resemble. In particular, they cite the uniformity of the till, the lack of supraglacial sediment in moraines, local derivation of homogeneous diamicton, characteristics of multiple channel fills, orientation of pebbles plunging down ice or into channels, and pebble concentrations on contacts as evidence of a deforming bed.

The deforming bed model has also been incorporated into two-dimensional numerical simulations of ice-sheet behavior along the Lake Michigan lobe flow line (Clark *et al.*, 1996; Jensen *et al.*, 1995, 1996) by assuming no sliding and a viscoplastic behavior of subglacial sediment. Hard bed and deforming bed were two end member conditions used, and strength properties of the subglacial sediments were assumed to be similar to properties of the present day till measured by Vela (1994). As predicted by Boulton & Jones (1979) (the "bowler hat" model), the ice surface on the up-ice, hard-bed end of the profile has a steeper ice-surface gradient and higher driving stress than the deforming soft bed. Modeling of sediment flux indicates that there would be sufficient velocity to produce the large sediment fluxes estimated by Johnson *et al.* (1991) only under a fairly narrow range of subglacial pore pressures. An important contribution of the modeling is the suggestion that the rapid ice-margin fluctuations documented in the southern part of the Lake Michigan lobe might be controlled by slight changes in subglacial sediment viscosity as opposed to being directly driven by climate change. The physical behavior of southern LIS lobes is still poorly known and recent numerical reconstructions of the LGM LIS by Marshall *et al.* (2002) fail to reproduce the known configuration of the southern margin because present numerical models do not reproduce all the lobe-scale processes of a real ice sheet.

Temperature Conditions at the Glacier Bed

There appear to have been distinct differences in the nature of the glacier bed across the southern LIS (Colgan *et al.*, 2003; Mickelson *et al.*, 1983) in addition to the "hard bed/soft bed" differences described above. In the far southern area (Ohio, Indiana, and Illinois), and in younger readvance deposits around the Great Lakes, the landscape is dominated by wide moraines with low internal relief. Between the moraines are flat till plains with only a few low flutes locally present (Hansel & Johnson, 1999). North of about the latitude of Chicago (Fig. 1), the end moraines have higher relief, and drumlins dominate the landscape between moraines. Tunnel channels (or valleys) are abundant to the north and rare, or absent, farther south. Drumlins and tunnel channels occur in areas where there was a soft bed as well as a hard bed, so some other explanation for the huge differences in landscape must be involved.

An explanation for this remarkable difference in subglacial landscape is likely the presence or absence of permafrost during ice advance, and therefore basal ice temperatures in the marginal zone of the ice sheet. The last twenty years have seen growth in our knowledge of the extent and effects of permafrost along the south margin of the LIS. The most convincing evidence of permafrost includes fossil ice-wedge casts and ice-wedge polygons seen in aerial photographs (Péwé, 1983). Such features are relatively common in Ohio (Konen, 1995), Illinois (Johnson, 1990), Wisconsin (Clayton *et al.*, 2001), Minnesota (Mooers, 1990b), Iowa (Walters, 1994), and in the Dakotas (Clayton *et al.*, 1980) and have been dated at between 21,000 and 16,000 ^{14}C yr B.P. Baker *et al.* (1986) showed that a boreal forest/tundra transition zone was present in southeast Iowa at between 18,090 and 16,710 ^{14}C yr B.P. based on abundant plant and animal fossils preserved in a silt-filled swale. In the east, an extensive permafrost zone is suggested by numerous periglacial phenomena in the Appalachian Plateau and Ridge and Valley province just south of the glacial border (Clark & Ciolkosz, 1988).

Ice-wedge casts on till surfaces in the southern parts of the Lake Michigan, Saginaw, and Huron lobes (Fig. 2) (Johnson, 1990) indicate the presence of permafrost, but it seems likely this was established during retreat from the glacial maximum, not during advance. Numerous radiocarbon dates on *Picea* wood indicate a lack of continuous permafrost during advance to the maximum. Ice-wedge casts are widespread farther north in Wisconsin and presumably in the remainder of the northern part of the U.S. covered by the LIS (Péwé, 1983). Recently, a two-dimensional model has been adapted to the topography and reconstructed climate of the Green Bay lobe in eastern Wisconsin. The climate record used to drive the advancing ice is a speleothem record from the vicinity of St. Louis (MSL on Fig. 1) (Dorale *et al.*, 1998). The model clearly demonstrates the development of permafrost in front of the advancing ice and that the permafrost slowly disappears after being covered by ice, taking several thousand years to do so (Cutler *et al.*, 2000, 2001). The model is at present being adapted to other southern LIS lobes, but the use of models for anything but the most crude representations of what actually took place is limited by our lack of good climate information. Although there are several fossil-derived climate records in the southern LIS area that extend back into or beyond the late Wisconsin (Baker *et al.*, 1986; Birks, 1976; Curry & Baker, 2000), they only constrain temperature somewhat and provide almost no information about precipitation. So far, the only semi-quantative information on precipitation is what is generated by global climate models.

It seems likely that distribution of drumlins and tunnel channels, and the internal relief of end moraines are closely tied to the difference in permafrost history from north to south. Although there are some small drumlins that may be depositional features, many drumlins show evidence that widespread subglacial erosion carved them out of pre-existing sediments by freeze-on to the base of the ice, by streaming subglacial sediment, or flowing water (Attig *et al.*, 1989; Boyce & Eyles, 1991; Colgan, 1999a; Newman & Mickelson, 1994; Whittecar & Mickelson, 1979). The drumlins in the northern part of the area covered by the southern LIS may have formed during disappearance of the frozen bed beneath the ice sheet, which would have produced an inhomogeneous bed allowing the drumlins to form by differential erosion and deposition (Cutler *et al.*, 2000; Stanford & Mickelson, 1985). Colgan & Mickelson (1997; Colgan, 1999a) demonstrated that advances of the Green Bay lobe that produced drumlins had steeper ice-surface profiles than advances that did not produce them, suggesting that driving stress and therefore bed resistance, was higher in the drumlin forming areas than in areas of flat till plains.

It has also been suggested that the permafrost wedge near the glacier margin was instrumental in the development of tunnel valleys or channels, although there is continuing debate about the genesis of these features. Wright (1973) suggested that tunnel valleys drained water catastrophically from beneath a wet bed glacier that had a frozen bed zone around its edge. Much the same explanation for these features was used by Patterson (1994), Clayton *et al.* (1999), and Cutler *et al.* (2002). Mooers (1989), however, has argued that the tunnel valleys in the area covered by the Superior lobe (Fig. 2) are valleys that were cut by relatively small, non-catastrophic flows that eroded their banks as channels migrated laterally, producing a wide channel under wet bed conditions. Neither of these interpretations of tunnel channel formation requires the huge, widespread, sheet flows of water that are discussed below.

The nature of internal relief in moraines also increases from south to north in the area covered by the southern LIS. The internal relief apparently is related to the thickness of sediment that accumulated on the ice surface during still stands of the ice margin. It has been postulated that this thick supraglacial sediment was a result of a frozen bed near the ice margin which caused compressive flow and upward movement of sediment within the ice. Subsequent melting out would have produced a thick sediment cover over stagnant ice that later collapsed (Ham & Attig, 1996).

Extent and Thickness of a Subglacial Water Layer

The presence of subglacial water, its abundance, and its importance in forming large-scale subglacial landforms has been much debated in the last 40 years. Although probably accepted by only a minority of researchers, the idea that large subglacial floods (megafloods) occurred under much of the southern LIS has been argued on theoretical grounds (Shoemaker, 1992a, b, 1999) and from field evidence (Shaw, 1989; Shaw *et al.*, 1989; Shaw & Gilbert, 1990; Shaw & Sharpe, 1987). These papers interpret drumlins and a variety of other streamlined and transverse forms as bed forms developed by a sheet of subglacial water flowing rapidly toward the margin for a relatively short time. In southern Ontario, a series of interdrumlin channels, called tunnel channels by Brennand & Shaw (1994), appear to be distinctly different than tunnel channels described by Clayton *et al.* (1999). Brennand & Shaw (1994) suggest that the channels were eroded by waning flows of the thick water layer as the glacier bed came back into contact with the landsurface. The question remains: Were there deep-water sheet flows from beneath the southern LIS or were these catastrophic flows confined to tunnel channels? Clearly there were catastrophic floods of subglacial water, but most evidence suggests that the flows were channelized in tunnels or valleys (Clayton *et al.*, 1999; Pair, 1997) and were not widespread water layers tens-of-meters thick.

Other Aspects of Ice Dynamics

Along with the concepts of gentle ice-surface slopes and deforming beds, the idea that advances of lobes of the southern LIS were surges or longer lasting ice streams has been argued. Unfortunately the chronology of advances and retreats, even where relatively well controlled by radiocarbon age determinations, is not detailed enough to accurately determine advance and retreat rate in most places. Clayton *et al.* (1985) estimated glacier margin advance and retreat rates during deglaciation in the southwestern area covered by the LIS of about 2 km/yr, somewhat higher than 0.7 km/yr estimated by Mickelson *et al.* (1981). These rapid rates are calculated for the later stages of deglaciation, after about 14,000 ^{14}C yr B.P. It has been suggested by many that as climate warmed during deglaciation, bed conditions changed, and that there were numerous retreats and readvances in a relatively short time, indicating very active, fast-flowing ice. This change in bed conditions appears to have migrated northward between 14,000 and 11,000 ^{14}C yr B.P., producing a different landform record than earlier advances (Mickelson *et al.*, 1983; Mooers, 1990a, b) when temperatures were colder.

Another indicator of former surges in the southwestern part of the ice sheet is the sediment and landform record. Thrust masses (Bluemle & Clayton, 1984) and thick supraglacial sediment (usually indicated by high-relief hummocky topography) indicate compressive flow and upward movement of sediment that was melted out at the ice surface. These features are common in modern glaciers that have surged and have been used as an indicator of surging in the southwest part of the LIS where clayey till is present (Clayton *et al.*, 1985; Colgan *et al.*, 2003). Evans *et al.* (1999) have attributed similar features, farther to the north in Alberta, Canada, to surging of the ice sheet by analogy with Icelandic glaciers. Ice streams feeding the Des Moines and James lobes (Fig. 2) have also been proposed for the southwestern part of the LIS by Patterson (1998).

The origin of high-relief hummocky topography, which is widespread in the area covered by the southern LIS, continues to be an issue of debate. The traditional view that it reflects thick supraglacial sediment (Clayton & Moran, 1974; Gravenor & Kupsch, 1959) has been challenged by resurrecting an older alternate hypothesis (Stalker, 1960) that suggests that hummocks are pressed forms developed on a soft deforming bed (Eyles *et al.*, 1999a, b). It has also been proposed that they were eroded by a megaflood beneath the ice (Munro-Stasiuk & Sjogren, 1999).

Chronology and Climate History

Since publication of *Quaternary of the United States* (Wright & Frey, 1965) there have been many advances in our understanding of the timing of events along the southern margin of the LIS. These advances include knowledge of the number and timing of pre-Wisconsin glaciations, the extent of an early Wisconsin glaciation, and details of the timing of late Wisconsin advances.

By the late 1960s and early 1970s it was becoming clear that the classic four-fold North American Stage terminology had become inadequate to describe the complexity of pre-Wisconsin glacial deposits (e.g. Dort, 1966; Reed & Dreezan, 1965). During the 1970s, fission-track dating applied to Yellowstone ashes interbedded between glacial tills of the central plains, showed that the terms Nebraskan and Kansan had little stratigraphic meaning, since five pre-Illinoian tills were documented, at least one older than 2 million years (Boellstorff, 1973, 1978). In Iowa and Nebraska, some tills previously classified as Kansan were proven to be older than those called Nebraskan (Boellstorff, 1973, 1978). During the 1980s, paleomagnetism (Easterbrook & Boellstorff, 1984) further helped to define the till stratigraphy in the classic type areas of the Nebraskan and Kansan sediments now informally referred to as pre-Illinioan (Hallberg, 1986) or early to middle Pleistocene (Richmond & Fullerton, 1986). As it stands now, at least six till units are recognized in the central plains region and tephrochronology, paleomagnetism, and paleopedology are increasingly being used to shed light on the pre-Illinoian record of the southern LIS (e.g. Aber, 1991; Colgan, 1999b; Guccione, 1983; Rovey & Keane, 1996).

The interpretation of younger Illinoian and Wisconsin stratigraphy has also experienced change in the last two decades. Tills formerly thought to be early Wisconsin have recently been reinterpreted as Illinoian in many key locations where they were first described, such as Illinois, Indiana, and Ohio (Fig. 1) (Clark, 1992b; Clark *et al.*, 1993; Goldthwait, 1992). These conclusions are based on reinterpretations of loess records and paleosols (Curry, 1989), and new dating techniques such as thermoluminescence and amino-acid methods (Miller *et al.*, 1992; Szabo, 1992). Because absolute dates are few and these methods are still problematic, many of these reinterpretations need to be tested. Debate has also focused on the extent of the early Wisconsin (OIS 4) LIS near Lake Ontario with some arguing that ice remained north of the lake (Eyles & Eyles, 1993), and others arguing that ice was grounded in the lake and advanced south of the lake (Dreimanis, 1992; Hicock & Dreimanis, 1989, 1992b). In New England (Fig. 1), tills older than late Wisconsin have been interpreted as both Illinoian (Newman *et al.*, 1990; Oldale & Coleman, 1992) and early Wisconsin (Colgan & Rosen, 2001; Stone & Borns, 1986). Unfortunately, the problem of the extent of an early Wisconsin advance is and will continue to be plagued by a lack of reliable dating methods that are effective beyond the 50,000-year range of radiocarbon. Hopefully future investigations with new dating methods will help solve the uncertainty in the extent of the early Wisconsin LIS.

The late Wisconsin history of ice advance and retreat continues to be refined with more radiocarbon dates, and new radiocarbon methods. A recent compilation of radiocarbon dates shows that the LGM southern LIS margin continues to be the best-dated ice margin in the world (Dyke *et al.*, 2002). Accelerator mass spectrometry (AMS) has allowed for smaller sample sizes and dates in areas where wood samples are rare (e.g. Maher *et al.*, 1998), but with smaller sample sizes come the problems of contamination and reworking of older materials. Tree-ring and coral-based calibration of radiocarbon dates has greatly improved the accuracy of dates younger than ~20,000 ^{14}C yr B.P. (Stuvier & Reimer, 1993; Stuiver *et al.*, 1998a, b). As was suggested by Mickelson *et al.* (1983), much less is known about the initial advance of the ice margin to its LGM position than is known about its subsequent retreat.

Much of the late Wisconsin chronology detailed by Mickelson *et al.* (1983) remains the same. Ice advanced into the northern U.S. about 26,000 ^{14}C yr B.P., yet the LGM extent was reached at different times in different places. Lobes in the Great Lakes and New England regions reached their maximum well before 21,000 ^{14}C yr B.P. Rapid decay of the ice sheet began after 14,500 ^{14}C yr B.P. Although lobes to the west of the Great Lakes also advanced before 21,000 ^{14}C B.P., they reached their maximum extent at about 14,000 ^{14}C yr B.P., out of phase with the rest of the ice-sheet margin (Hallberg & Kemmis, 1986). Readvance of lobes, some perhaps as surges, are recorded all along the southern LIS margin at 13,000 ^{14}C yr B.P., 11,800 ^{14}C yr B.P., and

9800 ^{14}C yr B.P. Ice retreated out of the northern U.S. shortly after 9800 ^{14}C yr B.P.

Advances in our knowledge of ice sheet chronology have occurred in New England also. In southern New England the date of the earliest advance is recorded by the youngest dates (~21,750 ^{14}C yr B.P.) found in the ice-thrust moraines of Long Island (Sirken & Stuckenrath, 1980). Retreat of the ice began at about 15,600 ^{14}C yr B.P. based on the number of varve years recorded in Glacial Lake Hitchcock and AMS dates at the north end of the lake (Ridge & Larsen, 1990; Ridge *et al.*, 2001). Numerous dates in coastal Maine have continued to refine ice-retreat history, retreat rates, and sea-level history (see papers in Retelle & Weddle, 2001).

In the Great Lakes regions, new dates have refined our knowledge of the deglaciation chronology (Attig *et al.*, 1985; Ekberg *et al.*, 1993; Hansel & Johnson, 1996) and revisions to the long established stratigraphic classification in Illinois have been adopted (Johnson *et al.*, 1997; Karrow *et al.*, 2000). The Lake Michigan lobe may have advanced to a maximum position as early as ~26,000 ^{14}C yr B.P. along its northwest edge (Hansel & Johnson, 1996), and then, subsequently retreated and readvanced to near its LGM position at ~22,500, 18,500, 17,500, and 15,500 ^{14}C B.P. Hansel & Johnson (1992) produced a time-distance diagram for the axis of the Lake Michigan lobe that is reproduced with slight modification in Fig. 3. It shows slightly different maximum advance times. Mickelson *et al.* (1983) assumed that glacial retreats and readvances along the southern LIS margin were synchronous because they could see no clear evidence otherwise. Lowell *et al.* (1999) have recently shown that most of the glacial retreats in this region were synchronous based on analyses of numerous radiocarbon dates. Lowell *et al.* (1999) and Clark *et al.* (2001) also show that advances and retreats of the southern LIS margin can be correlated with climate events recorded in the Greenland ice cores. This suggests that both the southern LIS and the Greenland Ice Sheet were responding to changes in North Atlantic climate. Although their results are convincing, as in the past, it has been extremely difficult to prove synchrony because of the inherent uncertainty in the radiocarbon method (as much as 1500 years in samples older than 10,000 years).

Cosmogenic isotope methods have recently been applied to late Wisconsin landscapes in New England (Larsen, 1996), the Great Lakes region (Colgan *et al.*, 2002), and along the southwestern LIS margin (Jackson *et al.*, 1997, 1999). Larsen (1996) dated a boulder resting on the LGM end moraine in New Jersey and estimated its exposure time as about 21,500 yr, consistent with the radiocarbon chronology. Colgan *et al.* (2002) sampled striated bedrock overridden by the Green Bay lobe in Wisconsin and found that ice began to retreat from its LGM margin probably well before 17,000 yr ago. They also found that bedrock near the ice margin contained inherited isotopes that made cosmogenic dating near (within 30 km) the margin impossible. Jackson *et al.* (1997, 1999) showed

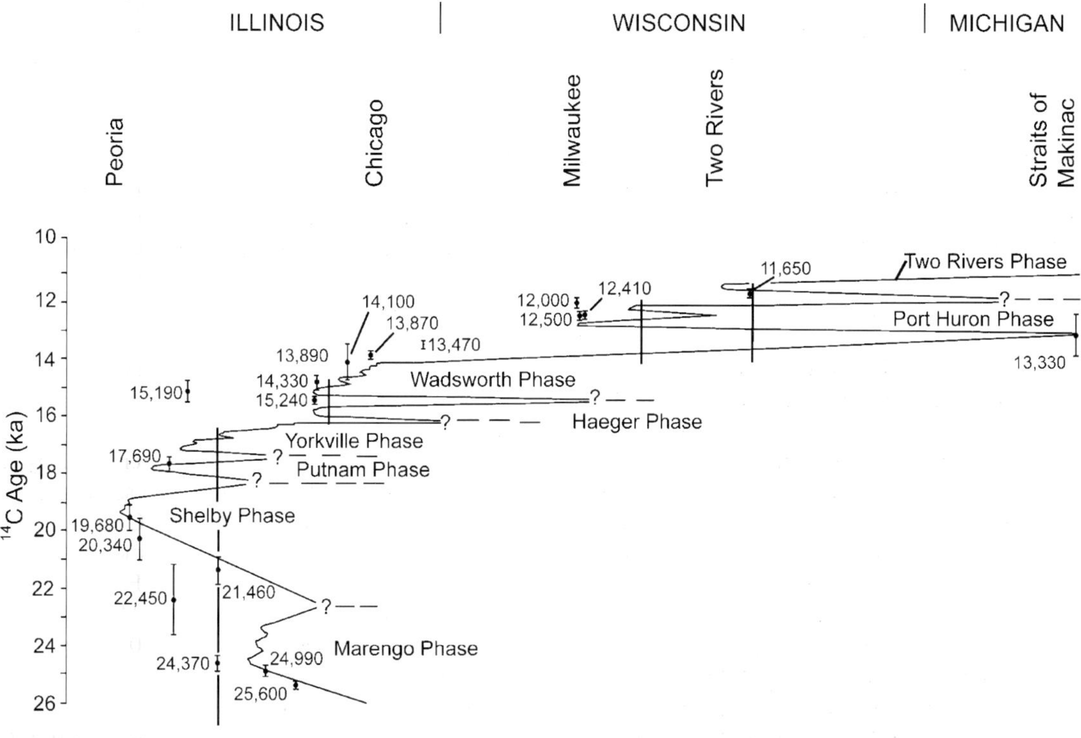

Fig. 3. Time-distance diagram showing glacial phases during the late Wisconsin glaciation in the Lake Michigan lobe. Representative radiocarbon age control (not calender years) is shown. Locations of the flow line and points on the flow line are shown in Fig. 2. Modified from Hansel & Johnson (1992).

that the southwestern LIS margin reached its greatest extent during the LGM sometime before 18,000 yr ago. These methods hold the promise of more closely limiting the age of glacial events if the problems of inheritance can be overcome, and uncertainty in isotope production rates can be reduced.

The amount of information about global climate history has exploded in the last 40 years. The deep-ocean drilling program of the late 1960s to the present has provided a detailed record of glacial and interglacial stages over the last 2.4 million years (e.g. Broecker & van Donk, 1970; Shackelton & Opdyke, 1973). The variations in ice sheet size and sea level history have been tied to changes in solar insolation driven by changes in Earth's orbit (e.g. Hays *et al.*, 1976). Ice cores from both Antarctica and Greenland have also produced detailed records with both excellent resolution and age control (e.g. Stuvier & Grootes, 2000). These records suggest that climate is driven by changes in solar insolation, internal ice sheet dynamics, changes in ocean currents, and changes in atmospheric composition (Mayewski *et al.*, 1997).

It is also apparent, based on the ocean record, that the LIS has discharged massive numbers of icebergs about every 5000–7000 years into the North Atlantic (Broecker *et al.*, 1992; Heinrich, 1988). Workers along the southern margin of the LIS have correlated local ice-lobe behavior to Heinrich events (Mooers & Lehr, 1997; Mullins *et al.*, 1996). Rapid drawdown in ice over Hudson Bay may have shifted ice divides and caused retreats along the southern LIS margin (Mooers & Lehr, 1997). It is clear from what we know about the chronology that ice retreated in many southern lobes shortly after the three youngest Henrich events ~21,000 (H2), 14,500 (H1), and 11,000 yr ago (H0 or Younger Dryas). These new studies that link southern LIS behavior to that of other ice sheets and major climate events is beginning to illuminate how the LIS responded to and influenced global climate. Future work along the LIS will continue to try to link the advances and retreats of lobes to both internal and external forcing mechanisms.

Conclusions

Glacial mapping and new data sources have led to a renewed interest in Quaternary mapping. Applied research helps to support new mapping initiatives. Glacial geology has become much more diverse, more oriented to global issues, and more connected to modern studies in paleoclimatology, glaciology, and sedimentology than in the past. Glacial sedimentology has flowered and embraced modern-process analogs and work in modern glacial environments.

Glacial geologists have used new information about the physics of modern ice sheets and glaciers to found a new discipline called paleoglaciology. This field has led the way in producing new models of past ice sheets. The late Wisconsin southern LIS consisted of thin, gently sloping lobes with low driving stresses. The southernmost lobes had a wet bed to the margin and surges were probably common. Ice lobes in lowlands may have been fed by ice streams. Farther north, ice advanced over permafrost and had a frozen bed near the margin. Not until after the late glacial maximum did ice warm to the margin. There are profound differences in the distribution and character of landforms such as moraines, drumlins and tunnel channels and likely there were differences in subglacial processes, with a soft deforming bed occurring in places, and sliding dominating in others.

There is general agreement that the pre-Wisconsin stratigraphic record is much more complex than thought in 1965. The terms "Nebraskan" and "Kansan" are no longer used, and the term "pre-Illinoian" is used in their place. The extent of early Wisconsin and mid-Wisconsin ice is now thought to have been less extensive than previously interpreted. A wide array of global climate records shows that the LIS responded to climate changes and may have also caused changes in climate because of discharges of meltwater and icebergs into the North Atlantic. Heinrich events may be correlated to major retreats of the southern LIS margin at about 21,000, and 14,500 ^{14}C yr B.P.

Acknowledgments

We thank Ardith Hansel and Thomas Lowell for their comments and constructive ideas on the manuscript. Richard Berg and Donald Luman also provided helpful comments, especially on the GIS and mapping discussion. Cornelia Winguth, Ben Laabs, Hans Hinke, and Steve Kostka kindly read and corrected near final drafts of the manuscript. We apologize to the many scientists who have contributed to our knowledge of the LIS during the last 40 years, but who are not cited here. Space limitations required that only representative references be cited.

References

Aber, J.S. (1991). The glaciation of northeastern Kansas. *Boreas*, **20**, 297–314.

Aber, J.S., Bluemle, J.P., Brigham-Grette, J., Dredge, L.A., Sauchyn, D.J. & Ackerman, D.L. (1995). Glaciotectonic map of North America. 1:6,500,000. *Geological Society of America, Maps and Charts Series, MCH079.*

Alley, R.B. (1991). Deforming-bed origin for southern Laurentide till sheets? *Journal of Glaciology*, **37**, 67–76.

Alley, R.B., Blankenship, D.D., Bentley, C.R. & Rooney, S.T. (1986). Deformation of till beneath Ice Stream B, West Antarctica. *Nature*, **322**, 8921–8929.

Alley, R.B., Blankenship, D.D., Bentley, C.R. & Rooney, S.T. (1987). Till beneath Ice Stream B, 3. Till deformation: Evidence and implications. *Journal of Geophysical Research*, **92**(B9), 8921–8929.

Anderson, L. & Sollid, J.L. (1971). Glacial chronology and glacial geomorphology in the marginal zones of the glaciers, Midtdalsbreen and Nigardsbreen, south Norway. *Norsk Geologisk Tidsskrift*, **25**, 1–38.

Andrews, J.T. (1987). The late Wisconsin glaciation and deglaciation of the Laurentide Ice Sheet. *In*: Ruddiman, W.F. & Wright, H.E., Jr. (Eds), *North America and Adjacent*

Oceans During the Last Deglaciation. Boulder, Colorado, Geological Society of America, 13–37.

Attig, J.W., Clayton, L. & Mickelson, D.M. (1985). Correlation of late Wisconsin glacial phases in the western great lakes area. *Geological Society of America Bulletin*, **96**, 1585–1593.

Attig, J.W., Mickelson, D.M. & Clayton, L. (1989). Late Wisconsin landform distribution and glacier- bed conditions in Wisconsin. *Sedimentary Geology*, **62**, 399–405.

Baker, R.G., Rhodes, R.S., II, Schwert, D.P., Ashworth, A.C., Frest, T.J., Hallberg, G.R. & Janssens, J.A. (1986). A full-glacial biota from southeastern Iowa, U.S.A. *Journal of Quaternary Science*, **1**, 91–107.

Beget, J.E. (1986). Modeling the influence of till rheology on the flow and profile of the Lake Michigan Lobe, southern Laurentide Ice Sheet, U.S.A. *Journal of Glaciology*, **32**, 235–241.

Bentley, C.R. (1987). Antarctic ice streams: A review. *Journal of Geophysical Research*, **92**, 8843–8858.

Birks, H.J.B. (1976). Late-Wisconsinan vegetational history at Wolf Creek, central Minnesota. *Ecological Monographs*, **46**, 395–429.

Bluemle, J.P. & Clayton, L. (1984). Large-scale glacial thrusting and related processes in North Dakota. *Boreas*, **13**, 279–299.

Boellstorff, J. (1973). Tephrochronology, petrology, and stratigraphy of some Pleistocene deposits in the central plains, U.S.A. [Ph.D. thesis]. Lousiana State University, 197 pp.

Boellstorff, J. (1978). A need for redefinition of the North American Pleistocene stages. *Transactions of the Society of the Gulf coast Association of Geological Societies*, **28**, 65–74.

Boulton, G.S. (1970a). On the deposition of subglacial and melt-out tills at the margins of certain Svalbard glaciers. *Journal of Glaciology*, **9**, 231–245.

Boulton, G.S. (1970b). On the origin and transport of englacial debris in Svalbard Glaciers. *Journal of Glaciology*, **9**, 213–229.

Boulton, G.S. (1971). Till genesis and fabric in Svalbard, Spitzbergen. *In*: Goldthwait, R.P. (Ed.), *Till, a Symposium* (pp. 41–72). Ohio State University Press.

Boulton, G.S. & Clark, C.D. (1990). A highly mobile Laurentide Ice Sheet revealed by satellite images of glacial lineations. *Nature*, **346**, 813–817.

Boulton, G.S. & Hindmarsh, R.C.A. (1987). Sediment deformation beneath glaciers: Rheology and geological consequences. *Journal of Geophysical Research*, **92**, 9059–9082.

Boulton, G.S. & Jones, A.S. (1979). Stability of temperate ice caps and ice sheets resting on beds of deformable sediments. *Journal of Glaciology*, **24**, 29–43.

Boyce, J.I. & Eyles, N. (1991). Drumlins carved by deforming till streams below the Laurentide Ice Sheet. *Geology*, **19**, 787–790.

Brennand, T.A. & Shaw, J. (1994). Tunnel channels and associated landforms, south-central Ontario: Their implications for ice-sheet hydrology. *Canadian Journal of Earth Science*, **31**, 505–522.

Broecker, W.S. & van Donk, J. (1970). Insolation changes, ice volumes and the ^{18}O record in deep-sea cores. *Reviews of Geophysics and Space Physics*, **8**, 169–197.

Broecker, W.S., Bond, G., McManus, J., Klar, M. & Clark, E. (1992). Origin of the North Atlantic's Heinrich events. *Climate Dynamics*, **6**, 265–273.

Brown, S.E, Bleuer, N.K., O'Neal, M.A., Olejnik, J. & Rupp, R. (2001). Glacial terrain explorer. Indiana Geological Survey Open-File Studies, OD-08, CD-Rom.

Brown, S.E. & Jones, H. (1999). Glacial Terrains of the Stroh, Indiana 7.5-Minute Quadrangle. Indiana Geological Survey Open-File Studies, OFS99-12, 4 pl.

Clark, P.U. (1992a). Surface form of the southern Laurentide Ice Sheet and its implications to ice-sheet dynamics. *Geological Society of America Bulletin*, **104**, 595–605.

Clark, P.U. (1992b). The last interglacial-glacial transition in North America: Introduction. *In*: Clark, P.U. & Lea, P.D. (Eds), *The last interglacial-glacial transition in North America*. Geological Society of America, Special Paper 270, 1–12.

Clark, P.U. (1994). Unstable behavior of the Laurentide Ice Sheet over deforming sediment and its implications for climate change. *Quaternary Research*, **41**, 19–25.

Clark, P.U. (1997). Sediment deformation beneath the Laurentide Ice Sheet. *In*: Martini, I.P. (Ed.), *Late glacial and postglacial environmental changes*. Quaternary, Carboniferous-Permian, and Proterozoic, 81–97. New York, Oxford University Press.

Clark, M.G. & Ciolkosz, E.J. (1988). Periglacial Geomorphology of the Appalachian Highlands and interior highlands south of the glacial border – A review. *Geomorphology*, **1**, 191–220.

Clark, P.U., Clague, J.J., Curry, B.B., Dreimanis, A., Hicock, S., Miller, G.H., Berger, G.W., Eyles, N., Lamonthe, M., Miller, B.B., Mott, R.J., Oldale, R.N., Stea, R.R., Szabo, J.P., Thorleifson, L.H. & Vincent, J.S. (1993). Initiation and development of the Laurentide and Cordelleran Ice Sheets following the last interglaciation. *Quaternary Science Reviews*, **12**, 79–114.

Clark, P.U., Licciardi, J.M., MacAyeal, D.R. & Jensen, J.W. (1996). Numerical reconstruction of a soft-bedded Laurentide Ice Sheet during the last glacial maximum. *Geology*, **23**, 679–682.

Clark, P.U., Marshall, S.W., Clarke, G.K.C., Licciardi, J.W. & Teller, J.T. (2001). Freshwater forcing of abrupt climate change during the last glaciation. *Science*, **293**, 283–286.

Clarke, G.K.C. (1987). Fast Glacier Flow: Ice streams, surging, and tidewater glaciers. *Journal of Geophysical Research*, **92**, 8835–8841.

Clarke, G.K.C., Collins, S.G. & Thompson, D.E. (1984). Flow, thermal structure, and subglacial conditions of a surge-type glacier. *Canadian Journal of Earth Sciences*, **21**, 232–240.

Clayton, L. (1964). Karst topography on stagnant glaciers. *Journal of Glaciology*, **5**, 107–112.

Clayton, L. (2001). Pleistocene geology of Waukesha County. *Wisconsin Geological and Natural History Survey Bulletin*, **99**, 33 pp. Plus map at 1:100,000 and cross sections.

Clayton, L., Attig, J.W. & Mickelson, D.M. (1999). Tunnel channels in Wisconsin. *In*: Mickelson, D.M. & Attig, J.W. (Eds), *Glaciers Past and Present, Geological Society of America Special Paper 337*, 69–82.

Clayton, L., Attig, J.W. & Mickelson, D.M. (2001). Effects of late Pleistocene permafrost on the landscape of Wisconsin, USA. *Boreas*, **30**, 173–188.

Clayton, L. & Moran, S.R. (1974). A glacial process-form model. *In*: Coates, D.R. (Ed.), *Glacial Geomorphology*. Binghampton, NY, State University of New York at Binghamton, 89–120.

Clayton, L., Moran, S.R. & Bluemle, J.P. (1980). *Geologic map of North Dakota*. U.S. Geological Survey Map.

Clayton, L., Mickelson, D.M. & Attig, J.W. (1989). Evidence against pervasively deformed bed material beneath rapidly moving lobes of the southern Laurentide Ice Sheet. *Sedimentary Geology*, **62**, 203–208.

Clayton, L., Teller, J.T. & Attig, J.W. (1985). Surging of the southwestern part of the Laurentide Ice Sheet. *Boreas*, **14**, 235–241.

CLIMAP (1976). The surface of ice age Earth. *Science*, **191**, 1131–1137.

CLIMAP (1984). The last interglacial ocean. *Quaternary Research*, **21**, 123–224.

COHMAP (1988). Climatic changes of the last 18,000 years: Observations and model simulations. *Science*, **241**, 1043–1052.

Coleman, S.M., Clark, J.A., Clayton, L., Hansel, A.K. & Larsen, C.E. (1989). Deglaciation, lake levels, and meltwater discharge in the Lake Michigan basin. *Quaternary Science Reviews*, **13**, 879–890.

Colgan, P.M. (1999a). Reconstruction of the Green Bay lobe, Wisconsin, United States, from 26,000 to 13,000 radiocarbon years B.P. *In*: Mickelson, D.M. & Attig, J.W. (Eds), *Glaciers Past and Present, Geological Society of America Special Paper 337*, 137–150.

Colgan, P.M. (1999b). Early middle Pleistocene Glaciation (780,000 to 610,000 B.P.) of the Kansas city area, northwestern Missouri, USA. *Boreas*, **28**, 477–489.

Colgan, P.M. & Mickelson, D.M. (1997). Genesis of streamlined landforms and flow history of the Green Bay lobe, Wisconsin, U.S.A. *Sedimentary Geology*, **111**, 7–25.

Colgan, P.M. & Rosen, P.S. (2001). Quaternary history of the Boston Harbor Islands, Massachusetts. *In*: Bailey, R.H. & West, D.R. (Eds), Geological Society of America, 2001 Meeting in Boston, U.S.A., Field Trip Guidebook, I-1–20.

Colgan, P.M., Bierman, P.R., Mickelson, D.M. & Caffee, M. (2002). Variation in glacial erosion near the southern margin of the Laurentide Ice Sheet, south-central Wisconsin, USA: Implications for cosmogenic dating of glacial terrains. *Geological Society of America Bulletin*, **114**, 1581–1591.

Colgan, P.M., Mickelson, D.M. & Cutler, P.M. (2003, in press). Glacial landsystems of the southern Laurentide Ice Sheet. *In*: Evans, D. (Ed.), *Glacial Landsystems*. Edwin Allen.

Curry, B.B. (1989). Absence of Altonian Glaciation in Illinois. *Quaternary Research*, **31**, 1–13.

Curry, B.B. & Baker, R.G. (2000). Palaeohydrology, vegetation, and climate since the late Illinois episode (~130 ka) in south-central Illinois. *Palaeogeography, Palaeoclimatology, Palaeoecology*, **155**, 59–81.

Curry, B.B., Berg, R.C. & Vaiden, R.C. (1997). Geologic mapping for environmental planning, McHenry County, Illinois, *Illinois Geological Survey Circular, 559*, 76 pp.

Cutler, P.M., Colgan, P.M. & Mickelson, D.M. (2002). Sedimentologic evidence for outburst floods from the Laurentide Ice Sheet margin in Wisconsin, USA: Implications for tunnel-channel formation. *Quaternary International*, **90**, 23–40.

Cutler, P.M., MacAyeal, D.R., Mickelson, D.M., Parizek, B.R. & Colgan, P.M. (2000). A numerical investigation of ice-lobe-permafrost interaction around the southern Laurentide Ice Sheet. *Journal of Glaciology*, **46**, 311–325.

Cutler, P.M., Mickelson, D.M., Colgan, P.M., MacAyeal, D.R. & Parizek, B.R. (2001). Influence of the great lakes on the dynamics of the southern Laurentide Ice Sheet: Numerical Experiments. *Geology*, **29**, 1039–1042.

DeSimone, D.J. & Dethier, D.P. (1992). *Surficial Geology of the Pownal and North Pownal area, Vermont*, **64**, 2 plates, scale 1:24000.

Dorale, J., Edwards, R.L., Gonzalez, L.A. & Ito, E. (1998). Mid-continent oscillations in climate and vegetation from 75 to 25 ka: A speleothem record from Crevice Cave, southeast Missouri, USA. *Science*, **282**, 1871–1874.

Dort, W., Jr. (1966). Nebraskan and Kansan Stades: Complexity and importance. *Science*, **154**, 771–772.

Dreimanis, A. (1992). Early Wisconsinan in the north-central part of the Lake Erie basin: A new interpretation. *In*: Clark, P.U. & Lea, P.D. (Eds), *The last interglacial-glacial transition in North America*. Boulder, Colorado, Geological Society of America Special Paper 270, 109–118.

Dyke, A.S. & Prest, V.K. (1987). Late Wisconsinan and Holocene history of the Laurentide Ice Sheet. *Geographie Physique et Quaternaire*, **41**, 237–263.

Dyke, A.S., Andrews, J.T., Clark, P.U., England, J.H., Miller, G.H., Shaw, J. & Veillette, J.J. (2002). The Laurentide and Innuitian ice sheets during the Last Glacial Maximum. *Quaternary Science Reviews*, **21**, 9–31.

Easterbrook, D.J. & Boellstorff, J. (1984). Paleomagnetism and chronology of early Pleistocene tills in central United States. *In*: Mahaney, W.C. (Ed.), *Correlation of Quaternary Chronologies*. Norwich, England Geo Books, 73–90.

Ekberg, M.P., Lowell, T.V. & Stuckenrath, R. (1993). Late Wisconsin glacial advance and retreat patterns in southwestern Ohio, USA. *Boreas*, **22**, 189–204.

Eschman, D.F. & Mickelson, D.M. (1986). Correlation of glacial deposits of the Huron, Lake Michigan, and Green Bay lobes in Michigan and Wisconsin. *Quaternary Science Reviews*, **5**, 53–58.

Evans, D.J.A., Lemmen, D.S. & Rea, B.A. (1999). Glacial landsystems of the southwest Laurentide Ice Sheet; modern Icelandic analogues. *Journal of Quaternary Science*, **14**, 673–691.

Eyles, C.H. & Eyles, N. (1993). Sedimentation in a large lake: A reinterpretation of the late Pleistocene stratigraphy of the Scarborough Bluffs, Ontario, Canada. *Geology*, **11**, 146–152.

Eyles, N., Boyce, J.I. & Barendregt, R.W. (1999a). Hummocky moraine; sedimentary record of stagnant Laurentide Ice Sheet lobes resting on soft beds. *Sedimentary Geology*, **123**, 163–174.

Eyles, N., Boyce, J.I. & Barendregt, R.W. (1999b). Hummocky moraine; sedimentary record of stagnant Laurentide Ice Sheet lobes resting on soft beds; reply. *Sedimentary Geology*, **129**, 169–171.

Eyles, N., Eyles, C.H. & Miall, A.D. (1983). Lithofacies types and vertical profile models: An alternative approach to description of glacial diamict and diamict sequences. *Sedimentology*, **30**, 393–410.

Flint, R.F. (1971). *Glacial and Quaternary geology*. New York, Wiley and Sons, 892 pp.

Fullerton, D.S. (1986). Stratigraphy and correlation of glacial deposits from Indiana to New York and New Jersey. *Quaternary Science Reviews*, **5**, 23–38.

Fullerton, D.S. & Colton, R.B. (1986). Stratigraphy and correlation of glacial deposits on the Montana plains. *Quaternary Science Reviews*, **5**, 69–82.

Goldthwait, R.P. (1974). Rates of formation of glacial features in Glacier Bay, Alaska. *In*: Coates, D.R. (Ed.), *Glacial Geomorphology*. State University of New York, Binghamton, 163–185.

Goldthwait, R.P. (1992). Historical overview of early Wisconsin glaciation. *In*: Clark, P.U. & Lea, P.D. (Eds), *The Last Interglacial-Glacial Transition in North America*. Boulder, Colorado, Geological Society of America Special Publication 270, 13–18.

Gravenor, C.P. & Kupsch, W.O. (1959). Ice-disintegration features in western Canada. *Journal of Geology*, **67**, 48–64.

Grimley, D.A. (2002). Surficial geology map, Elsah 7.5 minute Quadrangle (Illinois portion), Jersey and Madison Counties, Illinois. Illinois Geological Survey, Geological Quadrangle, Elsah-SG, scale 1:24,000.

Guccione, M. (1983). Quaternary sediments and their weathering history in north central Missouri. *Boreas*, **12**, 217–226.

Gustavson, T.C. & Boothroyd, J.C. (1987). A depositional model for outwash, sediment sources, and hydrologic characteristics, Malaspina Glacier, Alaska: A modern analog of the southeastern margin of the Laurentide Ice Sheet. *Geological Society of America Bulletin*, **99**, 187–200.

Hallberg, G.R. & Kemmis, T.J. (1986). Stratigraphy and correlation of the glacial deposits of the Des Moines and James lobes and adjacent areas in North Dakota, South Dakota, Minnesota and Iowa. *Quaternary Science Reviews*, **5**, 65–68.

Hallberg, G.R. (1986). Pre-Wisconsin glacial stratigraphy of the central plains region in Iowa, Nebraska, Kansas, and Missouri. *In*: Sibrava, V., Bowen, D.Q. & Richmond, G.M. (Eds), *Quaternary Glaciations in the Northern Hemisphere*. Oxford, Pergamon Press, 11–16.

Ham, N.R. & Attig, J.W. (1996). Ice wastage and landscape evolution along the southern margin of the Laurentide Ice Sheet, north-central Wisconsin. *Boreas*, **25**, 171–186.

Hansel, A.K. & Johnson, W.H. (1987). Ice-marginal sedimentation in a late Wisconsinan end moraine complex, northeastern Illinois, U.S.A. *In*: van der Meer, J.M.J. (Ed.), *Tills and Glaciotectonics*. Rotterdam, A.A. Balkkema, 97–104.

Hansel, A.K. & Johnson, W.H. (1992). Fluctuations of the Lake Michigan lobe during the late Wisconsin subepisode. *Geological Survey of Sweden*, Series Ca 81, 133–144.

Hansel, A. & Johnson, H.K. (1996). Wedron and mason groups. Lithostratigraphic reclassification of deposits of the Wisconsin Episode, Lake Michigan lobe area. *Illinois Geological Survey*, Bulletin 104, 116 pp.

Hansel, A.K. & Johnson, W.H. (1999). Wisconsin Episode Glacial Landscape of Central Illinois: A Product of Subglacial Deformation Processes? *In*: Mickelson, D.M. & Attig, J.W. (Eds), *Glacial Processes Past and Present, Geological Society of America Special Paper* 337, 121–135.

Hansel, A.K., Johnson, W.H. & Socha, B.J. (1987). Sedimentological characteristics and genesis of basal tills at Wedron, Illinois. *In*: Kujansu, R. & Saarnisto, M. (Eds), *INQUA, Till Symposium, Finland 1985*. Geological Survey of Finland Special Paper 3, 11–21.

Harris, K.L. & Luther, M.R. (1991). Surface geology of the goose river map area. *North Dakota Geological Survey*, Atlas Series 14-A1, scale 1:250,000.

Harrison, W. (1958). Marginal zones of vanished glaciers reconstructed from preconsolidation-pressure values of overridden silts. *Journal of Glaciology*, **66**, 72–95.

Hays, J.D., Imbrie, J. & Shackleton, N.J. (1976). Variations in earth's orbit: Pacemaker of the ice ages. *Science*, **194**, 1121–1132.

Heinrich, H. (1988). Origin and consequences of cyclic ice rafting in the northwest Atlantic Ocean during the past 130,000 years. *Quaternary Research*, **29**, 142–152.

Hicock, S.R. & Dreimanis, A. (1989). Sunnybrook drift indicates a grounded early Wiscosinan glacier in the Lake Ontario basin. *Geology*, **17**, 169–172.

Hicock, S.R. & Dreimanis, A. (1992a). Deformation till in the great lakes region; implications for rapid flow along the south-central margin of the Laurentide Ice Sheet. *Canadian Journal of Earth Sciences*, **29**, 1565–1579.

Hicock, S.R. & Dreimanis, A. (1992b). Sunnybrook drift in the Toronto area, Canada: Reinvestigation and Reinterpretation. *In*: Clark, P.U. & Lea, P.D. (Eds), *The last interglacial-glacial transition in North America*. Boulder, Colorado, Geological Society of America Special Paper 270, 163–170.

Hobbs, H.C. (1995). *Geologic atlas of Rice County*. Minnesota. Minnesota Geological Survey, County Atlas Series, C-9, Part A, Scale 1:100,000.

Hooke, R.L. (1970). Morphology of the ice-sheet margin near Thule, Greenland. *Journal of Glaciology*, **9**, 303–324.

Hooke, R.L. (1973). Flow near the margin of the Barnes Ice Cap, and the development of ice-cored moraines. *Geological Society of America Bulletin*, **84**, 3929–3948.

Hooyer, T.S. & Iverson, N.R. (2001). Diffusive mixing between shearing granular layers: Constraints on bed deformation from till contacts. *Journal of Glaciology*, **46**, 641–651.

Hooyer, T.S. & Iverson, N.R. (2003, in press). Flow mechanism of the Des Moines Lobe of the Laurentide Ice Sheet. *Journal of Glaciology*.

Hughes, T.J., Denton, G.H. & Fastook, J.L. (1985). The Antarctic Ice Sheet: An analog for Northern Hemisphere paleo-ice sheet? *In*: Woldenberg, M.J. (Ed.), *Models in Geomorphology*. Boston, Allen & Unwin, 25–72.

Jackson, L.E., Jr., Phillips, F.M. & Little, E.C. (1999). Cosmogenic ^{36}Cl dating of the maximum limit of the Laurentide Ice Sheet in southwestern Alberta. *Canadian Journal of Earth Sciences*, **36**, 1347–1356.

Jackson, L.E., Jr., Phillips, F.M., Shimamura, K. & Little, E.C. (1997). Cosmogenic ^{36}Cl dating of the Foothills Erratics Train, Alberta, Canada. *Geology*, **25**, 195–198.

Jensen, J.W., Clark, P.U., MacAyeal, D.R., Ho, C. & Vela, J.C. (1995). Numerical modeling of advective transport of saturated deforming sediment beneath the Lake Michigan Lobe, Laurentide Ice Sheet. *In*: Harbor, J.M. (Ed.), *Glacial Geomorphology; Process and Form Development*. Amsterdam, Netherlands, Elsevier, 157–166.

Jensen, J.W., MacAyeal, D.R., Clark, P.U., Ho, C.L. & Vela, J.C. (1996). Numerical modeling of subglacial sediment deformation; implications for the behavior of the Lake Michigan Lobe, Laurentide Ice Sheet. *Journal of Geophysical Research, B, Solid Earth and Planets*, **101**, 8717–8728.

Johnson, W.H. (1986). Stratigraphy and correlation of glacial deposits of the lake Michigan lobe prior to 14 ka B.P. *Quaternary Science Reviews*, **5**, 17–22.

Johnson, W.H. (1990). Ice-wedge casts and relict patterned ground in central Illinois and their environmental significance. *Quaternary Research*, **33**, 51–72.

Johnson, W.H. & Hansel, A.K. (1990). Multiple glacigenic sequences at Wedron, Illinois. *Journal of Sedimentary Petrology*, **60**, 26–41.

Johnson, W.H., Hansel, A.K. & Stiff, B.J. (1991). Glacial transport rates, late Wisconsin Lake Michigan lobe in central Illinois. Implications for transport mechanisms and ice dynamics [abs.]. *Geological Society of America Abstracts with Program*, **23**, 5, A-61.

Johnson, W.H., Hansel, A.K., Bettis, E.A., III, Karrow, P.F., Larson, G.J., Lowell, T.V. & Schneider, A.F. (1997). Late Quaternary temporal and event classifications, great lakes region, North America. *Quaternary Research*, **47**, 1–12.

Jopling, A.V. & McDonald, B.C. (Eds) (1975). Glaciofluvial and glaciolacustrine sedimentation. *Society of Economic Paleontologists and Mineralogists, Special Publication*, **23**, 320 pp.

Kamb, B., Raymond, C.F., Harrison, W.D., Englehardt, H., Echelmeyer, K.A., Humphrey, N., Brugman, M.M. & Pfeffer, T. (1985). Glacier surge mechanism: 1982–1983 surge of Variegated Glacier, Alaska. *Science*, **227**, 469–479.

Karrow, P.F., Dreimanis, A. & Barnett, P.J. (2000). A proposed diachronic revision of Late Quaternary time-stratigraphic classification in the eastern and northern great lakes area. *Quaternary Research*, **54**, 1–12.

Konen, M.E. (1995). Morphology and distribution of polygonal patterned ground and associated soils in Drake and Miami Counties, Ohio [M.S. Thesis]. The Ohio State University, Columbus, 200 pp.

Larsen, P.L. (1996). In-situ production rates of cosmogenic ^{10}Be and ^{26}Al over the past 21,500 years determined from the terminal moraine of the Laurentide Ice Sheet, north-central New Jersey [M.S. thesis]. Burlington, University of Vermont, 129 pp.

Lasemi, Z. & Berg, R.C. (2001). Three-dimensional geologic mapping: A pilot program for resource and environmental assessment in the Villa Grove Quadrangle, Douglas County, Illinois. *Illinois Geological Survey Bulletin*, **106**, 117 pp.

Lawson, D.E. (1979). A comparison of the pebble orientations in ice and deposits of the Matanuska, Alaska. *Journal of Geology*, **87**, 629–645.

Lawson, D.E. (1981). Distinguishing characteristics of diamictons at the margin of the Matanuska Glacier, Alaska. *Annals of Glaciology*, **2**, 78–84.

Lowell, T.V., Hayward, R.K. & Denton, G.H. (1999). Role of climate oscillations in determining ice-margin position. *In*: Mickelson, D.M. & Attig, J.W. (Eds), *Glacial Processes Past and Present, Geological Society of America Special Paper* 337, 193–203.

Maher, L.J., Jr., Miller, N.G., Baker, R.G., Curry, B.B. & Mickelson, D.M. (1998). Paleobiology of the sand beneath the Valders diamicton at Valders, Wisconsin. *Quaternary Research*, **49**, 208–221.

Marshall, S.J., James, T.S. & Clarke, G.K.C. (2002). North American Ice Sheet reconstructions at the Last Glacial Maximum. *Quaternary Science Reviews*, **21**, 175–192.

Mathews, W.H. (1974). Surface profiles of the Laurentide Ice Sheet in its marginal areas. *Journal of Glaciology*, **13**, 37–43.

Matsch, C.L. & Schneider, A.F. (1986). Stratigraphy and correlation of the glacial deposits of the glacial lobe complex in Minnesota and Wisconsin. *Quaternary Science Reviews*, **5**, 59–64.

Mayewski, P.A., Meeker, L.D., Twickler, M.S., Whitlow, S.I., Yang, Q., Lyons, W.B. & Prentice, M. (1997). Major features and forcing of high latitude Northern Hemisphere atmospheric circulation over the last 110,000 years. *Journal of Geophysical Research*, **102**, 345–366.

Mickelson, D.M., Acomb, L.A. & Bentley, C.R. (1981). Possible mechanisms for the rapid advance and retreat of the lake Michigan lobe between 13,000 and 11,000 years ago. *Annals of Glaciology*, **2**, 185–186.

Mickelson, D.M., Clayton, L., Fullerton, D.S. & Borns, H.W., Jr. (1983). The late Wisconsin glacial record of the Laurentide Ice Sheet in the United States. *In*: Porter, S.C. (Ed.), *The Late Pleistocene, United States*. Minneapolis, University of Minnesota Press, 3–37.

Miller, B.B., McCoy, W.D., Wayne, W.J. & Brockman, C.S. (1992). Ages of the Fairhaven and Whitewater tills in southwestern Ohio and southeastern Indiana. *In*: Clark, P.U. & Lea, P.D. (Eds), *The Last Interglacial-Glacial Transition in North America*. Geological Society of America Special Paper 270, 89–98.

Mooers, H.D. (1989). Drumlin formation: A time transgressive model. *Boreas*, **18**, 99–107.

Mooers, H.D. (1990a). A glacial-process model: The role of spatial and temporal variations in glacier thermal regime. *Geological Society of America Bulletin*, **102**, 243–251.

Mooers, H.D. (1990b). Ice-marginal thrusting of drift and bedrock: Thermal regime, subglacial aquifers, and glacial surges. *Canadian Journal of Earth Sciences*, **27**, 849–862.

Mooers, H.D. & Lehr, J.D. (1997). Terrestrial record of Laurentide ice sheet reorganization during Heinrich events. *Geology*, **25**, 987–990.

Muller, E.H. & Cadwell, D.H. (1986). Surficial geologic map of New York, Finger Lakes sheet: Albany, New York State Geological Survey, scale 1:250,000.

Mullins, H.T. & Hinchey, E.J. (1989). Erosion and infill of the New York Finger Lakes: Implications for Laurentide Ice Sheet deglaciation. *Geology*, **17**, 622–625.

Mullins, H.T., Hinchey, E.J., Wellner, R.W., Stephens, D.B., Anderson, W.T., Jr., Dwyer, T.R. & Hine, A.C. (1996). Seismic stratigraphy of the Finger Lakes: A continental record of Heinrich event H-1 and Laurentide ice sheet instability. *In*: Mullins, H.T. & Eyles, N. (Eds), *Subsurface Geologic Investigations of New York Finger Lakes: Implication for Quaternary Deglaciation and Environmental Change*. Boulder, Colorado, Geological Society of America, Special Paper 311, 1–35.

Munro-Stasiuk, M.J. & Sjogren, D.B. (1999). Hummocky moraine sedimentary record of stagnant Laurentide Ice Sheet lobes resting on soft beds; discussion. *Sedimentary Geology*, **129**, 165–168.

Newman, W.A., Berg, R.C., Rosen, P.S. & Glass, H.D. (1990). Pleistocene stratigraphy of the Boston Harbor drumlins, Massachusetts. *Quaternary Research*, **34**, 148–159.

Newman, W.A. & Mickelson, D.M. (1994). Genesis of Boston Harbor drumlins, Massachusetts. *Sedimentary Geology*, **91**, 333–343.

Oldale, R.N. & Coleman, S.M. (1992). On the age of the penultimate full glaciation of New England. *In*: Clark, P.U. & Lea, P.D. (Eds), *The Last Interglacial-Glacial Transition in North America*. Geological Society of America Special Paper 270, 163–170.

Pair, D.L. (1997). Thin film, channelized drainage, or sheetfloods beneath a portion of the Laurentide Ice Sheet; an examination of glacial erosion forms, northern New York State, USA. *Sedimentary Geology*, **111**, 199–215.

Patterson, C.J. (1994). Tunnel-valley fans of the St. Croix moraine, east-central Minnesota, U.S.A. *In*: Warren, W.P. & Croot, D.G. (Eds), *Formation and Deformation of Glacial Deposits*. Rotterdam, A.A. Balkema, 59–87.

Patterson, C.J. (1998). Laurentide glacial landscapes: The role of ice streams. *Geology*, **26**, 643–646.

Péwé, T.L. (1983). The periglacial environment in North America during Wisconsin time. *In*: Wright, H.E. & Porter, S.C. (Eds), *Late Quaternary Environments of the United States*. *In*: Porter, S.C. (Ed.), *The Late Pleistocene*. Minneapolis, Minnesota Press, **1**, 157–189.

Piotrowski, J.A., Mickelson, D.M., Tulaczyk, S., Krzyszkowski, D. & Junge, F.W. (2001). Were deforming subglacial beds beneath past ice sheets really widespread? *Quaternary International*, **86**, 139–150.

Raymond, C., Johannesson, T., Pfeffer, T. & Sharp, M. (1987). Propagation of a glacier surge into stagnant ice. *Journal of Geophysical Research*, **92**, 9037–9049.

Reed, E.C. & Dreezan, V.H. (1965). Revision of the classification of the Pleistocene deposits in Nebraska. *Nebraska Geological Survey Bulletin*, **23**, 65 pp.

Retelle, W.J. & Weddle, T.K. (2001). Deglacial history and relative sea-level changes. Northern New England and Adjacent Canada: Geological Society of America, Special Paper 351, 292 pp.

Richmond, G.M. & Fullerton, D.S. (1986). Summation of Quaternary glaciations in the United States. *In*: Sibrava, V., Bowen, D.Q. & Richmond, G.M. (Eds), *Quaternary Glaciations in the Northern Hemisphere*. Oxford, Pergamon Press, 183–196.

Ridge, J.C. & Larsen, F.D. (1990). Reevaluation of Antev's New England varve chronology and new radiocarbon dates of sediments from glacial Lake Hitchcock. *Geological Society of America Bulletin*, **102**, 889–899.

Ridge, J.C., Canwell, B.A., Kelly, M.A. & Kelley, S.Z. (2001). Atmospheric ^{14}C chronology for late Wisconsinan deglaciation and sea-level change in eastern New England using varve and paleomagnetic records. *In*: Retelle, W.J. & Weddle, T.K. (Eds), *Deglacial History and Relative Sea-Level Changes, Northern New England and Adjacent Canada*. Geological Society of America Special Paper 351, 171–190.

Ridky, R.W. & Bindschadler, R.A. (1990). Reconstruction and dynamics of the Late Wisconsin "Ontario" ice dome in the Finger Lakes region, New York. *Geological Society of America Bulletin*, **102**, 1055–1064.

Rovey, C.W. & Keane, W.F. (1996). Pre-Illinoian tills in north-central Missouri. *Quaternary Research*, **45**, 17–29.

Shackelton, N.J. & Opdyke, N.D. (1973). Oxygen isotope and paleomagnetic stratigrpahy of eequatorial Pacific core V28–238: Oxygen isotope temperatures and ice volumes on a 105 ka-year time scale. *Quaternary Research*, **3**, 39–55.

Shaw, J. (1989). Drumlins, subglacial metltwater floods, and ocean responses. *Geology*, **17**, 853–856.

Shaw, J. & Gilbert, R. (1990). Evidence for large scale subglacial meltwater flood events in southern Ontario and northern New York State. *Geology*, **18**, 1169–1172.

Shaw, J., Kvill, D. & Rains, B. (1989). Drumlins and catastrophic subglacial floods. *Sedimentary Geology*, **62**, 177–202.

Shaw, J. & Sharpe, D.R. (1987). Drumlin formation by subglacial meltwater erosion. *Canadian Journal of Earth Science*, **24**, 2316–2322.

Shoemaker, E.M. (1992a). Water sheet outburst floods from the Laurentide Ice Sheet. *Canadian Journal of Earth Sciences*, **29**, 1250–1264.

Shoemaker, E.M. (1992b). Subglacial floods and the origin of low relief ice sheet lobes. *Journal of Glaciology*, **38**, 105–112.

Shoemaker, E.M. (1999). Subglacial water-sheet floods, drumlins and ice-sheet lobes. *Journal of Glaciology*, **45**, 201–213.

Shreve, R.L. (1985). Late Wisconsin ice-surface profile calculated from esker paths and types, Katahdin esker system, Maine. *Quaternary Research*, **23**, 27–37.

Sibrava, V., Bowen, D.Q. & Richmond, G.M. (Eds) (1986). Quaternary Glaciations of the Northern Hemisphere. *Quaternary Science Reviews*, **5**, 514 pp.

Sirken, L. & Stuckenrath, R. (1980). The Portwashingtonian warm interval in the northern Atlantic Coastal Plain. *Geological Society of America Bulletin*, **91**, 332–336.

Soller, D.R. (1998). Map showing the thickness and character of Quaternary sediments in the glaciated United States east of the Rocky Mountains: W.S. Geological Survey, Miscellaneous Investigations Series, Map I-1970 A and B, Scale 1:1,000,000.

Soller, D.R. & Packard, P.H. (1998). Digital representation of a map showing the thickness and character of Quaternary sediment in the glaciated United States east of the Rocky Mountains.U.S. Geological Survey Digital Data Series DDS-38.

Stalker, A. M. (1960). Ice-pressed drift forms and associated deposits in Alberta. *Geological Survey of Canada Bulletin*, **57**, 38 pp.

Stanford, S.D., Harper, D.P. & Stone, B.D. (1998). Surficial Geology of the Hamburg Quadrangle, Sussex County, New Jersey: New Jersey Geological Survey, Geologic Map Series 98–1, Scale 1 to 24,000.

Stanford, S.D. & Mickelson, D.M. (1985). Till fabric and deformational structures in drumlins near Waukesha, Wisconsin, U.S.A. *Journal of Glaciology*, **31**, 220–228.

Stone, B.D. & Borns, H.W., Jr. (1986). Pleistocene glacial and interglacial stratigraphy of New England, Long Island, and adjacent Georges Bank and Gulf of Maine. *Quaternary Science Reviews*, **5**, 39–52.

Stuvier, M. & Grootes, P.M. (2000). GISP2 oxygen isotope ratios. *Quaternary Research*, **53**, 277–284.

Stuvier, M. & Reimer, P.J. (1993). Extended ^{14}C database and revised CALIB radiocarbon calibration program. *Radiocarbon*, **35**, 215–230.

Stuiver, M., Reimer, P.J., Bard, E., Beck, J.W., Burr, G.S., Hughen, K.A., Kromer, B., McCormac, F.G., v. d. Plicht, J. & Spurk, M. (1998a). INTCAL98 Radiocarbon age calibration 24,000–0 cal BP. *Radiocarbon*, **40**, 1041–1083.

Stuiver, M., Reimer, P.J. & Braziunas, T.F. (1998b). High-precision radiocarbon age calibration for terrestrial and marine samples. *Radiocarbon*, **40**, 1127–1151.

Szabo, J. (1992). Reevaluation of early Wisconsinan stratigraphy of northern Ohio. *In*: Clark, P.U. & Lea, P.D. (Eds), *The Last Interglacial-Glacial Transition in North America*. Geological Society of America Special Paper 270, 99–107.

Thelien, G.P., & Pike, R.J. (1989). *Digital shade relief map of the conterminous United States*. United States Geological Survey, scale 1:2,000,000.

Thompson, W.B. (1999). Surficial Geology of the Fryeburg 7.5' Quadrangle, Cumberland County, Maine: Augusta, Maine Geological Survey, Surficial Geology Report 99–8, 20 pp. and accompanying map, 99–7 (1:24,000).

Totten, S.M. (1988). Glacial geology of Medina County, Ohio. *Ohio Geological Survey Report of Investigations*, **141**, 38 pp., plus color map (1 inch equals about 1 mile).

Vela, J.C. (1994). Rheological testing of subglacial till material [M.S. Thesis]. Pullman, Washington State University, 189 pp.

Walters, J.C. (1994). Ice-wedge casts and relict polygonal patterned ground in north-east Iowa, U.S.A. *Permafrost and Periglacial Processes*, **5**, 269–282.

Whillans, I.M., Bolzan, J. & Shabtaie, S. (1987). Velocity of ice streams B and C, Antarctica. *Journal of Geophysical Research*, **92**, 8895–8902.

Whittecar, G.R. & Mickelson, D.M. (1979). Composition, internal structure, and a hypothesis of formation for drumlins, Waukesha County, Wisconsin, U.S.A. *Journal of Glaciology*, **22**, 357–371.

Wright, H.E., Jr. (1973). Tunnel valleys, glacial surges, and subglacial hydrology of the Superior Lobe, Minnesota. *Geological Society of America Memoir*, **136**, 251–276.

Wright, H.E. (1983). *The late quaternary of the United States*. Minneapolis, University of Minnesota Press, 2 vols, 277 plus 407 pp.

Wright, H.E., Jr. & Frey, D.G. (1965). *The Quaternary of the United States*. New Jersey, Princeton University Press, 922 pp.

The Cordilleran Ice Sheet

Derek B. Booth[1], Kathy Goetz Troost[1], John J. Clague[2] and Richard B. Waitt[3]

[1] *Department Earth & Space Sciences, University of Washington, Box 531310, Seattle, WA 98195, USA (206)543-7923 Fax (206)685-3836.*
[2] *Department of Earth Sciences, Simon Fraser University, Burnaby, British Columbia, Canada*
[3] *U.S. Geological Survey, Cascade Volcano Observatory, Vancouver, WA, USA*

Introduction

The Cordilleran Ice Sheet, the smaller of two great continental ice sheets that covered North America during Quaternary glacial periods, extended from the mountains of coastal south and southeast Alaska, along the Coast Mountains of British Columbia, and into northern Washington and northwestern Montana (Fig. 1). To the west its extent would have been limited by declining topography and the Pacific Ocean; to the east, it likely coalesced at times with the western margin of the Laurentide Ice Sheet to form a continuous ice sheet over 4,000 km wide. Because most of the marginal environments of the Cordilleran Ice Sheet were not conducive to preserving an extensive depositional record, much of our understanding of this ice sheet has come from limited areas where preservation is good and access unencumbered, notably along its lobate southern margin in northern Washington State and southern British Columbia.

Arrival of geologists into Puget Sound late in the 19th century initiated study of the Cordilleran Ice Sheet. The landscape displayed unmistakable evidence of past glaciations, but a sporadic sequence of deposits along valley walls and coastal bluffs only hinted at a long and intricate history of ice-sheet occupations. By the mid-20th century, extensive field studies had developed a framework for Pacific Northwest Quaternary history. Evidence of four glaciations, summarized by Crandell (1965) and detailed by Armstrong *et al.* (1965), Mullineaux *et al.* (1965), and Crandell (1963), followed the precedent from the American Midwest: four continental-scale glaciations, correlated across broad regions. In the Pacific Northwest, the youngest ice-sheet glaciation (Fraser) was constrained by radiocarbon dates and correlated with the Wisconsin glaciation of the mid-continent. Earlier glaciations (given the local names *Salmon Springs*, *Stuck*, and *Orting*) were identified only in the southeastern Puget Lowland. Crandell (1965) suggested that they spanned early through late Pleistocene time.

In the latter part of the 20th century, improved understanding of global and regional stratigraphy, and emphasis on geomorphic processes, have provided a new context for studies of the Cordilleran Ice Sheet. These advances are the topics of this chapter. The record of global warming and cooling recorded in deep-sea cores shows that there were many glaciations during the Quaternary Period, not just four. Global perspectives on past sea-level variations prove critical to understanding tidewater glacier systems like the southwestern part of the Cordilleran Ice Sheet. New dating techniques yield crude but consistent chronologies of local and regional sequences of alternating glacial and nonglacial deposits. These dates secure correlations of many widely scattered exposures of lithologically similar deposits and show clear differences among others.

Besides improvements in geochronology and paleoenvironmental reconstruction (i.e. glacial geology), glaciology provides quantitative tools for reconstructing and analyzing any ice sheet with geologic data to constrain its physical form and history. Parts of the Cordilleran Ice Sheet, especially its southwestern margin during the last glaciation, are well suited to such analyses. The approach allows interpretation of deposits and landforms at the now-exposed bed of the former ice sheet, and it also suggests likely processes beneath other ice sheets where reconstructions are less well-constrained.

Finally, expressions of the active tectonics of western North America are now widely recognized across the marginal zone of the Cordilleran Ice Sheet. Such conditions were little appreciated at mid-century. Only since the 1980s have the extent and potential influence of recent tectonics on the landscape of western Washington been appreciated. The regional setting for repeated glaciations owes much of its form to those tectonic influences; conversely, deformation and offset of ice-sheet deposits may be critical in unraveling the Quaternary expression of the region's tectonics.

Perhaps the greatest development in recent study of the Cordilleran Ice Sheet, especially its southwestern boundary, has been the scientific attention focused on this region – not only by geoscientists but also by resource managers, land-use planners, and the general public. In the last several decades, this glacial landscape has become a region of rapid population growth. In part because of these social pressures, the level of scientific study here has rapidly increased, which will likely render the story of the Cordilleran Ice Sheet presented in this synoptic paper even more quickly outdated than its predecessors.

Chronology and the Stratigraphic Record

Quaternary Framework

More than one hundred years after Bailey Willis published "Drift Phenomena of Puget Sound" (1898), geologists continue efforts to identify and correlate the Quaternary stratigraphic units across the area episodically covered by the southern part of the Cordilleran Ice Sheet (Fig. 1). Nearly

DEVELOPMENTS IN QUATERNARY SCIENCE
VOLUME 1 ISSN 1571-0866
DOI:10.1016/S1571-0866(03)01002-9

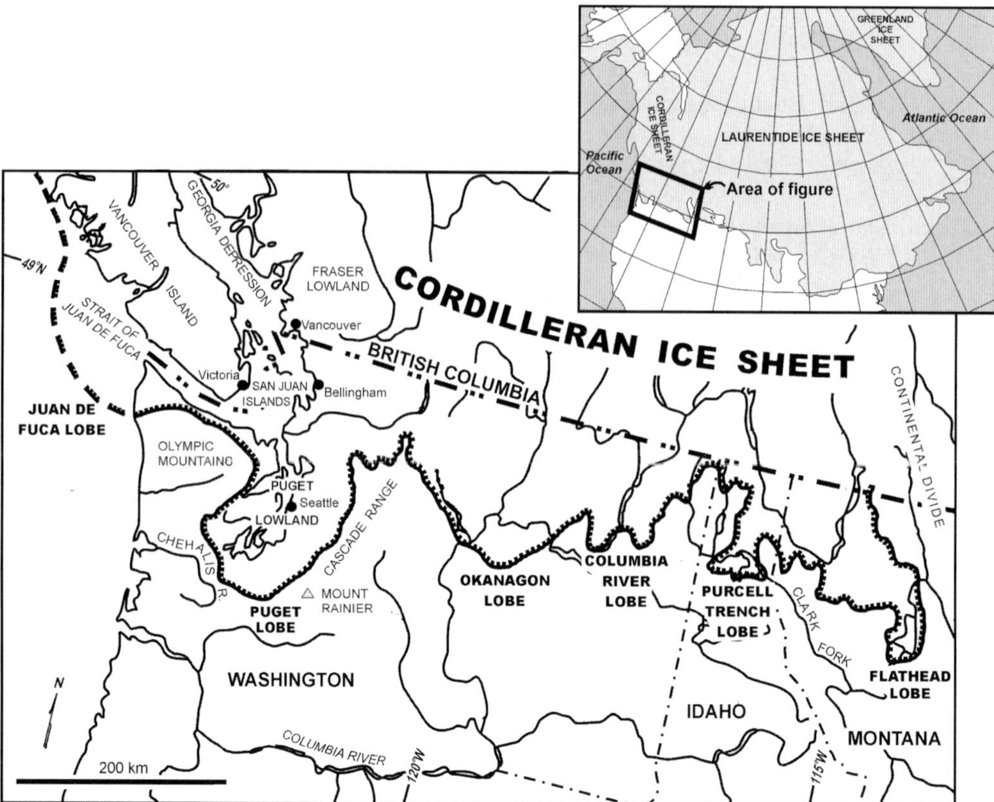

Fig. 1. Map of southern extent and lobes of the latest Pleistocene advance of the Cordilleran Ice Sheet in Washington and British Columbia.

a half century of field investigations in the southern Puget Lowland (Armstrong *et al.*, 1965; Crandell *et al.*, 1958; Mullineaux *et al.*, 1965; Noble & Wallace, 1966; Waldron *et al.*, 1962) and in the northern Puget Lowland (Clague, 1981; Easterbrook, 1986, 1994) show that ice sheets have advanced south into the lowlands of western Washington at least six times. The global climatic template of the marine-isotope record illustrates the likely number and frequency of glacier advances. It suggests that the current half-dozen known glacier advances do not include every advance into the region in the last 2.5 million years. The last three ice advances correlate with marine oxygen isotope stages (MIS) 2, 4, and probably 6 (Fig. 2). The most recent advance was the Fraser glaciation, discussed later in this chapter.

Little is known about the climate in the lowlands of southern British Columbia and western Washington during most of the Pleistocene. Recent research has focused on either MIS 2 (Hansen & Easterbrook, 1974; Heusser, 1977; Heusser *et al.*, 1980; Hicock *et al.*, 1999; Mathewes & Heusser, 1981; Whitlock & Grigg, 1999), MIS 2 and 3 (Barnosky, 1981, 1985; Grigg *et al.*, 2001; Troost, 1999), or MIS 5 (Heusser & Heusser, 1981; Muhs *et al.*, 1994; Whitlock *et al.*, 2000). From these studies we know climate during MIS 3 was cooler than today and sea level was lower. The climate of MIS 5 was similar to today's, with marine deposits commonly found slightly above and up to 60 m below modern sea level (Shackelton *et al.*, 1990).

Recognition of nonglacial environments in the depositional record is essential to unraveling the chronology here. The present Puget Lowland may be a useful analog for earlier nonglacial periods. Areas of nondeposition, soil formation, or minor upland erosion dominate most of the lowland (Fig. 3). Sediment is only accumulating in widely separated river valleys and lake basins, and in Puget Sound. Were the present lowland again invaded by glacier ice, it would bury a complex and discontinuous nonglacial stratigraphic record. Thick sedimentary sequences would pinch out abruptly against valley walls. Sediment deposited in valleys could be 100 m lower than coeval upland sediment or organic-rich paleosols. Thus, the thickness and lateral continuity of nonglacial sediment of any one nonglacial interval will be highly variable owing to the duration of the interval, subsidence and uplift rates, and the altitude and surface topography of fill left by the preceding glacier incursion (Troost, 1999).

West of the Cascade Range, Cordilleran glaciations were typified by the damming of a proglacial lake in the Puget Sound basin, the spreading of an apron of outwash, deep subglacial scouring and deposition of till, formation of large recessional outwash channels, formation of ice-contact terrain, and deposition of glaciomarine drift in the northern lowland. Glacial periods were marked by a change to cold-climate vegetation and increased deposition and erosion. Thick glaciomarine, glaciolacustrine, and outwash deposits accumulated in proglacial and subglacial troughs, capped

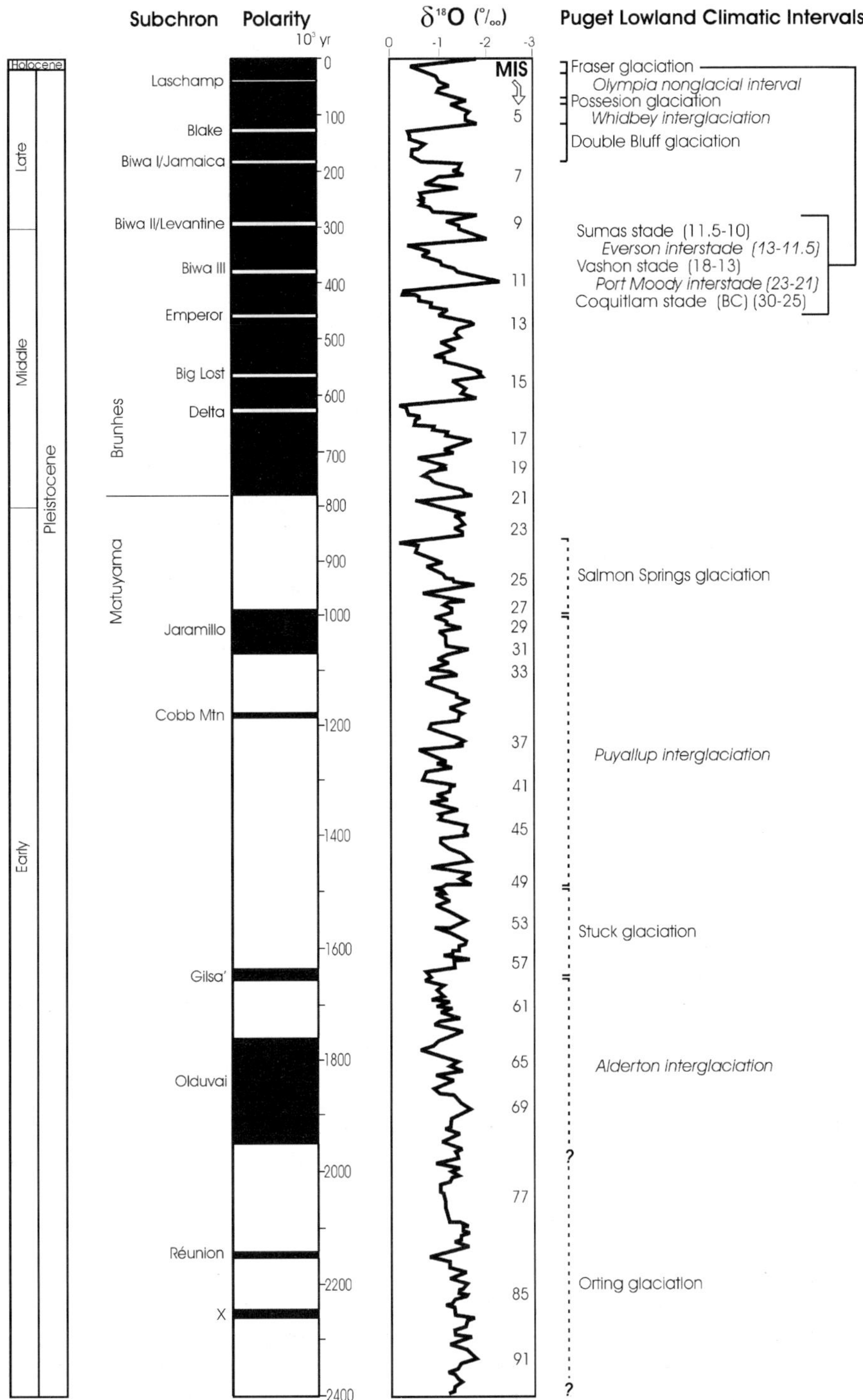

Fig. 2. Comparison of the marine oxygen-isotope curve stages (MIS) using the deep-sea oxygen-isotope data for ODP677 from Shackelton et al. (1990), global magnetic polarity curve (Barendregt, 1995; Cande & Kent, 1995; Mankinnen & Dalrymple, 1979), and ages of climatic intervals in the Puget and Fraser lowlands. Ages for deposits of the Possession glaciation through Orting glaciation from Easterbrook et al. (1981), Easterbrook (1986), Blunt et al. (1987), and Easterbrook (1994). Additional ages for deposits of the Puyallup Interglaciation from R.J. Stewart (pers. comm., 1999). Ages for the Olympia nonglacial interval from Armstrong et al. (1965), Mullineaux et al. (1965), Pessl et al. (1989), and Troost (1999). Ages for the Coquitlam stade from Hicock & Armstrong (1985); ages for the Port Moody interstade from Hicock & Armstrong (1981). Ages for the Vashon stade from Armstrong et al. (1965) and Porter & Swanson (1998). Ages for the Everson interstade from Dethier et al. (1995) and Kovanen & Easterbrook (2001). Ages for the Sumas stade from Clague et al. (1997), Kovanen & Easterbrook (2001), and Kovanen (2002).

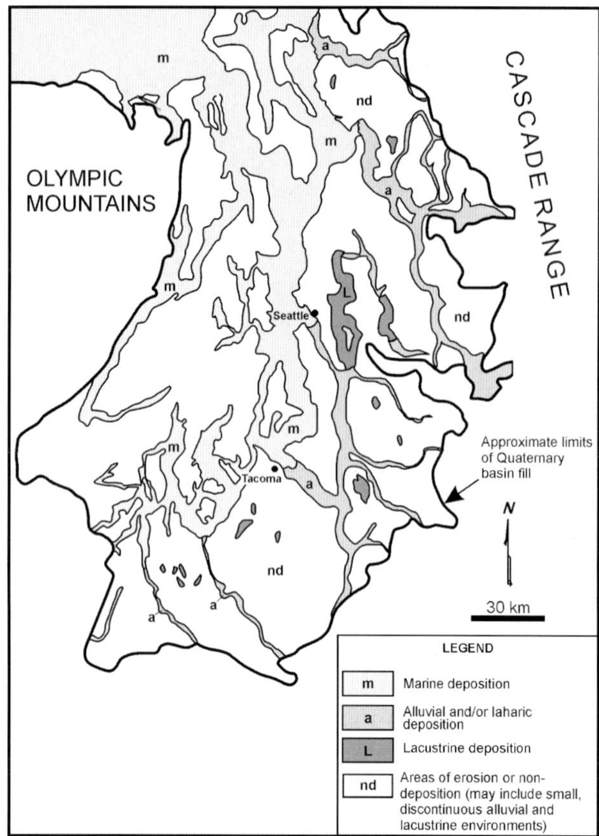

Fig. 3. Modern Puget Lowland depositional environments, providing one example of the extent of deposition during interglacial periods. Most of the land area is either erosional or non-depositional (except for minor upland soil formation). Modified from Borden & Troost (2001).

intermittently by subglacial till, predominantly of meltout origin. Likewise, subglacial drainage carved deep erosional troughs subsequently filled with postglacial volcanic debris flows and alluvium. Thus, there are many unconformities and buried topographies in the stratigraphic record.

Sediment lithology helps differentiate glacial from nonglacial deposits, given that source areas for glacial deposits are usually other than the headwaters of the current streams. This technique, although not new, is finding renewed use for proposing late Pleistocene glacier readvances (Kovanen & Easterbrook, 2001) and for interpreting bulk geochemistry analyses of central lowland deposits (Mahoney *et al.*, 2000).

Tectonic Setting

Plate movement of western North America governs the structural setting of the southwestern margin of the Cordilleran Ice Sheet. The Juan de Fuca plate (JDF) moves northeast and subducts beneath the North America plate at about 4 cm per year (Fig. 4a). From strike-slip plate movement farther south and crustal extension across the Basin and Range province, a series of crustal blocks between northern Oregon and southern British Columbia are colliding against the relatively fixed buttress of Canada's Coast Mountains (Wells *et al.*, 1998). The region is shortening N-S by internal deformation of the blocks and by reverse faulting along block boundaries.

Both the bedrock and overlying Quaternary sediment in the Puget Lowland have been deformed by faults and folds as a result of this tectonic activity. The Seattle fault is one of several active structures of the Puget Lowland showing displacement in the last 10,000 years. It separates the Seattle basin from the Seattle uplift, two of the structural blocks involved in the shortening in Oregon and Washington (Fig. 4b). Its displacement history embraces about 8 km of south-side-up movement since mid-Tertiary time (Johnson *et al.*, 1994; Pratt *et al.*, 1997), including 7 m of uplift during a great earthquake 1,100 years ago (Atwater & Moore, 1992; Bucknam *et al.*, 1992). This fault may have moved several times in the last 15,000 years; episodic movement throughout the Quaternary is likely, although not yet documented. Current investigations suggest that a similar fault may pass west-northwest near Commencement Bay at Tacoma (Brocher *et al.*, 2001). Other faults trending east-west or southeast-northwest cross the glaciated lowlands both north and south of the Seattle fault (Johnson *et al.*, 1996, 2001; Pratt *et al.*, 1997), with likely displacements of meters to tens of meters, thereby complicating interpretation of the Quaternary stratigraphic record.

Evidence of Pre-Fraser History and Depositional Environments

Puget Lowland

Abundant but fragmentary evidence of pre-Fraser glacial and interglacial deposition in the Puget Lowland exists in many geologic units named and described at type sections (Table 1, Figs 2 and 5). Because the evidence is scattered and discontinuous, reconstructions of pre-Fraser depositional environments and climate are sparse. Only the two latest nonglacial periods (MIS 3 and 5) are well known through abundant organic-bearing sediments and good exposures.

Evidence of nonglacial deposition during MIS 3 (broadly coincident with the Olympia nonglacial interval, defined by Armstrong *et al.*, 1965) has been found in bluff exposures and boreholes across the Puget Lowland. These deposits accumulated between about 70,000 yr ago and 15,000 ^{14}C yr B.P.; although a time-stratigraphic unit, Olympia deposits also have a defined type section at Fort Lawton in Seattle (Mullineaux *et al.*, 1965). During MIS 3, most of the lowlands of Washington were ice-free, allowing for subaerial deposition and weathering. Deposits of the Olympia nonglacial interval (named informally the Olympia beds in western Washington and the Cowichan Head Formation in southwestern British Columbia) consist of peat, tephra, lahars, mudflows, lacustrine, and fluvial deposits (Fig. 6). Dozens of radiocarbon dates from this interval confirm nonglacial conditions from about 15,000 ^{14}C yr B.P. to beyond the limit of radiocarbon dating (ca. 40–45,000 ^{14}C yr B.P.) (Borden &

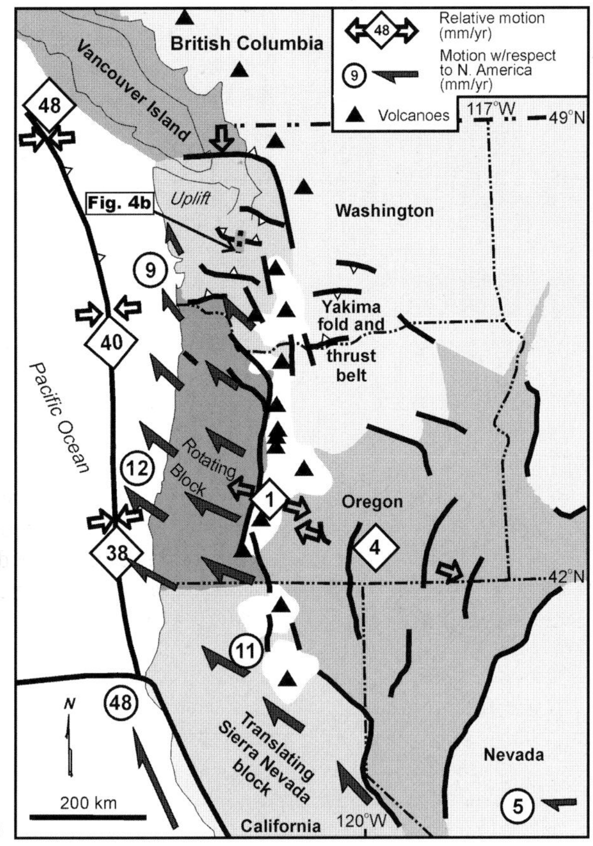

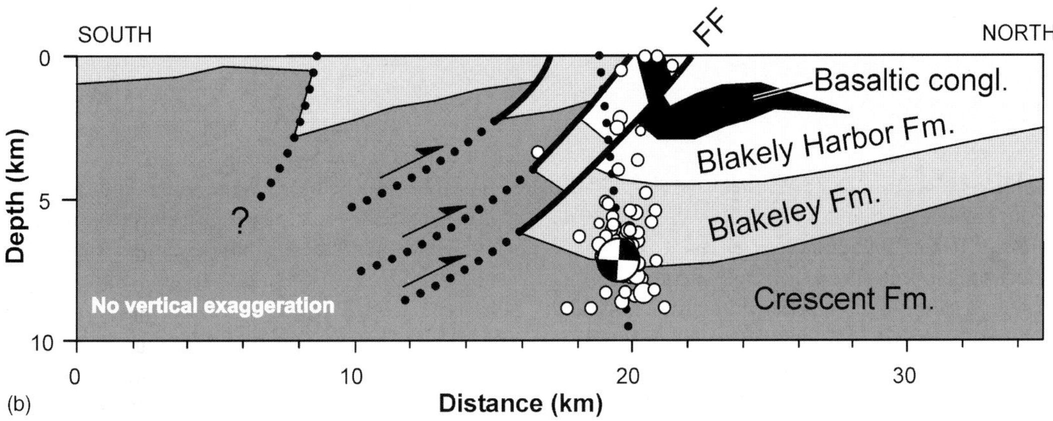

Fig. 4. Crustal blocks and major structures in the Puget Lowland, showing the north-verging compressional motion and the resultant displacement across the Seattle fault zone. Fig. 4a shows the relative motion of the western United States as transferred to western Washington (modified from Wells et al., 1998). Fig. 4b interprets the Seattle fault zone as a series of south-dipping reverse faults (FF = frontal fault; modified by B. Sherrod [USGS] from Johnson et al., 1994).

Troost, 2001; Clague, 1981; Deeter, 1979; Hansen & Easterbrook, 1974; Troost, 1999). Paleoecological analyses in the Puget Lowland indicate a wide range of paleoenvironments during the Olympia interval. Many locations of Olympia beds yield well-preserved pollen preservation with a predominance of pine and spruce; freshwater diatomites suggesting clear, shallow lakes and large littoral areas; and macrofossils including mammoth teeth and tusks, *Pinus contorta* type cones and needles, branches, leaf prints, and in situ tree roots (Troost, 2002).

As expected with deposition during nonglacial periods, Olympia beds vary in thickness, elevation, grain size, and composition over short distances. Topographic relief on the basal unconformity of the Olympia beds near Tacoma exceeds 230 m, 60 m of which lies below modern sea level. The thickest exposures of Olympia beds (>25 m) include multiple

Table 1. Sources of age data for Puget Lowland stratigraphic units.

Name (Climatic Intervals in Italics)	Type Section Location	Reference for Nomenclature	Reported Age (in 10^3 Years)	Type of Date	Location	Reference for Age	Comment
Sumas glaciation	Near Sumas, Canadian side	Armstrong (1957)	na	na	na	na	na
Sumas Drift		Easterbrook (1963)	11.3–10.0 and pre 11.9	^{14}C yr B.P.	Aldergrove–Fort Langley and Chilliwack R. valley, B.C.	Clague *et al.* (1997)	New dates
			11.5–10.0	^{14}C yr B.P.	Multiple locations, Fraser Lowland and Nooksack valley	Kovanen & Easterbrook (2001), Kovanen (2002)	Compilation and new dates; table of 69 dates on the Sumas interval
Sumas stade		Armstrong *et al.* (1965)	na	na	na	na	na
Everson Glaciomarine Drift, *Everson interstade*	Upstream of Everson, on the Nooksack R.	Armstrong *et al.* (1965)	13.0–11.0	^{14}C yr B.P.	Type section	Armstrong *et al.* (1965)	New dates
			13.0–11.5	^{14}C yr B.P.	Whidbey Is. to Campbell R.	Kovanen & Easterbrook (2001)	Compilation and new dates
			13.6–11.3	^{14}C yr B.P.	Northern Puget Lowland	Dethier *et al.* (1995)	New dates
	Southeast of Cedarville on the Nooksack R.	Easterbrook (1963)	na	na	na	na	Includes the Kulshan glaciomarine drift, Deming Sand, and Bellingham glaciomarine drift
Vashon till, *Vashon glaciation*	Vashon Island	Willis (1898)	>13.5	na	na	Rigg & Gould (1957)	Youngest limiting age
Vashon Drift, *Vashon stade*		Armstrong *et al.* (1965)	25.0–13.5	^{14}C yr B.P.	Multiple, Strait of Georgia to Lake Washington	Armstrong *et al.* (1965)	New dates
			18.0–13.0	^{14}C yr B.P.	Fraser Lowland	Kovanen & Easterbrook (2001)	Compilation
			16.0–13.5	^{14}C yr B.P.	Seattle, Bellevue, Issaquah	Porter & Swanson (1998)	Compilation and new dates
Steilacoom Gravel	Steilacoom plains	Willis (1898), Bretz (1913), Walters & Kimmel (1968)	Younger than 13.5	^{14}C yr B.P.	Ft. Lewis, Tacoma	Borden & Troost (2001)	Multiple, young, sub–Vashon dates
Esperance Sand Member of Vashon Drift	Fort Lawton, Seattle	Mullineaux *et al.* (1965)	15.0–13.5; 15.0–14.5	^{14}C yr B.P.	Seattle; Issaquah	Mullineaux *et al.* (1965), Porter & Swanson (1998)	Limiting ages
Lawton Clay Member of Vashon Drift	Fort Lawton, Seattle	Mullineaux *et al.* (1965)	15.0–13.5; 15.0–14.5	^{14}C yr B.P.	Seattle; Issaquah	Mullineaux *et al.* (1965), Porter & Swanson (1998)	Limiting ages

Port Moody nonglacial deposits	Port Moody	Hicock *et al.* (1982)	23.0–21.0	^{14}C yr B.P.		Hicock & Armstrong (1981)	New dates
Port Moody interstade					na	Hicock & Armstrong (1985)	Interstade informally introduced
Coquitlam Drift	Coquitlam–Port Moody	Hicock (1976)	21.7–18.7	^{14}C yr B.P.	Type section	Hicock & Armstrong (1981)	New dates
Coquitlam stade		Hicock & Armstrong (1985)	30.0–25.0	^{14}C yr B.P.	Multiple locations	Hicock & Armstrong (1985)	Compilation of 52 dates
			26.0–17.8	^{14}C yr B.P.	Multiple locations	Clague (1980), Armstrong *et al.* (1985)	Equivalent to Evans Creek stade?
Evans Creek Drift, *Evans Creek stade*	Carbon River valley, near mouth of Evans Creek	Crandell (1963)	25.0–15.0	^{14}C yr B.P.	Type section	Armstrong *et al.* (1965)	Alpine glaciation in Cascade Range
		Armstrong *et al.* (1965) (Crandell)	na	na	na	na	na
Olympia interglaciation	Fort Lawton	Armstrong *et al.* (1965)	35.0–15.0	^{14}C yr B.P.	Fort Lawton and multiple locations in WA and BC	Armstrong *et al.* (1965), Troost (1999)	Compilation and new dates; may be partly equivalent to Quadra sediments at Point Grey in Vancouver (Armstrong & Brown, 1953)
			24.0–15.0	^{14}C yr B.P.	Fort Lawton and West Seattle	Mullineaux *et al.* (1965)	New dates
Olympia beds		Minard & Booth (1988)	>45–13.5	^{14}C yr B.P.	Multiple locations around Seattle and Tacoma	Troost (1999), Borden & Troost (2001)	New dates
Possession Drift	Possession Point, Whidbey Island	Easterbrook *et al.* (1967)	80	Amino acid	Multiple locations	Easterbrook & Rutter (1981)	New dates
Whidbey Formation	Double Bluff, Whidbey Island	Easterbrook *et al.* (1967)	107–96, avg = 100	Amino acid	Multiple locations	Easterbrook & Rutter (1981)	New dates
			151–102	Thermo–luminescence		Easterbrook (1994)	

Table 1. (Continued)

Name (Climatic Intervals in Italics)	Type Section Location	Reference for Nomenclature	Reported Age (in 10^3 Years)	Type of Date	Location	Reference for Age	Comment
Double Bluff Drift	Double Bluff, Whidbey Island	Easterbrook *et al.* (1967)	250–150	Amino acid	Type section	Easterbrook & Rutter (1982)	New dates
			178–111	Amino acid		Blunt *et al.* (1987)	na
			291–177	Thermo–luminescence		Easterbrook *et al.*, 1992	na
Salmon Springs Drift	Near Sumner	Crandell *et al.* (1958)	1000	Inferred, based on Lake Tapps	Type section	Easterbrook (1994)	Reversely magnetized (Easterbrook, 1986)
Lake Tapps Tephra	Near Sumner	Crandell (1963), Easterbrook & Briggs (1979)	840	Fission track	3 locations	Easterbrook & Briggs (1979)	Correlation of other locations to type section based on chemistry
			1000	Fission track	Multiple locations	Westgate *et al.* (1987)	na
Puyallup interglaciation, Puyallup Sand	Near Alderton	Willis (1898)	1690–1640	Laser–argon	Type section	Easterbrook *et al.* (1992)	New date; reversely magnetized (Easterbrook, 1986)
Puyallup Formation		Crandell *et al.* (1958)	na	na	na	na	na
Stuck Drift	Near Alderton	Crandell *et al.* (1958)	Close to 1600	Based on bounding ages	Type section	Easterbrook (1994)	Reversely magnetized (Easterbrook, 1986)
Alderton Formation	Near Alderton	Crandell *et al.* (1958)	2400–1000, avg = 1600	Laser–argon	Type section	Easterbrook (1994)	Reversely magnetized (Easterbrook, 1986)
Orting Gravel, Orting Drift	Orting	Willis (1898), Crandell *et al.* (1958)	2000 (?)	Inferred	Type section	Easterbrook (1986), Easterbrook *et al.* (1988)	Reversely magnetized (Easterbrook, 1986)

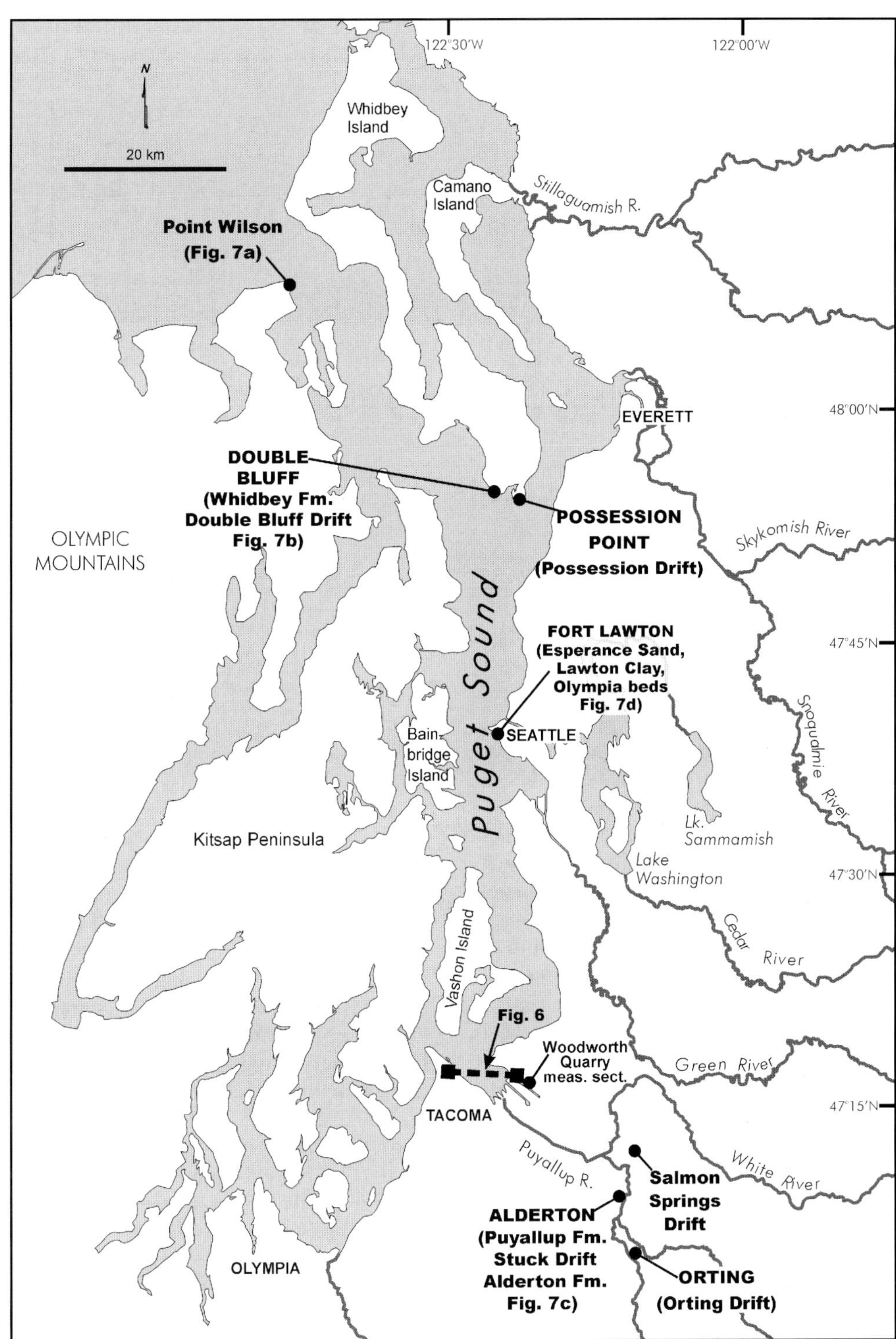

Fig. 5. Locations of type sections for the recognized pre-Fraser stratigraphic units in the Puget Lowland. Locations of cross section of Fig. 6 and measured sections in Fig. 7 are also shown. The Olympia nonglacial interval was first defined by Armstrong et al. (1965) with its type section at Fort Lawton (Mullineaux et al., 1965). The Possession Drift, Whidbey Formation, and Double Bluff units were named and described by Easterbrook et al. (1967, 1981). The Salmon Springs and older drifts were first described by Willis (1898) and formally named by Crandell et al. (1958).

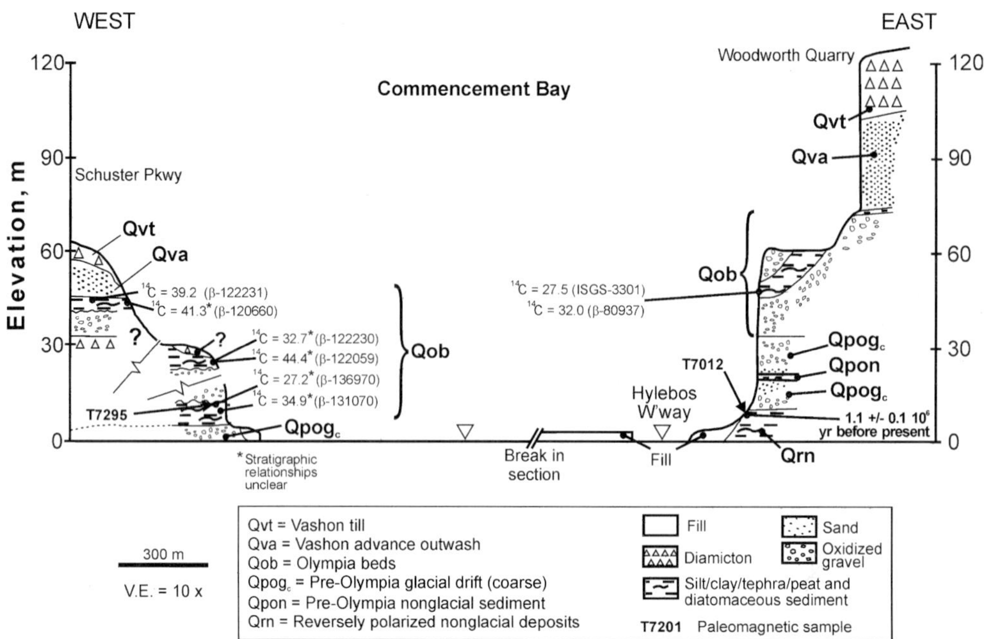

Fig. 6. East-west cross section through Commencement Bay near Tacoma, showing radiocarbon dates and topographic relief within the Olympia beds (unit Qob). Reversely magnetized nonglacial volcanic-rich deposits yield a zircon fission-track age of 1.1 × 10^6 years (modified from Troost et al., 2003). Unlabeled numbers are ^{14}C ages in 10^3 ^{14}C yr B.P.

tephra, lahar, peat, and diatomite layers (Troost *et al.*, 2003). At least five discontinuous Olympia-age tephras and lahars have been identified near Tacoma, with source areas including Mt. St. Helens and Mt. Rainier. Freshwater diatomites and in situ tree roots reveal lacustrine and forested environments across the lowland. Mastodons, mammoths, and bison roamed the Puget and Fraser lowlands during this nonglacial interval (Barton, 2002; Harrington *et al.*, 1996; Plouffe & Jette, 1997).

The next-oldest Pleistocene sediment in the Puget Lowland is the Possession Drift, probably related to glaciation during MIS 4 (Easterbrook, 1994) (Fig. 7a). The ice sheet responsible for this drift may have been less extensive than during MIS 2, according to reconstructions of global temperature. Away from the type section on Whidbey Island, pre-Fraser glacial deposits cannot be uniquely correlated with Possession Drift without age control. Thermoluminesence dating may prove most useful in this age range (Easterbrook, 1994), with preliminary results suggesting localities of Possession-age outwash south of the type section (Easterbrook, 1994; Mahan *et al.*, 2000).

The Whidbey Formation and its counterpart in British Columbia, the Muir Point Formation, correlate with MIS 5, the youngest full interglacial interval of the Pleistocene record. Climate was similar to that of today, with sea level perhaps slightly above today's level and as much as 60 m lower (Easterbrook, 1994; Easterbrook *et al.*, 1967). At its type section (Easterbrook, 1994) (Fig. 7b), the Whidbey Formation includes silt, sand, gravel, ash, and diatomite. On Whidbey Island, extensive sand deposits may be deltaic in origin. Like Olympia nonglacial deposits, sedimentary layers surely vary in thickness and composition over short distances; relief on the upper surface of the Whidbey Formation probably resembles today's landscape relief. Difficulties in dating sediments of this age, however, provide few constraints on the paleotopography from this time.

Still older mid- and early-Pleistocene deposits in the Puget Lowland include the Double Bluff Drift (Easterbrook, 1994) (Fig. 7b) and various unnamed glacial and interglacial deposits in the interval from 250,000 to 780,000 years ago, the existence of which are anticipated from climatic fluctuations expressed by the marine isotope record. Recent chronological and stratigraphic correlation efforts have begun to identify deposits in this age range and to confirm the presence of pre-Fraser deposits at locations away from their

→

Fig. 7. Measured sections at pre-Fraser localities on and near Whidbey Island and in the Puyallup River valley (reproduced from Easterbrook, 1994), and at Fort Lawton in Seattle. Fig. 7a depicts both the Possession Drift and the Whidbey Formation at Point Wilson. Fig. 7b shows the lithologies noted at the type locality of the Double Bluff Drift. Fig. 7c depicts the stratigraphic relationships between the Puyallup Formation, Stuck Drift, and Alderton Formation at the Alderton type locality; black dots depict reversely polarized samples. Fig. 7d shows the modern exposure at the type locality for deposits of the Olympia nonglacial interval, and for the Lawton Clay Member and Esperance Sand Member (the latter now generally mapped as Vashon advance outwash) of the Vashon Drift (Mullineaux et al., 1965).

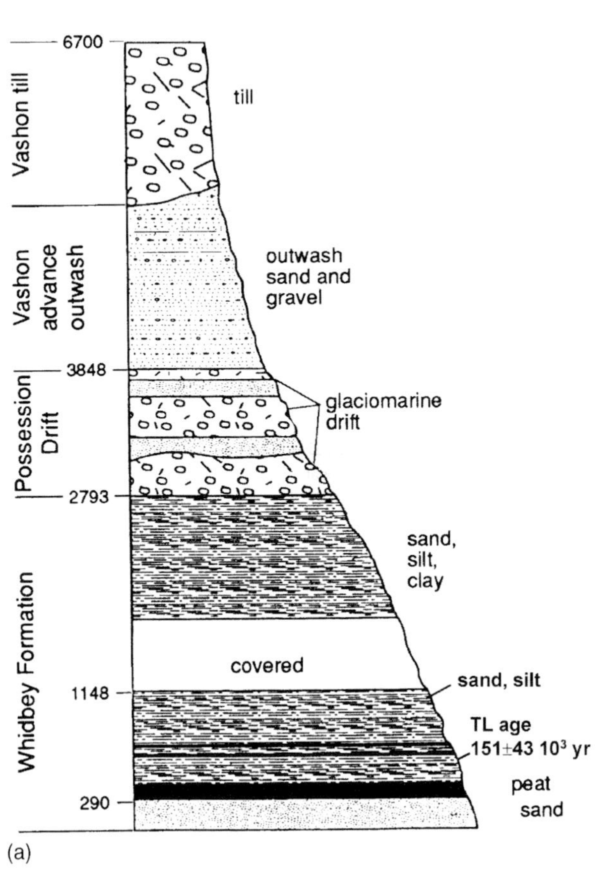
6700
3848
2793
1148
290
Vashon till
Vashon advance outwash
Possession Drift
Whidbey Formation
till
outwash sand and gravel
glaciomarine drift
sand, silt, clay
covered
sand, silt
TL age
151±43 10³ yr
peat
sand

(a)

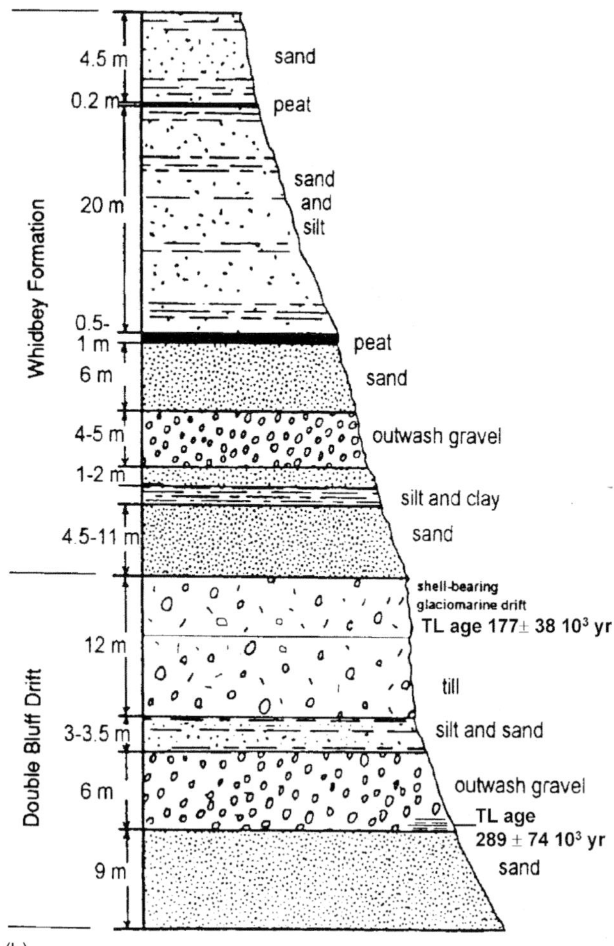
Whidbey Formation
Double Bluff Drift
4.5 m
0.2 m
20 m
0.5-1 m
6 m
4-5 m
1-2 m
4.5-11 m
12 m
3-3.5 m
6 m
9 m
sand
peat
sand and silt
peat
sand
outwash gravel
silt and clay
sand
shell-bearing glaciomarine drift
TL age 177± 38 10³ yr
till
silt and sand
outwash gravel
TL age
289 ± 74 10³ yr
sand

(b)

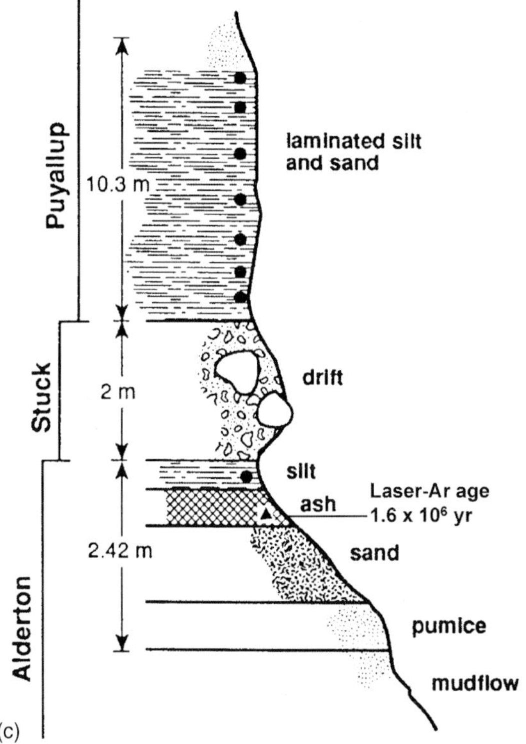
Puyallup
Stuck
Alderton
10.3 m
2 m
2.42 m
laminated silt and sand
drift
silt
ash
Laser-Ar age
1.6 x 10⁶ yr
sand
pumice
mudflow

(c)

Esperance Sand
(includes transitional zone with Lawton Clay)
Lawton Clay
Olympia beds
Landslide debris
Debris cones

(d)

Fig. 7.

type sections (Hagstrum *et al.*, 2002; Mahan *et al.*, 2000; Troost *et al.*, 2003). The oldest pre-Fraser deposits, about 1 million years old and older, are the Salmon Springs Drift, Puyallup Formation, Stuck Drift, Alderton Formation, and Orting Drift (Crandell, 1963; Westgate *et al.*, 1987) (Fig. 7c).

Eastern Washington

Discontinuous drift extending beyond the limits of Fraser-age drift in the Pend Oreille, Columbia, and Little Spokane valleys has stones that are highly weathered or deeply penetrated by cracks, has a slightly argillic soil, and overlies granite and gneiss bedrock that is highly decayed, even to grus. These characteristics indicate that the drift is pre-Fraser in age. Direct dating of pre-Fraser sediments is poor, but radiocarbon dates in Canada have been interpreted as denying the existence of an ice sheet between 65,000 and 25,000 yr B.P. (Clague, 1980), consistent with nonglacial conditions west of the Cascade Range during this time. The weathering of the drift and surrounding bedrock in places is so strong as to suggest an age very much older than late Wisconsin – equivalent to MIS 6 (160,000–130,000 years ago) or older. In northeastern Washington and adjacent Idaho, however, there is no objective basis for Richmond's (1986, Chart 1) assignment of any of these deposits to particular time intervals.

Probably there were several pre-Wisconsin Cordilleran ice-sheet glaciations in eastern Washington and farther east in Idaho and Montana. Glacial Lake Missoula and great floods from it are possible only when the Purcell Trench lobe advances far enough south (to 48°10′ N) to dam the Clark Fork of the Columbia. In southern Washington, deposits resembling Fraser-age Missoula-flood gravel bars but thickly capped by calcrete deeply underlie some of these Fraser deposits. One such gravel was dated to between 200,000 and 400,000 Th/U yr ago and another to before 780,000 Th/U yr ago (Bjornstad *et al.*, 2001).

Chronology of the Fraser Glaciation

The Cordilleran Ice Sheet most recently advanced out of the mountains of British Columbia about 25,000 ^{14}C yr B.P. It flowed west onto the continental shelf, east into the intermontaine valleys of British Columbia where it probably merged with the western edge of the Laurentide Ice Sheet, and south into the lowlands of Washington State (Fig. 8, Table 1). In southern British Columbia and western Washington the Puget lobe filled the Fraser Lowland and the Puget Lowland between the Olympic Mountains and Cascade Range. The Juan de Fuca lobe extended east along the Strait of Juan de Fuca to terminate some 100 km west of Washington's present coast. Several ice lobes east of the Cascade Range expanded south down the Okanogan Valley and down other valleys farther east. The Fraser-age ice-sheet maximum on both sides of the Cascade Range was broadly synchronous (Waitt & Thorson, 1983). It approximately coincided with the maximum advance of some parts of the Laurentide Ice Sheet in central North America at about 14,000 ^{14}C yr ago. but lagged several thousand years behind the culminating advance of most of the Laurentide Ice Sheet (Lowell *et al.*, 1999; Mickelson *et al.*, 1983; Prest, 1969).

Northern Puget Lowland/Southern Fraser Lowlands

The Fraser glaciation began about 25,000 ^{14}C yr B.P. with the expansion of alpine glaciers in the Coast Mountains of British

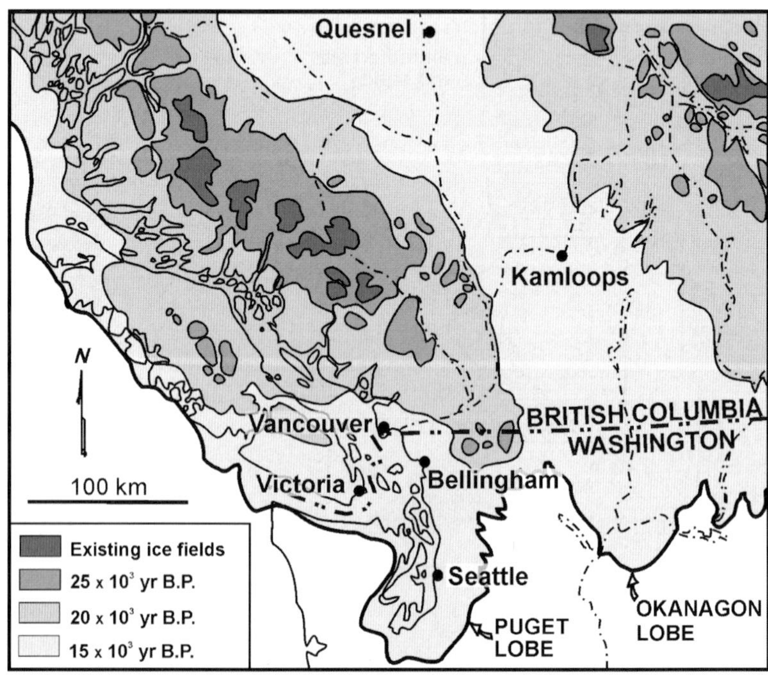

Fig. 8. Growth of the Cordilleran Ice Sheet during the Fraser glaciation (from Clague, 1981).

Columbia, the Olympic Mountains, and the Cascade Range of Washington. Glaciers in the Coast Mountains coalesced to form piedmont ice lobes that reached the Fraser Lowland of British Columbia about 21,000 ^{14}C yr B.P. during the Coquitlam stade (Hicock & Armstrong, 1981). The Coquitlam stade correlates with the Evans Creek stade of Washington, an early alpine phase of the Fraser glaciation in the Cascade Range (Armstrong *et al.*, 1965).

The Coquitlam stade was followed by a period of climatic amelioration that lasted from about 19,000 to 18,000 ^{14}C yr B.P. – the Port Moody interstade of Hicock & Armstrong (1985). The Port Moody interstade was in turn followed by the late Wisconsin advance of the Cordilleran Ice Sheet during the Vashon stade (Armstrong *et al.*, 1965). The Puget lobe advanced into northern Washington about 17,000 yr B.P. (Clague, 1981; Easterbrook, 1986) and retreated rapidly from its maximum position around 14,000 yr B.P. (Clague, 1981; Easterbrook, 1986; Porter & Swanson, 1998).

The Vashon stade was followed by a period of rapid and extensive glacier retreat (Everson interstade) that ended with a resurgence of the southwestern margin of the Cordilleran Ice Sheet in the Fraser Lowland about 12,000 ^{14}C yr B.P. (Sumas stade) (Clague *et al.*, 1997; Kovanen, 2002; Kovanen & Easterbrook, 2001). Several advances separated by brief periods of retreat apparently marked the Sumas stade. The final advance(s) occurred 11,000 ^{14}C yr B.P. or shortly thereafter. Soon after 10,500 ^{14}C yr B.P., the Cordilleran Ice Sheet rapidly disappeared from the lowlands.

Central Puget Lowland

Rates of ice-sheet advance and retreat are well constrained in the central Puget Lowland. The Puget lobe advanced to the latitude of Seattle by about 14,500 ^{14}C yr B.P. (17,590 cal yr B.P.) and to its maximum by 14,000 ^{14}C yr B.P. (16,950 cal yr B.P.) (Porter & Swanson, 1998). The ice apparently remained near its maximum position only a few hundred years and then rapidly retreated. It retreated past Seattle by 13,600 ^{14}C yr B.P. (16,575 cal yr B.P.) (Porter & Swanson, 1998) (Fig. 9). Glacial lakes, including Lake Russell, formed south of the retreating ice front, draining through a spillway to the Chehalis River (Bretz, 1913). The lakes coalesced into one lake, Lake Bretz (Lake Leland of Thorson, 1980), which enlarged northward as the ice front retreated until a northern spillway was uncovered. Further backwasting allowed sea water to enter the lowland from the Strait of Juan de Fuca. Glaciomarine drift and other marine deposits accumulated in the northern lowland where land had not yet rebounded from isostatic depression. This interstade – named the Everson by Armstrong *et al.* (1965) – ended about 12,000 ^{14}C yr B.P. Isostatic rebound raised the glaciomarine and marine deposits above sea level between about 13,500 and 11,300 ^{14}C yr B.P. (Dethier *et al.*, 1995).

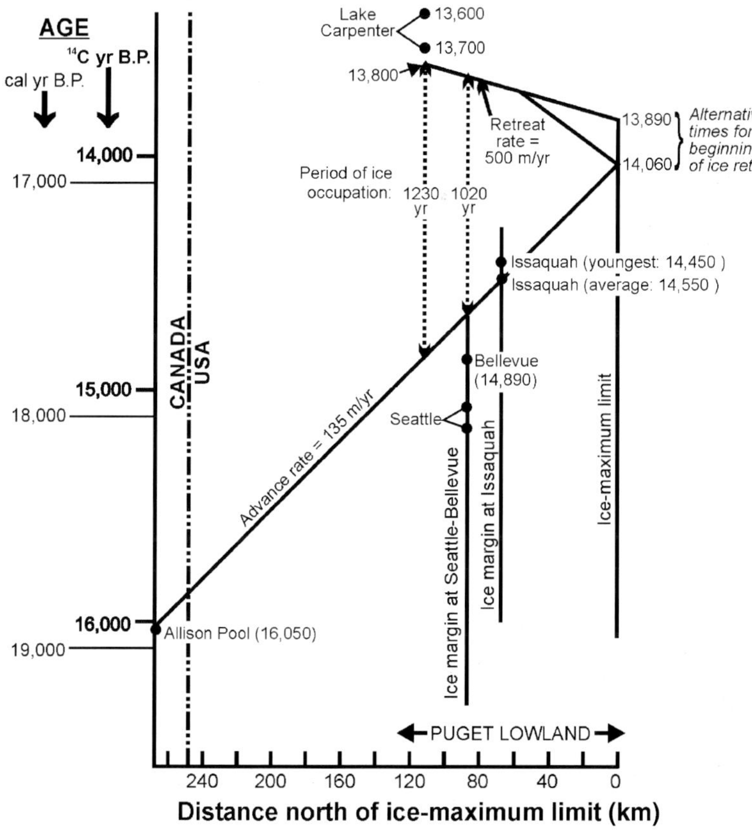

Fig. 9. Rates of Puget lobe advance and retreat in the Puget Lowland during the Vashon stade (modified from Porter & Swanson, 1998). Rapid advance and retreat are required to honor the limiting radiocarbon dates from Lake Carpenter, Seattle, Bellevue, and Issaquah. Maximum ice-sheet extent could have persisted at most a few hundred years.

Eastern Washington

In contrast to the tight age constraints west of the Cascade Range, limits on the Fraser maximum east of the Cascades and Coast Mountains are broad. They include a date of 17,240 ^{14}C yr B.P. for proglacial advance outwash, 100 km north of the ice limit, followed by advance to the glacier maximum, then a retreat of at least 80 km by 11,250 ^{14}C yr B.P., judged partly on the distribution of Glacier Peak tephra layer G (Clague *et al.*, 1980; Mehringer *et al.*, 1984; Porter, 1978). Lake Missoula flood deposits, interbedded with varves of glacial Lake Columbia that contain detrital wood dated 14,490 ^{14}C yr B.P., suggest that the Purcell Trench lobe blocked the Clark Fork for 2000–3000 yr and reached its maximum extent about 15,000 ^{14}C yr B.P. (Atwater, 1986).

Sea-Level Record

Changing sea levels greatly altered the shorelines of the Pacific Northwest. Variations in relative sea level, ranging from 200 m above present sea level to more than 100 m below, are the integrated result of eustasy, isostasy, and tectonism. These phenomena are difficult to assess separately, however, because eustasy and isostasy are interdependent and because the eustatic component has proven particularly difficult to quantify.

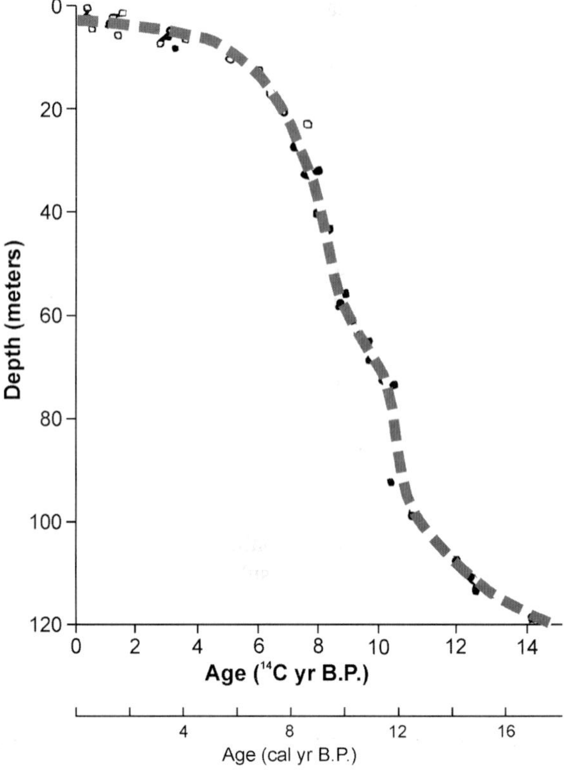

Fig. 10. Eustatic sea-level curve based on dating of shallow-water corals at Barbados (after Fairbanks, 1989).

Eustasy

Global Record

Eustatic sea-level changes are global and are caused mainly by changes in volume of ocean water. Fluctuating continental glaciers are the most important cause of eustatic sea-level change on the time scale of concern here – sea level falls when ice sheets grow and rises when they shrink. Seawater also decreases in volume as it cools, which further lowers sea level during glaciations.

The growth and decay of large ice sheets during the Pleistocene caused sea level to fluctuate by 120–140 m (Fairbanks, 1989; Lambeck *et al.*, 2000, 2002; Peltier, 2002; Yokoyama *et al.*, 2000). Estimates of sea-level lowering during the last glaciation (MIS 2) derive from fossil corals in Barbados, New Guinea, and Tahiti (Bard *et al.*, 1990a, b, 1993, 1996; Chappell & Polach, 1991; Fairbanks, 1989) and from more recent sediment cores taken from the Sunda Shelf (Hanebuth *et al.*, 2000) and Northwest Shelf of Australia (Yokoyama *et al.*, 2000). Eustatic sea-level changes have also been estimated from variations in the oxygen-isotope composition of air in bubbles trapped in the Greenland and Antarctica ice sheets (Dansgaard *et al.*, 1971; Epstein *et al.*, 1970; Grootes *et al.*, 1993; Johnsen *et al.*, 1972; Jouzel *et al.*, 2002; Lorius *et al.*, 1985; Petit *et al.*, 1999) and in foraminifera in ocean sediment (Chapman & Shackleton, 1999; Chappell & Shackleton, 1986; Lea *et al.*, 2002; Shackleton, 1987; Waelbroeck *et al.*, 2002). Numeric modeling and geologic data (summaries in Clark & Mix, 2002) provide equivalent sea-level lowering of 118–130 m for the volume of ice locked in glaciers at the last glacial maximum.

Eustatic sea level rose after about 18,000 ^{14}C yr B.P. as ice sheets in the Northern Hemisphere began to decay. Sea-level rise accelerated after about 15,000 ^{14}C yr B.P. and remained high until about 7000 ^{14}C yr B.P. when the Laurentide Ice Sheet had largely disappeared (Fig. 10; Fairbanks, 1989). Rates of eustatic sea-level rise were exceptionally high between about 11,000 and 10,500 ^{14}C yr B.P. and between 9000 and 8000 ^{14}C yr B.P. After 7000 ^{14}C yr B.P., the rate of eustatic sea-level rise sharply decreased, and by 4000 ^{14}C yr B.P. sea level was within 5 m of the present datum.

Regional Expression

It is difficult to disentangle the eustatic and glacio-isostatic components of the sea-level record in Washington and British Columbia. Isostatic depression and rebound dominate the late Pleistocene sea-level record in peripheral areas of the former Cordilleran Ice Sheet, but these effects decrease with distance beyond the ice margin. Estuaries in southwestern Washington record a mostly eustatic response, with the river valleys in this area drowned by rising sea level when ice sheets melted. In southwestern British Columbia and the northern Puget

Lowland, in contrast, relative sea level during deglaciation was higher than at present because these areas were isostatically depressed during the last glaciation.

Isostasy

Global Record

The growth and decay of ice sheets, and thus changes in global sea level, redistributed mass on the Earth's surface. Ice sheets depressed the crust beneath them, but just beyond their margins the crust warped as a "forebulge" (Walcott, 1970). Melting ice sheets reversed the process: the forebulge migrated back towards the former center of loading to cause uplift there.

Water transfer from oceans to ice sheets unloaded the seafloor; the opposite happened during deglaciation. These hydro-isostatic adjustments opposed the direction of glacio-isostatic adjustments. Continental shelves rose when seawater was removed and they subsided again when melting ice sheets returned water to the oceans.

Regional Expression

Expanding glaciers during the early part of the Fraser glaciation progressively depressed the land surface of southwestern British Columbia and northwestern Washington (Clague, 1983). This depression started beneath the Coast Mountains, where glaciers first grew. As glaciers continued to advance, the area of crustal subsidence expanded beneath coastal areas. Subsidence probably exceeded the eustatic fall in sea level as ice sheets grew between 25,000 and 15,000 ^{14}C yr B.P. (Chappell *et al.*, 1996; Lambeck *et al.*, 2002; Shackleton, 1987; Waelbroeck *et al.*, 2002). If so, relative sea level in the region rose during this period. The relative rise in sea level controlled deposition of thick bodies of advance outwash (the Quadra Sand in British Columbia and the Esperance Sand in western Washington) on braided floodplains and deltas, and in littoral environments (Clague, 1976). As the Puget lobe reached its limit near the city of Olympia, the region to the north was isostatically depressed. The depression was greatest beneath the Strait of Georgia and Fraser Lowland and decreased south along the Puget Lowland.

The height of the uppermost shorelines that formed during deglaciation gives some limits on isostatic depression. Marine deltas near Vancouver lie 200 m above present sea level (Clague *et al.*, 1982). With eustatic sea level-100 m at the time the highest shorelines formed (Fairbanks, 1989), local glacio-isostatic depression must have exceeded 300 m. The depression was actually larger, because the Cordilleran Ice Sheet had thinned before the highest shorelines formed, and thus rebound had started already.

The modern altitudes of the late-glacial marine limit display the variable isostatic influence of the Cordilleran Ice Sheet. The marine limit is highest around the Strait of Georgia and in the Canadian part of the Fraser Lowland, and it declines west and south (Clague *et al.*, 1982; Dethier *et al.*, 1995; Mathews *et al.*, 1970). From about 125 to 150 m above sea level (asl) near Bellingham, it drops to 70 m asl west of Victoria, below 50 m asl on the west coast of Vancouver Island at Tofino, and probably below 50 m asl near the entrance to Juan de Fuca Strait. The marine limit decreases south of Bellingham to about 35 m asl at Everett. At the heads of the British Columbia mainland fiords to the north, the marine limit is fairly low because these areas remained ice-covered until isostatic rebound was well along (Clague & James, 2002; Friele & Clague, 2002).

Isostatic uplift rates can be inferred from a variety of shoreline data. Proglacial lakes covered southern and central Puget Lowland during deglaciation (Fig. 11), the lakes

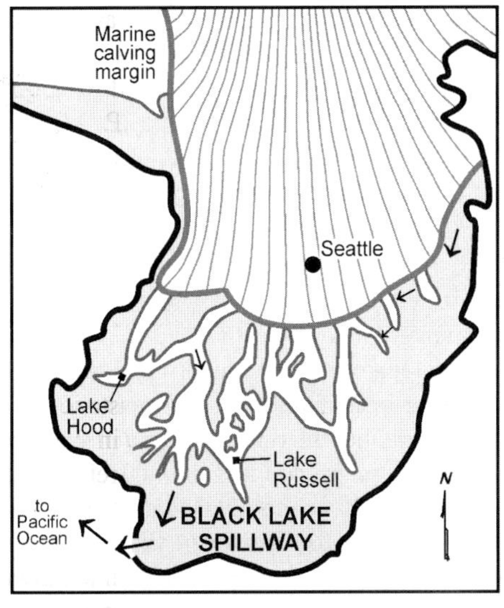

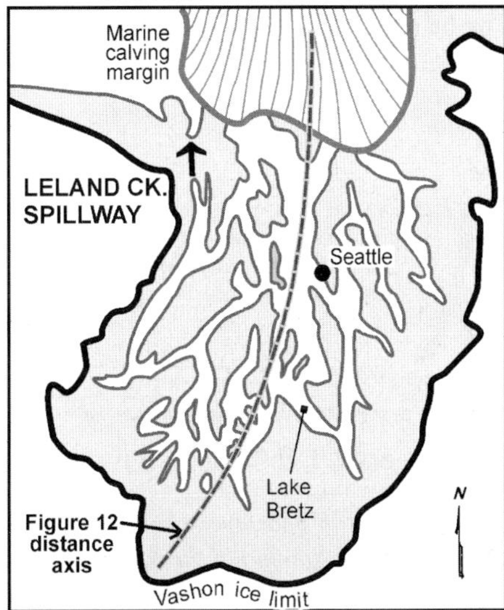

Fig. 11. Paleogeographic maps showing the maximum extents of Lake Russell and Lake Bretz (modified from Thorson, 1989, Fig. 2). Arrows show spillway locations controlling local and regional lake altitudes.

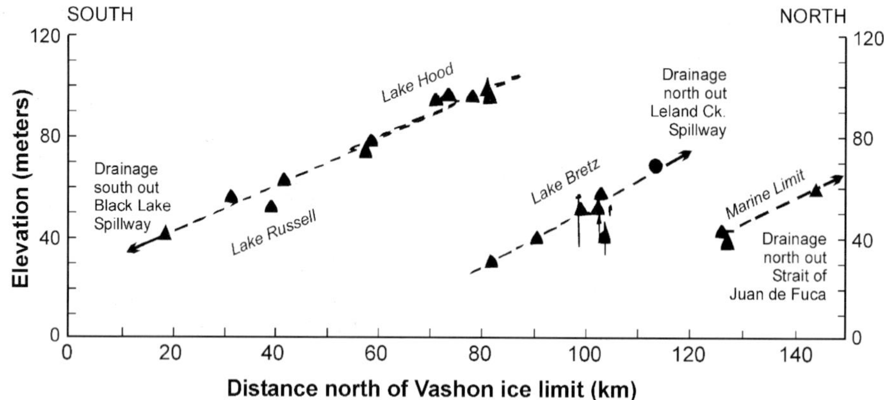

Fig. 12. North-south profile of shoreline features (delta tops) associated with Lake Russell, Lake Hood (confluent with Lake Russell), Lake Bretz, and the marine limit near Discovery Bay on the south side of the Strait of Juan de Fuca Strait (modified from Thorson, 1989, Fig. 5; with additional data from Dethier et al., 1995).

were dammed to the north by the retreating Puget lobe. The last lake drained when the Puget lobe retreated north of Port Townsend and marine waters entered Puget Sound. Differential isostatic rebound warped the shorelines of these proglacial lakes (Fig. 12) – shorelines of Lake Russell-Hood are tilted up to the north at 0.85 m/km; the tilt of Lake Bretz shorelines is 1.15 m/km (Thorson, 1989). Most uplift in the Fraser Lowland and on eastern Vancouver Island occurred in less than 1000 years (Clague *et al.*, 1982; Mathews *et al.*, 1970), as inferred from the shoreline tilt data and relative sea-level observations. These data underlie a postglacial rebound model of the Cordilleran Ice Sheet (Clague & James, 2002; James *et al.*, 2000) that predicts low mantle viscosities ($<10^{20}$ Pa s).

Besides rapid rebound, low mantle viscosities in this region are responsible for nearly complete glacio-isostatic uplift by the early Holocene (Clague, 1983). Relative sea level was lower 8000–9000 ^{14}C yr B.P. than it is today, by at least 15 m at Vancouver (Mathews *et al.*, 1970) and by perhaps as much as 50 m in Juan de Fuca Strait (Hewitt & Mosher, 2001; Linden & Schurer, 1988). Evidence for lower sea levels includes submerged spits, deltas, and wave-truncated surfaces on the floor of Juan de Fuca Strait, and buried terrestrial peats found well below sea level in the Fraser Lowland. Sea level seems to have tracked global eustatic sea-level rise thereafter (Clague *et al.*, 1982; Mathews *et al.*, 1970), except on the west coast of Vancouver Island where sea level was several meters higher in the middle Holocene than now (Clague *et al.*, 1982; Friele & Hutchinson, 1993). Tectonic uplift probably caused this anomaly (see below).

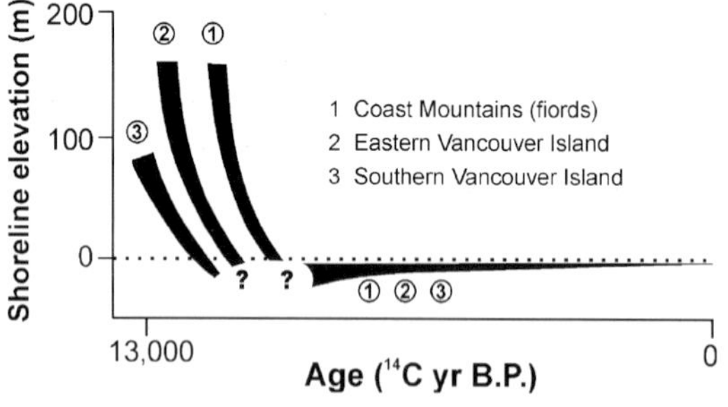

Fig. 13. Generalized patterns of sea-level change on south coast of British Columbia since the end of the last glaciation (Clague & James, 2002, Fig. 8; modified from Muhs et al., 1987, Fig. 10). Deglaciation and isostatic rebound occurred later in the southern Coast Mountains than on Vancouver Island. Line widths display range of uncertainty.

Isostatic uplift occurred at different times in southwestern British Columbia and northwestern Washington as the Cordilleran Ice Sheet retreated. Regions that deglaciated first rebounded earlier than those deglaciated later (Fig. 13). Glacio-isostatic response to deglaciation varied across the region, showing that the lithosphere responded non-uniformly as the ice sheet decayed (Clague, 1983).

Tectonics

Trends in elevations of the late-glacial marine limit and the patterns of sea-level change summarized above show that much of the crustal deformation is isostatic. Slippage on reactivated faults, however, may have caused some of the observed deformation, analogous to recognized movement on some faults in the Puget Lowland later in the Holocene (Bucknam *et al.*, 1992; Johnson *et al.*, 1996, 2001). As yet, no such late-glacial or early postglacial fault movements have been documented unequivocally.

Late Quaternary sea-level change in the coastal Pacific Northwest also includes a component of aseismic tectonic deformation, but the rates of such vertical motions are at least an order-of-magnitude less than those of late-glacial eustatic and glacio-isostatic sea-level change and so cannot be isolated from those signals. However, the much slower changes in late Holocene sea level may include a significant component of aseismic tectonic deformation, which may partly explain the late Holocene regression on the west coast of Vancouver Island (Clague *et al.*, 1982; Friele & Hutchinson, 1993).

Physical Behavior of the Cordilleran Ice Sheet

The Puget lobe of the Cordilleran Ice Sheet during the last glaciation provides an exceptional opportunity to examine the connection between glacier physics and the geomorphic products of the glacier system. Such an approach to interpreting the deposits and landforms of glaciated terrain has been widely applied only in the last several decades. The Puget lobe is not necessarily typical of every continental ice lobe, having a strong maritime influence. However, it is particularly well-constrained, with good age control, clearly recognized boundaries, moderately definitive source area, and good expression of its topographic effects and sedimentary deposits.

Ice-Sheet Reconstruction

By applying a height-mass balance curve (Porter *et al.*, 1983) over the reconstructed boundaries and surface altitudes of the Puget lobe, an ELA between 1200 and 1250 m balances the ice sheet (Booth, 1986). The ice flux peaks at the ELA, while meltwater flow increases monotonically downglacier (Fig. 14). The contribution to ice velocity from internal ice deformation (Paterson, 1981), based on reconstructed ice thickness and surface slope, is less than 2% of the total flux (Booth, 1986). Thus, basal sliding must account for virtually all of the predicted motion, several hundred meters per year over nearly the entire area of the lobe. From lobe dimensions, the calculated basal shear stress of the ice ranges between 40 and 50 kPa (Booth, 1986; Brown *et al.*, 1987). This value is low by the standards of modern valley glaciers but typical of ice streams and large modern ice lobes (Blankenship *et al.*, 1987; Mathews, 1974; Paterson, 1981), whose sliding velocities are also hundreds of meters per year. The system was thus one of rapid mass transport under a rather low driving stress across a bed of mainly unconsolidated sediment.

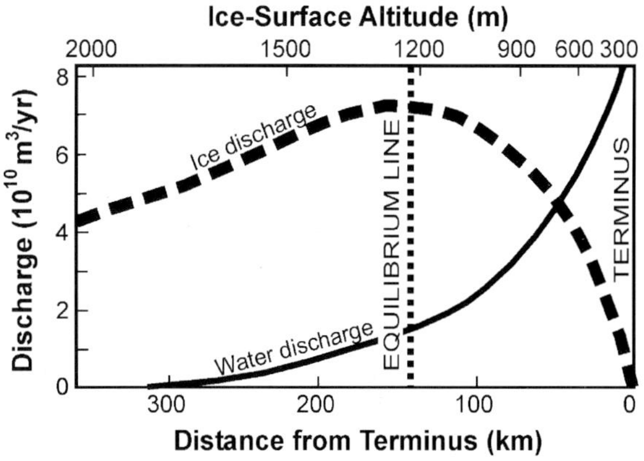

Fig. 14. Pattern of ice and water fluxes along the Puget lobe, reconstructed at ice-maximum conditions (from Booth, 1986).

Average pore-water pressures across the glacier bed closely approached the ice overburden, because so much water cannot be quickly discharged (Booth, 1991a), even through an extensive subglacial tunnel system. Thus, the ice loading of bed sediments was low except near the margins, and the strength of the sediments correspondingly poor; shearing and streamlining would have been widespread. The modern landscape amply testifies to these processes (Fig. 15).

Meltwater

The Puget Lowland basin became a closed depression once the ice advanced south past the entrance of the Strait of Juan de Fuca and blocked the only sea-level drainage route. Lacustrine sediment (e.g. Lawton Clay Member of the Vashon Drift; Mullineaux *et al.*, 1965) accumulated in ice-dammed lakes, followed by fluvial outwash (Esperance Sand Member of the Vashon Drift) that spread across nearly all of the

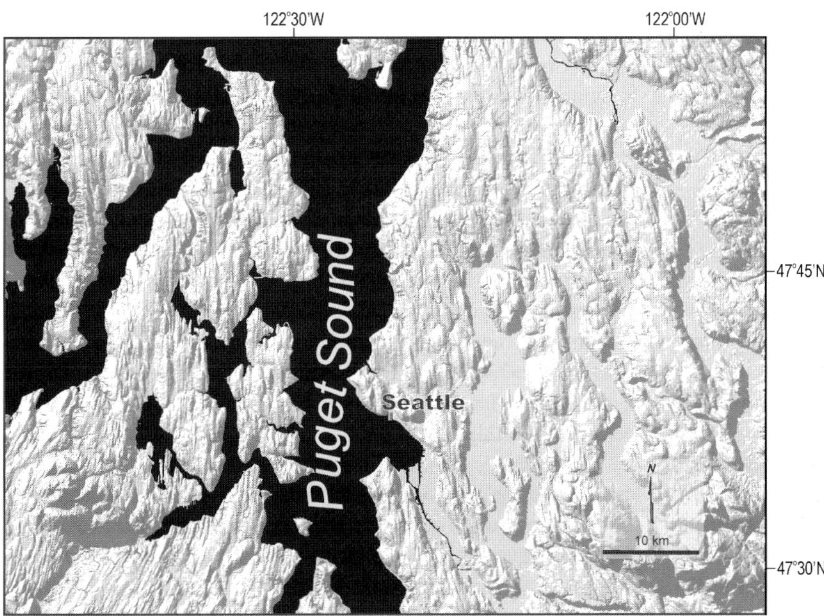

Fig. 15. Shaded topographic view of the central Puget Lowland, showing strongly streamlined landforms from the passage of the Puget lobe ice sheet during the Vashon stade. Modern marine waters of Puget Sound in black; city of Seattle is in the south-central part of the view. Nearly all streamlined topography is underlain by deposits of the last glaciation.

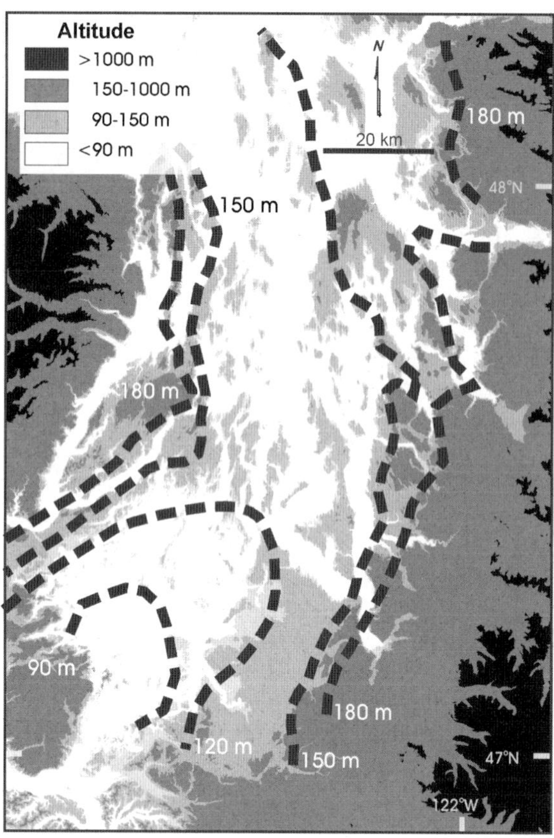

Fig. 16. Topography of the Puget Lowland, from U.S. Geological Survey 10-m digital elevation model. Contours show generalized topography of the great Lowland fill (modified from Booth, 1994), as subsequently incised by both subglacial channels and modern river valleys. Its modern altitude lies between 120 and 150 m across most of the lowland, reconstructed from the altitude of modern drumlin tops.

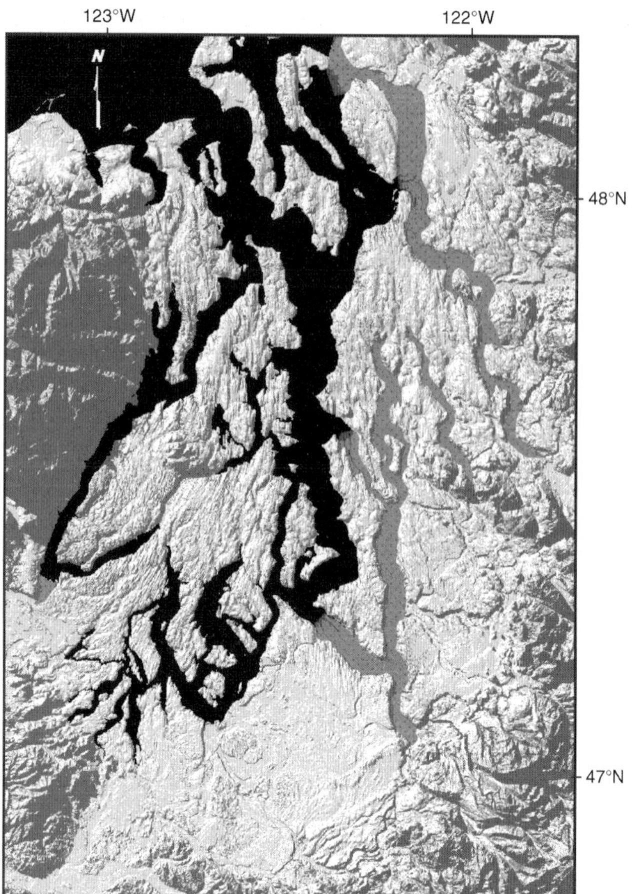

Fig. 17. Shaded topography of the Puget Lowland from U.S. Geological Survey 10-m digital elevation model, displaying the major subglacial drainage channels of the Puget lobe. Most are now filled by marine waters (black), with others by late-glacial and Holocene alluvium and mudflows (dark gray stipple).

Puget Lowland. The outwash must have prograded as deltas like those that formed during ice recession (Thorson, 1980). With the greater time available during ice advance, however, sediment bodies coalesced into an extensive outwash plain in front of the ice sheet (e.g. Boothroyd & Ashley, 1975), named the "great Lowland fill" by Booth (1994) (Fig. 16). With continued ice-sheet advance and outwash deposition, this surface ultimately would have graded to the basin outlet in the southern Puget Lowland. Crandell *et al.* (1966) first suggested that this deposit might have been continuous across the modern arms of Puget Sound; Clague (1976) inferred a correlative deposit (Quadra Sand) filled the Georgia Depression farther north.

The fill's depositional history lasted 2000–3000 years. Outwash of the ice-sheet advance did not inundate the Seattle area until shortly before 15,000 ^{14}C yr B.P. (Mullineaux *et al.*, 1965). Deposition may have begun a few thousand years earlier, but accumulation would have been slow until advancing ice blocked drainage out of the Strait of Juan de Fuca (about 16,000 ^{14}C yr B.P.). Although late in starting, deposition across the entire lowland must have been complete before the ice maximum at about 14,000 ^{14}C yr B.P. (Porter & Swanson, 1998) because basal till of the overriding ice sheet caps the great Lowland fill almost everywhere.

Incised up to 400 m into the fill (and the overlying till) are prominent subparallel troughs (Fig. 17), today forming one of the world's great estuarine systems. These troughs were once thought to result from ice tongues occupying a preglacial drainage system (Willis, 1898), preserving or enhancing a topography of fluvial origin. This scenario is impossible, however, because impounded proglacial lakes would have floated the ice tongues and precluded any bed contact or ice erosion. Incision by subaerial channels is impossible because the lowest trough bottoms almost 300 m *below* the southern outlet of the Puget Lowland basin, and Holmes *et al.* (1988) report seismic-reflection data that suggest that the troughs were excavated during ice occupation to more than twice their current depth. Thus, troughs must have been excavated after deposition of the great Lowland fill. Yet the troughs must predate subaerial exposure of the glacier bed during ice recession, because many of the eroded troughs are still mantled on their flanks with basal till (e.g. Booth, 1991b) and filled with deposits of recessional-age lakes (Thorson, 1989). Thus, the troughs were formed primarily (or exclusively) by subglacial processes and probably throughout the period of ice occupation, chiefly by subglacial meltwater (Booth & Hallet, 1993). A similar inference explains Pleistocene glacier-occupied troughs and tunnel valleys of similar dimensions and relief elsewhere in the Northern Hemisphere: Germany (Ehlers, 1981), Nova Scotia (Boyd *et al.*, 1988), New York (Mullins & Hinchey, 1989), Ontario (Shaw & Gilbert, 1990), and Minnesota (Patterson, 1994).

Missoula Floods

During several glaciations in the late Pleistocene, the Cordilleran Ice Sheet invaded Columbia River drainage and temporarily deranged it. The Purcell Trench lobe thwarted the Clark Fork of the Columbia to dam glacial Lake Missoula (Fig. 18) with volumes of as much as 2500 km^3 – as much water as Great Lakes Erie and Ontario together contain today. Stupendous floods from the lake swept the north and central part of the Columbia Plateau to carve a plexus of scabland channels as large as river valleys.

In the 1920s, J Harlen Bretz argued an astonishing idea: the Channeled Scabland originated by enormous flood (Bretz, 1923, 1925, 1928a, b, 1929, 1932). His scablands evidence included gigantic water-carved channels, great dry cataracts (Fig. 19), overtopped drainage divides, and huge gravel bars. But with no known water source, skeptics in the 1930s–1940s tried to account for the scabland channels by mechanisms short of cataclysmic flood, such as by sequential small floods around many huge ice jams. Then Pardee (1942) revealed giant current dunes and other proof of a colossal outburst of glacial Lake Missoula. Thus, a source for Bretz's great flood had been found. In the 1950s, Bretz himself vindicated his old story (Bretz *et al.*, 1956). Baker (1973) showed that Bretz's

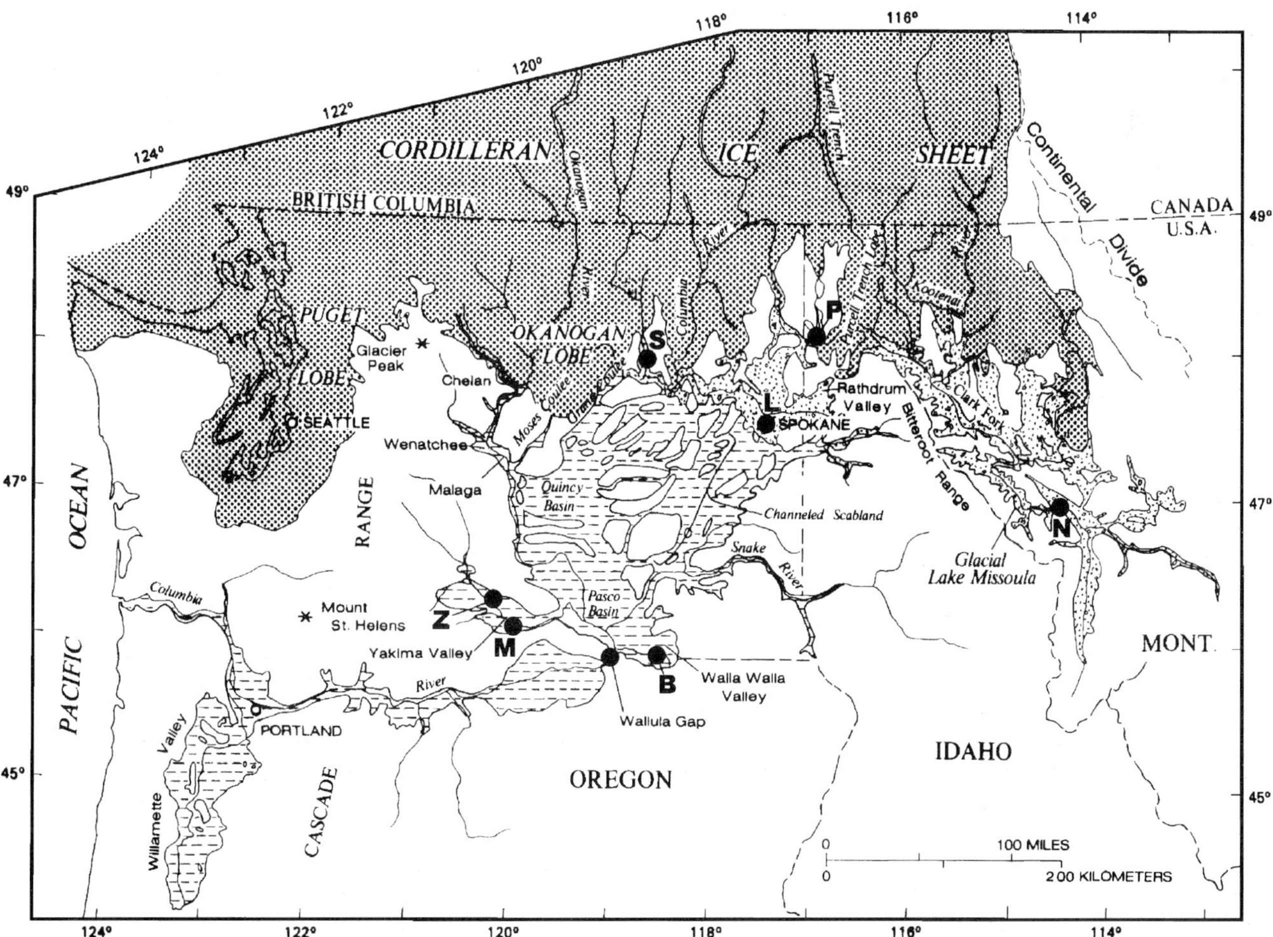

Fig. 18. Map of Columbia River valley and tributaries. Dark cross-hatching shows maximum extent of Cordilleran Ice Sheet; fine stipple pattern shows maximum area of glacial Lake Missoula east of Purcell Trench ice lobe and maximum extent of glacial Lake Columbia east of Okanogan lobe. Dashed-line pattern shows area that was swept by the Missoula floods in addition to these lakes. Large dots indicate key localities: B, Burlingame ravine; L, Latah Creek; M, Mabton; N, Ninemile Creek; P, Priest valley; S, Sanpoil valley; Z, Zillah. From Waitt (1985, Fig. 1). Relations at sites B, P, and N shown schematically on Fig. 21.

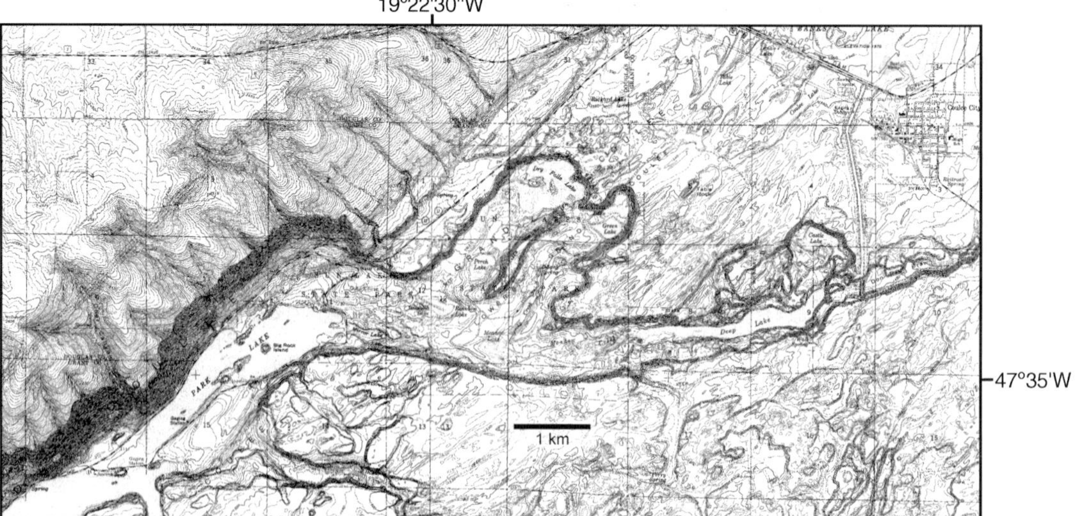

Fig. 19. Topographic map of Great Cataract Group, including Dry Falls in Grand Coulee (center of map). From U.S. Geological Survey 7.5-minute Park Lake and Coulee City quadrangles. Contour interval 10 ft. Land-grid squares (Township sections) are 1 mile (1.6 km) on a side. Top is north.

observations were in accord with principles of open-channel hydraulics. Bretz's old heresy now wore respectable clothes.

In the high-velocity, high-energy scabland reaches, one great flood eroded evidence of any earlier ones. But the waters also backflooded up tributary valleys and quietly deposited suspended load there in transitory hydraulic ponds. Within stacks of rhythmic beds in southern Washington (Fig. 20), the Mount St. Helens "set-S" ash couplet (14,000 ^{14}C yr B.P.) lies atop a floodlaid bed identical to other beds in these sections. This, and other evidence, shows that each graded bed is the deposit of a separate great flood. Numerous sites across the region tell a similar story of scores of separate floods (Atwater, 1984, 1986; Waitt, 1980, 1984, 1985, 1994). All together there were probably 95–100 Missoula floods during the last glaciation.

In northeastern Washington and Idaho, glacial lakes dammed along the Cordilleran ice margin (Fig. 18) accumulated sand-silt-clay varves. These beds are interrupted by many thick, coarse floodlaid beds. The numbers of varves indicate periods of six decades to a few years between successive floods (Atwater, 1984, 1986; Waitt, 1984, 1985). The only water body big and high enough to flood these glacial lakes was Lake Missoula. The sediment of Lake Missoula itself comprises dozens of fining-upward varve sequences, each the record of a gradually deepening then swiftly emptying lake (Chambers, 1971; Waitt, 1980). Fig. 21 relates the deposits across the region. The rhythmic beds of southern Washington record the floods, Lake Missoula bottom sediment records interflood periods, and the northern lake deposits record both.

East of the Cascade Range, the Fraser-age Cordilleran Ice Sheet is bracketed in time by preglacial dates as young as 17,200 ^{14}C yr B.P. and postglacial dates as old as 11,000 ^{14}C yr B.P. in southern British Columbia, 150–300 km north of the ice limit (Clague, 1981, 1989). Dammed at the ice terminus, Lake Missoula existed less than half this period. Fewer than 2500 varves are known from Lake Missoula bottom sediment or between Missoula-flood beds in other glacial lakes (Atwater, 1986; Chambers, 1971).

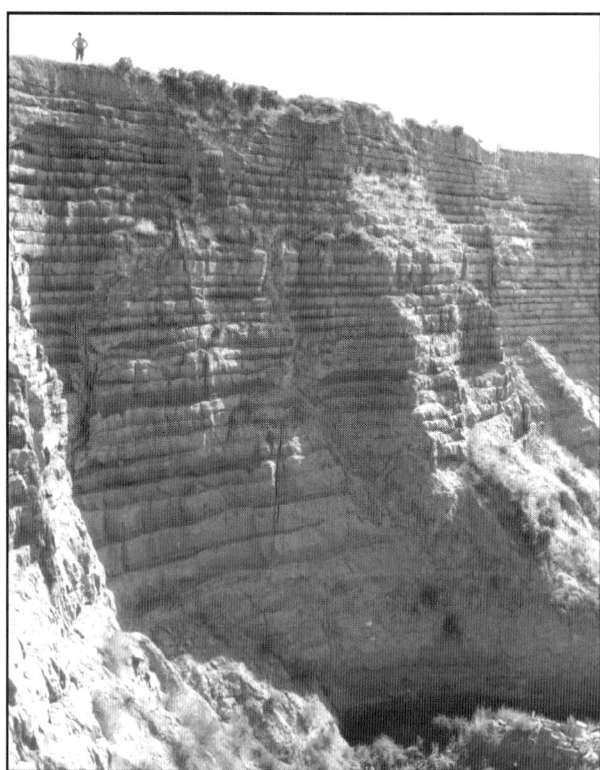

Fig. 20. Rhythmically bedded Missoula-backflood deposits at Burlingame ravine, Walla Walla valley (site B of Fig. 18). Each graded bed is the deposit of a separate flood.

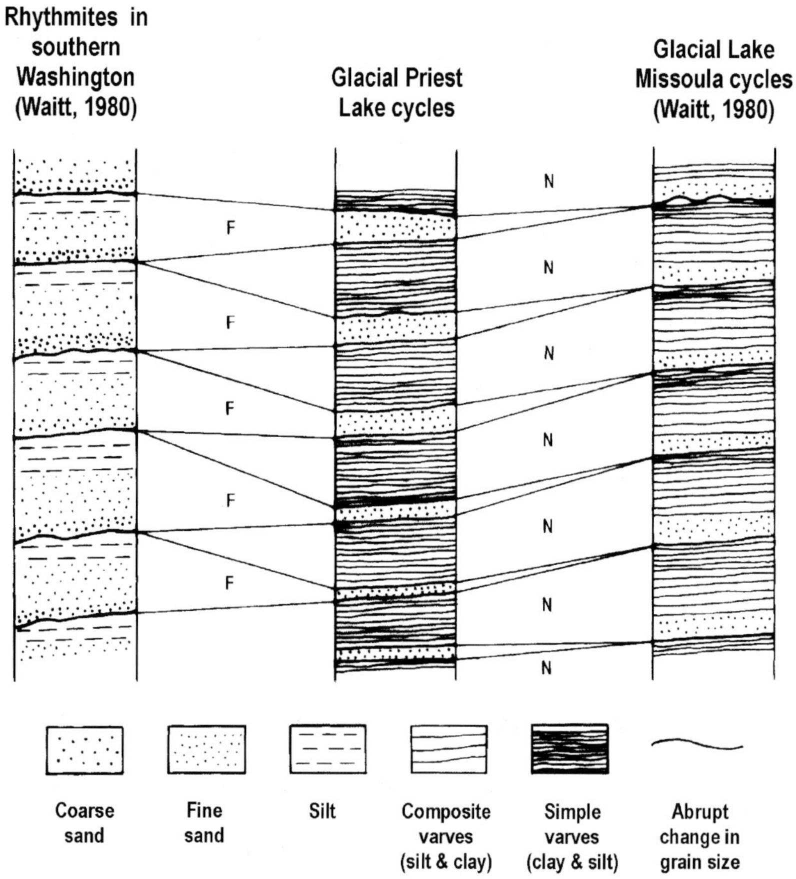

Fig. 21. Inferred relations between rhythmites in southern Washington, northern Washington and Idaho, and western Montana (Lake Missoula). These columns schematically represent sedimentary motif at sites B, P, and N of Fig. 18. From Waitt (1985, Fig. 17).

Radiocarbon ages and proxy ages further limit the age of the floods. Atwater (1986, Fig. 17) dated a wood fragment at 14,490 ^{14}C yr B.P. in the lower-middle of the Missoula-flood sequence in Sanpoil Valley. In Snake Valley, 21 Missoula-backflood couplets (Waitt, 1985) overlie gravel of the great flood from Lake Bonneville (ca. 14,500 ^{14}C yr B.P.; Oviatt *et al.*, 1992). The 14,000-yr-B.P. Mount St. Helens ash couplet overlies at least 28 giant-flood rhythmites in southern Washington and underlies eleven (Waitt, 1980, 1985). After these giant floods came several dozen smaller Missoula floods (Waitt, 1994). Organic matter within and below Missoula flood deposits in the Columbia gorge yielded three dates between 15,000 and 13,700 ^{14}C yr B.P. (O'Connor & Waitt, 1995). The 11,250 ^{14}C yr B.P. Glacier Peak tephra (Mehringer *et al.*, 1984) postdates ice-sheet retreat in northern Washington and Montana (Waitt & Thorson, 1983). These various limits suggest that glacial Lake Missoula existed for 2000 years or so during the period 15,700–13,500 ^{14}C yr B.P.

The controlling Purcell Trench ice dam became progressively thinner during deglaciation. Shallower lake levels were required to destabilize the smaller ice dam. Floods from the lake thus became smaller and more frequent. The average period between floods indicated by varves is about 30 years. At the glacial maximum it was much longer, and late during deglaciation it was much shorter. Atwater's varve counts (1986) detail a near-continuous record of the Missoula floods. The period between floods was about 50 years at the glacial maximum and during deglaciation decreased successively to 30, 20, and fewer than 10 years.

A recurring discharge every few decades or years suggests that glacial Lake Missoula emptied by a recurring hydraulic instability that causes glacier-outburst floods (jökulhlaups) from modern Icelandic glaciers (Waitt, 1980). As the water deepens against the ice dam, it buoys the lakeward end of the dam. Subglacial drainage occurs when the hydrostatic pressure of water from the lake exceeds the ice overburden pressure at the glacier bed (Bjornsson, 1974; Clarke *et al.*, 1984; Waitt, 1985). Drainage begins, and ice tunnels enlarge swiftly. The tunnel roof collapses; the whole lake drains. Glacier flow then repairs the damage, and within months the lake basin begins to refill.

The peak flow of Missoula floodwater down 10-km-wide Rathdrum Valley as modeled by O'Connor & Baker (1992, Figs 7 and 8) was at least 17 million m^3/s. More recent modeling suggests peak discharge almost twice that (Waitt *et al.*, 2000). During deglaciation the thinning ice dam fails at progressively shallower lake levels. Calculations suggest the Missoula floods ranged in peak discharge from as much as 30 million to as little as 200,000 m^3/s (Waitt, 1994). The largest were the Earth's grandest freshwater floods. Even a lake volume only one-third of maximum sufficed for a mighty flood down the Channeled Scabland and Columbia valley.

Summary

Advances in both global and regional understanding of Quaternary history, deposits, and geomorphic processes have brought new information and new techniques to characterize the growth, decay, and products of the Cordilleran Ice Sheet during the Pleistocene. Ice has advanced south into western Washington at least six times, but the marine-isotope record suggests that these are but a fraction of the total that entered the region in the last 2.5 million years. Several glacial and interglacial deposits are likely in the interval from 780,000 to 250,000 years ago but are not yet formally recognized. Growth and decay of large ice sheets during the Pleistocene have also caused sea level to fall and rise about 120–140 m, with strong influence on the tidewater margins of the Cordilleran Ice Sheet, as did progressive depression of the land surface as glaciers expanded during each glaciation. During the most recent (Fraser) glacier advance, local glacio-isostatic depression exceeded 270 m. Subsequent postglacial rebound of the Earth's crust, recorded in detail by proglacial lake shorelines, was rapid.

Reconstruction of the Puget lobe of the Cordilleran Ice Sheet during the last glacial maximum requires basal sliding at rates of several hundred meters per year, with pore-water pressures nearly that of the ice overburden. Landforms produced during glaciation include an extensive low-gradient outwash plain in front of the advancing ice sheet, a prominent system of subparallel troughs deeply incised into that plain and carved mainly by subglacial meltwater, and widespread streamlined landforms. At the southeastern limit of the ice sheet, the Purcell Trench lobe dammed glacial Lake Missoula to volumes as much as 2500 km^3, which episodically discharged as much as 30 million m^3/s. Scores of great floods swept across the Channeled Scablands of eastern Washington at intervals of typically a few decades, carving scabland channels as large as great river valleys. Modern geomorphic analysis of them confirms one of the region's early theories of wholesale development of landscape by the Cordilleran Ice Sheet.

Acknowledgments

We are indebted to our colleagues, and our predecessors, for the wealth of information on the Cordilleran Ice Sheet that we have summarized in this chapter. We also thank Doug Clark and David Dethier for their assistance as reviewers.

References

Armstrong, J.E. (1957). Surficial geology of the New Westminster map-area, British Columbia: Canada Geological Survey Paper 57–5, 25 pp.

Armstrong, J.E. & Brown, W.L. (1953). Ground-water resources of Surrey Municipality, British Columbia: Canada Geol. Survey Water-Supply Paper 322, 48 pp.

Armstrong, J.E., Crandell, D.R., Easterbrook, D.J. & Noble, J.B. (1965). Late Pleistocene stratigraphy and chronology in southwestern British Columbia and northwestern Washington. *Geological Society of America Bulletin*, **76**, 321–330.

Atwater, B.F. (1984). Periodic floods from glacial Lake Missoula into the Sanpoil arm of glacial Lake Columbia, northeastern Washington. *Geology*, **12**, 464–467.

Atwater, B.F. (1986). Pleistocene glacial-lake deposits of the Sanpoil River valley, northeastern Washington: U.S. Geological Survey Bulletin 1661, 39 pp.

Atwater, B.F. & Moore, A.L. (1992). A tsunami about 1000 years ago in Puget Sound, Washington. *Science*, **258**, 1614–1617.

Baker, V.R. (1973). Paleohydrology and sedimentology of Lake Missoula flooding in eastern Washington: Geological Society of America Special Paper 144, 79 pp.

Bard, E., Arnold, M., Fairbanks, R.G. & Hamelin, B. (1993). 230^{Th}–234^{U} and ^{14}C ages obtained by mass spectrometry on corals. *Radiocarbon*, **35**, 191–199.

Bard, E., Hamelin, B., Arnold, M., Montaggioni, L.F., Cabrioch, G., Faure, G. & Rougerie, F. (1996). Deglacial sea-level record from Tahiti corals and the timing of global meltwater discharge. *Nature*, **382**, 241–244.

Bard, E., Hamelin, B. & Fairbanks, R.G. (1990a). U/Th ages obtained by mass spectrometry in corals from Barbados: sea level during the past 130,000 years. *Nature*, **346**, 456–458.

Bard, E., Hamelin, B., Fairbanks, R.G. & Zindler, A. (1990b). Calibration of the ^{14}C timescale over the past 30000 years using mass spectrometric U-Th ages from Barbados corals. *Nature*, **345**, 405–410.

Barendregt, R.W. (1995). Paleomagnetic dating methods. *In*: Rutter, N.W. & Catto, N.R. (Eds), *Dating methods for Quaternary deposits*. Geological Association of Canada, GEOtext2, 308 pp.

Barnosky, C.W. (1981). A record of late Quaternary vegetation from Davis Lake, southern Puget Lowland, Washington. *Quaternary Research*, **16**, 221–239.

Barnosky, C.W. (1985). Late Quaternary vegetation near Battle Ground lake, southern Puget trough, Washington. *Geological Society of America Bulletin*, **96**, 263–271.

Barton, B.R. (2002). On the distribution of Late Pleistocene mammoth remains from Seattle and King County, Washington State. *In*: Program and Abstracts, 17th Biennial Meeting, Anchorage, AK, American Quaternary Association, p. 16.

Bjornsson, H. (1974). Explanation of jökulhlaups from Grímsvötn, Vatnajökull, Iceland. *Jökull*, **24**, 1–26.

Bjornstad, B.N., Fecht, K.R. & Pluhar, C.J. (2001). Long history of pre-Wisconsin, ice age cataclysmic floods – evidence from southeastern Washington. *Journal of Geology*, **109**, 695–713.

Blankenship, D.D., Bentley, C.R., Rooney, S.T. & Alley, R.B. (1987). Till beneath Ice Stream B: 1 – Properties derived from seismic travel times. *Journal of Glaciology*, **92**(B9), 8903–8911.

Blunt, D.J., Easterbrook, D.J. & Rutter, N.W. (1987). Chronology of Pleistocene sediments in the Puget Lowland, Washington. *In*: Schuster, J. (Ed.), *Selected papers on the geology of Washington*. Washington Division of Geology and Earth Resources Bulletin 77, 321–353.

Booth, D.B. (1986). Mass balance and sliding velocity of the Puget lobe of the Cordilleran ice sheet during the last glaciation. *Quaternary Research*, **29**, 269–280.

Booth, D.B. (1991a). Glacier physics of the Puget lobe, southwest Cordilleran ice sheet. *Géographie Physique et Quaternaire*, **45**, 301–315.

Booth, D.B. (1991b). Geologic map of Vashon and Maury Islands, King County, Washington: U.S. Geological Survey Miscellaneous Field Investigations Map MF-2161, scale 1: 24,000.

Booth, D.B. (1994). Glaciofluvial infilling and scour of the Puget Lowland, Washington, during ice-sheet glaciation. *Geology*, **22**, 695–698.

Booth, D.B. & Hallet, B. (1993). Channel networks carved by subglacial water – observations and reconstruction in the eastern Puget Lowland of Washington. *Geological Society of America Bulletin*, **105**, 671–683.

Boothroyd, J.C. & Ashley, G.M. (1975). Processes, bar morphology, and sedimentary structures on braided outwash fans, northeastern Gulf of Alaska. *In*: Jopling, A.V. & McDonald, B.C. (Eds), *Glaciofluvial and glaciolacustrine environments*. Society of Economic Paleontologists and Minerologists Special Publication No. 23, 193–222.

Borden, R.K. & Troost, K.G. (2001). Late pleistocene stratigraphy in the south-central Puget Lowland, West-Central Pierce County, Washington. Olympia, Washington State Department of Natural Resources, Report of Investigation 33, 33 pp.

Boyd, R., Scott, D.B. & Douma, M. (1988). Glacial tunnel valleys and Quaternary history of the outer Scotian shelf. *Nature*, **333**, 61–64.

Bretz, JH. (1913). Glaciation of the Puget Sound region. *Washington Geological Survey Bulletin No. 8*, 244 pp.

Bretz, JH. (1923). The channeled scabland of the Columbia plateau. *Journal of Geology*, **31**, 617–649.

Bretz, JH. (1925). The spokane flood beyond the channeled scablands, *Journal of Geology* **33**, 97–115, 236–259.

Bretz, JH. (1928a). Bars of Channeled Scabland. *Geological Society of America Bulletin*, **39**, 643–702.

Bretz, JH. (1928b). The Channeled Scabland of eastern Washington. *Geographical Review*, **18**, 446–477.

Bretz, JH. (1929). Valley deposits immediately east of the Channeled Scabland of Washington. *Journal of Geology*, **37**, 393–427, 505–541.

Bretz, JH. (1932). The grand coulee. *American Geographical Society Special Publication*, **15**, 89.

Bretz, JH., Smith, H.T.U. & Neff, G.E. (1956). Channeled Scabland of Washington – new data and interpretations. *Geological Society of America Bulletin*, **67**, 957–1049.

Brocher, T.M., Parsons, T., Blakely, R.J., Christensen, N.I., Fisher, M.A., Wells, R.E. & the SHIPS Working Group (2001). Upper crustal structure in Puget Lowland, Washington – results from 1998 Seismic Hazards Investigation in Puget Sound. *Journal of Geophysical Research*, **106**, 13,541–13,564.

Brown, N.E., Hallet, B. & Booth, D.B. (1987). Rapid soft bed sliding of the Puget glacial lobe. *Journal of Geophysical Research*, **92**(B9), 8985–8997.

Bucknam, R.C., Hemphill-Haley, E. & Leopold, E.B. (1992). Abrupt uplift within the past 1700 years at southern Puget Sound, Washington. *Science*, **258**, 1611–1614.

Cande, S.C. & Kent, D.V. (1995). Revised calibration of the geomagnetic polarity time scale for the late Cretaceous and Cenozoic. *Journal of Geophysical Research*, **100**, 6093–6095.

Chambers, R.L. (1971). Sedimentation in glacial Lake Missoula [M.S. thesis]. Missoula, University of Montana, 100 pp.

Chapman, M.R. & Shackleton, N.J. (1999). Global ice-volume fluctuations, North Atlantic ice-rafting events, and deep-ocean circulation changes between 130 and 70 ka. *Geology*, **27**, 795–798.

Chappell, J., Omura, A., Esat, T., McCulloch, M., Pandolfi, J., Ota, Y. & Pillans, B. (1996). Reconciliation of late Quaternary sea levels derived from coral terraces at Huon Peninsula with deep sea oxygen isotope records. *Earth and Planetary Science Letters*, **141**, 227–236.

Chappell, J. & Polach, H. (1991). Post-glacial sea-level rise from a coral record at Huon Peninsula, Papua New Guinea. *Nature*, **349**, 147–149.

Chappell, J. & Shackleton, N.J. (1986). Oxygen isotopes and sea level. *Nature*, **324**, 137–140.

Clague, J.J. (1976). Quadra Sand and its relation to the late Wisconsin glaciation of southwest British Columbia. *Canadian Journal of Earth Sciences*, **13**, 803–815.

Clague, J.J. (1980). Late Quaternary geology and geochronology of British Columbia, Part 1 – radiocarbon dates. Geological Survey of Canada Paper 80–13, 28 pp.

Clague, J.J. (1981). Late Quaternary geology and geochronology of British Columbia, Part 2. Geological Survey of Canada Paper 80–35, 41 pp.

Clague, J.J. (1983). Glacio-isostatic effects of the Cordilleran ice sheet, British Columbia, Canada. *In*: Smith, D.E. & Dawson, A.G. (Eds), *Shorelines and Isostay*. London, Academic Press, 321–343.

Clague, J.J. (1989). Quaternary geology of the Canadian cordillera. *In*: Fulton, R.J. (Ed.), *Quaternary geology of Canada and Greenland*. Geology of Canada, **1**, pp. 17–95 (Vol. 1 also printed as *Geological Society of America, Geology of North America*, **K-1**).

Clague, J.J., Armstrong, J.E. & Mathewes, W.H. (1980). Advance of the late Wisconsinin Cordilleran ice sheet in southern British Columbia since 22,000 yr BP. *Quaternary Research*, **13**, 322–326.

Clague, J.J., Harper, J.R., Hebda, R.J. & Howes, D.E. (1982). Late Quaternary sea levels and crustal movements, coastal

British Columbia. *Canadian Journal of Earth Sciences*, **19**, 597–618.

Clague, J.J. & James, T.S. (2002). History and isostatic effects of the last ice sheet in southern British Columbia. *Quaternary Science Reviews*, **21**, 71–87.

Clague, J.J., Mathewes, R.W., Guilbault, J.P., Hutchinson, I. & Ricketts, B.D. (1997). Pre-Younger Dryas resurgence of the southwestern margin of the Cordilleran ice sheet, British Columbia, Canada. *Boreas*, **26**, 261–278.

Clark, P.U. & Mix, A.C. (Eds) (2002). Ice sheets and sea levels at the Last Glacial Maximum. *Quaternary Science Reviews*, **21**, 454 pp.

Clarke, G.K.C., Mathews, W.H. & Pack, R.T. (1984). Outburst floods from glacial Lake Missoula. *Quaternary Research*, **22**, 289–299.

Crandell, D.R. (1963). Surficial geology and geomorphology of the Lake Tapps quadrangle, Washington. U.S. Geological Survey Professional Paper 388A, 84 pp.

Crandell, D.R. (1965). The glacial history of western Washington and Oregon. *In*: Wright, H.E., Jr. & Frey, D.G. (Eds), *The Quaternary of the United States*. Princeton University Press, 341–353.

Crandell, D.R., Mullineaux, D.R. & Waldron, H.H. (1958). Pleistocene sequence in the southeastern part of the Puget Sound Lowland, Washington. *American Journal of Science*, **256**, 384–397.

Crandell, D.R., Mullineaux, D.R. & Waldron, H.H. (1966). Age and origin of the Puget Sound trough in western Washington. U.S. Geological Survey Professional Paper 525-B, B132–B136.

Dansgaard, W., Johnsen, S.J., Clausen, H.B. & Langway, C.C. (1971). Climatic record revealed by the Camp Century ice core. *In*: Turekian, K.K. (Ed.), *Late Cenozoic glacial ages*. Yale University Press, Hartford, 37–56.

Deeter, J.D. (1979). Quaternary geology and stratigraphy of Kitsap County, Washington [M.S. thesis]. Bellingham, Western Washington University, Department of Geology, 243 pp.

Dethier, D.P., Pessl, F., Jr., Keuler, R.F., Balzarini, M.A. & Pevear, D.R. (1995). Late Wisconsinan glaciomarine deposition and isostatic rebound, northern Puget Lowland, Washington. *Geological Society of America Bulletin*, **107**, 1288–1303.

Easterbrook, D.J. (1963). Late Pleistocene glacial events and relative sea level changes in the northern Puget Lowland, Washington. *Geological Society of America Bulletin*, **74**, 1465–1484.

Easterbrook, D.J. (1986). Stratigraphy and chronology of Quaternary deposits of the Puget Lowland and Olympic Mountains of Washington and the Cascade Mountains of Washington and Oregon. *Quaternary Science Reviews*, **5**, 145–159.

Easterbrook, D.J. (1994). Chronology of pre-late Wisconsin Pleistocene sediments in the Puget Lowland, Washington. *In*: Lasmanis, R. & Cheney, E.S., conveners, *Regional geology of Washington State*. Washington Division of Geology and Earth Resources Bulletin 80, 191–206.

Easterbrook, D.J., Berger, G.W. & Walter, R. (1992). Laser argon and TL dating of early and middle Pleistocene glaciations in the Puget Lowland, Washington. *Geological Society of America Abstracts with Programs*, **24**, 22.

Easterbrook, D.J. & Briggs, N.D. (1979). Age of the Auburn reversal and the Salmon Springs and Vashon glaciations in Washington. *Geological Society of America Abstracts with Programs*, **11**, 76–77.

Easterbrook, D.J., Briggs, N.D., Westgate, J.A. & Gorton, M. (1981). Age of the Salmon Springs glaciation in Washington. *Geology*, **9**, 87–93.

Easterbrook, D.J., Crandell, D.R. & Leopold, E.B. (1967). Pre-Olympia Pleistocene stratigraphy and chronology in the central Puget Lowland, Washington. *Geological Society of America, Bulletin*, **78**, 13–20.

Easterbrook, D.J., Roland, J.L., Carson, R.J. & Naeser, N.D. (1988). Application of paleomagnetism, fission-track dating, and tephra correlation to lower Pleistocene sediments in the Puget Lowland, Washington. *In*: Easterbrook, D.J. (Ed.), *Dating Quaternary sediments*. Boulder, Colorado, Geological Society of America, Special Paper 227, 165 pp.

Easterbrook, D.J. & Rutter, N.W. (1981). Amino acid ages of Pleistocene glacial and interglacial sediments in western Washington. *Geological Society of America Abstracts with Programs*, **13**, 444.

Easterbrook, D.J. & Rutter, N.W. (1982). Amino acid analyses of wood and shells in development of chronology and correlation of Pleistocene sediments in the Puget Lowland, Washington. *Geological Society of America Abstracts with Programs*, **14**, 480.

Ehlers, J. (1981). Some aspects of glacial erosion and deposition in northern Germany. *Annals of Glaciology*, **2**, 143–146.

Epstein, S., Sharp, R.P. & Gow, A.J. (1970). Antarctic ice sheet: stable isotope analyses of Byrd station cores and interhemispheric climatic implications. *Science*, **168**, 1570–1572.

Fairbanks, R.G. (1989). A 17,000-year glacio-eustatic sea level record: influence of glacial melting dates on the Younger Dryas event and deep ocean circulation. *Nature*, **342**, 637–642.

Friele, P.A. & Clague, J.J. (2002). Readvance of glaciers in the British Columbia Coast Mountains at the end of the last glaciation. *Quaternary International*, **87**, 45–58.

Friele, P.A. & Hutchinson, I. (1993). Holocene sea-level change on the central west coast of Vancouver Island, British Columbia. *Canadian Journal of Earth Sciences*, **30**, 832–840.

Grigg, L.D., Whitlock, C. & Dean, W.E. (2001). Evidence for millennial-scale climate change during Marine Isotope stages 2 and 3 at Little Lake, western Oregon, USA. *Quaternary Research*, **56**, 10–22.

Grootes, P.M., Stuiver, M., White, J.W.C., Johnsen, S. & Jouzel, J. (1993). Comparison of oxygen isotope records from GISP2 and GRIP Greenland ice cores. *Nature*, **366**, 552–554.

Hagstrum, J.T., Booth, D.B. & Troost, K.G. (2002). Magnetostratigraphy, paleomagnetic correlation, and deformation of Pleistocene deposits in the south-central Puget Lowland, Washington. *Journal of Geophysical Research*, **107** (B4), 10.1029/2001JB000557, paper EPM 6, 14 pp.

Hanebuth, T., Stattegger, K. & Grootes, P.M. (2000). Rapid flooding of the Sunda Shelf: A late-glacial sea-level record. *Science*, **288**, 1033–1035.

Hansen, B.S. & Easterbrook, D.J. (1974). Stratigraphy and palynology of late Quaternary sediments in the Puget Sound region. *Geological Society of America Bulletin*, **86**, 587–602.

Harrington, C.R., Plouffe, A. & Jetté, H. (1996). A partial bison skeleton from Chuchi Lake, and its implications for the Middle Wisconsinin environment of Central British Columbia. *Géographie Physique et Quaternaire*, **50**, 73–80.

Heusser, C.J. (1977). Quaternary palynology of the Pacific slope of Washington. *Quaternary Research*, **8**, 282–306.

Heusser, C.J. & Heusser, L.E. (1981). Palynology and paleotemperature analysis of the Whidbey Formation, Puget Lowland, Washington. *Canadian Journal of Botany*, **18**, 136–149.

Heusser, C.J., Heusser, L.E. & Streeter, S.S. (1980). Quaternary temperatures and precipitation for the northwest coast of North America: Nature, v. 286, p. 702–704.

Hewitt, A.T. & Mosher, D.C. (2001). Late Quaternary stratigraphy and seafloor geology of eastern Juan de Fuca Strait, British Columbia and Washington. *Marine Geology*, **177**, 295–316.

Hicock, S.R. (1976). Quaternary geology – Coquitlam-Port Moody area, British Columbia [M.Sc. Thesis]. Vancouver, University British Columbia, 114 pp.

Hicock, S.R. & Armstrong, J.E. (1981). Coquitlam Drift – a pre-Vashon Fraser glacial formation in the Fraser Lowland, British Columbia. *Canadian Journal of Earth Sciences*, **18**, 1443–1451.

Hicock, S.R. & Armstrong, J.E. (1985). Vashon drift – definition of the formation in the Georgia Depression, southwest British Columbia. *Canadian Journal of Earth Sciences*, **22**, 748–757.

Hicock, S.R., Hebda, R.J. & Armstrong, J.E. (1982). Lag of the late-Fraser glacial maximum in the Pacific Northwest – pollen and macrofossil evidence from western Fraser Lowland, British Columbia. *Canadian Journal of Earth Sciences*, **19**, 2288–2296.

Hicock, S.R., Lian, O.B. & Mathewes, R.W. (1999). 'Bond Cycles' recorded in terrestrial Pleistocene sediments of southwestern British Columbia, Canada. *Journal of Quaternary Science*, **14**, 443–449.

Holmes, M.L., Sylvester, R.E. & Burns, R.E. (1988). Postglacial sedimentation in Puget Sound – the container, its history, and its hazards: Seattle, WA. Program with abstracts, Puget Sound Research Meeting, Puget Sound Water Quality Authority.

James, T.S., Clague, J.J., Wang, K. & Hutchinson, I. (2000). Postglacial rebound at the northern Cascadia subduction zone. *Quaternary Science Reviews*, **19**, 1527–1541.

Johnson, S.Y., Dadisman, S.V., Mosher, D.C., Blakely, R.J. & Childs, J.R. (2001). Active tectonics of the Devils Mountain fault and related structures, northern Puget Lowland and eastern Strait of Juan de Fuca region, Pacific Northwest: U.S. Geological Survey Professional Paper 1643, 45 pp.

Johnsen, S.Y., Dansgaard, W., Clausen, H.B. & Langway, C.C. (1972). Oxygen isotope profiles through the Antarctic and Greenland ice sheets. *Nature*, **235**, 429–434.

Johnson, S.Y., Potter, C.J. & Armentrout, J.M. (1994). Origin and evolution of the Seattle fault and Seattle basin, Washington. *Geology*, **22**, 71–74.

Johnson, S.Y., Potter, C.J., Armentrout, J.M., Miller, J.J., Finn, C. & Weaver, C.S. (1996). The southern Whidbey Island fault: an active structure in the Puget Lowland Washington. *Geological Society of America Bulletin*, **108**, 334–354.

Jouzel, J., Hoffmann, G., Parrenin, F. & Waelbroeck, C. (2002). Atmospheric oxygen 18 and sea-level changes. *Quaternary Science Reviews*, **21**, 307–314.

Kovanen, D.J. (2002). Morphologic and stratigraphic evidence for Allerod and Younger Dryas age glacier fluctuations of the Cordilleran Ice Sheet, British Columbia, Canada, and northwest Washington, USA. *Boreas*, **31**, 163–184.

Kovanen, D.J. & Easterbrook, D.J. (2001). Late Pleistocene, post-Vashon, alpine glaciation of the Nooksack drainage, North Cascades, Washington. *Geological Society of America Bulletin*, **113**, 274–288.

Lambeck, K., Yokoyama, Y., Johnston, P. & Purcell, A. (2000). Global ice volumes at the Last Glacial Maximum. *Earth and Planetary Sciences Letters*, **181**, 513–527.

Lambeck, K., Yokoyama, Y. & Purcell, T. (2002). Into and out of the last glacial maximum sea-level changes during oxygen isotope stages 3 and 2. *Quaternary Science Reviews*, **21**, 343–360.

Lea, D.W., Martin, P.A., Pak, D.K. & Spero, H.J. (2002). Reconstructing a 350 ky history of sea level using planktonic Mg/Ca and oxygen isotope records from a Cocos Ridge core. *Quaternary Science Reviews*, **21**, 283–293.

Linden, R.H. & Schurer, P.J. (1988). Sediment characteristics and sealevel history of Royal Roads Anchorage, Victoria, British Columbia. *Canadian Journal of Earth Sciences*, **25**, 1800–1810.

Lorius, C., Jouzel, J., Ritz, C., Merlivat, L., Barkov, N.I., Korotkevitch, Y.S. & Kotlyakov, V.M. (1985). A 150,000-year climatic record from Antarctic ice. *Nature*, **316**, 591–596.

Lowell, T.V., Hayward, R.K. & Denton, G.H. (1999). Role of climate oscillations in determining ice-margin position – Hypothesis, examples, and implications. *In*: Mickelson, D.M. & Attig, J.W. (Eds), *Glacial processes, past and present*. Geological Society of America Special Paper 337, 193–203.

Mahan, S.A., Booth, D.B. & Troost, K.G. (2000). Luminescence dating of glacially derived sediments – a case study for the Seattle Mapping Project: Vancouver, British Columbia, Abstracts with Programs, 96th Annual Meeting Cordilleran Section, Geological Society of America, A-27.

Mahoney, J.B., Brandup, J., Troost, K.G. & Booth, D.B. (2000). Geochemical discrimination of episodic glaciofluvial sedimentation, Puget Lowland, Washington: Vancouver, British Columbia, Abstracts with Programs, 96th Annual Meeting Cordilleran Section, Geological Society of America, A-27.

Mankinnen, E.A. & Dalrymple, G.B. (1979). Revised geomagnetic polarity time scale for the interval 0–5 m.y. b.p. *Journal of Geophysical Research*, **84**, 615–626.

Mathewes, R.W. & Heusser, L.E. (1981). A 12,000-year palynological record of temperature and precipitation trends in the southwestern British Columbia. *Canadian Journal of Botany*, **59**, 707–710.

Mathews, W.H. (1974). Surface profile of the Laurentide ice sheet in its marginal areas. *Journal of Glaciology*, **13**, 37–43.

Mathews, W.H., Fyles, J.G. & Nasmith, H.W. (1970). Postglacial crustal movements in southwestern British Columbia and adjacent Washington state. *Canadian Journal of Earth Sciences*, **7**, 690–702.

Mehringer, P.J., Jr., Sheppard, J.C. & Foit, F.F. (1984). The age of Glacier Peak tephra in west-central Montana. *Quaternary Research*, **21**, 36–41.

Mickelson, D.M., Clayton, L., Fullerton, D.S. & Borns, H.W., Jr. (1983). The late Wisconsin glacial record of the Laurentide ice sheet in the United States. *In*: Wright, H.E., Jr. & Porter, S.C. (Eds), *The Quaternary of the United States*. University of Minnesota Press, **1**, 3–37.

Minard, J.M. & Booth, D.B. (1988). Geologic map of the Redmond 7. 5' quadrangle, King and Snohomish Counties, Washington: U.S. Geological Survey Miscellaneous Field Investigations Map MF-2016, scale 1: 24,000.

Muhs, D.R., Kennedy, G.L. & Rockwell, T.K. (1994). Uranium-series ages of marine terrace corals from the Pacific coast of North America and implications for last-interglacial sea level history. *Quaternary Research*, **42**, 72–87.

Muhs, D.R., Thorson, R.M., Clague, J.J., Mathews, W.H., McDowell, P.F. & Kelsey, H.M. (1987). Pacific Coast and Mountain System. *In*: Graf, W.L. (Ed.), *Geomorphic systems of North America*. Geological Society of America, Centennial, **2**, 517–581.

Mullineaux, D.R., Waldron, H.H. & Rubin, M. (1965). Stratigraphy and chronology of late interglacial and early Vashon time in the Seattle area. Washington: U.S. *Geological Survey Bulletin* 1194-O, O1-O10.

Mullins, H.T. & Hinchey, E.J. (1989). Erosion and infill of New York Finger Lakes: Implications for Laurentide ice sheet deglaciation. *Geology*, **17**, 622–625.

Noble, J.B. & Wallace, E.F. (1966). Geology and groundwater resources of Thurston County, Washington. Washington Division of Water Resources Water-Supply Bulletin, **10**, 254.

O'Connor, J.E. & Baker, V.R. (1992). Magnitudes and implications of peak discharges from glacial Lake Missoula. *Geological Society of America Bulletin*, **104**, 267–279.

O'Connor, J.E. & Waitt, R.B. (1995). Beyond the channeled Scabland – field trip to Missoula flood features in the Columbia, Yakima, and Walla Walla valleys of Washington and Oregon. *Oregon Geology*, **57**, 51–60, 75–86, 99–115.

Oviatt, C.G., Currey, D.R. & Sack, D. (1992). Radiocarbon chronology of Lake Bonneville, eastern Great Basin, USA. *Palæogeography, Palaeoclimatology, Palæoecology*, **99**, 225–241.

Pardee, J.T. (1942). Unusual currents in glacial Lake Missoula, Montana. *Geological Society of America Bulletin*, **53**, 1569–1599.

Paterson, W.S.B. (1981). *The physics of glaciers*. Oxford, Pergamon Press, 380 pp.

Patterson, C.J. (1994). Tunnel-valley fans of the St. Croix moraine, east-central Minnesota, USA. *In*: Warren, W.P. & Croot, D.G. (Eds), *Formation and deformation of glacial deposits – proceedings of the meeting of the Commission on the Formation and Deformation of Glacial Deposits*. Dublin, Ireland, May 1991: Rotterdam, Balkema, p. 69–87.

Peltier, W.R. (2002). On eustatic sea level history: Last Glacial Maximum to Holocene. *Quaternary Science Reviews*, **21**, 377–396.

Pessl, F., Jr., Dethier, D.P., Booth, D.B. & Minard, J.P. (1989). Surficial geology of the Port Townsend 1:100,000 quadrangle, Washington. U.S. Geological Survey Miscellaneous Investigations Map I-1198F.

Petit, J.R. *et al.* (1999). Climate and atmospheric history of the past 420,000 years from the Vostok Ice Core, Antarctica. *Nature*, **399**, 429–436.

Plouffe, A. & Jette, H. (1997). Middle Wisconsinan sediments and paleoecology of central British Columbia sites at Necoslie and Nautley rivers. *Canadian Journal of Earth Sciences*, **34**, 200–208.

Porter, S.C. (1978). Glacier Peak tephra in the North Cascade range, Washington – Stratigraphy, distribution, and relationship to late-glacial events. *Quaternary Research*, **10**, 30–41.

Porter, S.C., Pierce, K.L., & Hamilton, T.D. (1983). Late Wisconsin mountain glaciation in the western United States. *In*: Wright, H.E., Jr. (Ed.), *Late Quaternary environments of the United States*. Minneapolis, University of Minnesota Press, **1**, 71–111.

Porter, S.C. & Swanson, T.W. (1998). Radiocarbon age constraints on rates of advance and retreat of the Puget lobe of the Cordilleran ice sheet during the last glaciation. *Quaternary Research*, **50**, 205–213.

Pratt, T.L., Johnson, S.Y., Potter, C.J., Stephenson, W.J. & Finn, C.A. (1997). Seismic reflection images beneath Puget Sound, western Washington State – the Puget Lowland thrust sheet hypothesis. *Journal of Geophysical Research, B, Solid Earth and Planets*, **102**, 27,469–27,489.

Prest, V.K. (1969). Retreat of recent and Wisconsin ice in North America: Geological Survey of Canada Map 1257A, scale 1: 5,000,000.

Richmond, G.M. (1986). Tentative correlations of deposits of the Cordilleran ice-sheet in the northern Rocky Mountains. *Quaternary Science Reviews*, **5**, 129–144.

Rigg, G.B. & Gould, H.R. (1957). Age of Glacier Peak eruption and chronology of postglacial peat deposits in Washington and surrounding areas. *American Journal of Science*, **255**, 341–363.

Shackleton, N.J. (1987). Oxygen isotopes, ice volume and sea level. *Quaternary Science Reviews*, **6**, 183–190.

Shackelton, N.J., Berger, A. & Peltier, W.R. (1990). An alternative astronomical calibration of the lower Pleistocene time-scale based on ODP site 677a. *Transactions of the Royal Society of Edinburgh, Earth Sciences*, **81**, 251–261.

Shaw, J. & Gilbert, R. (1990). Evidence for large-scale subglacial meltwater flood events in southern Ontario and northern New York State. *Geology*, **18**, 1169–1172.

Thorson, R.M. (1980). Ice sheet glaciation of the Puget Lowland, Washington, during the Vashon stade (late Pleistocene). *Quaternary Research*, **13**, 303–321.

Thorson, R.M. (1989). Glacio-isostatic response of the Puget Sound area, Washington. *Geological Society of America Bulletin*, **101**, 1163–1174.

Troost, K.G. (1999). The Olympia nonglacial interval in the southcentral Puget Lowland, Washington [M.S. thesis]. Seattle, University of Washington, 123 pp.

Troost, K.G. (2002). Summary of the Olympia nonglacial interval (MIS3) in the Puget Lowland, Washington. Corvallis, Oregon, Abstracts with Programs, 98th Annual Cordilleran Section Meeting, Geological Society of America, p. A109.

Troost, K.G., Booth, D.B. & Borden, R.K. (2003). Geologic map of the Tacoma North 7.5-minute quadrangle, Washington. U.S. Geological Survey Miscellaneous Field Investigation, scale 1: 24,000 (in press).

Waelbroeck, C., Labeyrie, L., Michel, E., Duplessy, J.C., McManus, J.F., Lambeck, K., Balbon, E. & Labracherie, M. (2002). Sea-level and deep water temperature changes derived from benthic foraminifera isotope records. *Quaternary Science Reviews*, **21**, 295–305.

Waitt, R.B. (1980). About forty last-glacial Lake Missoula jökulhlaups through southern Washington. *Journal of Geology*, **88**, 653–679.

Waitt, R.B. (1984). Periodic jökulhlaups from Pleistocene glacial Lake Missoula – new evidence from varved sediment in northern Idaho and Washington. *Quaternary Research*, **22**, 46–58.

Waitt, R.B. (1985). Case for periodic, colossal jökulhlaups from Pleistocene glacial Lake Missoula. *Geological Society of America Bulletin*, **96**, 1271–1286.

Waitt, R.B. (1994). Scores of gigantic, successively smaller Lake Missoula floods through Channeled Scabland and Columbia valley. *In*: Swanson, D.A. & Haugerud, R.A. (Eds), *Geologic field trips in the Pacific Northwest*. Seattle, Department of Geological Sciences, University of Washington, **1** (Chapter 1K), 88 pp.

Waitt, R.B., O'Connor, J.E. & Harpel, C.J. (2000). Varying routings of repeated colossal jökulhlaups through the channeled scabland of Washington, USA [abst]. *Orkustofnun Rept. OS-2000/036, Reykjavík, Extremes of the Extremes Conference*, 27.

Waitt, R.B. & Thorson, R.M. (1983). The Cordilleran ice sheet in Washington, Idaho, and Montana. *In*: Porter, S.C. & Wright, H.E., Jr. (Eds), *Late-Quaternary environments of the United States*. University of Minnesota Press, **1**, 53–70.

Walcott, R.I. (1970). Isostatic response to loading of the crust in Canada. *Canadian Journal of Earth Sciences*, **7**, 716–726.

Waldron, H.H., Liesch, B.A., Mullineaux, D.R. & Crandell, D.R. (1962). Preliminary geologic map of Seattle and vicinity, Washington. *U.S. Geological Survey Miscellaneous Investigations Map I-354*.

Walters, K.L. & Kimmel, G.E. (1968). Ground-water occurrence in stratigraphy of unconsolidated deposits, central Pierce County, Washington. *Washington Department of Water Resources Water-Supply Bulletin*, **22**, 428 pp.

Wells, R.E., Weaver, C.S. & Blakely, R.J. (1998). Fore-arc migration in Cascadia and its neotectonic significance. *Geology*, **26**, 759–762.

Westgate, J.A., Easterbrook, D.J., Naeser, N.D. & Carson, R.J. (1987). Lake Tapps tephra – an early Pleistocene stratigraphic marker in the Puget Lowland, Washington. *Quaternary Research*, **28**, 340–355.

Whitlock, C. & Grigg, L.D. (1999). Paleoecological evidence of Milankovitch and sub-Milankovitch climate variations in the western U.S. during the late Quaternary. *In*: Webb, R.S., Clark, P.U. & Keigwin, L.D. (Eds), *The roles of high and low latitudes in millennial-scale global climate change*. American Geophysical Union, 227–241.

Whitlock, C., Sarna-Wojcicki, A.M., Bartlein, P.J. & Nickmann, R.J. (2000). Environmental history and tephrostratigraphy at Carp Lake, southwestern Columbia basin, Washington, USA. *Paleogeography, Paleoclimatology, Palaeoecology*, **155**, 7–29.

Willis, B. (1898). Drift phenomena of Puget Sound. *Geological Society of America Bulletin*, **9**, 111–162.

Yokoyama, Y., Lambeck, K., De Deckker, P.P.J., Johnson, P. & Fifield, L.K. (2000). Timing of the last glacial maximum from observed sea-level minima. *Nature*, **406**, 713–716.

Controls, history, outbursts, and impact of large late-Quaternary proglacial lakes in North America

James T. Teller

Department of Geological Sciences, University of Manitoba, Winnipeg, Man., Canada R3T 2N2

Introduction

During recent decades, there has been an increasing recognition that lacustrine records provide one of the most continuous and responsive archives of environmental change. These can be used to establish past hydrological, biological, and climatic changes and to serve as baselines that guide our predictions of future change. Our understanding of individual proglacial lakes and, importantly, their relation with other lakes and other systems has grown enormously. Thousands of new ^{14}C dates, some on millimeter-sized seeds and organic fragments, plus the development of luminescence, amino acid, and paleomagnetic dating and a host of other radiometric and relative dating techniques (e.g. Bradley, 1999; Last & Smol, 2001a), have brought a great improvement in the resolution and correlation of events. Recognition that ^{14}C years are not the direct equivalent of calendar years has led to an improved understanding of the link between cause and effect. Coring and analytical techniques have improved (e.g. Last & Smol, 2001a, b). A new vitality in the study of pollen and plant microfossils, diatoms and siliceous microfossils, pigments, ostracodes, insects, cladocera, and other organics (see Smol *et al.*, 2001a, b; Warner, 1990), some of which were barely studied in the paleolimnological record 40 years ago, have provided important new insight into paleolake conditions and lake history. The use of stable isotope ratios, such as $\delta^{18}O$ and $\delta^{13}O$ in shelled organisms and cellulose has added valuable paleoenvironmental information about water temperature, chemistry, and source, and other ratios and isotopes such as δD, ^{10}Be, and ^{36}Cl are promising to provide even more information in our quest to understand these lakes (e.g. Gosse & Phillips, 2001). The past four decades have seen a revolution in Quaternary studies, and proglacial lakes have contributed significantly to our understanding.

This chapter will briefly review the controls on proglacial lake systems and present a capsule of our current understanding of the history and interrelation of the largest proglacial lakes in North America, specifically those that fringed the Laurentide Ice Sheet (LIS). Many other proglacial lakes formed in individual valleys dammed by the LIS and by Cordilleran ice, as well as in intermontane basins of western North America (see chapters by Benson and by Booth *et al.*), but these are not included in this review because they were comparatively small. The literature on proglacial lakes has grown exponentially in recent decades, as has the interest in them. Although more than a century of data gathering and interpretation of North American proglacial lakes has provided a wealth of knowledge about the postglacial history of the continent, our understanding of the interactive nature of these systems has only recently matured. Given the vast region covered by proglacial lakes in North America (Fig. 1), much of it in remote regions of Canada, much more is needed before we fully understand the complex story. The role of *sub*glacial water storage and its release (e.g. Shaw, 1996, 2002; Shoemaker, 1991, 1992) is an important, but contentious, issue that needs further study, especially because of its potential relationship to ice dynamics and the impact it may have had on linked proglacial lakes and hydrological systems beyond the ice margin. The last part of this chapter will explore some new developments in our understanding of the relationship of large proglacial lakes to global-scale changes in oceans and climate.

Controls on the Depth and Extent of Ice-Marginal Lakes

Proglacial lakes along the LIS in North America covered a total of more than 2 million km^2 (Fig. 1). In some areas, such as the Great Lakes basins, glaciation only led to expansion and contraction of existing bodies of water. In many cases, however, proglacial lakes formed in regions that had not been closed basins. During glaciation, northward and eastward drainage from the continent was impeded, because the LIS formed in high latitudes and eventually expanded into lowland areas such as the Hudson Bay basin. Thus, most new lakes formed north of the divide between the Hudson Bay-Arctic watershed and the Great Lakes-Mississippi River watershed (see Fig. 1). South of this divide, where runoff was only occasionally impeded by the LIS, proglacial lakes formed in pre-existing basins. Thus, as the LIS expanded southward, or retreated northward, lakes formed and changed their extent and depth in response to: (1) the location of the ice margin; (2) the elevation of the lowest point around the lake margin (which served as the overflow outlet); (3) differential isostatic rebound; and (4) the topography of the region south of the ice margin which helped control the outline of the lake. All of these controlling parameters changed through time.

Because the LIS retreated down slope, new, lower overflow routes were periodically available, each resulting in a drop in level of the proglacial lake. As discussed later, this occurred abruptly and sometimes re-routed large areas of runoff. Large volumes of water were transferred from one basin to another at these times. During glacial advances, there were rapid rises in lake level as ice closed outlets and diverted runoff from one drainage basin to another. Teller (1995a) describes the main topographic thresholds in North America

DEVELOPMENTS IN QUATERNARY SCIENCE
VOLUME 1 ISSN 1571-0866
DOI:10.1016/S1571-0866(03)01003-0

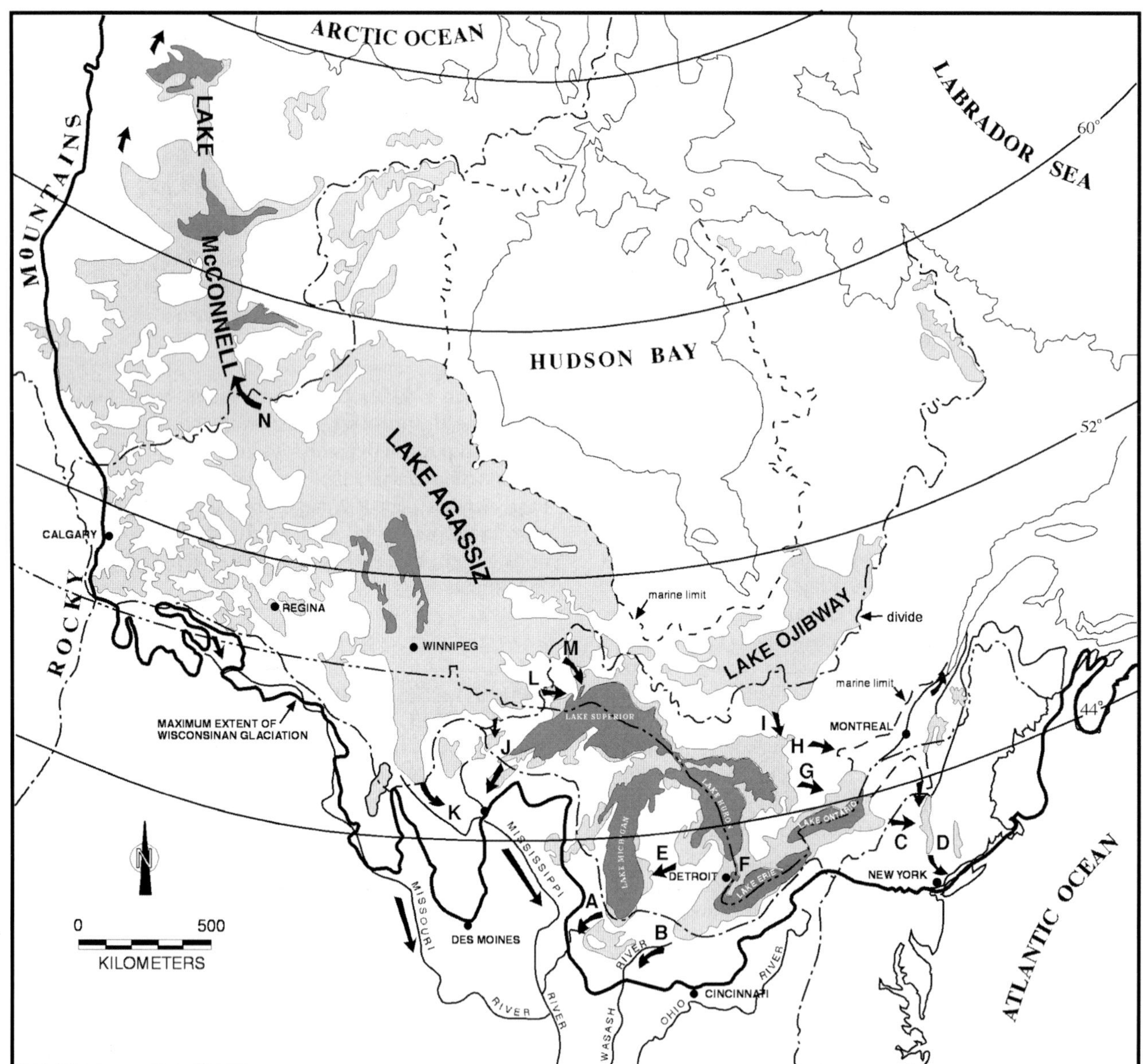

Fig. 1. Total area covered by large proglacial lakes in North America (grey) during the last retreat of the LIS (after Teller, 1987, Fig. 2). Major continental divides shown by dash-dot lines. Names of the major lakes are shown in the region where they formed. Major overflow routes from lakes are shown by arrows; letters identify their names as follows: A = Chicago outlet, B = Wabash River Valley, C = Mohawk Valley, D = Hudson Valley, E = Grand River Valley, F = Port Huron outlet, G = Fenelon Falls outlet, H = North Bay outlet, I = Temiskaming outlet, J = Duluth outlet, K = Minnesota River Valley, L = Kaminiskwia outlet, M = eastern Agassiz outlets, N = Clearwater outlet. Extent of proglacial lakes in Hudson Bay Lowland and St. Lawrence Lowland are not shown where lacustrine sediments are now buried by marine sediment.

that led to continental-scale amalgamations and subdivisions of proglacial lake basins. These critical topographic thresholds were located in (Fig. 2): (1) the Hudson Bay Lowland that leads to the North Atlantic Ocean; (2) two areas in the St. Lawrence Lowland; (3) Straits of Mackinac at the eastern end of the Superior basin; (4) Eastern Agassiz outlets into Lake Superior basin; and (5) Clearwater-Mackenzie River valley to the Arctic Ocean. These thresholds commonly controlled changes in lake level and were key in the history of North American proglacial lakes and the routing of glacial runoff from the continent to the oceans.

Equally important in controlling the depth and extent of proglacial lakes was differential isostatic rebound, which was greatest in the northern region where ice was thickest, because it controlled the elevation of lake outlets and the relative elevation of the lake floor across the basin. This resulted in the southward transgression of lakes south of the isobase (line of equal isostatic rebound) through the outlet,

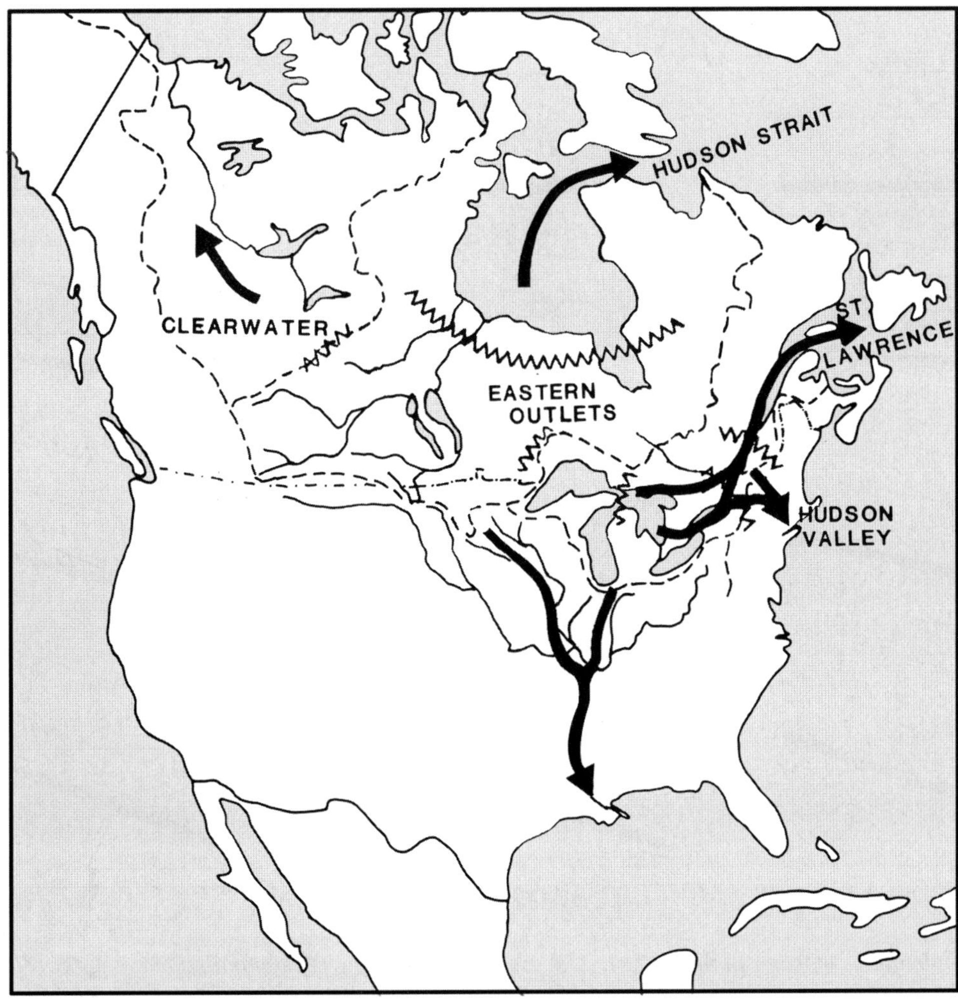

Fig. 2. Critical paleohydrological thresholds (serrated lines) associated with the LIS. These helped control proglacial lake development and major overflow routes shown by arrows (after Teller, 1995a). Major modern divides indicated by dashed lines.

and regression of the lake to the north of that isobase. Even today there is a southward shift of large water bodies such as Lake Winnipeg (e.g. Lewis *et al.*, 2001) because this differential rebound continues.

The combined controls of differential isostatic rebound and variable outlet location on lake level, beach formation, and distribution of offshore sediment has only been appreciated in recent years; this can be illustrated by a simple bathtub model. In Fig. 3A–C, overflow from the basin is shown through three different outlets: (1) southern end; (2) northern end; and (3) "middle." Because overflow from most of North America's great glacial lakes involved outlets in at least two different locations (e.g. Larsen, 1987; Lewis & Anderson, 1989), and because these outlets eroded during use, the histories of lake levels and beach formation were very complex. Figure 4 is a series of bathtub cartoons that show an example of the complexity that occurred, which is similar to what occurred during the early history of Lake Agassiz. Note that after outlets change, waters in any given area may change from a regressing to a transgressing mode (or vice versa) (Teller, 2001). In short, proglacial lake levels continually, and often rapidly, fluctuated, transgressing and/or regressing throughout their history.

Beaches and wave-cut cliffs formed in response to these changes, and most *large* beaches around proglacial lakes outline the maximum point of transgression during any lake stage; lake levels never remained stable and lower beaches were destroyed as water levels rose (Teller, 2001). In areas where the lake was regressing, only smaller beaches formed, mainly as a result of storm events when sediment supply was sufficient. These relict strandlines are today higher in elevation in the north than the south, with a difference of tens of meters in basins of the southern Great Lakes and >150 m across the northern basins of lakes McConnell, Agassiz, and Ojibway. Because the rate of rebound was greater during the early stages of deglaciation, older beaches have a steeper slope than do younger beaches along a specific isobase; this can be used as a relative age indicator in a proglacial lake basin.

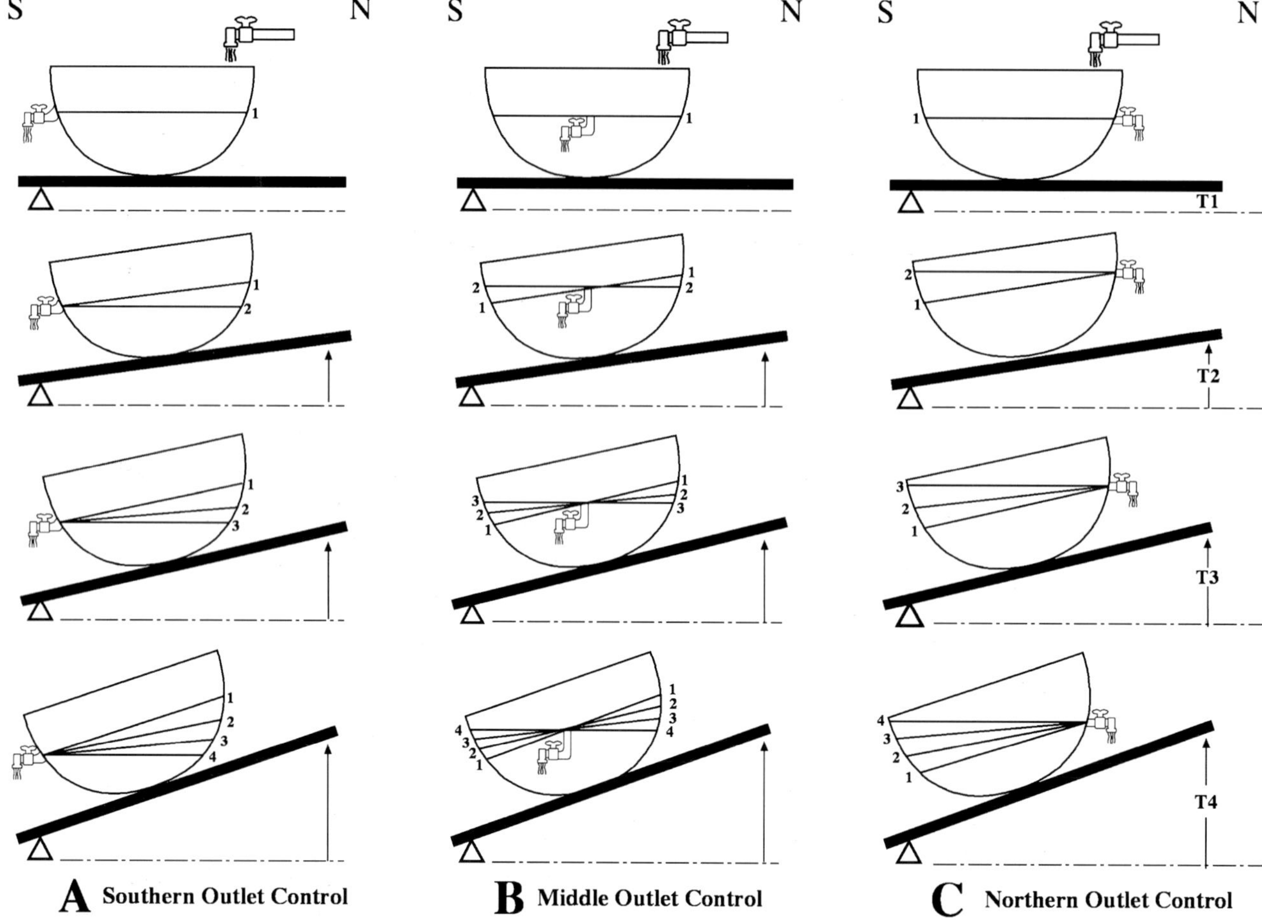

Fig. 3. Cartoons of bathtubs of water at times (T) 1, 2, 3, and 4, experiencing differential rebound and overflow through three different outlets. Strandlines are numbered in sequence of formation; for simplicity, beach curvature due to non-linear differential rebound is not shown (after Larsen, 1987; Teller, 2001). (A) Overflow from southern end of basin. Lake regresses everywhere through time. (B) Overflow from "middle" of basin. Through time, lake transgresses to south of outlet, regresses to north. (C) Overflow from northern end of basin. Lake transgresses upslope throughout basin.

A Brief History of Large Proglacial Lakes in North America

The largest proglacial lakes in North America during the last deglaciation were Lake McConnell, Lake Agassiz, Lake Ojibway, and the various lakes in the Great Lakes basins (Fig. 1). The record of older glaciations and proglacial lakes is incomplete, and their ages are not well constrained. Based on the magnetostratigraphic record, Barendregt & Irving (1998) concluded that the LIS was not as extensive during the early Pleistocene. If true, the LIS would not have provided a continuous barrier across the Hudson Bay Lowland, nor have expanded far enough west to impound drainage in the Mackenzie River Valley to the Arctic Ocean, and ancestors of glacial Lakes Agassiz, Ojibway and McConnell would not have formed. As a result, diversion of meltwater into the Mississippi River and Great Lakes from the vast region of Canada between the Rocky Mountains and Great Lakes would not have occurred, resulting in a dramatically different hydrological scenario from that of the last (and best known) glacial period, as well as a notably different role of North American fresh waters in ocean systems and climate (see later section).

The stratigraphic record of the last deglaciation is complex, and the chronology is not well controlled in some areas. Changes in proglacial lakes along the margin of the LIS were both frequent and rapid. Lakes periodically merged with others, sometimes forming bodies of water that spanned half a continent. At times, events in the upper end of the drainage system were responsible for re-directing millions of square kilometers of runoff from one lake basin to another. At other times, isostatic rebound or glacial advance deprived "downstream" basins of that water. Basin linkages were often abrupt, resulting in huge outbursts of overflow from one basin to the other. Now that we recognize this interrelationship and appreciate the magnitude and rapidity of the hydrological links, it is essential to examine the *whole* proglacial system along the LIS to understand how it worked. This is a challenge, because interpreted changes in one basin must be discussed (and tempered) with interpreted changes in hydrologically linked basins, even though these lakes may be

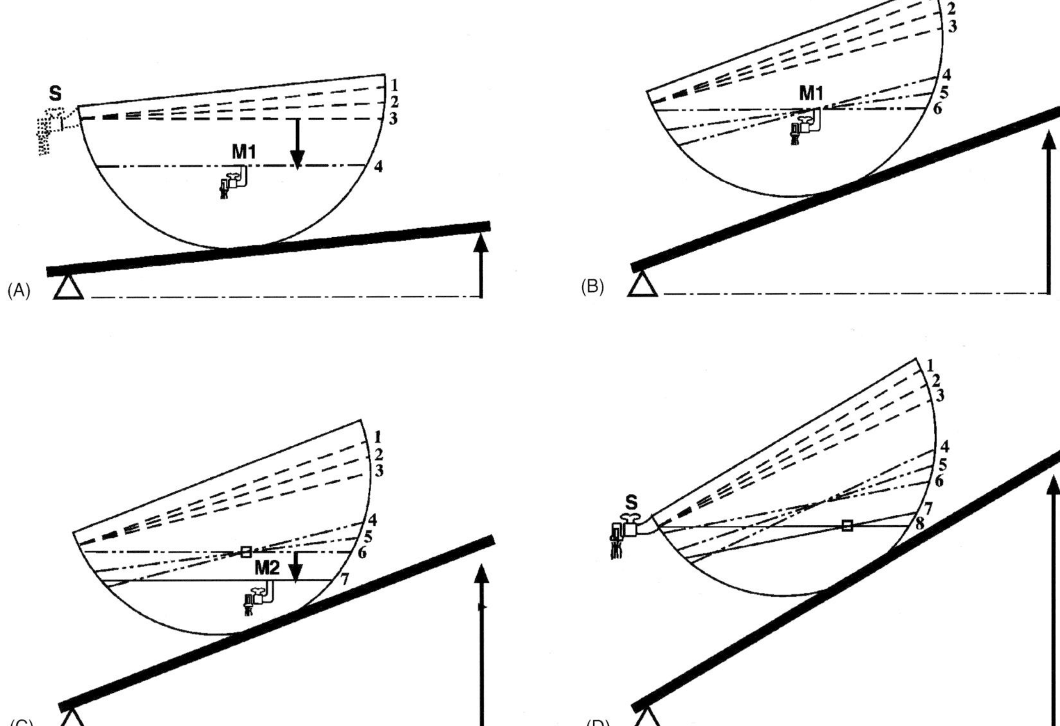

Fig. 4. Hypothetical development of lake levels through time as outlet changes from south to "middle" and back to south in basin. Sequence begins after southward drainage is abandoned, having formed as shown in Fig. 3A. Note that most large proglacial basins in North America experienced more outlet changes than shown here and that outlet erosion would also have impacted on lake level. (A) Overflow through southern outlet (S) ends after regressive lake levels (1, 2, 3) formed. Outlet then shifts to "middle" of basin (M1) and lake abruptly drops to level 4. No new beaches formed. (B) Overflow through outlet M1, with waters transgressing to south and regressing to north of outlet (levels 5, 6); beaches south of outlet reworked upslope; no distinct beaches formed north of outlet. (C) New lower outlet opens in "middle" of basin and overflow shifts to M2, with lake abruptly dropping to level 7; beach abandoned at transgressive maximum of level 6. (D) Lake transgresses to southern end of basin (level 8) and begins to flow out through outlet S, abandoning the now-higher outlet M2. As differential rebound continues, the lake will decline in level (regress) below level 8 throughout the basin.

thousands of kilometers apart. Perhaps our biggest advance in recent decades has been this effort to understand the whole system, even though there are still many unresolved issues and, undoubtedly, will be many surprises ahead.

Although the focus of this chapter is on large proglacial lakes, many smaller lakes developed along the ice margin as it retreated down slope (Teller, 1987). Many of these formed across the Canadian Prairies, commonly in north- or east-draining valleys and lowlands, and can be seen in Fig. 1 west of Lakes Agassiz and McConnell. Individual lakes typically were several thousand km^2 in extent (although some were $>10{,}000\,km^2$) and survived only a century or two (Teller & Kehew, 1994); many drained catastrophically into Lake Agassiz (e.g. Kehew, 1982). These lakes were integral in the development of postglacial drainage. Their abrupt drainage scoured deep valleys, led to the domino-like drainage of other small lakes downstream, and deposited flood fans in those basins (Kehew & Lord, 1986).

A number of syntheses of the history of North American proglacial lake systems have been published in the past few decades, many in (or as part of) books and special journal issues, including Prest (1970), Clayton & Moran (1982), Mickelson *et al.* (1983), Teller & Clayton (1983), Karrow & Calkin (1985), Teller (1987, 1995b), Dyke & Prest (1987), Fulton (1989), Lewis & Anderson (1989), Teller & Kehew (1994), Licciardi *et al.* (1999), and Teller & Leverington (pers. comm., 2003); these syntheses provide hundreds of additional references. The following synthesis is taken from these and other papers published mainly in the past 20 years.

Speculation About the History of Lakes During Advance of the LIS

Accumulation of continental ice over the upland areas of eastern Canada sometime after 110,000 years ago (marine oxygen isotope stage (MIS) 5e) probably was gradual, and did not initially result in the formation of large proglacial lakes. Not until ice thickened and expanded into the lower St. Lawrence Valley were major drainage basins impounded. Ice first advanced across southern Quebec into the St. Lawrence Lowland about 80,000–75,000 years ago

(Lamothe *et al.*, 1992; Occhietti, 1989), and may have extended into the Ontario basin and northern New England, eventually becoming extensive in the Great Lakes (see Clark, 1992). It is not known whether drainage west of the Great Lakes was impounded by this advance.

Subaerial exposure and a warmer climate in the Hudson Bay Lowland followed. Extensive glaciolacustrine sediment, dated by thermoluminescence at 40,000 years in Manitoba, suggests that runoff to Hudson Bay was again dammed by readvancing Late Wisconsinan ice (Thorleifson *et al.*, 1992); a final push of the LIS to its maximum extent across all proglacial lake basins occurred after 30,000 years ago (Dyke *et al.*, 2002).

As the Last Glacial Maximum approached, the LIS blocked a series of drainages (Fig. 2). The following speculative sequence of blockage during glacial advance illustrates the controls on glacial lakes (cf. Teller, 1995a):

(1) When ice first invaded the lower St. Lawrence Valley, east of Montreal, a lake would have formed in the western part of the lowland, re-routing overflow from the Great Lakes southward through the Lake Champlain and Hudson Valleys (CV and HV in Fig. 5; D in Fig. 1).
(2) Further ice advance across the Lake Champlain-Hudson outlet forced waters to deepen and expand westward into the Ontario basin, which overflowed through the Mohawk Valley (MV in Fig. 5; C in Fig. 1).
(3) Expansion of the LIS over the eastern Great Lakes forced water levels throughout those basins to rise and seek outlets into the Mississippi River basin via the Wabash and Illinois River Valleys (B and A in Fig. 1).
(4) Amalgamation of the Keewatin (western) and Labrador (eastern) ice centers eventually ponded waters in the Hudson Bay Lowland. Depending on the configuration of the ice margin, one or more ice-marginal lakes formed across the lowest terrain of Quebec, Ontario, Manitoba, and Saskatchewan, north of the continental divide (see Fig. 1); these were ancestors of Lakes Agassiz and Ojibway, which formed during the retreating phase of the LIS. Initially, overflow would have been from the Ojibway basin into the Ottawa River Valley (I and H in Fig. 1) but, as the eastern (Ojibway) and western (Agassiz) regions became separated by the ice, western overflow would have been diverted into the Superior basin (L and M in Fig. 1) and through the Clearwater outlet in the northwestern corner of the Agassiz basin (N in Fig. 1). Isostatic adjustment and outlet erosion would have combined with ice-margin location to dictate the extent and depth of waters.
(5) The westward expansion of LIS to the Mackenzie Mts. of the Yukon Territory in northwestern Canada impounded drainage of the >1.5 million km^2 Mackenzie River drainage basin, forming the predecessor of glacial Lake McConnell; overflow was west across the mountains to the Pacific Ocean (Lemmen *et al.*, 1994).
(6) After ice blocked outlets L, M, and N from the Agassiz basin (Fig. 1), causing waters to rise further, overflow was directed south through the Minnesota River Valley (K in Fig. 1) into the Mississippi River. To the west, lakes overflowed across the divide into the Missouri River drainage basin.

Eventually, all of these proglacial lake basins were completely overridden by the LIS, bringing this history of lakes during ice advance to an end. The subsequent northward retreat of ice is much better documented and, in a general way, followed a reverse pattern of lake evolution to that during the advancing stages, with lakes following on the heels of the retreating LIS, declining in level as new lower outlets were opened and amalgamating with (or overflowing into) other ice-marginal lakes as ice retreated down slope.

History and Interrelationship of Lakes Formed During Ice Retreat, Before 13,000 ^{14}C yr B.P.

The first lakes to form along the LIS margin during its retreat were in the southern parts of the Great Lakes' basins. The ice margin fluctuated throughout this period, so lakes formed, were overridden, and reformed on numerous occasions. As new and lower outlets were opened, lake levels fell but immediately began to transgress across the newly emergent lake floor because of differential isostatic rebound. Figure 6 shows the fluctuating water level in the Huron basin interpreted by Lewis *et al.* (1994), and illustrates the frequent and rapid changes that occurred throughout the Great Lakes. Although these fluctuations were not always synchronous, because they were controlled by the cryodynamics of different ice lobes, there was an overall similarity of expansion and contraction during deglaciation. Ice-margin retreats (interstades) alternated with readvances (stades), resulting in an overall rise and fall pattern of lake levels. In later stages, climate may have played a role (Blasco, 2001; Lewis *et al.*, 2002).

As shown in Fig. 7A, the Erie and southern Huron basins contained ice-marginal lakes during the Erie Interstade, perhaps between about 16,000 and 15,000 ^{14}C yr B.P.; ice retreated enough in the Ontario basin to allow overflow at this time through the Mohawk River Valley (M of Fig. 7A; MV of Fig. 5) to the Atlantic Ocean via the Hudson River Valley of New York state (C and D in Fig. 1). This outlet and these basins were overridden during the Port Bruce Stade (Fig. 7B). The subsequent retreat of ice led to a series of lakes in these basins and in the western Ontario basin during the Mackinaw Interstade (Fig. 7C, D, and E), as well as in the Michigan basin. Outflow from glacial Lake Maumee in the Erie basin first occurred through the Wabash River Valley of Indiana (W in Fig. 7C; B in Fig. 1) and included at least one catastrophic outburst (Fraser & Bleuer, 1988). When the ice margin retreated far enough, proglacial lakes in the Michigan basin received overflow from the Huron basin via the Imlay and Grand River Valley channels across the state of Michigan (Kehew, 1993) (Fig. 7D; E in Fig. 1). Retreat north of the isostatically depressed Fenelon Falls outlet (F in Fig. 7E; G in

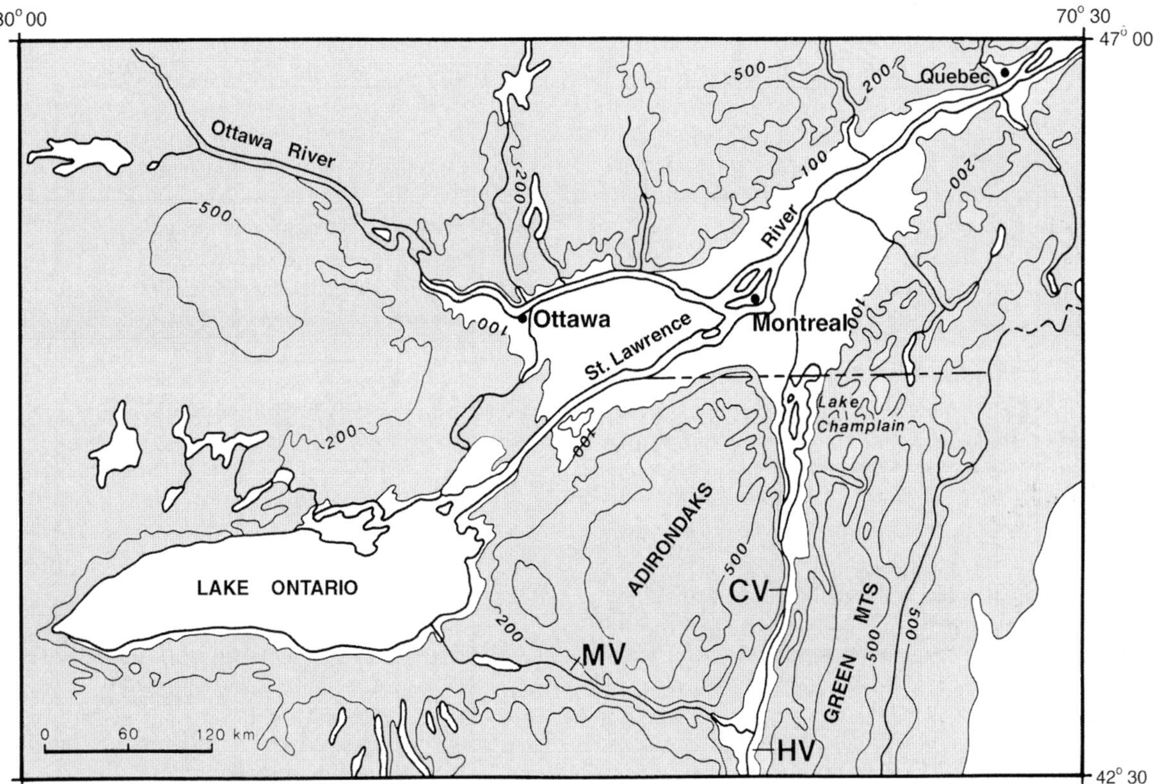

Fig. 5. Topography of the St. Lawrence Valley Lowland and adjacent area (elevations in m and shaded above 100 m), showing the Mohawk Valley (MV), Lake Champlain Valley (CV), and Hudson Valley (HV) routes for overflow from the glacial Great Lakes (after Gadd, 1988, Fig. 1).

Fig. 1) and the Ontario basin led to low lake levels in Erie and Huron basins, and overflow was again routed east through the Mohawk River Valley (M in Fig. 7E; also, see Fig. 6). These lake basins were not completely overridden during the next (Port Huron Stade) advance at 13,000 ^{14}C yr B.P.

History and Interrelationship of Lakes Formed After 13,000 ^{14}C yr B.P.

The Port Huron readvance closed the eastern overflow route, displaced waters in the Huron and Ontario basins, and raised lake levels in the Erie basin (Fig. 7F). The subsequent ice retreat (Two Creeks Interstade) once more initiated overflow from the Huron, Erie, and Ontario basins through the Mohawk valley system (Fig. 7G), producing generally low lake levels (e.g. Hansel *et al.*, 1985; Lewis *et al.*, 1994). Waters in the Michigan basin, which had previously overflowed south into the Mississippi River basin, spilled into the Huron basin after ice retreated north of the Straits of Mackinac. Glacial Lake St. Lawrence formed in the western St. Lawrence Valley shortly after 12,000 ^{14}C yr B.P. by the amalgamation of several smaller lakes (e.g. Rodrigues & Vilks, 1994). Other relatively small lakes formed as ice retreated from the northward-sloping valleys south of the St. Lawrence such as the Hudson and Connecticut River Valleys (Ashley, 1975; Connally & Sirkin, 1973; Fullerton, 1980).

Within a few hundred years after the next (Greatlakean) ice advance, which re-dammed waters in the Michigan basin and drowned the well-known Two Creeks forest bed, the ice margin again began to retreat. Overflow was progressively routed from the Huron basin through lower, more northern avenues, the Port Huron, Fenelon Falls, and North Bay outlets (F, G, and H in Fig. 1); the northern outlet carried water to the Ottawa River Valley and the Champlain Sea. Because of isostatic depression in the area of their outlets, lake levels in the Michigan, Huron, and Erie basins generally remained low for much of the next 5000 years (e.g. Figs 6 and 8) (Lewis *et al.*, 1994). During this time, ice retreated far enough to allow glacial Lake St. Lawrence to drain and the Champlain Sea to penetrate westward in the St. Lawrence Valley to the Ontario basin (Anderson & Lewis, 1985), bringing the history of ice-marginal lakes in the St. Lawrence Valley to an end (Rodrigues & Vilks, 1994).

The Superior basin was first deglaciated between 11,000 and 10,000 ^{14}C yr B.P., but ice invaded the basin again during the Marquette advance at around 10,000 ^{14}C yr B.P. (Lowell *et al.*, 1999). The Agassiz basin remained largely filled by ice until after 11,700 ^{14}C yr B.P. (Fenton *et al.*, 1983). Overflow from the early stages of lakes in the Superior and Agassiz basins was south into the upper Mississippi River drainage system. Lake Barlow (Fig. 8), and its successor Lake Ojibway, trapped between the retreating LIS and the continental divide, expanded into the eastern Hudson Bay Lowland (Fig. 1).

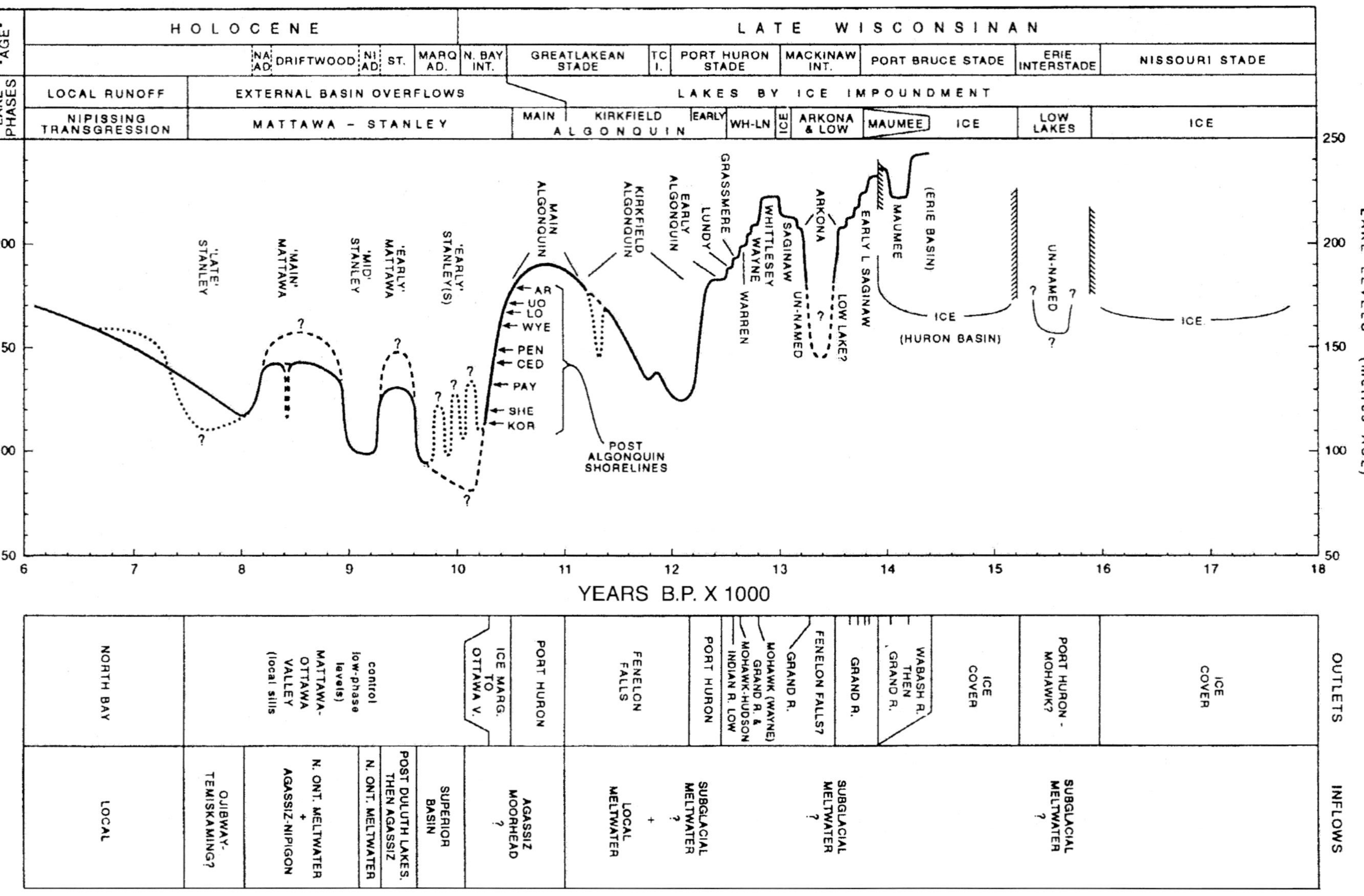

Fig. 6. Lake level fluctuations in the Huron basin, showing the relationship to inflow (mainly meltwater and Lake Agassiz) and the outlets used (see Fig. 1 for geographic locations) (Lewis et al., 1994, Fig. 14). Names of stades and interstades, and various Great Lakes' stages are shown; ages are in ^{14}C years.

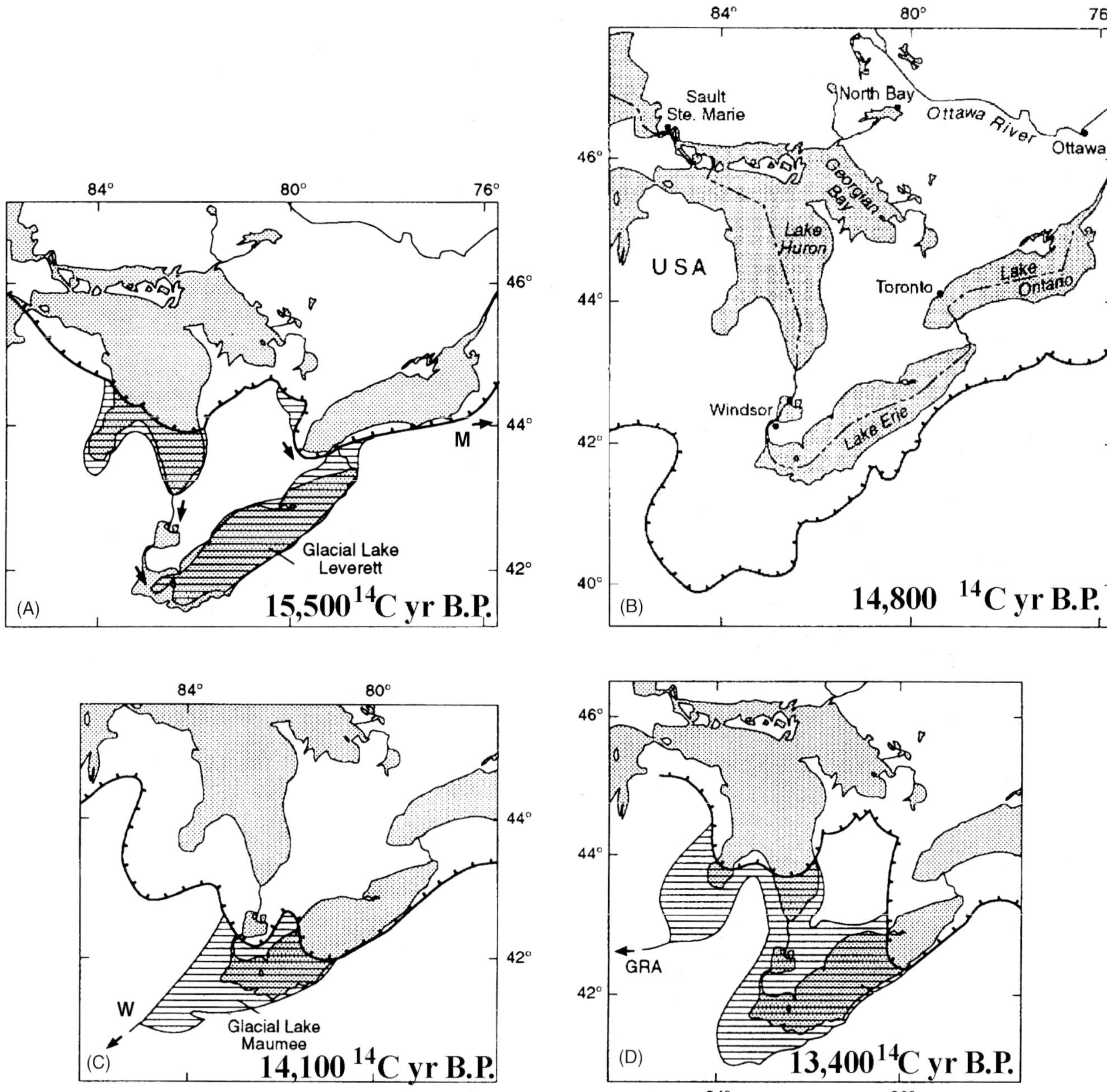

Fig. 7. Maps showing 7 snapshots of water bodies in the eastern glacial Great Lakes, resulting from advance and retreat of the LIS (hachured line) and differential isostatic rebound (after Lewis et al., 1994, Figs 3–5). Arrows show routing of overflow from lakes. Ages in ^{14}C yr B.P. (A) Maximum extent of lakes during the Erie Interstade at 15,500 ^{14}C yr B.P. (B) Extent of ice during the Port Bruce Stade about 14,800 ^{14}C yr B.P. (C) Highest phase of Lake Maumee about 14,100 ^{14}C yr B.P. at the start of the Mackinaw Interstade, which overflowed southwest via the Wabash River (W) in the Ohio River basin. (D) Continued ice retreat led to lake expansion and overflow west via the Grand River Valley (GRA) into the Michigan basin, which then spilled south into the Mississippi River basin via the Chicago outlet. (E) Continued ice retreat during Mackinaw Interstade opened the Fenelon Falls outlet (F) from the Huron basin and the Mohawk Valley route (M) from the Ontario basin, lowering lake levels and isolating these basins from one another about 13,200 ^{14}C yr B.P. (F) Reduction in extent of Great Lakes during the Port Huron Stade readvance about 13,000 ^{14}C yr B.P., when overflow was forced back into the Mississippi River basin. (G) Two Creeks interstade, showing low levels of the Huron, Erie, and Michigan basins with deeper water in the Ontario basin and in confluent glacial Lake St. Lawrence about 11,900 ^{14}C yr B.P.; overflow was east through the Mohawk Valley.

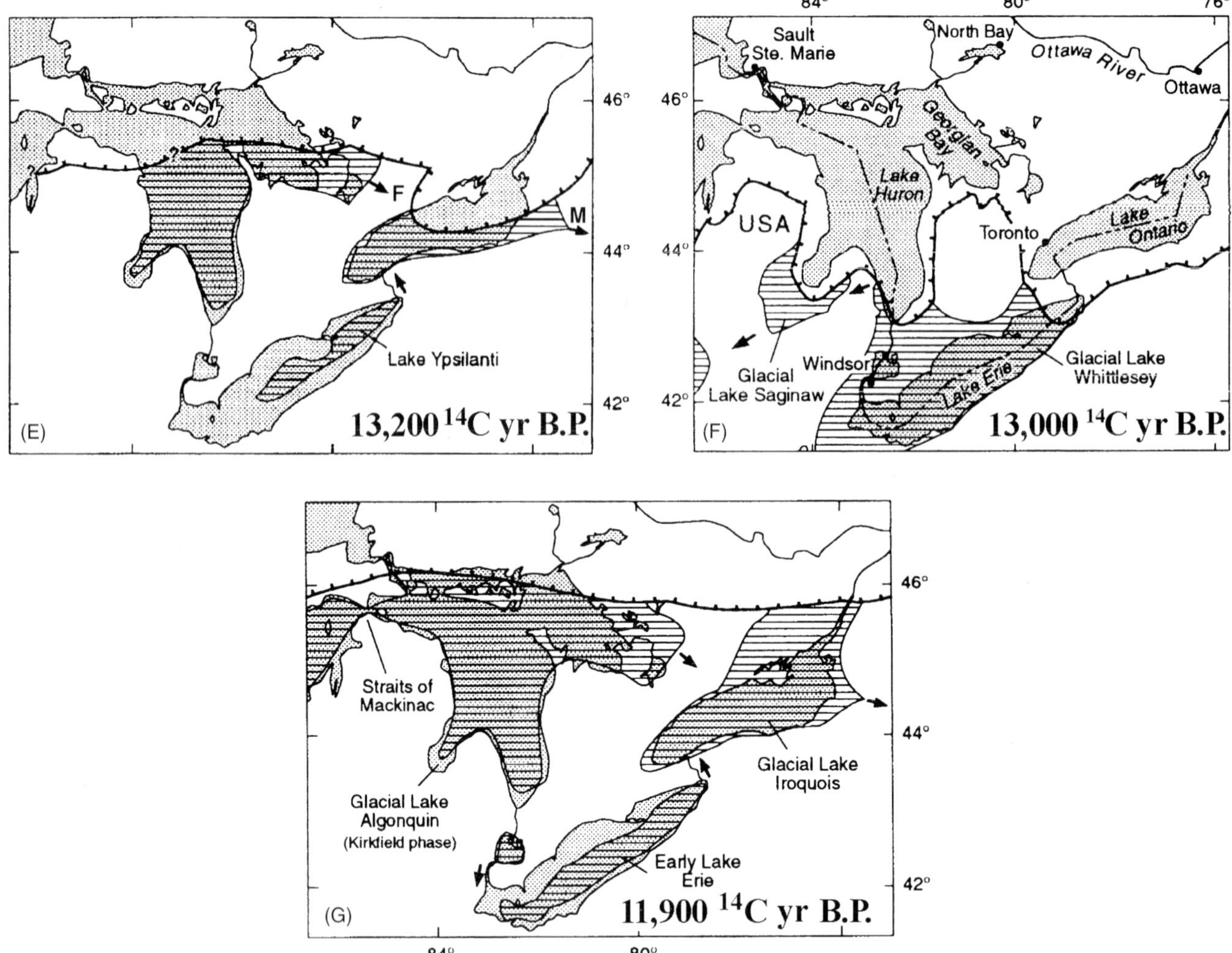

Fig. 7. (Continued)

As the LIS continued to retreat, other topographic thresholds (Fig. 2) came into play to control depth, distribution, and interconnection of North America's largest proglacial lakes. Differential isostatic rebound continued to cause southward transgression of these water bodies everywhere south of their outlets; the lake level changes of all proglacial lakes reflect the interplay of these factors and resulted in a saw-toothed lake-level history, like that shown for Lake Agassiz in Fig. 9.

Although there were several aborted starts to Lake Agassiz in the southern end of the Red River Valley, terminated when ice re-advanced to displace the lake (e.g. Clayton & Moran, 1982; Fenton *et al.*, 1983), early overflow was south, mainly though the Minnesota River Valley and into the Mississippi River at Minneapolis (K in Fig. 1) until ice retreated several hundred kilometers north, where a low point on the basin rim led east into the Superior basin. The level of Lake Agassiz abruptly dropped by about 110 m (Leverington *et al.*, 2000) when this eastern outlet was opened around 10,900 ^{14}C yr B.P., producing a low-water period (the Moorhead phase; Fig. 9) with an extensive subaerially exposed and vegetated lake floor that was progressively inundated by transgressing waters as differential isostatic rebound uplifted the outlet (e.g. Bajc *et al.*, 2000; Fenton *et al.*, 1983). This eastward diversion of Agassiz overflow generally corresponds to a relatively high lake level in the Huron basin (Fig. 6). Lake Agassiz eventually transgressed to the southern outlet again, but waters soon fell as a different outlet in the northwestern corner of the basin was opened (N in Fig. 1). Although details of events during the period from 10,100 to 9400 ^{14}C yr B.P. have been debated (see Teller, 2001), most overflow from Lake Agassiz, which was the world's largest proglacial lake at this time, was through the northwestern outlet to the Mackenzie River Valley and to the Arctic Ocean (Fisher & Smith, 1994; Smith & Fisher, 1993); a readvance about 9900 ^{14}C yr B.P. severed the link between Lake Agassiz and the northwestern outlet, forcing the lake to rise and overflow briefly through the southern outlet (Fig. 9) (Teller, 2001). This coincided with the Marquette re-advance into the Superior basin, which forced that lake to overflow west into the Agassiz basin and deposit a distinctive series of 24 red varves in northwestern Ontario (Zoltai, 1961).

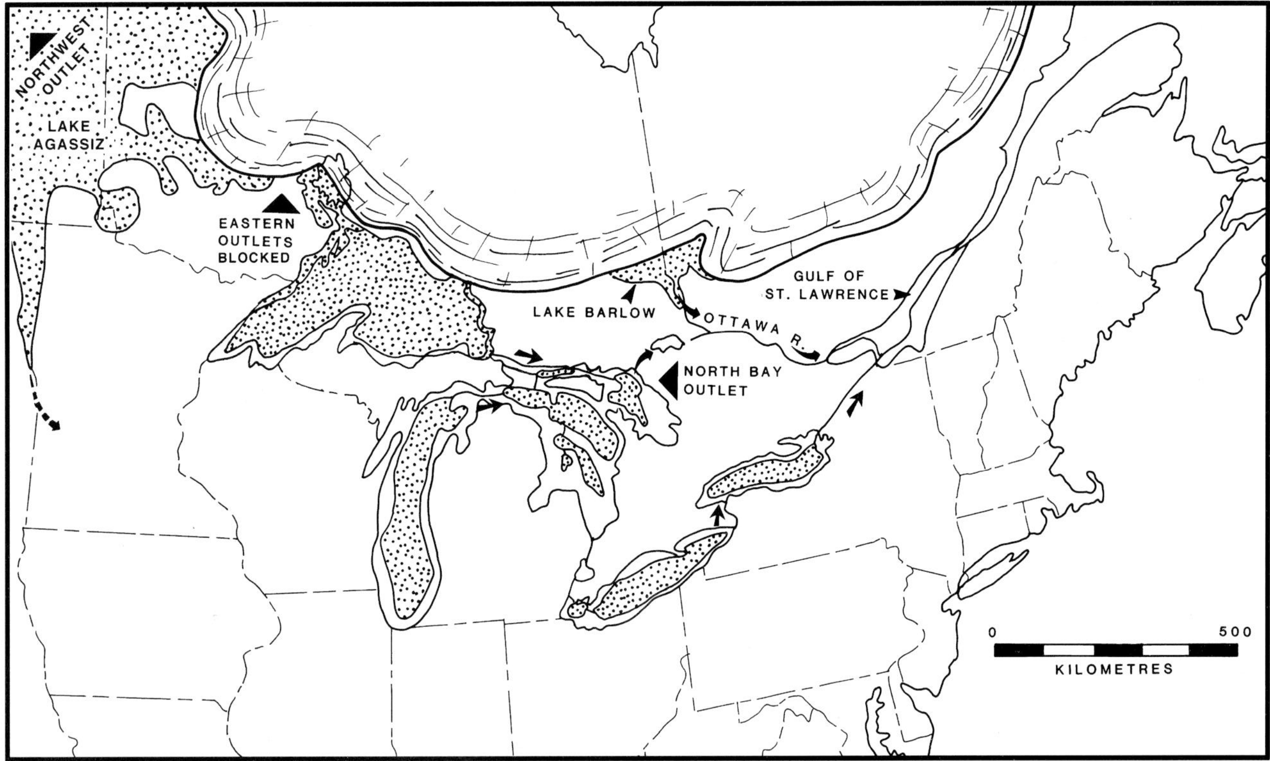

Fig. 8. Proglacial lakes along the LIS about 9500 ^{14}C yr B.P., just before the eastern outlets of Lake Agassiz were reopened, showing the lower levels of the Great Lakes resulting from their isostatically depressed outlets and the loss of overflow from the Agassiz basin (Teller, 1987, Fig. 20).

The fluctuating level of Lake Agassiz produced a sawtooth pattern (Fig. 9), dropping abruptly when new outlets opened and rising more gradually afterward (south of the outlet) as a result of differential rebound. The changing bathymetry and extent of Lake Agassiz, including the volume of flood outbursts that occurred when lower outlets were deglaciated, is shown and discussed by Leverington *et al.* (2000, 2002) and Teller & Leverington (in press).

Overflow and catastrophic outbursts from Lake Agassiz have been correlated with lake level rises and changes in the sedimentary and isotopic records in the Superior, Michigan, Huron, and Erie basins (e.g. Colman *et al.*, 1994; Farrand & Drexler, 1985; Lewis *et al.*, 1994; Rea *et al.*, 1994; Teller, 1985; Teller & Mahnic, 1988; Tinkler *et al.*, 1992; Tinkler & Pengelly, 1995). It was suggested by Kehew (1993) that the influx of Agassiz waters during its Moorhead phase (ca. 10,900–10,100 ^{14}C yr B.P.) may have forced the level of Lake Michigan to rise and overflow through the Chicago outlet. Lewis *et al.* (1994) suggest a linkage of Agassiz overflow to increasing lake level in the Huron basin (see Fig. 6) and a correlation of enriched (higher) δ^{18}O values in ostracode shells to the dilution of Huron basin waters by Agassiz waters that had a high precipitation/meltwater ratio; Rea *et al.* (1994) weave this into a history of the paleohydrology of the Great Lakes system. Colman *et al.* (1994), however, correlate lake level rise and *depleted* δ^{18}O values in the Michigan basin with the influx of Lake Agassiz overflow. Regardless, it is clear that the addition of the >2 million km^2 Agassiz drainage basin would have impacted the Great Lakes, especially at times when catastrophic outbursts through its eastern outlets introduced thousands of km^3 into the upsteam end of the Great Lakes in only a few months or years (see later discussion). When Agassiz flow was cut off, such as during the Marquette readvance into the Superior basin around 10,000 ^{14}C yr B.P., lake levels in the Great Lakes declined (Figs 6 and 8). Following that, meltwater continued to enter the Great Lakes from the Agassiz basin and from remnant ice until about 8000 ^{14}C yr B.P., when the LIS retreated far enough north of the divide with the Hudson Bay basin to allow waters to overflow from the eastern (Ojibway) part of this vast proglacial lake (I of Fig. 1).

During the period between about 10,000 and 9000 ^{14}C yr B.P., another great proglacial lake formed along the northwestern side of the LIS, Lake McConnell (Fig. 1). Trapped in the isostatically depressed area of the Mackenzie River basin, between the LIS and the Mackenzie Mountains, this lake changed its depth and configuration in response to ice marginal fluctuations, outlet change and incision, and isostatic rebound. Lemmen *et al.* (1994) and Smith (1994) summarize the history of this giant and complex lake, which eventually covered an area of 215,000 km^2. During its life, it received inflow from the Lake Agassiz basin when that lake overflowed through its northwestern outlet (N of Fig. 1) between about 10,100 and 9400 ^{14}C yr B.P. (Fig. 9). The

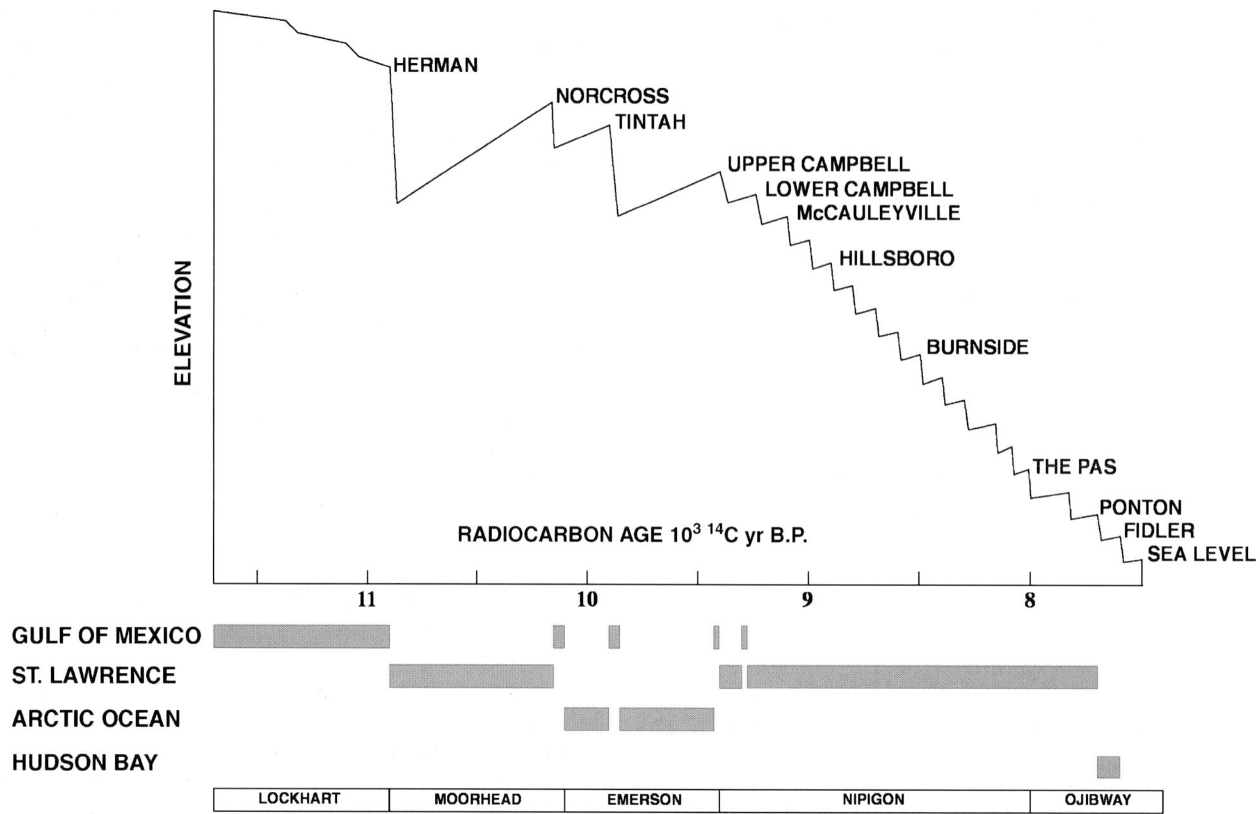

Fig. 9. Schematic of relative changes in level of Lake Agassiz south of its outlet from 11,700 to 7700 ^{14}C yr B.P., showing rising (transgressing) stages that resulted from differential isostatic rebound after each abrupt decline in lake level when new lower outlets were opened. Variable routing of overflow to oceans shown by bars. Names of selected beaches and the five main stages of the lake are shown.

eventual disappearance of glacial Lake McConnell, after reaching its maximum extent 10,500–10,000 ^{14}C yr B.P. (Lemmen *et al.*, 1994), occurred gradually, as differential rebound raised the depressed basin, causing it to drain through the Mackenzie River into the Arctic Ocean. Some of North America's largest modern lakes, Lakes Athabasca, Great Slave, and Great Bear, are remnants on the floor of this lake.

After about 9400 ^{14}C yr B.P., when ice had retreated from the Superior basin, overflow from Lake Agassiz was directed eastward through a series of 17 progressively lower channels into the Nipigon-Superior basin (Leverington & Teller, 2003; Teller & Thorleifson, 1983) until about 8000 ^{14}C yr B.P., producing the stair-step series of beaches seen in Fig. 9. When ice receded north of the Great Lakes-Hudson Bay divide, Agassiz overflow began to by-pass the Great Lakes, entering glacial Lake Ojibway, which had been expanding along the retreating LIS in the eastern Hudson Bay Lowland (Fig. 1); overflow from this giant ice-marginal lake was through the Kinojévis outlet into the Ottawa River-St. Lawrence Valley system (near I in Fig. 1).

Lakes Barlow (Fig. 8) and Ojibway (Fig. 1) formed north of the Great Lakes as ice retreated down slope into the Hudson Bay basin, overflowing through the Ottawa River Valley and into the Champlain Sea (St. Lawrence Valley, Fig. 8) (Veillette, 1994; Vincent & Hardy, 1979). Northward expansion was interrupted by several glacial surges into the lake (the Cochrane surges). After the merger with Lake Agassiz about 8000 ^{14}C yr B.P., this lake survived another few centuries, until ice in Hudson Bay thinned enough so it no longer provided a barrier to outflow to Hudson Strait and the North Atlantic Ocean. The catastrophic final drainage of this giant lake appears to have had an impact on oceans (see later discussion) (Barber *et al.*, 1999; Teller *et al.*, 2002).

After about 7700 ^{14}C yr B.P., the water level in the Huron basin fell below its outlet as climate warmed, possibly resulting in a closed basin and more saline environment for several hundred years (Blasco, 2001) (see Fig. 6). In general, throughout this late-glacial period, lakes in the Great Lakes' basins transgressed southward, although where the outlet lay on an isobase that extended through the "middle" of the basin rather than along its northern side, such as in the Superior basin, waters transgressed on to the southern shore, but receded from the northern side (see Fig. 3B). The Nipissing transgression brought all the western Great Lakes to a common level about 6500 ^{14}C yr B.P., and waters eventually reached the Chicago outlet, diverting overflow southward into the Mississippi River system (e.g. Eschman & Karrow, 1985; Hansel *et al.*, 1985; Larsen, 1987). This transgression also reached the Port Huron outlet from the Huron basin to the Erie basin, which led to erosion and eventual capture of overflow from the western Great Lakes by the eastern Great Lakes system by about 4700 ^{14}C yr B.P. (Lewis *et al.*, 1994). The complex history of the Great Lakes is summarized in Karrow & Calkin (1985), Larsen (1987), Teller (1987), Lewis & Anderson (1989), and Lewis *et al.* (1994).

The Impact of Large Glacial Lakes on Climate and Oceans

In recent years there has been an increasing recognition of the role that proglacial lakes played in influencing climate and in bringing about climate change. One aspect relates to the thermal capacity of water, which makes a lake slow to respond to seasonal warming and cooling; lake temperatures fluctuate less than do those on land and tend to moderate climate and increase cloudiness and precipitation in the region around it. As a result, ecosystems around large lakes tend to be different. The early stages of most North America proglacial lakes probably were turbid and cold, and lake productivity (and organic content in the sediment) was low. In their review, Lewis *et al.* (1994) note that waters in the ice-marginal Great Lakes were cold, with a mean temperature no higher than 3.6 °C, based on 112 taxa of aquatic and terrestrial mollusks; these authors and Moore *et al.* (2000) used variations in the δ^{18}O of ostracodes to identify the relative contributions of meltwater and precipitation runoff during the history of proglacial waters in the Huron basin.

The "lake effect" on land adjacent to large proglacial lakes produced contrasting climate, ecosystems, and isotopic signatures to those a few tens of kilometers away (e.g. Norton & Boisenga, 1993). Hu *et al.* (1997) found a shift to more depleted δ^{18}O values in a small lake in northern Minnesota from 10,200 to 9100 ^{14}C yr B.P., which coincided with the time when Lake Agassiz was expanding. Even though there was no increase in *Picea* at this time, which typically reflects cooler conditions, they interpreted that the isotopic depletion was due to decreased summer temperatures and the addition of precipitation derived from the isotopically light waters of nearby Lake Agassiz. Dean *et al.* (2002) suggest that the shift in climate signal about 7300 ^{14}C yr B.P. in the varved sequence of a small lake just south of the Agassiz basin reflects a change in atmospheric circulation that correlates with the final drainage of Agassiz. Hostetler *et al.* (2000) used an atmospheric GCM to model the likely effect of Lake Agassiz on the region, concluding that, during an expansive lake stage, there were cooler temperatures and *less* precipitation, because there was less evaporation from the cooler lake than from the land, and because the cool lake induced anticyclonic winds that blocked moisture-bearing winds from outside the region. During expanded lake stages they conclude that the suppressed precipitation would have contributed to retreat of the LIS.

Whenever new lower outlets were deglaciated and proglacial lake waters were drawn down, the influence of those lakes on regional climate would have been reduced. However, the climate around basins receiving these waters may have been influenced. Although the impact of small influxes of new water probably had a minimal effect on the depth of a lake and its overall nature, large influxes, such as the catastrophic outbursts from Lake Agassiz into the Great Lakes, may have briefly raised lake levels by several tens of meters and affected their sedimentary record, as noted before. Lewis & Anderson (1992) even suggested that the rise in *Picea* and shift in δ^{18}O and δ^{13}C around the Erie basin about 11,000–10,500 ^{14}C yr B.P. reflects cooling related to the first influx of Lake Agassiz overflow during that period. Rodrigues & Vilks (1994) identified an ostracode and isotopic signal in sediments of the Champlain Sea in the western St. Lawrence Valley that resulted from the diversion of Agassiz overflow into the Great Lakes.

The abrupt transfers of water from one lake basin to another had significant hydrological impacts on lakes, rivers, and oceans, as well as on climate. Of course there also was a morphological impact on many valleys receiving additional runoff or draw down from a lake basin; Teller & Thorleifson (1983) and Fisher & Smith (1994) describe some of the dramatic geomorphic and sedimentological changes that occurred as a result of outbursts from Lake Agassiz. The re-direction of overflow from one basin to another and the simple opening of a lower overflow route may produce such changes. In both cases there would have been a large initial, short-term increase in outflow (producing a flood hydrograph). The arrows in Fig. 1 show many of the new channels and routes for lake overflow around the LIS, which may have experienced these hydrological events. As previously discussed, the diversions of overflow were related to isostatic rebound, glacial advance, and glacial retreat. Some of the diversions of lake overflow were permanent, some lasted for centuries, and some last for only a few years.

The record of some drainage diversions in glaciated parts of North America is found in sediments of the Gulf of Mexico, and Leventer *et al.* (1982), Broecker *et al.* (1985), and others used this to suggest how the routing of North American meltwater might have impacted on ocean circulation. The potential role of Lake Agassiz overflow on ocean circulation and related climate was proposed by Broecker *et al.* (1989), who suggested that the diversion of Agassiz overflow from the Gulf of Mexico to the North Atlantic about 11,000 ^{14}C yr B.P. may have led to the Younger Dryas cooling. Although the impact of these fresh waters on ocean circulation continues to be debated, many GCM models indicate that a relatively small addition of fresh water to specific places in the North Atlantic Ocean, particularly at certain "transitional" stages of circulation, can shut down the production of North Atlantic Deep Water (NADW) (e.g. Fanning & Weaver, 1997; Manabe & Stouffer, 1997; Rind *et al.*, 2001); as discussed by Teller *et al.* (2002), several outbursts from Lake Agassiz exceeded those modeled values. As a result, thermohaline circulation and the transfer of heat into high latitudes of the North Atlantic were reduced, leading to climate cooling. The flux of freshwaters needed to alter NADW is related to the length of time over which this flux continued and to the specific nature of ocean circulation at the time of the freshwater addition.

The potential impact of fresh water additions to the North Atlantic Ocean and the close chronological relationship of major late-glacial climate changes to both short outbursts of overflow from Lake Agassiz and the diversion of North American drainage from one ocean to another have led Clark *et al.* (2001) and Teller *et al.* (2002) to conclude that those freshwater fluxes triggered or contributed to global climate change during the last period of deglaciation. In turn, these global climate changes are recorded in some pollen and isotopic records in small lakes in the Great Lakes and Agassiz region (Yu & Wright, 2001), although the degree to which these are related to global climate forcing from the North Atlantic region or reflect the control on climate by the proglacial lake system itself remains uncertain.

Four major late-glacial climate changes have been related to the influx of freshwaters from North America; these were superposed on the overall global warming, and are:

(1) The Older Dryas cold period. This cooling occurred between about 16,500 and 13,000 ^{14}C yr B.P. It spans the Erie Interstade in North America, in which the first large eastward overflow of freshwater from the Great Lakes to the North Atlantic Ocean (the Heinrich 1 event of ice rafting in the North Atlantic) was initiated. It also includes the Mackinaw Interstade, which led to the second period of Great Lakes outflow to the North Atlantic. This suggests a link between cooling and outflow from the Great Lakes and glacial re-advance, represented by the Port Huron Stade and by iceberg production (Clark *et al.*, 2001).
(2) The Younger Dryas cold period. This abrupt cooling at about 11,000–10,000 ^{14}C yr B.P. coincides with the largest outburst of water from Lake Agassiz and the accompanying 800-year diversion of its overflow into the Great Lakes and North Atlantic.
(3) The Preboreal Oscillation. Shortly after the end of the Younger Dryas, a relatively small cooling in the North Atlantic region occurred, which coincides in time with an outburst from Lake Agassiz to the Arctic Ocean (Fisher *et al.*, 2002; Teller *et al.*, 2002) and the draining of the Baltic Ice Lake into the North Atlantic (Björck *et al.*, 1996).
(4) The 8200 cal yr B.P. cold event. Following the draining of 163,000 km^3 of fresh water from Lake Agassiz-Ojibway about 7700 ^{14}C yr B.P. (8400 cal yr B.P.), a small cooling was recorded in the isotopic signal of the Greenland ice sheet, suggesting a link between these events (Barber *et al.*, 1999; Teller *et al.*, 2002). The 200-year difference between the age of this drainage, estimated by Barber *et al.* (1999), and the cooling may reflect (a) a two-step final draw down of Lake Agassiz, where complete drainage of the western part of this huge lake was delayed by 200 years (Leverington *et al.*, 2002), or it may indicate that the ocean carbonate reservoir effect used by Barber *et al.* (1999) to estimate the age was too small (Teller & Leverington, pers. comm., 2003). Only because the global ocean was in a more stable interglacial mode did this large freshwater flux not have a larger impact on climate.

Thus, not only did large proglacial lakes respond to climate and the bounding ice sheet, but they also influenced the ice and climate, both on a regional scale and, via the oceans, on a global scale. The chronological coincidence of large changes in proglacial lakes, the climate record in oceans and on adjacent continents, and the Greenland isotopic record supports a cause-effect linkage. Specifically, large fluxes of water from Lake Agassiz, along with changes in the routing of overflow from lakes around the LIS, slowed thermohaline circulation in the North Atlantic Ocean which, in turn, led to climate cooling.

Acknowledgments

This synthesis is the product of years of research by many people. My own research has been generously supported by the Natural Sciences and Engineering Research Council of Canada. My thanks to the many Quaternarists who have shared their expertise with me over the years. Reviews of this chapter by Mike Lewis, Tim Fisher, and Alan Kehew have been helpful in improving the final product. Thanks to Allen Patterson for his help in preparing the illustrations.

References

Anderson, T.W. & Lewis, C.F.M. (1985). Postglacial water-level history of the Ontario basin. *In*: Karrow, P.F. & Calkin, P.E. (Eds), *Quaternary evolution of the Great Lakes*. Geological Association of Canada Special Paper 30, 231–253.

Ashley, G.M. (1975). Rhythmic sedimentation in glacial Lake Hitchcock, Massachusetts-Connecticut. *In*: Jopling, A.V. & McDonald, B.C. (Eds), *Glaciofluvial and glaciolacustrine sedimentation*. Society of Economic Paleontologists and Mineralogists Special Publication 23, 304–320.

Bajc, A.F., Schwert, D.P., Warner, B.G. & Williams, N.E. (2000). A reconstruction of Moorhead and Emerson Phase environments along the eastern margin of glacial Lake Agassiz. Rainy River basin, northwestern Ontario. *Canadian Journal of Earth Sciences*, **37**, 1335–1353.

Barber, D.C., Dyke, A., Hillaire-Marcel, C., Jennings, A.E., Andrews, J.T., Kerwin, M.W., Bilodeau, G., McNeely, R., Southon, J., Morehead, M.D. & Gagnon, J.-M. (1999). Forcing of the cold event of 8200 years ago by catastrophic drainage of Laurentide lakes. *Nature*, **400**, 344–348.

Barendregt, R. & Irving, E. (1998). Changes in the extent of North American ice sheets during the late Cenozoic. *Canadian Journal of Earth Sciences*, **35**, 504–509.

Björck, S., Kromer, B., Johnsen, S., Bennike, O., Hammarlund, D., Lemdahl, G., Possnert, G., Rasmussen, T.L., Hammer, C.U. & Spurk, M. (1996). Synchronized terrestrial-atmospheric deglacial records around the North Atlantic. *Science*, **274**, 1155–1160.

Blasco, S.M. (2001). Geological history of Fathom Five National Marine Park over the past 15,000 years. *In*: Parker, S. & Munawar, M. (Eds), *Ecology, culture, and conservation of a protected area*. Fathom Five National Park, Canada, 45–62. Ecovision World Monograph Series, Backhuys Publ., Leiden, Netherlands.

Bradley, R.S. (1999). *Paleoclimatology: Reconstructing climates of the Quaternary* (2nd ed.). Harcourt/Academic Press, Burlington, Mass., 610 pp.

Broecker, W.S., Kennett, J., Flower, B., Teller, J.T., Trumbore, S., Bonani, G. & Wolfli, W. (1989). Routing of meltwater from the Laurentide Ice Sheet during the Younger Dryas cold episode. *Nature*, **341**, 318–321.

Broecker, W.S., Peteet, D. & Rind, D. (1985). Does the ocean-atmosphere have more than one stable mode of operation? *Nature*, **315**, 21–25.

Clark, P.U. (1992). The last interglacial-glacial transition in North America: introduction. *In*: Clark, P.U. & Lea, P.D. (Eds), *The last interglacial-glacial transition in North America*. Geological Society of America, Special Paper 270, 1–11.

Clark, P.U., Marshall, S.J., Clarke, G.K.C., Hostetler, S.W., Licciardi, J.M. & Teller, J.T. (2001). Freshwater forcing of abrupt climate change during the last glaciation. *Science*, **293**, 283–287.

Clayton, L. & Moran, S.R. (1982). Chronology of Late Wisconsinan glaciation in middle North America. *Quaternary Science Reviews*, **1**, 55–82.

Colman, S.M., Clark, J.A., Clayton, L., Hansel, A.K. & Larsen, C.E. (1994). Deglaciation, lake levels, and meltwater discharge in the Lake Michigan basin. *Quaternary Science Reviews*, **13**, 879–890.

Connally, G.G. & Sirkin, L.A. (1973). Wisconsin history of the Hudson-Champlain Lobe. *In*: Black, R.F., Goldthwait, R.P. & Willman, H.B. (Eds), *The Wisconsin Stage*. Geological Society of America Memoir 136, 47–69.

Dean, W.E., Forester, R.M. & Bradbury, J.P. (2002). Early Holocene change in atmosphere circulation in the Northern Great Plains: an upstream view of the 8.2 ka cold event. *Quaternary Science Reviews*, **21**, 1763–1775.

Dyke, A.S., Andrews, J.T., Clark, P.U., England, J.H., Miller, G.H., Shaw, J. & Veillette, J.J. (2002). The Laurentide and Innuitian ice sheets during the Last Glacial Maximum. *Quaternary Science Reviews*, **21**, 9–31.

Dyke, A.S. & Prest, V.K. (1987). Late Wisconsinan and Holocene history of the Laurentide Ice Sheet. *Géographie physique et Quaternaire*, **41**, 237–263.

Eschman, D.F. & Karrow, P.F. (1985). Huron basin glacial lakes: a review. *In*: Karrow, P.F. & Calkin, P.E. (Eds), *Quaternary evolution of the Great Lakes*. Geological Association of Canada Special Paper 30, 79–93.

Fanning, A.F. & Weaver, A.J. (1997). Temporal-geographical meltwater influences on the North Atlantic conveyor: implications for the Younger Dryas. *Paleoceanography*, **12**, 307–320.

Farrand, W.R. & Drexler, C.W. (1985). Late Wisconsinan and Holocene history of the Lake Superior basin. *In*: Karrow, P.F. & Calkin P.E. (Eds), *Quaternary evolution of the Great Lakes*. Geological Association of Canada Special Paper 30, 17–32.

Fenton, M.M., Moran, S.R., Teller, J.T. & Clayton, L. (1983). Quaternary stratigraphy and history in the southern part of the Lake Agassiz basin. *In*: Teller, J.T. & Clayton, L. (Eds), *Glacial Lake Agassiz*. Geological Association of Canada, Special Paper 26, 49–74.

Fisher, T.G. & Smith, D.G. (1994). Glacial Lake Agassiz: its northwest maximum extent and outlet in Saskatchewan (Emerson Phase). *Quaternary Science Reviews*, **13**, 845–858.

Fisher, T.G., Smith, D.G. & Andrews, J.T. (2002). Preboreal Oscillation caused by a glacial Lake Agassiz flood. *Quaternary Science Reviews*, **21**, 873–878.

Fraser, G.S. & Bleuer, N.K. (1988). Sedimentological consequences of two floods of extreme magnitude in the late Wisconsinan Wabash Valley. *In*: H.E. Clifton (Ed.), *Sedimentological consequences of convulsive events*. Geological Society of America, Special Paper 229, 111–125.

Fullerton, D.S. (1980). Preliminary correlation of post-Erie Interstadial events (16,000–10,000 radiocarbon years before present), central and eastern Great Lakes region, and Hudson, Champlain, and St. Lawrence Lowlands, United States and Canada. *U.S. Geological Survey Professional Paper*, 1089, 52 pp.

Fulton, R.J. (Ed.) (1989). Quaternary geology of Canada and Greenland. *Geological Survey of Canada*. Geology of Canada, No. 1, 839 pp.

Gadd, N.R. (1988). The basin, the ice, the Champlain Sea. *In*: N.R. Gadd (Ed.), *The Late Quaternary development of the Champlain Sea basin*. Geological Association of Canada Special Paper 35, 15–24.

Gosse, J.C. & Phillips, F.M. (2001). Terrestrial in situ cosmogenic nuclides: theory and application. *Quaternary Science Reviews*, **20**, 1475–1560.

Hansel, A.K., Mickelson, D.M., Schneider, A.F. & Larsen, C.E. (1985). Late Wisconsinan and Holocene history of the Lake Michigan basin. *In*: Karrow, P.F. & Calkin, P.E. (Eds), *Quaternary evolution of the Great Lakes*. Geological Association of Canada Special Paper 30, 39–53.

Hostetler, S.W., Bartlein, P.J., Clark, P.U., Small, E.E. & Solomon, A.M. (2000). Simulated interactions of proglacial Lake Agassiz with the Laurentide ice sheet 11,000 years ago. *Nature*, **405**, 334–337.

Hu, F.S., Wright, H.E., Jr., Ito, E. & Lease, K. (1997). Climatic effects of glacial Lake Agassiz in the midwestern United States during the last deglaciation. *Geology*, **25**, 207–210.

Karrow, P.R. & Calkin, P.E. (Eds) (1985). *Quaternary Evolution of the Great Lakes*. Geological Association of Canada Special Paper 30.

Kehew, A.E. (1982). Catastrophic flood hypothesis for the origin of the Souris spillway, Saskatchewan and North Dakota. *Geological Society of America Bulletin*, **93**, 1051–1058.

Kehew, A.E. (1993). Glacial lake outburst erosion of the Grand Valley, Michigan, and impacts on glacial lakes in the Lake Michigan basin. *Quaternary Research*, **39**, 36–44.

Kehew, A.E. & Lord, M.L. (1986). Origin and large-scale erosional features of glacial-lake spillways in the northern Great Plains. *Geological Society of America Bulletin*, **97**, 162–177.

Kehew, A.E. & Teller, J.T. (1994). History of late glacial runoff along the southwestern margin of the Laurentide ice sheet. *Quaternary Science Reviews*, **13**, 859–877.

Lamothe, M., Parent, M. & Shilts, W.W. (1992). Sangamonian and early Wisconsinan events in the St. Lawrence Lowland and Appalachians of southern Quebec. *In*: Clark, P.U. & Lea, P.D. (Eds), *The last interglacial-glacial transition in North America*. Geological Society of America, Special Paper 270, 171–184.

Larsen, C.E. (1987). *Geological history of glacial Lake Algonquin and the Upper Great Lakes*. U.S. Geological Survey Bulletin **B1801**, 36 pp.

Last, W.M. & Smol, J.P. (Eds) (2001a). *Basin analysis, coring, and chronological techniques*. Vol. 1 of "Tracking environmental change using lake sediments". Kluwer Academic Publishers, Dordrecht, Netherlands, 548 pp.

Last, W.M. & Smol, J.P. (Eds) (2001b). *Physical and geochemical methods*. Vol. 2 of "Tracking environmental change using lake sediments", Kluwer Academic Publishers, Dordrecht, Netherlands, 504 pp.

Lemmen, D.S., Duk-Radkin, A. & Bednarski, J.M. (1994). Late glacial drainage systems along the northwestern margin of the Laurentide Ice Sheet. *Quaternary Science Reviews*, **13**, 805–825.

Leventer, A., Williams, D.F. & Kennett, J.P. (1982). Dynamics of the Laurentide Ice Sheet during the last deglaciation: evidence from the Gulf of Mexico. *Earth and Planetary Science Letters*, **59**, 11–17.

Leverington, D.W., Mann, J.D. & Teller, J.T. (2000). Changes in the bathymetry and volume of glacial Lake Agassiz between 11,000 and 9300 ^{14}C yr B.P. *Quaternary Research*, **54**, 174–181.

Leverington, D.W., Mann, J.D. & Teller, J.T. (2002). Changes in the bathymetry and volume of glacial Lake Agassiz between 9200 and 7600 ^{14}C yr B.P. *Quaternary Research*, **57**, 244–252.

Leverington, D.W. & Teller, J.T. (2003). Paleotopographic reconstructions of the eastern outlets of glacial Lake Agassiz. *Canadian Journal of Earth Sciences*, **40**, 1259–1278.

Lewis, C.F.M. & Anderson, T.W. (1989). Oscillations of levels and cool phases of the Laurentian Great Lakes caused by inflows from glacial Lakes Agassiz and Barlow-Ojibway. *Journal of Paleolimnology*, **2**, 99–146.

Lewis, C.F.M. & Anderson, T.W. (1992). Stable isotope (O and C) and pollen trends in eastern Lake Erie, evidence for a locally induced climatic reversal of Younger Dryas age in the Great Lakes basin. *Climate Dynamics*, **6**, 241–250.

Lewis, C.F.M, Blasco, S.M. & Coakley, J.P. (2002). Severe dry climate impact on the Laurentian Great Lakes indicated by early to middle Holocene lake closure. Internat. Assoc. Great Lakes Research, Abstracts 45th Annual Conference, Winnipeg, p. 72.

Lewis, C.F.M., Forbes, D.L., Todd, B.J. et al. (2001). Uplift-driven expansion delayed by middle Holocene desiccation in Lake Winnipeg, Manitoba, Canada. *Geology*, **29**, 743–746.

Lewis, C.F.M., Moore, T.C., Jr., Rea, D.K., Dettman, D.L., Smith, A.J. & Mayer, L.A. (1994). Lakes of the Huron basin: their record of runoff from the Laurentide Ice Sheet. *Quaternary Science Reviews*, **13**, 891–922.

Licciardi, J.M., Teller, J.T. & Clark, P.U. (1999). Freshwater routing by the Laurentide Ice Sheet during the last deglaciation. *In*: Clark, P.U., Webb, R.S. & Keigwin, L.D. (Eds), *Mechanisms of Global Climate Change at Millennial Time Scales*. American Geophysical Union, Monograph, **112**, 171–202.

Lowell, T.V., Larson, G.J., Hughes, J.D. & Denton, G.H. (1999). Age verification of the Gribben forest bed and the Younger Dryas advance of the Laurentide Ice Sheet. *Canadian Journal of Earth Sciences*, **36**, 383–393.

Manabe, S. & Stouffer, R.J. (1997). Coupled ocean-atmosphere model response to freshwater input: comparison to Younger Dryas event. *Paleoceanography*, **12**, 321–336.

Mickelson, D.M., Clayton, L., Fullerton, D.S. & Borns, H.W. (1983). The late Wisconsin glacial record of the Laurentide Ice Sheet in the United States. *In*: Wright, H.E. (Ed.), *Late Quaternary environments of the United States*. University of Minnesota Press, Minneapolis, 3–37.

Moore, T.C., Walker, J.C.G., Rea, K.K., Lewis, C.F.M., Shane, L.C.K. & Smith, A.J. (2000). Younger Dryas interval and outflow from the Laurentide Ice Sheet. *Paleoceanography*, **15**, 4–18.

Norton, D.C. & Boisenga, S.J. (1993). Spatiotemporal trends in lake effect and continental snowfall in the Laurentian Great Lakes, 1951–1980. *Journal of Climate*, **6**, 1943–1956.

Occhietti, S. (1989). Quaternary geology of St. Lawrence Valley and adjacent Appalachian subregion. *In*: Fulton, R.J. (Ed.), *Quaternary geology of Canada and Greenland*. Geological Survey of Canada, Geology of Canada No. 1, 350–388.

Prest, V.K. (1970). Quaternary geology of Canada. *In*: Douglas, R.J.W. (Ed.), *Geology and economic minerals of Canada*. Geological Survey of Canada, Economic Geology Report 1, 676–764.

Rea, D., Moore, T., Anderson, T., Lewis, M., Dobson, D., Dettman, D., Smith, A. & Mayer, L. (1994). Great Lakes Paleohydrology: complex interplay of glacial meltwater, lake levels, and sill depths. *Geology*, **22**, 1059–1062.

Rind, D., deMenocal, P., Russell, G., Sheth, S., Collins, D., Schmidt, G. & Teller, J. (2001). Effects of glacial meltwater in the GISS coupled atmosphere-ocean model: Part I. North Atlantic Deep Water response. *Journal of Geophysical Research*, **106**, 27,335–27,354.

Rodrigues, C.G. & Vilks, G. (1994). The impact of glacial lake runoff on the Goldthwait and Champlain Seas: the relationship between glacial Lake Agassiz runoff and the Younger Dryas. *Quaternary Science Reviews*, **13**, 923–944.

Shaw, J. (1996). A meltwater model for Laurentide subglacial landscapes. *In*: McCann, S.B. & Ford, D.C. (Eds), Geomorphology sans frontière. J. Wiley, Chichister, 182–226.

Shaw, J. (2002). The meltwater hypothesis for subglacial bedforms. *Quaternary International*, **90**, 5–22.

Shoemaker, E.M. (1991). On the formation of large subglacial lakes. *Canadian Journal Earth Sciences*, **28**, 1975–1981.

Shoemaker, E.M. (1992). Water sheet outburst floods from the Laurentide Ice Sheet. *Canadian Journal Earth Sciences*, **29**, 1250–1264.

Smith, D.G. (1994). Glacial Lake McConnell: paleogeography, age, duration, and associated river deltas, Mackenzie River basin, western Canada. *Quaternary Science Reviews*, **13**, 829–843.

Smith, D.G. & Fisher, T.G. (1993). Glacial Lake Agassiz: the northwestern outlet and paleoflood. *Geology*, **21**, 9–12.

Smol, J.P., Birks, H.J.B., Last, W.M. (Eds) (2001a). *Terrestrial, algal, and siliceous indicators.* Vol. 3 of "Tracking environmental change using lake sediments". Kluwer Academic Publishers, Dordrecht, Netherlands, 371 pp.

Smol, J.P., Birks, H.J.B. & Last, W.M. (Eds) (2001b). *Zoological indicators.* Vol. 4 of "Tracking environmental change using lake sediments", Kluwer Academic Publishers, Dordrecht, Netherlands, 220 pp.

Teller, J.T. (1985). Lake Agassiz and its influence on the Great Lakes. *In*: Karrow, P. & Calkin, P.E. (Eds), *Quaternary evolution of the Great Lakes.* Geological Association of Canada, Special Paper 30, 1–16.

Teller, J.T. (1987). Proglacial lakes and the southern margin of the Laurentide Ice Sheet. *In*: Ruddiman, W.F. & Wright, H.E. (Eds), *North America and adjacent oceans during the last deglaciation.* Geological Survey of America, Decade of North American Geology Vol. K-3, 39–69.

Teller, J.T. (1995a). The impact of large ice sheets on continental paleohydrology. *In*: Gregory, K., Baker, V. & Starkel, L. (Eds), *Global Continental Paleohydrology.* J. Wiley & Sons, New York, pp. 109–129.

Teller, J.T. (1995b). History and drainage of large ice-dammed lakes along the Laurentide Ice Sheet. *Quaternary International*, **28**, 83–92.

Teller, J.T. (2001). Formation of large beaches in an area of rapid differential isostatic rebound: the three-outlet control of Lake Agassiz. *Quaternary Science Reviews*, **20**, 1649–1659.

Teller, J.T. & Clayton, L. (1983). An introduction to glacial Lake Agassiz. *In*: Teller, J.T. & Clayton, L. (Eds), *Glacial Lake Agassiz.* Geological Association of Canada, Special Paper 26, 3–5.

Teller, J.T. & Kehew, A.E. (Eds) (1994). Late glacial history of proglacial lakes and meltwater runoff along the Laurentide Ice Sheet. *Quaternary Science Reviews*, **13**, 795–981.

Teller, J.T. & Leverington, D.W. (in press). Glacial Lake Agassiz: a 5000-year history of change and its relationship to the isotopic record of Greenland. *Geological Society of American Bulletin.*

Teller, J.T. & Mahnic, P. (1988). History of sedimentation in the northwestern Lake Superior basin and its relation to Lake Agassiz overflow. *Canadian Journal of Earth Sciences*, **25**, 1660–1673.

Teller, J.T. & Thorleifson, L.H. (1983). The Lake Agassiz – Lake Superior connection. *In*: Teller, J.T. & Clayton, L. (Eds), *Glacial Lake Agassiz.* Geological Association of Canada, Special Paper 26, 261–290.

Teller, J.T., Leverington, D.W. & Mann, J.D. (2002). Freshwater outbursts to the oceans from glacial Lake Agassiz and their role in climate change during the last deglaciation. *Quaternary Science Reviews*, **21**, 879–887.

Thorleifson, L.H., Wyatt, P.H., Shilts, W.W. & Nielsen, E. (1992). Hudson Bay lowland Quaternary stratigraphy: evidence for early Wisconsinan glaciation centered in Quebec. *In*: Clark, P.U. & Lea, P.D. (Eds), *The last interglacial-glacial transition in North America.* Geological Society of America, Special Paper 270, 207–221.

Tinkler, K.J. & Pengelly, J.W. (1995). Great Lakes response to catastrophic inflows from Lake Agassiz: some simulations of a hydraulic geometry for chained lake systems. *Journal of Paleolimnology*, **13**, 251–266.

Tinkler, K.J., Pengelly, J.W., Parkins, W.G. & Terasmae, J. (1992). Evidence for high water levels in the Erie basin during the Younger Dryas chronozone. *Journal of Paleolimnology*, **7**, 215–234.

Veillette, J.J. (1994). Evolution and paleohydrology of glacial lakes Barlow and Ojibway. *Quaternary Science Reviews*, **13**, 945–971.

Vincent, J.-S. & Hardy, L. (1979). The Evolution of Glacial Lakes Barlow and Ojibway, Quebec and Ontario. *Geological Survey of Canada Bulletin*, **316**, 18 pp.

Warner, B.G. (Ed.) (1990). Methods in Quaternary Ecology. *Geoscience Canada Reprint Series 5, Geological Association of Canada*, 170 pp.

Yu, Z. & Wright, H.E. (2001). Response of interior North America to abrupt climate oscillations in the North Atlantic region during the last deglaciation. *Earth Science Reviews*, **52**, 333–369.

Zoltai, S.C. (1961). Glacial history of part of northwestern Ontario. *Proceedings Geological Association of Canada*, **13**, 61–83.

Pleistocene glaciations of the Rocky Mountains

Kenneth L. Pierce

U.S. Geological Survey, Northern Rocky Mountains Science Center, Montana State University, Bozeman, Montana, 59717-3492, USA; kpierce@usgs.gov

Introduction

This chapter presents the status of Rocky Mountain glacial studies in 1965 and progress from that time to the present. The Rocky Mountains and the adjacent Basin and Range of the United States consist of about 100 ranges distributed in a northwest trending belt 2000 km long and 200–800 km wide. Glaciation created much of the grandeur of the high parts of these ranges. Figure 1 shows the extent of late Pleistocene glaciers as well as a measure of snowline altitude across the western U.S.

Early in the 20th century, some noteworthy regional Rocky Mountain glacial studies included: the Uinta and Wasatch Mountains of Utah (Atwood, 1909), western Wyoming (Blackwelder, 1915), the San Juan Mountains of Colorado (Atwood & Mather, 1932), western Montana (Alden, 1953), and southern Rocky Mountains (Ray, 1940).

For simplicity in communication, we use the following regional terms and their probable correlations with marine oxygen-isotope stages (MIS): Pinedale (Late Wisconsin, MIS 2), early Wisconsin (MIS 4), Bull Lake (now thought to be largely MIS 6 and perhaps 5d), and pre-Bull Lake (pre-MIS 6).

Status in 1965

Blackwelder (1915) named the Bull Lake and Pinedale glaciations for moraines on the east and west sides of the Wind River Range. In glaciated valleys of the Rocky Mountains, researchers generally distinguished a younger set of moraines (Pinedale) from an older set 5–10% further downvalley (Bull Lake). In 1965, G.M. Richmond summarized the status of glacial studies in the Rocky Mountains in two publications: (1) an INQUA chapter on the glacial geology (Richmond, 1965); and (2) a guidebook and 16-day INQUA field trip through much of the Northern and Middle Rocky Mountains (Richmond *et al.*, 1965). Also in 1965, Crandall (1967) mapped and described the glacial sequence on the Wallowa Mountains of eastern Oregon. In 1965, non-quantitative morphology was the prime basis for distinguishing and correlating Bull Lake and Pinedale moraines. Richmond (1965) noted Bull Lake moraines were bulky with smooth slopes, did not retain lakes, and were less bouldery than Pinedale moraines. Bull Lake moraines were commonly subdivided into early and late stades. In contrast, Pinedale moraines were described as steep, irregular, having kettles commonly with ponds, and studded with numerous relatively unweathered boulders. Pinedale moraines were commonly subdivided into early, middle, and late stades (Richmond, 1965). Moraines of the early and middle stade were distinguished at and near the terminus, whereas the late stade moraines were 25–75% farther upvalley from the terminal moraines to the valley heads.

In 1965, Rocky Mountain glacial subdivisions and correlations were closely linked with those of the mid-continent. The Bull Lake and Pinedale glaciations were correlated with the early and late Wisconsin respectively (Richmond, 1965). Curiously, Leverett (1917) had correlated the Bull Lake with the Illinoian Glaciation of the mid-continent that he defined in 1899, but this correlation was not accepted in 1965. Three pre-Bull Lake glaciations were correlated with the Illinoian, Kansan, and Nebraskan Glaciations (Richmond, 1965). Also in 1965, erratic boulders and diamictons well beyond or above moraines of Pinedale and Bull Lake age had been noted at many sites in the Rocky Mountains; these were attributed to an older glaciation vastly more extensive than the Bull Lake or Pinedale.

Soil development was then emerging as the primary relative-age method (Birkeland, 1964; Morrison, 1965; Richmond, 1962). Pinedale deposits have an "immature zonal soil" with "B-horizons 0.3–0.6 m thick, that display very little illuviation and weak to moderate structural development" (Richmond, 1965). In contrast, Bull Lake deposits have "mature zonal soils" with "B-horizons 0.3–1.2 m thick, with sufficient illuvial clay to be slightly plastic . . . and moderately developed subangular blocky structure."

The 1965 synthesis was primarily descriptive; glacial-geologic sequences were correlated according to succession and the general appearance of moraines. Pleistocene ELA's were estimated to be about 1000 m lower than present ELAs and both ELAs and moraines decreased in altitude northward through the Rocky Mountains (Richmond, 1965).

Advances since 1965

The next major synthesis of Rocky Mountain glaciations was by Porter *et al.* (1983). It included chronology, ELA patterns, basal shear stress, types of glaciers based on valley slope and glacier thickness, and contrasting mass balance between ranges based on the modern snowpack at the Pleistocene ELA. This study was followed by Richmond (1986a), who summarized glacial extents, stratigraphy, and chronology for glaciated ranges in the Rocky Mountains. Additionally Richmond (1986b) presented a detailed chronology of Yellowstone. Madole (1976) summarized the Colorado Front Range. For the Great Basin west of the Rockies,

DEVELOPMENTS IN QUATERNARY SCIENCE
VOLUME 1 ISSN 1571-0866
DOI:10.1016/S1571-0866(03)01004-2

PUBLISHED BY ELSEVIER B.V.

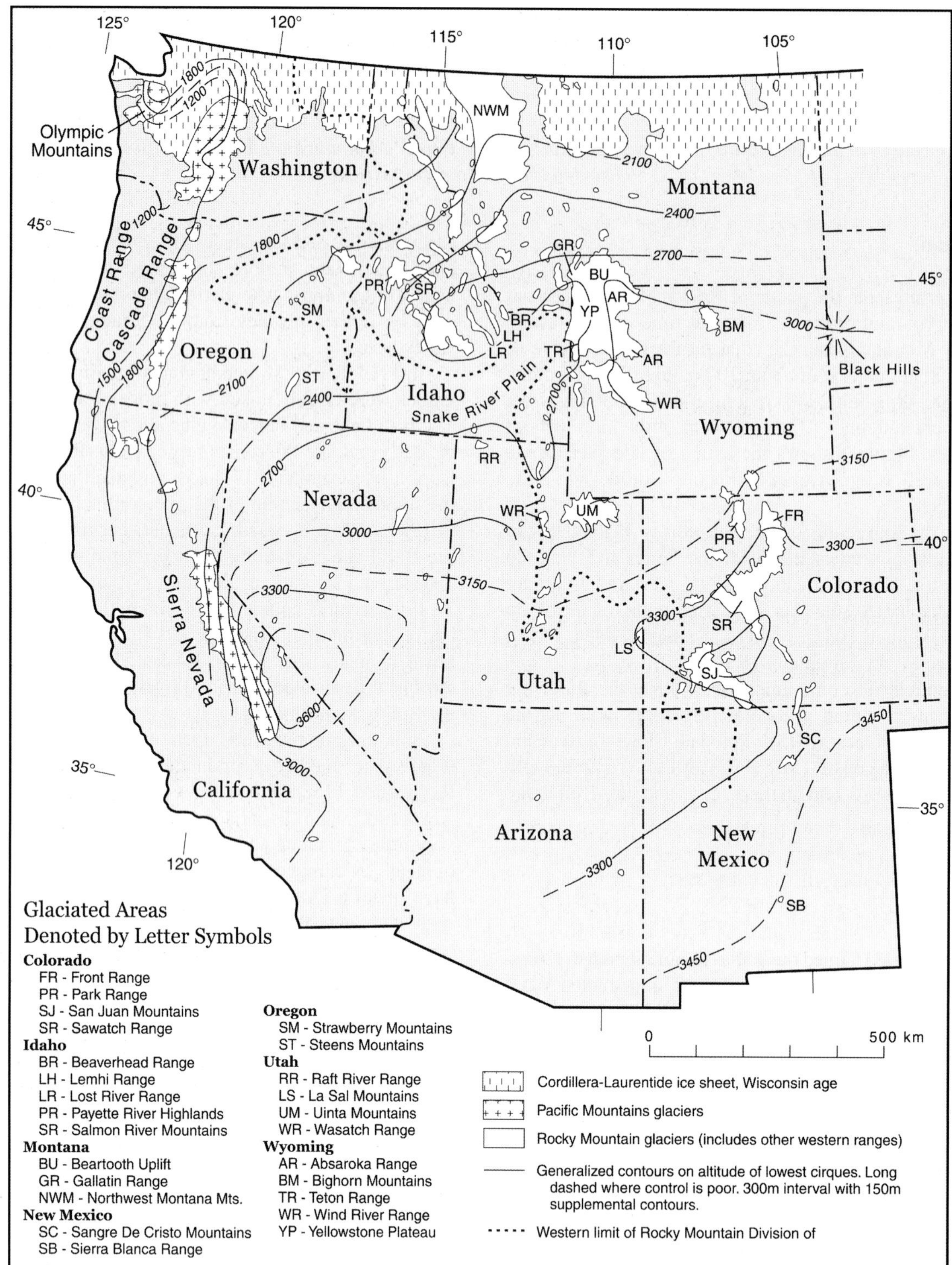

Fig. 1. Map showing extent of Late Wisconsin glaciers in the Rocky Mountains and other mountains in the western United. States (from Porter et al., 1983). Contours in meters are on the lowest cirque floors. This measure of late Pleistocene snowline increases in altitude both inland from the Pacific Ocean as moisture decreases, and southward as temperature increases (Hammond, 1965).

Wayne (1984) made a detailed study of the Ruby-Humboldt area, and Osborn & Bevis (2001) achieved an extensive overview including mapping and sequence descriptions and mapping for many ranges.

Quaternary Geochronologic Studies

Dating and correlation of glacial and other Quaternary deposits depend on a variety of dating techniques, as recently described in Noller *et al.* (2000). Dating techniques may be classified (Colman & Pierce, 2000) as: (1) relative age (chemical, biologic, geomorphic methods that include soils and weathering rinds, some of which can be refined to calibrated age; (2) numerical age (sidereal, isotopic, and radiogenic); and (3) correlated-age methods.

Selected Relative-Age Methods

Relative-age methods are broadly applicable and are useful in distinguishing deposits in a sequence. Correlation between sequences is better established if variables such as climate and lithology can be controlled. Methods may be combined; for example, Miller (1979) measured six relative age parameters on a glacial moraine sequence and applied statistical clustering techniques to define two ages of Pinedale moraines, two ages of Bull Lake moraines, and one age of pre-Bull Lake moraine.

Soil development. After 1965, soils became a primary basis for distinguishing glacial and other deposits in the Rocky Mountains, and other western mountains, where Pete Birkeland, his students and his text (1974, 1999) played a central role. Studies of his students include: (1) soils along a Rocky Mountain glacial transect (Shroba, 1977; Shroba & Birkeland, 1983); moraine-terrace transects in Colorado (Nelson & Shroba, 1998; Netoff, 1977); (2) the moraine sequence in the Wallowa Lake area, Oregon (Fig. 1, northeast of Strawberry Mountains) as well as part of the Sierra Nevada (Burke, 1979; Burke & Birkeland, 1979); (3) a terrace sequence near the Wyoming-Montana border (Reheis, 1987); (4) a toposequence in central Idaho from moraine crest to foot slope on Bull Lake and Pinedale glacial moraines (Berry, 1987). Some other important contributions were the formulation of a soil development index by Harden (1982), and quantitative studies of soils in the type areas of the Bull Lake and Pinedale glaciations (Dahms, 1991; Hall & Shroba, 1993; Swanson, 1985).

Weathering rinds and carbonate coats. Motivated by the results of Porter (1975), Colman & Pierce (1981) measured weathering rinds on basalt and andesite in the western U.S. and found that rind thicknesses systematically increased with stratigraphic sequence of moraines. They calibrated this method based on dating of the West Yellowstone Bull Lake as MIS 6. For the glacial succession near McCall, Idaho, weathering rinds show a clear difference between moraines correlated with MIS 2, 4, and 6. But for the corresponding soils, clay increase expressed as grams/cm^2 shows little change between moraines correlated with MIS 4 and MIS 6 based on weathering rinds. In areas of calcic soils (pedocals), carbonate coats on the undersides of stones increase in thickness with time. Based on Uranium-series ages, carbonate coats in central Idaho accumulate at ~0.6 mm/10^3 yr (Pierce, 1985; Pierce & Scott, 1982).

Numerical age methods. Few numerical ages existed in 1965 for glacial deposits in the Rocky Mountains. The establishment of marine oxygen isotope stages provided a reference standard for the large number of late Cenozoic glacial-interglacial oscillations and potentially associated glacial advances. The AGU Handbook (Noller *et al.*, 2000) describes many numerical-age techniques useful to dating glacial deposits, including radiocarbon and luminescence that can be applied to glacially related deposits such as bogs and loess. For Rocky Mountain and other glacial sequences, cosmogenic dating (Gosse & Phillips, 2001) is producing a numerical chronology that is based on the accumulation of isotopic changes in morainal boulders based on the duration of surface exposure to cosmic rays, provided one can account for surface erosion of the boulders and possible exhumation of the boulders. It has yielded ages that suggest a younger age than indicated by radiocarbon dating for the Pinedale glaciation and a possible MIS 5d correlation for type Bull Lake deposits.

Pre-Bull Lake Glacial Deposits

We use the designation pre-Bull Lake for glacial deposits that are older than MIS 6 (or 8?) and include deposits as old as the late Pliocene. Most deposits once attributed to very large pre-Bull Lake glaciations are now interpreted to have non-glacial origins. Primary reasons are lack of till fabric and glacial striations, and their satisfactory explanation as deeply weathered fluvial gravel. In addition, many of these deposits are located well outside Pinedale and Bull Lake glaciated areas; this requires an earlier, much more extensive glaciation generated during a much more severe glacial climate. However, the marine isotope record shows no glacial intervals much more severe than MIS 2 or 6, and suggests that pre-Bull Lake glaciations were not much more extensive than those of the Bull Lake or Pinedale.

Richmond (1965 and references therein) defined three pre-Bull Lake glacial deposits on the east side of the Wind River Range, from oldest to youngest: Washakie Point, Cedar Ridge, and Sacagawea Ridge. Later work establishes that only the Sacagawea Ridge has a glacial origin. At the type section for the Cedar Ridge and Washakie Point deposits, Hall & Jaworowski (1999) conclude that the only pre-Bull Lake glacial till present is the Sacagawea Ridge till. Chadwick *et al.* (1997) also did not confirm the presence of Cedar Ridge or Washakie Point glacial till. Moraines and associated outwash terraces of the Sacagawea Ridge glaciation (Richmond & Murphy, 1989) are somewhat older than the Lava Creek Ash dated at 640,000 ± 2000 yr (Christiansen, 2001), and probably correlate with MIS 16 (Chadwick *et al.*, 1997).

Three extensive pre-Bull Lake glaciations were also named and defined from diamictons of inferred glacial origin in the La Sal Mountains of Utah (Richmond, 1962). Shroder & Sewell (1985) concluded that these diamictons are extensive mass-movement deposits. They found the total area glaciated was only one twentieth (5%) the area mapped as glaciated in pre-Bull Lake time by Richmond (1962).

Madole (1982) concluded that upland bouldery deposits on the Colorado Front Range previously considered to be glacial were Tertiary deposits of non-glacial origin. For the southern Rocky Mountains, Scott (1975) concluded that all pre-Bull Lake glacial deposits are close to Bull Lake and Pinedale moraines, and that bouldery deposits once attributed to an extensive icecap glaciation are weathered Cenozoic fluvial gravels. In Jackson Hole, deposits of the "ghost" glaciation (Love, 1977) are either outlying deposits of no necessary glacial origin, or are part of the Munger (Bull Lake) glaciation of Pierce & Good (1992).

The best evidence for multiple pre-Bull Lake glaciations is exposed in sections just east of the Rocky Mountains near the US-Canada border. A succession of pre-Bull Lake diamictons identified as glacial by striations, till fabrics, stone shapes, and erratics from the Rocky Mountains are separated by soils but do not require a much more extensive glaciation than the Bull Lake (Karlstrom, 2000 and references therein; Fullerton *et al.*, 2003). Karlstrom (2000) identifies, from youngest to oldest, the following: two glaciations early in the Brunhes normal Chron (0–0.78 myr ago), at least three glacial events during the Matuyama reversed Chron (0.78–2.6 myr ago), and two events with normal polarity either in the Gauss Chron (2.6–3.6 myr ago), or possibly the Reunion or Olduvai events (2.23–2.20 or 1.93–1.76 myr ago).

Terrace sequences, which include pre-Bull Lake terraces, represent climatically modulated cycles of erosion followed by lateral planation and deposition. However, few terraces are tied directly to pre-Bull Lake glacial moraines. With age control on one or more terraces, the ages of other terraces can be approximated by incision rates (Reheis, 1987; R.C. Palmquist, written comm., 1989; Chadwick *et al.*, 1997). Locally such terrace sequences have age control provided by one or more volcanic ashes, such as the 640,000-yr-old Lava Creek ash in Verdos Alluvium roughly 100 m above drainage in the Colorado Piedmont (Scott, 1975).

Bull Lake Glaciation

Combined K-Ar and Obsidian Hydration Dating

In 1965, Bull Lake moraines were widely considered to be early Wisconsin in age (Richmond, 1965). However, the Bull Lake moraines near West Yellowstone are clearly older than the West Yellowstone rhyolite flow (Pierce *et al.*, 1976; Richmond, 1986b; Waldrop, 1975), which is best dated as 122,300 ± 2200 yr old (Obradovich, 1992, and spoken comm., Sept. 2002, based on the three older sanidine ages selected because complete degassing is a primary concern). Thickness of hydration rinds on glacial pressure cracks in obsidian clasts from the Bull Lake moraines at West Yellowstone are calibrated by hydration thicknesses on cooling cracks of dated rhyolite flows. A plot through time shows that the Bull Lake glacial cracking is ~30,000 yr older than the 122,000-yr-old West Yellowstone flow, and is ~40,000 yr younger than the 183,000 ± 3000 yr old Obsidian Cliff flow. Thus the age is about 150,000–140,000 yr (Pierce *et al.*, 1976) and correlates with the later part of MIS 6 (190,000–130,000 yr ago, Martinson *et al.*, 1987). A younger, less extensive glacial margin is indicated by the unusual embayed and perlitic eastern margin of the West Yellowstone rhyolite flow (Christiansen, 2001, p. G44). This recessional(?) glacial margin is about 15 km east of the Bull Lake terminus. Christiansen (2001, p. G46) estimates the K-Ar ages have a geologic uncertainty of ±10,000 yr. Thus, the age of the West Yellowstone flow may be from ~135,000 to 110,000 yr, and the glacier related to the embayed flow margin may date from late MIS 6 time, during recession from Bull Lake moraines, *or* a separate advance during MIS 5d time.

Wave-cut bluffs eroded into the Bull Lake end moraines at The Narrows of Hebgen Lake near West Yellowstone expose a 10-cm ash bed between imbricate thrusts of till. The ash had an apparent K-Ar age of 481,000 yr (Obradovich, 1992; Richmond, 1986b). Richmond (1986b) defined two glaciations based on position relative to the ash: the 610,000–481,000-yr-old "Till of Horse Butte" below the ash and the 481,000–399,000-yr-old "Till of Hebgen Lake" above the ash. J.D. Obradovich (spoken comm., 2002) and I conclude that contamination of the feldspar concentrate is likely. The ash chemistry is similar to the 162,000 ± 2000 yr old tuff of Bluff Point (G.A. Izett, written comm., 1990; Obradovich, 1992). I interpret the till thrust above the 162,000 yr old ash to be the same as the surface Bull Lake till and consistent with late MIS 6 time, and the till beneath the glacially thrusted ash to possibly be older than 162,000 yr and perhaps of an early MIS 6 age.

Cosmogenic Exposure Dating of Bull Lake Deposits

At the type Bull Lake on the east side of the Wind River Range, boulder exposure ages (^{36}Cl supplemented by ^{10}Be; Phillips *et al.*, 1997) from a sequence of 15 Bull Lake moraines (Chadwick *et al.*, 1997) yield the following ages and suggested MIS correlations:

Moraine Group & No.		Moraine Dated	Cosmogenic Age (10^3 yr)	MIS
D	XII–VX	XIII	95–120	5d
D		XII	95–120	5d
C	IX–XI	IX	100–130	5d
B	IV–VIII	(*not dated*)	>130	6
A	I–III	II–III	>130	6

Phillips *et al.* (1997) caution that in addition to laboratory analysis, additional uncertainty is about 10–15%.

Contrasting cosmogenic ages on boulders on the *same* Bull Lake moraines on the west side of the Wind River Range are illustrated and described in Fig. 2. Gosse & Phillips (2001) show that both ^{10}Be and ^{26}Al ages increase with increasing resistance to erosion (Fig. 2A), and conclude that these Bull Lake moraines are ~150,000 yr old. For the same moraines (Fig. 2B), Phillips *et al.* (1997) find an age of 120,000–100,000 yr, an age 20–35% younger than the ^{10}Be and ^{26}Al ages, but similar to the ^{36}Cl ages on Group C and D across the range at Bull Lake. In light of the U-series dating discussed next, it seems likely that the ^{36}Cl ages are too young, on both sides of the range.

U-series dating of terraces along the Wind River in the type area of the Bull Lake glaciation (Blackwelder, 1915) produces ages older than the cosmogenic ages of the nearby correlative moraines. Multiple analyses determining ^{230}Th/U ages on micro-stratigraphic layering of the carbonate coats on stones from soils on the terraces yield the following ages (Sharp *et al.*, 2003):

Terrace	Age (10^3 yr)	MIS	Glacial Unit
WR2	55 ± 8.6	4	Early Pinedale correlative
WR3	150 ± 8.3	6	Late Bull Lake
WR4	167 ± 6.4	6	Early Bull Lake

Based on these ages, the early and late Bull Lake terraces at the type area correlate with early and late MIS 6.

In the Colorado Front Range, ^{10}Be and ^{26}Al ages on only 2(?) boulders in Bull Lake moraines are 122,000 ± 26,000 yr and those from a Bull Lake terrace are 136,000 ± 28,000 yr, both with no correction for erosion of boulder surface (Schildgen *et al.*, 2002; Dethier *et al.*, 2003).

Discussion of Age and Correlation of the Bull Lake Glaciation

In the West Yellowstone area using combined K-Ar and obsidian hydration dating, Pierce *et al.* (1976) determined a MIS-6 age (Illinoian) for the Bull Lake Glaciation, revising the widely accepted early Wisconsin correlation (MIS 4?). Next, for Bull Lake moraines of the Wind River Range primarily based on ^{36}Cl ages, Phillips *et al.* (1997) dated Groups A and B as MIS 6 and Groups C and D as MIS 5d. However, dating the firmest boulders, Gosse & Phillips (2001) determined a mid-MIS-6 ^{10}Be age on the same moraine that Phillips *et al.* (1997) determined a MIS 5d ^{36}Cl age. Finally, U-series ages by Sharp *et al.* (2003) date the type terrace of the Bull Lake glaciation (WR3) as late MIS 6, and an older Bull Lake terrace (WR4) as early MIS 6, arguing against a MIS 5d age for the Bull Lake.

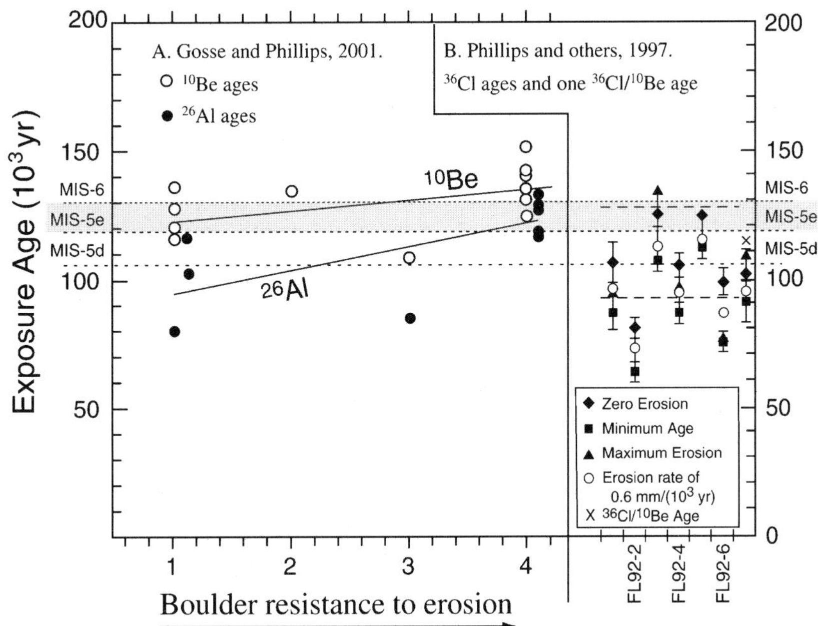

Fig. 2. Plots showing differences between two cosmogenic-dating studies on boulders from the same Bull Lake moraine on west side of Wind River Range near Fremont Lake. These differences illustrate the current difficulty in using exposure ages to distinguishing MIS 5d from MIS 6. (a) Plot of ages in order of increasing resistance to erosion and difficulty in sampling, from (1) weathered plagioclase porphyroblastic granodiorite to (4) unweathered granite. Both ^{10}Be and ^{26}Al exposure ages increase with increasing resistance of boulders to erosion. Gosse & Phillips (2001) conclude these moraines are ~150,000 yr old (MIS 6). (b) Plot of ^{36}Cl ages supplemented by ^{36}Cl/^{10}Be age for the same moraine as in Fig. 2A (Phillips et al., 1997). Phillips et al. (1997) conclude "the distribution of ^{36}Cl ages as quite similar to that for the Bull Lake IX moraine at Bull Lake, giving limits of 120 to 100 × 10^3 yr." This age range would suggest correlation with MIS 5d, although the left side of this figure as well as U-series dating (see text) indicate a MIS 6 age.

The ice-contact West Yellowstone flow is either recessional Bull Lake or a younger advance. Its age of 122,300 ± 2200 yr favors correlation with MIS 5d, although recession from the Bull Lake moraines (MIS 6) is permissible with the time scale of Martinson *et al.* (1987). However, if Winograd *et al.*'s (1997) dates at Devils Hole in southern Nevada are correct, and the last interglaciation (MIS 5e) in the western U.S. began 142,000 yr ago and lasted until 120,000 yr ago, then it follows that correlation of the ice-contact flow with MIS 5d is implied.

An extensive mountain glaciation during MIS 5d appears plausible in that a major minimum in solar insolation for the Northern Hemisphere culminated about 110,000 yr ago (Berger, 1978), and cool summers combined with enhanced precipitation from warm oceans then could have fostered the expansion of mountain glaciers. For Mono Basin moraines in Bloody Canyon of the Sierra Nevada, Phillips *et al.* (1990) advocate a MIS 5d age, although this age is appears to be out of stratigraphic sequence (see Kaufman *et al.*, this volume). Glaciation during MIS 5d at high latitudes is advocated for the Lake Baikal area, Siberia (Karabanov *et al.*, 1998). One implication of the assignment of some Bull Lake advances to MIS 5d is that glaciers more extensive than the Pinedale glaciers advanced and retreated in only ~12,000 yr (the MIS-5d age-span of Martinson *et al.*, 1987). Studies of ice cores in Antarctica and Greenland indicate it has taken ~10,000 yr *or longer* to accumulate the upper 1 km of ice in these areas of relatively slow accumulation. The Pinedale icecap on the Yellowstone Plateau built up to a thickness of a kilometer *after* an advance from the adjacent mountains (Good & Pierce, 1996). Also, the Wind River Pinedale icecap on the east side of the range locally was 600 m thick (William Locke, Earth Sciences web site at Montana State University, 2002, http://www.homepage.montana.edu/~ueswl/winds.html). Thus, there is a question whether adequate time exists between warm periods MIS 5c and 5e to build up icecaps thicker than the Pinedale, and then to deposit multiple, large-volume moraines. If this did occur in the 12,000-yr span of MIS 5d, it demonstrates the rapidity of icecap buildup, moraine deposition, and recession.

Lake-sediment records from the U.S. have not been interpreted to show a major glaciation in MIS 5d comparable in magnitude with MIS 6 or 2. Such records have been studied from Lake Bonneville (Oviatt *et al.*, 1999), Owens Lake, California (Bischoff *et al.*, 1997, using the timescale of Bischoff & Cummings, 2001; Litwin *et al.*, 1999), or from the mid-continent (Zhu & Baker, 1995). For Clear Lake, California, Adam (1988) shows that MIS 5d consisted of three cold intervals, each lasting only 1400–2400 yr. They were separated by two warm intervals of 3100–4700-yr duration. The oxygen-isotope record from Devils Hole, Nevada (Winograd *et al.*, 1997) also shows MIS 5d as shorter than MIS 2, 4, and 6. Whitlock *et al.* (2000; and spoken comm., 2002) concluded from pollen studies of cores from Carp Lake, western Washington, that MIS 5d was cool and humid, and not as severe than MIS 2. The MIS-5d landscape was covered by open forest.

In conclusion, early and late Bull Lake moraines have been distinguished in many Rocky Mountain areas (Richmond, 1965), locally on the basis of an intervening soil. Distinction as early and late Bull Lake may correlate either with: (1) MIS 6 and 5d respectively, as suggested for the Wind Rivers (Phillips *et al.*, 1997; Chadwick *et al.*, 1997); or (2) early and late MIS 6 (190,000–170,000 and 150,000–130,000 yr ago) as indicated by: (a) Sharp *et al.* (2003) for the Wind River areas (b) Fullerton *et al.* (2003) for early and late Bull Lake in the Glacier Park area, and (c) this paper for the West Yellowstone area. For different places, such contrasting correlations may be valid and show that different areas have contrasting surviving successions of Bull Lake-like moraines, a topic extensively developed in Gillespie & Molnar (1995). Nevertheless, I consider surviving Bull Lake moraines of MIS 5d yet to be established, where whereas those of MIS 6 are quite credible.

MIS 4 or Early Wisconsinan Glaciation

Based on weathering-rind thickness, Colman & Pierce (1981, 1986, 1992) found three successions in the western U.S. had early Wisconsin moraines (~60,000–70,000 yr old), including the Rocky Mountain sequence at McCall, Idaho. Other information also favors glaciation during MIS 4. On the east side of the Wind River Range, an outwash(?) terrace (WR2) between the Pinedale (WR1) and late Bull Lake (WR3) has a ^{230}Th/U age on carbonate coats of 55,000 ± 8600 yr (Sharp *et al.*, 2003), suggesting an early Wisconsin glacial advance, but with moraines subsequently overridden by younger glaciers. For Owens Lake, eastern California, abundant rock flour signifies Sierran glaciation from 78,000 to 66,000 yr ago (Bischoff & Cummings, 2001). A relatively deep lake in the Bonneville Basin, Utah dates 59,000 ± 5000 yr, and probably represents cooler conditions near the MIS 4/3 boundary (Kaufman *et al.*, 2001). A high stand of Summer Lake, southeastern Oregon dates between 89,000 and 50,000 yr, a period that includes MIS 4 (Cohen *et al.*, 2000). A loess-buried soil section in southern Jackson Hole records glacial(?) loess deposition ~65,000–75,000 yr ago based on TL ages, ^{10}Be accumulation, soil development, and minimum ^{14}C ages (K.L. Pierce, unpub. data). At Carp Lake, Washington, a cool-humid interval from 58,000 to 43,000 yr ago and cool-dry interval from 72,000 to 58,000 yr ago may represent glaciation in MIS 4 and early MIS 3 time, bracketed by warmer intervals (Whitlock *et al.*, 2000). Thus, an early Wisconsin (MIS 4) glacial advance probably has occurred in the Rocky Mountains, but surviving moraines are not recognized (see Gillespie & Molnar, 1995), excepting for the McCall, Idaho area.

Numerical Ages of Pinedale Deposits

Ages of the Pinedale (last) glaciation based on calibrated (cal) radiocarbon and obsidian-hydration dating are greater than new ages based on cosmogenic methods (Fig. 3). Fig. 3B shows the extent of Pinedale glaciation based on a 1983 compilation of ^{14}C and obsidian-hydration ages (Porter *et al.*, 1983, Fig. 4–28). Between Fig. 3A and B, the correction of ^{14}C age (right) to calibrated age (left) is show by the sloping

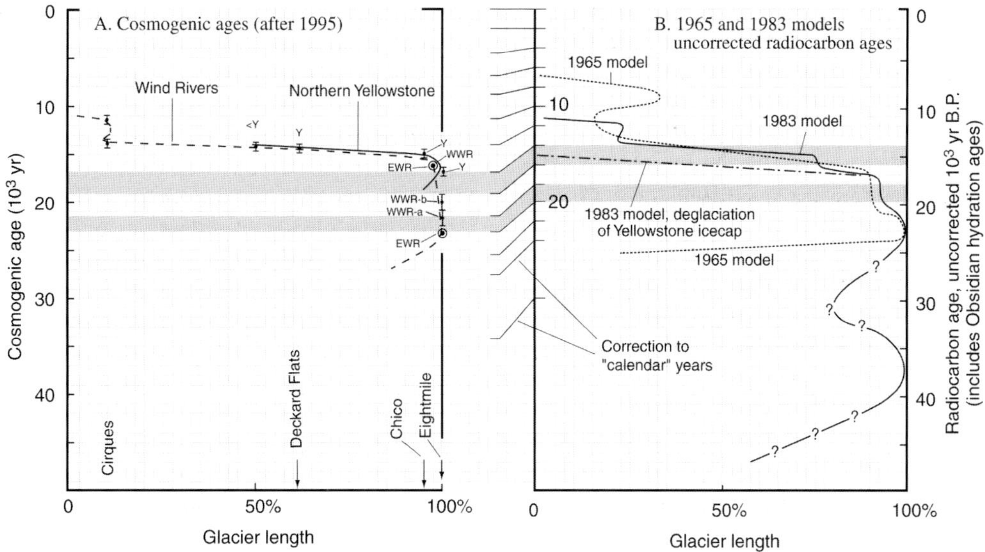

Fig. 3. Development of chronologies of the last (Pinedale) glaciation of the Rocky Mountains showing age vs. percent full-glacial length. A. Current cosmogenic dating for Wind River Range (Gosse et al., 1995a, b) and the Yellowstone icecap (Licciardi et al., 2001). EWR and WWR-east and west Wind River Range with WWR-a, original terminal moraine age of Goss and WWR-b, age calculated by J.M. Licciardi (written comm., 2003) using current scaled production rates; Y-Yellowstone. B. Plots on radiocarbon time scale showing 1965 model and 1983 compilation (both from Porter et al., 1983). For 1983 compilation of mountain glaciers and icecaps, control points are plotted and indexed to three tables (Porter et al., 1983, Figs 4–28, Tables 4–6 to 4–8). The center column between A and B shows the increase in age between radiocarbon ages (right) to calibrated ages (left); the shaded bands highlight this age offset. The cosmogenic chronology dates recession from near the terminal moraine position about 15,000 yr ago, whereas the calibrated ^{14}C compilation indicates an age nearer 20,000 cal yr B.P.

lines and is based on Stuiver & Reimer (1993, Calib. program) and Bard *et al.* (1998). Figure 3B shows the Pinedale glacial maximum occurred 25,000 to 20,000 ^{14}C yr B.P. (~23,400 to 29,200 cal yr B.P.) and was followed by rapid recession of ice-caps by 15,000 ^{14}C yr B.P. (~18,000 cal yr B.P.) and recession of valley glaciers to near cirque positions (shown as a small readvance) by 12,000 ^{14}C cal yr B.P. The 1983 model shows near complete deglaciation about 5000 yr earlier than the 1965 model (Fig. 3B, dotted line).

More recent ^{14}C ages also support the plot shown in Fig. 3B, with relatively old ages for the Pinedale maximum, and rapid recession to the cirques. For two glacially dammed lakes in the Colorado Front Range, Rosenbaum & Larson (1983) report the four oldest ages on finely disseminated plant debris were between 22,000 and 23,200 ($\pm$ ~1000) ^{14}C yr B.P. (27,000 to 25,700 cal yr B.P.). These newer ^{14}C ages are consistent with the antiquity of ^{14}C ages from the Colorado Front Range that were used in plotting Fig. 3B: (1) organic matter concentrated from sediments in a moraine-dammed lake dates to 22,400 ^{14}C yr B.P. (Madole, 1980), and (2) a section of interbedded till and peat indicates that the Pinedale glaciation lasted from >30,000 to 13,700 ^{14}C yr B.P. (Nelson *et al.*, 1979). In the Yellowstone-Grand Teton area, Whitlock (1993) obtained basal ages from cores in sediments deposited during the Pinedale recession. These ages were as old as 14,580 $\pm$ 150, 17,160 $\pm$ 210, 16,040 $\pm$ 220, 15,640 $\pm$ 150 and 14,490 $\pm$ 700 ^{14}C yr B.P., which would indicate glacial recession there by 20,000–19,000 cal yr B.P.

Cosmogenic Surface Exposure Ages of Pinedale Deposits

Cosmogenic ages for the Wind Rivers and Yellowstone produce ages for the Pinedale glacial maximum and recession that are younger than the ^{14}C based time-distance plot (Fig. 3). For the type Pinedale moraines on the west side of the Wind River Range, Gosse *et al.* (1995a) determined a terminal moraine age of 21,700 $\pm$ 700 ^{10}Be yr. They also found that moraine building continued near the terminus for ~6000 yr, or until ~15,800 ^{10}Be yr ago. Similarly, on the east side of the Wind River Range, Phillips *et al.* (1997) measured ^{36}Cl ages supplemented by ^{10}Be ages for boulders on the three outer Pinedale moraines, and concluded the moraines are between 23 and 16,000 yr old.

From the values determined by Gosse *et al.* (1995a), recalculation of the *mean* of each moraine group using current scaled production rates yields the following ages for the type Pinedale moraines (J.M. Licciardi, written comm., 2003):

Moraine or Group	Age ($\times 10^3$ ^{10}Be yr)
Terminal Pinedale	20.1 $\pm$ 1.0
Recessional group	17.6 $\pm$ 0.8
Difference	~2.5

In contrast, for the Pinedale sequence of the northern Yellowstone outlet glacier, Licciardi *et al.* (2001) analyzed a goodly number of boulders on each deposits and determined the following younger exposure ages:

Deposit (Number of Boulders Dated)	Cosmogenic Ages ($\times 10^3$ yr)
Eightmile terminal moraines (8)	16.5 ± 0.4 ^{3}He
Eightmile terminal moraines (9)	16.2 ± 0.3 ^{10}Be
Chico recessional moraines (8)	15.7 ± 0.5 ^{10}Be
Deckard Flats readjustment (10) (Yellowstone Plateau icecap no longer contributing)	14.0 ± 0.4 ^{10}Be
Late glacial outburst flood (6)	13.7 ± 0.5 ^{10}Be

These cosmogenic ages for the Eightmile (Pinedale) terminal moraines of the Yellowstone outlet glacier are about 3000–4000 ^{10}Be yr younger than the 20,100 ^{10}Be yr ages on outermost Pinedale terminal moraines on both sides of the Wind River Range (Fig. 3). This is probably a real age difference that can be explained by either: (1) the time interval required for the progressive buildup of the Yellowstone icecap; or (2) a climatic difference related to nearness to the Laurentide ice sheet (J.M. Licciardi, written comm., 2003). The 3000–4000 ^{10}Be yr contrast between the Yellowstone and the Wind Rivers in Pinedale terminal moraines shows that subdivision into early and middle stades does not necessarily indicate correlation from place to place.

For the Wallowa Range in eastern Oregon, the following are averages of cosmogenic ages on boulders on close-spaced moraines in the Wallowa Lake area (TTO, TTY, WTO) and in the glacial cirques (GL) (Licciardi, 2000; written comm., 2003):

Moraine (Number of Boulders Dated)	Age ($\times 10^3$ ^{10}Be yr)
TTO & some TTY (9)	21.1 ± 0.4
WTO & some TTY (9)	17.0 ± 0.3
GL (4), near cirques	10.2 ± 0.6

These ages are quite similar to those of the Wind Rivers.

From a Pinedale moraine site in Colorado Front Range, a few boulders yield an exposure age of $16,800 \pm 3400$ ^{10}Be and ^{26}Al yr B.P. (Dethier *et al.*, 2003). This age is also more than 5000 yr younger than corrected ^{14}C ages for moraine-dammed lakes in the same area.

The rate of retreat from the terminal-moraines areas to the cirques was remarkably fast (Fig. 3). For the Wind Rivers, glacial retreat of 33 km occurred at an average rate of 7.4 m/yr to moraines dated $12,100 \pm 500$ ^{10}Be yr in the cirque basins (Gosse *et al.*, 1995b). Retreat of the northern Yellowstone outlet glacier also occurred at the rapid rate of ~30 m/yr, given the 15,700 ^{10}Be yr for Chico recessional moraines and 13,700 ^{10}Be yr for flood deposits that originated more than 60 km upvalley (Licciardi *et al.*, 2001). For the Glacier National Park area (Carrara, 1987), much of deglaciation had occurred by about 12,000 ^{14}C yr B.P. (14,000 cal yr B.P., Carrara, 1995), also suggesting rapid retreat.

Discussion of Pinedale Ages

The cosmogenic ages for moraines of the northern Yellowstone outlet glacier are thousands of years younger than the Yellowstone chronology based on ^{14}C, obsidian-hydration (Pierce *et al.*, 1976), and U-series ages (Sturchio *et al.*, 1994), although none of non-cosmogenic methods used directly dated moraine deposition. As discussed previously, the calibrated ^{14}C ages for deglaciation of the Yellowstone Plateau are as old as 20,000–19,000 cal yr B.P. (Millspaugh *et al.*, 2000; Porter *et al.*, 1983; Whitlock, 1993; see discussion in Licciardi *et al.*, 2001). Either the cosmogenic ages are too young, or the carbon samples contained a large fraction of old carbon. Many of the carbon samples dating deglaciation were of finely disseminated organic carbon, which could be contaminated by older carbon. But some samples such as a peaty mud dating ~15,800 cal yr B.P. ($13,140 \pm 700$ ^{14}C yr B.P.; Porter *et al.*, 1983, W-2285) from near the center of the Yellowstone Plateau ice cap consisted of recognizable plant fragments and is inconsistent with the $14,000 \pm 400$ ^{10}Be yr age of the Deckard Flats readjustment which predates the complete deglaciation of the Yellowstone Plateau.

Obsidian-Hydration Age of Pinedale Glaciation

The obsidian-hydration dating for Yellowstone assumes that under a constant temperature the square of hydration thickness is a linear function of age and that the Yellowstone obsidians sampled hydrate at the same rate (Pierce *et al.*, 1976). Hydration was measured on pressure cracks resulting from glacial grinding *up to* the time of moraine deposition. Age calibration for obsidian in Bull Lake moraines is based on their hydration thickness compared to that on bracketing 122,000- and 183,000-yr-old rhyolite flows (Pierce *et al.*, 1976). But no bracketing rhyolite flows exists for the Pinedale moraines, and age estimation is particularly sensitive to both the difference in soil temperature between late Pleistocene and Holocene time *and* the present soil temperature difference between the Yellowstone Plateau and the West Yellowstone Basin (Pierce *et al.*, 1976). An age younger than the ~30,000–25,000 yr obsidian-hydration age calculated for the Pinedale moraines near West Yellowstone (Pierce *et al.*, 1976) would result either if (1) the Pleistocene-Holocene temperature difference was larger than the estimated 6 °C *or* (2) the temperature difference between the Yellowstone Plateau and West Yellowstone basin was larger than the estimated 0.5 °C, *or* (3) both.

Moraines Attributed to Younger Dryas

The younger Dryas (YD) was a dramatic cooling episode between 11,000 and 10,000 ^{14}C yr B.P. ($12,800 \pm 200$ and

11,500 ± 300 cal yr B.P.) in Europe. It has been attributed to a dramatic southward extension of cold North Atlantic water. At several localities in the Rocky Mountains, dating studies correlate a minor readvance in or near high cirques with the YD. In the core of the Wind River Range, the Inner Titcomb Lakes moraine dates 13,800 ± 600 to 11,400 ± 500 ^{10}Be yr (Gosse *et al.*, 1995b; see Fig. 3). Calculation of the *mean* age using the latest scaled production rates yields a *mean* age of 12, 600 ± 500 ^{10}Be yr (J.M. Licciardi, written comm., 2003). The Titcomb Lakes moraines correlate with the nearby Temple Lake moraines dated by sediment changes in nearby lake sediments as between 13,800 ± 900 and 11,800 ± 700 cal yr B.P. (Davis *et al.*, 1998).

In the Colorado Front Range, the Triple Lakes cirque moraines of Benedict (1985) have minimum ^{14}C ages just over and under ~10,000 ^{14}C yr B.P. (11,200 cal yr B.P.) indicating a latest Pleistocene age and suggesting they are candidates for the YD event (Davis, 1987). Also in the Colorado Front Range, lake sediments reflecting a nearby glacier advance date between 13,200 and 11,100 cal yr B.P. (Menounos & Reasoner, 1997). In Colorado Mountains, pollen spectra suggest a YD cooling event between 13,500 and 12,900 cal yr B.P. in Black Mountain Lake and 13,600 and 12,900 cal yr B.P. in Sky Pond (Reasoner & Jodry, 2000). In the southern Sangre de Cristo Range of New Mexico, lake sediments date an YD advance about 11,800 to 11,00 cal yr B.P. (Armour *et al.*, 2002).

Other Rocky-Mountain Glaciation Studies

Modeling Glacial Flow

Basal shear stress, mass balance, and glacial flow have been reconstructed for some late Pleistocene glaciers in the Rocky Mountains. For the glacial geologic reconstruction of the northern Yellowstone outlet glacier, basal shear stress averaged 1.2 bars for strongly extending reaches (converging flow lines), 1.0 bars for 11 uniform reaches, and 0.8 bars for 16 strongly compressing reaches (Pierce, 1979). Assuming precipitation similar to present, best estimates of mass balance yielded an annual accumulation of 2.8 km^3 above the ELA of 2850 m and annual loss of 3.3 km^3 below the ELA. Based on these parameters for a cross-section downvalley from the ELA, annual ice discharge was 2.7 km^3, of which 10% is modeled by flow within the glacier and the remaining 90% is attributed to basal sliding (Pierce, 1979).

On the mountains of northwestern Montana, a large ice cap built up on a complex mountain topography that includes several ranges. Based on terminal moraine positions, divide crossings, and nunataks, Locke (1995) modeled this ice sheet assuming basal shear stress values near 1 bar to produce a contour map of the ice surface. His model revealed much about the sources and non-sources of the multiple glacial lobes that radiate outward from this glacial source area. For some canyons draining the east side of Wind River Range icecap, W. Locke (written comm., Earth Sciences web site at Montana State University, 2002) calculates that the basal shear stress was as high as 8 bars. Possible factors for such high values are: (1) the need to funnel accumulation from a central icecap into narrow, steep canyons; and (2) low subglacial water pressure due to the draining of the glacial-bed water into permeable limestones high above the base level of the Wind River Basin.

Sedimentology of Glacial Deposits

On the southern margin of the San Juan Mountains, Colorado, glacial system, Johnson & Gillam (1995) found that end moraines are primarily of debris-flow sediment interbedded with sandy stream sediment. They conclude that "existing moraines were built rapidly (in 10 yr to a few tens of years)" but that outwash deposits indicate that the glacier stood at its terminus for a long time.

Pleistocene Climates

The contrast between present and glacial climates and departures of Pleistocene climate patterns from present can be made by comparison of late Pleistocene ELA's with modern snowpack patterns. For the Great Basin, Zielinski & McCoy (1987) located anomalies in the distribution of modern snowpack at late Pleistocene ELA's, in particular an anomaly in the NW Great Basin which appears to have been considerably wetter in Pleistocene times than present; this difference was also noted by Porter *et al.* (1983). Leonard (1984) found that contours on late Pleistocene ELA's for the San Juan Mountains in southern Colorado closely resembled modern snowpack pattern, suggesting little change in atmospheric circulation patterns. For the Sawatch Range in central Colorado, late Pleistocene mean summer temperatures are estimated to have been 7–9 °C colder (Brugger & Goldstein, 1999). Using modern climatic conditions at Pleistocene ELAs, Leonard (1989) found that Pleistocene glaciers could be maintained with either: (1) no change in total precipitation and 8.5 °C colder in summer, or (2) about 10–13 °C colder using Mears (1987) estimate of regional cooling and at least a 44% reduction in fall-through-spring precipitation. Locke (1990) used Mears (1987) estimate of regional cooling of 10 °C to infer an ~25% decrease in last glacial precipitation in Montana; Murray & Locke (1989) used ice-flow theory to independently estimate a similar drying.

Hostetler & Clark (1997), using a nested modeling strategy, determined that at 18,000 yr ago climatic conditions varied across the western U.S. Glaciers in the southern Rockies were sustained through decreases of 9–12 °C with little change in precipitation, whereas glaciers in the northern Rockies existed under relatively cold-summer and dry-winter conditions resulting from easterly anticyclonic wind flow off the Laurentide Ice Sheet. However, no evidence for this easterly flow was found by Muhs & Bettis (2000) based on late Wisconsin loess distribution, or by Locke (1990) or Gillespie (1991) based on changes in glacial ELA patterns in the northern Rocky Mountains and Sierra Nevada respectively.

Modeling of the climatic response in Western North America to the changes associated with sudden lowering of the Laurentide Ice Sheet in a Heinrich Event show a complex response that in different areas and times reinforce, cancel, or reverse local changes and thus make difficult simple correlations in the region (Hostetler & Bartlein, 1999).

Possible Uplift and Subsidence

Belts of uplift and subsidence may explain different patterns in the relative extent of Bull Lake and Pinedale glaciations near Yellowstone. The normal pattern is that Pinedale glaciers were ~90% as long as Bull Lake glaciers. But for both the Greater Yellowstone glacial system and nearby independent valley glaciers *on the western and southern* sides of Yellowstone, Pinedale glaciers are less than 80% and locally 60% the length of Bull Lake glaciers; whereas, *on the northern and eastern* sides of Yellowstone, Pindeale glaciers were >100% the length of Bull Lake glaciers and overran Bull Lake moraines (Pierce & Morgan, 1992). An explanation of this pattern is that the terrain to the northeast is rising and that to the southwest is subsiding. These areas appear to be on the currently uplifting (NE) and subsiding (SW) slopes of the "bow wave" of the Yellowstone hot spot. Compared to Bull Lake time, areas to the northeast became relatively higher in Pinedale time whereas those to the southwest became relatively lower (Pierce & Morgan, 1992).

Conclusions and Recommendations for Future Study

In their provocative paper "Asynchronous maximum advances of mountain and continental glaciers" Gillespie & Molnar (1995) cite much evidence that mountain glaciation did not proceed in lock step with either continental glaciation or its proxy, the marine oxygen isotope record. More dating may establish such variation for the Rocky Mountains. Global climate models (Kutzbach & Guetter, 1986) suggest that glacial-anticyclonic circulation weakened westerly flow and resulted in air cooler and drier than present, particularly for the northern Rocky Mountains. But Locke (1990) observed that the gradient of maximum glacial ELAs exactly parallels that of modern ELAs, indicating a similar rather than contrasting climatic pattern. In particular Locke noted no maximum glacial lowering of ELAs due to upslope precipitation on the east side of ranges that would be associated with such easterly flow. Assuming the Kutzbach & Guetter model results would suggest that differences in timing and relative magnitude of late Pleistocene glaciation is likely southward through the Rocky Mountains with greater distance from the continental ice sheet. Glacial culminations might follow storm tracks located at some distance south of the expanding or contracting continental ice sheet, a topic addressed by Licciardi (2000; written comm., 2003). More precisely dated glacial and lacustrine records may reveal patterns in such non-parallelism from south to north (colder) or east to west (wetter) throughout the Rocky Mountains.

Cosmogenic dating of Pinedale glaciers on both sides of the Wind River Range suggest a Pinedale culmination near the last glacial maximum 21,000 yr ago, but fail to reveal surviving older moraines of MIS 2, 3, or 4 age. Although imprecise, weathering rind thicknesses, some radiocarbon ages, and obsidian hydration suggest Wisconsin glacial advances in the 50,000–35,000 and 70,000–60,000 yr range (Colman & Pierce, 1981; Pierce *et al.*, 1976; Porter *et al.*, 1983). A 55,000-yr-old Wind River terrace (Sharp *et al.*, 2003) probably reflects early Wisconsin (MIS 4) glaciation. Records of glacial flour in lake sediments of the Rocky Mountains similar to that for the Sierra Nevada (Bischoff & Cummings, 2001) and the Cascades (Rosenbaum & Reynolds, 2003) could reveal much about the glacial record both preserved or subsequently overridden in the end moraine record, including helping resolve the relative magnitude of glacial activity in MIS 3, 4, 5d, and 6. For the many glacial cycles documented by the MIS record older than MIS 6, only the Sacagawea Ridge moraines of MIS 16 are well correlated with a particular MIS.

The potential distinction of moraines of MIS 4, 5d and MIS 6 may present interesting contrasts. The conference on the last interglaciation (Kukla *et al.*, 2002 and references therein) presents current knowledge about MIS 5d and its contrasts with MIS 5e and 5c. A warm ocean, a major low in northern Hemisphere solar insulation, and glacial expansion particularly at high latitudes, accompanied MIS 5d.

What were the glacial and climatic conditions along a north-south transect through the Rockies? In MIS 2, climate was cold and probably drier in the northern Rockies (Whitlock *et al.*, 2000), and MIS 6 was probably similar. Was MIS 5d a different character of glaciation than MIS 2 and 6 in the Rockies? If MIS 6 was similar to MIS 2, why are Bull Lake moraines (MIS 6 or MIS 5e) more bulky? The cross-feed between data for Pleistocene glaciations across the western mountains and climatic models such as Hostetler & Clark (1997) and Hostetler & Bartlein (1999) will enhance understanding in both subject areas.

Cosmogenic dating can benefit by refinement in calibration and changes in cosmic-ray flux through time. For boulders on deposits ~100,000 yr and older, the history of the boulder needs to be better understood including erosion of the boulder surface, emergence of the boulder from the eroding moraine, deposition and deflation of loess, and burial by snow.

Acknowledgments

I thank Pete Birkeland, Bill Locke, Joe Licciardi, Linda Pierce, and Nicole Davis for helpful reviews and editing. In the preparation of this paper, I benefited from discussions with Peter Clark, Bill Locke, Alan Gillespie, Steve Colman, Dave Fullerton, Joe Licciardi, Warren Sharp, Darrell Kaufman, Dave Dethier, Cathy Whitlock, Pete Birkeland, Dennis Dahms, Ralph Shroba, Eric Leonard, John Good, Don Easterbrook, John Obradovich, and Joe Rosenbaum.

References

Adam, D.P. (1988). Correlations of the Clear Lake, California, core CL-73–4 pollen sequence with other long climatic records. *Geological Society of America Special Paper 214*, 81–95.

Alden, W.C. (1953). Physiography and glacial geology of western Montana and adjacent areas. *U.S. Geological Survey Professional Paper 231*, 200 pp.

Armour, J., Fawcett, P.J. & Geissman, J.W. (2002). 15 k.y. paleoclimatic and glacial record from northern New Mexico. *Geology*, **30**, 723–726.

Atwood, W.W. (1909). Glaciation of the Uinta and Wasatch Mountains. *U.S. Geological Survey Professional Paper 61*, 96 pp.

Atwood, W.W. & Mather, K.F. (1932). Physiography and Quaternary Geology of the San Juan Mountains, Colorado. *U.S. Geological Survey Professional Paper 166*, 176 pp.

Bard, E., Arnold, M., Hamlin, B., Tisnerat-Laborde, N. & Cabioch, G. (1998). Radiocarbon calibration by means of mass-spectrometric ^{230}Th/^{234}U and ^{14}C ages of corals – An updated database including samples from Barbados, Mururoa, and Tahiti. *Radiocarbon*, **40**, 1085–1092.

Benedict, J.B. (1985). Arapaho Pass, Glacial geology and Archeology of the Crest of the Colorado Front Range. Center for Mountain Archeology Research Report No. 3, Ward Colorado, 197 pp.

Berger, A. (1978). Long-term variations of caloric insolation resulting from the Earth's orbital elements. *Quaternary Research*, **9**, 139–167.

Berry, M.E. (1987). Morphological and chemical characteristics of soil catenas on Pinedale and Bull Lake moraine slopes in the Salmon River Mountains, Idaho. *Quaternary Research*, **28**, 210–225.

Birkeland, P.W. (1964). Pleistocene glaciation of the northern Sierra Nevada, north of Lake Tahoe, California. *Journal of Geology*, **72**, 810–825.

Birkeland, P.W. (1974). *Pedology, weathering and geomorphological research.* Oxford University Press, 285 pp.

Birkeland, P.W. (1999). *Soils and Geomorphology*. Oxford University Press, New York, 430 pp.

Bischoff, J.L. & Cummings, K. (2001). Wisconsin Glaciation of the Sierra Nevada (79,000–15,000 yr B.P.) as recorded by rock flour in sediments of Owens Lake, California. *Quaternary Research*, **55**, 14–24.

Bischoff, J.L., Menking, K.M., Fitts, J.P. & Fitzpatrick, J.A. (1997). Climatic oscillations 10,000–155,000 yr B.P. at Owens Lake, California, reflected in glacial flour rock abundance and lake salinity in core OL-92. *Quaternary Research*, **48**, 313–325.

Blackwelder, E. (1915). Post-Cretaceous history of the mountains of central western Wyoming. *Journal of Geology*, **23**, 97–117, 193–217, 307–340.

Brugger, K.A. & Goldstein, B.S. (1999). Paleoglacier reconstruction and late Pleistocene equilibrium-line altitudes, southern Sawatch Range, Colorado. *Geological Society of America Special Paper 337*, 103–112.

Burke, R.M. (1979). Multiparameter relative dating (RD) techniques applied to morainal sequences along the eastern Sierra Nevada, California, and the Wallowa Lake area, Oregon. University of Colorado Ph.D. dissertation, 166 pp.

Burke, R.M. & Birkeland, P.W. (1979). Reevaluation of multiparameter relative dating techniques and their application to the glacial sequence along the eastern escarpment of the Sierra Nevada, California. *Quaternary Research*, **11**, 21–51.

Carrara, P.E. (1987). Late Quaternary glacial and vegetative history of the Glacier National Park region, Montana. *U.S. Geological Survey Bulletin 1902*, 64 pp.

Carrara, P.E. (1995). A 12,000-year radiocarbon date of deglaciation from the Continental Divide of northwestern Montana. *Canada Journal of Earth Sciences*, **32**, 1303–1307.

Chadwick, O.A., Hall, R.D. & Phillips, F.M. (1997). Chronology of Pleistocene glacial advances in the central Rocky Mountains. *Geological Society of America Bulletin*, **109**, 1443–1452.

Christiansen, R.L. (2001). The Quaternary and Pliocene Yellowstone Plateau Volcanic Field of Wyoming, Idaho, and Montana: U.S. Geological Survey Professional Paper 729-G, 145 pp.

Cohen, A.S., Palacios-Fest, M.R., Negrini, R.M., Wigand, P.E. & Erbes, D.B. (2000). A paleoclimate record for the past 250,000 years from Summer Lake, Oregon, USA: II. Sedimentology, paleontology, and geochemistry. *Journal of Paleolimnology*, **24**, 151–182.

Colman, S.M. & Pierce, K.L. (1981). Weathering rinds on andesitic and basaltic stones as a Quaternary age indicator, Western United States. *U.S. Geological Survey Professional Paper 1210*, 56 pp.

Colman, S.M. & Pierce, K.L. (1986). The glacial sequence near McCall, Idaho – weathering rinds, soil development, morphology, and other relative-age criteria. *Quaternary Research*, **25**, 25–42.

Colman, S.M. & Pierce, K.L. (1992). Varied records of early Wisconsin Alpine glaciation in the Western United States derived from weathering-rind thicknesses. *In*: Clark, P.U. & Lea, P.D. (Eds), *The last interglacial-glacial transition in North America*. Geological Society of America Special Paper 270, 269–278.

Colman, S.M. & Pierce, K.L. (2000). Classifications of Quaternary geochronologic methods. *In*: Noller, J.S., Sowers, J.M. & Lettis.W.R. (Eds), *Quaternary Geochronology, Methods and Applications*. AGU Reference Shelf, **4**, 2–5.

Crandall, D.R. (1967). Glaciation at Wallowa Lake, Oregon. *U.S. Geological Survey Professional Paper 575-C*, C124–C153.

Dahms, D.E. (1991). Eolian sedimentation and soil development on moraine catenas of the Wind River Mountains, west-central Wyoming. Ph.D. Dissertation, University of Kansas, Lawrence, 340 p.

Davis, P.T. (1987). Late Pleistocene age for type Triple Lakes moraines, Arapaho Cirque, Colorado Front Range. *Geological Society of America Abstracts with Programs*, **19**(5), 270.

Davis, P.T., Gosse, J.C., Romito, M., Sorenson, C., Klein, J., Dahms, D., Zielinski, G. & Jull, A.J.T. (1998). Younger Dryas age for type Titcomb Basin and type Temple Lake moraines, Wind River Range, Wyoming, USA. *Geological Society of America Abstract with Programs*, **30**, A-66.

Dethier, D.P., Benedict, J.B., Birkeland, P.W., Caine, N., Davis, P.T., Madole, R.F., Patterson, P. Price, A.B., Schildgen, T.F. & Shroba, R.R. (2003). Quaternary stratigraphy, geomorphology, soils, and alpine archeology in an alpine to plains transect, Front Range of Colorado. *In*: D. Easterbrook (Ed.), Field Trip of the INQUA Congress, 2003, Geological Society of America, in press.

Fullerton, D.S., Colton, R.B., Bush, C.A. & Straub, A.W. (2003). Spatial and temporal relationships of Laurentide continental glaciations and mountains glaciations on the northern Plains in Montana and northwestern North Dakota, with implications for reconstruction of the configurations of vanished ice sheets: U.S. Geological Survey Miscellaneous Geologic Investigations Map, with text.

Gillespie, A. (1991). Testing a new climatic interpretation for the Tahoe glaciation. *In*: Hall, C.A., Jr., Doyle-Jones, V. & Widawski, B. (Eds), *Natural History of Eastern California and High-Altitude Research*. White Mountain Research Station Symposium, 383–398.

Gillespie, A. & Molnar, P. (1995). Asynchronous maximum advances of mountain and continental glaciers. *Reviews of Geophysics*, **33**, 311–364.

Good, J.M., & Pierce, K.L. (1996). Interpreting the Landscapes of Grand Teton and Yellowstone National Parks, Recent and Ongoing Geology: Grand Teton National History Association, 58 pp., 57 illus., Third printing 2002, with additional revisions.

Gosse, J.C., Klein, J., Evenson, E.B., Lawn, B. & Middleton, R. (1995a). Beryllium-10 dating of the duration and retreat of the last Pinedale glacial sequence. *Science*, **268**, 1329–1333.

Gosse, J.C., Evenson, E.B., Klein, J., Lawn, B. & Middleton, R. (1995b). Precise cosmogenic ^{10}Be measurements in western North America: Support for a global Younger Dryas cooling event. *Geology*, **23**, 877–880.

Gosse, J.C. & Phillips, F.M. (2001). Terrestrial cosmogenic nuclides: Theory and application. *Quaternary Science Reviews*, **20**, 1475–1560.

Hall, R.D. & Jaworowski, C. (1999). Reinterpretation of the Cedar Ridge section, Wind River Range, Wyoming: Implications for the glacial chronology of the Rocky Mountains. *Geological Society of America Bulletin*, **111**, 1233–1249.

Hall, R.D. & Shroba, R.R. (1993). Soils developed in the glacial deposits of the type areas of the Pinedale and Bull Lake Glaciations, Wind River Range, Wyoming, USA. *Arctic and Alpine Research*, **25**, 368–373.

Hammond, E.H. (1965). Land surface form. *U.S. Geological Survey, National Atlas*, sheet 61.

Harden, J.W. (1982). A quantitative index of soil development from field descriptions: Examples from a chronosequence in central California. *Geoderma*, **28**, 1–28.

Hostetler, S.W. & Bartlein, P.J. (1999). Simulation of potential responses of regional climate and surface processes in western North America to a canonical Heinrich event. *In*: Clark, P.U., Webb, R.S. & Keigwin, L.D. (Eds), *Mechanisms of Global Climate Change at Millennial Time Scales*. American Geophysical Union Geophysical Monograph, **112**, 313–328.

Hostetler, S.W. & Clark, P.U. (1997). Climate controls of western U.S. glaciers at the last glacial maximum. *Quaternary Science Reviews*, **16**, 505–511.

Johnson, M.D. & Gillam, M.L. (1995). Composition and construction of late Pleistocene end moraines, Durango, Colorado. *Geological Society of America Bulletin*, **107**, 1241–1253.

Karabanov, E.B., Prokopenko, A.A., Williams, D.F. & Colman, S.M. (1998). Evidence from Lake Baikal for Siberian glaciation during oxygen isotope substage 5d. *Quaternary Research*, **50**, 46–55.

Karlstrom, E.T. (2000). Fabric and origin of multiple diamictons within the pre-Illinoian Kennedy Drift of Waterton-Glacier International Peace Park, Alberta, Canada and Montana, USA. *Geological Society of America Bulletin*, **112**, 1496–1506.

Kaufman, D.S., Forman, S.L. & Bright, J. (2001). Age of the Cutler Dam Alloformation Late Pleistocene, Bonneville Basin, Utah. *Quaternary Research*, **56**, 322–334.

Kukla, G.J. et al. (2002). Last interglacial climates. *Quaternary Research*, **58**, 2–13.

Kutzbach, J.E. & Guetter, P.J. (1986). The influence of changing orbital patterns and surface boundary conditions on climate simulations for the past 18,000 years. *Journal of Atmospheric Sciences*, **43**, 1726–1759.

Leonard, E.M. (1984). Late Pleistocene equilibrium-line altitudes and modern snow accumulation patterns, San Juan Mountains, Colorado, USA. *Arctic and Alpine Research*, **16**, 65–76.

Leonard, E.M. (1989). Climatic change in the Colorado Rocky Mountains based on modern climate at late Pleistocene equilibrium-lines. *Arctic and Alpine Research*, **21**, 245–255.

Leverett, F. (1917). Glacial formations in the western United States (abst). *Geological Society of America Bulletin*, **28**, 143–144.

Licciardi, J.M. (2000). Alpine Glacier and Pluvial Lake Records of Late Pleistocene Climate Variability in the Western Unites States: Ph.D. Dissertation, Oregon State University, Corvallis, Oregon, 155 p.

Licciardi, J.M., Clark, P.U., Brook, E.J., Pierce, K.L., Kurz, M.D., Elmore, D. & Sharma, P. (2001). Cosmogenic ^{3}He and ^{10}Be Chronologies of the late Pinedale Northern Yellowstone Ice Cap, Montana, USA. *Geology*, **29**, 1095–1098.

Litwin, R.J., Smoot, J.P., Durika, N.J. & Smith, G.I. (1999). Calibrating Late Quaternary terrestrial climate signals: radiometrically dated pollen evidence from the southern Sierra Nevada, USA. *Quaternary Science Reviews*, **18**, 1151–1171.

Locke, W.W. (1995). Modelling of icecap glaciation of the northern Rocky Mountains of Montana. *Geomorphology*, **14**, 123–130.

Locke, W.W. (1990). Late Pleistocene glaciers and climate of western Montana, USA. *Arctic and Alpine Research*, **22**, 1–13.

Love, J.D. (1977). Summary of upper Cretaceous and Cenozoic stratigraphy, and the tectonic and glacial events in Jackson Hole, Northwestern Wyoming: Wyoming Geological Association, Guidebook, Twenty-Ninth Annual Field Conference, 585–593.

Madole, R.F. (1980). Glacial Lake Devlin and the chronology of Pinedale Glaciation of the east slope of the Front Range, Colorado. U.S. Geological Survey Open-Rile Report 80–725, 32 pp.

Madole, R.F. (1976). Glacial geology of the Front Range, Colorado. *In*: W.C. Mahaney (Ed.), *Quaternary Stratigraphy of North America*. Dowden, Hutchinson, and Ross, Stroudsburg, Pennsylvania, 297–318.

Madole, R.F. (1982). Possible origins of till-like deposits near the summit of the Front Range in North-Central Colorado. *U.S. Geological Survey Professional Paper 1243*, 31 pp.

Martinson, D.G., Pisias, N.J., Hays, J.D., Imbrie, J., Moore, T.C., Jr. & Shackleton, N.J. (1987). Age dating and the orbital theory of the ice ages: development of a high resolution 0 to 300,000-year chronostratigraphy. *Quaternary Research*, **27**, 1–29.

Mears, B., Jr. (1987). Late Pleistocene periglacial wedge sites in Wyoming. *Geological Survey of Wyoming*, Memoir No. 3, 77 pp.

Menounos, B. & Reasoner, M.A. (1997). Evidence of cirque glaciation in the Colorado Front Range during the Younger Dryas Chronozone. *Quaternary Research*, **48**, 38–47.

Miller, C.D. (1979). A statistical method for relative-age dating of moraines in the Sawatch Range, Colorado. *Geological Society of America Bulletin*, **90**, 1153–1164.

Millspaugh, S.H., Whitlock, C. & Bartlein, P.J. (2000). Variations in fire frequency and climate over the past 17000 years in central Yellowstone National Park. *Geology*, **28**, 211–214.

Morrison, R.B. (1965). Quaternary geology of the Great Basin. *In*: Wright, H.D., Jr. & Frey, D.G. (Eds), *The Quaternary of the United States*. Princeton University Press, Princeton, New Jersey, 265–285.

Murray, D.R. & Locke, W.W., III (1989). Dynamics of the late Pleistocene Big Timber Glacier, Crazy Mountains, Montana, USA. *Journal of Glaciology*, **35**, 183–190.

Muhs, D.R. & Bettis, E.A., III (2000). Geochemical variations in Peoria Loess of western Iowa indicate paleowinds of mid-continental North America during last glaciation. *Quaternary Research*, **53**, 49–61.

Nelson, A.R., Millington, A.C., Andrews, J.T. & Nichols, H. (1979). Radiocarbon-dated upper Pleistocene glacial sequence, Fraser Valley, Colorado Front Range. *Geology*, **7**, 410–414.

Nelson, A.R. & Shroba, R.R. (1998). Soil relative dating of moraine and outwash-terrace sequences in the northern part of the upper Arkansas Valley, central Colorado, USA. *Arctic and Alpine Research*, **30**, 349–361.

Netoff, D.I. (1977). Soil clay mineralogy of Quaternary deposits in two Front Range-Piedmont transects, Colorado. Ph.D. dissertation, University of Colorado, Boulder, 169 pp.

Noller, J.S., Sowers, J.M. & Lettis, W.R. (Eds) (2000). Quaternary geochronology, methods and applications. *AGU Reference Shelf*, **4**, 582 pp.

Obradovich, J.D. (1992). Geochronology of the Late Cenozoic Volcanism of Yellowstone National Park and adjoining areas, Wyoming and Idaho. U.S. Geological Survey Open File Report 92–408, 45 pp.

Osborn, G. & Bevis, K. (2001). Glaciation of the Great Basin of the western United States. *Quaternary Science Reviews*, **20**, 1377–1410.

Oviatt, C.G., Thompson, R.S., Kaufman, D.S., Bright, J. & Forester, R.M. (1999). Reinterpretation of the Burmeister Core, Bonneville Basin, Utah. *Quaternary Research*, **52**, 180–184.

Phillips, F.M., Zreda, M.G., Gosse, J.C., Klein, J., Evenson, E.B., Hall, R.D., Chadwick, O.A. & Sharma, P. (1997). Cosmogenic ^{36}Cl and ^{10}Be ages of Quaternary glacial and fluvial deposits of the Wind River Range, Wyoming. *Geological Society of America Bulletin*, **109**, 1453–1463.

Phillips, F.M., Zreda, M.G., Smith, S.S., Elmore, D., Kubik, P.W. & Sharma, P. (1990). Cosmogenic Chlorine-36 chronology for glacial deposits at Bloody Canyon, eastern Sierra Nevada. *Science*, **248**, 1529–1532.

Pierce, K.L. (1979). History and dynamics of glaciation in the northern Yellowstone National Park area. *U.S. Geological Survey Professional Paper 729 F*, 91 pp.

Pierce, K.L. (1985). Quaternary history of movement on the Arco segment of the Lost River fault, central Idaho. *In*: Stein, R.S. & Bucknam, R.C. (Eds), Proceedings of Workshop XXVII on the Borah Peak, Idaho, earthquake. U.S. Geological Survey Open-File Report 85–290, pp. 195–206.

Pierce, K.L. & Good, J.D. (1992). Field guide to the Quaternary geology of Jackson Hole, Wyoming. U.S. Geological Survey Open-File Report 92–504, 49 pp.

Pierce, K.L. & Morgan, L.A. (1992). The track of the Yellowstone hot spot–volcanism, faulting and uplift. *In*: Link, P.K., Kuntz, M.A. & Platt, L.W. (Eds), Regional geology of eastern Idaho and western Wyoming. *Geological Society of America Memoir*, **179**, 1–53.

Pierce, K.L., Obradovich, J.D. & Friedman, I. (1976). Obsidian hydration dating and correlation of Bull Lake and Pinedale glaciations near West Yellowstone, Montana. *Geological Society of America Bulletin*, **87**, 703–710.

Pierce, K.L. & Scott, W.E. (1982). Pleistocene episodes of alluvial-gravel deposition, southeastern Idaho. *In*: Bonnichsen, B. & Breckenridge, R.M. (Eds), *Cenozoic geology of Idaho*. Idaho Bureau of Mines and Geology Bulletin 26, 685–702.

Porter, S.C. (1975). Weathering rinds as a relative-age criterion: Application to sub-division of glacial deposits in the Cascade Range. *Geology*, **3**, 101–104.

Porter, S.C., Pierce, K.L. & Hamilton, T.D. (1983). Late Pleistocene glaciation in the Western United States. *In*: Porter, S.C. (Ed.), *The Late Pleistocene*, Vol. 1, *of*: Wright, H.E., Jr. (Ed.), *Late Quaternary Environments of the United*

States. Minneapolis, Minn., University of Minnesota Press, 71–111.

Ray, L.L. (1940). Glacial chronology of the southern Rocky Mountains. *Geological Society of America Bulletin*, **51**, 1851–1917.

Reasoner, M.A., & Jodry (2000). Rapid response of alpine timberline vegetation to the Younger Dryas climatic oscillation in the Colorado Rocky Mountains, USA. *Geology*, **28**, 51–54.

Reheis, M.C. (1987). Soils in granitic alluvium in humid and semiarid climates along Rock Creek, Carbon County, Montana. *U.S. Geological Survey Bulletin* 1590-D, 71 pp.

Richmond, G.M. (1962). Quaternary stratigraphy of the La Sal Mountains, Utah. *U.S. Geological Survey Professional Paper 324*, 135 pp.

Richmond, G.M. (1965). Glaciation of the Rocky Mountains. *In*: Wright, H.D., Jr. & Frey, D.G. (Eds), *The Quaternary of the United States*. Princeton University Press, Princeton, New Jersey, 217–230.

Richmond, G.M. (1986a). Stratigraphy and correlation of glacial deposits of the Rocky Mountains, the Colorado Plateau, and the Ranges of the Great Basin. *In*: Sibrava, V., Bowen, D.Q. & Richmond, G.M. (Eds), *Quaternary Glaciations in the northern Hemisphere: Quaternary Science Reviews*, **5**, 99–127.

Richmond, G.M. (1986b). Stratigraphy and chronology of glaciations in Yellowstone National Park. *In*: Sibrava, V., Bowen, D.Q. & Richmond, G.M. (Eds), *Quaternary Glaciations in the northern Hemisphere: Quaternary Science Reviews*, **5**, 83–98.

Richmond, G.M., Fryxell, R., Montagne, J. & Trimble, D.E. (1965). Northern and Middle Rocky Mountains, Guidebook for field Conference E, VIIth INQUA Congress, Nebraska academy of Sciences, Lincoln, Nebraska, 129 pp.

Richmond, G.M. & Murphy, J.F. (1989). Preliminary Quaternary geologic map of the Dinwoody Lake area, Fremont County, Wyoming. *U.S. Geological Survey Open File Report 89–435*.

Rosenbaum, J.G. & Larson, E.E. (1983). Paleomagnetism of two late Pleistocene lake basins in Colorado: an evaluation of detrital remanent magnetization as a recorder of the geomagnetic field. *Journal of Geophysical Research*, **88**, 10,611–10,624.

Rosenbaum, J.G. & Reynolds, R.L. (2003). Record of Late Pleistocene glaciation and deglaciation in the southern Cascade Range: II. Flux of glacial flour in a sediment core from Upper Klamath Lake, Oregon, Paleolimnology, in press.

Schildgen, T., Dethier, D.P., Bierman, P. & Caffee, M. (2002). ^{26}Al and ^{10}Be dating of late Pleistocene and Holocene fill terraces: a record of fluvial deposition and incision, Colorado Front Range. *Earth Surface Processes and Landforms*, **27**, 773–787.

Scott, G.R. (1975). Cenozoic surfaces and deposits in the southern Rocky Mountains. *Geological Society of America Memoir*, **144**, 227–247.

Sharp, W., Ludwig, K.R., Chadwick, O.A., Amundson, R. & Glaser, L.L. (2003). Dating fluvial terraces by 230Th/U on pedogenic carbonate, Wind River Basin, Wyoming. *Quaternary Research*, **59**, 139–150.

Shroba, R.R. (1977). Soil development in Quaternary tills, rock glacier deposits, and taluses, southern and central Rocky Mountains. Ph.D. dissertation, University of Colorado, Boulder.

Shroba, R.R. & Birkeland, P.W. (1983). Trends in Late-Quaternary soil development in the Rocky Mountains and Sierra Nevada of the western United States. *In*: Porter, S.C. (Ed.), *The Late Pleistocene*, Vol. 1, of: Wright, H.E., Jr. (Ed.), *Late Quaternary Environments of the United States*. Minneapolis, Minnesota, University of Minnesota Press, 145–156.

Shroder, J.F. & Sewell, R.E. (1985). Mass movement in the La Sal Mountains, Utah. *In*: Christenson, G.E., Oviatt, C.G., Shroder, J.F. & Sewell, R.E. (Eds), *Contributions to the Quaternary Geology of the Colorado Plateau*. Utah Geological and Mineral Survey, Special Studies 65, 85 pp.

Stuiver, M. & Reimer, P.J. (1993). Extended ^{14}C data base and revised CALIB 3.0 ^{14}C age calibration program. *Radiocarbon*, **35**, 215–230.

Sturchio, N.C., Pierce, K.L., Morrell, M.T. & Sorey, M.L. (1994). Uranium-series ages of travertines and timing of the last glaciation in the northern Yellowstone area, Wyoming-Montana. *Quaternary Research*, **41**, 265–277.

Swanson, D.K. (1985). Soil catenas on Pinedale and Bull Lake moraines, Willow Lake, Wind River Mountains, Wyoming. *Catena*, **12**, 329–342.

Waldrop, H. A. (1975). Surficial geologic map of the West Yellowstone quadrangle, Yellowstone National Park and adjoining area, Montana, Wyoming, and Idaho. *U.S. Geological Survey Miscellaneous Investigations Series Map I-648* scale 1: 62,500.

Wayne, W.J. (1984). Glacial chronology of the Ruby Mountains – East Humboldt Range, Nevada. *Quaternary Research*, **21**, 286–303.

Whitlock, C. (1993). Postglacial vegetation and climate of Grand Teton and southern Yellowstone National Parks. *Ecological Monographs*, **63**(2), 173–198.

Whitlock, C., Sarna-Wojcicki, A.M., Bartlein, P.J. & Nickmann, R.J. (2000). Environmental history and tephrostratigraphy at Carp Lake, southwestern Columbia basin, Washington, USA. *Palaeogeography, Palaeoclimatology, and Palaeoecology*, **155**, 7–29.

Winograd, I.J., Landwehr, J.M., Ludwig, K.R., Coplen, T.B. & Riggs, A.C. (1997). Duration and structure of past four interglaciations. *Quaternary Research*, **48**, 141–154.

Zielinski, G.A. & McCoy, W.D. (1987). Paleoclimatic implications of the relationship between modern snowpack and late Pleistocene equilibrium-line altitudes in the mountains of the Great Basin, western USA. *Arctic and Alpine Research*, **19**, 127–134.

Zhu, H. & Baker, R.G. (1995). Vegetation and climate of the last glacial-interglacial cycle in southern Illinois, USA. *Journal of Paleolimnology*, **14**, 337–354.

Quaternary alpine glaciation in Alaska, the Pacific Northwest, Sierra Nevada, and Hawaii

Darrell S. Kaufman[1], Stephen C. Porter[2] and Alan R. Gillespie[2]

[1] *Department of Geology, Northern Arizona University, Flagstaff, AZ 86011, USA; Darrell.Kaufman@nau.edu*
[2] *Quaternary Research Center, University of Washington, Seattle, WA 98195, USA; scporter@u.washington.edu, alan@ess.washington.edu*

Introduction

This chapter deals with mountain glaciation in the North American Cordillera and Hawaii, exclusive of the Rocky Mountains (Pierce, this volume) and Canada. We focus on a few critical areas in each of these regions where research since 1965 has produced significant new results that advance our understanding of the extent, chronology, and dynamics of mountain glaciers, and enhance the paleoclimatic inferences that can be drawn from them.

Alaska

Unlike other high-latitude areas of North America, much of Alaska was never glaciated (Fig. 1). Even more land area lay exposed to the arid Pleistocene climate during intervals when sea level was lower than present. Despite the vastness of its unglaciated area, Alaska's mountainous terrain generated a mass of glacier ice on a par with all the rest of the western United States combined. The largest expanse of glaciers comprised the coalescent ice caps and piedmont lobes that extended from the Alaska Range to the Gulf of Alaska and from the southeastern panhandle to the Aleutian Islands. This amalgamation formed the western extension of the North American Cordilleran Ice Sheet, and it contained most of the glacier ice in Alaska. The ice caps that grew in the Brooks Range, a northern extension of the Rocky Mountain system in northern Alaska, and the Ahklun Mountains in southwestern Alaska were the only other major centers of ice accumulation in the state. Lower uplands across the state supported hundreds of smaller valley glaciers; most notable are the small ranges of Seward Peninsula and the Yukon-Tanana Upland. In all, glaciers once covered about 1,200,000 km^2 of Alaska and its adjacent continental shelf; during late Wisconsin time, the area was 727,800 km^2 (Manley & Kaufman, 2002). Presently, 74,700 km^2 of Alaska is covered by ice, or 4.9% of the state; most of the present volume of glacier ice is in the coastal ranges proximal to moisture sources around the Gulf of Alaska.

Because most of Alaska was not glaciated, mountain glaciers were free to expand onto adjacent lowlands where they left a rich record of moraines and morphostratigraphically related glacial-geologic features. Evidence for the extent of glaciers around the Gulf of Alaska is now submerged and obscured, but a succession of moraines is preserved along most mountain fronts. Evidence for multiple glacier fluctuations is also preserved in successions of glacially influenced deposits of lacustrine (e.g. Lake Atna: Ferrians, 1963; Lake Noatak: Hamilton, 2001), marine (e.g. Yakataga Formation: Plafker & Addicott, 1976; Hagemeister Island: Kaufman *et al.*, 2001a), fluvial (e.g. Epigurak Bluff: Hamilton *et al.*, 1993), and eolian systems (loess and sandsheets: Begét, 2001; Lea & Waythomas, 1990; and dunes: Carter, 1981; Mann *et al.*, 2002). The glacial history is often better dated in these depositional settings where volcanic products and organic material are more commonly preserved. The available age control shows that Alaska has a glacial record as long as that of anywhere else in the Northern Hemisphere. The geochronology also shows that glaciers fluctuated on time scales ranging from tens of thousands of years to decades, consistent with other records of global climatic changes. The ages of Holocene (e.g. Calkin *et al.*, 2001) and late Wisconsin (e.g. Porter *et al.*, 1983) glacier fluctuations are best documented. The ages and correlations of glacier deposits of the next-older (penultimate) ice advance are beyond the range of ^{14}C dating and have remained controversial.

Considering the vastness of the glaciated area in Alaska, the diversity of its glacial systems, and the widespread impact of glaciers on the non-glaciated regions, a single review cannot include a full account of Alaskan glacial geology. Instead, this section presents a brief summary of the major developments in Alaskan glacial geology prior to 1990, then highlights some of the progress in the last 10 years, since the last state-wide summaries of the Pliocene-Pleistocene (Hamilton, 1994) and Holocene (Calkin, 1988) glacial chronologies.

Status in 1965

Prior to the widespread use of helicopters, glacial-geologic research in Alaska was focused in the south-central part of the state, particularly in areas accessible by road (Péwé, 1965). The first attempt at a state-wide summary of Pleistocene glacier extents was by Péwé (1953). This effort led to the state-wide surficial map (Karlstrom *et al.*, 1964) and its derivative map of glacier extents (Coulter *et al.*, 1965), both published about the time of the VII INQUA Congress.

By the mid-1960's, the overall distribution of Quaternary glaciers in Alaska was known generally. Fifteen local Quaternary glacial sequences had been studied and the major

DEVELOPMENTS IN QUATERNARY SCIENCE
VOLUME 1 ISSN 1571-0866
DOI:10.1016/S1571-0866(03)01005-4

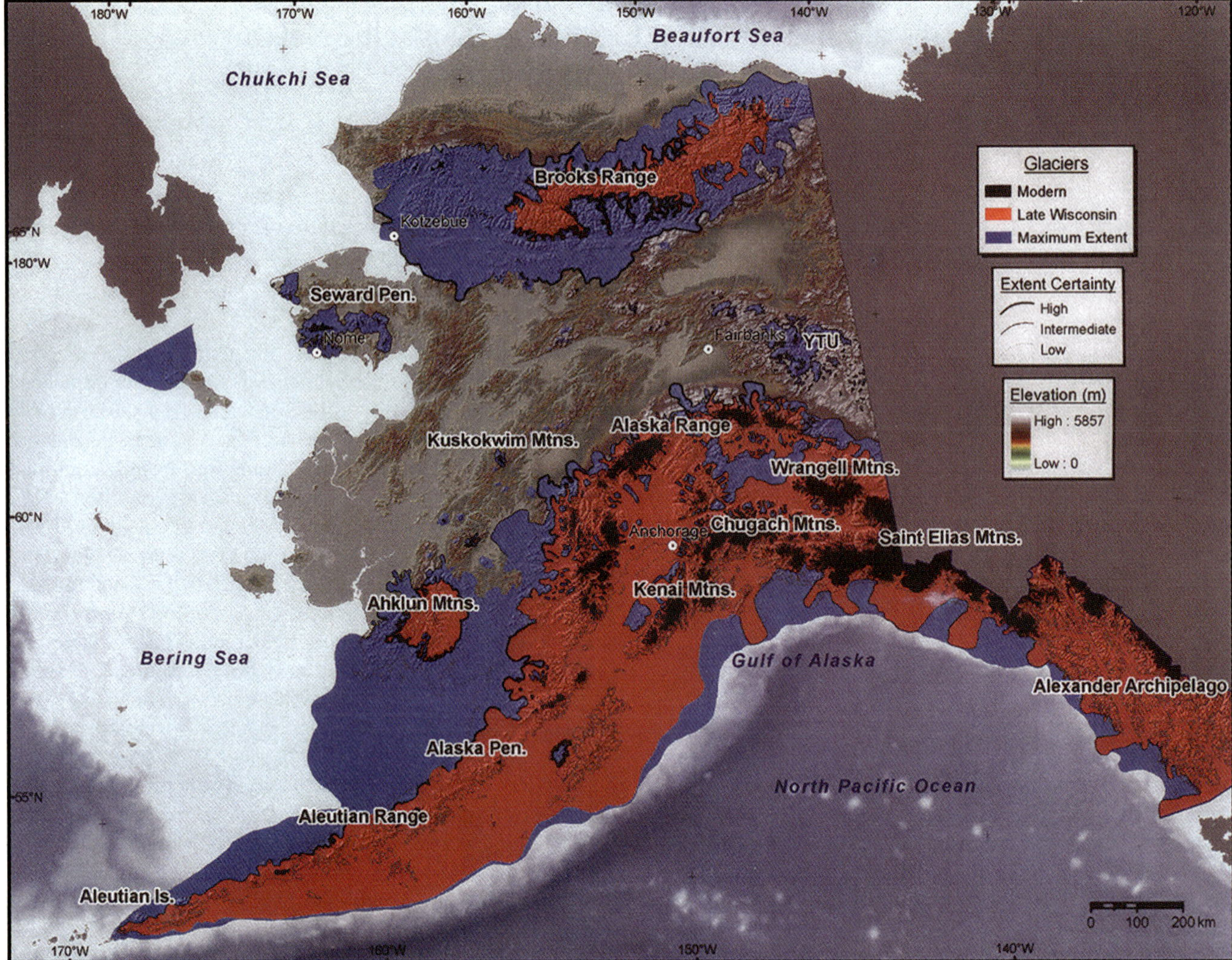

Fig. 1. Pleistocene maximum, late Wisconsin, and modern glacier extent across Alaska. Image from the Alaska Paleoglacier Atlas website (Manley & Kaufman, 2002).

advances in each sequence were tentatively correlated (Péwé *et al.*, 1965). Drift was subdivided into three principal units: Wisconsin, Illinoian, and pre-Illinoian, based mainly on semi-quantitative relative-weathering criteria and comparison with the mid-continent region. About half of the local sequences included a two-fold subdivision of the Wisconsin glaciation (early and late) and many authors recognized evidence of multiple advances during the early Holocene. Drift interstratified with marine deposits around Nome, Kotzebue (Hopkins *et al.*, 1965), and Cook Inlet (Karlstrom, 1964), provided additional age control based on the now-discredited Th/U dating of molluscs, and a few ^{14}C ages had been determined; these were some of the first ^{14}C analyses (ca. 1953) ever made. The dramatic difference between the limited extent of late Wisconsin glaciers and the vast extent of pre-Wisconsin ice was identified in most areas, and the exceptionally long record of glacial-marine sedimentation dating back to the latest Miocene had been recognized around the Gulf of Alaska (Miller, 1957).

Glacial-Geologic Research Between 1965 and 1990

Péwé (1975) reviewed the Quaternary geology of Alaska. His comprehensive synthesis included the first and only statewide compilation of glaciation thresholds. By the 1980s widespread application of ^{14}C dating had greatly refined the chronology of late Wisconsin glacier fluctuations; in Alaska; the technique was then, and has been since, most extensively applied to deposits of the Cordilleran Ice Sheet (Hamilton & Thorson, 1983) and in the Brooks and northern Alaska ranges (Porter *et al.*, 1983). These studies demonstrated that the most recent glaciation in Alaska occurred between 24,000 and 11,500 years ago and was therefore broadly synchronous with late Wisconsin glaciation elsewhere in North America.

Hamilton *et al.* (1986) complied the most complete collection of glacial-geologic studies in Alaska yet published. The volume includes detailed reports on ten different glaciated regions of the state, including: central Brooks Range (Hamilton, 1986), Seward Peninsula (Kaufman &

Hopkins, 1986), Yukon-Tanana Upland (Weber, 1986), Nenana River valley (Thorson, 1986), Beaver Mountains and northwestern Alaska Range (Kline & Bundtzen, 1986), Alaska Peninsula (Detterman, 1986), Aleutian Islands (Thorson & Hamilton, 1986), Cook Inlet (Schmoll & Yehle, 1986), Gulf of Alaska (Molnia, 1986), and southeast Alaska (Mann, 1986). Readers interested in the glacial geology of these regions are referred to this volume. The regional studies employed a wide range of approaches, but nearly all included ^{14}C ages or other geochronology, descriptive accounts of relative-weathering features, and paleoclimatological inferences. The authors recognized new complexities in the glacial sequences resulting from their detailed stratigraphic and geochronological studies; these were commensurate with their emerging understanding of Quaternary global climate changes provided by marine oxygen isotopes. About the same time, evidence for Holocene glaciation in Alaska was reviewed by Calkin (1988) and modern glaciers were discussed by Krimmel & Meier (1989).

Hamilton's (1994) review of Pleistocene glaciation of Alaska is the most up-to-date and complete synthesis. Based largely on the 1986 regional summaries, and updated with information from 109 additional papers published after 1986. Hamilton integrated evidence for late Cenozoic glaciation in 15 regions of the state and correlated glacial deposits within six broad age categories. Although the extent of glaciers during late Wisconsin time was relatively well known by the time of the review, the ages and correlations of glacier deposits of the next-older (penultimate) ice advance were controversial. Previous studies generally inferred that the penultimate advance was younger than the last interglaciation (i.e. early Wisconsin *s.l.*), but tephrostratigraphic, paleoecologic, pedogenic, and thermoluminescence evidence from the southern and central parts of Alaska suggested that the penultimate drift might predate the last interglaciation. Recognizing the likelihood that the drift might be of different ages in different places, Hamilton (1994) avoided the term "early Wisconsin" in favor of "penultimate" for glacier advances beyond the range of ^{14}C dating regardless of whether they predated or postdated the last interglaciation, and he assigned the next older drift, known to predate the last interglaciation, to the middle Pleistocene. Several new studies (see below) have amassed geochronological evidence favoring an early Wisconsin (*s.l.*) age for the penultimate drift.

Progress During the Last Decade

With the recent retirement and death of several prominent Alaskan Quaternary geologists, research into the glacial geology of Alaska has slowed during the last decade. In the recently published volume on the Quaternary Paleoenvironments of Beringia (Elias & Brigham-Grette, 2001), for example, only five of 35 chapters are devoted to the glacial geology of Alaska. Nonetheless, significant progress has been made as the research has shifted from the first-generation, regional mapping (mainly by the U.S. Geological Survey) to technique- and hypothesis-driven investigations at smaller scales (mainly by university scientists and their students). Areas of active glacial-geologic research include the Brooks Range, Ahklun Mountains, and Pacific coastal mountains.

Brooks Range

Building on decades of systematic, surficial-geologic mapping across the central Brooks Range, Hamilton (2002) recently completed a detailed glacial-geologic study of the Itkillik-Sagavanirktok River area. He identified six distinct late Pleistocene moraine sets, the most complete subdivision of the Ikillik (Wisconsin) glaciation yet discovered in the Brooks Range. The glacial history of the western and eastern sectors of the Brooks Range remains uncertain, however. In the west, Hamilton's (2001) recent study of the stratigraphic record exposed in the Noatak River basin has revealed evidence for multiple expansions of glaciers from the DeLong Mountains that dammed a succession of proglacial lakes. New photo-interpretive mapping and ^{14}C dating (T.D. Hamilton, unpub. data) indicate that glaciers were considerably less extensive (by an order of magnitude) in the DeLong Mountains during the late Wisconsin than has been depicted in previous state-wide compilations. Restricted ice in the DeLong Mountains during the late Wisconsin is consistent with the near absence of glaciers in the Baird Mountains south of the Noatak River (D.S. Kaufman, unpubl. data).

Ahklun Mountains

The Ahklun Mountains of southwestern Alaska supported the largest center of Pleistocene glaciers outside the Cordilleran Ice Sheet and the Brooks Range. Research during the last decade has clarified the age and extent of multiple Quaternary glacier advances. Amino acid analysis of mollusc shells from glaciomarine sediment in coastal exposures of northeastern Bristol Bay, combined with $^{40}Ar/^{39}Ar$ dating of lava from a tuya eruption, indicates that extensive piedmont glaciers emanating from an ice cap centered over the Ahklun Mountains advanced south across the present coast as many as four times during the middle Pleistocene (Kaufman *et al.*, 2001a). The youngest glaciers to reach the coast formed ice-thrust ridges containing glacially influenced marine sediment in Nushagak Bay area (Lea, 1990) and ice-contact stratified drift in the Togiak Bay area (Kaufman *et al.*, 2001b). Luminescence, amino acid, paleoecologic, and tephrostratigraphic evidence shows that the drift is younger than the last interglaciation (Kaufman *et al.*, 1996, 2001a, b; Manley *et al.*, 2001). The glacial-geologic evidence for thin, low-gradient glacier ice indicates a limited glacial-isostatic effect; this, combined with the available geochronology and the long distance to the edge of the shallow continental shelf, suggests that the glaciomarine sediment was deposited during periods of high eustatic sea level. Similar evidence from elsewhere in central Beringia (Brigham-Grette *et al.*, 2001; Huston *et al.*, 1990; Pushkar *et al.*, 1999) indicates that glaciers attained their maximum Pleistocene extent prior to the buildup of Northern Hemisphere ice sheets. The transitions between the interglaciations of marine oxygen-isotope stages (MIS) 11 and 5 and the subsequent glacial intervals are likely times when high sea level, warm sea-surface temperatures, and decreasing summer insolation conspired to generate the largest volumes of glacier ice in Alaska. The expansion of ice over high-latitude landmasses may have had an important positive feedback in the climate system during the onset of global glaciations.

The first published exposure ages for moraines in Alaska (Briner & Kaufman, 2000; Briner *et al.*, 2001) and ^{14}C ages from lake-sediment cores (Kaufman *et al.*, in press) clarify the ages of late Pleistocene glaciations in the Ahklun Mountains. No evidence for an extensive ice advance during MIS 6 has yet been discovered. Instead, outlet glaciers in the southwestern part of the mountain range advanced beyond the present coast and reached their maximum late Pleistocene extent ~60,000 years ago. Glaciers attained their maximum late Wisconsin extent ~24,000–20,000 cal yr B.P. when they terminated more than 60 km upvalley from their early Wisconsin limits. They then experienced a series of fluctuations as summer insolation increased, sea level rose, and ocean-atmospheric circulation shifted to its interglacial mode. The most dramatic readvance culminated at the end of the Younger Dryas interval (Briner *et al.*, 2002), consistent with emerging paleoenvironmental evidence for cooling around the state at that time (e.g. Bigelow & Edwards, 2001; Hu *et al.*, 2002; Mann *et al.*, 2001). In the Brooks Range (Hamilton, 1986) and Cook Inlet region (Reger & Pinney, 1996), however, prominent readvances occurred ~1500 to 1000 years before the Younger Dryas, and no Younger Dryas advances have been recognized in these two regions. Glaciers retreated during an interval of early Holocene warmth, then reformed in the highest elevations of the Ahklun Mountains beginning ca. 3400 cal yr B.P. (Levy *et al.*, in press).

The North Pacific Coast

Motivated by continued interest in biotic exchanges between the old and new worlds, Mann & Hamilton (1995) recently reviewed the paleogeography of the North Pacific coast since the last glacial maximum. They summarized evidence from around the southern margin of the Cordilleran Ice Sheet for time-transgressive glacier fluctuations and for several major climatic transitions between about 26,000 and 10,000 cal yr B.P. The understanding of the glacial history of the Cook Inlet region has been improved recently with work by Schmoll *et al.* (1999), Reger & Pinney (1996), and Reger *et al.* (1995). On the Alaska Peninsula, recent work by Wilson & Weber (2001) has attempted to correlate multiple Pleistocene drift units, and to understand the interaction of volcanic and glacier activity. These efforts have been frustrated, however, by differential glacier response of different source areas (Stilwell & Kaufman, 1996), and by the recognition that many of the moraines surrounding Bristol Bay are glacially tectonized ridges that do not necessarily record climatically significant ice-marginal positions (Kaufman & Thompson, 1998).

Recent studies of the dendrochronology, lichenometry, and moraine geomorphology of the coastal mountains rimming the northern Gulf of Alaska (Barclay *et al.*, 2001; Calkin *et al.*, 2001; Wiles & Calkin, 1994; Wiles *et al.*, 1999) and the Wrangell Mountains (Wiles *et al.*, 2002) provide the most detailed and geographically most extensive record of late Holocene glaciation in the state. Neoglacial expansions of many glaciers took place by about 4000–3500 cal yr B.P. Glaciers retreated by ~2000 cal yr B.P. before expanding again during the Little Ice Age advances of the 13th, 15th, middle 17th, and second half of the 19th centuries A.D.

On-Going Efforts

Three decades after the last Alaska-wide compilation of glacial geology (Coulter *et al.*, 1965), a collaborative effort has produced a new synthesis of reconstructed Pleistocene glacier extents (Manley & Kaufman, 2002). The Alaska PaleoGlacier (APG) Atlas integrates the results of glacial-geologic studies from 26 publications and 42 source maps into a Geographic Information System (GIS) targeted for a scale of 1:1,000,000. Maps and several GIS layers are available online (http://instaar.colorado.edu/QGISL/ak_paleoglacier_atlas). The APG Atlas is part of a larger effort led by the INQUA Commission on Glaciation to create a global GIS database of Pleistocene glacier extents. The atlas depicts several glaciated massifs that were not previously recognized by Coulter *et al.* (1965), mainly in the Yukon-Koyokuk region and the Kuskokwim Mountains. On-going spatial analysis based on more detailed digital mapping is aimed at reconstructing equilbribum-line altitudes (ELAs) and their paleoclimatic forcing using Pleistocene valley and cirque glaciers across Alaska (e.g. Manley & Kaufman, 1999).

Despite substantial progress since 1965, our understanding of the Alaskan glacial record is hindered by major gaps in ground-based mapping and geochronologic control. Recently, for example, interpretations made from satellite images led to the inference that a major Pleistocene ice sheet covered much of central Beringia (the "Beringian Ice Sheet"; Grosswald, 1998). New research in far eastern Russia (e.g. Gualtieri *et al.*, 2000), and previous studies (summarized by Brigham-Grette, 2001), refute the existence of this former ice sheet. New research is needed to address: (1) the age of the penultimate glaciation; (2) teleconnections to the global record of rapid climatic changes; (3) relation of the Alaskan glacial record to that in adjacent regions of the Yukon Territory (e.g. Froese *et al.*, 2000; Westgate *et al.*, 2001) and western Beringia (e.g. Glushkova, 2001; Heiser & Rousch, 2001); and (4) the role of sea level, atmospheric moisture, continentality, and other physiographic effects in controlling regional-scale response of glaciers to climate forcing.

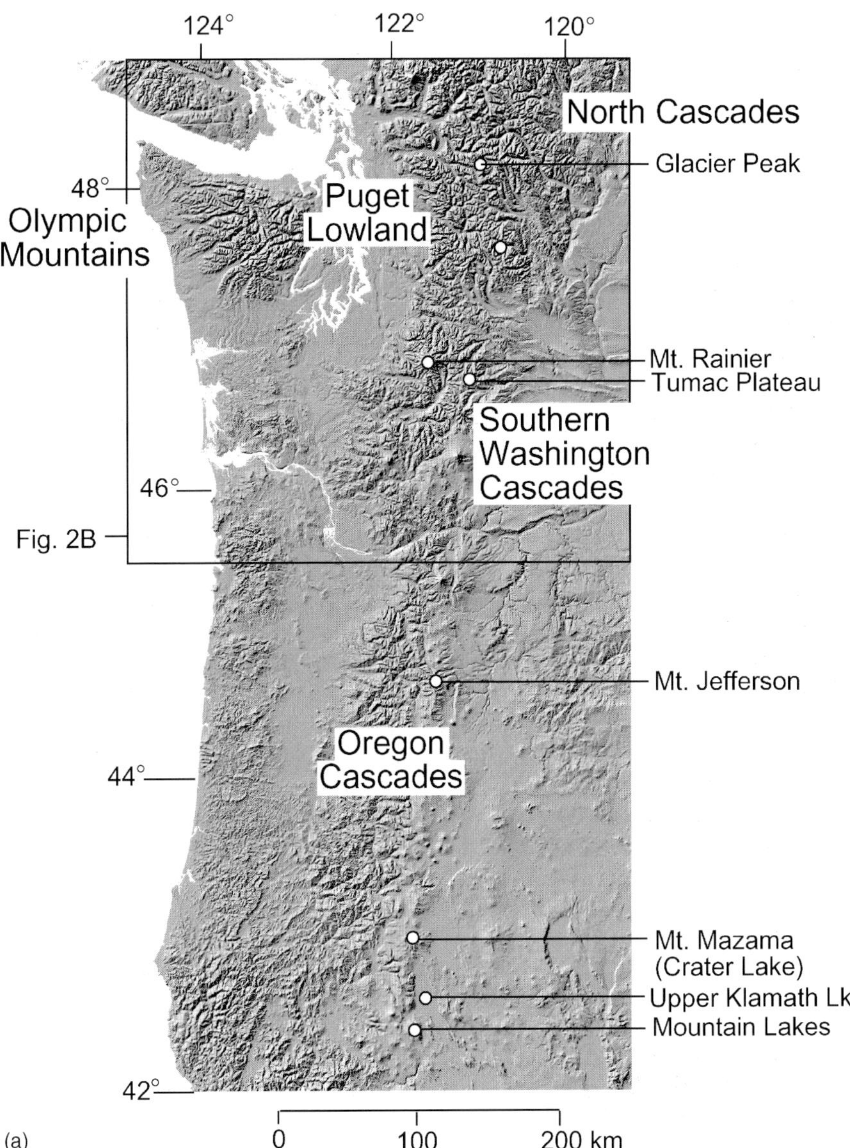

Fig. 2. (a) Shaded relief map of Olympic Mountains and Cascade Range of Washington and Oregon showing localities mentioned in the text. (b) Shaded relief map of western Washington showing localities mentioned in the text.

Cascade Range and Olympic Mountains, Washington and Oregon

The glaciated Cascade Range and Olympic Mountains of Washington and Oregon (Fig. 2) contain a wealth of data bearing on the Quaternary climatic and environmental history of the Pacific Northwest. The record of Pleistocene alpine glaciation is juxtaposed to a long record of ice-sheet glaciation found in adjacent lowlands, and terrain adjacent to hundreds of modern glaciers contains evidence of Holocene glacier variations.

During their greatest Pleistocene advance, alpine glaciers in the Washington Cascade Range and Olympic Mountains terminated as much as 70–80 km from their sources. During the last glaciation, the largest glaciers were only half as long. In the Oregon Cascades, glacier tongues terminated 10–30 km from ice fields that mantled the range crest.

The glaciated region encompasses a wide range of environments resulting from strong longitudinal and altitudinal climatic gradients. Cool, moist climates of western Washington and high snowfall zones on the western flank and crest of the Cascades and Olympics contrast with rainshadow conditions and drier climate farther east. Not surprisingly, radiocarbon dating of events during the last glaciation and the Holocene is largely restricted to the wetter flanks and crests of the mountains. Only recently have other dating methods become available that have the potential of developing a chronological framework spanning a significant part of the glacial history of these ranges.

Status of Alpine Glacial Studies in 1965

Pleistocene Glaciation in the Cascade Range

In his review of the Quaternary glacial record of the Puget Lobe of the Cordilleran Ice Sheet and the adjacent Cascade Range, Crandell (1965) recognized four Pleistocene drift units. The two youngest, Salmon Springs and Fraser, he

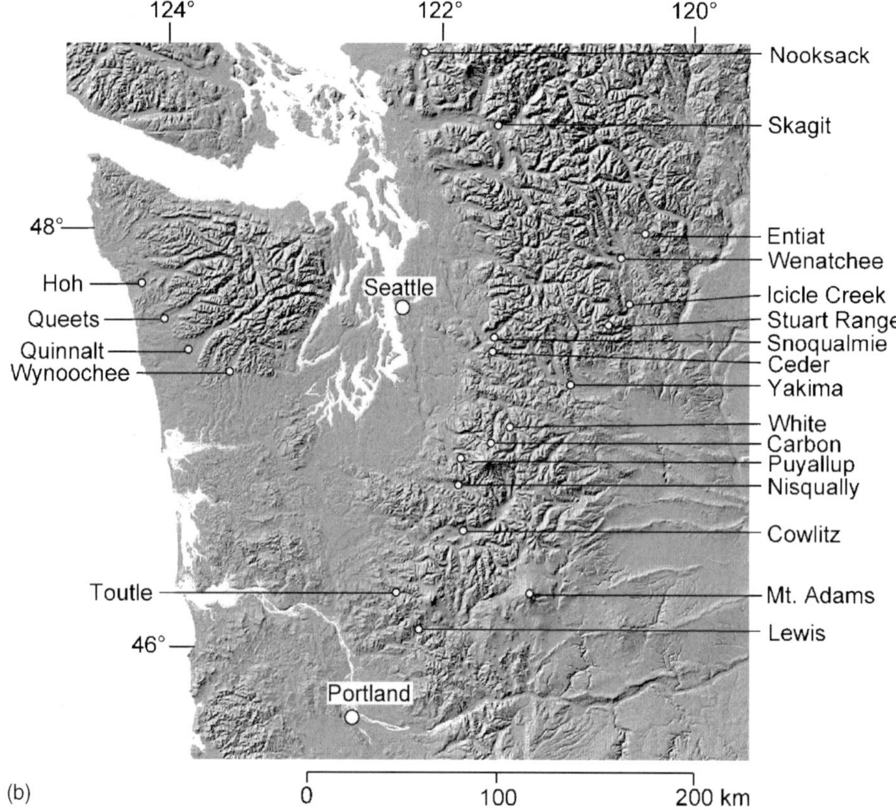

Fig. 2. (Continued)

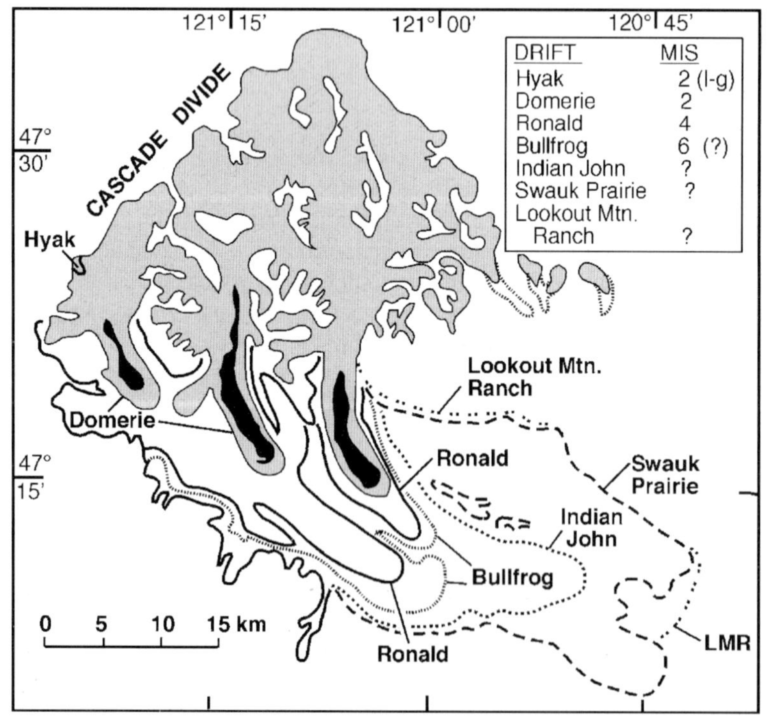

Fig. 3. Map of glaciated region of upper Yakima River drainage basin showing extent and inferred ages of Pleistocene drifts.

inferred to be of early to middle Wisconsin age and late Wisconsin age, respectively. Crandell divided the alpine glacial record into pre-Salmon Springs (?), Salmon Springs (?), Fraser, and Neoglacial drifts. Evidence of multiple glaciations had been found in valleys of the Entiat, upper Wenatchee, upper Yakima, Puyallup, Carbon, and White rivers, and at Mount Mazama (Fig. 2). Along much of the western range front in Washington, alpine limits were overlapped and obscured by drift of the Puget Lobe.

In 1965, the general advance-retreat chronology of the Puget Lobe was based on constraining ^{14}C dates, but age limits for alpine chronologies were based mainly on dated Holocene tephra layers. Crandell (1965) inferred that Evans Creek drift near Mount Rainier (ca. 20,000 ^{14}C years old) predated the maximum (Vashon) advance of the Puget Lobe (ca. 15,000–13,000 ^{14}C yr B.P.). According to prevailing opinion, by the time of the Vashon advance, the largest alpine glaciers had greatly shrunk in size or even disappeared.

Crandell (1965) summarized previous studies of the Oregon Cascades, noting that ice had covered the High Cascades at least once in late Pleistocene time. He also reported evidence of more-extensive glaciation in the west-draining valleys, as well as in the North Santiam River basin.

Pleistocene Glaciation on the Olympic Peninsula

Crandell (1964, 1965) focused his reconnaissance study of the glacial record on the southwestern part of the peninsula where glaciers spread across low terrain and advanced toward the Pacific coast. He recognized three drifts of likely Wisconsin age (early, middle, and late), as well as one or more of pre-Wisconsin age, distinguishing them on the basis of extent and contrasts in weathering.

Holocene Glaciation in the Cascades

In 1965, Crandell & Miller (1965) proposed a chronology for post-Hypsithermal (middle Holocene) glacier advances on Mt. Rainier based on limiting tephra ages and dendrochronology. Moraines of the Burroughs Mountain advance are overlain by tephra layers Wn (AD 1480) and C (ca. 2200 ^{14}C yr B.P.), but are younger than layer Yn (ca. 3300 ^{14}C yr B.P.). Moraines of Garda age were deposited during the Little Ice Age and were dated by tree-ring measurements (Sigafoos & Hendricks, 1972). An initial advance began in the late 12th or early 13th century and culminated variously in the mid-14th century to the mid-19th century.

Post-1965 Studies of Pleistocene Glaciation

Washington Cascades

Mapping and Relative-Age Control. During the past 30 years, most of the major Cascade valleys have been mapped and attempts made to derive a chronology of glaciation based largely on relative-age criteria (Fig. 2A and B and Table 1). As a result, limits of the last glaciation are well-delineated in major drainages. In most valleys evidence of two or more glaciations has been recognized, and in the most-thoroughly studied examples, evidence of at least nine ice advances has been documented.

Table 1. Post-1965 glacial-geologic studies in the Cascade Range and Olympic Mountains.[a]

Region	Study
Washington Cascades	
Western Drainages	
Mt. Baker	Thomas *et al.* (2000)
Nooksack	Kovanen & Easterbrook (2001)
Skagit	Heller (1980)
Glacier Peak and vicinity	Begét (1981), Davis & Osborn (1987)
North Cascades	Miller (1969), Porter (1978)
Tolt	Knoll (1967)
Middle Fork Snoqualmie	Williams (1971)
South Fork Snoqualmie	Porter (1976)
Cedar	Hirsch (1975)
Mt. Rainier and vicinity	Crandell & Miller (1965, 1974), Burbank (1981), Heine (1998)
White Pass	Clayton (1983)
Toutle	Hyde (1973)
Lewis	Mundorf (1984), Hammond (1987)
Eastern Drainages	
Mt. Adams and vicinity	Hopkins (1976)
Yakima	Porter (1976)
Wenatchee	
Ingalls	Hopkins (1966)
Icicle	Waitt *et al.* (1982), Swanson & Porter (1997, 2000)
Chumstick	Merrill (1966)
Upper Wenatchee	Nimick (1977)
Entiat	Thorp (1985)
Olympic Mountains	
Wynoochee	Carson (1970)
Quinnault	Moore (1965)
Queets	Thackray (2001)
Hoh	Thackray (2001)
Oregon Cascades	
Metolius	Scott (1977)
Mt. Jefferson and vicinity	Scott (1977)
Mountain Lakes	Carver (1973)

[a] Numerous unpublished manuscripts by William A. Long on various aspects of Quaternary glaciation in the Cascades and Olympics are housed in the library of the Quaternary Research Center, University of Washington.

Earliest-Dated Cascade Glaciation. The earliest evidence of glaciation that is at least partially controlled by radiometric ages is found southeast of Mt. Rainier in the southern Cascades, where Clayton (1983) found an alpine till near Penoyer Creek on the Tumac Plateau beneath a basalt flow with a K/Ar age of 1.75 ± 0.35 myr. Another till along the South Fork of Clear Creek, lies beneath a basaltic-andesite flow having a K/Ar age of 0.65 ± 0.08 myr. The regional extent of these tills is unknown.

Glaciation of Mt. Rainier Volcano. In their exhaustive study of Quaternary glaciation on and near Mt. Rainier volcano, Crandell & Miller (1974) recognized the deposits of four Pleistocene glaciations and two Holocene ice advances. At least one early glaciation is inferred from an intracanyon lava flow dating between $600{,}000 \pm 60$ K/Ar yr (for a mineral separate) and 325,000 K/Ar yr (whole-rock sample) that flowed down a glaciated valley. The largest glacier of the Wingate Hill glaciation extended 105 km from the mountain. Wingate Hill drift was differentiated from the subsequent Hayden Creek drift on the basis of weathering characteristics. The last glaciation (Fraser) is represented by the Evans Creek till, deposited by glaciers as much as 64 km long. A late-Fraser advance (McNeeley) left moraines far upvalley from their Evans Creek limit. No radiometric dates are available for these drifts, but Crandell and Miller inferred that the Evans Creek advance preceded the Vashon advance of the Puget Lobe. The McNeeley advance occurred prior to ca. 8850 ^{14}C years ago, the age of Rainier tephra layer *R* that mantles McNeeley moraines.

Glaciation of Upper Yakima River Drainage. The glacier system that occupied the upper Yakima River drainage was one of the longest in the Cascades and its record one of the most detailed (Fig. 3). Porter (1976) mapped the extent of eight glacier advances and determined their relative age based on various weathering parameters. ^{36}Cl ages for moraines of the Domerie advance, which impound three large lakes in the major tributary valleys, cluster in two groups, averaging ca. $23{,}200 \pm 1000$ and $16{,}300 \pm 1600$ years. Only a few preliminary ages are available for the next-older (pre-Domerie) Ronald and Bullfrog moraines, but both apparently are older than 50,000 ^{36}Cl years (T.W. Swanson, unpublished data). Three still-older drifts (Indian John, Swauk Prairie, and

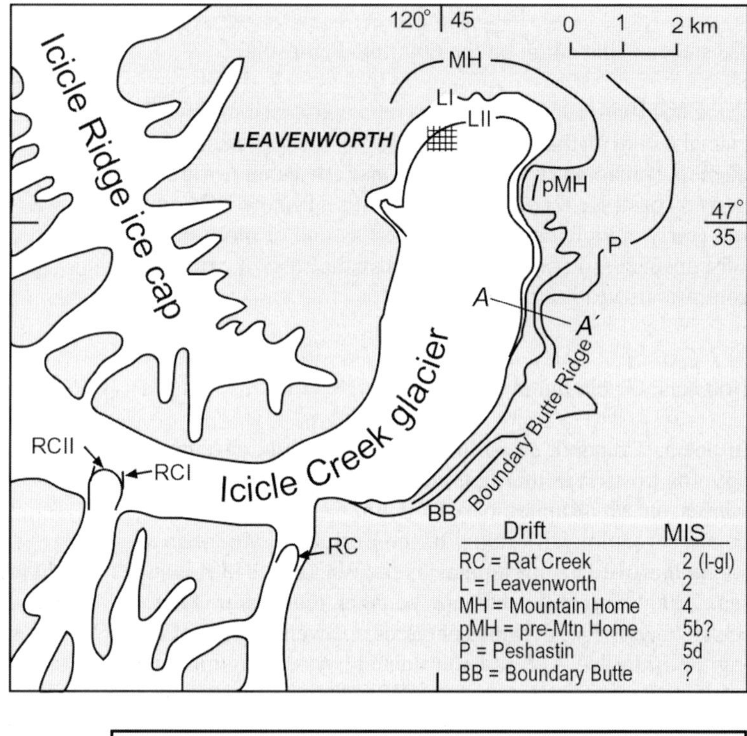

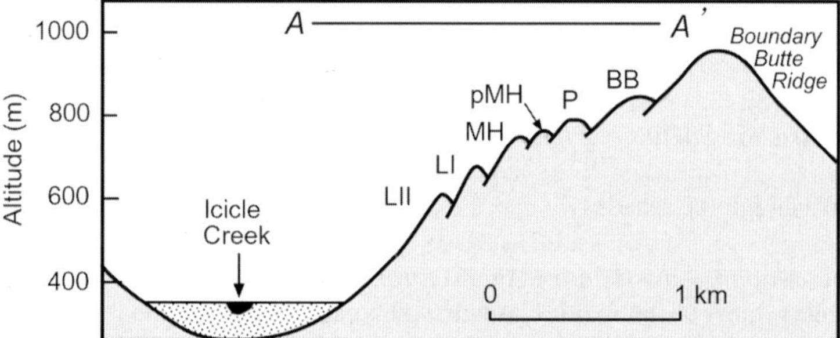

Fig. 4. Map of Leavenworth in the Wentachee River drainage basin showing extent and ages of Pleistocene moraines of Icicle Creek glacier.

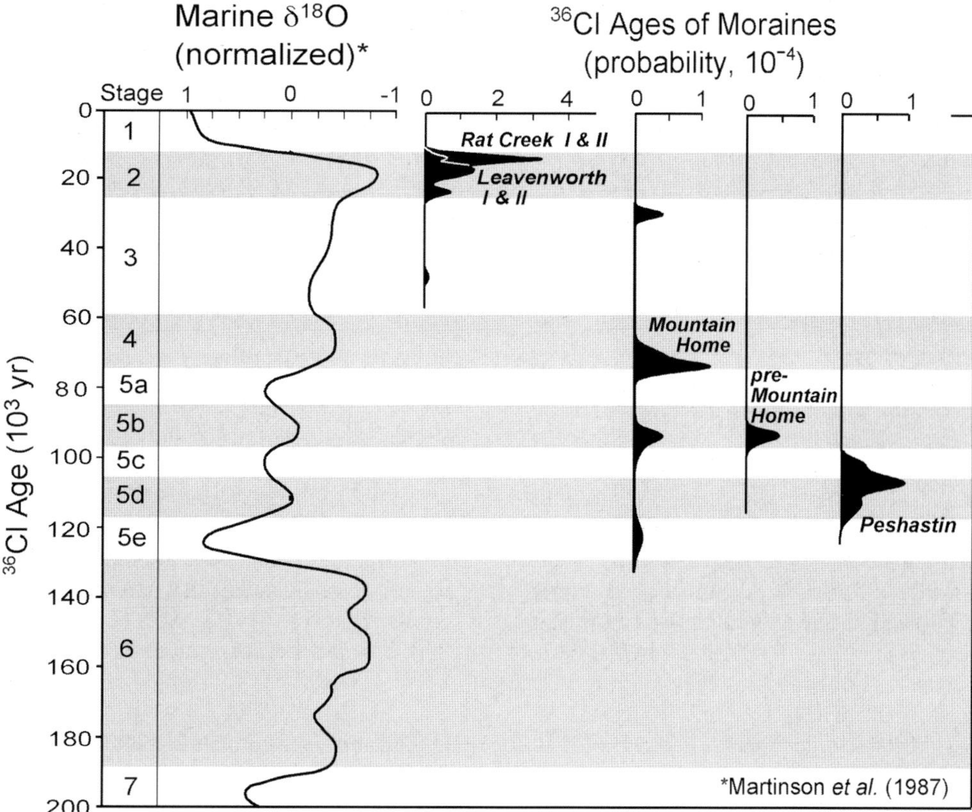

Fig. 5. Cosmogenic isotope (^{36}Cl) ages of moraine systems in Icicle Creek valley compared with the standard marine oxygen-isotope record (Martinson et al., 1987). The ^{36}Cl curves represent the collective probability distribution of individual samples ($\pm 1\sigma$).

Lookout Mountain) have not been dated, but likely are early-late to middle Pleistocene in age.

Glaciation of Icicle Creek Drainage. Right-lateral moraines of the Icicle Creek glacier, which occupied the next major valley north of the Yakima River, record seven advances that reached to or beyond the town of Leavenworth (Fig. 4). Two additional late-glacial moraines are found in tributary valleys upriver. The original sequence proposed by Page (1939) has been expanded with the discovery of five additional moraine systems. More than 100 ^{36}Cl ages have been obtained for moraine boulders and provide a chronology of late Pleistocene glacier variations for four of these moraines (Figs 5 and 6) (Swanson & Porter, 1997, and unpublished data). Prior to this study, the only age control was the presence of Mazama tephra (ca. 6800 ^{14}C yr B.P.) on late-glacial moraines of the Rat Creek advance (= Stuart glaciation of Page, 1939). Two moraines of the last glaciation (Leavenworth I and II) have mean ^{36}Cl ages of 20,000–18,000 and 18,000–15,000 years, respectively. Ages for the next older drift (Mountain Home) range from 77,000 to 71,000 years. Nine boulders on the Peshastin-age moraines range from 112,000 to 103,000 years and average 107,700 years. The oldest, strongly weathered drift (Boundary Butte) has not yet been dated. In addition, two late-glacial (late-Leavenworth) moraines have been dated (see below).

Time of Maximum Late Pleistocene Advance. Moraines marking the greatest expansion of ice in the mountains of western conterminous United States have often been inferred to predate the last interglaciation and correlate with MIS 6 or earlier isotope stages. The ^{36}Cl ages for Peshastin drift in the Icicle Creek drainage, however, imply an early last-glaciation age (i.e. equivalent to MIS 5d; Swanson & Porter, 2000). If this chronology is correct, it raises a question as to whether moraines marking maximum ice extent in other mountain ranges may also have formed early in the last glacial cycle. The difficulty is that surface exposure ages for moraines this old can be seriously affected by boulder weathering and moraine degradation, therefore requiring careful sample selection and interpretation. Nevertheless, these old moraines, heretofore largely "undatable" except by using relative-age criteria, may now be amenable to dating in carefully selected areas.

Late-Glacial Ice Advances. Was there an advance of Cascade glaciers during the Younger Dryas Stade (11,000–10,000 ^{14}C yr B.P.; ca. 12,900–11,600 cal yr B.P.)? This question has motivated several investigations during the past two decades. Moraines of presumed late-glacial age have been found in almost every glaciated valley investigated, but dating them has proved difficult. Often the only age control is the age of overlying Mazama tephra. Three recent studies

Fig. 6. Large granodiorite boulder of Rat Creek II moraine being sampled for ^{36}Cl dating.

provide closer-limiting ages based on radiocarbon and ^{36}Cl surface-exposure dating.

Late-glacial moraines in the Yakima River and Icicle Creek valleys have been dated by the ^{36}Cl method. In the former valley, two post-Domerie (i.e. late-glacial) moraines at Snoqualmie Pass have mean ^{36}Cl ages of 14,100 ± 500 and 12,700 ± 800 yr. The age of the younger moraine is consistent with a minimum limiting radiocarbon date from a bog inside the moraine of 11,050 ± 50 ^{14}C yr B.P. (ca. 12,600 cal yr B.P.) (Porter, 1976). In the Icicle Creek drainage, two late-glacial moraines (Rat Creek I and Rat Creek II) have mean ^{36}Cl boulder ages of 14,000–13,000 and 13,000–12,000 years, respectively. In each of these drainages the paired moraines are closely nested.

Kovanen & Easterbrook (2001) reported that three forks of the Nooksack valley in the western North Cascades contained long glacier systems during late-glacial time. Logs in a lateral moraine of the Nooksak Middle Fork glacier, lying 3 km beyond the terminus of the Derning Glacier have ages of 10,680 ± 70 and 10,500 ± 70 ^{14}C years. However, outwash heading at a moraine in the lower North Fork valley, some 50 km from the glacier source area at Mt. Shuksan (2782 m) and Mt. Baker (3285 m), overlies glacialmarine drift dated 11,910 ± 80 yr and contains charcoal dated 10,605 ± 70 and 10,790 ± 80 ^{14}C yr B.P., implying an ice tongue of exceptional length during Younger Dryas time.

The proposed exceptional extent of the Nooksack Valley glaciers at a time when glaciers along much of the crest of the Cascades were confined to valley heads and cirques is puzzling. Only on the north side of the Stuart Range, the source area of the type Rat Creek glacier, were late-glacial valley glaciers unusually long (ca. 8–10 km). Presumably this was because of their high accumulation area (1800–2200 m), as well as the potential for enhanced nourishment related to avalanching from steep valley slopes onto glacier surfaces. Mount Rainier, which is more than 1000 m higher than Mt. Baker and receives comparable record-high precipitation, supported long outlet glaciers during full-glacial time, but late-glacial ice was restricted in extent (see below).

A record of late-glacial glacier variations in Mt. Rainier National Park is based on mapping of moraines and radiocarbon dating of lake and bog sediments associated with the moraines (Heine, 1998). The outer of two moraines (McNeeley 1) in cirques and valley heads was deposited before 11,320 ± 60 ^{14}C yr B.P., and therefore predates the Younger Dryas interval. Moraines of a subsequent readvance (McNeeley 2) are overlain by layer R tephra (8850 ^{14}C yr B.P.). Basal organic matter at one site gives a minimum age of 9140 ± 100 ^{14}C yr B.P. for this advance. Detrital sediment layers in peat bogs that are inferred to be outwash from McNeeley 2 moraines are bracketed by ages of 10,080 ± 60 and 9120 ± 80 ^{14}C yr B.P. at one site, and between 9850 ± 60 and 8990 ± 60 ^{14}C yr B.P. at another site. These ages imply a post-Younger Dryas age for the advance. Heine therefore inferred that glaciers in this area were less extensive during the Younger Dryas interval than either before or after, possibly due to colder temperatures and diminished precipitation at that time.

These contrasting local moraine chronologies within the same mountain range, reflecting: (1) moderate advance; (2) exceptional advance; and (3) retreat during Younger Dryas time, raise questions about interpretations of the geology, the chronology, and the paleoclimatic environment, that will require additional research to resolve.

A recent study in southern British Columbia <100 km north of the international boundary bears on this question. Friele & Clague (2002) have reported evidence of a glacier readvance in the Squamish valley at the end of the last glaciation. Wood in till deposited by the glacier is 10,000–10,700 ^{14}C years old (ca. 12,500–12,900 cal yr), consistent with a Younger Dryas age.

Olympic Mountains

Two studies of glaciation in the Olympics have been reported since 1965. Carson (1970) established a relative sequence of glacial events in the Wynoochee River drainage of the southern Olympics, where he recognized two alpine drifts that predated the Salmon Springs glaciation (inferred at that time to be the penultimate glaciation), as well as one equated with the Salmon Springs. Drift of the last glaciation was divided into six phases, based on moraines and terraces.

Thackray's (2001) study of glacial landforms and thick stratigraphic sections exposed along the seacoast and the Hoh and Queets river valleys of the western Olympic Peninsula showed that Middle and Late Pleistocene glaciers advanced at least six times toward the Pacific coastal lowland. Abundant organic matter in the drifts permitted dating of deposits younger than ca. 50,000 yr B.P. The major glacier expansions are Hoh Oxbow 1 (beginning ca. 42,000–35,000 ^{14}C yr B.P.), Hoh Oxbow 2 (ca. 30,800–26,300 ^{14}C yr B.P.), Hoh Oxbow 3 (ca. 22,000–19,300 ^{14}C yr B.P.), Twin Creeks 1 (19,100–18,300 ^{14}C yr B.P.), and Twin Creeks 2 (undated).

Oregon Cascades

The glacial record of the Cascade volcanoes has been a primary glacial-geologic focus in Oregon. Two studies, in the northern and southern sectors of the range, demonstrate the general pattern and chronology of glaciations.

Mount Jefferson. Scott's (1977) study of Mt. Jefferson disclosed evidence of multiple drifts that he assigned to three glaciations based on relative-age criteria. The last glaciation (Cabot Creek) included two ice advances, Suttle Lake (last glacial maximum) and Canyon Creek (late-glacial). Multiple Suttle Lake moraines imply several stadial events. Canyon Creek moraines lie in valley heads and cirques, and are mantled with Mazama tephra (ca. 6800 ^{14}C yr B.P.). Moraines representing two phases of this glacial episode were detected in some cirques. Two older drifts (Jack Creek and Abbott Butte) preceded the pre-Cabot Creek interglaciation. The former is the oldest drift with preserved moraines. Scott inferred that Jack Creek drift correlates with MIS 4 or 6, and that Abbott Creek drift likely is Middle or Early Pleistocene in age.

Mountain Lakes Wilderness. Carver (1973) studied the glacial record in a region of ca. 1000 km^2 in the southern Oregon Cascades, focusing especially on the Mountain Lakes Wilderness area. Weathering parameters and morphology were used to characterize and correlate six drifts found in five valleys that radiate from the crest of a basaltic-andesitic volcano. At least four early drifts (unnamed, Winema, Moss Creek, Verney) predate a weathering interval preceding deposition of two drifts of the last glaciation (Waban, Zephyr Lake). Carver inferred that these younger drifts correlate with Evans Creek and McNeeley deposits, respectively, at Mt. Rainier. The Verney drift may correlate with Hayden Creek at Mt. Rainier, and the Moss Creek drift with Wingate Hill drift. Zephyr Lake drift predates the Mazama tephra, and likely was deposited close to the time of the Pleistocene-Holocene transition.

A recent study of sediment cores from nearby Upper Klamath Lake has provided a chronology for the Waban glaciation (J.G. Rosenbaum and R.L. Reynolds, pers. comm., 2002). Glacial rock flour produced by Cascade glaciers west of the lake, including those of the Mountain Lakes volcano, accumulated in the lake. Magnetic and grain-size data spanning the last ca. 37,000 yr, regarded as proxies for sediment flux, display a prominent peak between ca. 19,200 and 17,800 cal yr B.P. This peak is inferred to correlate with the outermost Waban moraines and to record the last glacial maximum.

Holocene Glacier Advances

Early Holocene (?) Advance

The occurrence of moraines near valley heads, and/or distal to Neoglacial moraines, that are mantled with Mazama tephra (ca. 6800 ^{14}C yr B.P) has led to speculation about their ages. Those who regard the tephra as providing a close minimum age suggest that the moraines are early Holocene in age, whereas others regard such a limiting age as permissive of a late-glacial advance. Purported evidence of an early Holocene advance of glaciers has been reported from sites at Mount Rainier and in the North Cascades.

At Mount Rainier National Park, an advance of cirque glaciers between ca. 9800 and 8950 ^{14}C yr B.P. has been suggested by Heine (1998), based on radiocarbon ages bracketing inferred meltwater sediment in lakes and bogs downstream from cirque glaciers.

Near Glacier Peak, in the North Cascades, moraines lying just beyond Neoglacial moraines are overlain by Mazama tephra. The drift overlies Glacier Peak tephra (ca. 11,250 ^{14}C yr B.P.), and charcoal from the purported till has an age of of 8300–8400 ^{14}C yr B.P (Begét, 1981); however, Davis & Osborn (1987) have questioned the interpretation of the sediment from which the dated samples were obtained. In the Enchantment Lakes basin, moraines lying just beyond Neoglacial moraines are mantled by Mazama tephra (Waitt *et al.*, 1982), and at Mount Baker, scoria from a local eruption (8420 ± 70 ^{14}C yr) predates moraines that are overlain by Mazama tephra (Thomas *et al.*, 2000). The scoria overlies adjacent moraines that may be equivalent to those of the inferred early Holocene advance at Mount Rainier.

Neoglacial Ice Advances

Washington Cascades. Miller (1969) used lichenometry and dendrochronology to date moraines in the Dome Peak area of the North Cascades. An early advance (ca. 4900 ^{14}C years ago) was inferred from ^{14}C-dated wood exposed by recent retreat of South Cascade Glacier. Evidence of an advance at that time is rarely seen in North America, for later advances apparently were more extensive. The oldest moraine of Chickamin glacier dates to the 13th century or earlier, but the main Little Ice Age moraine limits for most glaciers date to the 16th and 19th to 20th centuries. No moraines comparable to the Burroughs Mountain moraines at Mt. Rainier were found.

Burbank (1981) expanded on the work of Crandell & Miller (1965) by using a growth curve developed at Mt. Rainier for the lichen *Rhizocarpon geographicum* (Porter, 1981) to date moraines of four glaciers on the volcano. He showed that these glaciers built moraines in the early and middle 16th century, early, middle and late 17th century, early and middle, and late 18th century, early middle and late 19th century, and early 20th century. Moraines also have been dated by dendrochronology to the 13th and 15th centuries, although tephra stratigraphy suggests that they may instead predate tephra layer C (ca. 2200 ^{14}C yr B.P.).

Olympic Mountains. Outermost Neoglacial moraines of Blue Glacier in the Olympic Mountains remain poorly dated. The earliest dated moraine remnant likely was constructed in the mid-17th century, and the maximum late Little Ice Age ice limit dates to the early 19th century, possibly just prior to 1815 (Heusser, 1957). Spicer (1989) synthesized subsequent retreat of the glacier between 1815 and 1982 by assembling a discontinuous photographic record of the terminus.

Oregon Cascades. Scott (1977) mapped Neoglacial moraines on Three-Fingered Jack volcano and on Mt. Jefferson, where two phases were recognized. Some of the moraines are ice-cored. The oldest tree cored on the moraines is 90 years old, and *R. geographicum* thalli are $<$20 mm, implying that the glacier advance(s) occurred late during the Little Ice Age. Two Neoglacial periglacial phases in the southern Oregon Cascades are represented by rock glaciers, protalus ramparts, and block fields, but age control is limited (Carver, 1973). *Rhizocarpon* thalli on deposits of the early phase reach a diameter of 100 mm; those of the younger phase are $<$25 mm.

The present recession of glaciers in a time of warming climate may lead to exposure of terrain that has remained ice-covered for hundreds – even thousands – of years. This should afford an opportunity for obtaining datable samples to assess the extent of glaciers during major episodes of retreat, including the early to middle Holocene warm period prior to Neoglaciation.

Reconstructed Equilibrium-Line Altitudes

In several studies, estimates of the ELAs of former glaciers have been reconstructed, generally along west-east transects parallel to precipitation gradients. Most of these studies used the accumulation-area ratio method (AAR) (Porter, 2001), or a variant of it. Insufficient data are available to assess the three-dimensional full-glacial ELA surface, but this was possible for late-glacial time in a few areas where multiple small valley or cirque glaciers were clustered. Where modern glaciers exist, ELA depression (ΔELA) was also obtained.

The reconstructed full-glacial ELA of (Domerie) glaciers in the upper Yakima River valley lay close to 1100 m, and the ΔELA, relative to ELA's of nearby modern glaciers, was ca. 700 m (Porter *et al.*, 1983). The earlier Bullfrog ELA was about 100 m lower. Hurley (1996) made a study of late-glacial (Hyak/Rat Creek) ELAs for a population of 25 glaciers, based on the AAR and median altitude methods, along a west-to-east transect across the Cascades at this latitude. He showed that the regional ELA surface rose from 900 m at the western margin of the range to 1840 m in the east. Late-glacial ΔELAs averaged ca. 600 m, some 300 m higher than Domerie ELAs.

Near Mt. Rainier, the full-glacial ΔELA was estimated by Heine (1998) to be ca. 950 m. Late-glacial to early Holocene (McNeeley) ELAs were ca. 400–500 m below recent (late Holocene) ELAs.

Burbank (1981) calculated the average ELA of the late 18th to early 19th centuries for four valley glaciers on Mt. Rainier and arrived at a value of 1945 ± 65 m. His estimates were made relative to a modern steady-state value of 2105 ± 115 m. The average rise in ELA has therefore been ca. 160 m. However, he noted considerable variation among individual glaciers (e.g. 60–300 m); those showing the least change have a cover of supraglacial debris.

In the Oregon Cascades, reconstructed ELAs of Suttle Lake glaciers at Mt. Jefferson suggest a ΔELA of ca. 950 m during the last glaciation (Scott, 1977). The late-glacial Canyon Creek ELA depression was ca. 700–750 m. In the Mountain Lakes Wilderness, reconstructions by Carver (1973; AAR 0.65 ± 0.1) indicated that Varney Creek ELAs were 75–150 m lower than those of Waban glaciers, and that Zephyr Lake ELAs were 300–400 m higher. Total ELA depression could not be calculated because no glaciers now exist in the area.

Sierra Nevada, California

Small modern glaciers are sheltered in cirques high among the ridges and peaks of the Sierra Nevada. The modern climatic snowline near 37°N was estimated at 4500 m elevation by Flint (1957, p. 47), 600 m higher than the crest. However, during the maximum Pleistocene advances, the ELA was depressed to about ~3000 m, ~800 m lower than the ELA for modern glaciers (Gillespie, 1991; Warhaftig & Birman, 1965), and large Pleistocene glaciers and ice caps existed between 36.4° and 39.7°N (Fig. 7).

In the Sierra Nevada, the advance and retreat of glaciers are especially sensitive to changes in summer temperature and winter precipitation. At present, the Sierra Nevada receives moisture mainly from winter low-pressure systems from the Pacific Ocean guided by the jet stream. Mean annual

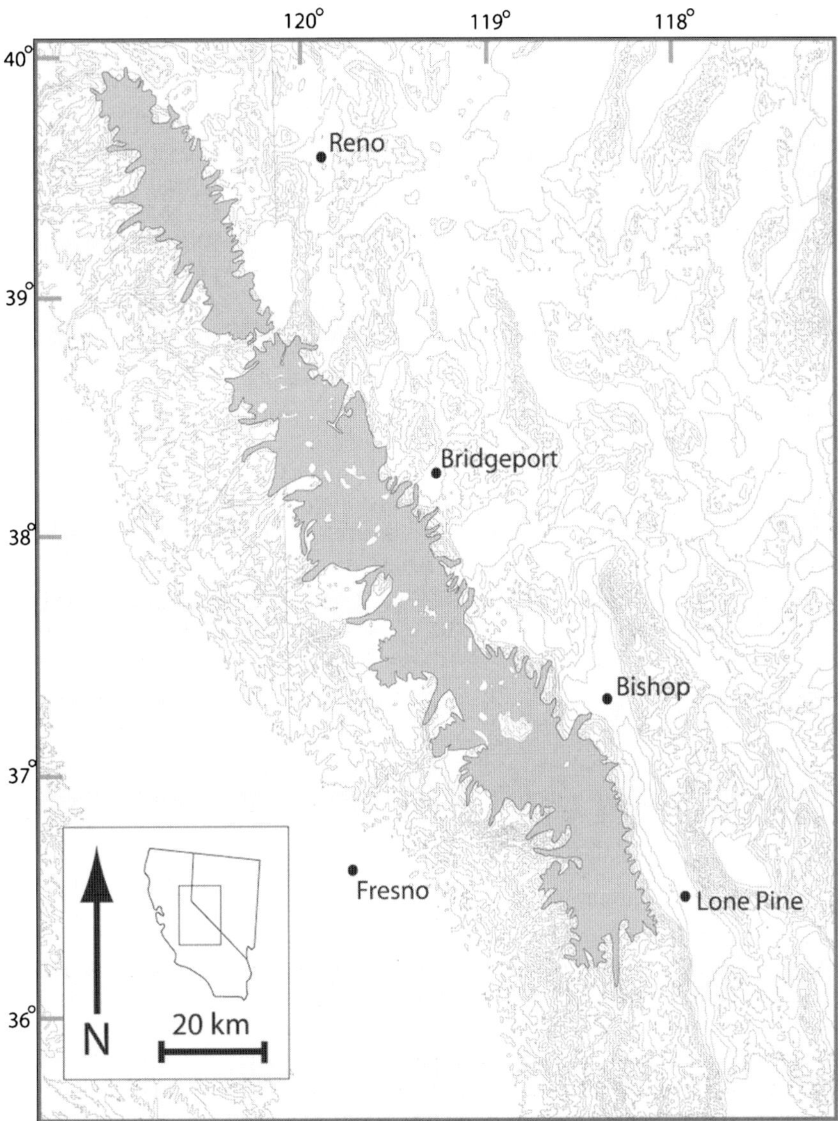

Fig. 7. Extent of late Pleistocene glaciers in the Sierra Nevada (courtesy Paul Zehfuss).

precipitation on the eastern slope of the Sierra Nevada at ~37°N ranges from ~100 cm/yr at the crest to ~25 cm/yr at the range front (Danskin, 1998); east of the crest it is controlled by a strong rainshadow. During Pleistocene glaciations, the jet stream shifted south so that precipitation increased and temperatures were reduced, although not necessarily in phase. Antevs (1938, 1948) first proposed this modern view of the ice-age climate, with Sierra Nevada precipitation controlled by intensified winter Pacific storms, driven farther south than today due to the influence of the Cordilleran and Laurentide ice sheets.

The modern framework of the Sierra Nevada glaciations was well understood by 1965 (Warhaftig & Birman, 1965, Table 1, p. 308), but numerical age control was weak or lacking. Major advances since INQUA VII have focussed on refinements to the glacial sequence, especially using numerical dating, and on inference of glacial history from lake-sediment cores. In addition to Warhaftig & Birman (1965), reviews of Sierra Nevada glaciations may be found in Porter *et al.* (1983) and Fullerton (1986).

Refinements to the Glacial Sequence

Major areas of activity have included: (1) additions to the list of recognized pre-MIS 6 glaciations; (2) reorganization of and controversy over MIS 3–6 glacial history; and (3) new insights into post-Tioga glacial history. Most of the research has taken place in the eastern Sierra Nevada. The improvements to the glacial history are based on continued exploration and mapping, new methods of numerical dating, and extraction and analysis of sediment cores from lakes and bogs. The recognized glaciations of the Sierra Nevada are summarized by Fullerton (1986, Chart 1, Table 1); those discussed in this Chapter are listed in Table 2.

Pliocene and Early Pleistocene Glaciations

The early record of glacial landforms and deposits is incomplete, and our understanding of this period remains sketchy. The most significant advances in our understanding

Table 2. Recognized glaciations of the Sierra Nevada, their ages, and east-side ELAs relative to Tioga and interpolated to 37°.

Glaciation	Mean Age (ka)[a]	ΔELA (m)[b]	Age References
Matthes (Little Ice Age)	0.6–0.1[c]	480 ± 25	Wood (1977), Stine (1994)
Recess Peak	14.2–13.1[c]	335 ± 20	Clark (1997)
Tioga (retreat)	15–14	125 ± 30	James *et al.* (2002), Clark & Gillespie (1997)
Tioga (start)	21–20[c]	0	Clark *et al.* (2003)
	<25[c]	0	Bursik & Gillespie (1993)
Tioga ("Tioga 2–4")	25–16	0	Phillips *et al.* (1996)[d]
Tenaya ("Tioga 1")	31	−45 ± 15	Phillips *et al.* (1996)[d], Benson *et al.* (1996)
	32[c]		Bursik & Gillespie (1993)
Tahoe II[††]	50–42	−95 ± 10	Phillips *et al.* (1996)
Tahoe I[††]	?	∼−100?	
Casa Diablo	126–62	?	Bailey *et al.* (1976)
pre-Tahoe (Bloody Canyon)[e]	220–140	?	Phillips *et al.* (1990), discussed in Gillespie *et al.* (2003)
Mono Basin	80–60	−195 ± 50	Phillips *et al.* (1990)[d]
Walker Creek[e]	∼550	?	Clark (1968); A. Sarna-Wojcicki, quoted in Gillespie *et al.* (2003)
Sherwin	∼820	∼−200	Sharp (1968), Birkeland *et al.* (1980), Nishiizumi *et al.* (1989)
Lower Rock Creek	∼920	?	Sharp (1968), Birkeland *et al.* (1980)
McGee	2700–1500	?	Huber (1981), Dalrymple (1963, 1964)

[a] Ages are shown as ranges or approximate values. Uncertainties may be found in text and/or references.
[b] Elevation differences are relative to the lowest Tioga ELA (∼3040 ± 150 m) and interpolated to 37°N (Gillespie, 1991). Uncertainties are ±1σ and are random and do not include the systematic uncertainties of ∼150 m. Accuracy depends critically on assignment of moraines to the correct glaciation. Regression was done on ≤70 glaciated drainages.
[c] 10^3 cal yr B.P.
[d] Revised (see discussions in James *et al.*, 2002, and Phillips *et al.*, 2001).
[e] Nomenclature from Gillespie *et al.* (2003).

involve the Sherwin till. Sharp (1968) demonstrated that the till underlies the Bishop Tuff (759,000 years: Sarna-Wojcicki *et al.*, 2000), a normally polarized regional marker bed that provides the best age constraint for the Bruhnes-Matuyama magnetic reversal. R.P. Sharp (1968; pers. comm., 1976) estimated that upon burial, the till had been weathered about as much as exposed Tahoe till is today, requiring ∼50,000 to 100,000 years. Other studies support this estimate. From the development of the paleosol on the buried till, Birkeland *et al.* (1980) estimated an age at burial of ∼50,000 years, and Nishiizumi *et al.* (1989) analyzed cosmogenic nuclides to estimate ∼67,000 to 53,000 years. Thus, the Sherwin glaciation probably occurred ∼820,000 years ago (i.e. late Early Pleistocene).

Sharp (1968) discovered an older till underneath the Sherwin till on Little Rock Creek. Birkeland *et al.* (1980) estimated from relative dating that the paleosol on this buried till represented ∼100,000 years of development. A similar till with a comparable paleosol underlies till mapped as Sherwin by R.P. Sharp (unpublished data, 1979) near the downvalley end of the left-lateral moraine of Big Pine Creek, ∼50 km to the south.

Both the Sherwin and the older till were deposited in the vicinity of late Pleistocene moraines. This suggests that both tills are significantly younger than the oldest tills of the Sierra Nevada, such as the Pliocene (?) McGee Till, which is found on the shoulder of McGee Mountain ∼800 m or more above the late Pleistocene moraines of McGee Creek.

A number of diamictons are locally preserved on old erosion surfaces now at or near the crest of the Sierra Nevada, and some of these diamictons may be glacial drift (Gillespie, 1982). Therefore, it seems likely that the long gap between the McGee and Sherwin glaciations does not indicate a million-year-long interglaciation; instead, it is an artifact of poor preservation of glacial evidence.

A few "tills" lie stratigraphically between the Sherwin and middle Pleistocene Mono Basin and Tahoe tills. One is from the West Walker River, east of Sonora Pass (Blackwelder, 1931; Clark, 1967, 1968). A roadcut near U.S. Highway 395 exposes fluvial gravel of Wheeler Flats that contains a tephra identified as Rockland Ash-Tuff by Sarna-Wojcicki *et al.* (1985), now dated at ∼550,000 years (A.M. Sarna-Wojcicki, quoted in Gillespie *et al.*, 2003). If the gravel is glaciofluvial, as seems likely, the tephra may date an unnamed, pre-Tahoe glaciation. Mathieson & Sarna-Wojcicki (1982) found the Rockland ash in a similar stratigraphic relation in the Mohawk Valley in the northern Sierra Nevada.

Fullerton (1986) discussed a till from Reds Meadow, near Devils Postpile, that overlies the Bishop Tuff and underlies an andesite, the age of which has been very loosely constrained by K/Ar to 650,000 ± 350,000 years. This till also is a candidate for a post-Sherwin, pre-Tahoe advance. Curry (1971) and Sharp (1972) presented evidence for other tills of

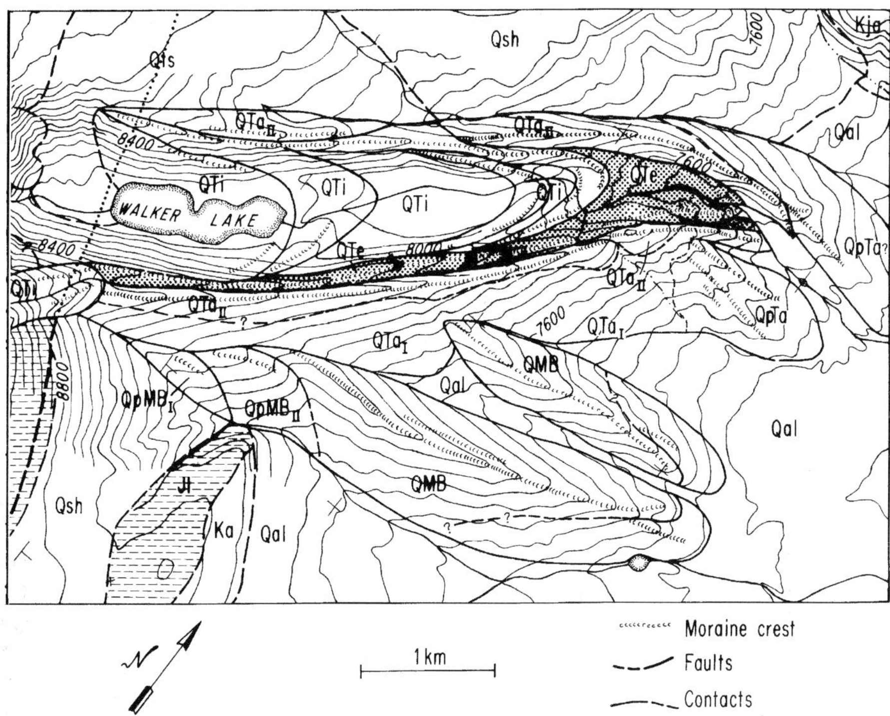

Fig. 8. Map of the Bloody Canyon moraines (after Gillespie, 1982). Jl, Ka, and Kja are plutonic rocks. Qal is Quaternary alluvium and Qls is colluvium. Qsh is Sherwin till of Sharp & Birman (1963). $QpMB_I$ and $QpMB_{II}$ are pre-Mono Basin moraines; QMB is Mono Basin moraines. QpTa is the oldest set of moraines (pre-Tahoe) along Walker Creek ("older Tahoe" of Phillips et al., 1990); QTa_I and QTa_{II} are the Tahoe I and II moraines (Gillespie, 1982). QTe (shaded for clarity) is the Tenaya moraine, and Qti are the undifferentiated Tioga moraines. Contour interval is 24 m (80 ft).

this same general period from Rock Creek and the Bridgeport Basin.

Gillespie (1982) described two pre-Mono Basin moraines at Bloody Canyon (Fig. 8), the type area of Mono Basin till. This finding together with those cited above suggests that the interval between the Mono Basin and Sherwin glaciations was marked by a number of ice advances, many to about the same maximum positions. In the absence of accurate and precise numerical dates, correlation of tills and elucidation of the glacial history from this interval remains problematical.

Mono Basin and Tahoe Glaciations

The Mono Basin and Tahoe moraines are well-preserved, yet their history is not well-understood. Age estimates for the Tahoe glaciation range from early MIS 6 to MIS 3, and even the relative age of the Mono Basin and Tahoe glaciations is controversial.

One key area of debate has been the partitioning of the glacial record into stades and glaciations. The distinction is only partly semantic. The larger issue involves the length of time that glaciers were at or near their maximum limits, and the length of time the Sierra Nevada was largely ice-free between advances. Although a number of interpretive schemes have been advanced (cited by Fullerton, 1986), it seems unlikely that definitive answers will be forthcoming until the records of glacial and lacustrine sediments for the Sierra Nevada have been integrated.

The current status of this debate is that multiple advances have been recognized for many of the glacial periods named by Blackwelder (1931), who noted that there were commonly two Tioga moraines in Sierra Nevada valleys. Burke & Birkeland (1979) used relative dating to regroup the moraines at Bloody Canyon into multiple advances, or stades, within only two glaciations, the Tioga and Tahoe. They regarded the Tenaya and Mono Basin glaciations of Sharp & Birman (1963) as early stades of the Tioga and Tahoe glaciations, respectively. Gillespie (1982) subdivided the Tahoe glaciation there (Fig. 8), and suggested that the two "Tahoe" stades were not necessarily related to the same glaciation, a conclusion supported by the soil-development data of Birkeland & Burke (1988) (R.M. Burke, pers. comm., 1998). There is no general consensus concerning the best classification, and the frustration in dealing with this thorny issue is evident in Fullerton's (1986) attempt to rationalize the Middle and Late Pleistocene glacial histories set forth by several researchers.

Direct, numerical dating of the glacial landforms and drift should cast the glacial history into sharp focus, yet this is not yet the case. The best dates for the Mono Basin and Tahoe glaciations are from ^{36}Cl cosmogenic exposure ages for boulders from moraine crests at Bloody Canyon (Phillips *et al.*, 1990, 1996). These dates have been revised downward as production-rate estimates have been refined. The age for the Mono Basin moraine at Bloody Canyon, its type area, has been reduced from ~103,000 ^{36}Cl years in Phillips *et al.* (1990) to ~80,000–60,000 ^{36}Cl years at present (Phillips *et al.*, 2001, revised as discussed in James *et al.*, 2002). The current "best estimate" for the "younger Tahoe" (Tahoe II) moraine is now ~50,000–42,000 ^{36}Cl years, reduced from ~60,000 ^{36}Cl years. The best dates for the "older Tahoe" (pre-Tahoe I, post-Mono Basin: Fig. 8) moraines at Bloody Canyon are ~220,000–140,000 ^{36}Cl years (Phillips *et al.*, 1990) (i.e. MIS 6). Although these older Tahoe values have not been adjusted downward, it is clear that they are greater than the ^{36}Cl ages for the Mono Basin moraines ~1 km away, and Phillips *et al.* (1990) regarded the older Tahoe as predating the Mono Basin glaciation. These findings contradict the field relations among the moraines, from which a clear stratigraphic sequence is inferred, with both Tahoe I and Tahoe II tills overlying Mono Basin till. It is noteworthy that only a small number of boulders have been dated (8 for Mono Basin, 5 for older Tahoe), and scatter among dates for each till is large. The dating studies have cast in sharp focus the conflicts between stratigraphic and chronologic analyses, and the current strengths and deficiencies in each.

A number of other tills have been identified from the eastern Sierra Nevada that may date from the same general period as the Tahoe and Mono Basin tills. One is the pre-Tahoe Casa Diablo till near the town of Mammoth Lakes, weathered to a similar degree as nearby Tahoe till (Burke & Birkeland, 1979; Birkeland *et al.*, 1980). Although it is interbedded with basalt flows that should afford a good dating opportunity, K/Ar analyses by Curry (1971) and by Bailey *et al.* (1976) are in substantial disagreement, constraining the age of the till either to ~453,000–288,000 or 126,000–62,000 years, respectively. Fullerton (1986) pointed out that the absence of 185,000-year-old quartz latite boulders in the Casa Diablo till, and their presence in the nearby Tahoe till, suggests that the Casa Diablo till may pre-date MIS 6. However, recent field work shows that the quartz latite boulders do occur in the Casa Diablo till, and the dates of Bailey *et al.* (1976) are now generally accepted as the age range for the till (R.M. Burke, pers. comm., 2003). Another till from the same general age range as the Casa Diablo till occurs in the oldest moraine in Sawmill Canyon, Inyo County. It, too, is also interbedded with basalt flows, but their dates only loosely constrain the moraine to ~465,000–130,000 years ago (Gillespie *et al.*, 1984).

Late Pleistocene Glaciations

The glacial history of the Sierra Nevada becomes a little clearer for the post-Tahoe period. Only two post-Tahoe glaciations have been proposed: Blackwelder's (1931) Tioga, and Sharp & Birman's (1963) Tenaya. Even so, the number of stades is variously counted, and the existence of the Tenaya as a glaciation separate from the Tahoe has been debated.

At Bloody Canyon, Sharp & Birman (1963) counted two Tenaya moraines; Burke & Birkeland (1979) grouped them with Sharp & Birman's (1963) Tioga moraines for a three-stade Tioga glaciation (Fullerton, 1986, Fig. 2a). From ^{36}Cl cosmogenic exposure ages at Bloody Canyon and other canyons, Phillips *et al.* (1996) inferred four separate Tioga stades ranging in age from 25,000 to 14,000 ^{36}Cl years, revised for the changes in production rates as discussed above. Their dates did not resolve the Tenaya as a separate glaciation at Bloody Canyon.

Based on ^{14}C ages of ostracodes in Mono Lake sediments interbedded with basaltic ash at nearby June Lake, Bursik & Gillespie (1993) inferred an age of $<25{,}200 \pm 2500$ cal yr B.P. for the maximum Tioga advance, which overrode one cinder cone. They inferred an age of 31,700 cal yr B.P. or more for the Tenaya moraine, through which an eruption may have occurred while ice was present. The two cinder cones are the only sources that have been discovered for the ash in the lake sediments. Bursik & Gillespie (1993) regarded the Tenaya as a separate advance. Clark *et al.* (2003) cored Grass Lake Bog, south of Lake Tahoe, and dated a sharp transition from Tioga glacial drift to underlying non-glacial sediments at 21,130–19,850 cal yr B.P. This may be the best date for the beginning of the Tioga maximum advance.

James *et al.* (2002) obtained ^{10}Be and ^{26}Al cosmogenic exposure ages for the Tioga glaciation and suggested that its maximum in the South Fork of the Yuba River was $\sim 18{,}600 \pm 1200$ years ago. Basal lake sediment elsewhere on the western slope of the Sierra Nevada yielded a minimum date for the beginning of Tioga retreat of $15{,}570 \pm 820$ ^{14}C yr B.P. (Wagner *et al.*, 1982, cited in Fullerton, 1986). James *et al.*'s (2002) cosmogenic data show that the Tioga glaciers retreated rapidly from the middle elevations of the Yuba River 15,000–14,000 years ago. Clark & Gillespie (1997) are in agreement that the Tioga glaciers vanished entirely or were restricted to cirques during this same interval.

Post-Tioga Advances

The Hilgard glaciation, proposed by Birman (1964), was regarded as a very late Tioga advance by Birkeland *et al.* (1976), and as a recessional standstill by M. Clark (pers. comm., 1988). Thus, the Hilgard advance is no longer regarded as a separate glaciation, and the Recess Peak advance is the oldest post-Tioga advance in the Sierra Nevada for which evidence has been discovered.

As first described by Birman (1964), Recess Peak moraines are restricted to the vicinity of Pleistocene cirques. Because of their fresh character, most early workers concluded that the Recess Peak moraines were Neoglacial in age, and constructed within the past 2000–3000 years (Birman, 1964; Curry, 1969; Scuderi, 1987). However, soil work by Yount *et al.* (1982) suggested that Recess Peak deposits were early Holocene or older. Firm numerical constraints on the moraines from sediment coring of nearby lakes demonstrated that the Recess Peak advance began by ~14,200 cal yr B.P. and ended before ~13,100 cal yr B.P. (Clark, 1997). Plummer's (2002) ^{36}Cl cosmogenic ages of $12{,}600 \pm 1300$ years (production rate uncertainties included) overlap Clark's (1997) age range. Therefore, the Recess Peak advance in the Sierra Nevada preceded the North Atlantic Younger Dryas event.

Curry (1971) recognized two Neoglacial advances having lichenometric ages equivalent to ~1100 and 970 ^{14}C yr B.P., and coring of the Conness Lakes indicates that Neoglaciation began by ~3200 ^{14}C yr B.P. (3400 cal yr B.P.) (Konrad & Clark, 1998). Curry (1971) dated the Matthes advances at 620–55 ^{14}C yr B.P. The absence of a ~700 ^{14}C yr B.P. (~630 cal yr B.P.) tephra blanketing Matthes moraines, as well as dendrochronology (Wood, 1977), indicate that Matthes glaciers reached their maximum positions after that eruption. Stine (1994) identified two droughts at about A.D. 1112–900 and 1350–1250, just before the onset of the Matthes advances in the Sierra Nevada.

The absence of any moraines between the Recess Peak and Matthes moraines, as well as the absence of any outwash deposits between 13,100 and 3400 cal yr B.P., indicates that no significant glacier advances in the Sierra Nevada occurred during that timespan, including during the Younger Dryas interval (Clark & Gillespie, 1997). If glaciers were present in the Sierra Nevada during Younger Dryas time, they must have been smaller than Recess Peak glaciers and largely restricted to cirques.

Numerical Dating of Glaciations

Rigorous dating in the Sierra began with K/Ar dating of lava flows interbedded with till (Dalrymple, 1963, 1964) and ^{14}C dating of latest Pleistocene sediments in bogs that could be related stratigraphically to glacial deposits (Adam, 1966, 1967). K/Ar dating, however, was problematic and opportunistic, because the glacial deposits and landforms themselves could not be dated. For example, Gillespie *et al.* (1984) obtained high-precision ^{40}Ar-^{39}Ar dates for basalt flows at Sawmill Canyon (Inyo Country), but in the end could only conclude that the Hogsback moraine was less than 119,000 ± 3000 years old and must postdate MIS 6, a conclusion already reached by Burke & Birkeland (1979) on the basis of soil development.

The Sierra Nevada provided an early opportunity to estimate production rates for cosmogenic ^{10}Be and ^{26}Al, based on analyses of bedrock exposed by retreat of the Tioga glaciers, assumed to have been 11,000 years ago (Nishiizumi *et al.*, 1989). Later, ^{14}C dates from lake cores demonstrated that the area actually was deglaciated by ~15,500 cal yr B.P. (Clark, 1997). As a result of this and similar findings elsewhere, estimated production rates for these isotopes were reduced and now give ages compatible with those obtained using other techniques.

Phillips *et al.* (1990, 1996) are the primary researchers responsible for the effort to obtain cosmogenic ages for Sierra Nevada moraines. Their most notable work has been in Bloody Canyon, where ^{36}Cl ages for Tenaya and Tioga agree closely with ^{14}C ages, as discussed above. Their ages are consistent with field observations that the latest Pleistocene moraines and boulders appear little eroded. The cosmogenic ages for the older Tahoe and Mono Basin moraines present a significant challenge to the field stratigraphy, however.

The Mono Basin moraines at Bloody Canyon appear to be buried by both Tahoe I and Tahoe II moraines (Fig. 8), and thus must be older. Figure 9 clearly shows that the dates

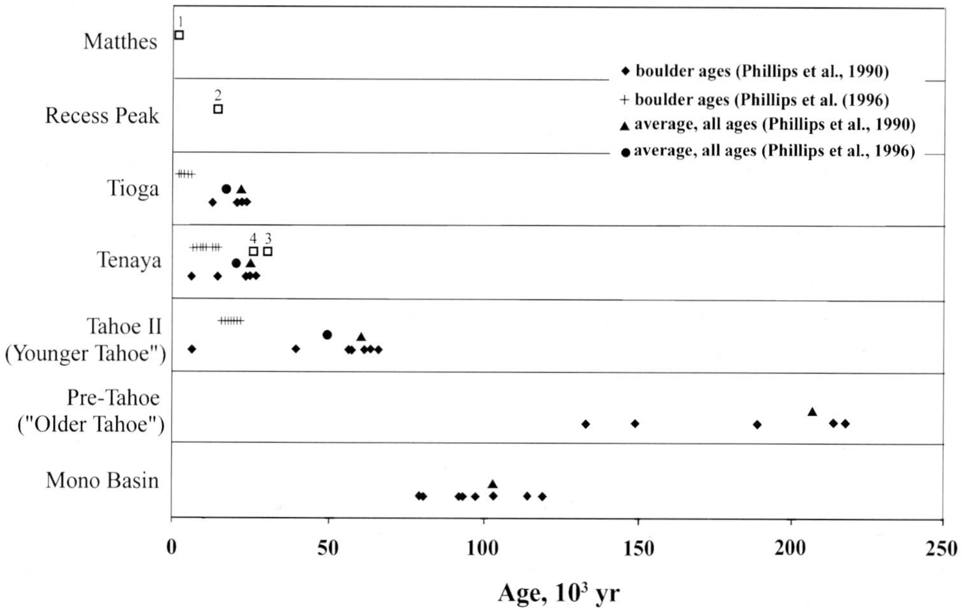

Fig. 9. ^{36}Cl dates (Phillips et al., 1990, 1996) for Bloody Canyon moraines with the following independent age control (courtesy of R.M. Burke): (1) 100–600 cal yr B.P. (Clark & Gillespie, 1997; Konrad & Clark, 1998; Yount et al., 1982). (2) 13,100–14,200 cal yr B.P. (11,200–12,200 ^{14}C yr) (Clark, 1997). (3) 30,000 cal yr B.P. (Bursik & Gillespie, 1993). (4) >25,000 on ^{10}Be and ^{26}Al (A.R. Gillespie, unpub. data).

for Tahoe II are indeed younger than Mono Basin. However, Phillips *et al.*'s (1990) "older Tahoe" appears to be older than Mono Basin, not younger. This finding appears to violate the stratigraphic order and, if correct, requires that the Bloody Canyon glaciers changed paths several times in order to explain the positions of the moraines.

The "older Tahoe" moraine dated by Phillips *et al.* (1990) is not demonstrably the same as the Tahoe I till, seen to bury the Mono Basin moraine in Fig. 8. Re-examination of field relations shows that the Tahoe I moraine buries the "older Tahoe" moraines. Thus, it is possible that Phillips *et al.* are correct in their dating and geologic interpretation. However, this requires that the large difference in soil characteristics between the Tahoe I and Tahoe II moraines (Birkeland & Burke, 1988) represent only a few thousand years during MIS 3, which appears unlikely.

A more likely possibility is that the age of the sharp-crested Mono Basin moraines is underestimated. Hallet & Putkonen (1994) showed by modeling that cosmogenic exposure ages for old, eroded moraines underestimate landform ages. They further concluded that even apparent age reversals, such as observed at Bloody Canyon, were possible if boulder erosion was also considered.

Although the ages, and even the relative age relations, among the older moraines of Bloody Canyon remain unresolved, direct dating of landforms less than ~50,000 years old seems to be a reality. This remarkable ability will surely yield important discoveries in the near future. It seems prudent, however, to regard cosmogenic ages for older landforms with skepticism until further developments explain the observed discrepancies.

Lake-Sediment Cores

Runoff and sediment from Sierra Nevada glaciers enter lakes fed by rivers draining the mountains. Sediment cores extracted from these lakes record climate and erosion, integrated over the watershed. Given the eroded nature of many older glacial deposits in the steep mountains, and the absence of deposits of small glaciers that were overridden by younger, larger ones, it is reasonable to turn to lake-sediment cores for a more complete record. Smith (1983) analyzed a core from Searles Lake, which intermittently received water from Owens River. Smith's study established the periods of major overflow from Owens Valley, upriver, and showed the promise of sediment cores for analyzing the eastern Sierra Nevada glacial history. Owens Lake and Mono Lake, which received runoff from the Sierra Nevada during all or most of the Quaternary Period, offer a more complete record, and soon attracted attention.

During times of decreased water flow, Owens Lake was closed, saline, alkaline, and biologically productive. During times of increased water flow, it was fresh and biologically unproductive. During glaciations, rock flour in the lacustrine sediment increased, total organic carbon decreased, and magnetic susceptibility peaked (Bischoff *et al.*, 1997). Therefore, oscillations in the chemical, mineralogical, and magnetic attributes of sediment cores from Owens Lake can provide clues to the climate and glacier response in the Sierra Nevada headwaters.

Detailed analysis of Owens Lake Core OL-92 focused on upper part of the core containing the Tahoe and Tioga record (MIS 6 to MIS 2). Bischoff *et al.* (1997) analyzed oscillations of lake salinity and rock flour content back to 155,000 years ago with a 1500-year resolution. They discovered evidence for two glacier advances during MIS 6 and three during MIS 4–2. The two MIS-6 stades occurred 155,000–146,000 and 140,000–120,000 years ago.

Rock-flour influx during the later MIS-6 stade terminated abruptly 118,000 years ago, which Bischoff *et al.* (1997) interpreted as marking the onset of the last interglaciation. However, $\delta^{18}O$ data from the same core show a gradual transition after ~140,000 years ago, parallel with isotopic evidence from carbonate deposited by groundwater near Death Valley (Winograd *et al.*, 1992).

The global glacial-interglacial transition occurred about 128,000 years ago, according to orbitally tuned data from North Atlantic marine sediment cores (Martinson *et al.*, 1987). Thus, large glaciers in the Sierra appear to have remained active for thousands of years after the regional climate began to change, and even after strong changes in the global climate and shrinking of the high-latitude ice sheets.

Bischoff & Cummins (2001) found evidence in the rock-flour data for four brief periods of glacier activity during the interval 80,000–53,000 years ago. Thus, from the end of MIS-5 and throughout MIS 4 there were brief glacier advances. The carbonate and rock-flour records are not perfectly correlated. Bischoff & Cummins (2001) concluded that at least some glacier advances began during dry, cold conditions. However, the activity of the glaciers increased during wet periods. For example, the MIS-2 LGM was a time of enhanced precipitation, as well as low temperatures.

Dates for the MIS 4–2 stades correspond well to the ^{36}Cl cosmogenic dates of Phillips *et al.* (1996) for boulders from Sierra Nevada moraines. Based on rock-flour data, Bischoff & Cummins (2001) suggested that brief glacier advances occurred ~47,000 and 41,000 years ago, and a longer one (the Tioga glaciation) from 30,500 to 15,500 years ago. High-resolution magnetic susceptibility data of Benson *et al.* (1998a) for Mono Lake indicate that the Tioga glaciation was characterized by four stades, and did not begin in earnest until 24,500 years ago (Fig. 10). The magnetic susceptibility and total organic carbon data revealed an additional 12 stades in the interval 52,500–31,000 years ago, at the base of the studied core section. Many of these stadial advances must have occurred in the vicinity of the Sierra Nevada crest, where moraine or drift preservation would be unlikely.

High-frequency fluctuation of Sierra Nevada glaciers is strongly supported by data from Mono Lake. Mono Lake $\delta^{18}O$ data for 35,400–12,900 ^{14}C yr B.P. show the same three scales of climatic oscillation as the Owens Lake data: Milankovitch, Heinrich, and Dansgaard-Oeschger (Benson *et al.*, 1996, 1998a, b). Benson *et al.* (1998a) interpreted their data to show that glacier activity in the Sierra Nevada, which peaked at times of low summer insolation, was synchronous with cold periods in the North Atlantic.

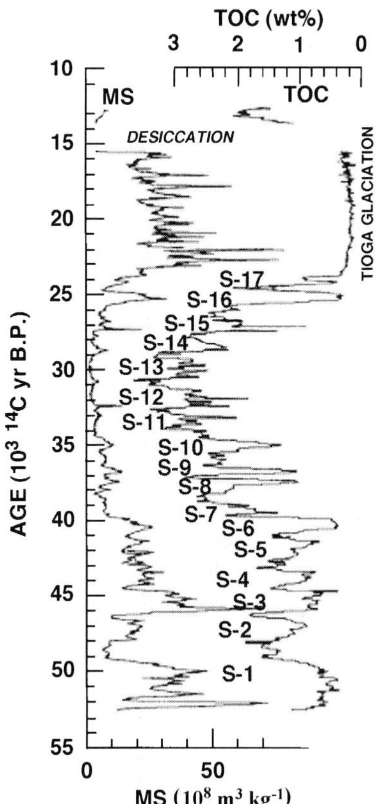

Fig. 10. Magnetic susceptibility (MS) and total organic carbon (TOC) from Owens Lake core OL-90. MS increases and TOC decreases with glacier activity (note that TOC values increase from right to left). The core shows evidence of 17 stades (S-1 through S-17) between 52,500 and 18,500 ^{14}C yr B.P., the start of the Tioga Glaciation (after Benson et al., 1998a, Fig. 10).

Equilibrium-Line Altitude

ELAs calculated using the accumulation area ratio (AAR) method and the highest-lateral moraine technique for paleoglaciers in 70 eastern Sierra Nevada valleys between 36.5° and 39.5°N are summarized in Table 2 (Gillespie, 1991). The modern ELA is taken to be 3860 m at 37°N (Burbank, 1991) based on selected glaciers, but the "true" value may be 100–200 m (e.g. Meierding, 1982) higher. If Flint's (1957) estimated climatic snowline is accepted, the difference would be even greater. Because firn limits rise to the east, the ELAs obtained with the AAR method are higher than they would be if there was no precipitation gradient.

At 37°N, the Tioga ELAs were ~820 m lower than for modern, sheltered cirque glaciers, or 920–1020 m (or even 1360 m) below the modern climatological ELA. They rose, on average, 3.1 ± 0.2 m/km toward the south. ELAs for earlier late Pleistocene glaciers were 45–195 m lower than for the Tioga glaciers. Tenaya ELAs were 45 m lower than Tioga, and Tahoe II were 95 m lower. Even for Sherwin glaciers, where evidence is best preserved in Bridgeport Basin (Sharp, 1972), ELAs were perhaps no more than 100 m below those of the Tahoe glaciers (Gillespie *et al.*, 2003). The overall spatial pattern of ELA depression is consistent with the generalized regional patterns for cirque elevations reported by Porter *et al.* (1983).

That ELA depressions should be increasingly greater for those older glaciations for which moraines were not obliterated by younger glaciers is not surprising. What is surprising is that the ELAs dropped time and again to within 10–20% of their MIS-2 values, even though the pattern of sea-level variation inferred from marine cores suggests that the high-latitude ice sheets were much larger during MIS 2 (Tioga) than during MIS 3–4 (Tenaya, Tahoe II) (e.g. Martinson *et al.*, 1987). Gillespie & Molnar (1995) emphasized this discrepancy, but Shackleton (2000) pointed out that the sawtooth history of ice volume inferred from the marine cores was due, more than previously suspected, to effects of cold water. Consequently, James *et al.* (2002) suggested that the record of mountain and high-latitude glaciations was more similar than Gillespie & Molnar (1995) suspected. Nevertheless, there does appear to be some fundamental, possibly climatic, limit to maximum ELA depressions over the last 800,000 years.

Glaciation of Hawaii

In the chain of volcanoes that comprise the Hawaiian Islands, only Mauna Kea (4206 m) on the island of Hawaii has an unequivocal record of multiple glaciation (Porter, 1979a) (Fig. 11). Haleakala volcano (3052 m) on Maui probably was glaciated repeatedly during the middle Pleistocene (Moore

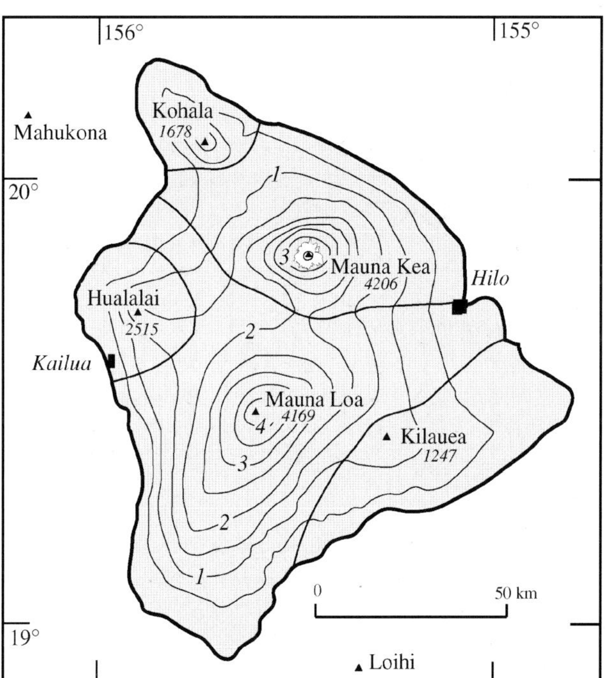

Fig. 11. Map of the island of Hawaii showing boundaries of the five main volcanoes. The Makanaka ice cap at last glacial maximum is shown on upper slopes of Mauna Kea. Altitude contour lines are in km.

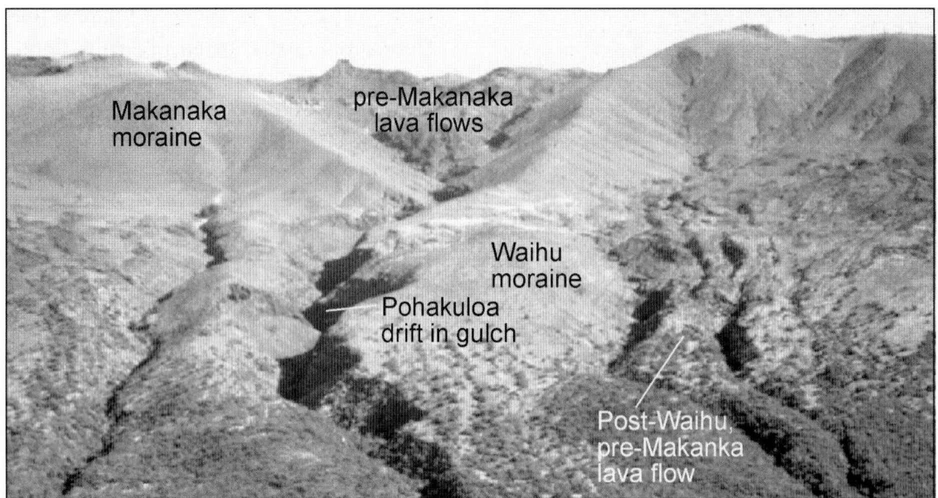

Fig. 12. Southern slope of Mauna Kea near Pohakuloa gulch showing Makanaka and Waihu moraines and interstratified lava flows.

et al., 1993), but details of its glacial history have yet to be elaborated. Mauna Loa (4169 m), Hawaii's second-highest summit, likely had a small ice cap during the last glacial maximum, but if so, the record lies buried beneath Holocene lavas that mantle the upper slopes.

The limit of the last (Makanaka) ice cap (ca. 70 km^2) on Mauna Kea is marked by a discontinuous terminal moraine or moraine complex as well as the upper limits of erratic boulders on cinder cones that lie within the glacial limit (Porter, 1979b) (Fig. 12). Upslope from the lobate outer moraines, striated bedrock surfaces are interspersed with loose bouldery drift.

Beyond the Makanaka drift limit, and below pre-Makanaka lava flows, remnants of an older moraine system are found on the southern flank of the volcano (Fig. 1). Although the area of this (Waihu) ice cap was greater than the Makanaka ice cap, its extent can only be estimated (ca.

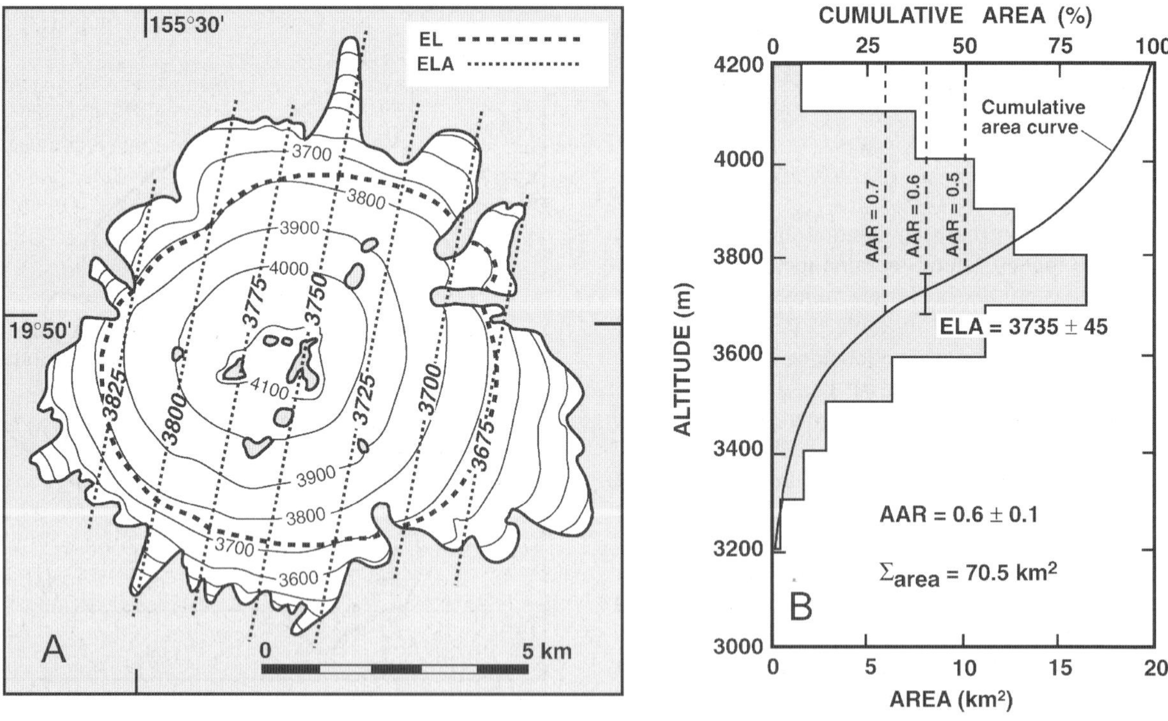

Fig. 13. (A) Reconstructed Makanaka ice cap showing topography, eastward-sloping gradient of the equilibrium line (EL), and the changing altitude (m) of the equilibrium line (ELA) around the mountain. (B) Histogram and cumulative curve showing altitudinal distribution of glacier area and the mean ELA based on an accumulation-area ratio (AAR) of 0.6 ± 0.1.

150 km^2). Still-older drift (Pohakuloa) is exposed mainly in several gullies on the southern slope of the mountain beneath pre-Waihu lavas.

The chronology of Mauna Kea glaciation is based largely on K/Ar dates of associated lava flows, ^{36}Cl surface-exposure ages of several moraine boulders, and several limiting radiocarbon ages. Due to the low K content of sub-Pohakuloa lavas, K/Ar dates have large error ranges, but those for younger alkalic lavas are smaller. The dates indicate that Pohakuloa drift likely correlates with MIS 6, Waihu drift with MIS 4, and Makanaka drift with MIS 2 (Wolfe *et al.*, 1997). Two Makanaka moraine boulders have ^{36}Cl ages of 18,900 and 20,300 years (Dorn *et al.*, 1991). A ^{36}Cl date of 14,700 years near the summit of the mountain and radiocarbon dates from the base of a lacustrine section in Lake Waihu (3968 m) provide minimum ages of close to 15,000 years for deglaciation of the summit.

The Makanaka ELA sloped gently toward the SSE and lies, on average, ca. 470 m lower than the present summit (corrected for isostatic subsidence; Porter, 1979b) (Fig. 13). The Waihu ELA can be reconstructed only on the southern flank of the volcano, where it lay ca. 675 m below the summit during that glaciation. The volcano is now too low to sustain a glacier. If the ELA of the last glacial maximum lay close to the modern July freezing isotherm (ca. 4715 m), then full-glacial ELA depression was about 935 ± 135 m.

Challenges for the Future

Whereas the extent of Late Quaternary glaciers is now reasonably well known in much of the western Cordillera, some major gaps and uncertainties remain. Extensive detailed mapping and study of glacial deposits in many Alaskan ranges, the Oregon Cascades, and the western Sierra Nevada remain to be done. Most extant chronologies still rely on minimum limiting (Holocene) radiocarbon dates and/or tephra layers. Cosmogenic isotope dating methods have yet to be applied widely, but hold considerable promise for dating and correlating Late Pleistocene moraines throughout the western mountains.

The stratigraphic resolution and dating of Middle and Early Pleistocene glaciations also remain a major problem. In many drainages, multiple pre-Late Pleistocene moraines exist, but they are poorly studied and none are closely dated. Relative-age criteria offer poor resolution for these old degraded deposits, and dating them by cosmogenic isotope methods at present seems problematic. The best hope for accurate age estimates and limits, especially for drift older than ca. 50,000 years, is further dating of till-bracketing lava flows, which are known from a few localities in Washington, the Sierra Nevada, Alaska, and Hawaii.

Detailed studies of Holocene moraines have been carried out in a few key areas in Alaska and west of the Rocky Mountains, but in many areas the existing chronology is tentative and incomplete. Without more data, it will not be possible to assess whether Neoglacial advances were synchronous along the crests of the Cascades and the Sierra Nevada. Because climate of the Cordillera varies both latitudinally and longitudinally, alpine glaciers may have responded more to local climate than to average regional climate.

References

Adam, D.P. (1966). Osgood Swamp (C-14) date A-545. *Radiocarbon*, **8**, 10.

Adam, D.P. (1967). Late Pleistocene and Recent palynology in the central Sierra Nevada. *In*: Cushing, E.J. & Wright, H.E., Jr. (Eds), *Quaternary Paleoecology; INQUA Congress VII, Proc. 7*. Yale University Press, 275–300.

Antevs, E. (1938). Postpluvial climatic variations in the southwest. *American Meteorological Society Bulletin*, **19**, 190–193.

Antevs, E. (1948). The Great Basin, with emphasis on glacial and post-glacial times – Climatic changes and pre-white man. *Bulletin of the University of Utah Biology Series*, **38**, 168–191.

Bailey, R.A., Dalrymple, G.B. & Lanphere, M.A. (1976). Volcanism, structure, and geochronology of Long Valley caldera, Mono County, California. *Journal of Geophysical Research*, **81**, 725–744.

Barclay, D.J., Calkin, P.E. & Wiles, G.C. (2001). Holocene history of Hubbard Glacier in Yakutat Bay and Russell Fiord, southern Alaska. *Geological Society of America Bulletin*, **113**, 8–402.

Begét, J.E. (1981). Early Holocene glacier advance in the North Cascade Range, Washington. *Geology*, **9**, 409–413.

Begét, J.E. (2001). Continuous late Quaternary Proxy climate records from loess in Beringia. *Quaternary Science Reviews*, **20**, 499–508.

Benson, L.V., Burdett, J.W., Kashgarian, M., Lund, S.P., Phillips, F.M. & Rye, R.O. (1996). Climatic and hydrologic oscillations in Owens Lake basin and adjacent Sierra Nevada, California. *Science*, **274**, 746–751.

Benson, L.V., May, H.M., Antweiler, R.C. & Brinton, T.I. (1998a). Continuous lake-sediment records of glaciation in the Sierra Nevada between 52,600 and 12,500 ^{14}C yr B.P. *Quaternary Research*, **50**, 113–127.

Benson, L.V., Lund, S.P., Burdett, J.W., Kashgarian, M., Rose, T.P., Smoot, J.P. & Schwartz, M. (1998b). Correlation of Late-Pleistocene lake-level oscillations in Mono Lake, California, with North Atlantic climate events. *Quaternary Research*, **49**, 1–10.

Bigelow, N.H. & Edwards, M.E. (2001). A 14,000 yr paleoenvironmental record from Windmill Lake, central Alaska: Late glacial and Holocene vegetation change in the Alaska Range. *Quaternary Science Reviews*, **20**, 203–216.

Birkeland, P.W., Burke, R.M., & Yount, J.C. (1976). Preliminary comments on late Cenozoic correlations on the Sierra Nevada. *In*: Mahoney, W.C. (Ed.), *Quaternary Stratigraphy of North America*. Stroudsburg, PA, Dowden, Hutchinson and Ross, 283–295.

Birkeland, P.W. & Burke, R.M. (1988). Soil catena chronosequences on eastern Sierra Nevada moraines, California, USA. *Arctic and Alpine Research*, **20**, 473–484.

Birkeland, P.W., Burke, R.M. & Walker, A.L. (1980). Soils and sub-surface weathering features of Sherwin and pre-Sherwin glacial deposits, eastern Sierra Nevada, California. *Geological Society of America Bulletin*, **91**, 238–244.

Birman, J.H. (1964). Glacial geology across the crest of the Sierra Nevada, California. *Geological Society of America Special Paper*, 75, 80 pp.

Bischoff, J.L. & Cummins, K. (2001). Wisconsin glaciation of the Sierra Nevada (79,000–15,000 yr B.P.) as recorded by rock flour in sediments of Owens Lake, California. *Quaternary Research*, **55**, 14–24.

Bischoff, J.L., Meisling, K.M., Fitts, J.P. & Fitzpatrick, J.A. (1997). Climatic oscillations 10,000–155,000 yr B.P. at Owens Lake, California reflected in glacial rock flour abundance and lake salinity in Core OL-92. *Quaternary Research*, **48**, 313–325.

Blackwelder, E. (1931). Pleistocene glaciation in the Sierra Nevada and Basin Ranges. *Geological Society of America Bulletin*, **42**, 865–922.

Brigham-Grette, J. (2001). New perspectives on Beringian Quaternary paleogeography, stratigraphy and glacial history. *Quaternary Science Reviews*, **20**, 15–24.

Brigham-Grette, J., Hopkins, D.M., Ivanov, V.F., Basilyan, A.E., Benson, S.L., Heiser, P.A. & Pushkar, V.S. (2001). Last interglacial (isotope stage 5) glacial and sea-level history of costal Chukotka Peninsula and St. Lawrence Island, western Beringia. *Quaternary Science Reviews*, **20**, 419–436.

Briner, J.P., Kaufman, D.S., Werner, A., Caffee, M., Levy, L., Kaplan, M.R. & Finkel, R.C. (2002). Glacier readvance during the late glacial (Younger Dryas?) in the Ahklun Mountains, southwestern Alaska. *Geology*, **30**, 679–682.

Briner, J.P. & Kaufman, D.S. (2000). Late Pleistocene glacial history of the southwestern Ahklun Mountains, Alaska. *Quaternary Research*, **53**, 13–22.

Briner, J.P., Swanson, T.W. & Caffee, M. (2001). Late Pleistocene cosmogenic ^{36}Cl glacial chronology of the southwestern Ahklun Mountains, Alaska. *Quaternary Research*, **56**, 148–154.

Burbank, D.W. (1981). A chronology of Late Holocene glacier fluctuations on Mount Rainier, Washington. *Arctic and Alpine Research*, **13**, 369–386.

Burbank, D. (1991). Late Quaternary snowline reconstructions for the southern and central Sierra Nevada, California: Reassessment of the "Recess Peak glaciation". *Quaternary Research*, **36**, 294–306.

Burke, R.M. & Birkeland, P.W. (1979). Re-evaluation of multiparameter relative dating techniques and their application to the glacial sequence along the eastern escarpment of the Sierra Nevada, California. *Quaternary Research*, **11**, 21–51.

Bursik, M.I. & Gillespie, A.R. (1993). Late Pleistocene glaciation of Mono Basin, California. *Quaternary Research*, **39**, 24–35.

Calkin, P.E. (1988). Holocene glaciation of Alaska (and adjoining Yukon Territory, Canada). *Quaternary Science Reviews*, **7**, 159–184.

Calkin, P.E., Wiles, G.C. & Barclay, D.J. (2001). Holocene coastal glaciation of Alaska. *Quaternary Science Reviews*, **20**, 449–461.

Carson, R.J. III (1970). Quaternary Geology of the south-central Olympic Peninsula, Washington [M.S. thesis]. Seattle, University of Washington.

Carter, L.D. (1981). A Pleistocene sand sea on the Alaskan Arctic Coastal Plain. *Science*, **211**, 381–383.

Carver, G.A. (1973). Glacial geology of the Mountain Lakes Wilderness and adjacent parts of the Cascade Range, Oregon [Ph.D. thesis]. Seattle, University of Washington.

Clark, D.H. (1997). A new alpine lacustrine sedimentary record from the Sierra Nevada: Implications for Late-Pleistocene paleoclimate reconstructions and cosmogenic isotope production rates [abst]. *EOS, Transactions American Geophysical Union*, **78**(46), F249.

Clark, D.H. & Gillespie, A.R. (1997). Timing and significance of late-glacial and Holocene glaciation in the Sierra Nevada, California. *Quaternary International*, **38/39**, 21–38.

Clark, D.H., Housen, B.A. & Filippelli, G.M. (2003). A deep, pre-Late Wisconsinan bog in the central Sierra Nevada, California. *Arctic, Antarctic, and Alpine Research* (in press).

Clark, M.M. (1967). Pleistocene glaciation of the drainage of the West Walker River, Sierra Nevada, California [Ph.D. thesis]. Stanford, Stanford University, 170 pp.

Clark, M.M. (1968). Pleistocene glaciation of the upper West Walker drainage, Sierra Nevada, California (abst). *Geological Society of America Special Paper 115*, 317.

Clayton, G.A. (1983). Geology of the White Pass Area, south-central Cascade Range, Washington [M.S. thesis]. Seattle, University of Washington.

Coulter, H.W., Hopkins, D.M., Karlstrom, T.N.V., Péwé, T.L., Wahrhaftig, C. & Williams, J.R. (1965). Map showing extent of glaciations in Alaska. *U.S. Geological Survey Map I-415*, 1:2,500,000.

Crandell, D.R. (1964). Pleistocene glaciations of the southwestern Olympic Peninsula: *U.S. Geological Survey Professional Paper, 501-B*, B135–B139.

Crandell, D.R. (1965). The glacial history of western Washington and Oregon. *In*: Wright, H.E., Jr. & Frey, D.G. (Eds), *The Quaternary of the United States*. Princeton, Princeton University Press, 341–353.

Crandell, D.R. & Miller, R.D. (1965). Post-Hypsithermal glacier advances at Mount Rainier, Washington. *U.S. Geological Survey Professional Paper, 501-D*, D110–D114.

Crandell, D.R. & Miller, R.D. (1974). Quaternary stratigraphy and extent of glaciation in the Mount Rainier region, Washington, U.S. Geological Survey Professional Paper 847, 59 pp.

Curry, R.R. (1969). Holocene climatic and glacial history of the central Sierra Nevada, California. *In*: Schumm, S.A. & Bradley, W.C. (Eds), *United States Contributions to Quaternary Research*. Geological Society of America Special Paper 123, 1–47.

Curry, R.R. (1971). Glacial and Pleistocene history of the Mammoth Lakes Sierra – A geologic Guidebook. Missoula, University of Montana, Department of Geology Geological Series Publication No. 11, 49 pp.

Dalrymple, G.B. (1963). Potassium-argon ages of some Cenozoic volcanic rocks of the Sierra Nevada, California. *Geological Society of America Bulletin*, **74**, 379–390.

Dalrymple, G.B. (1964). Cenozoic chronology of the Sierra Nevada, California. *University of California Publications in Geological Science*, **47**, 41 pp.

Danskin, W.R. (1998). Evaluation of the hydrologic system and selected water-management alternatives in the Owens Valley, California. *U.S. Geological Survey Water Supply Paper 2370*, 175 pp.

Davis, P.T. & Osborn, G. (1987). Age of pre-Neoglacial cirque moraines in the central North American Cordillera. *Géographie physique et Quaternaire*, **41**, 365–375.

Detterman, R.L. (1986). Glaciation of the Alaska Peninsula. *In*: Hamilton, T.D., Reed, K.M. & Thirson, R.M. (Eds), *Glaciation in Alaska – The geologic record*. Alaska Geological Society, Anchorage, 151–170.

Dorn, R.I., Phillips, F.M., Zreda, M.G., Wolfe, E.W., Jull, A.J.T., Donahue, D.J., Kubik, P.W. & Sharma, P. (1991). Glacial chronology of Mauna Kea, Hawaii, as constrained by surface-exposure dating. *National Geographic Research and Exploration*, **7**, 456–471.

Elias, S.A. & Brigham-Grette, J. (Eds) (2001). Beringian Paleoenvironments. *Quaternary Sciences Reviews*, **20**, 574 pp.

Ferrians, O.J., Jr. (1963). Glacolacustrine diamicton deposits in the Copper River Basin, Alaska. Short Papers in Geology and Hydrology 1963. *U.S. Geological Survey Professional Paper 475C*, 121–125.

Flint, R.F. (1957). *Glacial and Pleistocene Geology*. Wiley, New York, 553 pp.

Friele, P.A. & Clague, J.J. (2002). Younger Dryas readvance in Squamish River valley, southern Coast Mountains, British Columbia. *Quaternary Science Reviews*, **21**, 1925–1933.

Froese, D.G., Barendregt, R.W., Enkin, R.J. & Baker, J. (2000). Paleomagnetic evidence for multiple Late Pliocene-Early Pleistocene glaciations in the Klondike area, Yukon Territory. *Canadian Journal of Earth Sciences*, **37**, 863–877.

Fullerton, D.S. (1986). Chronology and correlation of glacial deposits in the Sierra Nevada, California. *In*: Sibrava, V., Bowen, D.Q. & Richmond, G.M. (Eds), Quaternary Glaciations in the Northern Hemisphere, *Quaternary Science Reviews*, **5**, 161–169.

Gillespie, A.R. (1982). Quaternary Glaciation and Tectonism in the Southeastern Sierra Nevada, Inyo County, California [Ph.D. thesis]. Pasadena, CA, Caltech, 695 pp.

Gillespie, A.R. (1991). Testing a new climatic interpretation for the Tahoe glaciation. *In*: Hall, Jr., Doyle-Jones, V. & Widawski, B. (Eds), *Natural History of Eastern California and High-Altitude Research, Proceedings of the White Mountain Research Station Symposium*, **3**, 383–398.

Gillespie, A.R. & Molnar, P. (1995). Asynchronism of maximum advances of mountain and continental glaciations. *Reviews of Geophysics*, **33**, 311–364.

Gillespie, A.R., Clark, D.H., Clark, M.C. & Burke, R.M. (2003). Field Trip B-3, Mountain Glaciations of the Sierra Nevada. *In*: Easterbrook, D.J. (Ed.), *XVI INQUA Congress Field Trip Guidebook* (in press).

Gillespie, A.R., Huneke, J.C. & Wasserburg, G.J. (1984). Eruption age of a ~100,000-year-old basalt from ^{40}Ar-^{39}Ar analysis of partially degassed xenoliths. *Journal of Geophysical Research*, **89**, 1033–1048.

Glushkova, O. Yu. (2001). Geomorphological correlation of late Pleistocene glacial complexes of western and eastern Beringia. *Quaternary Science Reviews*, **20**, 405–418.

Grosswald, M.G. (1998). Late-Weichselian ice sheets in Arctic and Pacific Siberia. *Quaternary International*, **45/46**, 3–18.

Gualtieri, L., Glushkova, O. & Brigham-Grette, J. (2000). Evidence for restricted ice extent during the last glacial maximum in the Koryak Mountains of Chukotka, for eastern Russia. *Geological Society of America Bulletin*, **112**, 1106–1118.

Hallet, B. & Putkonen, J. (1994). Surface dating of dynamic landforms: young boulders on aging moraines. *Science*, **265**, 937–940.

Hamilton, T.D. & Thorson, R.M. (1983). The Cordilleran Ice Sheet in Alaska. *In*: Porter, S.C. (Ed.), *Late Quaternary environments of the United States, 1, The Late Pleistocene*, 38–52. University of Minnesota Press, Minneapolis.

Hamilton, T.D. (1986). Late Cenozoic glaciation of the central Brook Range. *In*: Hamilton, T.D., Reed, K.M. & Thorson, R.M. (Eds), *Glaciation in Alaska – The geologic record*. Alaska Geological Society, Anchorage, 9–50.

Hamilton, T.D. (1994). Late Cenozoic glaciation of Alaska. *In*: Plafker, G. & Berg, H.C. (Eds), *The Geology of Alaska*. The Geology of North America, v. G-1, Geological Society of America, Boulder, Colorado, 813–844.

Hamilton, T.D. (2001). Quaternary glacial, lacustrine, and fluvial interactions in the western Noatak basin, northwest Alaska. *Quaternary Science Reviews*, **20**, 371–391.

Hamilton, T.D. (2002). Surficial Geology of the Dalton Highway (Itkillik-Sagavanirktok Rivers) area, southern Arctic Foothills, Alaska: Alaska Department of Natural Resources, Division of Geological and Geophysical Surveys Professional Report, **121**, 32 pp.

Hamilton, T.D., Ashley, G.M., Reed, K.M. & Schweger, C.E. (1993). Late Pleistocene vertebrates and other fossils from Epiguruk, northwestern Alaska. *Quaternary Research*, **39**, 381–389.

Hamilton, T.D., Reed, K.M. & Thorson, R.M. (Eds) (1986). *Glaciation in Alaska – The geologic record, Alaska*. Geological Society, Anchorage, 265 pp.

Hammond, P.E. (1987). Lone Butte and Crazy Hills: subglacial volcanic complexes, Cascade Range, Washington. *In*: Hill, M.L. (Ed.), *Centennial Field Guide Volume 1*. Cordilleran Section of the Geological Society of America, Boulder, CO, Geological Society of America, pp. 339–344.

Heine, J.T. (1998). Extent, timing, and climatic implications of glacier advances Mount Rainier, Washington, USA, at the Pleistocene/Holocene transition. *Quaternary Science Reviews*, **17**, 1139–1148.

Heiser, P.A. & Rousch, J.J. (2001). Pleistocene glaciation in Chukotka Russia: moraine mapping using satellite synthetic aperture radar (SAR) imagery. *Quaternary Science Reviews*, **20**, 393–404.

Heusser, C.J. (1957). Variations of Blue, Hoh, and White glaciers during recent centuries. *Arctic*, **10**, 139–150.

Hirsch, R.M. (1975). Glacial geology and geomorphology of the upper Cedar River watershed, Cascade Range, Washington [M.S. thesis]. Seattle, University of Washington.

Hopkins, D.M., MacNeil, F.S., Merklin, R.L. & Petrov, O.M. (1965). Quaternary correlations across Bering Strait. *Science*, **147**, 1107–1114.

Hopkins, K.D. (1966). Glaciation of Ingalls Creek valley, east-central Cascade Range, Washington [M.S. thesis]. Seattle, University of Washington.

Hopkins, K.D. (1976). Geology of the south and east slopes of Mount Adams volcano, Cascade Range, Washington [Ph.D. thesis]. Seattle, University of Washington.

Hu, F.S., Lee, B.Y., Kaufman, D.S., Yoneji, S., Nelson, D. & Henne, P. (2002). Response of tundra ecosystem in southwestern Alaska to Younger-Dryas climatic oscillation. *Global Change Biology*, **8**, 1156–1163.

Huber, N.K. (1981). Amount and timing of Cenozol uplift and tilt of the central Sierra Nevada, California – Evidence from the upper Sanjuaquin River. *U.S. Geological Survey Professional Paper 1197*, 28 pp.

Hurley, T.M. (1996). Late-glacial equilibrium-line altitudes of glaciers in the southern North Cascade Range, Washington [M.S. thesis]. Seattle, University of Washington.

Huston, M.M., Brigham-Grette, J. & Hopkins, D.M. (1990). Paleogeographic significance of middle Pleistocene glaciomarine deposits on Baldwin Peninsula, northwest Alaska. *Annals of Glaciology*, **14**, 111–114.

Hyde, J.H. (1973). Late Quaternary volcanic stratigraphy, south flank of Mount St. Helens, Washington [Ph.D. thesis]. Seattle, University of Washington.

James, L.A., Harbor, J., Fabel, D., Dahms, D. & Elmore, D. (2002). Late Pleistocene glaciations in the northwestern Sierra Nevada, California. *Quaternary Research*, **57**, 409–419.

Karlstrom, T.N.V. (1964). Quaternary geology of the Kenai lowland and glacial history of the Cook Inlet region, Alaska. *U.S. Geological Survey Professional Paper 443*, 69 pp.

Karlstrom, T.N.V. & others (1964). Surficial geology of Alaska. *U.S. Geological Survey Miscellaneous Geologic Investigations Map I-357*, 1:1,584,000.

Kaufman, D.S. & Hopkins, D.M. (1986). Glacial history of the Seward Peninsula. *In*: Hamilton, T.D., Reed, K.M. & Thorson, R.M. (Eds), *Glaciation in Alaska – The geologic record*. Alaska Geological Society, Anchorage, 51–78.

Kaufman, D.S., Forman, S.L., Lea, P.D. & Wobus, C.W. (1996). Age of pre-late-Wisconsin glacial-estuarine sedimentation, Bristol Bay, Alaska. *Quaternary Research*, **45**, 59–72.

Kaufman, D.S. & Thompson, C.H. (1998). Re-evaluation of pre-late-Wisconsin glacial deposits, lower Naknek valley, southwestern Alaska. *Arctic and Alpine Research*, **30**, 142–153.

Kaufman, D.S., Hu, F.S., Briner, J.P., Werner, A., Finney, B.P. & Gregory-Eave, I. (in press). A ~33,000 year record of environmental change from Arolik Lake, Ahklun Mountains, Alaska. *Paleolimnology*.

Kaufman, D.S., Manley, W.F., Forman, S.L., Hu, F.S., Preece, S.J., Westgate, J.A. & Wolfe, A.P. (2001b). Paleoenvironment of the last interglacial-to-glacial transition, Togiak Bay, southwestern Alaska. *Quaternary Research*, **55**, 190–202.

Kaufman, D.S., Manley, W.F., Forman, S.L. & Layer, P. (2001a). Pre-late-Wisconsin glacial history, coastal Ahklun Mountains, southwestern Alaska – New amino acid, thermoluminescence, and ^{40}Ar/^{39}Ar results. *Quaternary Science Reviews*, **20**, 337–352.

Kline, J.T. & Bundtzen, T.K. (1986). Two glacial records from west-central Alaska. *In*: Hamilton, T.D., Reed, K.M. & Thorson, R.M. (Eds), *Glaciation in Alaska – The geologic record*. Alaska Geological Society, Anchorage, 123–150.

Knoll, K.M. (1967). Surficial geology of the Tolt River area, Washington [M.S. thesis]. Seattle, University of Washington.

Konrad, S.K. & Clark, D.H. (1998). Evidence for an early Neoglacial glacier advance from rock glaciers and lake sediments in the Sierra Nevada, California, USA. *Arctic and Alpine Research*, **30**, 272–284.

Kovanen, D.J. & Easterbrook, D.J. (2001). Late Pleistocene, post-Vashon, alpine glaciation of the Nooksack drainage, North Cascades, Washington. *Geological Society of America Bulletin*, **113**, 274–288.

Krimmel, R.M. & Meier, M.F. (1989). Glaciers and glaciology of Alaska. *In*: *28th International Geological Congress Field Trip Guidebook T301*. American Geophysical Union, 61 pp.

Lea, P.D. (1990). Pleistocene glacial tectonism and sedimentation on a macrotidal piedmont coast, Ekuk Bluffs, southwestern Alaska. *Geological Society of America Bulletin*, **102**, 1230–1245.

Lea, P.D. & Waythomas, C.F. (1990). Late-Pleistocene eolian sand sheets in Alaska. *Quaternary Research*, **34**, 269–281.

Levy, L.B., Kaufman, D.S. & Werner, A. (in press). Neoglaciation recorded in proglacial Waskey Lake, Ahklun Mountains, southwestern Alaska. *The Holocene*.

Manley, W.F. & Kaufman, D.S. (1999). Using GIS to quantify ELA's and winter precipitation anomalies for Pleistocene valley glaciers: A case study from southwestern Alaska. *EOS*, **80**(46), F320.

Manley, W.F. & Kaufman, D.S. (2002). Alaska PaleoGlacier Atlas: Institute of Arctic and Alpine Research (INSTAAR), University of Colorado, http://instaar.colorado.edu/QGISL/ak_paleoglacier_atlas, 1.

Manley, W.F., Kaufman, D.S. & Briner, J.P. (2001). Late Quaternary glacial history of the southern Ahklun Mountains, southeast Beringia – Soil development, morphometric, and radiocarbon constraints. *Quaternary Science Reviews*, **20**, 353–370.

Mann, D.H. & Hamilton, T.D. (1995). Late Pleistocene and Holocene paleoenvironments of the north Pacific coast. *Quaternary Science Reviews*, **14**, 441–471.

Mann, D.H. (1986). Wisconsin and Holocene glaciation of southeast Alaska. *In*: Hamilton, T.D., Reed, K.M. & Thorson, R.M. (Eds), *Glaciation in Alaska – The geologic record*. Alaska Geological Society, Anchorage, 237–265.

Mann, D.H., Heiser, P.A. & Finney, B.P. (2002). Holocene history of the Great Kobuk sand dunes, northwestern Alaska. *Quaternary Science Reviews*, **21**, 709–731.

Mann, D.H., Peteet, D.M., Reanier, R.E. & Kunz, M.L. (2001). Responses of an arctic landscape to late glacial and early Holocene climatic changes, the importance of moisture. *Quaternary Science Reviews*, **21**, 997–1021.

Martinson, D.G., Pisia, N.G., Hayes, J.D., Imbrie, J., Moore, T.C., Jr. & Shackleton, N.J. (1987). Age dating and the orbital theory of the ice ages development of a high-resolution 0 to 300,000-year chronostratigraphy. *Quaternary Research*, **27**, 1–29.

Mathieson, S.A. & Sarna-Wojcicki, A.M. (1982). Ash layer in Mohawk Valley, Plumas County, California, correlated with the 0.45 M.Y.-old Rockland ash – Implications for the glacial and lacustrine history of the region [abst]. *Geological Society of America Abstracts with Program*, **14**(2), 184.

Meierding, T.C. (1982). Late Pleistocene glacial equilibrium-line altitudes in the Colorado Front range: A comparison of methods. *Quaternary Research*, **18**, 289–310.

Merrill, D.E. (1966). *Glacial geology of the Chiwaukum Creek drainage basin and vicinity*. Seattle, University of Washington.

Miller, C.D. (1969). Chronology of Neoglacial moraines in the Dome Peak area, North Cascade Range, Washington. *Arctic and Alpine Research*, **1**, 49–66.

Miller, D.J. (1957). Geology of the southeastern part of the Robinson Mountains, Yakataga district, Alaska. *U.S. Geological Survey Oil and Gas Inventory Map OM-187.*

Molnia, B.F. (1986). Glacial history of the northeastern Gluf of Alaska, a synthesis. *In*: Hamilton, T.D., Reed, K.M. & Thorson, R.M. (Eds), *Glaciation in Alaska – The geologic record*. Alaska Geological Society, Anchorage, 219–236.

Moore, J.G., Porter, S.C. & Mark, R. (1993). Glaciation of Haleakala volcano, Hawaii [abst]. *Geological Society of America Abstracts with Programs*, **25**, 123–124.

Moore, J.L. (1965). Surficial Geology of the Southwestern Olympic Peninsula [M.S. thesis]. Seattle, University of Washington.

Mundorf, M.J. (1984). Glaciation in the lower Lewis River Basin, southwestern Cascade Range, Washington. *Northwest Science*, **58**, 269–281.

Nimick, D.A. (1977). Glacial geology of Lake Wenatchee and vicinity, Washington [M.S. thesis]. Seattle, University of Washington.

Nishiizumi, K., Winterer, E.L., Kohl, C.P., Klein, J., Middleton, R., Lal, D. & Arnold, J.R. (1989). Cosmic ray production rates of ^{10}Be and ^{26}Al in quartz from glacially polished rocks. *Journal of Geophysical Research*, **94**, 17907–17915.

Page, B.M. (1939). Multiple alpine glaciation in the Leavenworth area, Washington. *Journal of Geology*, **47**, 785–815.

Péwé, T.L. (1975). The Quaternary Geology of Alaska. *U.S. Geological Survey Professional Paper 385*, 145 pp.

Péwé, T.L. (Ed.) (1965). Guidebook to the Quaternary geology of central and south-central Alaska. International Association for Quaternary Research VIIth Congress Field Conference F, 141 pp.

Péwé, T.L., Hopkins, D.M. & Giddings, J.L. (1965). Quaternary geology and archeology of Alaska. *In*: Wright, H.E., Jr. & Frey, D.G. (Eds), *The Quaternary of the United States*, Princeton University Press, Princeton, New Jersey, 355–374.

Péwé, T.L. (Ed.) (1953). Multiple glaciation in Alaska – A progress report. *U.S. Geological Survey Circular 289*, 13 pp.

Phillips, F.M., Stone, W.D. & Fabryka-Martin, J.T. (2001). An improved approach to calculating low-energy cosmic-ray neutron fluxes at the land/atmosphere interface. *Chemical Geology*, **175**, 689–701.

Phillips, F.M., Zreda, M.G., Benson, L.V., Plummer, M.A., Elmore, D. & Sharma, P. (1996). Chronology for fluctuations in late Pleistocene Sierra Nevada glaciers. *Science*, **274**, 749–751.

Phillips, F.M., Zreda, M.G., Smith, S.S., Elmore, D., Kubik, P.W. & Sharma, P. (1990). Cosmogenic chlorine-36 chronology for glacial deposits at Bloody Canyon, eastern Sierra Nevada. *Science*, **248**, 1529–1532.

Plafker, G. & Addicott, W.O. (1976). Glaciomarine deposits of Miocene through Holocene age in the Yakataga Formation along the Gulf of Alaska margin, Alaska. *In*: Miller, T.P. (Ed.), *Recent and Ancient Sedimentary Environments in Alaska*. Alaska Geological Society, Anchorage, Q1–Q23.

Plummer, M. (2002). Paleoclimatic conditions during the last deglaciation inferred from combined analysis of pluvial and glacial records [Ph.D. thesis]. Socorro, New Mexico Institute of Mining and Technology.

Porter, S.C. (1976). Pleistocene glaciation in the southern part of the North Cascade Range, Washington. *Geological Society of America Bulletin*, **87**, 61–75.

Porter, S.C. (1978). Glacier Peak tephra in the North Cascade Range, Washington: stratigraphy, distribution, and relationship to late-glacial events. *Quaternary Research*, **10**, 30–41.

Porter, S.C. (1979a). Quaternary stratigraphy and chronology of Mauna Kea, Hawaii: a 380,000-yr record of mid-Pacific volcanism and ice-cap glaciation. *Geological Society of America Bulletin*, 90, Part I: Summary, 609–611; Part II: Complete article, 980–1093.

Porter, S.C. (1979b). Hawaiian glacial ages. *Quaternary Research*, **12**, 161–187.

Porter, S.C. (1981). Lichenometric studies in the Cascade Range of Washington: establishment of *Rhizocarpon geographicum* growth curves at Mount Rainier. *Arctic and Alpine Research*, **13**, 11–23.

Porter, S.C. (2001). Snowline depression in the tropics during the last glaciation. *Quaternary Science Reviews*, **20**, 1067–1091.

Porter, S.C., Pierce, K.L., & Hamilton, T.D. (1983). Late Wisconsin mountain glaciation in the western United States. *In*: Porter, S.C. (Ed.), *Late Quaternary environments of the United States, Volume 1, The Late Pleistocene*, 71–111. University of Minnesota Press, Minneapolis.

Pushkar, V.S., Roof, S.R., Cherepanova, M.V., Hopkins, D.M. & Brigham-Grette, J. (1999). Paleogeographic and paleoclimatic significance of diatoms from middle Pleistocene marine and glaciomarine deposits on

Baldwin Peninsula, northwestern Alaska. *Palaeogeography, Palaeoclimatology, and Palaeoecology*, **152**, 67–85.

Reger, R.D. & Pinney, D.S. (1996). Late Wisconsin glaciation of the Cook Inlet region with emphasis on Kenai lowland and implications for early peopling. *In*: Davis, N.Y. & Davis, W.E. (Eds), *The Anthropology of Cook Inlet: Proceedings from a Symposium*. Cook Inlet Historical Society, Anchorage, 5–23.

Reger, R.D., Combellick, R.A. & Brigham-Grette, J. (1995). Late-Wisconsin events in the upper Cook Inlet region, southcentral Alaska. *In*: Combellick, R.A. & Tannian, F. (Eds), *Short notes on Alaska geology 1995*. Alaska Division of Geological and Geophysical Surveys Professional Report 117, 33–45.

Sarna-Wojcicki, A.M., Meyer, C.E., Bowman, H.R., Hall, N.T., Russell, P.C., Woodward, M.J. & Slate, J.L. (1985). Correlation of the Rockland ash bed, a 400,000-year-old stratigraphic marker in northern California and western Nevada, and implications for middle Pleistocene paleogeography of central California. *Quaternary Research*, **23**, 236–257.

Sarna-Wojcicki, A.M., Pringle, M.S. & Wijbrans, J. (2000). New $^{40}Ar/^{39}Ar$ age of the Bishop Tuff from multiple sites and sediment rate calibration for the Matayama-Bruhnes boundary. *Journal Geophysical Research*, **105**, 21,431–21,433.

Schmoll, H.R. & Yehle, L.A. (1986). Pleistocene glaciation of the upper Cook Inlet basin. *In*: Hamilton, T.D., Reed, K.M. & Thorson, R.M. (Eds), *Glaciation in Alaska – The geologic record*. Alaska Geological Society, Anchorage, 193–218.

Schmoll, H.R., Yehle, L.A. & Updike, R.G. (1999). Summary of Quaternary geology of the Municipality of Anchorage, Alaska. *Quaternary International*, **60**, 3–36.

Scott, W.E. (1977). Quaternary glaciation and volcanism, Metolius River area, Oregon. *Geological Society of America Bulletin*, **88**, 113–124.

Scuderi, L.A. (1987). Glacier variations in the Sierra Nevada, California, as related to a 1200-year tree-ring chronology. *Quaternary Research*, **27**, 220–231.

Sharp, R.P. (1968). Sherwin Till – Bishop Tuff relationship, Sierra Nevada, California. *Geological Society of America Bulletin*, **79**, 351–364.

Sharp, R.P. (1972). Pleistocene glaciation, Bridgeport Basin. *Geological Society of America Bulletin*, **83**, 2233–2260.

Sharp, R.P. & Birman, J.H. (1963). Additions to the classical sequence of Pleistocene glaciations, Sierra Nevada, California. *Geological Society of America Bulletin*, **74**, 1079–1086.

Sigafoos, R.S. & Hendricks, E.L. (1972). Recent activity of glaciers of Mount Rainier, Washington. *U.S. Geological Survey Professional Paper 387-B*, 24 pp.

Smith, G.I. (1983). Core KM-3, a surface-to-bedrock record of Late Cenozoic sedimentation in Searles Valley, California. *U.S. Geological Survey Professional Paper 1256*, 1–24.

Spicer, R.C. (1989). Recent variations of Blue Glacier, Olympic Mountains, Washington, USA. *Arctic and Alpine Research*, **21**, 1–21.

Stilwell, K.B. & Kaufman, D.S. (1996). Late Wisconsin glacial history of the Iliamna/Naknek/ Brooks Lake area, southwestern Alaska. *Arctic and Alpine Research*, **28**, 475–487.

Stine, S. (1994). Extreme and persistent drought in California and Patagonia during mediaeval time. *Nature*, **369**, 546–549.

Swanson, T.W. & Porter, S.C. (1997). Cosmogenic isotope ages of moraines in the southeastern North Cascade Range: Seattle, Pacific Northwest Friends of the Pleistocene Field Excursion Guidebook.

Swanson, T.W. & Porter, S.C. (2000). 36-Cl evidence for maximum late Pleistocene glacier extent in the Cascade Range during marine isotope substage 5d [abst]. *Geological Society of America Abstracts with Program*, **32**, 472.

Thackray, G.D. (2001). Extensive early and middle Wisconsin glaciation on the western Olympic Peninsula, Washington, and the variability of Pacific moisture delivery to the northwestern United States. *Quaternary Research*, **55**, 257–270.

Thomas, P.A., Easterbrook, D.J. & Clark, P.U. (2000). Early Holocene glaciation on Mount Baker, Washington State, USA. *Quaternary Science Reviews*, **19**, 1043–1046.

Thorp, P.W. (1985). *The Glacial Geology of the Entiat Valley in the Eastern North Cascade Range, Washington, USA*. London, Goldsmith's Company, London, England, p. 59.

Thorson, R.M. (1986). Late Cenozoic glaciation of the northern Nenana River valley. *In*: Hamilton, T.D., Reed, K.M. & Thorson, R.M. (Eds), *Glaciation in Alaska – The geologic record*. Alaska Geological Society, Anchorage, 99–122.

Wagner, D.L., Jennings, C.W., Bedrossian, T.L., Bortugno, E.J. & Saucedo, G.J. (1982). Sacramento Quadrangle, California: *California Division of Mines and Geology Regional Geologic Map Series*, scale 1:250,000.

Waitt, R.B., Jr., Yount, J.C. & Davis, P.T. (1982). Regional significance of an early Holocene moraine in Enchantments Lakes Basin, North Cascade Range, Washington. *Quaternary Research*, **17**, 191–210.

Warhaftig, C. & Birman, J.H. (1965). The Quaternary of the Pacific mountain system in California. *In*: Wright, H.E., Jr. & Frey, D.G. (Eds), *The Quaternary of the United States*. Princeton University Press, 299–340.

Weber, F.R. (1986). Glacial geology of the Yukon-Tanana Upland. *In*: Hamilton, T.D., Reed, K.M. & Thorson, R.M. (Eds), *Glaciation in Alaska – The geologic record*. Alaska Geological Society, Anchorage, 79–98.

Westgate, J.A., Preece, S.J., Froese, D.G., Walter, R.C., Sandhu, A.S. & Schweger, C.E. (2001). Dating early and middle (Reid) Pleistocene glaciations in central Yukon by tephrochronology. *Quaternary Research*, **56**, 335–356.

Wiles, G.C. & Calkin, P.E. (1994). Late Holocene, high-resolution glacial chronologies and climate, Kenai Mountains, Alaska. *Geological Society of America Bulletin*, **106**, 281–303.

Wiles, G.C., Jacoby, G.C., Davi, N.K. & McAllister, R.P. (2002). Late Holocene glacial fluctuations in the Wrangell Mountains, Alaska. *Geological Society of America Bulletin*, **114**, 896–908.

Wiles, G.C., Barclay, D.J. & Calkin, P.E. (1999). Tree-ring dated Little Ice Age histories of maritime glaciers from western Prince William Sound, Alaska. *The Holocene*, **9**, 163–173.

Williams, V.S. (1971). Glacial geology of the drainage basin of the Middle Fork of the Snoqualmie River. M.S. thesis, University of Washington.

Wilson, F.H. & Weber F.R. (2001). Quaternary geology of the Cold Bay and False Pass Quadrangles. *In*: Gough, L.P. & Wilson, F.H. (Eds), Geologic studies in Alaska by U.S. Geological Survey, 1999. *U.S. Geological Survey Professional Paper 1633*, 51–71.

Winograd, I.J., Coplen, T.B., Landwehr, J.M., Riggs, A.C., Ludwig, K.L., Szabo, B.J., Kolesar, P.T. & Revesz, K.M. (1992). Continuous 500,000-year climate record from vein calcite in Devils Hole, Nevada. *Science*, **258**, 255–260.

Wolfe, E.W., Wise, W.S. & Dalrymple, G.B. (1997). The geology and petrology of Mauna Kea volcano, Hawaii – a study of postshield volcanism. *U.S. Geological Survey Professional Paper 1557*, 129 pp.

Wood, S.H. (1977). Distribution, correlation, and radiometric dating of late Holocene tephra, Mono and Inyo craters, eastern California. *Geological Society of American Bulletin*, **88**, 89–95.

Yount, J.C., Birkeland, P.W. & Burke, R.M. (1982). Holocene glaciation, Mono Creek, central Sierra Nevada, California [abst]. *Geological Society of America Abstracts with Program*, **14**, 246.

Coupling ice-sheet and climate models for simulation of former ice sheets

Shawn J. Marshall[1], David Pollard[2], Steven Hostetler[3] and Peter U. Clark[4]

[1] *Department of Geography, University of Calgary, Calgary AB, Canada, Tel: 403-220-4884; Fax: 403-282-6561; shawn.marshall@ucalgary.ca*
[2] *Pennsylvania State University, University Park, PA, USA*
[3] *United States Geological Survey, Corvallis, OR, USA*
[4] *Department of Geosciences, Oregon State University, Corvallis, OR, USA*

Introduction

Many important intervals in Earth's paleoclimatic record are distinguished by periods of climatic, environmental, and biotic changes involving ice sheets. Some examples are late-Proterozoic Snowball Earth events between 750 and 590 myr ago (Hoffman & Schrag, 2000), initiation of the Antarctic ice sheet and its subsequent oscillations during the Oligocene, ~34–30 myr ago (Kennett, 1977), orbital-scale Northern Hemisphere glacial cycles of the last 3 myr (Imbrie *et al.*, 1993), and millennial scale climate changes embedded in glacial cycles (Bond *et al.*, 1993). Numerical models of the Earth system are needed to examine these global-scale events, including ice, ocean, atmosphere, and land surface processes and the biogeochemical, hydrological, and energy exchanges that couple them. Due to the inherently different time scales of various Earth system components, however, it is neither feasible nor desirable to simulate all components of the Earth system simultaneously. Instead, numerical models require asynchronous coupling strategies, with explicit prognostic models of the long-term components integrated continuously for 10^4–10^7 years and updated periodically by simulations of the short-term components.

In this chapter we review the development of coupled climate and ice-sheet models over the last two decades, discuss the current technical and physical capabilities of models, and identify future work for developing a better understanding of ice-climate events that have punctuated Earth history. To place the demands and needs for ongoing model development into context, we begin with a summary of late Cenozoic climate events and the role of ice sheets in long-term climate change. This review illustrates the complex behavior of the climate system and the modeling challenges posed by the observations. Recent recognition of extreme events like Snowball Earth and of abrupt climate changes occurring on millennial timescales is helping to spur the development of increasingly sophisticated coupled models for paleoclimate studies.

Overview of Ice Sheets and Late Pleistocene Climate Change

The initiation of Northern Hemisphere glaciation during the late Miocene culminated a long-term global cooling trend through much of the Cenozoic that had previously been marked by intermittent and rapid growth intervals of the East and West Antarctic ice sheets. Records of benthic oxygen isotopes ($\delta^{18}O_b$) and ice-rafted debris (IRD) reveal subsequent intensification of Northern Hemisphere glaciation beginning ~3.25 myr ago with the large Northern Hemisphere ice sheets (Laurentide, Eurasian, Cordilleran) first becoming established between 2.7 and 2.55 myr ago (Maslin *et al.*, 1998; Shackleton *et al.*, 1984). Unlike the permanent, post-Miocene Antarctic ice sheet system, however, these large Northern Hemisphere ice sheets deglaciated on a regular basis, initially on a 41,000-yr schedule, followed by the transition during the middle Pleistocene (~1 myr ago) to a 100,000-yr schedule. Statistical analyses of $\delta^{18}O_b$ records further demonstrate that large fluctuations in the volume of Northern Hemisphere ice sheets occurred at the same periodicities (100,000, 41,000, and 23,000 yr) as the orbital parameters (eccentricity, obliquity, precession) that control the seasonal and latitudinal distribution of insolation at high northern latitudes, providing critical support for the Milankovitch hypothesis as being the "pacemaker of the ice ages" (Hays *et al.*, 1976).

Ice sheets play an important role in the global climate system through their influence on the planetary albedo, the hydrological cycle, and atmospheric and oceanic circulation (Clark *et al.*, 1999). Their large topographic size presents an orographic barrier to atmospheric circulation, whereas their high albedo contributed as much as 60% of the total radiative forcing during glaciation maxima (Hewitt & Mitchell, 1997; Rind, 1987). Atmospheric and thermal effects are transmitted downwind, influencing ocean and continental surface temperatures, with attendant changes in such feedbacks as sea-ice extent and monsoonal strength. Finally, ice sheets affect the hydrological cycle through their radiative forcing, their effects on atmospheric circulation, and their capacity as the largest readily exchangeable reservoir of freshwater on Earth. The latter affect is particularly prominent in its ability to modulate the release of freshwater to the important sites of modern deepwater formation: the North Atlantic Ocean and the Southern Ocean bordering the Antarctic continent.

Aspects of ice-sheet history that define the key issues for modeling ice sheet-climate interactions include the origin of the 100,000-yr cycle, mechanisms of millennial-scale variability, and the size of ice sheets during the Last Glacial Maximum. The 100,000-yr ice-sheet cycle dominates the $\delta^{18}O_b$ record of the last 1 myr. In most cases, the cycle is asymmetric, with long (~90,000 yr), fluctuating growth phases and rapid (~10,000 yr) terminations (Fig. 1). In contrast, the

DEVELOPMENTS IN QUATERNARY SCIENCE
VOLUME 1 ISSN 1571-0866
DOI:10.1016/S1571-0866(03)01006-6

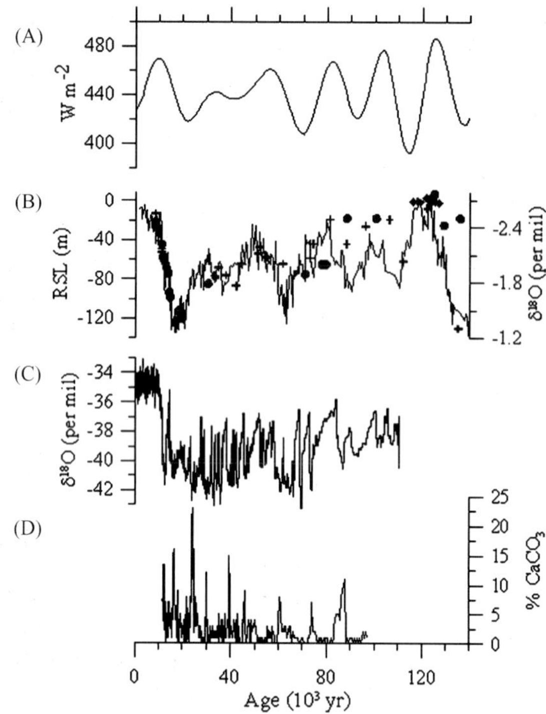

Fig. 1. (A) July (midmonth) insolation for 65°N for the last 140,000 yr (Berger & Loutre, 1991). (B) Marine $\delta^{18}O$ record (solid line) (Linsley, 1996) and relative sea-level data for the last 140,000 yr. Solid circles are samples from Barbados (Bard et al., 1990; Gallup et al., 2002), plus symbols are samples from New Guinea (Chappell et al., 1996), and diamonds are samples from Australia (Stirling et al., 1998). (C) The $\delta^{18}O$ record from the GISP2 ice core for the past 100,000 yr (Grootes et al., 1993). (D) Record of ice-rafted carbonate in North Atlantic core VM23–81 for the past 90,000 yr (Bond et al., 1999).

main effect of eccentricity variations on insolation is to modulate the amplitude of precession forcing, with the resulting 100,000-yr amplitude in orbital forcing being much too small to explain the large ice-volume response at this period. These observations indicate either a nonlinear response of ice sheets to eccentricity forcing or internal ice-sheet dynamics that give rise to a quasi-100,000-yr oscillation period that may or may not be paced by orbital forcing (Imbrie *et al.*, 1993). In either case, through their large influence on climate, ice sheets may become a dominant mechanism in driving the 100,000-yr climate cycle (Clark *et al.*, 1999; Imbrie *et al.*, 1993).

Modeling the origin of the 100,000-yr ice-sheet cycle and its abrupt terminations remains a challenge. Many models have successfully reproduced the ~23,000- and 41,000-yr cycles in direct response to orbital forcing, but modeling the 100,000-yr cycle, particularly the terminations, has required additional physical processes such as iceberg calving (Pollard, 1983), isostatic compensation (Peltier, 1987), ocean and CO_2 feedbacks (Tarasov & Peltier, 1997), or ice-sheet thermodynamics (MacAyeal, 1992; Marshall & Clark, 2002).

Embedded within the orbital-scale changes in ice-sheet size are millennial-scale variations that are too short and frequent to be explained by orbital forcing (Fig. 1). Such changes are identified by IRD records (Bond *et al.*, 1993), moraines (Clark, 1994), and sea-level records (Fairbanks, 1989) (Fig. 1) that indicate changes in the flux of ice and continental runoff to the ocean. High-resolution climate records reveal large and abrupt climate changes occurring at the same times as changes in the flux of water from ice sheets (Fig. 1), indicating either a common response to some climate forcing or a central role for ice sheets in causing millennial-scale climate change. In particular, ice sheets may force or amplify millennial-scale climate change through the release of freshwater to sites of deepwater formation, causing changes in sea-surface temperatures and deepwater formation that are transmitted through the atmosphere and ocean to cause regional or global climate change (Clark *et al.*, 2002).

The two dominant modes of millennial-scale ice-sheet variability are Dansgaard-Oeschger (D/O) cycles and Heinrich events. D/O cycles have an approximate (and intermittent) spacing of ~1500 yr, and each cycle has an IRD event associated with it (Bond *et al.*, 1999). The cause of D/O cycles remains under investigation, with possible mechanisms ranging from ice-sheet forcing through episodic release of fresh water (Clark *et al.*, 2001; Ganopolski & Rahmstorf, 2001; Schmittner *et al.*, 2002) to a periodic forcing that is amplified by stochastic processes such as ice-sheet forcing (Alley *et al.*, 2001). Heinrich events represent an abrupt release of icebergs from the Laurentide Ice Sheet (LIS) through Hudson Strait to the North Atlantic Ocean, causing a near shutdown of North Atlantic Deepwater (NADW) formation (Alley & Clark, 1999; Broecker *et al.*, 1992). Modeling Heinrich events has proven to be difficult (MacAyeal, 1993; Marshall & Clarke, 1997b), particularly with respect to identifying a climate mechanism to trigger them (Clarke *et al.*, 1999). Moreover, it is likely that not all Heinrich events are alike, with some having different source areas (Andrews *et al.*, 1994; Bond *et al.*, 1992), and some having significantly different sea-level signals (Chappell, 2002).

The Last Glacial Maximum (LGM) is defined as the most recent interval of Earth history when global ice volume reached a maximum. The $\delta^{18}O_b$ record, which integrates the contribution of individual ice sheets into a global signal, indicates that the LGM occurred 21,000 yr ago. The LGM remains an important topic for paleoclimate modeling because it represents a time when global climate was dramatically different than that of today, and thus provides a useful test of a climate model's sensitivity to change (Mix *et al.*, 2001).

Ice sheets represent one of the primary prescribed boundary conditions for modeling LGM climate. Their significance to understanding climate sensitivity is demonstrated by estimates of their contribution to the radiative forcing of the LGM that range as high as 60% (Broccoli, 2000; Hewitt & Mitchell, 1997). Nevertheless, there is longstanding debate over reconstructing the dimensions of the LGM ice sheets, which adds some level of uncertainty to modeling LGM climate. Far-field sea-level (Yokoyama *et al.*, 2000) and $\delta^{18}O_b$ records provide only an integrated record of ice volume; these are important for constraining the global volume, but provide no details on the spatial distribution of ice mass. Terrestrial geologic evidence identifies the former spatial distribution of former ice masses, but reveals only limited information on ice thickness. Modeling former ice sheets thus remains the principal means of partitioning the global ice volume constraint into its spatial components.

There are two modeling strategies used for reconstructing former ice masses. The first involves inverting relative sea level (RSL) records to determine the distribution of mass that best explains isostatic changes recorded by the RSL changes (Lambeck *et al*., 1998; Peltier, 1994, 1996; Shennan *et al*., 2002). Because most RSL records from formerly glaciated regions do not begin until well after the LGM, however, they provide little constraint on LGM loading. The second strategy, which involves using physically based ice-sheet models, is the basis of the present chapter.

The Environmental Processes of the Ice Age: Land, Oceans, Glaciers (EPILOG) program is revisiting the CLIMAP Project's landmark effort to provide a comprehensive reconstruction of LGM boundary conditions (Mix *et al*., 2001). Recent publication of a series of papers addressing geochemical, geological, and geophysical strategies has substantially narrowed the uncertainties in constraining LGM ice volume and its spatial distribution (Clark & Mix, 2002). LGM ice-sheet margins were reasonably close to those postulated in the CLIMAP "maximum" model. The magnitude and spatial distribution of ice load was substantially different, however, with the range of estimates (from ~118 to 135 m of eustatic equivalent) bounding the minimum CLIMAP model of 127 m. Because this ice was distributed over a larger area than the CLIMAP minimum model, however, it requires that the LGM ice sheets were thinner than those of the CLIMAP model. This observational constraint continues to pose a challenge to glaciological and climatological reconstructions of the Quaternary ice sheets. We consider these challenges in the general context of coupled ice-sheet with climate modeling below.

Development of Coupled Ice-Sheet/Climate Models

Prior to development of accessible computing infrastructure in the 1960s, there was considerable theoretical understanding of glacier dynamics, glacier-climate interactions (i.e. controls on glacier mass balance), global atmospheric circulation, and synoptic meteorology. Early theories of Nye (1952, 1959) and Robin (1955) laid out the fundamentals of ice-sheet dynamics, the underpinnings of the ice-sheet models that are in use today. Weertman (1961a) calculated simplified analytical profiles for isothermal, steady-state ice sheets.

Decades earlier, the theoretical foundations of atmospheric dynamics were set out by the father-and-son Bjerknes team in Bergen, Norway. Vilhelm Bjerknes introduced the "primitive equations" that describe atmospheric circulation, but meteorologists of the time had little ability to predict future weather or climate change in response to boundary forcing. Richardson (1922) had tried to integrate the equations of motion to predict the weather for a small area of Europe during World War I, using a mechanical calculator, but with poor results, due in part to including excessive physics. The calculation techniques that Richardson developed – division of space into grid cells, and finite difference solutions of differential equations – were the same ones employed by the first generations of climate models 30 years later. Richardson worked with simplified versions of Bjerknes' primitive equations in order to reduce the calculations required to a level where manual solution could be contemplated. Still, this task remained so large that Richardson's first attempt to calculate weather for a single eight-hour period took six weeks and ended in failure.

Charney *et al.* (1950) used early computers in the late 1940s and early 1950s to produce the first global circulation models with dynamics, but no long-term diabatic energy terms such as radiation. This was the start of a long development of general circulation models, first used for climate studies with other components included (land, snow, ocean) in the 1960s. Simple energy-balance models came to the fore in the late 1960s when Budyko (1969) and Sellers (1969) independently used them to demonstrate collapse to 100% global ice cover with only a few percent reduction in the solar constant. Linking of climate and ice-sheet models soon followed through the work of Budd (1969). This ushered in the era of numerical modeling in climate and ice sheet studies, a field which continues to evolve rapidly to this day. This evolution and the current frontiers in coupled climate-ice-sheet modeling are the primary concerns of this section.

Ice-Sheet Models

Huybrechts' (1990) landmark development of a three-dimensional ice-sheet model for coupled simulation of ice dynamics and thermodynamics has paved the way for a number of comparable ice-sheet models developed over the past decade. The recently completed European Ice-Sheet Modeling Initiative (EISMINT) model intercomparison featured 11 participants with fully-operational thermo-mechanical ice-sheet models, most based on Huybrechts' formulation (Payne *et al*., 2000). Coupling ice dynamic and thermodynamic systems into a thermomechanical framework was an important breakthrough because ice deformation is strongly dependent on ice temperature, with 1000-fold variations in the effective viscosity of ice over the range of temperatures found in ice sheets. Perhaps more importantly, it is the only way to rigorously simulate basal ice-sheet temperature, a primary control in enabling basal flow.

Huybrechts (1990) brought together the framework for ice dynamic simulation laid out by Mahaffy (1976) and the thermodynamic equations of Jenssen (1977) and asynchronously coupled the dynamics and temperature simulations to allow thermal feedbacks on ice stiffness (inverse effective viscosity),

$$B(T) = EB_0 \exp\left(-\frac{Q}{RT}\right) \Sigma_2^{(n-1)/2}, \tag{1}$$

where B_0 and n are coefficients in Glen's flow law for ice (Glen, 1958; Paterson, 1994), Q is creep activation energy, R is the ideal gas law constant, T is absolute temperature in the ice, and Σ'_2 is the second invariant of the deviatoric stress tensor in the ice, σ',

$$\Sigma'_2 = \tfrac{1}{2}\sigma'_{ij}\sigma'_{ji}. \tag{2}$$

The parameter E in Eq. (1) is a flow enhancement factor that is conventionally introduced into Glen's flow law to empirically encompass the effects of crystal anisotropy and impurities (e.g. dust) on bulk ice deformation (Fisher, 1987; Paterson, 1991). Ice deformation following Glen's flow law is implemented in thermomechanical ice-sheet models through calculation of strain rates ε_{ij} as a function of the deviatoric stress regime in the ice,

$$\varepsilon_{ij} = B(T)\sigma'_{ij} \tag{3}$$

In simplified terms, the standard Glen-law formulation in Eqs (1)–(3), with $n = 3$, gives ice deformation (strain rates) in proportion to the third power of the deviatoric stress. For grounded inland ice that is well-coupled with the substrate (i.e. not sliding over the bed), vertical shear strain is the dominant deformation regime in the ice. Eq. (3) can then be written in terms of the vertical gradients in the horizontal velocity (u, v) and the vertical shear stress,

$$\begin{aligned} \frac{\partial u}{\partial z} &= 2B(T)\sigma'_{\lambda z} = 2B(T)\frac{\rho g H}{R\cos\theta}\frac{\partial h_\mathrm{I}}{\partial \lambda}, \\ \frac{\partial v}{\partial z} &= 2B(T)\sigma'_{\theta z} = 2B(T)\frac{\rho g H}{R}\frac{\partial h_\mathrm{I}}{\partial \theta}. \end{aligned} \tag{4}$$

We have introduced a three-dimensional spherical Earth co-ordinate system with z upwards, longitude λ, latitude θ, and Earth radius R. Velocity components (u, v) are zonal and meridional, respectively. The shear stress is calculated from the product of ice density, ρ, gravitational acceleration, g, ice thickness, H, and ice surface slope, $\partial_j h_\mathrm{I}$.

Eq. (4) can be vertically integrated to give the average vertical ice velocities

$$\begin{aligned} u(z) &= u_\mathrm{b} - 2(\rho^\mathrm{I} g)^n \|\partial_j h_\mathrm{I}\|^{n-1}\frac{1}{R\cos\theta}\frac{\partial h_\mathrm{I}}{\partial\lambda} \\ &\quad\times \int_{h_\mathrm{b}}^{h_\mathrm{I}} B(T)(h_\mathrm{I} - z)^n\,\mathrm{d}z, \\ v(z) &= v_\mathrm{b} - 2(\rho^\mathrm{I} g)^n \|\partial_j h_\mathrm{I}\|^{n-1}\frac{1}{R}\frac{\partial h_\mathrm{I}}{\partial\theta} \\ &\quad\times \int_{h_\mathrm{b}}^{h_\mathrm{I}} B(T)(h_\mathrm{I} - z)^n\,\mathrm{d}z. \end{aligned} \tag{5}$$

where (u_b, v_b) are basal ice velocities, representing the sum of decoupled sliding over the bed and subglacial sediment deformation. Eq. (5) allows ice fluxes to be modeled as a function of local surface slope and ice thickness. Conservation of mass in ice sheets gives the governing equation for simulation of ice-sheet thickness evolution,

$$\frac{\partial H}{\partial t} = -\frac{\partial}{\partial x_j}(\bar{v}_j H) + b, \tag{6}$$

where b is the ice-equivalent mass balance rate (accumulation minus ablation rates) and $\partial_j(\bar{v}_j H)$ denotes the horizontal divergence of ice flux.

The three-dimensional distribution of ice temperature, for use in (5) via (1), is simulated from the energy balance at each point,

$$\frac{\partial T}{\partial t} = -v_k\frac{\partial T}{\partial x_k} + \kappa(T)\frac{\partial^2 T}{\partial z^2} + \frac{1}{\rho c}\left(\frac{\partial T}{\partial z}\right)^2 + \frac{\Phi}{\rho c}, \tag{7}$$

which includes the three-dimensional advection and vertical diffusion of heat, as well as strain heating produced by internal deformation of ice, $\Phi = \sigma_{ij}\varepsilon_{ij}$. Horizontal temperature diffusion is neglected in ice-sheet models because vertical temperature gradients are typically three orders of magnitude higher. In Eq. (7), $v_k(\lambda, \theta, z, t)$ is the three-dimensional ice velocity field and c, k, and κ are the heat capacity, thermal conductivity, and thermal diffusivity of ice.

Glen's flow law for ice deformation is the prevailing rheology used in ice-sheet modeling, although Peltier *et al.* (2000) have proposed a low-exponent flow law ($n = 1.7$) for ice flow under low-stress regimes. The proposed ice rheology offers a potential explanation for the relatively thin, low-sloping character of the Laurentide Ice Sheet that is suggested by geophysical inversions (Peltier, 1994, 1996). Geomorphological reconstructions also suggest a relatively thin, multi-domed ice sheet in North America (Dyke & Prest, 1987; Shilts, 1980), characterized by lobate, low-sloping southern outlets (e.g. Clark, 1992; Clayton *et al.*, 1985; Mathews, 1974) and active ice streams draining the ice sheet on the North Atlantic and Arctic margins (e.g. Bond *et al.*, 1993; Clark & Stokes, 2001; Dyke, 1999; Heinrich, 1988; MacAyeal, 1993).

It is uncertain whether exotic ice rheology is needed to explain the low Laurentide Ice Sheet volume, as this is also consistent with large-scale basal flow, as observed in present-day West Antarctica. Broad regions of the Laurentide Ice Sheet were underlain by deformable sediment and may have experienced pervasive basal flow, producing the thin and mobile ice sheet that is suggested by the geological record (Boulton *et al.*, 1985; Fisher *et al.*, 1985). This possibility is supported by the only model studies to date that have directly addressed the question (Clark *et al.*, 1996; Jenson *et al.*, 1996).

Although ice-sheet models have had good success in simulating the internal shear deformation that dominates flux in ice sheets that are well-coupled with their bed (e.g. East Antarctica, Greenland), modeling of basal flow remains a challenge. The controls of basal flow involve subglacial hydrology, roughness elements (basal pinning points or "sticky spots"), and sediment dynamics, with complex and subgrid-scale governing physics. In the case of West Antarctica, these basal flow controls produce subgrid-scale ice streams that drain over 90% of the ice sheet (Alley *et al.*, 1987; Paterson, 1994). Even in this extremely well-studied ice mass, ice-sheet models have yet to successfully simulate ice dynamics, typically predicting excessively thick ice.

Progress is being made on coupled sheet-stream-shelf dynamics (Hulbe & MacAyeal, 1999; Hulbe & Payne, 2001; Payne, 1999), and this is a major focus for the next generation of ice-sheet models. Significant progress will require the model capacity to simulate longitudinal stress/strain as well as vertical shear deformation, as this is the primary ice deformation mechanism for ice masses with fluxes dominated by basal flow (ice streams, surge lobes). Longitudinal stress coupling is also the physical mechanism by which fast-flowing outlets tap into and draw down interior regions of the ice sheet. The physics are well understood (e.g. Blatter, 1995; Kamb & Echelmeyer, 1986) but simulation of longitudinal stress and strain is challenging because the governing dynamics are non-local. This limitation is primarily technical, however, and we anticipate continental ice-sheet models that are better able to simulate higher-order ice mechanics in the near future.

This will also help to improve basal-flow modeling. In ice-sheet modeling studies, basal ice velocity is typically parameterized as a function of gravitational driving stress (e.g. Marshall *et al.*, 2000; Payne *et al.*, 2000). This may be representative of the physical controls of basal sliding via regelation over small-scale bedrock obstacles (Paterson, 1994), but it probably has no bearing on the physical controls of basal flow rates in ice streams and surge lobes, where velocity is either uncorrelated or negatively correlated with gravitational driving stress (e.g. Hulbe *et al.*, 2000). A higher-order stress solution will allow basal flow velocities to be calculated based on a physically consistent momentum balance at the bed (MacAyeal, 1989; MacAyeal *et al.*, 1995; Marshall & Clarke, 1997a).

Improved ice mechanics is only part of the challenge in capturing the influence of basal flow dynamics in ice sheets. As noted above, the temporal and spatial controls of basal flow processes are complex and, in many cases, operate on scales that are substantially less than the ca. 20-km resolution of current ice-sheet models. Ice streams on West Antarctica's Siple Coast are discrete, channelized entities that appear to interact through competition for ice and basal water supply in a shared source region (Anandakrishnan & Alley, 1997), giving rise to dynamical switching between active and inactive states (e.g. Anandakrishnan *et al.*, 2001). Subglacial thermal evolution also plays a role in the temporal regulation of ice stream activity, by enabling basal flow via high water pressures and subglacial sediment deformation or failure (Kamb, 1991; Tulaczyk *et al.*, 2000a, b). Widespread basal flow will not occur in the absence of subglacial water to decouple the ice from the bed or weaken the underlying sediment. Pervasive warm-based conditions are therefore needed to support basal melting and the accumulation and storage of pressurized subglacial water.

Most ice-sheet modeling studies to date employ a simple thermal switch for basal flow activation (e.g. MacAyeal, 1993; Marshall & Clarke, 1997b; Marshall *et al.*, 2000; Payne, 1998). Ice slides over the bed or sediment deforms wherever the ice sheet is warm-based. While ice needs to be warm-based to permit fast basal flow, this is more of a prerequisite than a switch in Nature. Much of the East Antarctic Ice Sheet is warm-based, for instance, as well as the marginal regions of Greenland Ice Sheet and the entire area of Icelandic ice caps. These ice masses have regions of subglacial flow, but the spatial patterns of basal motion are much more complex than the maps of warm vs. cold-based ice.

Subglacial geology plays a role in dictating where basal flow is possible, but the primary temporal switch will involve the subglacial water system. Simulation of subglacial hydrology is another emerging focus in glaciological modeling, and the first coupled models of ice dynamics and ice-sheet hydrology are just becoming available (Arnold & Sharp, 2002; Flowers & Clarke, 2002; Flowers *et al.*, 2003; Johnson & Fastook, 2002). While parameterically unwieldy because of the uncertainties in subglacial hydrological processes, this development should provide improved representation of the controls of basal flow and ice-stream dynamics.

Coupled Ice-Sheet Climate Modeling

Glaciological models are improving through the introduction of additional physical processes, much the way climate models have become more complex and comprehensive as additional coupled process models are built in (e.g. land surface schemes, dynamic sea ice, etc.). Atmospheric general circulation models (AGCMs) have evolved from early representations of the troposphere as a single viscous fluid layer over the Earth. The earliest glaciological models were simpler yet, treating the coupled climate-ice evolution as systems of coupled differential equations or one-dimensional (zonally averaged) dynamical systems.

Climate-ice-sheet models of this kind were first applied to the Quaternary ice ages in the 1970s. Climate was represented either by a snowline pattern vs. height and latitude (Oerlemans, 1980; Weertman, 1976), or by a seasonal energy-balance climate model (Birchfield *et al.*, 1982; Pollard, 1978). These climate representations responded to Milankovitch orbital forcing and current ice-sheet size, providing the mass balance to drive explicit 2-D models of Northern Hemispheric ice sheets through the last 10^5–10^6 years. The energy-balance model is run typically every few hundred years to update the climate, which is then used to integrate the ice-sheet model forward for the next 100- or 1000-year interval. [Some ice-age studies used energy-balance climate models alone with snow extent but no explicit ice-sheet physics (North *et al.*, 1983; Suarez & Held, 1979); the dichotomy between small ice-cap instability found in those studies (also North, 1984) due to snow-albedo feedback, and small ice-sheet instability due to vertical ice-sheet geometry (Abe-Ouchi & Blatter, 1993; Weertman, 1961b) remains largely unappreciated to this day (Maqueda *et al.*, 1998).

Throughout the 1980s and 1990s, models with climate and explicit ice sheets were used extensively for Quaternary studies (Berger *et al.*, 1999; Deblonde & Peltier, 1991, 1993; Pollard, 1983), along with more conceptual zero-dimensional models of all components (Le Treut & Ghil, 1983; Matteucci, 1989; Paillard, 1995; Saltzman & Verbitsky, 1993). More recently, the EBM-ice sheet genre has been applied to earlier glaciations of the Paleozoic and Precambrian (Hyde *et al.*, 1999, 2000). Huybrechts & T'Siobbel (1995) and Tarasov & Peltier (1997, 1999) applied this type of model to Northern Hemispheric Quaternary glaciations using a 3-D ice-sheet model, an EBM predicting climate "anomalies" (see below) from modern observed climatology. The use of energy balance climate models with their simple diffusive heat transport can arguably capture the effects of orbitally induced variations in summertime melt on ice sheets, but have little hope of capturing variations in snowfall patterns.

Beginning in the 1990s, comprehensive global climate models have been used to predict the mass balance on ice sheets. The computer time required by GCMs prohibited long-term integrations, and most studies investigated "snapshot" simulations of climate and the ice-sheet mass balance for particular times with prescribed ice sheets: for the present (Genthon & Krinner, 2001; Ohmura *et al.*, 1996; Thompson & Pollard, 1997) and the Last Glacial Maximum (Fabre *et al.*, 1997, 1998; Hall *et al.*, 1996; Pollard & Thompson, 1997a; Pollard *et al.*, 2000a; Ramstein *et al.*, 1997). Since typical GCM horizontal scales are several 100 km, they are inadequate to resolve the steep topography around ice-sheet margins that is important in determining ablation; hence, some of these studies used straightforward techniques to derive the mass balance on prescribed ice-sheet surfaces on a much finer ice-sheet grid from the predicted GCM climate. These techniques have included:

- Horizontal interpolation from the GCM grid to a much finer (~10 s of km) ice-sheet grid.
- Simple corrections to air temperature for the difference in GCM topography and the fine-grid ice topography, assuming a constant lapse rate. In some cases, a similar vertical correction to precipitation and downward infrared radiation flux.
- If only the monthly means of GCM temperature and other meteorological variables are saved, then the (re)imposition of a standard diurnal cycle in calculating mass balance using a column snow-ice surface model, or a more empirical degree-day method.
- Crude corrections for the refreezing of meltwater, along the lines of Pfeffer *et al.* (1991), for instance.
- For paleo and future studies, the use of predicted climate anomalies, whereby the *changes* in temperature and/or precipitation from the present predicted by the GCM are superimposed on modern observed climatology. In this way, it is hoped that GCM climate errors remain the same across periods, and so cancel to yield a more accurate climate for the period in question.
- A correction to net mass balance on nascent ice due to sub-grid topographic variability has been proposed, allowing for the extra melting in deep valleys not resolved by the model grid (Marshall & Clarke, 1999a; Pollard & Thompson, 1997b; Walland & Simmonds, 1996), but has been used in relatively few long-term studies to date.

Some impressive results have been obtained with GCMs for present-day Greenland and Antarctica, and for future changes in their mass balance due to anthropogenic change in the coming centuries (Wild *et al.*, pers. comm., 2003).[1] However, there is wide scatter between GCMs in the mass balances of modern and past ice sheets, due to the strong sensitivity of summer melt to air temperatures over the ablation zones (Pollard *et al.*, 2000a), suggesting that some ice-related results using GCMs may be model dependent.

Current State of Coupled Modeling

Since the late 1990s, a number of groups have begun to make use of GCMs in long-term ice-age simulations. Two basic strategies have been used: (1) GCM snapshots; and (2) a matrix of GCM simulations over the major forcing factors.

Snapshot Method

As in the EBM-ice age studies mentioned above, a GCM is run at several 1000-yr intervals using the current orbit and ice-sheet size, to provide the mass-balance forcing over the next few 1000 yr of ice-sheet integration. Each snapshot requires several decades of GCM integration to "spin up" the upper ocean and to average out interannual variability. For instance, Charbit *et al.* (2002) have used this technique to simulate the last Northern Hemispheric deglaciation from 21,000 yr ago to the present. However, the computational expense of full GCMs has limited this technique to relatively short intervals.

Marshall & Clarke (1999b) drove their Northern Hemispheric ice-sheet model over the last 10^5 years, with seasonal climate provided by a weighted average of GCM snapshots of the present and the Last Glacial Maximum (LGM). The weighting is prescribed from the observed $\delta^{18}O$ time series in the GRIP ice core from Summit, Greenland. Figure 2 shows a typical glacial cycle simulation with this model, in this case driven by the GISP2 $\delta^{18}O$ isotope record (Fig. 1c; Grootes *et al.*, 1993), with climate fields for LGM and present-day from Vettoretti *et al.* (2000). The $\delta^{18}O$ chronology is converted to an index of glaciation through model simulations of the last glacial cycle in Greenland (Marshall & Cuffey, 2000), correcting for the effect of elevation changes in the Summit region.

Figure 2(a and b) plot air temperature over the glacial cycle for the entire North American grid and for a sample point in interior North America (Churchill, Manitoba, near the Laurentide Ice Sheet divide at LGM (58°N, 94.5°W), to illustrate the manner in which the GISP2 record paces the climate history, while local temperature differences are a function of local ice thickness (elevation) and climate model reconstructions at LGM. Figure 2(c and d) show average precipitation and mass balance rates over the model grid, and Fig. 2(e and f) depict ice volume and area through the glacial cycle. Figure 2(g and h) plot integrated hydrological/mass-

[1] Wild, M., Calanca, P., Scherrer, S.C. & Ohmura, A. Effects of polar ice sheets on sea-level in high resolution greenhouse scenarios. Submitted to *Journal of Geophysical Research*.

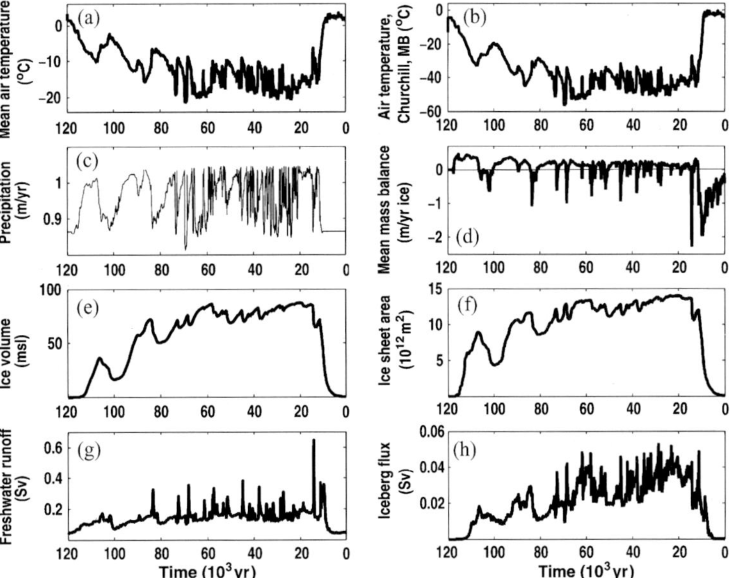

Fig. 2. Glacial cycle simulation in North America under GISP2 climate forcing: (a) average air temperature over the model grid (35°–85°N, 165°–45°W), (b) local air temperature at Churchill, Manitoba, near the Laurentide Ice Sheet divide (58°N, 94.5°W), (c) average precipitation rate over the model grid, m/yr, (d) average ice-sheet mass balance, m/yr ice equivalent, (e) ice sheet volume, m eustatic sea-level equivalent, (f) ice-sheet area, 10^{12} m^2, (g) freshwater runoff to the North Atlantic from meltwater and rainfall, Sv (10^6 m^3/s), and (h) total iceberg flux to the North Atlantic, Sv.

balance variables for North America: total freshwater runoff and total iceberg flux to the North Atlantic, in Sv (10^6 m^3/s). Figure 3 shows modeled climate fields (air temperature, precipitation, and mass balance) and ice-sheet surface topography at LGM.

This climate forcing methodology combines elements of both snapshot and matrix techniques, but has the drawback of imposing the climate fluctuations recorded in the Summit, Greenland ice cores, rather than allowing the model to freely simulate millennial and ice-age variability. This effectively forces a glacial cycle in the simulations, by driving the climate towards glacial and interglacial conditions. A preferred approach is to let climate freely evolve in response to changing ice-sheet conditions, through climate model snapshots that are updated as frequently as possible. This is computationally prohibitive, but Charbit *et al.* (2002) have productively explored this approach for asynchronously coupled ice sheet-GCM simulations from LGM to the present.

Matrix Method

Given that: (1) there are a limited number of degrees of freedom in the important long-term determinants of climate (e.g. orbit and ice sheet); and (2) each determinant is more or less one-dimensional (summer insolation at mid latitudes, total ice volume), then a "matrix" of generic GCM/RCM simulations can be built up, with orbits ranging over hot-medium-cold summers, and with prescribed ice sheets ranging over none-intermediate-maximum sizes. At any point in a long-term ice-sheet simulation, mass balance can be interpolated from elements of this matrix based on the current orbit and overall ice-sheet overall ice-sheet size. Even at 3 × 3 "resolution" with extreme cold-medium-extreme warm summer orbits and no-small-large ice-sheet size, the matrix approach can produce viable simulations of the last 10^5 years of Laurentide Ice Sheet variations (Pollard *et al.*, 2000b), with at least a crude capability to track the snowfall maxima on southern ice-sheet flanks for intermediate ice sizes.

An example of the differences between the matrix elements is shown in Fig. 4, emphasizing the effects of orbital variations alone on summer temperatures, similarly to Phillips & Held (1994). However, the obvious drawbacks are that: (1) other long-term determinants of climate may be important, such as the deep oceans for millennial variability; and (2) ice-sheet and orbit are not one-dimensional and/or need many more than three cases each to describe their effect on climate. These drawbacks may be ameliorated by using matrices with more dimensions and elements, but then the number of canonical GCM simulations required may approach that required by the simpler snapshot technique.

A combination snapshot/matrix techique has been used by DeConto & Pollard (2003, in press) in their coupled GCM-ice sheet study of Antarctic initiation in the early Oligocene. First, coupled GCM-ice sheet integrations are performed over 40,000 yr, driving a 3-D Antarctic ice-sheet model by running GCM snapshots every 10,000 yr, using an idealized orbital sequence of 20,000-yr precession, 40,000-yr obliquity and 80,000-yr eccentricity cycles. Several 40,000-yr integrations are performed, each with a different specified amount of atmospheric CO_2 in the GCM, saving all GCM snapshot climates from each integration. This library of GCM climates is then used to drive the ice-sheet model over much longer intervals of 10^7 yr, assuming the same idealized repeating orbital cycle, and weighting between the 40,000-yr integrations depending on the current atmospheric CO_2 value that is prescribed to have a slow downward trend over the 10^7-yr run. Although this does not fully capture ice-sheet climate feedbacks, the ice variations in the long-term runs are similar to those in the 40,000-yr integrations, especially around each chosen CO_2 value. Some results of snapshot and matrix techniques for long-term simulations of the Antarctic ice sheet in the early Cenozoic are shown in Figs 5 and 6.

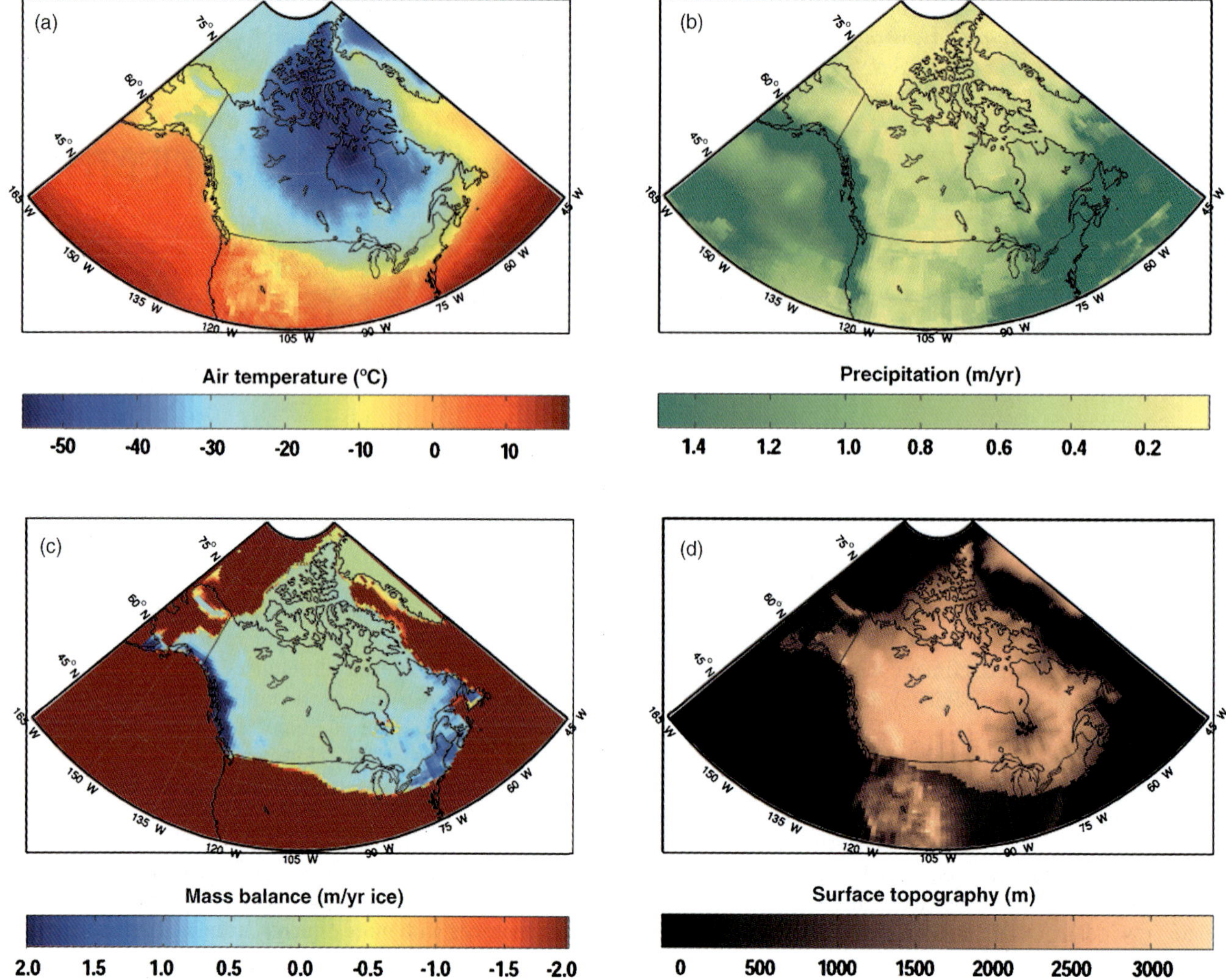

Fig. 3. Modeled North American climate fields and surface topography at 21,000 yr B.P., under GISP2 climate forcing with climate fields from the CCC AGCM (Vettoretti et al., 2000): (a) air temperature (°C), (b) precipitation (m/yr), (c) mass balance (m/yr ice), and (d) surface topography (m).

Alternative Methods: Reduced Climate Models

The matrix method attempts to address the primary shortcoming of the snapshot method: the inconsistency of modeled climate fields with evolving ice sheet climate-system feedbacks (e.g. topography and albedo fields). One can also expect different ice-sheet morphologies in the situation where the ice sheet is advancing or decaying on both millennial and orbital time scales (e.g. Marshall & Clark, 2002), with associated climate feedbacks that are likely to be substantial. It is desirable to have more tightly linked ice-sheet climate feedbacks in order to properly capture the evolving spatial patterns of surface temperature and snow accumulation as the ice sheets evolve. Reduced models of atmospheric dynamics that direct attention to moisture advection under changing boundary conditions could offer important insights (e.g. Hulton *et al.*, 2002; Purves & Hulton, 2000a, b). Such models can be integrated in truly coupled fashion with the ice sheet evolution, to ease the interpolation between infrequent GCM snapshots. Each GCM update (e.g. at 10,000-yr intervals during ice-sheet growth in the Pleistocene) could provide boundary conditions for the reduced model. The approach by Hulton *et al.* (2002) in Patagonian ice sheet reconstructions is an excellent example of progress in this direction, and we anticipate further progress with a blend of snapshot, matrix, reduced modeling, and comprehensive regional climate modeling, discussed below.

Regional Climate Modeling

The use of GCM output to drive ice-sheet models provides model-derived climatologies that reflect unparameterized internal and external forcing of the climate system. However, important regional climate forcing and feedbacks of ice sheet responses (e.g. precipitation and temperature gradients along the margins and elevation effects) may not be resolved by GCMs at the relatively smooth spatial resolution (order

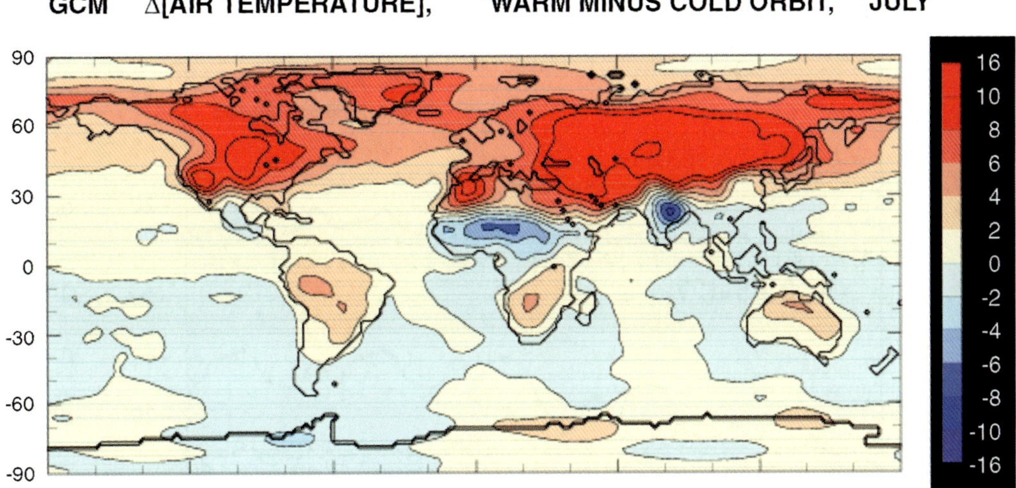

Fig. 4. Differences in summer (July) 2-m air temperatures between GCM climates with two different orbits yielding extreme Northern Hemispheric summers (warm minus cold). The GCM is Genesis v2 (Thompson & Pollard, 1997), using present-day boundary conditions except: obliquity, eccentricity, perihelion = 0.05, *270°, 24.5° for the "warm" orbit, and 0.05, 90°, 22.5° for the "cold" orbit. (Precession here is defined as prograde degrees from perihelion to Northern Hemispheric vernal equinox.) Similar and more extensive results are shown in Phillips & Held (1994, Fig. 9).*

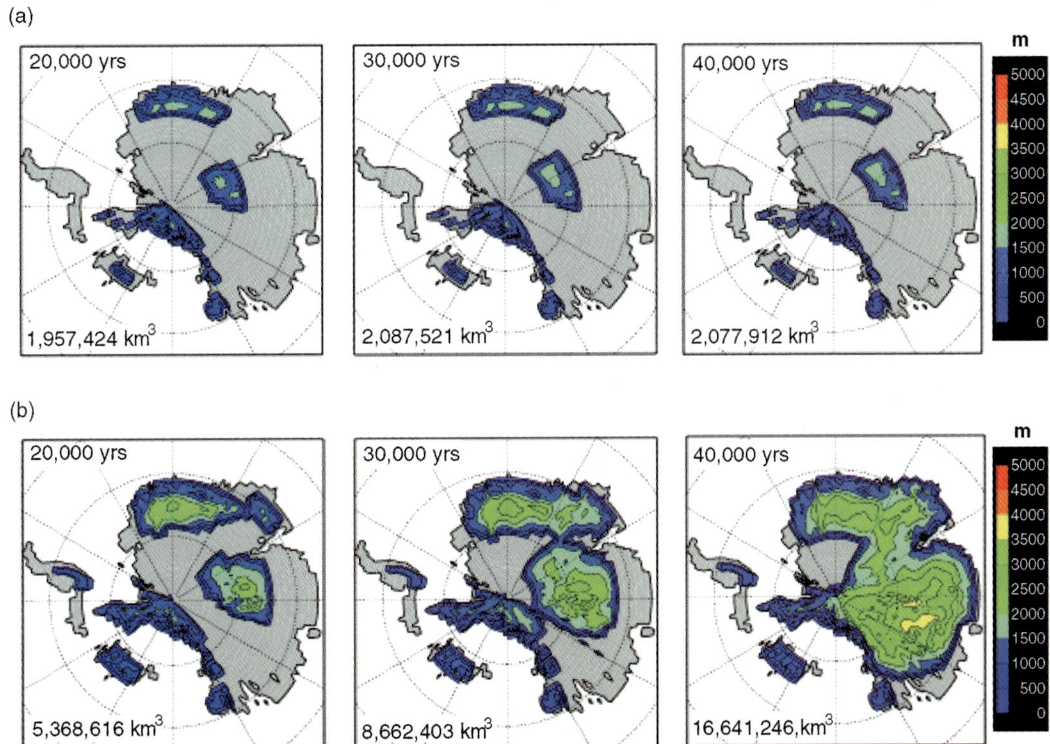

Fig. 5. Ice-sheet thickness over Cenozoic Antarctica at 10,000-yr intervals, from a "snapshot" simulation in which a global climate model is run every 10,000 yr, updating the climate used to drive a dynamic ice-sheet model continuously for 400,000 yr (from DeConto & Pollard, 2003, in press). Top panels (a): with atmospheric CO_2 levels in the GCM set to three times the present value. Bottom panels (b): with atmospheric CO_2 set to two times the present value. Total ice volumes are shown in each panel.

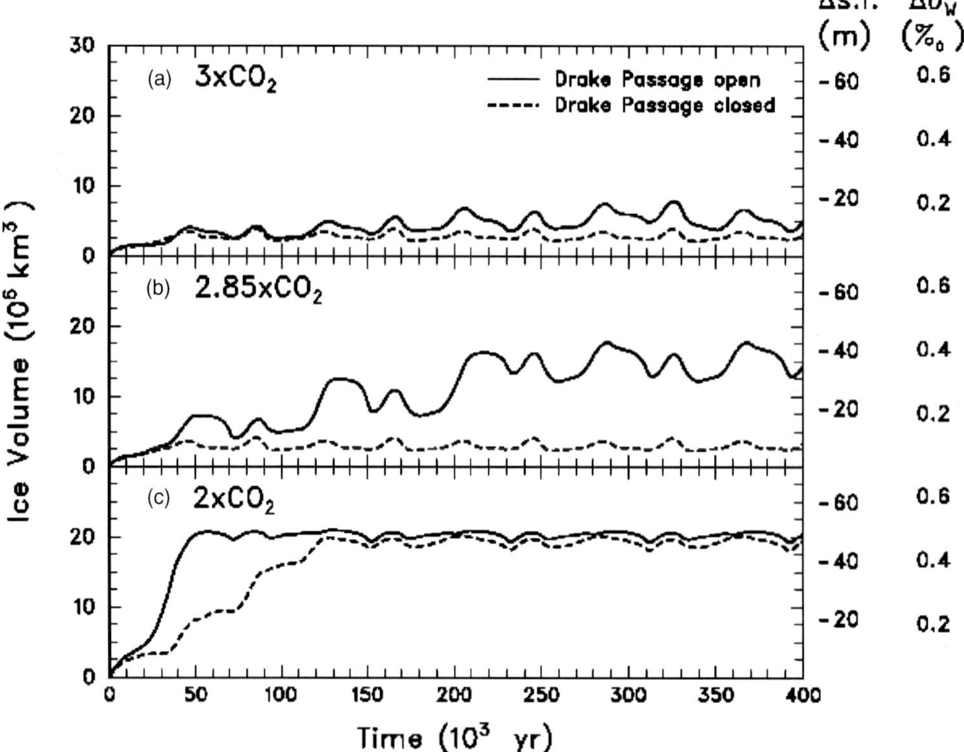

Fig. 6. Ice-sheet volumes for Cenozoic Antarctica, from a "matrix-method" simulation using stored GCM climates interpolated to drive a dynamic ice-sheet model continuously for 400,000 yr with idealized orbital forcing (DeConto & Pollard, 2003, in press). Solid curves show results using GCM climates with an open Drake Passage, and dashed curves show results with a GCM modification representing a closed Drake Passage. Each panel shows results for a different atmospheric CO_2 level (see DeConto & Pollard, 2003, in press).

of several 100 km) that are typically used in paleoclimate experiments. Lack of sufficient resolution may be particularly important in modeling abrupt, millennial-scale variations in ice sheets that involve dynamics of the ice margin. Resolving regional climate detail requires climate model resolutions on the order of 100 km or less, but present computer limitations make it impractical to run global climate models at such a fine resolution for extensive paleoclimate simulations such as those described in the matrix technique.

As an alternative, GCM simulations can be downscaled to finer resolutions with targeted applications of a regional climate model (RCM). Regional climate models are applied over a limited area (e.g. ice sheet) with lateral boundary conditions (e.g. surface pressure, SSTs and vertical fields of temperature, humidity, winds along the RCM boundary) derived from GCM history tapes. The synoptic-scale atmospheric circulation in the GCM is thus imparted to the RCM which simulates finer scale climatic features and that reflect the influences of regional forcings and feedbacks such as those induced by topography, coastlines, ice sheets and water bodies (Hostetler *et al.*, 1999, 2000). The finer-scale RCM fields can used to drive an ice-sheet model.

Regional climate models are computationally intensive and, as is the case with high-resolution global models, it is not practical to run them for long simulations. The models can be run for a decade or so to provide a suitable climatology for ice-sheet modeling. For the matrix technique, application of an RCM would be targeted to a subset of the various GCM runs comprising the matrix elements. By way of example, we have conducted a set of RCM simulations that are intended to assess the sensitivity of the LIS to a canonical Heinrich Event (H2). The GCM driving fields are derived from a series of GCM sensitivity experiments (Hostetler *et al.*, 1999) that include a nominal LGM, an LGM with a lowered LIS over Hudson Bay (Licciardi *et al.*, 1999 ice lowering to represent iceberg discharge into the North Atlantic) and all other boundary conditions the same as those of the LGM, and an LGM simulation with the lowered LIS and warmer SSTs in the North Atlantic representing post-discharge warming associated with reestablished thermohaline circulation. The RCM used in the experiments is the paleoclimate version of the NCAR RegCM2 (Hostetler *et al.*, 2000); the simulations were run for 15 yr using 12-hour boundary conditions from the GCM experiments.

Figure 7 depicts sample January and July air temperature and precipitation fields over the North American ice sheets from the RegCM2 regional climate model at 100-km horizontal resolution. The RegCM2 simulations for the LGM display considerable regional detail including steep temperature gradients along the margin of the LIS and

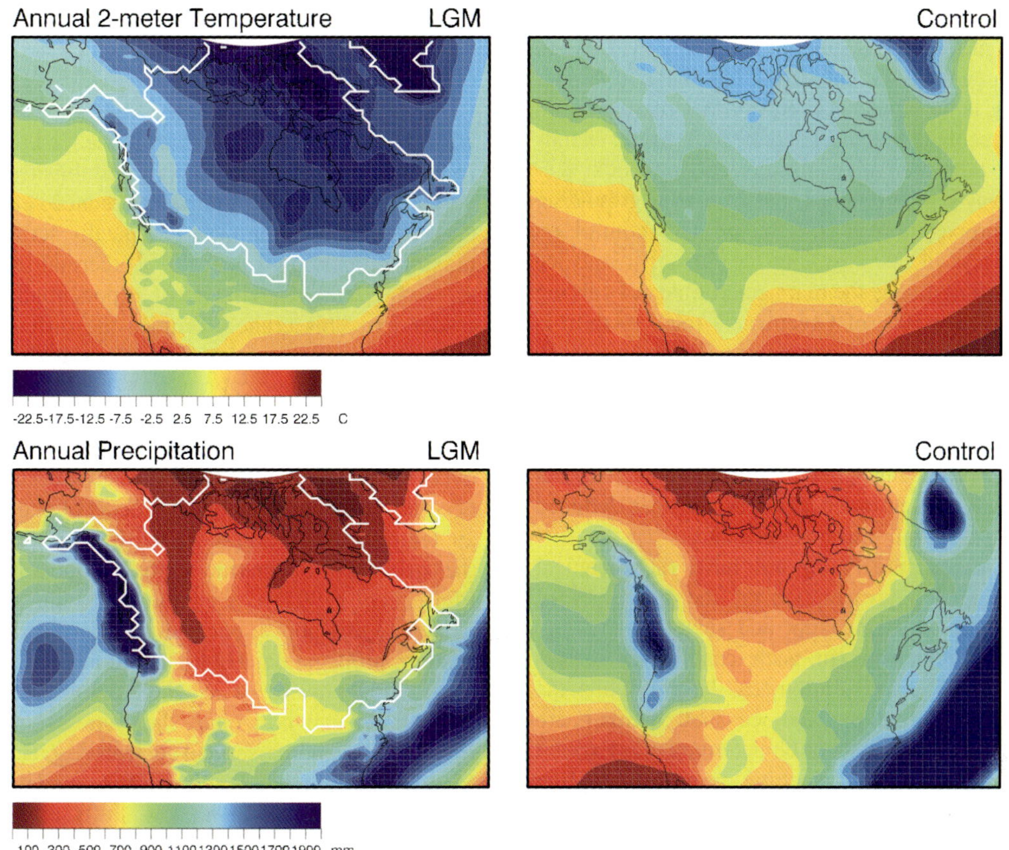

Fig. 7. Annual average (15-yr) surface air temperature and precipitation for LGM and control simulated with the RegCM2 regional climate model. The white lines indicate outlines of the continental ice sheets used as boundary conditions in the LGM simulation. For these simulations, the model was run on a 100 by 100 km horizontal grid with 14 vertical levels.

regional precipitation patterns that are induced thermally and topographically by the ice sheet. The high-resolution RCM output thus provides a climatology that reflects regional forcings that can be used to drive an ice-sheet model.

The Ice-Sheet Climate Interface: Mass-Balance Modeling

Given perfect knowledge of atmospheric conditions and ice-sheet topography, there is still uncertainty in estimating ice sheet mass balance, $b = a - m$, where a is the net annual snow/ice accumulation and m is the net annual mass loss through ablation. Each variable is typically measured in m/yr ice-equivalent. Snow and ice ablation occur through iceberg calving, surface melting, sublimation, and basal melting. Sources of accumulation are primarily meteoric, although "internal accumulation" occurs through the refreezing of surface meltwater that percolates into the snow or firn. This can be an important supplement to mass balance above the glacier or ice sheet equilibrium line. We summarize the methods being used to estimate each of these mass balance terms in ice-sheet models and discuss the associated uncertainties.

Marine Ablation

In present-day Antarctica, the dominant ice-sheet ablation mechanisms are iceberg calving and basal melting beneath floating ice shelves, as air temperatures are too cold to produce surface melting on most of the continent. The controls of ice shelf wastage through both calving and sub-shelf melting are complex and are not well quantified. Calving involves fracture generation and propagation processes and is known to be dependent on ice thickness and temperature (hence, tensile strength), water depth, tidal forcing, coastal/embayment geometry, and the flux of ice across the grounding line. There also appears to be a close relationship with spring/summer air temperatures, via the extent of the summer melt season (Doake & Vaughan, 1991; Vaughan & Doake, 1996). Scambos *et al.* (2000) demonstrate the role of water-filled crevasses in weakening overall ice-shelf competence, through forcing of vertical crack propagation.

Melting beneath ice shelves is difficult to quantify and is known to be spatially variable, ranging from cm/yr to several m/yr. Melt rates are controlled by ocean water temperature, density, and sub-shelf bathymetry, with much of the spatial variability associated with vertical convection cells in sub-shelf waters (Jacobs *et al.*, 1992, 1996; Jenkins *et al.*, 1997).

Ice shelves that drain the Greenland Ice Sheet and Arctic icefields in Canada and Russia are modest in scale relative to their Antarctic counterparts, but appear to behave in similar fashion, calving large tabular icebergs. Other marine-based outlets of the Greenland Ice Sheet and the high Arctic islands are more akin to tidewater valley glaciers, calving blocky icebergs from floating glacier tongues or at the marine grounding line. Fracture propagation is still the primary control on the calving process in this case, but the flux of inland ice across the grounding line is the main determinant of calving rates. Approximately 50% of Greenland's total ablation is believed to occur through iceberg calving (Paterson, 1994), with surface melt making up the remaining 50%. Recent interferometric studies in northern Greenland ice shelves indicate that basal melting beneath ice shelves is actually the dominant ablation mechanism in northern Greenland (Rignot *et al.*, 1997). Melt rates of order 20 m/yr have been deduced beneath major outlets in northeast Greenland.

Current ice-sheet models do not have the technical capacity for explicit representation of the processes of iceberg calving, ice shelf breakup, and sub-shelf melting. The governing mechanisms in each case have a finer scale than present model resolution (ca. 20 km), while the governing physics for calving and ice shelf breakup are not fully understood and may not be deterministic. Explicit modeling of ice shelf basal melting requires coupling with a regional-scale ocean model (e.g. Beckmann *et al.*, 1999; Hellmer & Olbers, 1989). This is technically feasible but has not yet been attempted in the ice-sheet modeling community.

From the perspective of modeling ice-sheet mass balance, it may be sufficient for many studies to neglect the details of marine-based ablation and accept that all ice to cross the grounding line ablates – the mechanism is unimportant. This is probably acceptable for first-order ice-sheet reconstructions (e.g. Last Glacial Maximum reconstructions for North America and Eurasia), and it is essentially valid if one is not interested in the detailed timing of ice loss in coastal regions, as the marine ablation mechanisms discussed above will effectively dispose of ice to cross the grounding line within a relatively short time frame, ca. 10^3 yr. Concerns with neglecting the details of marine ablation arise if one is interested in reconstructions of near-shore paleoenvironments or marine-triggered ice-sheet instability (e.g. tidewater glacier or ice-shelf collapse; North Atlantic Heinrich events).

Zweck & Huybrechts (2003) give a summary of current methods of portraying marine ablation mechanisms in continental ice-sheet models. In the simplest treatments, all ice to cross the grounding line or a pre-defined bathymetric contour (e.g. 400-m water depth) is simply removed. Zweck & Huybrechts introduce a slight variation on modern bathymetric controls, allowing time-varying water depth controls on calving rates. This builds in both the bathymetric and sea level influences on marine ice extent. Other studies parameterize calving losses, m_C as a function of water depth (Brown *et al.*, 1983), ice thickness, and ice temperature, proxy for ice strength/stiffness (e.g. Marshall *et al.*, 2000):

$$m_C = k_0 \exp\left[\frac{T - T_0}{T_C}\right] H H_W, \qquad (8)$$

where H_W is the water depth, H is the ice thickness, and T is ice temperature. The parameter k_0 establishes overall calving vigour, while T_0 and T_C introduce reduced calving rates with decreasing temperature. T_0 is typically set to 273 K, giving maximum calving rates for isothermal ice ($T = 273$ K; $m_C = k_0 H H_W$). Calving rates exponentially decreases for colder ice. This crudely mimics the difference in calving rates observed in maritime tidewater environments vs. polar environments, where calving rates are low enough to permit ice-shelf development.

This calving model permits floating/shelf ice to expand over the continental shelf or shelf break, but it is not necessarily closer to the truth; effects of oceanic circulation, air temperature (crevasse-forced fracture propagation), and the geometry of marine embayments are not captured. Pfeffer *et al.* (1997) have explored a calving parameterization similar to that above, but including an explicit treatment of fracture propagation. High-resolution ice-shelf models that include the physics of ice-shelf deformation (longitudinal stress/strain and horizontal shear stress) are better able to simulate these processes and controls (e.g. Hulbe *et al.*, 1998; MacAyeal *et al.*, 1995, 1996). These models are difficult to couple with inland ice models. Fully articulated ice-shelf models have not yet been coupled with continental ice-sheet models for the simulation of former ice sheets, although this technical step is imminent. Hulbe & MacAyeal (1999) have formulated the relevant coupled physics in a model of the Ross Sea sector of the West Antarctic Ice Sheet.

Surface Ablation

Surface melting is the dominant ablation mechanism for contemporary mid-latitude and subpolar ice masses, as well as high Arctic icefields. Surface melt also occurs over most of the Greenland Ice Sheet, but its high-altitude limits the amount of summer meltwater production in northern and interior regions of the ice sheet. Surface melting in ablation zones of southern Greenland is extensive, up to a few m/yr. Melt rates are as high as 10 m/yr on maritime icefields in Iceland, Svalbard, and western North America (e.g. Björnsson, 1979). Similar annual melt rates are found in the ablation zones of mid-latitude alpine ice masses (e.g. in the Rockies, the Andes, and the Alps).

The physics of snow and ice melt are well understood; melt rates m are governed by local energy balance,

$$\rho_{S/I} L m = Q_V(1 - \alpha) + Q_{IR}^{\downarrow} - Q_{IR}^{\uparrow} + Q_H + Q_L - Q_C + Q_P - Q_R \qquad (9)$$

where $\rho_{S/I}$ is the snow or ice density, L is the latent heat of fusion for water, α is the surface albedo, Q_H and Q_L are the sensible and latent heat fluxes, Q_V is incoming solar radiation, $Q_{IR}^{\downarrow}$ and $Q_{IR}^{\uparrow}$ are incoming and outgoing infrared radiation,

Q_C is the flux of energy conducted into the snow or ice, Q_P is the advective energy delivered by precipitation, and Q_R is the advective energy associated with meltwater runoff from the snow or ice surface. All fluxes have units W/m^2. Q_H, Q_L, and Q_P are assumed to be positive for heat transfer from the atmosphere to the snow/ice, while Q_C and Q_R are energy losses from the surface-atmosphere interface. Latent heat exchange occurs through sublimation, deposition, and evaporation of surface water during the melt season. The precipitation source term, Q_P, accounts for heat transferred to the snow or ice surface as rainwater of temperature T_P cools to 0 °C,

$$Q_P = \rho_W c_W P T_P, \qquad (10)$$

where ρ_W and c_W are the density and specific heat capacity of water and P is the precipitation rate. Rainfall temperature T_P is measured in °C.

Further elaborations of the terms in the energy balance can be found in Marks *et al.* (1999), Arnold *et al.* (1996), and Cline (1997a, b). These controls of surface melt are amenable to field measurement, making them reasonably well-understood, but they are difficult to quantify in a spatially distributed model due to their dependence on local meteorological conditions. The turbulent heat fluxes, Q_H and Q_L, for instance, are governed by local wind, humidity, and surface roughness properties, which exhibit substantial spatial and temporal variability. Other processes such as albedo evolution and snowpack hydrology are spatially complex and difficult to quantify in large-scale models.

The meteorological data demands and the spatial variability of governing processes make it difficult to apply a rigorous energy balance model to surface melt modeling in ice-sheet models. As a gross but tractable simplification, temperature-index models are widely used for estimation of surface melt (e.g. Fabre *et al.*, 1997, 1998; Huybrechts *et al.*, 1991; Letréguilly *et al.*, 1991; Tarasov & Peltier, 1999). These models make use of observations that air temperature is a strong indicator of radiative and heat energy available for melting, and parameterize meltwater generation, m, as a function of positive degree days (PDD),

$$m = d_{S/I}\,\text{PDD}, \qquad (11)$$

where $d_{S/I}$ is the degree-day melt factor for snow or ice, an empirically determined coefficient that differs for snow and ice to reflect the higher albedo of snow. Degree-day factors have units m water-equivalent melt production per °C. PDD is a measure of the integrated heat energy in excess of the melting point over the time interval of interest. Surface melt estimation from Eq. (11) is convenient because temperature (hence, PDD) is the only governing meteorological variable and is relatively amenable to spatial interpolation/extrapolation. A climate model temperature prediction for a region or synoptic grid cell can be distributed over a region through application of atmospheric-temperature lapse rates.

In ice-sheet modeling, with time steps of 1–10 yr, either monthly or mean annual temperatures are applied in estimating net annual melt through temperature-index models. Where monthly temperature fields are available, monthly mean temperatures are used to calculate PDD (Braithwaite, 1995; Braithwaite & Olesen, 1989),

$$\text{PDD} = \frac{\tau_m}{\sigma_m\sqrt{2\pi}} \int_0^{\infty} T \exp\left[-\frac{(T-\bar{T}_m)^2}{2\sigma_m^2}\right] dT, \qquad (12)$$

where τ_m is the length of the month, T is air temperature, $\bar{T}_m$ is monthly mean air temperature, and σ_m is the standard deviation in monthly temperature, ideally derived from hourly values in order to capture the critical influence of daily temperature maxima.

If annual mean climatology is applied, a sinusoidal temperature function is created to represent the annual temperature cycle, with knowledge of the summer temperature maximum, T_{max}, and the annual average, $\bar{T}_A$ (Reeh, 1991):

$$T_d(t) = \bar{T}_A - (T_{max} - \bar{T}_A)\cos\left(\frac{2\pi t}{\tau}\right), \qquad (13)$$

where T_d is the daily temperature, t is time in days, and τ is the length of the year (365.24 days). January 1 is taken to be $t = 0$. Net annual positive degree-days are then calculated through the integration

$$\text{PDD} = \frac{1}{\sigma_d\sqrt{2\pi}} \int_0^{\infty} T \left\{ \int_0^{\tau} \exp\left[-\frac{(T-T_d)^2}{2\sigma_d^2}\right] dt \right\} dT, \qquad (14)$$

where σ_d represents the standard deviation in daily temperature and T_d is taken from (13). Equation (14) is numerically integrated over the length of the year, τ.

Monthly or annual PDD go first towards melting of this year's snow accumulation, where available. If there is no remaining snow from the annual accumulation, it is assumed that melting has penetrated to the previous year's firn or ice surface, and any remaining melt energy is channeled into ice melt (e.g. Marshall *et al.*, 2000). Degree-day factors $d_{S/I}$ most commonly used in ice-sheet modeling are taken from Greenland Ice Sheet studies, although it is recognized that these values may not be appropriate for the extremely different radiation regime of the mid-latitude Pleistocene ice sheets. Even within Greenland, different sets of degree-day factors appear preferable in southern and northern sectors of the ice sheet (e.g. Bøggild *et al.*, 1994; Lefebre *et al.*, 2002).

Some fraction of surface melt percolates into the snowpack or pools on the surface and refreezes, particularly in the spring when overnight and snowpack temperatures are well below 0 °C. This refreezing is difficult to quantify, as a proper treatment of the process requires two- or three-dimensional modeling of surface runoff, meltwater percolation in the snowpack, and the detailed vertical temperature evolution of the snow and firn. The customary approach in ice-sheet models is to crudely estimate a fraction of the total melt that refreezes, of order 60% in the ice-sheet accumulation area. Refreezing values are much less in ablation zones or in alpine settings, of order 2–5% (Jóhannesson *et al.*, 1995). A study by Janssens & Huybrechts (2000) explored explicit modeling of the refreezing process in Greenland, concluding that differences from the assumption of 60% refreezing had only minor impacts on the modeled mass balance. However, the extent of refreezing will vary in different climatic and topographic settings, as well as within the melt season, so physically based models of the process are clearly preferable for application in different environments (e.g. ice-sheet ablation vs. accumulation zones; polar vs. mid-latitudes; steep, confined outlets vs. gently-sloping lobes).

The technical step of improving snow hydrology and melt rate calculations in ice-sheet models is within reach, and more physically based process models are likely to become the glaciological modeling standard over the next few years. This advance will be driven in part by the current effort to improve coupling between sophisticated atmospheric and ice-sheet models. It should be possible to move towards a more physical energy balance approach, including snow hydrology, following similar developments in AGCM simulation of present-day seasonal snowpack evolution (e.g. Marshall & Oglesby, 1994; Marshall *et al.*, 2001; Yang *et al.*, 1999).

An intermediate step towards the goal of more portable, physically based melt models is to include the effects of solar radiation in heat-index modeling. This can be done through a parameterization of the form

$$m = d_{S/I}\,\mathrm{PDD} + f_{S/I}(t) Q_V H_{\mathrm{PDD}}, \qquad (15)$$

where Q_V is incoming solar radiation and $f_{S/I}(t)$ is a radiation index for snow/ice melt (Cazorzi & Dalla Fontana, 1996; Hock, 1999). H_{PDD} is a Heavyside function that is equal to 1 for PDD > 0 and equal to 0 if PDD < 0. This precludes melting with sub-zero temperatures. Our radiation index $f_{S/I}(t)$ is time-dependent to allow the effects of changing surface albedo to be parameterized, although this has not yet been formulated in combined temperature-radiation index melt models. This relationship can be empirically calibrated through field observations, and it has important advantages over pure temperature-index modeling. In particular, incorporation of incoming solar radiation allows the effects of latitude and time of year (length of day, solar zenith effects), as well as terrain effects (shading, aspect) to be explicitly and objectively built into the melt model.

For distributed modeling, where spatial estimation of variables is required over a much larger area than is amenable to monitoring, Hock (1999) used potential direct solar radiation for Q_V in (15). Potential direct radiation can be calculated for any place and time, with shading effects incorporated through digital elevation models, as this is entirely governed by geometry. Pollard (1983) employed a surface melt parameterization of this type for Quaternary ice-sheet simulation, including potential direct solar radiation and surface temperature (rather than PDD) modeled by a global energy-balance model.

If sufficient insights about meteorological conditions are available, the radiation index $f_{S/I}$ could be made spatially variable, $f_{S/I}(\lambda, \theta, t)$, to parameterize the effects of cloud cover as well as changing albedo. This has not yet been attempted in ice-sheet models, but it is a promising avenue to pursue until a full energy balance becomes practical. Insights into regional and seasonal cloud conditions can be derived from paleoclimate simulations for different ice-sheet configurations, or more crudely based on local/regional precipitation rates.

Accumulation

Spatial-temporal reconstruction of past accumulation rates is a huge challenge in Quaternary ice-sheet simulations. There are two parts to the problem: (i) estimation of paleoprecipitation rates; and (ii) estimation of the fraction of precipitation to fall as snow. The former is a challenge because precipitation patterns have a complex dependence on synoptic circulation conditions and orographic forcing. Both synoptic conditions and surface topography change as ice sheets advance over the continents and global land and ocean surface conditions shift in the glacial world. Last Glacial Maximum atmospheric conditions have been extensively simulated (e.g. Fabre *et al.*, 1997, 1998; Ramstein *et al.*, 1997; Vettoretti *et al.*, 2000), revealing the impact of the Laurentide Ice Sheet in splitting the polar jet stream over North America and creating cool, wet conditions in southeastern North America. The latter observation is a result of both circulation shifts and orographic forcing of the southern Laurentide margin.

The region of intensified precipitation probably tracks the southern boundary of the ice sheet before and after the LGM. Climate models are just beginning to be applied to other temporal snapshots, with different ice-sheet configurations and climatic boundary conditions, so precipitation patterns at other periods during the last glacial cycle are a matter of speculation. General inferences can be made based on other paleoclimatic indicators. For instance, lake records in western North America and interior Alaska indicate that valley bottoms in these areas were largely ice-free during much of the last glacial period, an observation that requires more arid conditions than today. No ice-sheet model or couple ice-sheet climate simulation has been able to capture this yet; all studies to date predict a thick and pervasive Cordilleran Ice Sheet through most of the last glacial cycle (e.g. Marshall *et al.*, 2002; Tarasov & Peltier, 1997). Poorly resolved model topography in this region is part of the problem (Marshall & Clarke, 1999a), but it is also likely that important hydrological-cycle feedbacks associated with glacial climate are missing in these models. This is also true for parts of central and northeastern Asia, where models typically develop ice sheets in regions that are generally believed to have been ice-free at LGM.

High-resolution AGCM simulations focusing on the polar regions have had good success in modeling precipitation patterns over contemporary ice sheets (e.g. Bromwich *et al.*, 2001). No climate models that include advective atmospheric dynamics have ever been applied to glacial cycle integrations, however, due to the computational demands of long integrations. The principal approach to date has been to adopt present-day spatial precipitation patterns, locally scaled to reflect atmospheric aridity that accompanies cold glacial climates and elevation-desert effects (e.g. Tarasov & Peltier, 1997),

$$P(\lambda, \theta, t) = P(\lambda, \theta, 0) \exp[\beta_P (T_A(\lambda, \theta, t) - T_A(\lambda, \theta, 0))], \qquad (16)$$

where β_P is a parameter that introduces the decreasing moisture availability under cooling that one would expect in a Clasius-Clapeyron world. Glacial-period aridity in the polar regions is evident in ice-core accumulation-rate reconstructions in Greenland (Alley *et al.*, 1993), but as Kapsner *et al.* (1995) point out, changes in moisture supply are more complex than a simple thermodynamic control. Climate models suggest that this is particularly true at mid-latitudes (e.g. Hostetler *et al.*, 2000; Vettoretti *et al.*, 2000; Fig. 7). By fixing precipitation patterns to modern conditions, the effects of shifts in storm tracks and orographic forcing under different synoptic and topographic conditions are not incorporated. However, it has been defensibly argued that there is no objectively preferable alternative to (16), short of fully elaborated atmospheric dynamics.

The approach in Marshall *et al.* (2000, 2002) is to perturb modern observed precipitation rates locally as a function of the ratio of LGM vs. modern modeled precipitation, permitting some areas to be more wet and others more arid during the glaciation, as a function of AGCM-simulated patterns (e.g. Fig. 7). A harmonic combination of modern and LGM precipitation rates is adopted as a function of the Greenland ice-core glacial index introduced above.

This approach is helpful for LGM simulations but has severe deficiencies at other times during the glacial evolution. Greenland ice core $\delta^{18}O$ values approaching full glacial conditions at times other than LGM result in imposition of an LGM precipitation regime, even when the ice sheet is far from its southern limit. The regional precipitation patterns that are imposed by this treatment are certain to be incorrect. For instance, the increased precipitation rates seen over the southeastern Laurentide Ice Sheet lobes at LGM are a result of orographic forcing as well as anticyclonic circulation over the ice sheet (the Laurentide high), which forces the polar jet stream to the south and draws moist Atlantic air northwestwards, creating strong frontal precipitation. This precipitation belt will not be spatially fixed, but is likely to track the southern Laurentide margin as the ice sheet waxes and wanes. AGCM snapshots with different ice-sheet configurations/paleoclimate snapshots are required to capture this.

Given spatially distributed paleo-precipitation rates at a particular time, an arbitrary temperature cutoff, typically $T_A < 0\,^\circ C$ or $T_A < 1\,^\circ C$, is frequently used to estimate the fraction of precipitation to fall as snow, f_{snow}. This is a poor treatment with monthly or mean annual climatology, as it prohibits snowfall at mean values slightly above the threshold, while diurnal and daily variability will obviously give extensive periods with sub-freezing temperatures. Degree-day methodology offers a recommended alternative to estimate the fraction of precipitation to fall as snow. With monthly mean temperatures as in Eq. (12),

$$f_{snow} = \frac{1}{\sigma_m\sqrt{2\pi}}\int_{-\infty}^{T_S} \exp\left[-\frac{(T-\bar{T}_m)^2}{2\sigma_m^2}\right] dT, \quad (17)$$

where T_S is a threshold temperature for snowfall, taken to be $1\,^\circ C$ in Jóhannesson *et al.* (1995). With annual climatology as in Eqs (13) and (14), the equivalent integration is

$$f_{snow} = \frac{1}{\sigma_d\sqrt{2\pi}}\int_{-\infty}^{T_S} \frac{1}{\tau}\int_0^{\tau} \exp\left[-\frac{(T-\bar{T}_d)^2}{2\sigma_d^2}\right] dt\, dT. \quad (18)$$

In model implementations, atmospheric temperature lapse rates again become important, as for surface melt calculations.

Model Resolution Considerations for Mass Balance Modeling

Model resolution is an important technical consideration for simulation of both ablation and accumulation rates in coupled ice sheet-climate models. For marine calving parameterizations, the grounding zone transition between inland ice and floating ice is not well-resolved at the ca. 20-km resolution of ice-sheet models. Surface melt calculation is also known to be dependent on model resolution, as ice sheet ablation zones are narrow, relatively steep regions on the ice-sheet margins. These regions are not well-resolved in AGCMs, leading to substantial uncertainty in AGCM-driven melt calculations (Glover, 1999). High-resolution AGCMs have now been applied to Greenland, with improved results in modeling ice-sheet mass balance (Wild & Ohmura, 2000). These models are computationally expensive to run for a number of paleoclimatic time slices, but would be extremely helpful in capturing the ice-sheet topography that is critical for melt modeling. Alternatively, more coarse global AGCMs can be used to drive regional climate model simulations over the regions of interest.

Climate downscaling offers another avenue to take large-scale model fields and interpolate them to the scale of relevance for ice-sheet mass balance (Marshall & Clarke, 1999a; Thompson & Pollard, 1997). This approach requires assumption of atmospheric temperature lapse rates and parameterization of subgrid precipitation distribution, so contains inherent uncertainties. These uncertainties become more challenging with full energy-balance modeling, as winds, radiation fields, etc., all require subgrid interpolation (Glover, 1999). Substantial improvements are possible through physically based meteorological downscaling though, and ongoing efforts in the atmospheric modeling community should yield improvements in mass-balance modeling. It is practical to assume that climate models will never reach the resolution of interest for some phenomena of interest (e.g. km-scale hydrological and land surface processes; mountain glacier distributions). A combination of physically based downscaling strategies and regional/high-resolution climate models needs to be pursued.

Summary and Priorities for Model Development

Climate and ice-sheet models continue to improve, both in terms of model physics and technical capabilities. Considerable progress is being made in coupling ocean and atmosphere general circulation models, in regional climate modeling, and in simplified but comprehensive Earth/climate-system models (e.g. Ganopolski *et al.*, 2001; Petoukhov *et al.*, 2000; Weaver *et al.*, 1998, 2001). Receding computational limitations offer the possibility of further increases in model sophistication and resolution in the years ahead. However, new insights into the rich and complex behaviour of the climate system are arguably outpacing technical and technological improvements. It has become clear that fully coupled atmosphere-ocean-cryosphere-land surface models will be needed to address a number of paleoclimatic puzzles, particularly with respect to millennial climate variability.

Furthermore, the degree of coupling that is needed to explore centennial- and millennial-scale climate-system shifts is more intimate than previously imagined. Ideally, atmosphere-ocean-cryosphere-land surface models need to be simultaneously integrated for several millennia to explore Heinrich events or Dansgaard-Oeschger cycles in depth (e.g. Meissner *et al.*, 2002; Schmittner *et al.*, 2002). This is not feasible for AGCMs in the foreseeable future, so simplified climate-system models and clever asynchronous AGCM coupling strategies are needed. From the perspective of modeling glacial cycles, it will be a helpful first step to drive ice-sheet models with 10,000-year AGCM or OAGCM snapshots, something that has yet to be done. However, we already need to be thinking beyond this to explore the role of ice sheets in millennial-scale climate variability.

The matrix method discussed here, or some combination of matrix and snapshot techniques, offer avenues for improved paleoclimate reconstruction. Both techniques have the potential to give climatic fields that are more representative of concurrent ice sheet and orbital conditions than climate forcings that are currently in use. This will be an important step for allowing free exploration of ice-climate interactions and feedbacks through a glacial cycle. We also suggest that regional climate model simulations be carried out for each temporal snapshot or matrix element, for improved mass balance calculation over the ice sheets. Continued development of simpler climate models that are fully coupled with dynamic ice-sheet models offers an alternative and complementary approach to a suite of CPU-intensive OAGCM simulations (Purves & Hulton, 2000a; Schmittner *et al.*, 2002; Weaver *et al.*, 2001).

Both approaches – OAGCM snapshot/matrix methods and reduced models of the climate system – offer the possibility of free internal simulation of ice-sheet/climate feedbacks. This is important because current glacial cycle simulations that use GRIP/GISP climatology cannot address the climate dynamics that give rise to glacial cycles. Simulations with energy-balance models make it clear that ocean conditions and atmospheric greenhouse gas concentrations are essential internal climate variables for the 10^5-yr cycle (Deblonde & Peltier, 1993). These variables are not yet internally (freely) modeled in glacial cycle simulations. To derive a more complete understanding of the climate dynamics that give rise to glaciations and deglaciations, we anticipate a need for both OGCMs and carbon cycle modeling. The latter requires ocean and land surface schemes that treat CO_2 as an internal prognostic variable.

Further advances are also needed in ice-sheet modeling in order to address the puzzles posed by Heinrich events, Dansgaard-Oeschger cycles, and the evidence for thin LGM ice sheets. Current models do not adequately represent basal-flow and ice-stream dynamics, limiting their ability to emulate ice-sheet collapse or the complex southern lobe structure of the Laurentide Ice Sheet. The latter is key to meltwater routing, with implications for deepwater formation and D-O cycles (Clark *et al.*, 2001). The concerted effort to better understand West Antarctic Ice Sheet dynamics and the development of subglacial hydrological models should lead to improvements in the next generation of ice-sheet models, to help address millennial-scale variability in ice-sheet/climate models.

Acknowledgments

We thank the National Science Foundation Earth System History Program for support of the project Modeling Ice Sheet Evolution on Orbital and Millennial Time Scales (ATM-9905535). SJM receives addition support from the Climate System History and Dynamics Program, a research network sponsored by the Natural Sciences and Engineering Research Council of Canada. We thank an anonymous reviewer for helpful comments.

References

Abe-Ouchi, A. & Blatter, H. (1993). On the initiation of ice sheets. *Annals of Glaciology*, **18**, 203–207.

Alley, R.B., Anandakrishnan, S. & Jung, P. (2001). Stochastic resonance in the North Atlantic. *Paleoceanography*, **16**, 190–198.

Alley, R.B., Blankenship, D.D., Bentley, C.R. & Rooney, S.T. (1987). Till beneath ice stream B, 3, Till deformation: Evidence and implications. *Journal of Geophysical Research*, **92**(B9), 8921–8929.

Alley, R.B. & Clark, P.U. (1999). The deglaciation of the northern hemisphere: A global perspective. *Annual Reviews of Earth and Planetary Sciences*, **27**, 149–182.

Alley, R.B., Meese, D.A., Shuman, C.A., Gow, A.J., Taylor, K.C., Grootes, P.M., White, J.W.C., Ram, M., Waddington, E.D., Mayewski, P.A. & Zielinski, G.A. (1993). Abrupt increase in Greenland snow accumulation at the end of the Younger Dryas event. *Nature*, **362**, 527–529.

Anandakrishnan, S. & Alley, R.B. (1997). Stagnation of ice stream C, West Antarctica by water piracy. *Geophysical Research Letters*, **24**(3), 265–268.

Anandakrishnan, S., Alley, R.B., Jacobel, R.W. & Conway, H. (2001). The flow regime of ice stream C and hypotheses concerning its recent stagnation. *In*: Alley, R.B. & Bindschadler, R.A. (Eds), *The West Antarctic Ice Sheet: Behavior and Environment*. AGU Antarctic Research Series, **77**, 283–294.

Andrews, J.T., Erlenkeuser, H., Tedesco, K., Aksu, A.E. & Jull, A.J.T. (1994). Late Quaternary (stage 2 and 3) meltwater and Heinrich events, northwest Labrador Sea. *Quaternary Research*, **41**, 26–34.

Arnold, N.S. & Sharp, M.J. (2002). Flow variability in the Scandinavian ice sheet: Modeling the coupling between ice sheet flow and hydrology. *Quaternary Science Reviews*, **21**, 485–502.

Arnold, N.S., Willis, I.C., Sharp, M.J., Richards, K.S. & Lawson, M.J. (1996). A distributed surface energy-balance model for a small valley glacier. I. Development and testing for Haut Glacier d'Arolla, Valais, Switzerland. *Journal of Glaciology*, **42**(140), 77–89.

Bard, E., Hamelin, B. & Fairbanks, R.G. (1990). U-Th ages obtained by mass spectrometry in corals from Barbados: Sea level history during the past 130,000 years. *Nature*, **346**, 456–458.

Beckmann, A., Hellmer, H.H. & Timmermann, R. (1999). A numerical model of the Weddell Sea: Large scale circulation and water mass distribution. *Journal of Geophysical Research*, **104**(C10), 23,375–23,391.

Berger, A. & Loutre, M.F. (1991). Insolation values for the climate of the last 10 million years. *Quaternary Science Reviews*, **10**, 297–318.

Berger, A., Xi, X.S. & Loutre, M.F. (1999). Modeling Northern Hemisphere ice volume over the last 3 Ma. *Quaternary Science Reviews*, **18**, 1–16.

Birchfield, G.E., Weertman, J. & Lunde, A.T. (1982). A model study of the role of high-latitude topography and the climate response to orbital insolation anomalies. *Journal of the Atmospheric Sciences*, **39**, 71–87.

Björnsson, H. (1979). Glaciers in Iceland. *Jökull*, **29**, 74–80.

Blatter, H. (1995). Velocity and stress fields in grounded glaciers: A simple algorithm for including deviatoric stress gradients. *Journal of Glaciology*, **41**(138), 333–344.

Bøggild, C., Reeh, N. & Oerter, H. (1994). Modeling ablation and mass-balance sensitivity to climate change of Storstrømmen, Northeast Greenland. *Global and Planetary Change*, **9**, 79–90.

Bond, G.C., Broecker, W., Johnsen, S., McManus, J., Labeyrie, L., Jouzel, J. & Bonani, G. (1993). Correlations between climate records from North Atlantic sediments and Greenland ice. *Nature*, **365**, 143–147.

Bond, G.C., Heinrich, H., Broecker, W., Labeyrie, L., McManus, J., Andrews, J., Huon, S., Jantschik, R., Clasen, S., Simet, C., Tedesco, K., Klas, M., Bonani, G. & Ivy, S. (1992). Evidence of massive discharges of icebergs into the North Atlantic ocean during the last glacial period. *Nature*, **360**, 245–249.

Bond, G.C., Showers, W., Elliot, M., Evans, M., Lotti, R., Hajdas, I., Bonani, G. & Johnsen, S. (1999). The North Atlantic's 1–2 kyr climate rhythm: Relation to Heinrich events, Dansgaard/Oeschger cycles and the Little Ice Age. *In*: Clark, P.U., Webb, R.S. & Keigwin, L.D. (Eds), *Mechanisms of Global Climate Change at Millennial Timescales: Geophysical Monograph 112*. Washington, DC, American Geophysical Union, 35–58.

Boulton, G.S., Smith, G.D., Jones, A.S. & Newsome, J. (1985). Glacial geology and glaciology of the last mid-latitude ice sheets. *Geological Society of London Journal*, **142**, 447–474.

Braithwaite, R.J. (1995). Positive degree-day factors for ablation on the Greenland ice sheet studied by energy-balance modeling. *Journal of Glaciology*, **137**(41), 153–160.

Braithwaite, R.J. & Olesen, O.B. (1989). Calculation of glacier ablation from air temperature, West Greenland. *In*: Oerlemans, J. (Ed.), *Glacier Fluctuations and Climatic Change*. Kluwer Academic, Dordrecht, 219–233.

Broccoli, A.J. (2000). Tropical cooling at the last glacial maximum: An atmosphere-mixed layer ocean model simulation. *Journal of Climate*, **13**, 951–976.

Broecker, W.S., Bond, G., McManus, J., Klas, M. & Clark, E. (1992). Origin of the North Atlantic's Heinrich events. *Climate Dynamics*, **6**, 265–273.

Bromwich, D.H., Chen, Q., Bai, L., Cassano, E.N. & Li, Y. (2001). Modeled precipitation variability over the Greenland ice sheet. *Journal of Geophysical Research*, **106**(D24), 33,891–33,908.

Brown, C.S., Sikonia, W.G., Post, A., Rasmussen, L.A. & Meier, M.F. (1983). Two calving laws for grounded iceberg-calving glaciers. *Annals of Glaciology*, **4**, 295.

Budd, W.F. (1969). The dynamics of ice masses. *ANARE Scientific Reports*, **108**, 212 pp.

Budyko, M.I. (1969). The effect of solar radiation variations on the climate of the Earth. *Tellus*, **21**, 611–619.

Cazorzi, F. & Dalla Fontana, G. (1996). Snowmelt modeling by combining temperature and a distributed radiation index. *Journal of Hydrology*, **181**, 169–187.

Chappell, J. (2002). Sea level changes forced ice breakouts in the Last Glacial cycle: New results from coral terraces. *Quaternary Science Reviews*, **21**, 1229–1240.

Chappell, J., Omura, A., Esat, T., McCulloch, M., Pandolfi, J., Ota, Y. & Pillans, B. (1996). Reconciliation of late Quaternary sea levels derived from coral terraces at Huon Peninsula with deep sea oxygen isotope records. *Earth and Planetary Science Letters*, **141**, 227–236.

Charbit, S., Ritz, C. & Ramstein, G. (2002). Simulations of Northern Hemisphere ice-sheet retreat: Sensitivity to physical mechanisms involved during the Last Deglaciation. *Quaternary Science Reviews*, **21**, 243–265.

Charney, J.G., Fjörtoft, R. & von Neumann, J. (1950). Numerical integration of the barotropic vorticity equation. *Tellus*, **2**, 237–254.

Clark, C.D. & Stokes, C.R. (2001). Extent and basal characteristics of the M'Clintock Channel Ice Stream. *Quaternary International*, **86/1**, 81–101.

Clark, P.U. (1992). Surface form of the southern Laurentide Ice Sheet and its implications to ice-sheet dynamics. *Geological Society of America Bulletin*, **104**, 595–605.

Clark, P.U. (1994). Unstable behaviour of the Laurentide Ice Sheet over deforming sediment and its implications for climate change. *Quaternary Research*, **41**, 19–25.

Clark, P.U., Alley, R.B. & Pollard, D. (1999). Northern Hemisphere ice-sheet influences on global climate change. *Science*, **286**, 1104–1111.

Clark, P.U., Licciardi, J.M., MacAyeal, D.R. & Jenson, J.W. (1996). Numerical reconstruction of a soft-bedded

Laurentide Ice Sheet during the Last Glacial Maximum. *Geology*, **24**, 679–682.

Clark, P.U., Marshall, S.J., Clarke, G.K.C., Hostetler, S.W., Licciardi, J.W. & Teller, J.T. (2001). Freshwater forcing of abrupt climate change during the last glaciation. *Science*, **293**, 283–287.

Clark, P.U. & Mix, A.C. (2002). Ice sheets and sea level of the Last Glacial Maximum. *Quaternary Science Reviews*, **21**, 1–8.

Clark, P.U., Pisias, N.G., Stocker, T.S. & Weaver, A.J. (2002). The role of the thermohaline circulation in abrupt climate change. *Nature*, **415**, 863–869.

Clarke, G.K.C., Marshall, S.J., Hillaire-Marcel, C., Bilodeau, G. & Veiga-Pires, C. (1999). A glaciological perspective on Heinrich events. *In*: Clark, P.U., Webb, R.S. & Keigwin, L.D. (Eds), *Mechanisms of Global Climate Change at Millennial Timescales: Geophysical Monograph 112*. American Geophysical Union, Washington, D.C., 243–262.

Clayton, L., Teller, J.T. & Attig, J.W. (1985). Surging of the southwestern part of the Laurentide Ice Sheet. *Boreas*, **14**, 235–241.

Cline, D.W. (1997a). Effect of seasonality of snow accumulation and melt on snow surface energy exchanges at a continental alpine site. *Journal of Applied Meteorology*, **36**, 22–41.

Cline, D.W. (1997b). Snow surface energy exchanges and snowmelt at a continental, midlatitude Alpine site. *Water Resources Research*, **33**(4), 689–701.

Deblonde, G. & Peltier, W.R. (1991). Simulations of continental ice-sheet growth over the last glacial-interglacial cycle: Experiments with a one-level seasonal energy-balance model including realistic geography. *Journal of Geophysical Research*, **96**(D5), 9189–9216.

Deblonde, G. & Peltier, W.R. (1993). Pleistocene Ice Age scenarios based upon observational evidence. *Journal of Climate*, **6**, 709–727.

DeConto, R.M. & Pollard, D. (2003). Rapid Cenozoic glaciation of Antarctica induced by declining atmospheric CO_2. *Nature*, **421**, 245–249.

DeConto, R.M. & Pollard, D. (2003). A coupled climate-ice-sheet modeling approach to the early Cenozoic history of the Antarctic ice sheet. *Palaeogeography, Palaeoclimatology, Palaeoecology* (in press).

Doake, C.S.M. & Vaughan, D.G. (1991). Rapid disintegration of the Wordie ice shelf in response to atmospheric warming. *Nature*, **350**, 328–330.

Dyke, A.S. (1999). The last glacial maximum and the deglaciation of Devon Island: Support for an Innuitian Ice Sheet. *Quaternary Science Reviews*, **18**, 393–420.

Dyke, A.S. & Prest, V.K. (1987). Late Wisconsinan and Holocene history of the Laurentide Ice Sheet. *Géographie physique et Quaternaire*, **41**, 237–264.

Fabre, A., Ramstein, G., Ritz, C., Pinot, S. & Fournier, N. (1998). Coupling an AGCM with an ISM to investigate the ice sheets mass balance at the Last Glacial Maximum. *Geophysical Research Letters*, **25**(4), 531–534.

Fabre, A., Ritz, C. & Ramstein, G. (1997). Modeling of Last Glacial Maximum ice sheets using different accumulation parameterizations. *Annals of Glaciology*, **24**, 223–229.

Fairbanks, R.G. (1989). A 17,000-year glacio-eustatic sea level record: Influence of glacial melting rates on the Younger Dryas event and deep ocean circulation. *Nature*, **342**, 637–642.

Fisher, D.A. (1987). Enhanced flow of Wisconsin ice related to solid conductivity through strain history and recrystallization. *International Association of Hydrological Sciences*, **170**, 45–51.

Fisher, D.A., Reeh, N. & Langley, K. (1985). Objective reconstructions of the late Wisconsinan Laurentide Ice Sheet and the significance of deformable beds. *Géographie physique et Quaternaire*, **39**, 229–238.

Flowers, G.E., Björnsson, H. & Pálsson, F. (2003). New insights into the subglacial and periglacial hydrology of Vatnajökull, Iceland, from a distributed physical model. *Journal of Glaciology* (in press).

Flowers, G.E. & Clarke, G.K.C. (2002). A multicomponent coupled model of glacier hydrology, 1, Theory and synthetic examples. *Journal of Geophysical Research*, **107**, doi: 10.1029,2001JB001122.

Gallup, C.D., Cheng, H., Taylor, F.W. & Edwards, R.L. (2002). Direct determination of the timing of sea level change during Termination II. *Science*, **295**, 310–313.

Ganopolski, A., Petoukhov, V., Rahmstorf, S., Brovkin, V., Claussen, M., Eliseev, A. & Kubatzki, C. (2001). CLIMBER-2: A climate system model of intermediate complexity. Part II: Model sensitivity. *Climate Dynamics*, **17**, 735–751.

Ganopolski, A. & Rahmstorf, S. (2001). Rapid changes of glacial climate simulated in a coupled climate model. *Nature*, **409**, 153–158.

Genthon, C. & Krinner, G. (2001). Antarctic surface mass balance and systematic biases in general circulation models. *Journal of Geophysical Research*, **106**, 20,653–20,664.

Glen, J.W. (1958). The flow law of ice: A discussion of the assumptions made in glacier theory, their experimental foundations and consequences. *International Association of Hydrological Sciences*, **47**, 171–183.

Glover, R.W. (1999). Influence of spatial resolution and treatment of orography on GCM estimates of the surface mass balance of the Greenland Ice Sheet. *Journal of Climate*, **12**, 551–563.

Grootes, P.M., Stuiver, M., White, J.W.C., Johnsen, S.J. & Jouzel, J. (1993). Comparison of oxygen isotope records from GISP2 and GRIP Greenland ice cores. *Nature*, **366**, 552–554.

Hall, N.M.J., Valdes, P.J. & Dong, B. (1996). The maintenance of the last great ice sheets: A UGAMP GCM study. *Journal of Climate*, **9**, 1004–1019.

Hays, J.D., Imbrie, J. & Shackleton, N.J. (1976). Variations in the Earth's orbit: Pacemaker of the Ice Ages. *Science*, **194**, 1121–1132.

Heinrich, H. (1988). Origin and consequences of cyclic ice-rafting in the northeast Atlantic Ocean during the past 130,000 years. *Quaternary Research*, **29**, 141–152.

Hellmer, H.H. & Olbers, D. (1989). A two-dimensional model for the thermohaline circulation under an ice shelf. *Antarctic Science*, **1**(4), 325–336.

Hewitt, C.D. & Mitchell, J.F.B. (1997). Radiative forcing and response of a GCM to ice age boundary conditions: Cloud feedback and climate sensitivity. *Climate Dynamics*, **13**, 821–834.

Hock, R. (1999). A distributed temperature-index ice- and snowmelt model including potential direct solar radiation. *Journal of Glaciology*, **45**(149), 101–111.

Hoffman, P.F. & Schrag, D.P. (2000). Snowball Earth. *Scientific American* (January), 68–75.

Hostetler, S.W., Bartlein, P.J., Clark, P.U., Small, E.E. & Solomon, A.M. (2000). Simulated influences of Lake Agassiz on the climate of central North America 11,000 years ago. *Nature*, **405**, 334–337.

Hostetler, S.W., Clark, P.U., Bartlein, P.J., Mix, A.C. & Pisias, N.J. (1999). Atmopheric transmission of North Atlantic Heinrich events. *Journal of Geophysical Research*, **104**(D4), 3947–3952.

Hulbe, C.L., Joughin, I., Morse, D. & Bindschadler, R. (2000). Tributaries to West Antarctic ice streams: Characteristics deduced from numerical modeling of ice flow. *Annals of Glaciology*, **31**, 184–190.

Hulbe, C.L. & MacAyeal, D.R. (1999). A new thermodynamical numerical model of coupled ice sheet, ice stream, and ice shelf flow. *Journal of Geophysical Research*, **104**(B11), 25,349–25,366.

Hulbe, C.L. & Payne, A.J. (2001). The contribution of numerical modeling to our understanding of the West Antarctic Ice Sheet. *Antarctic Research Series*, **77**, 201–219.

Hulbe, C.L., Rignot, E. & MacAyeal, D.R. (1998). Comparison of ice-shelf creep flow simulations with ice-front motion of Filchner-Ronne Ice Shelf, Antarctica, detected by SAR interferometry. *Annals of Glaciology*, **27**, 182–186.

Hulton, N.R.J., Purves, R.S., McCulloch, R.D., Sugden, D.E. & Bentley, M.J. (2002). The Last Glacial Maximum and deglaciation in southern South America. *Quaternary Science Reviews*, **21**, 233–241.

Huybrechts, P. (1990). A 3-D model for the Antarctic Ice Sheet: A sensitivity study on the glacial-interglacial contrast. *Climate Dynamics*, **5**, 79–92.

Huybrechts, P., Letréguilly, A. & Reeh, N. (1991). The Greenland ice sheet and greenhouse warming. *Palaeogeography, Palaeoclimatology, Palaeoecology*, **89**(4), 399–412.

Huybrechts, P. & T'Siobbel, S. (1995). Thermomechanical modeling of northern hemisphere ice sheets with a two-level mass-balance parameterisation. *Annals of Glaciology*, **21**, 111–117.

Hyde, W., Crowley, T.J., Baum, S.K. & Peltier, W.R. (2000). Neoproterozoic 'snowball Earth' simulations with a coupled climate-ice-sheet model. *Nature*, **405**, 425–429.

Hyde, W., Crowley, T.J., Tarasov, L. & Peltier, W.R. (1999). The Pangean ice age: Studies with a coupled climate-ice-sheet model. *Climate Dynamics*, **15**, 619–629.

Imbrie, J., Berger, A., Bogle, E.A., Clemens, S.C., Duffy, A., Howard, W.R., Kukla, G., Kutzbach, J., Martinson, D.G., McIntyre, A., Mix, A.C., Molfino, B., Morley, J.J., Peterson, L.C., Pisias, N.G., Prell, W.L., Raymo, M.E., Shackleton, N.J. & Toggweiler, J.R. (1993). On the structure and origin of major glaciation cycles. 2. The 100,000-year cycle. *Paleoceanography*, **8**, 699–736.

Jacobs, S.S., Hellmer, H.H., Doake, C., Jenkins, A. & Frolich, R. (1992). Melting of ice shelves and the mass balance of Antarctica. *Journal of Glaciology*, **38**(130), 375–387.

Jacobs, S.S., Hellmer, H.H. & Jenkins, A. (1996). Antarctic ice sheet melting in the Southeast Pacific. *Geophysical Research Letters*, **23**(9), 957–960.

Janssens, I. & Huybrechts, P. (2000). The treatment of meltwater retention in mass-balance parameterisations of the Greenland Ice Sheet. *Annals of Glaciology*, **31**, 133–140.

Jenkins, A., Vaughan, D., Jacobs, S.S., Hellmer, H.H. & Keys, H. (1997). Glaciological and oceanographic evidence of high melt rates beneath Pine Island Glacier, West Antarctica. *Journal of Glaciology*, **43**(143), 114–121.

Jenson, J.W., MacAyeal, D.R., Clark, P.U., Ho, C.L. & Vela, J.C. (1996). Numerical modeling of subglacial sediment deformation: Implications for the behaviour of the Lake Michigan Lobe, Laurentide Ice Sheet. *Journal of Geophysical Research*, **101**(B4), 8717–8728.

Jenssen, D. (1977). A three-dimensional polar ice-sheet model. *Journal of Glaciology*, **18**, 373–389.

Jóhannesson, T., Sigurdsson, O., Laumann, T. & Kennett, M. (1995). Degree-day glacier mass-balance modeling with application to glaciers in Iceland, Norway, and Greenland. *Journal of Glaciology*, **41**, 345–358.

Johnson, J. & Fastook, J.L. (2002). Northern Hemisphere glaciation and its sensitivity to basal melt water. *Quaternary International*, **95–96**, 65–74.

Kamb, B. (1991). Rheological nonlinearity and flow instability in the deforming bed mechanism of ice stream motion. *Journal of Geophysical Research*, **96**, 585–595.

Kamb, B. & Echelmeyer, K. (1986). Stress-gradient coupling in glacier flow: I. Longitudinal averaging of the influence of ice thickness and surface slope. *Journal of Glaciology*, **32**(111), 267–284.

Kapsner, W.R., Alley, R.B., Shuman, C.A., Anandakrishnan, S. & Grootes, P.M. (1995). Dominant influence of atmospheric circulation on snow accumulation in Greenland over the past 18,000 years. *Nature*, **373**, 52–54.

Kennett, J.P. (1977). Cenozoic evolution of Antarctic glaciation, the Circum-Antarctic Ocean, and their impact on global paleoceanography. *Journal of Geophysical Research*, **82**, 3843–3860.

Lambeck, K., Smither, C. & Johnston, P. (1998). Sea-level change, glacial rebound and mantle viscosity for northern Europe. *Geophysical Journal International*, **134**, 102–144.

Le Treut, H. & Ghil, M. (1983). Orbital forcing, climatic interactions, and glaciation cycles. *Journal of Geophysical Research*, **88**, 5167–5190.

Lefebre, F., Gallee, H., van Ypersele, J.-P. & Huybrechts, P. (2002). Modelling of large-scale melt parameters with a regional climate model in South-Greenland during the 1991 melt season. *Annals of Glaciology*, **35**, 391–397.

Letréguilly, A., Reeh, N. & Huybrechts, P. (1991). The Greenland Ice Sheet through the last glacial-interglacial cycle.

Palaeogeography, Palaeoclimatology, Palaeoecology, **90**, 385–394.

Licciardi, J.M., Teller, J.T. & Clark, P.U. (1999). Freshwater routing by the Laurentide Ice Sheet during the last deglaciation. *In*: Clark, P.U., Webb, R.S. & Keigwin, L.D. (Eds), *Mechanisms of Global Climate Change at Millennial Timescales, AGU Geophysical Monograph 112*, 177–201.

Linsley, B.K. (1996). Oxygen-isotope record of sea level and climate variations in the Sulu Sea over the past 150,000 years. *Nature*, **380**, 234–237.

MacAyeal, D.R. (1989). Large-scale ice flow over a viscous basal sediment: Theory and application to Ice Stream B, Antarctica. *Journal of Geophysical Research*, **94**(B4), 4071–4087.

MacAyeal, D.R. (1992). Irregular oscillations of the West Antarctic ice sheet. *Nature*, **359**, 29–32.

MacAyeal, D.R. (1993). Binge/purge oscillations of the Laurentide ice sheet as a cause of the North Atlantic's Heinrich events. *Paleoceanography*, **8**, 775–784.

MacAyeal, D.R., Bindschadler, R.A. & Scambos, T.A. (1995). Basal friction of ice stream E, West Antarctica. *Journal of Glaciology*, **41**, 247–262.

MacAyeal, D.R., Rommelaere, V., Huybrechts, P., Hulbe, C.L., Determann, J. & Ritz, C. (1996). An ice-shelf model test based on the Ross Ice Shelf. *Annals of Glaciology*, **23**, 46–51.

Mahaffy, M.W. (1976). A three-dimensional numerical model of ice sheets: Tests on the Barnes Ice Cap, Northwest Territories. *Journal of Geophysical Research*, **81**(B6), 1059–1066.

Maqueda, M., Willmott, A.J., Bamber, J.L. & Darby, D.S. (1998). An investigation of the small ice cap instability in the Southern Hemisphere with a coupled atmosphere-sea ice-ocean-terrestrial ice model. *Climate Dynamics*, **14**, 329–352.

Marks, D., Domingo, J., Susong, D., Link, T. & Garen, D. (1999). A spatially distributed energy balance snowmelt model for application in mountain basins. *Hydrological Processes*, **16**, 1935–1959.

Marshall, S. & Oglesby, R.J. (1994). An improved snow hydrology for GCMs. Part I: Snow cover fraction, albedo, grain size, and age. *Climate Dynamics*, **10**, 21–37.

Marshall, S., Oglesby, R.J. & Nolin, A.W. (2001). Effect of western U.S. snow cover on climate. *Annals of Glaciology*, **32**, 82–86.

Marshall, S.J. & Clark, P.U. (2002). Basal temperature evolution of the North American Ice Sheets and implications for the 100-kyr glacial cycle. *Geophysical Research Letters*, **29**(18), doi: 10.1029/2002GL015192.

Marshall, S.J. & Clarke, G.K.C. (1997a). A continuum mixture model of ice stream thermomechanics in the Laurentide Ice Sheet 1. Theory. *Journal of Geophysical Research*, **102**, 20,599–20,614.

Marshall, S.J. & Clarke, G.K.C. (1997b). A continuum mixture model of ice stream thermomechanics in the Laurentide Ice Sheet 2. Application to the Hudson Strait Ice Stream. *Journal of Geophysical Research*, **102**, 20615–20638.

Marshall, S.J. & Clarke, G.K.C. (1999a). Ice sheet inception: Subgrid hypsometric parameterization of mass balance in an ice-sheet model. *Climate Dynamics*, **15**, 550–553.

Marshall, S.J. & Clarke, G.K.C. (1999b). Modeling North American freshwater runoff through the last glacial cycle. *Quaternary Research*, **52**, 300–315.

Marshall, S.J. & Cuffey, K.M. (2000). Peregrinations of the Greenland Ice Sheet divide in the last glacial cycle: Implications for central Greenland ice cores. *Earth and Planetary Science Letters*, **179**, 73–90.

Marshall, S.J., James, T.S. & Clarke, G.K.C. (2002). North American ice sheet reconstructions at the last glacial maximum. *Quaternary Science Reviews*, **21**, 175–192.

Marshall, S.J., Tarasov, L., Clarke, G.K.C. & Peltier, W.R. (2000). Glaciological reconstruction of the Laurentide Ice Sheet: Physical processes and modeling challenges. *Canadian Journal of Earth Sciences*, **37**, 769–793.

Maslin, M.A., Li, X.S., Loutre, M.-F. & Berger, A. (1998). The contribution of orbital forcing to the progressive intensification of Northern Hemisphere glaciation. *Quaternary Science Reviews*, **17**, 411–426.

Mathews, W.H. (1974). Surface profiles of the Laurentide ice sheet in its marginal areas. *Journal of Glaciology*, **13**, 37–43.

Matteucci, G. (1989). Orbital forcing in a stochastic resonance model of the late-Pleistocene climatic variations. *Climate Dynamics*, **3**, 179–190.

Meissner, K.J., Schmittner, A., Wiebe, E.C. & Weaver, A.J. (2002). Simulations of Heinrich Events in a coupled ocean-atmosphere-sea ice model. *Geophysical Research Letters*, **29**(14), 16:1–16:3.

Mix, A.C., Bard, E. & Schneider, R. (2001). Environmental Processes of the Ice Age: Land, Ocean, Glaciers (EPILOG). *Quaternary Science Reviews*, **20**, 627–657.

North, G.R. (1984). The small ice cap instability in diffusive climate models. *Journal of the Atmospheric Sciences*, **41**, 3390–3395.

North, G.R., Mengel, J. & Short, D. (1983). Simple energy balance model resolving the seasons and the continents: Application to the astronomical theory of the ice ages. *Journal of Geophysical Research*, **88**, 6576–6586.

Nye, J.F. (1952). The mechanics of glacier flow. *Journal of Glaciology*, **2**(12), 82–93.

Nye, J.F. (1959). The motion of ice sheets and glaciers. *Journal of Glaciology*, **3**(26), 493–507.

Oerlemans, J. (1980). Model experiments on the 100,000-yr glacial cycle. *Nature*, **287**, 430–432.

Ohmura, A., Wild, M. & Bengtsson, L. (1996). A possible change in mass balance of Greenland and Antarctic ice sheets in the coming century. *Journal of Climate*, **9**, 2124–2135.

Paillard, D. (1995). The hierarchical structure of glacial climatic oscillations: Interactions between ice-sheet dynamics and climate. *Climate Dynamics*, **11**, 162–177.

Paterson, W.S.B. (1991). Why ice-age ice is sometimes "soft". *Cold Regions Science and Technology*, **20**, 75–98.

Paterson, W.S.B. (1994). *The physics of glaciers* (3rd ed.). Elsevier Science Ltd, New York.

Payne, A.J. (1998). Dynamics of the Siple Coast ice streams, West Antarctica: Results from a thermomechanical ice-sheet model. *Geophysical Research Letters*, **25**(16), 3173–3176.

Payne, A.J. (1999). A thermomechanical model of ice flow in West Antarctica. *Climate Dynamics*, **15**, 115–125.

Payne, A.J., Huybrechts, P., Abe-Ouchi, A., Calov, R., Fastook, J., Greve, R., Marshall, S.J., Marsiat, I., Ritz, C., Tarasov, L. & Thomassen, M.P.A. (2000). Results from the EISMINT Phase 2 simplified geometry experiments: The effects of thermomechanical coupling. *Journal of Glaciology*, **46**, 227–238.

Peltier, W.R. (1987). Glacial isostasy, mantle viscosity, and Pleistocene climatic change. *In*: Ruddiman, W.F. & Wright, H.E., Jr. (Eds), *North America and Adjacent Oceans During the Last Deglaciation: The Geology of North America*. Geological Society of America, Boulder, CO, **K-3**, 155–182.

Peltier, W.R. (1994). Ice age paleotopography. *Science*, **265**, 195–201.

Peltier, W.R. (1996). Mantle viscosity and ice-age ice sheet topography. *Science*, **273**, 1359–1364.

Peltier, W.R., Goldsby, D.L., Kohlstedt, D.L. & Tarasov, L. (2000). Ice-age ice-sheet rheology: Constraints from the Last Glacial Maximum form of the Laurentide ice sheet. *Annals of Glaciology*, **30**, 163–176.

Petoukhov, V., Ganopolski, A., Brovkin, V., Claussen, M., Eliseev, A., Kubatzki, C. & Rahmstorf, S. (2000). CLIMBER-2: A climate system model of intermediate complexity. Part I: Model description and performance for present climate. *Climate Dynamics*, **16**, 1–17.

Pfeffer, W.T., Dyurgerov, M., Kaplan, M., Dwyer, J., Sassolas, C., Jennings, A., Raup, B. & Manley, W. (1997). Numerical modeling of late glacial Laurentide advance of ice across Hudson Strait: Insights into terrestrial and marine geology, mass balance, and calving flux. *Paleoceanography*, **12**, 97–110.

Pfeffer, W.T., Meier, M.F. & Illangasekare, T.H. (1991). Retention of Greenland runoff by refreezing: Implications for projected future sea-level change. *Journal of Geophysical Research*, **96**, 22,117–22,124.

Phillips, P.J. & Held, I.M. (1994). The response to orbital perturbations in an atmospheric model coupled to a slab ocean. *Journal of Climate*, **7**, 767–782.

Pollard, D. (1978). An investigation of the astronomical theory of the ice ages using a simple climate-ice-sheet model. *Nature*, **272**, 233–235.

Pollard, D. (1983). A coupled climate-ice-sheet model applied to the Quaternary ice ages. *Journal of Geophysical Research*, **88**(C12), 7705–7718.

Pollard, D., Clark, P., Hostetler, S. & Marshall, S. (2000b). Driving ice-sheet models using a generic matrix of GCM climates [abs]: Environmental Processes of the Ice Age: Land, Oceans, Glaciers (EPILOG) 2000 Workshop: Ice Sheets and Sea Level of the Last Glacial Maximum, Mt. Hood, OR, October 1–5, 2000.

Pollard, D. & PMIP Participating Groups (2000a). Comparisons of ice-sheet surface mass budgets from Paleoclimate Modeling Intercomparison Project (PMIP) simulations. *Global and Planetary Change*, **24**, 79–106.

Pollard, D. & Thompson, S.L. (1997a). Climate and ice-sheet mass balance at the last glacial maximum from the GENESIS version 2 global climate model. *Quaternary Science Reviews*, **16**, 841–864.

Pollard, D. & Thompson, S.L. (1997b). Driving a high-resolution dynamic ice-sheet model with GCM climate: Ice-sheet initiation at 116 Kyr BP. *Annals of Glaciology*, **25**, 296–304.

Purves, R.S. & Hulton, N.R.J. (2000a). Experiments in linking regional climate, ice-sheet models, and topography. *Journal of Quaternary Science*, **15**, 369–375.

Purves, R.S. & Hulton, N.R.J. (2000b). A climatic-scale precipitation model compared to the UKCIP baseline climate. *International Journal of Climatology*, **20**, 1809–1821.

Ramstein, G., Fabre, A., Pinot, S., Ritz, C. & Joussaume, S. (1997). Ice-sheet mass balance during the Last Glacial Maximum. *Annals of Glaciology*, **25**, 145–152.

Reeh, N. (1991). Parameterization of melt rate and surface temperature on the Greenland Ice Sheet. *Polarforschung*, **59**, 113–128.

Richardson, L.F. (1922). *Weather prediction by numerical process*. Cambridge University Press, Cambridge, UK.

Rignot, E.J., Gogineni, S.P., Krabill, W.B. & Ekholm, S. (1997). North and northeast Greenland ice discharge from satellite radar interferometry. *Science*, **276**, 934–937.

Rind, D. (1987). Components of the Ice Age circulation. *Journal of Geophysical Research*, **92**, 4241–4281.

de Robin, G.Q. (1955). Ice movement and temperature distribution in glaciers and ice sheets. *Journal of Glaciology*, **2**(18), 523–532.

Saltzman, B. & Verbitsky, M. (1993). Multiple instabilities and modes of glacial rhythmicity in the Plio-Pleistocene: A general theory of late Cenozoic climatic change. *Climate Dynamics*, **9**, 1–15.

Scambos, T.A., Hulbe, C.L., Fahnestock, M.A. & Bohlander, J. (2000). The link between climate warming and break-up of ice shelves in the Antarctic Peninsula. *Journal of Glaciology*, **46**(154), 516–530.

Schmittner, A., Yoshimori, M. & Weaver, A.J. (2002). Instability of glacial climate in a model of the ocean-atmosphere-cryosphere system. *Science*, **295**, 1489–1493.

Sellers, W.D. (1969). A global climatic model based on the energy balance of the Earth-atmosphere system. *Journal of Applied Meteorology*, **8**, 392–400.

Shackleton, N.J., Backman, J., Zimmerman, H., Kent, D.V., Hall, M.A., Roberts, D.G., Schnitker, D., Baldauf, J.G., Desprairies, A., Homrighausen, R., Huddlestun, P., Keene, J.B., Kaltenback, A.J., Krumsiek, K.A.D., Morton, A.C., Murray, J.W. & Westeberg-Smith, J. (1984). Oxygen isotope calibration of the onset of ice-rafting and history of glaciation in the North Atlantic region. *Nature*, **307**, 620–623.

Shennan, I., Peltier, W.R., Drummond, R. & Horton, B. (2002). Global to local scale parameters determining relative sea-level changes and the post-glacial isostatic adjustment of Great Britain. *Quaternary Science Reviews*, **21**, 397–408.

Shilts, W.W. (1980). Flow patterns in the central North American ice sheet. *Nature*, **286**, 213–218.

Stirling, C.H., Esat, T.M., Lambeck, K. & McCulloch, M.T. (1998). Timing and duration of the Last Interglacial: Evidence for a restricted interval of widespread reef growth. *Earth and Planetary Science Letters*, **160**, 745–762.

Suarez, M.J. & Held, I.M. (1979). The sensitivity of an energy balance climate model to variations in the orbital parameters. *Journal of Geophysical Research*, **84**, 4825–4836.

Tarasov, L. & Peltier, W.R. (1997). Terminating the 100 kyr ice age cycle. *Journal of Geophysical Research*, **102**, 21,665–21,693.

Tarasov, L. & Peltier, W.R. (1999). Impact of thermomechanical ice sheet coupling on a model of the 100 kyr ice age cycle. *Journal of Geophysical Research*, **104**(D8), 9517–9545.

Thompson, S.L. & Pollard, D. (1997). Ice-sheet mass balance at the last glacial maximum from the GENESIS version 2 global climate model. Greenland and Antarctic mass balances for present and doubled CO_2 from the GENESIS version-2 global climate model. *Journal of Climate*, **10**, 871–900.

Tulaczyk, S., Kamb, B. & Engelhardt, H. (2000a). Basal mechanics of Ice Stream B.I. Till mechanics. *Journal of Geophysical Research*, **105**, 463–481.

Tulaczyk, S., Kamb, B. & Engelhardt, H. (2000b). Basal mechanics of Ice Stream B. II. Plastic-undrained-bed model. *Journal of Geophysical Research*, **105**, 483–494.

Vaughan, D.G. & Doake, C.S.M. (1996). Recent atmospheric warming and retreat of ice shelves on the Antarctic Peninsula. *Nature*, **379**(6563), 328–331.

Vettoretti, G., Peltier, W.R. & McFarlane, N.A. (2000). Global water balance and atmospheric water vapour transport at Last Glacial Maximum. Climate simulations with the CCCma atmospheric general circulation model. *Canadian Journal of Earth Sciences*, **37**, 695–723.

Walland, D.J. & Simmonds, I. (1996). Sub-grid scale topography and the simulation of Northern Hemisphere snow cover. *International Journal of Climatology*, **16**, 961–982.

Weaver, A.J., Eby, M., Fanning, A.F. & Wiebe, E.C. (1998). The climate of the Last Glacial Maximum in a coupled atmosphere-ocean model. *Nature*, **394**, 847–853.

Weaver, A.J., Eby, M., Wiebe, E.C., Bitz, C.M., Duffy, P.B., Fanning, A.F., Holland, M.M., MacFadyen, A., Saenko, O., Schmittner, A., Wang, H. & Yoshimori, M. (2001). The UVic earth system climate model: Model description, climatology and applications to past, present and future climates. *Atmosphere-Ocean*, **39**(4), 361–428.

Weertman, J. (1961a). Equilibrium profile of ice caps. *Journal of Glaciology*, **3**(30), 953–964.

Weertman, J. (1961b). Stability of ice-age ice sheets. *Journal of Geophysical Research*, **66**, 3783–3792.

Weertman, J. (1976). Milankovitch solar radiation variations and ice age ice sheet sizes. *Nature*, **261**, 17–20.

Wild, M. & Ohmura, A. (2000). Change in mass balance of polar ice sheets and sea level from high-resolution GCM simulations of greenhouse warming. *Annals of Glaciology*, **30**, 197–203.

Yang, Z.-L., Dickinson, R.E., Hahmann, A.N., Niu, G.Y., Shaikh, M., Gao, X., Bales, R.C., Sorooshian, S. & Jinet, J.M. (1999). Simulation of snow mass and extent in global climate models. *Hydrological Processes*, **13**, 2097–2113.

Yokoyama, Y., Lambeck, K., Deckker, P., Johnson, P. & Fifield, K. (2000). Timing for the maximum of the Last Glacial constrained by lowest sea-level observations. *Nature*, **406**, 713–716.

Zweck & Huybrechts (2003). Modeling the marine extent of Northern Hemisphere ice sheets during the last glacial cycle. *Annals of Glaciology*, **37** (in press).

Permafrost process research in the United States since 1960

Bernard Hallet[1], Jaakko Putkonen[1], Ronald S. Sletten[1] and Noel Potter Jr.[2]

[1] *Quaternary Research Center, University of Washington, Box 351360, Seattle, WA 98195, USA; hallet@u.washington.edu*
[2] *Dickinson College, PO Box 1773, Carlisle, PA 17013, USA; pottern@dickinson.edu*

Introduction

Herein, we review significant scientific advances in understanding physical and chemical aspects of periglacial processes and the geomorphology of permafrost areas that have resulted from studies primarily conducted by United States researchers. We do not attempt to provide a comprehensive review but rather seek to briefly present the more significant advances in terms of new approaches, methodology, or insight gained into key processes. Because of space limitations, certain permafrost topics, particularly those that have been relatively lightly studied in the United States can only be acknowledged. These include pingos (Mackay, 1988a, b, 1990, 1998), palsas (Seppala, 1994), thaw lakes and thermokarsts (Dallimore *et al.*, 2000; Osterkamp *et al.*, 2000), and permafrost coastlines. For a comprehensive summary of fossil periglacial features in the United States, the reader is referred to Péwé (1983) and Washburn (1980).

We highlight advances in permafrost research from various sources that are widely dispersed in the scientific and engineering literature. Only a few publications are devoted entirely to permafrost topics; they include proceedings volumes from the international conferences on Permafrost held every 5 years since 1963, and the journal *Permafrost and Periglacial Processes* debuting in 1990. In addition to refereed journal articles, excellent reviews and considerable research have been published in reports by the United States Geological Survey and the Cold Regions Research and Engineering Laboratory (CRREL). *Geocryology* by A.L. Washburn (1980), which provided a thorough review of the field at that time, still stands out as the standard reference work on periglacial geomorphology.

Permafrost and Climate Change

Permafrost underlies approximately 25% of the world's land surface; it is widespread in high latitude and altitude regions, but is expected to diminish rapidly (Fig. 1) (Zhang *et al.*, 1999). Considerable permafrost also lies under the Arctic Ocean, on the northern continental shelves of North America and Eurasia; it is known as subsea or offshore permafrost. In this section we discuss three principal scientific aspects of terrestrial permafrost that are of special interest in view of the ongoing changes in climatic conditions: impact of climate change on permafrost, archives of climate change in permafrost, and cryosphere/climate change feedbacks.

Impact of Climate Change on Permafrost

Models and observations for contemporary greenhouse-induced climate change generally predict that warming will be substantially greater in high-latitude regions than the global average (Budyko & Izrael, 1992; IPCC, 2001; Maxwell & Barrie, 1989; Roots, 1989) although a recent analysis suggests that warming in the Arctic is not statistically greater than the average for the northern hemisphere (Polyakov *et al.*, 2002). The mean annual air temperature in the Arctic in 2071–2100 is predicted to be 4–6 °C higher than today (IPCC, 2001), and is expected to alter the surface energy balance and the distribution of permafrost profoundly, especially where soil temperatures are close to 0 °C (Nelson *et al.*, 1993; Osterkamp & Romanovsky, 1999; Riseborough & Smith, 1993). Indeed, recent environmental observations in the Arctic confirm the warming and show that it occurs primarily in the winter, which is attributed, at least partially, to increasing greenhouse aerosols (Serreze & Hurst, 2000). The 20th-century Arctic appears to be the warmest of the past 400 yr (Serreze *et al.*, 2000). In central Alaska, if temperatures in the upper 60 m of permafrost continue to rise at the current rate, 0.1 °C/yr, the upper 10 m of ice-bearing permafrost is projected to thaw within a century due to both air warming and increased insulation from snow (Osterkamp & Romanovsky, 1999). Boundaries between continuous permafrost, discontinuous permafrost, and no permafrost will shift several hundred kilometers to the north of their contemporary positions in both North America and Eurasia. The area underlain by permafrost is projected to decrease by 25–44%, a reduction of $6.3–11.3 \times 10^6$ km^2 (Anisimov & Nelson, 1996), and the area affected by rain on snow events and the consequent substantial soil warming will increase by 40% (Putkonen & Roe, 2003).

Specific effects of macro-scale climate change on permafrost are likely to be complex because of the nature of the interactions between climate, microclimate, and surface and ground thermal conditions (Goodrich, 1982; Nixon & Taylor, 1998; Zhang *et al.*, 1996). Changes in the thickness of the active layer are of particular interest because they have diverse and far-reaching implications as all hydrologic, geomorphic, pedogenic, chemical, and biological processes are sharply focused in this surface layer. The anticipated increase in active-layer thickness will also have direct societal consequences, aggravating problems associated with frost heave and differential thaw settlement (Nelson *et al.*, 2001). These changes are likely to damage and significantly increase the maintenance costs of houses, roads, airports, and other structures; hinder farming through thermokarst

DEVELOPMENTS IN QUATERNARY SCIENCE
VOLUME 1 ISSN 1571-0866
DOI:10.1016/S1571-0866(03)01007-8

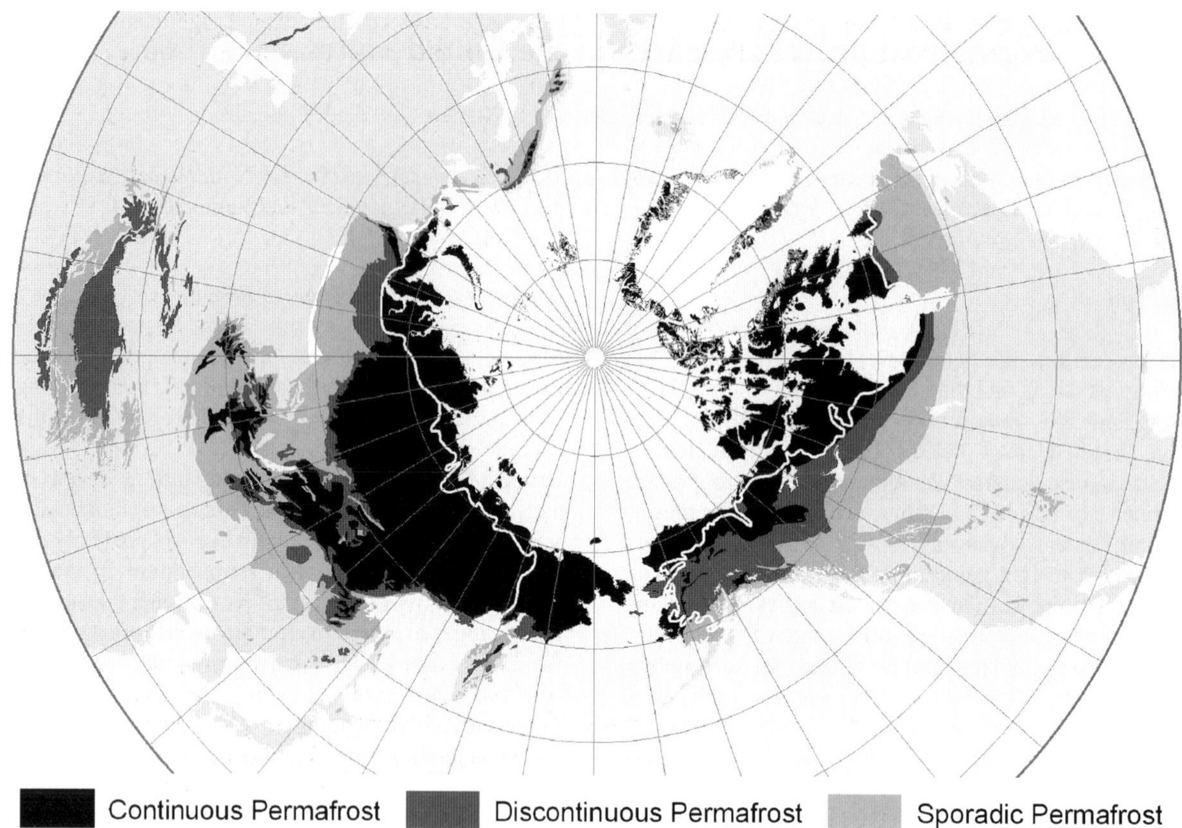

Fig. 1. Map of permafrost in the northern hemisphere indicating continuous, discontinuous, and sporadic permafrost, as well as treeline (modified from Brown et al., 1997).

formation; and increase slope instability problems (Judge & Pilon, 1983; Polar Research Board, 1983). Anisimov *et al.* (1997) estimate that by 2050, the Arctic warming will have caused a 20–30% increase in active-layer thickness for most of the permafrost areas in the northern hemisphere, with the largest relative increases in northernmost locations. Although a substantial range of permafrost response to global warming has been inferred, there is a general consensus that permafrost will degrade and the active layer will thicken.

An important trend in current research is the integration of knowledge and models developed in the permafrost community with general climate models to address the coupling between climate and permafrost. As our understanding of the effects of the climate change on aspects of permafrost improves, increasing research attention focuses on the feedbacks between various components of the climate/permafrost system (e.g. snow and plant cover; soil moisture and related changes in soil insulation). It is likely that the simple picture of degrading permafrost will become richer; some areas may experience surprising deviations from general trends in permafrost conditions, and unexpected behavior may emerge from the complex interactions between climate and permafrost, which both affect and are affected by the vegetation and snow cover. Tracking the annual change in the active layer depth is the focus of the international Circumpolar Active Layer Monitoring (CALM) network (Anisimov & Nelson, 1996; Brown *et al.*, 1997, 2000; Nelson *et al.*, 1998a, b).

Archives of (100–100,000 year) Climate Change in Permafrost

Abandoned drill holes in northern Alaska are rich archives of integrated past surface temperatures (Lachenbruch & Marshall, 1986; Lachenbruch *et al.*, 1982) that clearly record substantial warming (2–3 °C) of the top of the permafrost through the 20th century and shorter-term cyclic variations with similar amplitude (Osterkamp & Romanovsky, 1996; Osterkamp *et al.*, 1994). This warming is largely consistent with the few available instrumental records of generally increasing air temperature through the 20th century. However, geothermal records from other parts of North America reflect large spatial heterogeneity in ground thermal regime (e.g. Wang & Allard, 1995), and the interpretation of borehole temperature in terms of climate change is generally ambiguous because the temperature of permafrost is a complex function of air temperature, modulated by snow, plant cover, and active-layer processes. Temporal changes in the density, depth or structure of snow (Osterkamp & Romanovsky, 1999; Zhang *et al.*, 1996), active-layer depth, soil water/ice content, and rain-on-snow events (Putkonen, 1998), vegetation (Chapin *et al.*, 2000) could alter permafrost temperatures, independent of changes in air temperature.

Short-term (~100 yr) changes in atmospheric forcing primarily affect the maximum depth of the active layer and the temperature of the upper 100 m of permafrost. In contrast, the large glacial interglacial shifts in Quaternary climate should affect both the depth of permafrost and its temperature throughout (Harrison, 1991). For thick permafrost (~500 m), the depth adjustment lags the climatic forcing by about 20,000 yr. The permafrost thickness near Prudhoe Bay, Alaska probably varied by only about ±10% (±60 m) over the last glacial cycle, assuming surface temperatures were 6–8 °C lower than today (Osterkamp & Gosink, 1991). Similar modeling results suggest that the formation time of 1500 m-thick permafrost is of the order of millions of years (Lunardini, 1995). The evolution of permafrost in the vast regions formerly occupied by Quaternary ice sheets, which profoundly alter the energy exchange at the top of permafrost, is being examined in the latest ice-sheet models in which the thermal and dynamic evolution of the ice sheets is coupled to the evolution of the permafrost (Koutnik & Marshall, 2002).

Cryosphere/Climate Change Feedbacks

The degradation of permafrost, and decrease in snow cover induced by climate change can exacerbate global climate change due to the interactions between carbon and water cycling, and associated changes in the land-atmosphere energy and moisture exchange through a major portion of the earth's land area. Over the Holocene, the Arctic has been a major sink for carbon through the accumulation of plant material in soils and large peat bogs. Peat covers 3.5×10^6 km^2 in the boreal and subarctic peatlands with an average thickness of 2.3 m; its dry mass is estimated as 4.55×10^{14} kg, about a third of the total world pool of soil carbon (Gorham, 1991). Studies of contemporary carbon flux through the ground surface in northern Alaska and other regions show, however, that carbon is now being released to the atmosphere at rates up to about one tenth of those released globally by fossil-fuel usage (Oechel & Vourlitis, 1994), through increased drainage of the Arctic soils, higher winter soil temperature, and deepening seasonal thaw. Estimates of long-term carbon dynamics of the entire Arctic and feedback on climate change depend on projected shifts in tundra types across the Arctic (Serreze & Hurst, 2000; Smith & Riseborough, 1996). Replacing polar deserts with semi-deserts or semi-deserts with wet sedge tundra, for example, would increase the magnitude of net carbon exchange by 50–100% and would likely increase carbon sequestration across the vast landscapes of Canada, Russia, and Scandinavia with feedbacks to surface energy balances, albedo, and climate over vast regions (Chapin III *et al.*, 2000; Christensen & Kuhry, 2000).

Projected changes in the Arctic can also feedback on the climate system in other ways that are difficult to quantify. On land, carbon exchange between soil and the atmosphere over large areas can be affected significantly by numerous landscape processes that include enhanced thermokarst activity, soil creep, cryoturbation, and coastal cliff erosion (e.g. Jorgenson *et al.*, 2001). The warming of permafrost, primarily below sea level, can further contribute to global warming through the release of methane that is trapped in massive quantities within and below the permafrost, largely as a clathrate hydrate (Buffett, 2000). Recent studies of large and abrupt climate changes that may have been caused by the rapid release of this effective greenhouse gas from the vast reservoir of frozen methane in ocean sediments highlights the potentially critical role of this component of the Earth climate system (Kennett *et al.*, 2003; Nisbet, 2002).

Profound changes in hydrology are likely to results from the large anticipated warming of the permafrost in the Arctic. Such changes are important not only because they affect the sequestration of atmospheric CO_2 in Arctic soils but they are also likely to alter the freshwater input into the Arctic Ocean which feeds into the North Atlantic. The Atlantic freshwater budget is of special interest because it seems to play a key role in simulation of rapid climate change of the Quaternary including Dansgaard-Oeschger and Heinrich events, that can be triggered by relatively small perturbations in freshwater flux (Ganopolski & Rahmstorf, 2001). Moreover, this freshwater budget is getting increasing attention because of its potential impact on ocean stability and the global thermohaline circulation. Simulations by Stocker & Schmittner (1997) showed that when the fossil-fuel burning rate exceeds a particular threshold, the modern climate could flip to another mode with regional cooling over the North Atlantic. According to Bard's (2002) excellent review of recent, abrupt climate change research "The cooling is associated with a strong decrease in the Atlantic overturning, reminiscent of the scenario that prevailed during the Heinrich events." How close we are to such a flip is an urgent problem, particularly in view of the rapid freshening of North Atlantic water that has been observed over the last few decades (Dickson *et al.*, 2002).

On a much longer time scale ($>10^6$ yr), atmospheric CO_2 is modulated by the rate of weathering of silicate minerals, which constitutes another linkage between terrestrial processes and atmospheric CO_2. In the Arctic, it is generally expected that minerals dissolve slowly due to low temperature, sparse vegetation, and the scarcity of unfrozen water in winter. A richer picture emerges, however, from geochemical investigations on large high-latitude rivers such as the Lena, Omoloy, Yana, Indigirka, Kolyma, and Anadyr draining the Siberian Arctic (Edmond *et al.*, 1996; Gordeev & Sidorov, 1993; Huh & Edmond, 1999; Huh *et al.*, 1998a, b); and the Mackenzie (Reeder *et al.*, 1972), Fraser (Cameron *et al.*, 1995), and Yukon draining the Rocky Mountains. While solute fluxes and CO_2 consumption rates in the Siberian Arctic are less than those estimated in tropical areas (Amazon-Orinoco), the Mackenzie, Yukon, and Fraser Rivers have solute and CO_2 fluxes comparable to the tropical rivers (Huh & Edmond, 1999; Huh *et al.*, 1998a, b). This suggests that silicate weathering in permafrost areas is an important component of the long term carbon budget despite the low temperatures. An increase in thaw depths due to climate warming would lead to an increase of High Arctic CO_2 consumption by accelerating weathering as access to reactive surfaces and precipitation increases.

Fundamental Permafrost Processes

Heat Transport

The exchange of energy between the permafrost and atmosphere is confounded by an ensemble of factors, mentioned earlier, that modify the transfer of thermal energy from the snow surface to the soil beneath the snow pack. Considerable effort has been devoted to observing and modeling active-layer temperatures, thermal properties and heat-transfer processes (Hinkel & Outcalt, 1994; Hinzman *et al.*, 1998; Kane *et al.*, 1991; McKay *et al.*, 1998; Osterkamp & Gosink, 1991; Osterkamp & Romanovsky, 1996, 1999; Osterkamp *et al.*, 1994; Putkonen, 1998; Romanovsky & Osterkamp, 1995; Zhang & Osterkamp, 1993; Zhang *et al.*, 1996, 1997). The monitoring of thermal and other ground conditions has been greatly facilitated by recent advances in the electronic industry that have brought sophisticated data loggers and sensor technology within reach of the broad earth-science community. The main components of the micrometeorological surface energy balance are now routinely monitored by researchers in the field, along with soil temperatures, soil moisture content (time-domain reflectometry type probes), thermal properties and fluxes (soil heat flux plate, thermal conductivity, and heat capacity probes), and snow characteristics. Increasing amounts of data are available in public archives (International Permafrost Association, 1998).

The temporal evolution of soil thermal profiles can be simulated well utilizing simple conduction models that use realistic soil thermal properties varying with soil moisture, temperature, and depth. Non-conductive heat-transfer processes (Hinkel & Outcalt, 1993, 1994; Roth & Boike, 2001), involving vapor transport for example, appear important in some cases but there is no consensus about the circumstances under which they are significant in the soil energy balance.

Large integrated research initiatives in the Arctic (Arctic System Science/Land-Air-Ice Interactions) have recently provided much needed impetus for the permafrost community to expand from the traditional plot level research to regional observational networks (Brown *et al.*, 2000; Shiklomanov & Nelson, 2002) and regional permafrost thermal models (Hinzman *et al.*, 1998) to provide data products necessary for integrated regional climate and ecological models (Lynch *et al.*, 1999; McGuire *et al.*, 2000).

Freezing in Porous Media and Frost Heaving

The recurrent expansion and contraction of soils and rocks induced by ice growth and melting are fundamental to the principal periglacial processes, including size sorting, patterned ground formation, solifluction, and frost weathering. Contrary to popular belief, the 9% volumetric expansion associated with the water/ice phase transition has essentially nothing to do with frost heaving in soils. In fact, a number of experiments have demonstrated frost heaving by materials that contract upon freezing (Taber, 1929), most recently using Helium (Dash, 1992) and Argon (Zhu *et al.*, 2000). For all practical purposes, frost heaving in nature is due to the addition of water to freezing soils as water is thermodynamically drawn to the freezing front from nearby areas (e.g. Black & Hardenberg, 1991; Henry, 2000).

Frost Heaving

Remarkable insight about frost-heaving phenomena was gained long ago from simple experiments by Taber (1929, 1930). More recently, two forms of frost heaving in soils have been recognized. Primary heaving occurs due to relatively rapid freezing along a simple ice/water interface under minimal confining pressure; its most common manifestation in nature is the appearance of needle ice at the surface of moist soils during a cool clear night. Needle ice growth requires subfreezing temperatures at the soil surface, ample moisture, and soil sufficiently permeable to permit rapid moisture migration to the freezing front, yet fine-grained enough to inhibit ice growth significantly within the soil (Outcalt, 1971). This inhibition arises largely from the depression of the freezing point due to the ice/water interface curvature for ice grains confined to small pores (Cahn *et al.*, 1992; Everett, 1961; Miller, 1973). The physics of needle ice growth has much in common with the observed tendency for an upward-propagating freezing front to reject small non-buoyant particles in the water (Gilpin, 1980; Rempel *et al.*, 2001; Rempel & Worster, 2001).

Secondary heaving (Miller, 1973) generally occurs throughout a freezing soil mass; it is characterized by the relatively slow growth of ice-lenses, also known as segregation ice, fueled by the flow of water through a frozen fringe and by the migration of this ice fringe through the soil caused by temperature gradients (O'Neill & Miller, 1985; Rempel *et al.*, 2003). Secondary frost heaving has been studied extensively because of its geotechnical and geomorphic importance. Several recent quantitative models succeed in accounting for the known dependence of heave rate on temperature gradients and restraining pressure, and for the thickness and spatial arrangement of ice lenses (Gilpin, 1980; O'Neill & Miller, 1985). The model of O'Neill & Miller (1985) is the standard reference analyzing the physical basis of secondary heaving; it has been the subject of further mathematical analysis (Fowler & Krantz, 1994; Krantz & Adams, 1996; Rempel *et al.*, 2003). Although these models have been strikingly successful in simulating the salient aspects of the complex frost-heaving phenomenon they are highly idealized and remain to be verified rigorously.

Unfrozen Water and Solutes

The presence of unfrozen water in frozen soils is central to frost-heaving, and related frozen soil processes, including the transport of water and solutes, the rate and type of biochemical processes, and the persistence of biological activity through the winter (e.g. Schimel & Clein, 1996). The amount of unfrozen water also strongly influences the rheology and other mechanical properties of frozen soils. The dependence of this water on soil temperature, texture and mineralogy, and water composition, has long been the subject of attention within the permafrost community, as highlighted in the comprehensive review by Anderson & Morgenstern (1973). In clays, considerable water remains unfrozen to very low temperatures, as much as 20% at $-50\,^{\circ}$C (Low *et al.*, 1968; Smith & Tice, 1988; Tice *et al.*, 1984). This unfrozen water cannot be attributed to supercooling; rather it reflects the anomalously low free energy of water existing in thin films between ice and soil particles relative to bulk water. The free energy can also be lowered by solutes, and hence the amount of water remaining unfrozen increases with the solute content of the pore water, as measured by Tice *et al.* (1984) and Smith & Tice (1988). A chemical equilibrium model utilizing Pitzer equations to estimate ion activity in aqueous electrolyte solutions from -63 to $+25\,^{\circ}$C (Marion & Grant, 1994; Mironenko *et al.*, 1997) and more recent work (Grant & Sletten, 2002; Marion, 2001, 2002; Marion & Farren, 1999; Mironenko *et al.*, 2001) permit modeling how solute-rich waters develop as solutions freeze, leaving the solutes in decreasing volumes of remaining water. Strong solute enrichment in freezing bulk aqueous solutions (Gross, 1967; Gross *et al.*, 1977; Harrison & Tiller, 1963; Jaccard & Levi, 1961; Malo & Baker, 1968; Sletten, 1993) significantly lowers the freezing point. Ionic exclusion in solutions contained within porous media tends to be less effective than in bulk solutions, and it varies considerably among experiments (Leung & Carmichael, 1982; Wilson & Vinson, 1983).

Within the last decade, diverse advances in condensed matter physics dealing with ice surfaces and interfaces, and more generally with phase changes in confined spaces, have contributed considerably to the understanding of liquid water and freezing in porous media. Theoretical insights into the pre-melting phenomena, the occurrence in most materials of liquid films on surfaces and interfaces at temperatures distinctly below their bulk melting temperature, have shed considerable light on diverse aspects of ice in the environment (Dash *et al.*, 1995; Wettlaufer & Dash, 2000). The amount of unfrozen water in saturated porous media depends primarily on the curvature of the ice-water interface, van der Waals interactions between solid surfaces and nearby water, and solutes (Cahn *et al.*, 1992; Gay *et al.*, 1992; Wettlaufer *et al.*, 1997). The migration of this water along free-energy gradients and associated frost heaving have been examined closely (Rempel *et al.*, 2001; Wettlaufer, 2001; Wettlaufer & Worster, 1995; Wettlaufer *et al.*, 1997). This theoretical work has been closely connected with a suite of experimental studies using powerful new techniques, including time-domain reflectometry (Fu, 1993), atomic force microscopy (Pittenger *et al.*, 2001), optical interferometry to examine frost heaving on a single interface (Wilen & Dash, 1995), and small-scale experimental studies of frost heave (Dash, 1992; Zhu *et al.*, 2000). This collection of papers provides a powerful unifying theoretical and experimental underpinning for fundamental processes underlying much of permafrost research in the language of modern physics.

Freezing in Rock and Frost Weathering

Ice growth in rocks also occurs with migration of water to freezing centers much like segregation ice grows in soils, and it physically weathers the rock, as has been long recognized (Taber, 1950; Walder & Hallet, 1986). This should not be surprising because experiments with soils in a stiff apparatus connected to a water reservoir at atmospheric pressure have shown that ice growth in this hydrologically open system can generate pressures exceeding 20 MPa (Radd & Oertle, 1973). Pressures of this magnitude inside rock cracks are ample to fracture any rock. Indeed, experiments using acoustic emissions to record and locate fracture propagation events show that these events are not associated with the nominal freezing temperature of essentially $0\,^{\circ}$C (Hallet *et al.*, 1991); most of the fracture activity occurs at distinctly lower temperatures between -3 and $-6\,^{\circ}$C, in accord with theoretical predictions (Walder & Hallet, 1985). The 9% water-to-ice expansion, which is generally assumed to cause frost weathering, is undeniable but may rarely be significant in nature because rocks are seldom nearly saturated (Hall, 1986), and because the rapid freezing required to build up high transient pore pressures are not common, particularly at a significant depth below the rock surface. Accordingly, the long-standing notion that the frequency and amplitude (intensity) of freeze-thaw cycles are the dominant environmental determinant of frost weathering should be questioned. These cycles may, nevertheless, be important not because they cross $0\,^{\circ}$C, but because large temperature gradients arise as the rock is brought into the critical sub-zero temperature range (generally -3 and $-6\,^{\circ}$C) favorable for segregation ice growth. These ideas pave the way to a more fundamental understanding of the dependence of frost weathering on lithology and climate, present (e.g. Anderson, 1998) and past. They offer an attractive rationale for resolving problematic issues that have arisen in studies of frost weathering in the context of Quaternary soils and deposits, which reflect relatively intense frost action under former climates. They are also likely to lead to a sounder understanding of cirque development, rock-fall activity, and the production of rock debris that accumulates on talus and glaciers in alpine areas.

Soils and Patterned Ground

The physical and chemical processes reviewed above produce soil and landscape features that are unique to, or at least characteristic of, permafrost or frost influenced areas. Recurrent ice growth and thaw in soils of cold regions produce a rich diversity of patterns in soils that are defined by local relief, soil cracks, segregation of mineral material according to size, or differences in vegetation. These patterns can be subdivided into sorted patterns, which range in size from 0.1

to 3 m (Fig. 2) and are defined by spatial differentiation of soil material according to grain size, and non-sorted patterns. Among the latter, the most prominent variety is polygonal patterned ground on a scale of 10–30 m. In this section we review advances in understanding characteristic pedological features and patterned ground structures characteristic of periglacial areas.

Soil Processes

Polar soils and the primary soil-forming process have been systematically described for the Arctic (Tedrow, 1977) and Antarctic (Campbell & Claridge, 1987). The cold, arid conditions typical of Antarctica lead to slow soil development and salt accumulation that have been compared to desert soils (Claridge & Campbell, 1982). Soils in the Artic (Tedrow & Brown, 1962; Ugolini, 1986) and Antarctic (Bockheim & Ugolini, 1990) tend to be progressively less weathered poleward. The dominant pedogenic processes in these periglacial regions are the formation of oxide coatings, particle redistribution, salt accumulation, and physical weathering; conversion of primary silicate minerals to clay minerals is of secondary importance. Bockheim (1979, 1980, 1990) has used various soil-development indices (i.e. salt accumulation, oxidation) for relative-age dating of geomorphic features in Antarctica. Soil development is often mediated by cryoturbation which disrupts normal soil profile development (Ugolini *et al.*, 1973).

The role of freeze-thaw on the physical chemistry of soil profiles was reviewed by Marion (1995). Ionic exclusion during freezing of soil waters leads to high ionic concentrations and high unfrozen water contents at subzero temperatures, as discussed above. High solute concentrations can induce mineral precipitation (Sletten, 1987), and significant unfrozen water content may enable significant organic matter decomposition and below ground respiration while the soil is at subzero temperatures, especially during the transitional period before and after the summer season (Fahnestock *et al.*, 1998, 1999). Based on field studies conducted during the summer, Artic ecosystems appeared to be transitioning from a carbon sink to a carbon source. It is now recognized that significant, possibly even higher carbon efflux from the soil to the atmosphere occurs during the autumn to spring period when the temperatures are below 0 °C and any flux measurments that do not include this cold period underestimate the total carbon release from these systems (Oechel *et al.*, 1997). Unfrozen water during these periods is likely to be the principal factor controlling carbon flux from periglacial soils; direct studies of the linkage between unfrozen water and carbon flux are underway. Another factor that appears to control the soil carbon budget is the soil acidity, which appears to be primarily affected by eolian input of carbonate dust and the thickness of organic matter over mineral soil (Bockheim *et al.*, 1998; Walker *et al.*, 1998).

Overall physical and chemical weathering rates have been assessed in high-Arctic catchment basins by measuring

Fig. 2. Sorted stripes spaced 120–150 mm apart extend downslope for tens of meters on a ≅15° slope near the summit of Mauna Kea, Hawaii; coarse-grained stripes are comprised of pebbles 20–30 mm across. This type of patterned ground arises from frequent soil displacements due to needle ice and ice-lens growth on cold clear nights; it has been modeled numerically (Werner & Hallet, 1993).

dissolved and particulate transport (e.g. Hasholt & Hagedorn, 2000). They found that basins dominated by sedimentary and basaltic bedrock have 2–5 higher silicate denudation rates than high-grade metamorphic rocks; furthermore, absolute chemical denudation rates of soluble elements can exceed physical denudation due to redeposition within the catchment. Solute transport appears highly episodic occurring in part during storm events, as noted both in translocation of organics and metals in soil horizons (Stoner & Ugolini, 1988). Changes in Sr isotope composition in soil water collected over the season document the dependency of water composition on season. Abrupt changes in Sr isotope ratios during heavy rain events accompanied by mobilization of Al and Fe reflect the accumulation of relatively unstable metal phases and their sporadic release during extreme weather conditions (Hagedorn & Sletten, 2001).

Sorted Patterned Ground

Field and Laboratory Studies

A long history of observations has shed considerable light on patterned ground, as thoroughly reviewed by Washburn (1980, 1997). Since Washburn's *Geocryology* was published in 1980, only a few substantial field studies have been conducted. They involve detailed examination of active patterned ground, featuring high resolution mapping and extensive observations of displacements of soil and stones over periods of years (Hallet *et al.*, 1988; Walters, 1988; Washburn, 1989), systematic studies of the developmental stages of patterned ground (Washburn, 1997), and extensive use of electronic instrumentation for automated measurements of key properties and processes in active patterned ground, including temperature, soil heave and heaving strain, soil rotation, pore pressure, and total soil pressure (e.g. Hallet, 1998). Measurements confirm earlier observations that, in particularly active sorted circles, the periodic heaving and settling of the soil leads to residual long-term circulatory soil motion at rates of approximately 1 cm/yr (e.g. Hallet & Prestrud, 1986) with radial divergence from circle centers, much as was envisioned by Pissart (1990).

Laboratory studies of patterned ground formation have not been very successful in simulating patterned ground presumably because of the difficulty of speeding up a process that typically takes centuries or millennia in nature. However, one key aspect of the problem has been elucidated: the upfreezing process, in which larger stones move generally upward relative to the finer-grained soil matrix when soil is frozen from the top down (Anderson, 1988b). Kaplar (1976) produced a film that vividly illustrates the upfreezing process, as well as the distinct inflation of the soil due to ice-lens growth. After repeated freeze-thaw cycles, upfreezing results in an accumulation of rocks at the ground surface directly underlain by a soil layer depleted in stones. The thickness of this layer is dictated by the initial stone content of the soil and the maximum depth to which frost-induced volumetric changes of soil are significant in terms of both frequency and magnitude (Anderson, 1988a). This upfreezing process has much in common with the "kinetic sieving" phenomenon in which coarse particles in granular mixtures gradually move upward as the mixture is shaken because of the likelihood that small grains can settle further than larger ones (Rosato *et al.*, 1987). The repeated freeze-thaw alternations substitute for "shaking" as the growth and decay of ice displace stones in random directions with an upward bias.

More generally, a fresh perspective on patterned ground can be gained from the numerous laboratory studies of shaking granular materials largely within the last decade, which reveal a rich assortment of emergent behaviors including spatial self-organization, size segregation, convection, and mounding, all of whom have tantalizing analogs in freezing soils (e.g. Gallas *et al.*, 1996; Hill & Kakalios, 1994). Falcon *et al.* (1999) provide new support for the notion proposed by Berg (1990) that soil could mound spontaneously due to vigorous seismic shaking to form Mima mounds. Their origin, however, remains elusive in spite of a wealth of imaginative ideas, with suggested mechanisms ranging from biological to periglacial (Washburn, 1988).

Theoretical Studies

A major boost in understanding the spontaneous emergence and the dynamics of patterned ground comes from novel computer simulations of discrete particle motion in soil subjected to freeze/thaw activity (e.g. Fig. 3; Kessler *et al.*, 2001; Kessler & Werner, 2003; Werner & Hallet, 1993). Guided by field observations, Kessler *et al.* (2001) modeled the development and evolution of sorted patterns in frozen ground starting from a laterally uniform layer of fine-grained soil overlain by layer of coarser-grained material. The model shows that initially small, random variations in layer thickness grow and lead to the formation of diapir-like rising domains of fine-grained material that closely resemble the soil plugs described by Washburn (1997). With time, characteristic arrays of sorted circles and stripes arise spontaneously in the simulations, complete with realistic microtopography, subsurface architecture, and patterns of sorting (Fig. 3). Moreover, the average paths of soil particles is convection-like, with active boundary layers of relatively rapid soil displacement near the surface and periphery of sorted circles, much as observed in active sorted circles (Hallet *et al.*, 1988; Pissart, 1990).

A central component of the model is the explicit inclusion of the feedbacks between soil texture and heat flow in the slow non-linear dynamics of mixtures of fine-grained soil and rock particles that are freezing and thawing recurrently. On the one hand, textural segregation is a function of the heat flow because the pattern of upfreezing and soil compaction is largely dictated by the orientation of ice lenses that tend to parallel the freezing isotherm. On the other hand, the heat flow is a sensitive function of texture because latent heat release upon freezing is high for fine-grained material that retains substantial moisture, and decreases strongly with increasing grain size. The latest published models take into account slope effects and lateral confinement, which in essence represents

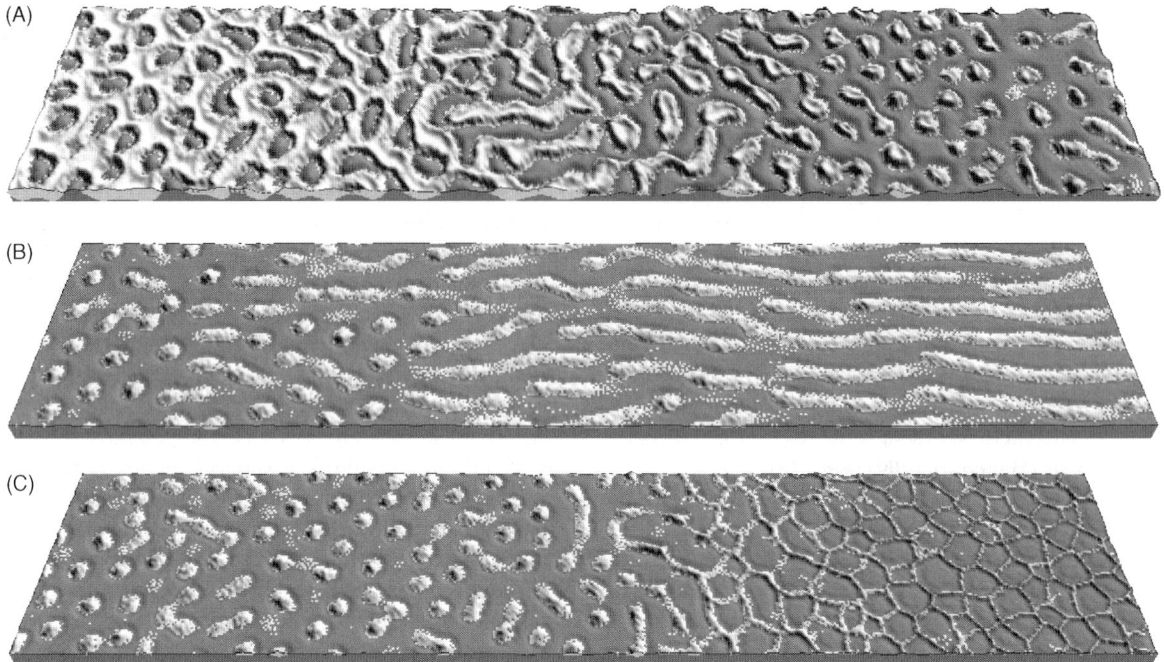

Fig. 3. Three-dimensional perspective view of numerically simulated sorted patterned ground after 500 iterations representing freeze-thaw cycles. They show pattern transitions and the influence of key model variables that vary from left to right as follows (gray stone domains and brown soil domains). (A) Stone concentration decreases. (B) Hillslope gradient increases from 0° to 30°. (C) Lateral confinement increases. Images provided courtesy of Mark Kessler.

the lateral push of freezing fine-grained domains on stony borders (Kessler & Werner, 2003), they are powerful tools for probing the origin and dynamics of patterned ground, its rich geometric complexity, and its sensitivity to environmental factors and soil properties. Figure 3 illustrates the striking range of distinct patterns that emerge spontaneously from the models as a function of controlling parameters, which are specified by Kessler & Werner (2003). The models are especially useful because past physical simulations have not been productive, and hence, they merit careful validation and further development though detailed comparison of pertinent field observations with model assumptions, parameterization, and results.

These recent simulations of patterned ground formation supersede other theoretical work on patterned ground, and circumvent the limitations of other approaches that require unrealistic idealizations or that address only one facet of patterned ground formation. A notable example of the latter is the intriguing notion introduced and developed in considerable mathematical detail by W. Krantz and coworkers (Gleason *et al.*, 1986, 1988; Ray *et al.*, 1983). They argued that many of the characteristics of patterned ground are indirect manifestations of percolative pore water convection through thawed soil during the embryonic stage of pattern development. Remarkably close agreement between model predictions of convection cell geometry and considerable data from diverse areas on the diameter of individual sorted polygons and depth of distinct sorting seems to provide compelling evidence for convection in the active layer (Gleason *et al.*, 1986). Hallet (1990) has stressed, however, that the pattern geometry is not diagnostic, and argued that it reflects convection of the soil rather than the water through the soil. Water convection in soil patterns probably does not occur and, if it does, it is likely incidental. No data exist showing thermally driven water convection sufficiently vigorous to affect the heat transport significantly in any realistic soil that is prone to patterning. Moreover the buoyancy forces involved are so weak that they are only capable of driving convection in very coarse sand or gravel that offer little resistance to the flow (Palm & Tveitereid, 1977), and yet patterned ground is usually associated with much finer grained material, orders of magnitude less permeable than coarse sand.

More recently Krantz & Adams (1996), Fowler & Noon (1997), and Peterson & Krantz (1998) have examined patterned ground initiation from a very promising perspective. They have conducted a rigorous mathematical analysis of the stability of secondary frost heaving to explore the early stages of patterned ground formation, and examined how initial perturbations on the ground surfaces can be selectively amplified, leading to pattern initiation. Using a rather different approach, Hallet & Waddington (1992) have stressed the potential importance of relatively large buoyancy forces that can arise naturally in thawing soils due to vertical gradients in soil bulk density caused by differences in water content. Following a period of frost heaving and attendant decrease in soil density, the soil thaws and consolidates. As the upper part of the soil consolidates first, the freshly thawed soil at the surface will tend to be densest particularly if the frozen soil is ice-rich and sufficiently fine-grained for the consolidation process to be slow due to the low permeability. This unstable density profile could drive diapirism in easily deformable fine-grained soil during much of the thaw season. However, Hallet & Waddington's (1992) analysis should also be viewed with caution because of the extreme idealization of soil as a continuum behaving as porous linear viscous fluid.

Non-Sorted Patterns: Polygons

Remarkably regular polygonal patterns have long been recognized in permafrost regions. Individual polygons are typically defined by five or six intersecting cracks commonly highlighted by ridges of soil warped up by the progressive growth of wedges of ice or sand that fill the cracks. Ice wedges grow in permafrost when surface meltwater infiltrates in the frozen soil and freezes; where meltwater is scarce, the cracks tend to fill with sand, as is typical in Antarctica. The resulting patterns are known as ice-wedge polygons or sand-wedge polygons, respectively.

In a classic monograph, which still stands out as exemplary after four decades, Lachenbruch (1962) examined the mechanics underlying these and similar crack patterns that form in response to tension resulting from a decrease in volume due to thermal contraction, desiccation, chemical reaction, or phase change. Differential contraction at the surface induces horizontal tensile stresses that tend to peak at the surface and cause fractures to form and propagate generally downward. As cracks form, they locally relieve tension in the surficial material in a region whose horizontal dimension scales with the crack depth. Progressively, cracks intersect to form polygons that "attain such a size that zones of stress relief of neighboring cracks are superposed at the polygon center so as to keep the stress there below the tensile strength" (Lachenbruch, 1962). The size and configuration of polygons depend upon the rheological behavior of the medium and the rate of induced volume change.

This seminal theoretical work provided the foundation for two recent sets of studies. Through novel two-dimensional numerical simulations, Plug & Werner (2001, 2002) study how the fracture networks self-organize through interactions between thermal stresses, fracture, and the evolving pattern of ice wedges. The cracks bounding polygons typically curve to join pre-existing cracks at right angles due to the influence of these early cracks on the stress field. This modeled fracture network geometry is very similar to that observed in drying mud (Kargel *et al.*, 1996; Weinberger, 1999, 2001) and other desiccating, contracting material (Shorlin *et al.*, 2000). The simulated networks of fractures and ice wedges in frozen ground display rich dynamics, including nonlinear dependence on climatic forcing that is particularly sensitive to severe cooling events (Plug & Werner, 2002). One component of the simulation, the development of ridges at the edges of polygons, was treated in a cursory fashion as it seemed of secondary interest for the focus of the study on the emergent network properties. This could be problematic, however, as troughs between paired ridges (Péwé, 1962; Sletten *et al.*, 2003) generally define the boundaries between polygons in nature, in contrast with the single ridges in the simulation. Importantly, the troughs could affect the timing and location of subsequent cracking because they tend to accumulate snow, which provides an effective thermal barrier that curtails thermal stresses in the winter. Hence, a promising objective for further modeling would be to simulate the polygon microrelief more realistically.

Mellon (1997) also started with the theoretical foundation developed by Lachenbruch (1962) and examined the conditions under which permafrost would crack on the surface of Mars. Interest in this type of study is increasing as high-resolution images (Fig. 4) show that patterned ground is widespread on Mars (Malin & Edgett, 2000, 2001; Seibert & Kargel, 2001). These images have also generated renewed interest in understanding how patterned ground on earth reflects the age of landscape surfaces and subsurface conditions, and in particular, the presence of buried ice (Sletten *et al.*, 2003). Recent spectroscopic evidence show hydrogen presumably due to the presence of ground ice (Boynton *et al.*, 2002) and digital elevation models suggest the past presence of large networks of lakes and rivers (Kramer *et al.*, 2003).

In addition to the theoretical work, polygonal patterned ground has been the subject of considerable field studies, hence more is known about this type of patterned ground than any other. Outstanding among the field studies are those of Black (1976), Black & Berg (1963), Péwé (1974), and Mackay (1974, 1993, 2000; Mackay & Burn, 2002); they have elucidated the initiation and growth of ice wedges and sand wedges. In addition, considerable paleoenvironmental data have also been extracted from the isotope geochemistry of ice wedges (Meyer *et al.*, 2002; Vasil'chuk *et al.*, 2000) with a broad spectrum of temporal dimensions ranging from 10 yr (e.g. Allard & Kasper, 1998) to 10^5 yr (Hamilton *et al.*, 1998) most recently using accelerator mass spectrometry for ^{14}C ages (Vasil'chuk *et al.*, 2000). Older ice wedges or their remains, and sand wedges have been used to infer past climate through the Quaternary (e.g. Johnson, 1990; Péwé, 1983) and the distant geologic past (e.g. Maloof *et al.*, 2002; Williams, 1994).

Landscape Features and Processes

Solifluction and Soil Creep

Among a number of field studies of solifluction and soil creep in diverse settings two stand out because of their thorough nature. Benedict (1970) examined in detail the rates, processes, and climatic significance of downslope soil movement in an alpine region of Colorado. In one of the longest studies to date (>20 yr), Price (1991) documented how the rate of downslope soil motion varied both with depth and position along a slope with distinct solifluction lobes in the Yukon. With the exception of important laboratory studies by Washburn *et al.* (1978), little experimental work has been conducted in the United States, in contrast with the considerable advances made in this area in Europe (Harris *et al.*, 1993).

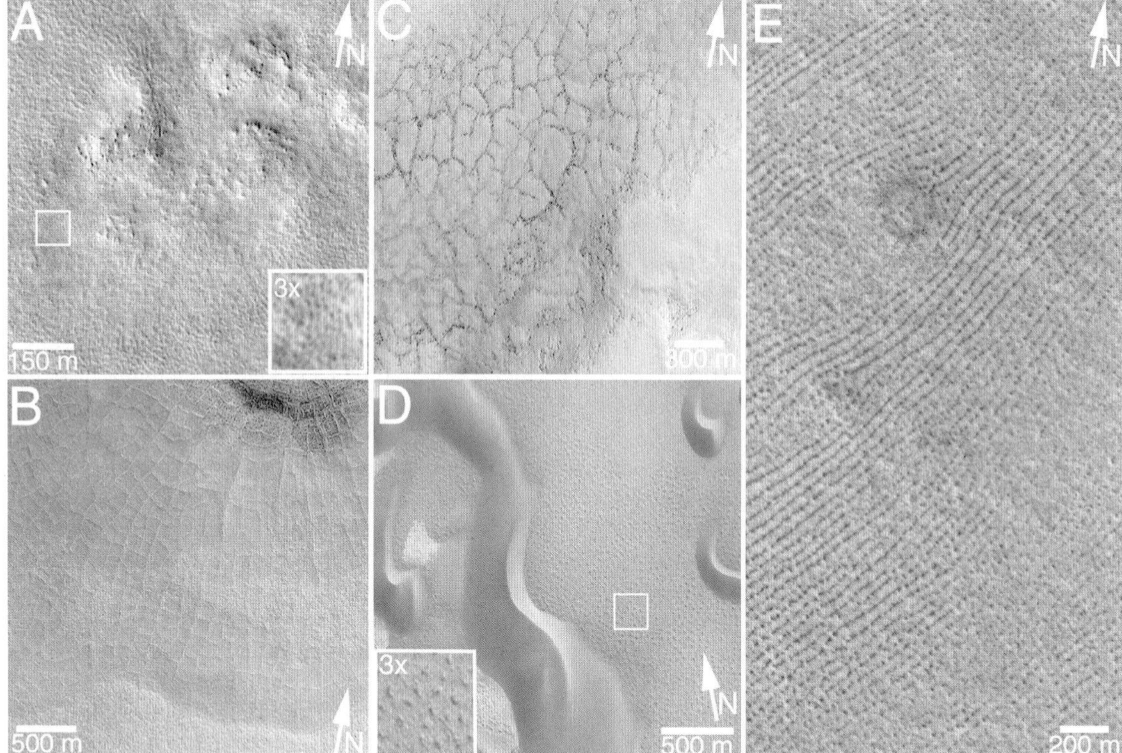

Fig. 4. Diverse types of patterned ground on Mars from the Mars orbiter camera. They range from hummocky, dimpled terrain (A and D), to polygonal ground (B and C). Note concentration of large boulders along the polygon boundaries the boundaries in C, similar to sorted polygons on earth. The stripe pattern in Fig. E has a strong resemblance with stone stripes, but instead of having ~0.15-m spacing as in Fig. 2, the spacing is ≅50 m. All images are illuminated from the left except for (D), which is illuminated from the upper right. (A) Patterned ground/polygons; M03–07241. (B) Polygons; M00–00602. (C) Polygons outlined by dark boulder; M19–01493. (D) "Basketball" textured surface of evenly-spaced mounds amid north polar dunes; M01–00063. (E) Striped pattern of mounds on northern plains; M02–04009. M.C. Malin and K.S. Edgett, Mars Global Surveyor Mars Orbiter Camera: Interplanetary cruise though primary mission. Journal of Geophysical Research, 106(E10), p. 23,552, 2001. Copyright 2001 American Geophysical Union. Reproduced by permission of American Geophysical Union.

The rate of soil creep sets the tempo for material transfer from hillslopes to fluvial systems over major portions of landscapes, and hence the rates of landscape evolution and delivery of terrigenous material to the oceans. Published data suggest that relatively rapid creep is associated with colder climates (Matsuoka, 2001). This tendency could exacerbate global climate change: a global drop in temperature, for example, could result in increased sequestration of atmospheric CO_2 as creep rates increase through two distinct pedogenic processes: accelerated mineral weathering due to increased mechanical refreshing of mineral surfaces and subaerial delivery of organic carbon stored in soils to rivers and long-lived reservoirs.

Rock Glaciers

Rock glaciers are common alpine features that occur in and emanate from cirques and valley walls. They are significant agents of debris transport. Haeberli (1985), for example, estimates that rock glaciers in the Alps cause about 20% of the total downslope movement of rock debris there. In their classic study of rock glaciers in the Alaska Range, Wahrhaftig & Cox (1959) concluded that rock glaciers consist dominantly of rock debris cemented by interstitial ice that could originate as: (1) glacial ice from compacted snow; (2) frozen meltwater or rain; or (3) frozen groundwater.

In his detailed study of Galena Creek rock glacier (GCRG) in Wyoming, Potter (1972) found clean ice containing minimal debris beneath a 1-m-thick coarse debris mantle. Based on the fabric of the ice, thin silty layers within the ice, and mass-balance of both ice and debris fed to the rock glacier, he concluded that the ice originated from the small snowfield at the head of the rock glacier. Thus, GCRG, with the classic morphology consisting of ridges and furrows and a steep front at the angle of repose, is in essence a debris-covered glacier rather than a rock glacier whose ice originated by the freezing of subsurface water. Clark *et al.* (1998) use the terms "glacigenic" and "periglacial" to distinguish the different origins. Barsch (1987), after a brief visit to GCRG, challenged the presence of ice of glacial origin there, mainly on the basis of seismic velocities. In that and subsequent papers

Barsch (1988, 1996) and others (e.g. Haeberli & Vonder Muhll, 1996) concluded that rock glaciers with ice of glacigenic origin do not exist. Since 1987, the subject of the origin of the ice in rock glaciers has been contentious, with GCRG a distinct lightening rod. A response to Barsch is in Potter *et al.* (1998), and the controversy is reviewed by Clark *et al.* (1998).

Studies of GCRG resumed in the 1990s, and with the analyses of oxygen and deuterium isotopes in a 9.5-m core of clean ice with thin silt bands, it is clear that the silt bands represent summer surfaces formed on the snow-field in the cirque and, hence, that the ice is of glacial origin (Clark *et al.*, 1996; Steig *et al.*, 1998). Several dates on organic matter from the ice in GCRG extend back to more than 2000 yr. The stratigraphy of the ice and dates suggest that the ice in glacigenic rock glaciers might yield a climate record at mid-latitudes (Clark *et al.*, 1996). The dates, high-precision velocity data, and temperature from a borehole through GCRG have been used to calibrate a two-dimensional, steady-state, flow-line model of GCRG (Konrad & Humphrey, 2000; Konrad *et al.*, 1999).

Studies of rock glaciers are now clearly moving from morphological descriptions to detailed studies of their dynamics through detailed surveys (Bucki & Echelmeyer, 2003; Kaab & Vollmer, 2000), surface and borehole geophysics (Arenson *et al.*, 2002; Haeberli *et al.*, 1988, 1999), and theoretical models (Loewenherz *et al.*, 1989; Olyphant, 1983). Recently satellite-based synthetic-aperture radar interferometry has been used to measure movement on the order of a few cm on a rock glacier in Beacon Valley, Antarctica (Rignot *et al.*, 2002). Giardino & Vick (1985) have reviewed the engineering hazards of rock glaciers.

It is now evident that rock glaciers have a variety of internal compositions and the ice within them can originate in various ways. Indeed, Elconin & LaChapelle (1997) have recently described naturally exposed sections of a rock glacier in Alaska, in which it is clear that the debris and ice content and its origin vary within a single rock glacier. As Potter (1972) proposed long ago, there is probably a continuum between glacigenic and periglacial rock glaciers reflecting largely the relative fluxes of ice and debris. The controls on rock-glacier type and their thermal and dynamic states, and the cause for the morphological distinction between classic rock glaciers and debris covered glaciers, which are typically marked by ablation pits and thermokarst features, are all subjects of continuing research. An impending report by the Task Force on Rock Glacier Dynamics established by the International Permafrost Association and the International Commission on Snow and Ice will provide an update on these topics, among others.

Concluding Remarks

The study of permafrost has evolved from a descriptive science based mostly on field observations and limited temperature measurements in the summer to a quantitative science capitalizing on advances in understanding fundamental principles in condensed matter physics, non-linear dynamics, soil physics, geochemistry, and on technological advances that make it possible to measure precisely soil properties and to monitor continuously key physical and biogeochemical processes year-round, including the important times of phase transitions in the spring and autumn.

In addition to well accepted theoretical models for heat and mass flow in the soil, recent quantitative models have been developed for a few key phenomena – frost heaving, chemical evolution of freezing pore water, contraction cracking, and patterned ground formation – but the models need to be rendered more realistic and require validation. We share the firm belief that significant advances in understanding most periglacial phenomena require the close integration of field and theoretical studies, and will greatly benefit from ideas and techniques in other disciplines within the geosciences and beyond. Much is to be gained from cross-fertilization between scientists and engineers as well. For example, the transfer of energy through a surface layer of coarse rock debris is central to understanding the thermal regime of rock glaciers. It is, however, poorly understood and little scientific insight has yet been gained from the extensive studies of a very similar engineering problem associated with the use of artificial rock cover to avoid warming or melting permafrost. Other examples include applying what has been learned about the physics of frost weathering to the engineering study of frost resistance of concrete and other porous materials, and exploring parallels between freezing in soils and similar phenomena considered in chemical engineering and medical science in connection with the freezing of food products for storage and the freezing of tissues and organs for medical purposes.

Research in permafrost processes is being re-energized and evolving as a result of the recent widespread recognition of the interdependency of physical, chemical, and biological processes in the active layer, the sensitivity of the upper permafrost to ongoing climate change, and importantly, the potential for changes in the polar regions to affect the global climate. The latter provides substantial impetus to expand the spatial scale of field studies from plot level to the circumarctic, and to examine the exchange of energy, water, and greenhouse gases between the atmosphere, polar lands, and oceans with a resolution compatible with that of Global Circulation Models. The growing interest in fresh-water input to the Arctic Ocean and role of the terrestrial Arctic in the global CO_2 budget, which are both sensitively dependent on groundwater conditions, also highlights the need to improve understanding of how current change in climate, vegetation and permafrost landscape in the Arctic will affect the hydrology of this vast region. A number of major national and international initiatives are aimed at these important issues.

Acknowledgments

This paper reflects the results of countless discussions and many collaborative efforts involving a large number of talented and knowledgeable individuals. We wish to thank in particular A.L. Washburn whose expertise, encouragement and interest have fostered our collective work on periglacial processes. We acknowledge the considerable support we have received for permafrost research from the National Science Foundation and U.S. Army Research Office. We thank M. Malin and M. Kessler for permission to use their figures, and are grateful for the constructive reviews by J. Brown and an anonymous reviewer.

References

Allard, M. & Kasper, J.N. (1998). Temperature conditions for ice-wedge cracking; field measurements from Salluit, northern Quebec. *In*: Permafrost; Seventh International Conference, Proceedings, Yellowknife, NWT, Canada, pp. 5–12.

Anderson, D.W. & Morgenstern, N.R. (1973). Physics, chemistry, and mechanics of frozen ground: A review. *In*: Permafrost; Second International Conference, Proceedings, Yakutsk, USSR, pp. 257–288.

Anderson, R.S. (1998). Near-surface thermal profiles in alpine bedrock: implications for the frost weathering of rock. *Arctic and Alpine Research*, **30**(4), 362–372.

Anderson, S.P. (1988a). Upfreezing in sorted circles, Western Spitsbergen. *In*: Permafrost; Fifth International Conference, Proceedings, Trondheim, Norway, pp. 666–671.

Anderson, S.P. (1988b). The upfreezing process: Experiments with a single clast. *Geological Society of America Bulletin*, **100**, 609–621.

Anisimov, O.A. & Nelson, F.E. (1996). Permafrost distribution in the Northern Hemisphere under scenarios of climatic change. *Global and Planetary Change*, **14**(1–2), 59–72.

Anisimov, O.A., Shiklomanov, N.I. & Nelson, F.E. (1997). Global warming and active-layer thickness: results from transient general circulation models. *Global and Planetary Change*, **15**(3–4), 61–77.

Arenson, L., Hoelzle, M. & Springman, S. (2002). Borehole deformation measurements and internal structure of some rock glaciers in Switzerland. *Permafrost and Periglacial Processes*, **13**(2), 117–135.

Bard, E. (2002). Climate shock: abrupt changes over millennial time scales. *Physics Today*, **55**, 32–38.

Barsch, D. (1987). The problem of the ice-cored Rock glacier. *In*: Giardino, J.R., Shroder, J.F., Jr. & Vitek, J.D. (Eds), *Rock Glaciers*. Boston, MA, United States, Allen & Unwin, pp. 45–53.

Barsch, D. (1988). Rockglaciers. *In*: Clark, M.J. (Ed.), *Advances in Periglacial Geomorphology*. Chichester, United Kingdom, Wiley, pp. 69–90.

Barsch, D. (1996). *Rockglaciers*. Berlin, Springer-Verlag, 331 pp.

Benedict, J.B. (1970). Downslope soil movement in a Colorado alpine region; rates, processes, and climatic significance. *Arctic and Alpine Research*, **2**(3), 165–226.

Berg, A.W. (1990). Formation of Mima mounds; a seismic hypothesis. *Geology (Boulder)*, **18**(3), 281–284.

Black, P.B. & Hardenberg, M.J. (1991). *Historical perspectives in frost heave research; the early works of S. Taber and G. Beskow*. U.S. Army Corps of Engineers Cold Regions Research and Engineering Laboratory, 91–23, 169 pp.

Black, R.F. (1976). Periglacial features indicative of permafrost: Ice and soil wedges. *Quaternary Research*, **6**, 3–26.

Black, R.F. & Berg, T.E. (1963). Patterned ground in Antarctica. *In*: Permafrost: First International Conference, Proceedings, Washington, DC, pp. 121–128.

Bockheim, J.G. (1979). Relative age and origin of soils in eastern Wright valley, Antarctica. *Soil Science*, **128**(3), 142–152.

Bockheim, J.G. (1980). Solution and use of chronofunctions in studying soil development. *Geoderma*, **24**, 71–85.

Bockheim, J.G. (1990). Soil development rates in the Transantarctic Mountains. *Geoderma*, **47**, 59–77.

Bockheim, J.G. & Ugolini, F.C. (1990). A review of pedogenic zonation in well-drained soils of the southern circumpolar region. *Quaternary Reseach*, **34**, 47–66.

Bockheim, J.G., Walker, D.A. & Everett, L.R. (1998). Soil carbon distribution in nonacidic and acidic tundra of arctic Alaska. *In*: Lal, R., Kimble, J.M., Follett, R.F. & Stewart, B.A. (Eds), *Advances in Soil Science*. Boca Raton, CRC Press, pp. 143–155.

Boynton, W.V., Feldman, W.C., Squyres, S.W., Prettyman, T.H., Bruckner, J., Evans, L.G., Reedy, R.C., Starr, R., Arnold, J.R., Drake, D.M., Englert, P.A.J., Metzger, A.E., Mitrofanov, I., Trombka, J.I., d' Uston, C., Wanke, H., Gasnault, O., Hamara, D.K., Janes, D.M., Marcialis, R.L., Maurice, S., Mikheeva, I., Taylor, G.J., Tokar, R. & Shinohara, C. (2002). Distribution of hydrogen in the near surface of Mars: Evidence for subsurface ice deposits. *Science*, **297**(5578), 81–85.

Brown, J., Ferrians, O.J., Jr., Heginbottom, J.A. & Melnikov, E.S. (1997). Circum-Arctic map of permafrost and ground-ice conditions: U.S. geological survey, scale 1:10,000,000.

Brown, J., Hinkel, K.M. & Nelson, F.E. (2000). The circumpolar active layer monitoring (CALM) program: Research designs and initial results. *Polar Geography*, **24**(3), 163–258.

Bucki, A. & Echelmeyer, K. (2003). The flow of Fireweed Rock Glacier. *Journal of Glaciology*, in press.

Budyko, M.I. & Izrael, Y.A. (1992). *Anthropogenic climatic change, hydrometeoizdat, Leningrad* (English edition). University of Arizona Press [*in Russian*] 405 pp.

Buffett, B.A. (2000). Clathrate hydrates. *Annual Review of Earth and Planetary Sciences*, **28**, 477–507.

Cahn, J.W., Dash, J.G. & Fu, H. (1992). Theory of ice premelting in monosized powders. *Journal of Crystal Growth*, **123**(1–2), 101.

Cameron, E.M., Hall, G.E.M., Veizer, J. & Krouse, H.R. (1995). Isotopic and elemental hydrogeochemistry of a major river system; Fraser River, British Columbia, Canada. *Chemical Geology*, **122**(1–4), 149–169.

Campbell, I.B. & Claridge, G.G.C. (1987). Antarctica: Soils, weathering processes and environment. *Developments in Soil Science*, **16**, Amsterdam, Elsevier, 368 pp.

Chapin, F.S., McGuire, A.D., Randerson, J., Pielke, R., Baldocchi, D., Hobbie, S.E., Roulet, N., Eugster, W.,

Kasischke, E., Rastetter, E.B., Zimov, S.A. & Running, S.W. (2000). Arctic and boreal ecosystems of western North America as components of the climate system. *Global Change Biology*, **6**(Suppl. 1), 211–223.

Chapin, F.S., III, Eugster, W., McFadden, J.P., Lynch, A.H. & Walker, D.A. (2000). Summer differences among Arctic ecosystems in regional climate forcing. *Journal of Climate*, **13**(12), 2002–2010.

Christensen, J.H. & Kuhry, P. (2000). High-resolution regional climate model validation and permafrost simulation for the East European Russian Arctic. *Journal of Geophysical Research Atmospheres*, **105**(D24), 29647–29658.

Claridge, G.C.G. & Campbell, I.B. (1982). A comparison between hot and cold desert soils and soil processes. *In*: Yalon, D.H. (Ed.), *Aridic Soils and Geomorphic Processes: Catena supplement 1*. Braunschweig, pp. 1–28.

Clark, D.H., Steig, E.J., Potter, N., Jr., Fitzpatrick, J., Updike, A.B. & Clark, G.M. (1996). Old ice in rock glaciers may provide long-term climate records. *EOS, Transactions, American Geophysical Union*, **77** (23, 217) 221–222.

Clark, D.H., Steig, E.J., Potter, N., Jr. & Gillespie, A.R. (1998). Genetic variability of rock glaciers: Geografiska Annaler Series A. *Physical Geography*, **80**(3–4), 175–182.

Dallimore, A., Schroder Adams, C.J. & Dallimore, S.R. (2000). Holocene environmental history of thermokarst lakes on Richards Island, Northwest Territories, Canada: Thecamoebians as paleolimnological indicators. *Journal of Paleolimnology*, **23**(3), 261–283.

Dash, J.G. (1992). Frost heave in helium and other substances. *Journal of Low Temperature Physics*, **89**, 277.

Dash, J.G., Fu, H. & Wettlaufer, J.S. (1995). The premelting of ice and its environmental consequences. *Reports on Progress in Physics*, **58**(1), 115–167.

Dickson, B., Yashayaev, I., Meincke, J., Turrel, B., Dye, S. & Holfort, J. (2002). Rapid freshening of the deep North Atlantic Ocean over the past four decades. *Nature*, **416**, 832–837.

Edmond, J.M., Palmer, M.R., Measures, C.I., Brown, E.T. & Huh, Y. (1996). Fluvial geochemistry of the eastern slope of the northeastern Andes and its foredeep in the drainage of the Orinoco in Colombia and Venezuela. *Geochimica et Cosmochimica Acta*, **60**(16), 2949–2976.

Elconin, R.F. & LaChapelle, E.R. (1997). Flow and internal structure of a rock glacier. *Journal of Glaciology*, **43**(144), 238–244.

Everett, D.H. (1961). The thermodynamics of frost damage to porous surfaces. *Transactions of the Faraday Society*, **57**, 1541–1551.

Fahnestock, J.T., Jones, M.H., Brooks, P.D., Walker, D.A. & Welker, J.M. (1998). Winter and early spring CO_2 efflux from tundra communities of northern Alaska. *Journal of Geophysical Research*, **103**(D22), 29023–29027.

Fahnestock, J.T., Jones, M.H. & Welker, J.M. (1999). Wintertime CO_2 efflux from arctic soils: Implications for annual carbon budgets. *Global Biogeochemical Cycles*, **13**(3), 775–779.

Falcon, É., Kumar, K., Bajaj, K.M.S. & Bhattacharj, J.K. (1999). Heap corrugation and hexagon formation of powder under vertical vibrations. *Physical Review E*, **59**(5), 5716–5720.

Fowler, A.C. & Krantz, W.B. (1994). Generalized secondary frost heave model. *SIAM Journal of Applied Mathematics*, **54**(6), 1650–1675.

Fowler, A.C. & Noon, C.G. (1997). Differential frost heave in seasonally frozen soils. *In*: Proceedings of the International Symposium on Physics, Chemistry and Ecology of Frozen Soils, Fairbanks, Alaska, pp. 247–252.

Fu, H. (1993). Investigation of the role of surface melting of ice in frozen media [Ph.D. thesis]: University of Washington, 153 pp.

Gallas, J.A.C., Herrmann, H.J., Poschel, T. & Sokolowski, S. (1996). Molecular dynamics simulation of size segregation in three dimensions. *Journal of Statistical Physics*, **82**(1–2), 443–450.

Ganopolski, A. & Rahmstorf, S. (2001). Rapid changes of glacial climate simulated in a coupled climate model. *Nature*, **409**, 153–158.

Gay, J.M., Suzanne, J., Dash, J.G. & Fu, H. (1992). Premelting of ice in exfoliated graphite: a neutron diffraction study. *Journal of Crystal Growth*, **125**(1–2), 33–41.

Giardino, J.R. & Vick, S.G. (1985). The engineering geology hazards of rock glaciers. *Bulletin of the Association of Engineering Geologists*, **22**(2), 201–216.

Gilpin, R.R. (1980). A model for the prediction of ice lensing and frost heave in soils. *Water Resources Research*, **16**, 918–930.

Gleason, K.J., Krantz, W.B. & Caine, N. (1988). Parametric effects in the filtration free convection model for patterned ground. *In*: Permafrost; Fifth International Conference, Proceedings, Trondheim, Norway, pp. 349–354.

Gleason, K.J., Krantz, W.B., Caine, N., George, J.H. & Gunn, R.D. (1986). Geometrical aspects of sorted patterned ground in recurrently frozen soil. *Science*, **232**(4747), 216–220.

Goodrich, L.E. (1982). The influence of snow cover on the ground thermal regime. *Canadian Geotechnical Journal*, **19**, 421–432.

Gordeev, V.V. & Sidorov, I.S. (1993). Concentrations of major elements and their outflow into the Laptev Sea by the Lena River. In: Second international symposium on the Biochemistry of model estuaries; estuarine processes in global change, Amsterdam, Netherlands, Elsevier, pp. 33–45.

Gorham, E. (1991). Northern peatlands: role in the carbon cycle and probable responses to climatic warming. *Ecological Applications*, **1**(2), 182–195.

Grant, S.A. & Sletten, R.S. (2002). Calculating capillary pressures in frozen and ice-free soils below the melting temperature. *Environmental Geology*, **42**(2–3), 130–136.

Gross, G.W. (1967). Ion distribution and phase boundary potentials during the freezing of very dilute ionic solutions at a uniform rate. *Journal of Colloid and Interface Science*, **25**, 270–279.

Gross, G.W., Wang, P.M. & Humes, K. (1977). Concentration dependent solute redistribution at the ice water boundary. III. Spontaneous convection. Chloride solutions. *The Journal of Chemical Physics*, **67**(1), 5240–5264.

Haeberli, W. (1985). Creep of mountain permafrost: Internal structure and flow of Alpine rock glaciers. *Mitteilungen der Versuchanstalt fur Wasserban, Hydrologie, und Glaziologie (Zurich)*, **77**, 142.

Haeberli, W., Huder, J., Keusen, H.R., Pika, J. & Roethlisberger, H. (1988). Core drilling through rock glacier-permafrost. *In*: Permafrost; Fifth International Conference, Proceedings, Trondheim, Norway, pp. 937–942.

Haeberli, W., Kaab, A., Wagner, S., Vonder Muhli, D., Geissler, P., Haas, J.N., Glatzel Mattheier, H. & Wagenbach, D. (1999). Pollen analysis and ^{14}C age of moss remains in a permafrost core recovered from the active rock glacier Murtel-Corvatsch, Swiss Alps: Geomorphological and glaciological implications. *Journal of Glaciology*, **45**(149), 1–8.

Haeberli, W. & Vonder Muhll, D. (1996). On the characteristics and possible origins of ice in rock glacier permafrost. *Zeitschrift fur Geomorphologie, Supplementband*, **104**, 43–57.

Hagedorn, B. & Sletten, R.S. (2001). Spatial and temporal variability of Sr isotope ratios in permafrost-affected soils, NE-Greenland, Zackenberg Station. *In*: 4th International Symposium on Applied Isotope Geochemistry, Pacific Grove, CA, pp. 148–150.

Hall, K. (1986). The utilization of the stress intensity factor (K_{IC^*}) in a model for rock fracture during freezing; an example from Signy Island, the maritime Antarctic. *British Antarctic Survey Bulletin*, **72**, 53–60.

Hallet, B. (1990). Self organization in freezing soils: from microscopic ice lenses to patterned ground. *Canadian Journal of Physics*, **68**, 842–852.

Hallet, B. (1998). Measurement of soil motion in sorted circles, western Spistbergen *In*: Permafrost; Seventh International Conference, Proceedings, Yellowknife, Canada, pp. 415–420.

Hallet, B., Anderson, S.P., Stubbs, C.W. & Gregory, E.C. (1988). Surface soil displacement in sorted circles. *In*: Permafrost; Fifth International Conference, Proceedings, Trondheim, Norway, pp. 770–775.

Hallet, B. & Prestrud, S. (1986). Dynamics of periglacial sorted circles in Western Spitsbergen. *Quaternary Research*, **26**, 81–99.

Hallet, B. & Waddington, E.D. (1992). Buoyancy forces induced by freeze-thaw in the active layer: implications for diapirism and soil circulation. *In*: Dixon, J.C. & Abrahams, A.D. (Eds), *Periglacial Geomorphology*, Proceedings 22nd annual symposium in geomorphology, Binghamton, 1991, Wiley, pp. 251–279.

Hallet, B., Walder, J.S. & Stubbs, C.W. (1991). Weathering by segregation ice growth in microcracks at sustained sub-zero temperatures: verification from an experimental study using acoustic emissions. *Permafrost and Periglacial Processes*, **2**, 283–300.

Hamilton, T.D., Craig, J.L. & Sellmann, P.V. (1998). The Fox permafrost tunnel; a late Quaternary geologic record in central Alaska. *Geological Society of America Bulletin*, **100**(6), 948–969.

Harris, C., Gallop, M. & Coutard, J.P. (1993). Physical modelling of gelifluction and frost creep; some results of a large-scale laboratory experiment. *Earth Surface Processes and Landforms*, **18**(5), 383–398.

Harrison, J.D. & Tiller, W.A. (1963). Controlled freezing of water. *In*: Kingery, W.D. (Ed.), *Ice and Snow*. Cambridge, MA, MIT Press, pp. 215–225.

Harrison, W.D. (1991). Permafrost response to surface temperature change and its implications for the 40,000-yr surface temperature history at Prudhoe Bay, Alaska. *Journal of Geophysical Research, B, Solid Earth and Planets*, **96**(2), 683–695.

Hasholt, B. & Hagedorn, B. (2000). Hydrology and geochemistry of river-borne material in a high arctic drainage system, Zackenberg, Northeast Greenland. *Arctic Antarctic and Alpine Research*, **32**(1), 84–94.

Henry, K.S. (2000). *A review of the thermodynamics of frost heave*. U.S. Army Corps of Engineers Engineer Research and Development Center, ERDC/CRREL Technical Report TR-00–16, 19 pp.

Hill, K.M. & Kakalios, J. (1994). Reversible axial segregation of binary mixtures of granular materials. *Physical Review E*, 49, 5, pt. A, R3610–3613.

Hinkel, K.M. & Outcalt, S.I. (1993). Detection of nonconductive heat transport in soils using spectral analysis. *Water Resources Research*, **29**(4), 1017–1023.

Hinkel, K.M. & Outcalt, S.I. (1994). Identification of heat-transfer processes during soil cooling, freezing & thaw in central Alaska. *Permafrost and Periglacial Processes*, **5**, 217–235.

Hinzman, L.D., Goering, D.J. & Kane, D.L. (1998). A distributed thermal model for calculating soil temperature profiles and depth of thaw in permafrost regions. *Journal of Geophysical Research*, **103**(D22), 28,975–28,991.

Huh, Y. & Edmond, J.M. (1999). The fluvial geochemistry of the rivers of eastern Siberia; III, Tributaries of the Lena and Anabar draining the basement terrain of the Siberian Craton and the Trans-Baikal Highlands. *Geochimica et Cosmochimica Acta*, **63**(7–8), 967–987.

Huh, Y., Panteleyev, G., Babich, D., Zaitsev, A. & Edmond, J.M. (1998a). The fluvial geochemistry of the rivers of eastern Siberia; II, Tributaries of the Lena, Omoloy, Yana, Indigirka, Kolyma & Anadyr draining the collisional/accretionary zone of the Verkhoyansk and Cherskiy ranges. *Geochimica et Cosmochimica Acta*, **62**(12), 2053–2075.

Huh, Y., Tsoi, M.Y., Zaitsev, A. & Edmond, J.M. (1998b). The fluvial geochemistry of the rivers of eastern Siberia; I, Tributaries of the Lena River draining the sedimentary platform of the Siberian Craton. *Geochimica et Cosmochimica Acta*, **62**(10), 1657–1676.

International Permafrost Association, Data and Information Working Group, comp. (1998). *Circumpolar Active-Layer Permafrost System (CAPS), version 1.0*. CD-ROM available form National Snow and Ice Data Center, nsidc@kryos.colorado.edu. Boulder, CO, NSIDC, University of Colorado at Boulder.

IPCC (Intergovernmental Panel on Climate Change) (2001). *Climate change 2001: The scientific basis*. Cambridge, Cambridge University Press.

Jaccard, R.C. & Levi, L. (1961). Ségrégation d'impurctés dans la glace. *Zeitschrift fur Angewandte Mathematik und Physik*, **12**, 70–77.

Johnson, W.H. (1990). Ice-wedge casts and relict patterned ground in central Illinois and their environmental significance. *Quaternary Research*, **33**(1), 51–72.

Jorgenson, M.T., Racine, C.H., Walters, J.C. & Osterkamp, T.E. (2001). Permafrost degradation and ecological changes associated with a warming climate in central Alaska. *Climatic Change*, **48**(4), 551–579.

Judge, A. & Pilon, J. (1983). Climate change and geothermal regime. *In*: Permafrost; Fourth international conference, proceedings, pp. 137–138.

Kaab, A. & Vollmer, M. (2000). Surface geometry, thickness changes and flow fields on creeping mountain permafrost: Automatic extraction by digital image analysis. *Permafrost and Periglacial Processes*, **11**(4), 315–326.

Kane, D.L., Hinzman, L.D. & Zarling, J.P. (1991). Thermal response of the active layer to climatic warming in a permafrost environment. *Cold Regions Science and Technology*, **19**(2), 111–122.

Kaplar, C.W. (1976). Stone migration by freezing of soil *In*: King, C.A.M. (Ed.), *Periglacial processes*. Stroudsburg, PA, Dowden, Hutchinson & Ross, Inc., pp. 44–45.

Kargel, J.S., Schreiber, J.F., Jr. & Sonett, C.P. (1996). Mud cracks and dedolomitization in the Wittenoom Dolomite, Hamersley Group, Western Australia. *Global and Planetary Change*, **14**(1–2), 73–96.

Kennett, J.P., Cannariato, K.G., Hendy, I.L. & Behl, R.J. (2003). *Methane hydrates in Quaternary climate change: The Clathrate Gun hypothesis*. American Geophysical Union Special Publication, 216 pp.

Kessler, M.A., Murray, A.B., Werner, B.T. & Hallet, B. (2001). A model for sorted circles as self-organized patterns. *Journal of Geophysical Research Solid Earth*, **106**(B7), 13287–13306.

Kessler, M.A. & Werner, B.T. (2003). Self-organization of sorted patterned ground. *Science*, **299**(5605), 380–383.

Konrad, S.K. & Humphrey, N.F. (2000). Steady-state flow model of debris-covered glaciers (rock glaciers). *IAHS Publication*, **264**, 255–263.

Konrad, S.K., Humphrey, N.F., Steig, E.J., Clark, D.H., Potter, N., Jr. & Pfeffer, W.T. (1999). Rock glacier dynamics and paleoclimatic implications. *Geology (Boulder)*, **27**(12), 1131–1134.

Koutnik, M.R. & Marshall, S. (2002). Modeling subglacial permafrost evolution. *In*: *EOS Transaction of the American Geophysical Union*, p. 83(47): Abstract C12A-1007.

Kramer, M.G., Potter, C.S., Des Marais, D. & Peterson, D. (2003). New insights on Mars: A network of ancient lakes and discontinuous river segments. *EOS, Transactions, American Geophysical Union*, **84**, 1, 1,6.

Krantz, W.B. & Adams, K.E. (1996). Application of a fully predictive model for secondary frost heave. *Arctic and Alpine Research*, **28**(3), 284–293.

Lachenbruch, A.H. (1962). Mechanics of thermal contraction cracks and ice-wedge polygons in permafrost, Special Paper-Geological Society of America: Boulder, CO, Geological Society of America, 69 pp.

Lachenbruch, A.H. & Marshall, B.V. (1986). Changing Climate: Geothermal Evidence from permafrost in the Alaskan Arctic. *Science*, **234**, 689–696.

Lachenbruch, A.H., Sass, J.H., Marshall, B.V. & Moses, T.H., Jr. (1982). Permafrost, Heat Flow & the Geothermal Regime at Prudhoe Bay, Alaska. *Journal of Geophysical Research*, **87**(B11), 9301–9316.

Leung, W.K.S. & Carmichael, G.R. (1982). Solute redistribution during normal freezing. *Water, Air & Soil Pollution*, **21**, 141–150.

Loewenherz, D.S., Lawrence, C.J. & Weaver, R.L. (1989). On the development of transverse ridges on rock glaciers. *Journal of Glaciology*, **35**(121), 383–391.

Low, P.F., Anderson, D.M. & Hoekstra, P. (1968). Some thermodynamic relationships for soils at or below the freezing point; [Part] 1, Freezing point depression and heat capacity. *Water Resources Research*, **4**(2), 379–394.

Lunardini, V.J. (1995). *Permafrost formation time*. U.S. Army Corps of Engineers Cold Regions Research & Engineering Laboratory, CRREL Special Report 95–8.

Lynch, A.H., Bonan, G.B., III, F.S.C. & Wu, W. (1999). Impact of tundra ecosystems on the surface energy budget and climate of Alaska. *Journal of Geophysical Research*, **104**, D6, 6647–6660.

Mackay, J.R. (1974). Ice-wedge cracks, Garry Island, Northwest Territories. *Canadian Journal of Earth Sciences*, **11**(10), 1366–1383.

Mackay, J.R. (1988a). The birth and growth of Porsild Pingo, Tuktoyaktuk Peninsula, District of Mackenzie. *Arctic*, **41**(4), 267–274.

Mackay, J.R. (1988b). Pingo collapse and paleoclimatic reconstruction. *Canadian Journal of Earth Sciences*, **25**(4), 495–511.

Mackay, J.R. (1990). Seasonal growth bands in pingo ice. *Canadian Journal of Earth Sciences*, **27**(8), 1115–1125.

Mackay, J.R. (1993). The sound and speed of ice-wedge cracking, Arctic Canada. *Canadian Journal of Earth Sciences*, **30**(3), 509–518.

Mackay, J.R. (1998). Pingo growth and collapse, Tuktoyaktuk Peninsula area, western arctic coast, Canada: A long-term field study. *Géographie Physique et Quaternaire*, **52**(3), 271–323.

Mackay, J.R. (2000). Thermally induced movements in ice-wedge polygons, western arctic coast: A long-term study. *Géographie Physique et Quaternaire*, **54**(1), 41–68.

Mackay, J.R. & Burn, C.R. (2002). The first 20 yr (1978–1979 to 1998–1999) of ice-wedge growth at the Illisarvik experimental drained lake site, western Arctic coast, Canada. *Canadian Journal of Earth Sciences*, **39**(1), 95–111.

Malin, M.C. & Edgett, K.S. (2000). Evidence for recent groundwater seepage and surface runoff on Mars. *Science*, **288**(5475), 2330–2335.

Malin, M.C. & Edgett, K.S. (2001). Mars global surveyor Mars orbiter camera: Interplanetary cruise through primary mission. *Journal of Geophysical Research*, **106**(E10), 23,429–23,570.

Malo, B.A. & Baker, R.A. (1968). Cationic concentration by freezing. *In*: Gould, R.F. (Ed.), *Trace Organics in*

Water: American Chemical Society Advances in Chemistry. Washington, DC, American Chemical Society, pp. 396.

Maloof, A.C., Kellogg, J.B. & Anders, A.M. (2002). Neoproterozoic sand wedges: crack formation in frozen soils under diurnal forcing during a snowball Earth. *Earth and Planetary Science Letters*, **204**(1–2), 1–15.

Marion, G.M. (1995). *Freeze-thaw processes and soil chemistry*. U.S. Army Cold Regions Research and Engineering Laboratory, CRREL Special Report 95–12.

Marion, G.M. (2001). Carbonate mineral solubility at low temperatures in the Na-K-Mg-Ca-H-Cl-SO_4-OH-HCO_3-CO_3-CO_2-H_2O system. *Geochimica et Cosmochimica Acta*, **65**(12), 1883–1896.

Marion, G.M. (2002). A molal-based model for strong acid chemistry at low temperatures (<200 to 298 K). *Geochimica et Cosmochimica Acta*, **66**(14), 2499–2516.

Marion, G.M. & Farren, R.E. (1999). Mineral solubilities in the Na-K-Mg-Ca-Cl-SO_4-H_2O system: A re-evaluation of the sulfate chemistry in the Spencer-Moller-Weare model. *Geochimica et Cosmochimica Acta*, **63**(9), 1305–1318.

Marion, G.M. & Grant, S.A. (1994). *FREZCHEM: A chemical-thermodynamic model for aqueous solutions at subzero temperatures*. Cold Regions Research & Engineering Laboratory, U.S. Army Corps of Engineers, CRREL Special Report 94–18.

Matsuoka, N. (2001). Solifluction rates, processes and landforms: a global review. *Earth Science Reviews*, **55**(1–2), 107–134.

Maxwell, J.B. & Barrie, L.A. (1989). Atmospheric and climatic change in the Arctic and Antarctic. *Ambio*, **1**, 42–49.

McGuire, A.D., Clein, J.S., Melillo, J.M., Kicklighter, D.W., Meier, R.A., Vorosmarty, C.J. & Serreze, M.C. (2000). Modelling carbon responses of tundra ecosystems to historical and projected climate: sensitivity of pan-Arctic carbon storage to temporal and spatial variation in climate. *Global Change Biology*, **6**(S1), 141–159.

McKay, C.P., Mellon, M.T. & Friedmann, E.I. (1998). Soil temperatures and stability of ice-cemented ground in the McMurdo Dry Valleys, Antarctica. *Antarctic Science*, **10**(1), 31–38.

Mellon, M. (1997). Small-scale polygonal features on Mars: seasonal thermal contraction cracks in permafrost. *Journal of Geophysical Research*, **102**(E11), 25617–25628.

Meyer, H., Dereviagin, A., Siegert, C., Schirrmeister, L. & Hubberten, H.-W. (2002). Palaeoclimate reconstruction on Big Lyakhovsky Island, north Siberia – hydrogen and oxygen isotopes in ice wedges. *Permafrost and Periglacial Processes*, **13**(2), 91–105.

Miller, R.D. (1973). Soil freezing in relation to pore water pressure and temperature *In*: Permafrost: The North American Contribution to the Second International Conference, Yakutsk, Siberia, pp. 344–352.

Mironenko, M.V., Boitnott, G.E., Grant, S.A. & Sletten, R.S. (2001). Experimental determination of the volumetric properties of NaCl solutions to 253 K. *Journal of Physical Chemistry B*, **105**(41), 9909–9912.

Mironenko, M.V., Grant, S.A., Marion, G.M. & Farren, R.E. (1997). *FREZCHEM2: A chemical thermodynamic model for electrolyte solutions at subzero temperatures*. U.S. Army Cold Regions Research and Engineering Laboratory, CRREL Report 97–5.

Nelson, F.E., Anisimov, O.A. & Shiklomanov, N.I. (2001). Subsidence risk from thawing permafrost. *Nature*, **410**, 889–890.

Nelson, F.E., Hinkel, K.M., Shiklomanov, N.I., Mueller, G.R. & Miller, L.L. (1998a). Active-layer thickness in north central Alaska: Systematic sampling, scale, and spatial autocorrelation. *Journal of Geophysical Research*, **103**(22), 28963–28973.

Nelson, F.E., Lachenbruch, A.H., Woo, M.K., Koster, E.A., Osterkamp, T.E., Gavrilova, M.K. & Guodong, C. (1993). Permafrost and changing climate. *In*: Permafrost; Sixth International Conference, Proceedings, Beijing, China, pp. 987–1005.

Nelson, F.E., Outcalt, S.I., Brown, J., Shiklomanov, N.I. & Hinkel, K.M. (1998b). Spatial and temporal attributes of the active-layer thickness record, Barrow, Alaska, USA. *In*: Permafrost; Seventh International Conference, Proceedings, Yellowknife, Canada, pp. 797–802.

Nisbet, E.G. (2002). Have sudden large releases of methane from geological reservoirs occurred since the Last Glacial Maximum, and could such releases occur again? *Philosophical Transactions of the Royal Society London, Series A*, **360**(1793), 581–607.

Nixon, F.M. & Taylor, A.E. (1998). Regional active layer monitoring across the sporadic, discontinuous and continuous permafrost zones, MacKenzie Valley, Northwestern Canda. *In*: Permafrost; Seventh International Conference, Proceedings, Yellowknife, NWT, Canada, pp. 815–820.

Oechel, W.C., Vourlitis, G. & Hastings, S.J. (1997). Cold season CO_2 emission from arctic soils. *Global Biogeochemical Cycles*, **11**, 163–172.

Oechel, W.C. & Vourlitis, G.L. (1994). The effects of climate change on Arctic tundra ecosystem. *Trends in Ecological Evolution*, **9**, 324–329.

Olyphant, G.A. (1983). Computer simulation of rock-glacier development under viscous and pseudoplastic flow. *Geological Society of America Bulletin*, **94**(4), 499–505.

O'Neill, K. & Miller, R.D. (1985). Exploration of a rigid ice model of frost heave. *Water Resources Research*, **21**, 281–296.

Osterkamp, T.E. & Gosink, J.P. (1991). Variations in permafrost thickness in response to changes in paleoclimate. *Journal of Geophysical Research*, **96**(B3), 4423–4434.

Osterkamp, T.E. & Romanovsky, V.E. (1996). Characteristics of changing permafrost temperatures in the Alaskan Arctic, USA. *Arctic and Alpine Research*, **28**(3), 267–273.

Osterkamp, T.E. & Romanovsky, V.E. (1999). Evidence for warming and thawing of discontinuous permafrost in Alaska. *Permafrost and Periglacial Processes*, **10**(1), 17–37.

Osterkamp, T.E., Viereck, L., Shur, Y., Jorgenson, M.T., Racine, C., Doyle, A. & Boone, R.D. (2000). Observations of thermokarst and its impact on boreal forests in Alaska, U.S.A. *Arctic Antarctic, and Alpine Research*, **32**(3), 303–315.

Osterkamp, T.E., Zhang, T. & Romanovsky, V.E. (1994). Evidence for a cyclic variation of permafrost temperatures in northern Alaska. *Permafrost and Periglacial Processes*, **5**, 137–144.

Outcalt, S.I. (1971). An algorithm for needle ice growth. *Water Resources Research*, **7**, 394–400.

Palm, E. & Tveitereid, M. (1977). On patterned ground and free convection. *Norsk Geologisk Tidsskrift*, **31**, 145–148.

Peterson, R.A. & Krantz, W.B. (1998). A linear stability analysis for the inception of differential frost heave. *In*: Permafrost Seventh International Conference Proceedings, pp. 883–890.

Péwé, T.L. (1962). Age of moraines in Victoria Land, Antarctica. *Journal of Glaciology*, **4**(31), 93–100.

Péwé, T.L. (1974). Geomorphic processes in polar deserts *In*: Smiley, T.L. & Zumberge, J.H. (Eds), *Polar Deserts and Modern Man*. Tucson, University of Arizona Press.

Péwé, T.L. (1983). The periglacial environment in North America during Wisconsin time. *In*: Porter, S.C. (Ed.), *The late Pleistocene*. Minneapolis, University of Minnesota Press, pp. 157–189.

Pissart, A. (1990). Advances in periglacial geomorphology. *Zeitschrift fur Geomorphologie*, **79**, 119–131.

Pittenger, B., Fain, S.C., Cochran, M.J., Donev, J.M.K., Robertson, B.E., Szuchmacher, A. & Overney, R.M. (2001). Premelting at ice-solid interfaces studied via velocity-dependent indentation with force microscope tips. *Physical Review B*, **63**(13), 134102 (1–15).

Plug, L.J. & Werner, B.T. (2001). Fracture networks in frozen ground. *Journal of Geophysical Research B: Solid Earth*, **106**(5), 8599–8613.

Plug, L.J. & Werner, B.T. (2002). Nonlinear dynamics of ice-wedge networks and resulting sensitivity to severe cooling events. *Nature*, **417**(6892), 929–933.

Polar Research Board (1983). *Permafrost research: an assessment of future needs*. Committee on Permafrost, Polar Research Board, Commission on Physical Sciences, Mathematics, and Resources, National Research Council: National Academy Press.

Polyakov, I., Akasofu, S., Bhatt, U., Colony, R., Motoyoshi, M., Makshtas, A., Swingley, C., Walsh, D. & Walsh, J. (2002). Polar amplification. *EOS, Transactions, American Geophysical Union*, **83**(47), 547–548.

Potter, N., Jr. (1972). Ice-cored rock glacier, Galena Creek, Northern Absaroka Mountains, Wyoming. *Geological Society of America Bulletin*, **83**(10), 3025–3058.

Potter, N., Jr., Steig, E.J., Clark, D.H., Speece, M.A., Clark, G.M. & Updike, A.B. (1998). Galena Creek rock glacier revisited-new observations on an old controversy. *Geografiska Annaler Series A: Physical Geography*, **80**(3–4), 251–265.

Price, L.W. (1991). Subsurface movement on solifluction slopes in the Ruby Range, Yukon Territory, Canada; a 20-yr study. *Arctic and Alpine Research*, **23**(2), 200–205.

Putkonen, J. & Roe, G. (2003). Rain on snow events impact soil temperatures and affect ungulate survival. *Geophysical Research Letters*, **30**(4), 1188, doi: 10.1029/2002GL016326.

Putkonen, J.K. (1998). Soil thermal processes and physical properties, Ny Ålesund, Northwest Spitsbergen. *Polar Research*, **17**, 165–179.

Radd, F.J. & Oertle, D.H. (1973). Experimental pressure studies of frost heave mechanisms and the growth-fusion behavior of ice. *In*: Permafrost; North American Contribution, Second International Conference, Proceedings, Washington, DC, pp. 377–384.

Ray, R.J., Krantz, W.B., Caine, T.N. & Gunn, R.D. (1983). A mathematical model for patterned ground; sorted polygons and stripes, and underwater polygons *In*: Permafrost; Fourth International Conference, Proceedings, pp. 1036–1041.

Reeder, S.W., Hitchon, B. & Levinson, A.A. (1972). Hydrogeochemistry of the surface waters of the Mackenzie River drainage basin, Canada; I, Factors controlling inorganic composition. *Geochimica et Cosmochimica Acta*, **36**(8), 825–865.

Rempel, A.W., Wettlaufer, J.S. & Worster, M.G. (2001). Interfacial premelting and the thermomolecular force: Thermodynamic buoyancy. *Physical Review Letters*, **87**(8), 088501-1–4.

Rempel, A.W., Wettlaufer, J.S. & Worster, M.G. (2003). Premelting dynamics in a continuum model of frost heave. *Journal of Fluid Mechanics*, in press.

Rempel, A.W. & Worster, M.G. (2001). Particle trapping at an advancing solidification front with interfacial-curvature effects. *Journal of Crystal Growth*, **223**(3), 420–432.

Rignot, E., Hallet, B. & Fountain, A. (2002). Rock glacier surface motion in Beacon Valley, Antarctica, from synthetic-aperture radar interferometry. *Geophysical Research Letters*, **29**(12), 13,494–13,497.

Riseborough, D.W. & Smith, M.W. (1993). Modeling permafrost response to climate change and climate variability. *In*: Proceedings, Fourth International Symposium on Thermal Engineering & Science for Cold Regions, pp. 179–187.

Romanovsky, V.E. & Osterkamp, T.E. (1995). Interannual variations of the thermal regime of the active layer and near-surface permafrost in northern Alaska. *Permafrost and Periglacial Processes*, **6**(4), 313–335.

Roots, E.F. (1989). Climate change: High latitude regions. *Climatic Change*, **15**(1–2), 223–253.

Rosato, A., Strandburg, K.J., Prinz, F. & Swendsen, R.H. (1987). Why the Brazil nuts are on top: size segregation of particulate matter by shaking. *Physical Review Letters*, **58**(10), 1038–1040.

Roth, K. & Boike, J. (2001). Quantifying the thermal dynamics of a permafrost site near Ny-Alesund, Svalbard. *Water Resources Research*, **37**(12), 2901–2914.

Schimel, J.P. & Clein, J.S. (1996). Microbial response to freeze-thaw cycles in tundra and taiga soils. *Soil Biology & Biochemistry*, **28**(8), 1061–1066.

Seibert, N.M. & Kargel, J.S. (2001). Small-scale Martian polygonal terrain: Implications for liquid surface water. *Geophysical Research Letters*, **28**(5), 899–902.

Seppala, M. (1994). Snow depth controls palsa growth. *Permafrost and Periglagcial Processes*, **5**(4), 283–288.

Serreze, M.C. & Hurst, C.M. (2000). Representation of mean Arctic precipitation from NCEP-NCAR and ERA reanalyses. *Journal of Climate*, **13**(1), 182–201.

Serreze, M.C., Walsh, J.E., Chapin, I.F.S., Osterkamp, T., Dyurgerov, M., Romanovsky, V., Oechel, W.C., Zhang, J.T. & Barry, R.G. (2000). Observational evidence of recent change in the northern high-latitude environment. *Climatic Change*, **46**(1–2), 159–207.

Shiklomanov, N.I. & Nelson, F.E. (2002). Active-layer mapping at regional scales: a 13-year spatial time series for the Kuparuk region, north-central Alaska. *Permafrost and Periglacial Processes*, **13**, 219–230.

Shorlin, K.A., de Bruyn, J.R., Graham, M. & Morris, S.W. (2000). Development and geometry of isotropic and directional shrinkage-crack patterns. *Physical Review E*, **61**(6), 6950–6957.

Sletten, R.S. (1987). Aragonite formation in polar soils [Thesis (M.S.E.) thesis]. University of Washington, 88 pp.

Sletten, R.S. (1993). Laboratory simulation of dolomite and limestone dissolution: rates, yield, and mineralogy of fine-grained residue. *In*: Permafrost; Sixth International Conference, Proceedings, Beijing, China, pp. 580–585.

Sletten, R.S., Hallet, B. & Fletcher, R.C. (2003). Resurfacing time of terrestrial surfaces by the formation and maturation of polygonal patterned ground. *Journal of Geophysical Reasearch-Planets*, **108**(E4), 8044, doi: 10.1029/2002JE001914.

Smith, M.W. & Riseborough, D.W. (1996). Permafrost monitoring and detection of climate change. *Permafrost and Periglacial Processes*, **7**, 301–309.

Smith, M.W. & Tice, A.R. (1988). Measurement of the unfrozen water content of soils; a comparison of NMR and TDR methods. *In*: Permafrost; Fifth International Conference, Proceedings, Trondheim, Norway, pp. 473–477.

Steig, E.J., Fitzpatrick, J.J., Potter, N., Jr. & Clark, A.D.H. (1998). The geochemical record in rock glaciers. *Geografiska Annaler Series A: Physical Geography*, **80**(3–4), 277–286.

Stocker, T.F. & Schmittner, A. (1997). Influence of CO_2 emission rates on the stability of the thermohaline circulation. *Nature*, **388**, 862–865.

Stoner, M.G. & Ugolini, F.C. (1988). Arctic pedogenesis: 2. Threshold-controlled subsurface leaching episodes. *Soil Science*, **145**(1), 46–51.

Taber, S. (1929). Frost heaving. *Journal of Geology*, **37**, 428–461.

Taber, S. (1930). The mechanics of frost heaving. *Journal of Geology*, **38**, 303–317.

Taber, S. (1950). Intensive frost action along lake shores. *American Journal of Science*, **248**(11), 784–793.

Tedrow, J.C.F. (1977). *Soils of the popular landscapes*. New Brunswick, NJ, Rutgers University Press, 638 pp.

Tedrow, J.C.F. & Brown, J. (1962). Soils of the northern Brooks Range, Alaska: Weakening of the soil-forming potential at high arctic altitudes. *Soil Science*, **93**, 254–261.

Tice, A.R., Yuanlin, Z. & Oliphant, J.L. (1984). *The effect of soluble salts on the unfrozen water contents of the Lanzhou, P.R.C. silt*. U.S. Army Cold Regions Research and Engineering Laboratory, CREEL Report 84–16.

Ugolini, F.C. (1986). Pedogenic zonation in the well-drained soils of the arctic regions. *Quaternary Research*, **26**, 100–120.

Ugolini, F.C., Bockheim, J.G. & Anderson, D.A. (1973). Soil development and patterned ground evolution in Beacon Valley, Antarctica. *In*: Permafrost: The North American Contribution to the Second International Conference, Yakutsk, Siberia, pp. 246–254.

Vasil'chuk, Y.K., Jvan der Plicht, J., Jungner, H., Sonninen, E. & Vasil'chuk, A.C. (2000). First direct dating of Late Pleistocene ice-wedges by AMS. *Earth and Planetary Science Letters*, **179**(2), 237–242.

Wahrhaftig, C.A. & Cox, A.V. (1959). Rock glaciers in the Alaska Range. *Geological Society of America Bulletin*, **70**(4), 383–436.

Walder, J.S. & Hallet, B. (1985). A theoretical model of the fracture of rock during freezing. *Geological Society of America Bulletin*, **96**(3), 336–346.

Walder, J.S. & Hallet, B. (1986). The physical basis of frost weathering: toward a more fundamental and unified perspective. *Arctic and Alpine Research*, **18**(1), 27–32.

Walker, D.A., Auerbach, N.A., Bockheim, J.G., Chapin, F.S., Eugster, W., King, J.Y., McFadden, J.P., Michaelson, G.J., Nelson, F.E., Oechel, W.C., Ping, C.L., Reeburg, W.S., Regli, S., Shiklomanov, N.I. & Vourlitis, G.L. (1998). Energy and trace-gas fluxes across a soil pH boundary in the arctic. *Nature*, **394**(6692), 469–472.

Walters, J.C. (1988). Observations of sorted circle activity, central Alaska. *In*: Permafrost; Fifth International Conference, Proceedings, Trondheim, Norway, pp. 885–891.

Wang, B. & Allard, M. (1995). Recent climatic trend and thermal response of permafrost at Salluit, northern Quebec, Canada. *Permafrost and Periglacial Processes*, **6**, 221–234.

Washburn, A.L. (1980). *Geocryology: a survey of periglacial processes and environments*. New York, Wiley, 406 pp.

Washburn, A.L. (1988). *Mima Mounds, an evaluation of proposed origins with special reference to the Puget Lowlands*. State of Washington, Department of Natural Resources, Division of Geology and Earth Resources, Report of Investigations, 53 pp.

Washburn, A.L. (1989). Near-surface soil displacement in sorted circles, Resolute area, Cornwallis Island, Canadian High Arctic. *Canadian Journal of Earth Sciences*, **26**(5), 941–955.

Washburn, A.L. (1997). Plugs and plug circles: A basic form of patterned ground, Cornwallis Island, Arctic Canada – origin and implications. *Memoir* – Geological Society of America, 190.

Washburn, A.L., Burrous, C. & Rein, R., Jr. (1978). Soil deformation resulting from some laboratory freeze-thaw experiments. *In*: Permafrost; Third international conference, proceedings Edmondton, Alberta, pp. 756–762.

Weinberger, R. (1999). Initiation and growth of cracks during desiccation of stratified muddy sediments. *Journal of Structural Geology*, **21**(4), 379–386.

Weinberger, R. (2001). Evolution of polygonal patterns in stratified mud during desiccation: The role of flaw distribution and layer boundaries. *Geological Society of America Bulletin*, **113**(1), 20–31.

Werner, B.T. & Hallet, B. (1993). Numerical simulation of self-organized stone stripes. *Nature*, **361**(6408), 142–145.

Wettlaufer, J.S. (2001). Dynamics of ice surfaces. *Interface Science*, **9**(1–2), 117–129.

Wettlaufer, J.S. & Dash, J.G. (2000). Melting below zero. *Scientific American (International Edition)*, **282**(2), 50–53.

Wettlaufer, J.S. & Worster, M.G. (1995). Dynamics of premelted films: frost heave in a capillary. *Physical Review E*, **51**(5), 4679–4689.

Wettlaufer, J.S., Worster, M.G. & Wilen, L.A. (1997). Premelting dynamics: Geometry and interactions. *Journal of Physical Chemistry B*, **101**(32), 6137–6141.

Wilen, L. & Dash, J.G. (1995). Frost heave dynamics at a single crystal interface. *Physical Review Letters*, **74**(25), 5076–5079.

Williams, G.E. (1994). The enigmatic Late Proterozoic glacial climate: an Australian perspective. *In*: Deynoux, M., Miller, J.M.G., Domack, E.W., Eyles, N., Fairchild, I. & Young, G.M. (Eds), *Earth's Glacial Record*. Cambridge, Cambridge University Press, pp. 146–164.

Wilson, R.C. & Vinson, T.S. (1983). *Solute redistribution and freezing rates in a coarse-grained soil with saline pore-water*. Oregon State University, Transportation Research Report 83–15.

Zhang, T., Barry, R.G., Knowles, K., Heginbottom, J.A. & Brown, J. (1999). Statistics and characteristics of permafrost and ground-ice distribution in the Northern Hemisphere. *Polar Geography*, **23**(2), 132–154.

Zhang, T. & Osterkamp, T.E. (1993). Changing climate and permafrost temperatures in the Alaskan Arctic. *In*: Permafrost; Sixth International Conference, Proceedings, pp. 783–788.

Zhang, T., Osterkamp, T.E. & Stamnes, K. (1996). Influence of the depth hoar layer of the seasonal snow cover on the ground thermal regime. *Water Resources Research*, **32**(7), 2075–2086.

Zhang, T., Osterkamp, T.E. & Stamnes, K. (1997). Effects of climate on the active layer and permafrost on the north slope of Alaska, USA. *Permafrost and Periglacial Processes*, **8**(1), 45–67.

Zhu, D.M., Vilches, O.E., Dash, J.G., Sing, B. & Wettlaufer, J.S. (2000). Frost heave in argon. *Physical Review Letters*, **85**(23), 4908–4911.

Quaternary sea-level history of the United States

Daniel R. Muhs[1], John F. Wehmiller[2], Kathleen R. Simmons[1] and Linda L. York[3]

[1] *U.S. Geological Survey, MS 980, Box 25046, Federal Center, Denver, CO 80225, USA*
[2] *Department of Geology, University of Delaware, Newark, DE 19716, USA*
[3] *U.S. National Park Service, Southeast Regional Office, 100 Alabama St. S.W., Atlanta, GA 30303, USA*

Introduction

In the past 30 years, there have been tremendous advances in our understanding of Quaternary sea-level history, due directly to developments in Quaternary dating methods, particularly uranium-series disequilibrium and amino acid racemization. Another reason for this progress is that coastline history can now be tied to the oxygen-isotope record of foraminifera in deep-sea cores. Furthermore, both records have been linked to climate change on the scale of glacial-interglacial cycles that are thought to be forced by changes in Earth-Sun geometry, or "orbital forcing" (Milankovitch, 1941).

Prior to about 1965, much less was known about the Quaternary sea-level record of the United States. Knowledge of the Quaternary history of U.S. coastlines at that time has been summarized by Richards & Judson (1965) for the Atlantic coast, Bernard & LeBlanc (1965) for the Gulf coast, Wahrhaftig & Birman (1965) for the Pacific coast, Péwé *et al.* (1965) for Alaska, and Curray (1965) for the continental shelves. On many segments of U.S. coastlines, Quaternary shorelines had not even been mapped adequately. Although marine terrace maps were available for small reaches of coast (e.g. the California studies of Alexander, 1953; Vedder & Norris, 1963; Woodring *et al.*, 1946), much larger areas had not been studied in sufficient detail for understanding sea-level history. The oxygen-isotope composition of foraminifera in deep-sea sediments had been explored (Emiliani, 1955) and the cores partially dated (Rosholt *et al.*, 1961), but the relations of this record to ice volume and the sea-level record were not understood. At that time, no reliable dating method beyond the range of radiocarbon was fully developed; some of the first attempts at U-series dating of marine fossils had just been published (Broecker & Thurber, 1965; Osmond *et al.*, 1965).

The direct dating of emergent marine deposits is possible because U is dissolved in ocean water but Th and Pa are not. Certain marine organisms, particularly corals (but not mollusks), co-precipitate U directly from seawater during growth. All three of the naturally occurring isotopes of U, ^{238}U and ^{235}U (both primordial parents), and ^{234}U (a decay product of ^{238}U), are therefore incorporated into living corals. ^{238}U decays to ^{234}U, which in turn decays to ^{230}Th. The parent isotope ^{235}U decays to ^{231}Pa. Thus, activity ratios of $^{230}Th/^{234}U$, $^{234}U/^{238}U$, and $^{231}Pa/^{235}U$ can provide three independent clocks for dating the same fossil coral (e.g. Edwards *et al.*, 1997; Gallup *et al.*, 2002). Until the 1980s, U-series dating was done by alpha spectrometry. Most workers since that time have employed thermal-ionization mass spectrometry (TIMS) to measure U-series nuclides, which has increased precision, requires much smaller samples, and can extend the useful time period for dating back to at least ~500,000 yr.

Because corals are not found in all marine deposits, amino-acid racemization has provided a complementary method for geochronology of coastal records (see review by Wehmiller & Miller, 2000). The basis of amino-acid geochronology is that proteins of living organisms (such as marine mollusks) contain only amino acids of the *L* configuration. Upon the death of an organism, amino acids of the *L* configuration convert to amino acids of the *D* configuration, a process called racemization. Racemization is a reversible reaction that results in increased *D*/*L* ratios in a fossil through time until an equilibrium ratio (1.00–1.30, depending on the amino acid) is reached. Racemization kinetics are nonlinear and are a function of environmental temperature history and taxonomy. Amino-acid methods are best applied to fossil mollusks, which occur on virtually every coastline of the world. Although numerical age estimates from amino-acid ratios are still tentative, the technique provides a valuable correlation tool that works best when combined with numerical methods such as U-series or ^{14}C dating.

The sea-level record of Quaternary glacial and interglacial periods is reflected in the oxygen-isotope composition (relative amounts of ^{16}O and ^{18}O) of foraminifera in deep-sea sediments. Oxygen-isotope compositions of foraminifera are a function of both water temperature and the oxygen isotope composition of ocean water at the time of shell formation. Foraminifera precipitate shells with more ^{18}O in colder water. The oxygen isotope composition of ocean water is a function of the mass of glacier ice on land, because glacier ice is enriched in ^{16}O. Thus, relatively heavy (^{18}O-enriched) oxygen isotope compositions in foraminifera reflect glacial periods, whereas light compositions (^{16}O-enriched) reflect interglacial periods (Fig. 1). Two difficulties arise with the intepretation of the oxygen-isotope record for sea-level studies, however: (1) it is difficult to decouple the ice-volume (or sea-level) component of oxygen-isotope composition from the water temperature component; and (2) deep-sea sediments can rarely be dated directly with any precision.

Coastal landforms and deposits provide an independent record of sea-level history. Constructional reefs or wave-cut terraces (Fig. 2) form when they can keep up with rising sea level (reefs) or when they can be cut into bedrock during a stable high sea stand (wave-cut terraces). Thus, emergent marine deposits, either reefs or terraces, on a tectonically

DEVELOPMENTS IN QUATERNARY SCIENCE
VOLUME 1 ISSN 1571-0866
DOI:10.1016/S1571-0866(03)01008-X

PUBLISHED BY ELSEVIER B.V.

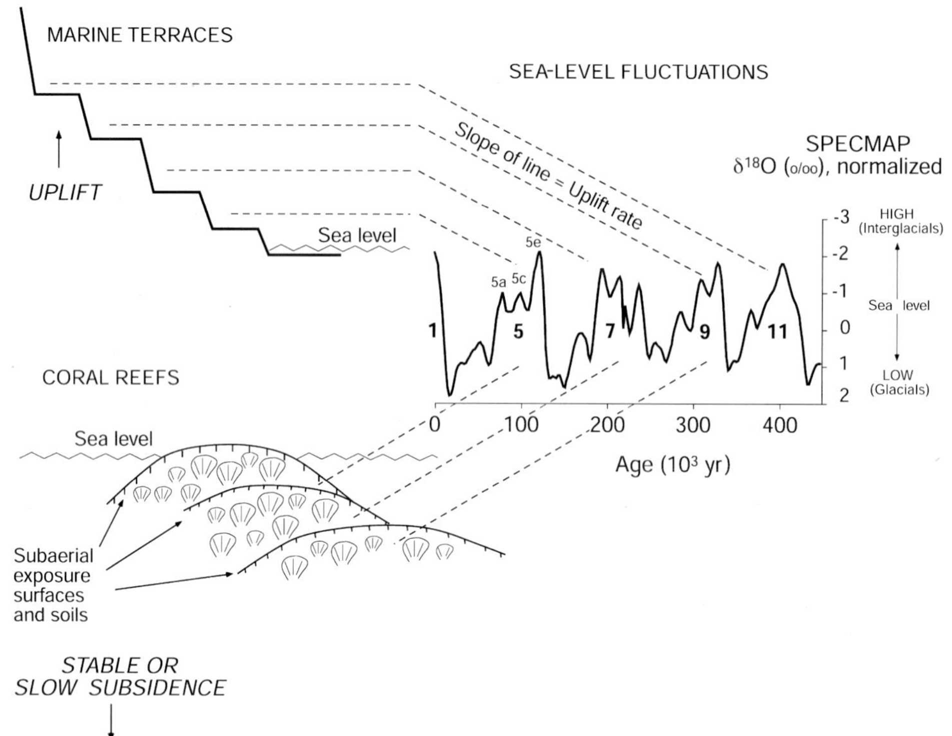

Fig. 1. Diagram of hypothetical coastlines showing relations of oxygen isotope records in foraminifera of deep-sea sediments to emergent reef or wave-cut terraces on an uplifting coastline (upper) and a tectonically stable or slowly subsiding coastline (lower). Emergent marine deposits record interglacial periods. Oxygen-isotope data shown are from the SPECMAP record (Imbrie et al., 1984).

active, uplifting coastline record interglacial periods (Fig. 1). On a tectonically stable or slowly subsiding coast, reefs will be emergent only from sea-level stands that were higher than present (Fig. 1). Paleo-sea levels can thus be determined from tectonically stable coastlines or even uplifting coastlines, if reasoned models of uplift rate can be made. Low stands of sea that occurred during glacial periods may also form reefs or wave-cut terraces, but such features will be offshore, whether on rising, stable or subsiding coasts. Coastal landforms have two advantages over the oxygen isotope record for sea-level history: (1) if corals are present, they can be dated directly; and (2) depending on the tectonic setting, estimates of paleo-sea level can be made.

A new framework for sea-level studies began with a seminal paper on U-series dating that linked the deep-sea oxygen isotope record with uplifted interglacial coral terraces of Barbados (Broecker *et al.*, 1968). These two independently dated records were, in turn, linked to the Milankovitch (1941), or astronomical (orbital forcing) theory of climate change. In the past three decades, geologists have compared a fragmentary geomorphic record (terraces) with a nearly complete record (deep-sea sediments) of Quaternary cycles of glaciations and interglaciations (Aharon & Chappell, 1986; Bender *et al.*, 1979; Bloom *et al.*, 1974; Chappell, 1974a; Chappell & Shackleton, 1986; Dodge *et al.*, 1983; Ku, 1968; Ku *et al.*, 1990; Mesolella *et al.*, 1969; Veeh & Chappell, 1970). Most of these pioneering studies concentrated on the spectacular flights of tectonically uplifted interglacial coral terraces on the islands of Barbados and New Guinea. Nevertheless, the coastlines of the United States also have abundant interglacial terraces and reefs and many localities have now been studied in detail (Fig. 3).

There have been far fewer studies of sea-level lowering on the coasts of the United States during glacial periods. The magnitude of sea-level lowering during the last glacial period has been debated for more than 160 years (see review in Bloom, 1983a). In fact, Bloom (1983a) points out that even at the time of his own review, little progress had been made in estimating the amount of last-glacial sea-level lowering since the studies of the previous century. However, in the 20 years since Bloom's (1983a, b) review, much progress has been made in estimating the magnitude of sea-level lowering during the last glacial period and subsequent sea-level rise in the Holocene. Many of the most recent estimates of sea-level lowering during the last glacial period are summarized in Fleming *et al.* (1998), Yokoyama *et al.* (2001) and Clark & Mix (2002).

The geologic record of late-glacial and Holocene sea-level history is complicated by differential glacio-hydro-isostatic responses of the Earth's crust. Models by Walcott (1972), Chappell (1974b), Clark *et al.* (1978), Nakata & Lambeck (1989), Mitrovica & Peltier (1991), and Peltier (1994, 1996, 1999, 2002) suggest that different sea-level records should be expected on continental coasts and mid-oceanic islands

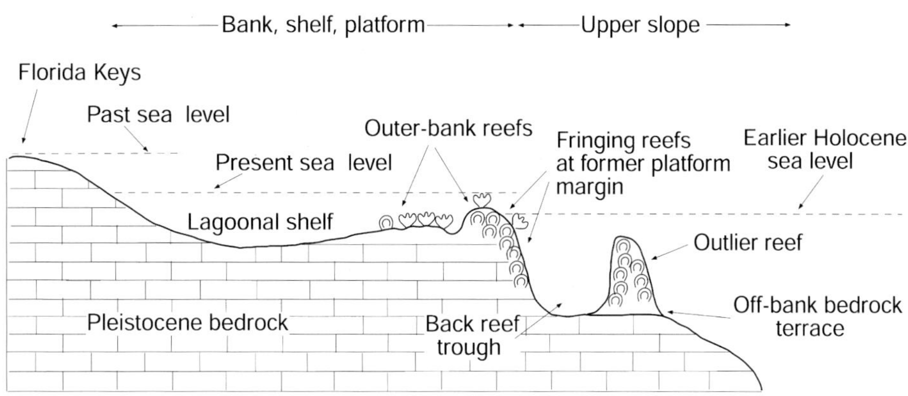

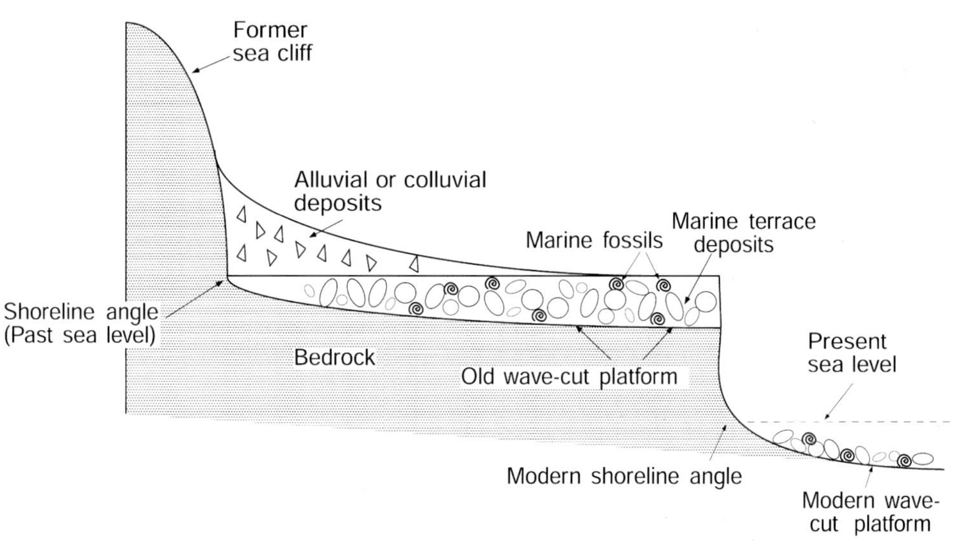

Fig. 2. Cross-sections showing idealized geomorphic and stratigraphic expression of coastal landforms and deposits found on low-wave-energy carbonate coasts of Florida and the Bahamas (upper) and high-wave-energy rocky coasts of Oregon and California (lower). Florida-Bahamas configuration redrawn from Lidz et al. (1997); Oregon-California type redrawn from Muhs (2000).

at different latitudes. During glacial periods, continental ice sheets depress the Earth's crust, with an associated forebulge at their margins. As the ice recedes, formerly glaciated regions experience rebound, or crustal uplift. As sea level rises during deglaciation, the additional water mass in the ocean basins causes crustal loading and depression at continental margins. Chappell (1974b) and Nakata & Lambeck (1989) proposed that the same meltwater loading on ocean floors would bring about mantle flow (below the sea floor) towards islands of some minimum size distant from former ice sheets. Thus, such islands would experience uplift and should have evidence of emergent Holocene shorelines. Some of the glacio-hydro-isostatic models of Holocene sea-level history have been tested on tropical islands of the United States.

In this paper, we review some of the accomplishments in understanding Quaternary sea-level fluctuations as recorded on the coastlines of the United States. It is necessarily an incomplete review, in part because of space limitations and in part because much of the research of the past three decades has focused on specific time periods. Thus, much of our emphasis is on the sea-level record of the last interglacial complex (defined informally here as those high sea stands represented by all of oxygen isotope stage 5). Nevertheless, we touch on progress made in understanding the U.S. record of pre-late Quaternary sea-level stands, as well as sea-level changes during the last glacial period and the Holocene. Because the sea-level record in the U.S. differs geographically, our discussion is arranged in that manner (Fig. 3).

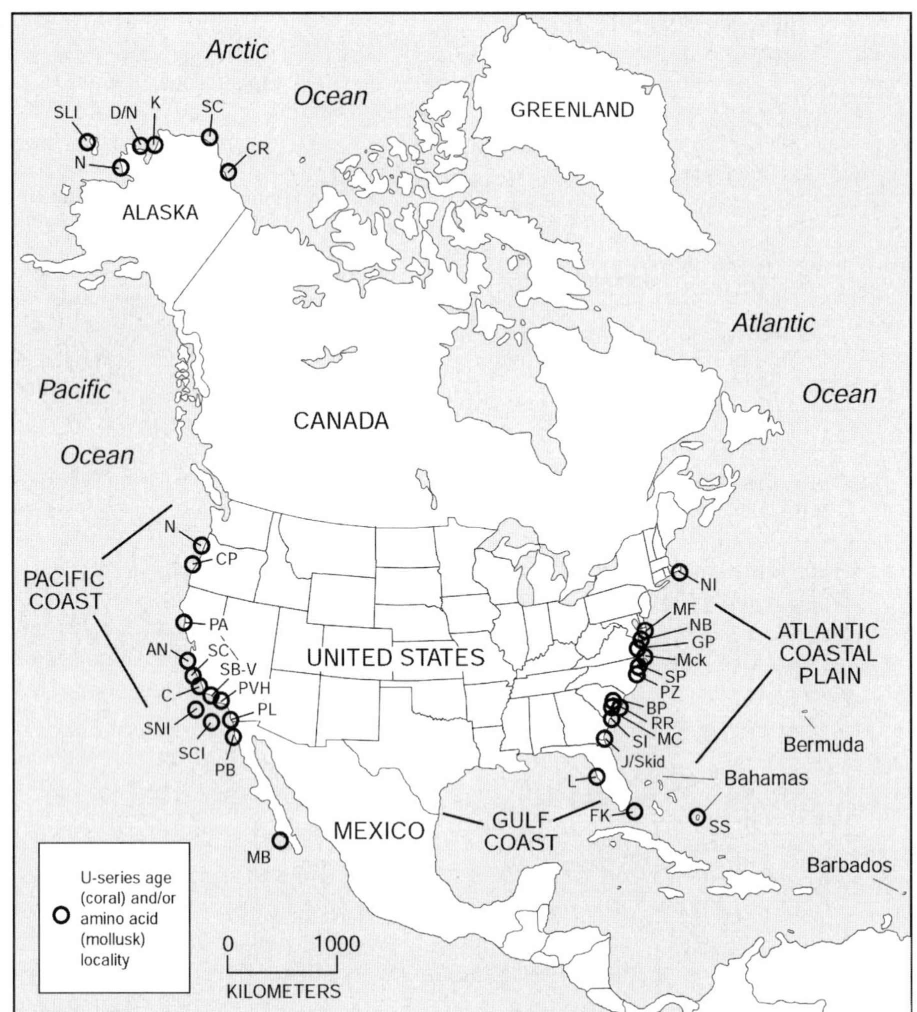

Fig. 3. Map of North America showing localities referred to in the text. Abbreviations: CR, Colville River; SC, Skull Cliff; K, Krusenstern; D/N, Deering/Nugnugaluktuk; SLI, St. Lawrence Island; N, Nome; NE, Newport; CP, Coquille Point; PA, Point Arena; AN, Point Año Nuevo; SC, Point Santa Cruz; C, Cayucos; SB-V, Santa Barbara-Ventura area; SNI, San Nicolas Island; SCI, San Clemente Island; PVH, Palos Verdes Hills; PL, Point Loma; PB, Punta Banda; MB, Magdalena Bay; NI, Nantucket Island; MF, Matthews Field; NB, Norris Bridge; GP, Gomez Pit,; Mck, Moyock; SP, Stetson Pit; PZ, Ponzer; BP, Berkeley Pit; RR, Rifle Range Pit; MC, Mark Clark Pit; SI, Scanawah Island; J/Skid, Jones Pit, Skidway Island; FK, Florida Keys; SS, San Salvador Island.

Florida

Southern Florida, and particularly the Florida Keys island chain (Fig. 4), is an important area for sea-level history because the region is tectonically stable, records of past sea-level stands are abundant, and materials suitable for dating by both uranium-series and radiocarbon methods are available. Because this platform represents long-term carbonate sedimentation on a passive continental margin, late Quaternary marine deposits on the Florida Keys have not experienced significant uplift, subsidence, or tectonic deformation.

Islands of the Florida Keys are important for Quaternary sea-level history because stratigraphic studies indicate that multiple episodes of reef growth and carbonate rock formation have taken place there (Halley *et al.*, 1997). The upper (northeastern) keys are composed of the Key Largo Limestone, a coral-dominated Quaternary carbonate rock (Coniglio & Harrison, 1983; Harrison & Coniglio, 1985; Hoffmeister *et al.*, 1967; Hoffmeister & Multer, 1968). The lower (southwestern) keys are composed of the Miami Limestone, a dominantly oolitic marine carbonate rock. Important sea level records have also been found offshore from the Florida Keys (Lidz *et al.*, 1991, 1997). Stratigraphic studies by Enos & Perkins (1977) and Multer *et al.* (2002) show that the Key Largo Limestone consists of five distinct, coral-dominated Quaternary limestones, from oldest to youngest, Q1 through Q5 (Fig. 5). The five units are separated by discontinuity surfaces, recognizable by subaerial laminated crusts (calcretes), root structures, solution surfaces and either soils or soil breccias.

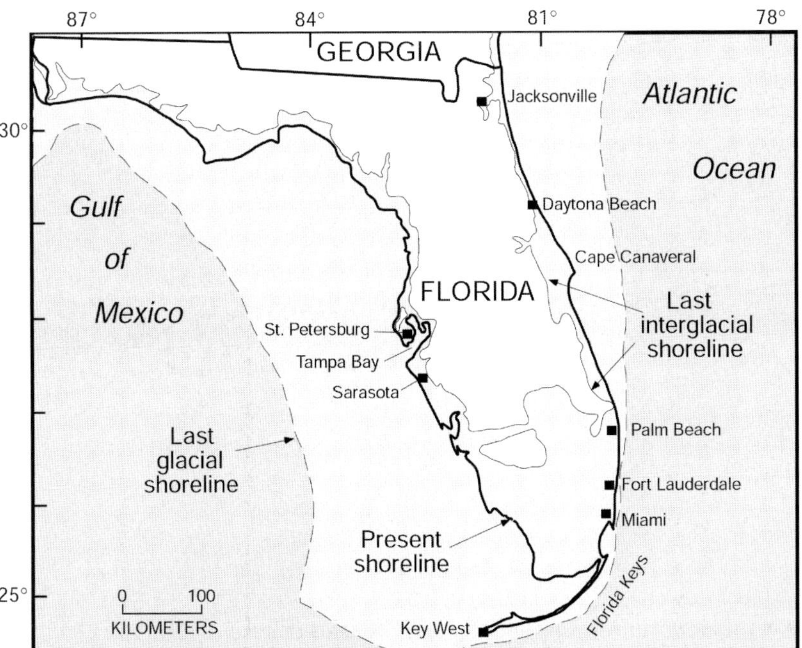

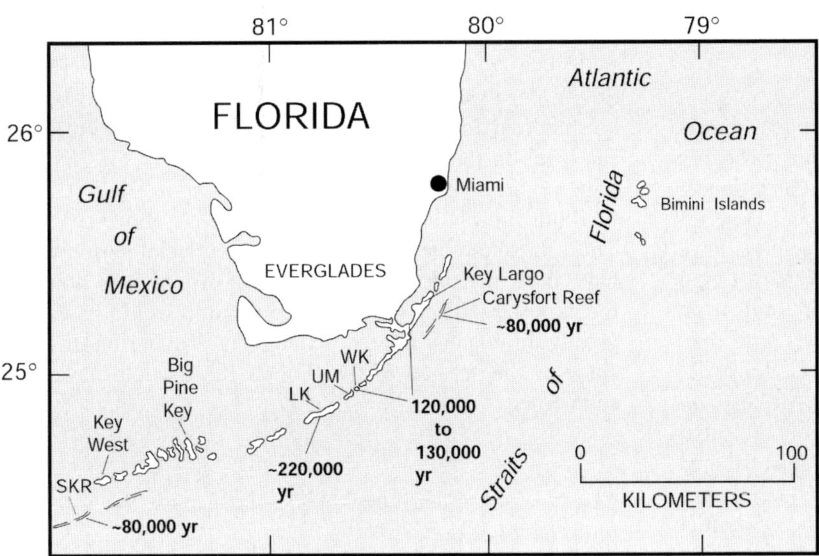

Fig. 4. Upper: map of the State of Florida, showing the modern, last glacial (~21,000 yr), and last-interglacial (~120,000 yr) shorelines. Estimated position of last-interglacial shoreline generated using 1:250,000-scale topographic maps and an assumed +6 m sea-level position, relative to present; estimated position of last-glacial shoreline generated using 1:750,000 topographic map and an assumed −120 m sea-level position, relative to present. Lower: detail of southern Florida, including the Florida Keys, and U-series ages of emergent or shallow-submerged Pleistocene reefs. Abbreviations: WK, Windley Key, UM, Upper Matecumbe Key; LK, Long Key; SKR, Sand Key Reef. Age data from Ludwig et al. (1996), Toscano & Lundberg (1999), Fruijtier et al. (2000), Multer et al. (2002) and this paper.

Florida Keys: Mid-Pleistocene Interglacial High Sea Stands

Recent studies by Multer *et al.* (2002) and the present authors have yielded some fragmentary evidence of the timing and magnitude of sea-level rise during interglacial high sea stands of the mid Pleistocene. Multer *et al.* (2002) reported U-series ages of ~370,000 yr for a coral (*Montastrea annularis*) from the Q3 unit drilled from a locality called Pleasant Point in Florida Bay. Although unrecrystallized, this coral showed clear evidence of open-system conditions and the age is probably closer to 300,000–340,000 yr ago, if the open-system model of Gallup *et al.* (1994) is correct. Nevertheless, the age and elevation of this reef suggest that sea level was close to present during marine oxygen isotope stage (MIS) 9, in agreement with recent data from tectonically stable Henderson Island in the South Pacific (Stirling *et al.*, 2001).

The Q4 unit on the Florida Keys has not received much attention from geochronologists. We collected samples of near-surface *Montastrea annularis* corals in quarry spoil piles that may date to this unit on Long Key. Analysis of a single sample shows a U content of 2.7 ppm, a $^{230}Th/^{232}Th$ ratio of ~14,000, a back-calculated initial $^{234}U/^{238}U$ value of 1.190 and an apparent age of 235,000 ± 4000 yr. The higher-than-modern initial $^{234}U/^{238}U$ value indicates a probable bias to an older age of ca. 7000 yr; thus, the true age may be closer to ~220,000–230,000 yr. If so, these data suggest that sea level also stood near present during MIS 7, the penultimate interglacial period.

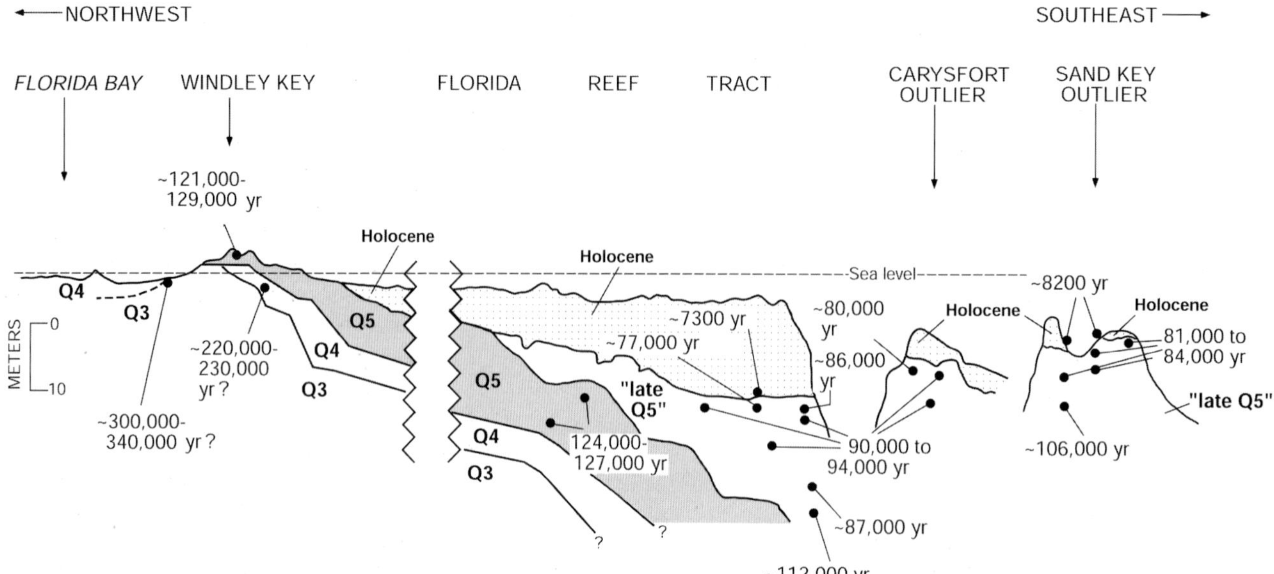

Fig. 5. Composite cross-section of the Florida Keys from northwest to southeast and U-series ages of corals from Quaternary reefs. Redrawn from Multer et al. (2002).

Florida Keys: The Last Interglacial Period

Early studies showed that the Miami Limestone and the uppermost part of the Key Largo Limestone (the Q5 unit) probably date to the peak of the last interglacial period (Broecker & Thurber, 1965; Osmond *et al.*, 1965). More recent U-series studies by Fruijtier *et al.* (2000) using TIMS methods have confirmed these age estimates, at least for the Key Largo Limestone. These workers reported new ages for corals from Windley Key, Upper Matecumbe Key, and Key Largo that, when corrected for high initial $^{234}U/^{238}U$ values (Gallup *et al.*, 1994), range from about ~130,000 to 121,000 yr. Multer *et al.* (2002) also showed, through an extensive series of cores, that the Q5 unit slopes downward to the southeast but can be traced laterally offshore (Fig. 5). They reported TIMS U-series ages of ~124,000 and ~127,000 yr for corals from the Q5 unit at water depths of ~16 and ~22 m, respectively. All these ages are in agreement with other records of peak last-interglacial corals (e.g. Chen *et al.*, 1991; Edwards *et al.*, 1997; Gallup *et al.*, 1994; Muhs *et al.*, 2002a, b).

The Florida Keys contain some of the best evidence that sea level during the peak of the last interglacial period must have been higher than present, because the region is tectonically stable. The youngest (Q5) unit of the Key Largo Limestone at Windley Key is 3–5 m above present sea level, on Grassy Key it is 1–2 m above sea level, and on Key Largo it is 3–4 m above sea level. Based on the elevations of the Key Largo Limestone at Windley Key and Key Largo and optimum growth depths of corals in the formation (Shinn *et al.*, 1989; Stanley, 1966), sea level during the last interglacial period (MIS 5e) must have been at least 5–8 m above present. An inference of a higher-than-present sea level on the Florida Keys during the last interglacial period is consistent with data from other tectonically stable coastlines (Muhs, 2002). For example, U-series ages of ~120,000–130,000 yr are reported for coral-bearing marine deposits ~5 m above sea level on the tectonically stable islands of the Bahamas (Chen *et al.*, 1991) and Bermuda (Muhs *et al.*, 2002b). Much of south Florida stands only a few meters above sea level. Thus, the shoreline during the last interglacial period would have been considerably landward of where it is now and the exposed part of the Florida peninsula would have been greatly diminished in areal extent (Fig. 4). During this high stand of the sea, all of the Florida Keys would have been submerged.

Florida Keys: A Sea-Level High Stand Late in the Last Interglacial Period

Seaward of the lower Florida Keys (Fig. 4), there are outlier reef tracts called Sand Key and Carysfort reef that have been described in detail by Lidz *et al.* (1991, 1997). Ludwig *et al.* (1996) and later Toscano & Lundberg (1999) reported TIMS U-series ages of corals from the crest of Sand Key reef, at depths of about 11–14 m, that range from about 80,000 to 84,000 yr. Similar ages were reported by Toscano & Lundberg (1999) for Carysfort Reef, another outlier reef seaward of Key Largo (Figs 4 and 5). Thus, these reefs formed during MIS 5a. The issue of how high sea level was at the time is difficult to resolve, because none of the ~80,000-yr-old corals from the crest of Sand Key reef is a shallow-water species. The main species from Sand Key, *Montastrea annularis*, has a depth range of −3 to −80 m. Toscano & Lundberg (1999) did succeed in recovering *Acropora palmata*, a strictly shallow-water species, from Carysfort reef at water depths of −15.2 m

(~85,000 yr) and −15.5 m (~92,000 yr). Because *Acropora palmata* grows within 5 m of the sea surface (Lighty *et al.*, 1982), sea level could have been no lower than ~20 m below present at these times. These data provide some measure of the possible amount of sea-level lowering during MIS 5b and the start of MIS 5a (Fig. 1).

Sea Level During the Last Glacial Period and the Holocene: Caribbean Islands and Florida

One of the best records of last-glacial sea-level lowering and postglacial sea-level rise comes from a series of submerged reefs off Barbados, studied by Fairbanks (1989) and Bard *et al.* (1990). Paleo-sea levels for such reefs can be estimated using the shallow-water, reef-crest coral *Acropora palmata* because: (1) *A. palmata* almost always occurs in waters shallower than ~5 m (Lighty *et al.*, 1982); (2) its branching form allows easy recognition of whether or not it is still in growth position; (3) it grows rapidly enough that it can keep pace with a rising sea level (Buddemeier & Smith, 1988); and (4) it can be dated by both U-series and radiocarbon methods. The submerged Barbados reefs indicate that during the last-glacial maximum, sea level was ~120 m lower than present, close to the estimate made by Shepard (1973), who used a worldwide average of the depth of the continental shelf-continental slope boundary. The sea-level curves given by Fairbanks (1989) and Bard *et al.* (1990) show two periods of relatively rapid sea-level rise due to rapid ice melting. The dual dating using U-series and radiocarbon has allowed calibration of radiocarbon ages older than what was possible using tree-ring data. Modeling efforts have shown that the Barbados submerged reefs may be one of the best records of eustatic sea-level rise since the last glacial maximum (Milne *et al.*, 2002; Peltier, 2002). Based on the estimate of ~120-m sea-level lowering during the last glacial period, the paleogeography of Florida would have been drastically different from present, with a Gulf Coast shoreline considerably farther to the west (Fig. 4).

Studies of *Acropora palmata*-dominated reefs off the mainland and islands of the U.S. have allowed an extension of the Barbados sea-level curve into the mid- and late Holocene (Fig. 6). Submerged reefs have been identified off the Florida mainland, the Florida Keys, Puerto Rico, and St. Croix (Lidz *et al.*, 1991, 1997; Lighty *et al.*, 1978, 1982; Ludwig *et al.*, 1996; Toscano & Lundberg, 1998). Although other Holocene sea-level curves have been made (Bloom, 1983b; Kidson, 1982), the advantage of the Florida and Puerto Rico-St. Croix records is that, as with Barbados, they utilize *Acropora palmata*-dominated reefs. For the reasons given above, this coral is probably more robust, both for dating and as a sea-level indicator, than other records that utilize shell beds or terrestrial peat/marine sediment contacts. Macintyre (1988) pointed out that the Florida reef record is limited to the early Holocene. About 7000–8000 cal yr B.P., flooding of the Florida continental shelf would have brought about soil erosion and increased turbidity that may have terminated the growth of *Acropora palmata*-dominated reefs. However, the early Holocene Florida record overlaps the Barbados record and is in broad agreement with it. Younger submerged reefs have been studied off Puerto Rico and St. Croix (Lighty *et al.*, 1982; Macintyre, 1988; Macintyre *et al.*, 1983). The combined records of the Florida

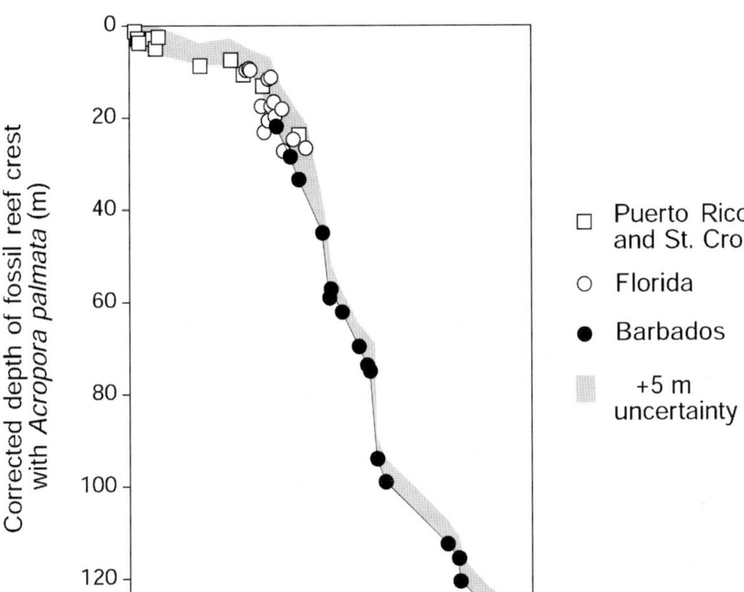

Fig. 6. Graph showing sea-level rise from the last-glacial maximum to the present based on elevations of dated specimens of the shallow-water coral Acropora palmata *from submerged reefs off Barbados (Fairbanks, 1989; Bard et al., 1990), Florida (Ludwig et al., 1996; Toscano & Lundberg, 1998), and Puerto Rico-St. Croix (Lighty et al., 1978, 1982; Macintyre et al., 1983). Barbados points are based on U-series ages and are corrected for tectonic uplift (Fairbanks, 1989). Some Florida reefs are U-series dated; others and all Puerto Rico-St. Croix reefs are dated by radiocarbon and have been converted to calendar years using Stuiver et al. (1998).*

and Puerto Rico-St. Croix reefs allow extension of the Barbados record into the Holocene and up to present sea level (Fig. 6).

Hawaiian Islands

The Hawaiian Islands contain some of the best records of Quaternary sea-level fluctuations, both onshore and offshore. As with the Florida Keys, Hawaiian marine deposits have the advantage of recording paleo-sea levels accurately with *in situ* reefs that contain corals suitable for U-series dating. Stearns (1978), who studied marine deposits on the Hawaiian Islands for more than 40 years, provided a summary of the geologic record of Quaternary sea-level fluctuations. Since the time of that summary, however, new studies have provided additional insight into sea-level history.

Unlike Florida, Oahu has experienced slow uplift over much of the mid-to-late Quaternary, an idea proposed originally by Moore (1970). Volcanic loading on the "big island" of Hawaii results in a compensatory, upward lithospheric flexure on distant islands such as Oahu, Molokai, and Lanai. For example, deposits of the last-interglacial Waimanalo Limestone on Oahu are, in places, several meters higher than the estimated +6 m position for this high sea stand (Muhs & Szabo, 1994). Veeh (1966), in a now-classic study, assumed that mid-plate Pacific islands such as Oahu would be ideal "dipsticks" for estimating paleo-sea levels. However, it is now apparent that marine deposits on the Hawaiian Islands, like the Cook Islands (Woodroffe *et al.*, 1991), need to be considered in light of at least modest Quaternary uplift.

Pre-last Interglacial High Stands of Sea

When Stearns (1978) summarized his career-long views on the sea-level history of the Hawaiian Islands, he was correct that records of older, pre-last-interglacial high stands of sea exist on the islands. On Oahu, southeast of Kaena Point (Fig. 7), a marine deposit ~30 m above sea level was designated as the "Kaena" shoreline by Stearns (1978). Muhs & Szabo (1994) observed these and other +30 m Kaena high stand deposits on Oahu and confirmed their general elevations. The sedimentology of these deposits and their limited, but similar elevations at a minimum of three localities on Oahu suggest that they are not deposits left by a tsunami from a submarine landslide of the sort that has been hypothesized for Lanai and other islands by Moore & Moore (1984, 1988). Szabo & others (1994), using TIMS methods, dated a coral from the ~30-m-high deposit at Kaena Point and reported a ^{230}Th/^{238}U age of 532,000 (+130,000/−70,000) yr, with an initial ^{234}U/^{238}U value that would permit interpretation of a closed-system history. Hearty (2002) reported a TIMS U-series age on a coral from the same deposit of 529,000 (+47,000/−35,000) yr (analyzed by R.L. Edwards & H. Cheng, University of Minnesota), in excellent agreement with the age reported by Szabo *et al.* (1994). Hearty (2002) chose to reject both ages on the basis of whole-rock amino acid data, calibrated to ~120,000-yr-old deposits on Oahu (also dated by U-series methods). Nevertheless, it seems simpler to interpret the two U-series ages as representing deposits of a high sea stand around 500,000–600,000 yr, elevated to ~30 m as a result of a modest (0.05–0.06 m/10^3 yr) long-term uplift rate. Such an uplift rate is consistent with that calculated for the ~120,000 yr deposits on Oahu (Muhs & Szabo, 1994).

Two recent studies have documented that deposits representing at least some part or parts of the penultimate interglacial complex (MIS 7) are found on the Hawaiian Islands. On the leeward (west) coast of Oahu, a nearshore terrace has been identified that slopes down to ~20 m depth. Corals recovered from cores taken at water depths of ~10 m on this terrace date to about 220,000–240,000 yr (Sherman *et al.*, 1999). Because the coral being dated, *Porites lobata*, lives at depths ranging from the intertidal zone down to −60 m, the depth data imply that sea level during some part of the penultimate interglacial complex must have been within 10 m of the present level and could have been higher. On the island of Lanai, Rubin *et al.* (2000) reported U-series ages of emergent coral clasts found in what appear to be California-style marine-terrace deposits. One of two age clusters (~196,000 to ~230,000 yr) dates to the penultimate interglacial period. However, Lanai, like Oahu, is probably undergoing slow uplift. Thus, the Hawaiian Islands have a record of the penultimate interglacial period, but precise timing and sea level position are not yet understood.

The Waimanalo Limestone and the Last Interglacial Period

The island of Oahu has one of the richest records of the last interglacial period in the marine deposits known as the Waimanalo Limestone (Fig. 7). This formation consists of a lower reef facies, with growth-position corals, overlain by a sand and gravel facies that contains coral clasts. Veeh (1966) was the first investigator to show that corals from this reef limestone dated to the peak of the last interglacial period. Ku *et al.* (1974) conducted an extensive study of corals from the Waimanalo Limestone and used both ^{230}Th/^{234}U and ^{231}Pa/^{235}U dating methods. Concordance between the two methods in this study is excellent and shows that the last interglacial high sea stand on Oahu could have begun as early as ~137,000 yr ago and lasted until ~112,000 yr ago. A later study by Muhs & Szabo (1994), also using alpha spectrometry, shows a range of ages from ~138,000 to ~120,000 yr. Szabo *et al.* (1994) and Muhs *et al.* (2002a) analyzed Oahu corals by TIMS with improved uncertainties (usually 1000 yr) and reported a range of ages between 134,000 and 113,000 yr, with most between 125,000 and 115,000 yr. Thus, the Waimanalo Limestone of Oahu has consistently shown evidence of a long last-interglacial period (Fig. 8).

A long last-interglacial period, with sea level at or above present for 10,000 to 20,000 yr, does not agree with the

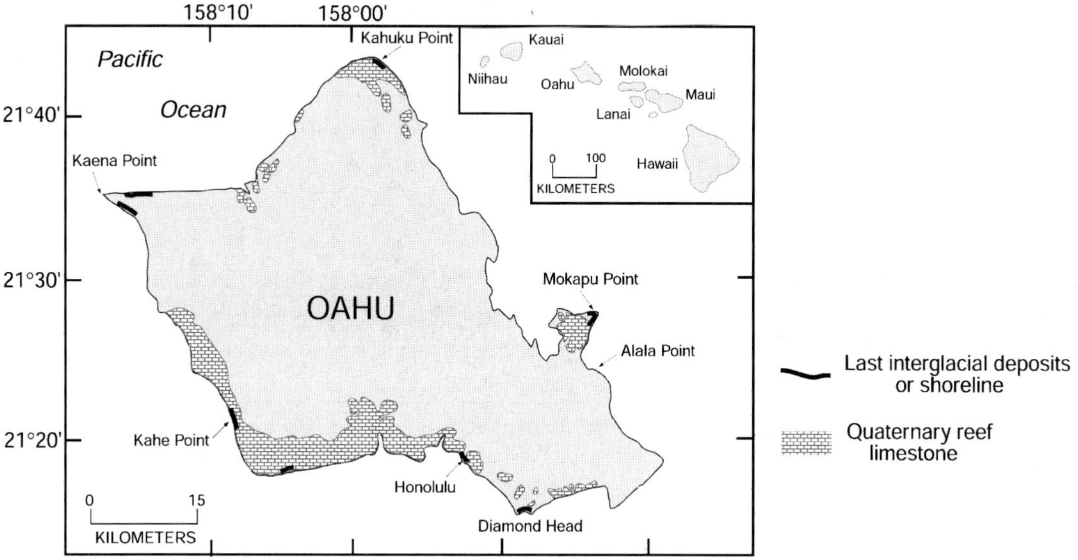

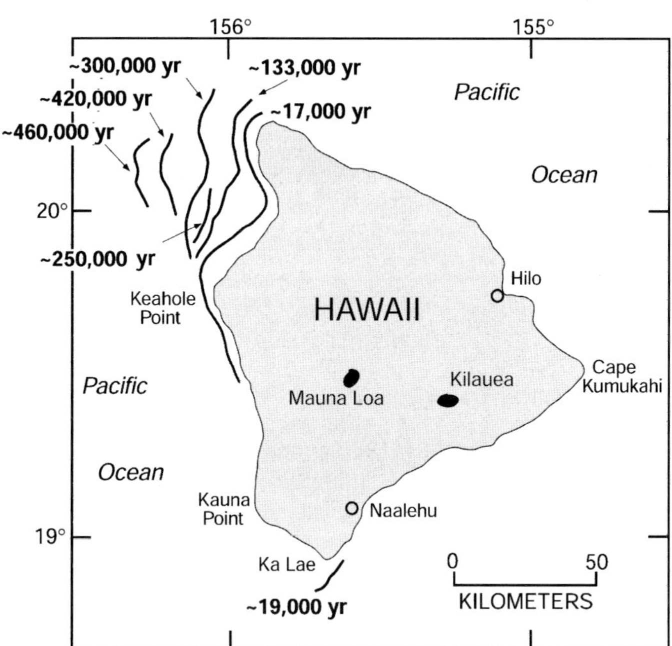

Fig. 7. Upper: map showing distribution of Quaternary limestone deposits (patterned areas) on the island of Oahu, Hawaii (from Stearns, 1974) and approximate locations of the last-interglacial Waimanalo shoreline (bold lines), based on U-series ages in Ku et al. (1971), Sherman et al. (1993), Szabo et al. (1994), Muhs & Szabo (1994) and Muhs et al. (2002a). Lower: map of the island of Hawaii, showing submerged coral reef crests and their U-series ages (data from Ludwig et al., 1991).

SPECMAP estimates (Martinson *et al.*, 1987) of the duration of this low-ice-volume period (Muhs, 2000). However, a long last interglacial period is in agreement with the Devils Hole, Nevada, oxygen isotope record, also dated by U-series methods (Winograd *et al.*, 1997). Furthermore, the early start of the last interglacial high sea stand, as recorded on Oahu, is in agreement with other recent data (Gallup *et al.*, 2002; Henderson & Slowey, 2000) that suggest this warm period preceded the peak of summer insolation in the Northern Hemisphere (Fig. 8). The distribution of coral ages and stratigraphy given in Ku *et al.* (1974), Szabo *et al.* (1994), Muhs & Szabo (1994) and Muhs *et al.* (2002a) also do not require or support the concept of a two-phase high stand at the peak of the last interglacial period, as proposed by Stearns (1978) and Sherman *et al.* (1993).

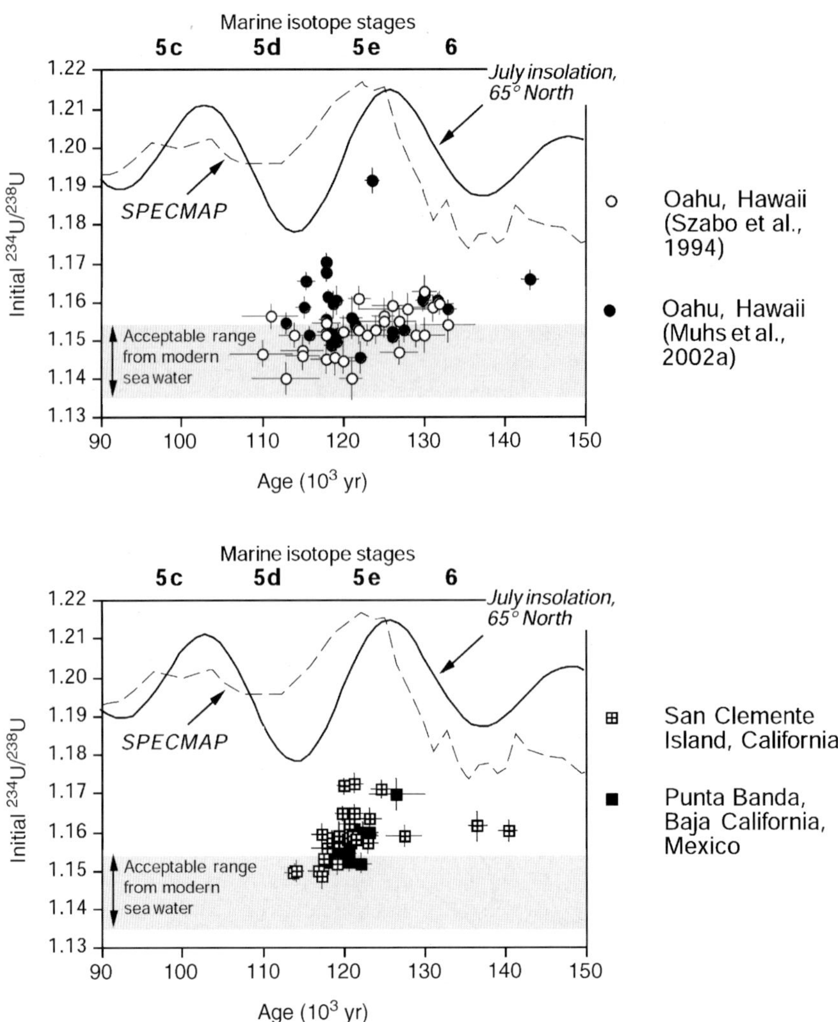

Fig. 8. Records of the duration of the last interglacial period illustrated by range in ages of corals from the Waimanalo Formation of Oahu, Hawaii and the Pacific Coast of North America. Shown for comparison are curves of July insolation at 65 N (from Berger & Loutre, 1991) and the SPECMAP record of oxygen isotopes in deep-sea cores (Martinson et al., 1987). Oahu data from Szabo et al. (1994) and Muhs et al. (2002a); San Clemente Island and Punta Banda data from Muhs et al. (2002b).

The ~120,000-yr-old deposits on Oahu have fossil mollusks that provide important information on last-interglacial marine paleotemperatures. Two key fossil localities studied by Kosuge (1969) and dated by Szabo *et al.* (1994) and Muhs *et al.* (2002a) show that numerous extralimital Indo-Pacific mollusks are present in Waimanalo Limestone deposits. These taxa indicate warmer-than-present waters off Oahu during the peak of the last interglacial period.

Origin of High-Elevation Marine Deposits on Lanai

Fossiliferous marine deposits on the island of Lanai, at elevations of up to 365 m, were interpreted by Stearns (1938) as representing a eustatic high stand of sea, a concept that he retained 40 years later (Stearns, 1978). However, a eustatic high stand of sea at this elevation is unlikely because there is not enough global ice on the continents for such a sea-level rise, a fact recognized by Stearns (1978, p. 13) himself. As mapped by Moore & Moore (1988), most fossiliferous marine deposits on Lanai occur at elevations of ~100 m or less, although some occur as high as ~155 m. Moore & Moore (1984, 1988) proposed that these deposits, as well as marine deposits at high elevations on Molokai and Maui, were deposited by a tsunami generated by a submarine landslide on an offshore scarp. U-series analyses of corals in the deposit on Lanai at elevations of 115–155 m gave ages of 101,000 to 134,000 yr, suggesting deposition at some time during the last interglaciation. Rubin *et al.* (2000) challenged the tsunami hypothesis. Corals in the gravels they studied date both to the last interglacial period (~130,000 to ~136,000 yr) and the penultimate interglacial period (~196,000 to ~230,000 yr). These workers proposed that the marine sediments were deposited and reworked by a combination of marine, fluvial and mass-movement processes, combined with slow uplift, as on Oahu.

Sea-Level Low Stands Recorded on Hawaii

Rapidly subsiding coastlines can have important records of sea-level low stands. The island of Hawaii, situated on an active hot spot, is subsiding due to ongoing volcanic loading (Moore & Fornari, 1984). As a result, submerged coral reefs occur at depths of 150 m to more than 1300 m (Fig. 7). U-series dating shows that the −150 m reef dates to the last deglacial period, about 14,000–16,000 yr ago (Moore *et al.*, 1990). Because sea level was no lower than about −120 m during the last glacial period (Fairbanks, 1989), the reef crest at −150 m demonstrates a subsidence rate of at least 2 m/10^3 yr. U-series ages of older submerged reefs show an average subsidence rate of about 2.6 m/10^3 yr for at least the past 475,000 yr (Ludwig *et al.*, 1991). These studies demonstrate that the reefs grew in shallow water but were drowned by rising sea level at the close of successive glacial periods. Subsidence lowered each just-drowned reef, ultimately to form a submerged, stairstep-like sequence of reefs, analogous to emergent, uplifted reefs.

Holocene Sea-Level History on the Hawaiian Islands

The Hawaiian Islands have been used to test glacio-hydro-isostatic models of sea-level history. Higher-than-present relative sea levels in the Holocene, followed by sea-level fall, are predicted for many low-latitude Pacific islands based on glacio-hydro-isostatic models (Clark *et al.*, 1978; Mitrovica & Peltier, 1991; Nakata & Lambeck, 1989; Walcott, 1972). Stearns (1978) reported that there was a record of a higher-than-present Holocene sea-level stand on several of the Hawaiian Islands, based on radiocarbon ages of emergent marine deposits. This proposal was challenged by Ku *et al.* (1974), Easton & Olson (1976), and Bryan & Stephens (1993) who maintained that there had not been a higher-than-present sea level on the Hawaiian Islands since the last interglacial period, ~120,000 yr ago. Detailed stratigraphic studies with good age control on the islands of Oahu and Kauai show that *relative* sea level was indeed higher than present during the mid-to-late Holocene (Calhoun & Fletcher, 1996; Fletcher & Jones, 1996; Grossman & Fletcher, 1998). This finding is consistent with data from many other tropical Pacific islands (Grossman *et al.*, 1998) and indicates that Holocene sea-level histories will differ from region to region (Bloom, 1983b).

Pacific Coast

Introduction: The Nature of the Pacific Coast Record

Quaternary sea-level fluctuations have left a record on the coasts of California and Oregon in the form of emergent marine terraces. Spectacular flights of multiple marine terraces, forming a stairstep-like landscape, are found in California at Santa Cruz, the Palos Verdes Hills, San Clemente Island, and San Nicolas Island, and in southern Oregon near Coquille Point (Figs 3 and 9). Marine terraces, unlike constructional reef terraces of the tropics, are erosional landforms, although a veneer of marine sediment, sometimes fossiliferous, is generally present (Fig. 2). This is important for understanding the timing of sea-level stands recorded by marine terraces, because erosional landforms need not form at the same time as constructional landforms during a given sea-level high stand. On a tropical, constructional-reef coast, coral growth may keep pace with a rising sea and record the early part, as well as the peak, of a high sea stand. In contrast, on an erosional coast, the early part of a sea-level high stand may be characterized by platform cutting; fossils left behind on the platform may date to the peak of the sea-level stand or even the early part of regression (Bradley & Griggs, 1976).

One problem that had long puzzled early researchers in California was why marine terraces occurred at different elevations in different places, if they were all due to the same sea-level high stands. Alexander (1953) provided the explanation when he showed that marine and stream terraces in central California formed as a result of sea-level high stands superimposed on a tectonically uplifting coast. Thus, flights of marine terraces from place to place could indeed have formed during the same succession of interglacial high sea stands, but differ in their elevations because of differing local uplift rates. Research conducted in the past three decades has confirmed Alexander's (1953) general model (Anderson & Menking, 1994; Grant *et al.*, 1999; Hanson *et al.*, 1992; Kelsey, 1990; Kelsey & Bockheim, 1994; Kelsey *et al.*, 1996; Kern & Rockwell, 1992; Lajoie *et al.*, 1979, 1991; Merritts & Bull, 1989; Muhs *et al.*, 1990, 1992a, 2002b; Rockwell *et al.*, 1989, 1992; Wehmiller & Belknap, 1978; Wehmiller *et al.*, 1977).

Sea-Level History Before the Last Interglacial Period

Several localities along the Pacific Coast, such as the Palos Verdes Hills, San Nicolas Island, and San Clemente Island (Figs 3 and 9), have fossil-bearing, high-elevation terraces. Few of them, however, have been studied for dating purposes. Some of these terraces can be correlated from locality to locality using amino acid and Sr-isotope methods (Kennedy *et al.*, 1982; Ludwig *et al.*, 1992a; Wehmiller *et al.*, 1977), but numerical ages are lacking.

Vedder & Norris (1963) reported corals in deposits of many of the older terraces on San Nicolas Island. We collected unaltered (100% aragonite) specimens of the solitary coral *Balanophyllia elegans* from the 10th terrace (inner edge elevation of ~240 m) on this island (LACMIP loc. 10626; Fig. 9). These results (Table 1) are presented here in order to demonstrate the potential for dating older high sea stands on the Pacific Coast. Laboratory methods for these analyses follow Ludwig *et al.* (1992b); half-lives used for age calculations are those of Cheng *et al.* (2000). Four of eight corals analyzed show evidence of closed-system conditions (initial $^{234}U/^{238}U$ values close to modern seawater, U contents similar to modern corals, and no evidence of inherited ^{230}Th). Because the corals are near the upper limit for TIMS U-series dating, uncertainties are relatively high, but the individual ages (451,000 ± 29,000; 596,000 ± 96,000; 498,000 ± 35,000; and 526,000 ± 41,000 yr) suggest a marine terrace age of perhaps ~500,000 yr. This age, along with ages of the lower two terraces on San Nicolas Island (Muhs *et al.*, 1994, 2002c), permit a possible correlation of intermediate terraces with other interglacial intervals found in the oxygen isotope record (Fig. 10). If these correlations are correct, San Nicolas Island

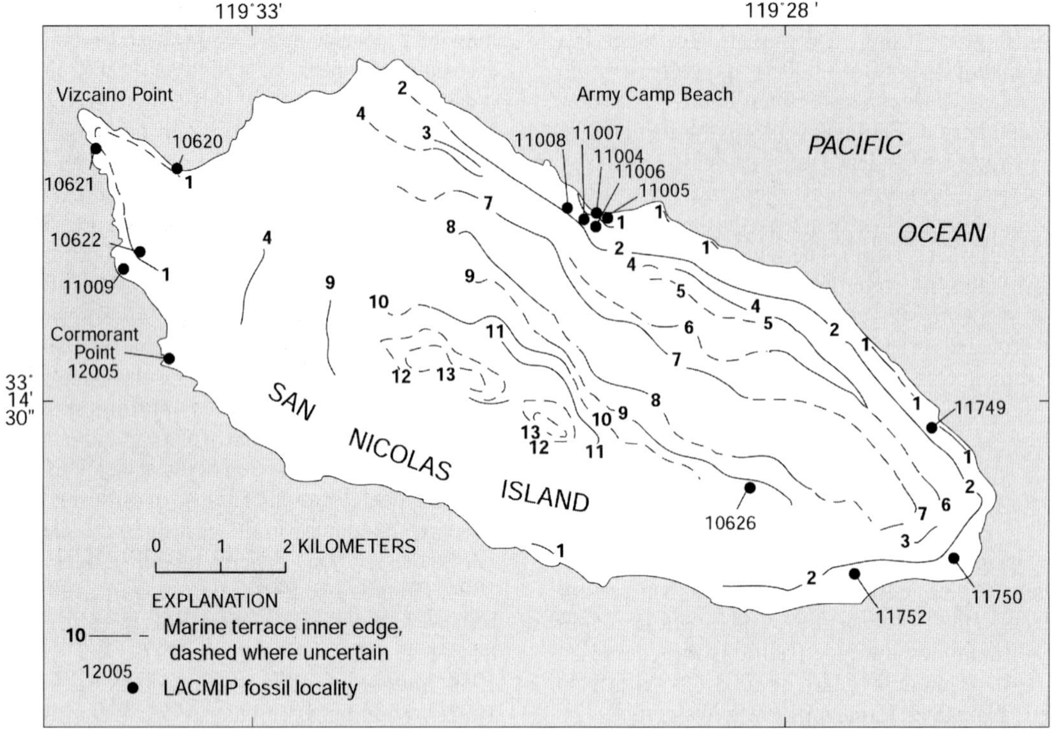

Fig. 9. Map of San Nicolas Island, California, showing inner edges of marine terraces and fossil localities (LACMIP = Los Angeles County Museum, Invertebrate Paleontology locality). Terrace inner edge data from Vedder & Norris (1963), except for the 1st terrace, which was mapped mostly by Muhs et al. (1994). U-series ages of fossil corals from the 10th terrace are from LACMIP loc. 10626 (see Table 1).

has much of the record of interglacial high sea stands of the past half-million years. Other fossil-bearing, high-elevation terraces elsewhere on the Pacific Coast (e.g. the Palos Verdes Hills) may have similar sea-level records that could be dated.

Aminostratigraphy and U-Series Ages of the Last Interglacial Period on the Pacific Coast

A major effort was made in the 1970s to link marine terraces all along the Pacific Coast with global high sea stands. This was accomplished by showing that terraces could be correlated laterally using amino acid isochrons that capitalized on the regional temperature gradient of this north-south-trending coastline (Kennedy *et al.*, 1982; Lajoie *et al.*, 1979; Wehmiller & Belknap, 1978; Wehmiller *et al.*, 1977). Early U-series dating of solitary corals recovered from low terraces at Cayucos, San Nicolas Island, and Point Loma (near San Diego) showed that the California coast has terraces that date to the last interglacial period (Ku & Kern, 1974; Valentine & Veeh, 1969; Veeh & Valentine, 1967). The age of these terraces, around 120,000 yr, is similar to what was being reported during the

Table 1. U and Th concentrations, isotopic activity ratios and ages of corals (Balanophyllia elegans) from LACMIP loc. 10626, 10th marine terrace, San Nicolas Island, CA[a]

Sample	U (ppm)	Th (ppm)	$^{234}U/^{238}U$ AR	±	$^{230}Th/^{238}U$ AR	±	$^{230}Th/^{232}Th$ AR	$^{230}Th/^{238}U$ Age (10^3 yr)	± (10^3 yr)	$^{234}U/^{238}U$ Init AR	±
SNI-19-A	3.23	0.0103	1.0452	0.0025	1.0499	0.0036	997	451	29	1.1619	0.0097
SNI-19-B	3.60	0.0116	1.0263	0.0022	1.0372	0.0030	980	596	96	1.1417	0.0311
SNI-19-C	3.22	0.0147	1.0389	0.0019	1.0476	0.0029	698	498	35	1.1588	0.0118
SNI-19-D	3.07	0.0062	1.0429	0.0014	1.0634	0.0024	1603	n.d.	n.d.	n.d.	n.d.
SNI-10-A	3.43	0.0135	1.0389	0.0018	1.0686	0.0027	825	n.d.	n.d.	n.d.	n.d.
SNI-10-B	3.20	0.0063	1.0418	0.0019	1.0701	0.0026	1637	n.d.	n.d.	n.d.	n.d.
SNI-10-C	2.78	0.0071	1.0341	0.0017	1.0395	0.0026	1228	526	41	1.1508	0.0139
SNI-10-D	3.16	0.0097	1.0668	0.0019	1.0939	0.0035	1078	705	208	1.4899	0.2800

[a] AR = activity ratio. Errors given are two-sigma. Ages calculated using half-lives given in Cheng *et al.* (2000). "n.d.," not determined.

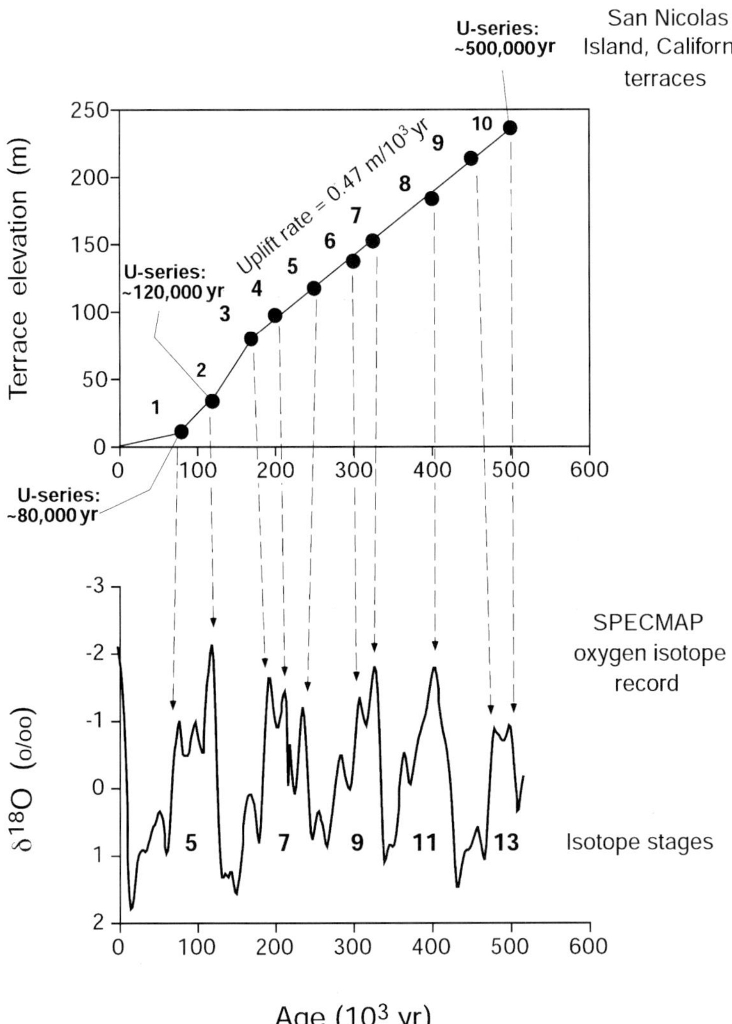

Fig. 10. Plot of terrace elevations as a function of terrace age, San Nicolas Island, California, and possible correlations with the deep-sea oxygen isotope record. Ages of the 1st, 2nd, and 10th terraces are based on U-series analyses of fossil corals; ages of other terraces are based on an assumed long-term uplift rate of ~0.47 m/1000 yr from the age, elevation and an assumed near-present paleo-sea level for the 10th terrace. U-series data are from Muhs et al. (1994, 2002c) for the 1st and 2nd terraces and Table 1 of this paper for the 10th terrace. Terrace elevations are from Muhs et al. (1994) for the 1st and 2nd terraces and from Vedder & Norris (1963) for all other terraces. Oxygen isotope data are from the SPECMAP oxygen isotope curve of Imbrie et al. (1984).

late 1960s for reef terraces on tropical coastlines (Broecker *et al.*, 1968; Ku, 1968; Mesolella *et al.*, 1969; Veeh, 1966; Veeh & Chappell, 1970). The three coral-bearing localities in California provided numerical age control that allowed correlation of the ~120,000-yr-old high sea stand over hundreds of kilometers of the Pacific Coast (Fig. 11). Furthermore, lower amino acid ratios showed that one or more late, last-interglacial (~80,000 or ~100,000 yr) and mid-Wisconsin (~30,000–60,000 yr) high stands were likely present at many localities. Even lower amino acid ratios showed that Holocene terraces were present in areas with high uplift rates, near Ventura and Santa Barbara (Fig. 3), where the "big bend" in the San Andreas fault zone results in a compressional tectonic style. Later U-series dating of corals has confirmed the mid-Wisconsin, ~80,000 and ~120,000 yr age estimates of many localities that were correlated by amino acid methods (Muhs *et al.*, 1990, 1994, 2002b; Stein *et al.*, 1991; Trecker *et al.*, 1998). Radiocarbon dating has confirmed the Holocene terrace ages (Lajoie *et al.*, 1979; Sarna-Wojcicki *et al.*, 1987).

The aminostratigraphy for the Pacific Coast generated by Wehmiller, Kennedy and Lajoie resolved another problem in Quaternary sea-level history and paleoclimatology. Some low-elevation marine terrace localities in California were observed to have extralimital southern ("warm") species in their fossil faunas and others had only extralimital northern ("cool") species (Addicott, 1966; Valentine, 1958, 1961; Vedder & Norris, 1963; Woodring *et al.*, 1946). Still other localities had mixtures of extralimital southern and northern mollusks. Resolution of the apparent contradiction of low terraces with cool species and other low terraces with warm species came with a combination of U-series dating, aminostratigraphy, and detailed faunal analyses (Kennedy *et al.*, 1982; Muhs *et al.*, 2002b; Wehmiller *et al.*, 1977). Those terrace localities with warm species, "neutral" species (i.e. those with faunas similar to those of today) or mixtures of warm and cool species date to the ~120,000-yr-old high sea stand (Fig. 11). Localities with slightly lower amino acid ratios, which correlate to the ~80,000-yr-old high sea stand, have faunas that contain some cool-water species but do not have warm-water species (Kennedy *et al.*, 1982).

High-precision, TIMS U-series dating of marine terrace corals has been accomplished for a number of localities in

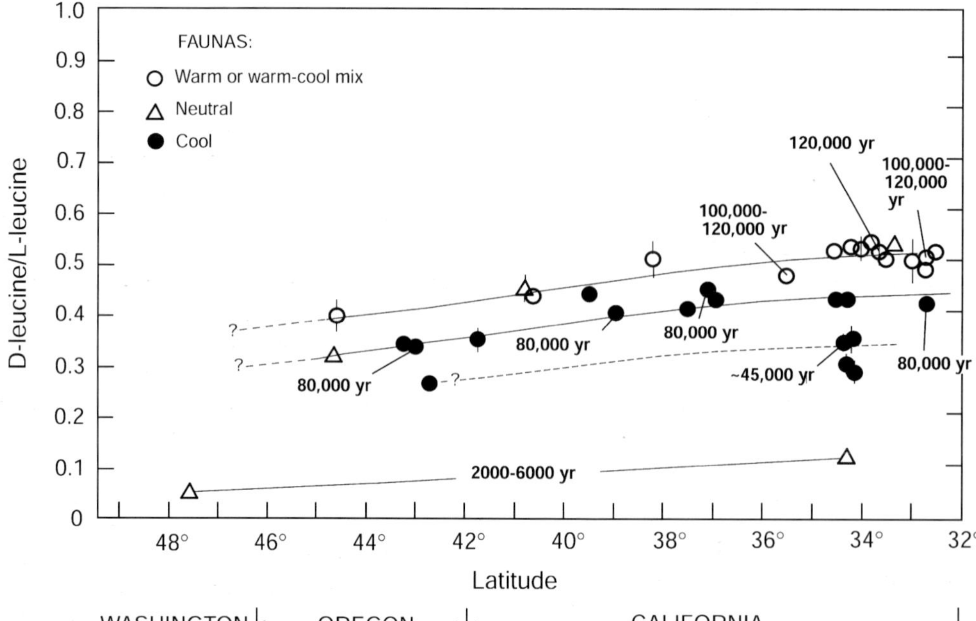

Fig. 11. Ratios of the amino acids D-leucine to L-leucine in the fossil bivalves Protothaca and Saxidomus from marine terrace deposits on the Pacific Coast of the United States, shown as a function of latitude (redrawn in modified form from Kennedy et al., 1982). Lines connect geographically proximal localities that are thought to be correlative based on similar, but northward-decreasing amino acid ratios and similar faunal zoogeographic aspects. Also shown are U-series ages on Pleistocene corals from localities where they are present (data from Muhs et al., 1990, 1994, 2002a; Grant et al., 1999; Trecker et al., 1998), and radiocarbon ages on Holocene mollusks (Lajoie et al., 1979; Sarna-Wojcicki et al., 1987).

California and Baja California (Muhs *et al.*, 2002b). Results show that the last interglacial high sea stand on the Pacific Coast lasted at least 9000 yr, from about 123,000 to ~114,000 yr ago (Fig. 8). In the Bahamas, Barbados, and Hawaii, the same high sea stand is recorded as early as ~130,000 yr (Chen *et al.*, 1991; Gallup *et al.*, 1994, 2002; Muhs *et al.*, 2002a). The difference in timing between the tropical island and Pacific Coast records supports a model proposed by Bradley & Griggs (1976). In this model, the early period of a high sea stand on the Pacific Coast is represented by an interval of terrace platform cutting. At the same time, vigorous upward reef growth is occurring along coasts in the tropics. Fossil deposition on the Pacific Coast follows the period of platform cutting. Thus, on the Pacific Coast, most corals date to either the regressional phase of a high sea stand or just before it, whereas tropical corals date as far back as the period when sea level was still rising.

Recent studies show that the ~80,000-yr-old sea-level stand on the Pacific Coast may have been of similar duration to the ~120,000-yr-old high stand (Muhs *et al.*, 2002c). New U-series ages of solitary corals from Coquille Point, Oregon and numerous localities in California (Point Arena, Point Año Nuevo, Point Santa Cruz, Palos Verdes Hills, and San Nicolas Island; Fig. 3) show that this high stand could have begun around 86,000–84,000 yr ago and lasted until ~76,000 yr ago, a duration of 8000–10,000 yr. This range of ages is similar to that of corals from the Southampton Formation of Bermuda (Muhs *et al.*, 2002a) and corals from the U.S. Atlantic Coastal Plain (see below). Similar records from all three coasts (Pacific, Atlantic, and Bermuda) indicate that this was probably a sea-level stand of a longer duration than has been interpreted from the deep-sea oxygen isotope record (Martinson *et al.*, 1987).

Sea-Level Positions ~100,000 and ~80,000 yr Ago on the Pacific Coast

Both ~100,000 and ~120,000-yr-old corals are found in terrace deposits at Cayucos and Point Loma, California (Muhs *et al.*, 2002b; Stein *et al.*, 1991). This finding provides evidence that the ~100,000-yr-old high stand likely reoccupied at least part of the ~120,000-yr-old terrace. Such a mechanism could explain the mixture of extralimital southern and northern species of mollusks found at both Cayucos and Point Loma. The northern species could date to the ~100,000-yr-old high sea stand whereas the southern species could date to the ~120,000-yr-old high sea stand (Muhs *et al.*, 2002b). At Cayucos, the shoreline angle of the terrace is only about 7–8 m above sea level (Fig. 12), indicating that little or no uplift has occurred in the past ~120,000 yr. Thus, Muhs *et al.* (2002b) interpreted these data to indicate that the ~100,000-yr-old high sea stand must have been close to present. At Punta Banda, Baja

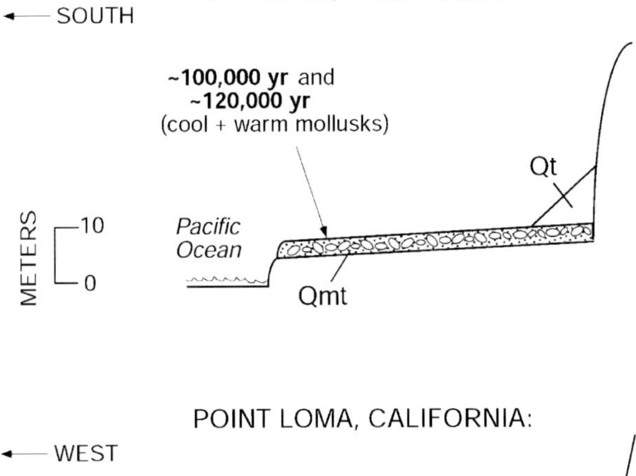

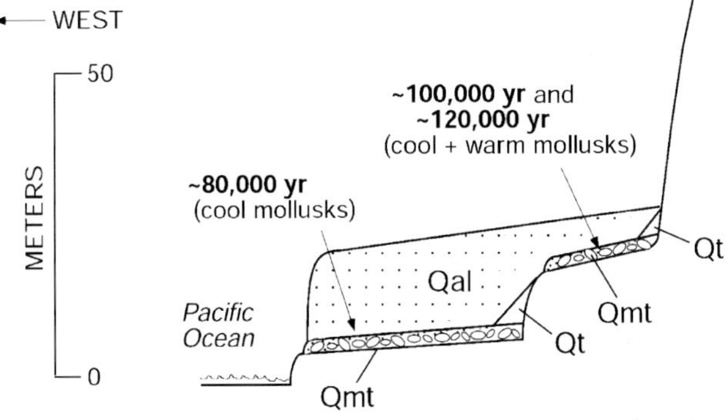

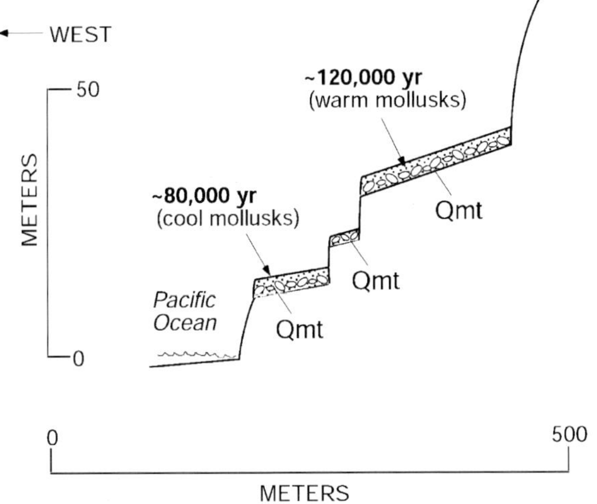

Fig. 12. Shore-normal profiles of low terraces at Cayucos and Point Loma, California and Punta Banda, Baja California, showing average U-series ages of individual corals and faunal thermal aspects. Abbreviations for deposits: Qal, alluvium; Qt, talus and other colluvial deposits; Qmt, marine terrace deposits. From data in Muhs et al. (2002b).

California, where the uplift rate is greater, terrace reoccupation did not take place, and there is no apparent mixing of southern and northern faunas (Fig. 12). A terrace, inferred to represent the ~100,000-yr-old high sea stand, is found between the dated 80,000 and 120,000 yr terraces at Punta Banda.

Terraces dating to the ~80,000-yr-old high sea stand have now been reported for localities from southern Oregon to northern Baja California (Muhs *et al.*, 1990, 1994, 2002b, c). At many of these localities, higher terraces are dated directly or indirectly to the ~120,000-yr-old high sea stand. The elevation of the 80,000 yr terrace can be plotted against the uplift rate (calculated using the elevation of the ~120,000 yr terrace) for a number of localities that span a range of uplift rates. If the plot is linear, the slope of the regression equation fitted to the data yields the approximate age (~80,000 yr) and the Y-axis intercept yields the paleo-sea level. We performed these calculations for the Pacific Coast of North America, the Huon Peninsula of New Guinea, Barbados, and Hateruma Island in the Ryukyu island chain of Japan (Fig. 13). The Pacific Coast, Hateruma Island and New Guinea data suggest that sea level ~80,000 yr ago was within 6 m of present. A widely cited sea-level curve derived from the oxygen isotope record suggests a level position significantly lower than present, about −25 m, at ~80,000 yr ago (Shackleton, 1987). The difference in estimates of sea level position between the terrace record and the oxygen isotope record is equivalent, in ice-volume terms, to approximately three Greenland ice sheets. A new sea-level curve derived from oxygen isotope data indicates that sea level could have been near present ~80,000 yr ago (Shackleton, 2000), which is in better agreement with the terrace-derived sea level estimate presented here. Nevertheless, this recent estimate is based on a single measurement and more data are needed (N.J. Shackleton, written comm. to D.R. Muhs, 2003).

Alaska

Emergent Quaternary marine terraces and their associated deposits and fossils have long been recognized in Alaska. One of the classic study areas has been the Nome coastal plain on the Seward Peninsula (Figs 1 and 14), where Hopkins *et al.* (1960) and Hopkins (1967) reported three wave-cut benches, overlain by fossiliferous sand and gravel that resemble "California-style" marine terraces (Fig. 2). Unlike California, however, the Nome coastal plain is mantled with glacial deposits left by ice that advanced from mountain ranges farther north on the Seward Peninsula (Kaufman, 1992). Emergent marine terraces are also found elsewhere on the Seward Peninsula and on the Arctic Coastal Plain of Alaska (Fig. 14).

Mid-Pleistocene Sea-Level History

As with other coastlines of the United States, mid-Pleistocene marine deposits have been found in Alaska. A marine deposit, representing what has been called the "Anvilian" marine transgression, occurs extensively along the Seward Peninsula and Arctic Ocean coast of Alaska. It occurs landward of "Pelukian" (last interglacial) marine deposits and is found

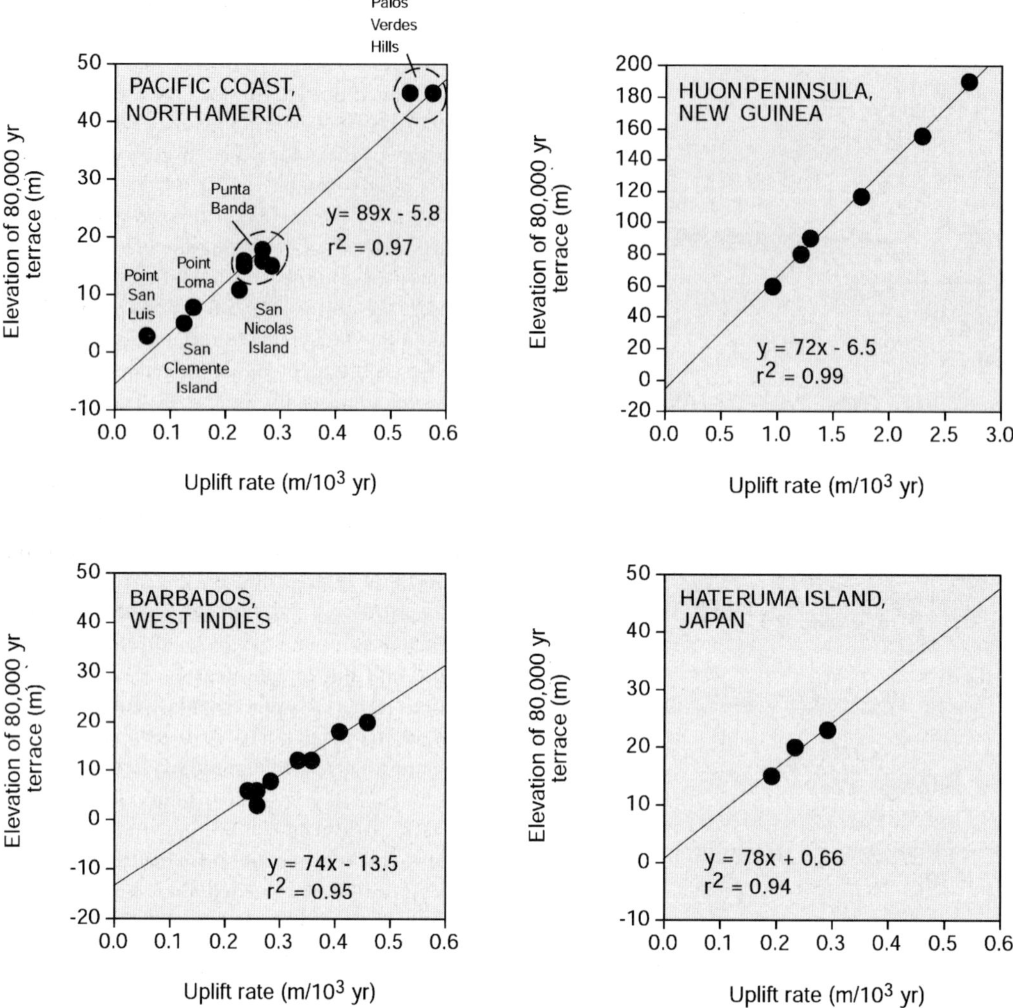

Fig. 13. Plots showing elevations of ~80,000-yr-old marine terraces or coral reefs as a function of uplift rate derived from ~120,000-yr-old marine terraces or reefs for four regions with differing uplift rates. Pacific Coast age and elevation data from Rockwell et al. (1989), Hanson et al. (1992), Kern & Rockwell (1992), and Muhs et al. (1992a, 1994, 2002b, c); New Guinea data from Bloom et al. (1974); Barbados data from Broecker et al. (1968), Matthews (1973), and Taylor & Mann (1991); Hateruma Island data from Ota & Omura (1992).

at altitudes of up to 22 m (Kaufman *et al.*, 1991). Amino acid ratios in mollusks presented by Kaufman & Brigham-Grette (1993) show that it is easily distinguishable from last-interglacial Pelukian deposits, but it is younger than deposits thought to be of Pliocene age (Fig. 15). Kaufman *et al.* (1991) reported that basaltic lava overlies drift of the Nome River glaciation, which in turn overlies Anvilian marine deposits. An average of several $^{40}Ar/^{39}Ar$ analyses on the lava yields an age of $470,000 \pm 190,000$ yr. Within the broad limits permitted by this age and using reasonable rates of epimerization of marine mollusks, Kaufman *et al.* (1991) proposed that the Anvilian marine transgression dates to ~400,000 yr ago and is correlative with a major interglacial period recorded by oxygen isotope stage 11.

The high elevation (up to ~22 m) of the Anvilian marine deposits in a tectonically stable region suggests that sea level at that time could have been higher than present and even higher than during the last interglacial period. Recent studies from Bermuda, the Bahamas, the Cariaco Basin and the Netherlands Antilles, localities that are either tectonically stable or only slowly uplifting, suggest that sea level could have been 20–25 m above present during a ~400,000-yr-old sea-level stand (Hearty *et al.*, 1999; Lundberg & McFarlane, 2002; Poore & Dowsett, 2001). A 20–25-m-high sea-level stand would require loss of all of the Greenland ice sheet, all of the West Antarctic ice sheet, and at least part of the East Antarctic ice sheet during a Quaternary interglacial period of exceptional warmth. Anvilian marine deposits contain no extralimital northern species of mollusks, but do contain at least six extralimital southern or southward-ranging species, including two that are now confined to lower latitudes in the Atlantic Ocean (Hopkins *et al.*, 1960, 1974). Thus, the

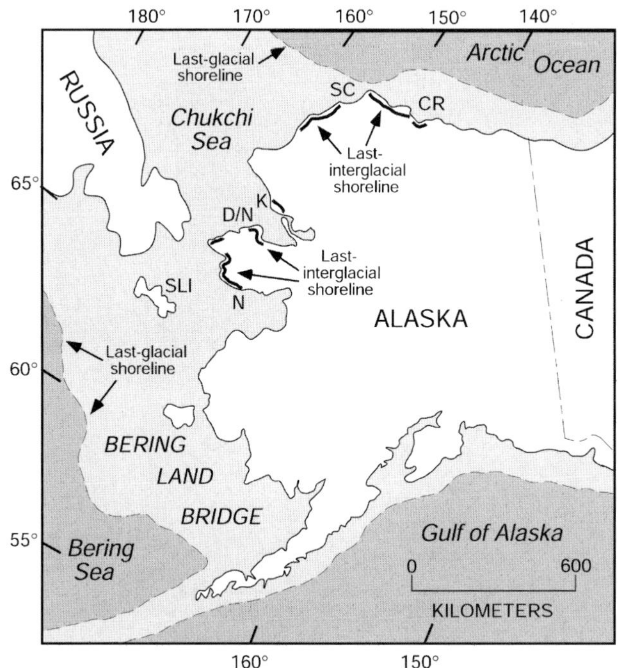

Fig. 14. Map of Alaska and adjacent regions showing last interglacial (~120,000 yr) and last-glacial (~20,000 yr) shorelines and localities referred to in text. Last interglacial shoreline is from Brigham-Grette & Hopkins (1995); last-glacial shoreline is approximate and is based on an assumed 120 m sea-level-lowering during the last glacial period (Fairbanks, 1989). Abbreviations: CR, Colville River; SC, Skull Cliff; K, Krusenstern; D/N, Deering/Nugnugaluktuk; SLI, St. Lawrence Island; N, Nome.

Anvilian marine transgression may record a mid-Pleistocene high sea level that was significantly higher than present during a time of warmer climate.

Last Interglacial Sea-Level History

The lowest marine terrace deposit on the Nome coastal plain, also found on the Arctic Coastal Plain of Alaska (Fig. 14), formed during what has been called the Pelukian transgression. Hopkins (1973) correlated the Pelukian transgression with the peak of the last interglacial period, substage 5e of the oxygen isotope record. Analyses of Pelukian marine mollusks have yielded non-finite radiocarbon ages, indicating that the deposits are more than ~40,000 yr old, but amino acid ratios (Fig. 15) indicate that they are late Pleistocene (Brigham-Grette & Hopkins, 1995; Goodfriend *et al.*, 1996). On tectonically stable and unglaciated portions of the coasts of the Seward Peninsula and Arctic Coastal Plain, Pelukian marine deposits are found at altitudes of less than 10 m, consistent with the inferred paleo-sea level position of ~120,000-yr-old deposits on tectonically stable coasts in middle and low latitudes. These various lines of evidence strongly support a correlation of Pelukian marine deposits to the peak of the last interglacial period, ~120,000 yr ago. Pelukian marine deposits, like their ~120,000 yr counterparts in mid-latitudes, contain a number of extralimital southern species of mollusks, indicating water temperatures higher than those of the present during the last interglacial period (Brigham-Grette & Hopkins, 1995; Hopkins *et al.*, 1960). Some mollusks found in Pelukian marine deposits at Nome and on St. Lawrence Island indicate that winter sea ice did not extend south of Bering Strait during the last interglacial period, a significant (~800 km) northward retreat of its present southern wintertime limit (Brigham-Grette & Hopkins, 1995).

Sea-Level History of the Last Glaciation

During the last glacial period, when sea level was ~120 m lower than present, Alaska experienced a dramatic change in coastal geography compared to most of the rest of North America. The present shelf areas of the Bering and Chukchi Seas off western Alaska and eastern Siberia are shallow, generally less than 100 m. Full-glacial sea-level lowering connected North America and Asia, creating the Bering Land Bridge (Fig. 14). During sea-level low stands of earlier glacial periods, this land bridge provided the conduit by which animals migrated from Asia to North America, possibly as long ago as early Pleistocene time (Guthrie & Matthews, 1971). During the last glacial period, the increased land area would have affected the climate of interior Alaska profoundly, increasing continentality and aridity. The increased aridity may in part explain the last-glacial herb tundra-dominated pollen record of interior Alaska (Ager & Brubaker, 1985). It has long been thought that the Bering Land Bridge, during last-glacial time, was the land corridor for the earliest human migrations from Asia into North America. A new hypothesis, however, suggests that humans may have arrived in North America via watercraft along the now-submerged last-glacial coastline south of the Bering Land Bridge (Dixon, 2001).

Atlantic Coast

Introduction: The Nature of the Atlantic Coast Record

The U.S. Atlantic Coastal Plain (Fig. 16) contains a rich record of Quaternary sea level change. With the exception of the carbonate deposits in southern Florida, these records are represented by a combination of erosional and clastic depositional features. Because of the low relief of this region, most of these records are found at elevations of less than 15 m, often as laterally adjacent landforms or as superposed deposits of variable thickness and continuity. Surface expression of Pleistocene sea levels is usually seen as a series of scarps, terraces, or paleoshorelines, rarely with more than 5 m of relief. From the earliest work (Richards & Judson, 1965 and references therein), many authors have interpreted the Atlantic Coastal Plain in terms of named terraces or terrace formations, as

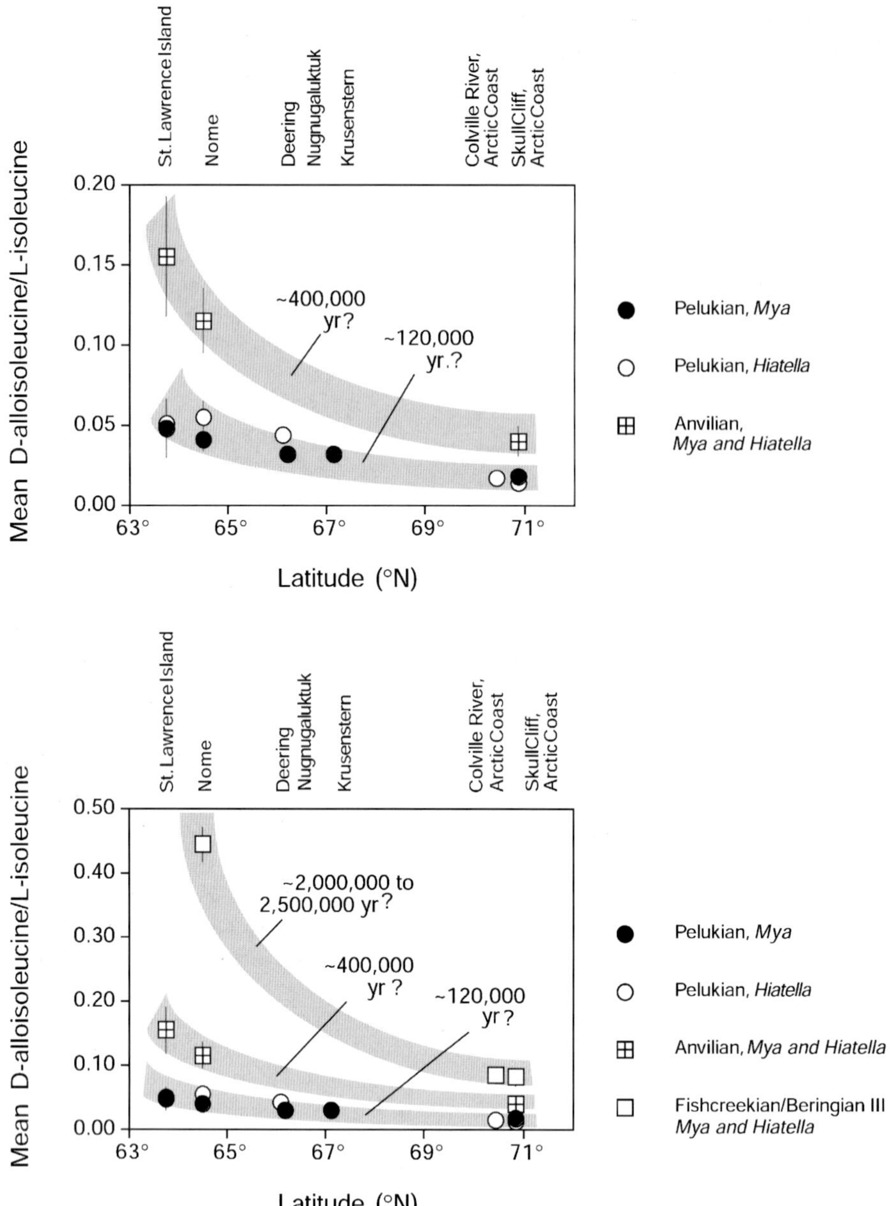

Fig. 15. Mean ratios of the amino acids D-alloisoleucine to L-isoleucine in fossil bivalves from marine deposits of Alaska, shown as a function of latitude. Upper panel is expansion of the vertical scale of the lower panel, showing only the youngest two ages of deposits. Data from Kaufman (1992), Kaufman & Brigham-Grette (1993), and Brigham-Grette & Hopkins (1995).

reviewed by Colquhoun *et al.* (1991). Several additional examples of Atlantic coast Quaternary sequences can be found in Nummendal *et al.* (1987) and Fletcher & Wehmiller (1992). U-series ages have contributed to an understanding of the complex history of these terraces and terrace formations, but it has proven remarkably difficult to identify features that are clearly of early last interglacial age (~125,000 yr or MIS 5e), a traditional reference point for many Quaternary coastal sequences. In addition to U-series methods, amino acid racemization, radiocarbon, and occasionally paleomagnetic methods have been used for geochronologic studies of Coastal Plain units. Middle and early Pleistocene units have often been correlated or assigned ages based on biostratigraphy (Ward *et al.*, 1991), but these approaches rarely have the age resolution that is required to assign deposits to specific ice volume minima of the marine isotopic record.

Colquhoun *et al.* (1991) divided the Atlantic Coastal Plain into upper, middle, and lower segments, each segment bounded by scarps. The lower Coastal Plain contains the Pleistocene coastal record, with late Pleistocene (<450,000 yr in the system of Colquhoun) coastal units mapped seaward of several major scarps, including the prominent Suffolk

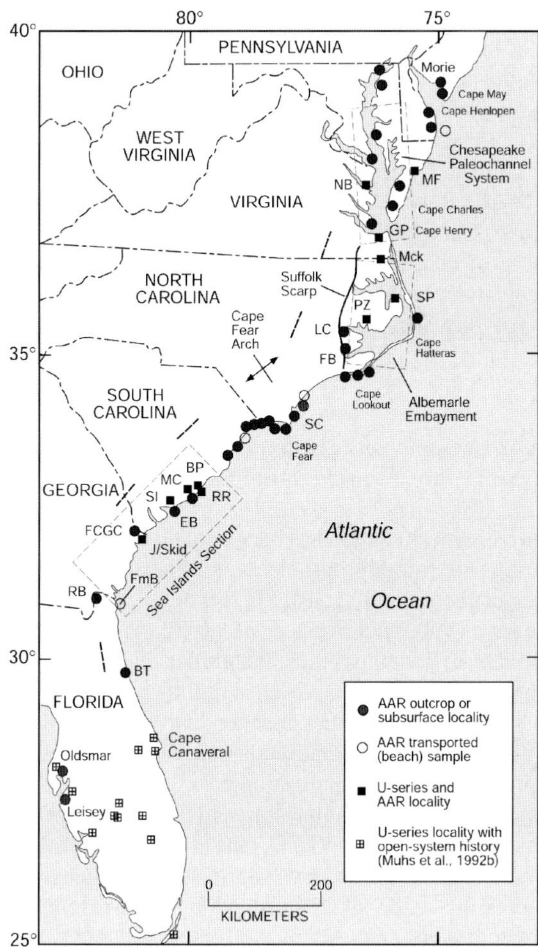

Fig. 16. Map of the U.S. Atlantic Coastal Plain. Prominent structural or geomorphic regions discussed in text (Chesapeake Paleochannel System, Albemarle Embayment, Cape Fear Arch, and Sea Islands Section). Localities with U-series or paired U-series/AAR data are identified by solid squares. Other AAR or U-series localities (e.g. Muhs et al., 1992b) are shown as indicated. AAR data from beach or shelf samples also identified separately. Locality abbreviation or relevant reference: Morie (O'Neal & McGeary, 2002; O'Neal et al., 2000); MF, Matthews Field VA (Szabo, 1985; Groot et al., 1990); NB, Norris Bridge, VA (Szabo, 1985; Groot et al., 1990); GP, Gomez Pit, VA (Mirecki et al., 1995); Mck, Moyock, NC (Cronin et al., 1981); SP, Stetson Pit NC (York et al., 1989); PZ, Ponzer NC (Cronin et al., 1981; Szabo, 1985; Wehmiller et al., 1988); LC, Lee Creek Mine, NC; FB, Flanner Beach, NC (McCartan et al., 1982); SC, Snows Cut, NC (Dockal, 1995); BP, Berkeley Pit SC; RR, Rifle Range Pit, SC; MC, Mark Clark Pit, SC (McCartan et al., 1982); SI, Scanawah Island, SC (McCartan et al., 1982); EB Edisto Beach, SC; FCGC, Forest City Gun Club, GA; J/Skid, Jones Pit, Skidway I, GA (Hulbert & Pratt, 1998); RB, Reids Bluff GA (Belknap, 1979; Huddleston, 1988); FrnB, Fernandina Beach, FL (Wehmiller et al., 1995); BT, Bon Terra, FL (R. Mitterer, 1975, 1995 pers. comm. to J.F. Wehmiller); Oldsmar (Olsmar 1 and 2, Karrow et al., 1996); Leisey (Jones et al., 1995).

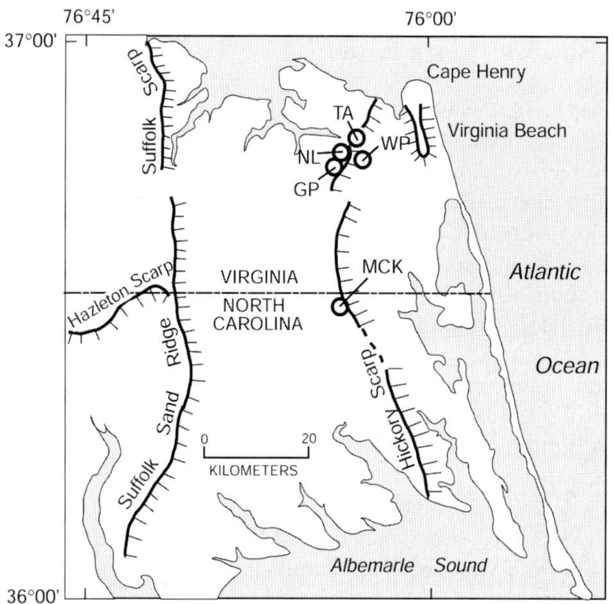

Fig. 17. Coastal plain of southeastern Virginia, with geomorphic features (from Oaks et al., 1974, their Fig. 7). See Mixon et al. (1982) for additional nomenclature on geomorphology. Locations of commercial excavations where collections for U-series coral dating or AAR mollusk analysis were made as follows: TA, Toy Avenue Pit; WP, Womack Pit; NL, New Light Pit; GP, Gomez Pit; MCK, Moyock Pit.

Scarp in North Carolina (Fig. 17). Pliocene and older units are found landward of the Surry Scarp, which separates the lower and middle Coastal Plain segments.

Holocene depositional environments of the Atlantic Coastal Plain are useful analogs for the Pleistocene records, which, in turn, have been influenced by the pre-Quaternary geologic framework. Hayes (1994) and Riggs *et al.* (1995) noted that variability in structure, antecedent topography, sediment supply, wave and tidal regimes, and Coastal Plain gradient have all affected both Holocene and Pleistocene depositional systems. South of the glacial limit in northern New Jersey, these systems include: (1) major estuaries associated with mid-Atlantic rivers (Delaware and Chesapeake bays); (2) the broad low relief Coastal Plain of North Carolina, with an extensive late Quaternary record of estuarine and shallow marine deposits; (3) the Cape Fear Arch, where the Coastal Plain gradient is steep and the Quaternary record is thin; and (4) the Sea Island section of South Carolina, Georgia, and northern Florida, where the topographic gradient is low and coast-parallel Holocene barrier-lagoon complexes are "welded" to a Pleistocene island core. The modern regional geomorphology demonstrates that older features (barriers or shorelines) exist seaward of younger back-barrier or estuarine environments, so it is likely that similar juxtapositions are preserved in the Pleistocene record. Similarly, records of Holocene sea-level rise on the Atlantic coast have been influenced by the underlying Pleistocene (or Tertiary) framework as well as regional postglacial tectonic adjustment (Peltier, 1999).

Uranium-Series Dating of the Atlantic Coastal Plain Quaternary Coastal Record

The earliest published applications of U-series methods to Atlantic coast sites involved alpha-spectrometric analyses of reef-building corals from central and southern Florida (Broecker *et al.*, 1965; Osmond *et al.*, 1965), reviewed above. Most Atlantic Coastal Plain U-series studies have employed ahermatypic corals from higher-latitude sites, particularly in southeastern Virginia and central South Carolina (Figs 3, 16 and 17) (Mixon *et al.*, 1982; Oaks *et al.*, 1974; Szabo, 1985). Cronin *et al.* (1981) discussed many of the paleoenvironmental implications of these ages, and McCartan *et al.* (1982) discussed the relation of the U-series results to lithostratigraphic and other independent chronologic information.

Advanced analytical methods (U-series dating by TIMS) have recently been applied to additional fossil coral collections from Virginia, South Carolina, and Georgia, using fresh samples collected in direct association with mollusk specimens used for racemization analyses (Simmons *et al.*, 1997; Wehmiller *et al.*, 1997; York *et al.*, 1999). There is a high degree of reliability for these samples because: (1) U contents of the corals are within the range that is typical for modern, colonial corals (2–3 ppm); (2) $^{230}Th/^{232}Th$ activity values are high, indicating little or no "inherited" ^{230}Th (a problem that had plagued previous studies: see Szabo, 1985); and (3) back-calculated initial $^{234}U/^{238}U$ values are within the range of modern sea water, indicating probable closed-system history with respect to U and its long-lived daughter products. The distinguishing feature of the both the new TIMS U-series results and the earlier alpha-spectrometry results is that the ages cluster in a time range (65,000–85,000 yr ago, roughly correlative with substage 5a) that would not be expected for emergent units on a stable or subsiding margin, if interpreted in terms of sea level records based on isotopic or tropical coral reef eustatic models (e.g. Lambeck & Chappell, 2001).

Only three sites on the Atlantic Coastal Plain north of Florida have yielded coral ages in the range of ~125,000 yr (MIS 5e). The few ages in this range that have been obtained, all from sites near Charleston or Myrtle Beach, South Carolina, are occasionally of questionable quality or do not unequivocally represent the age of their host unit (Hollin & Hearty, 1990; Szabo, 1985). On Nantucket Island, Massachusetts (Fig. 3), north of the Atlantic Coastal Plain proper, there is a single U-series age of ~130,000 yr on a coral from the glacially deformed Sankaty Sand (Oldale *et al.*, 1982). As with localities of this age in Hawaii, California, and Alaska, it contains a fauna that indicates warmer-than-present waters during this high sea stand. The absence of a prominent and well-dated ~125,000 yr unit on the Atlantic Coastal Plain, combined with the relatively large abundance of ~80,000 yr coral ages, remains unexplained in spite of many years of debate.

U-series (alpha-spectrometric) ages in the range of ~200,000 to 250,000 yr have been obtained from sites in North and South Carolina (Hollin & Hearty, 1990; Szabo, 1985) and a number of additional ages are at or near the limit of the $^{230}Th/^{238}U$ chronometer (Szabo, 1985). In some cases, the ages of these older samples (estimates range from 300,000 yr to >750,000 yr) have been derived from their $^{234}U/^{238}U$ values (Szabo, 1985). These oldest samples have been collected primarily along the Intracoastal Waterway near Myrtle Beach, South Carolina, from units known as the Socastee, Canepatch, and Waccamaw formations (Fig. 18). Because of concerns about sample reworking, stratigraphic nomenclature, and/or geochemical issues, the appropriate ages for the Socastee and Canepatch formations remain controversial (e.g. Hollin & Hearty, 1990).

Aminostratigraphy and Quaternary Chronology, Chesapeake Bay to Northern Florida

Amino acid racemization has been applied by several workers to various stratigraphic problems in either local or regional studies of the Atlantic coastal (Corrado *et al.*, 1986; Hollin & Hearty, 1990; McCartan *et al.*, 1982; Wehmiller, 1982; Wehmiller & Belknap, 1982; Wehmiller *et al.*, 1988, 1992). In Figs 18 through 20, we present a summary of critical results from sites that have either independent stratigraphic or U-series chronologic information that provides a framework for interpretation of the racemization data.

The racemization results plotted in Figs 18 and 19 are for a parameter identified as "VLPG," the numerical average of the D/L values of the amino acids valine, leucine, phenylalanine, and glutamic acid, all determined by high-resolution gas chromatographic analyses of well-preserved *Mercenaria*, a robust bivalve mollusk commonly found in Atlantic Coastal Plain Quaternary deposits (see chromatogram in Wehmiller & Miller, 2000). VLPG minimizes the potential variation in individual amino acid *D*/*L* values that can occasionally occur, although VLPG values are always within 3% of the observed values for *D*/*L* leucine, a common reference amino acid (e.g. Wehmiller & Belknap, 1982). A conceptual model for the relation of selected VLPG values to local stratigraphic sections is shown in Fig. 18, modified from McCartan *et al.* (1982). Because both stratigraphic and geochronologic interpretations of the units portrayed in Fig. 18 are debated or have been modified (e.g. Hollin & Hearty, 1990), and in some cases multiple aminozones are known to occur within a single named formation (Harris, 2000), there are optional interpretations that can be applied to individual sites in each region. Riggs *et al.* (1992; Fig. 2) further demonstrate the relation between local stratigraphy and geochronology for a superposed section in northeastern North Carolina.

In Fig. 19, the VLPG results are plotted vs. latitude, and each significant geomorphic or structural region of the Atlantic Coastal Plain is also identified so that racemization results can be discussed within these local frameworks. Figure 19 is plotted with VLPG values increasing downward to emphasize the superposed nature of many of the Atlantic Coastal Plain records. Results in Fig. 19 represent an ongoing effort to obtain new high-resolution chromatographic data for both earlier and more recent collections; earlier results

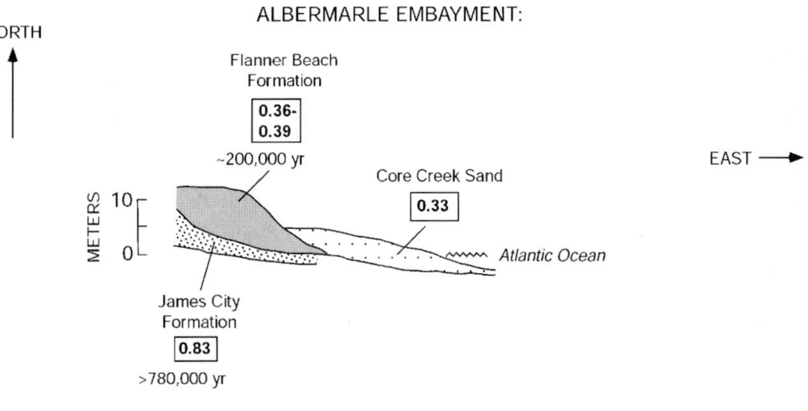

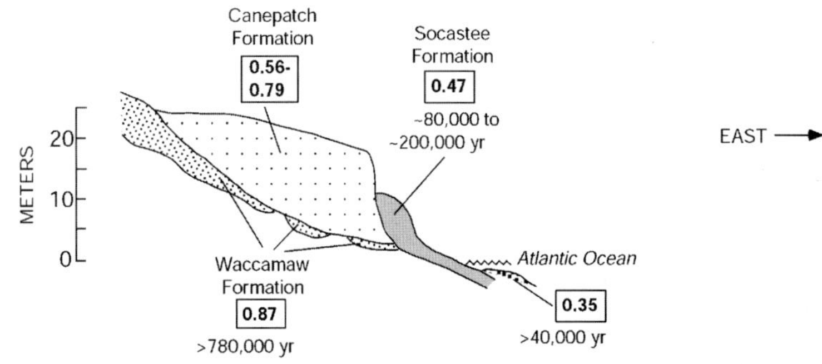

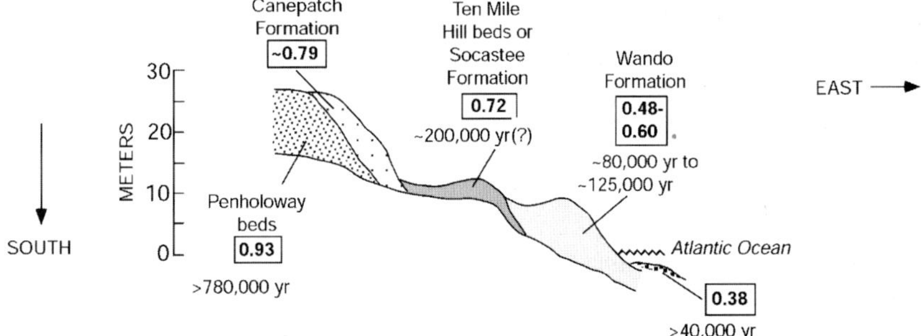

Fig. 18. Schematic cross sections of three Atlantic Coastal Plain sections, showing general relations of geomorphology, stratigraphy, and U-series or aminostratigraphic data. Sections modified from McCartan et al. (1982), with additional U-series data from Hollin & Hearty (1990). Ranges of U-series ages or VLPG values for a single formation identify situations where multiple ages may be found within a single mapped unit.

for many of these localities are published in Wehmiller *et al.* (1988).

Amino acid *D/L* (VLPG) values should always increase with increasing stratigraphic age in local regions or with increasing temperature for samples of equal age (Wehmiller & Miller, 2000). Results from Stetson and Gomez pits are among the best Coastal Plain examples of superposition (Fig. 19), as VLPG values cluster into three physically superposed aminozones at each site. Additionally, Fig. 19 demonstrates general trends of increasing *D/L* (VLPG) values with decreasing latitude or increasing temperature. However, the results in Fig. 19 raise a number of important issues regarding the potential utility of racemization for correlation or age estimation of Coastal Plain samples. These issues are probably related to geochemical, taphonomic, stratigraphic, and thermal factors (Corrado *et al.*, 1986; Hollin & Hearty, 1990; Wehmiller *et al.*, 1992, 2000a, b).

Five aminozones are specifically identified in Fig. 19, and others are cited below as necessary. The area labeled "$<8000\ ^{14}C$ yr" represents results for approximately 30 samples, with radiocarbon ages ranging to 7600 ^{14}C yr B.P., from inner shelf and beach sites (Wehmiller & York, 2001). The sloping band represents the upper limit of VLPG values for these Holocene samples. The shaded band labeled "$\sim$70,000 to $\sim$130,000 yr" is an aminozone that ranges from VLPG $\sim$0.30 at 36.5 N to $\sim$0.60 at 32 N and includes data for all sites where mollusk samples for racemization analysis were collected in close association with corals that yielded U-series ages between $\sim$130,000 and $\sim$70,000 yr ago. The shaded aminozone below this one represents racemization data associated with two North Carolina sites of this apparent age (Ponzer and Flanner Beach: McCartan *et al.*, 1982; Szabo, 1985) and the Socastee Formation at several sites along the Intracoastal Waterway near Myrtle Beach, South Carolina

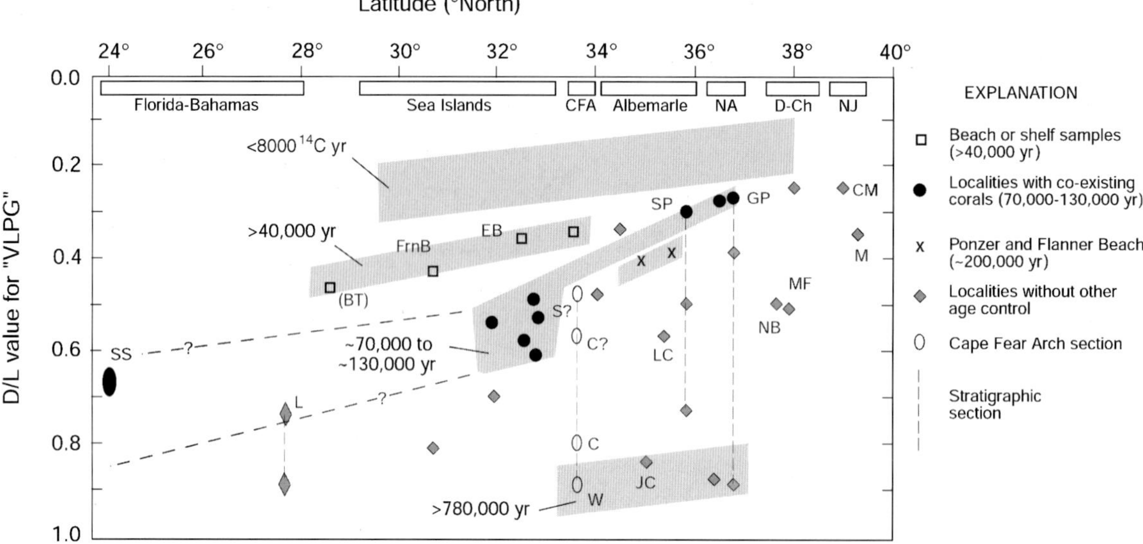

Fig. 19. Plot of VLPG vs. latitude for Atlantic Coastal Plain sites. VLPG is the average of the D/L values of valine, leucine, phenylalanine, and glutamic acid in Mercenaria. Stratigraphic or geomorphic provinces of the Coastal Plain are listed across the top of the figure. Province or locality abbreviations as in Fig. 16; CFA = Cape Fear Arch; NA = Norfolk Arch; D-Ch = Delmarva-Chesapeake. S, C, and W represent Socastee, Canepatch, and Waccamaw formations, respectively (see Fig. 18). X's represent ~200,000 yr "calibrated" results from the Albemarle Embayment, NC. Vertical dashed lines connect data points for local superposed sequences. Shaded regions represent aminozones associated with independent U-series ages as discussed in text. Open squares represent a beach or inner shelf aminozone that has infinite radiocarbon dates but appears younger than ~80,000 yr. The open square labeled "BT" represents data from Mitterer's (1975) Bon Terra site (pers. comm., R. Mitterer to JFW, 1992). Data point for San Salvador (SS) represents results from Chione specimens (Wehmiller et al., 1988; York et al., 2000a), increased by ~10% to make them comparable with the Mercenaria results plotted here.

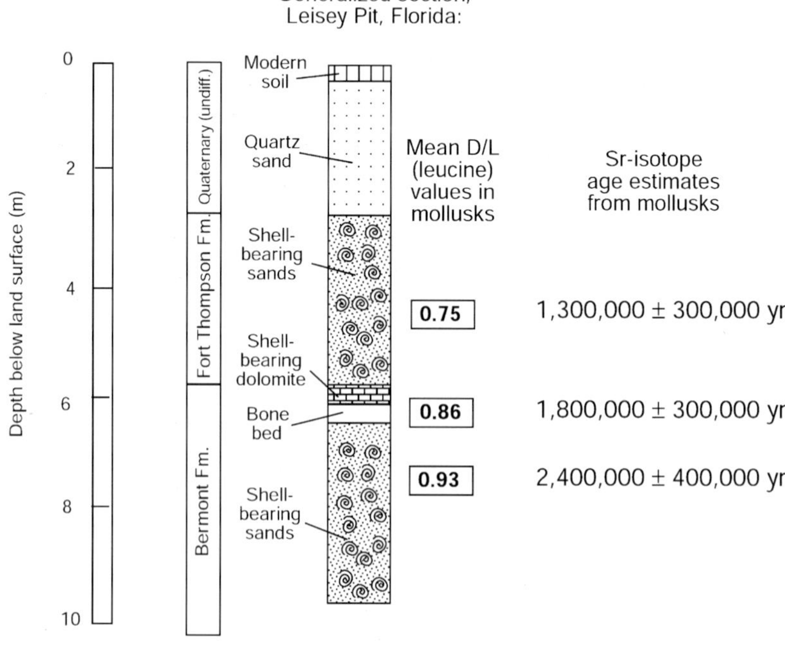

Fig. 20. Schematic summary of aminostratigraphic data and strontium isotopic age estimates for the Fort Thompson-Bermont section at Leisey Pit, Florida. Stratigraphic section modified from Jones et al. (1995); data from York et al. (2000a). Mean D/L leucine values have typical ranges of 5% for multiple samples.

(Fig. 15). The southward projection of this aminozone to latitude 32 N overlaps with the younger aminozone, and the southward projection of the "~70,000 to ~130,000 yr" aminozone does not include the well-calibrated results from San Salvador Island. These observations demonstrate a conflict that has been debated for over 20 years (reviewed by Wehmiller *et al.*, 1992). Hollin & Hearty (1990) resolved some of this conflict by rejecting ~200,000 yr ages as being from reworked samples, an interpretation that has major implications for all other aminostratigraphic age estimates in the region. Although theory (Wehmiller *et al.*, 1988) predicts that aminozones should rise and diverge with decreasing latitude (increasing temperature) (e.g. Figs 11 and 15, above, and Wehmiller, 1992), this principle does not hold for the U.S. Atlantic coast, even though the modern temperature gradient is a smooth function of latitude (Wehmiller *et al.*, 2000b). Correlation of calibrated racemization data from late Pleistocene Atlantic coast sites between Florida and Virginia instead presents a broad envelope that changes trend somewhere between 30 and 34°N (Corrado *et al.*, 1986; Hollin & Hearty, 1990; Wehmiller *et al.*, 1988).

Aminostratigraphic results from Leisey Pit, on the southeast margin of Tampa Bay, Florida (Figs 3 and 4) demonstrate the magnitude of potential conflicts with independently calibrated results. Figure 20, modified from Jones *et al.*, 1995) compares racemization data for the Leisey Pit section (from York *et al.*, 2000a) with associated strontium isotopic data (Jones *et al.*, 1995) from this same section. At Leisey Pit, a unit mapped as the Fort Thompson Formation occurs in the upper part of the section. The Fort Thompson Formation is often considered as "last interglacial" (cf. unit Q5, above) or specifically correlative with ~125,000 yr. U-series dated units elsewhere in Florida (DuBar *et al.*, 1991). In contrast, Sr-isotope age estimates for the Fort Thompson Formation at Leisey Pit are variable but are on the order of 1,000,000 years (Jones *et al.*, 1995; York *et al.*, 2000a). Aminostratigraphic and U-series data on mollusks (Karrow *et al.*, 1996) from nearby excavations at Oldsmar, Florida, at the north end of Tampa Bay, combined with amino acid data from Leisey Pit argue for an age of approximately 100,000 to 200,000 yr for the Fort Thompson Formation (Wehmiller *et al.*, 1999; York *et al.*, 2000a). Consequently, two vastly conflicting estimates for the age of the Fort Thompson Formation at Leisey Pit can be inferred, either "late Pleistocene" (<200,000 yr) or "middle to early Pleistocene" (~1,000,000 yr). The VLPG data from Leisey Pit (Fig. 19) are more consistent with the younger age estimate for the Fort Thompson Formation, especially an age around 200,000 yr when compared with the 125,000 yr calibrated results from San Salvador Island. Amino acid and Sr-isotope data are also available for the Bermont Formation, immediately underlying the Fort Thompson at Leisey (Fig. 20). These data are all stratigraphically consistent and the Sr-isotope data suggest an age of >1,500,000 yr for the Bermont Formation. The amino acid values for the Bermont are quite near equilibrium, so they can only be used to suggest ages greater than ~750,000 yr. Mitterer's (1975) original study of Florida aminostratigraphy identified similar results for central and southern Florida.

Because of the difficulties that can arise with broad correlations of the Atlantic coast racemization data over large latitude ranges, the following discussions focus on the relation of racemization results to stratigraphic or geomorphic sequences within each of the local regions identified in Figs 18–20. These localized results can be interpreted in terms of most probable chronologies for each region, using independent age control (isotopic or biostratigraphic) where available. Age estimates proposed in the following sections could easily change if additional calibrated results became available, or if alternative models for racemization kinetics were applied (e.g. York *et al.*, 1989).

Chesapeake Bay, Delmarva Peninsula, and Norfolk Arch

In the Chesapeake-Delmarva region, three Pleistocene aminozones are recognized (VLPG values of 0.26, 0.38, and 0.49), overlying Plio-Pleistocene units with near-equilibrium D/L values. At the southern end of Chesapeake Bay, the Gomez Pit section (ca. 15–20 m thick; land surface elevation ~+8 m) provides clear evidence for the physical superposition of these aminozones (Lamothe & Auclair, 1999; Mirecki *et al.*, 1995), the upper one (VLPG $= 0.26 \pm 0.02$) associated with ca. 80,000 yr U-series coral ages. This unit ("Gomez coral bed") occurs up to 7 m above present sea level, although the corals themselves occur between -1 and $+1$ m (Cronin *et al.*, 1981). The next older Gomez Pit aminozone (VLPG = 0.38), found in an extensive oyster bed at ca. -3 m, represents deposition roughly 300,000–350,000 yr ago, approximately equivalent to MIS 9 (Mirecki *et al.*, 1995). A still-older Pleistocene aminozone (VLPG = 0.49 – not plotted in Fig. 19 for GP), seen at several sites in the region, is seen at GP in shells that are likely reworked. At the base of the GP section are nearly racemic (VLPG = 0.9) shells from the late Pliocene Chowan River formation.

The three aminozones from southern Chesapeake Bay (VLPG = 0.26, 0.38, and 0.49) are also recognized in outcrop or limited subsurface samples from sites on the Delmarva Peninsula (Groot *et al.*, 1990). The three phases of deposition inferred from the racemization data correspond to major cut/fill sequences of the Susquehanna River paleochannel system (Colman & Mixon, 1988; Colman *et al.*, 1990; Oertel & Foyle, 1995). The Chesapeake paleochannel system represents the evolution of the ancestral Susquehanna, Potomac, Rappahanock, York, and James rivers. Earlier channels of these rivers passed through the region that is now the southern Delmarva Peninsula, which has migrated southward during the mid-to-late Pleistocene evolution of this drainage pattern. Chen *et al.* (1995) recognized the extension of this paleochannel network onto the inner shelf of southeastern Virginia, and Knebel & Circe (1988) identified comparable features in Delaware Bay. The VLPG value of 0.49 represents a unit that apparently fills the Exmore paleochannel, considered to be between 300,000 and 500,000 yr old, based on comparisons of racemization and U-Th data from the region. This aminozone is found at MF and NB (Figs 15 and 19), each site having yielded ambiguous U-series age estimates but with an age of >300,000 yr being most probable (Szabo, 1985; and pers. comm.), and also at other sites in the central Chesapeake Bay region. The two younger aminozones represent progressively younger channel-fill deposits that probably correspond with low-stand/high-stand cycles from ~300,000 yr through ~100,000 yr ago. These same two aminozones are also seen at exposures around the southern margins of Delaware Bay (CM, M on Fig. 19), where they interpreted to represent extensive late- or late-middle Pleistocene deposition (Groot *et al.*, 1990; O'Neal *et al.*, 2000).

Albemarle Embayment, Northeastern North Carolina

The thick (~50 m) Quaternary section preserved in the Albemarle Embayment preserves a racemization record spanning most of the Quaternary. The units of the Albemarle Embayment are preserved in both vertical and lateral sequences and record multiple cycles of cutting and filling during the Pleistocene (Boss *et al.*, 2002; Riggs *et al.*, 1992, 1995). Six Pleistocene aminozones are recognized in the sediments of the Albemarle Embayment, with approximate VLPG values of 0.3, 0.38, 0.49, 0.56, 0.72, and 0.83 (Fig. 19). The youngest of these (80,000 yr by U-series dating at Moyock and Stetson Pit) represents deposition during MIS 5. The next older aminozone records MIS 7, based on data from Ponzer and Flanner Beach (McCartan *et al.*, 1982; Szabo, 1985). Three older aminozones (VLPG 0.49, 0.56 and 0.72) are found deeper in the Stetson Pit section, near the center of the Albemarle Embayment, or in a mid-Pleistocene unit that is intermittently exposed at the Lee Creek Mine, ~10 km east of the Suffolk Scarp (Fig. 16; Ward *et al.*, 1991). Several of these aminozones are also seen in shallow core samples from the North Carolina inner shelf or in reworked beach specimens (Wehmiller *et al.*, 1995). A still-older aminozone (VLPG = 0.83) represents the James City Formation (early Pleistocene: Ward *et al.*, 1991). The oldest aminozone (VLPG =~ 0.90) in the region represents the Pliocene/early Pleistocene Chowan River formation, sampled in both NE North Carolina and in Gomez Pit. Collectively the Albemarle Embayment VLPG data define a series of aminozones that represent multiple cycles of deposition through the Pleistocene. These aminozones may, in fact, record only a small fraction of the numerous (≂15) transgressive records preserved in this region (Riggs *et al.*, 1992; York *et al.*, 1989).

The Albemarle VLPG values between ~0.50 and 0.70 fill a significant middle-to-early Pleistocene gap in the regional stratigraphy, as no named units in this interval are recognized (Soller & Mills, 1991). This aminostratigraphy may also place an upper limit on the age of the Suffolk Scarp in eastern North Carolina (mid-Pleistocene, ~350,000 to 500,000 yr, York *et al.*, 2000b), consistent with the chronology of Colquhoun *et al.* (1991). More rigorous calibration of this mid-Pleistocene aminostratigraphy would help to resolve some of the aminostratigraphic correlation problems identified by the results in Fig. 19.

The Suffolk Scarp, as a prominent boundary between late and early Pleistocene units, is a major geomorphic feature of the North Carolina Coastal Plain (Brill, 1996). The southward continuation of this boundary across the Cape Fear Arch suggests that late Pleistocene units (and their associated aminozones) seen in eastern North Carolina are infrequently observed in emergent units between Cape Lookout and Cape Romain (Fig. 16), mimicking the overall outcrop pattern of pre-Quaternary units across the Cape Fear Arch (Mixon & Pilkey, 1976; Riggs & Belknap, 1988; Snyder *et al.*, 1991; Wehmiller *et al.*, 1992; York & Wehmiller, 1992b).

Cape Fear Arch

Two shallow marine units, the Socastee and Canepatch formations, are exposed in the Intracoastal Waterway near the North Carolina-South Carolina border (Figs 16 and 18). Associated with these units are named barriers (the Myrtle Beach and Jaluco barriers, respectively, as mapped by DuBar *et al.*, 1980). The ages of these two formations are critical for any attempt to identify paleoshorelines or correlate aminozones between the north and south flanks of the Cape Fear Arch, a region where the entire preserved Cenozoic record is much thinner than in the Albemarle Embayment. Although the stratigraphic terminology, as applied to specific exposures on the Intracoastal Waterway, has varied (Colquhoun *et al.*, 1991; DuBar *et al.*, 1980; Hollin & Hearty, 1990; McCartan *et al.*, 1982; Owens, 1989; Wehmiller *et al.*, 1988), and the U-series chronology for these units is ambiguous (Szabo, 1985), most workers interpret the Socastee formation to be at least 200,000 yr old (~MIS 7) and the Canepatch formation to be as much as 500,000 yr old, the latter age being derived from biostratigraphic information and associated $^{234}U/^{238}U$ values in corals (Szabo, 1985). Hollin & Hearty (1990), however, interpreted the Socastee as 80,000 yr old (late MIS 5) and the Canepatch as 125,000 yr old (early MIS 5), an interpretation that appears consistent with aminostratigraphic comparison with the Sea Islands VLPG data summarized in Fig. 19. Hollin & Hearty (1990) emphasized the importance of using only the youngest U-series ages found in a unit because of the potential for mixing of fossils with different ages, even though the majority of the U-series ages from Intracoastal Waterway sites (and some from other sites) would then have to be rejected. Four aminozones (VLPG values of 0.47, 0.56, 0.79, and 0.87) are apparent in the results for the Cape Fear Arch region (Figs 18 and 19), the oldest (0.87) representing the Plio-Pleistocene Waccamaw formation. The three younger aminozones have been interpreted to represent three depositional events, with ages of >780,000 and ~460,000 yr recorded at different Canepatch sites, and ~200,000 yr recorded at Socastee sites (Corrado *et al.*, 1986; Wehmiller *et al.*, 1988). Racemization age estimates for the Socastee and Canepatch are sensitive to the choice of calibration samples and/or correlations from either the north or the south: older age estimates result from comparisons with results from the Albemarle Embayment, while younger age estimates result from comparisons with the Sea

Islands section of SC and GA. In spite of many years of study, serious ambiguities about the relation between racemization, U-series, and stratigraphic interpretations of the Socastee and Canepatch deposits remain unresolved.

Sea Islands Section

The Sea Islands section can be traced from Cape Romain, South Carolina to Cape Canaveral, Florida (Fig. 16). The typical thickness of the entire Quaternary section in the Sea Islands area is between 5 and 10 m, although both thinner and thicker sections are found over highs and lows in the underlying Tertiary deposits (Harris, 2000). Winker & Howard (1977) identified numerous Plio-Pleistocene paleoshorelines throughout this region.

Most of the U-series ages from the Sea Islands section cluster around 80,000 yr (MIS 5a: Wando Formation of McCartan *et al.*, 1984). However, two older coral U-series ages (~125,000–139,000 yr) have also been obtained from the Wando Formation (Szabo, 1985). Thus, it can be inferred that the "last interglaciation" in its broadest sense (all of MIS 5) may be recorded by this unit. Older (~200,000–250,000 yr) ages have come from a unit exposed inland from the Wando Formation localities (McCartan *et al.*, 1982; Szabo, 1985). Evidence for multiple aminozones within the Wando Formation (Corrado *et al.*, 1986; Harris, 2000; York *et al.*, 1999, 2001), raises the possibility of multiple ages of fossils within this unit, but there is no single site where multiple aminozones can be directly associated with one or more U-series ages. In fact, overlapping ranges of *D/L* values from sites with 80,000, 120,000, or 200,000 yr U-series ages (Corrado *et al.*, 1986; McCartan *et al.*, 1982; Wehmiller & Belknap, 1982) indicate that racemization data may not resolve these ages as well as has been claimed (Corrado *et al.*, 1986; Hollin & Hearty, 1990). Although these may be geochemical issues related to AAR (Wehmiller *et al.*, 1993), the complexity of the Sea Islands record itself is a potential cause of the variability in apparent ages inferred from the AAR data.

Three emergent aminozones (VLPG 0.53, 0.69, and 0.80) are recognized at the southern end of the Sea Island section. The youngest of these aminozones is associated with the 80,000 yr. U-series (TIMS) coral age at Skidaway Island (Figs 16 and 19), and the others may represent terrace formations that are probably of late or late-middle Pleistocene age (Colquhoun *et al.*, 1991; Huddleston, 1988; Markewich *et al.*, 1992). The Skidaway Island site is a good example of the Sea Island "model," as the dated samples come from a muddy unit exposed by shallow excavation near the center of this predominately Holocene island (Hulbert & Pratt, 1998).

A still-younger aminozone (VLPG $= 0.35$–0.42, radiocarbon age $>40{,}000$ ^{14}C yr B.P.) is found in reworked beach samples (EB and FrnB in Figs 16 and 19) as well as at beach and shelf sites near the North Carolina-South Carolina border (Fig. 19). The age of this aminozone is enigmatic, as it appears younger than 80,000 yr when compared with VLPG data from late Pleistocene emergent sites, but it is definitely older than ~40,000 ^{14}C yr B.P. (Wehmiller & York, 2001; Wehmiller *et al.*, 1995; York & Wehmiller, 1992a; York *et al.*, 2001). This same aminozone is inferred from the results of Mitterer (1975) for samples collected from an emergent outcrop near Bon Terra, Florida (R.M. Mitterer, pers. comm. to J.F. Wehmiller, 1995) (BT, Fig. 16). The Bon Terra collection is particularly important, as it most likely represents the Anastasia Formation, widely considered as of "last interglacial" age. Other data from northern Florida localities (Mitterer, 1975) are consistent with this four-fold amino acid zonation for the southern portion of the Sea Islands, although specific locality information for Mitterer's collections is not available.

Gulf Coast

North and west of the carbonate-rich records of central and southern Florida (discussed above), other coastal records from the Gulf Coast include those from the Florida panhandle extending west to southern Mississippi and those along the Louisiana-Texas coast (Fig. 3). Geochronological data for Pleistocene units in this entire region are limited, particularly for emergent sequences, the focus of this review. Seismic exploration of the thick Quaternary and pre-Quaternary records in the northern Gulf of Mexico, combined with extensive marine biostratigraphy, allow high-resolution chronology for portions of the offshore record, especially on the Texas-Louisiana shelf. DuBar *et al.* (1991) reviewed the characteristics and nomenclature of the onshore and offshore Quaternary record of the entire region. By analogy with the Atlantic Coastal Plain, it is quite likely that Gulf Coast Quaternary deposits are equally complex and variable.

In the panhandle region of northwest Florida (Fig. 4), the number of preserved Pleistocene shoreline features has been the subject of considerable debate (Donoghue & Tanner, 1992; Donoghue *et al.*, 1998; Otvos, 1992, 1995), as the location of "last interglacial" features remains in doubt. Additionally, the potential age range (late Pleistocene or most of the Pleistocene) of preserved coastal features is questioned. Mollusk samples from shallow excavations in Franklin County in the Florida panhandle have yielded racemization results that imply a late Pleistocene age when compared with other data from Florida and Georgia, but correlation of these results to a specific sea level transgression is not possible (Otvos & Wehmiller, unpublished data; Wehmiller *et al.*, 1996).

Pleistocene coastal features on the Texas coast include the Ingleside Barrier (interpreted as broadly of "last interglacial" age by Otvos & Howat, 1996), also referred to as the "Ingleside surface." The Ingleside surface is underlain by the Beaumont Clay, a complex unit that probably includes multiple ages of fluvial and nearshore units, perhaps representing the past 400,000 years of glacial-interglacial incision and valley filling (Blum & Price, 1998).

Records on the Texas shelf provide an important perspective on the origin and age of both submergent and emergent landforms in the region (Anderson *et al.*, 1996; Morton & Price, 1987; Rodriguez *et al.*, 2000; Suter *et al.*, 1987).

Valley incision, infilling, and delta construction are all recognized in various seismic studies, reinforced with core data. In one of the more provocative studies of the implications of these records for Pleistocene sea-level history, Rodriquez *et al.* (2000) identified a submerged escarpment and associated barrier features that indicate a shoreline at a depth (corrected for subsidence) of about -15 m. Based on seismic stratigraphic relations to other sequence boundaries in the record (interpreted to represent ~140,000 and ~20,000 yr, or "transitional MIS 6/5e" and "MIS 2" transgressive surfaces), this paleoshoreline is interpreted to be between 35,000 and 50,000 yr old (~MIS 3). Its depth is significantly shallower than would be predicted from most marine isotope or dated coral reef records (e.g. Lambeck & Chappell, 2001). Finite radiocarbon ages are consistent with this estimate (Rodriguez *et al.*, 2000), although ages in this range are near the limit of reliability. Racemization data for mollusks from these shelf sequences and from cores beneath the Ingleside Barrier are all internally consistent with the stratigraphic position of the samples (Wehmiller *et al.*, 2000a). Kinetic modeling is also consistent with an age of between 40,000 and 80,000 yr for the -15 m paleoshoreline (Wehmiller *et al.*, 2000a).

Elsewhere on the Gulf of Mexico shelf, Schroeder *et al.* (1995) used radiocarbon dating of oyster shells to produce a rough curve of postglacial sea-level rise and a more uncertain record for the time interval between about 27,000 and 40,000 ^{14}C yr B.P. Although many shell radiocarbon ages in this range can be challenged (Bloom, 1983a), some paired AAR/radiocarbon (AMS) analyses of massive and well-preserved shells from shelf/beach environments indicate that ages in this range may be valid (Wehmiller & York, 2001; Wehmiller *et al.*, 1995). Because all of the dated shells of Schroeder *et al.* (1995) were transported to some extent, a precise sea level curve cannot be constructed.

Summary: Atlantic and Gulf Coasts

Early Pleistocene to Last Interglacial Period

Age assignments for Atlantic coast marine units rely upon either biostratigraphic, radiometric, or suitably calibrated racemization modeling. All of these methods have significant uncertainties, particularly for the earlier Pleistocene, and racemization models are quite subjective, especially when used to correlate over broad latitude ranges. Nevertheless, for the Atlantic Coastal Plain, early to middle Pleistocene depositional records appear to be preserved throughout the region. Notable examples include the older parts of the Albemarle Embayment section in North Carolina and the Waccamaw formation in northeastern South Carolina. Late-middle Pleistocene, or pre-last interglacial units, include paleochannel fill in the Chesapeake Bay region, and U-series and racemization-dated units from the Albemarle and Cape Fear Arch regions. Determining the origin(s) and age(s) of the Suffolk Scarp would provide major insights into the history of sea level change on the Atlantic Coastal Plain. The Beaumont Clay on the Texas coast also appears to represent a complex of multiple units representing several transgressive-regressive cycles during the middle- to late Pleistocene.

Last Interglacial and Last Glacial Periods

Atlantic coast sites with U-series ages representing early MIS 5 (~125,000 yr) are rare north of the Florida Keys; where found, these units are at roughly the same elevation (all within ca. 6–8 m of present sea level) as those dated at 80,000 yr. These observations have been the focus of debate about the Atlantic Coastal Plain geochronology for nearly three decades. Nevertheless, a high sea level at ~80,000 yr is consistent with similar data from Bermuda (Muhs *et al.*, 2002b) and some data from the U.S. Pacific Coast (Fig. 13). Emergent shoreline features are found throughout the Gulf and Atlantic coastal plains, and in the absence of numerical geochronologic data, it is common to assign "last interglacial" ages to features found at elevations up to approximately $+8$ m.

The current elevations of emergent Atlantic coast paleoshorelines or shallow marine units U-series dated to late MIS 5 (80,000 yr) are higher than expected from marine isotope curves or dated tropical coral terrace records (Lambeck & Chappell, 2001; Shackleton, 1987). Records on the New Jersey, Maryland, and Texas shelves (Sheridan *et al.*, 2000; Rodriguez *et al.*, 2000; Toscano & York, 1992) also indicate sea levels during either late MIS 5 or MIS 3 that were higher than predicted from marine isotope curves. These observations suggest that major sections of the Atlantic and Gulf coasts, although considered as passive margin regions, may have been influenced by a combination of postglacial isostatic adjustment (through multiple glacial cycles), hydroisostasy, and shelf sediment loading.

Records of Holocene Sea-Level Rise, Atlantic and Gulf Coasts

Preserved records on continental shelves provide a general insight into the potential complexity of emergent records on any low gradient coastal plain. The youngest portions of these sequences represent the processes that occurred during last glacial low-stands and the rise in sea level since the last glacial maximum. In many cases, these are depositional records, preserved in areas of relatively large sediment supply. Examples include, but are not limited to, the inner shelves of New Jersey (Sheridan *et al.*, 2000), Maryland (Toscano, 1992; Toscano & York, 1992), North Carolina (Boss *et al.*, 2002; Riggs *et al.*, 1995), large estuaries (Colman *et al.*, 1992) and thick sections on the Texas shelf (Anderson *et al.*, 1996; Rodriguez *et al.*, 2000). Where sediment supply is more limited (such as the shelves of South Carolina and Georgia), extensive recycling of shelf sediments during multiple Quaternary transgressions is inferred from local stratigraphic sections where 1–2 m of preserved section may represent the entire Quaternary (Gayes *et al.*, 1992; Harris *et al.*, 2000; Heron *et al.*, 1984; Hine & Snyder, 1985; Macintyre *et al.*, 1978; Pilkey *et al.*, 1981).

Records of Holocene sea-level rise on the Atlantic and Gulf coasts are numerous, having been obtained in order to understand local or regional histories of coastal evolution or geophysical models of postglacial isostatic adjustment (e.g. Peltier, 1999). Major processes affecting these records include isostatic rebound in glaciated areas (Kelley *et al.*, 1992), steady subsidence associated with forebulge collapse in regions proximal to glacier advance (Kraft *et al.*, 1987), and, occasionally, rises of sea level to near present level during the middle Holocene in regions farther from glacial influence (Blum & Carter, 2000). Holocene depositional records that provide good models for Pleistocene sequences in a particular region are especially important. Good examples include channel erosion and filling in Chesapeake Bay (Colman *et al.*, 1990, 1992) and in Albemarle Sound (Riggs *et al.*, 1992, 1995), and Holocene coastal environments of the Florida peninsula (Davis *et al.*, 1992). In each of these regions, good analogies between preserved Pleistocene and Holocene records can be recognized.

Unresolved Issues on the Atlantic Coastal Plain

The following are three major unresolved geochronological issues related to the Pleistocene of the Atlantic Coastal Plain. Quite likely similar issues will arise for Gulf Coast sites as additional chronologic data (or samples for dating) become available for this region, but limited results prevent more specific conclusions.

(1) Identification of 125,000 yr erosional and depositional records using geomorphic and geochronologic analysis. Detailed mapping and dating in selected regions (such as the Cape Fear Arch and Albemarle Embayment) where both surface landforms and subsurface units can be clearly mapped is a high priority for establishing confidence in any of the Atlantic or Gulf coast chronologic records.

(2) Reconciling the elevation of the 80,000-yr-old units with the sea-level record from marine isotope and tropical coral reef records. The similar elevation of these units, and of possible 125,000-yr units (all within 6 m) from Virginia to Georgia (and also Bermuda), must be interpreted in the context of glacial-margin tectonics and hydroisostatic effects.

(3) Understanding the age-resolution capabilities of amino acid racemization methods, including assessing the reality and significance of the "offshore aminozone" found between North Carolina and Georgia. Inconsistencies in correlation of racemization data from region to region appear to contradict the stratigraphic coherence of racemizatoin data in local regions with common temperature and depositional histories. Continued refinement of racemization kinetic modeling by comparison with rigorous independent calibration is needed.

Summary

Much progress has been made in the past three decades in our understanding of the Quaternary sea-level history of U.S. coastlines. Two complimentary dating methods, U-series analyses of fossil corals and amino acid racemization of fossil mollusks, have generated hundreds of numerical and correlated ages. These data have allowed lateral correlation of marine deposits on the Pacific and Atlantic coasts and the Hawaiian Islands, as well as on the Bering Sea and Arctic Ocean coasts of Alaska.

Many of the sea-level high stands of the Quaternary are recorded in the reef record of the tectonically stable Florida Keys. Stratigraphic studies show that deposits of pre-last-interglacial high stands are present, although dating has yet to establish the precise timing of these deposits. Nevertheless, preliminary ages suggest that two high sea stands of the mid-Pleistocene are recorded on the Florida Keys, perhaps on the order of ~300,000–340,000 and 220,000–230,000 yr. Corals in reefs of two high sea stands of the last interglacial complex, the ~80,000 and ~120,000 yr stands, are present in this island chain. Offshore, Holocene reefs of the Florida Keys and mainland extend the Barbados record of sea-level rise since the last glacial period up to present sea level.

Reefs and coral-bearing marine deposits, both emergent and submergent, have been identified, mapped and dated in the Hawaiian Islands. On the island of Oahu, the Waimanalo Limestone was deposited during the peak of the last interglacial complex. Because Oahu is uplifting only very slowly, the range of U-series ages for this deposit is a good estimate of the duration of the peak of the last interglacial period. Results of recent high-precision dating indicate that the peak of the last interglacial period could have lasted at least 15,000 yr and possibly longer. Waimanalo Limestone deposits contain a significant number of extralimital southern mollusks, indicating warmer-than-present waters during the last interglacial period. The "big island" of Hawaii, unlike Oahu, is undergoing long-term Quaternary subsidence, due to volcanic loading. As a result, submerged reefs off Hawaii mark deglacial periods and yield a record of such events for more than 400,000 yr. The island of Lanai has long been the object of study because of high-elevation marine deposits. A novel hypothesis that these deposits were formed by a "giant wave" generated by a submarine landslide was proposed as a challenge to a decades-old concept that sea level was much higher than present. This hypothesis has in turn been challenged by new mapping and dating that suggest that the deposits were elevated due to lithospheric flexure, in response to the subsidence of the island of Hawaii, in turn induced by volcanic loading.

Multiple marine terraces on the Pacific Coast record a long history of sea-level high stands superimposed on long-term crustal uplift. Pre-last-interglacial high sea stands are recorded on the coast of California and several fossil-bearing localities in southern California hold promise for unraveling middle and early Pleistocene sea-level history. A high terrace on San Nicolas Island dates to ~500,000–600,000 yr ago and suggests that much of the sea-level history of the middle

Pleistocene may be recorded in a series of lower terraces. U-series analyses show that marine terraces from both the ~80,000- and ~120,000-yr-old sea-level high stands of the last interglacial complex, long recognized on tropical coasts, are also present on the Pacific Coast. The ~120,000 yr high sea stand lasted at least 8000–10,000 yr on the Pacific Coast, based on the range of individual coral ages. It was a major interglacial period that is distinguished by the presence of extralimital southern mollusks, indicating eastern Pacific Ocean water temperatures warmer than present. The ~80,000 yr high sea stand could also have been on the order of 8000–10,000 yr long, but it is distinguished by the presence of extralimital northern mollusks, indicating water temperatures cooler than present. Reworking of ~120,000 yr fossils during the ~100,000 yr high stand occurred on the Pacific Coast, based on U-series dating of individual corals. This reworking suggests that sea level at ~100,000 yr ago may have been close to that of today. "Capture" of the ~120,000 yr high sea stand during the ~100,000 yr high stand may also explain the mixture of extralimital southern and extralimital northern fossils in some deposits.

As with the coasts of California, Hawaii, and Florida, the peak of the last interglacial period is recorded on the Bering Sea and Arctic Ocean coasts of Alaska. Although corals have not been found in these deposits, amino acid ratios, non-finite radiocarbon ages, and terrace elevations all indicate that the "Pelukian" high sea stand corresponds to the peak of the last interglacial period. Similar to low-latitude and mid-latitude marine deposits of this high sea stand, Pelukian deposits of Alaska contain a number of extralimital southern species of mollusks, indicating warmer-than-modern water temperatures. In addition, an older high sea stand, that is probably on the order of ~400,000 yr old, may record another major interglacial period, also with warmer-than-modern water temperatures.

The Atlantic Coastal Plain of the eastern United States has been one of the most challenging regions for studies of past sea levels. Because this coast is on a passive continental margin, sea-level records are complex: long-term uplift does not "isolate" individual high stands of sea as distinct terraces in the manner that is typical for the Pacific Coast. Many transgressive-regressive cycles are preserved at roughly the same elevations, resulting in a highly complex stratigraphic record. Nevertheless, amino acid racemization studies have shown that there is a rich record of Quaternary sea-level fluctuations, at least as far back as the middle Pleistocene and more likely the early Pleistocene.

An unexpected result of recent studies on the Atlantic Coastal Plain is that, unlike the other coasts of the U.S., records of the peak last-interglacial high sea stand of ~120,000 yr ago have been remarkably difficult to find. The few U-series ages that have been reported for this high sea stand are equivocal. In contrast, the ~80,000 yr high sea stand is recorded as low-elevation (but emergent) marine deposits in Virginia, North Carolina, South Carolina and Georgia. Individual coral ages of this high sea stand are very similar to those on the Pacific Coast and indicate that the ~80,000 yr high stand could have lasted as long as ~10,000 yr. The presence of ~80,000 yr deposits at or even above sea level on a tectonically stable coast is a major challenge to widely held views that this sea stand was significantly lower than present sea level.

The longer-term Quaternary sea-level record of the Atlantic Coastal Plain is apparent in thick stratigraphic sequences, multiple aminozones ranging back to the early Pleistocene, and some independent chronologic control for units that date to >750,000 yr. Shore-parallel correlation of these deposits is also possible, as has been done on the Pacific Coast. The actual length of the Quaternary sea-level record in this region, as shown by amino acid racemization data, depends on the choice of calibration samples and on either a "short" or "long" time constant for racemization. Consequently, age estimates based on racemization results might vary by 100% or more. The "short" chronology suggests that the Atlantic Coastal Plain record spans only ~500,000 yr, whereas the "long" chronology would suggest a considerably greater part of the Quaternary is represented.

Certain results of sea-level studies on the coasts of the United States challenge the Milankovitch (1941) theory of climate change, as interpreted from the deep-sea oxygen isotope record. The duration of the peak of the last interglacial period (MIS 5e), when sea level was as high or higher than present, is interpreted from the oxygen isotope record to be on the order of only a few thousand years (Imbrie *et al.*, 1984; Martinson *et al.*, 1987). In contrast, the coral record from Oahu, Hawaii indicates that sea level as high or higher than present could have lasted 15,000–20,000 yr. The U-series ages from both Oahu and California also suggest that sea level was relatively high about 115,000 yr ago, a time that the oxygen isotope record would suggest that sea level was relatively low. U-series ages of ~80,000 yr for corals from the tectonically stable Atlantic Coastal Plain and the slowly uplifting Pacific Coast suggest that sea level at that time was near present, whereas the oxygen isotope record suggests that sea level was then well below present. The reasons for the discrepencies between the coastal sea-level record and the deep-sea oxygen isotope record are not understood, but provide an important challenge to future investigations on the coasts of the United States.

Acknowledgments

Work on sea-level history by Muhs and Simmons was supported by the Earth Surface Dynamics Program of the U.S. Geological Survey and is a contribution to the LITE (Last Interglacial: Timing and Environment) Project. Studies by Wehmiller and York were supported by the National Science Foundation. We thank Robert S. Thompson, Richard Z. Poore, and two anonymous referees for helpful reviews of an earlier version of this paper.

References

Addicott, W.O. (1966). Late Pleistocene marine paleoecology and zoogeography in central California. *U.S. Geological Survey Professional Paper* 523-C, pp. C1–C21.

Ager, T.A. & Brubaker, L. (1985). Quaternary palynology and vegetational history of Alaska. *In*: Bryant, V.M., Jr. & Holloway, R.G. (Eds), *Pollen records of Late-Quaternary North American sediments*. Dallas, Texas, American Association of Stratigraphic Palynologists Foundation, pp. 353–383.

Aharon, P. & Chappell, J. (1986). Oxygen isotopes, sea level changes and the temperature history of a coral reef environment in New Guinea over the last 10^5 years. *Palaeogeography, Palaeoclimatology, Palaeoecology*, **56**, 337–379.

Alexander, C.S. (1953). The marine and stream terraces of the Capitola-Watsonville area. *University of California Publications in Geography*, **10**, 1–44.

Anderson, J.B., Abdulah, K., Sarzalejo, S., Siringan, F. & Thomas, M.A. (1996). Late Quaternary sedimentation and high-resolution sequence stratigraphy of the east Texas shelf. *In*: De Batist, M. & Jacobs, P. (Eds), *Geology of siliciclastic shelf seas*. Geological Society Special Publication, 117, pp. 95–124.

Anderson, R.S. & Menking, K.M. (1994). The Quaternary marine terraces of Santa Cruz, California: Evidence for coseismic uplift on two faults. *Geological Society of America Bulletin*, **106**, 649–664.

Bard, E., Hamelin, B., Fairbanks, R.G. & Zindler, A. (1990). Calibration of the ^{14}C timescale over the past 30,000 years using mass spectrometric U-Th ages from Barbados corals. *Nature*, **345**, 405–410.

Belknap, D.F. (1979). Application of amino acid geochronology to stratigraphy of late Cenozoic marine units of the Atlantic coastal plain [Ph.D. thesis]. Newark, University of Delaware, 348 pp.

Bender, M.L., Fairbanks, R.G., Taylor, F.W., Matthews, R.K., Goddard, J.G. & Broecker, W.S. (1979). Uranium-series dating of the Pleistocene reef tracts of Barbados, West Indies. *Geological Society of America Bulletin*, **90**(Pt. I), 577–594.

Berger, A. & Loutre, M.F. (1991). Insolation values for the climate of the last 10 million years. *Quaternary Science Reviews*, **10**, 297–317.

Bernard, H.A. & LeBlanc, R.J. (1965). Résumé of the Quaternary geology of the northwestern Gulf of Mexico province. *In*: Wright, H.E., Jr. & Frey, D.G. (Eds), *The Quaternary of the United States*. Princeton, Princeton University Press, pp. 137–185.

Bloom, A.L. (1983a). Sea level and coastal morphology of the United States through the late Wisconsin glacial maximum. *In*: Porter, S.C. (Ed.), *Late Quaternary Environments of the United States*, Vol. 1, *The late Pleistocene*. University of Minnesota Press, Minneapolis, pp. 215–229.

Bloom, A.L. (1983b). Sea level and coastal changes. *In*: Wright, H.E., Jr. (Ed.), *Late Quaternary Environments of the United States*, Vol. 2, *The Holocene*. University of Minnesota Press, Minneapolis, pp. 42–51.

Bloom, A.L., Broecker, W.S., Chappell, J.M.A., Matthews, R.K. & Mesolella, K.J. (1974). Quaternary sea level fluctuations on a tectonic coast: New ^{230}Th/^{234}U dates from the Huon Peninsula, New Guinea. *Quaternary Research*, **4**, 185–205.

Blum, M.D. & Carter, A.E. (2000). Middle Holocene evolution of the central Texas coast. *Gulf Coast Association of Geological Societies Transactions*, **L**, 331–341.

Blum, M.D. & Price, D.M. (1998). Quaternary alluvial plain construction in response to interacting glacio-eustatic and climatic controls, Texas Gulf Coastal Plain. *In*: Shanley, K. & McCabe, P. (Eds), *Relative Roles of Eustasy, Climate, and Tectonism in Continental Rocks*: SEPM (Society for Sedimentary Geology) Special Publication, 59, pp. 31–48.

Boss, S.K., Hoffman, C.W. & Cooper, B. (2002). Influence of fluvial processes on the Quaternary geology framework of the continental shelf, North Carolina, USA. *Marine Geology*, **183**, 45–65.

Bradley, W.C. & Griggs, G.B. (1976). Form, genesis, and deformation of central California wave-cut platforms. *Geological Society of America Bulletin*, **87**, 433–449.

Brill, A.L. (1996). The Suffolk Scarp, a Pleistocene barrier island in Beaufort and Pamlico Counties, NC [M.S. thesis], Duke University, 154 pp.

Brigham-Grette, J. & Hopkins, D.M. (1995). Emergent marine record and paleoclimate of the last interglaciation along the northwest Alaskan coast. *Quaternary Research*, **43**, 159–173.

Broecker, W.S. & Thurber, D.L. (1965). Uranium-series dating of corals and oolites from Bahaman and Florida Key limestones. *Science*, **149**, 58–60.

Broecker, W.S., Thurber, D.L., Goddard, J., Ku, T.-L., Matthews, R.K. & Mesolella, K.J. (1968). Milankovitch hypothesis supported by precise dating of coral reefs and deep-sea sediments. *Science*, **159**, 297–300.

Bryan, W.B. & Stephens, R.S. (1993). Coastal bench formation at Hanauma Bay, Oahu. *Geological Society of America Bulletin*, **105**, 377–386.

Buddemeier, R.W. & Smith, S.V. (1988). Coral reef growth in an era of rapidly rising sea-level. *Coral Reefs*, **7**, 51–56.

Calhoun, R.S. & Fletcher, C.H., III (1996). Late Holocene coastal plain stratigraphy and sea-level history at Hanalei, Kauai, Hawaiian Islands. *Quaternary Research*, **45**, 47–58.

Chappell, J. (1974a). Geology of coral terraces, Huon Peninsula, New Guinea: A study of Quaternary tectonic movements and sea level changes. *Geological Society of America Bulletin*, **85**, 553–570.

Chappell, J. (1974b). Late Quaternary glacio- and hydro-isostasy on a layered Earth. *Quaternary Research*, **4**, 429–440.

Chappell, J. & Shackleton, N.J. (1986). Oxygen isotopes and sea level. *Nature*, **324**, 137–140.

Chen, J.H., Curran, H.A., White, B. & Wasserburg, G.J. (1991). Precise chronology of the last interglacial period: ^{234}U-^{230}Th data from fossil coral reefs in the Bahamas. *Geological Society of America Bulletin*, **103**, 82–97.

Chen, Z.-Q., Hobbs, C.H., III, Wehmiller, J.F. & Kimball, S.M. (1995). Late Quaternary paleochannel systems on the continental shelf, south of the Chesapeake Bay entrance. *Journal of Coastal Research*, **11**, 605–614.

Cheng, H., Edwards, R.L., Hoff, J., Gallup, C.D., Richards, D.A. & Asmerom, Y. (2000). The half-lives of uranium-234 and thorium-230. *Chemical Geology*, **169**, 17–33.

Clark, J.A., Farrell, W.E. & Peltier, W.R. (1978). Global changes in postglacial sea level: A numerical calculation. *Quaternary Research*, **9**, 265–287.

Clark, P.U. & Mix, A.C. (2002). Ice sheets and sea level of the Last Glacial Maximum. *Quaternary Science Reviews*, **21**, 1–7.

Colman, S.M. & Mixon, R.B. (1988). The record of major Quaternary sea-level changes in a large coastal plain estuary, Chesapeake Bay, eastern United States. *Palaeogeography, Palaeoclimatology, Palaeoecology*, **68**, 99–116.

Colman, S.M., Halka, J.P., Hobbs, C.H., III, Mixon, R.B. & Foster, D.S. (1990). Ancient channels of the Susquehanna River beneath Chesapeake Bay and the Delmarva Peninsula. *Geological Society of America Bulletin*, **102**, 1268–1279.

Colman, S.M., Halka, J.P. & Hobbs, C.H., III (1992). Patterns and rates of sediment accumulation in the Chesapeake Bay during the Holocene rise in sea level. *In*: Fletcher, C.H., III & Wehmiller, J.F. (Eds), *Quaternary Coasts of the United States: Marine and Lacustrine Systems*. SEPM (Society for Sedimentary Geology) Special Publication, 48, pp. 101–111.

Colquhoun, D.J., Johnson, G.H., Peebles, P.C., Huddleston, P.F. & Scott, T. (1991). Quaternary geology of the Atlantic coastal plain. *In*: Morrison, R.B. (Ed.), *Quaternary non-glacial geology: Conterminus U.S.* Boulder, Colorado, Geological Society of America, Geology of North America, K-2, pp. 629–650.

Coniglio, M. & Harrison, R.S. (1983). Facies and diagenesis of late Pleistocene carbonates from Big Pine Key, Florida. *Bulletin of Canadian Petroleum Geology*, **31**, 135–147.

Corrado, J.C., Weems, R.E., Hare, P.E. & Bambach, R.K. (1986). Capabilities and limitations of applied aminostratigraphy, as illustrated by analyses of *Mulinia lateralis* from the late Cenozoic marine beds near Charleston, South Carolina. *South Carolina Geology*, **30**, 19–46.

Cronin, T.M., Szabo, B.J., Ager, T.A., Hazel, J.E. & Owens, J.P. (1981). Quaternary climates and sea levels of the U.S. Atlantic Coastal Plain. *Science*, **211**, 233–240.

Curray, J.R. (1965). Late Quaternary history, continental shelves of the United States. *In*: Wright, H.E., Jr. & Frey, D.G. (Eds), *The Quaternary of the United States*. Princeton, Princeton University Press, pp. 723–735.

Davis, R.A., Jr., Hine, A.C. & Shinn, E.A. (1992). Holocene coastal development on the Florida peninsula. *In*: Fletcher, C.H., III & Wehmiller, J.F. (Eds), *Quaternary Coasts of the United States: Marine and Lacustrine Systems*. SEPM (Society for Sedimentary Geology) Special Publication, 48, pp. 193–212.

Dixon, E.J. (2001). Human colonization of the Americas: Timing, technology and process. *Quaternary Science Reviews*, **20**, 277–299.

Dockal, J.A. (1995). Documentation and evaluation of radiocarbon dates from the Cape Fear Coquina (Late Pleistocene) of Snows Cut, New Hanover County, North Carolina. *Southeastern Geology*, **35**, 169–186.

Dodge, R.E., Fairbanks, R.G., Benninger, L.K. & Maurrasse, F. (1983). Pleistocene sea levels from raised coral reefs of Haiti. *Science*, **219**, 1423–1425.

Donoghue, J.F., Stapor, F.W. & Tanner, W.F. (1998). Discussion of: Otvos, E.G. (1995). Multiple Pliocene-Quaternary marine highstands, northeast Gulf Coastal Plain-fallacies and facts. *Journal of Coastal Research*, **14**, 669–674.

Donoghue, J.F. & Tanner, W.F. (1992). Quaternary terraces and shorelines of the panhandle Florida region. *In*: Fletcher, C.H., III & Wehmiller, J.F. (Eds), *Quaternary Coasts of the United States: Marine and Lacustrine Systems*. SEPM (Society for Sedimentary Geology) Special Publication, 48, pp. 233–242.

DuBar, J.R., DuBar, S.S., Ward, L.W., Blackwelder, B.W., Abbot, W.H. & Huddleston, P.F. (1980). Cenozoic biostratigraphy of the Carolina outer coastal plain. *In*: Frey, R.W. (Ed.), *Excursions in Southeastern Geology*, 1, Field Trip 9, Geological Society of America Annual Meeting, Atlanta, Georgia (and American Geological Institute, Falls Church, Virginia), pp. 179–236.

DuBar, J.R., Ewing, T.E., Lundelius, E.L. Jr., Otvos, E.G. & Winker, C.D. (1991). Quaternary geology of the Gulf of Mexico coastal plain. *In*: Morrison, R.B. (Ed.), *Quaternary Non-Glacial Geology: Conterminus U.S.* Boulder, Colorado, Geological Society of America, Geology of North America, K-2, pp. 583–610.

Easton, W.H. & Olson, E.A. (1976). Radiocarbon profile of Hanauma Reef, Oahu, Hawaii. *Geological Society of America Bulletin*, **87**, 711–719.

Edwards, R.L., Cheng, H., Murrell, M.T. & Goldstein, S.J. (1997). Protactinium-231 dating of carbonates by thermal ionization mass spectrometry: Implications for Quaternary climate change. *Science*, **276**, 782–786.

Emiliani, C. (1955). Pleistocene temperatures. *Journal of Geology*, **63**, 538–578.

Enos, P. & Perkins, R.D. (1977). Quaternary sedimentation in south Florida. *Geological Society of America Memoir*, **147**, 198.

Fairbanks, R.G. (1989). A 17,000-year glacio-eustatic sea level record: influence of glacial melting rates on the Younger Dryas event and deep-ocean circulation. *Nature*, **342**, 637–642.

Fleming, K., Johnston, P., Zwartz, D., Yokoyama, Y., Lambeck, K. & Chappell, J. (1998). Refining the eustatic sea-level curve since the Last Glacial Maximum using far- and intermediate-field sites. *Earth and Planetary Science Letters*, **163**, 327–342.

Fletcher, C.H., III & Jones, A.T. (1996). Sea-level highstand recorded in Holocene shoreline deposits on Oahu, Hawaii. *Journal of Sedimentary Research*, **66**, 632–641.

Fletcher, C.H., III & Wehmiller, J.F. (Eds) (1992). *Quaternary coasts of the United States: Marine and lacustrine systems*. SEPM (Society for Sedimentary Geology) Special Publication, 48, pp. 1–450.

Fruijtier, C., Elliot, T. & Schlager, W. (2000). Mass-spectrometric ^{234}U-^{230}Th ages from the Key Largo Formation, Florida Keys, United States: Constraints on diagenetic age disturbance. *Geological Society of America Bulletin*, **112**, 267–277.

Gallup, C.D., Cheng, H., Taylor, F.W. & Edwards, R.L. (2002). Direct determination of the timing of sea level change during Termination II. *Science*, **295**, 310–313.

Gallup, C.D., Edwards, R.L. & Johnson, R.G. (1994). The timing of high sea levels over the past 200,000 years. *Science*, **263**, 796–800.

Gayes, P.T., Scott, D.B., Collins, E.S. & Nelson, D.D. (1992). A late Holocene sea-level fluctuation in South Carolina. *In*: Fletcher, C.H., III & Wehmiller, J.F. (Eds), *Quaternary Coasts of the United States: Marine and Lacustrine Systems*. SEPM (Society for Sedimentary Geology) Special Publication, 48, pp. 155–160.

Goodfriend, G.A., Brigham-Grette, J. & Miller, G.H. (1996). Enhanced age resolution of the marine Quaternary record in the Arctic using aspartic acid racemization dating of bivalve shells. *Quaternary Research*, **45**, 176–187.

Grant, L.B., Mueller, K.J., Gath, E.M., Cheng, H., Edwards, R.L., Munro, R. & Kennedy, G.L. (1999). Late Quaternary uplift and earthquake potential of the San Joaquin Hills, southern Los Angeles Basin, California. *Geology*, **27**, 1031–1034.

Groot, J.J., Ramsey, K.W. & Wehmiller, J.F. (1990). Ages of the Bethany, Beverdam & Omar Formations of southern Delaware. *Delaware Geological Survey Report of Investigations*, No. 47, 1–19.

Grossman, E.E. & Fletcher, C.H., III (1998). Sea level higher than present 3500 years ago on the northern main Hawaiian Islands. *Geology*, **26**, 363–366.

Grossman, E.E., Fletcher, C.H., III & Richmond, B.M. (1998). The Holocene sea-level highstand in the equatorial Pacific: Analysis of the insular paleosea-level database. *Coral Reefs*, **17**, 309–327.

Guthrie, R.D. & Matthews, J.V., Jr. (1971). The Cape Deceit fauna – early Pleistocene mammalian assemblage from the Alaskan Arctic. *Quaternary Research*, **1**, 474–510.

Halley, R.B., Vacher, H.L. & Shinn, E.A. (1997). Geology and hydrogeology of the Florida Keys. *In*: Vacher, H.L. & Quinn, T. (Eds), Geology and hydrogeology of carbonate islands. *Developments in Sedimentology*, **54**, Amsterdam, Elsevier, pp. 217–248.

Hanson, K.L., Lettis, W.R., Wesling, J.R., Kelson, K.I. & Mezger, L. (1992). Quaternary marine terraces, south-central coastal California: Implications for crustal deformation and coastal evolution. *In*: Fletcher, C.H., III & Wehmiller, J.F. (Eds), *Quaternary coasts of the United States: Marine and lacustrine systems*. SEPM (Society for Sedimentary Geology) Special Publication, 48, pp. 323–332.

Harris, M.S. (2000). Influence of a complex geologic framework on Quaternary coastal evolution: An example from Charleston, South Carolina [Ph.D. thesis]. Newark, University of Delaware, 330 pp.

Harris, M.S., Wehmiller, J.F., York, L.L. & Gayes, P.T. (2000). Quaternary evolution of the lower coastal plain and continental shelf near Charleston, South Carolina: stratigraphic construction, geomorphic expression & geochronology. *Geological Society of America Abstracts with Programs*, **32**, 2, 24.

Harrison, R.S. & Coniglio, M. (1985). Origin of the Pleistocene Key Largo Limestone, Florida Keys. *Bulletin of Canadian Petroleum Geology*, **33**, 350–358.

Hayes, M.O. (1994). The Georgia Bight barrier system. *In*: Davis, R.A., Jr. (Ed.), *Geology of Holocene Barrier Island Systems*. Berlin, Springer-Verlag, pp. 233–304.

Hearty, P.J. (2002). The Ka'ena highstand of O'ahu, Hawai'i: Further evidence of Antarctic ice collapse during the middle Pleistocene. *Pacific Science*, **56**, 65–81.

Hearty, P.J., Kindler, P., Cheng, H. & Edwards, R.L. (1999). A +20 m middle Pleistocene sea-level highstand (Bermuda and the Bahamas) due to partial collapse of Antarctic ice. *Geology*, **27**, 375–378.

Henderson, G.M. & Slowey, N.C. (2000). Evidence from U-Th dating against Northern Hemisphere forcing of the penultimate deglaciation. *Nature*, **404**, 61–66.

Heron, S.D., Moslow, T.F., Berelson, W.M., Herbert, J.R., Steele, G.A. & Susman, K.R. (1984). Holocene sedimentation of a wave-dominated barrier island shoreline: Cape Lookout, North Carolina. *Marine Geology*, **60**, 413–434.

Hine, A.C. & Snyder, S.W. (1985). Coastal lithosome preservation: Evidence from the shoreface and inner continental shelf off Bogue Banks, N.C. *Marine Geology*, **63**, 307–330.

Hoffmeister, J.E. & Multer, H.G. (1968). Geology and origin of the Florida Keys. *Geological Society of America Bulletin*, **79**, 1487–1502.

Hoffmeister, J.E., Stockman, K.W. & Multer, H.G. (1967). Miami Limestone of Florida and its Recent Bahamian counterpart. *Geological Society of America Bulletin*, **78**, 175–190.

Hollin, J.T. & Hearty, P.J. (1990). South Carolina interglacial sites and stage 5 sea levels. *Quaternary Research*, **33**, 1–17.

Hopkins, D.M. (1967). Quaternary marine transgressions in Alaska. *In*: Hopkins, D.M. (Ed.), *The Bering Land Bridge*. Stanford, California, Stanford University Press, pp. 451–484.

Hopkins, D.M. (1973). Sea level history in Beringia during the past 250,000 years. *Quaternary Research*, **3**, 520–540.

Hopkins, D.M., MacNeil, F.S. & Leopold, E.B. (1960). The coastal plain at Nome, Alaska: A late Cenozoic type section for the Bering Strait region. *In*: International Geological Congress, Report of the Twenty-First Session Norden, Part IV, *Chronology and Climatology of the Quaternary*, Copenhagen, Denmark, pp. 46–57.

Hopkins, D.M., Rowland, R.W., Echols, R.E. & Valentine, P.C. (1974). An Anvilian (early Pleistocene) marine fauna from western Seward Peninsula, Alaska. *Quaternary Research*, **4**, 441–470.

Huddleston, P.F. (1988). A revision of the lithostratigraphic units of the coastal plain of Georgia: The Miocene through Holocene. *Georgia Geological Survey Bulletin*, **104**.

Hulbert, R.C., III & Pratt, A.E. (1998). New Pleistocene (Rancholabrean) vertebrate faunas from coastal Georgia. *Journal of Vertebrate Paleontology*, **18**, 412–429.

Imbrie, J., Hays, J.D., Martinson, D.G., McIntyre, A., Mix, A.C., Morley, J.J., Pisias, N.G., Prell, W.L. & Shackleton, N.J. (1984). The orbital theory of Pleistocene climate: Support from a revised chronology of the marine δ^{18}O record.

In: Berger, A., Imbrie, J., Hays, J., Kukla, G. & Saltzman, B. (Eds), *Milankovitch and Climate: Understanding the Response to Astronomical Forcing*. Dordrecht: D. Reidel Publishing Company, pp. 269–305.

Jones, D.S., Mueller, P.A., Acosta, T. & Shuster, R.D. (1995). Strontium isotope stratigraphy and age estimates for the Leisey shell pit faunas, Hillsborough County, Florida: *Bulletin of the Florida Museum of Natural History. Biological Sciences*, **37**(pt. 1), 93–105.

Kaufman, D. (1992). Aminostratigraphy of Pliocene-Pleistocene high-sea-level deposits, Nome coastal plain and adjacent nearshore area, Alaska. *Geological Society of America Bulletin*, **104**, 40–52.

Kaufman, D. & Brigham-Grette, J. (1993). Aminostratigraphic correlations and paleotemperature implications, Pliocene-Pleistocene high-sea-level deposits, northwestern Alaska. *Quaternary Science Reviews*, **12**, 21–33.

Kaufman, D.S., Walter, R.C., Brigham-Grette, J. & Hopkins, D.M. (1991). Middle Pleistocene age of the Nome River glaciation, northwestern Alaska. *Quaternary Research*, **36**, 277–293.

Karrow, P.F., Morgan, G.S., Portell, R.W., Simons, E. & Auffenberg, K. (1996). Middle Pleistocene (early Rancholabrean) vertebrates and associated marine and non-marine invertebrates from Oldsmar, Pinellas County, Florida. *In*: Stewart, K.M. & Seymour, K.L. (Eds), *Palaeoecology and Palaeoenvironments of Late Cenozoic Mammals: Tributes to the Career of C.S. (Rufus) Churcher*. Toronto, University of Toronto Press, pp. 97–113.

Kelley, J.T., Dickson, S.M., Belknap, D.F. & Stuckenrath, R., Jr. (1992). Sea-level change and late Quaternary sediment accumulation on the southern Maine inner continental shelf. *In*: Fletcher, C.H., III & Wehmiller, J.F. (Eds), *Quaternary Coasts of the United States: Marine and Lacustrine Systems*. SEPM (Society for Sedimentary Geology) Special Publication, 48, pp. 23–34.

Kelsey, H.M. (1990). Late Quaternary deformation of marine terraces on the Cascadia subduction zone near Cape Blanco, Oregon. *Tectonics*, **9**, 983–1014.

Kelsey, H.M. & Bockheim, J.G. (1994). Coastal landscape evolution as a function of eustasy and surface uplift rate, Cascadia margin, southern Oregon. *Geological Society of America Bulletin*, **106**, 840–854.

Kelsey, H.M., Ticknor, R.L., Bockheim, J.G. & Mitchell, C.E. (1996). Quaternary upper plate deformation in coastal Oregon. *Geological Society of America Bulletin*, **108**, 843–860.

Kennedy, G.L., Lajoie, K.R. & Wehmiller, J.F. (1982). Aminostratigraphy and faunal correlations of late Quaternary marine terraces, Pacific Coast, USA. *Nature*, **299**, 545–547.

Kern, J.P. & Rockwell, T.K. (1992). Chronology and deformation of marine shorelines, San Diego County, California. *In*: Fletcher, C.H., III & Wehmiller, J.F. (Eds), *Quaternary Coasts of the United States: Marine and Lacustrine Systems*. SEPM (Society for Sedimentary Geology) Special Publication, 48, pp. 377–382.

Kidson, C. (1982). Sea level changes in the Holocene. *Quaternary Science Reviews*, **1**, 121–151.

Knebel, H.J. & Circe, R.C. (1988). Late Pleistocene drainage systems beneath Delaware Bay. *Marine Geology*, **78**, 285–302.

Kosuge, S. (1969). Fossil mollusks of Oahu, Hawaii Islands. *Bulletin of the National Science Museum [Tokyo, Japan]*, **12**, 783–794.

Kraft, J.C., Chrzastowski, M.J., Belknap, D.F., Toscano, M.A. & Fletcher, C.H., III (1987). The transgressive barrier-lagoon coast of Delaware: Morphostratigraphy, sedimentary sequences and responses to relative rise in sea level. *In*: Nummendal, D., Pilkey, O.H. & Howard, J.D (Eds), *Sea-Level Fluctuation and Coastal Evolution*. SEPM (Society for Sedimentary Geology) Special Publication, 41, pp. 129–143.

Ku, T.-L. (1968). Protactinium 231 method of dating coral from Barbados island. *Journal of Geophysical Research*, **73**, 2271–2276.

Ku, T.-L., Ivanovich, M. & Luo, S. (1990). U-series dating of last interglacial high sea stands: Barbados revisited. *Quaternary Research*, **33**, 129–147.

Ku, T.-L. & Kern, J.P. (1974). Uranium-series age of the upper Pleistocene Nestor terrace, San Diego, California. *Geological Society of America Bulletin*, **85**, 1713–1716.

Ku, T.-L., Kimmel, M.A., Easton, W.H. & O'Neil, T.J. (1974). Eustatic sea level 120,000 years ago on Oahu, Hawaii. *Science*, **183**, 959–962.

Lajoie, K.R., Kern, J.P., Wehmiller, J.F., Kennedy, G.L., Mathieson, S.A., Sarna-Wojcicki, A.M., Yerkes, R.F. & McCrory, P.A. (1979). Quaternary marine shorelines and crustal deformation, San Diego to Santa Barbara, California. *In*: Abbott, P.L. (Ed.), *Geological Excursions in the Southern California Area*. San Diego, Dept. of Geological Sciences, San Diego State University, pp. 3–15.

Lajoie, K.R., Ponti, D.J., Powell, C.L., II, Mathieson, S.A. & Sarna-Wojcicki, A.M. (1991). Emergent marine strandlines and associated sediments, coastal California; A record of Quaternary sea-level fluctuations, vertical tectonic movements, climatic changes, and coastal processes. *In*: Morrison, R.B. (Ed.), *Quaternary Nonglacial Geology; Conterminous U.S.* Boulder, Colorado, Geological Society of America, Boulder, Colorado, The Geology of North America, K-2, pp. 190–203.

Lambeck, K. & Chappell, J. (2001). Sea level change through the last glacial cycle. *Science*, **292**, 679–686.

Lamothe, M. & Auclair, M. (1999). A solution to anomalous fading and age shortfalls in optical dating of feldspar minerals. *Earth and Planetary Science Letters*, **171**, 319–323.

Lidz, B.H., Hine, A.C., Shinn, E.A. & Kindinger, J.L. (1991). Multiple outer-reef tracts along the south Florida bank margin: Outlier reefs, a new windward-margin model. *Geology*, **19**, 115–118.

Lidz, B.H., Shinn, E.A., Hine, A.C. & Locker, S.D. (1997). Contrasts within an outlier-reef system: Evidence for differential Quaternary evolution, south Florida windward margin, U.S.A. *Journal of Coastal Research*, **13**, 711–731.

Lighty, R.G., Macintyre, I.G. & Stuckenrath, R. (1978). Submerged early Holocene barrier reef south-east Florida shelf. *Nature*, **276**, 59–60.

Lighty, R.G., Macintyre, I.G. & Stuckenrath, R. (1982). *Acropora palmata* reef framework: A reliable indicator of sea level in the western Atlantic for the past 10,000 years. *Coral Reefs*, **1**, 125–130.

Ludwig, K.R., Muhs, D.R., Simmons, K.R., Halley, R.B. & Shinn, E.A. (1996). Sea level records at ~80 ka from tectonically stable platforms: Florida and Bermuda. *Geology*, **24**, 211–214.

Ludwig, K.R., Muhs, D.R., Simmons, K.R. & Moore, J.G. (1992a). Sr-isotope record of Quaternary marine terraces on the California coast and off Hawaii. *Quaternary Research*, **37**, 267–280.

Ludwig, K.R., Simmons, K.R., Szabo, B.J., Winograd, I.J., Landwehr, J.M., Riggs, A.C. & Hoffman, R.J. (1992b). Mass-spectrometric ^{230}Th-^{234}U-^{238}U dating of the Devils Hole calcite vein. *Science*, **258**, 284–287.

Ludwig, K.R., Szabo, B.J., Moore, J.G. & Simmons, K.R. (1991). Crustal subsidence rate off Hawaii determined from ^{234}U/^{238}U ages of drowned coral reefs. *Geology*, **19**, 171–174.

Lundberg, J. & McFarlane, D. (2002). Isotope stage 11 sea level in the Netherlands Antilles. *Geological Society of America Abstracts with Programs*, **34**, 6, 31.

Macintyre, I.G. (1988). Modern coral reefs of western Atlantic: New geological perspective. *The American Association of Petroleum Geologists Bulletin*, **72**, 1360–1369.

Macintyre, I.G., Pilkey, O.H. & Stuckenrath, R. (1978). Relict oysters on the United States Atlantic continental shelf: a reconsideration of their usefulness in understanding late Quaternary sea-level history. *Geological Society of America Bulletin*, **89**, 277–282.

Macintyre, I.G., Raymond, B. & Stuckenrath, R. (1983). Recent history of a fringing reef, Bahia Salina del Sur, Vieques Island, Puerto Rico. *Atoll Research Bulletin*, **268**, 1–9.

Markewich, H.W., Hacke, C.M. & Huddleston, P.F. (1992). Emergent Pliocene and Pleistocene sediments of southeastern Georgia: an anomalous, fossil-poor, clastic section. *In*: Fletcher, C.H., III & Wehmiller, J.F. (Eds), *Quaternary Coasts of the United States: Marine and Lacustrine Systems*. SEPM (Society for Sedimentary Geology) Special Publication, 48, pp. 173–192.

Martinson, D.G., Pisias, N.G., Hays, J.D., Imbrie, J., Moore, T.C., Jr. & Shackleton, N.J. (1987). Age dating and the orbital theory of the ice ages: Development of a high-resolution 0 to 300,000-year chronostratigraphy. *Quaternary Research*, **27**, 1–29.

Matthews, R.K. (1973). Relative elevation of late Pleistocene high sea level stands: Barbados uplift rates and their implications. *Quaternary Research*, **3**, 147–153.

McCartan, L., Lemon, E.M., Jr. & Weems, R.E. (1984). Geologic map of the area between Charleston and Orangeburg, South Carolina. U.S. Geological Survey Miscellaneous Investigations Series, Map I-1472, scale 1:250,000.

McCartan, L., Owens, J.P., Blackwelder, B.W., Szabo, B.J., Belknap, D.F., Kriausakul, N., Mitterer, R.M. & Wehmiller, J.F. (1982). Comparison of amino acid racemization geochronometry with lithostratigraphy, biostratigraphy, uranium-series coral dating, and magnetostratigraphy in the Atlantic coastal plain of the southeastern United States. *Quaternary Research*, **18**, 337–359.

Merritts, D. & Bull, W.B. (1989). Interpreting Quaternary uplift rates at the Mendocino triple junction, northern California, from uplifted marine terraces. *Geology*, **17**, 1020–1024.

Mesolella, K.J., Matthews, R.K., Broecker, W.S. & Thurber, D.L. (1969). The astronomical theory of climatic change: Barbados data. *Journal of Geology*, **77**, 250–274.

Milankovitch, M.M. (1941). *Canon of Insolation and the Ice Age Problem*. Beograd, Koniglich Serbische Akademie (English translation by the Israel Program for Scientific Translations, Jerusalem, Israel (1969)).

Milne, G.A., Mitrovica, J.X. & Schrag, D.P. (2002). Estimating past continental ice volume from sea-level data. *Quaternary Science Reviews*, **21**, 361–376.

Mirecki, J.E., Wehmiller, J.F. & Skinner, A. (1995). Geochronology of Quaternary coastal units, southeastern Virginia. *Journal of Coastal Research*, **11**, 1135–1144.

Mitrovica, J.X. & Peltier, W.R. (1991). On post-glacial geoid subsidence over the equatorial oceans. *Journal of Geophysical Research*, **96**, 20,053–20,071.

Mitterer, R.M. (1975). Ages and diagenetic temperatures of Pleistocene deposits of Florida based upon isoleucine epimerization in Mercenaria. *Earth and Planetery Science Letters*, **28**, 275–282.

Mixon, R.B. & Pilkey, O.H. (1976). Reconnaissance geology of the submerged and emerged Coastal Plain province, Cape Lookout area, North Carolina. U.S. Geological Survey Professional Paper, 859, pp. 1–45.

Mixon, R.B., Szabo, B.J. & Owens, J.P. (1982). Uranium-series dating of mollusks and corals, and age of Pleistocene deposits, Chesapeake Bay area, Virginia and Maryland. U.S. Geological Survey Professional Paper 1067-E, pp. 1–18.

Moore, G.W. & Moore, J.G. (1988). Large-scale bedforms in boulder gravel produced by giant waves in Hawaii. Geological Society of America Special Paper 229, pp. 101–110.

Moore, J.G. (1970). Relationship between subsidence and volcanic load, Hawaii. *Bulletin Volcanologique*, **34**, 562–576.

Moore, J.G. & Fornari, D.J. (1984). Drowned reefs as indicators of the rate of subsidence of the island of Hawaii. *Journal of Geology*, **92**, 752–759.

Moore, J.G. & Moore, G.W. (1984). Deposit from a giant wave on the island of Lanai, Hawaii. *Science*, **226**, 1312–1315.

Moore, J.G., Normark, W.R. & Szabo, B.J. (1990). Reef growth and volcanism on the submarine southwest rift zone of Mauna Loa, Hawaii. *Bulletin of Volcanology*, **52**, 375–380.

Morton, R.A. & Price, W.A. (1987). Late Quaternary sea-level fluctuations and sedimentary phases of the Texas coastal plain and shelf. *In*: Nummendal, D., Pilkey, O.H. & Howard, J.D. (Eds), *Sea-Level Fluctuation and Coastal Evolution*. SEPM (Society for Sedimentary Geology) Special Publication, 41, pp. 182–198.

Muhs, D.R. (2000). Dating marine terraces with relative-age and correlated-age methods. *In*: Noller, J.S., Sowers, J.M. & Lettis, W.R. (Eds), *Quaternary Geochronology,*

Applications and Methods. American Geophysical Union Reference Shelf, **4**, 434–446.

Muhs, D.R. (2002). Evidence for the timing and duration of the last interglacial period from high-precision uranium-series ages of corals on tectonically stable coastlines. *Quaternary Research*, **58**, 36–40.

Muhs, D.R., Kennedy, G.L. & Rockwell, T.K. (1994). Uranium-series ages of marine terrace corals from the Pacific coast of North America and implications for last-interglacial sea level history. *Quaternary Research*, **42**, 72–87.

Muhs, D.R., Kelsey, H.M., Miller, G.H., Kennedy, G.L., Whelan, J.F. & McInelly, G.W. (1990). Age estimates and uplift rates for late Pleistocene marine terraces: Southern Oregon portion of the Cascadia forearc. *Journal of Geophysical Research*, **95**, 6685–6698.

Muhs, D.R., Miller, G.H., Whelan, J.F. & Kennedy, G.L. (1992a). Aminostratigraphy and oxygen isotope stratigraphy of marine terrace deposits, Palos Verdes Hills and San Pedro areas, Los Angeles County, California. *In*: Fletcher, C.H., III & Wehmiller, J.F. (Eds), *Quaternary Coasts of the United States: Marine and Lacustrine Systems*. SEPM (Society for Sedimentary Geology) Special Publication, 48, pp. 363–376.

Muhs, D.R., Simmons, K.R., Kennedy, G.L. & Rockwell, T.K. (2002b). The last interglacial period on the Pacific Coast of North America: Timing and Paleoclimate. *Geological Society of America Bulletin*, **114**, 569–592.

Muhs, D.R., Simmons, K.R., Kennedy, G.L., Ludwig, K.R. & Groves, L.T. (2002c). A cool eastern Pacific Ocean at the close of the last interglacial complex, ~80,000 yr B.P. *Geological Society of America Abstracts with Programs*, **34**, 6, 130.

Muhs, D.R., Simmons, K.R. & Steinke, B. (2002a). Timing and warmth of the last interglacial period: New U-series evidence from Hawaii and Bermuda and a new fossil compilation for North America. *Quaternary Science Reviews*, **21**, 1355–1383.

Muhs, D.R. & Szabo, B.J. (1994). New uranium-series ages of the Waimanalo Limestone, Oahu, Hawaii: Implications for sea level during the last interglacial period. *Marine Geology*, **118**, 315–326.

Muhs, D.R., Szabo, B.J., McCartan, L., Maat, P.B., Bush, C.A. & Halley, R.B. (1992b). Uranium-series age estimates of corals from Quaternary marine sediments of southern Florida. Florida Geological Survey Special Publication, 36, pp. 41–50.

Multer, H.G., Gischler, E., Lundberg, J., Simmons, K.R. & Shinn, E.A. (2002). Key Largo Limestone revisited: Pleistocene shelf-edge facies, Florida Keys, USA. *Facies*, **46**, 229–272.

Nakata, M. & Lambeck, K. (1989). Late Pleistocene and Holocene sea-level change in the Australian region and mantle rheology. *Geophysical Journal International*, **96**, 497–517.

Nummendal, D., Pilkey, O.H. & Howard, J.D. (Eds) (1987). *Sea-level fluctuation and coastal evolution*. SEPM (Society for Sedimentary Geology) Special Publication, 41, pp. 1–267.

Oaks, R.Q., Coch, N.K., Sanders, J.E. & Flint, R.F. (1974). Post-Miocene shorelines and sea levels, southeastern Virginia. *In*: Oaks, R.Q., Jr. & DuBar, J.R. (Eds), *Post-Miocene Stratigraphy, Central and Southern Atlantic Coastal Plain*. Logan, Utah, Utah State University Press, pp. 53–87.

Oertel, G.F. & Foyle, A.M. (1995). Drainage displacement by sea-level fluctuation at the outer margin of the Chesapeake Seaway. *Journal of Coastal Research*, **11**, 583–604.

Oldale, R.N., Valentine, P.C., Cronin, T.M., Spiker, E.C., Blackwelder, B.W., Belknap, D.F., Wehmiller, J.F. & Szabo, B.J. (1982). Stratigraphy, structure, absolute age, and paleontology of the upper Pleistocene deposits at Sankaty Head, Nantucket Island, Massachusetts. *Geology*, **10**, 246–252.

O'Neal, M.L., Wehmiller, J.F. & Newell, W.L. (2000). Amino acid geochronology of Quaternary coastal terraces on the northern margin of Delaware Bay, southern New Jersey, USA. *In*: Goodfriend, G.A., Collins, M.J., Fogel, M.L., Macko, S.A. & Wehmiller, J.F. (Eds), *Perspectives in Amino Acid and Protein Geochemistry*. Oxford, Oxford University Press, pp. 301–319.

O'Neal, M.L. & McGeary, S. (2002). Late Quaternary stratigraphy and sea-level history of the northern Delaware Bay margin, southern New Jersey, USA: a ground penetrating radar analysis of composite Quaternary coastal terraces. *Quaternary Science Reviews*, **21**, 929–946.

Osmond, J.K., Carpenter, J.R. & Windom, H.L. (1965). Th^{230}/U^{234} age of the Pleistocene corals and oolites of Florida. *Journal of Geophysical Research*, **70**, 1843–1847.

Ota, Y. & Omura, O. (1992). Contrasting styles and rates of tectonic uplift of coral reef terraces in the Ryukyu and Daito Islands, southwestern Japan. *Quaternary International*, **15/16**, 17–29.

Otvos, E.G. (1992). Quaternary evolution of the Apalachicola coast, northeastern Gulf of Mexico. *In*: Fletcher, C.H., III & Wehmiller, J.F. (Eds), *Quaternary Coasts of the United States: Marine and Lacustrine Systems*. SEPM (Society for Sedimentary Geology) Special Publication 48, pp. 221–232.

Otvos, E.G. (1995). Multiple Pliocene-Quaternary marine highstands, northeast Gulf of Mexico coastal plain – fallacies and facts. *Journal of Coastal Research*, **11**, 984–1002.

Otvos, E.G. & Howat, W.E. (1996). South Texas Ingleside Barrier: coastal sediment cycles and vertebrate fauna: late Pleistocene stratigraphy revised. *Transactions of the Gulf Coast Association of Geological Societies*, **XLVI**, 333–344.

Owens, J.P. (1989). Geologic map of the Cape Fear region, Florence 1° × 2° quadrangle and northern half of the Georgetown 1° × 2° quadrangle, North Carolina and South Carolina. U.S. Geological Survey Miscellaneous Investigations Map I-1948-A, scale 1:250,000.

Pilkey, O.H., Blackwelder, B.W., Knebel, H.J. & Ayers, M.W. (1981). The Georgia embayment continental shelf: Stratigraphy of a submergence. *Geological Society America Bulletin*, **92**(Pt. I), 52–63.

Peltier, W.R. (1994). Ice age paleotopography. *Science*, **265**, 195–201.

Peltier, W.R. (1996). Mantle viscosity and ice-age ice sheet topography. *Science*, **273**, 1359–1364.

Peltier, W.R. (1999). Global sea level rise and glacial isostatic adjustment. *Global and Planetary Change*, **20**, 93–133.

Peltier, W.R. (2002). On eustatic sea level history: Last glacial maximum to Holocene. *Quaternary Science Reviews*, **21**, 377–396.

Péwé, T.L., Hopkins, D.M. & Giddings, J.L. (1965). The Quaternary geology and archaeology of Alaska. *In*: Wright, H.E., Jr. & Frey, D.G. (Eds), *The Quaternary of the United States*. Princeton, Princeton University Press, pp. 355–374.

Poore, R.Z. & Dowsett, H.J. (2001). Pleistocene reduction of polar ice caps evidence from Cariaco Basin marine sediments. *Geology*, **29**, 71–74.

Richards, H.G. & Judson, S. (1965). The Atlantic coastal plain and the Appalachian highlands in the Quaternary. *In*: Wright, H.E., Jr. & Frey, D.G. (Eds), *The Quaternary of the United States*. Princeton, Princeton University Press, pp. 129–136.

Riggs, S.R. & Belknap, D.F. (1988). Upper Cenozoic processes and environments of continental margin sedimentation: Eastern United States. *In*: Sheridan, R.E. & Grow, J.A. (Eds), *The Atlantic Continental Margin: U.S.* Boulder, Colorado, Geological Society of America, Geology of North America, I-2, pp. 131–176.

Riggs, S.R., York, L.L., Wehmiller, J.F. & Snyder, S.W. (1992). Depositional patterns resulting from high frequency Quaternary sea-level fluctuations in northeastern North Carolina. *In*: Fletcher, C.H., III & Wehmiller, J.F. (Eds), *Quaternary Coasts of the United States: Marine and Lacustrine Systems*. SEPM (Society for Sedimentary Geology) Special Publication, 48, pp. 141–153.

Riggs, S.R., Cleary, W.J. & Snyder, S.W. (1995). Influence of inherited geologic framework on barrier shoreface morphology and dynamics. *Marine Geology*, **126**, 213–234.

Rockwell, T.K., Muhs, D.R., Kennedy, G.L., Hatch, M.E., Wilson, S.H. & Klinger, R.E. (1989). Uranium-series ages, faunal correlations and tectonic deformation of marine terraces within the Agua Blanca fault zone at Punta Banda, northern Baja California, Mexico. *In*: Abbott, P.L. (Ed.), *Geologic Studies in Baja California*. Los Angeles, Pacific Section, Society of Economic Paleontologists and Mineralogists, pp. 1–16.

Rockwell, T.K., Nolan, J.M., Johnson, D.L. & Patterson, R.H. (1992). Ages and deformation of marine terraces between Point Conception and Gaviota, western Transverse Ranges, California. *In*: Fletcher, C.H., III & Wehmiller, J.F. (Eds), *Quaternary Coasts of the United States: Marine and Lacustrine Systems*. SEPM (Society for Sedimentary Geology) Special Publication, 48, pp. 333–341.

Rodriguez, A.B., Anderson, J.B., Banfield, L.A., Taviani, M., Abdulah, K. & Snow, J.N. (2000). Identification of a −15 m middle Wisconsin shoreline on the Texas inner continental shelf. *Palaeogeography, Palaeoclimatology, Palaeoecology*, **158**, 25–43.

Rosholt, J.N., Jr., Emiliani, C., Geiss, J., Koczy, F.F. & Wangersky, P.J. (1961). Absolute dating of deep-sea cores by the Pa^{231}/Th^{230} method. *Journal of Geology*, **69**, 162–185.

Rubin, K.H., Fletcher, C.H., III & Sherman, C. (2000). Fossiliferous Lana'i deposits formed by multiple events rather than a single giant tsunami. *Nature*, **408**, 675–681.

Sarna-Wojcicki, A.M., Lajoie, K.R. & Yerkes, R.F. (1987). Recurrent Holocene displacement on the Javon Canyon fault – A comparison of fault-movement history with calculated average recurrence intervals. U.S. Geological Survey Professional Paper 1339, pp. 125–135.

Schroeder, W.W., Schultz, A.W. & Pilkey, O.H. (1995). Late Quaternary oyster shells and sea-level history, inner shelf, northeast Gulf of Mexico. *Journal of Coastal Research*, **11**, 664–674.

Shackleton, N.J. (1987). Oxygen isotopes, ice volume and sea level. *Quaternary Science Reviews*, **6**, 183–190.

Shackleton, N.J. (2000). The 100,000-year ice-age cycle identified and found to lag temperature, carbon dioxide, and orbital eccentricity. *Science*, **289**, 1897–1902.

Shepard, F.P. (1973). *Submarine geology*. New York, Harper and Row, 517 pp.

Sheridan, R.E., Ashley, G.M., Miller, K.G., Waldner, J.S., Hall, D.W. & Uptegrove, J. (2000). Offshore-onshore correlation of upper Pleistocene strata, New Jersey coastal plain to continental shelf and slope. *Sedimentary Geology*, **134**, 197–207.

Sherman, C.E., Fletcher, C.H. & Rubin, K.H. (1999). Marine and meteoric diagenesis of Pleistocene carbonates from a nearshore submarine terrace, Oahu, Hawaii. *Journal of Sedimentary Research*, **69**, 1083–1097.

Sherman, C.E., Glenn, C.R., Jones, A.T., Burnett, W.C. & Schwarcz, H.P. (1993). New evidence for two highstands of the sea during the last interglacial, oxygen isotope substage 5e. *Geology*, **21**, 1079–1082.

Shinn, E.A., Lidz, B.H., Kindinger, J.L., Hudson, J.H. & Halley, R.B. (1989). Reefs of Florida and the Dry Tortugas: A guide to the modern carbonate environments of the Florida Keys and the Dry Tortugas. St. Petersburg, Florida, U.S. Geological Survey, 53 pp.

Simmons, K.R., Wehmiller, J.F., Krantz, D.E., Ludwig, K., Markewich, H.W., Rich, F. & Hulbert, R.C., Jr. (1997). TIMS U-series ages for Atlantic coastal plain corals suggest 80 ka sea level similar to the present. *EOS, Transactions American Geophysical Union*, **78**, 46, F788.

Snyder, S.W., Snyder, Stephen W., Riggs, S.R. & Hine, A.C. (1991). Sequence stratigraphy of Miocene deposits, North Carolina continental margin. *In*: Horton, J.W., Jr. & Zullo, V.A. (Eds), *The Geology of the Carolinas*. Knoxville, University of Tennessee Press, pp. 263–273.

Soller, D.R. & Mills, H.H. (1991). Surficial geology and geomorphology. *In*: Horton, J.W., Jr. & Zullo, V.A. (Eds), *The Geology of the Carolinas*. Knoxville, University of Tennessee Press, pp. 290–308.

Stanley, S.M. (1966). Paleoecology and diagenesis of Key Largo Limestone, Florida. *Bulletin of the American Association of Petroleum Geologists*, **50**, 1927–1947.

Stearns, H.T. (1938). Ancient shorelines on the island of Lanai, Hawaii. *Geological Society of America Bulletin*, **49**, 615–628.

Stearns, H.T. (1974). Submerged shorelines and shelves in the Hawaiian Islands and a revision of some of the eustatic emerged shorelines. *Geological Society of America Bulletin*, **85**, 795–804.

Stearns, H.T. (1978). Quaternary shorelines in the Hawaiian Islands. *Bernice P. Bishop Museum Bulletin*, **237**, 57.

Stein, M., Wasserburg, G.J., Lajoie, K.R. & Chen, J.H. (1991). U-series ages of solitary corals from the California coast by mass spectrometry. *Geochimica et Cosmochimica Acta*, **55**, 3709–3722.

Stirling, C.H., Esat, T.M., Lambeck, K., McCulloch, M.T., Blake, S.G., Lee, D.-C. & Halliday, A.N. (2001). Orbital forcing of the marine isotope stage 9 interglacial. *Science*, **291**, 290–293.

Stuiver, M., Reimer, P.J., Bard, E., Beck, J.W., Burr, G.S., Hughen, K.A., Kromer, B., McCormac, G., van der Plicht, J. & Spurk, M. (1998). INTCAL 98 Radiocarbon age calibration, 24,000–0 cal BP. *Radiocarbon*, **40**, 1041–1083.

Suter, J.R., Berryhill, H.L., Jr. & Penland, S. (1987). Late Quaternary sea-level fluctuations and depositional sequences, southwest Louisiana continental shelf. *In*: Nummendal, D., Pilkey, O.H. & Howard, J.D. (Eds), *Sea-Level Fluctuation and Coastal Evolution*. SEPM (Society for Sedimentary Geology) Special Publication, 41, pp. 199–219.

Szabo, B.J. (1985). Uranium-series dating of fossil corals from marine sediments of southeastern United States Atlantic Coastal Plain. *Geological Society of America Bulletin*, **96**, 398–406.

Szabo, B.J., Ludwig, K.R., Muhs, D.R. & Simmons, K.R. (1994). Thorium-230 ages of corals and duration of the last interglacial sea-level high stand on Oahu, Hawaii. *Science*, **266**, 93–96.

Taylor, F.W. & Mann, P. (1991). Late Quaternary folding of coral reef terraces, Barbados. *Geology*, **19**, 103–106.

Toscano, M.A. (1992). Record of oxygen isotope stage 5 on the Maryland inner shelf and Atlantic coastal plain – a post-transgressive-highstand regime. *In*: Fletcher, C.H., III & Wehmiller, J.F. (Eds), *Quaternary Coasts of the United States: Marine and Lacustrine Systems*. SEPM (Society for Sedimentary Geology) Special Publication, 48, pp. 89–100.

Toscano, M.A. & Lundberg, J. (1998). Early Holocene sea-level record from submerged fossil reefs on the southeast Florida margin. *Geology*, **26**, 255–258.

Toscano, M.A. & Lundberg, J. (1999). Submerged late Pleistocene reefs on the tectonically-stable S.E. Florida margin: high-precision geochronology, stratigraphy, resolution of Substage 5a sea-level elevation, and orbital forcing. *Quaternary Science Reviews*, **18**, 753–767.

Toscano, M.A. & York, L.L. (1992). Quaternary stratigraphy and sea-level history of the U.S. Middle Atlantic Coastal Plain. *Quaternary Science Reviews*, **11**, 301–328.

Trecker, M.A., Gurrola, L.D. & Keller, E.A. (1998). Oxygen-isotope correlation of marine terraces and uplift of the Mesa Hills, Santa Barbara, California, USA. *In*: Stewart, I.S. & Vita-Finzi, C. (Eds), *Coastal Tectonics*. London: Geological Society of London Special Publications, 146, pp. 57–69.

Valentine, J.W. (1958). Late Pleistocene megafauna of Cayucos, California and its zoogeographic significance. *Journal of Paleontology*, **32**, 687–696.

Valentine, J.W. (1961). Paleoecologic molluscan geography of the Californian Pleistocene. *University of California Publications in Geological Sciences*, **34**, 309–442.

Valentine, J.W. & Veeh, H.H. (1969). Radiometric ages of Pleistocene terraces from San Nicolas Island, California. *Geological Society of America Bulletin*, **80**, 1415–1418.

Vedder, J.G. & Norris, R.M. (1963). Geology of San Nicolas Island California. U.S. Geological Survey Professional Paper, 369, 65 pp.

Veeh, H.H. (1966). Th^{230}/U^{238} and U^{234}/U^{238} ages of Pleistocene high sea level stand. *Journal of Geophysical Research*, **71**, 3379–3386.

Veeh, H.H. & Chappell, J. (1970). Astronomical theory of climatic change: Support from New Guinea. *Science*, **167**, 862–865.

Veeh, H.H. & Valentine, J.W. (1967). Radiometric ages of Pleistocene fossils from Cayucos, California. *Geological Society of America Bulletin*, **78**, 547–550.

Wahrhaftig, C. & Birman, J.H. (1965). The Quaternary of the Pacific mountain system in California. *In*: Wright, H.E., Jr. & Frey, D.G. (Eds), *The Quaternary of the United States*. Princeton, Princeton University Press, pp. 299–340.

Walcott, R.I. (1972). Past sea levels, eustasy and deformation of the Earth. *Quaternary Research*, **2**, 1–14.

Ward, L.W, Bailey, R.H. & Carter, J.G. (1991). Pliocene and early Pleistocene stratigraphy, depositional history, and molluscan paleobiogeography of the coastal plain. *In*: Horton, J.W., Jr. & Zullo, V.A. (Eds), *The Geology of the Carolinas*. Knoxville, University of Tennessee Press, pp. 274–289.

Wehmiller, J.F. (1982). A review of amino acid racemization studies in Quaternary mollusks: stratigraphic and chronologic applications in coastal and interglacial sites, Pacific and Atlantic coasts, United States, United Kingdom, Baffin Island, and tropical islands. *Quaternary Science Reviews*, **1**, 83–120.

Wehmiller, J.F. (1992). Aminostratigraphy of Southern California Quaternary marine terraces. *In*: Fletcher, C.H., III & Wehmiller, J.F. (Eds), *Quaternary Coasts of the United States: Marine and Lacustrine Systems*. SEPM (Society for Sedimentary Geology) Special Publication, 48, pp. 317–321.

Wehmiller, J.F. & Belknap, D.F. (1978). Alternative kinetic models for the interpretation of amino acid enantiomeric ratios in Pleistocene mollusks: examples from California, Washington, and Florida. *Quaternary Research*, **9**, 330–348.

Wehmiller, J.F. & Belknap, D.F. (1982). Amino acid age estimates, Quaternary Atlantic coastal plain: comparison with U-series dates, biostratigraphy, and paleomagnetic control. *Quaternary Research*, **18**, 311–336.

Wehmiller, J.F. & Miller, G.H. (2000). Aminostratigraphic dating methods in Quaternary geology. *In*: Noller, J.S., Sowers, J.M. & Lettis, W.R. (Eds), *Quaternary*

Geochronology, Methods and Applications. American Geophysical Union Reference Shelf, **4**, 187–222.

Wehmiller, J.F. & York, L.L. (2001). Chronostratigraphic and paleoclimatic implications of paired radiocarbon/racemization analyses of Quaternary mollusks from the mid- and southeastern Atlantic coastal plain. *Geological Society of America Abstracts with Programs*, **33**, 6, 171.

Wehmiller, J.F., Belknap, D.F., Boutin, B.S., Mirecki, J.E., Rahaim, S.D. & York, L.L. (1988). A review of the aminostratigraphy of Quaternary mollusks from United States Atlantic Coastal Plain sites. Geological Society of America Special Paper, 227, pp. 69–110.

Wehmiller, J.F., York, L.L., Belknap, D.F. & Snyder, S.W. (1992). Aminostratigraphic discontinuities in the U.S. Atlantic coastal plain and their relation to preserved Quaternary stratigraphic records. *Quaternary Research*, **38**, 275–291.

Wehmiller, J.F., York, L.L., Krantz, D.E., & Gayes, P.T. (1993). Aminostratigraphy of the Wando Fm., Charleston, S.C.: When is an aminozone a valid chronostratigraphic unit? *Geological Society of America Abstracts with Programs*, **25**, 4, 76.

Wehmiller, J.F., York, L.L. & Bart, M.L. (1995). Amino acid racemization geochronology of reworked Quaternary mollusks on U.S. Atlantic coast beaches: Implications for chronostratigraphy, taphonomy, and coastal sediment transport. *Marine Geology*, **124**, 303–337.

Wehmiller, J.F., Otvos, E., Wingard, G.L. & Scott, T.M. (1996). Aminostratigraphy of the Gulf Coast Quaternary – approaches to correlation of aminozones between Atlantic and Pacific coast sites. *Geological Society of America Abstracts with Programs*, **28**, 2, 49.

Wehmiller, J.F., Krantz, D.E., Simmons, K.R., Ludwig, K.R., Markewich, H.W., Rich, F. & Hulbert, R.C., Jr. (1997). U.S. Atlantic Coastal Plain late Quaternary geochronology; TIMS U-series coral dates continue to indicate 80 kyr sea level at or above present. *Geological Society of America Abstracts with Programs*, **29**, 6, 346.

Wehmiller, J.F., York L.L., Jones, D.S. & Portell, R.W. (1999). Racemization isochrons for the U.S. Atlantic coastal plain Quaternary: independent calibration and geochemical implications of results from marginal marine units, central Florida. V.M. Goldschmidt Geochemistry Conference Proceedings Volume, Cambridge, Massachusetts, p. 321.

Wehmiller, J.F., Rodriquez, A.B., Anderson, J.B. & York, L.L. (2000a). Gulf of Mexico (east Texas inner shelf) Quaternary geochronology based on marine mollusk amino acid racemization data. *EOS, Transactions American Geophysical Union*, **81**, 48, F650.

Wehmiller, J.F., Stecher, H.A., III, York, L.L. & Friedman, I. (2000b). The thermal environment of fossils: effective ground temperatures (1994–1998) at aminostratigraphic sites, U.S. Atlantic coastal plain. *In*: Goodfriend, G.A., Collins, M.J., Fogel, M.L., Macko, S.A. & Wehmiller, J.F. (Eds), *Perspectives in Amino Acid and Protein Geochemistry*. Oxford, Oxford University Press, pp. 219–250.

Wehmiller, J.F., Lajoie, K.R., Kvenvolden, K.A., Peterson, E., Belknap, D.F., Kennedy, G.L., Addicott, W.O., Vedder, J.G. & Wright, R.W. (1977). Correlation and chronology of Pacific coast marine terrace deposits of continental United States by fossil amino acid stereochemistry – Technique evaluation, relative ages, kinetic model ages, and geologic implications. U.S. Geological Survey Open-File Report 77–680, 196 pp.

Winker, C.D. & Howard, J.D. (1977). Correlation of tectonically deformed shorelines on the southern Atlantic coastal plain. *Geology*, **5**, 123–127.

Winograd, I.J., Landwehr, J.M., Ludwig, K.R., Coplen, T.B. & Riggs, A.C. (1997). Duration and structure of the past four interglaciations. *Quaternary Research*, **48**, 141–154.

Woodring, W.P., Bramlette, M.N. & Kew, W.S.W. (1946). Geology and paleontology of Palos Verdes Hills, California. U.S. Geological Survey Professional Paper 207, 145 pp.

Woodroffe, C.D., Short, S.A., Stoddart, D.R., Spencer, T. & Harmon, R.S. (1991). Stratigraphy and chronology of late Pleistocene reefs in the southern Cook Islands, south Pacific. *Quaternary Research*, **35**, 246–263.

Yokoyama, Y., De Deckker, P., Lambeck, K., Johnston, P. & Fifield, L.K. (2001). Sea-level at the Last Glacial Maximum: Evidence from northwestern Australia to constrain ice volumes for oxygen isotope stage 2. *Palaeogeography, Palaeoclimatology, Palaeoecology*, **165**, 281–297.

York, L.L. & Wehmiller, J.F. (1992a). Molluscan aminostratigraphy of Pleistocene marine deposits offshore of Cape Fear, N.C. and Murrells Inlet, S.C. *Geological Society of America Abstracts with Programs*, **24**, 2, 74.

York, L.L. & Wehmiller, J.F. (1992b). Aminostratigraphic results from Cape Lookout, N.C. and their relation to the preserved Quaternary marine record of SE North Carolina. *Sedimentary Geology*, **80**, 279–291.

York, L.L., Wehmiller, J.F., Cronin, T.M. & Ager, T.A. (1989). Stetson Pit, Dare County, North Carolina: An integrated chronologic, faunal, and floral record of subsurface coastal sediments. *Palaeogeography, Palaeoclimatology, Palaeoecology*, **72**, 115–132.

York, L.L., Harris, M.S., Wehmiller, J.F. & Krantz, D.E. (1999). Implications of TIMS U-series dates for the Late Pleistocene sea level and aminostratigraphic record in the Coastal Plain of central South Carolina. *EOS, Transactions of the American Geophysical Union*, **80**, 46, F585.

York, L.L., Jones, D.S., Martin, E.E., Portell, R.W. & Wehmiller, J.F. (2000a). Comparison of strontium isotope and amino acid age estimates for Plio-Pleistocene mollusks, Central Florida. *Geological Society of America Abstracts with Programs*, **32**, 7, 20.

York, L.L., Thieler, E.R., Brill, A.L., Riggs, S.R. & Wehmiller, J.F. (2000b). Aminostratigraphic age estimate for the Suffolk Scarp, North Carolina Coastal plain. *Geological Society of America Abstracts with Programs*, **32**, 2, 85.

York, L.L., Doar, W.R., III & Wehmiller, J.F. (2001). Late Quaternary aminostratigraphy and geochronology of the St. Helena island area, South Carolina coastal plain. *Geological Society of America Abstracts with Programs*, **33**, 2, 26.

Western lakes

Larry Benson

U.S. Geological Survey, 3215 Marine St., Boulder, CO 80303, USA

Status of Western Lake Studies in 1965

Wright & Frey (1965) contained a chapter entitled the "Quaternary Geology of the Great Basin" in which Roger Morrison (1965) emphasized the use of soils as stratigraphic markers. He used these markers to correlate the rise and fall of Great Basin lakes (Fig. 1) with the advance and retreat of North American alpine and continental glaciers. This study was entirely outcrop-based and the correlations were made under the assumption that soils had formed synchronously across North America during relatively dry and warm periods.

Although not reflected in the Morrison (1965) chapter, new concepts in climate forcing and methods of age control had already emerged. Antevs (1948), nearly two decades previously, had suggested that maximum levels of Great Basin lakes were linked to the presence of a permanent ice sheet over North America. He hypothesized that the size of the ice sheet, combined with a permanent high-pressure area located over it, caused storm tracks associated with the polar jet stream (PJS) to be pushed south of their present-day average path over the northern Great Basin. In addition, Emiliani & Geiss (1957) had suggested that variation in insolation strongly influenced Pleistocene glacial-interglacial cycles.

Broecker and his colleagues (Broecker & Kaufman, 1965; Broecker & Orr, 1958; Broecker & Walton, 1959; Kaufman & Broecker, 1965) had already applied the relatively new radiocarbon (^{14}C) and uranium-series methods of age control to studies of carbonates precipitated from Lake Bonneville and Lake Lahontan (Fig. 1), and Flint & Gale (1958) and Stuiver (1964) had obtained ^{14}C ages of organic materials in sediment cores from Searles Lake, California (Fig. 1). These studies demonstrated that the last large lake cycle in each of these basins occurred between ~25,000 and ~10,000 ^{14}C yr B.P.

New Approaches and Technologies

Because of constraints on the length of this chapter, I will emphasize high-resolution studies of sediment cores taken from lakes located within the western United States as defined by the boundaries of Fig. 1. High-resolution will be taken to mean a resolution of 500 to 2000 yr for records $\leq$150,000 yr in length, a resolution of 100 to 500 yr for records $\leq$50,000 yr in length, a resolution of 5 to 100 yr for records $\leq$10,000 yr in length, and a resolution of $\leq$5 yr for records $\leq$1000 yr in length.

Sediment-based records that illustrate evolving concepts of climate change (e.g. solar forcing) will be emphasized. The work of those who broke new ground as well as the work of those who attempted to quantify the elements of climate change will be featured. Records that extend beyond the Last Interglaciation (e.g. Adam *et al.*, 1989; Cohen, 1996; Davis & Moutoux, 1998; Kowalewska & Cohen, 1998; Thompson, 1996), records that are salt-dominated (e.g. Jannik *et al.*, 1991; Li *et al.*, 1996; Lowenstein *et al.*, 1999; Smith, 1984), records that are discontinuous (e.g. Hooke, 1999; Wells *et al.*, 1987), and most outcrop-based records will not be discussed in this chapter.

Studies of western lakes and their sediments have improved in several ways since 1965, including: (1) recovery of continuous sediment records from the deepest areas of lake basins by coring; (2) improvements in age control; and (3) introduction and application of new proxies of climate change.

Sediment Coring

Coring of a few lake basins had occurred prior to 1965; e.g. Great Salt Lake, Utah (Eardley & Gvosdetsky, 1960), and Searles lake, California (Smith, 1962); however, the core sites were few and some were chosen for their economic rather than their climatic value. During the past two decades, scientists have sought to obtain continuous or nearly continuous records of climate change by coring the deepest areas of lake basins. Cores have been recovered from several presently dry lake basins including Owens Lake basin, California (Lund, 1996; Smith & Bischoff, 1997), Summer Lake basin, Oregon (Cohen *et al.*, 2000), and Estancia basin, New Mexico (Allen & Anderson, 2000) (Fig. 1). Numerous perennial lakes also have been cored, including Great Salt Lake, Utah (Spencer *et al.*, 1984), Clear Lake, California (Sims *et al.*, 1988), Carp Lake, Washington (Whitlock & Bartlein, 1997), and Pyramid Lake, Nevada (Benson *et al.*, 1997a, 2002) (Fig. 1).

Methods of Age Control

Methods of age control have been expanded and in many cases greatly improved since 1965.

Radiocarbon (^{14}C). In 1965, ~5 g of C was needed for ^{14}C dating. With the advent of accelerator mass spectrometry (AMS) in the late 1970s, very precise ^{14}C analyses could be performed on as little as 1 mg of C (Southon *et al.*, 1982). This enhancement in sensitivity has allowed the dating of terrestrial macrofossils, lacustrine shells, humic acids (e.g. Reasoner & Jodry (2000), pollen (Long *et al.*, 1992; Mensing & Southon, 1999), and the total inorganic or organic carbon fraction of lake sediments (e.g. Benson *et al.*, 1997a, b).

DEVELOPMENTS IN QUATERNARY SCIENCE
VOLUME 1 ISSN 1571-0866
DOI:10.1016/S1571-0866(03)01009-1

PUBLISHED BY ELSEVIER B.V.

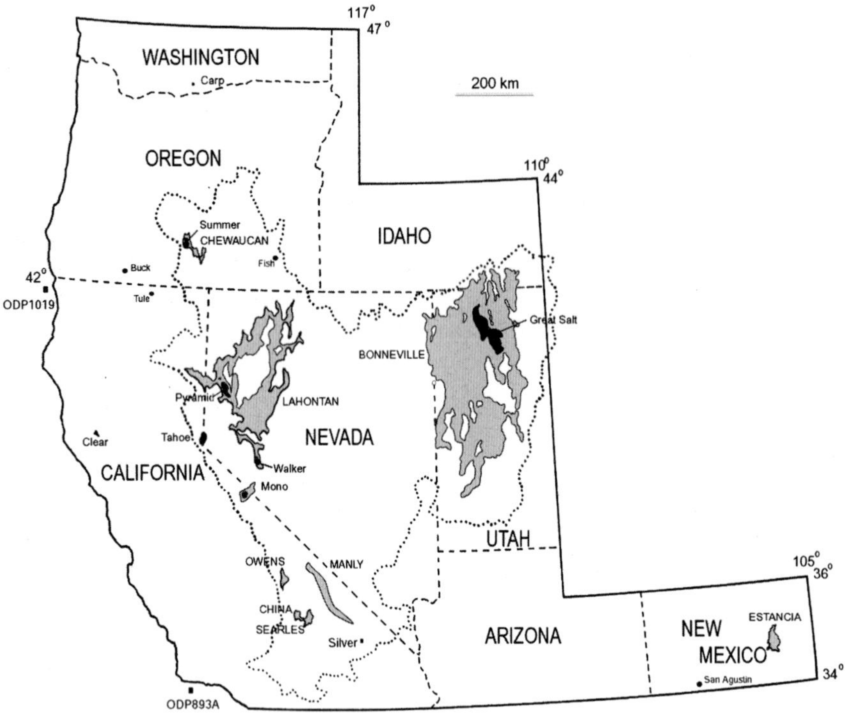

Fig. 1. Location map of lake and marine core sites cited in this paper. This map defines the boundary of the western United States for the purposes of this paper. Dotted line indicates the boundary of the Great Basin. Solid black areas indicate existing water bodies and gray areas indicate the extent of late-Pleistocene pluvial lakes.

Terrestrial macrofossils deposited in lake sediments generally yield reliable ^{14}C dates unless the time they have spent in the terrestrial environment is long compared to the time they have spent in the lacustrine environment. Unfortunately, deep-water sediments of large lakes such as Walker Lake, Pyramid Lake, and Great Salt Lake, usually lack terrestrial macrofossils. Other carbon sources often yield ^{14}C ages older than their date of deposition because of reworking of "old" forms of detrital carbon to the core site (e.g. pollen, Benson *et al.*, 2002) or because the carbon source was formed from a lake whose dissolved inorganic carbon (DIC) reservoir was not in equilibrium with atmospheric ^{14}C (reservoir effect) (Broecker & Walton, 1959).

Radiocarbon dates of organic and inorganic carbon-bearing materials that are now subaerially exposed offer a means of estimating lake-level histories of Great Basin paleolake systems. Such data have been obtained for Lake Bonneville, Utah (Oviatt *et al.*, 1992), Lake Lahontan, Nevada (Benson *et al.*, 1995, and references therein), and Mono Lake, California (Benson *et al.*, 1998a, and references therein) (Fig. 1). Two of these paleolake systems reached their highest levels between ~16,000 and ~14,000 cal yr B.P. (Fig. 2). Lake Bonneville's well-documented earlier decline at 17,000 cal yr B.P. resulted from the catastrophic downcutting of its basin margin and not from climate change. If the downcutting had not occurred, Lake Bonneville may have achieved its highstand at ~15,000 cal yr B.P. (dashed line in Fig. 2A).

Calibration of ^{14}C ages. Not only has the ^{14}C method been improved in terms of greater sensitivity and smaller sample size, there are now procedures for conversion of ^{14}C dates to calibrated ages. Stuiver *et al.* (1998) obtained ^{14}C ages of dendrochronologically dated wood samples (mostly 10-yr segments) back to 11,620 cal yr B.P. Hughen *et al.* (1998) used varved sediments from the Cariaco Basin to provide a new ^{14}C calibration data set for the period 14,500 to 10,000 cal yr B.P., and Bard *et al.* (1998) have used high-precision ^{14}C and ^{230}Th-^{234}U ages of marine corals to calibrate the ^{14}C time scale to ~23,500 cal yr B.P. In addition, Kitagawa & van der Plicht (1998) used 250 ^{14}C dates of terrestrial macrofossils from annually laminated sediments from Lake Suigetsu, Japan, to provide an atmospheric calibration for the past 45,000 cal yr B.P. Most recently, Schramm *et al.* (2000) have calibrated the ^{14}C time scale to 40,000 cal yr B.P. using high-precision ^{14}C and ^{230}Th-^{234}U ages of aragonites found in laminated sediments of Lake Lisan, Israel. Voelker *et al.* (1998) have correlated a ^{14}C-dated plankton-based marine ^{18}O record with the GISP2 ^{18}O record, providing a means of linking terrestrial ^{14}C records with the GISP2 record. In the present paper conversions from ^{14}C to calibrated ages have been done using, in order of preference, the work of Stuiver *et al.* (1998), Hughen *et al.* (1998), and Bard *et al.* (1998).

Uranium series. Uranium-series methods, have been greatly improved since the seminal study of Kaufman & Broecker (1965). For example, the work of Bischoff & Fitzpatrick (1991) and Luo & Ku (1991) have demonstrated that the correction for contamination by a single source of detrital Th is best accomplished using total sample dissolution and an isochron technique. Much of the unreliability in uranium-series age estimates may be due to multiple and often poorly understood sources of initial Th (Bischoff & Fitzpatrick, 1991; Lin *et al.*, 1996; Szabo *et al.*, 1996). If there are more than two sources of Th, the total dissolution and isochron approach may not succeed.

Uranium-series methods have been applied with some success in attempts to date carbonates from Lake Lahontan

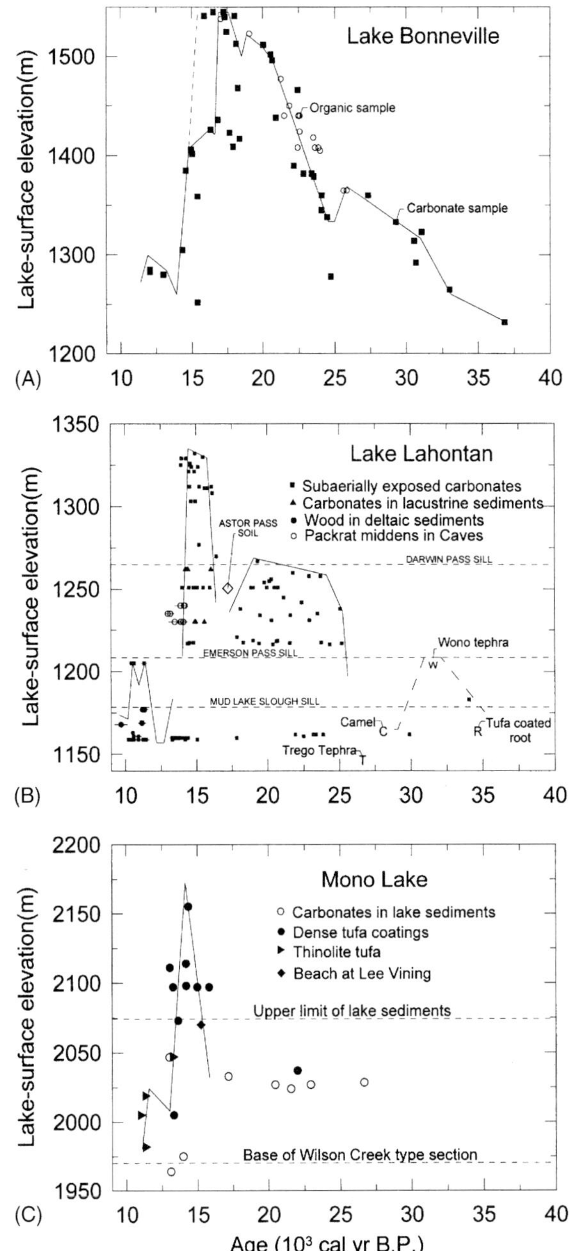

Fig. 2. Lake-level histories based on surficial materials for: (A) the Bonneville basin (Oviatt et al., 1992), (B) the Lahontan basin (Benson et al., 1995), and (C) the Mono Lake basin (Benson et al., 1998a).

(Bischoff & Fitzpatrick, 1991; Lao & Benson, 1988; Lin *et al.*, 1996; Szabo *et al.*, 1996) and to date salts and carbonates from Searles and Manly lakes (Bischoff *et al.*, 1985; Ku *et al.*, 1998; Lin *et al.*, 1998; Peng *et al.*, 1978) (Fig. 1). Uranium-series age estimates of salts and/or clays and organic materials contained within salts appear to yield reasonable values after correction for a single source of detrital ^{230}Th (Bischoff *et al.*, 1985; Ku *et al.*, 1998; Lin *et al.*, 1998; Peng *et al.*, 1978); i.e. most of the ages are in stratigraphic order.

Tephrochronology. Wilcox (1965) was one of the first to suggest the usefulness of volcanic ashes as stratigraphic markers in the western United States. However, it wasn't until the 1970s that analytical tools such as the electron microprobe (EMP) were available for chemical analysis of individual glass shards. Tephrochronology was applied to California lakes by Sarna-Wojcicki *et al.* (1988) and Adam *et al.* (1989) and to Great Basin lakes by Davis (1978, 1983). A recent review of tephrochronologic methods can be found in Sarna-Wojcicki (2000), and a recent application of tephrochronologic methods to lake-sediment age control can be found in Whitlock *et al.* (2000). New lake-sediment-based ages of two of the most widespread Great Basin tephras (Wono and Trego Hot Springs tephras) can be found in Benson *et al.* (1997a).

Paleomagnetic secular variation (PSV). At the time of the publication of the Wright & Frey (1965) volume, geophysicists were beginning studies of paleomagnetic reversals (e.g. Cox *et al.*, 1965) and had not developed detailed PSV (inclination, declination, intensity) records of the Earth's field. Since then, studies of lake sediments and archeological materials have helped create detailed PSV records (Lund & Banerjee, 1979, 1985a, b; Verosub, 1979; Verosub & Mehringer, 1984). Creer & Tucholka (1982, 1983) were among the first to suggest that type PSV curves could be applied to the dating of lake sediments and Lund (1996) was the first to demonstrate that distinctive field signatures in inclination and declination could be traced across North America without significant change in pattern.

During the last 25 yr, secular variation has been used in studies of several Great Basin lakes, including Mono Lake, California (Liddicoat & Coe, 1979; Lund *et al.*, 1988), Fish Lake, Oregon (Verosub *et al.*, 1986), Pyramid Lake, Nevada (Benson *et al.*, 2002), pluvial Lake Lahontan (Liddicoat, 1992, 1996; Liddicoat & Coe, 1997), Owens Lake, California (Li *et al.*, 2000), pluvial Lake Bonneville (Liddicoat & Coe, 1998), and Summer Lake, Oregon (Negrini *et al.*, 1984, 2000). Recently, Benson *et al.* (2003a) have shown that the geomagnetic feature, termed the Mono Lake excursion, has a date of $28{,}620 \pm 300$ ^{14}C yr B.P. ($\sim$32,400 GISP2 yr using the Voelker *et al.* (1998) conversion).

Methods for age control of young sediments. Various methods for age control of sediments <150 yr old have been applied to Great Basin lakes, including: ^{210}Pb and $^{239+240}$Pu concentrations in Mono Lake sediments (Jellison, 1996), ^{137}Cs and elemental Hg concentrations in Pyramid Lake (Benson *et al.*, 2002) and Lake Tahoe (Heyvaert *et al.*, 2000) sediments, and Pb concentrations in Owens Lake sediments (Li *et al.*, 2000).

Climate Change Proxies

Several organic and inorganic proxies of climate change are preserved in lake sediments. They can be roughly grouped into biological, chemical (including isotopic), and physical parameters. Changes in most parameters actually reflect variations in the chemical and physical nature of a lake, only indirectly indicating shifts in the local or regional climate system.

Biological proxies of climate change. Three biological proxies of climate change (diatoms, ostracodes, pollen) will be discussed. The discussion of pollen in this section of the paper will be confined to studies that indicate its response to lake-size change. In general, these proxies had not been applied in high-resolution studies of western lakes at the time of publication of the Wright & Frey (1965) volume.

Thompson, in Bradbury *et al.* (1989), was the first to show that the transition from relatively deep- to shallow-water conditions in a western lake (Walker Lake, Nevada), exposed large areas of playa sediment, providing expanded habitat for local desert vegetation such as greasewood (*Sarcobatus*). An increase in the concentration of greasewood pollen in the sediments of Walker Lake was, therefore, interpreted to reflect an increase in the area of the dry, salty playa that bordered the lake which was exposed during lake-level declines.

Bradbury (1987, 1992, 1997) was the first to use down-core changes in diatom species to infer changes in the salinity of western lakes. In his study of Walker Lake, Nevada, Bradbury (1987) demonstrated that transitions to shallow-water conditions were evidenced by abrupt decreases in the number of diatoms and in shifts from moderately saline (*C. quillensis*) to saline (*Navicula*) diatoms.

Forester pioneered the application of ostracode assemblages in studies of western lakes, including Great Salt Lake, Utah (Spencer *et al.*, 1984), Walker Lake, Nevada (Bradbury *et al.*, 1989), and the San Agustin basin, New Mexico (Markgraf *et al.*, 1984) (Fig. 1), to infer variations in the physical and chemical properties of these paleolake systems. More recently, Carter (1997) used ostracodes as indicators of changing salinity in the Owens Lake basin, and Allen & Anderson (2000) used ostracode species to infer salinity changes in Estancia basin (Fig. 1).

Chemical and physical proxies of alpine glaciation. Chemical analyses of cored sediments can be used to infer the advance and retreat of alpine glaciers in a lake's watershed as well as changes in the lake's hydrologic balance. Benson *et al.* (1996) produced the first continuous record of Sierra Nevada glacier oscillations using core-based data from Owens Lake Nevada (Fig. 1). They applied two glacial proxies, total organic carbon (TOC) and magnetic susceptibility (χ) (Fig. 3). They suggested that χ was a proxy for the magnetic component of glacial rock flour and that decreases in TOC were caused by decreases in biological productivity and dilution of the TOC fraction with rock flour. The results of this study were supported by independent ^{36}Cl age determinations made on Sierra Nevada moraines (Phillips *et al.*, 1996).

Bischoff *et al.* (1997) suggested that hydrologic closure of Owens Lake corresponded to interglacial conditions and that smectites dominated the clay-size fraction of sediments deposited during hydrologic closure. This led them to hypothesize that glacial rock flour dominated the clay-size fraction of Owens Lake when it overflowed and they suggested that Na, Ti, Mn, and Ba were excellent proxies of rock-flour abundance.

In the following year, Benson *et al.* (1998b) showed that diatoms flourished during interstadial dry periods and that organic carbon associated with their soft body parts and amorphous SiO_2 associated with their hard parts accounted for elevated values of TOC and decreased concentrations of

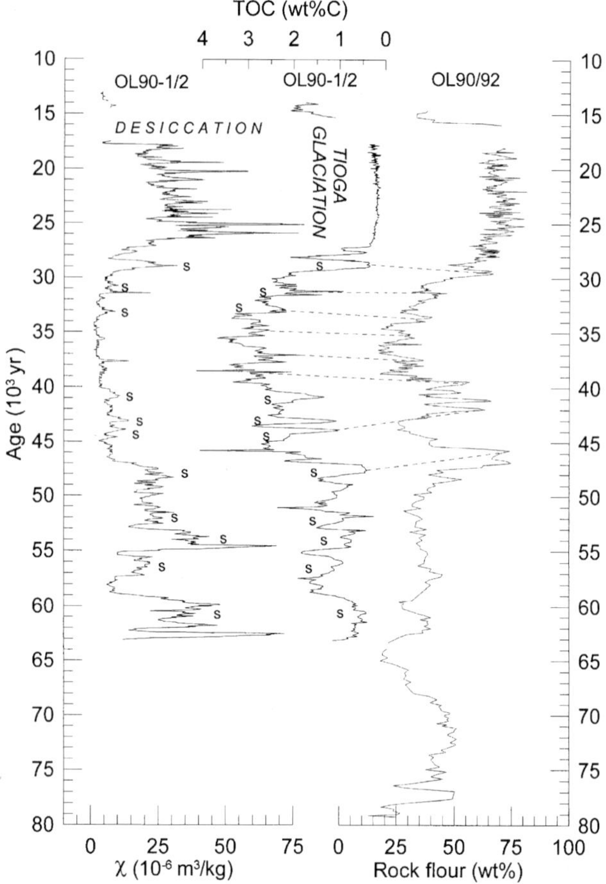

Fig. 3. Magnetic susceptibility (χ), TOC (Benson et al., 1998b), and rock flour (Bischoff & Cummins, 2001) glacial-proxy records for Owens Lake cores OL90–1/2 and OL92. "S" indicates the presence of an alpine stade.

Na, Ti, Mn, Ba, etc. Benson *et al.* (1998b) also demonstrated that numerous chemical elements contained in the clay-size fraction, which are common to the granitic Sierran bedrock, are suitable proxies of Sierran glacier advances. Elements such as Ti, whose mineral hosts have low solubilities, are probably the best and most durable indicators of glacier activity.

Bischoff & Cummins (2001) performed a high-resolution study of the mineral-weighted rock-flour component in sediments from Owens Lake core OL92 and from the lower part of core OL90. The record they produced spans the last 80,000 yr. A comparison of their rock-flour proxy of Sierra Nevada alpine glacial activity with the TOC and χ glacial proxies of Benson *et al.* (1996) indicates a high degree of correspondence in the younger part of the record (Fig. 3). Many of the offsets in the locations of peaks and troughs in the older parts of the records are due to differences in age models employed by the two groups of investigators. Bischoff & Cummins (2001) combined uranium-series dates of ostracodes from OL92 (Bischoff *et al.*, 1998) with uranium-series dates determined for saline minerals from a Searles Lake, California, core (Bischoff *et al.*, 1985) correlated to OL92, using pollen spectra (Litwin *et al.*, 1999). In contrast, the age model employed by Benson *et al.* (1996, 1998b) utilized ^{14}C dates (ranging from 37,000 to 12,000 ^{14}C yr B.P.) of the TOC fraction.

TOC data from Pyramid Lake core PLC92B and Owens Lake core OL90–1/2 (Benson *et al.*, 1998b) have also been used to define the beginning and end of the Tioga glacial stage. The Tioga appears to have commenced at 28,000 cal yr B.P. and ended by 15,000 cal yr B.P. (Fig. 4). Total inorganic carbon (TIC) can sometimes act as a proxy for glaciation (Fig. 4); however, its use as a glacial proxy is complicated by the effect of changes in the hydrologic balance on its concentration (see below).

Chemical proxies of change in the hydrologic balance. Five chemical proxies of change in the hydrologic balance have been applied to western lake sediments in the past several years, including: TIC concentrations, $\delta^{18}O$ values of the TIC fraction, $\delta^{18}O$ values of ostracode valves, and Mg/Ca and Sr/Ca ratios in ostracode valves.

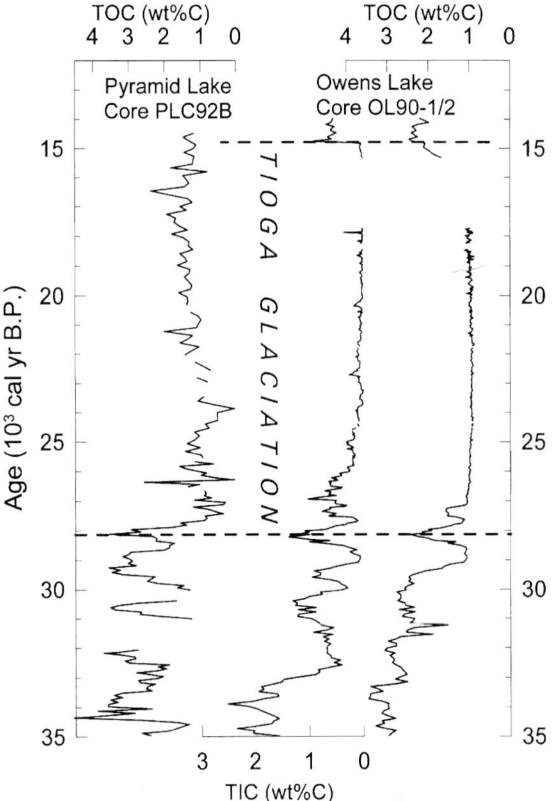

Fig. 4. TOC and TIC records from Pyramid Lake core PLC92B and Owens Lake core OL90–1/2. Low values of TOC were caused by reduced productivity, accompanying introduction of glacial rock flour to Owens Lake and by dilution of the TOC fraction by siliceous rock flour. The dashed lines enclose the Tioga glaciation. The gap in the Owens Lake record resulted from a desiccation that led to the removal of several cm of sediment by wind and water erosion. The PLC92B record does not extend to times younger than ~15,000 cal yr B.P. TOC values from another Pyramid Lake core (PLC97–3) fell sharply after ~15,000 cal yr B.P. (unpublished data of L. Benson).

The hydrologic state of a lake (open or closed) and, to a lesser extent, the relative size of a hydrologically closed lake can be determined from its down-core TIC distribution *in the absence of glacial runoff into the lake* (Benson, 1999; Benson *et al.*, 1996, 1997b, 1998a, 2002). During times of hydrologic closure, most of the dissolved calcium reaching Great Basin lakes precipitates within the lake basin as a carbonate mineral, usually calcite or aragonite. When a lake overflows, its saturation state (with respect to carbonate phases) decreases as the rate of overflow increases; thus, precipitation of $CaCO_3$ occurs infrequently when the climate is very wet.

Glaciation within a lake's watershed alters this simple process because rock flour input to the lake during glacier advances acts to dilute the TIC fraction (Fig. 4). Data presented in Bischoff & Cummins (2001) compares the distribution of Na_2O in the clay-size fraction (an alpine glacial proxy) with the distribution of TIC in Owens Lake sediment (Fig. 5). At the onset of marine oxygen isotope stage (MIS) 5, an abrupt decrease in Na_2O signals the termination of Sierran alpine glaciation. The abrupt increase in TIC may reflect either the return of Owens Lake to hydrologic closure, or it may reflect the cessation of glacial rock flour input to the Owens Lake basin. The glacial proxy (Na_2O) signals the onset of alpine glaciation at 77,000 yr; however, low TIC values do not indicate the establishment of rapid overflow of Owens Lake until 67,000 yr ago.

As summarized by Benson *et al.* (2002), for a closed-basin lake at hydrologic steady state,

$$\delta^{18}O_{lake} = \delta^{18}O_{in} - {}^{18}\alpha_{H_2O(W)\text{-}H_2O(v)}.$$

where $\delta^{18}O_{lake}$ is the steady state $\delta^{18}O$ value of lake water, $\delta^{18}O_{in}$ is the volume-weighted $\delta^{18}O$ value of discharge and on-lake precipitation, and ${}^{18}\alpha_{H_2O(W)\text{-}H_2O(v)}$ is the fractionation factor between lake water and water vapor produced during evaporation. The fractionation factor is a complicated function of air temperature, water temperature, humidity over the lake, wind speed, and the $\delta^{18}O$ value and humidity of advected air that passes over the lake (Benson & White, 1994). During the warm season when evaporation rates are high, a closed-basin desert lake may be assumed to create its own local climate; i.e. it generates the humidity and $\delta^{18}O$ values in the air mass overlying the lake surface (Benson & White, 1994). Under such conditions, variability in the fractionation factor is mostly due to changes in water temperature and amounts to $\sim 0.1‰\,^{\circ}C^{-1}$.

Because climate changes on all time scales, hydrologic and isotopic steady states are never fully achieved. During a wet period that results in rising lake levels, input of isotopically light river water exceeds loss of isotopically light evaporated water. As the volume of a closed-basin lake increases, $\delta^{18}O_{lake}$ decreases; and the faster the volume increase, the greater the rate of decrease in $\delta^{18}O_{lake}$. During a dry period, $\delta^{18}O_{lake}$ increases as the lake shrinks, reflecting the dominance of evaporation on the hydrologic and isotopic balances.

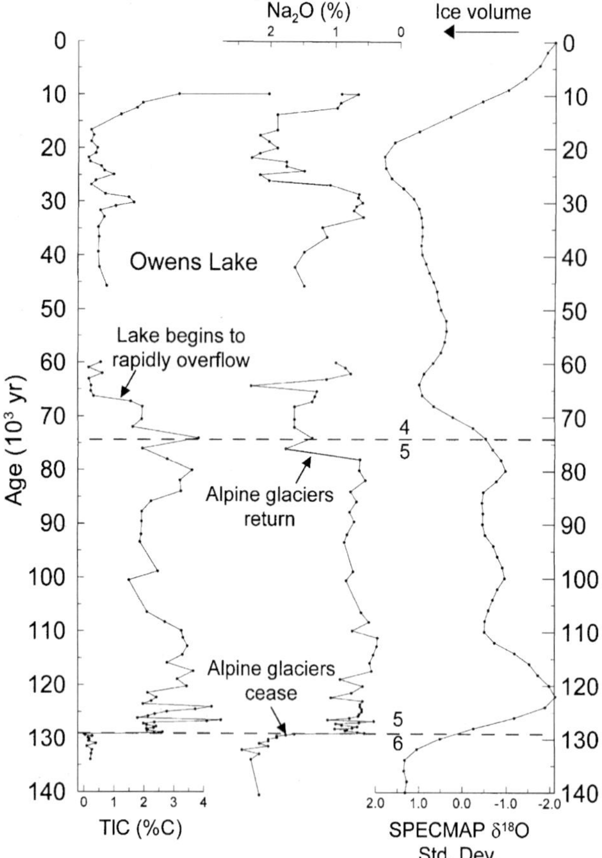

Fig. 5. TIC and Na_2O records from Owens Lake core OL90–2 plotted with the SPECMAP ^{18}O proxy for continental ice volume (Bischoff et al., 1997). Dashed horizontal lines indicate boundaries between marine oxygen isotope stages 4, 5, and 6. Note the general correspondence between low ice-volume values and relatively dry conditions (high TIC values) and the absence of alpine glaciers (low Na_2O values) in the Sierra Nevada during marine oxygen isotope stage 5. Note also that the return of alpine glaciers precedes the overflow of Owens Lake.

When a lake overflows, the $\delta^{18}O$ value of the overflowing lake water, at hydrologic steady state, is proportional to the ratio of the spill (V_{spill}) rate relative to lake volume (V_{lake}). This is because the residence time of water in the spilling lake decreases with increased rate of spill, lessening the effect of evaporation on the $^{18}O/^{16}O$ ratio of the spilling lake. The relationship between $\delta^{18}O_{lake}$ and V_{lake} cannot, therefore, be expressed as an equation of state; i.e. a simple constant relationship between these two parameters does not exist. For example, input of a unit volume of isotopically depleted river water will cause a greater negative shift in $\delta^{18}O_{lake}$ when lake volumes are small than when lake volumes are large.

Abrupt change in the hydrologic balance of a Great Basin lake is not the only process governing the value of $\delta^{18}O_{lake}$ and the $\delta^{18}O$ values of carbonate phases precipitated from lake water. The $\delta^{18}O$ values of precipitation falling in the watershed and on-lake precipitation are functions of condensation air temperature, and the $\delta^{18}O$ value of a carbonate precipitate decreases ~0.2‰ for every 1 °C increase in water temperature (Kim & O'Neil, 1997; O'Neil *et al.*, 1969).

Application of ostracode $\delta^{18}O$ values to lake-based studies of climate change has generally proven successful. Experiments performed by Xia *et al.* (1997a) using *Candona Rawsoni* juveniles showed that the $\delta^{18}O$ values of 15 °C cultures were ~2‰ higher than 25 °C cultures, a difference nearly identical to that expected for inorganic carbonates across a 10 °C temperature range. Recent work by Zeebe (1999) has shown that the oxygen-isotope fractionation factor between water and dissolved carbonate species ($H_2CO_3 + HCO_3^- + CO_3^{2-}$) decreases with increasing pH. If a carbonate solid is formed from a mixture of the carbonate species in proportion to their relative contribution to the total amount of dissolved inorganic carbonate (DIC), the $\delta^{18}O$ value of the solid will also decrease with increasing pH. Keetings *et al.* (2002) have used this concept to explain observed $\delta^{18}O$ enrichments in two species of ostracodes relative to theoretical values obtained using the temperature-based equations of Kim & O'Neil (1997).

The environment in which ostracodes molt, acquiring an isotopic signature, will change with changing lake volume. Deep-water temperatures in Pyramid Lake, Nevada, remain at ~6 °C, whereas surface-water temperatures range from 6 to 24 °C over the annual cycle (Benson, 1994). Some ostracodes prefer to live within a lake's hypolimnion where temperature is relatively stable and cold (Last *et al.*, 1994). If Pyramid Lake were to decrease in volume such that the base of the mixed layer (epilimnion) was at the bottom of Pyramid Lake, ostracodes that formerly inhabited the hypolimnion would experience an 18 °C range of temperature over the annual cycle. If ostracodes were to molt in the warmest season, the $\delta^{18}O$ value of their valves would be 3.6‰ less than the value acquired in the cold hypolimnion of a deep Pyramid Lake. Thus the $\delta^{18}O$ values of ostracode valves may reflect different environments having differing temperature regimes, whereas the $\delta^{18}O$ value of an inorganic carbonate, that precipitates from the mixed layer during the autumn of each year, will largely reflect annual changes in the hydrologic balance of the lake.

The application of ostracode Mg/Ca and Sr/Ca ratios to lake-based studies of climate change has met with some difficulty. Chivas *et al.* (1986) suggested that, for a simple closed-basin lake, ostracode Mg/Ca and Sr/Ca ratios should increase with increasing salinity. In a closed-basin lake, salinity increases with time and with decreasing lake volume. If precipitation of calcite or calcitic ostracodes occurs throughout these salinity increases, Ca is preferentially sequestered (relative to Mg and Sr) in the calcite phases, leading to increasing concentrations of Mg and Sr in the open waters of the closed-basin lake (Rayleigh fractionation). Unfortunately, changes in Mg/Ca and Sr/Ca ratios have sometimes been equated with changes in a lake's hydrologic balance as opposed to changes in its salinity. Changes in salinity can reflect changes in lake volume; however, increases in salinity will also occur over time even if lake volume remains constant.

Most closed-basin lakes are not simple, at least in terms of their chemical evolution. A change in the nature of the inorganic carbonate that precipitates in a lake (e.g. calcite, aragonite, monohydrocalcite, dolomite, strontianite) can affect the Mg/Ca or Sr/Ca ratio of lake water in a step-function manner. In addition, the formation of smectites in the sediments of saline closed-basin lakes can remove large quantities of Mg over time (Jones, 1986). Partitioning coefficients for Mg and Sr with respect to Ca may vary with the chemical composition of the lake. For example, a study of ostracode chemistry in two moderately saline Dakota lakes (Xia *et al.*, 1997b) indicated that the Mg/Ca partitioning coefficient decreased when the Mg/Ca ratio of the lake water reached high values.

A recent example of the application of ostracode minor element and isotopic ratios to the study of western paleolake systems is illustrated in Weimer (1997) who measured Mg/Ca and Sr/Ca ratios in *Limnocythere ceriotuberosa* shells, extracted every 2 cm from the Wilson Creek Formation, Mono Lake basin, California (Fig. 6). Ostracode elemental ratios for the period 40,000 to 24,000 cal yr B.P. have been plotted together with their $\delta^{18}O$ values (Benson, 2002) and the $\delta^{18}O$ values of the TIC fraction (Fig. 6) (Benson *et al.*, 1998a). Both $\delta^{18}O$ records look remarkably similar. The Mg/Ca and $\delta^{18}O$ records have many features in common between 40,000 and 30,000 cal yr B.P. but differ after 30,000 cal yr B.P. The Sr/Ca record looks somewhat similar to the Mg/Ca record but troughs in the Sr/Ca record often lag corresponding features in the Mg/Ca and $\delta^{18}O$ records.

The examples discussed above suggest that ostracode Mg/Ca and Sr/Ca records do not always afford simple interpretations of change in the hydrologic balance of a lake. Ostracode-based $\delta^{18}O$ records can, however, be as reliable as TIC-based $\delta^{18}O$ records.

Geophysical proxies of change in the hydrologic balance. Sediment magnetic properties have been used, albeit infrequently, to infer changes in the hydrologic balance of surface-water systems. In a study of sediments from Buck Lake, Oregon, Rosenbaum *et al.* (1996) used magnetic susceptibility to subdivide cored sediments into four zones. They found that zones of high magnetic susceptibility corresponded to pollen intervals that were indicative of cold, dry environments and that zones of low magnetic susceptibility corresponded to pollen intervals indicative of warm climate. Rosenbaum *et al.* (1996) suggested that magnetic variations arose from changes in peak runoff; i.e. during high flows, fresh volcanic rock fragments containing magnetite were transported to Buck Lake.

Benson *et al.* (2002) showed that magnetic susceptibility of surficial Pyramid Lake sediments decreased with increasing lake depth (distance from shore). They argued that most of the magnetic content resided in ferromagnetic magnetite and that magnetite, being denser than average sediment, was preferentially retained in shallow-water

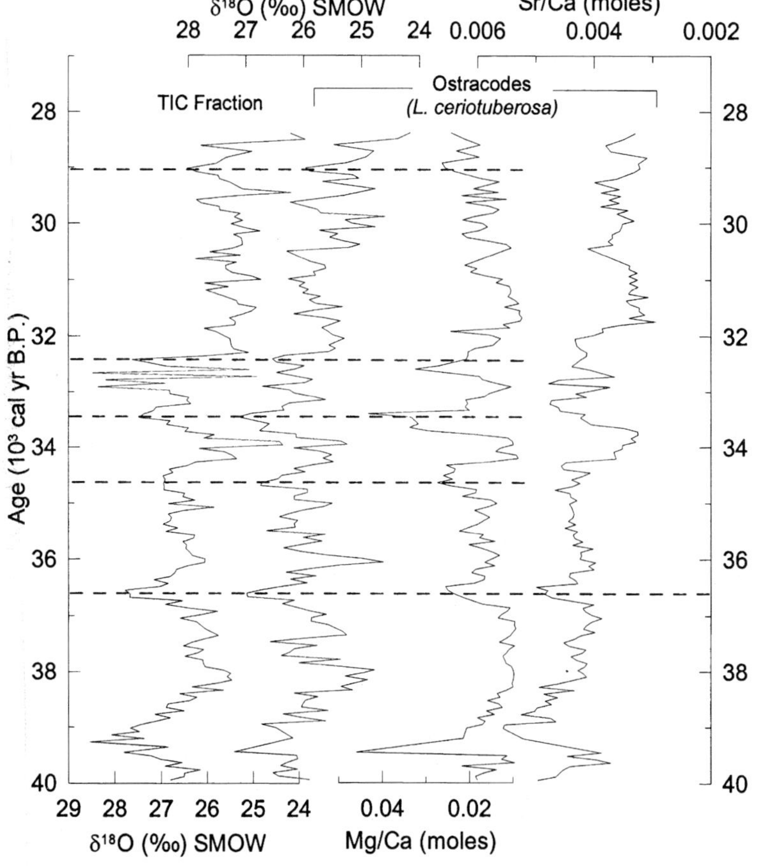

Fig. 6. $\delta^{18}O$ records from the Mono Lake Wilson Creek Formation for TIC (Benson et al., 1998a) and ostracode (unpublished data of L. Benson) fractions between 40,000 and 28,500 cal yr B.P. compared with Mg/Ca and Sr/Ca ratios of ostracode valves (Weimer, 1997). Dashed lines indicates times when three or more records indicate low-lake levels.

environments, whereas relatively non-magnetic clays were preferentially transported to deep-water sites. They applied this concept to Holocene-age cores, interpreting high-susceptibility intervals as indicative of shallow-water environments.

Zic *et al.* (2002) used isothermal remanent magnetization (IRM) to estimate the concentrations of ferromagnetic material in a core from Summer Lake, Oregon. They argued that low IRM values were associated with shallow-lake conditions, in which elevated biological productivity led to anoxic conditions which, in turn, favored the partial dissolution of magnetite grains. This concept was supported by TOC values obtained on samples from the upper part of the core where decreases in TOC (a proxy for productivity) were found to parallel increases in IRM (Zic, 2001).

Development and Application of Numerical Models of the Effect of Climate Change on the Thermal Structure, Hydrologic Balance, and Isotopic Composition of Lakes

One area of advancement since 1965 is the development of numerical models that describe changes in the thermal structure, hydrologic balance, and stable-isotopic composition of western lakes. Until the 1980s, arguments as to what controlled lake size (increased precipitation or decreased evaporation) were either qualitative or empirical in nature (e.g. Leopold, 1951; Mifflin & Wheat, 1979; Snyder & Langbein, 1962).

Benson (1981, 1986) derived a highly parameterized one-dimensional energy-balance model for the determination of the relative importance of temperature, humidity, solar insolation, and cloudiness on lake-surface evaporation rates. He found that change in air temperature had little effect on evaporation rates if water temperature also changed in the same direction, that changes in humidity resulted in small changes in evaporation rate, and that changes in insolation resulted in very small changes in evaporation rate. However, changes in the fractional distribution and type of sky cover led to large changes in evaporation rate.

In terms of processes that affect the hydrologic balance of large lake systems, Benson (1981) suggested that the presence of lake ice could significantly reduce evaporation rates. Benson & Thompson (1987a) further argued that the hydrologic balance of large Pleistocene lakes could be affected by lake-effect snowstorms with precipitation on their eastern margins being triggered by upslope conditions and/or thermal instability.

Benson & Paillet (1989) showed that the proper gage of lake response to change in the hydrologic balance was neither lake depth (level) nor lake volume, but rather surface area. In this study, they applied a hydrologic balance model (FILLSPIL) to the Lahontan basin and showed that lake levels in each of the different subbasins of the Lahontan system responded differently to the same change in hydrologic balance. However, the simulated change in lake surface area in each basin was shown to be proportional to the change in hydrologic balance.

Phillips *et al.* (1986) were the first to introduce a numerical model for simulation of the isotopic evolution of a western closed-basin lake. Several years later, Phillips *et al.* (1992, 1994) applied a variant of this model to approximate the late-Pleistocene surface areas of lakes in the paleo-Owens River system, including Owens, China, Searles, and Manly lakes (Fig. 1). This transient numerical model was based on equations initially presented by Gonfiantini (1965) and differed from the model described in Phillips *et al.* (1986) in being formulated in terms of a time derivative rather than a volume derivative.

Hostetler & Bartlein (1990) developed a one-dimensional model that simulated the thermal structure of and evaporation from a lake. Hostetler & Benson (1990) applied this model to paleo-Lake Lahontan. They showed that the shift to a jet-stream climatology (increased cloudiness and decreased temperature) during the late Pleistocene may have caused a 42% reduction in evaporation from lakes in the Lahontan basin.

Hostetler (1991) coupled a lake-ice model with the Lake Lahontan thermal model. Invoking reductions in mean annual air temperature and solar radiation associated with jet-stream cloud cover, ice cover lasting four months was simulated. In addition, the simulated evaporation rate was shown to be 60% lower than the historic mean.

Hostetler *et al.* (1994) coupled the thermal and lake-ice models to a high-resolution, regional climate model (RCM), nested within a general circulation model (GCM), to study the interaction of Great Basin Pleistocene lakes and the atmosphere. They found that lake-effect precipitation made up a substantial component of the input to Lake Bonneville, supporting the lake-effect and orographic hypotheses of Benson & Thompson (1987a).

Benson & White (1994) developed a model of the isotopic fractionation process that accompanies evaporation. This model differed from the work of Craig & Gorden (1965) in that the Benson & White (1994) model accounted for the effects of external air masses and wind speed on isotopic fractionation. Hostetler & Benson (1994) coupled the Benson & White (1994) fractionation model to the Hostetler & Bartlein (1990) lake-thermal model and successfully simulated the measured $\delta^{18}O$ evolution of Pyramid Lake between 1987 and 1989.

Benson & Paillet (2002) coupled the Benson & White (1994) fractionation model to a two-box hydrologic-balance model and were able to simulate the $\delta^{18}O$ evolution of Pyramid Lake between 1985 and 1994. They also were able to demonstrate the effect of hydrologic closure of Lake Tahoe on the $\delta^{18}O$ value of Pyramid Lake and the effect of step-function changes in the hydrologic balance on the $\delta^{18}O$ value of paleo-Owens Lake.

Application of the above models to western lakes has enabled a better understanding of their thermal, hydrologic, and isotopic responses to variables of climate change (e.g. air temperature, humidity, cloud cover, and wind speed). These models have proved invaluable in understanding (1) the nature of the climate system that led to the creation of pluvial lakes; (2) the structure of lake-atmosphere feedbacks associated with paleolake systems; and (3) the magnitude in lags in volume and $\delta^{18}O$ values of lakes relative to changes in climate forcing.

The Response of Western Lake Research to Shifts in the Prevailing Hypothesis of Climate Forcing

Prior to 1965, many paleoclimate researchers believed that the Pleistocene was a time during which four continental glaciations (Nebraskan, Kansan, Illinoian, and Wisconsin) had occurred in North America (e.g. Frye *et al.*, 1965). To some this concept suggested that four worldwide climate oscillations had occurred during the Quaternary Period, and they found it desirable to subdivide western alpine glaciations (e.g. Richmond, 1965) and lake-level oscillations (Morrison, 1965) into four corresponding periods. The outcrop-based forced-correlation approach to lake-level histories began to crumble when Scott *et al.* (1983) demonstrated that highstand lakes existed in the Bonneville Basin only during MIS 2 and 6, not during MIS 3 and 4 as suggested by Morrison (1965).

The Croll-Milankovitch Hypothesis

The Croll-Milankovitch hypothesis (Milankovitch, 1941) was adopted by researchers in the 1970s to explain the temporal evolution of their paleoclimate records. For example, Wijmstra & Van der Hammen (1974) and Woillard (1979) produced European pollen records that strongly resembled the marine $\delta^{18}O$ record of ice volume. In essence, the Croll-Milankovitch hypothesis states that changes in the Earth's orbital parameters control the distribution of solar insolation, and that past changes in the Earth's heat budget in northern latitudes led to the growth and decay of continental ice sheets.

The effect of changing solar insolation and continental ice volumes on western United States lacustrine climate records was first documented in a pollen record from northern California Adam (1988). Adam demonstrated that oak (*Quercus*) pollen from Clear Lake, California (Fig. 1) exhibited peaks in abundance between 125,000 and 75,000 yr ago, indicating that relatively warm and dry climates occurred frequently during MIS 5 when continental-ice volumes were minimal (Fig. 7C and F).

Heusser & Florer (1973) were the first to demonstrate that pollen deposited in marine continental-margin sediments provided stratigraphically controlled records of terrestrial vegetation. Two oak records from marine cores ODP 1019 and 893A, taken from the continental margin of northern and southern California (Heusser, 1995; Heusser *et al.*, 2000) (Fig. 1), indicated the occurrence of oak-rich assemblages at times during MIS 5 (Fig. 7B and D). On the millennial scale, oscillations of pollen in both lake and marine records do not appear to have occurred synchronously nor do they appear to correlate particularly well with summer insolation (45°N) or ice volume (marine $\delta^{18}O$) records (Fig. 7A and F). However, the pollen records suggest the frequent occurrence of a Mediterranean type of climate (wet winters and dry summers) during MIS 5 at times when continental ice volumes were minimal. More recently, Litwin *et al.* (1999) were able to correlate pollen zones from Owens Lake core OL92 with MIS 1 through 9.

The minimal glacier ice volume and relatively warm summer insolation characteristic of MIS 5 are also reflected in the glacial rock-flour record from Owens Lake, California (Bischoff *et al.*, 1997) (Fig. 1). The Na_2O glacial proxy indicates significantly reduced alpine glaciation occurred during MIS 5 when global ice volume was small and July solar insolation was high (Fig. 7E and F). In shorter, high-resolution records, the effect of solar insolation on alpine glacier advance and retreat has been demonstrated by Benson *et al.* (1998a) and Benson (1999) using TOC as a proxy for glacier activity in outcrop- and core-based studies of the Mono and Owens lake basins, California (Fig. 1).

In the Milankovitch hypothesis, solar insolation has a fundamental impact on the growth and decay of high-latitude continental ice sheets. As mentioned previously, Antevs (1948) suggested that maximum levels of Great Basin lakes were linked to the presence of a permanent ice sheet over North America through its effect on the mean position of the polar jet stream. Interest in this concept was renewed by experiments using atmospheric global climate models (Kutzbach & Guetter, 1986; Kutzbach & Wright, 1985; Manabe & Broccoli, 1985). These experiments demonstrated that glacial-age boundary conditions, including the size and shape of the Laurentide Ice Sheet (LIS) were sufficient to produce the effect that Antevs had hypothesized. Shortly thereafter, Benson & Thompson (1987b) invoked the polar jet-stream hypothesis to explain the rise and fall of Lake Lahontan. The hypothesis has continued to be used to explain the changing paleohydrologies of Great Basin lake systems, including Lake Lahontan, Lake Bonneville, Owens Lake, Mono Lake, Searles Lake, and Summer Lake (Negrini *et al.*, 2000; Oviatt, 1997; Palacios-Fest *et al.*, 1993).

Abrupt Millennial-Scale Climate Change

Impetus for the concept of millennial-scale abrupt change from studies of Greenland ice and marine sediments. During the past decade, high-resolution $\delta^{18}O$ data sets have been obtained from Greenland by the Greenland Ice Core Project (GRIP) (Johnsen *et al.*, 1992, 1997) and by the Greenland Ice Sheet Project 2 (GISP2) (Grootes & Stuiver, 1997). The detailed $\delta^{18}O$ records from these cores permit the recognition of 22 interstades within the last 100,000 yr (Dansgaard *et al.*, 1993). Stadial-interstadial oscillations have become known as Dansgaard-Oeschger (**DO**) events (Fig. 8).

Bond *et al.* (1993) were able to correlate sea-surface temperature (SST) proxy records from two North Atlantic sediment cores with the Greenland $\delta^{18}O$ air-temperature record. They noted that millennial-scale **DO** events were bundled into cooling cycles, each terminated by an abrupt shift from cold to warm temperatures. Exceptionally large discharges of icebergs known as Heinrich events (Bond *et al.*, 1992; Heinrich, 1988) occurred near some of the stadial terminations.

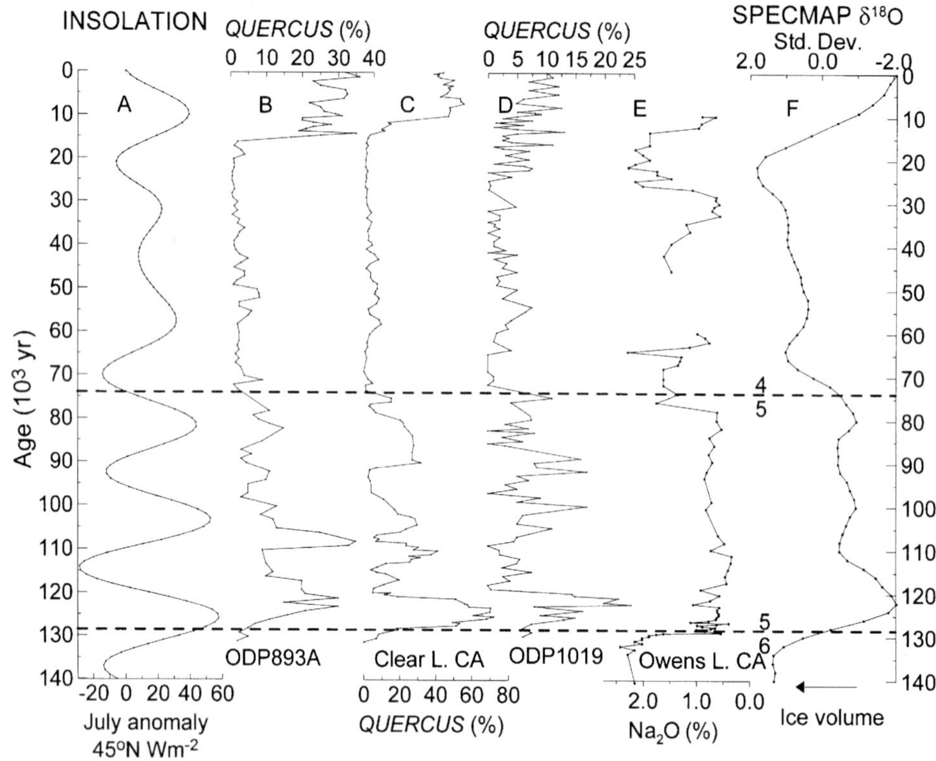

Fig. 7. Climate records from western lakes and eastern Pacific cores (see Fig. 1) compared with summer insolation at 45°N and the $\delta^{18}O$ proxy for continental ice volume. Dashed horizontal lines indicate boundaries between marine isotope stages 4, 5, and 6. (A) Summer insolation record. (B) Oak (Quercus) record from the Santa Barbara basin (Heusser et al., 2000). (C) Oak record from northern California (Adam, 1988). (D) Oak record from the northern California continental shelf (Heusser, 1995). (E) Sierra Nevada glacial proxy record (Na_2O) from Owens Lake, California (Bischoff et al., 1997). (F) SPECMAP marine $\delta^{18}O$ record. A and F adapted from Whitlock & Bartlein (1997). Note that the presence of a large continental ice sheet was accompanied by Sierran alpine glaciation and absence of a Mediterranean-type climate which is characteristic of present-day California.

During the past decade, abrupt millennial- and centennial-scale climate change has become a major focus of climate-oriented researchers. One question that remains concerns the spatial scale over which climate varies coherently on centennial and millennial time scales. Whereas climate change over the entire North Atlantic region appears to have occurred nearly synchronously, some scientists question whether centennial and millennial scales of climate change affected the Northern Hemisphere in a uniform and synchronous manner (e.g. Benson, 1999).

Perhaps the strongest evidence for the hemispheric nature of synchronous **DO** events comes from the work of Kennett and his colleagues. Behl & Kennett (1996) correlated laminated intervals in a Santa Barbara basin core (ODP 893, Fig. 1) with **DO** interstades, and Hendy & Kennett (1999) correlated planktonic $\delta^{18}O$ and assemblage changes in the same core to the GISP2 $\delta^{18}O$ record.

Examples of lake-based records of decadal- to millennial-scale abrupt climate change. Evidence for abrupt millennial-scale climate change in western lakes is common. As noted above, declines of Great Basin lakes from their highstand elevations appear to have occurred abruptly (Fig. 2). In addition, the oscillations recorded in the Owens and Mono lake sediments cores (Figs 3 and 6) are certainly millennial in scale.

Recently, Zic *et al.* (2002) correlated a Summer Lake proxy climate record with the GISP2 $\delta^{18}O$ record, stretching the isothermal remanent magnetization (IRM) record from the Summer Lake B&B core to match Heinrich & Dansgaard-Oeschger features in GISP2. The shape of the stretched Summer Lake IRM record (plotted on a log scale) is remarkable in its similarity to the shape of the GISP2 $\delta^{18}O$ record (Fig. 8). The two records have yet to be unequivocally demonstrated to be in phase, except between **DO-6** and **DO-7** (Fig. 8) where the Mono Lake Excursion (MLE) ties the Summer Lake record to the GISP2 record (Wagner *et al.*, 2000; Zic *et al.*, 2002).

The Summer Lake IRM record (Fig. 8) is primarily an indicator of the amount of magnetite in a sample. Zic *et al.* (2002) argued that low magnetite values occur when Lake Chewaucan (pluvial Summer Lake) was shallow and highly productive, leading to the selective dissolution of magnetite. This argument is supported by data from other cores taken in the Summer Lake basin that show that low IRM values

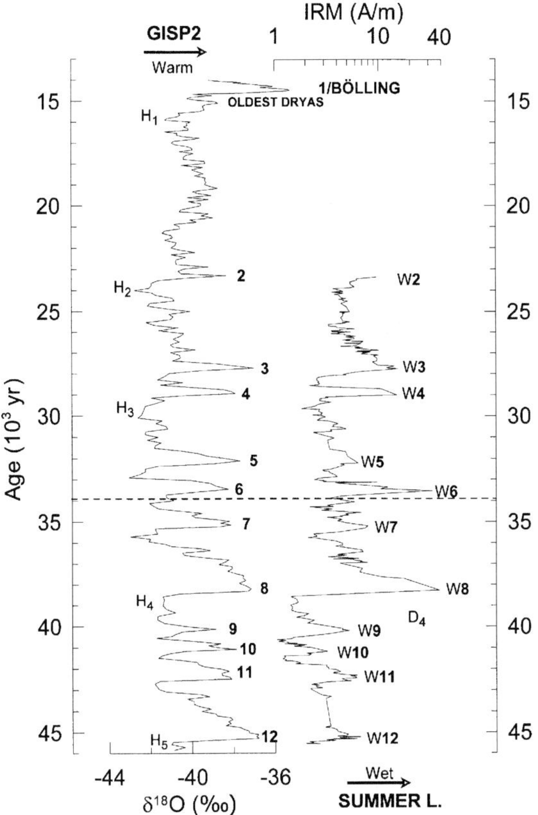

Fig. 8. Graph comparing the Summer Lake IRM record with the GISP2 $\delta^{18}O$ record. Heinrich events are indicated by $\boldsymbol{H_n}$*; Dansgaard-Oeschger events are indicated by a single number in bold; e.g.* **2***; lacustrine wet events are indicated by* **WN***. The location of the Mono Lake excursion (MLE) in the GISP2 and Summer Lake records is indicated by the thin dashed line. See Zic et al. (2002) for details of the age model used to construct this plot.*

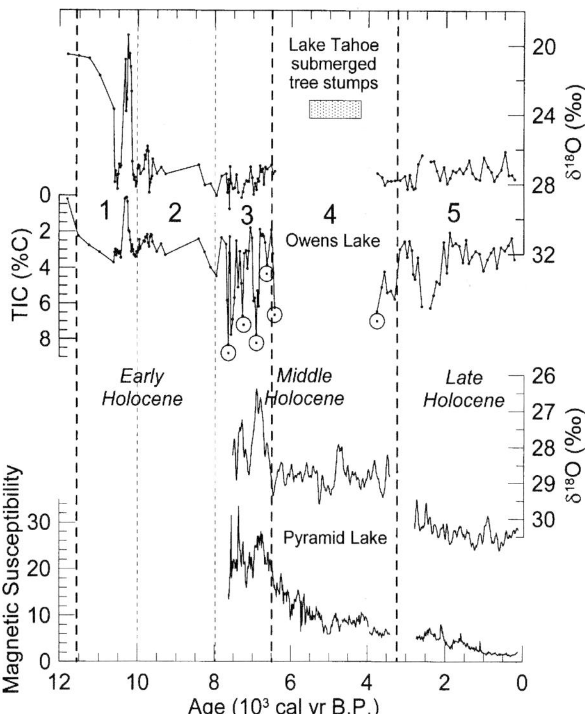

Fig. 9. TIC and $\delta^{18}O$ records from Owens Lake since 11,600 cal yr B.P. and $\delta^{18}O$ and magnetic susceptibility records from Pyramid Lake since 7630 cal yr B.P. Data used in the construction of the records were taken from Benson et al. (2002). The data sets indicate that the Holocene of the western Great Basin can be divided into five distinct climatic intervals (separated by thin and thick dashed lines). Circles with a dot at the center indicate the presence of shallow-water oolites. The rectangle in the upper part of the figure indicates the age range of tree stumps found submerged in Lake Tahoe, California (Lindstrom, 1990).

are associated with high TOC values, the latter indicating an increase in productivity (Negrini *et al.*, 2000). Thus, at least one climate record from the Great Basin tentatively suggests that wet periods in the western Great Basin may have been associated with interstadial warm periods in the North Atlantic region.

High-resolution records of decadal- to centennial-scale change in the hydrologic balance of western lakes have been obtained from both Pyramid and Owens lake basins for the Holocene Epoch (Fig. 9). These records suggest that the Holocene was characterized by five climatic intervals (Benson *et al.*, 2002). TIC and δ^{18}O records from Owens Lake indicate that the first interval (11,600 to 10,000 cal yr B.P.), in the early Holocene, was characterized by a drying trend that was interrupted by a brief (200-yr) wet oscillation centered at 10,300 cal yr B.P. This was followed by a second early-Holocene interval (10,000 to 8000 cal yr B.P.) during which relatively wet conditions prevailed. During the early part of the third Holocene interval (8000 to 6500 cal yr B.P.), high-amplitude oscillations in TIC in Owens Lake and δ^{18}O in Pyramid Lake indicate the presence of shallow lakes that underwent relatively strong oscillations in hydrologic balance. The shallow nature of Owens Lake is further attested to by the presence of shallow-water oolite deposits, and low levels of Pyramid Lake are confirmed by high magnetic susceptibility values (Benson *et al.*, 2002). During the fourth interval (6500 to 3500 cal yr B.P.), drought conditions dominated and Owens Lake desiccated. Pyramid Lake δ^{18}O values indicate that Lake Tahoe was not overflowing to the Truckee River and the existence of trees that grew below the sill level of Lake Tahoe during the fourth interval also implies the existence of a dry period (Lindstrom, 1990) (Fig. 9).

Owens Lake reformed ~3800 cal yr B.P. and experienced a marked increase in volume (negative shift in δ^{18}O) at 3200 cal yr B.P. The δ^{18}O record from Pyramid Lake indicates that Lake Tahoe rose to its sill level between 3200 and 2700 cal yr B.P. Thus the fifth Holocene interval began ~3200 cal yr B.P. Both the Owens and Pyramid lake records are consistent with the estimate of the onset of the late Holocene at ~3200 cal yr B.P.; i.e. the shift in δ^{18}O and leveling of the TIC and magnetic susceptibility records occur

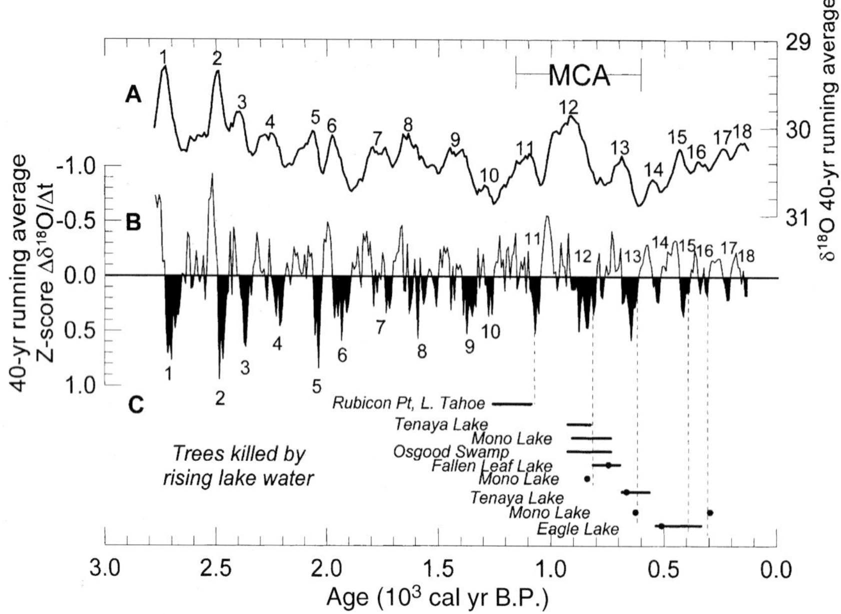

Fig. 10. Comparison of Pyramid Lake $\delta^{18}O$ record (panel A) and its derivative with respect to time (panel B) with ages of tree stumps killed by rising lake waters (panel C). A. 40-yr running average of $\delta^{18}O$ record from core PLC97–1. Minima in $\delta^{18}O$ are numbered 1–18. B. Smoothed derivative of normalized $\delta^{18}O$ record shown in panel A; positive values of the derivative, shown as solid areas below the zero line, indicate times when the lake was falling. Hydrologic droughts are numbered 1 through 18. MCA refers to the time interval occupied by the Medieval Climatic Anomaly (1150–600 cal yr B.P.). C. Ages of tree stumps killed by rising lake waters (Lawrence & Lawrence, 1961; Stine, 1990, 1994). Data for tree stumps at Rubicon Point and Fallen Leaf Lake from Benson et al. (2002). Solid dots indicate best estimates of drought termination. Horizontal lines reflect uncertainty in the calendar-age estimates of drought termination. Dashed vertical lines indicate possible correlation between core-based and stump-based drought records.

at ~3100 ± 200 cal yr B.P. However, Enzel *et al.* (1989) have documented evidence of an earlier playa lake (Silver Lake) in the Mojave Desert at 3916 + 160/−80 cal yr B.P., indicating that climate change may not have occurred synchronously across the entire western United States during the Holocene.

Continuous, high-resolution $\delta^{18}O$ records from Pyramid Lake, Nevada, indicate that, on average, oscillations in the hydrologic balance occurred about every 150 yr during the past 7630 cal yr (Fig. 10). The records are not stationary, and during the past 2740 yr, drought durations ranged from 20 to 100 yr with intervals between droughts ranging from 80 to 230 yr. Many droughts occurring within the last 1000 yr are also documented by the ages of currently submerged stumps of trees that grew at lower elevations when the lakes were shallower (Fig. 10).

In the West, the period between 1150 and 600 cal yr B.P. was generally warm, but neither consistently warm nor consistently dry (Hughes & Diaz, 1994). In the Sierra Nevada and White Mountains of California, this period, which has been termed the Medieval Climatic Anomaly (MCA), was found to contain two intervals of intense drought separated by an intervening wet interval (Leavitt, 1994; Stine, 1994). In the Pyramid Lake record, the MCA interval is punctuated by three droughts separated by two wet periods (Fig. 10), and the two younger droughts (#12 and #13) are equivalent to the intense droughts identified by Stine (1990) in his study of Mono Lake.

Benson *et al.* (2003b) developed a climate record for Mono Lake, California, with nearly annual resolution, for comparison with tree-ring based records of climate change and with Pacific Ocean indices of climate change (e.g. El Niño-Southern Oscillation (ENSO) and the Pacific Decadal Oscillation (PDO)). They first compared a tree-ring-based reconstruction of the PDO index (D'Arrigo *et al.*, 2001) with a coral-based reconstruction of Subtropical South Pacific SST (Linsley *et al.*, 2000), showing that a high degree of correlation existed between the two records for the past 300 yr (Fig. 11C and E). This indicated that the PDO was a pan-Pacific phenomena for at least the past few hundred years.

Benson *et al.* (2003b) also demonstrated that Mono Lake sediments recorded five major oscillations in hydrologic balance between A.D. 1700 and 1941 (Fig. 11A) which could be correlated with tree-ring-based oscillations of the Sierra Nevada snowpack (Fig. 11B and D). Extreme droughts in the records of Sierran wetness occurred during PDO maxima at A.D. 1710, 1770, 1850, and 1930, suggesting that the PDO exerts a strong influence on the climate of the western United States.

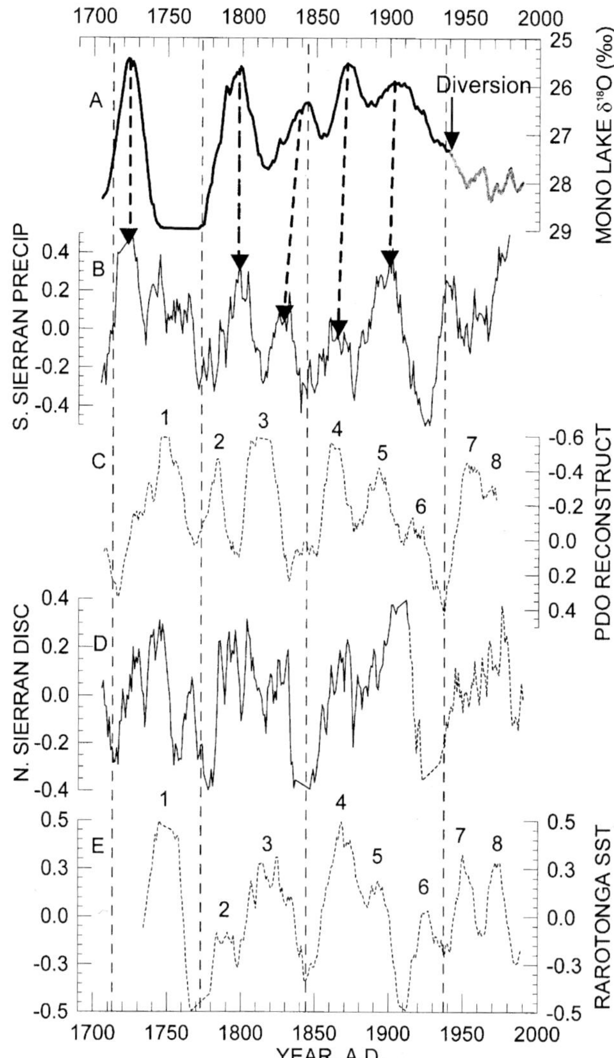

Fig. 11. Comparison of smoothed (15-yr running average) records of (A) Mono Lake $\delta^{18}O$ record (B) reconstructed Southern Sierran precipitation (Graumlich, 1993) (C) reconstructed PDO index (D'Arrigo et al., 2001) (D) reconstructed Northern Sierran discharge (Meko et al., 2001), and (E) detrended reconstructed Subtropical South Pacific SST (Linsley et al., 2000). Detrending was done using a 2nd-degree polynomial. All data sets except the $\delta^{18}O$ record were normalized by subtracting the mean from each value and dividing by the standard deviation. Arrows connect volume maxima of Mono Lake with precipitation maxima in the southern Sierra Nevada. The gray line in (A) indicates Mono Lake $\delta^{18}O$ values after diversion of water out of the Mono Lake basin. Numbers refer to PDO minima and central South Pacific gyre SST maxima. Thin vertical dashed lines connect PDO maxima with Sierran precipitation minima and depressed Subtropical South Pacific SST.

Concluding Remarks

Summary

During the past four decades, advances have been made in lake-based climate studies of the western United States. Sediment coring has allowed access to continuous records of climate change from the deepest parts of lake basins and major advances have been made in terms of age control. Various biological, chemical, and geophysical proxies of climate change have been developed, and the production of reasonably well-dated, high-resolution records of lake-size change and alpine glacier oscillations is now possible. The development of atmospheric, hydrologic, thermal, and isotopic climate models has allowed evaluation of the magnitudes and directions of climate variability and has permitted numerical experimentation with lake-atmosphere feedbacks.

Millennial-scale-resolution records of climate change from western lakes give support to the concept that changes in solar insolation affect Northern Hemisphere climate. Correlation of a Summer Lake centennial-scale climate record with the GISP2 record suggests that the northern Great Basin was anomalously wet during Dansgaard-Oeschger interstades. A centennial-scale Mono Lake climate record indicates that some western lakes may have experience low levels at times of Heinrich events, and centennial-scale climate records from the Owens Lake basin have provided a continuous history of Sierra Nevada glaciation for the past 140,000 yr.

A core-based record of climate change, with less-than-decadal resolution, from Pyramid Lake, Nevada, indicates that droughts of decadal- to centennial-scale occurred frequently in the Great Basin throughout the Holocene. During the mid-Holocene (6500 to 3500 cal yr B.P.), drought conditions caused Lake Tahoe to become hydrologically closed.

A 300-yr annually-resolved record of climate change from Mono Lake, California, indicates five major oscillations in the hydrologic balance that are correlative with tree-ring-based oscillations in Sierra Nevada precipitation. The oscillations in precipitation were found to correspond to changes in the sign of the PDO with extreme droughts occurring during PDO maxima.

Thoughts on the Future of Lake-Based Studies of Climate Change

Lakes are extremely useful in testing global, hemispheric, and regional models of climate forcing, given adequate age control and accurate and precise proxies of climate change. Age control for marine-, ice- and lake-based records is presently inadequate when comparing decadal- to millennial-scale oscillations in climate records. A possible way of solving this problem is to provide a means of correlation over wide geographic areas. Tephrochronology utilizes such an approach, but volcanic eruptions have occurred infrequently over the past 100,000 yr, and the areal distribution of a tephra is often not sufficient to link important climate records.

An approach that shows promise is that of magnetic-field variability. Two magnetic excursions, the Laschamp excursion and Mono Lake excursion, are important markers that have been found in both lake and marine sediments across the Northern Hemisphere (e.g. Denham & Cox, 1971; Levi & Karlin, 1989; Liddicoat *et al.*, 1982; Negrini *et al.*, 1984; Nowaczyk, 1997). The magnetic-intensity lows associated with these excursions have been linked to the GISP2 ice-core record via perturbations in the ^{36}Cl production rate (Wagner *et al.*, 2000).

In order to provide a means of correlation on at least a centennial time scale, a detailed history of changes in the Earth's magnetic field (intensity, inclination, and declination) should be constructed and referenced to phenomena such as Dansgaard-Oeschger oscillations, Heinrich events, meltwater pulses, etc. recorded in marine cores. Some progress in this area has already been made; e.g. Lund (1996) has created a set of Holocene PSV curves for North America, and Stoner *et al.* (1998, 2000) has linked palaeointensity records of North Atlantic Labrador Sea cores to the SPECMAP ^{18}O record. One area of concern that will have to be addressed is whether the magnetic field varies coherently over the entire Northern Hemisphere or whether some perturbations in the field are only regionally manifested.

Interdisciplinary, multiproxy approaches to studies of lake sediments have become common; however, integrated process-based approaches to surface-water systems have yet to be implemented. One of the reasons that marine-based studies have been so instrumental in developing concepts and mechanisms of climate change is that the oceans and atmosphere are linked by transfers of mass and energy. Knowledge regarding a process occurring in one part of the marine system is, therefore, often transferable to another part. Unfortunately, lakes are small and often somewhat unique systems; therefore, knowledge about chemical or biological phenomena in one lake system is often not transferable to another lake system.

For this reason, it is suggested that "archetypal" lakes should be selected as long-term study sites and the processes (chemical, physical, and biological) that influence selected climate indicators should be studied. Such a program would be similar to the Long Term Ecological Research (LTER) sites established by the National Science Foundation. However, the lakes program would be more comprehensive in terms of the types of processes that were selected for study. It would tend to focus on those processes that influence the history of climate proxy indicators as opposed to but not exclusive of those processes that influence the ecology/biology of lakes.

References

Adam, D.P. (1988). Correlations of the Clear Lake, California, core CL-73-4 pollen sequence with other long climate records. *In*: Sims, J.D. (Ed.), *Late Quaternary Climate, Tectonism, and Sedimentation in Clear Lake, Northern California Coast Ranges*. Geological Society of America Special Paper **214**, pp. 81–95.

Adam, D.P., Sarna-Wojcicki, A., Rieck, H.J., Bradbury, J.P., Dean, W.E. & Forester, R.M. (1989). Tulelake, California: The last 3 million years. *Palaeogeography, Palaeoclimatology, Palaeoecology*, **72**, 89–103.

Allen, B.D. & Anderson, R.Y. (2000). A continuous, high-resolution record of late Pleistocene climate variability from the Estancia basin, New Mexico. *Geological Society of America Bulletin*, **112**, 1444–1458.

Antevs, E. (1948). The Great Basin, with emphasis on glacial and post-glacial times-Climatic changes and pre-white man. *Bulletin of the University of Utah Biological Series*, **38**, 168–191.

Bard, E., Arnold, M., Hamelin, B., Tisnerat-Laborde, N. & Cabioch, G. (1998). Radiocarbon calibration by means of mass spectrometric $^{230}TH/^{234}U$ and ^{14}C ages of corals: An updated database including samples from Barbados, Mururoa and Tahiti. *Radiocarbon*, **40**, 1085–1092.

Behl, R.J. & Kennett, J.P. (1996). Brief interstadial events in the Santa Barbara basin, NE Pacific, during the past 60 kyr. *Nature*, **379**, 243–246.

Benson, L.V. (1981). Paleoclimatic significance of lake-level fluctuations in the Lahontan Basin. *Quaternary Research*, **16**, 309–403.

Benson, L.V. (1986). The sensitivity of evaporation rate to climate change–results of an energy balance approach. *In*: *U.S. Geological Survey Water Resources Investigation Report, 86-4148*, 77.

Benson, L.V. (1994). Stable isotopes of oxygen and hydrogen in the Truckee River-Pyramid Lake surface-water system. 1. Data analysis and extraction of paleoclimatic information. *Limnology and Oceanography*, **39**, 344–355.

Benson, L.V. (1999). Records of millennial-scale climate change from the Great Basin of the Western United States. *In*: Clark, P.U., Webb, R.S. & Keigwin, L.D. (Eds), *Mechanisms of Global Climate Change at Millennial Time Scales, American Geophysical Union Monograph 112*, 203–225.

Benson, L.V., Burdett, J.W., Kashgarian, M., Lund, S.P., Phillips, F.M. & Rye, R.O. (1996). Climatic and hydrologic oscillations in the Owens Lake basin and adjacent Sierra Nevada California. *Science*, **274**, 746–749.

Benson, L.V., Burdett, J.W., Lund, S.P., Kashgarian, M. & Mensing, S. (1997b). Nearly synchronous climate change in the Northern Hemisphere during the last glacial termination. *Nature*, **388**, 263–265.

Benson, L.V., Kashgarian, M. & Rubin, M. (1995). Carbonate deposition, Pyramid Lake subbasin, Nevada: 2. Lake levels and polar jet stream positions reconstructed from radiocarbon ages and elevations of carbonate deposits (tufas). *Palaeogeography, Palaeoclimatology, Palaeoecology*, **117**, 1–30.

Benson, L.V., Kashgarian, M., Rye, R.O., Lund, S.P., Paillet, F.L., Smoot, J., Kester, C., Mensing, S., Meko, D. & Lindstrom, S. (2002). Holocene multidecadal and multicentennial droughts affecting northern California and Nevada. *Quaternary Science Reviews*, **21**, 659–682.

Benson, L.V., Liddicoat, J.C., Smoot, S.P., Sarna-Wojcicki, A., Negrini, R.M. & & Lund, S. (2003a). The age of the

Mono Lake excursion. *Quaternary Science Reviews*, **22**, 135–140.

Benson, L.V., Linsley, B., Smoot, J., Mensing, S., Lund, S., Stine, S. & & Sarna-Wojcicki, A. (2003b). Influence of the Pacific Decadal Oscillation (PDO) on the climate of the Sierra Nevada, California and Nevada. *Quaternary Research*, **59**, 151–159.

Benson, L.V., Lund, S.P., Burdett, J.W., Kashgarian, M., Rose, T.P., Smoot, J. & Schwartz, M.D. (1998a). Correlation of late-Pleistocene lake-level oscillations in Mono Lake, California, with North Atlantic climate events. *Quaternary Research*, **49**, 1–10.

Benson, L.V., May, H.M., Antweiler, R.C., Brinton, T.I., Kashgarian, M., Smoot, J. & Lund, S.J. (1998b). Continuous lake-sediment records of glaciation in the Sierra Nevada between 52,600 and 12,500 ^{14}C yr B.P. *Quaternary Research*, **50**, 113–127.

Benson, L.V. & Paillet, F.L. (1989). The use of total lake-surface area as an indicator of climatic change: examples from the Lahontan basin. *Quaternary Research*, **32**, 262–275.

Benson, L.V. & Paillet, F.L. (2002). HIBAL: a hydrologic-isotopic-balance model for application to paleolake systems. *Quaternary Science Reviews*, **21**, 1521–1539.

Benson, L.V., Smoot, J., Kashgarian, M., Sarna-Wojcicki, A. & Burdett, J.W. (1997a). Radiocarbon ages and environments of deposition of the Wono and Trego hot springs tephra layers in the Pyramid Lake subbasin, Nevada. *Quaternary Research*, **47**, 251–260.

Benson, L.V. & Thompson, R.S. (1987a). The physical record of lakes in the Great Basin. *In*: Ruddiman, W.F. & Wright, H.E. (Eds), *North America and adjacent oceans during the last glaciation: Geological Society of America, The Geology of North America*, 241–260.

Benson, L.V. & Thompson, R.S. (1987b). Lake-level variation in the Lahontan Basin for the past 50,000 years. *Quaternary Research*, **28**, 69–85.

Benson, L.V. & White, J.W.C. (1994). Stable isotopes of oxygen and hydrogen in the Truckee River-Pyramid Lake surface-water system. 3. Source of water vapor overlying Pyramid Lake. *Limnology and Oceanography*, **39**, 1945–1958.

Bischoff, J.L., Bullen, T.D., Canavan I.V., Forester, R.M. (1998). A test of uranium-series dating of ostracode shells from the last interglaciation at Owens Lake, California, core OL-92. U.S. Geological Survey Open-File Report 98-132.

Bischoff, J.L. & Cummins, K. (2001). Wisconsin glaciation of the Sierra Nevada (79,000-15,000 yr B.P.) as recorded by rock flour in sediments of Owens Lake, California. *Quaternary Research*, **55**, 14–24.

Bischoff, J.L. & Fitzpatrick, J.A. (1991). U-series dating of impure carbonates: an isochron technique using total-sample dissolution. *Geochimica et Cosmochimica Acta*, **55**, 543–554.

Bischoff, J.L., Menking, K.M., Fitts, J.P. & Fitzpatrick, J.A. (1997). Climatic oscillations 10,000–155,000 yr B.P. at Owens Lake, California reflected in glacial rock flour abundance and lake salinity in Core OL-92. *Quaternary Research*, **48**, 313–325.

Bischoff, J.L., Rosenbauer, R.J. & Smith, G.I. (1985). Uranium-series dating of sediments from Searles Lake, California: differences between continental and marine climate records. *Science*, **227**, 1221–1222.

Bond, G., Broecker, W., Johnsen, S., McManus, J., Labeyrie, L., Jouzel, J. & Bonani, G. (1993). Correlations between climate records from North Atlantic sediments and Greenland ice. *Nature*, **365**, 143–147.

Bond, G., Heinrich, H., Broecker, W., Labeyrie, L., McManus, J., Andrews, J., Huon, S., Jantschik, R., Clasen, S., Simet, C., Tedesco, K., Klas, M., Bonani, G. & Ivy, S. (1992). Evidence for massive discharges of icebergs into the North Atlantic ocean during the last glacial period. *Nature*, **360**, 245–249.

Bradbury, J.P. (1987). Late Holocene diatom paleolimnology of Walker Lake, Nevada. *Archive fur Hydrobiologie*, **79**, 1–27.

Bradbury, J.P. (1992). Late Cenozoic lacustrine and climatic environments at Tule Lake, northern Great Basin USA. *Climate Dynamics*, **6**, 275–285.

Bradbury, J.P. (1997). A diatom record of climate and hydrology for the past 200 ka from Owens Lake, California with comparison to other Great Basin records. *Quaternary Science Reviews*, **16**, 203–219.

Bradbury, J.P., Forester, R.M. & Thompson, R.S. (1989). Late Quaternary paleolimnology of Walker Lake, Nevada. *Journal of Paleolimnology*, **1**, 249–267.

Broecker, W.S. & Kaufman, A. (1965). Radiocarbon chronology of Lake Lahontan and Lake Bonneville II, Great Basin. *Bulletin of the Geological Society of America*, **76**, 537–566.

Broecker, W.S. & Orr, P.C. (1958). Radiocarbon chronology of Lake Lahontan and Lake Bonneville. *Bulletin of the Geological Society of America*, **69**, 1009–1032.

Broecker, W.S. & Walton, A. (1959). The geochemistry of C^{14} in freshwater systems. *Geochimica et Cosmochimica Acta*, **16**, 15–38.

Carter, C. (1997). Ostracodes in Owens Lake core OL-92: alternation of saline and freshwater forms through time. *In*: Smith, G.I. & Bischoff, J.L. (Eds), *An 800,000-year Paleoclimatic Record from Core OL-92, Owens Lake, Southeast California*. Geological Society of America Special Paper **317**, 113–120.

Chivas, A.R., De Decker, P. & Shelley, J.M. (1986). Magnesium and strontium in non-marine ostracod shells as indicators of palaeosalinity and palaeotemperature. *Hydrolbiologia*, **143**, 135–142.

Cohen, A.S., Palacios-Fest, M.R., Negrini, R.M., Wigand, P.E. & Erbes, D.B. (2000). A paleoclimate record for the past 250,000 years from Summer Lake, Oregon, USA, II. sedimentology, paleontology and geochemistry. *Journal of Paleolimnology*, **24**, 151–181.

Cox, A., Doell, R.R. & Dalrymple, G.B. (1965). Quaternary paleomagnetic stratigraphy. *In*: Wright, H.E. & Frey, G.D. (Eds), *The Quaternary of the United States*. Princeton University Press, Princeton, 817–830.

Craig, H. & Gorden, L.I. (1965). Isotopic oceanography: deuterium and oxygen 18 variations in the ocean and marine atmosphere. *In*: Schink, D.R. & Corless, J.T. (Eds), *Marine Geochemistry.* University, Rhode Island, 277–374.

Creer, K.M. & Tucholka, P. (1982). Construction of type curves of geomagnetic secular variation for dating lake sediments from east central North America. *Canadian Journal of Earth Sciences*, **19**, 1106–1115.

Creer, K.M. & Tucholka, P. (1983). On the current state of lake sediment paleomagnetic research. *Geophysical Journal of the Royal Astronomical Society*, **74**, 223–238.

Dansgaard, W., Johnsen, S., Clausen, H.B., Dahl-Jensen, D., Gundestrup, N.S., Hammer, C.U., Hvidberg, C.S., Steffensen, J.P., Sveinbjornsdottir, A.E., Jouzel, J. & Bond, G. (1993). Evidence for general instability of past climate from a 250-kyr ice-core record. *Nature*, **364**, 218–220.

D'Arrigo, Villalba, R. & Wiles, G. (2001). Tree-ring estimates of Pacific decadal climate variability. *Climate Dynamics*, **18**, 219–224.

Davis, J.O. (1978). Quaternary tephrochronology of the Lake Lahontan area, Nevada and California. *Nevada Archeological Survey Research Paper no. 7*, 137.

Davis, J.O. (1983). Level of Lake Lahontan during deposition of the Trego Hot Springs tephra about 22,400 years ago. *Quaternary Research*, **19**, 312–324.

Davis, O.K. & Moutoux, T.E. (1998). Tertiary and Quaternary vegetation history of the Great Salt Lake, Utah, USA. *Journal of Paleolimnology*, **19**, 417–427.

Denham, C.R. & Cox, A. (1971). Evidence that the La schamp polarity event did not occur 13,300–30,400 years ago. *Earth Planetary Science Letters*, **13**, 181–190.

Eardley, A.J. & Gvosdetsky, V. (1960). Analysis of Pleistocene core from Great Salt Lake, UT. *Bulletin of the Geological Society of America*, **71**, 1323–1344.

Emiliani, C. & Geiss. (1957). On glaciations and their causes. *Geologische Rundschau*, **46**, 576.

Enzel, Y., Cayan, D.R., Anderson, R.Y. & Wells, S.G. (1989). Atmospheric circulation during Holocene lake stands in the Mojave Desert: evidence of regional climate change. *Nature*, **341**, 44–48.

Flint, R.F. & Gale, W.A. (1958). Stratigraphy and radiocarbon dates at Searles Lake, California. *American Journal of Science*, **256**, 689–714.

Frye, J.C., Willman, H.B. & Black, R.F. (1965). Outline of glacial geology of Illinois and Wisconsin. *In*: Wright, H.E. & Frey, D.G. (Eds), *The Quaternary of the United States* Princeton University Press, Princeton, 43–62.

Gonfiantini, R. (1965). Effetti isotopici nell'evaporazione di acque salate. Atti della Societa Toscana di Scienze Naturali Residente in Pisa, Memorie, Processi Verbali, Serie A **72**, 550–569.

Graumlich, L.J. (1993). A 1000-year record of temperature and precipitation in the Sierra Nevada. *Quaternary Research*, **39**, 249–255.

Grootes, P.M. & Stuiver, M. (1997). Oxygen 18/16 variability in Greenland snow and ice with 10^{-3}- to 10^{5}-year time resolution. *Journal of Geophysical Research*, **102**, 26455–26470.

Heinrich, H. (1988). Origin and consequences of cyclic ice rafting in the Northeast Atlantic Ocean during the past 130,000 years. *Quaternary Research*, **29**, 142–152.

Hendy, I.L. & Kennett, J.P. (1999). Latest Quaternary North Pacific surface-water responses imply atmosphere-driven climate instability. *Geology*, **27**, 291–294.

Heusser, C.J. & Florer, L.E. (1973). Correlation of marine and continental Quaternary pollen records from the Northeast Pacific and western Washington. *Quaternary Research*, **3**, 661–670.

Heusser, L. (1995). Pollen stratigraphy and paleoecologic interpretation of the 160-K.Y. record from Santa Barbara Basin, Hole 893A. *In*: Kennett, J.P., Baldaul, J. & Lyle, M. (Eds), *Proceedings of the Ocean Drilling Program Scientific Results*, 265–279.

Heusser, L., Lyle, M. & Mix, A. (2000). Vegetation and climate of the northwest coast of North America during the last 500K.Y.: high-resolution pollen evidence from the Northern California margin. *In*: Lyle, M., Koizumi, I., Richter, C. & Moore, T.C. (Eds), *Proceedings of the Ocean Drilling Program Scientific Results*, **167**, 217–226.

Heyvaert, A.C., Reuter, J.E., Slotton, D.G. & Goldman, C.R. (2000). Paleolimnological reconstruction of historical atmospheric lead and mercury deposition at Lake Tahoe, California-Nevada. *Environmental. Science and Technology*, **34**, 3588.

Hooke, R.L. (1999). Lake Manly(?) shorelines in the eastern Mojave Desert, California. *Quaternary Research*, **52**, 328–336.

Hostetler, S.W. (1991). Simulation of lake ice and its effect on the late-Pleistocene evaporation rate of Lake Lahontan. *Climate Dynamics*, **6**, 43–48.

Hostetler, S.W. & Bartlein, P.J. (1990). Simulation of lake evaporation with application to modeling lake-level variations at Harney-Malheur Lake. *Water Resources Research*, **26**, 2603–2612.

Hostetler, S.W. & Benson, L.V. (1990). Paleoclimatic implications of the high stand of Lake Lahontan derived from models of evaporation and lake level. *Climate Dynamics*, **4**, 207–217.

Hostetler, S.W. & Benson, L.V. (1994). Stable isotopes of oxygen and hydrogen in the Truckee River-Pyramid Lake surface-water system. 2. A predictive model of $\delta^{18}O$ and $\delta^{2}H$ in Pyramid Lake. *Limnology and Oceanography*, **39**, 356–364.

Hostetler, S.W., Giorgi, F., Bates, G.T. & Bartlein (1994). Lake-atmosphere feedbacks associated with paleolakes Bonneville and Lahontan. *Science*, **263**, 665–668.

Hughen, K.A., Overpeck, J.T., Lehman, S.J., Kashgarian, M., Southon, J.R. & Peterson, L.C. (1998). A new ^{14}C calibration data set for the last deglaciation based on marine varves. *Radiocarbon*, **40**, 483–494.

Hughes, M.K. & Diaz, H.F. (1994). Was there a "Medieval Warm Period", and if so, where and when? *Climatic Change*, **26**, 109–142.

Jannik, N.O., Phillips, F.M., Smith, G.I. & Elmore, D. (1991). A ^{36}Cl chronology of lacustrine sedimentation in the Pleistocene Owens River system. *Geological Society of America Bulletin*, **103**, 1146–1159.

Jellison, R. (1996). Organic matter accumulation in sediments of hypersaline Mono Lake during a period of changing salinity. *Limnology, Oceanography*, **41**, 1539–1544.

Johnsen, S., Clausen, H.B., Dansgaard, W., Fuhrer, K., Gundestrup, N.S., Hammer, C.U., Iversen, P., Jouzel, J., Stauffer, B. & Steffensen, J.P. (1992). Irregular glacial interstadials recorded in a new Greenland ice core. *Nature*, **359**, 311–313.

Johnsen, S., Clausen, H.B., Dansgaard, W., Gundestrup, N.S., Hammer, C.U., Andersen, U., Andersen, K.K., Hvidberg, C.S., Dahl-Jensen, D., Steffensen, J.P., Shoji, H., Sveinbjornsdottir, A.E., White, J.W.C., Jouzel, J. & Fisher, D. (1997). The $\delta^{18}O$ record along the Greeland Ice Core Project deep ice core and the problem of possible Eemian climate instability. *Journal of Geophysical Research*, **102**, 26397–26410.

Jones, B.F. (1986). Clay mineral diagenesis in lacustrine sediments. *In*: Mumpton, F. (Ed.), *Studies in Diagenesis*. U.S. Geological Survey Bulletin **1578**, 291–300.

Kaufman, A. & Broecker, W.S. (1965). Comparison of ^{230}Th and ^{14}C ages for carbonate materials from Lake Lahontan and Bonneville. *Journal of Geophysical Research*, **70**, 4039–4054.

Keetings, K.W., Heaton, T.H. & Holmes, J.A. (2002). Carbon and oxygen isotope fractionation in non-marine ostracodes: Results from a 'natural culture' environment. *Geochimica et Cosmochimica acta*, **66**, 1701–1711.

Kim, S.-T. & O'Neil, J. (1997). Equilibrium and nonequilibrium oxygen isotope effects in synthetic carbonates. *Geochimica et Cosmochimica acta*, **61**, 3461–3475.

Kitagawa, H. & van der Plicht, J. (1998). A 40,000-year varve chronology from Lake Suigetsu, Japan: extension of the ^{14}C calibration curve. *Radiocarbon*, **40**, 505–515.

Kowalewska, A. & Cohen, A.S. (1998). Reconstruction of palaeoenvironments of the Great Salt Lake Basin during the late Cenozoic. *Journal of Paleolimnology*, **20**, 381–407.

Ku, T-L., Luo, S., Lowenstein, T.K., Li, J. & Spencer, R.J. (1998). U-series chronology of lacustrine deposits in Death Valley, California. *Quaternary Research*, **50**, 261–275.

Kutzbach, J.E. & Guetter, P.J. (1986). The influence of changing orbital parameters and surface boundary conditions of climate simulations for the past 18,000 years. *Journal of Atmospheric Science*, **43**, 1726–1759.

Kutzbach, J.E. & Wright, H.E. (1985). Simulation of the climate of 18,000 yr B.P.: results for the North American/North Atlantic/European sector. *Quaternary Science Reviews*, **4**, 147–187.

Lao, Y. & Benson, L.V. (1988). Uranium-series age estimates and paleoclimatic significance of Pleistocene tufas from the Lahontan basin, California and Nevada. *Quaternary Research*, **30**, 165–176.

Last, W.M., Teller, J.T. & Forester, R.M. (1994). Paleohydrology and paleochemistry of lake Manitoba, Canada: the isotope and ostracode records. *Journal of Paleolimnology*, **12**, 269–282.

Lawrence, D.B. & Lawrence, E.G. (1961). Response of enclosed lakes to current glaciopluvial climatic conditions in middle latitude Western North America. *Annals of the New York Academy of Science*, **95**, 341.

Leavitt, S.W. (1994). Major wet interval in White mountains Medieval Warm Period evidenced in ^{13}C bristlecone pine tree rings. *Climate Change*, **26**, 299–308.

Leopold, L.H. (1951). Pleistocene climate in New Mexico. *American Journal of Science*, **249**, 152–168.

Levi, S. & Karlin, R. (1989). A sixty thousand year paleomagnetic record from the Gulf of California sediments: Secular variation late Quaternary excursions and geomagnetic implications. *Earth Planetary Science Letters*, **92**, 219–233.

Li, H.C., Bischoff, J.L., Ku, T., Lund, S.P. & Stott, L.D. (2000). Climate variability in east-central California during the past 1000 years reflected by high-resolution geochemical and isotopic records from Owens Lake sediments. *Quaternary Research*, **54**, 189–197.

Li, J., Lowenstein, T.K., Brown, C.B., Ku, T.L. & Luo, S. (1996). A 100ka record of water tables and paleoclimates from salt cores, Death Valley, California. *Palaeoecology Palaeogeography, Palaeoclimatology*, **123**, 179–203.

Liddicoat, J.C. (1992). Mono Lake excursion in Mono Basin, California, and at Carson Sink and Pyramid Lake, Nevada. *Geophysical Journal International*, **108**, 442–452.

Liddicoat, J.C. (1996). Mono Lake excursion in the Lahontan Basin, Nevada. *Geophysical Journal International*, **125**, 630–635.

Liddicoat, J.C. & Coe, R.S. (1979). Mono Lake geomagnetic excursion. *Journal of Geophysical Research*, **84**, 261–271.

Liddicoat, J.C., Lajoie, K.R., and Sarna-Wojcicki, A. (1982). Detection and dating of the Mono Lake excursion in the Lake Lahontan Sehoo Formation, Carson Sink Nevada. EOS, *Transactions of the American Geophysical Union*, **63**, 920.

Liddicoat, J.C. & Coe, R.S. (1997). Paleomagnetic investigation of Lake Lahontan sediments and its application for dating pluvial events in the northwestern Great Basin. *Quaternary Research*, **47**, 45–53.

Liddicoat, J.C. & Coe, R.S. (1998). Paleomagnetic investigation of the Bonneville Alloformation, Lake Bonneville, Utah. *Quaternary Research*, **50**, 214–220.

Lin, J.C., Broecker, W.S., Anderson, R.F., Hemming, S.R., Rubenstone, J. & Bonani, G. (1996). New ^{230}Th/U and ^{14}C ages from Lake Lahontan carbonates, Nevada, USA, a discussion of the origin of initial thorium. *Geochimica et Cosmochimica Acta*, **60**, 2817–2832.

Lin, J.C., Broecker, W.S., Hemming, S.R., Hajdas, I., Anderson, R.F., Smith, G.I., Kelley, M. & Bonani, G. (1998). A reassessment of U-Th and ^{14}C ages for late-glacial high-frequency hydrological events at Searles Lake, California. *Quaternary Research*, **49**, 11–23.

Linsley, B.K., Wellington, G.M. & Schrag, D.P. (2000). Decadal sea surface temperature variability in the Subtropical South Pacific from 1726 to 1997 A.D. *Science*, **290**, 1145–1148.

Lindstrom, S. (1990). Submerged tree stumps as indicators of Mid-Holocene aridity in the Lake Tahoe Basin.

Journal of California and Great Basin Anthropology, **12**, 146–157.

Litwin, R.J., Smoot, J.P., Durika, N.J. & Smith, G.I. (1999). Calibrating Late Quaternary terrestrial climate signals: radiometrically dated pollen evidence from the southern Sierra Nevada, USA. *Quaternary Science Reviews*, **18**, 1151–1171.

Long, A., Davis, O.K. & De Lanois, J. (1992). Separation and ^{14}C dating of pure pollen from lake sediments: nanofossil AMS dating. *Radiocarbon*, **34**, 557–560.

Lowenstein, T.K., Li, J., Brown, C., Roberts, S.M., Ku, T.-L., Luo, S. & Yang, W. (1999). 200 k.y. paleoclimate record from Death Valley salt core. *Geology*, **27**, 3–6.

Lund, S.P. (1996). A comparison of Holocene paleomagnetic secular variation records from North America. *Journal of Geophysical Research*, **101**, 8007–8024.

Lund, S.P. & Banerjee, S.K. (1979). Paleosecular variations from lake sediments. *Reviews of Geophysics and Space Physics*, **17**, 244–249.

Lund, S.P. & Banerjee, S.K. (1985a). Late quaternary paleomagnetic field secular variation from two Minnesota lakes. *Journal of Geophysical Research*, **90**, 803–825.

Lund, S.P. & Banerjee, S.K. (1985b). The paleomagnetic record of late Quaternary secular variation from Anderson Pond, Tennessee. *Earth and Planetary Science Letters*, **72**, 219–237.

Lund, S.P., Liddicoat, J.C., Lajoie, K.R., Henyey, T.L. & Robinson, S.W. (1988). Paleomagnetic evidence for long-term (10^4 year) memory and periodic behavior in the Earth's core dynamo process. *Geophysical Research Letters*, **15**, 1101–1104.

Luo, S. & Ku, T.-H. (1991). U-series dating: A generalized method employing total-sample dissolution. *Geochimica et Cosmochimica Acta*, **55**, 555–564.

Manabe, S. & Broccoli, A.J. (1985). The influence of continental ice sheets on the climate of an ice age. *Journal of Geophysical Research*, **90**, 2167–2190.

Markgraf, V., Bradbury, J.P., Forester, R.M., Singh, G. & Sternberg, R.S. (1984). San Agustin Plains, New Mexico age and paleoenvironmental potential reassessed. *Quaternary Research*, **22**, 336–343.

Meko, D.M., Therrell, M.D., Baisan, C.H. & Hughes, M.K. (2001). Sacramento River flow reconstructed to A.D. 869 from tree rings. *Journal American Water Resources Association*, **37**, 1029–1040.

Mensing, S.A. & Southon, J.R. (1999). A simple method to separate pollen for AMS radiocarbon dating and its application to lacustrine and marine sediments. *Radiocarbon*, **41**, 1–8.

Mifflin, M.D. & Wheat, M.M. (1979). Pluvial lakes and estimated pluvial climates of Nevada. *Nevada Bureau of Mines and Geology Bulletin*, **94**, 57.

Milankovitch, M.M. (1941). Canon of insolation and the ice-age problem: *In*: *Koniglich Servische Akademie, Beograd*. U.S. Department of Commerce and the National Science Foundation, Washington D.C.

Morrison, R.B. (1965). Quaternary geology of the Great Basin. *In*: Wright, H.E. & Frey D.G. (Eds), *The Quaternary of the United States*. Princeton University Press, Princeton, 265–286.

Negrini, R.M., Davis, J.O. & Verosub, K.L. (1984). Mono Lake geomagnetic excursion found at Summer Lake, Oregon. *Geology*, **12**, 464–643.

Negrini, R.M., Erbes, D.B., Faber, K., Herrera, A.M., Robers, A.P., Cohen, A.S., Wigand, P.E. & Foit, F.F. (2000). A paleoclimate record for the past 250,000 years from Summer Lake, Oregon, USA: I. Chronology and magnetic proxies for lake level. *Journal of Paleolimnology*, **24**, 125–149.

Nowaczyk, N.R. (1997). High-resolution magnetostratigraphy of four sediment cores from the Greenland Sea–II. Rock magnetic and relative palaeointensity data. *Geophysical Journal International*, **131**, 325–334.

O'Neil, J.R., Clayton, R.N. & Mayeda, T.K. (1969). Oxygen isotope fractionation in divalent metal carbonates. *Journal of Chemical Physics*, **50**, 5547–5558.

Oviatt, C.G. (1997). Lake Bonneville fluctuations and global climate change. *Geology*, **25**, 155–158.

Oviatt, C.G., Currey, D.R. & Sack, D. (1992). Radiocarbon chronology of Lake Bonneville, Eastern Great Basin, USA. *Palaeogeography, Palaeoclimatology Palaeoecology*, **99**, 225–241.

Palacios-Fest, M.R., Cohen, A.S., Ruiz, J. & Blank, B. (1993). Comparative paleoclimatic interpretations from nonmarine ostracodes using faunal assemblages, trace elements shell chemistry and stable isotope data. *In*: Swart, P.K., Lohman, K.C., McKenzie, J.A. & Savin, W. (Eds), *Climate Change in Continental Isotopic Records*. Geophysical Monograph, **78**, 179–190.

Peng, T.H., Goddard, J.G. & Broecker, W.S. (1978). A direct comparison of ^{14}C and ^{230}Th ages at Searles Lake California. *Quaternary Research*, **9**, 319–329.

Phillips, F.M., Campbell, A.R., Kruger, C., Johnson, P., Roberts, R.G. & Keyes, E. (1992). A reconstruction of the response of the water balance in western United States lake basins to climatic change. New Mexico Water Resources Research Institute Report **269**.

Phillips, F.M., Campbell, A.R., Smith, G.I. & Bischoff, J.L. (1994). Interstadial climatic cycles: a link between western North America and Greenland? *Geology*, **22**, 1115–1118.

Phillips, F.M., Person, M.A. & Muller, A.B. (1986). A numerical model for simulating the isotopic evolution of closed-basin lakes. *Journal of Hydrology*, **85**, 73–86.

Phillips, F.M., Zreda, M.G., Benson, L.V., Plummer, M.A., Elmore, D. & Sharma, P. (1996). Chronology for fluctuations in Late Pleistocene Sierra Nevada glaciers and lakes. *Science*, **274**, 749–751.

Reasoner, M.A. & Jodry, M.A. (2000). Rapid response of alpine timberline vegetation to the Younger Dryas climate oscillation in the Colorado Rocky Mountains, USA. *Geology*, **28**, 51–54.

Richmond, G.M. (1965). Glaciation of the Rocky Mountains. *In*: Wright, H.E. & Frey, D.G. (Eds), *The Quaternary of the United States*. Princeton University Press, Princeton, 217–230.

Rosenbaum, J., Reynolds, R., Adam, D.P., Drexler, J., Sarna-Wojcicki, A.M. & Whitney, G.C. (1996). Record of middle Pleistocene climate change from Buck Lake, Cascade Range, southern Oregon–Evidence from sediment magnetism, trace-element geochemistry, and pollen. *Bulletin of the Geological Society of America*, **108**, 1328–1341.

Sarna-Wojcicki, A. (2000). Tephrochronology. *In*: *Quaternary Geochronology: Methods and Applications*. American Geophysical Union Reference Shelf 4, 357–377.

Sarna-Wojcicki, A., Meyer, C.E., Adam, D.P. & Sims, J.D. (1988). Correlations and age estimates of ash beds in late Quaternary sediments of Clear Lake, California. *Geological Society of America Special Paper*, **214**, 141.

Schramm, A., Stein, M. & Goldstein, S.L. (2000). Calibration of the ^{14}C time scale to >40 ka by ^{234}U-^{230}Th dating of Lake Lisan sediments (last glacial Dead Sea. *Earth and Planetary Science Letters*, **175**, 27–40.

Scott, W.E., McCoy, W.D., Shroba, R.R. & Rubin, M. (1983). Reinterpretation of the exposed record of the last two cycles of Lake Bonneville, Western United States. *Quaternary Research*, **20**, 185–261.

Sims, J.D., Rymer, M.J. & Perkins, J.A. (1988). Late Quaternary deposits beneath Clear Lake, California; physical stratigraphy, age, and paleogeographic implications. *In*: Sims, J.D. (Ed.), *Late Quaternary Climate, Tectonism, and Sedimentation in Clear Lake, Northern California Coast Ranges*. Geological Society of America Special paper **214**, 21–44.

Smith, G.I. (1962). Subsurface stratigraphy of late Quaternary deposits, Searles Lake, California, a summary.U.S. *Geological Survey Professional Paper*, **450-C**, 65–69.

Smith, G.I. (1984). Paleohydrologic regimes in the southwestern Great Basin, 0–3.2 my Ago, compared with other long records of "global" climate. *Quaternary Research*, **22**, 1–17.

Smith, G.I. & Bischoff, J.L. (1997). Core OL-92 from Owens Lake: project rationale, geologic setting, drilling procedures, and summary. *In*: Smith, G.I. & Bischoff, J.L. (Eds), *An 800,000 Year Paleoclimatic Record from Core OL-92, Owens Lake, Southeast California Geological Society of America Special Paper* **317**, 1–8.

Snyder, C.T. & Langbein, W.B. (1962). The Pleistocene lake in Spring Valley, Nevada, and its climatic implications. *Journal of Geophysical Research*, **67**, 2385–2394.

Southon, J.R., Nelson, D.E., Korteling, R., Nowikow, I., Hammaren, E., McKay, J. & Burke, D. (1982). Techniques for the direct measurement of natural beryllium-10 and carbon-14 with a tandem accelerator. *In*: Currie, L.A. (Ed.), *Nuclear and Chemical Dating Techniques*. American Chemical Society, Washington, DC, 75–88.

Spencer, R.J., Baedecker, M.J., Eugster, H.P., Forester, R.M., Goldhaber, M.B., Jones, B.F., Kelts, K., McKenzie, J., Madsen, D.B., Rettig, S.L., Rubin, M. & Bowser, C.J. (1984). Great Salt Lake and precursors, Utah; the last 30,000 years. *Contributions to Mineralogy Petrology*, **86**, 321–334.

Stine, S. (1990). Late Holocene fluctuations of Mono Lake, eastern California. In: Meyers, P.A. & Benson, L.V. (Eds), *Special Issue Paleoclimates: the record form lakes, ocean and land*. Palaeogeography, Palaeoclimatology, Palaeoecology, **78**, 333–381.

Stine, S. (1994). Extreme and persistent drought in California and Patagonia during mediaeval time. *Nature*, **369**, 546–549.

Stoner, J.S., Channell, J.E.T. & Hillaire-Marcel (1998). A 200 ka geomagnetic chronostratigraphy for the Labrador Sea: Indirect correlation of the sediment record to SPECMAP. *Earth and Science Planetary Letters*, **159**, 165–181.

Stoner, J.S., Channell, J.E.T., Hillaire-Marcel & Kissel, C. (2000). Geomagnetic palaeointensity and environmental record from Labrador Sea core MD95-2024: global marine sediment and ice core chronostratigraphy for the last 110 kyr. *Earth and Science Planetary Letters*, **183**, 1651–177.

Stuiver, M. (1964). Carbon isotopic distribution and correlated chronology of Searles Lake sediments. *American Journal of Science*, **262**, 377–392.

Stuiver, M., Reimer, P.J. & Braziunas, T.F. (1998). High-precision radiocarbon age calibration for terrestrial and marine samples. *Radiocarbon*, **40**, 1127–1151.

Szabo, B.J., Bush, C.A. & Benson, L.V. (1996). Uranium-series dating of carbonate (tufa) deposits associated with Quaternary fluctuations of Pyramid Lake, Nevada. *Quaternary Research*, **45**, 271–281.

Thompson, R.S. (1996). Pliocene and early Pleistocene environments and climates of the western Snake River Plain, Idaho. *Marine Micropaleontology*, **27**, 141–156.

Verosub, K.L. & Mehringer, P.J. (1984). Congruent paleomagnetic and archeomagnetic records from the western United States: A.D. 750 to 1450. *Science*, **224**, 387–389.

Verosub, K.L., Mehringer, P.J. & Waterstraat, P. (1986). Holocene secular variation in western North America: paleomagnetic record from Fish Lake, Harney County, Oregon. *Journal of Geophysical Research*, **91**, 3609–3623.

Voelker, A.H., Sarnthein, M., Grootes, P.M., Erlenkeuser, H., Laj, C., Mazaud, A., Nadeau, M. & Schleicher, M. (1998). Correlation of marine ^{14}C ages from the Nordic seas with the GISP2 isotope record: implications for ^{14}C.calibration beyond 25 ka BP. *Radiocarbon*, **40**, 517–534.

Verosub, K.L. (1979). Paleomagnetism of carved sediments from western New England: variability of the paleomagnetic recorder. *Geophysical Research Letters*, **6**, 241–244.

Wagner, G., Beer, J., Laj, C., Kissel, C., Masarik, J., Muschele, R. & Synal, H. (2000). Chlorine-36 evidence for the Mono Lake event in the Summit GRIP ice core. *Earth and Planetary Science Letters*, **181**, 1–6.

Weimer, M.B. (1997). An examination of solute evolution and lake level fluctuations of Mono Lake, California, using minor element and oxygen isotope chemistries of the ostracod limnocythere ceriotuberosa. Unpublished Master of Science thesis, University of Southern Colorado.

Wells, S.G., McFadden, L.D. & Dohrenwend, J.C. (1987). Influence of late Quaternary climatic changes on geomorphic and pedogenic processes on a desert piedmont, Eastern Mojave Desert, California. *Quaternary Research*, **27**, 130–146.

Whitlock, C. & Bartlein, P.J. (1997). Vegetation and climate change in northwest America during the past 125 kyr. *Nature*, **388**, 57–61.

Whitlock, C., Sarna-Wojcicki, A., Bartlein, P.J. & Nickmann, R.J. (2000). Environmental history and tephrostratigraphy at Carp Lake, southwestern Columbia Basin, Washington, USA. *Palaeogeography, Palaeoclimatology, Palaeoecology*, **155**, 7–29.

Wijmstra, T.A. & Van der Hammen, T. (1974). The last interglacial-glacial cycle state of affairs of correlation between data obtained from the land and from the ocean. *Geologie en Mijnbouw*, **53**, 386–392.

Wilcox, R.E. (1965). Volcanic-ash chronology. *In*: Wright, H.E. & Frey, D.G. (Eds), *The Quaternary of the United States*. Princeton University Press, Princeton, 807–816.

Woillard, G.M. (1979). Grande Pile peat bog a continuous pollen record for the last 140,000 years. *Quaternary Research*, **9**, 1–21.

Wright, H.E. & Frey, D.G. (1965). *The Quaternary of the United States*. Princeton University Press, Princeton, 922 pp.

Xia, J., Engstrom, D.R. & Ito, E. (1997a). Geochemistry of ostracode calcite: Part 2. The effects of water chemistry and seasonal temperature variation on Candona rawsoni. *Geochimica et Cosmochimica Acta*, **61**, 383–391.

Xia, J., Ito, E. & Engstrom, D.R. (1997b). Geochemistry of ostracode calcite: Part 1. An experimental determination of oxygen isotope fractionation. *Geochimica et Cosmochimica Acta*, **61**, 377–382.

Zeebe, R.E. (1999). An explanation of the effect of seawater carbonate concentration on foraminiferal oxygen isotopes. *Geochimica et Cosmochimica Acta*, **63**, 2001–2007.

Zic, M. (2001). Sediment magnetism of the B&B core from Summer Lake, Oregon, USA: implications for regional and global millennial-scale climate change from 46 to 23 ka. Unpublished Master of Science thesis, California State University.

Zic, M., Negrini, R.M. & Wigand, P.E. (2002). Evidence of synchronous climate change across the northern hemisphere between 50 and 20 KA. *Geology* (in press).

Isotopic records from ground-water and cave speleothem calcite in North America

Jay Quade

Department of Geosciences, University of Arizona, Tuscon, AZ 85721, USA

Introduction

The purpose of this chapter is to provide an overview of stable isotope records of Quaternary paleoclimate from calcite formed in the presence of ground water descending through the vadose zone (speleothems in caves) and within the saturated zone (ground-water calcite from Devils Hole, southern Nevada). The Devils Hole record is arguably the most remarkable – and at the same time controversial – continental isotope record in the Americas. There is an urgent need to sample more such deposits, which probably coat the fractures of many carbonate aquifers, but will be hard to access. Globally, only the recently documented speleothems from West Jerusalem and Soreq Caves in Israel (Bar-Mathews *et al.*, 1999; Frumkin *et al.*, 1999) and Hulu Cave in China (Wang *et al.*, 2001) approach the Devils Hole record in terms of detail, and then only over the last one or two glacial cycles.

Speleothems are easier to sample than Devils Hole-type calcites but are also in many cases more complex to interpret. The level of complexity in interpretation has increased over time, especially in the past decade since the last major review by Gascoyne (1992) appeared in the literature. My overview of speleothems will be, of necessity, far from comprehensive, given the many published records. The focus will be more on developments in the last ten years, and on records that best represent the growing array of interpretations now applied to speleothem records. In the process I shall offer some cautionary views of my own, particularly with respect to the carbon isotope record in speleothems.

I will not treat isotope records in calcite formed at the terminus of the ground-water system, such as spring travertines. Enrichments in ^{13}C and ^{18}O due to CO_2 off-gassing and evaporation, respectively, are well documented in these settings (e.g. Amundson & Kelly, 1987; Dandurand *et al.*, 1982; Gonfiantini *et al.*, 1968; Usdowski *et al.*, 1979), greatly reducing the usefulness spring travertines in paleoclimatic reconstruction.

Types of Records and Mechanisms of Formation

Cave speleothems and Devil's Hole calcite can be viewed as two end-member expressions of a single ground-water flow continuum. The general equation for calcite dissolution and precipitation in this continuum is:

$$CaCO_3 + CO_2 + H_2O \rightarrow Ca^{2+} + 2HCO_3^- \quad (1)$$

All the situations under review here involve dissolution of carbonate-rich rocks somewhere along the ground-water flow path. Calcite dissolution is initiated in the vadose (recharge) zone in the presence of elevated soil pCO_2 (>350 ppmV) produced by plant respiration and decay. This dissolution process can continue below the vadose zone and into the saturated zone until significant levels of supersaturation with respect to calcite are reached, whereupon calcite precipitation occurs. Supersaturation also develops through: (1) off-gassing of dissolved CO_2 due to a reduction in ambient pCO_2 (as in caves and springs) or due to increased T; (2) evaporation, as in some caves and springs; and/or (3) through reduction of pressure, as with ascending ground water, perhaps such as at Devils Hole.

Soil solutions should undergo little or no change in temperature or pressure between the deep soil zone and caves at depths of generally ten's of meters. Therefore, soil solutions entering a cave must lose water or CO_2 or both in order for calcite to form. This in turn requires some advective or diffusive gas leakage to outside of the cave (Wigley & Brown, 1976). Most caves in temperate regions are at or near 100% humidity except in ventilated areas such as around entrances. Thus, CO_2 off-gassing is thought to be the cause of speleothem formation in caves from temperate regions. Humidity is typically <95% in caves in drier regions (e.g. Carlsbad Caverns; Ingraham *et al.*, 1990), and the possibility of evaporative effects on the $\delta^{18}O$ value of calcite makes them much less attractive for paleoclimatic reconstruction (e.g. Polyak & Asmerom, 2001).

Caves from which most speleothems have been sampled lie near (ten's of meters) the surface and receive recharge from small catchments (generally $<10\,km^2$). Because of some storage and mixing in the overlying vadose zone, seasonal variations in $\delta^{18}O$ of soil water are usually dampened out (e.g. Goede *et al.*, 1982; Harmon, 1979; Thompson *et al.*, 1976; Yonge *et al.*, 1985; but see also Ayalon *et al.*, 1998; Bar-Mathews *et al.*, 1995), whereas longer term (10–100 yr) changes in $\delta^{18}O$ value of soil water will probably be preserved in many speleothem records.

In sharp contrast, the Devils Hole calcites in southern Nevada lie toward the distal end of a large ($\sim$12,000 km^2) flow system that receives most of its recharge as winter/spring moisture in the Spring and Sheep Mountains >30 km away. Mixing and additions of water from such a large recharge area dampen out short-term variations, but apparently preserve them at a longer (=1000 yr) scale (Winograd *et al.*, 1992). The flow system is hosted largely by fractured, Paleozoic-age carbonate rocks. Ground-water calcites coat fractures just above, at, and below the top of the water table probably along

DEVELOPMENTS IN QUATERNARY SCIENCE
VOLUME 1 ISSN 1571-0866
DOI:10.1016/S1571-0866(03)01010-8

much of the distal portion of the aquifer (Riggs *et al.*, 1994; Szabo *et al.*, 1994). Access to these fracture coatings has only been possible at Devils Hole, where collapse of the roof of one large fracture has exposed the aquifer and permitted divers to enter (Winograd *et al.*, 1988), as well as at various locations in nearby Death Valley, where recent tectonism has exhumed other fracture coatings (Winograd *et al.*, 1985). Water at Devils Hole and nearby emergent springs is saturated with respect to calcite (Riggs *et al.*, 1994; Winograd & Pearson, 1976), perhaps abetted by a drop in pressure as the mildly thermal waters ascend from depths >1 km.

The $\delta^{18}O$ Record

Speleothems

In this section we will review how the paleoclimatic interpretation of the $\delta^{18}O$ record from speleothems has evolved in the last three decades in North America, with particular emphasis on the much studied speleothem records in Iowa and Missouri. The seminal works of Hendy & Wilson (1968), Hendy (1971), Schwarcz *et al.* (1976), and Harmon *et al.* (1978, 1979a) framed the basic approach of most studies prior to 1990 to reconstructing paleoclimate from speleothem isotope records. Three factors were and are seen to potentially control the $\delta^{18}O$ value of speleothem calcite (henceforth $\delta^{18}O_c$): (1) cave and air temperature (2) the $\delta^{18}O$ value of meteoric water (henceforth $\delta^{18}O_{mw}$), and (3) the kinetic effects of evaporation and rapid CO_2 off-gassing on the $\delta^{18}O$ value of soil and cave waters. In temperate caves, the effects of evaporation in soils on $\delta^{18}O_{mw}$ are generally viewed as negligible. This assumption and the strong correlation between the $\delta^{18}O_{mw}$ and the $\delta^{18}O$ of cave drip water (henceforth $\delta^{18}O_{dw}$) were confirmed in such studies as Schwarcz *et al.* (1976), Harmon (1979), Goede *et al.* (1982), and Yonge *et al.* (1985).

The kinetic effects of evaporation and rapid CO_2 off-gassing can fatally compromise a paleoclimate $\delta^{18}O$ record from speleothems. These processes can lead to significant enrichment in ^{18}O in the case of evaporation, and in both ^{13}C and ^{18}O in the case of rapid CO_2 off-gassing. Hendy (1971) devised several tests for these effects that have since been widely applied in speleothem research. Water will both evaporate and lose CO_2 as it drips from a hanging stalactite and then splashes onto and flows as a thin film down a stalagmite or flowstone. If pCO_2 is much lower in the cave than in the drip, off gassing of CO_2 can be so rapid that ^{13}C and ^{18}O are preferentially (kinetically) lost with little or no isotopic back-reaction with the solution. This process can be identified as a gradual enrichment (and covariance between) in ^{13}C and ^{18}O in $\delta^{18}O_c$ along a single growth layer. The effects of evaporation can be identified as an increase in $\delta^{18}O_c$ alone along a growth layer. In addition to these screening criteria, sampling from poorly ventilated parts of the cave where pCO_2 and humidity are high will reduce the risk of – but are no absolute insurance against – kinetic effects.

The general equation used by most in describing the controls on change (Δ) in the $\delta^{18}O_c$ value in speleothems can be developed as follows:

$$\Delta\delta^{18}O_c = \left(\frac{d\{\Delta_{c\text{-}w}\}}{dT}\right)\Delta T + \Delta(\delta^{18}O_{dw}) \tag{2}$$

$$\delta^{18}O_{dw} = \delta^{18}O_{mw} \tag{3}$$

$$\Delta(\delta^{18}O_{mw}) = \left(\frac{d\{\delta^{18}O_{mw}\}}{dT}\right)\Delta T + \Delta(\delta^{18}O_{sw}) \tag{4}$$

substituting Eqs (3) and (4) into Eq. (2) yields:

$$\Delta\delta^{18}O_c = \left(\frac{d\{\Delta_{c\text{-}w}\}}{dT}\right)\Delta T + \left(\frac{d\{\delta^{18}O_{mw}\}}{dT}\right)\Delta T + \Delta(\delta^{18}O_{sw}) \tag{5}$$

where

(1) $d\{\Delta_{c\text{-}w}\}/dT$ is the temperature-dependent fractionation (Kim & O'Neil, 1997) between calcite and water, or

$$\Delta_{c\text{-}w} = 1000\ln\alpha_{c\text{-}w} = \left(\frac{18.03\times10^3}{T}\right) - 32.42 \tag{6}$$

which is the same as approximately $-0.2‰/°C$ at $0\,°C$ and $-0.24‰/°C$ at $30\,°C$. Temperature (in °K) in $d\{\Delta_{c\text{-}w}\}/dT$ is assumed to be cave temperature, which is close to mean annual temperature ($\pm1\,°K$) in most caves provided that the cave is: (1) not so shallow that dripwater does not reach thermal equilibrium with the host rock; (2) so deep (>100 m) that rock temperature exceeds mean annual air temperature; or (3) sampled too closely to the cave entrance.

(2) $\Delta(\delta^{18}O_{sw})$ = changes in the $\delta^{18}O$ value of seawater. This amounts to $-0.08‰/10$ m of sea-level change, or $\sim-1.0\pm0.1‰$ across the last glacial/interglacial transition (Shrag *et al.*, 1996). This can be factored into evaluations of $\Delta\delta^{18}O_c$ provided that the age of the speleothem is known.

(3) $d\{\delta^{18}O_{mw}\}/dT$ = the dependence of $\delta^{18}O_{mw}$ on local air temperature variations. This dependence is a major imponderable in Eq. (5) and requires some careful consideration here. The most widely cited value for $d\{\delta^{18}O_{mw}\}/dT$ in the speleothem literature is $\sim+0.7‰/°C$ from Dansgaard (1964) (Fig. 1). Closer inspection of more extensive compilations of rainfall/T data shows that $\delta^{18}O_{mw}$ is a non-linear function of temperature (Yurtsever & Gat, 1981). Arctic stations display higher $d\{\delta^{18}O_{mw}\}/dT$ values of $+0.67$–$0.90‰/°C$, whereas low-latitude sites show lower values ($<+0.5‰/°C$). Continentality also has an influence, with island stations displaying much lower $d\{\delta^{18}O_{mw}\}/dT$ values of $+0.17‰/°C$ than continental sites (Rozanski *et al.*, 1993). For continental settings at temperate latitudes, the $d\{\delta^{18}O_{mw}\}/dT$ relation appears to be approximately linear at $+0.6\pm0.1‰/°C$ (Fig. 1). This relation is also followed by modern cave drip waters in the temperate U.S. (Yonge *et al.*, 1985).

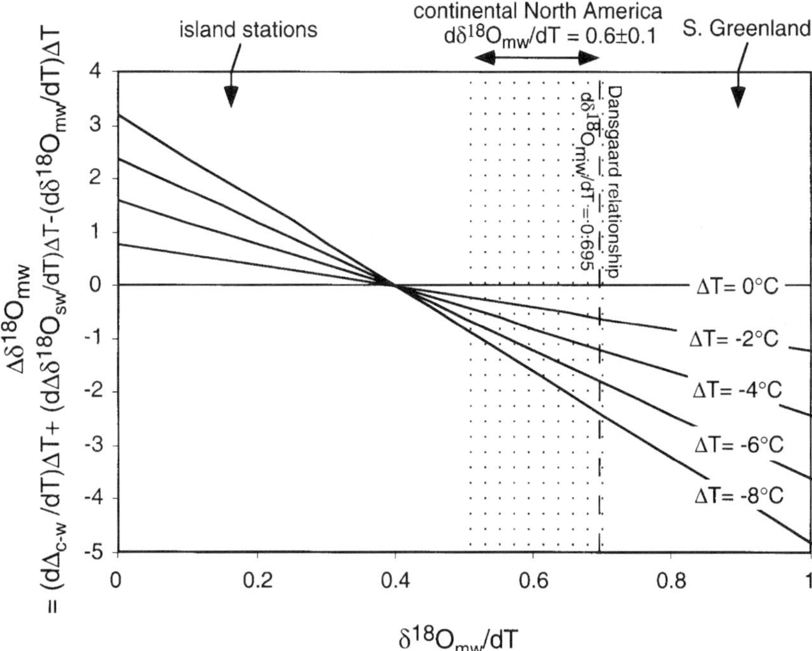

Fig. 1. A graphical depiction of Eq. (5) (see text) showing the dependence of $\Delta\delta^{18}O_c$ on $d\delta^{18}O_{mw}/dt$ as a function of total temperature change (ΔI The $d\delta^{18}O_{mw}/dt$ relation for continental North America today is $0.6 \pm 0.1/^\circ C$ (shaded zone), and is higher farther north (S. Greenland). As such, during colder times ($-\Delta I$, $\Delta\delta^{18}O_c$ should have been negative in North American speleothems. In fact, this pattern is only displayed by some records (see Table 1), suggesting that (1) $d\delta^{18}O_{mw}/dt$ was lower in glacial times than at present, or more plausibly, that (2) the dependence of $\Delta\delta^{18}O_c$ on ΔI expressed in Eq. (5) is overly simplistic. To construct this graph, I used $d\Delta_{c\text{-}w}/dT = 0.23‰/^\circ C$ and $d\Delta\delta^{18}O_{sw}/dT = 1.0‰/6\,^\circ C = 0.17‰/^\circ C$.

Paleoclimate Reconstructions

Most isotope studies of cave speleothems prior to 1985 focused on paleotemperature reconstruction. The notion was to measure the isotopic composition of paleo-water in fluid inclusions ($\delta^{18}O_{fiw}$) in speleothem calcite and to assume that:

$$\Delta\delta^{18}O_{fiw} = \left(\frac{d\{\delta^{18}O_{mw}\}}{dT}\right)\Delta T + \Delta(\delta^{18}O_{sw})$$

substituting this into Eq. (4) yields:

$$\Delta\delta^{18}O_c = \left(\frac{d\{\Delta_{c\text{-}w}\}}{dT}\right)\Delta T + \Delta\delta^{18}O_{fiw} \qquad (7)$$

However, O in the fluid inclusions exchanges isotopically with O in calcite, whereas H in fluid inclusions cannot. So, the global average relation between $\delta^{18}O_{mw}$ and δD_{mw} of $\delta D_{mw} = 8\delta^{18}O_{mw} + 10$ can be substituted into Eq. (7), assuming $\delta^{18}O_{fiw} = \delta^{18}O_{mw}$, as:

$$\Delta\delta^{18}O_c = \left(\frac{d\{\Delta_{c\text{-}w}\}}{dT}\right)\Delta T + \frac{\Delta(\delta D_{mw} - 10)}{8}$$

To actually solve for temperature at any time in the past, we can take Eq. (6) and rearrange it to solve for T, and as above, substitute $(\delta D_{mw} - 10)/8$ for $\delta^{18}O_{mw}$, to obtain:

$$T(^\circ C) = \left[\frac{18030}{1000\ln(\delta^{18}O_c + 1000/((\delta D_{mw} - 10)/8 + 1000)) + 32.42}\right] - 273.15 \qquad (8)$$

Harmon *et al.* (1979b) and Harmon & Schwarcz (1981) provided good examples of how Eq. (8) has been used in paleo-temperature reconstruction, using measurements from coexisting calcite and fluid inclusions from North America. The results proved problematic because some of the paleotemperature estimates came out implausibly low, in some cases below 0 °C. In hindsight it is clear that this approach to paleo-temperature estimation ran aground on several issues:

(1) The analysis assumed that $\delta D_{fiw} = \delta D_{dw}$. However, during the extraction step, either by crushing or low-temperature heating, inclusive water was fractionated. The exact mechanism is unclear but appears to involve adsorption of water into calcite crystal faces. The extent of this fractionation was found to be $+22.1 \pm 3.9‰$ from a variety of North American caves (Yonge *et al.*, 1985), a result similar to that found by Goede *et al.* (1986) from a Tasmanian speleothem. This fractionation effect alone results in underestimation of paleo-temperature by an average of ~12.5 °C, with an uncertainty of ±2.3 °C. Rather poor reproducibility of replicate fluid inclusion analyses is also apparent in the results of Schwarcz & Yonge (1983) and Winograd *et al.* (1985).

(2) The analysis assumed that δD_{fiw} and $\delta^{18}O_{fiw}$ values follow the global meteoric water line as defined by $\delta D_{mw} = 8\delta^{18}O_{mw} + 10$, where $d = 10‰$ (the deuterium excess), and $s = 8$. However, many studies have shown that d and s vary according to location, as they probably did in time (Harmon & Schwarcz, 1981). A reasonable range for these parameters in temperate North America today might be $10 \pm 2‰$ for d and 8 ± 1.5 for s. The choice of these ranges of values would introduce an additional uncertainty in paleo-temperature estimates at least ±7 °C.

(3) Typical analytical uncertainties for $\delta^{18}O_c$ are ±0.1‰, producing ±0.4 °C error. The total uncertainty for paleo-temperature estimation is therefore about ±10 °C.

Table 1. Apparent response of $\delta^{18}O_c$ from speleothem calcites to changes in temperature and $\delta^{18}O$ of vapor.

Temperature dominant		$\delta^{18}O_v$ dominant
$d\delta^{18}O_c/dt > 0$	$d\delta^{18}O_c/dt < 0$	$d\delta^{18}O_c/d\delta^{18}O_v > 0$
Cascade Cave (Gascoyne *et al.*, 1981)	Gardner's Gut and Waipuna Caves (Hendy & Wilson, 1968)	West Cave (Frumkin *et al.*, 1999)
Lynds Cave (Goede & Hitchman, 1983)	Norman Bone/Grapevine Cave (Thompson *et al.*, 1976)	Crag Cave (McDermott *et al.*, 2001)
Little Trimmer Cave (Goede *et al.*, 1986)	Caves in Bermuda, Kentucky, and W. Virginia (Harmon *et al.*, 1978)	Cold Water Cave (Denniston *et al.*, 1999)
Francome Cave (Goede *et al.*, 1990)	Cold Water Cave (Harmon *et al.*, 1979a)	Hulu Cave (Wang *et al.*, 2001)
Cold Water Cave (Dorale *et al.*, 1992)	Unnamed cave (Wilson *et al.*, 1979)	Burns *et al.* (2001)
Onondaga Cave (Denniston *et al.*, 2001)	Ease Gill Caverns (Gascoyne, 1992)	
Okshola and Stordalsgrotta Caves (Lauritzen, 1995)	Ingleborough Cave (Gascoyne, 1992)	
Lithofagus Cave (Lauritzen & Onac, 1999)	Lost John's Pothole (Gascoyne, 1992)	
Spannegal Cave (Spötl *et al.*, 2002)	Victoria Cave (Gascoyne, 1992)	
	Exhaleair and Nettlebed Caves (Hellstrom *et al.*, 1998)	
	Soreq Cave (Bar-Mathews *et al.*, 1999)	

For these reasons, little paleo-temperature estimation based on fluid inclusion analysis has been published since 1985. Although too imprecise for paleo-temperature reconstruction, the δD_{fiw} analyses have shown that the δD_{mw} was $\sim 12‰$ (or very roughly $-1.5‰$ in $\delta^{18}O_{mw}$) lower in the full-glacial period than today. Moreover, all but a few studies prior to 1990 found a negative relation between coexisting δD_{fiw} and $\delta^{18}O_c$ values through a stalagmite time series (Table 1). Since δD_{mw} is positively correlated with temperature, so must temperature be negatively correlated with $\delta^{18}O_c$. This relation was further confirmed in most speleothem data sets by the fact that $\delta^{18}O_c$ values are lower in cold (full glacial) times compared to values in modern calcite (Gascoyne, 1992).

Dorale *et al.* (1992) revisited the paleo-temperature issue at Cold Water Cave (first examined by Harmon *et al.*, 1979b) by studying a record from deep in the cave spanning much of the Holocene. Rather than relying in the δD value from fluid inclusions, Dorale *et al.* (1992) made a simplifying assumption to get at paleo-temperature (after Hendy & Wilson, 1968), that all variations in the $\delta^{18}O_{mw}$ in the past are T dependent and can be approximated as $d\{\delta^{18}O_{mw}\}/dT = 0.695‰/°C$ (from Dansgaard, 1964). Equation (5) can be solved iteratively for ΔT by inserting this term in the equation and by assuming that $\Delta(\delta^{18}O_{sw})$ is negligible for the Holocene. Using this approach, Dorale *et al.* (1992) interpreted the increase of nearly 2‰ in $\delta^{18}O_c$ during the mid-Holocene (Fig. 2b) to indicate a 2–3 °C *increase* in temperature.

This approach is also reflected in the interpretation of $\delta^{18}O_c$ results from Crevice Cave (Dorale *et al.*, 1998) and Onondoga Cave (Denniston *et al.*, 2001) in nearby Missouri. In both caves, decreases in $\delta^{18}O_c$ of 2‰ during the Younger Dryas (Denniston *et al.*, 2001) and 1.5‰ between 55,000 and 41,000 yr ago (Dorale *et al.*, 1998) were interpreted to result from temperature decreases.

The conundrum here is that most other caves studies up until 1990 (Gascoyne, 1992), as described previously and including Cold Water Cave (Harmon *et al.*, 1978, 1979a), concluded just the opposite, that increases in $\delta^{18}O_c$ values reflected a temperature *decrease*. Differing estimates of $d\{\delta^{18}O_{mw}\}/dT$ contributed to the differing interpretations (Fig. 1). Dorale *et al.* (1992, 1998) and Denniston *et al.* (2001) assumed the rather high, global average value for $d\{\delta^{18}O_{mw}\}/dT$ from Dansgaard (1964), high enough that $d\{\delta^{18}O_{mw}\}/dT > (d\{D_{c\text{-}w}\}/dT)\Delta T + \delta D^{18}O_{sw}$ in Eq. (5). Harmon *et al.* (1978) and Schwarcz & Yonge (1983), using constraints from δD_{dw}, concluded that $d\{\delta^{18}O_{mw}\}/dT$ must have been smaller, on the order of 0.3‰/°C and as a consequence that $(d\{\Delta_{c\text{-}w}\}/dT)\Delta T + \Delta(\delta^{18}O_{sw}) > (d\{\delta^{18}O_{sw}\}/dT)\Delta T$ in Eq. (5) (Fig. 1).

Denniston *et al.* (1999) obtained the most detailed Holocene $\delta^{18}O_c$ records from Cold Water Cave yet (Fig. 2b), and their interpretations potentially add a further complication to paleoenvironmental interpretation of speleothem $\delta^{18}O_c$ values. Three speleothems were analyzed from three widely spaced sites deep in the cave. One of the speleothems (1S) was the same as that analyzed by Dorale *et al.* (1992) in less detail, and the same clear increase in $\delta^{18}O_c$ was observed during the mid-Holocene. The three separate speleothem records show poor agreement (Fig. 2b), especially with respect to the strongly opposed shifts in $\delta^{18}O_c$ values in speleothem 1S and 2SS during the mid-Holocene, and the comparative lack of response in $\delta^{18}O_c$ values from speleothem 3L. Denniston *et al.* (1999) attributed the large discrepancies between records to the differing effects of

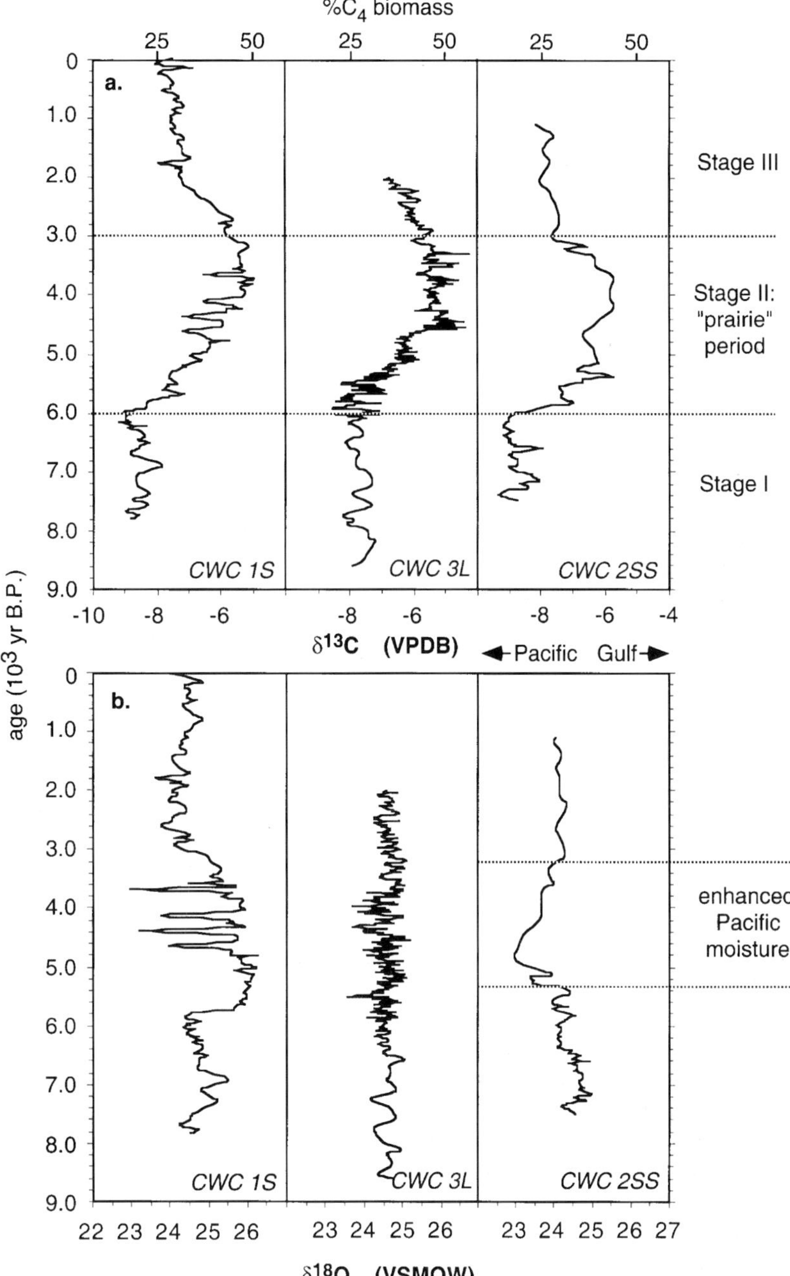

Fig. 2. Time (10^3 yr) vs. a. $\delta^{13}C_c$ values and b. $\delta^{18}O_c$ values from Cold Water Cave, Iowa (from Denniston et al., 1999). Estimated percent C_4 biomass shown in a. uses the soil-diffusion model of Cerling (1984) and Quade et al. (1989). The percent C_4 biomass (summer grasses) increases to near 50% in the mid-Holocene "prairie" period. Denniston et al. (1999) interpret the $\delta^{18}O_c$ values from CWC 2SS as the least effected by evaporation and therefore the most representative of changes in $\delta^{18}O_{mw}$ through time. They view the decrease in $\delta^{18}O_{mw}$ values in CWC 2SS during the mid-Holocene as indicating an increase in the fraction of moisture derived from Pacific vs. the Gulf of Mexico.

evaporation on recharge waters above each of the speleothem sites. For example, speleothem 1S shows the largest enrichments in ^{18}O during the mid-Holocene of the three records. And the 1S site underlies a south-facing slope most subject to evaporative enrichment of ^{18}O in soil water.

In contrast, speleothem 2SS underlies a north-facing slope and was therefore deemed to be the most useful for paleoenvironmental reconstruction due to a lack of evaporation effects. The results from 2SS show a ~1.8‰ increase in $\delta^{18}O_c$ from the early to mid-Holocene (Fig. 2b). Using the same assumptions as Dorale *et al.* (1992) did previously for interpretation of 1S, this 1.8‰ shift would imply 3–4 °C cooling during the mid-Holocene. However, no such cooling is indicated by other, independent records from the region. To explain the decrease in $\delta^{18}O_c$, Denniston *et al.* (1999) invoked an increase in the proportion of winter (with lower $\delta^{18}O_{mw}$ values) compared to summer rainfall (with higher $\delta^{18}O_{mw}$ values) during the mid-Holocene. In essence, Denniston *et al.* (1999), along with other recent publications described below, add another variable ($\delta^{18}O_v$) to Eq. (5):

$$\Delta\delta^{18}O_c = \left(\frac{d\{\Delta_{c\text{-}w}\}}{dT}\right)\Delta T + \left(\frac{d\{\delta^{18}O_{mw}\}}{dT}\right)\Delta T + \Delta(\delta^{18}O_{sw}) + \Delta(\delta^{18}O_v) \quad (9)$$

where $\delta^{18}O_v$ denotes the $\delta^{18}O$ value of water vapor that arrives at a site, and which is determined by a variety of factors not directly dependant on temperature, such as storm trajectory, seasonality of rainfall, and rainfall amount (but not sea-water $\delta^{18}O$ changes).

In summary, the five studies over the past 25 years of the $\delta^{18}O_c$ records from Iowa and Missouri have yielded up three very different paleoclimatic interpretations. Harmon *et al.* (1978, 1979a) combined $\delta^{18}O_c$ and $\delta^{18}O_{iw}$ results to interpret the higher $\delta^{18}O_c$ values during full-glacial period as indicating a *colder* cave temperatures during speleothem formation. From the same cave, Dorale *et al.* (1992) took higher $\delta^{18}O_c$ values to indicate *warmer* conditions in the mid-Holocene. Denniston *et al.* (1999), acting from a larger data set, invoked evaporative effects and changes in $\delta^{18}O_v$ responding to changes in storm track, not temperature, as the main determinants of $\delta^{18}O_c$. More fundamentally, the interpretations of Denniston *et al.* (1999) challenge a key assumption of nearly all previous speleothem studies, that in temperate caves $\delta^{18}O_{mw} = \delta^{18}O_{dw}$. For temperate caves this assumption was seemingly satisfied by the work of Thompson *et al.* (1976), Goede *et al.* (1982), Yonge *et al.* (1985), and others, although in drier regions such as the Judean Hills (mean annual precipitation ~0.5 m/yr) evaporative enrichment in ^{18}O of all drip waters in the vadose zone has been convincingly argued (Ayalon *et al.*, 1998; Bar-Mathews *et al.*, 1995). In the end, the causes for the lack of agreement in the $\delta^{18}O_c$ values between stalagmites at Cold Water Cave *could* be explained by the evaporation effects suggested by Denniston *et al.* (1999). Clearly, there is a need to rigorously test this explanation in a number of caves by systematically sampling temperate cave drip waters fed by recharge on slopes of different aspect.

Some Results From Outside North America

Other recent results from around the world reflect the same evolution in thinking in North America on how $\delta^{18}O_c$ results should be interpreted in the absence of independent constraints on $\delta^{18}O_{mw}$ (once thought to be provided by fluid-inclusion analysis). As one example, the record from Hulu Cave in eastern China spans 10,000–70,000 yr and shows a long-term oscillation in $\delta^{18}O_c$ values of 4–5‰ but also shorter term shifts of 1–2‰ that seem to correlate with Heinrich Events in the Greenland ice records (Wang *et al.*, 2001). $\delta^{18}O_c$ is positive during cold periods such as marine oxygen isotope stage (OIS) 2, the Heinrich Events, and the Younger Dryas. Whereas the short-term shifts in $\delta^{18}O_c$ might be explainable by changes in cave temperature alone, the longer term 4‰ shift between ~16,000 and 10,000 yr ago is not (Fig. 1). Barring any unforeseen kinetic fractionation effects in the cave, a major, temperature-independent shift in $\delta^{18}O_{mw}$ seems called for. Wang *et al.* (2001) invoke insolation-related oscillations in the strength of the East Asian summer (low $\delta^{18}O_{mw}$) and winter (high $\delta^{18}O_{mw}$) monsoon to explain the record, and down-play the role of temperature change. Burns *et al.* (2001) similarly interpreted the $\delta^{18}O_c$ speleothem record from Hoti Cave in Oman in terms of local changes in the strength of the Indian Monsoon. Both these instances fit with the strong correlation between rainfall amount and $\delta^{18}O_{mw}$ values at low latitudes (Hong Kong, $r^2 = 0.71$; New Delhi, $r^2 = 0.45$, Rozanski *et al.*, 1993).

Other recent cases for large, temperature-independent shifts in the $\delta^{18}O_v$ come from West Cave near Jerusalem studied by Frumkin *et al.* (1999) and Crag Cave in Ireland (McDermott *et al.*, 2001). The West Cave record extends almost continuously for the last 170,000 yr, and shows a very similar coarse structure to $\delta^{18}O$ results from ice and marine records for the same period. Glacial-age (OIS 2 and 6) $\delta^{18}O_c$ values are 3.5–5‰ more positive than interglacial values. This would appear to place the West Cave results in the same category as many other speleothem records, where $(d\{\Delta_{c\text{-}w}\}/dT)\Delta T + \Delta\delta^{18}O_{sw}$ dominates over the $(d\{\delta^{18}O_{mw}\}/dT)\Delta T$ term in Eq. (5). However, even assuming a very low $d\{\delta^{18}O_{mw}\}/dT = 0.3‰/°C$ would only produce a ~1‰ increase in glacial-age $\delta^{18}O_c$ (see Fig. 1, $\Delta T = -8$ °C line), well short of the observed 3.5–5‰ shift. Frumkin *et al.* (1999) therefore invoke a major increase in local glacial $\delta^{18}O_{sw}$, due to evaporative enrichment of the Mediterranean during glacial lowstands to largely explain the very low glacial-age $\delta^{18}O_c$ values in West Cave. In an analogous manner, McDermott *et al.* (2001) interpreted decreases in $\delta^{18}O_c$ values from Crag Cave in Ireland in terms of temperature decreases during the Holocene enhanced by decreases in $\delta^{18}O_v$ caused by North Atlantic meltwater surges.

As in North America, however, some researchers outside the U.S. continue to view changes in temperature, not in $\delta^{18}O_v$, as the main determinant of $\delta^{18}O_c$. One recent example comes from Soreq Cave, only 330 m higher and 10 km away from West Cave, which yielded a $\delta^{18}O_c$ record remarkably similar to that of West Cave and yet the paleoclimatic interpretation of Bar-Mathews *et al.* (1997, 1999) is totally different. The Soreq Cave record displays the same 3–3.5‰ decrease in $\delta^{18}O_c$ during the last deglaciation as in the West Cave record. However, Bar-Mathews *et al.* (1999) interpreted the shift largely in terms of temperature change, not changes in $\delta^{18}O_v$.

In summary, most speleothem studies prior to 1990 both within and outside North America documented higher $\delta^{18}O_c$ values during cold phases such as the last glacial period compared to values from modern speleothems. This response was viewed to indicate that the combined $(d\{\Delta_{c\text{-}w}\}/dT)\Delta T + \Delta\delta^{18}O_{sw}$ terms dominated over a relatively low $(d\{\delta^{18}O_{mw}\}/dT)\Delta T$ term for most continental settings. In contrast, most speleothem records documented in the last ten years show lower $\delta^{18}O_c$ values during cold phases. The reason for this contradiction remains unclear and only partly can reside in the improved quality of U-series dating using TIMS (thermal ionization mass spectrometry) compared to the less precise alpha-count dating of pre-1990. The more recent results from North America are consistent with the recognition that $d\{\delta^{18}O_{mw}\}/dT$ is much larger than once thought for inland continental settings. More recent studies also are recognizing that Eq. (5) is an unrealistically simplistic view of the determinants of $\delta^{18}O_c$ values in most caves, and that in many caves the former promise of speleothems in yielding paleotemperature estimates appears to be out of the question.

Devils Hole

Three important isotope records from ground-water calcites, two from Devils Hole, and one from nearby Death Valley, have come out of the Mojave Desert. The Death Valley record involved measurement of the δD values of inclusive water in ground-water calcites exhumed by recent tectonism along the east side of Death Valley (Winograd *et al.*, 1985). The record spans the last 2 million years and shows a large increase of 40‰ (much larger than the uncertainty on the δD_{fiw} measurements) in δD_{fiw} values. Winograd *et al.* (1985) attributed this to a decrease in the average value of $\delta^{18}O_v$ as the Sierra Nevada mountains were uplifted to the west.

The best-known ground-water calcite records in the region come from two cores obtained by SCUBA divers at Devils Hole. DH–2 spans 50,000 to 310,000 yr ago (Winograd *et al.*, 1988) and DH–11 50,000–550,000 yr ago (Fig. 3; Winograd *et al.*, 1992), based on multiple alpha and TIMS U-series dates from both cores. Neither record has any apparent depositional hiatuses. Further records filling in much of the last 80,000–19,000 yr have been obtained but are largely unpublished (Winograd *et al.*, 1996). Dates from the cores lag climatic events by the time required for water to travel from the recharge area to Devils Hole. Thus, dates from Devils Hole represent minimum ages (by $<$10,000 yr) for the recorded climate events. The results from DH–2 and DH–11 are nearly identical and henceforth can be discussed as a single "DH" record.

$\delta^{18}O_c$ (SMOW) values from Devils Hole fluctuate between +13.05 and 15.75‰ and display the same sawtooth pattern of marine benthic foram $\delta^{18}O_c$ records over the same time span (Fig. 3b and c). Some differences between the $\delta^{18}O_c$ records from DH and SPECMAP have been pointed out, such as the amplitude of termination III, decreasing (DH–11) vs. increasing (SPECMAP) peak heights during the first half of OIS 6, and the greater duration of interglacial periods in the DH records (Winograd *et al.*, 1992, 1997). Overall, however, the similarity in the form of the $\delta^{18}O_c$ curve from the Devils Hole record to marine (viewed as a proxy for ice volume changes) and ice records such as Vostok (viewed as a proxy for Antarctic air temperature) are striking.

How is the $\delta^{18}O_c$ record from Devils Hole being interpreted? The conditions of formation of the $\delta^{18}O_c$ record from Devils Hole make it a much simpler record to interpret than $\delta^{18}O_c$ records from most caves. First, the temperature of formation of the calcite has likely remained constant due to the deep circulation of water and the large volume of the flow system. Second, the temperature of the water (25–35 °C) is low enough and the volume of water large enough that changes in primary $\delta^{18}O_{mw}$ due to water-rock interaction are negligible. Thus, Winograd *et al.* (1988, 1992) have suggested that the main determinant of $\delta^{18}O_c$ in the Devils Hole records is the value of $\delta^{18}O_{mw}$ falling mainly in winter and spring on the recharge area to the Devils Hole/Ash Meadows. The $\delta^{18}O_{mw}$ value is in turn determined by $\Delta\delta^{18}O_{sw}$ and by changes in the average winter/spring temperature in cloud masses arriving over the recharge area. Following the lead of the speleothem literature, we can make a rough estimate of these temperature contrasts between glacial and interglacial periods from the Devils Hole record using Eq. (5). The amplitude of the $\delta^{18}O_c$ contrast varies from 2.2‰ for across Termination II to 1.7 across Termination III (Fig. 3b). The $(d\{\Delta_{c-w}\}/dT)\Delta T$ term in Eq. (5) can be dropped because the temperature change at the point of calcite formation is assumed to be invariant. Solving for ΔT, Eq. (5) becomes:

$$\Delta T(^{\circ}C) = \frac{\Delta\delta^{18}O_c - \Delta(\delta^{18}O_{sw})}{d\{\delta^{18}O_{mw}\}/dT}$$

Substituting $\Delta(\delta^{18}O_{sw}) = -1.0‰$ and $d\{\delta^{18}O_{mw}\}/dT = 0.6 \pm 0.1‰/^{\circ}C$, we obtain $\Delta T = 5.3 \pm 1.1\,^{\circ}C$ across Termination II and $4.5 \pm 1.1\,^{\circ}C$ across Termination III. Two key assumptions here are that all variation in $\delta^{18}O_{mw}$ is temperature dependent and that that dependence lies within $0.6 \pm 0.1‰/^{\circ}C$.

The one feature of the Devils Hole record that has sparked great controversy is the dating of glacial terminations II and III at 140,000 ± 3000 and 253,000 ± 3000 yr ago (2σ, respectively. These are minimum age estimates because of the time lag in the Devils Hole record mentioned before. In the orbitally tuned SPECMAP records these same terminations date to 128,000 ± 3000 and 244,000 ± 3000 yr, respectively (Fig. 3). The offset of $\geq$9000 years between age estimates for the terminations is non-trivial and cuts to the heart of the debate over the origin of the Ice Ages. For example, the orbitally tuned date of 128,000 ± 3000 yr for termination II from SPECMAP coincides with a pronounced summer insolation maximum 128,000 ± 1000 yr ago for the northern hemisphere (Fig. 3c and d), strongly supporting a causal link between maximum summer heating in the northern hemisphere and ice-sheet retreat. If, however, the Devils Hole chronology is correct, then the linkage between orbital forcing and waxing and waning of glaciation is non-linear.

Part of the controversy has pivoted on the quality of dating for the Devils Hole and both marine core and terrace deposits. Discussion of this debate here is beyond the scope of this paper. Suffice it to say, questions (Edwards & Gallup, 1993; Shackleton, 1993) regarding the original TIMS ^{230}Th–^{234}U–^{238}U dates from Devils Hole (Ludwig *et al.*, 1992, 1993) have largely abated (Imbrie *et al.*, 1993), particularly in the face of confirming dates using the ^{231}Pa–^{235}U decay series (Edwards *et al.*, 1997).

More serious are questions as to whether the Devils Hole $\delta^{18}O_c$ record represents a regional vs. more global record of temperature change. Winograd *et al.* (1992) have convincingly argued that local air temperature in the recharge area is the dominant control on the $\delta^{18}O_c$ value on the Devils Hole record. Herbert *et al.* (2001) contend that changes in air temperature in the region have not changed in phase with average global air temperature. The basis for this is their paleo-temperature estimates spanning multiple glacial and interglacial cycles using unsaturated alkenones from marine

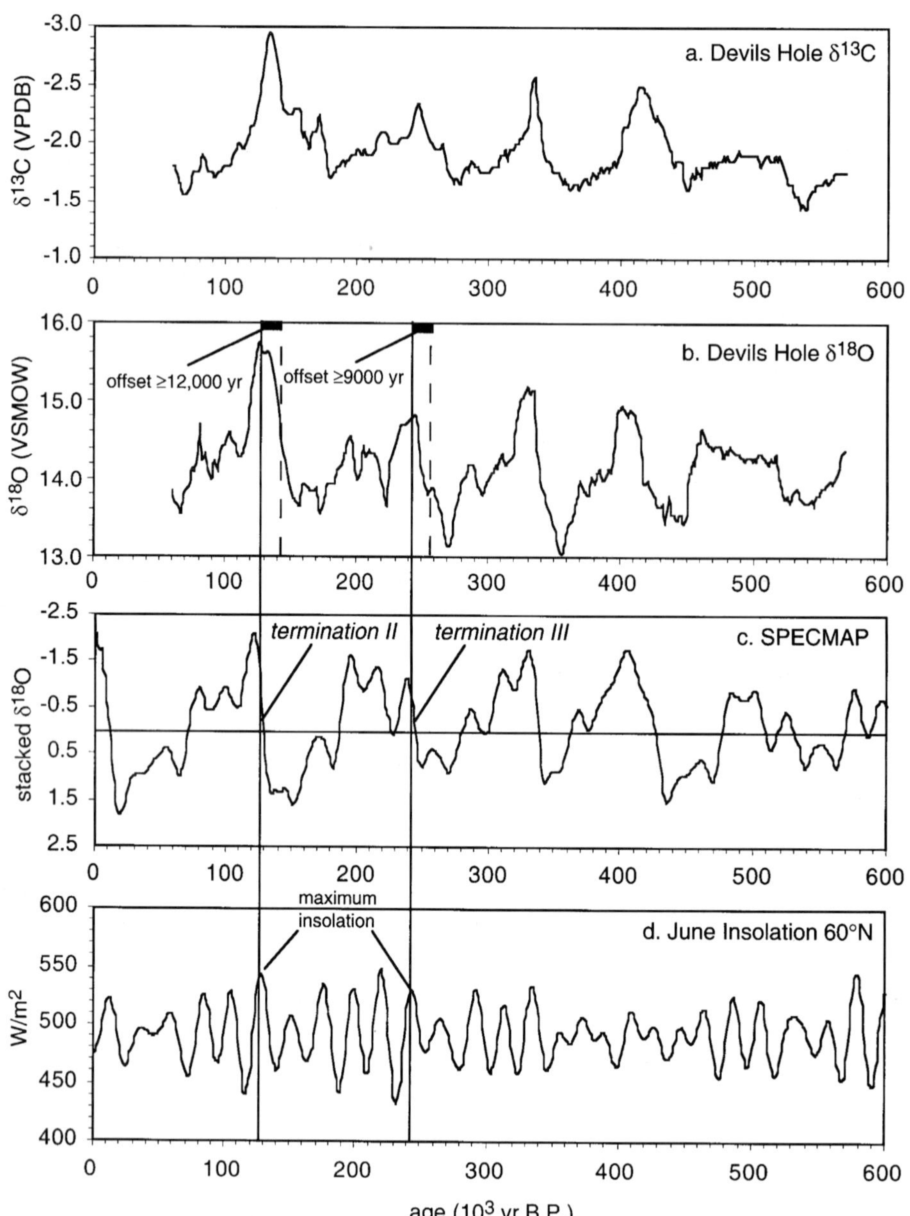

Fig. 3. Time (10^3 yr) vs. a. $\delta^{13}C_c$ (from Coplen et al., 1994) and b. $\delta^{18}O_c$ (from Winograd et al., 1992) from Devils Hole, c. stacked $\delta^{18}O_c$ values for benthic forams from SPECMAP compiled by Imbrie et al. (1984), and d. June insolation values in watts/meter for 60°N (from Berger & Loutre, 1991). Also shown are the contrasting ages of Terminations II and III for the DH–11$\delta^{18}O_c$ record (dashed vertical lines) compared to SPECMAP (solid vertical lines). Age offsets for the terminations are minimum estimates due to the ground-water travel lag-time in the DH–11 record.

sediments off the west coast of North America. Reconstructed sea-surface temperatures (SST) begin to increase on average 11,000 yr earlier than deglaciation, as reflected in the decrease in $\delta^{18}O_c$ values from co-occurring benthic forams (Fig. 4). Herbert *et al.* (2001) point out the strong correlation ($r = 0.70$–0.78) of their paleo-SST record with the $\delta^{18}O_c$ record from DH–11, in contrast to the lack of correlation ($r = -0.5$) between the DH–11 record and SPECMAP. This early warming of waters off the coast of California is then thought to produce (in some manner) warming of air-temperatures in southern Nevada, thus explaining the temporal lag of the DH records compared to SPECMAP. In essence, Herbert *et al.* (2001) suggest that Devils Hole records air-temperature changes caused by changes in local SST off the California Coast that are not correlated to global temperature changes.

Winograd (2002) have recently countered by pointing out that early SST warming during deglaciation is a feature of many marine records globally, not just a local feature of the easternmost north Pacific Ocean. Moreover, many (but not all) U-series dates from marine corals and marine sediments place the attainment of near-modern sea-level

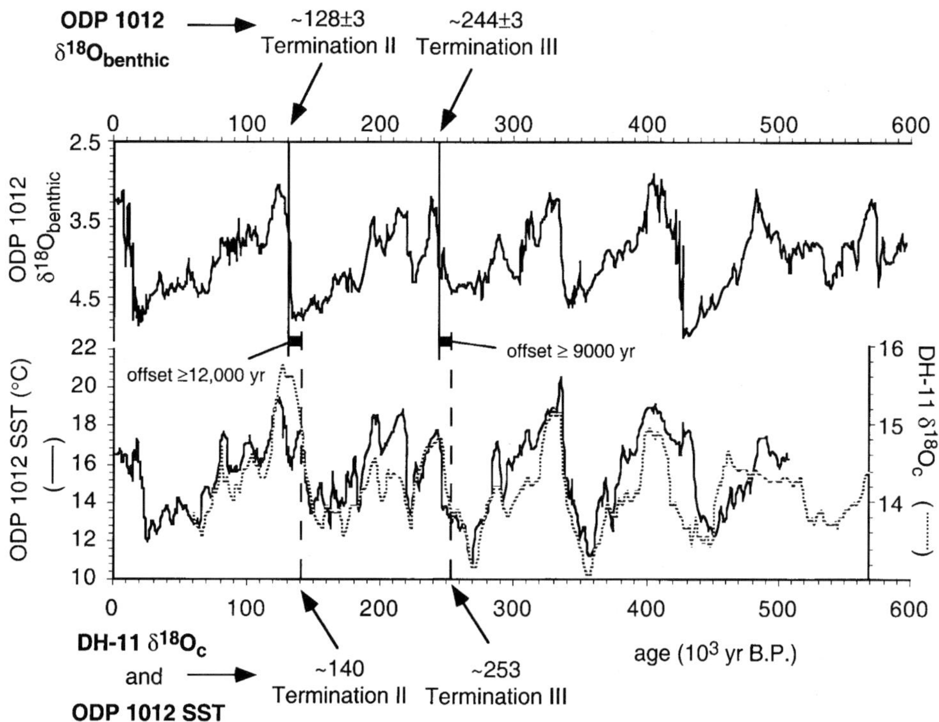

Fig. 4. Time (10^3 yr) vs. $\delta^{18}O_c$ values of benthic forams (upper graph) and reconstructed sea-surface temperature (SST, lower graph, solid line) from ODP 1012 taken off the northern California Coast (from Herbert et al., 2001), compared to $\delta^{18}O_c$ from core DH–11 (lower graph, dashed line) from Devils Hole, southern Nevada (Winograd et al., 1992). Ages for ODP 1012 are orbitally tuned (see Herbert et al., 2001; Shackleton, 2000) while the ages for Devils Hole are based in ^{230}Th–^{234}U–^{238}U ages uncorrected for ground-water travel times (from Ludwig et al., 1992). Also shown are the nearly synchronous ages of Terminations II and III for DH–11 and ODP 1012 SST's (dashed vertical lines), contrasted with the ages of the same terminations (solid vertical lines) based on ODP 1012 $\delta^{18}O_c$ values of benthic forams.

at 132,000–135,000 yr ago, inconsistent (too early) with orbitally "dated" $\delta^{18}O_c$ evidence for deglaciation from SPECMAP (Winograd *et al.*, 1992, with more recent support from Gallup *et al.*, 2002; Henderson & Slowey, 2000).

The whole issue of how globally representative the DH record is remains unclear, and awaits, in part, development of ground-water calcite and speleothem records outside of southwestern North America. Two such speleothem records have recently been published, but they do not appear to agree on the dating of termination II. The impressively dated SPA–52/11 speleothems from Spannegel Cave in Austria point to the onset of deglaciation before 135,000 ± 1200 yr ago (Spötl *et al.*, 2002), whereas the perhaps less robust results from a Norwegian speleothem point to an age for termination II of 128,000 ± 5000 yr (Lauritzen, 1995).

The $\delta^{13}C$ Record

The carbon isotopic composition of calcite ($\delta^{13}C_c$) in speleothems and ground-water calcites is initially determined by: (1) the $\delta^{13}C$ value of plant and atmospheric CO_2 in the soil zone; and later by (2) water-rock (limestone) interaction; and by (3) the causes and rates of calcite formation.

As for (1) above, the relation between the $\delta^{13}C$ value of CO_2 and $CaCO_3$ in soils has been carefully studied (Cerling, 1984; Cerling & Quade, 1993; Quade *et al.*, 1989). The $\delta^{13}C_c$ value in soils is determined by the $\delta^{13}C$ value of coexisting soil CO_2. The $\delta^{13}C$ of soil CO_2 is in turn determined by the relative contribution of CO_2 from the atmosphere, and from decay of and respiration by C_3, C_4, and CAM plants. Outside of deserts, the contribution of atmospheric CO_2 and of CAM plants (largely desert succulents) to soil CO_2 is negligible below about 10-cm soil depth. C_3 (average $\delta^{13}C = \sim -26‰$) plants include all trees, most shrubs, and grasses favored by cool growing seasons. C_3-dominated ecosystems occur where rain is year-round or confined to the cool/cold part of the year, including deserts. C_4 plants (average $\delta^{13}C = \sim -13‰$) are mainly grasses favored by a warm rainy season. They tend to dominate ecosystems with moderate (<1 m/year) rainfall but only where abundant rain falls in the summer.

Dissolution of limestone ((2) above) either in the soil or during infiltration or flow below the soil can modify the $\delta^{13}C$ value of carbon species in solution. The extent of this influence depends on how open the system is to back exchange with surrounding gaseous CO_2, and on how large the surrounding soil CO_2 reservoir is. Soil water stays in continuous contact with plant-derived CO_2, and thus C species in solution remain in isotopic equilibrium with a very large, isotopically unchanging reservoir of CO_2. However, as soil solutions infiltrate they may lose contact (become "closed") with respect to plant-derived CO_2; here the dissolution of carbonates along the flow path may gradually modify the $\delta^{13}C$ value of carbon in solution. The more extreme case of this is represented by Devils Hole discussed below, where ground-water flow through ten's of kilometers of fractured limestone has greatly modified the $\delta^{13}C$ value of C in solution. Caves represent a more intermediate case, although it is likely that the rate of infiltration to some caves is rapid, and exchange with local limestone is therefore negligible.

The causes and rate of precipitation of calcium carbonate ((3) above), as well as the pCO_2 and the $\delta^{13}C$ value of CO_2 in the surrounding atmosphere, are critical to the $\delta^{13}C_c$ value of the resultant calcium carbonate. In soils, carbonate precipitation is largely caused by gradual dewatering of the soil by plant evapotranspiration (see Eq. (1)). $CO_2(aq)$ will also gradually be lost, but it will remain in isotopic equilibrium with plant-derived CO_2. The resultant soil carbonate will be 13.4 (25 °C) to 16.4 (0 °C)‰ enriched in ^{13}C with respect to coexisting plant cover (Cerling & Quade, 1993), where soil respiration rates are moderate to high.

The carbon isotopic composition of speleothem calcite once received little attention in the literature, but in the last decade it has expanded greatly in papers both North America and abroad, largely in parallel with the increasing use of $\delta^{13}C_c$ from soils in paleovegetation reconstruction. There is a tendency in the recent speleothem literature (e.g. Bar-Mathews *et al.*, 1997, 1999; Baskaran & Krishnamurthy, 1993; Brook *et al.*, 1990; Denniston *et al.*, 1999, 2000, 2001; Dorale *et al.*, 1992, 1998) to treat the processes of calcite formation in caves like those in the soils described above, in other words, to interpret changes in the $\delta^{13}C_c$ value of speleothems in terms of changes in the proportion of C_3 and C_4 plants growing in soils in the recharge zone above the cave. Unfortunately, the behavior of carbon isotopes in most caves has not been measured in the same rigorous manner that it has been recently in soils. The single exception here is the recently opened Soreq Cave, where careful measurements (Bar-Mathews *et al.*, 1995) showed that the carbon species in the cave system do not follow the Cerling soil-diffusion model. In the absence of such measurements, a key test of the soil-diffusion model developed by Cerling (1984) can be applied to speleothems, in which the $\delta^{13}C$ value of carbonate is compared to that of co-existing soil organic matter. Measured soil values fall within the range of values predicted by diffusion-model calculations (Fig. 5). The same test can be applied to speleothems using a large $\delta^{13}C$ data set from British speleothems (Baker *et al.*, 1997; Gascoyne, 1992), where soil respiration rates are high (thus excluding atmospheric CO_2) and C_4 plants do not grow. In this comparison, most of the $\delta^{13}C_c$ from the speleothems fall outside the range predicted by the soil-diffusion model (Fig. 5). This pattern is also visible in the broad range in $\delta^{13}C_c$ values (−2 to −12‰) found in speleothems from Tasmania (Goede *et al.*, 1983, 1986, 1990), from Norway (−1.9 to −7.5‰; Lauritzen, 1995), and elevated values (<−7‰) from Cascade Cave in Vancouver (Gascoyne, 1992). Like Britain, all these areas are and probably always were vegetated by C_3 plants during the Quaternary.

Clearly, the $\delta^{13}C_c$ value from many speleothems is dependent on more than just the proportion of C_3 and C_4 plants growing above the cave, as originally modeled by Hendy (1971) and Dulinski & Rozanski (1990), discussed by Baker *et al.* (1997), and measured by Bar-Mathews *et al.* (1995). Other complicating processes include rapid off-gassing of CO_2, producing kinetic enrichments in both C and O isotopes, or slower CO_2 off-gassing, so that C species in solution and in the surrounding cave remain in isotopic equilibrium. In the latter case, the $\delta^{13}C$ value (but not the $\delta^{18}O$ value) of the solution will gradually increase as CO_2 (with a lower $\delta^{13}C$ value than dissolved C) is off-gassed. This

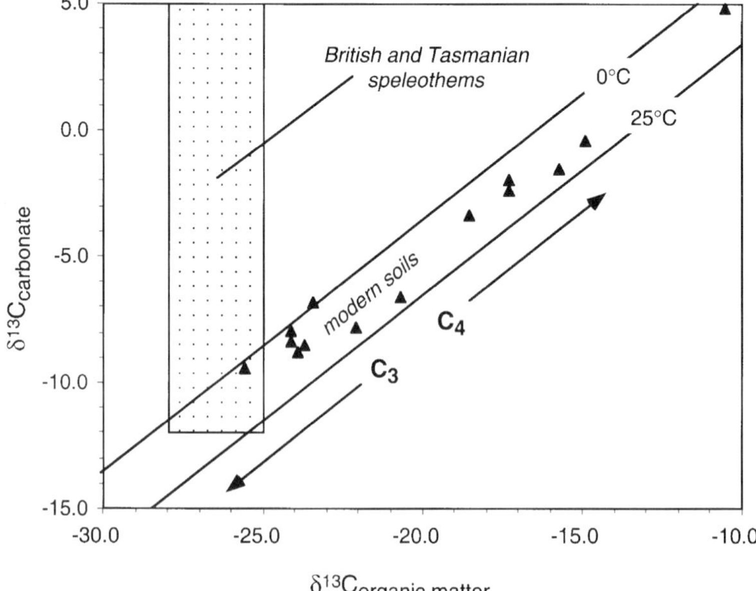

Fig. 5. $\delta^{13}C$ (VPDB) of soil organic matter vs. $\delta^{13}C$ (VPDB) of carbonates formed in modern soils (solid triangles; data from Cerling & Quade, 1993) and caves from Britain and Tasmania (shaded box; data from Baker et al., 1997, Gascoyne (1992), and Goede et al. (1983, 1986, 1990)). Lines depict the $\delta^{13}C_c/\delta^{13}C_{organic\ matter}$ relations predicted by the soil-diffusion model of Cerling (1984) and Quade et al. (1989) but using temperature-dependent carbon isotope fractionation factors from Romanek et al. (1992). The range of values for $\delta^{13}C_{organic\ matter}$ of British and Tasmanian soils is the observed range for non-desert C_3 plants. Note that the $\delta^{13}C_c/\delta^{13}C_{organic\ matter}$ relation from most speleothems fall well outside observed or modeled relation in modern soils globally.

process is the likely explanation for the up to 10‰ enrichment in $\delta^{13}C_c$ values between fast drips and pools in Soreq Cave, Israel (Bar-Mathews *et al.*, 1995). A third process suggested by Baker *et al.* (1997) is that cave solutions could off-gas or exchange with the atmosphere prior to entering the cave. Any three of the processes separately or combined in some manner could produce the enrichments in ^{13}C observed in British, Norwegian, Canadian, and Tasmanian speleothems.

This is not to say that some caves do not behave like a giant Cerling soil pore; but at least two conditions would must be met to make this so. First, pCO_2 in the cave would have to be high enough to slow the rate of CO_2 loss from solution, thus preventing kinetic enrichment of ^{13}C, and facilitating full isotopic exchange between C species (including cave CO_2). The slow rate of back-reaction between HCO_3 (in the cave drip) and cave CO_2 (Dulinski & Rozanski, 1990) may be particularly problematic in caves compared to soils. In caves, calcite precipitation appears to occur on the scale of minutes to hours, whereas in soils, dewatering (after a rainfall event) by plants happens more in the scale of days to weeks, ample time for back-reaction of $HCO_3(aq)$ and soil $CO_2(g)$.

Second, the $\delta^{13}C$ value of cave CO_2 with which the cave solution is back reacting has to be the same as that of soil CO_2 in the recharge zone. Alteration of the $\delta^{13}C$ value of cave CO_2 by off-gassing or by advection of atmospheric CO_2 (caves must and do "breathe" in order to precipitate calcium carbonate) would lead to a potential violation of this condition.

Speleothems

Paleovegetation reconstruction using $\delta^{13}C_c$ values from speleothems figure prominently in the speleothem studies from Iowa and Missouri. To their credit, researchers there are aware of many of the complicating issues surrounding carbon isotopes in caves mentioned above and have endeavored to obtain samples that would minimize such problems as rapid CO_2 off-gassing. For example, Dorale *et al.* (1992) and Denniston *et al.* (1999) both obtained samples from deep within caves where cave pCO_2 should be higher and humidity near 100%. Moreover, Dorale *et al.* (1998) and Denniston *et al.* (1999) replicated their records from a single cave, the best way at present to test if carbon isotopes actually record vegetation change in the vicinity of the cave. However, whether carbon isotopes in most caves actually behave as they do in soils is unproven and remains to be demonstrated rigorously with replication of records or better yet with careful soil and cave gas measurements.

Dorale *et al.* (1992) defined three stages in $\delta^{18}O_c$ and $\delta^{13}C_c$ values from speleothem 1S during the Holocene: stage I (8600 to ~6000 yr ago) where $\delta^{13}C_c$ values are a low at ~−9 to −8‰; stage II (~6000–3000 yr ago) where $\delta^{13}C_c$ values increase to −5‰: and stage III where $\delta^{13}C_c$ values gradually decline again <−8‰ (Fig. 2a). Unlike the $\delta^{18}O_c$ results discussed previously, these patterns were verified by more detailed sampling from three speleothems in Denniston *et al.* (1999), strongly supporting the idea that the $\delta^{13}C_c$ values are tracking some single external change, such as the $\delta^{13}C$ value of changing vegetation covering the cave. The simplest explanation is that C_4 grasses (prairie) expanded into the area during stage II. Dorale *et al.* (1992) see the prairie expansion as a response to increasing mid-Holocene temperature, based on the $\delta^{18}O_c$ evidence previously discussed. Denniston *et al.* (1999), however, showed that the $\delta^{18}O_c$ record from Cold Water Cave could not be interpreted simply in terms of paleotemperature change. Unfortunately, the alternative interpretation that Denniston *et al.* (1999) offer for the mid-Holocene decrease in $\delta^{18}O_c$ – that the proportion of Pacific-derived (winter/spring) storms increased with respect to Gulf-derived (summer) storms – would probably lead to a decrease, not increase in the proportion of C_4 grasses. In sum, I feel that the evidence for an increase in the fraction of C_4 biomass in the mid-Holocene above Cold Water Cave is well supported by the $\delta^{13}C_c$ values, but that the $\delta^{18}O_c$ evidence does not provide an unambiguous answer why these changes occurred.

A similar case for reconstructing changes in the fraction of C_3 to C_4 plants using $\delta^{13}C_c$ values from speleothems comes from Crevice Cave in Missouri (Dorale *et al.*, 1998) (Fig. 6) and from number of other caves in the region (Denniston *et al.*, 2000). As with Cold Water Cave, the strength of the evidence from Crevice Cave lies in the fact that the changes in $\delta^{13}C_c$ values are repeated in three different speleothems widely separated in the cave. Here Dorale *et al.* (1998) argue for a sharp expansion in the fraction C_4 biomass (i.e. prairie) between 70,000 and 55,000 yr ago based on increases in $^{13}C_c$ values from −9 to −2‰, followed by a nearly equivalent re-expansion of the C_3 biomass (forest) after 55,000 yr ago (Fig. 6). Likewise, duplication of shifts in $\delta^{13}C_c$ values from some (but not all) speleothem records studied by Denniston *et al.* (2000) in the Ozarks supports the idea that least some these records are tracking changes in the fraction C_3/C_4 biomass. Key assumptions underlying these vegetation reconstructions are that carbonate dissolution occurred under open-system conditions, that cave solutions are in full carbon isotopic equilibrium with cave CO_2, that $\delta^{13}C_{cave\,CO_2} = \delta^{13}C_{soil\,CO_2}$, and that grasses in the region are C_4, not C_3.

Other recent studies from Romania (Lauritzen & Onac, 1999) and Zaire (Brook *et al.*, 1990) make a less compelling case for using the $\delta^{13}C_{c\,value}$ from speleothems to track paleovegetation changes. The main weakness of these studies are that they rely on single speleothem records to reconstruct vegetation.

Devils Hole

Core DH–11 from Devils Hole has yielded a continuous record of change in the $\delta^{13}C$ value of ground water from ~550,000 to 50,000 yr ago (Coplen *et al.*, 1994) (Fig. 3a). $\delta^{13}C_c$ values vary between ~−1.6 and −2.9‰. The variation in $\delta^{13}C_c$ is highly inversely correlated ($r^2 = -0.75$) with $\delta^{18}O_c$ but with a lag in the $\delta^{18}O_c$ time series of 7000 yr. Thus, $\delta^{13}C_c$ values are highest during glacial periods (low $\delta^{18}O_c$). Obliquity (40,000 yr) and precession (23,000 yr) periodicities also are evident in the DH–11 $\delta^{13}C_c$ record.

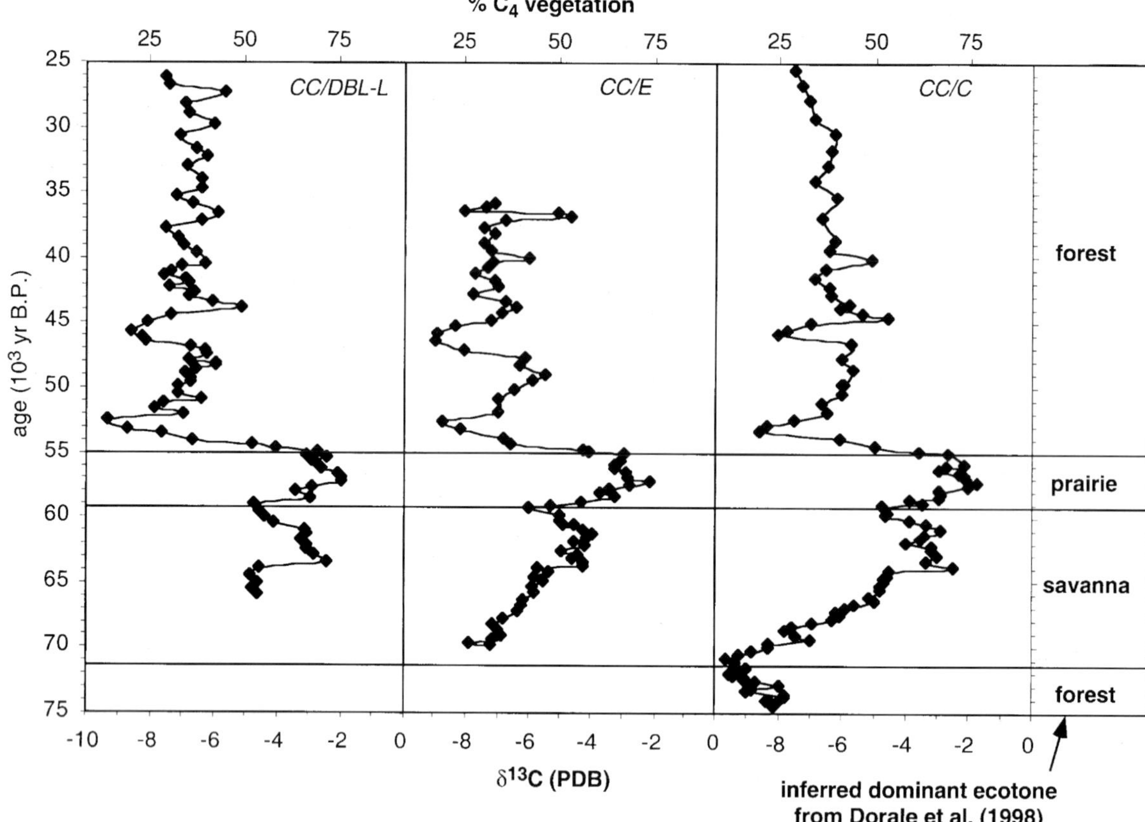

Fig. 6. $\delta^{13}C_c$ *vs. time* (10^3 *yr) from three separate speleothems in Crevice Cave, Missouri (data from Dorale et al., 1998). Dorale et al. (1998) reconstruct dominant vegetation ecotones (on right) from* $\delta^{13}C_c$ *values by assuming that grasses past and present around Crevice Water Cave are largely* C_4*, and by using values predicted by the soil-diffusion model of Cerling (1984). The proportion of* C_3 *to* C_4 *biomass predicted by that soil model are shown in top.* $\delta^{13}C_c$ *and* $\delta^{18}O_c$ *values co-vary in the Crevice Cave record, which Dorale et al. (1998) interpret in terms of temperature changes (higher temperature = greater* C_4 *biomass and higher* $\delta^{18}O_{mw}$*) in the region.*

The $\delta^{13}C_c$ values in DH–11 likely record fluctuations in the $\delta^{13}C$ value of dissolved inorganic carbon ($\delta^{13}C_{dic}$) in the aquifer. The $\delta^{13}C_{dic}$ value originates in the soil zone of the winter recharge areas in the adjacent mountain ranges. These recharge areas are divided into two very distinct zones. The upper elevations above the 0 °C isotherm are largely devoid of vegetation and the $\delta^{13}C_{dic}$ value of recharge waters should be heavily influenced by the atmosphere (i.e. $\delta^{13}C_{dic} > 0‰$). Below the 0 °C isotherm soils in the recharge area are thickly vegetated with C_3 plants, largely conifers, and $\delta^{13}C_{dic}$ should be close to $-11‰$ (Quade *et al.*, 1989). In addition, with the long ($\sim 10^3$–10^4 yr) travel times to Devils Hole, significant exchange with limestone would have both decreased the ^{14}C (thus making the water appear older than it is) and increased the ^{13}C content of the dissolved inorganic carbon in water.

In their conclusions, Coplen *et al.* (1994) offer two explanations for the distinct periodicity in the $\delta^{13}C_c$ record from DH–11. One is that the shifts in the $\delta^{13}C$ of atmospheric CO_2 drive changes in $\delta^{13}C_c$, with higher $\delta^{13}C$ values in the glacial times and lower ones in the interglacial periods. In support of this, some $\delta^{13}C_c$ records from marine plankton show a periodicity similar to that in DH–11, although the amplitude of most of these changes appears to be much too small ($<1.0‰$) to account for the $>1.3‰$ change in the $\delta^{13}C_c$ values in DH–11.

Another, perhaps more plausible explanation is that the periodicity in the $\delta^{13}C_c$ record from DH–11 reflects changes in the proportion of vegetated to non-vegetated areas in the recharge area (Coplen *et al.*, 1994). In glacial periods, the 0 °C isotherm (the main control on upper treeline elevation) would have dropped by 800–1000 m, thereby greatly increasing the proportion of bare or thinly vegetated soil in the recharge zone. The larger fraction atmospheric CO_2 in soil CO_2 would increase $\delta^{13}C_{dic}$, and in turn $\delta^{13}C_c$ in DH–11.

Where Isotope Studies from Speleothem and Ground-Water Calcite Stand

The trend away from interpreting the $\delta^{18}O_c$ record from speleothems in purely temperature terms, while probably realistic, reduces the allure of speleothems for paleoclimate reconstruction. Nonetheless, speleothems remain one of the most datable of Quaternary continental records, and other aspects of the deposits, such as the age distributions and growth rates (e.g. Ayliffe *et al.*, 1998; Hennig *et al.*, 1983; Musgrove *et al.*, 2001; Qian & Zhu, 2002; Spötl *et al.*, 2002), the apparent potential for annual or semi-annual banding (e.g. Baker *et al.*, 1993; Polyak & Asmerom, 2001; Shopov *et al.*, 1994, but see Betancourt *et al.*, 2002; Qian & Zhu, 2002), and the use of $\delta^{13}C_c$ values in paleovegetation reconstruction will continue to attract researchers to speleothem study. The prospects for the use of $\delta^{18}O_c$ values would be enhanced by development of a more isotopically reproducible method of fluid inclusion extraction, by testing for the evaporation effects in temperate caves suggested by Denniston *et al.* (1999), and by improvements in our understanding of the controls on the spatial variation of the $d\{\delta^{18}O_{mw}\}/dT$ dependence and of the slope and intercept in $\delta D = 8\delta^{18}O + 10$.

The paucity of rigorous measurement and modeling of the carbon isotope system in modern caves hampers the use of $\delta^{13}C_c$ values in paleovegetation reconstruction. I caution the speleothem community against the rush the over the last decade to use the $\delta^{13}C_c$ from speleothems to reconstruct shifts in the fraction C_3/C_4 biomass. As demonstrated by several studies, the use of $\delta^{13}C_c$ values to reconstruct the fraction C_3/C_4 biomass simply does not work in many and perhaps most cave situations. Relatively simple measurements of the $\delta^{13}C$ and $\delta^{14}C$ of coupled soil and cave CO_2 gas and dissolved C would go a long way toward testing how faithful a $\delta^{13}C_c$ record from a certain cavern of a cave is to the $\delta^{13}C$ of surface plant biomass above the cave. In the absence of such measurements, the best approach to convincingly reconstructing paleovegetation involves duplication of $\delta^{13}C_c$ records within caves from widely spaced speleothems in poorly ventilated portions of the cave, as in Dorale *et al.* (1998) and Denniston *et al.* (1999).

It is lamentable that more isotope records like that at Devils Hole have not been developed, not only as a goal in itself but as a way of resolving the current controversy surrounding interpretation of the DH record. In my view, ground-water calcites are probably widespread globally and coat fractures in the distal portion of many carbonate aquifers. Ground-water calcites obtained from drill cores extracted from carbonate aquifers may offer more widespread opportunities for sampling than those exactly like Devils Hole, where divers only entered the aquifer through a collapsed cavern. In the meantime, development of speleothem records outside the southwestern U.S., like that in West Cave in Israel (Frumkin *et al.*, 1999) or Spannagel Cave in Austria (Spötl *et al.*, 2002), seems to offer the most direct means of testing the "regionality" issues currently surrounding the Devils Hole record.

Acknowledgments

My thanks to T. Cerling, I. Winograd, J. Dorale, R. Denniston, and L. Gonzalez for discussions and comments recently or over the years, to J. Dorale, T. Herbert and L. Gonzalez for kindly sharing their data, and to L. Klasky, J. Betancourt, and the FEDEX office in Sucre, Bolivia for getting all those references to me during my sabbatical.

References

Amundson, R. & Kelly, E. (1987). The chemistry and mineralogy of a CO_2-depositing spring in the California Coast Range. *Geochimica et Cosmochimica Acta*, **51**, 2883–2890.

Ayalon, A., Bar-Mathews, M. & Sass, E. (1998). Rainfall-recharge within a karstic terrain in the eastern Mediterranean semi-arid region, Israel: $\delta^{18}O$ and δD characteristics. *Journal of Hydrology*, **207**, 18–31.

Ayliffe, L.K., Marianelli, P.C., Moriarty, K.C., Wells, R.T., McCulloch, M.T., Mortimer, G.E. & Hellstrom, J.C. (1998). 500 ka precipitation record from southeastern Australia: evidence for interglacial relative aridity. *Geology*, **26**, 147–150.

Baker, A., Ito, E., Smart, P.L. & McEwan, R.G. (1997). Elevated and variable values of ^{13}C in speleothems in a British cave system. *Chemical Geology*, **136**, 263–270.

Baker, A., Smart, P.L., Edwards, R.L. & Richards, D.A. (1993). Annual growth banding in a cave stalagmite. *Nature*, **304**, 518–520.

Bar-Mathews, M., Ayalon, A., Kaufman, A. & Wasserburg, G.J. (1997). Late Quaternary climate in the eastern Mediterranean region from the stable isotope analysis of speleothems of Soreq cave (Israel). *Quaternary Research*, **47**, 155–168.

Bar-Mathews, M., Ayalon, A., Kaufman, A. & Wasserburg, G.J. (1999). The eastern Mediterranean paleoclimate as a reflection of regional events: Soreq Cave, Israel. *Earth and Planetary Science Letters*, **166**, 85–95.

Bar-Mathews, M., Ayalon, A., Mathews, A., Sass, E. & Halicz, L. (1995). Carbon and oxygen isotope study of the active water-carbonate system in a karstic Mediterranean cave: implications for paleoclimate research in semiarid regions. *Geochimica et Cosmochimica Acta*, **60**, 337–347.

Baskaran, M. & Krishnamurthy, R.V. (1993). Speleothems as a proxy for the carbon isotope composition of atmospheric CO_2. *Geophysical Research Letters*, **20**(7), 603–606.

Berger, A. & Loutre, M.F. (1991). Insolation values for the climate of the last 10 million years. *Quaternary Sciences Review*, **10**(4), 297–317.

Betancourt, J.L., Grissino-Meyer, H.D., Salzer, M.W. & Swetnam, T.W. (2002). A test of "annual resolution" in stalagmites using tree rings. *Quaternary Research*, **58**, 197–199.

Brook, G.A., Burney, D.A. & Cowart, J.B. (1990). Desert palaeoenvironmental data from cave speleothems with examples from the Chihuahuan, Somali-Chalbi, and Kalahari Deserts. *Palaeogeography, Palaeoclimatology, Palaeoecology*, **76**, 311–329.

Burns, S.J., Fleitmann, D., Matter, A., Neff, U. & Mangini, A. (2001). Speleothem evidence from Oman for continental pluvial events during interglacial periods. *Geology*, **29**, 623–626.

Cerling, T.E. (1984). The stable isotopic composition of modern soil carbonate and its relationship to climate. *Earth and Planetary Science Letters*, **71**, 229–240.

Cerling, T.E. & Quade, J. (1993). Stable carbon and oxygen isotopes in soil carbonates. *In*: Swart, P., McKenzie, J.A. & Lohman, K.C. (Eds), *Continental Indicators of Climate*.

Proceedings of Chapman Conference, Jackson Hole, Wyoming, American Geophysical Union Monograph 78, 217–231.

Coplen, T.B., Winograd, I.J., Landwehr, J.M. & Riggs, A.C. (1994). 500,000-year stable carbon isotope record from Devils Hole, Nevada. *Science*, **263**, 361–364.

Dandurand, J.L., Gout, R., Hoefs, J., Menschel, G., Schott, J. & Usdowski, E. (1982). Kinetically controlled variations of major components and carbon and oxygen isotopes in a calcite-precipitating spring. *Chemical Geology*, **36**, 299–315.

Dansgaard, W. (1964). Stable isotopes in precipitation. *Tellus*, **4**, 43–468.

Denniston, R.F., Gonzalez, L.A., Asmerom, Y., Baker, R.G., Reagan, M.K. & Bettis, E.A. (1999). Evidence for increased cool season moisture during the middle Holocene. *Geology*, **27**, 815–818.

Denniston, R.F., Gonzalez, L.A., Asmerom, Y., Polyak, V.J., Reagan, M.K. & Saltzman, M.R. (2001). A high-resolution speleothen record of climatic variability at the Allerod-Younger Dryas transition in Missouri, central United States. *Palaeogeography, Palaeoclimatology, Palaeoecology*, **176**, 147–155.

Denniston, R.F., Gonzalez, L.A., Asmerom, Y., Reagan, M.K. & Recelli-Snyder, H. (2000). Speleothem carbon isotopic records of Holocene environments in the Ozarks Highlands, USA. *Quaternary International*, **67**, 21–27.

Dorale, J.A., Edwards, R.L., Ito, E. & Gonzalez, L.A. (1998). Climate and vegetation history of the midcontinent from 75 to 25 ka: a speleothen record from Crevice Cave, Missouri, USA. *Science*, **282**, 1871–1874.

Dorale, J.A., Gonzalez, L.A., Reagan, M.K., Pickett, D.A., Murrell & Baker, R.G. (1992). A high-resolution record of Holocene climate change in speleothem calcite from Cold Water Cave, northeast Iowa. *Science*, **258**, 1626–1630.

Dulinski, M. & Rozanski, K. (1990). Formation of $^{13}C/^{12}C$ isotope ratios in speleothems: a semi-dynamic model. *Radiocarbon*, **32**, 7–16.

Edwards, R.L., Cheng, H., Murrell, M.T. & Goldstein, S.J. (1997). Protactinium-231 dating of carbonates by thermal ionization mass spectrometry: implications for Quaternary climate change. *Science*, **276**, 782.

Edwards, R.L. & Gallup, C.D. (1993). Dating the Devils Hole calcite vein. *Science*, **259**, 1626.

Frumkin, A., Ford, D.C. & Schwarcz, H.P. (1999). Continental oxygen isotopic record of the last 170,000 years in Jerusalem. *Quaternary Research*, **51**, 317–327.

Gallup, C.D., Cheng, H., Taylor, F.W. & Edwards, R.L. (2002). Direct determination of the timing of sea level change during Termination II. *Science*, **295**, 310–313.

Gascoyne, M. (1992). Paleoclimatic determination from cave calcite deposits. *Quaternary Science Reviews*, **11**, 609–632.

Gascoyne, M., Ford, D.C. & Schwarcz, H.P. (1981). Late Pleistocene chronology and paleoclimate of Vancouver Island determined from cave deposits. *Canadian Journal of Earth Sciences*, **18**, 1643–1652.

Goede, A., Green, D.C. & Harmon, R.S. (1982). Isotopic composition of precipitation, cave drips, and actively forming speleothems at three Tasmanian cave sites. *Helictite*, **20**, 17–27.

Goede, A., Green, D.C. & Harmon, R.S. (1986). Late Pleistocene palaeotemperature record from a Tasmanian speleothem. *Australian Journal of Earth Sciences*, **33**, 333–342.

Goede, A. & Hitchman, M.A. (1983). Late Quaternary Climate change – Evidence from a Tasmanian speleothem. *In*: Vogel, J.C. (Ed.), *Late Cainozoic Palaeoclimates of the Southern Hemisphere*. Proceedings of an international symposium held by the South African Society for Quaternary Research, Swaziland, South Africa, Balkema, Rotterdam, 221–232.

Goede, A., Veeh, H.H. & Ayliffe, L.K. (1990). Late Quaternary palaeotemperature records for two Tasmanian speleothems. *Australian Journal of Earth Sciences*, **37**, 267–278.

Gonfiantini, R., Panichi, C. & Tongiorgi, E. (1968). Isotopic disequilibrium in travertine deposition. *Earth and Planetary Science Letters*, **5**, 55–58.

Harmon, R.S. (1979). An isotopic study of groundwater seepage in the central Kentucky karst. *Water Resources Research*, **15**(2), 476–480.

Harmon, R.S. & Schwarcz, H.P. (1981). Changes in the ^{2}H and ^{18}O enrichment of meteoric water and Pleistocene glaciation. *Nature*, **290**, 125–128.

Harmon, R.S., Schwarcz, H.P., Ford, D.C. & Koch, D.L. (1979a). Late Pleistocene paleoclimates of North America as inferred from stable isotope studies of speleothems. *Quaternary Research*, **9**, 54–70.

Harmon, R.S., Schwarcz, H.P., Ford, D.C. & Koch, D.L. (1979b). An isotopic temperature record for the late Wisconsin in NE Iowa. *Geology*, **7**, 430–433.

Harmon, R.S., Thompson, P., Schwarcz, H.P. & Ford, D.C. (1978). Late Pleistocene paleoclimates of North America as inferred from stable isotope studies of speleothems. *Quaternary Research*, **9**, 54–70.

Hellstrom, J., McCulloch, M. & Stone, J. (1998). A detailed 31,000-year record of climate and vegetation change from the isotope geochemistry of two New Zealand speleothems. *Quaternary Research*, **50**, 167–178.

Henderson, G.M. & Slowey, N.C. (2000). Evidence from U-Th dating against Northern Hemisphere forcing of the penultimate deglaciation. *Nature*, **404**, 61–66.

Hendy, C.H. (1971). The isotopic geochemistry of speleothems – I. The calculation of the effects of different modes of formation on the isotopic composition of speleothems and their applicability as palaeoclimatic indicators. *Geochimica et Cosmochinica Acta*, **35**, 801–824.

Hendy, C.H. & Wilson, A.T. (1968). Paleoclimate data from speleothems. *Nature*, **219**, 48–51.

Hennig, G.J., Grun, R. & Brunnacker, K. (1983). Speleothems, travertines, and paleoclimates. *Quaternary Research*, **20**, 1–29.

Herbert, T.D., Schuffert, J.D., Andreasen, D., Heusser, L., Lyle, M., Mix, A., Ravelo, A.C., Stott, L.D. & Herguera, J.C. (2001). Collapse of the California current during glacial maxima linked to climate change in land. *Science*, **293**, 71–76.

Imbrie, J., Hays, J.D., Martinson, D.G., McIntyre, A., Mix, A.C., Morley, J.J., Pisias, N.G., Prell, W.L. & Shackleton, N.J. (1984). *In*: Berger, A.L., Imbrie, J., Hays, J., Kukla, G. & Saltzman, B. (Eds), *Milankovitch and Climate, Part 1*. Reidel, Dordrecht, Netherlands, 269–305.

Imbrie, J., Mix, A.C. & Martinson, D.G. (1993). Milankovitch theory viewed from Devils Hole. *Nature*, **363**, 531–533.

Ingraham, N., Chapman, J.B. & Hess, J.W. (1990). Stable isotopes in cave pool systems: Carlsbad Cavern, New Mexico, USA. *Chemical Geology (Isotope Geoscience Section)*, **86**, 65–74.

Kim, S.-T. & O'Neil, J.R. (1997). Equilibrium and non-equilibrium oxygen isotope effects in synthetic carbonates. *Geochimica et Cosmochimica Acta*, **61**, 3461–3475.

Lauritzen, S.-E. (1995). High-resolution paleotemperature proxy record from the last interglaciation based on Norwegian speleothems. *Quaternary Research*, **43**, 133–146.

Lauritzen, S.-E. & Onac, B.P. (1999). Isotopic stratigraphy of a last intergalcial stalagmite from northwestern Romania: correlation with the deep-sea record and northern-latitude speleothem. *Journal of Cave and Karst Studies*, **61**(1), 22–30.

Ludwig, K.R., Simmons, K.R., Szabo, B.J., Winograd, I.J., Landwehr, J.M., Riggs, A.C. & Hoffman, R.J. (1992). Mass-spectrometric ^{230}Th–^{234}U–^{238}U dating of the Devils Hole calcite vein. *Science*, **258**, 284–287.

Ludwig, K.R., Simmons, K.R., Szabo, B.J., Winograd, I.J., Landwehr, J.M., Riggs, A.C. & Hoffman, R.J. (1993). Response to Edwards and Gallup. *Science*, **259**, 1626–1627.

McDermott, F., Mattey, D. & Hawkesworth, C. (2001). Centennial-scale Holocene climate variability revealed by a high-resolution speleothem δ^{18}O record from SW Ireland. *Science*, **294**, 1328–1331.

Musgrove, M., Banner, J.L., Mack, L.E., Combs, D.M., James, E.W., Cheng, H. & Edwards, R.L. (2001). Geochronology of late Pleistocene to Holocene speleothems from central Texas: implications for regional paleoclimate. *Geological Society of America Bulletin*, **113**, 1532–1543.

Polyak, V.J. & Asmerom, Y. (2001). Late Holocene climate and cultural changes in the Southwestern United States. *Science*, **294**, 148–151.

Qian, W. & Zhu, Y. (2002). Little Ice Age climate near Beijing, China, inferred from historical and stalagmite records. *Quaternary Research*, **57**, 109–119.

Quade, J., Cerling, T.E. & Bowman, J.R. (1989). Systematic variations in the carbon and oxygen isotopic composition of pedogenic carbonate along elevation transects in the southern Great Basin, USA. *Geological Society of America Bulletin*, **101**, 464–475.

Riggs, A.C., Carr, W.J., Kolesar, P.T. & Hoffman, R.J. (1994). Tectonic speleogenesis of Devils Hole, Nevada, and implications for hydrogeology and the development of long, continuous paleoenvironmental records. *Quaternary Research*, **42**, 241–254.

Romanek, C.S., Grossman, E.T. & Morse, J.W. (1992). Carbon isotopic fractionation in synthetic aragonite and calcite: effects of temperature and precipitation rate. *Geochimica et Cosmochimica Acta*, **56**, 419–430.

Rozanski, R., Araguas-Araguas, L. & Gonfiantini, R. (1993). Isotopic patterns in modern global precipitation. *In*: Swart, P., McKenzie, J.A. & Lohman, K.C. (Eds), *Continental Indicators of Climate*. Proceedings of Chapman Conference, Jackson Hole, Wyoming, American Geophysical Union Monograph 78, 1–36.

Shrag, D.P., Hampt, G. & Murray, D.W. (1996). The temperature and oxygen isotopic composition of the glacial ocean. *Science*, **272**, 1930–1932.

Schwarcz, H.P., Harmon, R.S., Thompson, P. & Ford, D.C. (1976). Stable isotope studies of fluid inclusions in speleothems and their paleoclimatic significance. *Geochimica et Cosmochimica Acta*, **40**, 657–665.

Schwarcz, H.P. & Yonge, C. (1983). Isotopic composition of palaeowaters as inferred from speleothem and its fluid inclusions. *In*: *Paleoclimates and Paleowaters: A Collection of Environmental Studies*, *STI/PUB* 1621. International Atomic Energy Agency (I.A.E.A.), Vienna, 115–133.

Shackleton, N.J. (1993). Last interglacial in Devils Hole. *Nature*, **362**, 596.

Shackleton, N.J. (2000). The 100,000-year ice-age cycle identified and found to lag temperature, carbon dioxide, and orbital eccentricity. *Science*, **289**, 1897–1902.

Shopov, Y.Y., Ford, D.C. & Schwarcz, H.P. (1994). Luminescent microbanding in speleothems: high-resolution chronology and paleoclimate. *Geology*, **22**, 407–410.

Spötl, C., Mangini, A., Frank, N., Eichstadter, R. & Burns, S.J. (2002). Start of the last interglacial period at 135 ka: evidence from a high alpine speleothem. *Geology*, **30**, 815–818.

Szabo, B.J., Kolesar, P.T., Riggs, A.C. & Winograd, I.J. (1994). Paleoclimatic inference from a 120,000-yr calcite record of water-table fluctuation in Browns Room of Devils Hole, Nevada. *Quaternary Research*, **41**, 59–69.

Thompson, P.T., Schwarcz, H.P. & Ford, D.C. (1976). Stable isotope geochemistry, geothemometey, and geochronology of speleothems from West Virginia. *Geological Society of America Bulletin*, **87**, 1730–1738.

Usdowski, E., Hoeffs, J. & Menschel, G. (1979). Relationship between ^{13}C and ^{18}O fractionation and changes in major element composition in a recent calcite-depositing spring – a model of chemical variations with inorganic $CaCO_3$ precipitation. *Earth and Planetary Science Letters*, **42**, 267–276.

Wang, Y.J., Cheng, H., Edwards, R.L., An, Z.S., Wu, J.Y., Shen, C.-C. & Dorale, J.A. (2001). A high-resolution absolute-dated late Pleistocene monsoon record from Hulu Cave, China. *Science*, **294**, 2345–2348.

Wigley, T.M.L. & Brown, M.C. (1976). Cave physics. *In*: Ford, T.D. & Cullingford, C.M.D. (Eds), *The Science of Speleology*. Academic Press, New York, 329–358.

Wilson, A.T., Hendy, C.H. & Reynolds, C.P. (1979). Short-term climate change and New Zealand temperatures during the last millennium. *Nature*, **279**, 315–317.

Winograd, I.J. (2002). Technical Comments: The California Current, Devils Hole, and Pleistocene climate. *Science*, **296**, 7a.

Winograd, I.J., Coplen, T.B., Landwehr, J.M., Riggs, A.C., Ludwig, K.R., Szabo, B.J., Kolesar, P.T. & Revesz, K.M. (1992). Continuous 500,000-year climate record from Devils Hole, Nevada. *Science*, **258**, 255–260.

Winograd, I.J., Coplen, T.B., Ludwig, K.R., Landwehr, J.M. & Riggs, A.C. (1996). High resolution $\delta^{18}O$ record from Devils Hole, Nevada, for the period 80–19 ka. *EOS, Transactions of the American Geophysical Union*, **77**(17), S169.

Winograd, I.J., Landwehr, J.M., Ludwig, K.R., Coplen, T.B. & Riggs, A.C. (1997). Duration and structure of the past four glaciations. *Quaternary Research*, **48**, 141–154.

Winograd, I.J. & Pearson, F.J. (1976). Major carbon-14 anomaly in a regional carbonate aquifer: possible evidence for megascale channeling, south central Great Basin. *Water Resources Research*, **12**, 11125–11143.

Winograd, I.J., Szabo, B., Coplen, T.B. & Riggs, A.C. (1988). A 250,000-year climatic record from Great Basin vein calcite: implications for Milankovitch theory. *Science*, **242**, 1275–1280.

Winograd, I.J., Szabo, B., Coplen, T.B., Riggs, A.C. & Kolesar, P.T. (1985). Two-million-year record of deuterium depletion in Great Basin ground waters. *Science*, **227**, 519–522.

Yonge, C.J., Ford, D.C., Gray, J. & Schwarcz, H.P. (1985). Stable isotope studies of cave seepage water. *Chemical Geology*, **58**, 97–105.

Yurtsever, Y. & Gat, J.R. (1981). Atmospheric waters. *In*: Ford, J.R. & Gonfiantini, R. (Eds), *Stable Isotope Hydrology: Deuterium and Oxygen in the Water Cycle*. International Atomic Energy Agency (I.A.E.A.), Vienna, Technical Report Series, **210**, 103–142.

Rivers and riverine landscapes

David R. Montgomery[1] and Ellen E. Wohl[2]

[1] *Department of Earth and Space Sciences, University of Washington, Seattle, WA, USA*
[2] *Department of Earth Resources, Colorado State University, Ft. Collins, CO, USA*

Introduction

The study of fluvial processes and sediment transport has a long history (e.g. Chézy, 1775; du Boys, 1879; Manning, 1891; Shields, 1936) before groundbreaking studies in the 1950s and 1960s established fundamental empirical aspects of hydraulic geometry and advanced understanding of the general processes governing river morphology and dynamics. Over the last 40 years fluvial geomorphology has grown from a focus primarily on studies of the mechanics and patterns of alluvial rivers to an expanded interest in mountain channels, the role of rivers in landscape evolution and as a geological force, and the relation of fluvial processes to aquatic and riparian ecology. An increasing emphasis on quantitative analysis and process models has forged new views of river networks as systems controlled by suites of processes, from landslide-dominated headwater valleys, to high-energy bedrock channels in mountains, lowland alluvial valleys, and estuarine channels. Key recent advances in understanding of rivers include: the processes and dynamics that lead to the development of different types of channels in different portions of a channel network; increased understanding of the fundamental coupling and interaction of rivers and tectonics; the influences of vegetation – both live and dead – on river processes and forms; and the role of riverine disturbance processes on ecological systems. In addition, advances in understanding the nature, extent, and legacies of post-glacial changes and human activities on rivers systems have increased knowledge of regional river systems. Increasingly, investigators are exploring the influences of fluvial processes on fields as diverse as the ecology of benthic macroinvertebrates and metamorphic petrology, as well as for practical efforts in conservation biology and watershed management. River restoration is emerging as an area of substantial societal investment, and presents a wealth of research opportunities in applied fluvial geomorphology.

Advances in Understanding

We cannot pretend even to attempt to review advances across the entire field of fluvial geomorphology in these few pages. Consequently, we will focus on a few topics we consider to have advanced fundamentally over the past several decades. Our review is biased and incomplete: we hope that these limitations help make it useful.

Types of Channels

Recognition that there are different types of river and stream channels is nothing new. In a paper based on his experiences with the U.S. Exploring Expedition from 1838 to 1842, J.D. Dana discussed fundamental differences between mountain channels and lowland rivers on islands of the South Pacific (Dana, 1850). Similarly, large differences in river patterns (e.g. braided, meandering, and straight) have been recognized and studied for decades. Although many fundamental aspects of river processes have been applied in the study of rivers worldwide, researchers have increasingly recognized that rivers also have distinctly regional character (e.g. rivers of the Colorado Plateau, Great Plains, Rocky Mountains, Cascades, and the coast ranges of the Pacific states). Broad variations in hydrology, geology, and vegetation impart a strong regional imprint to the morphology and dynamics of many river systems. Hydrologic regimes differ among arid, tropical, temperate, and polar regions; the geomorphic processes influencing mountain rivers differ from those in lowland regions; and the influences of vegetation reflect the dominance of forest, grassland, or shrub/scrub communities. Because different combinations of these fundamental regimes impart different characteristics to river systems in different regions, rivers are best understood in the context of their climatic and geomorphic setting, and disturbance history (Booth *et al.*, 2003; Buffington *et al.*, 2003; Montgomery, 1999; Montgomery & MacDonald, 2002).

Until recent decades research on mountain rivers and streams was eclipsed by a greater number of studies on lowland alluvial rivers. Recent work has advanced understanding of connections between process and form in mountain channel networks where reach-scale distinctions are apparent in both channel bed morphology and basin-wide relations between drainage area and slope. Montgomery & Buffington (1997) showed that different types of alluvial bed morphology in mountain channel reaches reflect the balance between transport capacity and sediment supply. Due to long-term differences in processes driving bedrock erosion, debris-flow-dominated colluvial channels and fluvial channels in upland bedrock valleys have different relations between drainage area and slope (Montgomery & Foufoula-Georgiou, 1993; Stock & Dietrich, 2003). These studies showed that different portions of mountain channel networks are controlled by different processes, with key distinctions between colluvial, bedrock, and alluvial channels. In the past several decades channel and

DEVELOPMENTS IN QUATERNARY SCIENCE
VOLUME 1 ISSN 1571-0866
DOI:10.1016/S1571-0866(03)01011-X

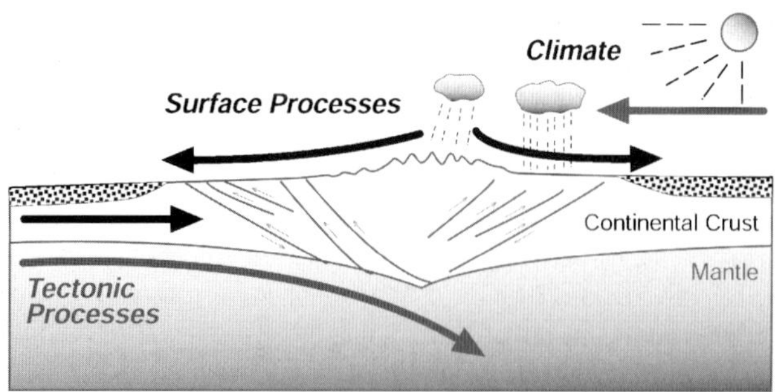

Fig. 1. Schematic illustration of relations between climate, tectonics, and erosion in shaping topography (after Willett, 1999).

floodplain classification systems have proliferated, particularly for use in the regulatory and river management arenas.

Rivers and Tectonics. Digital topography provides for quantitative, landscape-scale analyses that are modernizing the practice of geomorphology, and especially current investigations focused on relations between bedrock river incision, rock uplift, and landscape evolution (Fig. 1). The ability to analyze landforms quantitatively has been revolutionized by geographic information systems (GIS) and high-resolution topographic data. Increasing availability and resolution of digital terrain models for much of Earth's surface open new opportunities for studying the role of rivers in the evolution of particular landscapes and on interactions between rivers and tectonics. Recent interest has focused in particular on the role of rivers as a primary boundary condition on landscape evolution (Burbank *et al.*, 1996; Finlayson *et al.*, 2002; Montgomery & Brandon, 2002; Seidl & Dietrich, 1992; Whipple *et al.*, 1999) and the role of bedload cover and sediment transport in bedrock river incision (Sklar & Dietrich, 1998, 2001). Interest in the interaction of rivers and tectonics also focuses on the role of fluvial processes in maintaining steady-state orogens (Willett & Brandon, 2002) and in the dynamics of knickpoint-dominated systems (Seidl *et al.*, 1994). Research addressing spatial and temporal scales over which steady-state assumptions may be reasonable (Whipple, 2001) highlights interest in understanding the coupling of fluvial and tectonic processes.

The characteristic concave upward profiles of rivers have long been thought to reflect the downstream trade-off in erosion rates or transport capacity between increasing discharge and decreasing slope (Gilbert, 1877; Mackin, 1948). Models of river profile development predict exponential, logarithmic, or power function forms for steady-state river profiles (Snow & Slingerland, 1987), and deviations from expected trends are interpreted to reflect differences in either geologic or climate history, or spatial variability in erosion resistance, erosional processes, or rock uplift rates (Hack, 1957; Snow & Slingerland, 1990). Seeber & Gornitz (1983), for example, used the alignment of knickpoints on trans-Himalayan rivers to argue for active deformation along the Main Central Thrust between the Lesser and High Himalaya. Aided by digital elevation models (DEMs), Seidl *et al.* (1996) used the longitudinal profiles of rivers draining the flank of the south-east Australian escarpment to investigate the kinematics and pattern of escarpment retreat. Based on the spatial coincidence of distinct knickpoints at the head of major tributaries they inferred that long-term escarpment retreat at about 2 mm yr^{-1} was controlled by rock strength and fracturing more than by fluvial discharge or stream power. Hence, analyses of DEM-derived river profiles can help to elucidate the mechanisms behind the long-term evolution of both active and passive margins.

Many workers report that channel slope varies as an inverse power function of drainage area

$$S = cA^{-\theta} \tag{1}$$

where θ varies from 0.2 to 1.0 (Flint, 1974; Hack, 1957; Hurtrez *et al.*, 1999; Kirby & Whipple, 2001; Moglen & Bras, 1995; Snyder *et al.*, 2000; Tarboton *et al.*, 1989). Headwater channels prone to debris flows exhibit different values of θ than do downstream fluvial channels (Montgomery & Foufoula-Georgiou, 1993; Seidl & Dietrich, 1992), and plots of drainage area vs. channel slope have been used to characterize different portions of a river system dominated by different processes (Montgomery, 2001; Montgomery & Foufoula-Georgiou, 1993; Snyder *et al.*, 2000).

Over the past decade it has become common for the local erosion rate (E) to be modeled as a function of drainage area (A) and local slope (S) for detachment-limited channel incision

$$E = KA^m S^n, \tag{2}$$

where K is an empirical coefficient that incorporates climatic factors and bedrock erodibility, and m and n are thought to vary with different erosional processes. For the special case of steady-state topography, the local erosion rate at a distance x along the channel $E(x)$ everywhere equals the local rock uplift rate $U(x)$, and Eq. (2) can be rearranged to yield a relation between drainage area and slope

$$S = \left[\frac{U(x)}{K(x)}\right]^{1/n} A^{-(m/n)} \tag{3}$$

For spatially uniform rock uplift and lithology (i.e. $U(x)$ and $K(x)$ are constants), Eqs (1) and (3) imply that for steady-state topography $c = [U/K]^{1/n}$ and $\theta = m/n$. Models of detachment limited bedrock river incision based on both shear stress and unit stream power formulations hold that $m/n \approx 0.5$ (Whipple & Tucker, 1999).

Stock & Montgomery (1999) analyzed patterns of 13 rivers where initial river profiles of known age were compared with modern river profiles to constrain possible values of K and m/n. They found that for roughly half of the available examples the optimal m/n value ranged from 0.3 to 0.5, but that for the other half of the cases studied there was only a weak area dependence, with $m/n \approx 0.1$–0.2. They also found that K varied by at least five orders of magnitude among different lithologies, implying a huge range of potential time scales of landscape response to changes in climate or tectonic forcing. Hence, in many cases the assumption of steady state may be difficult to justify.

Lague *et al.* (2000) rearranged Eq. (3) to solve for the ratio of uplift to erodibility (i.e. $[U/K] = S^n A^{-[m-1]}$) using slopes and drainage areas derived from DEMs. By calibrating this ratio to drainage basins where the assumption of homogeneous uplift appeared reasonable, they evaluated differences in erodibility for areas underlain by different lithologies. They found a four-fold variation in erodibility between areas underlain by erodible and resistant lithologies. They also evaluated spatial patterns of rock uplift rate after normalizing to account for these lithological effects.

In a similar approach, Kirby & Whipple (2001) analyzed downstream variations in θ to evaluate longitudinal gradients in rock uplift in the Siwalik Hills in central Nepal. They found that by assuming the river profiles were in steady state, erosion rates predicted by their calibrated stream power parameters implied rock uplift rates similar to those modeled by Hurtrez *et al.* (1999) from empirical relations between erosion rate and local relief. Consequently, in some cases area-slope characteristics of river profiles may provide insight into spatial patterns (and perhaps rates) of rock uplift across active geological structures.

Roe *et al.* (2002), however, showed that feedback between orographically variable precipitation and discharge-driven river incision implies that $\theta \neq m/n$ for steady-state landscapes with strong orographic precipitation regimes. Moreover, if the sediment flux through the reach is an important factor in controlling the rate of bedrock river incision (Sklar & Dietrich, 1998), as indicated by recent flume experiments (Sklar & Dietrich, 2001), then Eqs (2) and (3) become less pertinent to the field problem. Hence, as noted by Snyder *et al.* (2000), care needs to be taken in trying to infer m/n from observations of θ.

Influences of Vegetation. Studies in the past several decades have established that vegetation is a major influence on channel morphology and patterns at scales from individual channel units (e.g. pools and bars) to entire valley bottoms. Both live and dead vegetation influence channel form and processes. Live trees and grasses can contribute substantially to bank strength, and large woody debris (logs and logjams) can cause both local erosion and deposition (Fig. 2). Surveys of pool frequency in forest channels revealed that the majority of pools are forced by flow around wood, and the frequency of wood obstructions controls pool spacing (Beechie & Sibley, 1997; Buffington *et al.*, 2002, 2003; Montgomery *et al.*, 1995). The importance of bank vegetation has been demonstrated in a number of studies, and has long been recognized (e.g. Schumm & Lichty, 1965). In a series of field experiments, Smith (1976) demonstrated that plant roots can provide the dominant component of stream bank strength. Millar (2000) recently showed that the strength contributed by bank vegetation can be significant enough to influence the transition from a meandering to a braided channel pattern. Such an influence is apparent in evidence for a rapid change from meandering to braided river morphology coincident with the global plant die-off 250 million years ago at the Permian/Triassic mass extinction event (Michaelsen, 2002; Ward *et al.*, 2000). The presence of grass or forest on river banks influences channel width, although whether channels widen or narrow depends on the type and geomorphic context of the channel (Davies-Colley, 1997; Stott, 1997; Trimble, 1997). Several studies have also investigated the role of vegetation on generating an anastomosing channel form by creating and maintaining local flow diversions that split a channel into a network of multiple channels (Collins *et al.*, 2002; Harwood & Brown, 1993; Tooth & Nanson, 1999). Such studies have broadened the range of scales over which vegetation is recognized as a primary influence on channel form.

Fig. 2. Stable logjam acting as bank revetment on the Queets River, Washington.

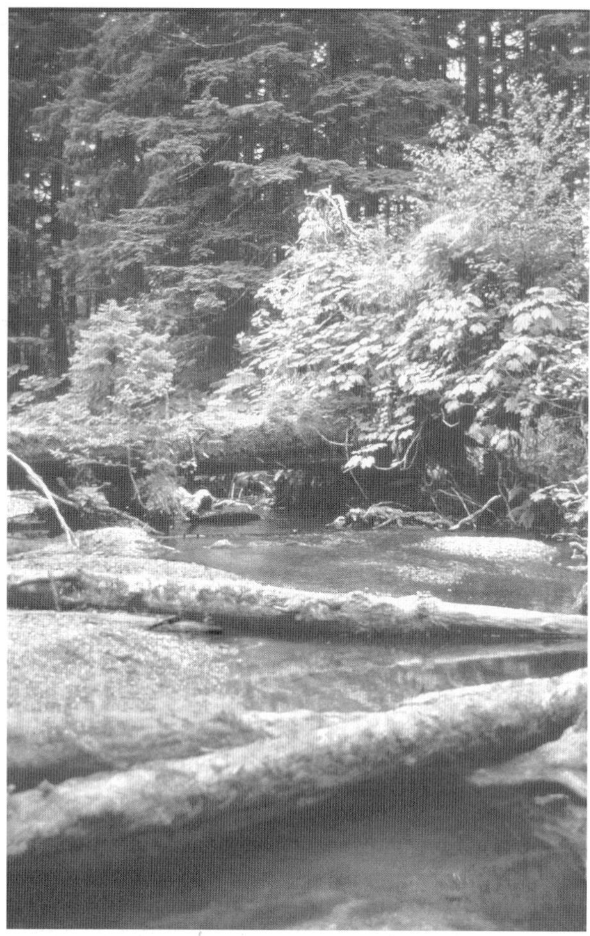

Fig. 3. Log-filled reach of a small Pacific Northwest stream.

In recent decades, the distribution of in-stream wood and the accumulation of logs into log jams has been studied extensively in Europe and North America (Abbe & Montgomery, 1996; Downs & Simon, 2001; Gregory *et al.*, 1985, 1993; Gregory & Davis, 1992; Gurnell & Sweet, 1998; Piégay, 1993; Piégay & Marston, 1998; Robison & Beschta, 1990). In addition, researchers have established that large wood affects many channel processes (Fig. 3). In particular, recent studies have shown that wood influences channel roughness (Buffington & Montgomery, 1999a, b; MacFarlane & Wohl, 2003; Manga & Kirchner, 2000; Shields & Gippel, 1995), bed-surface grain size (Buffington & Montgomery, 1999a, b; Lisle, 1995), pool formation (Abbe & Montgomery, 1996; Keller & Tally, 1979; Lisle, 1995; Montgomery *et al.*, 1995), channel-reach morphology (Keller & Swanson, 1979; Lisle, 1986; Montgomery *et al.*, 1996; Montgomery & Buffington, 1997; Nakamura & Swanson, 1993; Piégay & Gurnell, 1997), and the formation of valley-bottom landforms (Abbe & Montgomery, 1996; Collins *et al.*, 2002; Gurnell *et al.*, 2001). Some studies have shown that many of the geomorphic effects of wood in rivers arise from the influence of large stable wood as obstructions to flow and sediment transport (Abbe & Montgomery, 1996; Keller & Tally, 1979; Nakamura & Swanson, 1993). A number of workers have noted how the organization of wood and its effects on channels vary with position in the channel network (Abbe & Montgomery, 1996; Gurnell *et al.*, 2001; Keller & Swanson, 1979; Swanson *et al.*, 1982; Wallerstein *et al.*, 1997). Although many of the effects of wood occur at the scale of individual channel units (Bisson *et al.*, 1982), the integrated affects of these changes can alter channel properties at larger spatial scales of channel reaches and entire valley bottoms.

Disturbance Regimes. The characteristics and dynamics of stream habitat are recognized as providing a "geomorphic template" upon which aquatic ecosystems develop (Southwood, 1977). Disturbance regimes set by spatial and temporal variability in geomorphic processes capable of disrupting ecological systems or processes are viewed as a primary geomorphological control on stream ecosystems (Swanson *et al.*, 1988). The frequency, magnitude, and intensity of effects associated with a geomorphic process define its disturbance regime, and areas characterized by a similar disturbance regime define distinct process domains (Montgomery, 1999).

Episodicity is a fundamental characteristic of most geomorphic processes. Landslides and floods do not happen every day. The periodic nature of geomorphic phenomena has motivated ongoing examination of what controls the spatial and temporal scales over which steady-state assumptions may or may not apply. At the broad spatial and temporal scales of mountain range evolution, steady-state can be defined by constant exhumation rates over millions of years (Brandon *et al.*, 1998; Willett & Brandon, 2002). At this scale, individual storms may trigger catastrophic pulses of erosion but the long-term erosion rate is set by the rock uplift rate. The rise of mountains or periods of glaciation can act as disturbances over evolutionary time scales, but a single landslide can prove catastrophic to a local population of stream dwelling organisms. Although sediment transport in mountain streams is fundamentally episodic (Bunte & MacDonald, 1995), the time scales over which the integrated effects of discrete events can be meaningfully averaged depends on the nature of the analysis and the problem to be addressed. Incorporation of geomorphic processes into disturbance ecology, is reshaping understanding of aquatic and riparian ecosystem dynamics (e.g. Fausch *et al.*, 2002), in particular the role of periodic disturbances on the morphology and variability of mountain channel systems (e.g. Benda *et al.*, 1998).

Different kinds of organisms occupy different parts of a river system in part due to variations in the physical habitat template. Habitat characteristics and variability are influenced by the type, intensity, and frequency of disturbances. River systems exhibit both local and systematic downstream variability, as well as regional differences due to factors such as the geomorphic importance of small hydrologic events in a wet climate and the contrasting importance of rare events in arid climates. In mountain drainage basins, for example, headwater channels in confined valleys tend to be prone to high intensity, low frequency disturbances such as landslides and debris flows. After being scoured by debris flows such channels exhibit a temporal succession of habitat characteristics as material falls into the channel and gradually accumulates until the next scouring event. In contrast, lower-gradient alluvial channels in unconfined valley bottoms are frequently disturbed by lower intensity disturbances and by channel migration and avulsion. The style of disturbances in these different environments leads to differences in community structure and composition. Moreover, disturbance processes that may adversely impact local populations may be essential for creating and maintaining high-quality habitat in disturbance-prone environments. Consequently, understanding aquatic and riparian ecology may depend, in large part, on the integration of spatial and temporal disturbance processes and on their relation to the life history and distribution of particular organisms.

Post-Glacial Changes

Studies of post-glacial changes in river systems have improved understanding of regional river systems. In particular, legacies of Pleistocene glaciation on modern rivers have come to be appreciated. The oscillation between glacial and interglacial climates can result in sustained high sediment yields from rivers that never reach a steady state (Church & Ryder, 1972). Church & Slaymaker (1989), for example, showed how reworking of Pleistocene sediments still dominates the sediment budget for glaciated drainage basins in British Columbia.

Much of the progress in understanding post-glacial changes in riverine landscapes has been closely tied to advances in geochronology. Prior to the 1980s, late Quaternary geochronology was largely based on radiocarbon-dating or on relative dating using soils, stratigraphic position, rock weathering, or archeological context. Since the 1980s, numerical dating using cosmogenic isotopes, thermoluminescence, fission track, amino acid racemization, electron spin resonance and other techniques has been much more widely applied in Quaternary studies. These techniques have been especially useful in establishing chronologies for erosional or depositional episodes not directly associated with preservation of fossils. Many Holocene glacial chronologies for mountain ranges in the western U.S., for example, were originally based on radiocarbon ages from interbedded lake or marsh deposits. The use of cosmogenic isotopes to date glacial moraines directly has potential for improving the temporal resolution of glacial chronologies (e.g. Phillips *et al.*, 1990).

Coupled with advances in geochronology has been an increasingly quantifiable understanding of the episodicity of geomorphic change. The Pleistocene-Holocene transition was marked by enormous outburst floods from meltwaters ponded along the glacial margins. In the Channeled Scabland and northern margins of the Great Plains, these floods created landscapes that have been little modified by subsequent geomorphic processes (Baker & Nummedal, 1987; Kehew, 1993; Lord & Kehew, 1987). And, in regions as geologically and climatically diverse as the Appalachians and the Colorado Front Range, the Holocene was characterized at timescales of centuries to millennia by episodic geomorphic change driven by climatic variability. The relative importance of different styles of post-glacial change varied regionally across the United States, and these changes have left an imprint on modern river systems.

Puget Sound Rivers. The Puget Lobe of the Cordilleran Ice Sheet overran the Puget Sound about 17,000–16,000 cal yr B.P. (Porter & Swanson, 1998). As the Puget Lobe retreated northward at a rate of several hundred meters per year (Porter & Swanson, 1998), the melting ice exposed deeply incised valleys carved by sub-glacial streams. As river networks were re-established through a shifting network of spillways (Booth, 1994), some rivers came to occupy overdeepened subglacial meltwater troughs, whereas others were carved into the upland formed by the advance outwash. The modern character of Puget Sound rivers retains a legacy of these glacial origins (Booth *et al.*, 2003). Rivers flowing through sub-glacial meltwater troughs have aggraded during the Holocene, and were characterized historically by meandering channels (Fig. 4), some of which flowed through extensive valley bottom wetlands (Collins & Montgomery, 2001). In contrast, channels incised into advance outwash had few valley bottom wetlands and were characterized by an extensive network of anastomosing sloughs and side-channels (Collins & Montgomery, 2001). The distinction between these two contrasting styles of post-glacial history controlled the type and relative abundance of salmonid habitat in Puget Sound rivers at the time of Euro-American colonization.

Extensive post-glacial changes have also reshaped Puget Sound rivers. Post-glacial sea-level rise and isostatic rebound of up to 200 m in the North Sound have altered the extent of Holocene river valleys (Dethier *et al.*, 1995). Incision of rivers through the Holocene has altered the expanse of riverine valley bottoms (Beechie *et al.*, 2001). Immense mid-Holocene lahars from Glacier Peak created the extensive delta of the Skagit River (Dragovich *et al.*, 2000). In the 1980 eruption of Mount St. Helens, extensive lahars inundated the valley of the Toutle River, creating a broad valley flat (Fig. 5). Post-glacial establishment of forests further influenced Puget Sound rivers until historic clearing of snags and forest cover depleted in-stream wood and transformed the morphology of many Puget Sound rivers from complexes of anastomosing channels into relatively simple, single-thread meandering channels (Collins *et al.*, 2002). The post-glacial legacy has been one of extensive changes in Puget Sound rivers.

Mississippi River Drainage Basin. The retreat of the Laurentide ice sheet sent enormous volumes of meltwater flowing down the Mississippi River drainage network until the retreating ice sheet exposed the St. Lawrence and Hudson drainages. Recently-derived records of these meltwater floods come from $\delta^{18}O$ content in foraminifera (Joyce *et al.*, 1993) and grain-size variations of siliciclastic mud (Brown & Kennett, 1998) in the Gulf of Mexico, as well as from geomorphic evidence of channel cutting (Kehew & Lord, 1987; Knox, 1996) and alluvial fan deposition (Porter & Guccione, 1994). These records suggest that the ice sheet began to melt circa 14,000 ^{14}C yr B.P., with a meltwater megaflood from 12,600 to 12,000 ^{14}C yr B.P. (Brown & Kennett, 1998). Between 11,000 and 9500 ^{14}C yr B.P., a rapid decrease

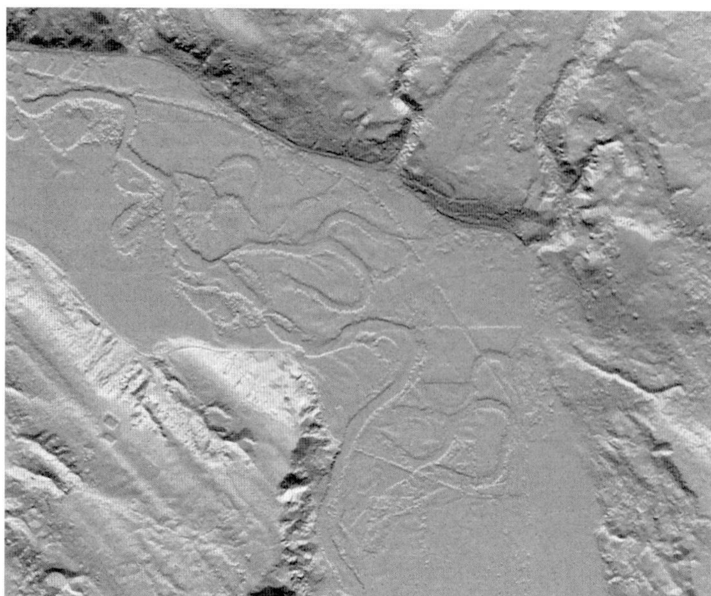

Fig. 4. Shaded relief map of LIDAR derived topography along the Snoqualmie River, Washington.

in discharge rate of the Mississippi drainage occurred as meltwater was directed eastward through the Hudson and St. Lawrence rivers (Broecker *et al.*, 1989; Teller, 1990).

The changes in water and sediment yield associated with the latest Pleistocene glaciation caused large changes in the Mississippi River. The river initially incised in response to lowered baselevel during the height of the Wisconsinan glaciation (Schumm & Brakenridge, 1987). As the ice retreated, rapid drainage development occurred in the newly exposed land at the northern margin of the drainage basin (Anderson, 1988), and the central and lower portions of the river experienced aggradation and enhanced lateral movement. The channel in the lower basin began to change from braided to meandering ca. 8800 ^{14}C yr B.P. (Baker, 1983).

The Holocene sedimentary record of the upper Mississippi River basin indicates fluctuations of $\pm 30\%$ of contemporary bankfull discharge, despite only modest changes in mean annual temperature and mean annual precipitation (Knox, 1993). During periods of larger floods (6000–5000 ^{14}C yr B.P., 3300–2000 ^{14}C yr B.P., A.D. 1450–1200), relatively rapid channel migration reworked or removed substantial amounts of valley-bottom alluvium (Knox, 1985). During periods of smaller floods (8000–6500 ^{14}C yr B.P., 5000–3300 ^{14}C yr B.P., and

Fig. 5. Lateral blast zone and lahar filled valley along the North Fork Toutle River at Mount St. Helens.

2000 ^{14}C yr B.P. to A.D. 1450), relatively slow lateral channel migration occurred and the channel and floodplain remained relatively stable (Knox, 1985). Overbank sedimentation on floodplains accelerated with the advent of agriculture in the region after A.D. 1820 (Knox, 1987).

The Holocene evolution of the lower Mississippi Valley has been a response to the effects of relative sea-level rise and variations in discharge and sediment delivery as driven by climate (Autin *et al.*, 1991). Individual rivers have alternately incised, aggraded, and changed their plan-view form during the Holocene (Autin, 1993), but regional stratigraphic records are not yet sufficient to determine whether these changes were broadly synchronous. Lobes of the Mississippi River delta, each approximately 30,000 km^2 and averaging 35 km thick, suggest that the delta's primary depositional site changes on average every 1500 years (Coleman, 1988).

Rivers of the Colorado Front Range. The eastern edge of the Colorado Rocky Mountains from the Wyoming border south to the Arkansas River drainage constitutes the Colorado Front Range. The Front Range is drained by the channels of the South Platte River, which begin as mountain rivers confined within narrow bedrock canyons, and continue beyond the mountain-front as piedmont cobble- and gravel-bed rivers before becoming sand-bed channels farther east on the Great Plains. The post-glacial history of rivers in this region represents that of many mountain ranges in the Intermountain West in that the riverine landscape reflects erosional and depositional episodes over various timescales.

The piedmont along the eastern base of the Colorado Front Range has four pediment surfaces, the oldest of which is Pliocene in age, and a younger set of five strath or fill terraces of late Pleistocene and Holocene age (Morrison, 1987). The chronology for these surfaces was largely established during the 1960s using radiocarbon and relative geochronologic methods (Scott, 1960, 1963). The larger episodes of incision have been hypothesized to represent climatic changes, and the younger surfaces may reflect late Quaternary glaciation (Morrison, 1987). Front Range glacial chronologies based on radiocarbon and relative dating methods suggest between two and four glacial episodes during the Holocene (Benedict, 1973; Birkeland *et al.*, 1971; Burke & Birkeland, 1983; Richmond, 1960). More recent dating of Pleistocene glaciations indicate that Bull Lake moraines along upper Boulder Creek, one of the drainages in the Front Range, have minimum average ^{10}Be and ^{26}Al ages of 101,000 $\pm$ 21,000 and 122,000 $\pm$ 26,000 yr. Pinedale moraines along Boulder Creek have average model ages of 16,900 $\pm$ 3500 yr and 17,500 $\pm$ 3600 yr (Dethier *et al.*, 2000). Fill terraces downstream from the moraines along Boulder Canyon represent Bull Lake, Pinedale, and Holocene surfaces (Schildgen & Dethier, 2000). Limited cosmogenic and radiocarbon dating and soil development suggest that these terraces correlate with the terraces on the piedmont. Since ~600,000 yr ago, net incision rates on the High Plains near Boulder Creek have been ~0.04 mm yr^{-1}, whereas rates in Boulder Canyon have averaged ~0.15 mm yr^{-1} since about 130,000 yr ago, suggesting that downcutting rates along the canyon have increased since early Pleistocene time (Dethier *et al.*, 2000).

In addition to erosional and depositional episodes driven by glacial and climatic change at timescales of thousands of years, rivers in the Colorado Front Range have undergone episodic change at timescales of hundreds of years as a result of hillslope instability and floods driven by precipitation and forest fires. Intense convective precipitation associated with summer thunderstorms can trigger slope mass movements and valley-bottom floods such as occurred in the Big Thompson River drainage during July 1976 (McCain *et al.*, 1979; Shroba *et al.*, 1979). Only these extreme floods, which recur at intervals of about 300–500 yr (Wohl, 2001), generate enough shear stress to overcome the high boundary resistance of the Front Range rivers. Such rare floods are thus very important in shaping valley and channel geometry.

Sierra Nevada Rivers. Once a matter considered decided, the topographic evolution of the Sierra Nevada is again controversial. According to classic studies of uplift of the Sierra Nevada, the range rose in post-Miocene or Pliocene time (Axelrod, 1957; Huber, 1981). Similarly, Wakabayashi & Sawyer (2001) argued that westward tilting, stream incision, and east-down normal and dextral faulting along the eastern escarpment of the range began ca. 5 myr ago. During these five million years, alpine glaciers repeatedly advanced and retreated, streams incised up to 1 km, and alluvial fan complexes developed along the mountain front. However, other recent studies, using new geochemical techniques, support the interpretation of little surface uplift in the Sierra Nevada since the early Tertiary (Chamberlain & Poage, 2000; House *et al.*, 1998).

Unlike the uplift history of the range, paleoclimate records of the Sierra Nevada have become less controversial in the past several decades. Pollen records indicate maximum glacial cooling of approximately 7–8 °C and, although precipitation inferences are less reliable, up to 2 m more annual precipitation during the glacial maximum (Adam & West, 1983). Radiocarbon and surface-exposure ages record multiple late Wisconsin advances in the Sierra Nevada (Osborn & Bevis, 2001), and glacial rock flour beneath Owens Lake suggests at least seven glacial advances between 84,000 and 15,000 yr ago (Bischoff & Cummins, 2001). Cosmogenic isotope ages from moraines in the eastern Sierra Nevada suggest that the transition from interglacial to full glacial conditions was rapid, with earlier glacial advances (ca. 200,000, 145,000, 115,000 yr ago) more extensive than later advances (ca. 65,000, 24,000, 21,000 yr ago) (Phillips *et al.*, 1990).

Pollen records from the Sierra Nevada indicate a drier climate 11,000–7000 yr ago, a slight increase in precipitation 7000–3000 yr ago, and establishment of the present cool-moist climate after 3000 yr ago (Anderson & Smith, 1994; Davis *et al.*, 1985). These climatic fluctuations have been associated with glacial advances during the latest Pleistocene, during an episode ca. 4000–3000 yr ago, and during the past several hundred years (Anderson & Smith, 1994). Sierra Nevada tree-ring records indicate that climate has remained relatively stable during the late Holocene (LaMarche, 1973) and that late Holocene hydrologic fluctuations are largely synchronous across the western United States (Earle, 1993).

An initial study if paleosalinity records from San Francisco Bay indicated no overall trends in the discharge of Sierra Nevada rivers during the past 2700 yr (Ingram *et al.*, 1996). However, average nonglacial erosion rates in the mountainous granitic terrain of the Sierra Nevada have varied by 2.5-fold during the Holocene (Riebe *et al.*, 2001). Spatial variability in erosion rates across the Sierra Nevada has been attributed to proximity to fault scarps and river canyons. Erosion rates and hillslope gradients are strongly correlated at sites close to scarps and canyons. These sites appear to have accelerated local baselevel lowering and catchment erosion rates that are up to 15-fold higher than those of sites more distant from scarps and canyons, where erosion rates are much more uniform and less sensitive to average hillslope gradient (Riebe *et al.*, 2000).

Rivers of the Sierra Nevada adjoin bouldery debris fans at the canyon mouths that commonly merge to form an alluvial apron. Relative fan size may reflect distribution of subsidence rates in the depositional basin (Whipple & Trayler, 1996), lithology and climate as these control both weathering rate and availability of unconsolidated material on canyon floors, and intense thunderstorm precipitation that generates sediment transport through flash floods and debris flows (Beaty, 1990; Bull, 1977). Alluvial fan deposition may be dominated by debris flows which determine both the structure of the channel network on the fan, and the long-term pattern of deposition on the fan suface (Whipple & Dunne, 1992). Alluvial fan deposition may also reflect sedimentation from outburst floods produced by failure of glacial moraines. Such fans are characterized by thick, unsorted, unstratified deposits that are a boulder-rich mix of clay to blocks deposited from noncohesive sediment gravity flows (Blair, 2001). Fans dominated by deposits from outburst floods lack the constituent levees, lobes and channel plugs, and alternating stacks of matrix-rich beds and washed gravel beds, present on fans dominated by debris flows (Blair, 2001).

Appalachian Rivers. The retreat of the Laramide ice sheet from the northeastern U.S. starting circa 16,000–15,000 ^{14}C yr B.P. was associated with rising baselevel for rivers draining to the Atlantic Ocean, glacial outburst floods from meltwater ponded along the margins of the ice sheet, and warming climate and associated changes in vegetation and weathering regime. Pollen and macrofossil records from the central Appalachians indicate an overall warming trend from 14,000 to ca. 7500–6000 yr ago (Kneller & Peteet, 1999; Webb *et al.*, 1993). The northward expansion of boreal and temperate trees during this period produced many ephemeral forest communities. Today the southern Appalachians contain the most diverse tree flora of the eastern region (Davis, 1983). From a geomorphic perspective, probably the most important point is that the Appalachians and the eastern U.S. remained forested throughout the post-glacial period, so that rivers have responded to storms and floods rather than to changes in vegetative cover (Knox, 1983). Recent research in the Appalachians has tended to focus on (i) hillslope instability and the evolution of alluvial fans; (ii) the role of large floods in shaping contemporary river landscapes; and (iii) Cenozoic river incision and landscape evolution.

The late Pleistocene was a period of intense mechanical weathering and denudation in the Appalachian highlands (Clark & Ciolkosz, 1988; Mills & Delcourt, 1991); late Pleistocene slope denudation rates were an order of magnitude higher than Holocene rates (Braun, 1989; Saunders & Young, 1983). Sediment generated from creep and solifluction was stored in mountain hollows and episodically delivered to the valley floor by debris flows that on average recurred about every 2500 yr (Eaton & McGeehin, 1997). Holocene warming terminated periglacial slope processes and reduced the rate of mechanical weathering. The reduction in sediment supply initiated stream incision through debris fans of late Pleistocene age, which resulted in lower fans of Holocene age (Eaton, 1999). Landforms such as block fields and boulder streams that are relicts from late Pleistocene colder climates are now being modified by Holocene processes (Braun, 1989; Delcourt & Delcourt, 1988; Gardner *et al.*, 1991).

Jacobson *et al.* (1989) emphasized two scales of temporal variation that influence hillslope instability and large floods in the central Appalachians: Quaternary climatic changes, and a higher frequency variation of rare events during the Holocene. The rare events arise from interactions between tropical storm paths and topographic barriers in the Appalachians. These interactions produce intense rainfall that may trigger hillslope mass movements and flooding. Topographically influenced flow concentration is the most important factor in determining relative slope stability throughout the region, but climate, lithology, geologic history and structure, and land use all exert an important influence (Mills *et al.*, 1987). Comparison of the minimum precipitation threshold necessary to trigger debris flows in various regions of the U.S. indicates that longer and more intense rainfall is necessary in the Blue Ridge than elsewhere in the country (Wieczorek *et al.*, 2000).

Hillslope instability in the Appalachians creates the alluvial fans that are the most prominent Cenozoic deposits in the region (Mills, 2000b). The instability also supplies debris that influences the evolution of channels and bottomlands (Miller, 1990; Mills *et al.*, 1987). The Appalachians contain freely meandering streams, ingrown meandering streams confined within asymmetrical bedrock walls, and straight rivers within symmetrical bedrock walls that have little or no net valley-floor alluviation (Brakenridge, 1987). Although Appalachian valley bottoms do not preserve alluvial cut and fill cycles like those common throughout the southwestern U.S., Appalachian rivers do appear to have episodically enhanced rates of lateral channel migration, cutbank erosion and convex bend sedimentation that produce fill terraces (Brakenridge, 1987). The causes of such periodic enhanced erosion remain unclear, but certainly the occurrence of large floods plays a role. The geomorphic role of a flood varies in relation to drainage size. During a widespread flood in 1985, small, steep drainages scoured extensively; drainages of 1–65 km^2 had mixed erosion and deposition with continuous reworking of the valley floor; and drainages larger than 100 km^2 had only localized, discontinuous reworking (Miller, 1990). On a reach scale, the location and severity of flood impacts reflect longitudinal variations in valley width and channel orientation more than average width (Miller, 1995; Miller & Parkinson, 1993). And during moderate floods, basin geologic characteristics modify the severity of flooding, whereas the discharge of extreme floods is more closely controlled by precipitation characteristics resulting from storm motion and topographic features (Jacobson *et al.*, 1989; Smith *et al.*, 1996).

Research into Cenozoic river incision and landscape evolution in the Appalachians began more than a century ago (Davis, 1889). John Hack (1960) developed the concept of dynamic equilibrium in landscape evolution to explain his observations in the Valley and Ridge of Virginia. From Davis onward, geomorphologists have proposed dynamic incision of rivers into an asymmetrical mountain range, superposition from now-eroded overlying rocks, and superposition of river cutting through asymmetrical folds and thrust plates to explain the riverine landscapes in the Appalachians. Episodes of increased sedimentation in Mesozoic and Cenozoic marine basins (Poag & Sevon, 1989) may reflect periods of tectonic and/or climatic change, or divide migration and capture (Harbor, 1996). Investigations of individual rivers, including Virginia's New River and those to the north, have provided evidence of divide migration (Bartholomew & Mills, 1991; Hack, 1973).

Longitudinal profiles of most Appalachian rivers include distinct convexities where the streams traverse the Fall Zone, in which resistant rocks of the Piedmont bend downward beneath erodible rocks of the Coastal Plain (Pazzaglia & Gardner, 1994). Individual river incision rates include 0.027 mm yr^{-1} for the New River, Virginia (Granger *et al.*, 1997), 0.056–0.063 mm yr^{-1} for the Cheat River, West Virginia (Springer *et al.*, 1997), and 0.006–0.010 mm yr^{-1} for the Susquehanna River, Pennsylvania (Pazzaglia *et al.*, 1998). The rate of river incision may have increased during the late Cenozoic (Mills, 2000a). The incision history and terrace record of the Susquehanna River are the best-studied in the Appalachians.

The longitudinal profile and Miocene-Pleistocene age terraces of the Susquehanna River suggest complex interactions among relative baselevel, long-term flexural isostatic processes, climate, and river grade. Pazzaglia & Gardner (1993) proposed that the Susquehanna attained and maintained a characteristic graded longitudinal profile, such that bedrock straths were continually cut during periods of relative baselevel stability, with a change in climate or baselevel causing river incision and the formation of strath terraces. The Susquehanna strath terraces converge at the river mouth, diverge through the Piedmont, and reconverge to the north. This terrace profile deformation records progressive and cumulative flexural upwarping of the Atlantic margin (Pazzaglia & Gardner, 1993).

Anthropogenically Induced Changes

Despite the substantial post-glacial changes that have occurred in rivers throughout the United States, changes associated with human activities during the past two centuries have been so widespread and intense that some river systems have been more dramatically altered during this period than during earlier Quaternary climate changes. Channelization and levees have contained floodwaters and altered channel form; flow regulation has reduced peak flows and increased base flows, completely removed flow from a river channel, or driven a complete change in channel form; mining has induced massive increases in sediment transport and channel instability; and a variety of human activities have so loaded rivers with sediments, toxic contaminants, and excess nutrients that aquatic and riparian communities are impoverished in species diversity (NRC, 1992). Research during the past two decades has increasingly focused on human impacts to rivers and river landscapes as investigators have recognized how pervasive and substantial such impacts may be, even in apparently little-altered rivers (Wohl, 2001).

Snagging, Levees, and Channelization. Aboriginal forests across much of the United States were cleared throughout the 19th and 20th centuries (Fig. 6). Clearing of snags from the large rivers of the eastern and midwestern states was a matter of great commercial and military importance to the expanding country (Hill, 1957). In the 50 years after the first snagboat was built in 1829 to remove logs from the Mississippi and Ohio Rivers, more than 800,000 snags were pulled from the

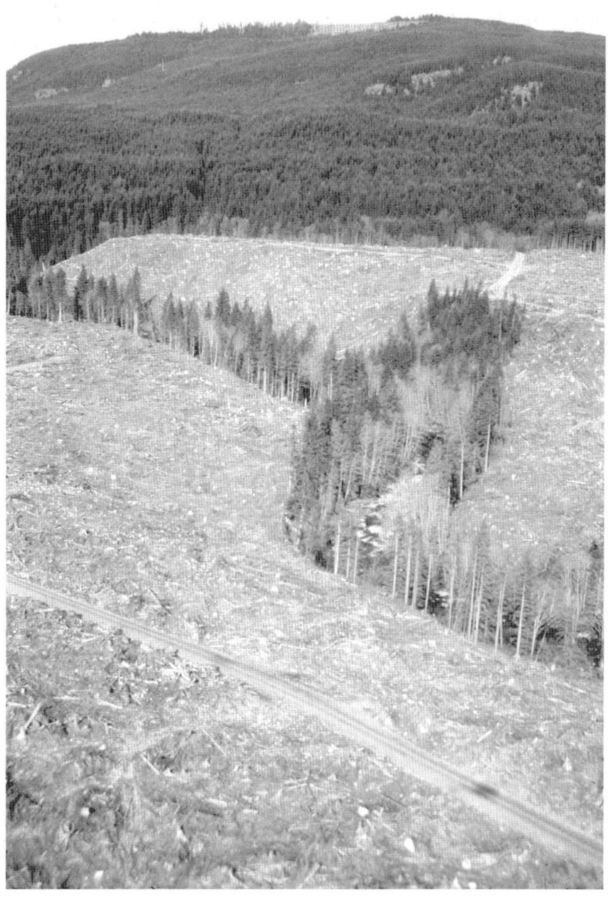

Fig. 6. Forested buffer strip left after logging along a stream in the watershed of the Tolt River, Washington.

lower Mississippi alone. Over time, snagging extended to rivers throughout the Southeast and Midwest, and Pacific Northwest (Collins *et al.*, 2002; Sedell & Froggatt, 1984), where rivers were snagged and massive log rafts dismantled even before the valley bottom forests could be logged. The average diameter of cottonwood and sycamore snags from the Mississippi and Red Rivers exceeded 1.5 m (Sedell *et al.*, 1982; Triska, 1984); those pulled from rivers in western Washington were as large as 5.3 m in diameter (Collins & Montgomery, 2001). Records of snagging operations suggest wood loading in large Pacific Northwest rivers 100 times greater than now (Sedell & Froggatt, 1984), a difference similar to that estimated by comparing present-day wood loading in a protected reach of the lower Nisqually River to cleared reaches of the Stillaguamish and Snohomish rivers (Collins *et al.*, 2002).

After large rivers were cleaned of snags and jams in New England and the Pacific Northwest, tributary streams were catastrophically cleared through the ubiquitous practice of splash damming, in which a dam-break flood was induced to transport logs to the larger rivers from where they were then rafted to market. Splash damming was common throughout the Northwest (Sedell & Duvall, 1985), Intermountain West (Wohl, 2001), Midwest, and Northeast (Sedell *et al.*, 1982). These torrents scoured sediment and wood from streambeds and banks and reduced roughness and obstructions to flow, leaving many channels scoured down to bedrock.

As access to rivers increased, agriculture rapidly spread across adjacent floodplains. Within a matter of decades floodplain sloughs were typically ditched, valley-bottom wetlands drained, and side channels plugged (Fig. 7). Human occupation and use of floodplains made flooding a problem, so extensive networks of levees were built and many rivers were straightened, or channelized, in order to prevent flooding. An estimated 25,000 miles of levees enclose more than 30,000 square miles of floodplain in the United States (NRC, 1992). On large rivers, extensive construction of levees and dikes diminished floodplain storage of water during floods, thereby creating greater flow depths (higher stages of water) for the same discharge volume (e.g. Criss & Shock, 2001; Sparks, 1995). Flood control efforts in many instances exacerbated flooding hazards by promoting occupation of flood-prone areas (Fig. 8).

Cattle Grazing. Paleochannels and alluvial stratigraphy along rivers in the arid and semiarid regions of the Southwest and the intermountain West record numerous episodes of channel incision and aggradation throughout the Quaternary (Patton & Boison, 1986). The episodic instability of these dryland rivers became a focus of geomorphic research following an episode of widespread channel incision during the 1880s and 1890s (Graf, 1983). A lively debate has since persisted as to whether such channel change is driven by climatic variability, land use, or an intrinsic cycle of filling and entrenchment. Authors attributing channel change to land use have noted how intensive grazing results in lower vegetation density and higher runoff and sediment yields from uplands, as well as vegetation removal, trampling of banks, and associated reductions in bank stability along channels (Cooke & Reeves, 1976; Dodge, 1902; Leopold, 1921; Thornthwaite *et al.*, 1941). Before the mid-1940s, most studies attributed channel incision primarily to grazing. Consensus then shifted toward climatic variability as the main trigger for channel incision or aggradation (Graf, 1988). Various investigators concluded that (a) increasing precipitation created higher mean discharge, leading to greater erosive capacity; (b) decreasing precipitation led to a decline in stabilizing vegetation and allowed knickpoints to form and propagate upstream; or (c) changes in storm frequency led to periods of larger floods (channel incision) or smaller floods (aggradation) (Bryan, 1925; Hack, 1939; Leopold, 1951, 1976; Love, 1979). Since the early 1970s, episodic channel incision and aggradation have also been attributed to the inherent episodicity of channel processes along ephemeral rivers, where gradual aggradation by flows not competent to move sediment completely through a channel network eventually produces over-steepening, leading to formation of a headcut and channel incision (Patton & Schumm, 1975; Schumm & Hadley, 1957).

Fig. 7. Urbanized channel in the Puget Lowland, Washington.

The relative importance of climatic variability, land use, and intrinsic channel processes in regulating channel incision and aggradation remains a subject of contention and certainly varies among channels. However, there is consensus that intensive grazing within the riparian zone alters channel form and conditions. The net effect of grazing is that the channel

Fig. 8. Snoqualmie River, Washington, in flood.

becomes wider and shallower, has a finer substrate, less pool volume, less overhead cover from vegetation, undercut banks, warmer water and lower dissolved oxygen, more unstable banks, and less habitat diversity (Kauffman & Krueger, 1984; Magilligan & McDowell, 1997; Platts & Nelson, 1985; Trimble & Mendel, 1995). Changes in channel characteristics associated with grazing in the riparian zone include reduced shading and input of organic matter and reduced bank stability due to removal of riparian vegetation (Platts & Nelson, 1985; Trimble & Mendel, 1995), bank compaction, increased runoff, decreased infiltration, increased erosion due to trampling (Kauffman & Krueger, 1984; Magilligan & McDowell, 1997), and excess nutrient loads, lower dissolved oxygen, algal blooms and eutrophication from animal excrement (Behnke & Zarn, 1976; Trimble & Mendel, 1995). Grazing in the riparian zone is widespread on public lands in the western United States, where it may be the single greatest threat to the integrity of aquatic habitat (Behnke & Zarn, 1976).

Flow Regulation

Changes in the flow regime – the magnitude, frequency, and duration of flow – along a river may result from diverse human activities including dams, diversions, groundwater withdrawal, and urbanization. Many of these activities, and the concomitant channel changes, are now ubiquitous across the United States, although the specific impacts vary by region. In some regions of North America dams have had a greater effect on rivers and aquatic ecosystems than Quaternary climate changes.

The effects of flow regulation depend on the associated changes in flow regime. The effects of dams, for example, depend on the purpose to which the dam was built. Dams operated for flood control, hydroelectric power generation, or water storage commonly decrease peak flows, result in strong fluctuations over 24-hr period (hydroelectric) or shift in timing of seasonal peak (water storage), decrease downstream sediment supply, leading to channel bed armoring or erosion, and channel instability, and may change water temperature and chemistry (Graf, 1996; Hirsch *et al.*, 1990; Williams & Wolman, 1984). Flow diversions may remove or add water to channel, and change the magnitude and timing of flow. Where water withdrawal is so pronounced that channel form (e.g. substrate grain-size, pool volume, width/depth ratio, flood conveyance), habitat, or recreational uses are impaired, efforts may be made to purchase or legislate some minimum volume of flow within the channel (instream flow), specified as an annual minimum or as minimum flows at various times during an annual hydrograph (Schleusener *et al.*, 1962; Stromberg & Patten, 1990). Groundwater withdrawals can lower the local or regional water table to the degree that in extreme cases base flow to a river may be reduced or even cause a perennial river to become intermittent or ephemeral. Water withdrawals may also cause channel incision, particularly if the withdrawal causes regional compaction or subsidence (Kondolf, 1996). Changes in flow due to increased impermeable surface in a drainage basin increases the magnitude and rate of runoff for a given precipitation input (Morisawa & LaFlure, 1979; Urbonas & Benik, 1995; Wolman, 1967). This commonly increases the peak flow of small to moderate floods, and makes flood hydrographs more flashy. Combined with a decrease in sediment yield, such changes increase channel erosion and instability.

The continental U.S. has 75,000 dams that together are capable of storing a volume of water almost equaling one year's mean runoff (Graf, 1999). The greatest impacts to river flow from dams occur in the Great Plains, Rocky Mountains, and the arid Southwest, where storage is almost 4 times the mean annual runoff (Graf, 1999). In these regions and elsewhere, such as the Columbia River basin, dams have been identified as a major impact on native fish species (Ligon *et al.*, 1995). Surface-water withdrawal for offstream uses such as irrigation is greatest in California, Idaho, Colorado, and the western Great Plains.

The Mississippi River presents an example of how Holocene patterns of water and sediment discharge have been altered by dams, diversions, and channel structures constructed during the 20th century. In the 208,000 km^2 of the upper Mississippi River basin, more than 13,000 km of levees have been built, and 65% of the original wetlands have been drained. By 1950, a system of 29 locks and dams had been built along the upper Mississippi River from St. Louis to Minneapolis, and a 2.7-m navigation channel had been dredged from St. Louis to Sioux City, Iowa (Watson & Biedenharn, 2000). Although the Mississippi River historically had the greatest water and sediment discharges in the U.S. (Meade *et al.*, 1990), the storage of large volumes of sediment behind more than 8000 dams (Graf, 1999) is causing various downstream impacts, including subsidence and erosion of the Mississippi delta and adjacent coastlines (Britsch & Dunbar, 1993).

Some of the most dramatic changes in riverine landscapes as a result of flow regulation have occurred in the western Great Plains. Rivers such as the South Platte, the North Platte, and the Arkansas, were broad, shallow, braided channels when people of European descent first described them in the 19th century (Eschner *et al.*, 1983; Williams, 1978). Phrases such as "a mile wide and an inch deep" and "too thick to drink but too thin to plow" were used to describe these rivers which flowed high with late spring-early summer snowmelt, then shrank back to very low flows by autumn. The channels had sparse riparian vegetation and warm, turbid water. Beginning at the end of the 19th century, reductions in the snowmelt flood peak, increased late summer base flow, and higher regional water tables resulting from flow regulation and extensive agricultural irrigation, facilitated the establishment of riparian trees. Within a few decades, braided channels that had been 450 m wide became 150-m-wide sinuous channels with densely vegetated islands and banks (Nadler & Schumm, 1981).

The flow regime of most U.S. rivers has been modified to some degree by human actions. Growing recognition that aquatic ecosystems can be finely adjusted to the hydrologic regime, and the channel features and habitats that it creates or sustains, is focusing attention on how to set flow regimes in regulated or managed channels so as to maintain channel form and ecological functions. Approaches to setting instream flows in managed or regulated rivers are generally based on habitat preferences that can be characterized in terms of flow depth or velocity. But as higher flows are generally required to form and maintain habitat, a major concern with such approaches for determining minimum flows in regulated rivers is that they specify flows needed to maintain the use of habitats but not the habitats themselves (Whiting, 2002). Although there is no simple way to determine the flows needed to maintain a channel, it has become clear that the closer the annual hydrograph is to the natural flow regime, the more likely it is that flow-habitat interactions will be ecologically effective and sustainable (Poff *et al.*, 1997; Whiting, 2002).

Mining. Mining may occur within the river corridor, as when placer metals disseminated through valley-bottom sediments are removed, or alluvial sand and gravel deposits are mined for construction aggregate. Mining may occur elsewhere within a watershed if lode metals disseminated through bedrock outcrops are mined, or if fossil fuels such as coal and oil are removed.

Impacts of mining on rivers generally vary with the type of mine. Placer mining decreases bed and bank stability, increases downstream sediment transport, and reduces water quality. Associated toxins such as mercury, increased sediment transport and channel instability can each stress or destroy aquatic and riparian organisms (Hilmes & Wohl, 1995; James, 1991, 1994, 1999; Van Haveren, 1991; Van Nieuwenhuyse & LaPerriere, 1986). Aggregate mining decreases bed and bank stability, and depressions created by mining may initiate headcut erosion, trap sediment and create sediment depletion downstream, or divert flow and cause lateral channel movement, which increases downstream sediment transport and thus reduces water quality and alters aquatic habitat (Bull & Scott, 1974; Chang, 1987; Kondolf, 1994, 1997; Lagasse *et al.*, 1980; Norman *et al.*, 1998). Lode mining may increase sediment yield to channels from tailings and from slope instability associated with deforestation, and may cause acid-mine drainage (Starnes & Gasper, 1995; Stiller, 2000; Wohl, 2001). Finally, strip-mining may completely alter topography and water and sediment yields, or even obliterate streams (as, for example, in mountain-top removal in West Virginia). In addition, contaminants in water used to mine or process fuels may completely alter the local river flow regime (e.g. coal-bed methane mining in the intermountain West) (Starnes & Gasper, 1995).

Placer mining in the continental U.S. occurred mainly in the intermountain West during the 1850s–1950s. California's Sierra Nevada, Colorado's Front Range, western Montana, western Nevada, and the Black Hills of South Dakota were among the regions with the most intense placer mining (Fig. 9). The massive amounts of sediment mobilized by mining activities, as well as associated changes in flow regime and introduction of toxic contaminants, continue to affect these rivers (Alpers & Hunerlach, 2000; Hilmes & Wohl, 1995; James, 1991, 1999; Stiller, 2000).

Aggregate mining for sand and gravel is widespread in the U.S. (Tepordei, 1987). In 1990, approximately 4200 companies mined 830 billion kg of sand and gravel from 5700 operations along rivers and floodplains (Meador & Layher, 1998). Nearly all of this material is used in construction, usually within 50–80 km of the mine. In-channel mining occurs as: (a) dry-pit mining in which a pit is excavated below the thalweg of a dry ephemeral stream; (b) wet-pit mining in which the pit extends below the water table; (c) bar skimming in which all the sediment in a gravel bar above an imaginary line sloping upwards from the summer water's edge is removed; or (d) safe yield in which extraction is limited to

Fig. 9. Hydraulic mining at the Malakoff diggings of the North Bloomfield Gravel Mining Company in the Sierra Nevada (Plate A from Whitney, 1880).

the removal of annual aggradation (Kondolf, 1994, 1997; Sandecki, 1989). Floodplain and terrace pit mining occur in fluvial deposits beyond the active channel. Instream mining commonly causes channel incision that may propagate up- and downstream from the mine, undermining structures such as bridges. Incision may also induce channel instability that changes substrate grain-size distribution and bedform configuration, and causes downstream siltation and reduced water quality. And incision may lower valley-bottom water tables and destroy riparian environments and hyporheic exchanges (Kondolf, 1997; Norman *et al.*, 1998; Sandecki, 1989).

Lode mining has focused on industrial metals such as iron in the upper Great Lakes, and precious metals such as silver and gold in the southern Appalachians and the West. Most active contemporary lode mining occurs in the western U.S., and the majority of this region's toxic waste sites are associated with historical or contemporary mining. Impacts to rivers adjacent to lode mining derive primarily from increased sediment yields and introduced toxic contaminants (Rampe & Runnells, 1989; Stiller, 2000; Stoughton & Marcus, 2000).

Coal has been mined extensively in the Appalachians, the upper and central Midwest and Great Lakes region, the intermountain West and the upper Great Plains. Oil in the continental U.S. has come primarily from the southern Great Plains and Gulf Coast, the Mississippi Valley, and parts of the northwestern Great Plains. Natural gas came from the Appalachians, the Mississippi Valley, and the southern and western Great Plains. Some of the most active contemporary mining for fossil fuels occurs in the western Great Plains (e.g. Wyoming), the South (e.g. Texas and Louisiana), and the Appalachians (e.g. West Virginia). The effects on rivers of such mining vary in association with the type of mining, but commonly involve increased sediment yields and the introduction of toxic contaminants.

Water Quality. Most rivers in the continental United States have contaminants or reduced water quality resulting from human activity. The Clean Water Act of 1972 set a "fishable" and "swimmable" goal for all waterways. Over the next twelve years the federal government contributed a third of the $310 billion spent to clean up surface waters. Spending on water pollution remained high in subsequent decades; in 1992, for example, the Environmental Protection Agency spent $2.9 billion, largely in the form of grants to states for sewage treatment plants. By 1994, such programs had reduced sewage in American rivers by 90% relative to 1970 (Vileisis, 1997). However, the U.S. did not, and has not, come close to meeting the 1985 target date of the Clean Water Act for fishable and swimmable waters everywhere.

Pollutants in streams represent have various sources, and present differing potential hazards to humans and other organisms. Elevated sediment loads derived from runoff from agricultural and otherwise managed lands can change channel substrate and form and destroy aquatic and riparian habitat (Waters, 1995). Excess nutrients (N, P) from fertilizers or sewage systems (animal waste, laundry detergent) can lead to excessive algal growth and low levels of dissolved oxygen, or introduce carcinogenic by-products of treatment in chlorinated drinking water (Graffy *et al.*, 1996; Steingraber, 1997). Trace elements (such as As, Cd, Cr, Cu, Pb, Hg, Ni, Se, Zn) can be introduced by atmospheric deposition (volcanic emissions; combustion of municipal solid waste and fossil fuels in coal- and oil-fired power plants; releases from metal smelters, automobile emissions, biomass burning), point source releases to surface water (municipal sewage sludge, effluent from coal-fired power plants, releases from industrial uses, acid-mine drainage), or nonpoint source releases (natural rock weathering, agricultural runoff of manure and artificial fertilizers, wear of automobile parts, irrigation return flow). Many trace elements are adsorbed to fine sediments, taken up by invertebrates, and passed through the food web. Individual elements may bioaccumulate (accumulate within the body of an organism) and biomagnify (concentrate as they are passed between organisms). Trace elements are commonly teratogens (cause developmental changes and abnormalities), mutagens (cause chromosomal changes) and carcinogens (cause cancerous growths) (Rice, 1999). Organochlorine compounds (pesticides, PCBs) from agricultural and municipal application of herbicides and insecticides; waste from electricity-generating facilities, and many common items (e.g. photocopy toner) in wastewater effluents and atmospheric fallout from incinerators, or that leach from landfills can have a range of behavior (mobility, persistence, toxicity) that varies widely among individual compounds. The worst-case scenarios are compounds such as DDT, the toxic breakdown products (DDE) of which are still present 30 years after the last application in the U.S., and which act as endocrine-disrupters in many species (Colborn *et al.*, 1997; Nowell *et al.*, 1999; USGS, 1999). The behavior and health effects of volatile organic compounds (VOCs) introduced into streams from nonpoint sources (primarily urban land surfaces and urban air) are variable and largely unknown, but this large group of more than 60 compounds includes diverse substances and carcinogens such as benzene, the solvent terachloroethylene, toluene, and chloroform (Lopes & Bender, 1998; Rathbun, 1998).

Sediment remains the most important river pollutant in terms of number of stream miles degraded (Waters, 1995), and excess fermentable organic wastes from human and animal sewage create locally significant impacts on water quality. The most insidious contaminants come from industrial and agricultural activities. The U.S. Geological Survey's National Water Quality Assessment (NAWQA) program, begun in 1991, provides a comprehensive index of national water conditions. The first phase of the program included 59 study units throughout the continental U.S., Alaska and Hawaii. Standardized sampling of surface and ground waters within these study units assessed water chemistry, streambed sediments, invertebrate and fish tissue, and stream habitat. Sample testing included analyses for 9 trace elements, 33 organochlorine compounds and 106 pesticides, 5 nutrients, and 60 volatile organic compounds. Every one of the 59 study units sampled had some type of contaminant that locally exceeded either drinking-water standards, or standards for the protection of aquatic life. Even forested watersheds with little direct land use contain residues of such synthetic compounds as DDT or PCBs, which reached the watershed from atmospheric sources, although the use of DDT was banned in the U.S. in 1972, and the use and manufacture of PCBs was banned in the U.S. in 1979. The cumulative impact of these contaminants impairs aquatic and riparian ecosystem functioning by reducing the diversity and abundance of organisms in river landscapes.

Summary of Anthropogenic Changes. Other than rives once overrun by glacial ice, anthropogenic changes exceed those due to Quaternary climate changes in many river systems. Hooke (1999, 2000) estimated that throughout much of the U.S., humans now move more sediment (average 31,000 kg yr^{-1} per capita) than do rivers. The net result of a diverse array of human activities is to move rivers and riverine landscapes toward increasing homogeneity and reduced environmental quality. We straighten and deepen channels; confine floodwaters and stop the processes maintaining floodplains and riparian corridors; increase sediment movement and alter channel bedforms and planform; reduce peak flows and increase base flows; and poison water and sediment with adsorbed toxics. All of this moves diverse, stable, functional river ecosystems toward a condition in which river form and process are so altered that channels begin to resemble irrigation canals or drainage culverts.

Emerging Research Directions

Much remains to be learned about rivers in general, and about the status, behavior, and history of particular river systems. In particular, we see four emerging areas of research interest as providing significant new opportunities: river restoration and rehabilitation, biogeomorphology, resistant-boundary channels, and interactions among tectonics, climate, and erosion.

River Restoration and Rehabilitation. River restoration programs around the United States aim to improve the quality of rivers for use by both humans and wildlife. Rivers are dynamic systems in which specific attributes are continually created, altered, and destroyed. Consequently, river restoration means not only reestablishing certain prior conditions but also reestablishing the processes that create those conditions. In contrast, river rehabilitation aims to improve river conditions but does not necessarily seek reestablishment of natural conditions and dynamics. Given the extensive historic changes to rivers, and the resulting constraints, most projects billed as "river restoration" actually achieve only a form of river rehabilitation.

Techniques being used to rehabilitate rivers include setting levees back away from channel banks to allow the channel to migrate within a proscribed corridor on its historical floodplain. Delineation of channel migration zones and erosion hazard zones also are starting to be used in regulatory arenas to account for the potential for channel movement that could affect long-term capital projects or impact the assumptions or objectives underpinning forest-harvest planning. Streamside buffer zones have been widely applied to forestlands and are now being adopted in some urbanized settings. The central importance of the natural flow regime in stream ecology also is becoming recognized in stream restoration and rehabilitation programs and projects.

Biogeomorphology. Disturbances are generally considered to negatively impact aquatic ecosystems, but catastrophic disturbances such as floods and landslides also can locally create or enhance aquatic and riparian habitat (Everest &

Fig. 10. Salmon swimming through the Ballard Locks in Seattle.

Meehan, 1981; Friedman & Auble, 2000a, b; Reeves *et al.*, 1995). Consequently, the net benefit or detriment to an aquatic population will depend on the type, frequency, and impact of a disturbance and its relative importance in creating essential characteristics for that species. To assess the relative role of disturbance processes on aquatic ecosystems and habitat, one must consider the net effect of habitat destruction and creation. Recognizing that an organism evolved in a dynamic environment does not necessarily imply that disturbance processes are essential for maintaining that organism. Understanding the effects of disturbance processes on populations of organisms requires understanding the full distribution of events and their effects across space and through time. For most systems and organisms such understanding remains in its infancy (Fig. 10).

Resistant-Boundary Channels. Resistant-boundary channels are those formed on bedrock or on very coarse clasts, such as rivers in mountain regions formed on boulders. These types of channels are not adequately described by the conceptual and mathematical models commonly applied to lower gradient alluvial rivers (Tinkler & Wohl, 1998; Wohl, 2000). Resistant-boundary channels tend to have steep gradients and rough boundaries that resist fluvial erosion. These channels have highly turbulent flow; supply-limited sediment transport; highly stochastic sediment movement that is difficult to parameterize and model; abrupt downstream variation in channel geometry; and episodic channel change restricted to the relatively high magnitude, low frequency flows that are capable of exceeding erosional thresholds along bedrock rivers and mountain rivers (Baker, 1988; Tinkler & Wohl, 1998; Wohl, 1998).

Bedrock rivers have increasingly been the subject of research because knowledge of processes and rates of bedrock channel incision is vital to quantitative modeling of landscape evolution. River channels are the conduits by which weathering products are removed from a drainage basin. This rate of removal influences the efficiency of landscape change. For example, Burbank *et al.* (1996) describe a balance between channel incision and hillslope profiles in the Indus River drainage basin such that, as a bedrock channel incises, adjacent hillslopes become over-steepened, triggering mass movements. The resultant influx of sediment to the channel decreases the rate of channel incision until the sediment has been transported downstream, at which time a new period of incision occurs. The rate and manner of bedrock channel erosion thus partly control hillslope stability and evolution of the entire catchment area.

Many recent studies have focused on quantifying the variables controlling erosional processes and channel geometry along bedrock channels, as well as the resultant long-term incision rate (Howard, 1998; Roe *et al.*, 2002; Seidl *et al.*, 1994; Snyder *et al.*, 2000; Stock & Montgomery, 1999; Whipple & Tucker, 1999). The findings begin to delineate the conditions under which specific channel incision regimes occur. Stock & Montgomery (1999), for example, have proposed that a weak dependence of incision rate on drainage area characterizes areas where abrupt baselevel fall produces channel incision primarily through knickpoint retreat. In contrast, rates of channel incision under conditions of stable baselevel depends strongly on drainage area. Recent studies also suggest that bedrock channel geometry responds consistently to the balance between hydraulic driving forces and substrate resisting forces, such that bedrock channel geometry may be predictable in some circumstances (Montgomery & Gran, 2001; Wohl & Merritt, 2001).

Investigations have increasingly focused on mountain rivers in recognition that these headwater stream segments produce a disproportionately large component of the sediment yield from a drainage basin (Milliman & Syvitski, 1992). However, adequate equations do not yet exist to describe hydraulics and sediment transport along mountain rivers. Steep gradients and large grain and form roughness promote non-logarithmic velocity profiles (Wiberg & Smith, 1991), localized critical and supercritical flow, and strongly three-dimensional flow in these rivers (Wohl, 2000). Recent work on the hydraulics of mountain rivers has attempted to: (a) predict flow resistance coefficients as a function of gradient, relative submergence, flow depth, particle size distribution, or other channel characteristics (Jarrett, 1990; Marcus *et al.*, 1992; Maxwell & Papanicolaou, 2001); (b) quantify the contribution of the components of grain and/or form roughness to total flow resistance (Bathurst, 1993; Curran & Wohl, 2003; Prestegaard, 1983); and (c) characterize cross-stream velocity and vertical velocity distribution and the associated forces of lift or shear stress exerted on the channel boundaries (Furbish, 1993; Wohl & Thompson, 2000).

Students of mountain channels have barely begun to understand how sediment in transport variously shields and scours river beds (e.g. Sklar & Dietrich, 1998, 2001). Spatial and temporal variations in hydraulics and bed substrate, and limited sediment supply, render bedload entrainment and transport equally difficult to characterize in mountain streams. Research has focused on (a) the grain-size distribution of channel-bed sediments (Buffington & Montgomery, 1999a, b; Ferguson & Paola, 1997; Nikora *et al.*, 1998; Wohl *et al.*, 1996; Wolcott & Church, 1991); (b) sediment entrainment (Johnston *et al.*, 1998) and the occurrence of equal mobility vs. selective entrainment (Kuhnle, 1992; Montgomery *et al.*, 1999; Wathen *et al.*, 1995; Wilcock, 1993); and (c) the mechanics of bedload transport (Gomez & Troutman, 1997; Thompson *et al.*, 1999; Wilcock *et al.*, 1996).

Other areas of recent focus in mountain rivers include bedforms and channel morphology, and longitudinal profile development. Flow energy and sediment supply likely interact along a continuum of channel-bed gradient to produce predictable trends in channel bedforms and channel morphology (Chin, 1998; Montgomery & Buffington, 1997), but conceptual models describing these trends are presently limited by a lack of field-based data describing how channel morphology varies as a function of potential controls. Similarly, a quantitative understanding of the controls on spatial distribution and relative importance of different processes of incision along mountain rivers requires further detailed, extensive field measurements of hillslope and channel processes in mountainous regions (Wohl, 2000).

Tectonics, Climate, and Erosion. An exciting area of active research focuses on coupling and feedback among climate, erosion, and tectonic processes. Over the past decade geologists have recognized that the development and evolution of geologic structures can depend on spatial gradients in the climate forcing that drives erosion (Avouac & Burov, 1996; Hoffman & Grotzinger, 1993; Horton, 1999). Models that couple geodynamics and surface processes in evolving and steady-state orogens, and which predict the resulting metamorphic gradients exposed at the surface, reflect the influence of spatial variability in surface erosion (Willett, 1999; Willett *et al.*, 1993, 2001). Development of mountain ranges strongly influences patterns of precipitation (Barros & Lettenmaier, 1994) and therefore patterns of erosional intensity, which in turn governs the development and evolution of topography. Steady-state river long profiles, for example, are influenced by orographic controls on precipitation patterns (Roe *et al.*, 2002). The role of fluvial erosion in reducing mass accumulation in mountain ranges is perhaps best illustrated by the converse cases where lack of rainfall, and therefore limited erosion, allows accumulation of enough mass to engage the mechanical limit to crustal thickening (Pope & Willett, 1998) and result in development of high plateaus like the Altiplano and Tibet (Montgomery *et al.*, 2001) (Fig. 11). Climate, erosion, and tectonics are thus coupled through feedbacks. How this leads to positive feedback between erosion and tectonics provides a fruitful avenue for further inquiry.

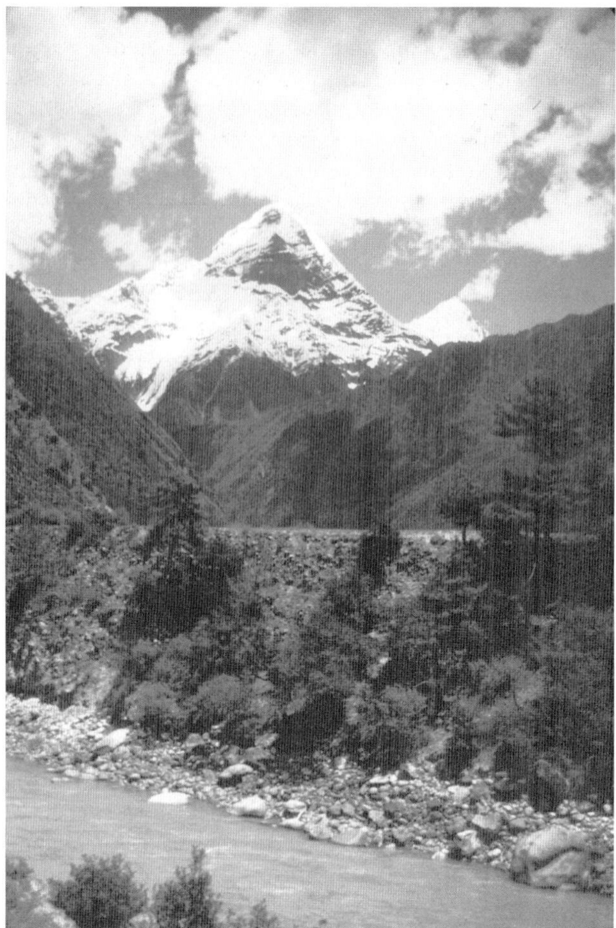

Fig. 11. High peak along a river near Namche Barwa in eastern Tibet.

Summary

The discipline of fluvial geomorphology has broadened as the science has developed over the past 40 years. Understanding of the fundamental aspects of rivers and the variability in fluvial processes and conditions due to regional differences in climate and geologic histories has advanced considerably. Connections to other disciplines are coming to the forefront of research, and offer exciting potential for further connections and for investigation of feedback between both system components and systems traditionally treated as separate domains. The investigation of resistant-boundary channels is an exciting and under-explored area of fluvial geomorphology that challenges us to rethink traditional assumptions based on alluvial channels. The dramatic influence of fluvial processes on aquatic and riparian ecology paves the way for collaboration with stream ecologists. Modern geobiology, encompassing the interactions of organisms and their environment, offers frontier opportunities to researchers willing to work at the interface between geomorphology and biology. Over vastly expanded temporal horizons the role of erosion as a geological process remains an area with exciting research potential. Finally, the pressing need for better integration of fluvial geomorphology (and fluvial geomorphologists) in river restoration and rehabilitation presents an important challenge to the research community.

Acknowledgments

We thank John Pitlick and John Buffington for their helpful comments on a draft manuscript.

References

Abbe, T.B. & Montgomery, D.R. (1996). Interaction of large woody debris, channel hydraulics and habitat formation in large rivers. *Regulated Rivers: Research & Management*, **12**, 201–221.

Adam, D.P. & West, G.J. (1983). Temperature and precipitation estimates through the last glacial cycle form Clear Lake, California, pollen data. *Science*, **219**, 168–170.

Alpers, C.N. & Hunerlach, M.P. (2000). Mercury contamination from historic gold mining in California. U.S. Geol. Survey Fact Sheet FS-061-00.

Anderson, R.C. (1988). Reconstruction of preglacial drainage and its diversion by earliest glacial forebulge in the upper Mississippi Valley region. *Geology*, **16**, 254–257.

Anderson, R.S. & Smith, S.J. (1994). Paleoclimatic interpretations of meadow sediment and pollen stratigraphies from California. *Geology*, **22**, 723–726.

Autin, W.J. (1993). Influences of relative sea-level rise and Mississippi River delta plain evolution on the Holocene Middle Amite River, southeastern Louisiana. *Quaternary Research*, **39**, 68–74.

Autin, W.J., Burns, S.A., Miller, B.J., Saucier, R.T. & Snead, J.I. (1991). Quaternary geology of the lower Mississippi valley. *In*: Morrison, R.B. (Ed.), *Quaternary Nonglacial Geology: Conterminous United States*. Geological Society of America, Boulder, CO, 547–582.

Avouac, J.-P. & Burov, E.B. (1996). Erosion as a driving mechanism of intracontinental mountain growth. *Journal of Geophysical Research*, **101**, 17747–17769.

Axelrod, D.I. (1957). Late Tertiary floras and the Sierra Nevadan uplift. *Geological Society of America Bulletin*, **68**, 19–46.

Baker, V.R. (1983). Late-Pleistocene fluvial systems. *In*: Porter, S.C. (Ed.), *Late-Quaternary Environments of the United States: The Late Pleistocene*. University of Minnesota Press, Minneapolis, **1**, 115–129.

Baker, V.R. (1988). Flood erosion. *In*: Baker, V.R., Kochel, R.C. & Patton, P.C. (Eds), *Flood Geomorphology*. Wiley, New York, 81–95.

Baker, V.R. & Nummedal, D. (Eds) (1987). *The channeled Scabland*. NASA, Washington, DC, 186 pp.

Barros, A.P. & Lettenmaier, D.P. (1994). Dynamic modeling of orographically produced precipitation. *Reviews of Geophysics*, **32**, 265–284.

Bartholomew, M.J. & Mills, H.H. (1991). Old courses of the New River: its late Cenozoic migration and bedrock control inferred from high-level stream gravels, southwestern Virginia. *Geological Society of America Bulletin*, **103**, 73–81.

Bathurst, J.C. (1993). Flow resistance through the channel network. *In*: Beven, K. & Kirkby, M.J. (Eds), *Channel Network Hydrology*. Wiley and Sons, Chichester, 69–98.

Beaty, C.B. (1990). Anatomy of a White Mountains debris-flow – the making of an alluvial fan. *In*: Rachocki, A.H. & Church, M. (Eds), *Alluvial Fans: A Field Approach*. Wiley, New York, 69–89.

Beechie, T.J., Collins, B.D. & Pess, G.R. (2001). Holocene and recent geomorphic processes, land use, and salmonid habitat in two North Puget Sound rivers bains. *In*: Dorava, J.M., Montgomery, D.R., Palcsak, B.B. & Fitzpatrick, F.A. (Eds), *Geomorphic Processes and Riverine Habitat*. American Geophysical Union, Water Science and Application, **4**, 37–54.

Beechie, T.J. & Sibley, T.H. (1997). Relationship between channel characteristics, woody debris, and fish habitat in northwestern Washington streams. *Transactions of the American Fisheries Society*, **126**, 217–229.

Behnke, R.J. & Zarn, M. (1976). Biology and management of threatened and endangered western trout. Gen. Tech. Report RM-GTR-28. U.S. Department of Agriculture, Forest Service, Rocky Mountain Forest and Range Experiment Station, Ft. Collins, CO.

Benda, L.E., Miller, D.J., Dunne, T., Reeves, G.H. & Agee, J.K. (1998). Dynamic landscape systems. *In*: Naiman, R.J. & Bilby, R.E. (Eds), *River Ecology and Management*. Springer-Verlag, New York, NY, 261–288.

Benedict, J.B. (1973). Chronology of cirque glaciation, Colorado Front Range. *Quaternary Research*, **3**, 584–599.

Birkeland, P.W., Crandell, D.R. & Richmond, G.M. (1971). Status of correlation of Quaternary stratigraphic units in the western conterminous United States. *Quaternary Research*, **1**, 208–227.

Bischoff, J.L. & Cummins, K. (2001). Wisconsin glaciation of the Sierra Nevada (79,000–15,000 yr BP) as recorded by rock flour in sediments of Owens Lake, California. *Quaternary Research*, **55**, 14–24.

Bisson, P.A., Nielson, J.L., Palmason, R.A. & Gore, L.E. (1982). A system of naming habitat types in small streams, with examples of habitat utilization by salmonids during low stream flow. *In*: Armantrout, N.B. (Ed.), *Acquisition and Utilization of Aquatic Habitat Information*. Western Division, American Fisheries Society, Portland, OR, 62–73.

Blair, T.C. (2001). Outburst flood sedimentation on the proglacial Tuttle Canyon alluvial fan, Owens Valley, California, USA. *Journal of Sedimentary Research*, **71**, 657–679.

Booth, D.B. (1994). Glaciofluvial infilling and scour of the Puget Lowland, Washington, during ice-sheet glaciation. *Geology*, **22**, 695–698.

Booth, D.B., Haugerud, R.A. & Troost, K.G. (2003). The geology of Puget Lowland rivers. *In*: Montgomery, D.R., Bolton, S., Booth, D.B. & Wall, L. (Eds), *Restoration of Puget Sound Rivers*. University of Washington Press, Seattle & London, 14–45.

Brakenridge, G.R. (1987). Fluvial systems in the Appalachians. *In*: Graf, W.L. (Ed.), *Geomorphic Systems of North America*. Geological Society of America, Boulder, Colorado, 37–46.

Brandon, M.T., Roden-Tice, M.K. & Garver, J.I. (1998). Late Cenozoic exhumation of the Cascadia accretionary wedge in the Olympic Mountains, northwest Washington State. *Geological Society of America Bulletin*, **110**, 985–1009.

Braun, D.D. (1989). Glacial and periglacial erosion of the Appalachians. *In*: Gardner, T.W. & Sevon, W.D. (Eds), *Appalachian Geomorphology*, 233–258.

Britsch, L.D. & Dunbar, J.B. (1993). Land loss rates: Louisiana coastal plain. *Journal of Coastal Research*, **9**, 324–338.

Broecker, W.S., Kennett, J.P., Flower, B.P., Teller, J.T., Trumbore, S., Bonani, G. & Wolfli, W. (1989). Routing of meltwater from the Laurentide ice sheet during the Younger Dryas cold episode. *Nature*, **341**, 318–321.

Brown, P.A. & Kennett, J.P. (1998). Megaflood erosion and meltwater plumbing changes during last North American deglaciation recorded in Gulf of Mexico sediments. *Geology*, **26**, 599–602.

Bryan, K. (1925). Date of channel trenching (arroyo cutting) in the arid southwest. *Science*, **62**, 338–344.

Buffington, J.M., Lisle, T.E., Woodsmith, R.D. & Hilton, S. (2002). Controls on the size and occurrence of pools in coarse-grained forest rivers. *River Research and Applications*, **18**, 507–531.

Buffington, J.M. & Montgomery, D.R. (1999a). Effects of hydraulic roughness on surface textures of gravel-bed rivers. *Water Resources Research*, **35**, 3507–3522.

Buffington, J.M. & Montgomery, D.R. (1999b). A procedure for classifying textural facies in gravel-bed rivers. *Water Resources Research*, **35**, 1903–1914.

Buffington, J.M., Woodsmith, R.D., Booth, D.B. & Montgomery, D.R. (2003). Fluvial processes in Puget Sound rivers and the Pacific Northwest. *In*: Montgomery, D.R., Bolton, S., Booth, D.B. & Wall, L. (Eds), *Restoration of Puget Sound Rivers*. University of Washington Press, Seattle & London, 46–78.

Bull, W.B. (1977). The alluvial-fan environment. *Progress in Physical Geography*, **1**, 222–270.

Bull, W.B. & Scott, K.M. (1974). Impact of mining gravel from urban stream beds in the southwestern U.S. *Geology*, **2**, 171–174.

Bunte, K. & MacDonald, L.H. (1995). Detecting changes in sediment loads: where and how is it possible? *In*: Osterkamp, W. (Ed.), *Effects of Scale on Interpretation and Management of Sediment and Water Quality*. International Association of Hydrological Sciences Publication No. 226, 253–261.

Burbank, D.W., Leland, J., Fielding, E., Anderson, R.S., Brozovic, N., Reid, M.R. & Duncan, C. (1996). Bedrock incision, rock uplift and threshold hillslopes in the northwestern Himalayas. *Nature*, **379**, 505–510.

Burke, R.M. & Birkeland, P.W. (1983). Holocene glaciation in the mountain ranges of the western United States. *In*: Wright, H.E., Jr. (Ed.), *Late-Quaternary environments of the United States*. The Holocene. University of Minnesota Press, Minneapolis, **2**, 3–11.

Chamberlain, C.P. & Poage, M.A. (2000). Reconstructing the paleotopography of mountain belts from the isotopic composition of authigenic minerals. *Geology*, **28**, 115–118.

Chang, H.H. (1987). Modelling fluvial processes in streams with gravel mining. *In*: Thorne, C.R., Bathurst, J.C. & Hey, R.D. (Eds), *Sediment Transport in Gravel-Bed Rivers*. Wiley, Chichester, 977–988.

Chézy, A.D. (1775). Memoire sur la vitesse de l'eau conduite dan une régole. Reprinted in *Annals des Ponts et Chaussées*, **60**, 1921.

Chin, A. (1998). On the stability of step-pool mountain streams. *Journal of Geology*, **106**, 59–69.

Church, M. & Ryder, J.M. (1972). Paraglacial sedimentation: A consideration of fluvial processes conditioned by glaciation. *Geological Society of America Bulletin*, **83**, 3059–3072.

Church, M. & Slaymaker, O. (1989). Disequilibrium of Holocene sediment yield in glaciated British Columbia. *Nature*, **337**, 452–454.

Clark, G.M. & Ciolkosz, E.J. (1988). Periglacial geomorphology of the Appalachian Highlands and Interior Highlands south of the glacial border – a review. *Geomorphology*, **1**, 191–220.

Colborn, T., Dumanoski, D. & Myers, J.P. (1997). Hormone impostors. *Sierra*, **82**, 28–35.

Coleman, J.M. (1988). Dynamic changes and processes in the Mississippi River delta. *Geological Society American Bulletin*, **100**, 999–1015.

Collins, B.D. & Montgomery, D.R. (2001). Importance of archival and process studies to characterizing presettlement riverine geomorphic processes and habitat in the Puget Lowland. *In*: Dorava, J.B., Montgomery, D.R., Palcsak, B. & Fitzpatrick, F. (Eds), *Geomorphic Processes and Riverine Habitat*. American Geophysical Union, Washington, DC, 227–243.

Collins, B.D., Montgomery, D.R. & Haas, A. (2002). Historic changes in the distribution and functions of large woody debris in Puget Lowland rivers. *Canadian Journal of Fisheries and Aquatic Sciences*, **59**, 66–76.

Cooke, R.U. & Reeves, R.W. (1976). *Arroyos and environmental change in the American South-West*. Clarendon Press, Oxford, 213 pp.

Criss, R.E. & Shock, E.L. (2001). Flood enhancement through flood control. *Geology*, **29**, 875–878.

Curran, J.H. & Wohl, E.E. (2003). Large woody debris and flow resistance in step-pool channels, Cascade Range, Washington. *Geomorphology*, **53**, 141–157.

Dana, J.D. (1850). On denudation in the Pacific. *American Journal of Science*, **9**(2), 48–62.

Davies-Colley, R.J. (1997). Stream channels are narrower in pasture than in forest, New Zealand. *Journal of Marine and Freshwater Research*, **31**, 599–608.

Davis, M.B. (1983). Holocene vegetational history of the eastern United States. *In*: Wright, H.E., Jr. (Ed.), *Late-Quaternary Environments of the United States*. The Holocene. University of Minnesota Press, Minneapolis, **2**, 166–181.

Davis, O.K., Anderson, R.S., Fall, P.L., O'Rourke, M.K. & Thompson, R.S. (1985). Palynological evidence for early Holocene aridity in the southern Sierra Nevada, California. *Quaternary Research*, **24**, 322–332.

Davis, W.M. (1889). The rivers and valleys of Pennsylvania. *National Geographic*, **1**, 183–253.

Delcourt, P.A. & Delcourt, H.R. (1988). Quaternary landscape ecology: relevant scales in space and time. *Landscape Ecology*, **2**, 23–44.

Dethier, D.P., Pessl, F., Jr., Deuler, R.G., Balzarini, M.A. & Pevear, D.R. (1995). Late Wisconsinan glaciomarine deposition and isostatic rebound, northern Puget Lowland, Washington. *Geological Society of America Bulletin*, **197**, 1288–1303.

Dethier, D.P., Schildgen, T.F., Bierman, P. & Caffee, M. (2000). The cosmogenic isotope record of late Pleistocene incision, Boulder Canyon, Colorado. *Geological Society of America Abstracts with Programs*, **32**, A-473.

Dodge, R.E. (1902). Arroyo formation. *Science*, **15**, 746.

Downs, P.W. & Simon, A. (2001). Fluvial geomorphological analysis of the recruitment of large woody debris in the Yalobusha River network, Central Mississippi, USA. *Geomorphology*, **37**, 65–91.

Dragovich, J.D., McKay, D.T., Jr., Dethier, D.P. & Beget, J.E. (2000). Holocene Glacier Peak lahar deposits in the lower Skagit River valley, Washington. *Geology*, **28**, 19–21.

du Boys, P. (1879). Le Rhône et les riviérs a lit affouillable, Annales des Ponts et Chaussées. *Mémoires et Documents, Series 5*, **18**, 141–195.

Earle, C.J. (1993). Asynchronous droughts in California streamflow as reconstructed from tree rings. *Quaternary Research*, **39**, 290–299.

Eaton, L.S. (1999). Debris flows and landscape evolution in the Upper Rapidan basin, Blue Ridge Mountains, central Virginia. Ph.D. dissertation, University of Virginia, 154 pp.

Eaton, L.S. & McGeehin, J.P. (1997). Frequency of debris flows and their role in long term landscape evolution in the central Blue Ridge, Virginia. *Geological Society of America Abstracts with Programs*, **29**, 410.

Eschner, T.R., Hadley, R.F. & Crowley, K.D. (1983). Hydrologic and morphologic changes in channels of the Platte River Basin in Colorado, Wyoming, and Nebraska: A historical perspective. U.S. Geological Survey Professional Paper 1277-A.

Everest, F.H. & Meehan, W.R. (1981). Forest management and anadromous fish habitat productivity. *In*: Sabol, K. (Ed.), *Transactions of the Fourty-sixth North American Wildlife and Natural Resources Conference*. Wildlife Management Institute, Washington, DC, 521–530.

Fausch, K.D., Torgersen, C.E., Baxter, C.V. & Li, H.W. (2002). Landscapes to riverscapes: Bridging the gap between research and conservation of stream fishes. *BioScience*, **52**, 1–16.

Ferguson, R.I. & Paola, C. (1997). Bias and precision of percentiles of bulk grain size distributions. *Earth Surface Processes and Landforms*, **22**, 1061–1077.

Finlayson, D., Montgomery, D.R. & Hallet, B.H. (2002). Spatial coincidence of rapid inferred erosion with young metamorphic massifs in the Himalayas. *Geology*, **30**, 19–222.

Flint, J.J. (1974). Stream gradient as a function of order, magnitude, and discharge. *Water Resources Research*, **10**, 969–973.

Friedman, J.M. & Auble, G.T. (2000a). Floods, flood control, and bottomland vegetation. *In*: Wohl, E.E. (Ed.), *Inland Flood Hazards: Human, Riparian, and Aquatic Communities*. Cambridge University Press, Cambridge, UK, 219–237.

Friedman, J.M. & Auble, G.T. (2000b). Floods, flood control, and bottomland vegetation. *In*: Wohl, E. (Ed.), *Inland Flood Hazards: Human, Riparian and Aquatic Communities*. New York, NY, Cambridge University Press, 219–237.

Furbish, D.J. (1993). Flow structure in a bouldery mountain stream with complex bed topography. *Water Resources Research*, **29**, 2249–2263.

Gardner, T.W., Ritter, J.B., Shuman, C.A., Bell, J.C., Sasowsky, K.C. & Pinter, N. (1991). A periglacial stratified slope deposit in the Valley and Ridge province of central Pennsylvania, USA: sedimentology, stratigraphy, and geomorphic evolution. *Permafrost and Periglacial Processes*, **2**, 141–162.

Gilbert, G.K. (1877). Geology of the Henry Mountains, U.S. Geographical and Geological Survey of the Rocky Mountain Region: Washington, DC, U.S. Government Printing Office, 160 pp.

Gomez, B. & Troutman, B.M. (1997). Evaluation of process errors in bed load sampling using a dune model. *Water Resources Research*, **33**, 2387–2398.

Graf, W.L. (1983). The arroyo problem – palaeohydrology and palaeohydraulics in the short term. *In*: Gregory, K.J. (Ed.), *Background to Palaeohydrology*. Wiley, Chichester, 279–302.

Graf, W.L. (1988). *Fluvial processes in dryland rivers*. Springer-Verlag, Berlin, 346 pp.

Graf, W.L. (1996). Geomorphology and policy for restoration of impounded American rivers: What is "natural"? *In*: Rhoads, B.L. & Thorn, C.E. (Eds), *The Scientific Nature of Geomorphology*. Wiley, New York, 443–473.

Graf, W.L. (1999). Dam nation: A geographic census of American dams and their large-scale hydrologic impacts. *Water Resources Research*, **35**, 1305–1311.

Graffy, E.A., Helsel, D.R. & Mueller, D.K. (1996). Nutrients in the nation's waters: identifying problems and progress. U.S.G.S. Fact Sheet FS-218-96.

Granger, D.E., Kirchner, J.W. & Finkel, R.C. (1997). Quaternary downcutting rate of the New River, Virginia, measured from differential decay of cosmogenic ^{26}Al and ^{10}Be in cave-deposited alluvium. *Geology*, **25**, 107–110.

Gregory, K.J. & Davis, R.J. (1992). Coarse woody debris in stream channels in relation to river channel management in woodland areas. *Regulated Rivers: Research and Management*, **7**, 117–136.

Gregory, K.J., Davis, R.J. & Tooth, S. (1993). Spatial distribution of coarse woody debris dams in the Lymington Basin, Hampshire, U.K. *Geomorphology*, **6**, 207–224.

Gregory, K.J., Gurnell, A.M. & Hill, C.T. (1985). The permanence of debris dams related to river channel processes. *Hydrological Sciences Journal*, **30**, 371–381.

Gurnell, A.M., Petts, G.E., Hannah, D.M., Smith, B.P.G., Edwards, P.J., Kollmann, J., Ward, J.V. & Tockner, K. (2001). Riparian vegetation and island formation along the gravel-bed Riume Tagliamento, Italy. *Earth Surface Processes and Landforms*, **26**, 31–62.

Gurnell, A.M. & Sweet, R. (1998). The distribution of large woody debris accumulations and pools in relation to woodland stream management in a small, low-gradient stream. *Earth Surface Processes and Landforms*, **23**, 1101–1121.

Hack, J.T. (1939). Late Quaternary history of several valleys of northern Arizona, a preliminary announcement. *Museum of Northern Arizona, Museum Notes*, **11**, 63–73.

Hack, J.T. (1957). Studies of longitudinal profiles in Virginia and Maryland. U.S. Geological Survey Professional Paper 294-B.

Hack, J.T. (1960). Interpretation of erosional topography in humid temperate regions. *American Journal of Science*, **258A**, 80–97.

Hack, J.T. (1973). Drainage adjustment in the Appalachians. *In*: Morisawa, M. (Ed.), *Fluvial Geomorphology*. Allen and Unwin, Boston, 51–69.

Harbor, D.J. (1996). Nonuniform erosion patterns in the Appalachian Mountains of Virginia. *Geological Society American Abstracts with Programs*, **28**, A116.

Harwood, K. & Brown, A.G. (1993). Fluvial processes in a forested anastomosing river: Flood partitioning and changing flow patterns. *Earth Surface Processes and Landforms*, **18**, 741–748.

Hill, F.G. (1957). *Roads, rails and waterways, the Army Engineers and early transportation*. University of Oklahoma Press, Norman, OK, 248 pp.

Hilmes, M.M. & Wohl, E.E. (1995). Changes in channel morphology associated with placer mining. *Physical Geography*, **16**, 223–242.

Hirsch, R.M., Walker, J.F., Day, J.C. & Kallio, R. (1990). The influence of man on hydrologic systems. *In*: Wolman, M.G. & Riggs, H.C. (Eds), *Surface Water Hydrology. Geological Society of America*. Boulder, 329–359.

Hoffman, P.F. & Grotzinger, J.P. (1993). Orographic precipitation, erosional unloading, and tectonic style. *Geology*, **21**, 195–198.

Hooke, R.L. (1999). Spatial distribution of human geomorphic activity in the United States: comparison with rivers. *Earth Surface Processes and Landforms*, **24**, 687–692.

Hooke, R.L. (2000). On the history of humans as geomorphic agents. *Geology*, **28**, 843–846.

Horton, B.K. (1999). Erosional control on the geometry and kinematics of thrust belt development in the central Andes. *Tectonics*, **18**, 1292–1304.

House, M.A., Wernicke, B.P. & Farley, K.A. (1998). Dating topography of the Sierra Nevada, California, using apatite (U-Th)/He ages. *Nature*, **396**, 66–69.

Howard, A.D. (1998). Long profile development of bedrock channels: interaction of weathering, mass wasting, bed erosion, and sediment transport. *In*: Tinkler, K.J. & Wohl, E.E. (Eds), *Rivers Over Rock: Fluvial Processes in Bedrock Channels*. American Geophysiology Union Press, Washington, DC, 297–319.

Huber, N.K. (1981). Amount and timing of late Cenozoic uplift and tilt of the central Sierra Nevada, California – Evidence from the upper San Joaquin River basin. *U.S. Geological Survey Porfessional Paper 1197*, 28 pp.

Hurtrez, J.-E., Lucazeau, F., Lavé, J. & Avouac, J.-P. (1999). Investigation of the relationships between basin morphology, tectonic uplift, and denudation from the study of an active fold belt in the Siwalik Hills, central Nepal. *Journal of Geophysical Research*, **104**, 12,799–12,796.

Ingram, B.L., Ingle, J.C. & Conrad, M.E. (1996). Stable isotope record of late Holocene salinity and river discharge in San Francisco Bay, California. *Earth and Planetary Science Letters*, **141**, 237–247.

Jacobson, R.B., Miller, A.J. & Smith, J.A. (1989). The role of catastrophic geomorphic events in Central Appalachian landscape evolution. *Geomorphology*, **2**, 257–284.

James, L.A. (1991). Incision and morphologic evolution of an alluvial channel recovering from hydraulic mining sediment. *Geological Society of America Bulletin*, **103**, 723–736.

James, L.A. (1994). Channel changes wrought by gold mining: northern Sierra Nevada, California. *In*: *Effects of Human-Induced Changes on Hydrologic Systems*. American Water Resources Association, Bethesda, 629–638.

James, L.A. (1999). Time and the persistence of alluvium: river engineering, fluvial geomorphology, and mining sediment in California. *Geomorphology*, **31**, 265–290.

Jarrett, R.D. (1990). Hydrologic and hydraulic research in mountain rivers. *Water Resources Bulletin*, **26**, 419–429.

Johnston, C.E., Andrews, E.D. & Pitlick, J. (1998). In situ determination of particle friction angles of fluvial gravels. *Water Resources Research*, **34**, 2017–2030.

Joyce, J.E., Tjalsma, L.R.C. & Prutzman, J.M. (1993). North American glacial meltwater history for the past 2.3 m.y.: Oxygen isotope evidence from the Gulf of Mexico. *Geology*, **21**, 483–486.

Kauffman, J.B. & Krueger, W.C. (1984). Livestock impacts on riparian ecosystems and streamside management implications . . . a review. *Journal of Range Management*, **37**, 430–438.

Kehew, A.E. (1993). Glacial-lake outburst erosion of the Grand Valley, Michigan, and impacts on glacial lakes in the Lake Michigan basin. *Quaternary Research*, **39**, 36–44.

Kehew, A.E. & Lord, M.L. (1987). Glacial-lake outbursts along the mid-continent margins of the Laurentide ice-sheet. *In*: Mayer, L. & Nash, D. (Eds), *Catastrophic Flooding*. Allen and Unwin, Boston, 95–120.

Keller, E.A. & Swanson, F.J. (1979). Effects of large organic material on channel form and fluvial processes. *Earth Surface Processes*, **4**, 361–380.

Keller, E.A. & Tally, T. (1979). Effects of large organic debris on channel form and fluvial processes in the coastal Redwood environment. *In*: Rhodes, D.D. & Williams, G.P. (Eds), *Adjustments of the Fluvial System*. Kendal-Hunt, Dubuque, IA, 169–197.

Kirby, E. & Whipple, K. (2001). Quantifying differential rock-uplift rates via stream profile analysis. *Geology*, **29**, 415–418.

Kneller, M. & Peteet, D. (1999). Late-glacial to early Holocene climate changes from a central Appalachian pollen and macrofossil record. *Quaternary Research*, **51**, 133–147.

Knox, J.C. (1983). Responses of river systems to Holocene climates. *In*: Wright, H.E., Jr. (Ed.), *Late-Quaternary Environments of the United States*. The Holocene. University of Minnesota Press, Minneapolis, **2**, 26–41.

Knox, J.C. (1985). Responses of floods to Holocene climatic change in the Upper Mississippi Valley. *Quaternary Research*, **23**, 287–300.

Knox, J.C. (1987). Historical valley floor sedimentation in the Upper Mississippi Valley. *Annals Association of American Geographers*, **77**, 224–244.

Knox, J.C. (1993). Large increases in flood magnitude in response to modest changes in climate. *Nature*, **361**, 430–432.

Knox, J.C. (1996). Late Quaternary upper Mississippi River alluvial episodes and their significance to the lower Mississippi River system. *Engineering Geology*, **45**, 263–285.

Kondolf, G.M. (1994). Geomorphic and environmental effects of instream gravel mining. *Landscape and Urban Planning*, **28**, 225–243.

Kondolf, G.M. (1996). A cross section of stream channel restoration. *Journal of Soil and Water Conservation*, **51**, 119–125.

Kondolf, G.M. (1997). Hungry water: effects of dams and gravel mining on river channels. *Environmental Management*, **21**, 533–551.

Kuhnle, R.A. (1992). Bed load transport during rising and falling stages on two small streams. *Earth Surface Processes and Landforms*, **17**, 191–197.

Lagasse, P.F., Winkley, B.R. & Simons, D.B. (1980). Impact of gravel mining on river system stability, Port, Coastal and Ocean Division. *American Society of Civil Engineers Journal of Waterway*, **106**, 389–404.

Lague, D., Davy, P. & Crave, A. (2000). Estimating uplift rate and erodibility from the area-slope relationship: Examples from Brittany (France) and numerical modelling. *Physics and Chemistry of the Earth (A)*, **25**, 543–548.

LaMarche, V.C. (1973). Holocene climatic variations inferred from treeline fluctuations in the White Mountains, California. *Quaternary Research*, **3**, 632–660.

Leopold, A. (1921). A plea for recognition of artificial works in forest erosion and control policy. *Journal of Forestry*, **19**, 267–273.

Leopold, L.B. (1951). Rainfall frequency: An aspect of climatic variation. *Trans. American Geophysiology Union*, **32**, 347–357.

Leopold, L.B. (1976). Reversal of erosion cycle and climatic change. *Quaternary Research*, **6**, 557–562.

Ligon, F.K., Dietrich, W.E. & Trush, W.J. (1995). Downstream ecological effects of dams. *BioScience*, **45**, 183–192.

Lisle, T.E. (1986). Stabilization of a gravel channel by large streamside obstructions and bedrock bends, Jacoby Creek, northwestern California. *Water Resources Research*, **97**, 999–1011.

Lisle, T.E. (1995). Effects of coarse woody debris and its removal on a channel affected by the 1980 eruption of Mount St. Helens, Washington. *Water Resources Research*, **31**, 1797–1808.

Lopes, T.J. & Bender, D.A. (1998). Nonpoint sources of volatile organic compounds in urban areas – relative importance of urban land surfaces and air. *Environmental Pollution*, **101**, 221–230.

Lord, M.L. & Kehew, A.E. (1987). Sedimentology and paleohydrology of glacial-lake outburst deposits in southeastern Saskatchewan and northwestern North Dakota. *Geological Society of America Bulletin*, **99**, 663–673.

Love, D.W. (1979). Quaternary fluvial geomorphic adjustments in Chaco Canyon, New Mexico. *In*: Rhodes, D.D. & Williams, G.P. (Eds), *Adjustments of the Fluvial System*. Kendall/Hunt, Dubuque, 277–308.

MacFarlane, W.A. & Wohl, E. (2003). The influence of step composition on step geometry and flow resistance in step-pool streams of the Washington Cascades. *Water Resources Research*, **39**(2), 1037, doi: 10.1029/2001WR001238.

Mackin, J.H. (1948). Concept of the graded river. *Bulletin of the Geological Society of America*, **59**, 463–512.

Magilligan, F.J. & McDowell, P.F. (1997). Stream channel adjustments following elimination of cattle grazing. *Journal of American Water Resources Association*, **33**, 867–878.

Manga, M. & Kirchner, J.W. (2000). Stress partitioning in streams by large woody debris. *Water Resources Research*, **36**, 2373–2379.

Manning, R. (1891). On the flow of water in open channels and pipes. *Transactions of the Institution of Civil Engineers of Ireland*, **20**, 161–207.

Marcus, W.A., Roberts, K., Harvey, L. & Tackman, G. (1992). An evaluation of methods for estimating Manning's n in small mountain streams. *Mountain Research and Development*, **12**, 227–239.

Maxwell, A.R. & Papanicolaou, A.N. (2001). Step-pool morphology in high-gradient streams. *International Journal of Sediment Research*, **16**, 380–390.

McCain, J.F., Hoxit, L.R., Maddox, R.A., Chappell, C.F. & Caracena, F. (1979). Storm and flood of July 31-August 1, 1976, in the Big Thompson River and Cache la Poudre

River basins, Larimer and Weld Counties, Colorado. Part A. Meteorology and hydrology in the Big Thompson River and Cache la Poudre River basins. U.S. Geological Survey Professional Paper 1115.

Meade, R.H., Yuzyk, T.R. & Day, T.J. (1990). Movement and storage of sediment in rivers of the United States and Canada. *In*: Wolman, M.G. & Riggs, H.C. (Eds), *Surface Water Hydrology*. Geological Society of America, Boulder, CO, 255–280.

Meador, M.R. & Layher, A.O. (1998). Instream sand and gravel mining. *Fisheries*, **23**(11), 6–12.

Michaelsen, P. (2002). Mass extinction of peat-forming plants and the effect on fluvial styles across the Permian – Triassic boundary, northern Bowen Basin, Australia. *Palaeogeography, Palaeoclimatology, Palaeoecology*, **179**, 173–188.

Millar, R.G. (2000). Influence of bank vegetation on alluvial channel patterns. *Water Resources Research*, **36**, 1109–1118.

Miller, A.J. (1990). Flood hydrology and geomorphic effectiveness in the central Appalachians. *Earth Surface Processes and Landforms*, **19**, 681–697.

Miller, A.J. (1995). Valley morphology and boundary conditions influencing spatial patterns of flood flow. *In*: Costa, J.E., Miller, A.J., Potter, K.W. & Wilcock, P.R. (Eds), *Natural and Anthropogenic Influences in Fluvial Geomorphology*. American Geophysical Union Press, Washington, DC, 57–81.

Miller, A.J. & Parkinson, D.J. (1993). Flood hydrology and geomorphic effects on river channels and flood plains: the flood of November 4–5, 1985, in the South Branch Potomac River Basin of West Virginia. *U.S. Geological Survey Bulletin 1981-E*, 96 pp.

Milliman, J.D. & Syvitski, J.P.M. (1992). Geomorphic/tectonic control of sediment discharge to the ocean: the importance of small mountainous rivers. *Journal of Geology*, **100**, 525–544.

Mills, H.H. (2000a). Apparent increasing rates of stream incision in the eastern United States during the late Cenozoic. *Geology*, **28**, 955–957.

Mills, H.H. (2000b). Controls on form, process, and sedimentology of alluvial fans in the central and southern Appalachians, southeastern U.S.A. *Southeastern Geology*, **39**, 281–313.

Mills, H.H., Brakenridge, G.R., Jacobson, R.B., Newell, W.L., Pavich, M.J. & Pomeroy, J.S. (1987). Appalachian mountains and plateaus. *In*: Graf, W.L. (Ed.), *Geomorphic Systems of North America*. Geological Society of America, Boulder, Colorado, 5–50.

Mills, H.H. & Delcourt, P.A. (1991). Quaternary geology of the Appalachian Highlands and Interior Low Plateaus. *In*: *Quaternary Nonglacial Geology*. Conterminous U.S. Geological Society of America, Boulder, CO, 611–628.

Moglen, G.E. & Bras, R.L. (1995). The effect of spatial heterogeneities on geomorphic expression in a model of basin evolution. *Water Resources Research*, **31**, 2613–2623.

Montgomery, D.R. (1999). Process domains and the river continuum. *Journal of the American Water Resources Association*, **35**, 397–410.

Montgomery, D.R. (2001). Slope distributions, threshold hillslopes and steady-state topography. *American Journal of Science*, **301**, 432–454.

Montgomery, D.R., Abbe, T.B., Peterson, N.P., Buffington, J.M., Schmidt, K. & Stock, J.D. (1996). Distribution of bedrock and alluvial channels in forested mountain drainage basins. *Nature*, **381**, 587–589.

Montgomery, D.R., Balco, G. & Willett, S. (2001). Climatic, tectonics and the morphology of the Andes. *Geology*, **29**, 579–582.

Montgomery, D.R. & Brandon, M.T. (2002). Non-linear controls on erosion rates in tectonically active mountain ranges. *Earth and Planetary Science Letters*, **201**, 481–489.

Montgomery, D.R. & Buffington, J.M. (1997). Channel reach morphology in mountain drainage basins. *Geological Society of America Bulletin*, **109**, 596–611.

Montgomery, D.R., Buffington, J.M., Smith, R., Schmidt, K. & Pess, G. (1995). Pool spacing in forest channels. *Water Resources Research*, **31**, 1097–1105.

Montgomery, D.R. & Foufoula-Georgiou, E. (1993). Channel network source representation using digital elevation models. *Water Resources Research*, **29**, 3925–3934.

Montgomery, D.R. & Gran, K.B. (2001). Downstream variations in the width of bedrock channels. *Water Resources Research*, **37**, 1841–1846.

Montgomery, D.R. & MacDonald, L.H. (2002). Diagnostic approach to stream channel assessment and monitoring. *Journal of the American Water Resources Association*, **38**, 1–16.

Montgomery, D.R., Panfil, M.S. & Hayes, S.K. (1999). Channel-bed mobility response to extreme sediment loading at Mount Pinatubo. *Geology*, **27**, 271–274.

Morisawa, M.E. & LaFlure, E. (1979). Hydraulic geometry, stream equilibrium and urbanization. *In*: Rhodes, D.D. & Williams, G.P. (Eds), *Adjustments of the Fluvial System*. Allen and Unwin, Boston, 333–350.

Morrison, R.B. (1987). Long-term perspective: changing rates and types of Quaternary surficial processes: Erosion-deposition-stability cycles. *In*: Graf, W.L. (Ed.), *Geomorphic Systems of North America*. Geological Society of America, Boulder, Colorado, 167–176.

Nadler, C.T. & Schumm, S.A. (1981). Metamorphosis of South Platte and Arkansas Rivers, eastern Colorado. *Physical Geography*, **2**, 95–115.

Nakamura, F. & Swanson, F.J. (1993). Effects of coarse woody debris on morphology and sediment storage of a mountain stream system in western Oregon. *Earth Surface Processes and Landforms*, **18**, 43–61.

National Research Council (NRC) (1992). *Restoration of Aquatic Ecosystems*. National Academy Press, Washington, DC, 552 pp.

Nikora, V.I., Goring, D.G. & Biggs, B.J.F. (1998). On gravel-bed roughness characterization. *Water Resources Research*, **34**, 517–527.

Norman, D.K., Cederholm, C.J. & Lingley, W.S. (1998). Flood plains, salmon habitat, and sand and gravel mining. *Washington Geology*, **26**, 3–20.

Nowell, L.H., Capel, P.D. & Dileanis, P.D. (1999). Pesticides in stream sediment and aquatic biota – distribution, trends, and governing factors. *Pesticides in the Hydrologic System Series*, **4**. CRC Press, Boca Raton, 1040 pp.

Osborn, G. & Bevis, K. (2001). Glaciation in the Great Basin of the western United States. *Quaternary Science Reviews*, **20**, 1377–1410.

Patton, P.C. & Boison, P.J. (1986). Processes and rates of formation of Holocene alluvial terraces in Harris Wash, Escalante River basin, south-central Utah. *Geological Society of American Bulletin*, **97**, 369–378.

Patton, P.C. & Schumm, S.A. (1975). Gully erosion, northwestern Colorado: A threshold phenomenon. *Geology*, **3**, 88–89.

Pazzaglia, F.J. & Gardner, T.W. (1993). Fluvial terraces of the lower Susquehanna River. *Geomorphology*, **8**, 83–113.

Pazzaglia, F.J. & Gardner, T.W. (1994). Late Cenozoic flexural deformation of the middle U.S. Atlantic margin. *Journal of Geophysical Research*, **99**(B6), 12143–12157.

Pazzaglia, F.J., Gardner, T.W. & Merritts, D.J. (1998). Bedrock fluvial incision and longitudinal profile development over geologic time scales determined by fluvial terraces. *In*: Tinkler, K.J. & Wohl, E.E. (Eds), *Rivers Over Rock: Fluvial Processes in Bedrock Channels*. American Geophysical Union Press, Washington, DC, 207–235.

Phillips, F.M., Zreda, M.G., Smith, S.S., Elmore, D., Kubik, P.W. & Sharma, P. (1990). Cosmogenic chlorine-36 chronology for glacial deposits at Bloody Canyon, eastern Sierra Nevada. *Science*, **248**, 1529–1532.

Piégay, H. (1993). Nature, mass and preferential sites of coarse woody debris deposits in the lower Ain Valley (Mollon Reach), France. *Regulated Rivers: Research and Management*, **8**, 359–372.

Piégay, H. & Gurnell, A.M. (1997). Large woody debris and river geomorphological pattern: examples from S.E. France and S. England. *Geomorphology*, **19**, 99–116.

Piégay, H. & Marston, R.A. (1998). Distribution of large woody debris along the outer bend of meanders in the Ain River, France. *Physical Geography*, **19**, 318–340.

Platts, W.S. & Nelson, R.L. (1985). Impacts of rest-rotation grazing on stream banks in forested watersheds in Idaho. *North America Journal of Fisheries Management*, **5**, 547–556.

Poag, C.W. & Sevon, W.D. (1989). A record of Appalachian denudation in postrift Mesozoic and Cenozoic sedimentary deposits of the U.S. middle Atlantic continental margin. *Geomorphology*, **2**, 303–318.

Poff, N.L., Allan, J.D., Bain, M.B., Karr, J.R., Prestegaard, K.L., Richter, B.D., Sparks, R.E. & Stromberg, J.C. (1997). The natural flow regime. *BioScience*, **47**, 769–784.

Pope, D.C. & Willett, S.D. (1998). A thermal-mechanical model for crustal thickening in the central Andes driven by ablative subduction. *Geology*, **26**, 511–514.

Porter, D.A. & Guccione, M.J. (1994). Deglacial flood origin of the Charleston Alluvial Fan, lower Mississippi alluvial valley. *Quaternary Research*, **41**, 278–284.

Porter, S.C. & Swanson, T.W. (1998). Radiocarbon age constraints on rates of advance and retreat of the Puget Lobe of the Cordilleran Ice Sheet during the last glaciation. *Quaternary Research*, **50**, 205–213.

Prestegaard, K.L. (1983). Bar resistance in gravel bed streams at bankfull stage. *Water Resources Research*, **19**, 472–476.

Rampe, J.J. & Runnells, D.D. (1989). Contamination of water and sediment in a desert stream by metals from an abandoned gold mine and mill, Eureka District, Arizona, USA. *Applied Geochemistry*, **4**, 445.

Rathbun, R.E. (1998). Transport, behavior, and fate of volatile organic compounds in streams. *U.S. Geological Survey Professional Paper 1589*, 160 pp.

Reeves, G.H., Benda, L.E., Burnett, K.M., Bisson, P.A. & Sedell, J.R. (1995). A disturbance-based ecosystem approach to maintaining and restoring freshwater habitats of evolutionarily significant units of anadromous salmonids in the Pacific Northwest. *In*: Nielsen, J.L. (Ed.), *Evolution and the Aquatic Ecosystem: Defining Unique Units in Population Conservation*. American Fisheries Society Symposium 17, 334–349.

Rice, K.C. (1999). Trace-element concentrations in streambed sediments across the conterminous United States. *Environmental Science and Technology*, **33**, 2499–2504.

Richmond, G.M. (1960). Glaciation of the east slope of Rocky Mountain National Park, Colorado. *Geological Society of America Bulletin*, **71**, 1371–1382.

Riebe, C.S., Kirchner, J.W., Granger, D.E. & Finkel, R.C. (2000). Erosional equilibrium and disequilibrium in the Sierra Nevada, inferred from cosmogenic ^{26}Al and ^{10}Be in alluvial sediment. *Geology*, **28**, 803–806.

Riebe, C.S., Kirchner, J.W., Granger, D.E. & Finkel, R.C. (2001). Minimal climatic control on erosion rates in the Sierra Nevada, California. *Geology*, **29**, 447–450.

Robison, E.G. & Beschta, R.L. (1990). Coarse woody debris and channel morphology interactions for undisturbed streams in southeast Alaska, USA. *Earth Surface Processes and Landforms*, **15**, 149–156.

Roe, G., Montgomery, D.R. & Hallet, B. (2002). Effects of orographic precipitation variations on the concavity of steady-state river profiles. *Geology*, **30**, 143–146.

Sandecki, M. (1989). Aggregate mining in river systems. *California Geology*, **42**, 88–94.

Saunders, I. & Young, A. (1983). Rates of surface processes on slopes, slope retreat and denudation. *Earth Surface Processes and Landforms*, **8**, 473–501.

Schildgen, T.F. & Dethier, D.P. (2000). Fire and ice: using isotopic dating techniques to interpret the geomorphic history of Middle Boulder Creek, Colorado. *Geological Society of America Abstracts with Programs*, **32**, A-18.

Schleusener, R.A., Smith, G.L. & Chen, M.C. (1962). Effect of flow diversion for irrigation on peak rates of runoff from watersheds in and near the Rocky Mountain foothills of Colorado. *International Association of Hydrologists Bulletin*, **7**, 53–61.

Schumm, S.A. & Brakenridge, G.R. (1987). River responses. *In*: Ruddiman, W.F. & Wright, H.E., Jr. (Eds), *North America and Adjacent Oceans During the Last Deglaciation*. Geological Society of America, Boulder, Colorado, 221–240.

Schumm, S.A. & Hadley, R.F. (1957). Arroyos and the semiarid cycle of erosion. *American Journal of Science*, **255**, 161–174.

Schumm, S.A. & Lichty, R.W. (1965). Time, space, and causality in geomorphology. *American Journal of Science*, **263**, 110–119.

Scott, G.R. (1960). Subdivision of the Quaternary alluvium east of the Front Range near Denver, Colorado. *Geological Society of America Bulletin*, **71**, 1541–1543.

Scott, G.R. (1963). Quaternary geology and geomorphic history of the Kassler Quadrangle, Colorado. *U.S. Geological Survey Professional Paper 421-A*, 70 pp.

Sedell, J.R. & Duvall, W.S. (1985). Water transportation and storage of logs. General Technical Report PNW-186, USDA Forest Service, Pacific Northwest Research Station, Portland, OR, 68 pp.

Sedell, J.R., Everest, F.H. & Swanson, F.J. (1982). Fish habitat and streamside management: Past and present. *In*: *Proceedings of the 1981 Convention of The Society of American Foresters, September 27–30, 1981*. Society of American Foresters, Bethesda, MD, Publication 82–01, 244–255.

Sedell, J.R. & Froggatt, J.L. (1984). Importance of streamside forests to large rivers: the isolation of the Willamette River, Oregon, USA, from its floodplain by snagging and streamside forest removal. *Verhandlungen-Internationale Vereinigung für Theoretifche und Angewandte Limnologie*, **22**, 1828–1834.

Seeber, L. & Gornitz, V. (1983). River profiles along the Himalayan arc as indicators of active tectonics. *Tectonophysics*, **92**, 335–367.

Seidl, M.A. & Dietrich, W.E. (1992). The problem of channel erosion into bedrock. *In*: Schmidt, K.-H. & de Ploey, J. (Eds), *Functional Geomorphology*. Catena Supplement 23, Catena-Verlag, Cremlingen-Destedt, 101–124.

Seidl, M.A., Dietrich, W.E. & Kirchner, J.W. (1994). Longitudinal profile development into bedrock: an analysis of Hawaiian channels. *Journal of Geology*, **102**, 457–474.

Seidl, M.A., Weissel, J.K. & Pratson, L.F. (1996). The kinematics and pattern of escarpment retreat across the rifted continental margin of SE Australia. *Basin Research*, **12**, 301–316.

Shields, A. (1936). Anwendung der Aehnlichkeitsmechanik und der Turbulenzforschung auf die Geschiebebewegung. *Mitteilungen der Preussischen Versuchsanstalt fur Wasserbau und Schiffbau*, **26**, 26.

Shields, F.D. & Gippel, C.J. (1995). Prediction of effects of woody debris removal on flow resistance. American Society of Civil Engineers. *Journal of Hydraulic Engineering*, **121**, 341–354.

Shroba, R.R., Schmidt, P.W., Crosby, E.J., Hansen, W.R. & Soule, J.M. (1979). Storm and flood of July 31 to August 1, 1976, in the Big Thompson River and Cache la Poudre River basins, Larimer and Weld Counties, Colorado. Part B. Geologic and geomorphic effects in the Big Thompson Canyon area, Larimer County. U.S. Geological Survey Professional Paper 1115.

Sklar, L. & Dietrich, W.E. (1998). River longitudinal profiles an dbedrock incision models; stream power and the influence of sediment supply. *In*: Tinkler, K.J. & Wohl, E.E. (Eds), *Rivers Over Rock: Fluvial Processes in Bedrock Channels*. Amercian Geophysical Union Geophysical Monograph 107, Washington, DC, 237–260.

Sklar, L. & Dietrich, W.E. (2001). Sediment and rock strength controls on river incision into bedrock. *Geology*, **29**, 1087–1090.

Smith, D.G. (1976). Effect of vegetation on lateral migration of anastomosed channels of a glacier meltwater river. *Geological Society of America Bulletin*, **87**, 857–860.

Smith, J.A., Baeck, M.L., Steiner, M. & Miller, A.J. (1996). Catastrophic rainfall from an upslope thunderstorm in the central Appalachians: The Rapidan storm of June 27, 1995. *Water Resources Research*, **32**, 3099–3113.

Snow, R.S. & Slingerland, R.L. (1987). Mathematical modeling of graded river profiles. *Journal of Geology*, **95**, 15–33.

Snow, R.S. & Slingerland, R.L. (1990). Stream profile adjustment to crustal warping: Nonlinear results from a simple model. *Journal of Geology*, **98**, 699–708.

Snyder, N.P., Whipple, K.X., Tucker, G.E. & Merritts, D.J. (2000). Landscape response to tectonic forcing: Digital elevation model analysis of stream profiles in the Mendocino triple junction region, northern California. *Geological Society of America Bulletin*, **112**, 1250–1263.

Southwood, T.R.E. (1977). Habitat, the templet for ecological strategies? *Journal of Animal Ecology*, **46**, 337–365.

Sparks, R.E. (1995). Need for ecosystem management of large rivers and their floodplains. *Bioscience*, **45**, 168–182.

Springer, G.S., Kite, J.S. & Schmidt, V.A. (1997). Cave sedimentation, genesis, and erosional history in the Cheat River Canyon, West Virginia. *Geological Society of America Bulletin*, **109**, 524–532.

Starnes, L.B. & Gasper, D.C. (1995). Effects of surface mining on aquatic resources in North America. *Fisheries*, **20**, 20–23.

Steingraber, S. (1997). *Living downstream: An ecologist looks at cancer and the environment*. Addison-Wesley Publishing, Reading, MA, 357 pp.

Stiller, D. (2000). *Wounding the West: Montana, mining, and the environment*. University of Nebraska Press, Lincoln, 212 pp.

Stock, J. & Dietrich, W.E. (2003). Valley incision by debris flows: Evidence of a topographic signature. *Water Resources Research*, **39**(4), 1089, doi: 10.1029/2001WR001057.

Stock, J.D. & Montgomery, D.R. (1999). Geologic constraints on bedrock river incision using the stream power law. *Journal of Geophysical Research*, **104**, 4983–4993.

Stott, T. (1997). A comparison of stream bank erosion processes on forested and moorland streams in the Balquhidder catchments, central Scotland. *Earth Surface Processes and Landforms*, **22**, 383–399.

Stoughton, J.A. & Marcus, W.A. (2000). Persistent impacts of trace metals from mining on floodplain grass communities along Soda Butte Creek, Yellowstone National Park. *Environmental Management*, **25**, 305–320.

Stromberg, J.C. & Patten, D.T. (1990). Riparian vegetation instream flow requirements: a case study from a diverted stream in the eastern Sierra Nevada, California, USA. *Environmental Management*, **14**, 185–194.

Swanson, F.J., Gregory, S.V., Sedell, J.R. & Campbell, A.G. (1982). Land-water interactions: The riparian zone. *In*: Edmonds, R.L. (Ed.), *Analysis of Coniferous Forest Ecosystems in the Western United States*. Hutchinson Ross, Stroudsburg, Pennsylvania, 267–291.

Swanson, F.J., Krata, T.K., Caine, N. & Woodmansee, R.G. (1988). Landform effects on ecosystem patterns and processes. *BioScience*, **38**, 92–98.

Tarboton, Bras, R.L. & Rodriguez-Iturbe, I. (1989). Scaling and elevation in river networks. *Water Resources Research*, **25**, 2037–2051.

Teller, J.T. (1990). Volume and routing of late-glacial runoff from the Southern Laurentide ice sheet. *Quaternary Research*, **34**, 12–23.

Tepordei, V.V. (1987). 1986 aggregate mining data. *Rock Products*, **90**, 25–31.

Thompson, D.M., Wohl, E.E. & Jarrett, R.D. (1999). Velocity reversals and sediment sorting in pools and riffles controlled by channel constrictions. *Geomorphology*, **27**, 229–241.

Thornthwaite, C.W., Sharpe, C.F.S. & Dosch, E.F. (1941). Climate of the southwest in relation to accelerated erosion. *Soil Conservation*, **6**, 298–304.

Tinkler, K.J. & Wohl, E.E. (1998). A primer on bedrock channels. *In*: Tinkler, K.J. & Wohl, E.E. (Eds), *Rivers Over Rock: Fluvial Processes in Bedrock Channels*. American Geophysical Union Press, Washington, DC, 1–18.

Tooth, S. & Nanson, G.C. (1999). Anabranching rivers on the Northern Plains of arid central Australia. *Geomorphology*, **29**, 211–233.

Trimble, S.W. (1997). Stream channel erosion and change resulting from riparian forests. *Geology*, **25**, 467–469.

Trimble, S.W. & Mendel, A.C. (1995). The cow as a geomorphic agent – a critical review. *Geomorphology*, **13**, 233–253.

Triska, F.J. (1984). Role of large wood in modifying channel morphology and riparian areas of a large lowland river under pristine conditions: a historical case study. *Verhandlungen-Internationale Vereinigung für Theorelifche und Angewandte Limnologie*, **22**, 1876–1892.

Urbonas, B. & Benik, B. (1995). Stream stability under a changing environment. *In*: Herricks, E.E. (Ed.), *Stormwater Runoff and Receiving Systems: Impact, Monitoring, and Assessment*. Lewis Publishers, Boca Raton, 77–101.

U.S. Geological Survey (1999). The quality of our nation's waters: nutrients and pesticides. *U.S. Geological Sciences Circular*, **1225**, 82 pp.

Van Haveren, B.P. (1991). Placer mining and sediment problems in interior Alaska. *In*: Fan, S.S. & Kuo, Y.H. (Eds), *Proceedings of the 5th Federal Interagency Sedimentation Conference*. Las Vegas, 10-69–10-73.

Van Nieuwenhuyse, E.E. & LaPerriere, J.D. (1986). Effects of placer gold mining on primary production in subarctic streams of Alaska. *Water Resources Bulletin*, **22**, 91–99.

Vileisis, A. (1997). *Discovering the unknown landscape: A history of America's wetlands*. Island Press, Washington, DC.

Wakabayashi, J. & Sawyer, T.L. (2001). Stream incision, tectonics, uplift, and evolution of topography of the Sierra Nevada, California. *Journal of Geology*, **109**, 539–562.

Wallerstein, N., Thorne, C.R. & Doyle, M.W. (1997). Spatial distribution and impact of large woody debris in northern Mississippi. *In*: Wang, C.C., Langendoen, E.J. & Shields, F.D. (Eds), *Proceedings of the Conference on Management of Landscapes Disturbed by Channel Incision*. University of Mississippi, Oxford, MS, 145–150.

Ward, P., Montgomery, D.R. & Smith, R. (2000). Altered river morphology in South Africa associated with the Permian-Triassic mass extinction. *Science*, **289**, 1740–1743.

Waters, T.F. (1995). Sediment in streams: sources, biological effects, and control. *American Fisheries Society Monograph 7*. American Fisheries Society, Bethesda, 251 pp.

Wathen, S.J., Ferguson, R.I., Hoey, T.B. & Werritty, A. (1995). Unequal mobility of gravel and sand in weakly bimodal river sediments. *Water Resources Research*, **31**, 2087–2096.

Watson, C.C. & Biedenharn, D.S. (2000). Comparison of flood management strategies. *In*: Wohl, E.E. (Ed.), *Inland Flood Hazards: Human, Riparian and Aquatic Communities*. Cambridge University Press, 381–393.

Webb, T., Bartlein, P.J., Harrison, S.P. & Anderson, K.H. (1993). Vegetation, lake levels, and climate in eastern North America for the past 18,000 years. *In*: Wright, H.E., Jr. (Ed.), *Global Climates since the Last Glacial Maximum*. University Minnesota Press, Minneapolis, 415–467.

Whipple, K.X. (2001). Fluvial landscape response time: how plausible is steady-state denudation? *American Journal of Science*, **301**, 313–325.

Whipple, K.X. & Dunne, T. (1992). The influence of debris-flow rheology on fan morphology, Owens Valley, California. *Geological Society of America Bulletin*, **104**, 887–900.

Whipple, K.X., Kirby, E. & Brocklehurst, S.H. (1999). Geomorphic limits to climate-induced increases in topographic relief. *Nature*, **401**, 39–43.

Whipple, K.X. & Trayler, C.R. (1996). Tectonic control of fan size: the importance of spatially variable subsidence rates. *Basin Research*, **8**, 351–366.

Whipple, K.X. & Tucker, G.E. (1999). Dynamics of stream-power river incision model: implications for height limits of mountain ranges, landscape response timescales, and research needs. *Journal Geophysical Research*, **104B**, 17661–17674.

Whiting, P.J. (2002). Streamflow necessary for environmental maintenance. *Annual Reviews of Earth and Planetary Science*, **30**, 181–206.

Whitney, J.D. (1880). The Auriferous Gravels of the Sierra Nevada of California. *Contributions to American Geology, Volume I, Memoirs of the Museum of Comparative Zoölogy*. University Press, Cambridge, 567 pp.

Wiberg, P.L. & Smith, J.D. (1991). Velocity distribution and bed roughness in high-gradient streams. *Water Resources Research*, **27**, 825–838.

Wieczorek, G.F., Morgan, B.A. & Campbell, R.H. (2000). Debris-flow hazards in the Blue Ridge of central Virginia. *Environmental and Engineering Geoscience*, **6**, 3–23.

Wilcock, P.R. (1993). Critical shear stress of natural sediments. *ASCE Journal of Hydraulic Engineering*, **119**, 491–505.

Wilcock, P.R., Barta, A.F., Shea, C.C., Kondolf, G.M., Matthews, W.V.G. & Pitlick, J. (1996). Observations of flow and sediment transport on a large gravel-bed river. *Water Resources Research*, **32**, 2897–2909.

Willett, S.D. (1999). Orogeny and orography: The effects of erosion on the structure of mountain belts. *Journal of Geophysical Research*, **104**, 28957–28981.

Willett, S.D., Beaumont, C. & Fullsack, P. (1993). Mechanical model for the tectonics of doubly vergent compressional orogens. *Geology*, **21**, 371–374.

Willett, S.D. & Brandon, M.T. (2002). On steady states in mountain belts. *Geology*, **30**, 175–178.

Willett, S.D., Slingerland, R. & Hovius, N. (2001). Uplift, shortening, and steady state topography in active mountain belts. *American Journal of Science*, **301**, 455–485.

Williams, G.P. (1978). The case of the shrinking channels – the North Platte and Platte Rivers in Nebraska. *U.S. Geological Survey Circular 781*.

Williams, G.P. & Wolman, M.G. (1984). Effects of dams and reservoirs on surface-water hydrology; changes in rivers downstream from dams. *U.S.G.S. Professional Paper 1286*, 83 pp.

Wohl, E. (2000). Mountain rivers. American Geophysical Union Press, Washington, DC, 320 pp.

Wohl, E.E. (1998). Bedrock channel morphology in relation to erosional processes. *In*: Tinkler, K.J. & Wohl, E.E. (Eds), *Rivers Over Rock: Fluvial Processes in Bedrock Channels*. American Geophysical Union Press, Washington, DC, 133–151.

Wohl, E.E. (2001). *Virtual rivers: Lessons from the mountain rivers of the Colorado Front Range*. Yale University Press, New Haven, 210 pp.

Wohl, E.E., Anthony, D.J., Madsen, S.W. & Thompson, D.M. (1996). A comparison of surface sampling methods for coarse fluvial sediments. *Water Resources Research*, **32**, 3219–3226.

Wohl, E.E. & Merritt, D.M. (2001). Bedrock channel morphology. *Geological Society of America Bulletin*, **113**, 1205–1212.

Wohl, E.E. & Thompson, D.M. (2000). Velocity characteristics along a small step-pool channel. *Earth Surface Processes and Landforms*, **25**, 353–367.

Wolcott, J. & Church, M. (1991). Strategies for sampling spatially heterogeneous phenomena: The example of river gravels. *Journal of Sedimentary Petrology*, **61**, 534–543.

Wolman, M.G. (1967). A cycle of sedimentation and erosion in urban river channels. *Geografiska Annaler*, **49A**, 385–395.

Landscape evolution models

Frank J. Pazzaglia

Department of Earth and Environmental Sciences, Lehigh University, 31 Williams, Bethlehem, PA 18015, USA; fjp3@lehigh.edu

Introduction

Geomorphology is the study of Earth's landforms and the processes that shape them. From its beginnings, geomorphology has embraced a historical approach to understanding landforms; the concept of evolving landforms is firmly ingrained in geomorphic thought. Modern process geomorphologists use the term landscape evolution to describe the interactions between form and process that are played out as measurable changes in landscapes over geologic as well as human time scales. More traditionally defined landscape evolution describes exclusively time-dependent changes from rugged youthful topography, through the rounded hillslopes of maturity, to death as a flat plain. Bridging the considerable gulf in these different views of landscape evolution is the task of a different paper altogether. Rather, this chapter provides some historical perspectives on landscape evolution, identifies the key qualitative studies that have moved the science of large-scale geomorphology forward, explores some of the new numeric models that simulate real landscapes and real processes, and provides a glimpse of future landscape evolution studies.

In 1965, the year of the last INQUA meeting held in the United States, thoughts on landscape evolution were still dominated by the classic, philosophy-based arguments of William Morris Davis, Lester King, and John Hack. Geomorphology, on the other hand, had become more quantitative and the number of process studies was growing rapidly (Leopold *et al.*, 1964). Systems theory was on still on the horizon, but a landmark paper by Schumm & Licthy (1965) had laid the framework for scaling the coming generation of process and physical modeling studies into the landscape context. In 1965, landscape evolution models were becoming stale; process studies were viewed as more scientifically rigorous and more directly applicable to human dimension problems. At the same time, plate tectonics was emerging and thought at the orogen scale enjoyed growth and acceptance in the structure-tectonics community. Decades later that community would begin questioning basic characteristics of active orogens such as: what limits mean elevation or mean relief of a landscape or what controls the rate that an orogen erodes? Orogen-scale geomorphology became relevant again to geologic and tectonic questions about the uplift and erosion of mountains. Particularly in the past decade, interests shared by the structure-tectonics community and the geomorphic community have inspired new thinking at the orogen scale and a new generation of landscape evolution models.

Landscape evolution models come in two basic flavors, qualitative and quantitative, that can be applied across a wide range of spatial and temporal scales. This chapter will primarily consider models that address large-scale landforms and processes over the graded and cyclic scales of Schumm & Licthy (1965) (spatial dimensions equivalent to an orogen or physiographic province, temporal dimensions equivalent to 10^4 to 10^6 years). This scale of observation is useful because it encompasses investigations common to physical geography, process geomorphology, paleoclimatology, and geodynamics. Qualitative models are well known to most students of geomorphology and they form the basis for the more quantitative approaches. Louis Agassiz (1840) is best figured as the grandfather of all landscape models. Agassiz's approach to understanding the impact of glaciation on landscapes forced him to think in terms of form and process as well as irreversible changes in the overall configuration of landforms as a function of time. Ironically, Agassiz's integrated approach diverged with the next generation of geomorphologists and landscape evolution models. By the late 19th century William Morris Davis had published his two seminal papers on the geographic cycle (Davis, 1889, 1899a) and Grove Karl Gilbert laid the foundation for process-oriented approaches with his influential chapter on land sculpture in his monograph on the Henry Mountains (Gilbert, 1877). The middle part of the 20th century saw the blossoming of physical analogue models and more aggressive pursuit of process-oriented studies, particularly with respect to hillslope hydrology (e.g. Horton, 1945) and fluvial geomorphology (e.g. Leopold *et al.*, 1964). Growth of interdisciplinary studies in the latter part of the 20th century allowed the qualitative and quantitative approaches to begin to find common ground, spurring a proliferation of numeric approaches (e.g. Willgoose *et al.*, 1991) and ultimately, the coupled geodynamic-surface process model (e.g. Beaumont *et al.*, 1992; Koons, 1989).

This chapter begins by defining terms and suggesting a taxonomy for types of landscape evolution models. It then discusses the classic qualitative paradigms of landscape evolution as a basis for explaining where the science is today. It follows with an exploration of physical models where the physical bases for geomorphic processes were first explored. Finally, it summarizes four different types of numeric landscape evolution models.

A Taxonomy of Landscape Evolution Models

Any organization of the various types of landscape evolution models necessarily begins with a definition of some terms. A landform is a feature of topography that exerts an influence

DEVELOPMENTS IN QUATERNARY SCIENCE
VOLUME 1 ISSN 1571-0866
DOI:10.1016/S1571-0866(03)01012-1

on, and is in turn shaped by surficial processes. A hillslope, a river valley, a sand dune, and a colluvial hollow are landforms. A landscape is an aggregate of landforms for a region. The evolution of a landscape here is not restricted to long-term changes in topographic metrics, but rather is more generally applied to describe any change of form-process interactions of constituent landforms. This chapter organizes landscape evolution models into three broad categories: qualitative, physical, and surficial process models.

Qualitative landscape models tend to describe the long-term changes in the size, shape, and relief of landforms over continental or subcontinental regions. They are not rooted in the principles of physics and they are strongly colored by the geography of their origin. Nevertheless, the qualitative models that endure are based on good, reproducible field observations that must represent a common suite of geodynamic and surficial processes that shape topography. Close inspection of the several main qualitative models reveals that they have much more in common then their often incorrectly celebrated differences. Qualitative models tend to be heavily skewed towards the influence of numerical age on the resulting landforms. For cyclic time and space scales in a decaying orogen setting, the age dependence is warranted. But for landscape evolution at the scale of individual watersheds, the resulting landforms many depend less on numerical age than on the time scale of response and adjustment to driving and resisting forces (e.g. Bull, 1991).

Physical models are scale representations of landforms. A flume is a good example of a physical model used extensively to understand channel form and process and there have been remarkable accomplishments in the understanding of natural channels based on flume studies (Schumm *et al.*, 1987). Well-known criticisms of physical models, based on the fact that it is difficult to correctly scale for material properties (Paola *et al.*, 2001; Schumm *et al.*, 1987), should not be viewed as dismissing the potential insights from physical experiments. For example, it would be easy to build a model river channel in a flume with scaled down grain size and density (coal dust) and a fluid other than water with a similarly scaled down viscosity (acetone). The problem lies in the fact that the resulting acetone-like fluid does not erode the coal dust substrate by any process resembling what happens in a real channel, not to mention that acetone volatilizes and coal dust has high electrostatic charges. By retaining sand and water for the flume, what is lost in the scaling is more than made up for by retaining similar processes of grain entrainment and erosion observed in natural channels.

Surface process models are rooted in the principles of physics and chemistry. Good surface process models are both inspired by and verified by field observations. Surficial process models can focus on a specific landform, such as a hillslope or river channel, they can explore the linkages between landforms, and/or they can consider the feedbacks between surface and tectonic processes. A example that focuses on a single landform is surface process modeling of the evolution of hillslopes in a humid environment. Field observations reveal these hillslopes to be concavo-convex in cross-section. Their regoliths originate near the slope crest and creep downslope. Evolution of the hillslope profile, if not the regolith itself, is elegantly modeled by diffusion where the diffusivity is a function of the regolith material properties. As outlined in detail below, the rate of diffusion is proportional to the slope, so form and process mutually adjust. Numeric surface process modeling opens the door for exploring the linkages between landforms and/or surficial and tectonic processes. Numeric models are constructed by determining the appropriate mathematical proxy for all, or at least the major processes acting on a landscape. Typically, the processes are landform specific such that one mathematical equation describes the dominant hillslope process, another describes regolith production, and another describes fluvial transport. The surface process model reconstructs long-term changes in a landscape by integrating simultaneously across these different process using a large number of time steps in a computer program. The simplest kinds of numeric models deal with mass balances and cannot truly reconstruct the complexity of a real landscape but rather attempt to track some metric of that landscape, such as mean elevation. More complex models are typically 1-, 2-, or 3-D finite difference or finite element code that actually attempt to build and shape a synthetic landscape. Numeric surface process models have been successfully linked with geodynamic models. Such linked models are used to understand the feedbacks between surficial and tectonic processes in active orogens.

Qualitative Models: Classic Paradigms of Long-Term Landscape Evolution

It has been said that the study of landscape evolution is first and foremost a study of hillslope form and process (Carson & Kirkby, 1972; Hooke, 2000). This is not to say that rivers do not play an important role in the shaping of landscapes or in limiting the overall rate of landscape evolution (Howard *et al.*, 1994; Whipple & Tucker, 1999). But across widely variable climates, rock types, and rates of rock uplift, rivers tend to respond relatively quickly to driving forces and their forms lack a memory of the changes in driving forces. Hillslopes respond more slowly on average and as a result, retain a richer memory of the changes in process and driving forces expressed in their forms, especially in landscapes underlain by rock types of different erodibility.

Observations from a large number of field studies conclude that most hillslopes generally have rounded convex summits and are separated from stream valleys at their base by shallow concave reaches. The degree of convexity or concavity, the linear separation between summits and the base, and average slope angle constitute the differences in overall form. Both the upslope convexity and downslope concavity are the result of transport-limited processes. The former results from weathering, creep, and rainsplash (Gilbert, 1909; Schumm, 1956) and the latter from hillslope retreat, not necessarily at the same angle, by surface wash and solution (Schumm, 1956; Schumm & Chorley, 1964), or from deposition. The remaining slope profile between the upper convexity and lower concavity is called the main slope and its form is primarily a result of weathering limited processes such as mass movement and surface wash. The main slope can either retreat parallel to itself, decline by laying back about a fixed hinge point at its base, or shorten by having the upper convexity extend downward to join the lower concavity. These three main behaviors vary with rock type, layering of rock types of variable resistance, and climate. The behaviors are common to the important qualitative models of long-term landscape evolution.

Base Level, Erosion, and Landscape Genesis – Powell (1875)

John Wesley Powell, a Civil War hero of the Battle of Shiloh in which he lost an arm, is the true father of the genetic principles of landscape evolution, many of which are incorrectly attributed to William Morris Davis. Powell's views on landscape evolution are vividly described in the accounts of his expeditions through the river gorges of the Rocky Mountains and Colorado Plateau (Powell, 1875). The concepts of base level and widespread erosion of great mountain ranges to low elevation and relief are the cornerstones of Powell's work. These ideas were a natural consequence of the features that Powell saw during his river trips: great gorges carved by rivers attempting to lower their gradients, great torrents of sediment-laden water resulting from material washing off steep hillslopes, and perhaps most influential, the Great Unconformity of the Grand Canyon. More than any other feature, the Great Unconformity forced Powell to appreciate the enormity of erosion that must have occurred across a once lofty mountain range to produce the strikingly horizontal boundary between deformed Precambrian and flat-lying Paleozoic rocks. That unconformity marked the beveling of a former world and it is this point that most influenced the peneplain concept of William Morris Davis. River gorges carved through variable rock types led Powell to consider a genetic classification of streams, hinging primarily on their broad tectonic setting. The concepts of antecedent, consequent, subsequent, and superimposed drainages have their origin in Powell's writings, but were greatly popularized and further defined by Gilbert (1877) and Davis (1889).

Form and Process – Gilbert (1877, 1909)

The work of Grove Karl Gilbert has had perhaps the most lasting impact on modern geomorphology (Yochelson, 1980). The core of Gilbert's work on landscape evolution was elegantly laid out in his monograph on the geology of the Henry Mountains, Utah (Gilbert, 1877). Chapter V in that monograph is entitled "Land Sculpture" and it is here that Gilbert begins the science of modern process geomorphology. The overall message in Chapter V is that there is an interaction between form and process, or in Gilbert's words, an equality of action. In this view, the landforms of the Henry Mountains reflect their underlying rock type, the tectonic processes that uplifted, emplaced (in the case of laccoliths) and deformed

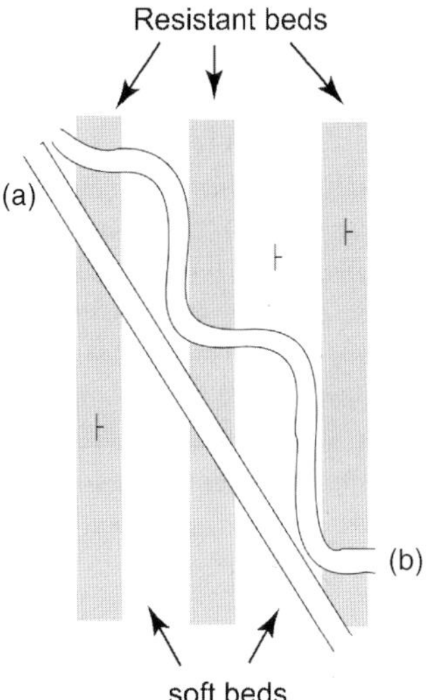

Fig. 1. Sketch, modified from Gilbert (1877), of monoclinal shifting. Gilbert noted that when channels traverse dipping sedimentary rocks of variable resistance (a), their courses adjust to maximize interaction with soft rocks and minimize interaction with hard rocks (b).

them, and the predominant surficial processes of weathering and transport. This represents a wholly modern view of process geomorphology and it is ironic that it lay largely dormant for several decades following Gilbert's monograph, only to be rediscovered a half a century ago (Hack, 1960).

Gilbert built his ideas on the foundation laid by John Wesley Powell (Powell, 1875) and on pioneering work of European engineers, particularly Du Buat (1786), Dausse (1857), Beardmore (1851) and Taylor (1851). Gilbert's genius was applying the results of these earlier studies to the long periods of time represented by landscape evolution. Particularly noteworthy are Gilbert's ideas about the processes and rates of erosion, land sculpture (landform evolution), and drainage adjustment (Fig. 1). For Gilbert, erosion encompasses mechanical and chemical weathering in soils as well as during transport. Erosion rates are most rapid where slopes are steep – a concept not fully popularized until Ahnert (1970). Because such erosion would produce a landscape with a single slope angle, it is clear that rock type must play a primary role also in determining erosion rates. Gilbert cast the role of climate in controlling erosion rates in the context of competing interests of effective precipitation and vegetation with semi-arid landscapes having the highest rates of erosion – an idea not fully appreciated until Langbein & Schumm (1958). Fluvial channels erode in proportion to their slope-discharge product, and sediment transport in a channel depends on the shear stress on the bed. Most impressively, Gilbert was the first to describe the concave-up profile of rivers as graded, improving upon the original definition by European engineers by attributing the graded condition to a balance between available discharge and capacity for adjacent reaches. Gilbert described drainage divides in the context of opposing and interacting graded

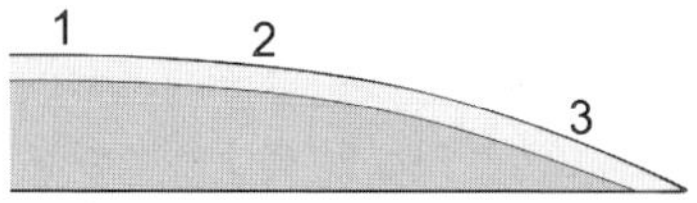

Fig. 2. Sketch of a hillslope profile from Gilbert (1909) showing the zone of creep (light shading) and two hillslope profiles (solid lines) separated by a period of erosion. Gilbert reasoned that the prism of material between 1 and 2 had to be carried past 2 during the period of erosion in the same way that the prism between 2 and 3 was carried past 3 over the same period of time. The quantity passing 2 and 3 is proportional to the distance to the summit (point 1) if the creeping layer is of uniform thickness and velocity of the creeping material is proportional to slope.

profiles. Divides are strictly fluvial features in this description and thus should migrate in the direction of their more gentle side. In this respect, Gilbert was perplexed as to why the "law of divides" seem to be violated in certain badlands where opposing slopes were not concave up, but rather distinctly convex. He surmised that it has something to do with a transition from hillslope processes at or very near the divide to fluvial processes further downslope. This hunch proved to be the correct answer which is ultimately laid out in a later paper (Gilbert, 1909) that describes modern understanding of transport and weathering limited hillslopes, creep as a hillslope process, and the basis for hillslope transport proportional to slope (Fig. 2). These contributions later form the basis for Horton's zone of no fluvial erosion (Horton, 1945).

Gilbert's findings were only slowly appreciated in terms of long-term changes in the landscape. Part of the problem is that Gilbert did not philosophically cast his observations into a temporal framework. He focused on current interactions between form and process with little regard for how those interactions would play out over long periods of time to change a landscape from its current form, to something different. Ironically, understanding the interactions between form and process and casting them into mathematical equations is the basis for numeric surface process modeling that has proven to be a powerful tool for predicting long-term landscape evolution.

The Geographic Cycle – Davis (1889, 1899a, 1932)

The most influential paradigm on long-term landscape evolution comes from the work of the America physical geographer William Morris Davis. Much has been written regarding the obvious differences between the Davisian approach with respect to Gilbert's process approach. There is a place for both approaches in modern views of long term landscape evolution. The Davisian approach provides the temporal framework and the recognition that all landscapes are palimpsests, overwritten by numerous tectonic and climatic processes (Bloom, 2002). A Pennsylvanian by birth, Davis summered with his family in the Pennsylvania Ridge and Valley, a landscape that would forever color his geomorphology views. Davis fully appreciated the best geologic interpretations of his day and used these to constrain his models of landscape evolution. He was also careful to point out that his model was idealized. He never intended concepts like an instantaneous, impulsive uplift, or completion of the geographic cycle to be attained in every case or to be taken and applied literally (Davis, 1899b). Unfortunately, these and other idealizations of his model have been taken literally by many outside of the field of geomorphology, leading to much confusion and stifling reconciliation of his ideas with modern process-oriented thought.

Among Davis's numerous publications, three define and apply the concept of the geographic cycle to a range of landscapes. The first paper is "The rivers and valleys of Pennsylvania" (Davis, 1889) which focuses on the development and evolution of the current drainage of Pennsylvania. Basic observations of accordant summits, wind gaps, and water gaps routing the master streams transverse to structure in the Ridge and Valley anchor the justifications for repeated landscape uplift and beveling. The second paper is "The geographic cycle" (Davis, 1899a) where the Davisian model for long term landscape evolution is described. The third paper is "Piedmont benchlands and Primärrümpfe" (Davis, 1932) where Davis confronts a major alternative paradigm to the geographic cycle and the formation of peneplains presented by the ideas of Walter Penck (1924).

The geographic cycle (Fig. 3) is a simple, but compelling treatment of how mean elevation and mean relief change as a landscape erodes. It also places a large emphasis on transport

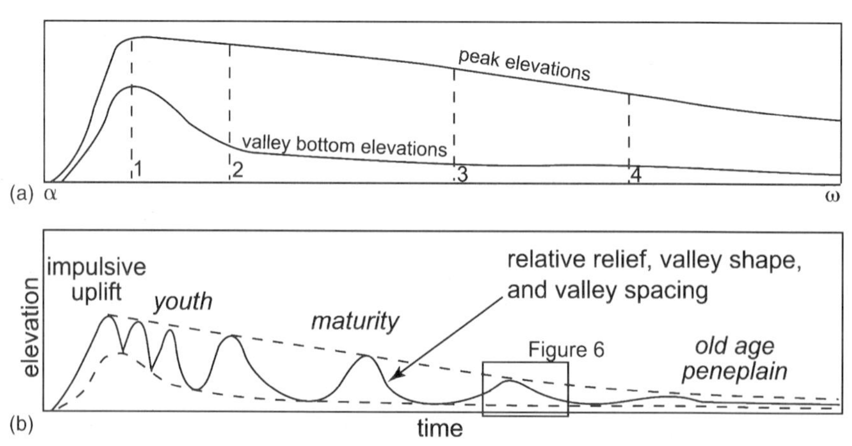

Fig. 3. The geographic cycle of Davis showing (a) a simplified version of the original figure from the 1899 paper and (b) a modified interpretation of the original figure (from Summerfield, 1991). Stages 1 through 4 in (a) refer to uplift, valley bottom deepening, valley bottom widening, and finally interfluve lowering. Upper line denotes mean elevation of interfluves and lower line, mean elevation of valley bottoms.

limited processes for the overall erosion and lowering of hillslopes. The geographic cycle has four major stages. In stage 1, a landscape is born by rapid, if not impulsive uplift of both rocks and the land surface above sea level. The uplifted landscape has a high mean elevation, but low mean relief. Erosion is initially concentrated in the river which leads to the carving of deep, narrow valleys. The landscape passes into stage 2 where it is youthful in appearance, it has both a relatively high mean elevation and high relief. Valleys rapidly lower to base level, at which point they begin to widen laterally. At the same time, hillslope erosion becomes important as interfluves are rounded. Hillslopes take on a concavo-convex form and weathered material is moved across this form to the river valley predominantly by creep. These conditions describe stage 3, the mature landscape. Both mean elevation and mean relief steadily decrease, and valleys continue to widen. Stage 4 is reached when valleys can no longer deepen or widen and all landscape lowering is accomplished by the progressive rounding and lowering of interfluves by hillslope creep. The completely beveled landscape resulting from the completion of the geographic cycle, the "penultimate plain," is called a peneplain. Davis envisioned the peneplain being formed at different rates depending on the size of the landscape being beveled and the relative erodibility of the rocks. Resistant islands of high standing topography are called monadnocks (for Mt. Monadnock in southern New Hampshire), and evidence of partial peneplains is common on softer rocks. Davis fully envisioned the difficulty in running the cycle to completion. He inferred that renewed tectonic uplift of a landscape occurs often enough to interrupt the cycle. This realization led to the preoccupation with finding peneplains and partial peneplains in the landscape, essentially an attempt to define an landscape's erosional stratigraphy. Time was the critical factor in determining a landscape's mean elevation and relief; no interactions between form and process or driving and resisting forces are considered. The concept of the peneplain proved the most controversial part of the geographic cycle (Davis, 1899b). Ironically, Davis never claimed ownership of the term. In his view, he was just giving a name to a process and a feature described by geologists before him, most notable, Powell's description of the Great Unconformity in his *Exploration of the Colorado River* (Powell, 1875).

The Geographic Cycle makes little reference to hillslope processes and what references do exist are considered together under the concept of agencies of removal. Here Davis suggests that the hillslopes are shaped by a myriad of processes that include surface wash, ground water, temperature change, freeze-thaw, chemical weathering, root and animal bioturbation. These collectively drive creeping regolith downslope. Relative amounts of weathered products and exposed bedrock are considered in the context of whether the slope is steep and youthful, or old and rounded.

An important idea that has survived from the geographic cycle is that slopes in tectonically active landscapes rapidly attain steep, almost straight profiles, then more slowly decay into concavo-convex, sigmoid-shaped profiles as tectonism wanes, valleys reach their base level of erosion, and divides are lowered. Davis also suggested several modifications to the geographic cycle model. Most notable are the ideas about cyclic erosion in arid landscapes (Davis, 1930). In this paper, Davis introduces elements of slope retreat into his model. From this model Bryan (1940) and later King (1953) describe and expand upon the role that retreating escarpments play in long-term landscape evolution. Unfortunately, Davis appears to abandon his view of slope form and process being a consequence of climate in his 1932 paper where he returns only to the ideas originally espoused in the Geographic Cycle as a defense against the ideas of Walter Penck (see below).

The triumph of the geographic cycle is that it was almost a century ahead of its time in showing how mean elevation and mean relief might evolve in a landscape. Only recently, with the realization that surficial and geodynamic processes are linked in orogenesis (Molnar & England, 1990), has the broader geologic community come to appreciate the lag times and feedbacks between rock uplift and erosion. The four stages of the geographic cycle resemble modern concepts about growing, steady-state, and decaying orogens (Hovius, 2000). Furthermore, the erosional response to impulsive uplift described by the geographic cycle is completely consistent with the observed flux of material from disturbed Earth systems across a wide range of space and time scales (Schumm & Rea, 1995). This flux follows a distinct exponential decay after an impulsive increase. Over cyclic time scales, uplift and the initial erosional response are in fact impulsive, and the landscape response in terms of lowering mean elevation and mean relief is most simply described by an exponential decay function. The tragedy of the geographic cycle is that individual parts of it have been accepted too literally. A prime example is equating meandering streams with Davis's wide, low gradient valleys of mature landscapes. Meandering channels do not necessairly indicate landscape maturity. Rather, meandering reflects the interaction of driving and resisting forces such as watershed hydrology, prevailing grain size, bank stability, and rock type erodibility in a fluvial system.

Slope Replacement – Penck (1924, 1953)

The most influential European geomorphologist in the early part of the 20th century and a direct challenger to the Davisian model was Walter Penck. Penck was influenced by field observations made in Germany, particularly the Rhine graben area, in northern Argentina, and to a lesser degree, in Africa. Like Davis, Penck envisioned long-term landscape evolution occurring in stages from youth to old age, when all relief in a landscape was beveled by erosion. And despite the assertions that Penck was the original proponent of slope retreat (King, 1953), he like Davis, believed in the flattening of slopes through time. Misunderstandings arose from his obscure writing style (Simons, 1962), and from Davis's incorrect translation of some of his work (Davis, 1932). But unlike Davis, Penck was more process oriented, actually making measurements in the field, and he focused his process studies on hillslopes.

Penck recognized concavo-convex hillslopes similar to what Davis was observing in eastern North America.

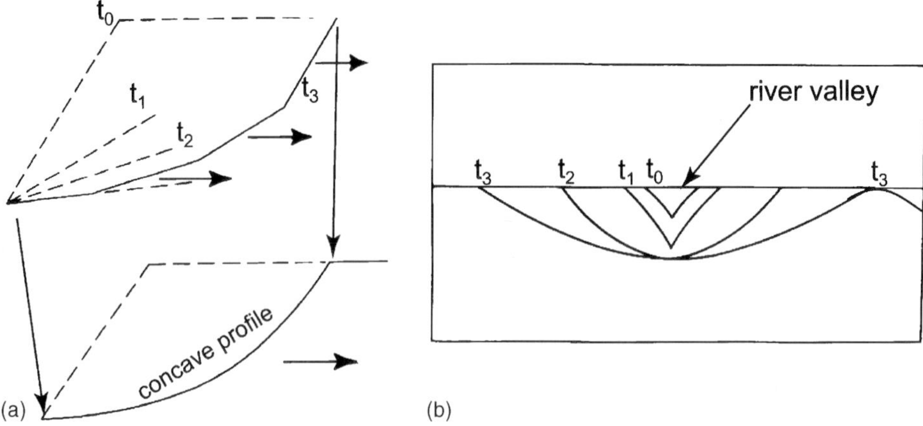

Fig. 4. The Penck model. (a) Diagram showing slope replacement (upper profile) caused by successive retreat of an infinite number of inclined valley bottoms leading to a smooth concave profile (lower concave profile). Penck envisioned an initially steep linear cliff (t_0), being replaced by progressively more gentle slope segments ($t_1 - t_3$). (b) Penck's original figure, misinterpreted by Davis (1932), illustrating Penck's cycle of erosion. Much like Davis's model, this diagram shows that Penck considered erosion to be concentrated in river valleys first (t_0), then extended to hillslopes ($t_1 - t_2$). If adjacent, ever-widening concave valley profiles meet at an interfluve (t_3), that interfluve becomes rounded and lowered (modified from Carson & Kirkby, 1972).

However, the Penckian model for their origin was altogether different. Penck is the father of slope replacement as a mechanism for hillslopes evolution (Fig. 4). Whereas Davis's hillslopes are transport-limited and always covered with a creeping regolith mantle, Penck's hillslopes are predominantly-weathering limited with little to no regolith cover, except, implicitly at their base where regolith transported downslope is allowed to accumulate. Originally steep and straight slope profiles weather parallel to themselves except for a small, step-like flattening at the base of the slope, the haldenhang, which is presumably controlled by the angle of repose of the hillslope debris. Retreat of the haldenhang results in an even lower-gradient basal slope called the abflachungshang. The process continues with successive abflachungshang retreating, each leaving a basal slope of lower angle than itself. The integrated result is a concave-up slope that has replaced the original steep, straight slope. Penck went on to propose mechanisms of how the upper part of the concave profile would flatten to produce an upper slope convexity using arguments similar to those used by Gilbert (1909).

The concept of waxing and waning slopes is not as much a part of Penck's original model as it is Davis's interpretation of it. But Penck's views on the effects of base level change on slope form are notable. Penck argued that given a stable base level, a hillslope will attain a characteristic profile determined by the erodibility of underlying rock. Steep slopes are underlain by hard rocks whereas gentle slopes are underlain by soft rocks. Furthermore, Penck was influenced by his studies in the Alpine foreland, where it appeared certain that orogeny was protracted and occurred at variable rates, rather than being impulsive as in the Davisian model. These observations led Penck to believe that base level fall, if linked to orogeny, starts slow, accelerates to a maximum (waxing), and decelerates back to stability (waning). Hillslopes adjust to that base level fall throughout the waxing and waning cycle. A fall in base level causes that segment of the hillslope immediately adjacent to the base level fall to steepen, and that steepened hillslope segment then causes the next segment upslope to steepen and so on. The result is the replacement of a gentle slope with a more steep one as the base level fall signal is translated upslope. If base-level fall is a one-time event, the steep hillslope segment propagates upslope, and in turn is replaced by a less gentle segment downstream. But if the base-level fall continues or accelerates, there is a downslope trend of progressively steeper slope segments, such that the hillslope becomes convex in profile. Alternatively, under base-level rise or long-term stability, there is a downslope shallowing of slope segment gradients and the hillslope is concave in profile. The Penckian model predicts convex hillslopes when base level is actively falling (waxing slopes) as part of a youthful landscape; the Davisian model predicts concavo-convex hillslopes where base level is stable and a mature landscape is slowly being beveled. In summary, Penck viewed base-level changes (uplift) as long-lived, rather than impulsive, and the overall landscape response to those changes as having short lag times.

Even though Penck's ideas on landscape evolution differed significantly from the Geographic Cycle, his landforms are still time-dependent features (Fig. 4). The Penck model holds that landscapes are born from a base level fall that affects a low relief landscape called a primärrümpf, resulting in stream incision and convex hillslopes. Acceleration in the rate of base-level fall increases hillslope convexity and results in the formation of benches (piedmottreppen) near drainage headwaters that have not been affect by the slope replacement process. Eventually, base-level fall decelerates and stops completely. Slopes are replaced to lower and lower

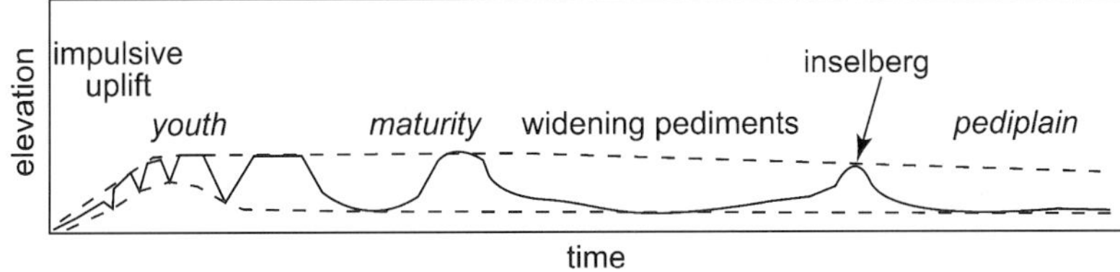

Fig. 5. The pediplanation model of King (1953) (modified from Summerfield, 1991). Penck's model stresses the development of concave-up slopes that retreat faster than interfluves lower, resulting in widespread pediments that coalesce into a pediplain. Note that mean elevation decreases in this model, but relief persists along escarpments or inselbergs.

declivities until the relatively straight profiles of base level stability retreat headward leaving a beveled plain called an endrümpf with large, residual inselbergs at the drainage divides.

Pediplanation – L.C. King (1953, 1967)

Drawing upon work conducted primarily in Africa, but versed on landscapes worldwide, Lester King championed a model for long term landscape evolution very similar to the geographic cycle, but differing in the dominant mode of hillslope evolution (King, 1953, 1967). Like Davis, King envisioned impulsive uplift and long response times of landscape adjustment. King never accepted the Davisian concavo-convex slope; instead, he favored Penck's view of concave hillslopes and slope replacement. In fact, King took the Penckian model of slope replacement literally, to conclude that the landscape assumes the form of a series of nested, retreating escarpments (Fig. 5). King called the low-gradient footslope extensions of the steep escarpments pediments and the flat beveled surface they leave in their retreating wake a pediplane. Such a restricted interpretation of the Penckian model is unwarranted because Penck made it clear that the rates of slope modification are proportional to gradient, a relationship that would continually produce slope replacement, not parallel retreat. In fact, King's model is based more on the earlier work of Kirk Bryan (1922, 1940) than it is on Penck.

The notion of a landscape dominated by pediments should not come as a surprise to a geomorphologist influenced by South African landscapes. But King was wholly convinced that these landforms are not restricted to just the arid and semi-arid climates of his homeland. Rather, he argued, Davis himself had described precisely the same features in New England where Mt. Monadnock, the supposed type locality of a remnant, rounded, concavo-convex hillslope has a concave-up pediment profile. To King's credit he built upon the earlier work of Penck in making measurements in the field and incorporating the process work on pediment formation from other noted geomorphologists including Kirk Bryan (1922). King proposed that once pediments form, that they persist indefinitely until consumed by younger retreating escarpments following renewed base level fall. The base level fall in King's model is inherently episodic because it occurs on passive margins, which at the time, were thought to respond isostatically to episodes of erosional unloading and offshore deposition (Schumm, 1963) – a concept viewed today as untenable.

King summarizes his pediplane model as well as his overall views on landscape evolution in fifty canons, published in the 1953 paper. Of these canons, the following are particularly insightful, and research continues on many of them: (No. 1) Landscape is a function of process, stage, and structure; the relative importance of these is indicated by their order. (No. 3) There is a general homology between all (fluvially sculpted) landscapes. The differences between landforms of humid-temperate, semiarid, and arid environments are differences only of degree. Thus, for instance, monadnocks and inselbergs are homologous. (No. 6) The most active elements of hillslope evolution are the free face and the debris slope. If these are actively eroded, the hillslope will retreat parallel to itself. (No. 9) When the free face and debris slope are inactive, the waxing slope becomes strongly developed and may extend down to met the waning slope. Such concavo-convex slopes are degenerate. (No. 42) Major cyclic erosion scarps retreat almost as fast as the knickpoints which travel up the rivers transverse to the scarp. Such scarps therefore remain essentially linear and lack pronounced re-entrants where they cross the rivers.

Dynamic Equilibrium – Hack (1960)

The simple, embodying concept of driving and resisting forces – the interaction between form and process proposed by Gilbert (1877) – was rediscovered and presented as an alternative to the time-dependent paradigms of landscape evolution by J.T. Hack of the U.S. Geological Survey (Hack, 1960). By his own admission, Hack had been carefully studying the same Appalachian landscape of Davis and making a conscious effort to seek alternatives to cyclic theories of landscape evolution. The alternative favored by Hack is that landscapes are in a state of dynamic equilibrium (Fig. 6). They are in equilibrium in the sense that given the same driving and resisting forces over a long period of time, a "time-independent" characteristic landscape will emerge. This is a landscape where the rivers and hillslopes are all "graded" and the processes acting on the interfluves and channel bottoms, although different, are lowering their respective parts of the landscape at the same rate. The landscape is dynamic in

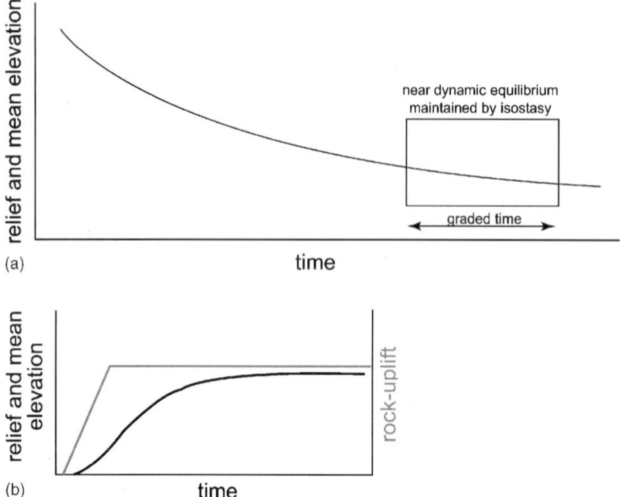

Fig. 6. Attainment of a near time-invariant relief and mean elevation of a dynamic equilibrium landscape (a) attained over graded time during a protracted period of decay (cyclic time) and (b) attained as a flux steady-state between the input of rocks by tectonic processes and output by erosion.

the sense that climate, tectonics, and rock type change as subaerial erosion progresses so there is a constant adjustment between the principal driving and resisting forces such that true equilibrium is probably only asymptotically reached. Strictly speaking, a truly characteristic or steady-state landscape cannot exist, except in a tectonically active setting where the mass flux in and out of the orogen is conserved (Pazzaglia & Knuepfer, 2001; Willett & Brandon, 2002). In a decaying orogen, mean elevation and mean relief must be reduced, as Davis surmised, because isostatic rebound of a low density crustal root only recovers ~80% of what is removed via erosion.

Hack's concept of dynamic equilibrium can be used to describe the origin of the same landscape that Davis worked in, but without cycles of erosion and peneplains. The Hack model predicts accordant summits among resistant rock types because the similar processes have established themselves on slopes of similar declivity, and on rocks sharing similar structure over a long period of time. Hack did not propose an erosional stratigraphy in landscapes. Rather, he proposed that all landscapes are essentially modern and a reflection of modern processes. Truly flat landscapes are not erosional, but rather aggradational. At the time, Hack's paper was only the most recent of a long list of cycle of erosion critics to point out the conspicuous global lack of peneplains at or near sea level.

Hack's contribution very much refocused the geomorphologic community back towards process and as such, helped to lay the foundation upon which modern analogue and surface process models are built. This is not to say that process geomorphology was absent before Hack. But his concept of dynamic equilibrium moved geomorphology back into the wider geologic arena of basic principle chemical, and biologic principles that govern landscape evolution.

Process Linkage – Bull (1991)

Several decades of process studies (e.g. Leopold *et al.*, 1964), watershed hydrology (Dunne, 1978; Horton, 1945), and the Hack tradition of dynamic equilibrium (Hack, 1960) collectively form the basis for understanding landscape evolution from the perspective of linked processes. This view of landscape evolution, colored by the dramatic effects of Quaternary climate change, is best represented in the textbook *Climatic Geomorphology* (Bull, 1991). Many geomorphologists in North America and elsewhere contributed to a modern understanding of process linkage. But it was Bill Bull, working in the American southwest and with colleagues abroad, particularly in Israel, who led this effort.

Process linkage refers to the direct and indirect ways in which individual components of a watershed mutually respond to an external driving force, the two most obvious being tectonics and climate. Implicit in the linkage among individual components is the realization that there may not be a unique response for a given external stimuli. Such non-unique responses arise because thresholds (Patton & Schumm, 1975) are embedded in most geomorphic processes, and because a large external stimulus is required to elicit the same response in multiple watersheds.

Process linkage has a particular focus on landscape change as a function of the creation and routing of sediment thorough a watershed. In this respect, it differs from other models of landscape evolution that focus on the evolution of hillslope forms. Climate change, an important control on sediment creation and routing affects watersheds at all spatial scales and has been particularly acute during Pleistocene glacial-interglacial cycles (Pederson *et al.*, 2000, 2001). Quaternary stratigraphy, like that of an alluvial fan, commonly reveals relatively brief pulses of deposition interspersed with longer periods of landscape stability and pedogenesis. The Bull model considers weathering, the production rate of regolith, the liberation of that regolith off hillslopes, the response time, thresholds, and equilibrium all interacting in the watershed as the primary controls on the resulting alluvial fan stratigraphy (Fig. 7). Although many factors including rock type, climate, seasonality, relief, and vegetation influence the hydrology and the unique response of a given watershed, a simple, representative scenario best illustrates Bull's model.

Particularly in arid and semi-arid climates, regolith production is thought to be maximized under relatively cool, moist climatic conditions when vegetative cover on the hillslope accelerates chemical and physical weathering. Stripping of that regolith occurs when a switch to drier, warmer conditions leads to the loss of the vegetative cover, leaving the regolith prone to gulling and removal by overland flow. Even without much loss of vegetation, changes such as an increase in precipitation seasonality or the lowering of infiltration rates in well-developed, mature soils can eventually provoke loss of a hillslope's regolith to increased overland flow. The time between the climatic perturbation and the beginning of hillslope stripping involves crossing a geomorphic threshold

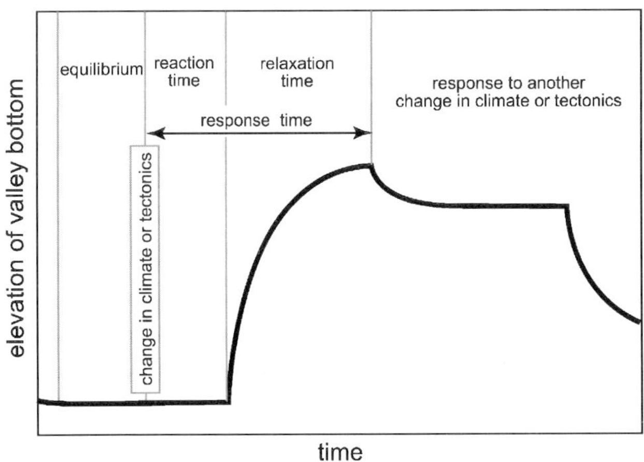

Fig. 7. Reaction time, relaxation time, and response times of the Bull (1991) process-response model (modified from Bull, 1991).

and is called the reaction time (Fig. 7). The time needed to achieve a new equilibrium condition between the new climate and hillslope process is called the relaxation time. The sum of the reaction time and relaxation time is the response time (Brunsden & Thornes, 1979). Whipple & Tucker (1999) described a similar definition of the response time as it applies to channel headwaters adjusting to changes in the rate of rock uplift.

The transition from the reaction to relaxation times of when the hillslopes begin to liberate regolith is affected by the overall climatic regime (arid, semi-arid, semi-arid and glaciated) of the watershed (Ritter *et al.*, 2000). That regolith is delivered to the fluvial system which moves it out of the watershed. The hydraulic geometry may change as sediment is routed through them, and the streams may aggrade or incise depending on the sediment flux from the hillslopes. Even so, the residence time of sediment in the fluvial system is typically short compared with the residence time of regolith on the hillslopes. The sediment is delivered to a basin where its depositional architecture retains clues about climatically influenced transport processes as well as the weathering processes in the source (Smith, 1994). Some watersheds have very short response times between the hillslopes, channels, and depositional basin where a direct temporal link can be established between the climatic event that affected the hillslopes and the depositional response far downstream (Fig. 7). In other cases, long response times between the hillslopes and depositional basin make it far more difficult to link a climate change to the depositional response.

Among the numerous studies that view process linkage as an important agent of landscape evolution, three are noteworthy in illustrating the relative influences of rock type (Bull & Schick, 1979), pedogenesis (Wells *et al.*, 1987), and climate-watershed hydrology (Meyer *et al.*, 1995). The Bull & Schick (1979) study is one of the first to show the response of two adjacent watersheds in an arid environment to the same climate change. The watersheds are underlain by different rock types, leading to different regoliths, soil infiltration rates, and response times to late Pleistocene and Holocene climate change. Some watersheds continue to deliver sediment from hillslopes as a response to late Pleistocene to Holocene climate change, so that alluvial fans at the mouths of these watersheds have a primarily aggradational history. In contrast, other watersheds cease liberating sediment from their hillslopes and the corresponding alluvial fans pass into a period of incision as the sediment supply wanes.

Wells *et al.* (1987) showed how the landscape evolution of an alluvial fan surface, including the location of fan deposition, is strongly controlled by the linked factors of soil genesis and how those soils control runoff. Alluvial fan deposition is initially affected by the late Pleistocene-Holocene climate change with the watershed hillslopes being the dominant source of sediment delivered to the fan. Subsequent Holocene fan deposition reflects a change in the source from the hillslopes, where runoff has been reduced by renewed colluvial mantle development, to the piedmont where runoff has been increased by development of progressively impermeable clay and calcic horizons in eolian soils.

Meyer *et al.* (1995) described the landscape evolution of rugged montane valleys in the Rocky Mountains, where liberation of sediment from hillslopes is influenced by large slope-clearing fires as well as climate. The fires recur at intervals that average approximately 200 years. Relatively dry climate over a period of years increases the chances of large, destructive, slope-clearing fires. Drier climatic conditions also correlate with relatively small winter snowpacks and intense summer convective storms. The hydrology associated with these relatively dry climatic conditions favors debris flow activity in burned regions during the summer. Sediment accumulates in debris fans along the valley margins and the fluvial system has flashy discharges in a braided, incised channel with a low base flow. Wet years, in contrast, are characterized by a climate with heavy winter snowpacks, depressed summer convective storm activity, and decreased numbers of fires and debris flows. The result is a meandering stream fed by the high, stable base flows and ample sediment supply recruited from the debris fans as the channel meanders across a widening valley floor. The transition between relative dry and relatively wet climatic conditions occurs at the millennial scale. In their long-term evolution, hillslopes and rivers in this setting are clearly linked, but there is a millennial-scale response time of the river system following the initial reaction time of the hillslope to the climatic perturbation.

Geodynamic and Surficial Process Feedbacks – Molnar & England (1990)

A new way of thinking about orogen-scale landscape evolution, in terms of surface processes that limit rates of tectonic deformation and the uplift of rocks originated with Molnar & England (1990) who linked the building of high-standing

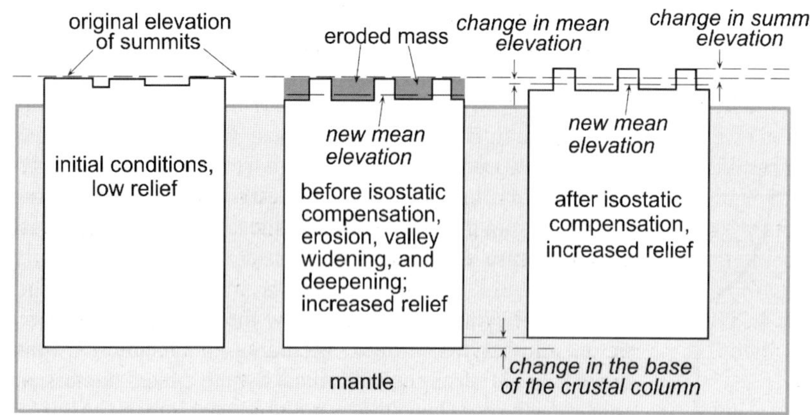

Fig. 8. Changes in mean elevation and mean local relief in response to erosion that deepens and widens valleys. Rocks are uplifted as an isostatic response to this erosion because the topography is compensated by a low-density crustal root protruding into the mantle. Despite rock uplift and surface uplift of the summits, the mean elevation of the landscape decreases as the crustal root is consumed (modified from Burbank & Anderson, 2001).

mountains to cool climates. The idea of coupled surficial and geodynamic processes is not new, extending back as least as far as the original considerations of isostasy and erosion (Ransome, 1896). More recently, King (1953) argued that nested, inset pediments were the result of epeirogenic upheavals caused by the unloading effects of retreating escarpments. Schumm (1963) forwarded a similar argument for uplift in the Appalachians caused by erosion crossing an isostatic threshold after which the land surface would rapidly rebound. These arguments of episodic uplift (or subsidence) due to erosion have largely been shown to be incorrect (Gilchrist & Summerfield, 1991), but the general idea that uplift and erosion are linked persists.

The central idea in coupled geodynamic-surficial processes is best illustrated by the simple Airy isostatic case of a mountain range where the topography is supported by a low-density crustal root protruding into the mantle (Fig. 8). Uniform erosion across the range results in isostatic rebound of the root proportional to the density difference between the mantle and the crust, but the land surface and mean elevation are lowered with respect to sea level. Airy isostatic rebound recovers approximately 80% of the mass removed by erosion; a column of rock 5 km thick must be consumed to reduce mean elevation by a kilometer. The shape of a mountain range responds dramatically to the rebound if the erosion rate is not uniform across the range. When erosion is concentrated in valleys and interfluves are lowered much more slowly than valley bottoms, local relief is increased and the resulting isostatic rebound pushes the interfluves (mountain peaks) higher even though mean elevation has been lowered (Fig. 8). The effect is muted below the flexural wavelength of the lithosphere (Small & Anderson, 1998) but it still may be an important source of recent surface uplift.

Feedbacks between surficial and geodynamic processes go beyond simple consideration of surface uplift by isostasy. The concept encompasses climatically controlled limits on mean elevation (Brozovic *et al.*, 1997), the width of orogens (Beaumont *et al.*, 1992), the metamorphic grade of rocks exposed in ancient orogens (Hoffman & Grotzinger, 1993), the dominant river long profiles and hillslope erosion processes (Hovius, 2000), the structure of convergent mountain belts (Willett, 1999), and a geomorphic throttle on orogenic plateau evolution (Zeitler *et al.*, 2001). Consideration of surficial-geodynamic feedbacks releases geomorphic thinking from viewing landscape evolution at the large scale as only reacting, rather than interacting to impulsive climatic or tectonic changes. At the scale of the orogen, landscapes evolve as a consequence of both tectonics and climate and the evolutionary path itself plays a role in defining the feedbacks between the two.

Physical Models

Physical models of landscape evolution existed in parallel with, but originally did not enjoy the spotlight focused on the philosophical approaches. A history of the rise and development of physical models is provided by Schumm *et al.* (1987). Daubrée (1879) was among the first to recognize that experiments can reveal basic relationships and general hypotheses to guide field studies. Early experiments focused on formation of drainage networks (Hubbard, 1907), the erosional evolution of an exhumed laccolith (Howe, 1901), the downslope movement of scree (Davisson, 1888a, b), fluvial transport of sediment (Gilbert, 1914, 1917), and behavior of channel meanders (Jaggar, 1908). Evolution of drainage networks within a given watershed were first simulated by Glock (1931) and Würm (1935, 1936).

There are three broad classes of physical models: (1) segments of unscaled reality, (2) scale models, and (3) analog models (Chorley, 1967). A segment of unscaled reality does not imply that the model is constructed at some scale very different than what is found in nature. Rather, it refers to a model that can actually be of a natural system, like the segment of an alluvial river channel that is considered representative of a broader range of river channels in form, process, and geography. Simple sand-bed flumes or stream tables also illustrate unscaled reality. There are particularly useful applied geomorphology practices where data from segments of unscaled reality have led to important policies in land management. An example widely used by agricultural engineers is the Universal Soil Loss Equation (Wischmeier & Smith, 1965), which was derived from both field studies and physical models of soil erosion. Commonly, space and time are substituted in segments of unscaled reality models so that processes that unfold over millions of years in large watersheds can be investigated over a period of years. For example, Ritter & Gardner (1993) linked dependence of infiltration recovery with channel development on mined landscapes in a physical model that plays out over a period of years to mimic the evolution of larger watersheds. In the process, the model provides insight as to how climate change affects basin hydrology.

Scale models make an attempt to reproduce natural landforms in such a way that ratios of significant dimensions and forces are equal to those found in nature. True scale models are difficult to construct because of the limited range of fluid and substrate material properties that are practically available to maintain ratios of dimension and force. Jurassic Tank at the University of Minnesota St. Anthony's Falls laboratory illustrates a scale model dedicated to the study of depositional landforms and their resulting underlying stratigraphy (Paola *et al.*, 2001).

Analogue models are designed to mimic a given natural phenomena without having to reproduce the same driving forces, processes, or materials. Examples include modeling the flow of glaciers using oatmeal (Romey, 1982) or modeling landslides using a flume filled with beans (Densmore *et al.*, 1997).

Physical models typically isolate one part of the geomorphic system, or even one part of a watershed and, as a result, rarely show the behavior of an alluvial channel, for example, in the context of broader landscape change. Fortunately, virtually all geomorphic processes produce erosion which physical models can reproduce as a general proxy for landscape change. Erosion, channel incision, and sediment transport, common to many physical models, are considered below.

Drainage Networks

Drainage networks have attracted a great deal of attention in the overall evolution of a watershed. Given the growing appreciation for the rate of fluvial incision and characteristic drainage density in limiting rates of rock uplift and erosion in tectonically active landscapes, the initiation and evolution of drainage networks remains a central pursuit in landscape evolution models. There are three main hypotheses regarding drainage network evolution: (1) network growth by overland flow (Horton, 1945); (2) headward growth of established channel heads (Howard, 1971; Smart & Moruzzi, 1971); and (3) dominance of rapidly elongated channels (Glock, 1931). The Glock model incorporates components of both overland flow and headward retreat. In this model, a trunk channel initiates, rapidly elongates, elaborates, and then abstracts, settling into a characteristic drainage density consistent with the prevailing infiltration characteristics of contributing hillslope. In fact, it is the evolution of infiltration characteristics on the hillslope that determines the speed at which the network progresses through the various stages. Predictions of the Glock model have been directly observed in field studies (Morisawa, 1964; Ritter & Gardner, 1993) and have been reproduced and refined with other physical models (Parker, 1977).

Hillslopes

Physical models are common in studies of hillslope form and process. Geological engineering and watershed hydrology literature is rich in examples of physical models designed to understand landslides, earthflows, and debris flows. The U.S. Geological Survey debris-flow experimental flume (e.g. Denlinger & Iverson, 2001; Iverson & Vallance, 2001) is an example of recent approaches that have greatly advanced the understanding of Coulomb and granular dispersion processes in hyperconcentrated and debris flows. However, it is typically more difficult to reproduce scale representatives of hillslopes in the laboratory, so many "physical" models are actually carefully controlled field monitoring studies designed to isolate one process acting over a limited area. The fall of Threatening Rock, a well-documented example of slab-failure processes, is one such field study that passes for a physical model (Schumm & Chorley, 1964). This study revealed that movement of the slab away from the cliff face accelerated exponentially approaching the moment of failure. The rates were fastest in the winter months, probably because of frost action. Studies of hillslope creep are another class of physical models, typically conducted in the field. Creep experiments consider a column or known shape of regolith or soil periodically measured with respect to fixed reference points. The measurements may include markers or lines on the surface of the regolith, or horizontal pins inserted in the free face of a Young Pit (Young, 1960).

Weathering and rock disaggradation in the hillslope environment are easily simulated by repeated mass and volume measurements of representative rock samples under different weathering environments. Rates of disintegration and mass loss depend on whether the rock sample remains exposed at the surface, or if it remains buried in a soil profile, exhumed only for the purpose of measurement. Such experiments may influence the resulting measured rates, but they do provide insights into processes. The classic experiment of this type was conducted on sandstone collected from cliff faces in the Colorado Plateau (Schumm & Chorley, 1966). The results from measurements over a period of two years, showed that granular disintegration, rather than fracturing into small pieces, was the dominant weathering process. Steep cliff faces are probably maintained in the Colorado Plateau as the rock disintegrates and is blown away; little accumulates against the cliff base.

There are at least two recent examples of small-scale physical models with applications to long-term hillslope evolution based on the study of avalanches. Field studies support creep processes to be dominant on low-gradient hillslopes or for the convex upper slope portion, whereas landslides dominate steep hillslopes or on the steep, straight main slope. A physical model using dried beans of various sizes and shapes (Densmore *et al.*, 1997) helps show the genesis and role of landslides on the steep main slope. The apparatus used in these experiments was a narrow flume made of clear acrylic walls 2.5 cm apart. The front of the flume could be lowered to simulate base-level fall. The flume was alternately filled with red, oval-shaped beans or white, more spherically-shaped beans. Base-level was lowered and the flux of beans shed from the flume was measured for each base-level lowering step. The beans are thought to represent strong, coherent blocks of rock with weak grain boundaries and as such, collectively approximate rock behavior where strength is a function of both zones of weakness and block anisotropy. Results showed that base level lowering caused a step to develop at the toe of the slope. As the step grew, the slope above it destabilized and was swept clear of a layer of beans in a slope-clearing landslide. The landslides happened at irregular intervals and their frequency depended on bean anisotropy. The slope clearing events, analogous to landslides, accounted for 70% of the total mass removed from the model hillslope. In real landscapes, the steep inner gorges in mountainous topography may be analogous to the lower step seen in the model. The applicability of this model to real landscapes has been questioned because subsequent experiments demonstrate that the narrow width of the flume may have influenced the initiation of the inner gorges and slope-clearing events (Aalto *et al.*, 1997). Nevertheless, the study does underscore the importance of both creep and landslide processes on hillslopes have creep and landslide components of mass removal.

Creep and landslide removal of mass from a hillslope has been treated as linear and non-linear diffusive processes respectively (Martin, 2000). Non-linear sediment transport from model hillslopes was investigated by a physical model (Roering *et al.*, 1999, 2001) consisting of a plexiglass box, with open ends, filled with a hill of sand. The sand was subjected to acoustic vibrations that caused the outermost layer of grains to vibrate, dilate, and creep downslope, simulating the biologic and freeze-thaw turbation of regolith on real hillslopes. A key result of the experiment is that the flux of sediment from this model hillslope approximate a linear diffusive behavior when gradients were low (<0.3). For slopes between 0.3 and 0.52, creep remained the dominant process, but the sediment flux varied non-linearly with slope. Landsliding become the dominant process at gradients greater than 0.52. These results suggest that the threshold of non-linear sediment flux from hillslopes is lower than the threshold gradient that triggers landsliding, and that creep cannot be wholly described by linear diffusion.

Alluvial and Bedrock Channels

Physical models of alluvial channels, beginning with studies like those of Thomson (1879), grew rapidly into numerous permanent and semi-permanent hydraulics labs in academic, engineering, and government facilities worldwide. The focus has traditionally been on understanding the processes and behaviors of alluvial channels (Schumm *et al.*, 1987). More recently, some researchers have focused on bedrock channels (Thompson & Wohl, 1998), a shift in focus that reflects the interest of bedrock channel behavior in tectonically active settings. Most physical models of channels investigate the mutually dependent variables of gradient, width, depth, flow velocity, and bed grain size as a function of discharge. Braided channels produced in physical models have similar, scaled hydraulic geometries to natural channels. In contrast, meandering channels produced in physical models lack symmetry with natural channels, particularly in meander amplitude and wavelength. The difference may be related to natural controls on bank stability, such as vegetation, that are difficult to reproduce in the scale models.

Physical models of bedrock channels have focused more on the processes and rates of channel bed erosion, than on channel geometries. Abrasion by coarse bedload is considered one of the key processes that erode natural bedrock channels (Whipple *et al.*, 2000). In a clever experiment to capture the effects of abrasion, modified ball mills subjected disks of variable erodibility to a swirling mixture of cobbles and water (Sklar & Dietrich, 2001). These experiments illustrated how an optimal amount of cobble tools maximized the abrasion. Too many tools in the mill allowed cobbles to protect the bed (disk) and to decrease in the rate of erosion. Too few tools resulted in infrequent interactions with the bed and again, a decrease in the rate of erosion. Large cobbles tend to erode the bed more than did small cobbles. These results have direct applications to natural systems where changes in climate, for example, modulate the sediment flux and discharge in a channel over 10^3 to 10^5 years time scales. Such changes in climate have long been suspected to limit the rates of vertical and horizontal channel incision. The Sklar & Dietrich (2001) experiments confirm that these incision rates are influenced by abrasion.

Process Sedimentology

Physical models dedicated to studying process sedimentology share many characteristics in common with the flume fluvial geomorphology approaches. A class of process sedimentology physical models is dedicated to synthesizing basin stratigraphy (Paola *et al.*, 2001). The main application of these studies for landscape evolution is developing the ability to deconvolve the record of base level, sediment supply, and subsidence from a resulting synthetic stratigraphy. Understanding the effects of sediment supply is particularly valuable in determining the spatial and temporal variations in erosional lowering of the source. In a source to sink approach, basin stratigraphy provides limits on what the processes and rates of erosion must be in the landscape.

Numeric Models

A numeric model describes landscape evolution by representing a geomorphic process, multiple processes, or landscape characteristic as mathematical expressions. A typical numeric strategy renders a landscape as a rasterized arrangement of cells whose elevation above a base level is determined by one or more mathematical equations that govern the addition or removal of mass from the cell.

Many iterations of solving the equations lead to changes in cell height that mimic real time-dependent changes in landscape elevation and relief. Numeric models provide alternatives to cumbersome physical models where scaling relationships may be violated. They also facilitate the focused investigation of an individual process or suite of processes. There is a certain elegance and level of objectivity to numeric approaches because the outcome of a good model is the result of independent equations based on physical laws. Lastly, and perhaps most importantly, numeric models can be used to test paradigms of landscape evolution and to predict future changes in the landscape, using the modern topography as a starting point. However, erroneous interpretations result if the boundary conditions processes are not properly identified and defined.

There are several different types of numeric models that have been developed over the past three decades. Their rapid expansion into studies of landscape evolution has been greatly enhanced by computers. There are at least five reviews or compilations of numeric landscape models from the past ten years: the *Tectonics and Topography* special volume published in the Journal of Geophysical Research (Merritts & Ellis, 1994), Koons (1995), Beaumont *et al.* (2000), *Landscape erosion and evolution modeling* (Harmon & Doe, 2001), and the *Steady-State Orogen* special volume published by the American Journal of Science (Pazzaglia & Knuepfer, 2001). The Beaumont *et al.* (2000) paper is particularly noteworthy for its attempt to categorize numeric models based on relative complexity. The organization and explanations of numeric models in this chapter borrows heavily upon these sources. The interested reader is directed to them for detailed descriptions of the models described herein.

The three major types of numeric landscape evolution models are surrogate models, multi-process models, and coupled geodynamic-surface process models. A surrogate model typically tracks a single metric of the landscape, such as mean elevation or mean relief, and does not discriminate among the myriad of processes that redistribute mass in real landscapes. The multi-process model attempts to isolate and represent all processes that contribute to the redistribution of mass with individual mathematical equations. A recent trend is to link a robust multi-process model with a geodynamic model of rock deformation in the Earth's crust, to yield a coupled geodynamic-surface process model. A true coupled model is one where the surface process model predicts lateral mass fluxes throughout the model landscape and the corresponding geodynamic model calculates the solid-Earth mass fluxes in response to the surface processes and as a consequence of tectonic forces or velocity fields (Beaumont *et al.*, 2000). Such coupled models are particularly intriguing because their results provide insights into the feedbacks between geodynamic and surficial processes.

All numeric models share the common goal of representing erosion of the landscape, either explicitly or implicitly as a key component of landscape evolution. Nearly all models treat erosion as proportional to local relief, mean elevation (Fig. 9), or local gradient and hydrology. There are separate justifications for all three. Erosion proportional to mean relief is supported by studies that show a correlation between river suspended sediment yields and relief (Ahnert, 1970). Erosion proportional to mean elevation is supported by Ruxton & McDougall (1967) and Ohmori (2000) among other studies, and a correlation between mean elevation and mean relief has been established for landscapes with well integrated drainages (Ohmori, 2000; Summerfield & Hulton, 1991). An important caveat to models that appeal to relief dependent erosion is that they can only be applied to decaying landscapes. Constructional or steady-state landscapes may exhibit no correlation between mean elevation, mean local relief, and rate of erosion. Erosion proportional to local gradient and hydrology is predicated on the field and lab studies that began with Gilbert (1877, 1909) and continues with experiments like those of Roering *et al.* (2001).

Another feature commonly shared by numeric models is the concept of landscape response time. This numeric formalization of the response time of Bull (1991) is typically derived from linear system behavior. Response times are particularly important in coupled geodynamic-surface process models because the tectonic forcing in these models is commonly treated as a step function. Linear system behavior treats the system response to a step-like forcing as an exponential and the response time (sometimes also called the characteristic time) is defined as $(1 - 1/e)$ where $e = 2.71828$. In other words, the response time indicates the amount of time needed to accomplish 63% of the landscape change or adjustment following a perturbation. Response times are sensitive to climate, model rock type, and the spatial-temporal scale of a model tectonic forcing. In real orogens that are well-drained, the response time scales approximately with the size of the orogen. Small orogens like Taiwan have short response times on the order of millions of years. Large orogens with orogenic plateaux like the Himalaya may have response times in excess of 100 million years. Linear system behavior is supported by real observations of how perturbed Earth systems liberate sediment as they seek new equilibria (Schumm & Rea, 1995). Flumes, watersheds, and orogens all show an exponential decay in sediment yield following the perturbation. In the case of a whole orogen, the response times are long enough that the simplest explanation for exponential reduction in sediment yield is a decrease in the mean elevation and mean relief of the source. At cyclic space and time scales, the Davisian cycle of erosion is probably an appropriate landscape evolution model. Analytical solutions to the landscape response time have been proposed by Whipple & Tucker (1999) and Whipple (2001).

Surrogate Models

The goal of the surrogate landscape model is to track a metric of the landscape, such as mean elevation above base level, or evolution of a single landscape component like a stream channel (Howard *et al.*, 1994; Whipple & Tucker, 1999), using a physically based equations or established functional relationships between a landscape metrics or processes. Erosion rate is commonly indexed to mean elevation (Fig. 9). Several early models use an elevation-erosion rate relationship to solve for

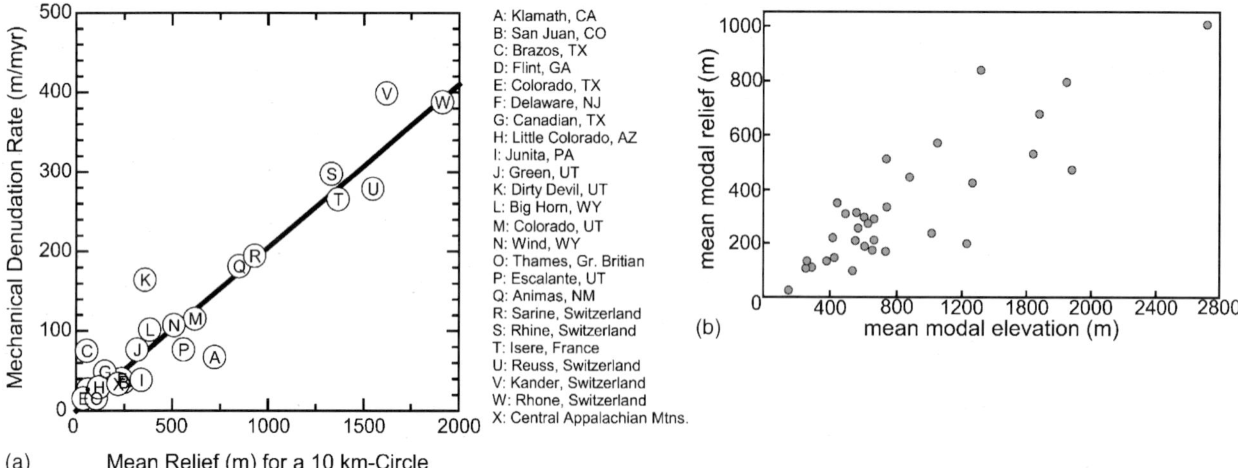

Fig. 9. (a) Correlation between mean local relief (measured in a circular window 10 km in diameter) and mean denudation rate, calculated from suspended sediment yield data from large, mid-latitude, well-drainage watersheds (modified from Pazzaglia & Brandon, 1996). (b) Correlation between mean elevation and mean local relief for well drained landscapes (modified from Summerfield & Hulton, 1991).

rock uplift, evolution of mean elevation, and flexural deformation of the lithosphere (Moretti & Turcotte, 1985; Stephenson, 1984; Stephenson & Lambeck, 1985). Such models have also simulated critical taper on an emergent accretionary wedge (Dahlen & Suppe, 1988), predicted the height limit of mountains (Ahnert, 1984; Slingerland & Furlong, 1989; Whipple & Tucker, 1999), and predicted the thermal structure of an eroding crustal column (Batt, 2001; Zhou & Stüwe, 1994).

Surrogate landscape models are not designed to capture the full range of changes in the landscape as it is eroded, but rather are useful in exploring the relative scaling relationships between landscape metrics and driving and resisting forces. For example, Pitman & Golovchenko (1991) justified an erosion proportional to mean elevation relationship to define the necessary and sufficient conditions for generating peneplains in the Appalachian landscape. Pazzaglia & Brandon (1996) followed a similar approach to reconstruct the evolution of mean elevation of the post-rift central Appalachians (Fig. 10a). The erosion rate was reconstructed from known quantities of sediment trapped in the Baltimore Canyon Trough, and a simple linear equation was solved for the general relationship between the flux of rocks into the mountain belt and erosional flux out of the belt. The results of a tectonic model, in which the flux of rocks into the belt was allowed to vary, but climate (rock erodibility) was held constant shows that three kilometers of rock would have to be fluxed through the belt in the past 20 myr to account for the recent offshore sediment volumes. Mean elevation of the range would have had to increase from near sea level before the Miocene to 1100 m, before diminishing to its present value of about 350 m. Tested by, for example, thermochronology, such predictions help place limits on the range of landscape responses that might be expected from a tectonic or climatic perturbation.

A less common surrogate approach is to apply a stochastic technique where landscape change is determined by the

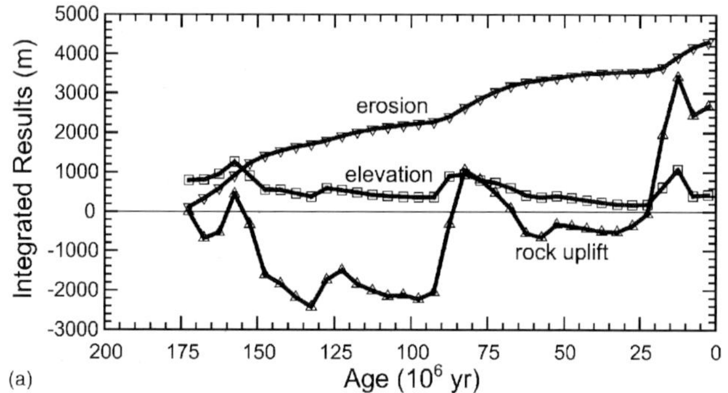

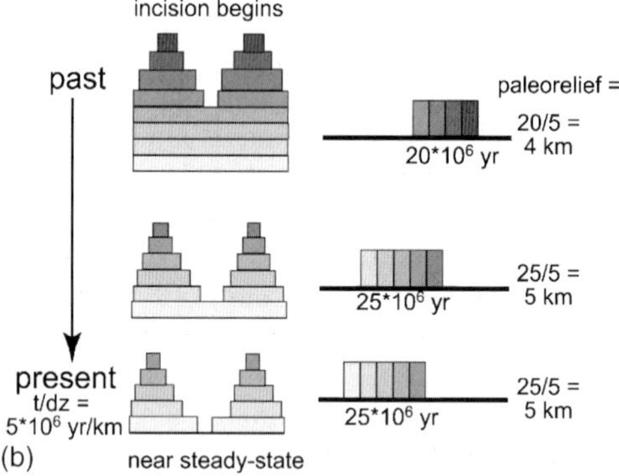

Fig. 10. (a) Reconstruction of rock uplift (source, triangles), mean elevation (squares), and total erosion (inverted triangles) for the post-rift Appalachian mountains (modified from Pazzaglia & Brandon, 1996). This simple single-process landscape model provides insights to the scaling relationships between mean elevation and rock uplift to explain the known volume of sediment delivered to offshore basins. (b) Illustration of detrital mineral age spectra during exhumation of the landscape to the steady-state relief condition (modified from Stock & Montgomery, 1996).

mean and variance of landscape slope elements (Craig, 1982, 1989). This particular type of model is one of the few that carefully incorporates map-scale considerations of the influence of structure, rock type, and stratigraphy on landscape evolution. One-kilometer-long slope elements called "draws" are constructed between adjacent cells in a DEM where the rock type underlying the cells is noted. The mean and variance of draws for any particular rock type pair is calculated and recorded. The slope of the draws is assumed to be well adjusted to the rock type and structure. The model can only be applied in settings where driving and resisting forces approach a dynamic equilibrium. Erosion is introduced as a random perturbation by lowering one cell. All adjacent cells are adjusted randomly by choosing a new draw that falls within the variance limits of all possible draw slopes developed between specific rock type pairs. The result is a diffusive rippling effect of slopes adjusting away from the perturbation. Relief can increase or decrease depending on the rock types encountered or uncovered. Erosion stops when the landscape has been lowered to base level.

Another type of surrogate model involves inversion techniques to estimate paleoelevation or paleorelief using thermochronology in the source and/or detrital grain ages (e.g. Bernet *et al.*, 2001; Brandon & Vance, 1992; Garver *et al.*, 1999; Stock & Montgomery, 1996; Fig. 10b). Conceptually these models assume a relief or elevation dependence on the exhumation of bedrock in the source. Mineral isochrons commonly increase in age with elevation especially where isotherms are a horizontally subdued reflection of topography. Measurable mineral age gradients are apparent in landscapes with moderate relief ($\sim$3 km) and erosion rates around 0.5 mm/yr. The numerical ages of the minerals depend on relief, the geotherm, and erosion rate. If erosion is uniform across the landscape and relief remains more or less steady as the landscape is unroofed, the vertical distribution of mineral ages in a depositional basin essentially reflects that unroofing process with a constant age gradient. If relief instead changes as the land erodes, the age gradients in the depositional basin also change. An inversion technique has been proposed to convert that change in age gradient to changes in relief (Fig. 10b; Stock & Montgomery, 1996). The technique requires high-precision single-grain ages, no change in mineral age during transport and deposition, mixing of the eroded sediment, and minimal sediment storage between source and sink.

Multi-Process Landscape Models

A multi-process landscape model considers two or more mutually interacting geomorphic processes that sculpt real landscapes. The model mathematically describes these processes, links the mathematical descriptions together in such a way that the continuity of mass is maintained, provides inputs of rock type, climate, and tectonics, and predicts the resulting landscape evolution (Slingerland *et al.*, 1994). Multi-process landscape models differ from the surrogate models in that the former actually strives to simulate the form and process of real landscapes, as opposed to simply exploring the scaling relationships among landscape metrics. Two generations of multi-process landscape models are recognized, the second borrowing heavily on the first. First generation multi-process models include attempts to mathematically capture bedrock weathering, hillslope creep, landsliding, fluvial transport of sediment, and channel initiation (Ahnert, 1976, 1987; Anderson & Humphrey, 1990; Armstrong, 1976; Chase, 1992; Gregory & Chase, 1994; Kirkby, 1986; Musgrave *et al.*, 1989; Willgoose *et al.*, 1991). Second-generation models specifically explore broader, more complex interactions among processes by introducing, for example, climatic perturbations (Tucker & Slingerland, 1997) or migration of drainage divides (Kooi & Beaumont, 1994).

The basic physical principle underlying all multi-process models is the conservation of mass. Mass rates into and out of model cells are driven by the geomorphic processes acting on the landscape (Fig. 11). These include but are not limited to bedrock weathering, hillslope creep, hillslope landsliding, glacial erosion, bedrock channel erosion, alluvial channel erosion, and sediment transport. A simple multi-process model collapses all hillslope processes into one mathematical equation that treats them collectively as diffusion. Similarly, it treats all fluvial transport processes collectively as advection, with sediment transport proportional to unit stream power (the discharge-slope product; Howard, 1994). More sophisticated approaches consider non-linear processes such as hillslope landsliding. A fine example of a robust multi-process landscape model, GOLEM, with examples of the various model inputs is discussed in detail in Slingerland *et al.* (1994).

Bedrock weathering is typically modeled as an alteration front that penetrates downward at a rate inversely proportional to the thickness of the regolith cover (Ahnert, 1976; Anderson & Humphrey, 1990; Armstrong, 1976; Rosenbloom & Anderson, 1994). The justification for this relationship is that soil genesis is a self-limiting process. Thick, well developed soils (or regoliths) tend to influence infiltration rates and generate runoff resulting in overland flow that strips the upper horizons. Regolith thickness approximates a steady state when the rate of stripping equals the rate of descent of the weathering front (Heimsath *et al.*, 1999; Pavich *et al.*, 1989).

Hillslope creep is typically modeled as a linear diffusive process where sediment transport is proportional to slope. Diffusion has attracted the greatest attention of all modeled geomorphic processes both within and outside the modeling community because of its simplicity and its apparent verification from slope degradation studies (e.g. Culling, 1960, 1963; Hanks *et al.*, 1984; Nash, 1980). On any uniformly eroding hillslope, it has been long known that sediment flux must increase systematically away from the hilltop (Gilbert, 1909). If the sediment flux is dependent on the slope, the combination of a slope-dependent transport law and the continuity equation leads to mathematical equation that looks like linear diffusion (Culling, 1960; Nash, 1980). However, this equation describes change in the *hillslope profile* and is not meant to describe the downslope transport of individual particles of sediment, a process that is probably non-diffusive. The diffusion constant or diffusivity in the

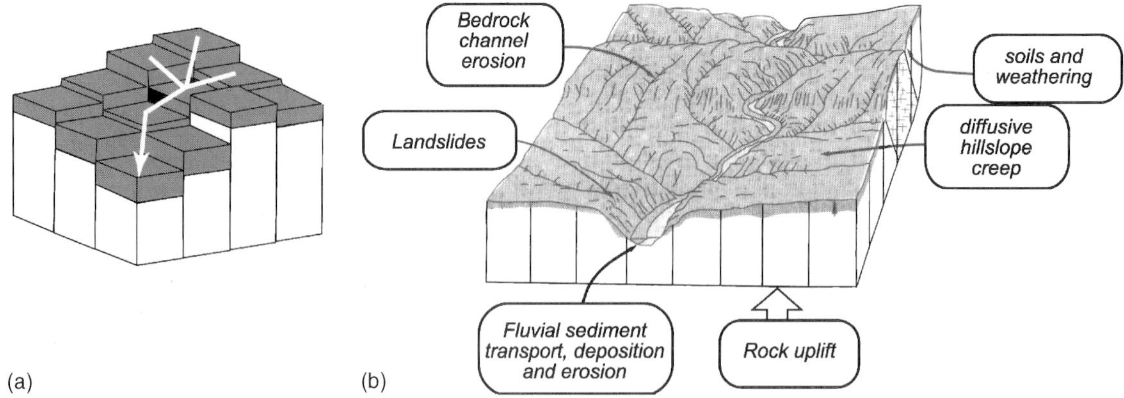

Fig. 11. Schematics showing (a) the arrangement of cells in a traditional 2-D, multi-process model where flow is directed down the steepest paths between rectilinear cells and (b) the major processes that are represented mathematically in model landscapes (both modified from Tucker & Slingerland, 1994). Most models now employ a triangulated network, rather than rectilinear cells.

diffusion equation is a scale-dependent parameter that implicitly includes climate and substrate characteristics.

The idea of simple hillslope diffusion is complicated by field observations and physical models of slope-clearing landslides as important agents for removing mass from hillslopes, especially in steep landscapes undergoing tectonic uplift. Multi-process models that account for landsliding typically assign a threshold slope angle, below which sediment is transported by diffusion, and above which, sediment is transported by slope-clearing landslides until the slope diminishes to a threshold angle (Kirkby, 1984). More sophisticated approaches consider potential failure planes as well as threshold angles (Densmore *et al.*, 1998).

Fluvial erosion and sediment transport have been approached by various multi-process models, most of which treat sediment transport in alluvial channels as proportional to the shear stress on the bed (Bagnold, 1977) and bedrock erosion proportional to unit stream power (Howard *et al.*, 1994), with channel width determined by the square root of the reach discharge. The models typically distinguish between alluvial and bedrock channel behavior based upon the presence or absence of sediment in a given model cell. A cell can experience both sediment transport and bedrock erosion if excess stream power remains after all of the sediment has been transported. Channel network development is a tricky consideration because the extent of channels – essentially the distinction between a model cell that operates as a channel vs. one that operates as a hillslope – can change through model time, especially if model precipitation is allowed to vary (Tucker & Slingerland, 1997). Changes in the total network channel length are expressed as changes in drainage density. One way to explore the sensitivity of precipitation changes on drainage density is to impose a shear stress threshold on the extent of the channel network (Tucker & Slingerland, 1994). A stochastic theory for erosion and sediment transport based on the Poisson pulse rainfall model (Eagleson, 1978) is another method for predicting changes in drainage density (Tucker & Bras, 1998; Tucker *et al.*, 2001).

Glacial erosion is a relatively new consideration in multi-process models, especially those exploring landscape evolution at high latitudes, through several glacial-interglacial cycles, or in high mountainous areas (Hallet *et al.*, 1996). Processes in glacial erosion include chemical weathering, subglacial stream erosion, abrasion by entrained rock, and bedrock plucking (Lliboutry, 1994). Abrasion, perhaps the most important of these, may scale linearly with basal sliding velocity, not with normal stress (Hallet, 1979; Harbor *et al.*, 1988). MacGregor *et al.* (2000) use basal abrasion and water-pressure fluctuations, which controls the bedrock quarrying rate, to model glacial erosion of alpine valleys. Both processes are thought to be maximized at the equilibrium line altitude (ELA). In the model, the longitudinal profiles of valleys are rapidly flattened, and tributary glaciers create steps down valley of the tributary junction by repeated glacial advances and retreats. The results are broadly consistent with the major landforms observed in glaciated valleys of the Sierra Nevada. A more sophisticated multi-process approach has been used to investigate the role of glaciation with respect to hillslope and fluvial erosion on a narrow, rapidly uplifting orogen likened to the Southern Alps of New Zealand (Braun *et al.*, 1999; Tomkin & Braun, 2002). The results illustrate the geomorphic importance of freezing at the bed of a glacier. Relief in the model landscape increases if glaciers are assumed to be frozen to their bed. Without frozen beds, relief in the model landscape decreases and erosion becomes concentrated in interfluves. Total erosion increases in a landscape suffering glacial interglacial cycles in comparison to landscapes that experience strictly fluvial erosion.

Quantifying the effects of climate change not involving glaciation on landscape evolution remains an important pursuit of multi-process models (Tucker & Bras, 2000; Tucker & Slingerland, 1997). For this application, models are being used to test many of the qualitative concepts in the process linkage approach to landscape evolution (Bull, 1991). The GOLEM model has been used to simulate landscape evolution in a watershed, including the ways that watersheds liberates sediment, under the influence of cyclic changes in model precipitation varied in such a way that it mimic glacial-interglacial climate swings of the Quaternary (Tucker & Slingerland, 1997). The watershed response and response time both depend on the direction of climate change. The model includes a parameter describing the land surface resistance to erosion by overland flow. This parameter is a proxy for the presence or absence of vegetative cover and acts to impose an erosion threshold. Increases in runoff intensity (increase in precipitation seasonality or storminess) or a decrease in vegetative cover lead to rapid expansion of drainage networks and delivery of sediment to the trunk channel that initially aggrades, then incises as a sediment pulse ends. In contrast, a decrease in runoff intensity (decrease in precipitation seasonality or storminess) or an increase in vegetation lead to a retraction of the channel network and relative stabilization in the trunk channel. Repeated cycles of runoff variability lead to highly variable denudation rates, reminiscent of the complex response phenomena of Schumm *et al.* (1987). However, most of the denudation appears concentrated during periods of increasing runoff intensity or loss of vegetative cover. These results are completely consistent with the Bull (1991) model for watershed response to late Pleistocene-to-Holocene climate change. Similarly, the Poisson pulse rainfall model (Tucker & Bras, 2000) suggests that climatic variability, such as increased seasonality in precipitation distribution throughout a year, causes watersheds to experience higher rates of erosion, develop greater drainage densities, and ultimately lose relief. In a parallel study in which fluvial incision was parameterized by the stream power law, climate change to overall wetter conditions reduces mean relief in a landscape (Whipple *et al.*, 1999).

Coupled Geodynamic-Surface Process Models

A coupled geodynamic-surface process model predicts topographic and landscape evolution in concert with the underlying deformation in the Earth's crust or lithosphere (Beaumont *et al.*, 2000). The most basic coupled model can be envisioned as a one-dimensional column of crust with a specified erosion law removing mass from its top and isostatic rebound occurring in response to that mass removal and subsequent adjacent deposition. Truly coupled models typically involve the full complexity of a multi-process landscape model and contain geometric or kinematic assumptions about deformation of crustal materials, including horizontal and lateral variations. The mechanics of tectonic processes must be built into the models if the dynamics of the coupled system are to be investigated. This necessity arises because the primary coupling occurs through the gravitational and buoyancy forces of topography, which influence the state of stress and deformation rate (Beaumont *et al.*, 2000). An important current limitation in the coupled model is the ability to match the resolution of a 2-D planform multi-process landscape model with a geodynamic model of equal resolution. Computational restrictions typically restrict the geodynamic model to be mated to the multi-process landscape model as a thin sheet (Ellis *et al.*, 1990). Alternatively, the geodynamic model is represented as a 2-D vertical section or panel through the planform landscape model.

One example of a coupled model was used to investigate qualitative concepts about long-term landscape evolution (Kooi & Beaumont, 1996). Though this model involves kinematic uplifts, the predicted surface erosion does not influence the uplift rate. In other words, the surface processes are coupled to gradients created by tectonic uplift, but feedbacks between them are not considered. The model represents hillslope processes as diffusion and fluvial transport is modeled as a network of one-dimensional rivers with the equilibrium sediment transport rate determined by local stream power. Discharge is fixed by equal precipitation spread across the model topography and by a routing routine that funnels water down the steepest pathways among and collected by model cells. No distinction is made between a hillslope model cell and a river model cell. Rather, the local sediment flux is determined by a length scaling parameter which determines sediment entrainment as a diffusive (hillslope) or advective (fluvial) process. The geodynamic model assumes linear system behavior, and kinematic symmetric uplifts. The landscape response time depends upon climate, rock type, and the spatial scale of the tectonic forcing. Soft rocks and high precipitation typically lower the response time. The ratio between the time scale of tectonic forcing and landscape response time is used to model the topographic response to different types of uplift (Fig. 12). If the time scale of the tectonic forcing is slower than the landscape response time, total erosion of the landscape closely matches tectonic uplift. The system maintains a flux steady-state (Willett & Brandon, 2002), the landscape is in a state of dynamic equilibrium though the time of tectonic uplift, and the topography has a characteristic relief and mean elevation over graded time spans (Fig. 12a). This type of landscape evolution closely approximates the qualitative landscape evolution model of Hack (1960). If the time scale of the tectonic forcing is the same as the landscape response time, maximum erosion of the landscape occurs after maximum tectonic forcing (Fig. 12b). In real landscapes, this behavior translates into rapid increase in relief, followed by establishment of a characteristic slope form, and finally a systematic decrease in erosion rates as slopes and interfluves are lowered. This type of landscape evolution approximates the qualitative model of Penck (1924). One of the most intriguing components of this model is the lag between maximum tectonic uplift and maximum erosion. The lag time in model years corresponds to millions or tens of millions of years in real landscapes, depending on the length scale of the uplift and the landscape response time. Lag times have important implications for dating orogenic events from a stratigraphic record. If the time scale of the tectonic forcing is much shorter than the landscape response time, the landscape cannot respond in kind and adjusts mainly to the long-wavelength components of the forcing (Fig. 12c). In real landscapes, this type of forcing is evident in river long profiles that have remained graded to mean sea level despite rapid glacio-eustatic fluctuations in the Quaternary. If the tectonic forcing is impulsive with respect to the landscape response, the landscape responds

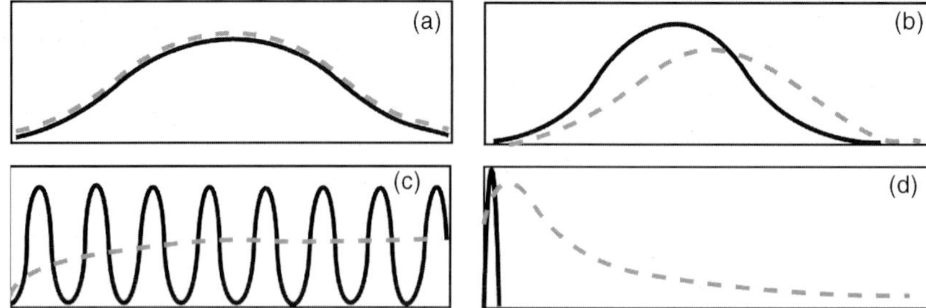

Fig. 12. The erosional response (dashed, gray line) to tectonic forcing (solid, black line) in a model landscape if the response time of the landscape is (a) faster than the time scale of the tectonic perturbation, (b) approximately the same as the time scale of the tectonic perturbation, (c) slower than the time scale of the tectonic perturbation, and (d) much slower than the tectonic perturbation which is impulsive (modified from Beaumont et al., 2000).

with an exponential decay in mean elevation and relief (Fig. 12d). This kind of landscape evolution approximates the qualitative landscape evolution model of Davis (1899a).

Coupled models are also used to investigate specific feedbacks between geodynamic and surficial processes (e.g. Barry, 1981; Beaumont *et al.*, 1992; Koons, 1989; Willett, 1999; Willett *et al.*, 2001). The emerging paradigm from these modeling efforts is that the macrogeomorphic profile of active orogens is determined by a feedback between the dynamics of the uplift and climate. The key variables influencing the feedback are the vergence direction for contractional orogens approximated as an orogenic wedge, and the prevailing direction of the storm tracks that deliver moisture to the orogen. Rock particle velocity vectors appear to be influenced by the direction of the moisture source for a modeled orogen that approximates the tectonic setting of the south island of New Zealand (Beaumont *et al.*, 1992; Fig. 13). The flux of rocks through the orogenic wedge increases with surface erosion and with precipitation derived from model west. A feedback that draws rocks up from depth results in a surface distribution of metamorphic grade similar to that observed along the Alpine fault.

Willett (1999), among others, has expanded upon these results by investigating the effects of orographic precipitation on the shapes of mountain ranges (Fig. 14). Model orogens with uniform precipitation exhibit asymmetric topography, with shallower slopes facing the subducting plate and maximum exhumation opposite the subduction vergence. Directional moisture modifies this basic topographic and exhumational asymmetry. If precipitation is delivered in the direction of subduction vergence, the zone of exhumation broadens and is maximized in the orogen interior. If precipitation is delivered opposite subduction vergence, exhumation is concentrated on the margin of the orogen and the topographic asymmetry can be reversed if the landscape response time is short. The Olympic Mountains of western Washington State represent a real landscape that approximates the former model landscape, and the New Zealand Alps represent a real landscape that approximates the latter model landscape. For New Zealand, the landscape response time, though rapid, is not rapid enough to effect a reversal in topographic asymmetry.

Subsequent coupled models have cast some doubt on the overall effectiveness of climate alone in determining the macrogeomorphic shape of orogens. Particularly where horizontal motions of rocks in a convergent setting are important, the ratio of the horizontal motion to vertical uplift and the processes that control drainage divide movement appear to exercise an overriding influence on orogen shape (Willett *et al.*, 2001; Fig. 15). These models challenge the traditional notions of what parts of the topography are fixed and which parts are moving as the landscape is denuded. For an orogen that approximates a flux steady state and also has significant horizontal advection of rock, the asymmetry in the range profile reflects a gradient the drainage divide can migrate against, and at a rate opposing the tectonic advection of rocks. The drainage divide can be envisioned as the crest of a topographic standing wave through which the rocks are advecting. The processes that influence drainage divide migration are well known, but some of them such as landslides, drainage capture, or glaciation may not be well simulated by diffusion which remains the most common mathematical process applied to divides in most surface process models (Densmore *et al.*, 1999).

Different Approaches

Coupled models are computationally intensive and as such present challenges to adequately capture the full range and interactions of geomorphic and geodynamic processes. For example, the spatial resolution that would truly represent river channels and their coupling to hillslopes is beyond the resolution of most DEM data as well as being computationally impractical (Stark & Stark, 2001). Some studies have begun to address these problems by employing efficient algorithms, from the field of computational geometry, that operate on arbitrarily discretized model cells (Braun & Sambridge, 1997). Other studies approximate fine scale model resolution with a parameterization scheme applied at a computationally feasible coarse model scale (Stark & Stark, 2001).

Braun & Sambridge (1997) adopt an approach where a 2-D planform multi-process landscape model operates on an irregular, self-refining grid. In essence, the model cells adapt in size and shape in response to the predominant, locally acting geomorphic process. Cell spacing shrinks

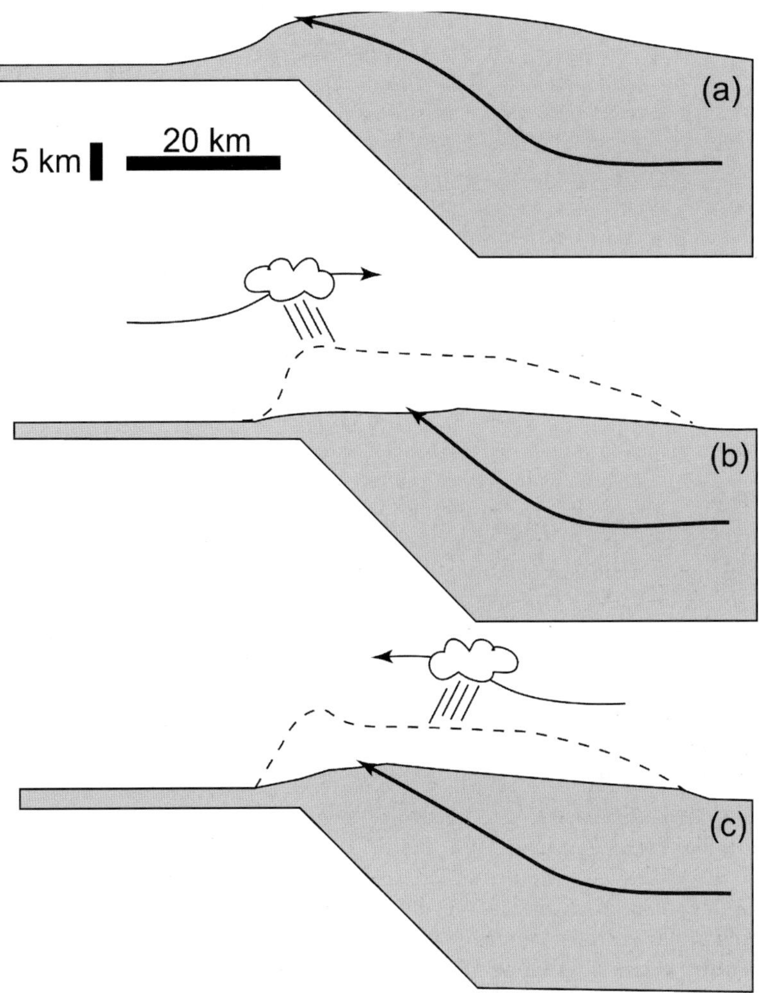

Fig. 13. Cross-sectional view of a model convergent orogen (orogenic wedge shaded). A representative particle path moving through the wedge is shown by the thick arrow. In all cases, there has been 16 km of shortening over 800,000 yr. Growth of topography above sea level is shown (a) with no erosion, (b) with the moisture source in the same direction of subduction, and (c) with the moisture source opposite the direction of subduction. In both (b) and (c), the section of eroded rock is shown by the area beneath the dashed line (modified from Beaumont et al., 1992).

along channels where fine discretization helps resolve the real interactions between channels and hillslopes, whereas cell spacing widens across hillslopes where diffusion does an adequate job modeling sediment flux. The irregular spatial discretization method removes the directional bias introduced by rectangular meshes that force streams to flow in four preferred directions. The method also makes it possible to solve problems involving complex geometries and/or boundary conditions, allows for self-adapting discretization, and allows for horizontal tectonic movements such as strike-slip faults (Braun & Sambridge, 1997).

Stark & Stark (2001) approached the problem of limited model discretization with a parameterization scheme. Most models implicitly account for parameterization in river networks, for example, when channels are considered to be in equilibrium which assumes that reach channel width varies as the square root of reach discharge. This particular parameterization may hold for modeled alluvial rivers as it is a common metric of alluvial rivers in nature, but the precise relationship between discharge and channel geometry for bedrock streams, if one is exists, remains poorly constrained (Montgomery & Gran, 2001; Snyder *et al.*, in press). Moreover, coupled models typically apply to mountainous topography where bedrock streams dominate. An alternative parameterization scheme is justified. The resulting parameterization is based on the assumption that channel disequilibrium is fundamental in overall landscape evolution. A scaling analysis of DEM data shows that channel slope and upstream drainage area can be corrected for pixel resolution. Stream power, for example, measured at the coarse pixel resolution contains enough topographic information to permit a sub-pixel description of the topography in a distributional sense, even if there is not enough information to explicitly reconstruct channel morphology. The proposed channelization index, a lumped parameter, integrates all factors at the sub-pixel level that influence the flow of sediment, but not water, through a channel reach. The coupled system equations describing the channelization-based model are solved by a novel approach of using a network of non-linear resistors and capacitors, rather than finite difference or self-adaptive finite element meshes. Electricity in the resistor-capacitor network has direct analogues to the coupled model parameters. Topography, for example, is mapped to node potential, equal to the potential difference across a nodal capacitor. The erosion rate is mapped to the rate of discharge of the capacitor. Hillslope sediment flux is modeled by the flow of current through a resistor. The channelization index is represented by resistor temperature. The resistor-capacitor network can contain on

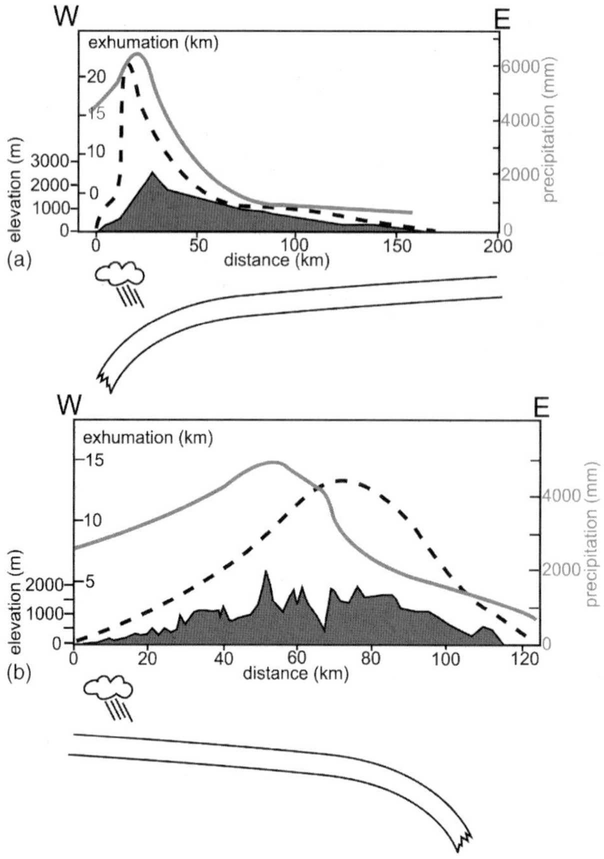

Fig. 14. Simplified, average elevation (dark shaded polygon), exhumation (dashed line), and precipitation (gray line) for model landscapes representing the (a) Southern Alps of New Zealand and (b) the Olympic Mountains of Washington State. Sketch beneath each graph shows the subduction direction and source of precipitation for each setting (modified from Willett, 1999).

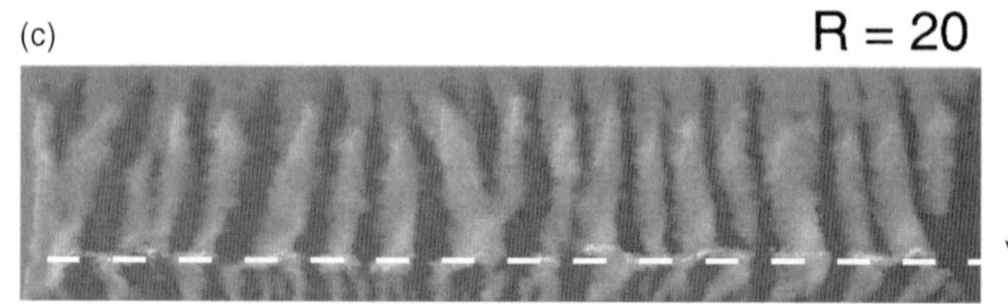

Fig. 15. Plan view of the effect of horizontal motion of rock in a convergent orogen on range asymmetry and location of the orogen's drainage divide (dashed white line). R, ratio of convergence velocity to uplift rate. In all cases, the model topography has reached a steady state so each divide may be viewed as a uniquely shaped standing wave through which rocks advect enroute to their erosion. (a) rocks lack horizontal motion and a more or less symmetrical divide forms. In (b) and (c), horizontal motion of rocks produces an asymmetric divide (modified from Willett et al., 2001).

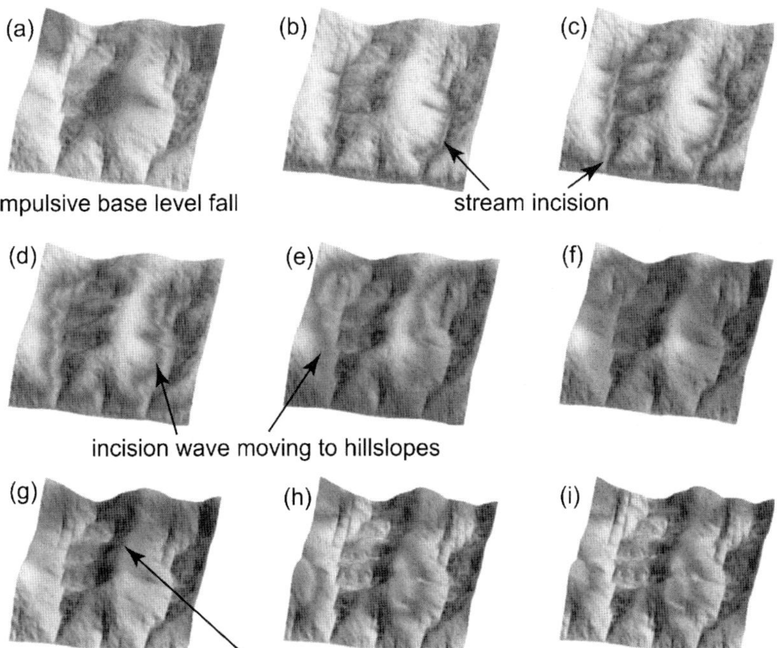

Fig. 16. Erosional response of a model landscape simulated with a channelized parameterization and a linked non-linear resistor-capacitor network. Erosion is shown by the transition from lighter to darker shading, individually scaled in each image. The landscape in (a) has the streams at their base level of erosion (grade) and limited erosion in the interfluves. An impulsive base level fall sends a wave of incision up the streams (b), but the hillslopes do not begin to completely respond until (e). By (f) and (g), the streams are once again at grade while the hillslopes continue to erode. In (h) and (i), the landscape returns to rates and distribution like those in (a).

the order of 10^4 nodes and remain numerically tractable. The model visually illustrates the landscape response to kinematic tectonics (Fig. 16). It is particularly useful for tracking a wave of erosion as it moves from the channel network out across the model hillslopes.

Concluding Synthesis: Landscape Evolution Models and Geomorphic Frontiers

Current understanding of landscape evolution is a rich amalgamation of qualitative models that emphasize description and in some cases philosophy, physical models typically geared towards understanding specific geomorphic processes, and numeric models in which coupling and feedbacks between Earth surface and geodynamic processes has begun to be revealed. Semantic disagreements aside, there currently is room for all of the major qualitative paradigms to be "correct" when the system, its boundary conditions, and time and space scales of observation are clearly defined.

This chapter neglects major recent strides in geochronology, such as cosmogenic dating. Also neglected is the search for a simple, but useful stream-power based river incision law. However, the dating of landforms and a greater understanding of the fundamental limits on local erosion will drive the development of new qualitative, physical, and numeric models in the foreseeable future.

The geomorphologic equivalent to the much sought unified theory of physical forces is a single landscape evolution model that can successfully explain the bewildering display of landforms and landscapes at all spatial and temporal scales as well successfully predict the time-dependent changes in that landscape. Geomorphology is no closer to its grand unified model than physics is to grand unified theory, but an intriguing convergence among field observations, the qualitative paradigms, and numeric approaches is emerging.

Increasingly complex and realistic landscape evolution models reveal interactions between geomorphic and tectonic processes that are not immediately obvious from field studies alone. A useful landscape evolution model is not necessarily one that can accurately replicate all components of a natural landscape, but rather, can guide the geomorphologist towards designing a field study to understand the complicated, and commonly non-linear feedbacks between tectonic and surficial processes. In this way, landscape evolution models become heuristic tools (Oreskes *et al.*, 1994) that allow the geomorphologist to progressively refine field based hypotheses and field experiments. For example, interest in the origin and landscape evolution of the Great Escarpments of the southern continents spawned a flurry of numeric models that can be used to directly test long-held beliefs on landscape antiquity and rates of landscape change (Fig. 17). In the qualitative model of King (1953), the Great Escarpments are thought to have originated as the rift flanks of the continental rifts that ultimately split Gondwanaland into separate continents 80–120 myr ago. High-standing, seaward facing escarpments formed at this time are thought to have receded inland tens to hundreds of kilometers, more or less as a minimally embayed wall, to their present locations. This and related models are testable with thermochronologic and cosmogenic dating, which would determine the record of long-term erosion and persistence of old land surfaces respectively. Results from these studies imply links among tectonics, base level fall, and parallel retreat of escarpments – links different from those envisioned by King (1953) and later investigations (e.g. Partridge & Maud, 1987). In summary, the thermochronologic data are consistent with moderate to low

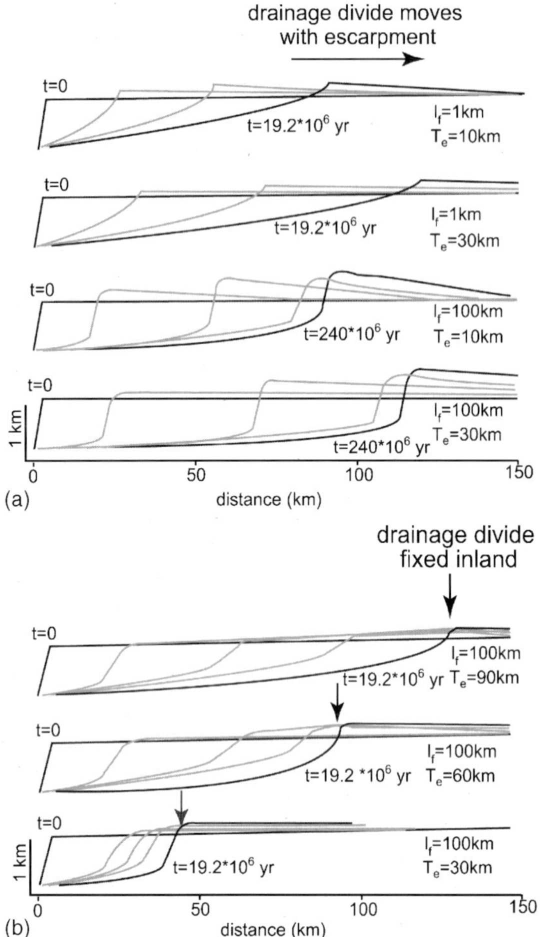

Fig. 17. Profiles showing escarpment evolution for a coupled model in which scarp retreat is influenced by (a) an originally flat, pre-rift topography and a drainage divide localized by flexure at the escarpment lip and (b) gentle seaward sloping pre-rift topography and a drainage divide located landward of the rift flank. In both models, lf denotes the erosion length scale, Te the effective elastic thickness of the lithosphere, and gray lines the intermediate profiles. Model (b) is consistent with thermochronologic and cosmogenic data (modified from Kooi & Beaumont, 1994).

amounts of erosion and landscape change during the syn-rift period, followed by substantial amounts of rapid erosion in the immediate post-rift period (Brown *et al.*, 2000). The thickness of crustal column eroded generally decreases inland approaching the current escarpment, but there is locally high spatial variability in the total amount of erosion both near the coast and the escarpment. Cosmogenic studies corroborate the thermochronologic data by demonstrating very slow current rates of escarpment retreat that have been apparently maintained over significant periods of time (Bierman & Caffee, 2001). Both data sets suggest that the Great Escarpments assumed their present expression rapidly, in a burst of major landscape evolution after rifting, they then settled down to relatively slow rates of landscape change and in fact are approximating a topographic steady state (Pazzaglia & Knuepfer, 2001).

Coupled geodynamic-surficial process models provide a viable explanation for these results concerning Great Escarpment evolution (Gilchrist *et al.*, 1994; Kooi & Beaumont, 1994; Tucker & Slingerland, 1994). The models show that high mean elevations are a necessary condition to make the escarpments. This idea is completely consistent with the notions King (1953) put forth for their origin. However, the numeric models also show that any Great Escarpment, when first formed, will rapidly embay as channels elongate headward into the continental interior. An escarpment defines, in essence, a drainage divide between opposing stream heads. Any process that favors the steep stream heads to breach the divide and capture the opposing more gently-sloped streams, as suggested in Gilbert's Law of Divides (Gilbert, 1877), will ultimate embay the escarpment and lead to its destruction as streams begin lowering the divide aggressively from both flanks. The numeric models show that rapid elongation of stream heads that embay and destroy the escarpment occurs if there is no coupled feedback to the flexural bending of the lithosphere in response to erosional unloading and sediment deposition offshore. If instead lithospheric flexure is considered, the long-wavelength arching of the margin pins the divide at a prescribed distance inland atop the flexural forebulge. Seaward-draining channels rapidly elongate headward to this position but have little incentive to go further. The result is rapid erosion seaward of the flexurally-defined drainage divide and consolidation of the Great Escarpment immediately seaward of the flexural forebulge. So in essence, the numeric models argue that the Great Escarpments were carved out of the seaward flank of a passive margin's flexural forebulge, which had a high-standing origin on a rift margin. The great, sweeping pediments at the foot of the Great Escarpments therefore were not formed as an escarpment retreated, but rather when the escarpment formed, base level stabilized, and has remained stable for tens of millions of years.

One of the frontiers of landscape evolution research, if they can be predicted, may include direct reconstruction the mean elevation of topography (Rowley *et al.*, 2001), as a global response to glacial climates (Hallet *et al.*, 1996; Zhang *et al.*, 2001), and to reconcile the present disparity of erosion rates from landscape of markedly similar relief and mean elevation. The latter issue pertains directly to the growing appreciation by geomorphologist, especially those that are not intimately familiar with the landscape of the Gondwanan continents, that long-term landscape erosion is neither uniform nor steady leading to the possibility of the preservation of landforms of great antiquity (Gunnell, 1998; Nott, 1994; Ollier & Pain, 1994; O'Sullivan *et al.*, 2000; Partridge & Maud, 1987; Twidale, 1976, 1985, 1997). There remains no good single explanation for why dramatically steep slopes on the Great Escarpments of passive margins erode at slow rates approaching 5 m/myr whereas equally steep slopes in tectonically active settings may erode at rates three orders of magnitude faster approaching 5000 m/myr. The answer may lie in the basic fact that landscapes erode more or less at the rate that they are experiencing base level fall, with all of the appropriate caveats for lag time and factors influencing landscape response times such as climate and rock type. If so, a new generation of landscape evolution models, qualitative and numeric, will have to be developed to simulate the nuances of base-level change. Such approaches may hold great promise in moving current understanding towards a unifying model of long-term landscape evolution.

Acknowledgments

The author is indebted to Josh Roering and Noah Snyder for carefully reading the first draft of this manuscript and providing thoughtful, constructive comments that ultimately made this chapter a better contribution to the volume. However, mistakes or misinterpretations of the many studies referenced herein are the sole responsibility of the author.

References

Aalto, R., Montgomery, D.R., Hallet, B., Abbe, T.B., Buffington, J.M., Cuffey, K.M. & Schmidt, K.M. (1997). A hill of beans. *Science*, **277**, 1911–1912.

Agassiz, L. (1840). Etudes sur les glaciers: Privately published, Neuchâtel.

Ahnert, F. (1976). Brief description of a comprehensive process-response model for landform development. *Zeitschrift für Geomorphologie Suppl.*, **25**, 20–49.

Ahnert, F. (1984). Local relief and the height limits of mountain ranges. *American Journal of Science*, **284**, 1035–1055.

Ahnert, F. (1970). Functional relationship between denudation, relief, and uplift in large mid-latitude drainage basins. *American Journal of Science*, **268**, 243–263.

Ahnert, F. (1987). Process-response models of denudation at different spatial scales. *Catena Supplement*, **10**, 31–50.

Anderson, R.S. & Humphrey, N.F. (1990). Interaction of weathering and transport processes in the evolution of arid landscapes. *In*: Cross, T.A. (Ed.), *Quantitative Dynamic Stratigraphy*. Prentice-Hall, Englewood Cliffs, NJ, pp. 349–361.

Armstrong, A.C. (1976). A three-dimensional simulation of slope forms. *Z. Geomorph. Suppl.*, **25**, 20–28.

Bagnold, R.A. (1977). Bed-load transport by natural rivers. *Water Resources Research*, **13**, 303–312.

Barry, R.G. (1981). *Mountain weather and climate*. London, Methuen, 313 pp.

Batt, G.E. (2001). The approach to steady-state thermochronological distribution following orogenic development in the southern Alps of New Zealand. *American Journal of Science*, **301**, 374–384.

Beardmore, N. (1851). *Manual of hydrology*. London, Waterlow and Sons, 384 pp.

Beaumont, C., Fullsack, P. & Hamilton, J. (1992). Erosional control of active compressional orogens. *In*: McClay, K.R. (Ed.), *Thrust Tectonics*. Chapman and Hall, New York, pp. 1–18.

Beaumont, C., Kooi, H. & Willett, S. (2000). Coupled tectonic-surface process models with applications to rifted margins and collisional orogens. *In*: Summerfield, M.A. (Ed.), *Geomorphology and Global Tectonics*. Chirchester, John Wiley and Sons, pp. 28–55.

Bernet, M., Zattin, M., Garver, J.I., Brandon, M.T. & Vance, J. (2001). Steady-state exhumation of the European Alps. *Geology*, **29**, 35–38.

Bierman, P.R. & Caffee, M. (2001). Slow rates of rock surface erosion and sediment production across the Namib desert and escarpment, Southern Africa. *American Journal of Science*, **301**, 326–358.

Bloom, A. (2002). Teaching about relict, no-analog landscapes. *Geomorphology*, **47**, 303–311.

Brandon, M.T. & Vance, J.A. (1992). Tectonic evolution of the Cenozoic Olympic subduction complex, Washington State, as deduced from Fission track ages for detrital zircons. *American Journal of Science*, **292**, 565–636.

Braun, J. & Sambridge, M. (1997). Modeling landscape evolution on geologic timescales: A new method based on irregular spatial discretation. *Basin Research*, **9**, 27–52.

Braun, J., Zwartz, D. & Tomkin, J.H. (1999). A new surface processes model combining glacial and fluvial erosion. *Annals of Glaciology*, **28**, 282–290.

Brown, R.W., Gallagher, K., Gleadow, A.J.W. & Summerfield, M.A. (2000). Morphotectonic evolution of the South Atlantic margins of Africa and South America. *In*: Summerfield, M.A. (Ed.), *Geomorphology and Global Tectonics*. Chirchester, John Wiley and Sons, pp. 255–281.

Brozovic, N., Burbank, D. & Meigs, A. (1997). Climatic limits on landscape development in the northwestern Himalaya. *Science*, **276**, 571–574.

Brunsden, D. & Thornes, J.B. (1979). Landscape sensitivity and change. *Transactions of the Institute of British Geographers*, **4**, 463–484.

Bryan, K. (1922). Erosion and sedimentation in the Papago Country, Arizona, with a sketch of the geology. *U.S. Geological Survey Bulletin* 730-B, pp. 19–90; (reprinted in: Adams, G.F. (Ed.), 1975, *Planation Surfaces; Peneplains, Pediplains, and Etchplains*. Dowden, Hutchinson and Ross, Stroudsburg, Pa, pp. 207–227).

Bryan, K. (1940). The retreat of slopes. *Annals of the Association of American Geography*, **30**, 254–268.

Bull, W.B. & Schick, A.P. (1979). Impact of climatic change on an arid watershed: Nahal Yael, southern Israel. *Quaternary Research*, **11**, 153–171.

Bull, W.B. (1991). *Geomorphic responses to climatic change*. Oxford University Press, New York, 326 pp.

Burbank, D.W. & Anderson, R.S. (2001). *Tectonic geomorphology*. Malden, Mass, Blackwell Science, 274 pp.

Carson, M.A. & Kirkby, M.J. (1972). *Hillslope form and process*. Cambridge University Press, London, 475 pp.

Chase, C.G. (1992). Fluvial landsculpting and the fractal dimension of topography. *Geomorphology*, **5**, 39–57.

Chorley, R.J. (1967). Models in geomorphology. *In*: Chorely, R.J. & Haggett, P. (Eds), *Models in Geography*. London, Methuen, pp. 59–96.

Craig, R. (1982). The ergodic principle in erosional models. *In*: Thorn, C.E. (Ed.), *Space and Time in Geomorphology*. London, Allen and Unwin, pp. 81–115.

Craig, R. (1989). Computing Appalachian geomorphology. *Geomorphology*, **2**, 197–207.

Culling, W.E.H. (1960). Analytical theory of erosion. *Journal of Geology*, **68**, 336–344.

Culling, W.E.H. (1963). Soil creep and the development of hillside slopes. *Journal of Geology*, **71**, 127–161.

Dahlen, F.A. & Suppe, J. (1988). Mechanics, growth, and erosion of mountain belts. *In*: Clark, S.P., Jr., Burchfiel, B.C. & Suppe, J. (Eds), *Processes in Continental Lithospheric Deformation*. Geological Society of America Special Paper 218, pp. 425–478.

Daubrée, A.G. (1879). *Etudes Synthetiques de Geologie Experimentale*. Paris, Dunod, 828 pp.

Dausse, M.F.B. (1857). Note sur un principe important et nouveau d'hydrologie. *Comptes Rendus de l'Académie des Sciences*, Paris, **44**, 756–766.

Davis, W.M. (1889). The rivers and valleys of Pennsylvania. *National Geographic Magazine*, **1**, 183–253.

Davis, W.M. (1899a). The geographical cycle. *Geography Journal*, **14**, 481–504.

Davis, W.M. (1899b). *Response to literal interpretations of the geographic cycle*.

Davis, W.M. (1930). Rock floors in arid and humid climates. *Journal of Geology*, **38**(1–27), 136–158.

Davis, W.M. (1932). Piedmont benchlands and Primärrumpfe. *Geological Society of America Bulletin*, **43**, 399–440.

Davisson, C. (1888a). Note on the movement of scree material. *Quarterly Journal of the Geological Society of London*, **44**, 232–238.

Davisson, C. (1888b). Second note on the movement of scree material. *Quarterly Journal of the Geological Society of London*, **44**, 825–826.

Denlinger, R.P. & Iverson, R.M. (2001). Flow of variably fluidized granular masses across three-dimensional terrain 2. Numerical predictions and experimental tests. *Journal of Geophysical Research*, **106**, 553–566.

Densmore, A.L., Anderson, R.S., Ellis, M.A. & McAdoo, B.G. (1997). Hillslope evolution by bedrock landslide. *Science*, **275**, 369–372.

Densmore, A.L., Ellis, M.A. & Anderson, R.S. (1998). Landsliding and the evolution of normal-fault-bounded mountains. *Journal of Geophysical Research*, **103**, 15,203–15,219.

Densmore, A.L., Ellis, M.A. & Anderson, R.S. (1999). Numerical experiments on the evolution of mountainous topography. *Journal of Geophysical Research*, **103**, 15,203–15,219.

Du Buat, L.G. (1786). Principes d'hydraulique. *Paris, De l'Imprimerie de Monsieur*, **1**, 453 pp.; **2**, 402 pp.

Dunne, T. (1978). Field studies of hillslope flow processes. *In*: Kirkby, M.J. (Ed.), *Hillslope Hydrology*. John Wiley and Sons, New York, pp. 227–294.

Eagleson, P.S. (1978). Climate, soil, and vegetation: 2. The distribution of annual precipitation derived from observed storm sequences. *Water Resources Research*, **14**, 713–721.

Ellis, S., Fullsack, P. & Beaumont, C. (1990). Incorporation of erosion into thin sheet numerical models of continental collision. *EOS, Transactions of the American Geophysical Union*, **71**, 1562.

Garver, J.I., Brandon, M.T., Roden-Tice, M.K. & Kamp, P.J.J. (1999). Exhumation history of orogenic highlands determined by detrital fission track thermochronology. *In*: Ring, U. *et al.* (Eds), *Exhumation Processes: Normal Faulting, Ductile Flow, and Erosion*. London, Geological Society Special Publication 154, pp. 283–304.

Gilbert, G.K. (1877). *Geology of the Henry Mountains (Utah)*. U.S. Geographical and Geological Survey of the Rocky Mountains Region, U.S. Government Printing Office, Washington D.C., 170 pp.

Gilbert, G.K. (1909). The convexity of hillslopes. *Journal of Geology*, **17**, 344–350.

Gilbert, G.K. (1914). *The transportation of debris by running water*. U.S. Geological Survey Professional Paper 86, 263 pp.

Gilbert, G.K. (1917). *Hydraulic mining debris in the Sierra Nevada*. U.S. Geological Survey Professional Paper 105, 154 pp.

Gilchrist, A.R., Kooi, H. & Beaumont, C. (1994). Post-Gondwana geomorphic evolution of southwestern Africa implications for the controls on landscape development from observations and numerical experiments. *Journal of Geophysical Research, B, Solid Earth and Planets*, **99**(6), 12,211–12,228.

Gilchrist, A.R. & Summerfield, M.A. (1991). Denudation, isostasy, and landscape evolution. *Earth Surface Processes and Landforms*, **16**, 555–562.

Glock, W.S. (1931). The development of drainage systems: A synoptic view. *Geography Review*, **21**, 475–482.

Gregory, K.M. & Chase, G.C. (1994). Tectonic and climatic significance of a late Eocene low-relief, high level geomorphic surface, Colorado. *Journal of Geophysical Research*, **99**, 20,141–20,160.

Gunnell, Y. (1998). The interaction between geological structure and global tectonics in multistoried landscape development: A denudation chronology of the South Indian shield. *Basin Research*, **10**, 281–310.

Hack, J.T. (1960). Interpretation of erosional topography in humid temperate regions. *American Journal of Science*, **258-A**, 80–97.

Hallet, B. (1979). A theoretical model of glacial abrasion. *Journal of Glaciology*, **17**, 209–222.

Hallet, B., Hunter, L. & Bogen, J. (1996). Rates of erosion and sediment evacuation by glaciers; a review of field data and their implications. *Global and Planetary Change*, **12**(1–4), 135–213.

Hanks, T.C., Bucknam, R.C., Lajoie, K.R. & Wallace, R.E. (1984). Modification of wave-cut and faulting-controlled landforms. *Journal of Geophysical Research*, **89**, 5771–5790.

Harbor, J.M., Hallet, B. & Raymond, C.F. (1988). A numerical model of landform development by glacial erosion. *Nature*, **333**, 347–349.

Harmon, R.S. & Doe, W.W. III (Eds) (2001). *Landscape erosion and evolution modeling*. New York, NY, Kluwer Academic/Plenum Publishers, 540 pp.

Heimsath, A.M., Dietrich, W.E., Nishiizumi, K. & Finkel, R.C. (1999). Cosmogenic nuclides, topography, and the spatial variation of soil depth. *Geomorphology*, **27**, 151–172.

Hoffman, P.F. & Grotzinger, J.P. (1993). Orographic precipitation, erosional unloading, and active tectonic style. *Geology*, **21**, 195–198.

Hooke, R.L.B. (2000). Toward a uniform theory of clastic sediment yield in fluvial systems. *Geological Society of America Bulletin*, **112**, 1778–1786.

Horton, R.E. (1945). Erosional development of streams and their drainage basins: Hydro-physical approach to quantitative morphology. *Geological Society of America Bulletin*, **56**, 275–370.

Hovius, N. (2000). Macro scale process systems of mountain belt erosion. *In*: Summerfield, M. (Ed.), *Geomorphology and Global Tectonics*. New York, John Wiley and Sons, pp. 77–105.

Howard, A.D. (1994). A detachment-limited model of drainage basin evolution. *Water Resources Research*, **30**, 2261–2285.

Howard, A.D. (1971). Simulation of stream networks by headward growth and branching. *Geogr. Anal.*, **3**, 29–50.

Howard, A.D., Dietrich, W.E. & Siedl, M.A. (1994). Modeling fluvial erosion on regional to continental scales. *Journal of Geophysical Research*, **99**, 13,971–13,986.

Howe, E. (1901). Experiments illustrating intrusion and erosion. *U.S. Geological Survey*, 21st annual report, Part 3, pp. 291–303.

Hubbard, G.D. (1907). Experimental physiography. *American Geographic Society Bulletin*, **39**, 658–666.

Iverson, R.M. & Vallance, J.W. (2001). New views of granular mass flows. *Geology*, **29**, 115–118.

Jaggar, T.A. (1908). Experiments illustrating erosion and sedimentation. *Bulletin of the Museum of Comparative Zoology Harvard College*, **49** (Geological Series 8), 285–305.

King, L.C. (1953). Canons of landscape evolution. *Geological Society of America Bulletin*, **64**, 721–752.

King, L.C. (1967). *The morphology of the earth* (2nd ed.). Edinburgh, Oliver and Boyd.

Kirkby, M.J. (1984). Modeling cliff development in South Wales Savigear re-reviewed. *Zeitschrift für Geomorphologie*, **28**, 405–426.

Kirkby, M.J. (1986). A two-dimensional simulation model for slope and stream evolution. *In*: Abrahams, A.D. (Ed.), *Hillslope Processes*. Boston, Allen and Unwin, pp. 203–222.

Kooi, H. & Beaumont, C. (1994). Escarpment evolution on high-elevation rifted margins insights derived from a surface processes model that combines diffusion, advection, and reaction. *Journal of Geophysical Research, B, Solid Earth and Planets*, **99**, 12191–12209.

Kooi, H. & Beaumont, C. (1996). Large-scale geomorphology classical concepts reconciled and integrated with contemporary ideas via a surface processes model. *Journal of Geophysical Research, B, Solid Earth and Planets*, **101**(2), 3361–3386.

Koons, P.O. (1989). The topographic evolution of collisional mountain belts: A numerical look at the southern Alps, New Zealand. *American Journal of Science*, **289**, 1044–1069.

Koons, P.O. (1995). Modeling the topographic evolution of collisional belts. *Annual Reviews of Earth and Planetary Sciences*, **23**, 375–408.

Langbein, W.B. & Schumm, S.A. (1958). Yield of sediment in relation to mean annual precipitation. *EOS Transactions, American Geophysical Union*, 1076–1084.

Leopold, L.B., Wolman, M.G. & Miller, J.P. (1964). *Fluvial processes in geomorphology*. W.H. Freeman and Company, San Francisco and London, 522 pp.

Lliboutry, L.A. (1994). Monolithologic erosion of hard beds by temperate glaciers. *Journal of Glaciology*, **40**, 433–450.

MacGregor, K.R., Anderson, R.S., Anderson, S.P. & Waddington, E.D. (2000). Numerical simulations of glacial-valley longitudinal profile evolution. *Geology*, **28**, 1031–1034.

Martin, Y. (2000). Modeling hillslope evolution: Linear and nonlinear transport relations. *Geomorphology*, **34**, 1–21.

Merritts, D.J. & Ellis, M. (Eds) (1994). *Tectonics and Topography: Special Issue of the Journal of Geophysical Research*, **99**, 12,135–12,315, 13,871–14,050, 20,063–20,321.

Meyer, G.A., Wells, S.G. & Jull, A.J.T. (1995). Fire and alluvial chronology in Yellowstone National Park: Climate and intrinsic controls on Holocene geomorphic processes. *Geological Society of America Bulletin*, **107**, 1211–1230.

Molnar, P. & England, P. (1990). Late Cenozoic uplift of mountain ranges and global climate change: Chicken or egg? *Nature*, **346**, 29–34.

Montgomery, D.R. & Gran, K.B. (2001). Downstream variations in the width of bedrock channels. *Water Resources Research*, **37**, 1841–1846.

Moretti, I. & Turcotte, D.L. (1985). A model for erosion, sedimentation, and flexure with applications to New Caledonia. *Journal of Geodynamics*, **3**, 155–168.

Morisawa, M.E. (1964). Development of drainages systems on an upraised lake floor. *American Journal of Science*, **262**, 340–354.

Musgrave, F.K., Kolb, C.E. & Mace, R.S. (1989). The synthesis and rendering of eroded fractal terrains. *Comp. Graph.*, **23**, 41–50.

Nash, D. (1980). Morphologic dating of degraded normal fault scarps. *Journal of Geology*, **88**, 353–360.

Nott, J. (1994). Long-term landscape evolution in the Darwin region and its implications for the origin of landsurfaces in the north of Northern Territory. *Australian Journal of Earth Sciences*, **41**, 407–415.

O'Sullivan, P.B., Gibson, D.L., Kohn, B.P., Pillans, B. & Pain, C. (2000). Long-term landscape evolution of the Northparkes region of the Lachlan fold belt, Australia: Constraints from fission track and paleomagnetic data. *Journal of Geology*, **108**, 1–16.

Ollier, C.D. & Pain, C.F. (1994). Landscape evolution and tectonics in southeastern Australia. *Journal of Australian Geology and Geophysics*, **15**, 335–345.

Ohmori, H. (2000). Morphotectonic evolution of Japan. *In*: Summerfield, M.A. (Ed.), *Geomorphology and Global Tectonics*. Chirchester, John Wiley and Sons, pp. 147–166.

Oreskes, N., Shrader-Frenchette, K. & Belitz, K. (1994). Verification, validation, and confirmation of numeric models in the Earth Sciences. *Science*, **263**, 641–646.

Paola, C., Mullin, J., Ellis, C., Mohrig, D.C., Swenson, J.B., Parker, G., Hickson, T., Heller, P.L., Pratson, L., Syvitski, J., Sheets, B. & Strong, N. (2001). Experimental stratigraphy. *GSAToday*, **11**, 4–9.

Parker, R.S. (1977). Experimental study of basin evolution and it hydrologic implications. Unpublished Ph.D. dissertation, Fort Collins, CO, Colorado State University, 331 pp.

Partridge, T.C. & Maud, R.R. (1987). Geomorphic evolution of southern Africa since the Mesozoic. *South African Journal of Geology*, **90**, 179–208.

Patton, P.C. & Schumm, S.A. (1975). Gully erosion, northwestern Colorado: A threshold phenomenon. *Geology*, **3**, 88–90.

Pavich, M.J., Leo, G.W., Obermeier, S.F. & Estabrook, J.R. (1989). Investigations of the characteristics, origin, and residence time of the upland mantle of the Piedmont of Fairfax County, Virginia: U.S. Geological Survey Professional Paper 1352, 114 pp.

Pazzaglia, F.J. & Brandon, M.T. (1996). Macrogeomorphic evolution of the post-Triassic Appalachian mountains determined by deconvolution of the offshore basin sedimentary record. *Basin Research*, **8**, 255–278.

Pazzaglia, F.J. & Knuepfer, P.L.K. (2001). The steady state orogen: Concepts, field observations, and models. *American Journal of Science*, **302**, 313–512.

Pederson, J., Smith, G.A. & Pazzaglia, F.J. (2001). Comparing the modern, Quaternary, and Neogene records of climate-controlled hillslope sedimentation in Southeast Nevada. *Geological Society of America Bulletin*, **113**, 305–319.

Pederson, J., Pazzaglia, F.J. & Smith, G.A. (2000). Ancient hillslope deposits; missing links in the study of climate controls on sedimentation. *Geology*, **28**, 27–30.

Penck, W. (1924). Die Morphologische Analyse (Morphological Analysis of Landforms): J. Engelhorn's Nachfolger, Suttgart, 283 p. English translation by Czech, H. & Boswell, K.C., London, 1953, St. Martin's Press, New York, 429 pp.

Pitman, W.C. & Golovchenko, X. (1991). The effect of sea level changes on the morphology of mountain belts. *Journal of Geophysical Research*, **96**, 6879–6891.

Powell, J.W. (1875). *Exploration of the Colorado River of the West*. Washington, D.C., U.S. Government Printing Office, 291 pp.

Ransome, F. L. (1896). *The great valley of California: A criticism of the theory of isostasy*. University of California Publications in Geological Sciences, pp. 371–428.

Ritter, J.B. & Gardner, T.W. (1993). Hydrologic evolution of drainage basins disturbed by surface mining, central Pennsylvania. *Geological Society of America Bulletin*, **105**, 101–115.

Ritter, J.B., Miller, J. & Husek-Wulforst, J. (2000). Environmental controls on the evolution of alluvial fans in Buena Vista Valley, north central Nevada, during late Quaternary time. *Geomorphology*, **36**, 63–87.

Roering, J.J., Kirchner, J.W. & Dietrich, W.E. (1999). Evidence for nonlinear diffusive sediment transport on hillslopes and implications for landscape morphology. *Water Resources Research*, **35**, 853–870.

Roering, J.J., Kirchner, J.W., Sklar, L.S. & Dietrich, W.E. (2001). Hillslope evolution by nonlinear creep and landsliding: An experimental study. *Geology*, **29**, 143–146.

Romey, W.D. (1982). Earth in my oatmeal. *EOS Transactions*, **63**, 162.

Rosenbloom, N.A. & Anderson, R.S. (1994). Hillslope and channel evolution in a marine terraced landscape, Santa Cruz, California. *Journal of Geophysical Research*, **99**, 14,013–14,029.

Rowley, D.B., Pierrehumbert, R.T. & Currie, B.S. (2001). A new approach to stable isotope-based paleoaltimetry; implications for paleoaltimetry and paleohypsometry of the High Himalaya since the late Miocene. *Earth and Planetary Science Letters*, **188**, 253–268.

Ruxton, B.P. & McDougall, I. (1967). Denudation rates in northeast Paupa from potassium-argon dating of lavas. *American Journal of Science*, **265**, 545–561.

Schumm, S.A. & Chorley, R.J. (1964). The fall of threatening rock. *American Journal of Science*, **262**, 1041–1054.

Schumm, S.A. & Chorley, R.J. (1966). Talus weathering and scarp recession in the Colorado Plateau. *Zeitschrift für Geomorphologie*, **10**, 11–36.

Schumm, S.A. & Rea, D.K. (1995). Sediment yield from disturbed earth systems. *Geology*, **23**, 391–394.

Schumm, S.A. (1963). Disparity between present rates of denudation and orogeny: U.S. Geological Survey Professional Paper 454-H.

Schumm, S.A. (1956). The role of creep and rain-wash on the retreat of badland slopes. *American Journal of Science*, **254**, 693–706.

Schumm, S.A. & Licthy, R.W. (1965). Time, space, and causality in geomorphology. *American Journal of Science*, **263**, 110–119.

Schumm, S.A., Mosley, M.P. & Weaver, W.E. (1987). *Experimental fluvial geomorphology*. Wiley Interscience, New York, 413 pp.

Simons, M. (1962). The Morphological Analysis of Landforms: A new review of the work of Walter Penck (1888–1923). *Transactions of the Institute of British Geographers*, **31**, 1–14.

Sklar, L.S. & Dietrich, W.E. (2001). Sediment and rock strength controls on river incision into bedrock. *Geology*, **29**, 1087–1090.

Slingerland, R. & Furlong, K.P. (1989). Geodynamic and geomorphic evolution of the Permo-Triassic Appalachian Mountains. *Geomorphology*, **2**, 23–37.

Slingerland, R.S., Harbaugn, J.W. & Furlong, K.P. (1994). *Simulating clastic sedimentary basins*. Englewood Cliffs, New Jersey, Prentice-Hall, 220 pp.

Small, E.E. & Anderson, R.S. (1998). Pleistocene relief production in Laramide mountain ranges, western United States. *Geology*, **26**, 123–126.

Smart, J.S. & Moruzzi, V.L. (1971). Random-walk model of stream network development. *Journal of Research Development, IBM*, **15**, 197–203.

Smith, G.A. (1994). Climatic influences on continental deposition during late-stage filling of an extensional basin, southeastern Arizona. *Geological Society of America Bulletin*, **106**, 1212–1228.

Snyder, N.P., Whipple, K.X., Tucker, G.E. & Merritts, D.J. (in press). Channel response to tectonic forcing: Analysis of stream morphology and hydrology in the Mendocino triple junction region, northern California. *Geomorphology*.

Snyder, N.P., Whipple, K.X., Tucker, G.E. & Merritts, D.J. (2000). Landscape response to tectonic forcing digital elevation model analysis of stream profiles in the Mendocino triple junction region, Northern California. *Geological Society of America Bulletin*, **112**, 1250–1263.

Stark, C.P. & Stark, G.J. (2001). A channelization model of landscape evolution. *American Journal of Science*, **301**, 486–512.

Stephenson, R. & Lambeck, K. (1985). Erosion-isostatic rebound models for uplift: An application to south-eastern Australia. *Geophysical Journal of the Royal Astronomical Society*, **82**, 31–55.

Stephenson, R. (1984). Flexural models of continental lithosphere based on long-term erosional decay of topography. *Geophysical Journal of the Royal Astronomical Society*, **77**, 385–413.

Stock, J.D. & Montgomery, D.R. (1996). Estimating paleorelief from detrital mineral age ranges. *Basin Research*, **8**, 317–327.

Summerfield, M.A. (1991). *Global Geomorphology*. Longman Scientific and Technical. Co-published by John Wiley and Sons Inc., New York, 537 pp.

Summerfield, M.A. & Hulton, N.J. (1994). Natural controls of fluvial denudation rates in major world drainage basins. *Journal of Geophysical Research*, **99**, 13,871–13,883.

Taylor, T.J. (1851). *An inquiry into the operation of running streams and tidal waters*. London, Longman, Brown, Green and Longmans, 119 pp.

Thomson, J. (1879). On the flow of water round river bends. *Proceedings of the Institute of Mechanical Engineers* (August 6).

Thompson, D. & Wohl, E. (1998). Flume experimentation and simulation of bedrock channel processes. *In*: Tinkler, K. & Wohl, E. (Eds), *Rivers Over Rocks*. American Geophysical Union Monograph 107, Washington, D.C., pp. 297–296.

Tomkin, J.H. & Braun, J. (2002). The influence of alpine glaciation on the relief of tectonically active mountain belts. *American Journal of Science*, **302**, 169–190.

Tucker, G.E., Catani, F., Rinaldo, A. & Bras, R.L. (2001). Statistical analysis of drainage density from digital terrain data. *Geomorphology*, **36**, 187–202.

Tucker, G.E. & Bras, R.L. (2000). A stochastic approach to modeling the role of rainfall variability in drainage basin evolution. *Water Resources Research*, **36**, 1953–1964.

Tucker, G. & Slingerland, R.L. (1994). Erosional dynamics, flexural isostasy, and long-lived escarpments: A numerical modeling study. *Journal of Geophysical Research*, **99**(B6), 12,229–12,243.

Tucker, G.E. & Slingerland, R.L. (1997). Drainage basin response to climate change. *Water Resources Research*, **33**, 2031–2047.

Twidale, C.R. (1976). On the survival of paleoforms. *American Journal of Science*, **276**, 77–95.

Twidale, C.R. (1985). Old land surfaces and their implications for models of landscape evolution. *Revue de Géomorphology Dynamique*, **34**, 131–147.

Twidale, C.R. (1997). The great age of some Australian landforms: Examples of, and possible explanations for, landscape longevity. In: Widdowson, M. (Ed.), *Palaeosurfaces: Recognition, Reconstruction, and Palaeoenvironmental Interpretation*. Geological Society of London Special Publication 120, pp. 13–23.

Wells, S.G., McFadden, L.D. & Dohrenwend, J.C. (1987). Influence of late Quaternary climatic changes on geomorphic and pedogenic processes on a desert piedmont, eastern Mojave Desert, California. *Quaternary Research*, **27**, 130–146.

Whipple, K.X. (2001). Fluvial landscape response time: How plausible is steady-state denudation? *American Journal of Science*, **301**, 313–325.

Whipple, K.X. & Tucker, G.E. (1999). Dynamics of bedrock channels in active orogens: Implications for the height limits of mountain ranges, landscape response time scales, and research needs. *Journal of Geophysical Research*, **104**, 17,661–17,674.

Whipple, K., Kirby, E. & Brocklehurst, S. (1999). Geomorphic limits to climate-induced increases in topographic relief. *Nature*, **401**, 39–43.

Whipple, K.X., Hancock, G.S. & Anderson, R.S. (2000). River incision into bedrock: Mechanics and relative efficacy of plucking, abrasion and cavitation. *Geological Society of America Bulletin*, **112**, 490–503.

Willett, S.D., Slingerland, R. & Hovius, N. (2001). Uplift, shortening, and steady state topography in active mountain belts. *American Journal of Science*, **301**, 455–485.

Willett, S.D. (1999). Orogeny and orography: The effects of erosion on the structure of mountain belts. *Journal of Geophysical Research*, **104**(B12), 28,957–28,981.

Willett, S.D. & Brandon, M.T. (2002). On steady states in mountain belts. *Geology*, **30**, 175–178.

Willgoose, G., Bras, R.L. & Rodriguez-Iturbe, I. (1991). Results from a model of river basin evolution. *Earth Surface Processes and Landforms*, **16**, 237–254.

Wischmeier, W.H. & Smith, D.D. (1965). *Predicting rainfall erosion losses from cropland, a guide to conservation planning*. U.S. Department of Agriculture, Agricultural Handbook, Washington, D.C., 58 pp.

Würm, A. (1935). Morphologische Analyse und Experiment Schichtstufenlandschaft. *Zeitschrift für Geomorphologie*, **9**, 1–24.

Würm, A. (1936). Morphologische Analyse und Experiment Hangentwicklung, Ernebnung, Piedmonttreppen. *Zeitschrift für Geomorphologie*, **9**, 58–87.

Yochelson, E.L. (Ed.) (1980). *The scientific ideas of G.K. Gilbert*. Geological Society of America Special Paper 183, 148 pp.

Young, A. (1960). Soil movement by denudational processes on slopes. *Nature*, **188**, 120–122.

Zeitler, P.K., Meltzer, A.S., Koons, P.O., Craw, D., Hallet, B., Chamberlain, C.P., Kidd, W.S.F. & Park, S.K. (2001). Erosion, Himalayan geodynamics, and the geomorphology of metamorphism. *GSA Today*, **11**, 4–9.

Zhang, P., Molnar, P. & Downs, W.R. (2001). Increased sedimentation rates and grain sizes 2–4 Myr ago due to the influence of climate change on erosion rates. *Nature*, **410**, 891–897.

Zhou, S. & Stüwe, K. (1994). Modeling of dynamic uplift, denudation rates, and thermomechanical consequences of erosion in isostatically compensated mountain belts. *Journal of Geophysical Research*, **99**, 13,923–13,940.

Eolian sediments

Alan J. Busacca[1], James E. Begét[2], Helaine W. Markewich[3], Daniel R. Muhs[4], Nicholas Lancaster[5] and Mark R. Sweeney[6]

[1] *Department of Crop and Soil Sciences, Washington State University, Pullman, WA 99164-6420, USA; busacca@wsu.edu*
[2] *Geophysical Institute and Department of Geology and Geophysics, University of Alaska, Fairbanks, AK 99775-5780, USA; ffjeb1@uaf.edu*
[3] *U.S. Geological Survey, 3039 Amwiler Road, Peachtree Business Center, Atlanta, GA 30360–2824, USA; helainem@usgs.gov*
[4] *U.S. Geological Survey, MS 980, Box 25046, Denver Federal Center, Denver, CO 80225-0046, USA; dmuhs@usgs.gov*
[5] *Division of Earth and Ecosystem Sciences, Desert Research Institute, UCCSN, 2215 Raggio Parkway, Reno, NV 89512, USA; nick@dri.edu*
[6] *Department of Geology, Washington State University, Pullman, WA 99164-2812, USA; sweeney@wsunix.wsu.edu*

Introduction

Quaternary loess and eolian sand cover vast areas of the United States (Fig. 1). In the past 35 years, there has been a keen interest in eolian deposits. Many studies have given new perspective on the ages, origin, and paleoclimatic significance of these deposits since the publication of the "Quaternary of the United States" in 1965 (Wright & Frey, 1965). Several of the advancements in eolian studies and new insights to regional eolian systems are outlined in this chapter.

The study of sand dunes spans dry mid-continent settings to more humid coastal areas, and major progress has been made in understanding dunes in all settings. The study by Hunter (1977) of modern sand dunes led to the identification of sedimentary structures that are distinctly eolian. This was a key advance in allowing the interpretation of ancient eolian sandstones based on the modern record, interpretations that in the past were met with much controversy. Utility of dune strata in paleoclimatic interpretation has been revolutionized by luminescence dating techniques. Luminescence ages derived from a dune-interdune sequence in Wyoming were paired with radiocarbon ages that not only validated the luminescence technique but also allowed one of the first studies that detailed paleoclimatic reconstruction in an eolian setting (Stokes & Gaylord, 1993). More recently, much work has been done to advance what is known about sedimentary dynamics and geomorphology related to sand dunes (Goudie *et al.*, 1999; Pye & Tsoar, 1990).

The recognition of some loess in North America as not only glacially derived but as "desert" or "non-glaciogenic" has expanded the scope of generation mechanisms of loessial silt and of eolian processes in general (Aleinikoff *et al.*, 1998, 1999; Mason, 2001; Tsoar & Pye, 1987). Controls on the distribution patterns of loess have been defined as influenced by variable winds and proximity to source (Handy, 1977), vegetation cover effects (Pye, 1996; Tsoar & Pye, 1987) and various controls exerted by topography (Mason *et al.*, 1999; Pye, 1996). Recent interest has been sparked in regional and global dust emissions related to loess formation as well as atmospheric loading that affects solar insolation and climate (Kohfeld & Harrison, 2001). The role of dust as additions to non-loessial soils has also generated interest in global biogeochemical cycles (Chadwick *et al.*, 1999).

In a more regional context of the United States, work on eolian deposits of the Great Plains (GP) has added rich detail about paleoenvironments, sediment provenance, and stratigraphy of sand deposits to a record where the basic stratigraphy of the loess component was already well studied by 1965. In Alaska, much progress has been made in understanding the significance of eolian sediments since the pioneering work of Péwé (1955), important because loess is the most widespread surficial deposit in the state. Recent work on the geochronology, stratigraphy, paleoecology, paleoclimatology, geophysical properties, and sedimentology of loess deposits in unglaciated central Alaska has revealed loess successions as old as 3.5 million years that contain a record of glacial-interglacial cycles similar to loess deposits in China and Europe.

In the deserts of the southwestern U.S., very little had been published up to 1965 about the history and paleoclimatic significance of eolian deposits, apart from pioneering investigations in Arizona by J.T. Hack (1941). Building on reconnaissance investigations of eolian deposits (e.g. Evans, 1961; Eymann, 1953), H.T.U. Smith was the first to recognize the significance and long history of past eolian activity in the Mojave Desert (Smith, 1967), and recent advances in dating have allowed periods of dune activity to be reconstructed.

In some cases, such as the Palouse and Snake River Plain of the Pacific Northwest U.S. (PNW) and the Atlantic Coastal Plain (ACP), very little was known in 1965 and progress has been made across all fronts, beginning with the pioneering work of Roald Fryxell (Richmond *et al.*, 1965). In the Palouse eolian system, for example, it is now thought that loess deposition occurred mainly during some interglaciations whereas soil development predominated during glaciations, an observation that counters the record documented in many other loess areas world wide. Ultimately, each eolian system has been driven by different controls and there are few synchronous units or soils that span the continent. Each region responded differently to glacial/interglacial conditions as a result of variable climate and vegetative cover, sediment character, and supply.

In the future, eolian studies aim to tackle issues such as the interaction between juxtaposed sand dune and loess systems like those in the Great Plains and the Palouse of

DEVELOPMENTS IN QUATERNARY SCIENCE
VOLUME 1 ISSN 1571-0866
DOI:10.1016/S1571-0866(03)01013-3

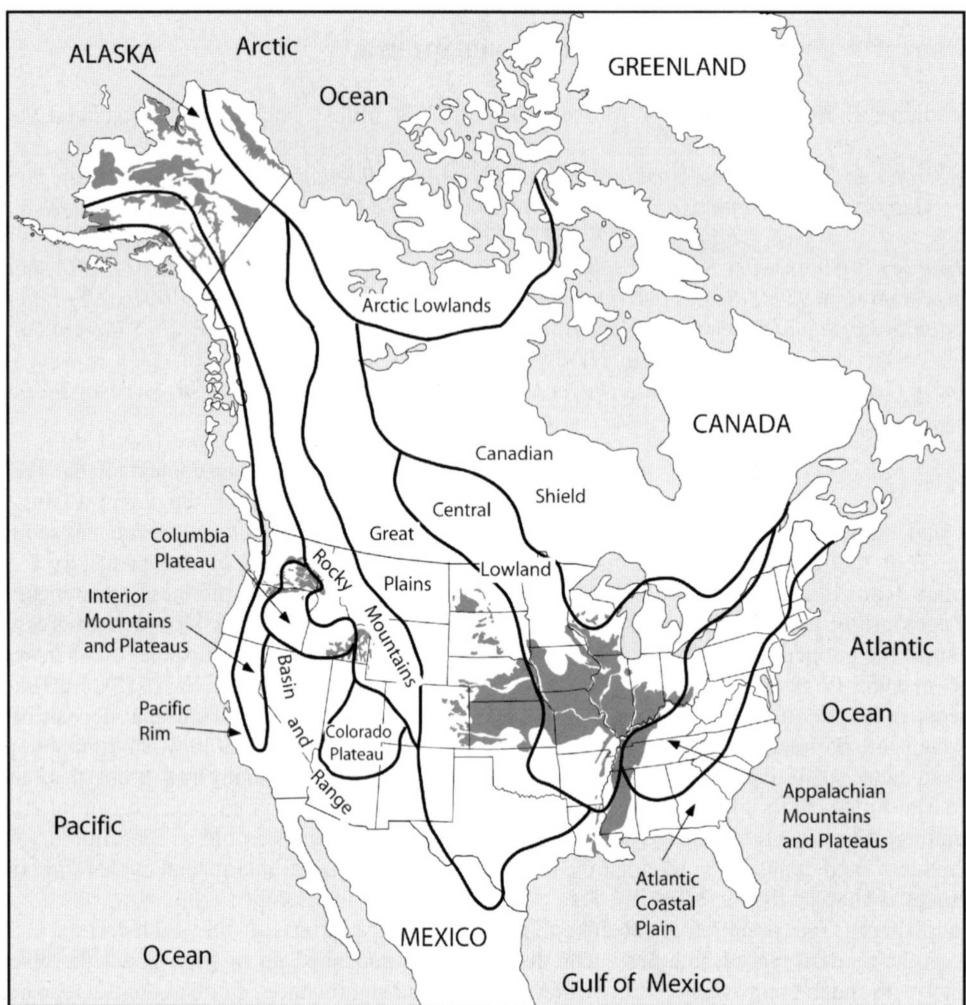

Fig. 1. Map showing physiographic provinces of North America and principal areas of loess discussed in text. Areas of eolian sand and some areas of loess not discussed or too small to show at the scale of this map not shown.

the Pacific Northwest. Stratigraphy of eolian successions and relation to paleoclimate fluctuations will also become better understood as luminescence techniques are more widely applied. Field studies paired with enhanced computer simulations will also aid in modeling eolian response to changing circulation patterns and climate fluctuations.

This review is selective and not exhaustive. We have attempted to highlight major advances since 1965 in selected areas; necessarily, a number of smaller areas of eolian deposits in the United States, such as areas of coastal dunes, are not discussed in this chapter.

Late Quaternary Loess and Eolian Sand in the Central U.S.

The majority of new findings for the central U.S. reported over the past four decades have been on Peoria Loess, which is the primary focus of this section. In this section, "Central Lowland" refers to states east and north of the Missouri River, within the drainage of the upper Mississippi River (Fig. 1). "Great Plains" refers to states east of the Rocky Mountains but west of ~95° west longitude.

Loess in the Central United States

Stratigraphy and Chronology

Loess is extensive across the central United States (Fig. 2), where four middle-to-late Quaternary loess units have been identified and correlated. These units are, from oldest to youngest: (1) the marine oxygen isotope stage 6 (MIS-6) Loveland Loess, (2) the mid-Wisconsin (MIS 3) Roxana Silt (in the Central Lowland) and Gilman Canyon Formation (in the Great Plains; about 30,000–50,000 yr ago), (3) the late Wisconsin (MIS 2) Peoria Loess, and (4) the Holocene (MIS 1) Bignell Loess, which is found only in the Great Plains. Radiocarbon ages indicate that in places the Gilman Canyon Formation could be as old as ~40,000 yr (Muhs *et al.*, 1999)

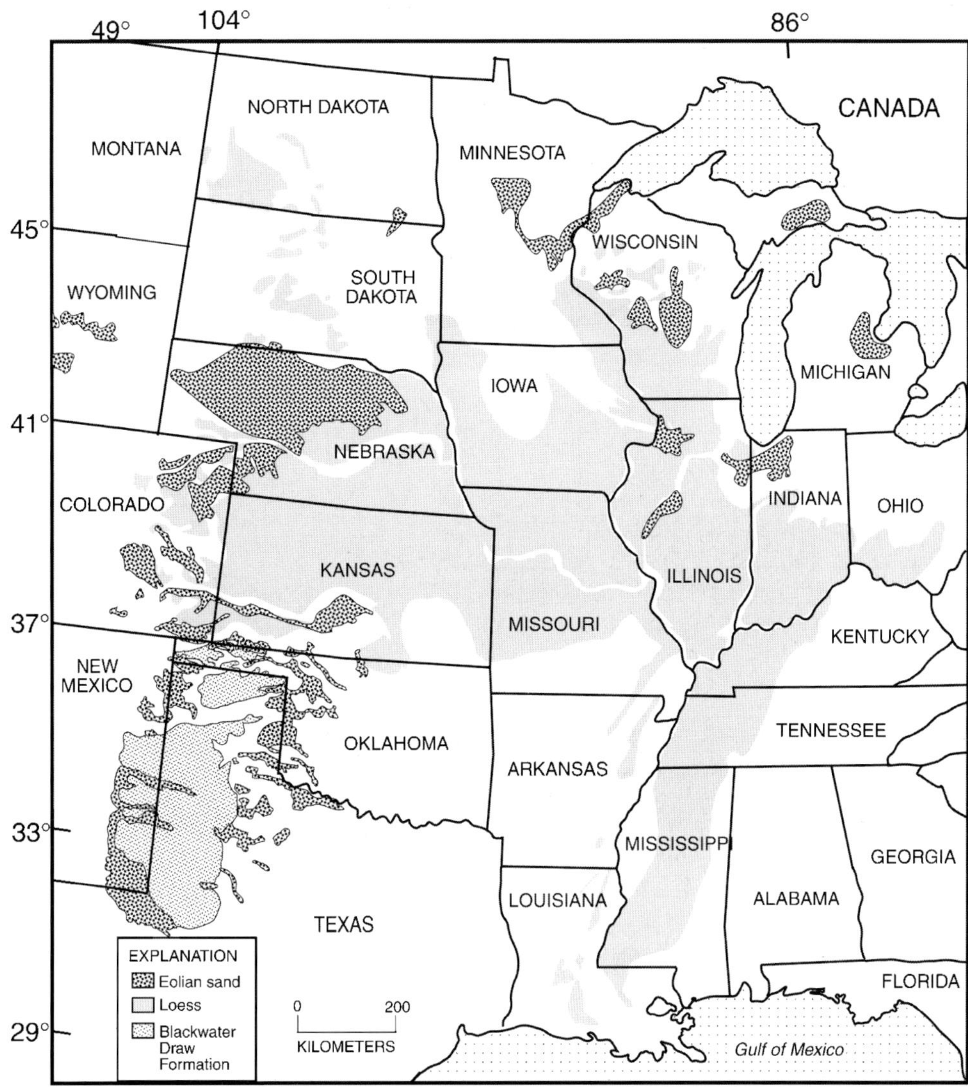

Fig. 2. Map of the central United States showing the distribution of eolian sand (from compilations in Muhs & Holliday, 1995, 2001), loess (modified from compilation in Muhs & Bettis, 2000), and the Pleistocene Blackwater Draw Formation, an eolian sheet sand (from Holliday, 1989).

and the equivalent Roxana Silt could be as old as ~55,000 yr (Leigh & Knox, 1994). Stratigraphic evidence indicates that the Gilman Canyon Formation could include more than one period of loess deposition (Maat & Johnson, 1996; Muhs *et al.*, 1999).

Despite the main focus on Peoria Loess over the past four decades, there have been some important new findings on the ages of pre-Peoria loesses in the central United States. In the Great Plains, thick exposures indicate that a long loess record, perhaps going back to the early Pleistocene, is present (e.g. Jacobs & Knox, 1994; May *et al.*, 1995, pp. 16–17). Possible early and middle Pleistocene loesses are undated, but approximate thermoluminescence ages have been measured for some pre-Late Pleistocene loess units. For example, although thermoluminescence (TL) ages older than ~100,000 yr are considered to be of questionable accuracy, estimates of ~160,000–100,000 yr for Loveland loess in Iowa and Nebraska support the notion that this unit dates to the penultimate glacial period (MIS 6) (Forman *et al.*, 1992; Maat & Johnson, 1996).

Many radiocarbon ages reported in the past 30 years demonstrate conclusively that Peoria Loess dates to the last glacial maximum (MIS 2), whether near glaciated terrain in the Central Lowlands or distant from the Laurentide Ice Sheet (LIS) on the Great Plains (Fig. 3). In the Central Lowland, radiocarbon ages of plant macrofossils, snails and organic extractions indicate that Peoria Loess deposition began sometime after ~25,000 ^{14}C yr B.P., and probably continued until about 11,000 ^{14}C yr B.P. (Curry & Follmer, 1992; Forman *et al.*, 1992; Grimley *et al.*, 1998; Leigh & Knox, 1994; Ruhe, 1969; Ruhe *et al.*, 1971; Wang *et al.*, 2000). In the Great Plains, similar ages have been reported (Johnson & Willey,

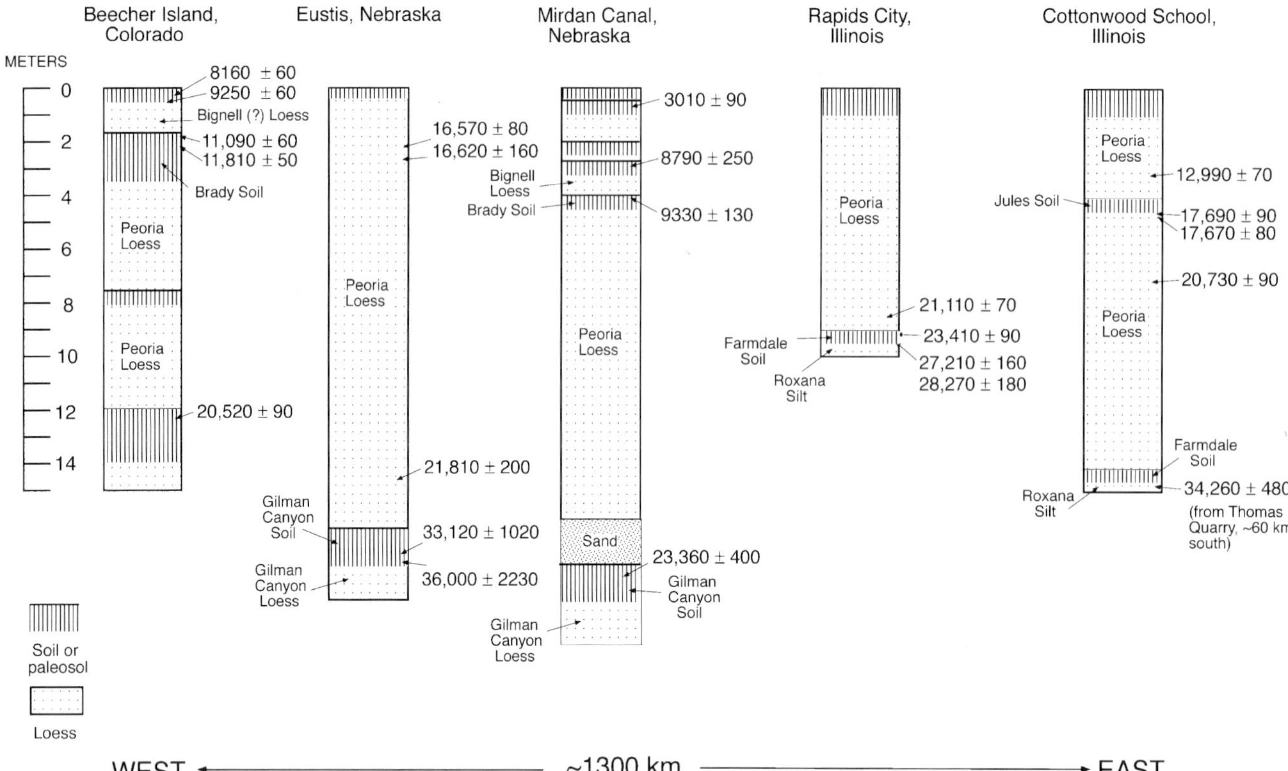

Fig. 3. Stratigraphy and radiocarbon ages (^{14}C yr B.P.) of recently studied, representative late Quaternary loess sections in the central United States. Beecher Island from Muhs et al. (1999; Eustis from Maat & Johnson (1996); Mirdan Canal from Mason & Kuzila (2000); Rapids City from Muhs et al. (2001a) and Cottonwood School from Grimley et al. (1998).

2000; Maat & Johnson, 1996; Martin, 1993; Muhs *et al.*, 1999; Wells & Stewart, 1987). The radiocarbon chronology agrees well with direct TL ages of the Peoria Loess itself, both in the Central Lowland and the Great Plains (Forman *et al.*, 1992; Maat & Johnson, 1996; Pye *et al.*, 1995).

During the last glacial maximum (MIS 2), loess deposition rates were neither spatially nor temporally constant. Ruhe *et al.* (1971) showed that maximum-limiting radiocarbon ages for Peoria Loess in Iowa become younger with distance to the east of the Missouri River. This suggests that while all loess deposition took place broadly within the last glacial period, areas close to glaciofluvial source valleys received sediment earlier than those farther away. Alternatively, the apparently younger ages of the basal zone at distal sites could reflect contamination by younger carbon if it remained in the rooting zone of the surface soil for a longer period than the basal zone at thick proximal sites. It is possible that dust deposition was widespread early on in Peoria Loess accumulation, but slow accumulation rates resulted in compressed sections at distal sites, thus making it difficult to resolve its maximum age (J. Mason, pers. Comm., 2003). In addition, loess sedimentation was not constant through time at any specific site. In Iowa, Illinois and Indiana, weakly developed paleosols occur within Peoria Loess, indicating brief periods of little or no sedimentation (Hayward & Lowell, 1993; Ruhe *et al.*, 1971; Wang *et al.*, 2000); no regional correlations of these have been attempted.

Sedimentology and Mineralogy

Studies from several localities also show that composition of Peoria Loess has varied through time. Frye *et al.* (1968), McKay (1979) and Grimley *et al.* (1998) showed that Peoria Loess in Illinois has distinct zonations of clay mineral types, dolomite content and magnetic susceptibility. These zonations reflect changes in source sediment provenance due to fluctuations of major lobes of the LIS Grimley (2000). In western Iowa, Muhs & Bettis (2000) showed that Peoria Loess has three distinct zones, based on particle size, mineralogy and geochemistry. The lower and middle zones reflect slow and rapid sedimentation, respectively, from the Missouri River valley; the upper zone is derived from a combination of local and distal sediment sources.

Provenance studies have shown that much loess in the central United States is nonglaciogenic. Although at small scales loess appears to be a continuous blanket across the region (Fig. 2), geochemical studies show that at subregional scales, loess has distinctly different compositions. Loess in Illinois is much higher in carbonate minerals (calcite and dolomite) than loess in Nebraska (Muhs & Bettis, 2000). The high-carbonate loess in Illinois is a direct reflection of its glaciogenic source, because the LIS traversed Paleozoic carbonate terrains. In contrast, Nebraska was much farther away from the LIS. Pb-isotopic compositions of silt-sized K-feldspars and U-Pb ages of individual silt-sized zircons show that the volcaniclastic siltstone of the Tertiary White River Group, exposed well away from the margins of the LIS, is a significant source of loess in Nebraska (Aleinikoff *et al.*, 1998). Transport directions inferred from thickness trends also show that much of Nebraska Peoria Loess cannot be glaciogenic (Mason, 2001). In still other areas, there may have been both glaciogenic and nonglaciogenic sources for loess. Peoria Loess in Minnesota and eastern Iowa was produced at least in part from erosion of pre-Wisconsin tills in a periglacial environment (Mason *et al.*, 1994). In eastern Colorado, the White River Group (a nonglacial source) alternated with sediments derived from the Front Range (probably from glaciofluvial sources) as a loess source during MIS 2 (Aleinikoff *et al.*, 1999). Thus, loess in the Central Lowland and Great Plains is not always simply a glacially derived sediment. Climatic conditions during a glacial period – aridity, sparsely vegetated landscapes and strong winds – may be critical factors in loess generation.

Paleoenvironments and Paleoclimate

Loess in the Central U.S. contains valuable records of paleoclimate. Although plant and large animal fossils are rare in loess deposits, terrestrial gastropods are common and yield important information about plant cover during loess fall. Ruhe (1969), Wells & Stewart (1987) and Rousseau & Kukla (1994) showed that loess deposits in Iowa, Kansas, and Nebraska all contain northern or western gastropods that today have habitats associated with boreal forest in Canada or coniferous forest in the Rocky Mountains. Carbon isotope compositions of loess and paleosol organic matter have yielded valuable information on the past distribution of C3 (forest or cool grassland) vs. C4 (warm grassland) plant communities in central U.S. loess sequences (Johnson & Willey, 2000; Muhs *et al.*, 1999; Wang *et al.*, 2000). The Bignell Loess may record mid-Holocene droughts on the Great Plains that generated large dust storms. A very complete record of mid-Holocene droughts, matched by a detailed pollen record of a dry, prairie-dominated period (Fig. 4), is found in eolian sediments accumulated at a high rate in Elk Lake, Minnesota (Dean, 1993).

Because loess is deposited by wind, it is a direct record of atmospheric circulation. Loess-derived paleowind reconstructions, summarized by Muhs & Bettis (2000) and Mason (2001), were dominantly from the west or northwest during Peoria time, as indicated by thickness, particle size and geochemical trends. In contrast, some computer-model simulations of atmospheric circulation show that dominant winds over the central United States during the last glacial period were from the northeast or east, due to the presence of a strong glacial anticyclone, in both winter and summer (Bartlein *et al.*, 1998; COHMAP Members, 1988). Recent models show seasonal changes in circulation patterns during the last glacial maximum, including easterly wind anomalies that were not as strong or prevalent as those predicted in earlier models (Whitlock *et al.*, 2001).

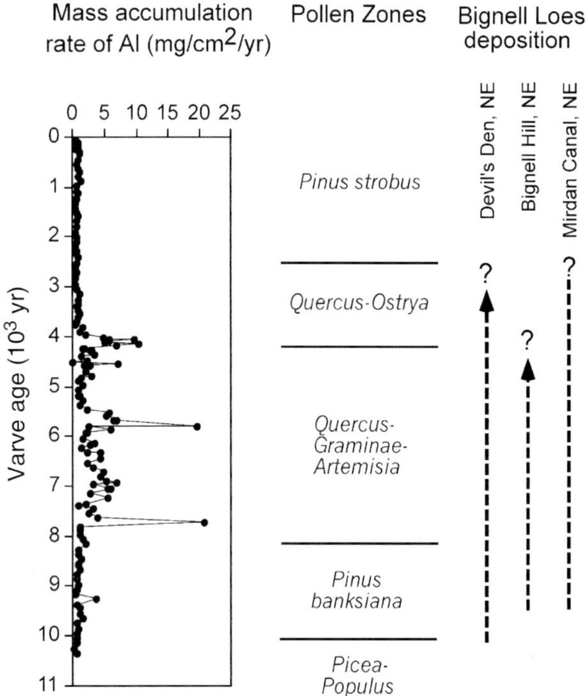

Fig. 4. Eolian record of mid-Holocene droughts in sediments of Elk Lake, Minnesota, based on mass accumulation rates of Al, which proxies for eolian particles (data from Dean, 1993). Shown for comparison are calendar-year ages of the possible accumulation periods of Holocene Bignell Loess from three localities in Nebraska (age data from Pye et al., 1995; Maat & Johnson, 1996; Johnson & Willey, 2000; Mason & Kuzila, 2000).

Eolian Sand in the Central United States

Stratigraphy and Chronology

The most extensive eolian sands in North America are in the central and southern Great Plains (Fig. 2). Landforms are commonly parabolic dunes and sand sheets (Arbogast, 1996; Holliday, 2001; Muhs & Holliday, 2001; Muhs *et al.*, 1996, 1997). In the Nebraska Sand Hills, however, a wide variety of landforms are found, including sand sheets, parabolic dunes, linear dunes, dome-like dunes, and barchanoid ridges (Ahlbrandt & Fryberger, 1980; Swinehart, 1990). An older sheet sand, the stabilized Blackwater Draw Formation, occurs in Oklahoma, Texas, and New Mexico and may span much of the Pleistocene (Holliday, 1989).

It was long assumed that dunes on the Great Plains, particularly the Nebraska Sand Hills, were last active during MIS 2 (see reviews in Muhs *et al.*, 1997; Wright *et al.*, 1985). There is, however, only limited evidence that eolian sand sheets and dunes were deposited then in Nebraska and Colorado (Forman *et al.*, 1995; Loope *et al.*, 1995; Madole, 1995; Muhs *et al.*, 1996; Swinehart & Diffendal, 1990; Wright *et al.*, 1985). One of the more remarkable findings of the past two decades has been that many dunes of the Great Plains are of late Holocene age. Both radiocarbon and

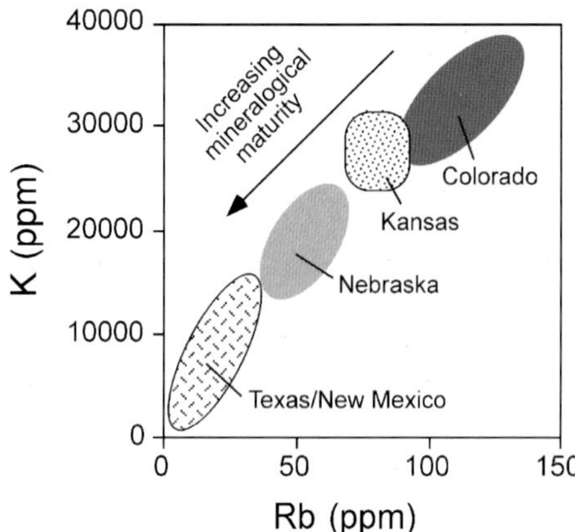

Fig. 5. Plots of K and Rb, which proxy for K-feldspar, in Holocene eolian sands of Great Plains dune fields, showing varying degrees of mineralogical maturity. Data from Muhs et al. (1996, 1997), Arbogast & Muhs (2000), and Muhs & Holliday (2001).

luminescence methods demonstrate that eolian sands over much of the Great Plains have been active in the past 3000 yr (Ahlbrandt *et al.*, 1983; Arbogast, 1996; Forman *et al.*, 1995; Holliday, 2001; Loope *et al.*, 1995; Madole, 1995; Muhs *et al.*, 1996, 1997; Stokes & Swinehart, 1997; Swinehart & Diffendal, 1990). These new observations show that Great Plains dune fields can be active under essentially modern climatic conditions. Furthermore, late Holocene sand deposition was episodic, as shown by the presence of multiple paleosols.

Sedimentology and Mineralogy

It was previously asserted that Great Plains dune fields were all derived from a single source, the late Tertiary Ogallala Group rocks, found over much of the region (Lugn, 1968). The composition of eolian sands in the Great Plains is now known to be highly variable, which in part reflects derivation from diverse sediment sources, such as alluvium, bedrock, and reworking of older eolian sheet sands (Arbogast, 1996; Muhs *et al.*, 1996, 1997; Muhs & Holliday, 2001; Swinehart, 1990). In addition, mineralogical and geochemical data show that dune fields also vary in their degree of mineralogical maturity (Fig. 5), or amount of quartz vs. feldspars (Arbogast & Muhs, 2000; Muhs & Holliday, 2001; Muhs *et al.*, 1997). A viable explanation for quartz enrichment in dunes compared to their source sediments is destruction of feldspars by ballistic impacts under strong winds (Dutta *et al.*, 1993). If this is the mechanism for mineralogical maturation, it is an important measure of the cumulative duration of eolian activity in a dune field's history.

Paleoenvironments and Paleoclimate

The history of eolian activity vs. stability is a valuable record of Holocene paleoclimate in the Great Plains. Although dunes in this region may become active under interglacial climate conditions, historical evidence suggests that dune reactivation is probably induced by a loss of vegetation cover, in turn related to decreased effective moisture (Muhs & Holliday, 1995). In addition, because eolian sediments are deposited by wind, they are direct records of circulation in the lower atmosphere. Late Holocene dunes in Nebraska, Colorado, New Mexico and Texas have, for the most part, similar orientations to modern sand-transporting winds (Madole, 1995; Muhs & Holliday, 2001; Muhs *et al.*, 1996, 1997, 2000a, b; Swinehart, 1990). These observations suggest that during droughts of the late Holocene, overall circulation differed little from that of the present, although there was probably a greater frequency of zonal (westerly) circulation (Muhs *et al.*, 1996). In addition, eolian activity in the Great Plains is now known to have affected other geomorphic systems. Studies by Loope *et al.* (1995) and Muhs *et al.* (2000) have shown that dune movement can create sand dams on rivers and streams. Eolian sand dams were generated in Holocene and possibly late glacial time in Nebraska.

Eolian Sediments in the Lower Mississippi Valley and the Southeastern U.S.

Loess in the Lower Mississippi Valley

Lyell (1847) is credited with first recording observations on the 20–30 m high loess bluffs along the Mississippi River in the lower Mississippi Valley (LMV), that part of the Mississippi Valley below the confluence of the Mississippi and Ohio rivers. Since then, researchers have addressed the origin, silt sources, and the relations between valley-train breadth and the thickness/extent of upland loess. They also have studied the relations of dominant and subsidiary winds to loess thickness and particle size, and loess chronostratigraphy. Rutledge *et al.* (1996) reviewed the early literature on LMV loess.

Loess Distribution

The Mississippi River received much of the sediment-rich meltwater from the LIS, placing it as one of the longest glacial meltwater rivers in the world. The presence of identifiable loess units along the entire length of the Mississippi alluvial valley (Fig. 6) suggests that this region was directly affected by glacial and proglacial meltwaters and strong west and northwest winds at least four times during the late Quaternary Period. LMV loess units are source-bordering deposits (Figs 2 and 6) with maximum thickness immediately east of the valley (Wascher *et al.*, 1947). Thickness and particle size of the Peoria Loess decrease with increasing distance from the river bluffs (Smith, 1942). The thickness of pre-Peoria loess appears to be related more to geomorphic position and degree of dissection than to distance from source area.

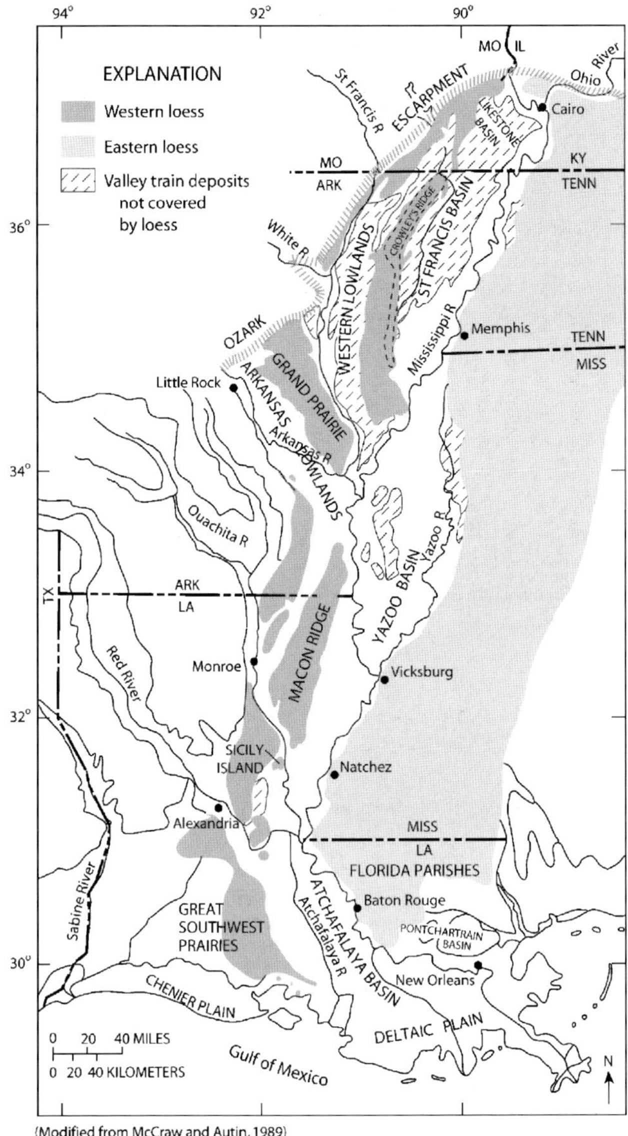

(Modified from McCraw and Autin, 1989)

Fig. 6. Distribution of loess in the lower Mississippi valley. Names on the map are locations referred to in cited references. MO: Missouri; ARK: Arkansas; LA: Louisiana; KY: Kentucky; TENN: Tennessee.

Stratigraphy and Chronology

Most stratigraphic studies of midcontinent loess have relied on a model associating loess deposition with glacial advances and nondeposition and pedogenesis with glacial retreat, resulting in loess/soil units. McKay & Follmer (1985) and Rutledge *et al.* (1990) used stratigraphic position, degree of soil development, and sedimentological/chemical criteria to correlate loess/soil units in the LMV with those in the central U.S. Thermoluminescence, isotopic (^{14}C and ^{10}Be), and amino acid racemization (AAR) age determinations have been used to establish chronostratigraphic correlations (Table 1) (Clark *et al.*, 1989; Johnson *et al.*, 1984; Markewich, 1993, 1994; Markewich *et al.*, 1998b; Millard & Maat, 1994; Miller *et al.*, 1985, 1986; Mirecki & Miller, 1994; Pye & Johnson, 1988; Rodbell *et al.*, 1997).

Dating of loess units in the LMV (north of 34°N, on Crowley's Ridge and east of the Mississippi River) supports correlation of the three youngest LMV loess/soil units with age-equivalent loess/soil units in the central U.S.: Peoria Loess/modern soil, Roxana Silt/Farmdale Soil, and Loveland Loess/Sangamon Soil (Table 1). In the LMV from 34°N south to the Gulf of Mexico, the correlation of the loess/soil sequences is unclear. Several factors contribute to the difficulty in constructing a chronostratigraphic framework. First, pre-Peoria loess units are more eroded, resulting in limited loess preservation and an erosional unconformity that locally represents all or parts of MIS 5 to 2. Second, the few available age and stratigraphic data conflict (see references above and in Table 1). Third, pedogenic characteristics of the Sangamon Soil change from southern Illinois to Mississippi, making identification based solely upon local characteristics virtually impossible without supporting TL or isotopic age data. Fourth, throughout the LMV, loess units older than MIS 6 are poorly preserved or represent only localized deposition.

Sedimentology and Mineralogy

LMV loess units are similar in particle size, mineralogy and chemistry (Table 2). There are differences, however. The Peoria Loess has greater areal extent and thickness ($\leq$20 m) than older units and "rhythmic" variations in coarse silt to fine silt ratio that may reflect bedding. The Roxana Silt has laminae or very thin beds with evidence of nematode- to crawfish-sized bioturbation, five-fold higher background ^{10}Be values, higher dithionite-citrate extractable iron values, more variable magnetic susceptibility values, lower carbonate content, and higher pyroxene, organic carbon and clay contents compared to the Peoria or Loveland Loess. The Sangamon Soil has high chroma/high value, distinct horizonation, strong angular blocky structure, many thick continuous clay films, high clay content, and characteristic clay mineralogy with minor amounts of a randomly interstratified smectite-kaolinite (D.A. Wysocki, unpublished data, 1999). Additional loess/soil data for Crowley's Ridge and Marianna can be found in Miller *et al.* (1985, 1986), Rutledge *et al.* (1990) and Markewich *et al.* (1998b).

Paleoenvironments and Paleoclimate

The chronostratigraphy of LMV loess generally supports the climatic sequence described by Beaudoin *et al.* (1999) for North America. The Crowley's Ridge and Mariana loesses suggest there were pre-Illinoian periods of loess deposition followed by periods of landscape stability, weathering and pedogenesis. Illinoian-age Loveland loess represents a period of at least 50,000 yr of cool arid climate from 190,000 to 120,000 TL yr ago (Table 1), with strong prevailing winds, resulting from advances of the LIS into the central Lowland

Table 1. Stratigraphic models of loess units in the lower Mississippi valley.

Time stratigraphic units (10^3 yr) and marine oxygen isotope stages (MIS)	^{10}Be accumulation periods (10^3 yr)	TL, AAR, and isotopic age estimate ranges (10^3 yr)[a]	Miller *et al.* (1986), Vicksburg Mississippi	Clark *et al.* (1989), Sicily Island	Clark *et al.* (1989), at Vicksburg Mississippi	Rutledge *et al.* (1996), Crowley's Ridge	Rodbell *et al.* (1997), Western Tennessee	Markewich *et al.* (1998a), Crowley's Ridge, Western Tennessee
Late Wisconsin MIS 2 (ca 25–10 ka)		^{14}C, AAR, and TL 25 to 9	Peoria Loess	Peoria Loess	Peoria Loess	Peoria Loess	Peoria Loess	Peoria Loess
Middle to Late Wisconsin		^{14}C 28–25				Farmdale Soil	Farmdale Soil	Farmdale Soil
MIS 3 (ca 60–25 ka)	25–30	^{14}C and TL 56–30	Basal mixed zone		Pre-Peoria loess	Roxana Silt	Roxana Silt	Roxana Silt
		TL 53					Sangamon Soil in Loveland Loess 3	
Early Wisconsin MIS 4 (ca 75–60 ka)		TL 70	Sicily Island paleosol				Unnamed paleosol in Loveland Loess 2	
		TL 95–75	Sicily Island loess with basal mixed zone			Sicily Island paleosol or Sangamon Soil		
MIS 5 (ca 130–70 ka)	50–80		Crowley's Ridge paleosol					Sangamon Soil
						Sicily Island loess or Loveland Loess		
Illinoian MIS 6 (ca 180–130)		TL 190–120	Crowley's Ridge loess		Loveland Silt (?)		Loveland 1	Loveland Loess
MIS 7						Crowley's Ridge paleosol		Crowley's Ridge paleosol?
Pre-Illinoian MIS 8?	50		Marianna	Sicily Island		Crowley's Ridge loess		Crowley's Ridge loess?
Pre-Illinoian						Marianna paleosol Marianna loess		Marianna paleosol? Marianna loess?

[a] Also includes data from Pye & Johnson (1988).

and lowstands of global sea level. An unnamed paleosol in the basal Loveland Loess suggests at least one period of nondeposition during the Illinoian. The Sangamon represents a 50,000–80,000 yr period of interglacial landscape stability and pedogenesis from 130,000 to 70,000 yr B.P. in a warm to hot climate with seasonal or longer periods of drought and temperature and precipitation extremes greater than at present. B horizon hues of the Sangamon Soil range from 2.5 to 7.5 yr. B-horizon structure is very-coarse prismatic (>50 cm) parting to moderate-strong subangular blocky or strong angular blocky. Argillic horizon texture is generally silty clay loam, but clay loam is not uncommon. In the argillic horizons, clay films (cutans) are everywhere present on ped faces. Downvalley trends in the pedologic data suggest that structure, clay films, color differences (between clay films and inside of peds), and mineralogy are interpreted to reflect a continuous gradient of paleoclimate. The presence of interstratified kaolinite-smectite indicates climatic conditions more seasonal than present. Mirroring the present climate of the LMV, down valley pedologic changes in the Sangamon Soil reflect increasing precipitation and temperature and decreasing seasonality with decreasing latitude. The anomalous age of 95,000–75,000 yr for Loveland/Sicily Island loess (Table 1) suggests short periods of aridity and strong prevailing winds during MIS 5.

In the LMV, the uppermost Loveland/Sicily Island and (or) the overlying Roxana silt and Peoria Loess, are everywhere eroded. Sangamon soil characteristics indicate that: (a) no A horizon has been preserved; and (b) the uppermost B horizon differs from locality to locality (such as, all or nearly all the B horizons are present, or the uppermost horizon represents a middle or lower B or a BC horizon). At several localities (generally large borrow pits) the Loveland/Sicily Island is observed to thin and (or) "pinch out" in the distance of a few hundred feet. Locally, where erosion has cut deep channels into or through loess units, the Peoria Loess lies directly on the Loveland/Sicily Island loess or underlying Pleistocene(?) gravel. This eroded/disconformable contact between the Loveland/Sicily Island and overlying Roxana silt or Peoria Loess suggests an unconformity that represents all of MIS 4, and south of 34°N, most of MIS 3.

As discussed in Pavich & Chadwick (this volume), deposition of Roxana Silt coincided with a period of rapid stadial/interstadial cycles (Boyle, 2000; Markewich *et al.*, 1998a). Sedimentological characteristics of Roxana Silt suggest rapidly oscillating climate conditions and slow, nearly continuous loess deposition, primarily in topographic positions favorable for organic carbon accumulation and rapid dissolution of carbonate (Markewich *et al.*, 1998a). Enigmatic features of the Roxana silt in the LMV include its apparent low deposition rate (net maximum thickness of $\leq$3 m in about 30,000 yr) and the lack of significant syndepositional pedogenesis (excepting the Farmdale and a few unnamed, minimally developed, intercalated paleosols). Roxana silt deposition was followed by a 2000–5000 yr cool, wet period of nondeposition characterized by organic carbon accumulation and minimal soil formation. Throughout the LMV, this period is represented by the Farmdale Soil.

The abrupt contact between Roxana silt and Peoria Loess suggests rapid climate change to maximum aridity and strong prevailing winds about 25,000 yr ago. Isotopic ages and the absence of paleosols in the Peoria Loess indicate nearly continuous loess deposition at rates much higher than during Roxana time.

Eolian Sand in the Atlantic Coastal Plain

Recent work on the eolian sand of the Atlantic Coastal Plain (ACP) has focused on determining the ages of inland dunes in Georgia and South Carolina (Brooks *et al.*, 2001; Ivester *et al.*, 2001; Leigh & Ivester, 1998; Zayac *et al.*, 2001). Markewich & Markewich (1994) give an overview of the distribution, relative ages, geomorphology, sedimentology, and paleoenvironment of extensive late Cenozoic dunes/sandsheets in the central and eastern parts of the ACP from North Carolina through Georgia, and a review of the available literature (to that date) for inland dunes of the ACP from New Jersey to Georgia.

Distribution of Sand Dunes and Sand Sheets

Quartz-rich sand dunes and sand sheets in the southeastern ACP are source-bordering deposits east and northeast of streams, rivers, and shallow circular to elliptical depressions called Carolina Bays (Fig. 7). The highest dunes are up to 25 m high and the most continuous dune fields cover 50 km^2. They border drainages, such as the Ohoopee and Satilla rivers, that have headwaters in the ACP. Smaller fields with dunes 3–10 m high are present along drainages such as the Altamaha, Savannah, and PeeDee rivers that have headwaters in the Piedmont and Blue Ridge. Dunes up to 3 m high associated with Carolina Bays are common in the Cape Fear and Black River valleys.

Younger dunes are commonly lower and are open parabolas with surfaces that have closed depressions. Older dunes are typically infilled and are higher, more areally extensive and have smooth upper surfaces. Sandsheets have planar, eastward sloping upper surfaces that extend from the front of a dune field. In many places, younger dunes have migrated across the landscape, infilling or covering older dunes, sandsheets, or the rims of Carolina Bays.

There are no active dunes/sheets in the southeastern ACP today; all are vegetated and stabilized. Most are found in the driest part of the region where present precipitation to evaporation ratio is less than 1.1:1. This suggests that conditions were at least as dry or drier than present, and that winds were stronger than present. Dune axes indicate formation by west and southwest winds. No data are available on the factors that controlled dune size. Data presented in Markewich & Markewich (1994) suggest that at most localities, modern winds do not reach even the lowest wind energies thought to be needed for dune formation (as defined by Fryberger & Dean, 1979).

Table 2. Particle size, mineralogy, and chemistry for some lower mississippi valley loesses.

Loess Unit or Paleosol	Particle Size (wt.%)		Mineralogy			Chemistry (wt.%)		
	Clay	Sand	Clay[a]	Silt[a]	Sand	Fe (DC/DBC)	OC	$CaCO_3$
Peoria Loess	<10	<1	S ≫ K > V	Q > D > KF > OW = PF = H = P (Pye and Johnson show PF > KF)		<0.5	<0.1	10–23
Farmdale paleosol	<20	<3	S > K > V	Q > KF > OW = PF = H = P		0.7–1.0	0.10–0.22	<2
Roxana Silt	<15	<5	S ≥ K > V = MI = Q	Q > KF > OW = PF = H = P		0.7–1.2	0.07–0.10	Tr
Sangamon paleosol	>20 < 35	<2	K > MI = S = G > V = HE = Q	Q > KF > OW = PF = H = P		2–4.5	0.02–0.10	N.D.
Loveland Loess	<20	<1	S ≥ K > MI = V = Q	Q > KF > PF = OW = H = P (Pye and Johnson show KF ≫ PF)	Mn concretions; quartz if subjacent unit is sand	<2	0.01–0.05	N.D.
Sicily Island paleosol	>17 < 33	<5	S > K > MI = V	N.D.		N.D.	N.D.	Locally 10–30 if thick and deeply buried or erosional remnant
Sicily Island loess	<17	<10	S ≫ K ≥ MI ≫ V	N.D.		N.D.	N.D.	N.D.
Crowley's Ridge paleosol	>20 < 50	<10	K ≫ S = MI > V	N.D.		N.D.	N.D.	N.D.
Crowley's Ridge loess	<10	<10	K > S > MI ≫ V	Q > KF > OW = PF = P		N.D.	N.D.	N.D.

Note: Data are from Miller *et al.* (1986) Pye & Johnson (1988), Rodbell *et al.* (1997), and Markewich *et al.* (1998b).
[a] Dolomite, D; goerthite, G; hematite, HE; hornblende, H; kaolinite, K; mica, MI; quartz, Q; plagioclase feldspar, PF; potassium feldspar, KF; other weatherable, OW; pyroxene, P; smectite, S; vermiculite, V; trace, Tr; Iron, Fe; dithionate-Citrate extractable, DC; dithionate-carbonate-citrate extractable, DCB; organic carbon, OC; calcium carbonate, $CaCO_3$; no data, N.D.

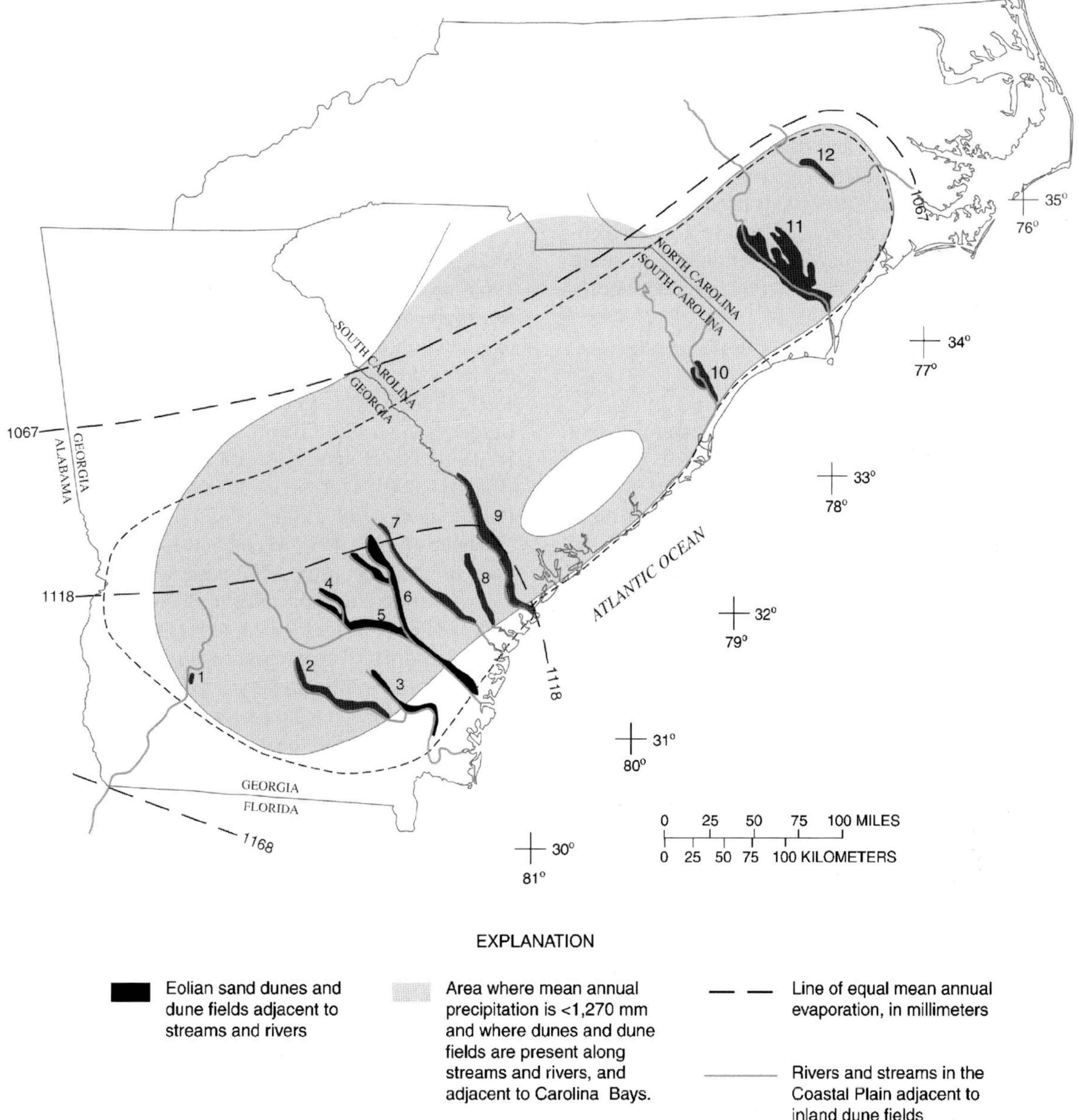

Fig. 7. Location of inland dune fields along rivers and streams in the Atlantic Coastal Plain of Georgia, South Carolina, and North Carolina. Drainages are (1) Flint River; (2) Satilla River; (3) Alabaha River; (4) Little Ocmulgee River; (5) Altamaha River; (6) Ohoopee and Little Ohoopee Rivers and Pendleton Creek; (7) Canoochee River; (8) Ogeechee River; (9) Savannah River; (10) Great Pee Dee River; (11) combined valleys of Cape Fear and Black Rivers; and (12) Neuse River. Dunes and dune fields present along smaller rivers and streams adjacent to Carolina Bays cannot be shown at this scale (modified from Markewich & Markewich, 1994).

Stratigraphy, Chronology, and Sedimentology

Surface morphology, soil development, and distribution in distinct belts suggest that dunes and sandsheets developed over several phases. Dunes typically overlie Pleistocene alluvium or older dunes and their associated organic deposits. Where dunes migrated onto the uplands, they overlie Tertiary fluvial and marine sediments. Optically stimulated luminescence (OSL) and ^{14}C age data for southeast Georgia dunes suggest major dune-forming periods from about 26,000 to 15,000 yr ago, at ca. 30,000 yr ago, ca. 45,000 yr ago, ca. 77,000 yr ago, and prior to 120,000 yr ago (Ivester *et al.*, 2001), with minor reworking of dunes and localized formation of dunes in the early Holocene (see references in Markewich & Markewich, 1994; Zayac *et al.*, 2001).

Most of the dune sands are massive, or at least show no evidence of bedding. This is probably due to extremely uniform grain size and mineralogy, and to deposition onto vegetated surfaces. The sand size fraction is moderately to well-sorted, symmetrical to fine skewed, >90% quartz, and has <1%

heavy minerals by weight. Medium and fine sand makes up ≥80 wt.% of the total sand. Silt and clay content varies with depth and source material. Dunes adjacent to Piedmont and Blue Ridge rivers typically have higher silt and clay contents than dunes adjacent to ACP rivers. The clay size fraction (<2 μm) is primarily quartz and kaolinite and represents either an original mineral fraction or subsequent airborne additions. Some of the <2 μm material is present as coatings on sand grains. Younger dunes generally have a more complex clay mineral assemblage than do older dunes, probably from greater weathering of the clay fraction in older dunes.

Paleoenvironments and Paleoclimate

Holocene dunes in the southeastern ACP were active during the Hypsithermal (ca. 8000–4000 yr ago), when a Midwest-type warm arid climate expanded eastward into the southeastern ACP (Delcourt *et al.*, 1980; Watts, 1971). Eolian activity at that time appears to have been restricted to reworking of older dunes along with the formation of small localized dunes. The concurrence of MIS-2 and MIS-6 or greater ages between dunes in the southeastern ACP and loess in the LMV suggests that periods of climates drier than at present and perhaps strong westerly winds were common when the LIS was present in the upper Mississippi valley. The similarity between the OSL ages of 45,000–29,000 yr on dunes and the TL ages of 55,000–28,000 yr on loess suggest similar climatic conditions in the LMV and ACP during MIS 3, when the LIS may have been restricted to Canada. Although few in number, OSL ages of 80,000–50,000 yr on dunes may correlate with the TL ages of 95,000–75,000 yr for LMV loess, supporting the hypothesis that there may have been periods in the LMV and ACP of climates that were drier than at present or perhaps with stronger prevailing westerly winds than at present during the MIS 5 to MIS 4 interglacial/glacial transition.

Eolian Sediments in Alaska

Loess in Alaska is found in broad zones bordering all the major river systems (Fig. 8), with the thickest and oldest deposits occurring in areas that were not glaciated, and thinner deposits, sometimes entirely of Holocene age, found in some glaciated areas. The deposition and preservation of Alaskan loess was strongly affected by local topography and microclimate, and the thickness of loess deposits changes dramatically over relatively short distances. Most loess deposits in central Alaska consist of a thin mantle of silt from a few meters to less than 10 m thick, but in areas adjacent to both the Tanana and Yukon Rivers extensive deposits of loess are more than 50–100 m thick. The thickest sections of loess typically contain both primary loess deposits and reworked, often soliflucted, silt beds. Scanning electron microscope studies of loess grains provide evidence of glacial crushing and abrasion, and dust flux measurements during modern dust storms confirm the glacial-fluvial origin of Alaskan loess and indicate that deposition is still continuing (Begét, 1990, 1996). Fields of ancient and stabilized sand dunes are also found in several areas adjacent to and downwind from braided, glacier-fed rivers that carry heavy sediment loads.

Loess in Alaska

Stratigraphy and Chronology

Studies of loess deposits in unglaciated central Alaska since 1965 have shown that Alaskan loess deposits in some areas are as much as 3.5 million years old, and contain a record of glacial-interglacial cycles similar to those preserved in the loess deposits in China and Europe (Begét, 1988, 1990, 1991, 1996, 2001; Begét & Hawkins, 1989; Begét *et al.*, 1990; Berger *et al.*, 1996; Edwards & McDowell, 1991; Hamilton *et al.*, 1988; Lagroix & Banerjee, 2002; McDowell & Edwards, 2001; Muhs *et al.*, 2001b, 2003a; Péwé *et al.*, 1997; Preece *et al.*, 1999; Sher *et al.*, 1997; Vlag *et al.*, 1999; Westgate *et al.*, 1990). Magnetic susceptibility profiling has been used to reconstruct the pattern of Pleistocene climate change in Alaska, especially through the last glacial cycle. Older loess sequences appear to be interrupted by significant unconformities. The pattern of magnetic susceptibility variation in Alaska is similar to that seen in loess from high latitude parts of Siberia, and in marine and lacustrine sediments, but is reversed compared to the glacial-interglacial pattern found in China (Begét, 2001; Begét *et al.*, 1990; Lagroix & Banerjee, 2002). Pedogenic processes that enrich soils with high-susceptibility iron-bearing minerals in temperate regions like China are weak or inactive in cold, high-latitude areas, where gleying is common and results in the dissolution and removal of iron-bearing minerals from soils (Liu *et al.*, 1999). The susceptibility record in central Alaskan loess is thought instead to reflect both primary, depositional differences caused by glacial-interglacial variations in storm frequency and wind intensity, with a secondary overprint controlled by pedogenic effects, including gleying (Begét, 2001). The fabric of magnetic anisotropy in central Alaskan loess has recently been shown to record paleowind directions and to indicate NW to SE winds predominating during glacial periods, and N to S winds during interglacial periods (Lagroix & Banerjee, 2002). This is consistent with the general distribution of loess deposits.

The application of modern dating techniques since 1965 has revolutionized the understanding of the age and stratigraphy of Alaskan loess deposits. The widespread Old Crow tephra has been identified throughout Alaska and adjacent Yukon Territory (Begét *et al.*, 1991a; Preece *et al.*, 1992, 1999; Westgate *et al.*, 1983, 1985). Westgate (1988, 1989) produced multiple isothermal-plateau fission-track dates of the Old Crow tephra, which yielded an averaged age of 140,000 ± 10,000 yr. This age is in broad agreement with recent TL age determinations of loess around the Old Crow tephra (Berger *et al.*, 1994, 1996). The Old Crow tephra is associated with the last interglacial forest bed and paleosols in loess across interior Alaska, typically occurring in loess just below or within the lowermost part of the last

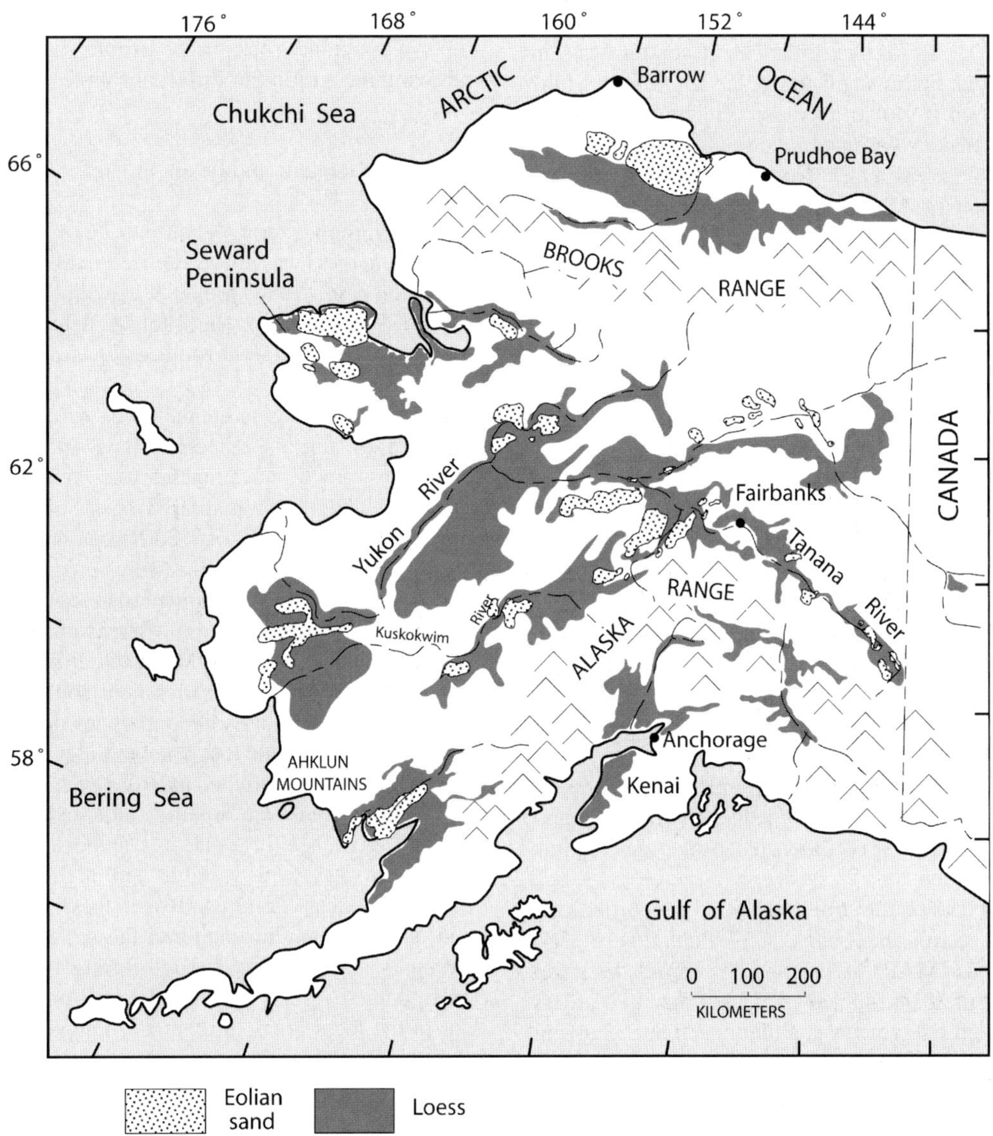

Fig. 8. Generalized distribution of deposits of loess and eolian sand in Alaska. Most loess deposits are found in lowland areas near river systems that drain glaciated mountain ranges. Loess distribution redrawn from Péwé (1975); eolian sand distribution from Lea & Waythomas (1990).

interglacial sequence (Hamilton & Brigham-Grette, 1991; Muhs *et al.*, 2001b, 2003a).

The 1.9-myr-old PA tephra, found at the base of thick loess sections at several places in central Alaska, has also been dated by the fission track method, and its age confirmed by magnetostratigraphy. The oldest loess underlies the PA tephra, and may be more than 3 million years old, based on the pattern of paleomagnetic reversals below the PA tephra (Westgate *et al.*, 1990).

Radiocarbon dating of wood, charcoal and humic acids has also been used widely since 1965 to determine the age of the upper portion of loess sections. Well-preserved charcoal, wood, and other plant macrofossils yield consistent and apparently reliable ages back to about 40,000 ^{14}C yr B.P. (Hamilton *et al.*, 1988; Muhs *et al.*, 2001b). Humic acid extracts (Abbott & Stafford, 1996) from soil organic matter in paleosols and loess have also recently been dated and are in reasonable agreement with other radiocarbon ages back to ca. 30,000 ^{14}C yr B.P. (McGeehin *et al.*, 2001). For older samples, humic acid ages are, however, probably minima (Muhs *et al.*, 2003a).

Cosmogenic ^{10}Be has been used in Alaskan loess sequences to estimate the time of accumulation. Beryllium-10, with a half-life of ca. 1.5 million years (Hofmann *et al.*, 1987), is formed in the atmosphere by interactions of N and O atoms with cosmic rays, and is quickly deposited on the Earth's surface where it is complexed with clay minerals. The abundance of ^{10}Be in a soil or paleosol, combined with an estimate of the ^{10}Be production rate, will yield the amount of time of accumulation since stabilization of a surface after deposition (Curry & Pavich, 1996; Markewich *et al.*, 1998a, b). This method of dating using "meteoric" ^{10}Be is distinguished from the method of measuring the amount of ^{10}Be produced in situ in quartz. Muhs *et al.* (2003a) measured ^{10}Be in central Alaskan loess and paleosols. Following Curry & Pavich (1996), Muhs *et al.* (2003a) used a global average ^{10}Be production rate of 1.3×10^6 atoms/cm^2/yr and a correction for inherited ^{10}Be to calculate age estimates. Concentrations of ^{10}Be are consistently higher in Alaskan paleosols than in enclosing loess beds, indicating longer periods of subaerial exposure and cosmogenic ^{10}Be accumulation from the atmosphere. Age estimates derived from ^{10}Be concentrations in paleosols and loess at several different sites are in reasonable agreement with those obtained on charcoal and plant remains back to the limit of radiocarbon dating, but ^{10}Be age estimates on loess just above the Old Crow tephra yield an age of ca. 57,000 yr, which is clearly too young based on the fission-track and TL age estimates. Similarly, the ^{10}Be age estimate of ca. 48,000 yr obtained for the Dome tephra is too young compared to its radiocarbon age of >55,900 yr and stratigraphic position within the last interglacial Eva Creek Forest Bed (Muhs *et al.*, 2001b, 2003a). The differences between the ^{10}Be and other age estimates for tephras suggest that the production rate assumed for ^{10}Be in Alaska may need to be re-evaluated, and perhaps also that one or more unconformities lie above the tephras. In spite of these problems, the ^{10}Be method holds much promise, and inventories of cosmogenic beryllium from Alaska loess sections have already provided estimates of sedimentation rate, the duration of paleosol formation, and independent age estimates of loess, paleosols, and tephras (Muhs *et al.*, 2003a).

Accumulated data on the stratigraphy and geochronology of Alaskan loess sections has enabled the calculation of mass accumulation rates (MAR) at several sites across the state. Holocene MARs of Alaskan loess span more than two orders of magnitude, from ~3 g/m^2/yr at a tundra site near Sagwon to ~1540 g/m^2/yr at a boreal forest site near Delta Junction (Muhs *et al.*, 2003a). In general, localities under boreal forest in central and south-central Alaska had Holocene rates of ~300–400 g/m^2/yr, whereas tundra localities in the Brooks Range, Seward Peninsula, and Ahklun Mountains had rates of less than 100 g/m^2/yr. Full glacial MARs from two localities in western Alaska were estimated at ~500 and 668 g/m^2/yr, but this record is short and data are needed from more full-glacial age sites (Muhs *et al.*, 2003a).

Sedimentology and Mineralogy

Muhs *et al.* (2003a) recently characterized the mineralogy and geochemistry of loess across central Alaska using both major and trace elements. Some local differences exist, but central Alaskan loess derived from glaciers in the Alaska Range and distributed along the Tanana River typically contains quartz, K-feldspar, plagioclase, mica, and small amounts of heavy minerals. Carbonate minerals are notably absent and unweathered loess has clay contents of only 3–6%. Concentrations of Fe_2O_3 and Al_2O_3 are surprisingly high, reflecting abundant maghemite and other Fe-rich minerals. In contrast, loess derived from limestones and other sedimentary rocks in the Brooks Range is calcium rich, allowing recognition of different source areas.

Paleoenvironments and Paleoclimate

The stratigraphic and paleoclimatic significance of paleosols in Alaskan loess was not recognized until the recent studies of Hamilton *et al.* (1988), Begét & Hawkins (1989), Begét *et al.* (1990), Begét (1990), Hamilton & Brigham-Grette (1991), Vlag *et al.* (1999) McDowell & Edwards (2001) and Muhs *et al.* (2001b). Unaltered loess in Alaska typically has 2.5Y Munsell color hues, whether moist or dry, whereas buried soil O or A horizons generally have 10 YR or 7.5 YR hues and almost always have much lower values and chromas that distinguish them from unaltered loess. Soil B horizons, where they occur, have 10 YR or 7.5 YR hues with chromas that are higher than unaltered loess (Muhs *et al.*, 2003a). Paleosols also have organic matter content of at least 1–2% or more and higher phosphorus contents, whereas unaltered loess generally has less than 0.5% OM. Muhs *et al.* (2001b) reported that zones with a P_2O_5 enrichment could be interpreted to be either buried tundra soils or perhaps minimally developed boreal forest soils, and that paleosols identified on the basis of color are typically enriched in P_2O_5, even when deeply buried and lacking significant organic matter (Muhs *et al.*, 2003a).

The existence of paleosols indicates that loess sedimentation rates were variable and at times slowed enough that pedogenesis dominated over loess accretion. Fine silt and clay is minor in unaltered loess, suggesting that wind strengths may be great enough during glacials periods that only coarse particles (coarse silts and some fine silts) settled out in the loess areas. In contrast, fine silt and clay is more abundant in paleosols, suggesting that during periods of soil formation loess deposition rates were low and took place under generally weaker winds (Muhs *et al.*, 2003a). Begét *et al.* (1990) proposed a similar model for Alaskan loess based on the systematic variations in magnetic susceptibility, which are a proxy for heavy mineral content and wind competence. In contrast, changes in loess particle size in the Central U.S. appear to be functions of transport distance, source, and pedogenesis (Muhs & Bettis, 2000; Ruhe, 1983).

Most Alaskan loess is associated with present or past occurrences of ice-rich permafrost (Fig. 9). In some instances, thick deposits of loess and reworked loess contain specimens of frozen Pleistocene megafauna, including mammoth, bear, and bison (Guthrie, 1990). Frozen Alaskan loess deposits also contain buried forest beds, consisting of horizons of logs, branches, leaves, and other biologic materials, as well as peat, beaver-chewed wood, and paleosols correlative with the last interglaciation (Edwards & McDowell, 1991; Muhs *et al.*, 2001b; Péwé *et al.*, 1997). As many as three older forest beds lie below deposits of the last interglaciation, the oldest of which is associated with the PA tephra at sites along both the Tanana and Yukon rivers. The 1.9 myr PA tephra occurs just below a forest bed at the Palisades of the Yukon, and just above ancient ice wedge casts, documenting the

Fig. 9. Typical loess deposits in central Alaska. (A) Hydraulic gold mining, carried out through most of the 20th century, has produced numerous exposures of loess in central Alaska. Streams of water sprayed on 50-m-thick cliffs of loess in a mine near Fairbanks thaw the frozen silts and then wash them away to uncover gold-bearing gravels at their base. (B) An 30-m-high exposure of loess along Birch Creek near the Yukon River in central Alaska contains a 8-m-thick body of massive ice in late Pleistocene loess near the top. Thawing of such ice bodies produces unconformities in Alaska loess sections.

earliest known occurrence of permafrost in Alaskan loess to the Pliocene-Pleistocene transition (Begét, 2000; Matheus *et al.*, 2000). The melting of massive ice bodies and the erosion of frozen loess in interglacial times have produced significant unconformities in many loess sections (Fig. 9).

The Alaskan Loess Record of the Last Glacial Cycle

Stratigraphic and geochronologic data obtained over the last two decades indicate that Alaskan loess was deposited from the very end of the Tertiary through the entire Quaternary, recording multiple cycles of loess deposition and soil development. Magnetic susceptibility sequences in Alaskan loess can be correlated with the main stages of the marine oxygen isotope record (Martinson *et al.*, 1987) from the present through the last glacial cycle. Some shorter events, like the Younger Dryas, are also identifiable in the loess record (Begét *et al.*, 1991b; Bigelow *et al.*, 1990). Efforts to characterize and subdivide long sequences of the Alaskan loess record more precisely and to match events in the loess record of Alaska with the stages and substages of the marine record, however, have been problematical.

Recent research has been concentrated on the loess record of the last full interglaciation, e.g. MIS 5, which lasted at least 50,000 yr, and was punctuated by at least three separate warm intervals at ca. 125,000, 100,000, and 80,000 yr ago. No single Alaskan section of loess and paleosols, however, appears to have a complete record of all three MIS 5 warm periods. The peak warmth of MIS 5 (substage 5e) is represented by the Eva Forest Bed, which consists of logs, branches, and an associated paleosol (Péwé *et al.*, 1997). At one site, the Old Crow tephra occurs in a paleosol and near the base of the Eva Forest Bed, which in turn is overlain by unaltered loess that contains the Dome tephra (Muhs *et al.*, 2001b). At a second loess site, near Halfway House, the Old Crow tephra is present but is overlain by a weak paleosol and the Dome tephra is absent. At a third loess site at Gold Hill, the Old Crow tephra is missing and the Dome tephra occurs in the upper part of a paleosol. In a fourth loess section along Chena Hot Springs Road, the Old Crow tephra is similarly missing and the Dome tephra occurs beneath a paleosol (Muhs *et al.*, 2003a). McDowell & Edwards (2001) reported that a paleosol developed in loess at Birch Creek contains the Old Crow tephra, but the warmest part of MIS 5 was recorded higher in the section in association with the Dome tephra. Similar inconsistencies exist among last-interglacial lacustrine and other non-loess sequences that contain tephras (Hamilton & Brigham-Grette, 1991). These problems in identifying and matching the substages of the last interglacial among different loess and non-loess sections would not have been apparent without the precise temporal ties provided by correlated tephras. The highly variable stratigraphic records are probably best explained either by local variability in microclimate or by unconformities within Alaskan loess sections (Muhs *et al.*, 2003a).

Loess correlated with MIS 4 is typically recorded by relatively thin, unweathered, somewhat coarser loess occurring above the sediments and paleosols correlative with MIS 5. MIS 4 loess in turn is overlain by alternating thin loess beds and minimally developed paleosols of the mid-Wisconsin period at sites along the Tanana River. The oldest paleosols in the mid-Wisconsin loess have been dated to ca. 42,000 ^{14}C yr B.P. At least two paleosols separated by a loess layer are found at several sites and are dated to 29,000–30,000 ^{14}C yr B.P. The loess record suggests that environmental conditions in central Alaska were quite variable at this time, consistent with Alaskan and Siberian pollen data from sites west of Alaska (Anderson & Lozhkin, 2001) and ice-core data from central Greenland far to the east (Grootes *et al.*, 1993; Meese *et al.*, 1997).

MIS-3 loess is overlain in many areas by 1–2 m of unaltered, slightly coarser loess correlative with the last glacial maximum, MIS 2, and by finer-grained Holocene loess. The Holocene loess contains several thin paleosols and, in some cases, abundant wood macrofossils in its upper parts. Many radiocarbon ages on wood and paleosols from multiple sites show that loess was deposited widely but episodically in central Alaska during the Holocene (Begét, 1990; Hamilton *et al.*, 1983; Muhs *et al.*, 2003a).

Sites in Alaska that span the last glacial cycle contain surprisingly little loess that can be firmly correlated with MIS 6, 4, or 2 (Muhs *et al.*, 2003a). This is in strong contrast to well-studied loess sites in China, Europe, and the mid-continental United States where glacial loess predominates and interglacial loess and interstadial loess is rare or even entirely absent. It is clear that glacier expansion and the dry, continental conditions that predominated in Alaska during glacial periods were conducive to the creation and dispersal of loess, so the minimal record of full-glacial loess may more likely be related to unfavorable conditions for the deposition and preservation of loess. An important factor may have been the disappearance of the boreal forest and its replacement with herbaceous tundra vegetation, which was less efficient at trapping windblown dust (Begét, 1988; Muhs *et al.*, 2003a). Higher velocity full-glacial winds may also have caused intermittent erosion of loess deposits (Thorson & Bender, 1985).

Sand Dunes and Sand Sheets in Alaska

In addition to loess deposits, extensive areas in Alaska are covered by both active and stabilized fields of sand dunes and local deposits of thin sand sheets, often associated with periglacial sand wedges (Lea & Waythomas, 1990; Péwé *et al.*, 1965). Several studies have focused on unique aspects of the sedimentology of these high-latitude sand deposits, including the interactions of windblown sand with snow and the preservation of footprints of Pleistocene megafauna (Lea, 1990, 1996).

Alaskan dunes are mobilized by summer winds, when snow cover is absent and the sediment unfrozen. The distribution of Pleistocene dunefields, the morphology of dunes, and the orientation of sand beds in dune deposits suggest that winds, particularly in late-glacial time, came predominantly from the northeast (Hopkins, 1982; Lea & Waythomas, 1990). Global climate simulation runs suggest, however, that full-glacial paleowinds in Alaska were from the south in winter and southwest in summer (Bartlein, 1997; Bartlein *et al.*, 1998). This conflict is difficult to explain. Loess derived primarily from the Tanana River in central Alaska has been dispersed mainly to north, requiring southerly winds. Also, Thorson & Bender (1985) showed that katabatic winds, derived from expanded glaciers in the Alaska and Brooks ranges, were restricted in their extent. It seems likely that the divergent paleowind reconstructions from sand deposits and loess, and conflicts with global climate modeling can be resolved by assuming the geologic data are reflecting both the regional-scale wind regimes and also local katabatic winds and other effects resulting from interactions between local topography and microclimate (Muhs *et al.*, 2003a).

Eolian Sediments in the Desert Southwest U.S.

This section summarizes current knowledge of Quaternary eolian history in the desert Southwest U.S. and examines the relationships among periods of eolian activity and stability and paleoclimates and paleohydrology. J.T. Hack conducted pioneering investigations in NE Arizona (Hack, 1941). H.T.U. Smith was the first to recognize the significance and long history of past eolian activity in the Mojave Desert (Smith, 1967). Little progress was made, however, on developing a chronology of Quaternary eolian activity in the region until the advent of luminescence dating and its application to studies of eolian deposits (Clarke, 1994; Clarke & Rendell, 1998; Clarke *et al.*, 1996; Rendell & Sheffer, 1996; Rendell *et al.*, 1994; Stokes & Breed, 1993; Wintle *et al.*, 1994). This approach was complemented on the Colorado Plateau by use of radiocarbon ages from archaeological sites in dune areas (e.g. McFadden & McAuliffe, 1997; Wells *et al.*, 1990) to develop chronologies of eolian activity.

Eolian Systems in Southwestern Deserts

The Desert Southwest includes the Chihuahuan, Sonoran, Colorado Plateau, Mojave, and Colorado deserts (Fig. 10). Sand deposits are comprised of sand sheets, dunes, and sand ramps, which are amalgamated accumulations of eolian sand, alluvial, and colluvial deposits on the slopes of desert mountains. In addition, silt- and clay-sized eolian materials are an important component of many soils in the region (e.g. McDonald *et al.*, 1995; McFadden *et al.*, 1987).

Mojave Desert

Eolian sediments and landforms are widespread throughout the Mojave Desert of southern California and adjacent areas of Nevada and Arizona (Fig. 11). Topographic control of winds and therefore sand transport has resulted in well-defined corridors of transport (Fig. 11) that are characterized by areas of active and inactive dunes and sand sheets, together with sand ramps (Zimbelman *et al.*, 1995). Geochemical and mineralogical studies indicate

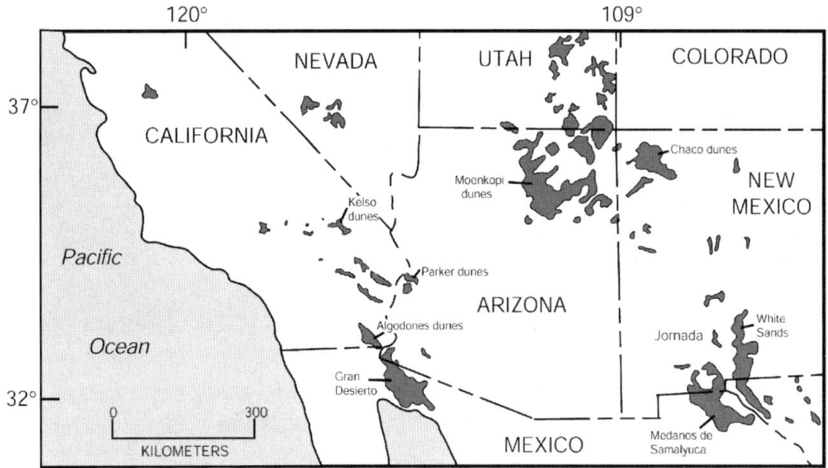

Fig. 10. Areas of eolian sand in the desert Southwest. Modified from Muhs & Zárate (2001).

several major (e.g. Mojave River) and many local sources of sediment and a general west to east transport of sand (Ramsey *et al.*, 1999; Sharp, 1966; Zimbelman & Williams, 2002). The Colorado River has provided sediment for several dune fields, including those near Parker, Arizona, the Algodones Dunes of southeastern California, and parts of the Gran Desierto sand sea of northern Mexico (Muhs *et al.*, 1995, 2003b).

The history of Late Quaternary eolian activity along the major sand transport corridors has been documented using TL and Infrared stimulated luminescence (IRSL) dating of dunes (Wintle *et al.*, 1994) and eolian deposits in sand ramps (Clarke & Rendell, 1998; Clarke *et al.*, 1996; Rendell & Sheffer, 1996). Because the sand ramps represent a long-term net accumulation of eolian, alluvial, and slope deposits with little reworking, they provide a long history of eolian processes in the Mojave Desert (Lancaster & Tchakerian, 2002). The results of these studies are summarized in Figs 12 and 13. By contrast, areas of dunes may be partially or completely reworked by each period of eolian activity and therefore often preserve a relatively short record of past eolian dynamics.

Following the interpretations of Rendell & Sheffer (1996), the earliest dated phase of eolian activity in the Mojave Desert appears to have spanned the period 20,000–30,000 yr ago (Fig. 12). A later phase of eolian accumulation extends from 15,000 or 20,000 to 7000 yr. Mid- and late-Holocene eolian deposits (4000–2000 yr ago and <2000 yr ago, respectively) occur in the Kelso Dunes transport system (Fig. 12; Clarke & Rendell, 1998). Interstratified paleosols in many sand ramps suggest one major and a number of minor periods of regional-scale eolian stability. The oldest dated of these periods occurred 20,000–15,000 yr ago, with shorter periods of soil formation around 14,000 and 4000 yr ago. The record of periods of eolian construction and stability in the Kelso dunes system can be compared to the lacustrine record from Lake Mojave (Fig. 12; Wells *et al.*, 2002). Periods of eolian stability generally coincide with the highest lake levels,

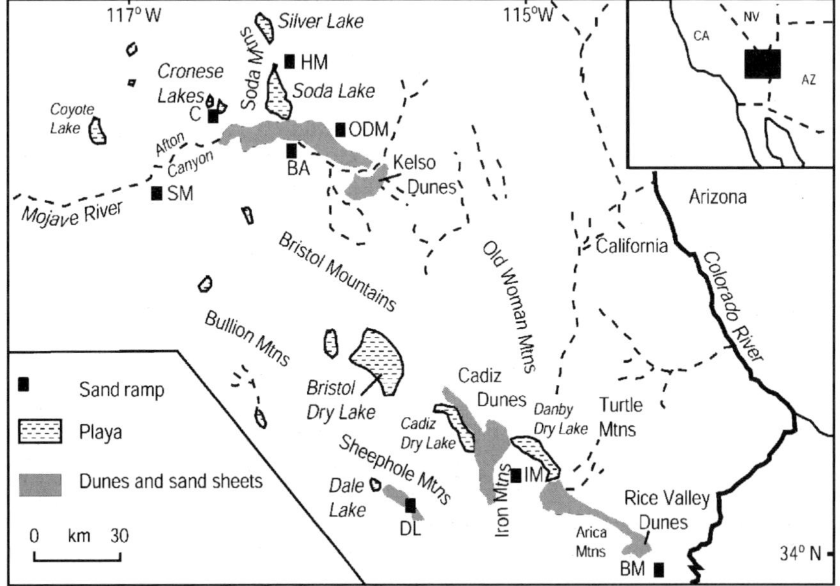

Fig. 11. Location of dunes and sand ramps in the central and eastern Mojave Desert (after Rendell & Sheffer, 1996). SM – Soldier Mountain; C – Cronese Basin; HM – Hanks Mountain; BA – Balch; ODM – Old Dad Mountain; DL – Dale Lake; IM – Iron Moutain; BM – Big Maria.

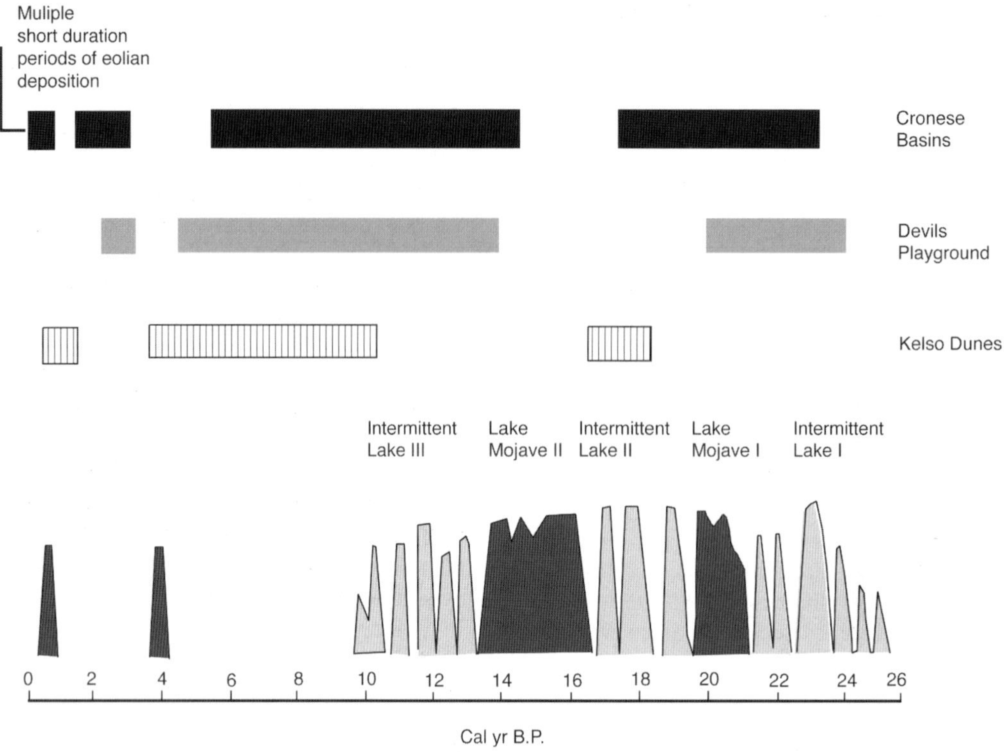

Fig. 12. Comparison of periods of eolian construction in the Kelso Dunes system (Striped bars), Cronese Basins (black bars), and Devils Playground (gray bars) with the level of lake Mojave. Data on lake levels from Wells et al. (2003). Luminescence ages from data in Rendell & Sheffer (1996), Wintle et al. (1994), and Rendell & Clarke (1998). Lake Mojave ^{14}C time scale has been calibrated to facilitate comparison with the luminescence ages.

whereas periods of eolian construction are linked to periods of fluctuating or dessicating lakes. This has been interpreted to indicate the strong influence of sediment supply on eolian processes in this region (Clarke & Rendell, 1998; Lancaster & Tchakerian, 2003).

Multiple short-duration episodes of late Holocene eolian accumulation are recorded close to modern sand source areas (Fig. 13) (Clarke & Rendell, 1998; Clarke *et al.*, 1996). IRSL ages for sediments in the Kelso Dunes sand transport system (including the Cronese basins) occur in clusters dating from 140 to 6700 yr ago (Fig. 13; Clarke & Rendell, 1998). Episodes of eolian deposition correlate well with the record of major floods in the Mojave River system (Enzel & Wells, 1997; Enzel *et al.*, 1992) and likely reflect the effect of enhanced sediment supply from the Mojave River (Clarke & Rendell, 1998). In addition, multiple periods of eolian activity and reworking of dunes likely occurred throughout the late Holocene, most notably around 1500 and 400–700 yr ago (Wintle *et al.*, 1994).

Colorado and Lower Sonoran Deserts

There are three main areas of eolian deposits in this region: (1) the Coachella Valley (Beheiry, 1967; Griffiths *et al.*, 2002; Wasklewicz & Meek, 1995); (2) the Algodones Dunes (McCoy *et al.*, 1967; Muhs *et al.*, 1995; Stokes *et al.*, 1997; Sweet *et al.*, 1988; Winspear & Pye, 1995) and (3) the Gran Desierto sand sea of northern Mexico (Lancaster, 1992, 1995). Although multiple generations of dunes have been identified in the Gran Desierto, chronological information is only available for the Algodones, where Stokes *et al.* (1997) identified periods of eolian activity >30,000 yr ago, 3100, and <400 yr ago. The latest period of eolian deposition apparently postdates the last major flooding of the Salton Trough by the Colorado River to form the youngest highstand of Lake Cahuilla (Waters, 1983).

Colorado Plateau Desert

Extensive areas of partly to mostly vegetated linear, parabolic, and barchanoid dunes with associated sand sheets occur in northeastern Arizona and adjacent areas of New Mexico and Utah (Fig. 10). The dunes were first described by Hack (1941) and are derived from underlying Mesozoic sedimentary rocks and Pleistocene to Recent alluvial deposits.

Soil-stratigraphic and geomorphic relations indicate eolian activity was widespread prior to the Holocene, in addition, several periods of more geographically restricted eolian activity occurred during the Holocene (McFadden & McAuliffe, 1994, 1997). In the area of the Petrified Forest National Park, Ellwein (1997) used soil-geomorphic relations to identify three periods of eolian deposition: Mid Pleistocene

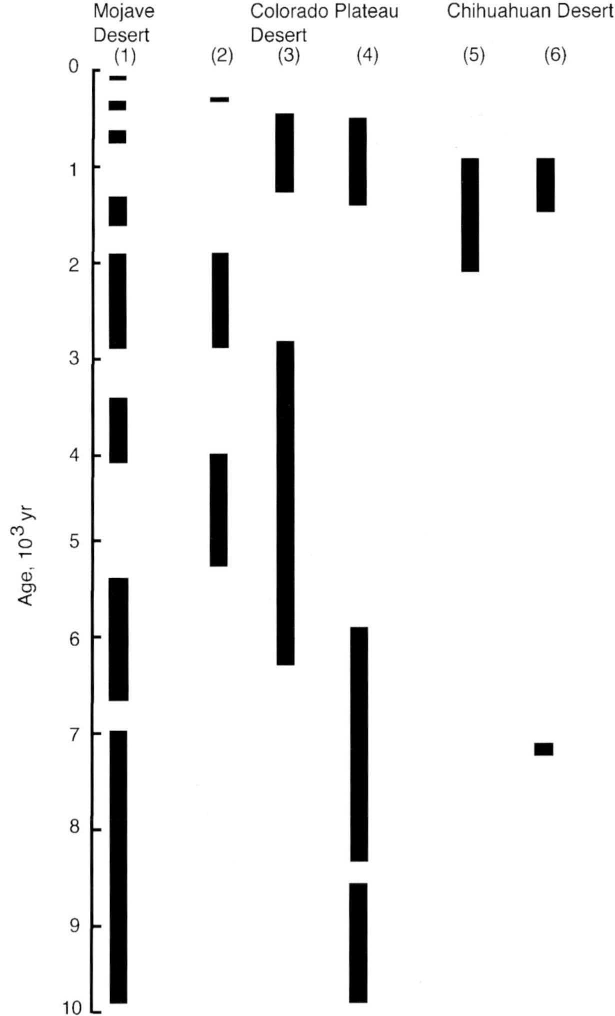

Fig. 13. Dated periods of Holocene eolian activity in the Southwest deserts; data from (1) Clarke & Rendell (1998); (2) Stokes & Breed (1993); (3) Wells et al. (1990); (4) Smith & McFaul (1997); (5) Gile (1999); (6) Wilkins & Currey (1999).

(75,000–200,000 yr ago); Mid Holocene (2000–8000 yr ago); and late Holocene (<1000 yr ago), summarized in Fig. 13.

In the southwestern part of the area, Stokes & Breed (1993) provide OSL ages for linear dunes on the Moenkopi Plateau. They identified three periods of eolian activity: (1) around 4,700 ± 600 yr; (2) between 2000 and 3000 yr ago; and (3) around 400 yr ago. In the Chaco Dunefield of northwestern New Mexico Wells *et al.* (1990), used ^{14}C dated materials in eolian units, dated cultural materials, and geomorphic and soil-stratigraphic relations to distinguish three main periods of eolian activity in this area: (1) prior to 14,000 yr ago and most likely between 14,000 and 19,000 yr ago; (2) between 2900 and 6400 yr ago, with the bulk of eolian activity occurring after 4500 yr ago; and (3) between 520 and 1370 yr ago. Smith & McFaul (1997) working in the same area provided numerous charcoal ^{14}C dates for eolian deposits. The calibrated dates fall into four main groups: 8600 to >10,900, 5900–8400, 2100–4400, and 580–1500 yr ago. The periods prior to 8600 and after 2200 yr ago were most active.

Chihuahuan Desert

Eolian deposits in the Chihuahuan Desert consist of gypsum dunes such as those at White Sands (McKee & Moiola, 1975) and quartz sands in the Rio Grande Rift and adjacent areas (Langford, 2000). The latter have a long history, summarized by Gile (1999), who identified four Pleistocene eolian depositional units in the Jornada area, with age estimates based on soil characteristics that range from mid- to early Pleistocene (Stage V calcic horizons) to late Pleistocene (Stage III calcic horizons). An extensive late Holocene unit with a Stage I calcic horizon is correlated with surfaces that have a calibrated ^{14}C age of 1020–2200 cal yr B.P. At White Sands, Langford (2003) recognized a extensive regional deflation event around 7800 cal yr B.P., similar to that identified in the Estancia Basin (Bachhuber, 1982). In the nearby Guadalupe Mountains, Wilkins & Currey (1999) mapped an extensive gypsum eolianite of inferred late Pleistocene age, capped by two quartz-rich eolian units of Holocene age that have calibrated ^{14}C ages of 7200 and 310–1580 cal yr B.P. The Holocene units are overlain by modern active dunes.

Relations to Paleoclimate and Paleohydrology

The dynamics of transport systems for eolian sediment on any time scale are determined by the relations between the supply, availability, and mobility of sediment of a size suitable for transport by wind (Kocurek & Lancaster, 1999). In turn, sediment supply, availability, and mobility are determined in large part by regional and local climate and vegetative cover. Interactions between these factors result in episodic input of sediment to the eolian system.

In the east-central Mojave Desert, sediment supply from the Mojave River system has largely determined rates and styles of episodes of eolian construction. Elsewhere, the influence of sediment supply is less clear, but played an important role in the development of the Algodones dunefield and Gran Desierto sand sea. In the eastern Mojave and the Colorado Plateau deserts, changes in sediment availability as a result of changes in precipitation that affected vegetation cover likely determined the timing of the many episodes of Holocene eolian activity. Reconstruction of the Holocene record for the desert southwest indicate that conditions have been close to the threshold for eolian activity throughout much of the Holocene (Fig. 13).

Eolian Sediments in the Pacific Northwest U.S.

Bodies of loess and eolian sand are prominent and widespread in two distinct areas of the Pacific Northwest U.S., the Columbia Plateau of eastern Washington and the Snake River Plain of southern Idaho (Fig. 14). Few studies had been completed in either area prior to 1965, although Kirk Bryan's paper titled "The Palouse Soil Problem" (Bryan, 1927) foreshadowed with prescience the major research questions of the origin of the loess. Roald Fryxell in Richmond *et al.* (1965) reviewed research on the loess of the Columbia Plateau up to 1965.

Loess on the Columbia Plateau

Distribution

The Columbia Plateau is the area of eastern Washington, northeast Oregon, and northern Idaho that lies between the Cascade Mountains to the west and the Bitterroot Mountains to the east. It is drained in part by the Columbia and Snake rivers and is underlain by the Miocene-age Columbia River Basalt. Loess covers >50,000 km^2 of the Columbia Plateau with deposits that are from a few meters to 75 m thick (Fig. 14). The loess on the plateau accumulated on the gently SW-sloping to locally warped surface of the Columbia River Basalt. The area of deepest and most continuous loess, which straddles the Washington-Idaho border, is termed the "Palouse." Palouse loess shares the Columbia Plateau with the Channeled Scabland, the huge complex of anastomosing flood coulees cut into the loess and basalts during cataclysmic outburst flooding from glacial Lake Missoula. The Palouse shares the plateau also with the sedimentary deposits from outburst flooding (Baker & Bunker, 1985). Research since 1965 suggests that fine facies of these flood sediments have been the major source of late Pleistocene deposits of Palouse loess (Busacca & McDonald, 1994; Sweeney *et al.*, 2002b). Other research further suggested that episodes of outburst floods accompanied perhaps six or more glacial maxima during the late, middle, and even early Pleistocene (Bjornstad *et al.*, 2001; McDonald & Busacca, 1988; Patton & Baker, 1978), providing a source for older Palouse loess deposits as well.

Stratigraphy and Chronology

The majority of field studies since 1965 have concentrated on late Pleistocene exposures; however, several deep roadcut exposures and one deep drill core have illuminated the older record. The great age of the Palouse loess is demonstrated, at least in a general way, by normal-over-reverse magnetic polarity zonation in several sites (Busacca, 1991; Foley, 1982; Kukla & Opdyke, 1980; Packer, 1979). For example, reversely magnetized loess occurs below 11 m in a 25-m-deep roadcut near the town of Washtucna, Washington (Busacca, 1991), implying deposition before the Brunhes Normal Magnetic Polarity Chron or about 790,000 yr ago. Drilling in 1996 recovered a 38-m core near the town of Winona, Washington that had normal polarity in the upper 28 m and reverse from 28 to 38 m (Busacca *et al.*, 1997, 1998). Normal-reverse-normal polarity signatures have also been documented (Busacca, 1991), suggesting that loess deposition may have begun by 1–2 myr ago.

More than 200 exposures of late Pleistocene loess on the Columbia Plateau were described and sampled from 1985 to 1993 (McDonald, 1987; McDonald & Busacca, unpublished data, 1985–1993), forming the basis of regional correlation of loess units, paleosols and distal tephras (Busacca *et al.*, 1992; McDonald & Busacca, 1988, 1990, 1992). Two major units of loess that span approximately the last 75,000 yr have been named L1 and L2 (Fig. 15; McDonald & Busacca, 1992). The modern surface soil is formed at the top of L1 and

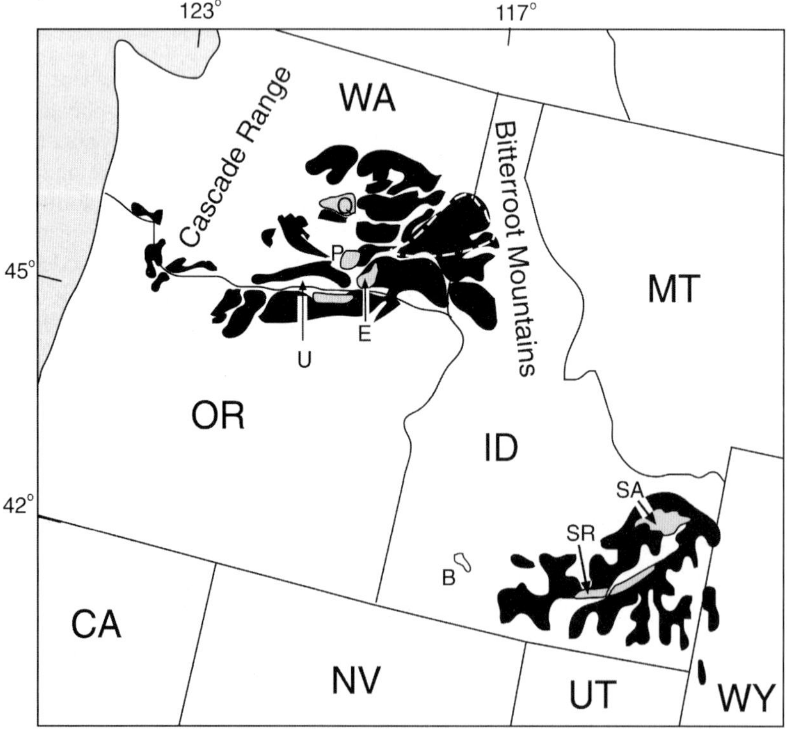

Fig. 14. Generalized distribution of deposits of loess and eolian sand in the Pacific Northwest United States. Black are areas of loess greater than approximately two meters thick, stipples are areas of eolian sand. U is the Umatilla Basin, which contains the Juniper Canyon dunes, E is Eureka Flat, with its dunes and sand sheet, P is the Pasco Basin, which contains the Hanford dunes, Smith Canyon dunes, and Juniper dunes (shown as one area of dunes at this scale), and Q is the Quincy Basin, which contains the Quincy dunes. B is the Bruneau dunes, SR is the area of sand sheet and thin dunes along the Snake River floodplain in SE Idaho, and SA is the St. Anthony dune field. Some other significant areas of sand dunes, loess and other eolian deposits in Oregon, NE California, and N Nevada are not discussed in text and are not shown. Modified from Muhs & Zárate (2001), Busacca & McDonald (1994), and Scott (1982).

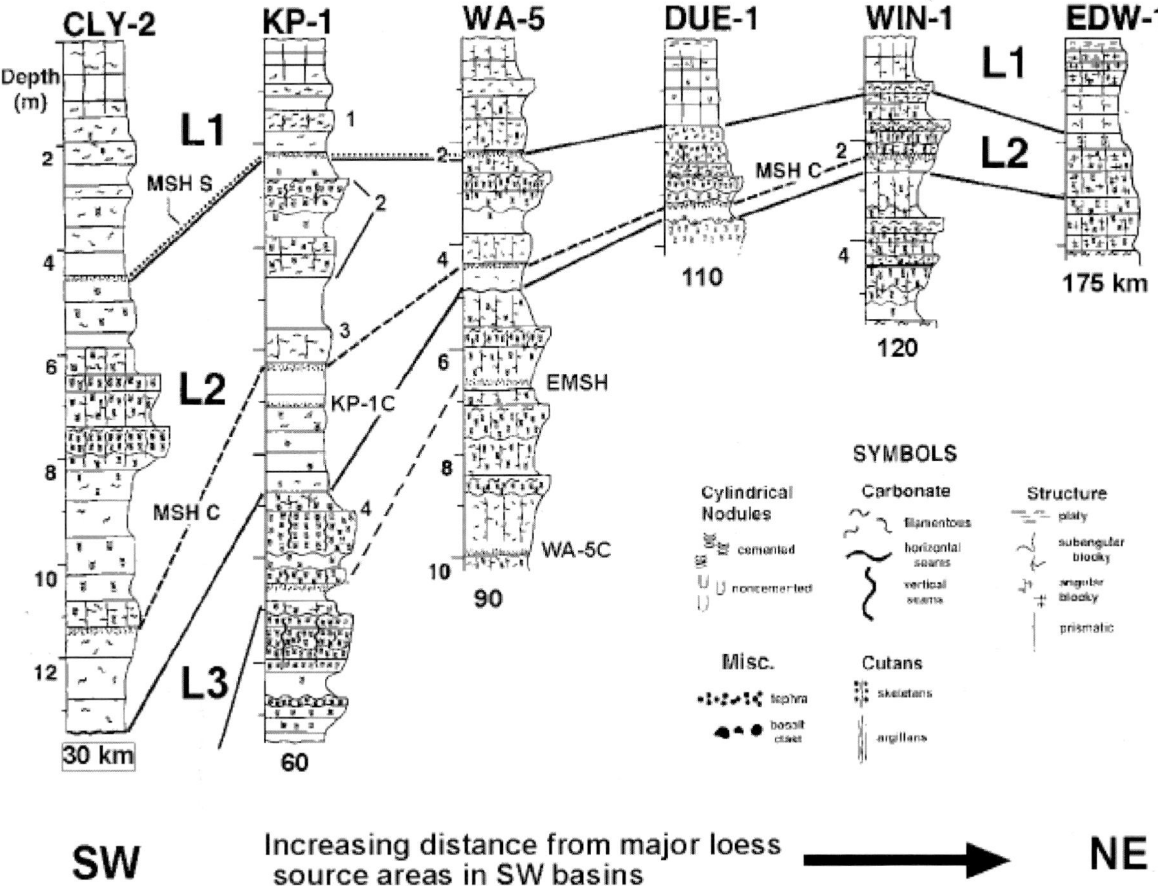

Fig. 15. Regional Correlation of Late Quaternary loess strata in the Palouse region showing thinning of loess units to the NE; loess units L1–L3; correlated tephras including Mount St. Helens (MSH) set S, MSH set C, tephra KP-1C, an earlier MSH tephra (EMSH), and tephra WA-5C. Correlated soil stratigraphic units are labeled to the right of the KP-1 section and include (1) Sand Hills Coulee Soil; (2) Washtucna Soil complex; (3) Old Maid Coulee Soil; (4) Devils Canyon Soil. Modified from McDonald & Busacca (1990, 1992).

four buried soil stratigraphic units have been named: (1) The weakly developed Sand Hills Coulee Soil occurs mid way through L1 at proximal sites; (2) The strongly developed Washtucna Soil (Fig. 15) has formed at the top of L2 across the Columbia Plateau; however, it is expressed as one or two buried soils depending on position in the loess depositional system (McDonald & Busacca, 1990) and its properties change with position in the regional bioclimatic gradient (Busacca, 1989); (3) The weakly developed Old Maid Coulee Soil occurs mid way through L2 at proximal sites; (4) The strongly developed Devils Canyon Soil occurs at the top of the next older loess unit (McDonald & Busacca, 1992).

Because the Palouse loess lies downwind of active volcanoes of the Cascade Range (Fig. 14), many layers of distal tephra are preserved in the loess. These have been very important in establishing stratigraphic correlations among sites. Busacca *et al.* (1992) fingerprinted tephras using an electron microprobe. Distal tephra layers have been correlated to, for example, Glacier Peak layers G and B (ca. 13,300 cal yr B.P.), Mount St. Helens set S tephra (MSH S; ca. 15,300 cal yr B.P.; TL age ca. 18,000 yr), Trego Hot Springs tephra (King *et al.*, 2001), and Mount St. Helens set C distal tephra (MSH C; TL age ca. 50,000–55,000 yr). Many other older layers of tephra, even including several in reversely magnetized loess, have been useful for local correlation among loess sites despite not being correlated to any known and dated eruptions. For example, the "EMSH" (pre-Set C MSH) layer and the WA-5C tephras underlie the Devils Canyon Soil at sites such as WA-5 and KP-1 (Fig. 15).

Materials suitable for radiocarbon dating are virtually unknown from Palouse loess. Early age estimates were based largely on tephra correlations to the radiocarbon chronology at volcanoes such as Mount St. Helens (Foley, 1982; Mullineaux, 1986). In the early 1990s, luminescence dating techniques were first applied to Palouse loess and Berger & Busacca (1995) reported useful TL ages back to 83,000 yr (Fig. 16). TL ages associated with MSH S and C tephras are older than calibrated ^{14}C ages for these layers, at

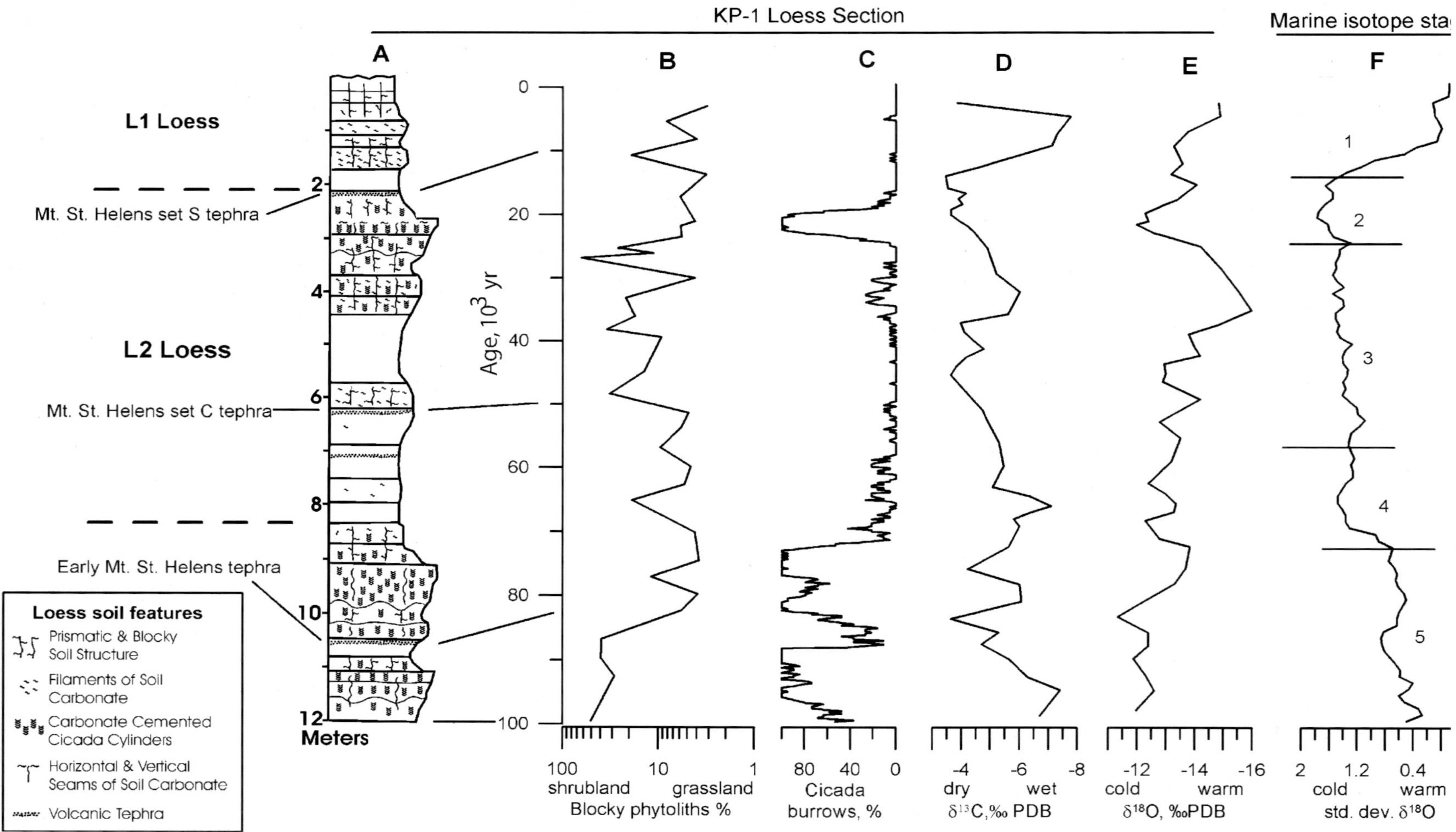

Fig. 16. Stratigraphy and paleoclimate proxies, KP-1 loess section. (A) Loess stratigraphy and features of buried soils. Age-depth model for loess (Blinnikov et al., 2001) was derived from luminescence age dating (Berger & Busacca, 1995). (B) Relative % of shrub (arid Artemisia) vs. grassland (moist perennial grasses) cover through time based on phytoliths preserved in the loess (Blinnikov et al., 2001). (C) Depth function of abundance of cicada burrows; increasing abundance a proxy for increasing Artemisia shrubland (from O'Geen & Busacca, 2001; O'Geen & Busacca, unpublished data). (D and E) Depth function of stable carbon and oxygen isotopes (respectively) extracted from pedogenic carbonates (Stevenson, 1997). Palouse native vegetation is ~100% C3 type. Carbon isotope signature is a proxy for paleoprecipitation. Oxygen isotope signature is a proxy for paleotemperature. (F) Depth function of stable oxygen isotopes from marine sediments, divided into major isotope stages, with odd numbers representing interglaciations and even numbers representing glaciations. Modified from Blinnikov et al. (2001).

about 18,000 and 50,000 yr, respectively. The 83,000 yr age is from loess immediately under the EMSH tephra (Fig. 16) and extrapolation of age vs. depth relations at the WA-5 site was used to predict an age of ca. 125,000 yr for the WA-5C tephra at 10 m in site WA-5 (Fig. 15; Berger & Busacca, 1995). More recently, loess from a number of additional sites was dated by both TL and IRSL methods (Richardson *et al.*, 1997, 1999), with generally similar age estimates for similar loess strata. The oldest age so far determined by luminescence methods on Palouse loess, 158,000 ± 16,800 yr, is from a site near Pendleton, Oregon (Richardson *et al.*, 1997). The oldest age of any kind determined by luminescence methods, 246,000 ± 29,000 yr, was on glass in a distal tephra in a deep layer of Palouse loess (Berger, 1991).

Sedimentology and Mineralogy

A detailed discussion of the mineralogy of the Palouse loess can be found in Busacca (1991). To summarize, as early as 1927, Kirk Bryan recognized that the mineralogy of the loess supported a source in the granitic, gneissic, and metamorphic terrains of the Cordillera to the north and east of the Columbia Plateau (Bryan, 1927) and he suggested that the ancestral Columbia and Snake Rivers had carried these minerals to the vicinity of their junction in southcentral Washington so that prevailing southwesterly winds could carry it into the Palouse. Not one but three potential sources for the loess occur on the Columbia Plateau: the late Miocene and Pliocene Ringold Formation (Packer, 1979), normal outwash from the Cordilleran Ice Sheet in the Columbia River system (Waitt, 1983), and Touchet Beds (Flint, 1938), quietwater or "slackwater" sediments from last-glacial giant outburst floods in the Channeled Scabland.

Several lines of evidence indicate that giant-flood slackwater sediments have been the principal source of Palouse loess. Fryxell & Cook (1964) reported that post-scabland-flood loess (L1 of McDonald & Busacca, 1992) can be traced directly into eroding outcrops of slackwater sediments. This is confirmed by more recent and ongoing work in the Umatilla Basin and Eureka Flat (Fig. 14; Sweeney *et al.*, 2000, 2001, 2002b). Also, the L1 and L2 loess units have a combined thickness of more than 13 m in proximal sites and thin to less than 3 m to the northeast away from major bodies of slackwater sediment in south-central Washington and north-central Oregon (Fig. 15; Busacca & McDonald, 1994). The regional patterns of L2 and L1 are consistent with independent evidence that two major episodes of Scabland flooding, one about 70,000–60,000 yr ago (MIS 4) and the other the last-glacial floods (MIS 2, ca. 20,000–15,000 TL yr ago) triggered the last two major cycles of loess deposition on the Columbia Plateau (Busacca & McDonald, 1994; McDonald & Busacca, 1988). Research in progress using geochemical matching of provenance indicates that late Pleistocene loess is derived from slackwater sediments (Sweeney *et al.*, 2002a).

Sand content of basal sediment in some loess units at source-proximal sites ranges up to 75% and diminishes upward to 25–35% (McDonald & Busacca, 1988, 1990). Such spikes in sand have been interpreted as being derived from outburst flood sediment that were remobilized by the wind following flood events. Sand sheets were initially formed and it is thought that these then translated upward into deposition of loess from suspension as climate changed (McDonald & Busacca, 1988; Sweeney *et al.*, 2002b). Based on TL dating, stratigraphy, and paleosols, the working model for the timing of major episodes of loess deposition suggests that deposition of at least the L2 and L1 loess units was initiated at the end of climatic cold phases and continued throughout much of the succeeding interstadial to interglacial intervals (McDonald & Busacca, 1998). For example, L2, whose deposition was triggered by outburst flooding during MIS 4, is dated to about 60,000–70,000 yr at its base and 20,000 yr at its top. L1, whose deposition was triggered by outburst flooding during MIS 2 appears to have been accumulating nearly without interruption from about 17,000 yr ago to the present. Intensive dating of L1 across a transect of sites is underway to test this view of uninterrupted deposition of L1 through the latest Pleistocene and Holocene. This timing, if verified, is strikingly different from that documented from many other loess deposits around the world.

Paleopedology and Paleoenvironments

The steep gradients of increasing precipitation and decreasing temperature with distance away from the rain shadow of the Cascade Range influence a number of properties of Palouse loess and of the paleosols contained in the loess. Four interacting factors are thought to account for much of the variation in properties of surface soils and individual buried soils that have been traced across the Columbia Plateau: (1) Mean parent material particle size decreases and clay content increases to the north, east, and southeast (away from source areas), affecting weathering rates (Busacca & McDonald, 1994; Marks, 1996). (2) The gradient of increasing present-day precipitation and of paleoprecipitation to the northeast, east, and southeast created a gradient of increasing weathering and leaching intensity with concomitant changes in pedogenic process across the gradient (Busacca *et al.*, 1994). (3) Loess units are thick in south-central Washington and thin considerably to the north and east so that buried soils increasingly overlap or weld in thinning loess beds (Busacca, 1989; Kemp *et al.*, 1998; McDonald & Busacca, 1990). (4) Episodes of rapid loess deposition that diluted or suppressed soil development occurred dominantly during interstades and interglaciations, whereas episodes of slow loess deposition and significant soil development often occurred during glacial stages (Tate, 1998).

Soils and paleosols in source-proximal loess are coarse textured, attenuated, and generally weakly to moderately developed, with <10% carbonates. Surface soils in this zone receive less than 250 mm mean annual precipitation (MAP) and are dominantly Xeric Haplocambids (Soil Survey Staff, 1999). Buried soils in this zone have horizons with biofabrics that have been interpreted to be ochric horizons (Tate, 1998) because they are thoroughly ramified by burrows of cicada nymphs (*Homoptera: Cicadidae*), and cambic and calcic horizons. These soils have been interpreted as paleo-Aridisols (Tate, 1998). Cicada burrows always are backfilled with soil, exhibit crescentic infill structures, and have characteristic diameter of 1–2 cm (O'Geen & Busacca, 2001). These burrows impart a distinctive structure to most paleosols when they become cemented that has been called "cylindrical structure" (Hugie & Passey, 1963). The Washtucna, Devils Canyon, and many other soils are dominated by these burrows, yet the Holocene soil formed in L1 loess has few if any of these burrows (Figs 15 and 16). Burrows have been cemented as the zone of secondary precipitation of pedogenic carbonates migrated upward in the accreting landscape (McDonald & Busacca, 1990; O'Geen & Busacca, 2001; O'Geen *et al.*, 2002). Carbonates in the Washtucna Soil at proximal to medial sites reach Stage IV morphology (Gile *et al.*, 1966) with laminar carbonate caps in less than 10,000–15,000 years because of this process, resulting in strongly cemented, visually distinct, but pedogenically weakly weathered soils (Busacca *et al.*, 1994).

Sequences of buried paleosols can have an alternation of horizons interpreted to have ochric and mollic-like biofabrics when examined in thin sections, reflecting an alternation from Aridisols during cold, dry glacial climates and Mollisols during warmer, moister interglaciations (Tate, 1998).

Research using thin-section micromorphology (King, 2000; Tate, 1998) demonstrated that several paleosol such as the Washtucna (MIS 2) and Devils Canyon (MIS 4) have suites of features, such as platy freeze-thaw fabrics, gypsum ghosts, and cicada burrow fabrics, that indicate formation during cool to cold, dry shrub-steppe conditions. For the Washtucna Soil, dating indicates that the deposition of L2 loess was episodic, allowing formation of the aggradational pedocomplex seen today as the lower and upper Washtucna Soils in proximal sites (Fig. 16). The rate had apparently slowed greatly by the onset of MIS 2, allowing the landscape to stabilize, forming the upper Washtucna Soil with its strong suite of cold-climate features (Tate, 1998).

Research on the ecology of cicadas combined with study of paleosol sequences suggest that an Artemisia-dominated periglacial steppe, which supported dense populations of sagebrush-dependent soil-dwelling cicada nymphs and the above-ground adult forms, was widespread on the Columbia Plateau during Washtucna-Soil time and during the development of other cold-phase soils (O'Geen, 1998; O'Geen & Busacca, 2001; Tate, 1998). Lacking a shrub component, the bunchgrass soils that developing during the Holocene did not support cicada nymphs, consequently the L1 loess lacks burrows (Figs 15 and 16; except that sage persisted in the driest core area of the plateau and there the L1 loess has scattered burrows).

Parallel research to deduce paleoclimate signals involved extraction and analysis of plant opal phytoliths to reconstruct plant community structure through time across the plateau (Blinnikov *et al.*, 2001, 2002) and analysis of stable oxygen and carbon isotopes in native plants, soil organic matter, and pedogenic carbonates (Stevenson, 1997). Phytoliths were extracted from samples of loess and paleosol horizons in the 12-m KP-1 section (Fig. 16) and several other sites on the bioclimatic transect (Blinnikov *et al.*, 2002). Plant identification was based on phytoliths extracted from a modern reference collection of plants and soils. Many dominant plants such as *Artemisia tridentata*, *Agropyron spicatum*, *Poa sandbergii*, and *Pinus ponderosa* contain characteristic morphotypes and thus could be identified in the paleorecord. Eight vegetation types, including four grassland types, three forest types, and *Artemisia* shrub steppe could be identified based on phytolith assemblages in soils. The fossil assemblages were directly compared with the modern assemblages by means of the squared-chord distance measure to establish the degree of similarity between modern analogues and the fossil samples. Based on the reconstruction at KP-1, the vegetation composition has undergone significant changes in the last 100,000 years. The modern *Agropyron-Festuca* assemblage was preceded by a late-Pleistocene assemblage that contained a greater proportion of *Stipa-Artemisia* tridentate phytoliths, implying colder and drier climate (Blinnikov *et al.*, 2001). Based on results from four sites on the bioclimatic transect into the Blue Mountains, grasslands and shrub steppe were more widespread at mid to high elevations during MIS 2 and 4, probably due to cold and dry conditions (Blinnikov *et al.*, 2002).

In a similar way, stable carbon isotopes were analyzed from soil organic matter and soil carbonates with respect to position along a C3-plant dominated climatic gradient (Stevenson, 1997). $\delta^{13}C$ from pedogenic carbonates ranged from -3.1 in the arid sites to -10.5 (PDB) in the wetter sites and were highly correlated with MAP ($r^2 = 0.90$, $p = 0.001$), estimated above-ground productivity, and soil organic carbon. This relationship may serve with some caution as a proxy of paleoproductivity, probably influenced principally by MAP. A depth plot of $\delta^{13}C$ for carbonates at KP-1 suggested that conditions were more arid across the Columbia Plateau during MIS 2, parts of MIS 3, and parts of MIS 5 (Fig. 16). Relationships also were developed by Stevenson (1997) between the $\delta^{18}O$ signature of meteoric precipitation, soil water, and modern pedogenic carbonates across the climatic gradient. $\delta^{18}O$ values for carbonates ranged from -14.13 to -11.38 per mil, becoming more enriched with decreasing temperature. Although the interactions and uncertainties are greater for oxygen isotopes in soil systems than for carbon isotopes, as a first approximation, the carbonate $\delta^{18}O$ values may serve as a paleothermometer (Fig. 16).

Sand Dunes on the Columbia Plateau

Dune fields such as the Hanford dunes, Smith Canyon dunes, Juniper dunes, Eureka Flat dunes, and Umatilla Basin dune field occur in south-central Washington and north-central Oregon (Gaylord & Stetler, 1994; Gaylord *et al.*, 1991, 1997, 1998, 2001; Schatz, 1996; Smith, 1992; Stetler, 1993; Stetler & Gaylord, 1996). Most of these dune fields are downwind of deflational flats of giant-flood deposits, have SW-NE orientations, and are upwind of or cap loess. The Quincy dune field on the northern plateau occupies more than 1000 km^2 (Fig. 14; Petrone, 1970). Sediments from the dunes are similar compositionally to outburst-flood deposits and Palouse loess (Busacca & McDonald, 1994; Gaylord *et al.*, 1991; Gaylord *et al.*, 2001).

Datable materials are rare in the dunes of the Columbia Plateau; however, Gaylord & Stetler (1994) and Gaylord *et al.* (2001) concluded based on ages inferred from interstratified tephra that dune activity on the Columbia Plateau peaked during the early and middle Holocene. Modern sand dunes on the plateau are primarily parabolic and blowout dune types, whose activity and form are controlled by stabilizing plant cover. Active dunes today tend to be in higher topographic positions subject to higher wind speeds and tend to have modified parabolic-barchan-barchanoid shapes.

Research underway on the Columbia Plateau emphasizes the need to understand the dynamic interactions between flood-sediment sources in basins, proximal sand dunes, and distal loess (Gaylord *et al.*, 2002; Sweeney *et al.*, 2001). The Columbia Plateau eolian system appears to have been controlled by three primary end-member factors: character of the source sediment, bioclimate (plant cover density and soil moisture), and topographic traps. In the Channeled Scabland system, flood-proximal sites contain coarse-grained outburst flood deposits. These have been reworked by prevailing SW-NE winds into extensive sand dunes accompanied only by thin loess, such as at the Quincy dunes (Fig. 14). In contrast, flood-distal sites contain coarse-depleted, fine-grained slackwater sediments. These have been reworked downwind into relatively thin, discontinuous sand dunes and sand sheets but this succeeds to thick loess even farther downwind. An example is Eureka Flat (Fig. 14). Thick loess deposits apparently have been generated where bioclimatic controls (plant cover density and soil moisture) along the SW-NE semiarid to subhumid climatic gradient have enhanced loessial accumulation (Sweeney *et al.*, 2002b). Topographic irregularities in and near sediment source basins have influenced loess accumulation by selectively trapping saltating sand, promoting aggradation downwind of thick units of suspended load silt, such as at Juniper Canyon, Oregon (Fig. 14; Sweeney *et al.*, 2001).

Loess on the Snake River Plain

Distribution and Thickness

On the eastern Snake River Plain in southern Idaho, loess deposits greater than 6 m thick cover several thousand km^2 adjacent to the Snake and other rivers (Fig. 14; Lewis & Fosberg, 1982). Loess on the plain reaches a maximum thickness of about 12 m and exceptionally to 36 m (Lewis & Fosberg, 1982) or even greater than 60 m (Malde, 1991; K.L. Pierce, pers. comm., 2002) in protected positions. Thickest loess is generally found to the south and southeast of the Snake River and is presumably derived largely from alluvial sediments on the Snake River Plain (Glenn *et al.*, 1983; Lewis *et al.*, 1975; Scott, 1982). Loess extends into the uplands of mountainous areas in far southeastern Idaho and southwestern Wyoming (Glenn *et al.*, 1983). Because of the great range in age of lava flows and other surfaces in the Snake River Plain, the age and thickness of loess preserved at any place is partly a function of the age of the landform on which it is deposited (Malde, 1991). Loess in the western Snake River Plain and Boise Valley is thin, reaching a thickness of only about 2 m on late middle Pleistocene and younger terrace surfaces there (Othberg, 1994).

Stratigraphy, Paleopedology, and Chronology

Stratigraphic studies of loess on the Snake River Plain by Pierce *et al.* (1982) led to the division of the upper part of the loess into two informal units named "loess A" and "loess B." Loess units thin and become finer grained to the east and southeast of the Snake River (Glenn *et al.*, 1983; Lewis *et al.*, 1975; Lewis & Fosberg, 1982; Pierce *et al.*, 1982), supporting the interpretation of a floodplain source and westerly or northwesterly wind direction for much of the loess.

The near absence of loess on Holocene basalt surfaces suggests that loess deposition has been minimal on the SRP over the last 12,000 yr. In part because of this, a surface soil developed in the top of loess unit A that is comparable to soils developed on latest Pleistocene deposits in similar climatic settings (Pierce *et al.*, 1982). Soils on loess unit A display calcic, argillic or both types of horizons depending on annual precipitation (Lewis & Fosberg, 1982). Stratigraphic relations to dated pumice layers, Bonneville Flood deposits, and dated lava flows suggest that loess A was deposited during part or all of MIS 4–2, or between about 70,000 and perhaps as late as 12,000 yr ago, because local accumulations of loess overlie 14,400-yr-old Bonneville Flood gravels. (Pierce *et al.*, 1982).

Loess unit B is capped by a strongly developed buried soil, named the Fort Hall geosol by McDole (1969; McDole *et al.*, 1973) that is more developed than the post-loess A soil and is therefore estimated to have formed over several tens of thousands of years (Pierce *et al.*, 1982). Loess unit B conformably overlies glacial deposits at sites near West Yellowstone and Jackson Hole that are correlated to MIS 6 or about 130,000 yr ago (K.L. Pierce, pers. Comm., 2002), Because loess unit A has a lower limiting age of about 70,000 yr ago, the strong buried soil on loess unit B arguably formed during MIS 5, the last full interglaciation. Several other dark bands and weak to moderate buried soils occur in loess A and B, suggesting periods of interrupted or slowing loess deposition.

Forman *et al.* (1993) used TL dating at a site well north of the Snake River to interpret an age of 40,000 to 10,000 yr ago for loess unit A. They associated its deposition with Pinedale glaciation and pluvial lake expansion. They interpret loess unit B to have been deposited from about 80,000–60,000 yr ago, and the well developed paleosol between the two units to have likely formed between 70,000 and 40,000 yr ago Based on the TL ages associated with loess B and that it is underlain in their research area by a lava flow with a K/Ar age of 95,000 $\pm$ 25,000 yr ago, Forman *et al.* (1993) reinterpreted loess B as representing an episode of dust deposition that occurred during MIS 4 rather than during the long-accepted MIS 6 interval. Older loesses occur on the Snake River Plain but lack of exposure has prevented detailed characterization or dating up to now. Loess on the north side of the Snake River near Twin Falls contains many loess units, and the

630,000-yr-old Lava Creek ash occurs in loess that is many tens of meters thick in the Gay Mines area southeast of Blackfoot, Idaho (K.L. Pierce, pers. comm., 2002).

Sand Dunes on the Snake River Plain

Eolian sands occur on the Snake River Plain almost wholly in the eastern part. They consist of thin sand sheets and deeper accumulations of stabilized dunes and small areas of active dunes (Coleman, 2002; Kuntz, 1979; Scott, 1982). Bruneau dunes (Fig. 14) lie to the south of Mountain Home, Idaho in the central part of the plain and cover an area of more than 100 km^2 up to 150 m thick.

Stabilized dunes and sand sheets form two areas on the eastern plain. The first covers a linear zone that extends more than 50 km on the north side of the Snake River from near Burley to American Falls, Idaho. This accumulation continues on the south side of the river for more than 100 km from American Falls to Idaho Falls, Idaho. This area of stabilized dunes and sand sheets ranges from less than 1 km to more than 20 km wide and in thickness from less than 1–7 m (Scott, 1982). These dunes apparently have not been dated but overlie and thus are at least partly younger than loess A (Malde, 1991), that is, Holocene in age. They appear to have been derived from deflation of MIS 2 glacial outwash along the Snake River and from terminal playas of rivers from the north such as the Big Lost River (Malde, 1991; Scott, 1982).

The second consists of stabilized to active dunes that cover more than 300 km^2 with sand up to 140 m thick (Coleman, 2002) on the northern margin of the Snake River Plain (Fig. 14) from the Little Lost River on the southwest about 100 km to the northeast to the town of St. Anthony, Idaho. The active dunes are called variously the Juniper Butte dunes and St. Anthony dunes. The sand in St. Anthony dunes was derived from sediments in glacial Lake Terreton and from the Henry's Fork of the Snake River (Coughlin & Gaylord, 1999; Scott, 1982). Active dunes in the complex consist of barchan, barchanoid ridge, seif, linear, parabolic, and blowout dunes (Coughlin, 2000; Coughlin & Gaylord, 1999). Active dunes trend about N20E, whereas stabilized dunes trend about N40E (Koscielniak, 1973). Recent work has documented a record of migrating dunes, dune damming of streams, flooding and stream avulsions, and eolian remobilization in several episodes from at least 4280 ^{14}C yr B.P. (Coleman, 2000; Gaylord *et al.*, 2000).

Summary

Great progress has been made in the study of eolian sediments in North America since 1965. Some of the more exciting finds include uncovering the long eolian record preserved in the extensive loess of Alaska, better understanding of sources and timing of loess and sand dunes in the central U.S., better understanding of loess stratigraphy in the lower Mississippi valley, enhanced resolution of dune activity in the desert southwest U.S. and its relation to paleoclimate and paleohydrology, and better understanding of source areas, timing, and controls on loess accumulation linked to catastrophic flooding and environmental change in the Palouse loess of the Pacific Northwest. This progress has been made possible by technological advancements such as luminescence geochronology and by the collaboration of many scientists whose expertise spans multiple disciplines, including sedimentology, geomorphology, geochronology, pedology, and ecology.

In the future, eolian researchers will tackle issues such as the poorly understood interactions between juxtaposed sand dune and loess systems like those in the Great Plains and the Palouse of the Pacific Northwest. Bodies of sand dunes and loess are often thought of as separate entities, but are sometimes linked by common sources and processes. Kocurek & Lancaster (1999) outlined a state theory for eolian sediment, defined by sediment supply, sediment availability, and transport capacity, which, if we could analyze these forcings simultaneously, might allow complete quantification of an eolian system. Understanding of eolian systems will continue to improve through such approaches, especially when attempting to calculate mass balances in open or closed eolian systems. Part of understanding mass balances lies in determination of source areas that may fluctuate in size, shape, and level of activity through time. Stratigraphic and geochemical fingerprinting techniques will help pinpoint sources and sinks. Once sources and sinks can be identified, the incorporation of field studies with enhanced computer simulations will aid in modeling eolian response to changing atmospheric circulation patterns and other climatic fluctuations. Attempts are being made to quantify natural vs. anthropogenic dust emissions in order to create global climate simulations that can forecast how dust emissions have affected climate in the past and how they may affect climate in the future.

Acknowledgments

The authors thank Vance Holliday and Joe Mason for thoughtful and constructive reviews.

References

Abbott, M.B. & Stafford, T.W., Jr. (1996). Radiocarbon geochemistry of modern and ancient Arctic lake systems, Baffin Island, Canada. *Quaternary Research*, **45**, 300–311.

Ahlbrandt, T.S. & Fryberger, S.G. (1980). Eolian deposits in the Nebraska Sand Hills. *U.S. Geological Survey Professional Paper*, 1120-A, 24 pp.

Ahlbrandt, T.S., Swinehart, J.B. & Maroney, D.G. (1983). The dynamic Holocene dune fields of the Great Plains and Rocky Mountain basins, U.S.A. *In*: M.E. Brookfield & T.S. Ahlbrandt (Eds), *Eolian Sediments and Processes*. New York, Elsevier, pp. 379–406.

Aleinikoff, J.N., Muhs, D.R. & Fanning, C.M. (1998). Isotopic evidence for the sources of late Wisconsin (Peoria) loess, Colorado and Nebraska: Implications for paleoclimate. *In*: Busacca, A.J. (Ed.), *Dust Aerosols, Loess Soils and Global Change*. Miscellaneous Publication No. MISC0190: Pullman, Washington, Washington State University College of Agriculture and Home Economics, pp. 124–127.

Aleinikoff, J.N., Muhs, D.R., Sauer, R.R. & Fanning, C.M. (1999). Late Quaternary loess in northeastern Colorado, II–Pb isotopic evidence for the variability of loess sources. *Geological Society of America Bulletin*, **111**, 1876–1883.

Anderson, P.M. & Lozhkin, A.V. (2001). The Stage 3 interstadial complex (Karginskii/middle Wisconsinan interval) of Beringia variations in paleoenvironments and implications for paleoclimatic interpretations. *Quaternary Science Reviews*, **20**, 93–125.

Arbogast, A.F. (1996). Stratigraphic evidence for late-Holocene aeolian sand mobilization and soil formation in south-central Kansas, U.S.A. *Journal of Arid Environments*, **34**, 403–414.

Arbogast, A.F. & Muhs, D.R. (2000). Geochemical and mineralogical evidence from eolian sediments for northwesterly mid-Holocene paleowinds, central Kansas, U.S.A. *Quaternary International*, **67**, 107–118.

Bachhuber, F.W. (1982). Quaternary history of the Estancia Valley, central New Mexico. *In*: Callender, J.F., Grambling, J.A. & Wells, S.G. (Eds), *Guidebook, Albuquerque Country II*. Socorro, New Mexico, New Mexico Geological Society, pp. 343–346.

Baker, V.R. & Bunker, R.C. (1985). Cataclysmic late Pleistocene flooding from glacial Lake Missoula: A review. *Quaternary Science Reviews*, **4**, 1–41.

Bartlein, P.J. (1997). Paleoclimatic variations in Beringia: Large-scale controls and regional responses in general circulation model simulations. *In*: Edwards, M.E., Sher, A. & Guthrie, R.D. (Eds), *Terrestrial Paleoenvironmental Studies in Beringia*. Fairbanks, Alaska, The Alaska Quaternary Center, University of Alaska, pp. 43–47.

Bartlein, P.J., Anderson, K.H., Anderson, P.M., Edwards, M.E., Mock, C.J., Thompson, R.S., Webb, R.S., Webb, T., III & Whitlock, C. (1998). Paleoclimate simulations for North America over the past 21,000 years: Features of the simulated climate and comparisons with paleoenvironmental data. *Quaternary Science Reviews*, **17**, 549–585.

Beaudoin, A., Davis, O., Delcourt, H., Delcourt, P., Richard, P.J.H. & Adams, J. (1999). Palaeovegetation Maps for North America: http://www.esd.ornl.gov/projects/qen/nercNORTHAMERICA.html

Begét, J.E. (1988). Tephras and sedimentology of frozen loess. *In*: Senneset, K. (Ed.), Proceedings 5th International Permafrost Conference. Trondheim, Tapir, Vol. 1, pp. 672–677.

Begét, J.E. (1990). Middle Wisconsin climate fluctuations recorded in central Alaskan loess. *Geographie Physique et Quaternaire*, **44**, 3–13.

Begét, J.E. (1991). Paleoclimatic significance of high latitude loess deposits. *In*: Weller, G. (Ed.), *International Conference on the Role of the Polar Regions in Global Change*. Fairbanks, Alaska, Geophysical Institute, University of Alaska, Vol. II, pp. 594–598.

Begét, J.E. (1996). Tephrochronology and paleoclimatology of the last interglacial-glacial cycle recorded in Alaskan loess deposits. *Quaternary International*, **34–36**, 121–126.

Begét, J.E. (2000). Integrating loess, glacial, and periglacial Quaternary history in central Alaska using tephrochronology. Troy Péwé Memorial Workshop, Fairbanks, Alaska, Abstracts and Program, p. 14.

Begét, J.E. (2001). Continuous Late Quaternary proxy climate records from loess in Beringia. *Quaternary Science Reviews*, **20**, 499–507.

Begét, J.E. & Hawkins, D.B. (1989). Influence of orbital parameters on Pleistocene loess deposition in central Alaska. *Nature*, **337**, 151–153.

Begét, J.E., Stone, D.B. & Hawkins, D.B. (1990). Paleoclimatic forcing of magnetic susceptibility variations in Alaskan loess during the Quaternary. *Geology*, **18**, 40–43.

Begét, J., Edwards, M., Hopkins, D. & Kukla, G. (1991a). Old Crow tephra found at the Palisades of the Yukon. *Quaternary Research*, **35**, 291–296.

Begét, J., Bigelow, N. & Powers, R. (1991b). Reply to the comment of C. Waythomas and D. Kaufmann on Bigelow, N., Begét, J. & Powers, R. (1990). Latest Pleistocene increase in wind intensity recorded in eolian sediments from central Alaska: Reply. *Quaternary Research*, **34**, 160–168.

Beheiry, S.A. (1967). Sand forms in the Coachella Valley, Southern California. *Annals of the Association of American Geographers*, **57**, 25–48.

Berger, G.W. (1991). The use of glass for dating volcanic ash by thermo-luminescence. *Journal of Geophysical Research*, **96**(B12), 19,705–19,720.

Berger, G.W. & Busacca, A.J. (1995). Thermoluminescence dating of late Pleistocene loess and tephra from eastern Washington and southern Oregon, and implications for the eruptive history of Mount St. Helens. *Journal of Geophysical Research*, **100**(B11), 22,361–22,374.

Berger, G.W., Pillans, B.J. & Palmer, A.S. (1994). Test of thermoluminescence dating of loess from New Zealand and Alaska. *Quaternary Science Reviews*, **13**, 309–333.

Berger, G.W., Péwé, T.L., Westgate, J.A. & Preece, S.J. (1996). Age of Sheep Creek tephra (Pleistocene) in central Alaska from thermoluminescence dating of bracketing loess. *Quaternary Research*, **45**, 263–270.

Bigelow, N., Begét, J. & Powers, R. (1990). Latest Pleistocene increase in wind intensity recorded in eolian sediments from central Alaska. *Quaternary Research*, **34**, 160–168.

Bjornstad, B.N., Fecht, K.R. & Pluhar, C.J. (2001). Long history of pre-wisconsin, ice age cataclysmic floods; evidence from southeastern Washington state. *Journal of Geology*, **109**, 695–713.

Blinnikov, M., Busacca, A. & Whitlock, C. (2001). A new 100,000-yr. phytolith record from the Columbia Basin, Washington, U.S.A. *In*: Meunier, J.D. & Colin, F. (Eds), *Phytoliths – Applications in Earth Science and Human History*. Dordrecht, A.A. Balkema, 380, pp. 27–55.

Blinnikov, M., Busacca, A. & Whitlock, C. (2002). Reconstruction of the late-Pleistocene grassland of the Columbia basin, Washington, USA, based on phytolith records in loess. *Palaeogeography, Palaeoclimatology, Palaeoecology*, **177**, 77–101.

Boyle, E.A. (2000). Is thermohaline circulation linked to abrupt stadial/interstadial transitions? *Quaternary Science Reviews*, **19**, 255–272.

Brooks, M.J., Taylor, B.E., Stone, P.A. & Gardner, L.R. (2001). Pleistocene encroachment of the Wateree River sand sheet into Big Bay on the Middle Coastal Plain of South Carolina. *Southeastern Geology*, **40**, 241–257.

Bryan, K. (1927). The Palouse soil problem. *U.S. Geological Survey Bulletin*, **790**, 21–46.

Busacca, A.J. (1989). Long Quaternary record in eastern Washington, U.S.A. interpreted from multiple buried paleosols in loess. *Geoderma Special Issue on Climatic and Lithostratigraphic Significance of Paleosols*, **45**, 105–122.

Busacca, A.J. (1991). Loess deposits and soils of the Palouse and vicinity. *In*: Morrison, R.B. (Ed.), *Quaternary Non-Glacial Geology of the United States*. Geological Society of America Geology of North America, Vol. K-2, pp. 216–228.

Busacca, A.J. & McDonald, E.V. (1994). Regional sedimentation of late Quaternary loess on the Columbia Plateau: Sediment Source Areas and Loess Distribution Patterns, Regional Geology of Washington State. *Washington Division of Geology and Earth Resources Bulletin*, **80**, 181–190.

Busacca, A.J., Nelstead, K.T., McDonald, E.V. & Purser, M.D. (1992). Correlation of distal tephra layers in loess in the Channeled Scabland and Palouse of Washington state. *Quaternary Research*, **37**, 281–303.

Busacca, A.J., Lilligren, S. & McDonald, E.V. (1994). Major-element analysis used to interpret lithology and pedogenesis in a bioclimatic sequence of paleosols, Palouse loess. *Agronomy Abstracts*, Madison, WI, American Society of Agronomy, p. 345.

Busacca, A.J., Verosub, K.L., McDonald, E.V. & Murphy, E.M. (1997). A 1 My environmental magnetic and pedologic record from Palouse Loess. *Agronomy Abstracts*, American Society of Agronomy, Annual Meeting, Anaheim, CA, p. 308.

Busacca, A.J., Verosub, K.L., McDonald, E.V., Murphy, E.M. (1998). *Magnetostratigraphy, environmental magnetism and pedology of a 1 My record of the Palouse loess*, Washington state. American Geophysical Union, Spring Meeting, Boston, MA, Abstract No. GP31A-07.

Chadwick, O.A., Derry, L.A., Vitousek, P.M., Huebert, B.J. & Hedin, L.O. (1999). Changing sources of nutrients during four million years of ecosystem development. *Nature*, **397**, 491–497.

Clarke, M.L. (1994). Infra-red stimulated luminescence ages from aeolian sand and alluvial fan deposits from the eastern Mojave Desert, California. *Quaternary Science Reviews*, **13**, 533–538.

Clarke, M.L. & Rendell, H.M. (1998). Climatic change impacts on sand supply and the formation of desert sand dunes in the south-west U.S.A. *Journal of Arid Environments*, **39**, 517–532.

Clarke, M.L., Richardson, C.A. & Rendell, H.M. (1996). Luminescence dating of Mojave Desert sands. *Quaternary Science Reviews*, **14**, 783–790.

Clark, P.U., Nelson, A.R., McCoy, W.D., Miller, B.B. & Barnes, D.K. (1989). Quaternary aminostratigraphy of Mississippi Valley loess. *Geological Society of America Bulletin*, **101**, 918–926.

COHMAP Members (1988). Climatic changes of the last 18,000 years: Observations and model simulations. *Science*, **241**, 1043–1052.

Coleman, A.J. (2002). Mid-to-Late Holocene eolian, alluvial, and lacustrine deposits and paleoclimate fluctuations from the St. Anthony dune field, southeastern Idaho [M.S. thesis]. Pullman, Washington State University.

Coughlin, J.J. (2000). Sedimentary and geomorphic development and climatic implications of the St. Anthony dune fields, Fremont County, southeastern Idaho [M.S. thesis]. Pullman, Washington State University, 83 pp.

Coughlin, J.J. & Gaylord, D.R. (1999). Sedimentary and geomorphic development and climatic implications of the St. Anthony dune field, Fremont County, Idaho. Abstracts with Programs, Geological Society of America, Vol. 31, 54 pp.

Curry, B.B. & Pavich, M.J. (1996). Absence of glaciation in Illinois during marine isotope stages 3 and 5. *Quaternary Research*, **46**, 19–26.

Curry, B.B. & Follmer, L.R. (1992). The last interglacial-glacial transition in Illinois: 123–25 ka. *In*: Clark, P.U. & Clark, P.D. (Eds), *The Last Interglacial-Glacial Transition in North America*. Geological Society of America Special Paper 270, pp. 71–88.

Dean, W.E. (1993). Physical properties, mineralogy, and geochemistry of Holocene varved sediments from Elk Lake, Minnesota, in Elk Lake, Minnesota: Evidence for rapid climate change in the north-central United States. *In*: Bradbury, J.P. & Dean W.E. (Eds), Geological Society of America Special Paper, pp. 135–157.

Delcourt, P.A., Delcourt, H.R., Brister, R.C. & Lackey, L.E. (1980). Quaternary vegetation history of the Mississippi Embayment. *Quaternary Research*, **13**, 111–132.

Dutta, P.K., Zhou, Z. & dos Santos, P.R. (1993). A theoretical study of mineralogical maturation of eolian sand. Geological Society of America Special Paper, 284, pp. 203– 209.

Edwards, M.E. & McDowell, P.F. (1991). Interglacial deposits at Birch Creek, northeast interior Alaska. *Quaternary Research*, **35**, 41–52.

Ellwein, A.L. (1997). Quaternary evolution of eolian landforms, soils, and landscapes of the Petrified Forest National Park, Arizona [M.S. thesis]. University of New Mexico, 176 pp.

Enzel, Y. & Wells, S.G. (1997). Extracting Holocene paleohydrology and paleoclimatology information from modern extreme flood events: An example from southern California. *Geomorphology*, **19**, 203–226.

Enzel, Y., Brown, W.J., Anderson, R.Y., McFadden, L.D. & Wells, S.G. (1992). Short-Duration Holocene Lakes in the Mojave River Drainage Basin, Southern California. *Quaternary Research*, **38**, 60–73.

Evans, J.R. (1961). Falling and climbing sand dunes in the Cronese ("Cat") Mountain area, San Bernadino County, California. *Journal of Geology*, **70**, 107–113.

Eymann, J.L. (1953). A study of sand dunes in the Colorado and Mojave Deserts [M.S. thesis]. University of Southern California.

Flint, R.F. (1938). Origin of the Cheney-Palouse scabland tract. *Geological Society of America Bulletin*, **49**, 461–524.

Foley, L.L. (1982). Quaternary chronology of the Palouse loess near Washtucna, Washington [M.S. thesis]. Bellingham, Western Washington University, 137 pp.

Forman, S.L., Bettis, E.A., III, Kemmis, T.J. & Miller, B.B. (1992). Chronologic evidence for multiple periods of loess deposition during the late Pleistocene in the Missouri and Mississippi River valley, United States: Implications for the activity of the Laurentide Ice Sheet. *Palaeogeography, Palaeoclimatology, Palaeoecology*, **93**, 71–83.

Forman, S.L., Smith, R., Hackett, W., Tullis, J. & McDaniel, P. (1993). Timing of late Quaternary glaciations in the Western United States based on the age of loess on the eastern Snake River Plain, Idaho. *Quaternary Research*, **40**, 30–37.

Forman, S.L., Oglesby, R., Markgraf, V. & Stafford, T. (1995). Paleoclimatic significance of Late Quaternary eolian deposition on the Piedmont and High Plains, Central United States. *Global and Planetary Change*, **11**, 35–55.

Fryberger, S.G. & Dean, G. (1979). Dune forms and wind regime. *In*: McKee, E.D. (Ed.), A study of global sand seas. *U.S. Geological Survey Professional Paper*, 1052, pp. 137–169.

Frye, J.C., Glass, H.D. & Willman, H.B. (1968). Mineral zonation of Woodfordian loesses of Illinois. *Illinois State Geological Survey Circular*, **427**, 1–44.

Fryxell, R. & Cook, E.F. (1964). A field guide to the loess deposits and Channeled Scablands of the Palouse area, eastern Washington, Pullman. *Washington State University Laboratory of Anthropology Report of Investigations*, **27**, 32.

Gaylord, D.R. & Stetler, L.D. (1994). Eolian-climatic thresholds and sand dunes at the Hanford Site, south-central Washington, USA. *Journal of Arid Environments*, **28**, 95–116.

Gaylord, D.R., Stetler, L.D., Smith, G.D. & Chatters, J.C. (1991). Holocene and recent eolian activity at the Hanford Site, Washington. American Geophysical Union Pacific Northwest Section Meeting, Richland, WA, Sept. 18–20, *Abstracts with Programs*.

Gaylord, D.R., Schatz, J.K., Coleman, A.J. & Foit, F.F., Jr. (1997). Holocene to recent sedimentary, stratigraphic, and geomorphic development of the Juniper dune field, Washington. Geological Society of America Annual Meeting, Salt Lake City, Utah, *Abstracts with Programs*, 29, A218 pp.

Gaylord, D.R., Foit, F.F., Schatz, J.K. & Coleman, A.J. (1998). Sand dune activity within sensitive agricultural lands, Columbia Plateau, USA. *In*: Busacca, A.J. (Ed.), *Dust Aerosols, Loess Soils and Global Change*, Miscellaneous Publication No. MISC0190. Pullman, Washington State University College of Agriculture and Home Economics, pp. 145–148.

Gaylord, D.R., Coughlin, J.J., Coleman, A.J., Sweeney, M.R. & Rutford, R.H. (2000). Holocene sand dune activity and paleoclimate from Sand Creek, St. Anthony dune field, Idaho. *Abstracts with Programs, Geological Society of America*, **32**, 10.

Gaylord, D.R., Foit, F.F., Jr., Schatz, J.K. & Coleman, A.J. (2001). Smith Canyon dune field, Washington, U.S.A.: relation to glacial outburst floods, the Mazama eruption, and Holocene paleoclimate. *Journal of Arid Environments*, **47**, 403–424.

Gaylord, D.R., Sweeney, M.R. & Busacca, A.J. (2002). Geomorphic development of a late Quaternary paired eolian sequence, Columbia Plateau, Washington. *Geological Society of America, Abstracts with Programs*, **34**, 245.

Gile, L.H. (1999). Eolian and associated pedogenic features of the Jornada Basin floor, southern New Mexico. *Soil Science Society of America Journal*, **63**, 151–163.

Gile, L.H., Peterson, F.F. & Grossman, R.B. (1966). Morphological and genetic sequences of carbonate accumulation in desert soils. *Soil Science*, **10**, 347–360.

Glenn, W.R., Nettleton, W.D., Fowkes, C.J. & Daniels, D.M. (1983). Loessial deposits and soils of the Snake and tributary valleys of Wyoming and eastern Idaho. *Soil Science Society of America Journal*, **47**, 547–552.

Goudie, A.S., Livingstone, I. & Stokes, S. (1999). *Aeolian environments, sediments and landforms*. Chichester, John Wiley and Sons, Ltd., 325 pp.

Griffiths, P.G., Webb, R.H., Lancaster, N., Kaehler, C.A. & Lundstrom, S.C. (2002). Long-term sand supply to Coachella Valley fringe-toed lizard (*Uma inornata*) habitat in the northern Coachella Valley, California. United States Geological Survey Water-Resources Investigations Report: Washington D.C.

Grimley, D.A. (2000). Glacial and nonglacial sediment contributions to Wisconsin episode loess in the central United States. *Geological Society of America Bulletin*, **112**, 1475–1495.

Grimley, D.A., Follmer, L.R. & McKay, E.D. (1998). Magnetic susceptibility and mineral zonations controlled by provenance in loess along the Illinois and central Mississippi River valleys. *Quaternary Research*, **49**, 24–36.

Grootes, P.M., Stuiver, M., White, J.W.C., Johnsen, S. & Jouzel, J. (1993). Comparison of oxygen isotope records from the GISP2 and GRIP Greenland ice cores. *Nature*, **366**, 552–554.

Guthrie, R.D. (1990). *Frozen fauna of the Mammoth Steppe: the story of Blue Babe*. Chicago, University of Chicago Press, 323 pp.

Hack, J.T. (1941). Dunes of the Western Navajo County. *Geographical Review*, **31**, 240–263.

Hamilton, T.D. & Brigham-Grette, J. (1991). The last interglaciation in Alaska: Stratigraphy and paleoecology of potential sites. *Quaternary International*, **10–12**, 49–71.

Hamilton, T.D., Ager, T.A. & Robinson, S.W. (1983). Late Holocene ice wedges near Fairbanks, Alaska, U.S.A.: Environmental setting and history of growth. *Arctic and Alpine Research*, **15**, 157–168.

Hamilton, T.D., Craig, J.L. & Sellmann, P.V. (1988). The Fox permafrost tunnel: A late Quaternary geologic record in central Alaska. *Geological Society of America Bulletin*, **100**, 948–969.

Handy, R.L. (1977). Loess distribution by variable winds. *Geological Society of America Bulletin*, **87**, 915–927.

Hayward, R.K. & Lowell, T.V. (1993). Variations in loess accumulation rates in the mid-continent, United States, as reflected by magnetic susceptibility. *Geology*, **21**, 821–824.

Hofmann, H.J., Beer, J., Bonani, G., Von Gunten, H.R., Raman, S., Suter, M., Walker, R.L., Wšlfi, W. & Zimmermann, D. (1987). ^{10}Be: Half-life and AMS standards. *Nuclear Instruments and Methods in Physics Research*, **B29**, 32–36.

Holliday, V.T. (1989). The Blackwater Draw Formation (Quaternary): A 1.4-plus m.y. record of eolian sedimentation and soil formation on the Southern High Plains. *Geological Society of America Bulletin*, **101**, 1598–1607.

Holliday, V.T. (2001). Stratigraphy and geochronology of upper Quaternary eolian sand on the Southern High Plains of Texas and New Mexico, United States. *Geological Society of America Bulletin*, **113**, 88–108.

Hopkins, D.M. (1982). Aspects of the paleogeography of Beringia during the late Pleistocene. *In*: Hopkins, D.M., Matthews, J.V., Jr., Schweger, C.E. & Young, S.B. (Eds), *Paleoecology of Beringia*. New York, Academic Press, pp. 3–28.

Hugie, V.K. & Passey, H.B. (1963). Cicadas and their effect upon soil genesis in certain soils in southern Idaho, northern Utah, and northeastern Nevada. *Soil Science Society of America Proceedings*, **27**, 78–82.

Hunter, R.E. (1977). Basic types of stratification in small eolian dunes. *Sedimentology*, **24**, 361–387.

Ivester, A.H., Leigh, D.S. & Godfrey-Smith, D.I. (2001). Chronology of inland eolian dunes on the Coastal Plain of Georgia, U.S.A. *Quaternary Research*, **55**, 293–302.

Jacobs, P.M. & Knox, J.C. (1994). Provenance and petrology of a long-term Pleistocene depositional sequence in Wisconsin's Driftless Area. *Catena*, **22**, 49–68.

Johnson, R.A., Pye, K. & Stipp, J.J. (1984). Thermoluminescence dating of southern Mississippi loess. American Quaternary Association, 8th Biennial Meeting, Boulder, Colorado, *Program and Abstracts*, **8**, 64.

Johnson, W.C. & Willey, K.L. (2000). Isotopic and rock magnetic expression of environmental change at the Pleistocene-Holocene transition in the central Great Plains. *Quaternary International*, **67**, 89–106.

Kemp, R.A., McDaniel, P.A. & Busacca, A.J. (1998). Genesis and relationship of macromorphology and micromorphology to contemporary hydrological conditions of a welded Argixeroll from the Palouse in NW U.S.A. *Geoderma*, **83**, 309–329.

King, M. (2000). Late Quaternary loess-paleosol sequences in the Palouse, Northwest USA: pedosedimentary and paleoclimatic significance [Ph.D. dissertation]. Royal Holloway, University of London. 231 pp.

King, M., Busacca, A.J., Foit, F.F., Jr. & Kemp, R.A. (2001). Identification of the Trego Hot Springs tephra in the Palouse, Washington State. *Quaternary Research*, **56**, 165–169.

Koscielniak, D.E. (1973). Eolian deposits on a volcanic terrain near Saint Anthony, Idaho [M.S. thesis]. Buffalo, State University of New York, 28 pp.

Kocurek & Lancaster (1999). Aeolian system sediment state; theory and Mojave Desert Kelso dune field example. *Sedimentology*, **46**, 505–515.

Kohfeld, K.E. & Harrison, S.P. (2001). DIRTMAP: The geological record of dust. *Earth-Science Reviews*, **54**, 81–114.

Kukla, G.J. & Opdyke, N.D. (1980). Matuyama loess at Columbia Plateau, Washington. *American Quaternary Association, Abstracts with Programs*, **6**, 122.

Kuntz, M.A. (1979). Geologic map of the Juniper Buttes area, eastern Snake River Plain, Idaho. U.S. Geological Survey Miscellaneous Investigations Map I-1115, scale 1:48,000.

Lagroix, F. & Banerjee, S.K. (2002). Paleowind directions from the magnetic fabric of loess profiles in central Alaska. *Earth and Planetary Science Letters*, **195**, 99–112.

Lancaster, N. (1992). Relations between dune generations in the Gran Desierto, Mexico. *Sedimentology*, **39**, 631–644.

Lancaster, N. (1995). Origin of the Gran Desierto Sand Sea: Sonora, Mexico: Evidence from dune morphology and sediments. *In*: Tchakerian, V.P. (Ed.), *Desert Aeolian Processes*. New York, Chapman and Hall, pp. 11–36.

Lancaster, N. & Tchakerian, V.P. (2003). Late Quaternary eolian dynamics, Mojave Desert, California. *In*: Enzel, Y., Wells, S.G. & Lancaster, N. (Eds), *Paleoenvironments and Paleohydrology of the Mojave and Southern Great Basin Deserts*. Boulder, CO, Geological Society of America.

Langford, R.P. (2000). Nabkha (coppice dune) fields of south-central New Mexico, U.S.A. *Journal of Arid Environments*, **46**, 25–41.

Langford, R.P. (2003). The Holocene history of the White Sands dune field and influences on eolian deflation and playa lakes. *Quaternary International*, **104**, 31–39.

Lea, P.D. (1990). Pleistocene periglacial eolian deposits in southwestern Alaska: sedimentary facies and depositional processes. *Journal of Sedimentary Petrology*, **60**, 582–591.

Lea, P.D. (1996). Vertebrate tracks in Pleistocene eolian sand-sheet deposits of Alaska. *Quaternary Research*, **45**, 226–240.

Lea, P.D. & Waythomas, C.F. (1990). Late-Pleistocene eolian sand sheets in Alaska. *Quaternary Research*, **34**, 269–281.

Leigh, D.S. & Ivester, A.H. (1998). Eolian dunes at Fort Stewart, Georgia: Geomorphology, age and archaeological site burial potential. Geomorphology Laboratory Research Report 1, University of Georgia, 108 pp.

Leigh, D.S. & Knox, J.C. (1994). Loess of the upper Mississippi Valley driftless area. *Quaternary Research*, **42**, 30–40.

Lewis, G.C. & Fosberg, M.A. (1982). Distribution and character of loess and loess soils in southeastern Idaho. *In*: Bonnichsen, B. & Breckenridge, R.M. (Eds), Cenozoic Geology of Idaho. Moscow, Idaho. *Idaho Bureau of Mines and Geology, Bulletin*, Vol. 26, pp. 705–716.

Lewis, G.C., Fosberg, M.A., McDole, R.E. & Chugg, J.C. (1975). Distribution and some properties of loess in south-central and south-eastern Idaho. *Soil Science Society of America Proceedings*, **39**, 1165–1168.

Liu, X.M., Hesse, P., Rolph, T. & Begét, J. (1999). Properties of magnetic mineralogy of Alaskan loess: Evidence for pedogenesis. *Quaternary International*, **62**, 93–102.

Loope, D.B., Swinehart, J.B. & Mason, J.P. (1995). Dune-dammed paleovalleys of the Nebraska Sand Hills: Intrinsic vs. climatic controls on the accumulation of lake and marsh sediments. *Geological Society of America Bulletin*, **107**, 396–406.

Lugn, A.L. (1968). The origin of loesses and their relation to the Great Plains in North America. *In*: Schultz, C.B. & Frye, J.C. (Eds), *Loess and Related Eolian Deposits of the World*. Lincoln, Nebraska, University of Nebraska Press, pp. 139–182.

Lyell, C. (1847). On the delta and alluvial deposits of the Mississippi, and other points in the geology of North America, observed in the years 1845–1846. *The American Journal of Science and Arts*, Second Series, III, May, 34–39.

Maat, P.B. & Johnson, W.C. (1996). Thermoluminescence and new ^{14}C age estimates for late Quaternary loesses in southwestern Nebraska. *Geomorphology*, **17**, 115–128.

Madole, R.F. (1995). Spatial and temporal patterns of late Quaternary eolian deposition, eastern Colorado, U.S.A. *Quaternary Science Reviews*, **14**, 155–177.

Malde, H.E. (1991). Quaternary geology and structural history of the Snake River plain, Idaho and Oregon. *In*: Morrison, R.B. (Ed.), *Quaternary Non-Glacial Geology of the United States*. Geological Society of America Geology of North America, Vol. K-2, pp. 251–281,

Markewich, H.W. (Ed.) (1993). Progress report on chronostratigraphic and paleoclimatic studies, middle Mississippi River Valley, eastern Arkansas and western Tennessee. *U.S. Geological Survey Open-File Report*, 93–273, 61 pp.

Markewich, H.W. (Ed.) (1994). Second progress report on chronostratigraphic and paleoclimatic studies, middle Mississippi River Valley, eastern Arkansas, western Tennessee, and northwestern Mississippi. *U.S. Geological Survey Open-File Report*, 94–208, 51 pp.

Markewich, H.W. & Markewich, W. (1994). An overview of Pleistocene and Holocene inland dunes in Georgia and the Carolinas – Morphology, distribution, age, and paleoclimate. *U.S. Geological Survey Bulletin*, 2069, 32 pp.

Markewich, H.W., Pavich, M.J. & Wysocki, D.A. (1998a). Stage 3 Climate/loess relations in the Lower Mississippi Valley, U.S.A. *In*: Busacca, A.J. (Ed.), *Dust Aerosols, Loess Soils, and Global Change*. Miscellaneous Publication No. MISC-0190. Pullman, Washington State University Department of Agriculture and Home Economics Miscellaneous Publication, pp. 159–162.

Markewich, H.W., Wysocki, D.A., Pavich, M.J., Rutledge, E.M., Millard, H.T., Jr., Rich, F.J., Maat, P.B., Rubin, M. & McGeehin, J.P. (1998b). Paleopedology plus TL, ^{10}Be, and ^{14}C dating as tools in stratigraphic and paleoclimatic investigations, Mississippi River Valley, U.S.A. *Quaternary International*, **51/52**, 143–167.

Marks, H.M. (1996). Characteristics of surface soils that affect windblown dust on the Columbia Plateau [M.S. thesis]. Pullman, Washington State University, 134 pp.

Martin, C.W. (1993). Radiocarbon ages on late Pleistocene loess stratigraphy of Nebraska and Kansas, central Great Plains, U.S.A. *Quaternary Science Reviews*, **12**, 179–188.

Martinson, D.G., Pisias, N.G., Hays, J.D., Imbrie, J., Moore, T.C., Jr. & Shackleton, N.J. (1987). Age dating and the orbital theory of the ice ages: Development of a high-resolution 0 to 300,000-year chronostratigraphy. *Quaternary Research*, **27**, 1–29.

Mason, J.A. (2001). Transport direction of Peoria loess in Nebraska and implications for loess sources on the central Great Plains. *Quaternary Research*, **56**, 79–86.

Mason, J.A., Nater, E.A. & Hobbs, H.C. (1994). Transport direction of Wisconsinan loess in southeastern Minnesota. *Quaternary Research*, **41**, 44–51.

Mason, J.A., Nater, E.A., Zanner, C.W. & Bell, J.C. (1999). A new model of topographic effects on the distribution of loess. *Geomorphology*, **28**, 223–236.

Mason, J.A. & Kuzila (2000). Episodic Holocene loess deposition in central Nebraska. *In*: Wolfe, S.A., Goodfriend, G.A. & Baker, R. (Eds), Holocene environmental change on the Great Plains of North America. *Quaternary International*, Vol. 67, pp. 119–131.

Matheus, P., Beget, J., Mason, O. & Gelvin-Reymiller, C. (2000). The Palisades site, central Yukon River, Alaska. Troy Péwé Memorial Workshop, Fairbanks, Alaska, *Abstracts and Program*, p. 18.

May, D., Swinehart, J.B., Loope, D. & Souders, V. (1995). Late Quaternary fluvial and eolian sediments: Loup River basin and the Nebraska Sand Hills. *In*: Flowerday, C.A. (Ed.), *Geologic Field Trips in Nebraska and Adjacent Parts of Kansas and South Dakota*, Guidebook No. 10: Conservation and Survey Division, University of Nebraska-Lincoln, pp. 13–31.

McCoy, F.W., Noeleberg, W.J. & Norris, R.M. (1967). Speculations on the origin of the Algodones Dunes, California. *Geological Society of America Bulletin*, **78**, 1039–1044.

McDole, R.E. (1969). Loess deposits adjacent to the Snake River flood plain in the vicinity of Pocatello, Idaho [Ph.D. dissertation]. Moscow, University of Idaho, 231 pp.

McDole, R.E., Lewis, G.C. & Fosberg, M.A. (1973). Identification of paleosols and the Fort Hall geosol in southeastern Idaho loess deposits. *Soil Science Society of America Proceedings*, **37**, 611–616.

McDonald, E.V. (1987). Correlation and interpretation of the stratigraphy of the Palouse loess of eastern Washington [M.S. thesis]. Pullman, Washington State University, 218 pp.

McDonald, E.V. & Busacca, A.J. (1988). Record of pre-late Wisconsin giant floods in the Channeled Scabland interpreted from loess deposits. *Geology*, **16**, 728–731.

McDonald, E.V. & Busacca, A.J. (1990). Interaction between aggrading geomorphic surfaces and the formation of a late Pleistocene paleosol in the Palouse loess of eastern Washington state. *Geomorphology*, **3**, 449–470.

McDonald, E.V. & Busacca, A.J. (1992). Late Quaternary stratigraphy of loess in the Channeled Scabland and Palouse regions of Washington State. *Quaternary Research*, **38**, 141–156.

McDonald, E.V. & Busacca, A.J. (1998). Unusual timing of regional loess sedimentation triggered by glacial outburst flooding in the Pacific Northwest U.S. *In*: Busacca, A.J. (Ed.), *Dust Aerosols, Loess Soils, and Global Change*, Miscellaneous Publication No. MISC0190. Pullman, Washington State University College of Agriculture and Home Economics, pp. 163–166.

McDonald, E.V., McFadden, L.D. & Wells, S.G. (1995). The relative influences of climate change, desert dust and lithologic control on soil-geomorphic processes on alluvial fans, Mojave Desert, California. *In*: Reynolds, R.E. & Reynolds, J. (Eds), *Ancient Surfaces of the East Mojave Desert*. San Bernardino, CA, San Bernardino County Museum Association, pp. 35–42.

McDowell, P.F. & Edwards, M.E. (2001). Evidence of Quaternary climatic variations in a sequence of loess and related deposits at Birch Creek, Alaska: implications for the Stage 5 climatic chronology. *Quaternary Science Reviews*, **20**, 63–76.

McFadden, L.D. & McAuliffe, J.R. (1994). Late Quaternary eolian landscapes and soils of the south central Colorado Plateau. EOS. *Transactions of the American Geophysical Union Supplement*, **74**, 44.

McFadden, L.D. & McAuliffe, J.R. (1997). Lithologically induced geomorphic responses to Holocene climatic changes in the southern Colorado Plateau, Arizona: A soil-geomorphic and ecologic perspective. *Geomorphology*, **19**, 303–332.

McFadden, L.D., Wells, S.G. & Jercinovich, M.J. (1987). Influences of eolian and pedogenic processes on the origin and evolution of desert pavements. *Geology*, **15**, 504–508.

McGeehin, J., Burr, G.S., Jull, A.J.T., Reines, D., Gosse, J., Davis, P.T., Muhs, D. & Southon, J.R. (2001). Stepped-combustion ^{14}C dating of sediment: A comparison with established techniques. *Radiocarbon*, **43**, 255–261.

McKay, E.D. (1979). Wisconsinan loess stratigraphy of Illinois. *In*: Follmer, L.R., McKay, E.D., Lineback, J.A. & Gross, D.L. (Eds), *Wisconsinan, Sangamonian, and Illinoian Stratigraphy in Central Illinois*, Guidebook 13. Illinois State Geological Survey, pp. 95–108.

McKay, E.D. & Follmer, L.R. (1985). A correlation of Lower Mississippi Valley loesses to the glaciated Midwest, Geological Society of America. *Abstracts with Programs*, **17**, 167.

McKee, E.D. & Moiola, R.J. (1975). Geometry and growth of the White Sands Dune Field, New Mexico. *United States Geological Survey Journal of Research*, **8**(1), 59–66.

Meese, D.A., Gow, A.J., Alley, R.B., Zielinski, G.A., Grootes, P.M., Ram, M., Taylor, K.C., Mayewski, P.A. & Bolzan, J.F. (1997). The Greenland Ice Sheet Project 2 depth-age scale: Methods and results. *Journal of Geophysical Research*, **102**, 26411–26423.

Millard, H.T. & Maat, P.B. (1994). Thermoluminescence dating procedures in use at the U.S. Geological Survey, Denver, Colorado. *USGS Open File Report*, 94–249, 112 pp.

Miller, B.J., Lewis, G.C., Alford, J.J. & Day, W.J. (1985). *Loesses in Louisiana and at Vicksburg, Mississippi*. Friends of the Pleistocene Field Trip Guidebook. Baton Rouge, Louisiana Agricultural Experiment Station, Louisiana State University, 126 pp.

Miller, B.J., Day, W.J. & Schumacher, B.A. (1986). *Loesses and loess-derived soils in the Lower Mississippi Valley*. Guidebook for Soil-Geomorphology Tour 28–30. Baton Rouge, Louisiana Agricultural Experiment Station, Louisiana State University, 144 pp.

Mirecki, J.E. & Miller, B.B. (1994). Aminostratigraphic correlation and geochronology of two Quaternary loess localities, central Mississippi Valley. *Quaternary Research*, **41**, 289–297.

Muhs, D.R. & Bettis, E.A., III (2000). Geochemical variations in Peoria loess of western Iowa indicate paleowinds of midcontinental North America during last glaciation. *Quaternary Research*, **53**, 49–61.

Muhs, D.R. & Holliday, V.T. (1995). Evidence of active dune sand on the Great Plains in the 19th century from accounts of early explorers. *Quaternary Research*, **43**, 198–208.

Muhs, D.R. & Holliday, V.T. (2001). Origin of late Quaternary dune fields on the Southern High Plains of Texas and New Mexico. *Geological Society of America Bulletin*, **113**, 75–87.

Muhs, D.R. & Zárate, M. (2001). Late Quaternary eolian records of the Americas and their paleoclimatic significance. *In*: Markgraf, V. (Ed.), *Interhemispheric Climate Linkages*. New York, Academic Press, pp. 183–216.

Muhs, D.R., Bush, C.A., Cowherd, S.D. & Mahan, S. (1995). Geomorphic and geochemical evidence for the source of sand in the Algodones dunes, Colorado Desert, southeastern California. *In*: Tchakerian, V.P. (Ed.), *Desert Aeolian Processes*. London, Chapman & Hall, pp. 37–74.

Muhs, D.R., Stafford, T.W., Jr., Cowherd, S.D., Mahan, S.A., Kihl, R., Maat, P.B., Bush, C.A. & Nehring, J. (1996). Origin of the late Quaternary dune fields of northeastern Colorado. *Geomorphology*, **17**, 129–149.

Muhs, D.R., Stafford, T.W., Jr., Swinehart, J.B., Cowherd, S.D., Mahan, S.A., Bush, C.A., Madole, R.F. & Maat, P.B. (1997). Late Holocene eolian activity in the mineralogically mature Nebraska Sand Hills. *Quaternary Research*, **48**, 162–176.

Muhs, D.R., Aleinikoff, J.N., Stafford, T.W., Jr., Kihl, R., Been, J., Mahan, S.A. & Cowherd, S.D. (1999). Late Quaternary loess in northeastern Colorado, I-Age and paleoclimatic significance. *Geological Society of America Bulletin*, **111**, 1861–1875.

Muhs, D.R., Swinehart, J.B., Loope, D.B., Been, J., Mahan, S.A. & Bush, C.A. (2000a). Geochemical evidence for an eolian sand dam across the North and South Platte Rivers in Nebraska. *Quaternary Research*, **53**, 214–222.

Muhs, D.R., Bettis, E.A., Jr., Been, J. & McGeehin, J. (2001a). Impact of climate and parent material on chemical weathering in loess-derived soils of the Mississippi River Valley. *Soil Science Society of America Journal*, **65**, 1761–1777.

Muhs, D.R., Ager, T.A. & Begét, J. (2001b). Vegetation and paleoclimate of the last interglacial period, central Alaska. *Quaternary Science Reviews*, **20**, 41–61.

Muhs, D.R., Ager, T.A., Bettis, E.A., III, McGeehin, J., Been, J.M., Begét, J.E., Pavich, M.J., Stafford, T.W., Jr. & Stevens, D.S.P. (2003a). Stratigraphy and paleoclimatic significance of late Quaternary loess-paleosol sequences of the last interglacial-glacial cycle in central Alaska. *Quaternary Science Reviews*, **22**, 1947–1986.

Muhs, D.R., Reynolds, R., Been, J. & Skipp, G. (2003b). Eolian sand transport pathways in the southwestern United

States: Importance of the Colorado River and local sources. *Quaternary International*, **104**, 3–18.

Mullineaux, D.R. (1986). Summary of pre-1980 tephra-fall deposits from Mount St. Helens, Washington State, U.S.A. *Bulletin of Volcanology*, **48**, 17–26.

O'Geen, A.T. (1998). Faunal paleoecology of Late Quaternary paleosols on the Columbia Plateau [M.S. thesis]. Pullman, Washington State University, 96 pp.

O'Geen, A.T. & Busacca, A.J. (2001). Faunal burrows as indicators of paleo-vegetation in eastern Washington, U.S.A. *Palaeogeography, Palaeoclimatology, Palaeoecology*, **169**, 23–37.

O'Geen, A.T., McDaniel, P.A. & Busacca, A.J. (2002). Cicada burrows as indicators of buried soils in arid and semiarid regions of the Pacific Northwest. *Soil Science Society of America Journal*, **66**, 1584–1586.

Packer, D.R. (1979). Paleomagnetism and age dating of the Ringold Formation and loess deposits in the State of Washington. *Oregon Geology*, **41**, 119–132.

Patton, P.C. & Baker, V.R. (1978). New evidence for pre-Wisconsin flooding in the Channeled Scabland of eastern Washington. *Geology*, **6**, 567–571.

Pavich, M.J. & Chadwick, O.A. (2003). Soils and the Quaternary Climate System. *In*: Gillespie et al. (Ed.) (this volume).

Petrone, A. (1970). The Moses Lake sand dunes [M.S. thesis]. Pullman, Washington State University, 89 pp.

Péwé, T.L. (1955). Origin of the upland silt near Fairbanks, Alaska. *Geological Society of America Bulletin*, **66**, 699–724.

Péwé, T.L., (1975). Quaternary geology of Alaska. U.S. Geological Survey Professional Paper 835, 145 pp.

Péwé, T.L., Hopkins, D.L. & Giddings, J.L. (1965). The Quaternary geology and archeology of Alaska. *In*: Wright, H.E., Jr. & Frey, D.G. (Eds), *The Quaternary of the United States*. Princeton, New Jersey, Princeton University Press, pp. 354–374.

Péwé, T.L., Berger, G.W., Westgate, J.A., Brown, P.M. & Leavitt, S.W. (1997). Eva Interglaciation Forest Bed, Unglaciated East-Central Alaska: Global Warming 125,000 Years Ago. Geological Society of America Special Paper 319, 54 pp.

Pierce, K.L., Fosberg, M.A., Scott, W.E., Lewis, G.C. & Colman, S.M. (1982). Loess deposits of southeastern Idaho: Age and correlation of the upper two loess units. *In*: Bonnichsen, W. & Breckenridge, R.M. (Eds), Cenozoic Geology of Idaho. Boise, Idaho. *Idaho Bureau of Mines and Geology Bulletin*, Vol. 26, pp. 717–725.

Preece, S.J., Westgate, J.A. & Gorton, M.P. (1992). Compositional variation and provenance of late Cenozoic distal tephra beds, Fairbanks area, Alaska. *Quaternary International*, **13/14**, 97–101.

Preece, S.J., Westgate, J.A., Stemper, B.A. & Péwé, T.L. (1999). Tephrochonology of late Cenozoic loess at Fairbanks, central Alaska. *Geological Society of America Bulletin*, **111**, 71–90.

Pye, K. (1996). The nature, origin and accumulation of loess. *Quaternary Science Reviews*, **14**, 653–667.

Pye, K. & Johnson, R. (1988). Stratigraphy, geochemistry, and thermoluminescence ages of Lower Mississippi Valley loess. *Earth Surface Processes and Landforms*, **13**, 103–124.

Pye, K. & Tsoar, H. (1990). *Aeolian sand and sand deposits*. London, Unwin Hyman.

Pye, K., Winspear, N.R. & Zhou, L.P. (1995). Thermoluminescence ages of loess and associated sediments in central Nebraska, USA. *Palaeogeography, Palaeoclimatology, and Palaeoecology*, **118**, 73–87.

Ramsey, M.S., Christensen, P.R., Lancaster, N. & Howard, D.A. (1999). Identification of sand sources and transport pathways at the Kelso Dunes, California using thermal infrared remote sensing. *Geological Society of America Bulletin*, **111**, 646–662.

Rendell, H., Lancaster, N. & Tchakerian, V.P. (1994). Luminescence dating of Late Pleistocene Aeolian Deposits at Dale Lake and Cronese Mountains, Mojave Desert, California. *Quaternary Science Reviews*, **13**, 417–422.

Rendell, H.M. & Clarke, M.L. (1998). Climate change impacts on sand supply and the formation of desert sand dunes in the south-west U.S.A. *Journal of Arid Environments*, **39**, 517–532.

Rendell, H.M. & Sheffer, N.L. (1996). Luminescence dating of sand ramps in the eastern Mojave Desert. *Geomorphology*, **17**, 187–198.

Richardson, C.A., McDonald, E.V. & Busacca, A.J. (1997). Luminescence dating of loess from the Northwest United States. *Quaternary Science Reviews (Quaternary Geochronology)*, **16**, 403–415.

Richardson, C.A., McDonald, E.V. & Busacca, A.J. (1999). A luminescence chronology for loess deposition in Washington State and Oregon, U.S.A. *Zeitschrift für Geomorphologie*, **116**, 77–95.

Richmond, G.M., Fryxell, R., Neff, G.E. & Weiss, P.L. (1965). The Cordilleran Ice Sheet of the northern Rocky Mountains and related Quaternary history of the Columbia Plateau. *In*: Wright, H.E., Jr. & Frey, D.G. (Eds), *The Quaternary of the United States*. Princeton, New Jersey, Princeton University Press, pp. 231–242.

Rodbell, D.T., Forman, S.L., Pierson, J. & Lynn, W.C. (1997). Stratigraphy and chronology of Mississippi Valley loess in western Tennessee. *Geological Society of America Bulletin*, **109**, 1134–1148.

Rousseau, D.-D. & Kukla, G. (1994). Late Pleistocene climate record in the Eustis loess section, Nebraska, based on land snail assemblages and magnetic susceptibility. *Quaternary Research*, **42**, 176–187.

Ruhe, R.V. (1969). *Quaternary landscapes in Iowa*. Ames, Iowa State University Press.

Ruhe, R.V. (1983). Depositional environment of late Wisconsin loess in the midcontinental United States. *In*: Wright, H.E., Jr. & Porter, S.C. (Eds), *Late-Quaternary Environments of the United States, The Late Pleistocene*. Minneapolis, University of Minnesota Press, Vol. 1, pp. 130–137.

Ruhe, R.V., Miller, G.A. & Vreeken, W.J. (1971). Paleosols, loess sedimentation and soil stratigraphy. *In*: Yaalon,

D.H. (Ed.), *Paleopedology – Origin, nature and dating of paleosols*. Jerusalem, Israel Universities Press, pp. 41–59.

Rutledge, E.M., West L.T. & Guccione, M.J. (1990). Loess deposits of northeast Arkansas. *In*: Guccione, M.J. & Rutledge, E.M. (Ed.), *Field Guide to the Mississippi Alluvial Valley, northeast Arkansas and southeast Missouri*. Fayetteville, Arkansas, Friends of the Pleistocene, South Central Cell, Geology Department, University of Arkansas, pp. 57–98.

Rutledge, E.M., Guccione, M.J., Markewich, H.W., Wysocki, D.A. & Ward, L.B. (1996). Loess stratigraphy of the Lower Mississippi Valley. *Engineering Geology*, **45**, 167–183.

Schatz, J.K. (1996). Provenance and depositional histories of the Smith Canyon and Sand Hills Coulee dunes, south-central, Washington [M.S. thesis]. Pullman, Washington State University, 87 pp.

Scott, W.E. (1982). Surficial geologic map of the eastern Snake River Plain and adjacent areas, areas, 111° to 115° W, Idaho and Wyoming. U.S. Geological Survey Miscellaneous Investigations Series Map I-1372.

Sharp, R.P. (1966). Kelso Dunes, Mohave Desert, California. *Geological Society of America Bulletin*, **77**, 1045–1074.

Sher, A.V. Edwards, M.E., Begét, J.E., Berger, G.W., Guthrie, R.D., Preece, S.J., Virina, E.I. & Westgate, J.A. (1997). The stratigraphy of the Russian Trench at Gold Hill, Fairbanks, Alaska. *In*: Edwards, M.E., Sher, A.V. & Guthrie, R.D. (Eds), *Terrestrial Paleoenvironmental Studies in Beringia*. Fairbanks, Alaska, The Alaska Quaternary Center, University of Alaska Museum, pp. 31–40.

Smith, G.D. (1942). Illinois loess – Variations in its properties and distribution, a pedologic interpretation. *University of Illinois Agricultural Experiment Station Bulletin*, **490**, 139–184.

Smith, G.D. (1992). Sedimentology, stratigraphy, and geoarcheology of the Tsulim site, on the Hanford Site, Washington [M.S. thesis]. Pullman, Washington State University, 169 pp.

Smith, G.D. & McFaul, M.D. (1997). Paleoenvironmental and geoarchaeologic implications of late Quaternary sediments and paleosols: North-central to southwestern San Juan basin, New Mexico. *Geomorphology*, **21**, 107–138.

Smith, H.T.U. (1967). *Past vs. present wind action in the Mojave Desert region, California*. U.S. Army Cambridge Research Laboratory, AFCRL-67-0683.

Soil Survey Staff (1999). Soil Taxonomy: USDA-NRCS. *Agriculture Handbook*, 436, U.S. Government Printing Office, Washington, D.C.

Stetler, L.D. (1993). Eolian dynamics and Holocene eolian sedimentation at the Hanford Site, Washington [Ph.D. dissertation]. Pullman, Washington State University, 181 pp.

Stetler, L.D. & Gaylord, D.R. (1996). Evaluating eolian-climatic interactions using a regional climate model from Hanford, Washington (USA). *Geomorphology*, **17**, 99–113.

Stevenson, B.A. (1997). Stable carbon and oxygen isotopes in soils and paleosols of the Palouse loess, eastern Washington state: modern relationships and applications for paleoclimatic reconstruction [Ph.D dissertation]. Fort Collins, Colorado State University, 128 pp.

Stokes, S. & Breed, C.S. (1993). A chronostratigraphic re-evaluation of the Tusayan Dunes, Moenkopi Plateau and Ward Terrace, Northeastern Arizona. *In*: Pye, K. (Ed.), *The Dynamics and Environmental Context of Aeolian Sedimentary Systems*. London, Geological Society, pp. 75–90.

Stokes, S. & Gaylord, D.R. (1993). Optical dating of Holocene dune sands in the Ferris dune field, Wyoming. *Quaternary Science Reviews*, **39**, 274–281.

Stokes, S. & Swinehart, J.B. (1997). Middle- and late-Holocene dune reactivation in the Nebraska Sand Hills, USA. *The Holocene*, **7**, 263–272.

Stokes, S., Kocurek, G., Pye, K. & Winspear, N.R. (1997). New evidence for the timing of aeolian sand supply to the Algodones dunefield and East Mesa area, southeastern California, U.S.A. *Palaeogeography, Palaeoclimatology, Palaeocology*, **128**, 63–75.

Sweeney, M.R., Busacca, A.J. & Gaylord, D.R. (2000). Eolian facies of the Palouse region, Pacific Northwest: clues to controls on thick loess accumulation. *Geological Society of America, Abstracts with Programs*, **32**, 22.

Sweeney, M.R., Gaylord, D.R. & Busacca, A.J. (2001). Geomorphic, sedimentologic and bioclimatic models for loess accumulation on the Columbia Plateau, Pacific Northwest. *Geological Society of America, Abstracts with Programs*, **33**, 317.

Sweeney, M.R., Busacca, A.J., Gaylord, D.R. & Zender, C. (2002a). Provenance of Palouse loess and relation to late Pleistocene glacial outburst flooding, Washington state. *EOS Trans. AGU*, 83, 47, Fall Meet. Suppl., Abstract H22B-0899.

Sweeney, M.R., Gaylord, D.R. & Busacca, A.J. (2002b). Changes in loess-dune deposition in response to climate fluctuations on the Columbia Plateau, Washington. *Geological Society of America, Abstracts with Programs*, **34**, 77.

Sweet, M.L., Nielson, J., Havholm, K. & Farralley, J. (1988). Algodones dune field of southeastern California: Case history of a migrating modern dune field. *Sedimentology*, **35**, 939–952.

Swinehart, J.B. (1990). Wind-blown deposits. *In*: Bleed, A. & Flowerday, C. (Eds), *An Atlas of the Sand Hills, Resource Atlas No. 5a*. University of Nebraska-Lincoln, pp. 43–56.

Swinehart, J.B. & Diffendal, R.F., Jr. (1990). Geology of the pre-dune strata. *In*: Bleed, A. & Flowerday, C. (Eds), *An Atlas of the Sand Hills, Resource Atlas No. 5a*. University of Nebraska-Lincoln, pp. 29–42.

Tate, T.A. (1998). Micromorphology of loessial soils and paleosols on aggrading landscapes on the Columbia Plateau [M.S. thesis]. Pullman, Washington State University, 192 pp.

Thorson, R.M. & Bender, G. (1985). Eolian deflation by ancient katabatic winds: A late Quaternary example from the north Alaska Range. *Geological Society of America Bulletin*, **96**, 702–709.

Tsoar, H. & Pye, K. (1987). Dust transport and the question of desert loess formation. *Sedimentology*, **34**, 139–153.

Vlag, P.A., Oches, E.A., Banerjee, S.K. & Solheid, P.A. (1999). The paleoenvironmental-magnetic record of the Gold Hill steps loess section in central Alaska. *Physics and Chemistry of the Earth*, **A24**, 779–783.

Waitt, R.B., Jr. (1983). Tens of successive, colossal Missoula floods at north and east margins of the Channeled Scabland. *U.S. Geological Survey Open-File Report*, 83–671, 29 pp.

Wang, H., Follmer, L.R. & Liu, J.C. (2000). Isotope evidence of paleo-El Nino-Southern Oscillation cycles in loess-paleosol record in the central United States. *Geology*, **28**, 771–774.

Wascher, H.L., Humbert, R.F. & Cady, J.G. (1947). Loess in the southern Mississippi Valley: Identification and distribution of the loess sheets. *Soil Science Society of America Proceedings*, **12**, 389–399.

Wasklewicz, T.A. & Meek, N. (1995). Provenance of aeolian sediment: The upper Coachella Valley, California. *Physical Geography*, **16**, 539–556.

Waters, M.R. (1983). Late Holocene lacustrine chronology and archaeology of ancient Lake Cahuilla, California. *Quaternary Research*, **19**, 373–387.

Watts, W.A. (1971). Postglacial and interglacial vegetation history of southern Georgia and central Florida. *Geological Society of America Bulletin*, **80**, 631–642.

Wells, P.V. & Stewart, J.D. (1987). Spruce charcoal, conifer macrofossils, and landsnail and small-vertebrate faunas in Wisconsinan sediments on the High Plains of Kansas. *In*: Johnson, W.C. (Ed.), *Quaternary Environments of Kansas. Kansas Geological Survey Guidebook Series 5*, pp. 129–140.

Wells, S.G., McFadden, L.D. & Schultz, J.D. (1990). Eolian landscape evolution and soil formation in the Chaco dune field, southern Colorado Plateau, New Mexico. *In*: Knuepfer, P.L.K. & McFadden, L.D. (Eds), *Soils and Landscape Evolution*. Proceedings, Symposium in Geomorphology, 21st, Binghamton, Elsevier, Amsterdam, pp. 517–546.

Wells, S.G., Brown, W.J., Enzel, Y., Anderson, R.Y. & McFadden, L.D. (2003). Late Quaternary geology and paleohydrology of pluvial Lake Mojave, southern California. *In*: Enzel, Y., Wells, S.G. & Lancaster, N. (Eds), *Paleoenvironments and Paleohydrology of the Mojave and Southern Great Basin Deserts*. Boulder, CO, Geological Society of America.

Westgate, J. (1988). Isothermal plateau fission-track age of the late Pleistocene Old Crow tephra, Alaska. *Geophysical Research Letters*, **15**, 376–379.

Westgate, J. (1989). Isothermal plateau fission-track ages of hydrated glass shards from silicic tephra beds. *Earth and Planetary Science Letters*, **95**, 226–234.

Westgate, J.A., Hamilton, T.D. & Gorton, M.P. (1983). Old Crow tephra: A new late Pleistocene stratigraphic marker across north-central Alaska and western Yukon Territory. *Quaternary Research*, **19**, 38–54.

Westgate, J.A., Walter, R.C., Pearce, G.W. & Gorton, M.P. (1985). Distribution, stratigraphy, petrochemistry, and palaeomagnetism of the late Pleistocene Old Crow tephra in Alaska and the Yukon. *Canadian Journal of Earth Sciences*, **22**, 893–906.

Westgate, J.A., Stemper, B.A. & Péwé, T.L. (1990). A 3 m.y. record of Pliocene-Pleistocene loess in interior Alaska. *Geology*, **18**, 858–861.

Whitlock, C., Bartlein, P.J., Markgraf, V. & Ashworth A.C. (2001). The mid-latitudes of North and South America during the last glacial maximum and early Holocene: Similar paleoclimate sequences despite differing large-scale controls. *In*: V. Markgraf (Ed.), *Interhemispheric Climate Linkages*, pp. 391–416.

Wilkins, D.E. & Currey, D.R. (1999). Radiocarbon chronology and $\delta^{13}C$ analysis of mid- to late-Holocen aeolian environments, Guadalupe Mountains National Park, Texas, U.S.A. *The Holocene*, **9**, 363–371.

Winspear, N.R. & Pye, K. (1995). Sand supply to the Algodones dunefield, south-eastern California, U.S.A. *Sedimentology*, **42**, 875–892.

Wintle, A.G., Lancaster, N. & Edwards, S.R. (1994). Infrared stimulated luminescence (IRSL) dating of late-Holocene aeolian sands in the Mojave Desert, California, USA. *The Holocene*, **4**, 74–78.

Wright, H.E., Jr. & Frey, D.G. (Eds) (1965). *The Quaternary of the United States*. Princeton, NJ, Princeton University Press, 922 pp.

Wright, H.E., Jr., Almendigner, J.C. & Gruger, J. (1985). Pollen diagram from the Nebraska Sandhills and the age of the dunes. *Quaternary Research*, **24**, 115–120.

Zayac, T., Rich, F.J. & Newsom, L. (2001). The paleoecology and depositional environments of the McClelland sandpit site, Douglas, Georgia. *Southeastern Geology*, **40**, 259–272.

Zimbelman, J.R. & Williams, S.H. (2002). Geochemical indicators of separate sources for eolian sands in the eastern Mojave Desert, California, and western Arizona. *Geological Society of America Bulletin*, **114**, 490–496.

Zimbelman, J.R., Williams, S.H. & Tchakerian, V.P. (1995). Sand transport paths in the Mojave Desert, southwestern United States. *In*: Tchakerian, V.P. (Ed.), *Desert Aeolian Processes*. New York, Chapman and Hall, pp. 101–129.

Soils and the Quaternary climate system

Milan J. Pavich[1] and Oliver A. Chadwick[2]

[1] *U.S. Geological Survey, Reston, VA, USA; mpavich@usgs.gov*
[2] *University of California at Santa Barbara, Santa Barbara, CA, USA; oac@geog.ucsb.edu*

Introduction: Quaternary Soil Interpretation

Our knowledge of Quaternary soils includes the recognition of dynamic equilibrium of soil with environment, the inoculation of geomorphology with that concept (Hack, 1960), and subsequent attempts to understand soils and geomorphology (Birkeland, 1999). The growing urgency of predicting landscape response to climate change highlights the issue of how soils have responded to past climate cycles. This review of soil research illustrates the complexity of extracting paleoenvironmental and paleoclimatic information from preserved soils. One problem is to figure out the best sites and methods for extracting that information.

Since the INQUA VII review of 1965, there have been major advances in pedology that further the goal of elucidating the climate system. These include the USDA Soil Taxonomy (Soil Survey Staff, 1974, 1999), soil landscape studies (Gile *et al.*, 1981; Ruhe *et al.*, 1967), and improved chronology of stratigraphic sequences containing soils. Ruhe (1965) summarized Quaternary paleopedology for the last North American INQUA volume. He highlighted the importance of paleosols, soils formed on past landscapes, to the interpretation of till and loess sequences in the Midwestern U.S. He argued that paleosols, including those buried by younger deposits and those exhumed after temporary burial, are an important aspect of reconstructing the history of Quaternary landscapes. Almost 40 years later, we are still debating the definition and interpretation of paleosols (e.g. Birkeland, 1999; Fenwick, 1985; the INQUA Commission on Paleopedology: http://fadr.msu.ru/inqua/discussions/; Mack *et al.*, 1993).

One of the major arguments made by Ruhe (1965) was the need for combined geomorphologic-pedologic approaches to solving problems related to soil property distribution across the Earth's surface. The same criteria for identifying soils are applicable to identifying paleosols (Fenwick, 1985); thus, the same processes apparently operate despite differences in geologic age. Ruhe argued that chemical and physical analyses are important to assessing environmental conditions responsible for paleosol genesis, dating horizons, and calculating residence times. For example, he pointed out that intensity (climate) and time can confuse the interpretation of past environments. He cited horizon color as a possible indicator of either intensity or time; thus, these must be assessed as independent variables. Analytical methods, some of which are noted in Table 1, have advanced significantly since 1965. In particular we are now able, in some circumstances, to measure age or residence time independent of temperature. Our improved ability to assess time independently in soils that contain climate information is of great significance, and our ability to link isotopes to atmospheric or biologic processes helps define intensity factors.

Integration of soil research with other Quaternary disciplines is critical to improving models of the Quaternary climate system. McFadden & Knuepfer (1990) correctly caution that we can't simply assume that soil forming intervals exist; soils may be polygenetic expressions of constantly changing environments. There is increasing evidence that the well-studied marine oxygen isotope record provides an important, but insufficient, basis for interpreting Quaternary climate change and its impact on terrestrial ecology and geomorphology. Spatial scale and surface processes are important in the regional to local responses to global-scale climate drivers (Bartlein, 1997; Entin *et al.*, 2000).

Previous reviews by Mahaney (1978), Bronger & Catt (1989), Catt & Bronger (1998) and Birkeland (1999) are essential resources for Quaternary scientists. We highlight examples showing how soil is related to climate and, in turn, the types of climate change information that can be derived from soil studies.

Soil Formation Pathways Related to Climate

One of the problems in formulating a model of soil development is that progressive development of soil properties is not simply related to single external variables such as rainfall (Runge, 1973; Simonson, 1978). Soils are extremely complex integrative systems that function as a boundary layer at the surface of the ice- and water-free portions of continents. Even if we can't solve the complex equation that expresses the interactions of these variables, the spatial pattern of soil orders indicates that soil processes work to establish dynamic equilibrium with energy and material fluxes at landscape surfaces; thus, there is a relation to climate. Distinct pathways of soil development appear to operate, and recur, over significantly long periods of geologic time (Retallack, 1990). One of the fundamental global patterns that we see in Fig. 1 is the correspondence of soil orders and climate as defined by the USDA Soil Taxonomy (Soil Survey Staff, 1999). Does the global soil pattern reflect climate boundaries that have been stable over the time scale of surface soil genesis? Does present climate represent average climate over the time scale of soil genesis? Once soil formation is initiated, is further development relatively insensitive to climate cycles? Do old, relict soils become unresponsive to climate change (Nettleton *et al.*, 1989)?

Models of the relations of soils to landscape (Birkeland, 1999; Bockheim, 1980; Yaalon, 1971, 1983) stress time as an

DEVELOPMENTS IN QUATERNARY SCIENCE
VOLUME 1 ISSN 1571-0866
DOI:10.1016/S1571-0866(03)01014-5

Table 1. Parameters and methods relevant to quantifying soil characteristics and chronology.

Soil Field Methods (Soil Survey Staff, 1999)
  Thickness
  Horizons
  Color
  Texture

Petrography and Mineralogy
  Micromorphologic Fabric (Bullock, 1985)
  Clay Mineralogy (Singer, 1980; Velde, 1992)

Chemistry Methods
  Major elements
  Trace elements
  Extractable Fe and Al (McKeague *et al.*, 1971; Schwertman, 1993)
  Stable Isotopes
    Carbon (Cerling & Wang, 1996)
    Oxygen (Cerling & Wang, 1996)

Magnetic Properties
  Susceptibility (Reynolds & King, 1995)

Chronologic methods
  Uranium series (Sharp *et al.*, 2003)
  ^{14}C (Amundson *et al.*, 1999; Scharpenseel, 1971; Wang *et al.*, 1996)
  ^{10}Be (Pavich & Vidic, 1993)
  Luminescence (Forman, 1989; Millard & Maat, 1994)

important process variable in Quaternary soils. On a broad scale, however, the soil order boundaries that are roughly parallel to latitude (Fig. 1) suggest that mean temperature and precipitation are important climatic variables affecting soil processes. Soil Taxonomy (Soil Survey Staff, 1999) includes 11 orders differentiated by diagnostic horizons, and numerous suborders that reflect soil climate or parent material. Thus, using a taxonomy based on diagnostic characteristics of soils, leads to a map that reflects climate.

Soil moisture boundaries have fluctuated through the Quaternary in response to glacial/interglacial cycles. For non-glaciated areas the dominant Quaternary variable, along with temperature, has been water balance. Both terrestrial and marine evidence support the argument for multiple glaciations associated with water and temperature cycles through the Quaternary (Winograd *et al.*, 1997). In some cases, moisture delivery to high latitudes increased preceding glacial maxima (Winograd, 2001) and decreased at maxima (Jouzel *et al.*, 1989). During glacial periods, many pluvial lakes and inland seas expanded greatly due to shifts in moisture patterns and boundaries (Smith & Street-Perrot, 1983).

Within the large geographic ranges of soil orders, there are significant spatial variations of seasonal and interrannual temperature and rainfall (Entin *et al.*, 2000). Mean monthly temperature and precipitation are commonly used to estimate soil water balance in specific profiles (Soil Survey Staff, 1999). Longer-term means must reflect the inter-annual variability of these properties, but we have few instrumental records that correspond to the duration of soil development. Minimum limits appear important to defining major order boundaries. The broad-scale pattern of soil distribution suggests that there are limiting conditions for specific pathways of soil development, and that once a pathway is initiated the profile tends to evolve to a characteristic state. This assertion is supported by evidence that soil diversity is low in areas of uniform parent material and climate (Hole & Campbell, 1985).

Pathway Models

Many physical, chemical, and biological processes affect the expression of $S = f(climate)$. At a global scale, however, there are mappable soil taxa defined by climate-related processes. Figure 2 shows the important divergent pathways of soil order development. Detailed treatment of soil genesis is found in Buol *et al.* (1997) among other texts. Birkeland (1999) provides a ranking of the importance of processes in soil orders. Soil characteristics change at different rates; thus, their relative differences can be related to time. Harden (1982) demonstrated that soil field properties can be used to index progressive soil development relative to parent materials. By deciphering the chemical and isotopic signatures of soil components, we can understand the progression of the soil mineral formation, and possibly draw conclusions about present and past climatic influences. It is difficult to interpret unambiguously the paleoclimatic significance of soil mineral assemblages. Singer (1980) has highlighted several critical assumptions that must be tested: (1) clay mineral formation is directly related to climatic parameters (2) once formed, clay minerals do not undergo further alteration, and (3) clay mineral assemblages are uniform throughout the weathering profile and if not, that the sampled profiles have not been truncated by physical erosion (Birkeland, 1999).

Rates of Soil Formation Under Warm/Wet and Warm/Dry Climates

The interpretation of paleoenvironmental conditions from soils is based, in part, on evidence that initiation of soil development occurs with onset of surface stabilization and that measurable change is rapid and systematic (Harden *et al.*, 1991a). Accumulation of organic material and dust can result in measurable soil properties in tens to hundreds of years, especially in coarse, permeable parent materials such as those studied by McFadden & Weldon (1987) and McFadden (1988). In addition, as shown in Fig. 3, the processes that occur in soils are increasingly non-linear through time and the response to input variations become increasingly difficult to decipher (Birkeland, 1999). Some soil properties appear to be "irreversible" and persistent. Chronosequence studies (Birkeland, 1999; Harden, 1982; Harden & Taylor,

Global Soil Regions

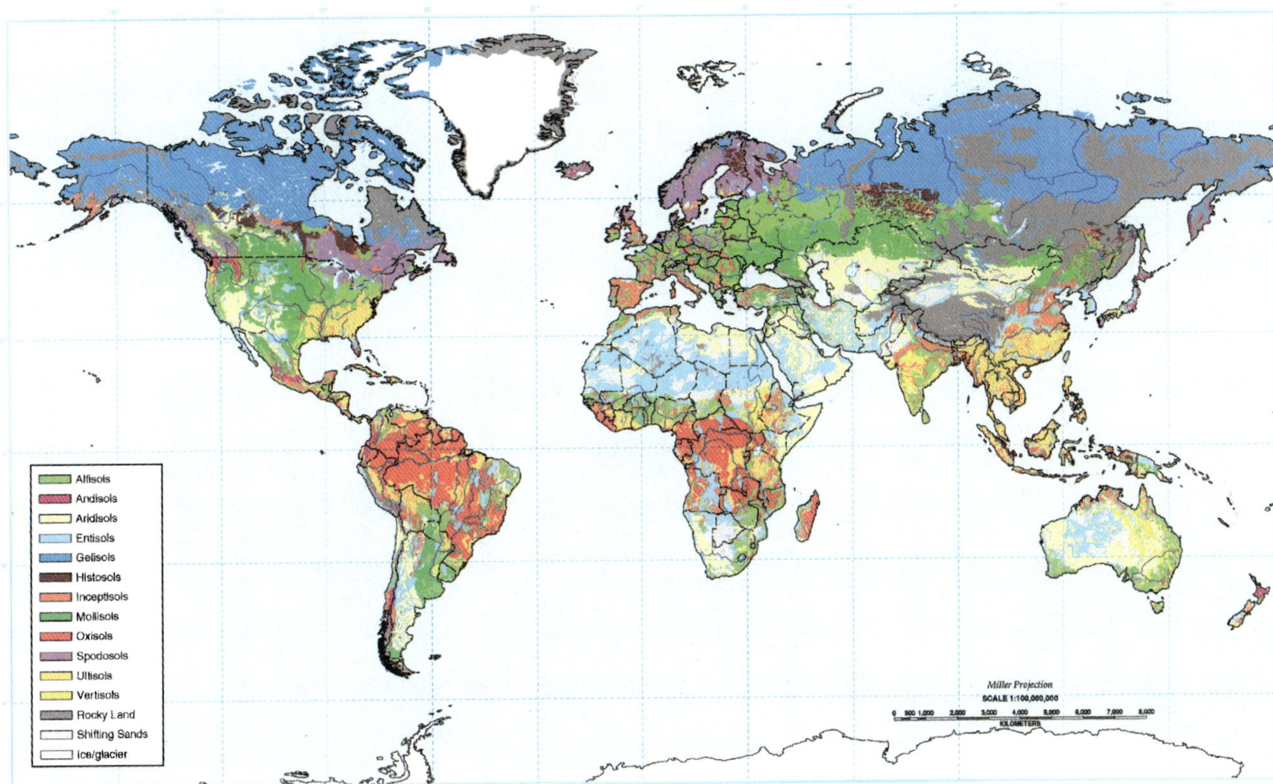

Fig. 1. Global soil-order map (Soil Survey Staff, 1999).

1983; Harden *et al.*, 1991; Markewich *et al.*, 1987, 1998; Vidic, 1998) show that many soil properties initially change very rapidly over decadal to 10,000-year periods (Fig. 3). Birkeland (1999) cited Richmond's (1962) and Morrison's (1964) arguments that, based on field criteria, it is difficult to distinguish soils that approach 10^5 yr residence time. Between 10^4 and 10^5 yr, soil properties change dramatically, but by 10^5 yr processes slow and the time scale of change is compressed. For instance, in humid regions, argillic horizons (Alfisols and Ultisols) show decreasing rates of change after 10^5 yr. In arid to semi-arid climates, carbonate-cemented horizons form in less than 10^5 yr (Gile *et al.*, 1981; Reheis *et al.*, 1992). In areas of significant dust input, calcic horizons in Aridisols can show continued development well beyond 10^5 yr (Machette, 1985). Vidic (1998) demonstrated that beyond 0.4 myr, laboratory soil parameters in several chronosequences are difficult to distinguish.

Importance of Thresholds

As emphasized above, soil formation rates are rarely linear for long time periods. McFadden & Weldon (1987) provided a very clear example of an extrinsic threshold created by the addition of organic matter and dust to gravelly parent material. The rapid transition to a less permeable soil environment creates a positive feedback that accelerates subsequent change. Thus, rapid rates of change can occur without any change in the external climate as intrinsic thresholds create step functions (Chadwick & Chorover, 2001; McFadden & Weldon, 1987; Muhs, 1984). Given the complexity of parent materials and dust inputs in arid environments, great caution should be exercised in correlations of soil properties or deduction of climate change without independent evidence (McFadden & Weldon, 1987). Another aspect of thresholds is that pathways of development can diverge under the same climate. For example Markewich *et al.* (1986) found that slight differences in parent material clay content determined whether Ultisols or Spodosols developed in South Carolina beach-barrier sediments. Without addition of fresh parent material, consumption of weatherable minerals in humid environments typically leads to an eventual deceleration in profile development. Recent applications of cosmogenic isotopes (Heimsath *et al.*, 2001) confirm that soil production is an inverse exponential function of soil depth providing a quantitative explanation of an important pedogenic and geomorphic threshold. Decreasing rates can be due in some cases to depletion of leachable bases (Chadwick *et al.*, in press), slowing of weathering front decent (Pavich & Vidic, 1993), and to decreasing permeability of argillic and calcic horizons (Buol *et al.*, 1997). This may take hundreds of thousands of years (McFadden & Weldon, 1987; Muhs, 1984). In some

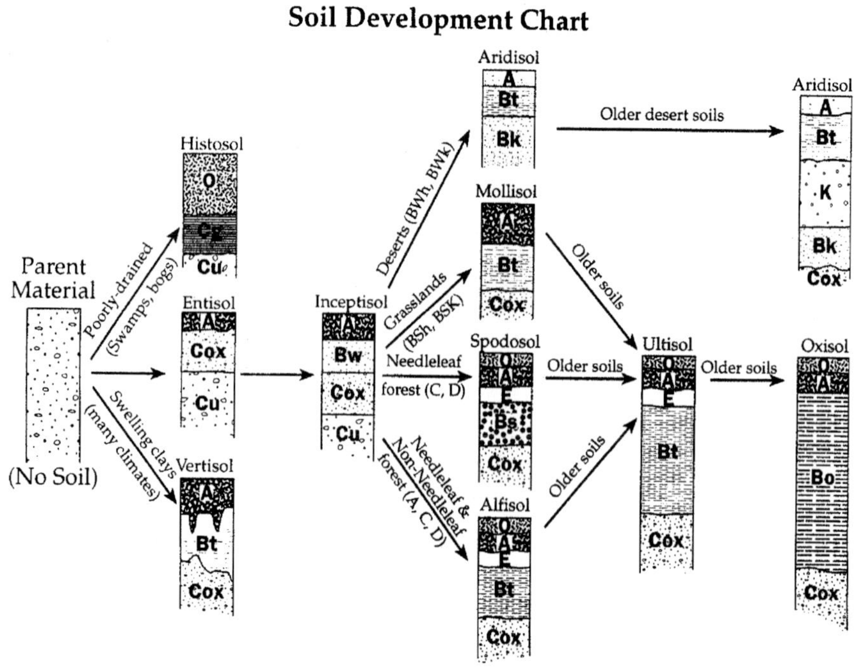

Fig. 2. Soil formation pathways. These schematic profiles are arranged in order of progressive development of soil horizons and properties. The soils discussed in this paper that change as functions of time and climate follow the pathways from parent material to Entisol to Inceptisol to those that are differentiated by precipitation, soil moisture, and vegetation (Aridisols to Alfisols). Letters in parentheses follow climate classification of Koppen (from D. Netoff, pers. comm., 1997 and Birkeland, 1999).

Soil Development Chart

O Horizon: Surface horizon dominated by organic matter (e.g., leaves) in various stages of decay

A Horizon: Surface horizon or beneath O horizon dominated by mineral matter, but with sufficient humus to darken color

E Horizon: A light-gray subsurface horizon that has been leached of pigments by organic acids produced by the decay of needleleaf litter

Bw Horizon: A young B horizon that has been slightly reddened by oxidation, but is not yet clay-enriched

Bs Horizon: Illuvial accumulation of amorphous organic matter – sesquioxide (Al_2O_3; Fe_2O_3) complexes

Bk Horizon: A light-colored B horizon due to the coating of mineral grains by carbonates

Bt Horizon: A 'mature' B horizon that is reddened by oxidation as well as clay-enriched

Bo Horizon: A deep red, highly weathered and leached B horizon rich in residual sesquioxides; usually restricted to very old soils in tropical climates

Cox Horizon: An oxidized horizon beneath a B or A horizon

K Horizon: Similar to a Bk horizon, but so greatly enriched in carbonates that the horizon is white

Cg Horizon: A 'gleyed' subsurface horizon that shows patchy zones of blue, gray, and green colors from reduced conditions; may be local areas of oxidized material

Cu Horizon: Unweathered parent material

environments, however, the rates of change decrease dramatically after the first few tens of thousands of years, particularly in environments of high rainfall and intense leaching (Chadwick *et al.*, 1999; Chadwick *et al.*, in press; Hotchkiss *et al.*, 2000). A deduction from the trend toward steady state profile is that soil stability also increases through time; that is, a negative threshold to further change increases. One way to express this qualitatively is by: *Change Probability = Magnitude of External Force/Threshold.*

Soils and soil-landscape relations, therefore, can approach a dynamic stability unless disrupted by erosion driven by tectonics, climate, or human land-use; or by rapid introduction of fresh parent material as in loess sequences. This dynamic stability provides a useful conceptual model to explain some of the Quaternary soil-landscape relations that we observe; numerous examples are presented in Birkeland (1999). Even if climate changes significantly, it is difficult to alter the progressive characteristics of a soil profile quickly (Olson, 1989); it appears more likely that physical processes such as erosion and burial lead to composite stratigraphies that include soils.

Soil Geomorphic Processes and Soil Preservation

Before presenting interpretations of climate change from soil stratigraphic studies, we need to emphasize that soil geomorphology is also an important approach to the interpretation of Quaternary soils. McFadden & Knuepfer (1990) argue that we should link soil investigations with geomorphic analysis of landscape evolution and that soil stratigraphic research can reach erroneous conclusions without an understanding of soil geomorphic relations. Soil stratigraphic studies have commonly assumed an external control on time boundaries, but those are not necessarily related to climate.

Preservation presents problems for both soil stratigraphy and soil geomorphology. The concept of paleosols requires that they exist in a stratigraphic framework; thus, they can

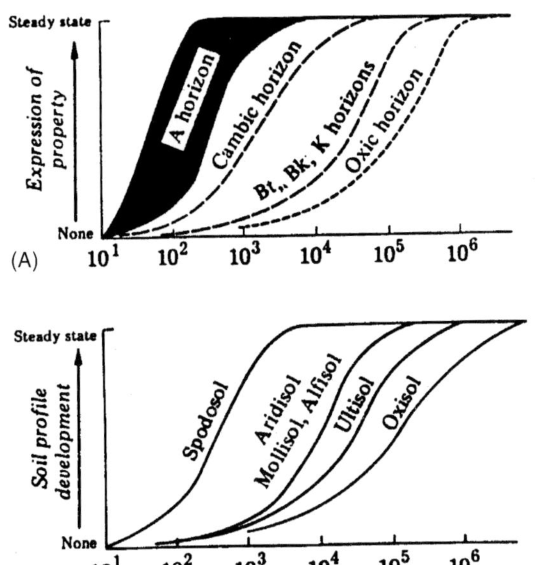

Fig. 3. Soil properties vs. time. These schematic curves represent the rates a which diagnostic horizons (A) and soil order profiles (B) approach steady states (from Birkeland, 1999).

potentially be bounded in time even if the upper bound is a subaerial geomorphic surface. Fenwick (1985) thought that the term paleosol should be restricted to buried soils that, by definition, would be bounded by the age of the parent material and the age of burial. Fenwick (1985) also pointed out that laterally extensive paleosols, or the past landscapes on which they formed, are rare. Without spatial boundaries for paleosols, it is difficult to reconstruct the spatial scale of the paleoclimate boundary; thus, we don't know if we're observing more than a micro-climate. Physical constraints on preservation can be related to non-linear geomorphic process responses to climate forcing; geomorphic systems such as fluvial valleys can amplify and respond non-linearly to relatively small climate changes or events (Knox, 1985). Valley deposits and their soils may be removed by erosion that is not simply related to climate. The Mississippi Valley has been the conduit of glacier ice and glacial meltwater through the Quaternary. The oldest loess and paleosols in the valley are restricted, however, to the last few glacial cycles (Markewich *et al.*, 1998). This, and the fact that Quaternary paleosols are not commonly preserved in glaciated areas, significantly restricts the geography of where we might find Quaternary paleoclimate evidence in soils. Chemical alteration after burial can also obscure records (Jenkins, 1985; Ruhe & Olson, 1980; Valentine & Dalrymple, 1976).

Interpretations of Climate State Changes from Soils

Despite the problems of thresholds and preservation, soils can be used to assess changes of climate state. Field identification and analytical verification of the existence of buried or relict soils have been carried out in many areas. The variability of Quaternary climate and climate-related sediments has provided initiation of new sites of soil formation. The variations of Quaternary climate have produced a complex and possibly unprecedented range of soil/parent material associations, paleosols, and buried soils. Some sites have been described and sampled repeatedly as analytical techniques improve.

Climate State Changes

Geologically there is evidence that climate has exhibited dramatic state changes, over time scales ranging from decades to hundreds of millions of years (Crowley, 1983; Rohling *et al.*, 2003). Even though the global climate system comprises an open system with many variables, there is evidence for characteristic states of the climate system (Crowley & North, 1991) and for global scale synchrony in state changes. Bartlein (1997) pointed out that most characteristic features of Quaternary climate are reversible and that they return to a previous general mean state. The stability of the mean is probably related to atmospheric humidity (which is, in part, dependent on atmosphere/land/vegetation interactions). For time intervals of 10^4 yr and greater, a mean climate state appears to be a valid concept for both humid and semi-arid environments. The importance of solar insolation to surface temperature variations is clear in modern instrumental records (Friis-Christensen & Lassen, 1991), and the existence of recurrent climate states in the Quaternary is apparently related to how the climate system has responded to orbital modulation of incident solar radiation. Crowley & North (1991) provide a good summary of how variation of incident solar radiation plays a role in controlling climate.

Despite the apparent success of the orbital cycle model in explaining the frequency variations in marine climate proxies, there are also important terrestrial state changes that do not correlate with the time scale for climate variation derived from orbital forcing models (Gillespie & Molnar, 1995; Sarthein *et al.*, 1986). Recently documented cycles of about 1500 yr (Campbell *et al.*, 1998) appear to be significant in terrestrial and some shallow marine stratigraphies (Poore *et al.*, 2003). Records in different media (e.g. ocean sediments, lake sediments, ice, loess, soils) are related to climate via different processes. Changes in processes such as glacier melting and sea-level rise, are not simple linear functions of orbitally controlled solar insolation (Bard *et al.*, 1996). Thus, spatially separate records often don't correlate in time or magnitude, and they often show that response to a specific regional change (e.g. the younger Dryas cooling of Europe) is neither globally immediate nor linear in magnitude. Terrestrial compilations (e.g. COHMAP, 1988) show regional complexity. By contrast with marine and ice records, terrestrial data sets are generally not synoptic and are not derived from standard methodologies. We stress the complexity of the terrestrial responses because over the past few decades the power of the solar insolation/ice volume model prompted correlation of the terrestrial glacial records to the marine oxygen isotope stratigraphy, such as in Sibrava *et al.* (1986). Correlations were done with little independent age control and are contradicted by examples from the last interglacial and montaine glacial records discussed later in this paper. Despite the complexities, and despite the possibility that future climate models will radically alter our thinking, it is still worthwhile to consider how soils relate to the significant glacial/interglacial cycles since 0.8 myr ago.

Preserved Soils as Indicators of Climate State Changes

The evidence cited above for a complex mosaic rather than a uniform cyclicity of climate indicates that much more work is needed on local to regional reconstruction of soil chronology. We have made significant progress since 1965 in such reconstructions using a variety of soil parameters. Birkeland (1999) cites organic content, clay content, clay mineralogy, color and iron, fragipan occurrence, calcic horizons, salt horizons, and chemical trends as possible indicators of paleoclimate. The interpretation of residence times of soils, rates of soil processes, and relation to external inputs of precipitation and dust has improved dramatically over the past few decades due to application of chemical, magnetic, isotopic and luminescence techniques (Table 1). Recent studies have shown that soils are, in some cases, very useful chronometers (Birkeland, 1999; Pavich & Vidic, 1993; Sharp *et al.*, 2003). While there has been very successful focus on the most recent interglacial periods, lack of preservation presents problems. Extraction of paleoclimate information from soils depends on our ability to invert the climate function, so that *inferred paleoclimate* $= f$ (*soil*). While we will cite specific examples of how climate parameters (e.g. mean annual temperature, water balance, $\delta^{18}O$ of rainfall) have been extracted from soils, we're faced with the problem of extrapolating geographically limited soils data. Climate is viewed as exhibiting global cycles, but we don't see global correlations of soil-climate relations as a practical goal until many more local studies are completed.

Importance of the Holocene Process Studies

As pointed out by Fenwick (1985), interpretation of paleosols depends on understanding modern analogs. Holocene and late Pleistocene parent materials provide opportunities to compare and contrast soil response to variations in temperature and precipitation. Holocene soils offer the best opportunity to measure and model water balance, and to develop process models for interpretation of longer-lived soils. There is increasing evidence that the transition from glacial to interglacial state was abrupt. Rapid sea level rises due to deglaciation are not simply synchronous with orbital forcing and appear to occur faster than insolation changes (Bard *et al.*, 1996). Moreover, the major meltwater pulse of termination I recorded in Barbados & Tahiti (Bard *et al.*, 1990, 1996) lags the initial ^{18}O change in ice and marine sediments by about 700 yr (14,000 vs. 14,700 cal yr B.P.). Although no simple correlation between continental warming, glacial melting, sea level rise, and solar insolation appears in one of the most detailed paleoclimate records available, there is no question that the termination of the LGM was rapid. In the past 10,000 years, since the end of the last glaciation, temperature has varied over an atypically narrow range (Broecker, 1997), a state conducive to agricultural use of soils and expansion of human populations (Lamb, 1995). Recent studies show important temporal variations in Holocene climate (Bradbury & Dean, 1993; Broecker, 1997; Keigwin, 1996) some of which are recorded in soils. The dominant factors in North America were the rapid recovery from the LGM and the solar insolation maximum that caused the Altithermal, first recognized by Antevs (1953). While the Altithermal appears in many records, its interpreted impacts vary depending on location. Sirocko *et al.* (1996) argued for connection of the strength of the Asian monsoon and solar insolation, but their evidence is that the monsoon shows a non-linear response to solar peaks, and a lag of several thousand years. The variation of temperatures and precipitation associated with the Altithermal is expressed in land-surface response to water balance. Mean states of precipitation and temperature that characterize broad physiographic regions have been documented from proxy records. Millennial data for parts of North America (Bartlein & Whitlock, 1993; Woodhouse & Overpeck, 1998) show decadal variations around mean states.

Soils developed on latest Pleistocene or Holocene parent materials, dominantly in glaciated areas or areas of active fluvial, lacustrine, or eolian deposition, tend to be weakly to moderately developed. In well-drained environments they are typically leached and oxidized, have colors distinct from their parent materials, and may show incipient argillic or calcic horizons. In arid to semi-arid environments, eolian processes and dust have been significant factors in soil development. One of the important Pleistocene to Holocene state changes in the southwestern U.S. was a decrease in precipitation. With warming and increase in evapotranspiration, effective moisture decreased significantly, having a major effect on Holocene soil properties (Chadwick *et al.*, 1995; Harden *et al.*, 1991b). Harden *et al.* (1991a) found linear rates of profile development through the Holocene, and similarity of rates over the southern Great Basin suggesting that climate has been relatively stable at a regional scale, probably due to the importance of dust flux in profile development. Differences of rates are small enough that soils can be correlated and approximately dated. They found that the Holocene/Pleistocene contrast was much greater than the geographical variation of rates and that linear Holocene rates could not be extended into the Pleistocene. They attribute rate changes to climate-related processes such as effective moisture, dust, and erosion in addition to intrinsic chemical processes. By contrast, McFadden & Weldon (1987) found in southern California that silt increased progressively up to about 8000 cal yr B.P., followed by a very rapid increase in clay content in late Pleistocene deposits. This acceleration is attributed to crossing a threshold for the production of clay related to the accumulation of hygroscopic fines. Thus, age of parent material rather than Pleistocene/Holocene climate change can explain the observed differences in soil properties across that age boundary. This study highlights that in Holocene investigations we have the opportunity to compare different climate settings at a regional scale. The identification of the importance of the dust component in Cajon Pass relied on comparison with data from the San Joaquin Valley (Harden, 1986). Because of the relatively good preservation of Holocene and Late Pleistocene soils, there are numerous opportunities to compare processes and rates not available in older records.

Wang *et al.* (1994) showed that isotopic records from soil carbonate provide information about Holocene climate trends in an arid environment. They found that carbon and oxygen isotopes indicate a warmer and drier Holocene climate relative to the late Pleistocene in the Mojave Desert. Curiously, ^{14}C ages from soil organic matter and carbonate are older than other age estimates for the alluvial fans in which the soils developed. McFadden *et al.* (1998) demonstrated that vesicular A horizons (Av's) have formed during the Holocene (as well as in previous interglaciations) due to specific environmental conditions in desert soils. The combination of gravelly surface and dust influx are critical to the formation of these diagnostic carbonate surface horizons. The accelerated dust supply after the Pleistocene-Holocene transition due to drier Holocene climate facilitated their formation. An important aspect of their work on Av's has been the application of thermoluminescence (TL) dating.

Holocene soils can also provide evidence for instability in geographic zones of narrow climate transition. A well-documented type of Holocene soil variability has been related to dune reactivation in semi-arid landscapes. Dune reactivation has occurred through the Holocene in the Great Plains (Holliday, 1990; Muhs, 1985), including in historical period (Muhs & Holliday, 1995). Stratigraphically, sand sheets and dunes alternate with soils, some containing Bt horizons. The alternations and correlations with other climate records show that variations in water-balance (drought) and wind velocities result in destabilization of surfaces and conversion of alluvium to eolian dunes and sand sheets. These complex stratigraphic relations occur in an area of transition from semi-arid to humid climate. By contrast, dune reactivation in the humid east is a less common process. Markewich & Markewich (1994) documented Pleistocene dune reactivation in Georgia. Holocene dune reactivation in Georgia occurred during the early Holocene solar maximum, but a connection to climate change is equivocal (Ivester *et al.*, 2001) possibly indicating that the threshold for response to drought conditions is significantly higher in a generally more humid region.

It is likely that along the increasing rainfall gradient, the frequency of response decreases with increasing mean precipitation, probably reflecting the importance of vegetation in stabilizing soil surfaces. Resistant thresholds may be weak in young soils and the observed responses of Holocene soils to climate may not yield process models that accurately represent the climate responses of older soils. The climate variable that has produced a response in a Holocene soil may not be of sufficient magnitude to produce a response by a last interglacial soil. Thus, mapping of Holocene parent materials in areas of complex stratigraphy is useful for assessing potential climate impacts. This is extremely important to the problem of predicting regional responses to future climate change (Arnold *et al.*, 1990).

Climosequences provide possible calibration of soil properties with quantified climate parameters. Recent work by Cerling & Wang (1996) demonstrates that soil carbonate can provide information about the carbon isotope composition of organic material, and oxygen isotopes related to local meteoric water provide trends related to climate gradients. Muhs *et al.* (2001) documented that latitudinal climate gradients along ca. 1000 km of the Mississippi valley affect the degree of Holocene chemical weathering in the Peoria loess. Schwertman *et al.* (1982) found evidence or soil color change associated with hematite along a several 100 km temperature gradient. At the local scale, topography is an important factor, as shown by playa soils (Eghbalm *et al.*, 1989) whose soil salt and water balances change progressively along topographic profiles (catenas) that determine water transport and evaporation. Since it is rare that we see complete exposures of older catenas (Birkeland, 1999) that give such complex information, it is important to recognize that part of a catena may give an incomplete picture of the climate prevalent over the broader landscape.

In summary, in some cases Holocene soils can provide valuable information about local to regional scale relations to climate. In areas of temperature or precipitation gradients, we can measure systematic relations of soil properties to climate. Although soil development is weak, and can be complicated by inputs of eolian dust, thresholds, and erosion, there are areas in which the systematic changes in soil properties fit an interpretation of stable climate. The previous interglaciation provides examples of longer-term soil development.

How Long was the Last Interglaciation?

One of the ongoing paleoclimate debates is about the length of the last interglaciation. One of the possible applications of soils is to test the degree and timing of climate-state change associated with previous interglaciations, some of which are possible analogs of future warming. Ruhe (1965) pointed out that pre-Wisconsinan paleosols are more strongly developed than Wisconsinan or Holocene soils in the Midwest. Time is probably an important factor for these differences. Since soil formation is dependent on temperature and moisture, the coldest intervals and driest environments are generally not conducive to rapid profile development. A simple binary model of Quaternary climate, which ignores atmospheric moisture variations, suggests that rates of soil formation should vary significantly through contrasting cold glacial and warm interglacial climate cycles. The Devils Hole record (Winograd *et al.*, 1997) shows that the beginning of termination II (the onset of the last interglaciation) was also rapid and also predated the rise in solar insolation calculated by Berger (1988). If the last two glacial terminations are characteristic of earlier terminations, then the glacial-to-interglacial transitions may be very short relative to the duration of the contrasting climate states. Initial soil formation following a termination may be under conditions similar to the mean climate of the interglaciation.

The last interglaciation provides the most detailed terrestrial information we have for the duration of interglacial periods. The variations in ^{18}O during marine oxygen isotope stage (MIS) 5 are complex (Fig. 4a), and possibly driven by solar insolation (Imbrie *et al.*, 1993). Muhs *et al.* (1994, 2002), Smith *et al.* (1997) and Bischoff (1998) discuss the complexity of the last interglaciation in various marine and continental settings, and whether it corresponds just to MIS 5e (lasting ~10,000 yr), or whether it lasted all of MIS 5 (~60,000 yr). For example, sea level can be identified and precisely dated using the uranium/thorium systematics of corals. Gallup *et al.* (1994) dated high sea levels on Barbados over the past 200,000 yr; they point out the persistence of high sea levels for 10,000 yr following insolation peaks. Dating by Szabo *et al.* (1994) and Muhs *et al.* (1994) also shows that following the peak interglaciation at MIS 5e, sea-level drop was not synchronous with marine ^{18}O changes. Muhs *et al.* (1994) found that Pacific coast sea levels dated at ~105,000 and ~80,000 yr ago were close to present sea level. The dated sea level records from the Atlantic and Pacific coasts show that the last interglacial sea level may have stayed high until about 80,000 yr ago, and perhaps until 60,000 yr ago. This is consistent with evidence that ice volume increases following the last interglaciation were not simply linked to orbital variations (Clark *et al.*, 1993), and that marine ^{18}O variation is not only related to ice volume (Cronin *et al.*, 1996). Devil's Hole (Winograd *et al.*, 1997) provides a continental record consistent with Vostok ice (Petit *et al.*, 1990) and sea level records, showing that the last interglaciation spanned at least 20,000 yr and probably 57,000 yr, significantly longer than the 12,000-yr duration estimated by CLIMAP Project Members (1984). Marine sediments from the North Atlantic (Oppo *et al.*, 1997) also show that MIS 5e, the earliest part of the interglaciation, lasted more than 20,000 yr. If the last interglaciation is characteristic, then time periods for Quaternary soil development are potentially long with respect to the rate of soil differentiation. Vidic *et al.* (2003) discuss the problem of resolving the duration of periods of soil formation in the Chinese loess through the last six interglaciations. The estimates of interglacial duration from marine oxygen isotope stages (Fig. 5a) are not consistent with the relative degrees of soil formation. This may be due to significant paleo-precipitation differences among interglaciations.

For soils, the importance of an interglacial duration depends on precipitation as well as temperature. Temperature variations may correlate well with glacial/interglacial cycles, but precipitation is not simply related to a global climate state. For some continental areas, such as the western U.S., interglaciations were warm and dry relative to wet and cool glacial intervals. Pluvial lake records, discussed in detail by Benson (this volume), provide evidence for major variations of atmospheric moisture transport on 10^4–10^5 yr timescales. One of the best-studied continental records is from the Owens Lake Basin, California. Bischoff (1998) found evidence for a closed lake between ~118,000 and ~53,000 yr ago in the Owens Lake basin, interpreted as the last interglaciation, contrasting with the wet, pluvial glacial periods. Menking *et al.* (1997) and Li *et al.* (1998) found geochemical evidence that short (<10,000-yr) wet periods punctuated the generally dry last interglaciation in the Owens system. This implies a long continental interglaciation in the Owens Valley with short excursions possibly related to variations in solar insolation. Pollen from Owens Lake (Litwin *et al.*, 1997) shows that within the last interglaciation, the relative percentage of junipers was high from ~120,000 to ~60,000 yr ago, while the percentage of pines fluctuated. The overall pattern supports the geochemical evidence that short wet periods occurred during the long, relatively dry interglaciation. By contrast to the Owens Valley record, the Grand Pile record from southern France (Woillard, 1978), which presumably reflects Atlantic moisture, shows a relatively long, wet interglaciation punctuated by short excursions indicating colder, drier intervals (Beaulieu & Reille, 1992).

The spatial variability of moisture is apparently related to the balance of dry, polar vs. wet, tropical air mass movements (Rohling *et al.*, 2003), which is influenced by a variety of processes. Kutzbach & Guetter (1986) discuss the changes in atmospheric circulation related to continental ice sheet expansion; jet stream diversion increased moisture supply to temperate, pluvial lakes and caused drying of some tropical wetlands. The prevalent growth of mountain glaciers in the period between ~100,000 and 50,000 yr ago (Gillespie & Molnar, 1995) suggests that in that period atmospheric circulation of water vapor was very different from the present (when most mountain glaciers are retreating) and at the LGM. Major changes in atmospheric transport of water probably also occurred going into the LGM about 35,000 yr ago when high-latitude fresh-water flux into the Arctic ocean increased (Poore *et al.*, 1999), the Laurentide Ice Sheet expanded (Winograd, 2001), and global sea level (Bard, 1996; Fairbanks, 1989) changed rapidly. The data presented by Gillespie & Molnar (1995) are consistent with other evidence that terrestrial records vary spatially due to air mass (e.g. jet stream) and atmospheric moisture variations. Thus, in periods conducive to soil formation, the spatial patterns of moisture can be extremely variable on millennial and shorter timescales. This presents major concerns about extrapolating from local scales to much broader paleoclimate and paleoenvironmental reconstruction.

The Last Interglaciation of the U.S. Mid-Continent

The mid-continent of North America provides detailed information about the last interglaciation and the transition from it to the LGM. Glacier advances and retreats were complex over the past 130,000 yr. Recent reviews (e.g. Clark *et al.*, 1993), show that Laurentide ice-sheet growth and decay was a highly complex process. Based on work by Bond *et al.* (1992) and Boulton & Clark (1990), Clark *et al.* (1993) argue for a thin and dynamic ice-sheet complex that was strongly influenced by regional as well as global climate. The recent soil stratigraphic information is consistent with a slow increase in Laurentide ice thickness since 110,000 yr ago (Hunt & Malin, 1998), and with short-lived expansions and contractions of Laurentide ice between ~110,000 and ~30,000 yr ago, culminating in a major expansion between ~25,000 and ~15,000 yr ago (Winograd, 2001). For the southern part of the Laurentide ice sheet, there is increasing evidence that despite a major ^{18}O peak defining MIS 4 in the marine record (Martinson *et al.*, 1987), ice did not cover as large an area as in MIS 6 or 2 (Curry, 1989; Curry & Follmer, 1992; Curry & Pavich, 1996) in North America.

Soil studies are critical to interpreting the last interglacial landscape. The soil stratigraphy of southern Canada and the Illinois lobe of the Laurentide Ice sheet is relatively well studied (Follmer *et al.*, 1979; Hall & Anderson, 2000; Olson, 1989). The Sangamon soil dominates the record of pre-Holocene soils. The Sangamon Soil clearly developed under warm, humid climate (Hall & Anderson, 2000; Markewich, 1993, 1994) and is much-better developed than any of the Holocene soils in the mid-continent. In well-drained environments, the Sangamon Soil is >1 m thick, red, and very clay-rich (detailed descriptions are found in Hall & Anderson, 2000; Markewich, 1993, 1994; Markewich *et al.*, 1998). The contrast with weakly developed Holocene soils is striking.

The Sangamon Soil paleosol is distributed from the western great plains through the Mississippi Valley (Curry & Follmer, 1992). This thick, clayey and often red soil is better developed than surface soils and has long been recognized as a remnant of interglacial conditions, occupying part of the time interval between the Late Wisconsinan (MIS 2) and Illinoian (MIS 6) Laurentide glaciations of the mid-continent. Boardman (1985a) argued from soil formation rates that the last interglacial soils of the U.S. and western Europe reflect a long period of landscape stability relative to Holocene soils. Based on clay accumulation, he argued that the Sangamon Soil from the midwestern U.S. formed over a period roughly an order of magnitude greater than the 10^4 yr of Holocene clay accumulation. Recent studies of the Sangamon Soil developed on till (Curry & Pavich, 1996) and loess (Markewich *et al.*, 1998) show that the Sangamon soil was exposed at the surface for >60,000 yr, between ~120,000 and ~55,000 yr ago. The duration of exposure of the Sangamon Soil, developed both on till and loess in the Mississippi valley, shows that MIS 4 did not produce a major stratigraphic break (Curry & Pavich, 1996; Markewich *et al.*, 1998). This result shows that the Mississippi valley did not experience glaciations similar to MIS 6 and 2 in the intervening interval. A similar duration for the last interglaciation has been documented along a transcontinental transect running from Europe to China. Frechen *et al.* (1997) dated loess using thermoluminescence and infrared stimulated luminesence and found a paleosol that formed from about 130,000 to 60,000 yr ago in loess of the penultimate glaciation. These results are inconsistent with large ice volume increases in MIS 5d and 5b as suggested by the Milankovitch model and Imbrie *et al.* (1984); rather, they are consistent with independent evidence for a long interglaciation, discussed earlier, or with the Sangamon Soil reaching a state of insensitivity to climate variability documented in isotopic or biologic records.

Shifts in atmospheric circulation that may have affected the Sangamon Soil development are recorded in more arid soils as well. Pedogenic carbonates in the Wind River basin provide evidence that interglacial soils there may have developed under moisture sources different from those of the Holocene (Amundson *et al.*, 1996). Their analyses of ^{18}O shows that a stronger monsoonal flow from the Gulf of Mexico contributed much more precipitation than in the recent past. The stability of interglacial soils may also be significant for interpreting the record of atmospheric dust. Dust transport to Greenland, Antarctica (De Angelis *et al.*, 1997; Petit *et al.*, 1990), and Devil's Hole (I.J. Winograd, pers. comm., 2002) increased during cold periods. The origin of that dust is quite likely the continental loess areas. Dust generated during drier, windier periods is increasingly recognized in ice records. The Antarctic Vostok core contains a record of dust concentration variations back to ~160,000 yr ago (Petit *et al.*, 1990). The independently "dated" dust record corresponds roughly to the SPECMAP time scale of glacial/interglacial variations (Imbrie *et al.*, 1984), but with some important offsets (Fig. 12 from Petit *et al.*, 1990). For example, the Vostok ice does not contain a dust peak at 110,000 yr ago, during the cold MIS 5d. Petit *et al.* (1990) attribute this to insufficient wind or high sea level, which is consistent with the sea-level records cited previously. Moreover, the dust concentrations in Vostok are minimal until <70,000 yr ago, much younger than the MIS 4 cold peak. Petit *et al.* (1990) also point out a possible correlation with the Chinese loess record; the period of minimal dust, from ~120,000 to ~70,000 yr ago, corresponds to a wet, soil-forming period at the Xifeng section. The possible connection between dust, loess (Porter *et al.*, 2001) and the strength of the Asian monsoon is very significant to identifying global scale wet/dry cycles (Rohling *et al.*, 2003). Heller *et al.* (1993) present evidence that transient periods 25,000 and 55,000 yr ago were significantly wetter than the early Holocene (Reynolds & King, 1995). These may have been transient climate state changes as discussed above. The increasing recognition of transient pulses in many climate records is also changing our view of climatic variability and is consistent with spatial and temporal variability of dust flux in the west and southwest (Birkeland, 1999).

The evidence from soils over large continental areas distant from ice margins is important because it shows that not all parts of landscape are highly responsive to climate fluctuations recorded in more sensitive proxy media such as the marine oxygen isotope record. In other cases, soils may provide evidence for climatic variability that is not well recorded in the marine record. The previously cited examples of Holocene water balance variations are not reflected in marine oxygen isotope records. These process details must be taken into consideration in the application of climate models to predictions of regional responses to climate change. For humid environments, time allows for progressive increases in thickness, clay mass, and loss of weatherable minerals. Soil properties more in equilibrium with their mean environment become less responsive to extremes of external variables. We again emphasize that soil changes require state changes of increasing magnitude or durations as internal thresholds increase. The transition into the LGM may be used to illustrate the significance of short, extreme climate state variations.

Did Landscape Respond to Stadial/Interstadial Variability?

While the evidence for a preservation of well-developed soils of the last interglaciation is well established, there is also evidence for alteration of soils by periglacial processes in the transition into the LGM. Bullock (1985) showed how soil micromorphology could be used to identify the impact of cold climate on soil fabric developed under milder conditions. That work illustrates that in some cases soil properties can remain relatively unmodified under a new climate regime. The mid-continent of the U.S. provides a possible example of the soil response to highly variable stadial/interstadial fluctuations. Work in the Mississippi valley (Curry & Pavich, 1996; Leigh, 1994; Markewich, 1993, 1994) demonstrates that an enigmatic silt unit, the Roxana, occupies the time period of MIS 3. The existence of a "non-glacial" silt deposit formed during MIS 3 is significant to the assumed linkage of loess with glaciation. A more useful concept may be the relation of loess to hydrologic conditions in continental areas. The Roxana Silt, a 1- to 7-m-thick loess is present from Minnesota to Mississippi and Iowa to Indiana. Age data (Leigh & Knox, 1993; Markewich, 1993, 1994) indicate deposition during a 30,000-yr period from ~55,000 to ~26,000 yr ago, thus representing most of MIS 3. The Roxana Silt appears to be conformably overlain by the Peoria Loess and lies disconformably on the Loveland Loess. The Farmdale paleosol marks the uppermost surface of the Roxana Silt; the Sangamon paleosol marks the uppermost surface of the Loveland Loess. The Sangamon paleosol is everywhere truncated and where the Roxana Silt is thin is commonly welded to an unnamed paleosol at the base of the Roxana Silt. Markewich *et al.* (1998) hypothesize that the unconformity between the Loveland Loess and the Roxana Silt may represent MIS 4 in the lower Mississippi valley. The Roxana Silt contrasts sharply with the Loveland and Peoria loesses, in color, mineral content, paleosol characteristics, and ^{10}Be content. In some exposures the Roxana Silt appears to be very thinly bedded and bioturbated. The bedding and the paucity of carbonate minerals (H.W. Markewich, unpublished data) suggest dissolution shortly after deposition. One possible explanation is that locally the Roxana Silt was deposited in a floodplain environment; another would be almost immediate redeposition from upslope to lowerslope positions. The comparatively high content of organic carbon, particularly in the Farmdale paleosol, suggests a cool suboxic environment. The comparatively high concentration of inherited ^{10}Be content suggests a source other than outwash. This hypothesis is supported by stratigraphic evidence suggesting that glacier ice was not present in the upper Mississippi valley in the early Wisconsin (Curry & Pavich, 1996). Preliminary mineral magnetic data have indicated variations in the amount, grain size, and type of magnetic minerals in the lower Mississippi valley. The high frequency variations in magnetic susceptibility may record short-term fluctuations in wind-speed, composition, or climate (Markewich, 1993; Rodbell *et al.*, 1997).

Evidence for high-magnitude climate variation during MIS 3 is present in numerous records. For example, MIS–3 records from Greenland ice (Broeker, 1997; Cuffy & Clow, 1997), the Sierra Nevada (Benson *et al.*, 1996; Phillips *et al.*, 1996), and North Atlantic Deep Water (Boyle, 2000; Oppo & Lehman, 1995) show high frequency/high magnitude fluctuations of climate proxies. Solar insolation was relatively stable from 60,000 to 25,000 yr ago, at a value intermediate between the insolation extremes of MIS 5e and 2. Despite the apparent solar insolation stability, Greenland ice shows an impressive record of interstadial variability. In addition to the extreme contrasts of full-glacial and interglacial climate, highly oscillatory periods with rapid, millennial-scale temperature fluctuations are recorded in ice records (Cuffy & Clow, 1997; Daansgard *et al.*, 1984). (Fig. 14 from Cuffy & Clow, 1997). Poore *et al.* (1999; in press) found evidence for reduced ice cover in the Arctic during this period. The onset of Heinrich events and high variability of mountain glaciers (Benson *et al.*, 1996) also occurred in MIS 3. Benson also presents evidence that Owens Lake overflowed between ~53,000 and 13,000 yr ago, indicating a significantly wetter climate than during the last interglaciation or the Holocene. Benson (1999; this volume) has summarized recent data on transient climatic events of the past 50,000 yr in the Great Basin of the western U.S. Evidence from Owens, Pyramid, and Mono lakes in California support a correlation between periods of lake-level fall and Heinrich and Younger Dryas periods of the North Atlantic. In the interval between 52,500 and 12,500 yr ago, Owens Lake overflowed much of the time (Benson *et al.*, 1996). In this interval most of the Heinrich event iceberg discharges into the North Atlantic occurred, and also numerous oscillations of Sierran glaciers. Benson (1999) identifies ~20 stadial/interstadial oscillations between 52,000 and 14,000 yr ago in the Sierra Nevada. Even if the number is not accurate, "...periods of greatest climatic (hydrologic) instability in the Owens basin were confined to periods of intermediate continental ice volume" (Benson, 1999). Such high-resolution correlation, if correct, implies a synchronous response of North Pacific and western continental climates with thermohaline circulation variations in the North Atlantic, as reflected in MIS 3 in Greenland ice. However, correlation at the millennial timescale over large areas is also uncertain. Benson (1999) points out that these Dansgaard/Oeschger and Heinrich events can be correlated in the North Atlantic, but correlations across the continent into Great Basin lacustrine records are unconvincing. Benson (this volume) also states that "...the number and durations of oscillations...recorded in sediments from the Owens, Pyramid and Mono basins are not the same as the number and durations of oscillations in the δ^{18}O documented in the GISP2 ice core."

The evidence for high-magnitude variability in MIS 3 supports the argument that a relatively major change of climate state was necessary to create an upper boundary for the Sangamon soil. For parent materials exposed over the past 50,000 yr in north temperate latitudes, strong climate change cycles have interrupted soil development. Thus, soils are less-well developed than the Sangamon Soil and their thresholds to change are lower. This is important not only to the record of variability in MIS 3, but also to future responses to possible climate change.

Importance of Dust and Loess in Global Processes

The mid-continent and semi-arid West provide numerous examples of the importance of eolian process and sediments to pedogenic processes. Dust is of increasing interest as a factor in global climate processes. While outside the scope of this volume, it is important to acknowledge the important research in Chinese loess. The 2.5-myr record (Liu *et al.*, 1985) of alterations of unweathered silt and paleosols on Chinese loess have been correlated with the marine oxygen isotope record (Heller & Liu, 1986) and with alternations in the strength of the Asian monsoon (An *et al.*, 1991). Magnetic stratigraphy has been particularly useful in deciphering the pedogenic processes in Chinese loess and their relation to atmospheric circulation (Reynolds & King, 1995). Soils are formed as dust deposition wanes. With the onset of more rapid dust deposition, soils are buried and become paleosols. Correlations across the plateau indicate that erosional truncation has not removed much of the soil record (Vidic *et al.*, 2003) and are evidence for widespread synchrony of variations in loess deposition rate. Even so, the correlation of paleosols across the Chinese loess plateau is not adequately tested. Vidic *et al.* (2003) show that there are major problems in correlating individual interglacial soils with the duration of interglaciations indicated by the marine oxygen isotope time scale. Independent dating of loess is needed. Outside of China it is not certain that loess stratigraphy shows a convincing correlation with the marine record (Busacca *et al.*, this volume; Markewich *et al.*, 1998).While the Chinese loess may correlate well with marine records, Mississippi valley loess is probably younger than 300,000 yr and does not correlate exactly with the cold peaks (even-numbered stages) in the marine record (Markewich *et al.*, 1998). There is evidence that global increases of atmospheric dust were periods of atmospheric cooling and drying, but silt deposits such as the Roxana silt in the Mississippi valley appear to have a complex origin related to periods of climate instability. One general aspect of Quaternary climate, highlighted by loess, is that atmospheric dust may be more common in cold, dry periods than in warmer climate periods (Muhs *et al.*, 2001). Addition of fresh dust to soils changes the parent mineral/secondary mineral relations and significantly increases soil fertility in highly depleted soils (Chadwick *et al.*, 1999; Simonson, 1995). Dust increases during the LGM may also have contributed to cooling of the tropics (Claquin *et al.*, 2003). Short periods, or events, of dust fall can have a significant impact on soil development (Chadwick & Davis, 1990; Machette, 1985). The identification and dating of dust events is important to understanding the details of a particular soil profile. A systematic program of research on eolian sediments is necessary for correlation of local, regional and global events, and for evaluating the possible primary impact of Asian monsoon processes on global climate. Evaluation of such hypotheses will require an increased research focus on the sources, transport, and fate of dust.

Future Challenges for Pedology in Landscape, Ecosystem and Climate Models

Compared to other earth sciences, pedology has suffered from lack of a unifying paradigm (Jacob & Nordt, 1991). Despite the need for information about Quaternary soil-landscape evolution, there has been no broad and systematic effort comparable to the Ocean Drilling Program. This is unfortunate since, despite numerous problems cited in this review, we feel that soils can provide information about local scale timing of environmental change that is different from and complementary to other climate proxies. The examples above demonstrate how field research supported by improved analytical techniques is critical to advancing our understanding of Quaternary landscapes. We are far from formulating a comprehensive, predictive model to explain soils and landscapes in terms of climate inputs. One of the problems in formulating models is that soils, like organisms, modify their environments. The rate curves (Fig. 3) suggest that positive feedbacks accelerate profile development early in pedogenesis. This is followed by a deceleration due to negative feedbacks that tend to stabilize the soil profile. The soil systems that develop, as reflected by the distributions of soil orders and suborders, can persist even though climate is unstable.

Because of the pedogenic processes that attempt to equilibrate with earth-surface variables, soils can be more stable than rocks at the surface. Evidence of soil stabilizing the landscape is seen in humid river valleys (Pavich & Vidic, 1993) and in broad interfluvial landscapes (Cremaschi, 1987). Eppes *et al.* (2002) demonstrated that in semi-arid southern California, development of soil carbonate horizons influences long-term geomorphic evolution independent of climate change. The increasing applications of cosmogenic surface-exposure dating (Bierman *et al.*, 2002) promise to revolutionize our ability to relate geomorphic surfaces to processes and climatic and tectonic forcing functions. There are limitations, however, in preservation of exposed rocks that constrain geomorphic reconstructions based on cosmogenic data. Thus, exposure dating of rocks may provide incomplete information about landscape stability and should be combined, where possible, with soils research.

Despite all of the cited caveats, there are important contributions of soil research to Quaternary studies. Foremost, soils provide critical information about the landscape response to climate variability. Humid and semi-arid climates have persisted over broad regions despite transient departures of temperature and precipitation; temporal cycles appear to oscillate between persistent boundary states. Thus, the persistence of soil characteristics over intervals longer than 10^5-yr climate cycles may be related to the recurrence of climates conducive to the formation of those properties, or to the irreversibility of threshold-related soil properties. If the latter, then soils may provide negative feedbacks that favor dynamic equilibrium and tend to stabilize geomorphic systems. The persistence of soil characteristics influences landscape evolution (Eppes *et al.*, 2002; McAuliffe, 1994) and, therefore, the stability of ecosystems.

The examples cited lead to the conclusion that we need to improve our understanding of terrestrial processes to improve the overall Quaternary climate model. Orbital forcing fails to explain the timing and magnitude of changes in many continental climate records (Clark *et al.*, 1993; Imbrie *et al.*, 1993). This view is supported by increasing numbers of investigations of climate variability on millennial and centennial time scales. While the concept of a variable oceanic conveyor of heat (Broecker *et al.*, 1985) may be the key global concept to emerge over the past decade in explaining non-Milankovitch variations (Charles, 1998), we also need better terrestrial models. We suggest that the Chinese loess/paleosol record is a critical area for understanding both the record and drivers of terrestrial climate. Rohling *et al.* (2003) point out that correlative responses in all parts of the Northern Hemisphere of millennial-scale records implies an atmospheric rather than an oceanic/thermohaline driver. The evidence that the Asian dust flux reveals a primary mode of atmospheric variability (Rohling *et al.*, 2003) is a compelling reason to pursue the hypothesis that variations in the relative strength of summer and winter monsoons over Asia play an important role in synchronizing Northern Hemisphere records. The identification and dating of eolian deposits should, therefore, receive more attention, combined with efforts to improve the chronology of Quaternary soils.

Soil Chronology and Rates

We agree with Birkeland's (1999) assertion that the relation of soil properties to time is perhaps the major challenge for Quaternary pedology. Various isotopic techniques have been applied to improving soil chronology (Table 1). Stratigraphic sequences can yield organic samples that provide burial dates (Ruhe, 1976; Markewich, 1993, 1994). ^{14}C dating has changed with the advent of accelerator mass spectrometry, but many problems persist. The depth dependence of radiocarbon dates documented by Sharpenseel (1971) is still a major obstacle to extracting simple age interpretations (Wang *et al.*, 1996). Wang *et al.* (1994) showed that carbonate coatings in soils can yield useful dates if the complexity of the soil CO_2 system is taken into account. Other cosmogenic isotopes, such as ^{10}Be, have also been applied to argillic horizons to estimate the duration of soil development (Pavich, 1986; Pavich & Vidic, 1993). Improved chronology would help evaluate the relative importance of linear and exponential rates and step-functions in soil development (Harden *et al.*, 1991a) as well as providing better time-stratigraphic control.

Variations in Soil-Water Balance

Many of the examples of soil response to climate involve variations in water balance. As mentioned previously, process models in young, Holocene soils are of great value. Runge (1973) emphasized water balance, organic matter production, and time in modeling soil development. He developed a model to explain differences in Holocene soils in loess related to drainage and water balance. McDonald *et al.* (1996) utilized numerical modeling of soil-water balance to evaluate Holocene climate records in carbonates in the Mojave Desert. Their results show that annual variations in rainfall can strongly influence the depth of carbonate precipitation, possibly obscuring the effects of Holocene climate trends. Their work highlights the potential for extraction of climate information from soil carbonate systems. Soil carbonate is particularly informative because of the opportunity to measure carbon and oxygen isotopes on the same samples. Soil carbonates are linked to climate through plant metabolism (Cerling & Wang, 1996). Deutz *et al.* (2001) demonstrated that stable carbon isotopes and radiocarbon ages can be used to decipher the environmental history of pedogenic carbonate since 25,000 yr ago in the semi-arid Rio Grande Rift. δ^{13}C values in carbonates indicated increasing C4 vegetation during the times of glacial climate, establishment of C3 vegetation by about 9000 yr ago, and a subsequent C4 vegetation increase since 6000 yr ago. δ^{18}O from soil carbonate provided information about moisture sources and seasonal water balance.

Soil moisture variations are not well documented and model simulations are imprecise (Pan *et al.*, 2001; Robock *et al.*, 1998). Soil moisture at temperate latitudes shows strong intra-annual variation, but tends to modulate interannual variations (Entin *et al.*, 2000). Soil hydrology also can influence slope evolution. An example of the influence of soil infiltration on slope development is presented in a model by Ergenzinger & Schmidt (1992). They demonstrate that soil infiltration has a strong influence on slope morphology, a relation that could produce a stabilizing feedback.

Soil Duration and Dynamic Equilibrium

Comparing the time to attain steady-state from chronosequence studies (Birkeland, 1999) (Fig. 3) with the timescale of global climate cycles (Fig. 4) (Martinson *et al.*, 1987), Spodosols, Mollisols, and Alfisols would have all approached steady states within the time available in warm, interglacial climates. This presents the general problem of whether our interpretation of soil boundaries and properties depends more on decelerating rates of change or on external, climatic forcing of the soil-forming processes. An important implication of observing decreasing rates of change is that soil pathways may progress toward steady state, or dynamic equilibrium, in which energy and material flows produce little or no additional change. It is unlikely that true steady state is ever achieved, but rates of change do decrease in many soil orders (Birkeland, 1999). Soil profile indices based on field criteria (Harden & Taylor, 1983) show the

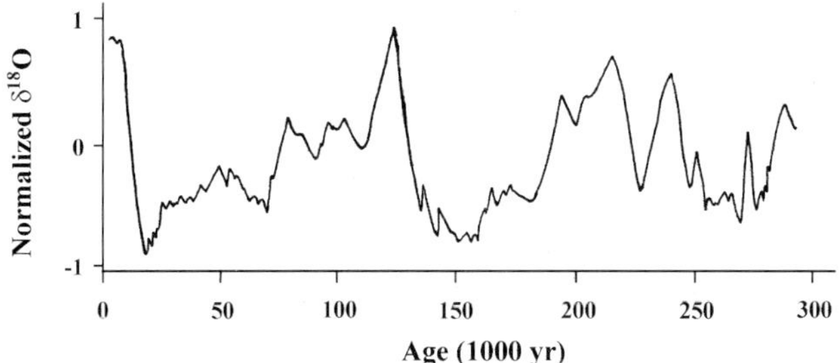

Fig. 4. Marine oxygen isotope climate record for the past 300,000 yr (Martinson et al., 1987).

applicability of this concept to large geographic areas. The evidence for approach toward steady states in soil chronosequences provides indirect evidence that soils are decreasingly sensitive to climate disturbances over time, thus possible obscuring the impacts of short-term increases in climate variability.

The persistence of preserved soils through climate cycles of various frequencies and magnitudes indicate that soils provide negative feedbacks that promote landscape stability and dynamic equilibrium. The systematic formation of secondary soil minerals, such as carbonates, in specific time intervals over very long timescales is providing information about long-term trends of atmospheric CO_2. Ekart & Carling (1999) discuss the record of atmospheric CO_2 based on $\delta^{13}C$ extracted from paleosol carbonates back to 400 myr ago. Quade & Cerling (1995) documented a shift in soil carbonate carbon isotopes during the Neogene. Their data from soil carbonate and organic matter indicate a shift from C3 to C4 vegetation in the northern Indian subcontinent associated either with intensification of the Asian monsoon or decrease of atmospheric pCO_2.

Soils data are needed at various timescales to understand the relation of the landscape stability to climate change. Rapid transitions, including the last glacial termination and sea level rise, are much more rapid than the gradual changes in soil properties discussed earlier. Deglaciation of the Laurentide ice sheet was slow relative to the rapid change of termination I; time stratigraphy shows that sea-level rise was abrupt, perhaps in one or two major pulses at about 13,500 cal yr B.P. (Bard *et al.*, 1996). Improved soils data would help in evaluating the role that soils and biota play in mediating the landscape response to abrupt transitions. The stability or inertia of soils is also important in considering the potential response of landscape to future climate changes. Boardman (1985b) stressed that landscape stability is more important to lengthy periods of soil development than paleoclimate. Therefore, what controls landscape stability? Since landscape stability is affected by climate via soils and other geomorphic processes (Bull, 1991), determining the range of temperature and precipitation associated with a particular soil stratigraphic unit is important. Markewich *et al.* (1998) argue that, based on chemistry and mineralogy, the Sangamon Soil in the Mississippi valley transect developed under climates ranging from weakly to stongly seasonal. The observations that soils in the U.S. and northern Europe formed over long last-interglacial periods, despite major fluctuations in the marine oxygen isotope record, and that possibly synchronous climate changes terminated the last-interglacial soil development about 55,000 yr ago over large areas, indicates that the insolation model alone does not explain an extremely important landscape/climate relation. If the marine record is dominated by ice volume and temperature, then temperature *per se* may not be as significant to landscape. Hydrologic changes may be more important than temperature to landscape stability. Thus, the understanding the relation of soil thresholds to hydrologic response may provide critical improvements in our landscape models.

Assessing Human Impacts

The growth of human population creates major impacts on landscape (Hooke, 2000). Soil erosion has long been associated with agriculture. In many areas human disturbance has caused acceleration of soil erosion rates (Trimble, 1999). One impact is that well-preserved surface soil profiles are becoming rare (Amundson, 1998). The increasing demand for quantifying human impacts (IPCC, 2001) requires that we improve our understanding of complex systems. The future impact on soil will be determined to a great extent by human disturbance and management of landscapes. Clearly, along with models of rapid climate changes related to thermohaline ocean circulation, which potentially could disrupt mid-latitude growing cycles (Broecker, 1997), we need to understand the long-term stability of agricultural soils and their susceptibility to being disrupted by short-term climate change. Are soils becoming more vulnerable because of human disturbance of what were dynamically equilibrated profiles?

Much work is needed to determine how soil stability models can be tied into broader models of landscape stability. This is of immediate concern for assessing impacts of Global Change (IPCC, 2001). Because we see evidence that climate can depart dramatically from the norm during the Quaternary and that soils persist through climate extremes, what general concepts should guide our research? Whipple (2001) highlights the problem of landscape response to climate oscillations: "... given that the timescale of climatic forcing is so short compared with system response time, it may reasonably be expected that rapid climatic fluctuations would have negligible impact on landscape form." This appears to be true for the fluvial profiles he modeled, as well as for well-developed argillic soils. In the semi-arid Central Asian climate, by contrast, landscape has been highly responsive to climate oscillations. Zhang *et al.* (1999) have documented variation of dust sources through Asian climate cycles. Zhang's detailed geochemical work on dust also highlights the problem of identifying sources and loads of atmospheric pollution due to human activities in a sensitive landscape. The relative impacts of land use and rapid climate fluctuations is only resolved with information about soil ages and processes in profiles over broad regions. Lack of response in terrestrial systems governed by thresholds may simply indicate that the energetics of climate change recorded in less-resistant systems (such as marine sediments, glacier ice, chemical precipitates, unvegetated floodplains, major river valleys, and semi-arid and arid lands with sparse vegetation) is not sufficient to alter the more resistant system. Conversely, for systems with significant thresholds, only the big events, rather than gradual change due to the cumulative effect of many moderate events (Wolman & Gerson, 1978), may be important.

Soil-System Models

Soil genesis models are rudimentary at best and their elaboration will require integration with climate models. The geologic evidence, some of which has been reviewed here, indicates that soil-climate relations, like other complex earth systems, represent a highly variable balance of complex processes (Mitchell, 1976; Nahon, 1991; Runge, 1973) that do not presently fit in a comprehensive theory or model. While Earth science is becoming more deductive and model-driven, soil research is still at an inductive stage. Soil science is relatively young (Jacob & Nordt, 1991; Yaalon & Berkowicz, 1997) and thus presents great potential for observation and hypothesis. The relatively recent recognition of erosion as an important component of steady-state orogenic systems (Whipple, 2001; Willet & Brandon, 2002) illustrates how geophysical hypotheses (e.g. isostatic balance) can evolve into models that incorporate data from many sub-disciplines of geology. We need more interdisciplinary collaboration in geomorphology and pedology.

Process models describing soil development are useful for portraying the interactions of complex variables as soils evolve toward steady states. Inherent in that evolution is the dominance of a few variables and the feedback between soil properties and processes. Runge (1973) emphasized the partitioning of runoff and infiltration of precipitation. Other than Simonson (1978), there have been few additional attempts to synthesize the array of soil information into simplifying genetic models.

Climate modeling is an iterative process in which the terrestrial impacts of dust production, albedo, and soil moisture on climate must be assessed, along with terrestrial impacts of climate variation on soils. In general, soils can potentially provide major insights into how geomorphic systems integrate oscillatory processes into stable forms. Complementing the now-voluminous studies of less-inertial systems, soils can provide information about lower-frequency climate cycles and about rapid state changes that create stratigraphic boundaries. While other proxies increasingly reveal complexity of climate, we have to consider which climate changes are really significant to altering soils and landscapes. One of the major enigmas about soils is that, despite very energetic climatic, physical, and biological processes working to disrupt fabric and horizon structure, they exhibit similar recurrent characteristics over broad spatial and long temporal scales. That astounding fact is reason enough to dig.

References

Amundson, R. (1998). *Soil preservation and the future of pedology*. 16th World Congress of Soil Science, Montpelier, France.

Amundson, R., Chadwick, O.A., Kendall, C., Wang, Y. & DeNiro, M. (1996). Isotopic evidence for shifts in atmospheric circulation patterns during the late Quaternary in mid-North America. *Geology*, **24**(1), 23–26.

An, Z.S., Kukla, G.J., Porter, S.C. & Xiao, J. (1991). Magnetic susceptibility evidence of monsoon variation on the Loess Plateau of central China during the last 130,000 years. *Quaternary Research*, **36**, 29–36.

Antevs, E. (1953). Geochronology of the deglacial and neothermal ages. *Journal of Geology*, **61**(3), 195–230.

Arnold, R.W., Szabolcz, I. & Targulian, V.O. (Eds) (1990). *Global soil change*. Laxenburg, Austria, International Institute For Applied Systems Analysis, 110 pp.

Bard, E., Hamelin, B., Arnold, M., Montaggioni, L., Cabioch, G., Faure, G. & Roughiere, F. (1996). Deglacial sea-level record from Tahiti corals and the timing of global meltwater discharge. *Nature*, **382**, 241–244.

Bard, E., Hamelin, B., Fairbanks, R.G. & Zindler, A. (1990). Calibration of the ^{14}C timescale over the past 30,000 years using mass spectrometric U-Th ages from Barbados corals. *Nature*, **345**, 405–410.

Bartlein, P.J. (1997). Past environmental changes: Characteristic features of Quaternary climate variations. *In*: Huntley, B., Cramer, W., Morgan, A.V., Prentice, H.C. & Allen, J.R.M. (Eds), *Past and Future Rapid Environmental Changes: The Spatial and Evolutionary Responses of Terrestrial Biota*. NATO ASI Series, Vol. 147, pp. 11–29.

Bartlein, P.J. & Whitlock, C. (1993). Paleoclimatic interpretation of the Elk Lake pollen record. *In*: Bradbury, J.P. & Dean, W.E. (Eds), The Paleoenvironmental History of Elk Lake, A 10,000 Year Record of Varved Sediments in Minnesota. *Geological Society of America Special Paper*, pp. 275–293.

Beaulieu, J.L. & Reille, M. (1992). The last climatic cycle at La Grande Pile (Vosges, France): A new pollen profile. *Quaternary Science Reviews*, **11**, 431–438.

Benson, L.V. (1999). Records of millennial-scale climate change from the Great Basin of the western United States. *In*: *Mechanisms of Global Climate Change at Millennial Time Scales*, American Geophysical Union Geophysical Monograph No. 112, pp. 203–225.

Benson, L.V., Burdett, J.W., Kashgarian, M., Lund, S.P., Phillips, F.M. & Rye, R.O. (1996). Climatic and hydrologic oscillations in the Owens Lake basin and Adjacent Sierra Nevada, California. *Science*, **274**, 746–748.

Berger, A. (1988). Milankovitch theory and climate. *Reviews of Geophysics*, **26**, 624–657.

Bierman, P.R., Caffee, M.W., Davis, P.T., Marsella, K, Pavich, M.J., Colgan, P., Mickelson, D. & Larsen, J. (2002). Understanding the rates and timing of earth surface processes using in-situ produced cosmogenic ^{10}Be. *In*: Grew, E.S. (Ed.), Beryllium: Mineralogy, petrology and geochemistry.

Reviews in Mineralogy and Geochemistry, Vol. 50, pp. 147–205.

Birkeland, P.W. (1999). *Soils and geomorphology* (3rd ed.). Oxford Univ. Press, 430 pp.

Bischoff, J.L. (Ed.) (1998). The last interglaciation at Owens Lake, California: Core OL–92. *U.S. Geological Survey, Open-File Report*, 98–132, 186 pp.

Boardman, J. (1985a). Comparison of soils in midwestern United States and western Europe with the interglacial record. *Quaternary Research*, **23**, 62–75.

Boardman, J. (Ed.) (1985b). *Soils and Quaternary landscape evolution*. Chichester, Wiley and Sons, 391 pp.

Bockheim, J.G. (1980). Solution and use of chronofunctions in studying soil development. *Geoderma*, **24**, 71–85.

Bond, G. *et al.* (1992). Evidence for massive discharges of icebergs into the North Atlantic ocean during the last glacial period. *Nature*, **360**, 245–249.

Boulton, G.S. & Clark, C.D. (1990). A highly mobile Laurentide ice sheet revealed by satellite images of glacial lineations. *Nature*, **346**, 813–817.

Boyle, E.A. (2000). Is ocean thermohaline circulation linked to abrupt stadial/interstadial transitions? *Quaternary Science Reviews*, **19**, 255–272.

Bradbury, J.P. & Dean, W.E. (Eds) (1993). The paleoenvironmental history of Elk Lake, a 10,000 year record of varved sediments in Minnesota. *Geological Society of America Special Paper*, 276.

Broecker, W. (1997). Will our ride into the greenhouse future be a smooth one? *GSA Today*, **7**, 1–7.

Broecker, W.S., Peteet, D.M. & Rind, D. (1985). Does the ocean-atmosphere system have more than one stable mode of operation? *Nature*, **315**, 21–26.

Bronger, A. & Catt, J.A. (1989). *Paleopedology: Nature and application of paleosols*. Catena Supplement, No. 16, 232 pp.

Bull, W.B. (1991). *Geomorphic responses to climatic changes*. Oxford University Press, New York, 326 pp.

Bullock, P. (1985). The role of micromorphology in soil studies. *In*: Boardman, J. (Ed.), *Soils and Quaternary Landscape Evolution*. Chichester, Wiley and Sons, pp. 45–68.

Buol, S.W., Hole, F.D., McCracken, R.J. & Southard, R.J. (1997). *Soil genesis and classification* (4th ed.). Ames, IA, Iowa State Univ. Press.

Busacca, A., Muhs, D.R., Markewich, H.W., Beget, J.E., Lancaster, N. & Sweeney, M.R. (2003). Eolian sediments. *In*: Gillespie A. *et al.* (Eds), This Volume.

Campbell, I.D., Campbell, C., Apps, M.J., Rutter, N.W. & Bush, A.B.G. (1998). Late Holocene ~1500 yr climatic periodicities and their implications. *Geology*, **26**(5), 471–473.

Catt, J.A. & Bronger, A. (1998). Reconstruction and climatic implications of paleosols. *Catena*, **34**(1–2), 207.

Cerling, T.E. & Wang, Y. (1996). Stable carbon and oxygen isotopes in soil CO_2 and soil carbonate: Theory, practice and application to some prairie soils of the upper Midwestern North America. *In*: Boutton, T.W. & Yamasaki, S.-I. (Eds), *Mass Spectrometry of Soils*. New York, Marcel Dekker, pp. 113–131.

Chadwick, O.A. & Chorover, J. (2001). The chemistry of pedogenic thresholds. *Geoderma*, **100**, 321–353.

Chadwick, O.A. & Davis, J.O. (1990). Soil forming intervals caused by eolian sediment pulses in the Lahontan Basin, northwestern Nevada. *Geology*, **18**, 243–246.

Chadwick, O.A. *et al.* (in press). The impact of climate on the biogeochemical functioning of volcanic soils. *Chemical Geology*.

Chadwick, O.A., Derry, L.A., Vitousek, P.M., Huebert, B.M. & Hedin, L.O. (1999). Changing sources of nutrients during four million years of ecosystem development. *Nature*, **391**, 491–497.

Chadwick, O.A., Nettleton, W.D. & Staidl, G.J. (1995). Soil Polygenesis as a function of Quaternary climate change, Northern Great Basin, USA. *Geoderma*, **68**, 1–26.

Charles, C. (1998). The ends of an era. *Nature*, **394**, 422–423.

Claquin, T., Roelandt, C., Kohfeld, K., Harrison, S., Tegen, I., Prentice, I., Balkanski, Y., Bergametti, G., Hansson, M., Mahowald, N., Rodhe, H. & Schulz, M. (2003). Radiative forcing of climate by ice-age atmospheric dust. *Climate Dynamics*, **20**(2/3), 193–202.

Clark, P.U. *et al.* (1993). Initiation and development of the Laurentide and Cordilleran Ice Sheets following the last interglaciation. *Quaternary Science Reviews*, **12**, 79–114.

CLIMAP Project Members (1984). The last interglacial ocean. *Quaternary Research*, **21**, 123–224.

COHMAP Members (1988). Climate changes of the last 18,000 years: Observations and model simulations. *Science*, **241**, 1043–1052.

Cremaschi, M. (1987). *Paleosols and vetusols in the central Po Plain (northern Italy). A study in Quaternary geology and soil development*. Milano, Unicopli, 316 pp.

Cronin, T. *et al.* (1996). Deep sea ostracode shell chemistry (Mg:Ca ratios) and late Quaternary Arctic Ocean history. *Geological Society of America Special Publication*. No. 111, pp. 117–134.

Crowley, T.J. & North, G.B. (1991). *Paleoclimatology*. New York, Oxford University Press, 339 pp.

Crowley, T.J. (1983). The geological record of climatic change. *Reviews of Geophysical Space Physics*, **21**, 828–877.

Cuffy, K.M. & Clow, G.D. (1997). Temperature, accumulation and ice sheet elevation in central Greenland through the last deglacial transition. *Journal of Geophysical Research*, **102**(C12), 26383–26396.

Curry, B.B. (1989). Absence of Altonian glaciation in Illinois. *Quaternary Research*, **31**, 1–13.

Curry, B.B. & Follmer, L.R. (1992). The last interglaicial-glacial transition in Illinois: 123–25 ka. *In*: Clark, P.U. & Lea, P.D. (Eds), The Last Interglacial-Glacial Transition in North America. *Geological Society of America Special Paper*, Vol. 270, pp. 71–88.

Curry, B.B. & Pavich, M.J. (1996). Absence of glaciation in Illinois during marine isotope stages 3 through 5. *Quaternary Research*, **46**, 19–26.

De Angelis, M., Steffensen, J.P., Legrand, M.R., Clausen, H.B. & Hammer, C.U. (1997). Primary aerosol (sea salt and soil dust) deposited in Greenland ice during the last climatic cycle: Comparison with east Antarctic records. *J. Geophysical Research Oceans*, **102**, 26681–26698.

Deutz, P., Montanez, I.P., Monger, H.P. & Morrison, J. (2001). Morphology and isotope heterogeneity of late Quaternary pedogenic carbonates: Implications for paleosol carbonates as paleoenvirnomental proxies. *Paleogeography, Paleoclimatology, Paleoecology*, **166**, 293–317.

Eghbalm, K., Southand, J. & Whittig, L.D. (1989). Dynamics of evaporik distribution in soils on a fan-playa transect in the Carrizo Plain, California, USA. *Soil Science Society of America Journal*, **53**(3), 898–903.

Ekart, D.D. & Cerling, T.E. (1999). A 400 million year carbon isotope record of pedogenic carbonate: Implications for atmospheric carbon dioxide. *American Journal of Science*, **299**, 805–817.

Entin, J.K., Robock, A., Vinnikov, K.Y., Hollinger, S.E., Liu, S. & Namkai, A. (2000). Temporal and spatial scales of observed soil moisture variations in the extratropics. *Journal of Geophysical Research*, **105**(11), 865–877.

Eppes, M.C., McFadden, L.D., Matti, J. & Powell, R. (2002). Influence of soil development on geomorphic evolution of landscapes: An example from the Transverse Ranges of California. *Geology*, **30**, 195–198.

Ergenzinger, P. & Schmidt, J. (1992). Slope geometry resulting from the spatial variation of soil permeability *In*: Schmidt, K.-H. & dePloey, J. (Eds), *Functional Geomorphology*, Catena Supplement 23, pp. 151–156.

Fairbanks, R.G. (1989). A 17,000–year glacio-eustatic sealevel record: Influence of glacial melting rates on the Younger Dryas event and deep-ocean circulation. *Nature*, **342**, 637–642.

Fenwick, I. (1985). Paleosols: Problems of recognition and interpretation. *In*: Boardman, J. (Ed.), *Soils and Quaternary Landscape Evolution*. Chichester, Wiley and Sons, pp. 3–21.

Follmer, L.R., McKay, E.D., Lineback, J.A. & Gross, D.L. (1979). Wisconsinan, Sangamonian and Illinoian stratigraphy of Central Illinois: Midwest Friends of the Pleistocene Field Conference. *Illinois State Geological Survey Guidebook*, **13**, 134.

Forman, S.L. (1989). Applications and limitations of thermoluminescence to dating Quaternary sediments. *Quaternary International*, **1**, 47–59.

Frechen, M., Horvath, E. & Gabris, G. (1997). Geochronology of middle and upper Pleistocene loess sections in Hungary. *Quaternary Research*, **48**, 291–312.

Friis-Christensen, E. & Lassen, K. (1991). Length of the solar cycle: An indicator of solar activity closely associated with climate. *Science*, **254**, 698–700.

Gallup, C.D., Edwards, G.L. & Johnson, R.G. (1994). The timing of high sea-levels over the past 200,000 years. *Science*, **263**, 796–800.

Gile, L., Hawley, J. & Grossman, R. (1981). Soils and geomorphology in the Basin and range area of southern New Mexico-guidebook to the desert project. *New Mexico Bureau of Mines and Mineral Resources Memoir*, **39**, 222.

Gillespie, A. & Molnar, P. (1995). Glacial asynchronism. *Reviews of Geophysics*, **33**, 311–364.

Hack, J. (1960). The interpretation of erosional topography in humid temperate regions. *American Journal of Science*, **258a**, 80–97.

Hall, R.D. & Anderson, A.K. (2000). Comparative development of Quaternary Paleosols of the central United States. *Paloegeography, Paleoclimatology, Paleoecology*, **158**(1–2), 109–145.

Harden, J.W. (1982). A quantitative index of soil development from field descriptions. Examples from a chronosequence in central California. *Geoderma*, **28**, 1–28.

Harden, J.W. & Taylor, E.M. (1983). A quantitative comparison of soil development in four climatic regimes. *Quaternary Research*, **20**, 342–359.

Harden, J.W., Taylor, E.M., Hill, C., Mark, R.K., McFadden, L.D., Reheis, M.C., Sowers, J.M. & & Wells, S.G. (1991a). Rates of soil development from four soil chronosequences in the southern Great Basin. *Quaternary Research*, **35**, 383–399.

Harden, J.W., Taylor, E.M., McFadden, L.D. & Rehies, M.C. (1991b). Calcic, gypsic and siliceous soil chronosequences in arid and semi-arid environments. *Soil Science Society of America Special Publication*, No. 26, pp. 1–16.

Heimsath, A.M., Dietrich, W.E., Nishizumi, K. & Finkel, R.C. (2001). Stochastic processes of soil production and transport: Erosion rates, topographic variation and cosmogenic nuclides in the Oregon coast range. *Earth Surface Processes and Landforms*, **26**, 531–552.

Heller, F. & Liu, T.S. (1986). Paleoclimatic and sedimentary history from magnetic susceptibility of loess in China. *Geophysical Research Letters*, **13**, 1169–1172.

Heller, F., Shen, C.D., Beer, J., Liu, T.S., Bronger, A., Suter, M. & Bonani, G. (1993). Quantitative estimations of pedogenic ferromagnetic formation in Chinese loess and paleoclimatic implications. *Earth Planetary Science Letters*, **184**, 125–139.

Hole, F.D. & Campbell, J.B. (1985). *Soil landscape analysis*. Totowa, N.J., Rowman and Allanheld, 196 pp.

Holliday, V.T. (1990). Soils and landscape evolution of eolian plains in the Southern High Plains of Texas and New Mexico. *In*: Knuepfer, P.L.K. & McFaddden, L.D. (Eds), *Soils and Landscape Evolution*, Elsevier, pp. 489–516.

Hooke, R.LeB. (2000). On the history of humans as geomorphic agents. *Geology*, **28**, 843–846.

Hotchkiss, S.C., Vitousek, P.M., Chadwick, O.A. & Price, J. (2000). Climate cycles, geomorphological change and the interpretation of soil and ecosystem development. *Ecosystems*, **3**, 322–333.

Hunt, A.G. & Malin, P.E. (1998). Possible triggering of Heinrich events by ice-load-induced earthquakes. *Nature*, **393**, 155–158.

Imbrie, J. *et al.* (1984). The orbital theory of Pleistocene climate: Support from a revised chronology of the marine ^{18}O record. *In*: Berger, A.L., Imbrie, Hays, Kukla, Saltzman, (Eds), *Milankovitch and Climate*, Pt. 1. Dordrecht, The Netherlands, D. Reidel, pp. 269–305.

Imbrie, J. *et al.* (1993). On the structure and origin of major glaciation cycles 2. The 100,000-year cycle. *Paleooceanography*, **8**, 699–735.

IPCC WGI, Climate Change (2001). The science of climate change, contribution of the WGI to the second assessment report of the Intergovernmental Panel on Climate Change. *In*: Houghton, J.T., Meira Filho, L.G., Callander, B.A., Harris, N., Kattenberg, A. & Maskell, K. (Eds). Cambridge University Press.

Ivester, A.H., Leigh, D.S. & Godfrey-Smith, D.I. (2001). Chronology of inland eolian dunes on the Coastal Plain of Georgia, USA. *Quaternary Research*, **55**, 293–302.

Jacob, J.S. & Nordt, L.C. (1991). Soil and landscape evolution: A paradigm for pedology. *Soil Science Society of America Journal*, **55**, 1194.

Jenkins, D.A. (1985). Chemical and mineralogical composition in the identification of paleosols. *In*: Boardman, J. (Ed.), *Soils and Quaternary Landscape Evolution*. Chichester, Wiley and Sons, pp. 23–43.

Jouzel *et al*. (1989). Global change over the last climatic cycle. *Quaternary International*, **2**, 15–24.

Keigwin, L. (1996). The little ice age and medeival warm period in the Sargasso sea. *Science*, **274**, 1504.

Knox, J.C. (1985). Responses of floods to Holocene climatic change in the upper Mississippi Valley. *Quaternary Research*, **23**, 287–300.

Kutzbach, J.E. & Guetter, P.J. (1986). The influence of changing orbital parameters and surface boundary conditions on climate simulations for the past 18,000 years. *Journal of Atmospheric Sciences*, **43**, 1726–1759.

Lamb, H.H. (1995). *Climate; present, past and future*. Methuen, London, 835 pp.

Leigh, D.S. (1994). Roxana silt of the upper Mississippi Valley: Lithology, source, and paleoenvironment. *Geological Society of America Bulletin*, **106**, 430–442.

Leigh, D.S. & Knox, J.C. (1993). AMS radiocarbon age of the Upper Mississippi Valley Roxana Silt. *Quaternary Research*, **39**, 282–289.

Li, H.C., Ku, T.L., Bischoff, J.L. & Stott, L.D. (1998). Climatic and hydrologic conditions in Owens Basin, California between 45 and 145 Ka as reconstructed from the high-resolution stable isotope records. *In*: Bischoff, J.L. (Ed.), *The Last Interglaciation at Owens Lake, California: Core OL–92. U.S. Geological Survey, Open-File Report*, 98–132, pp. 66–81.

Litwin, R.J., Adam, D.P., Frederiksen, N.V. & Woolfenden, W.B. (1997). An 800,000 year pollen record from Owens Lake, California: Preliminary Analyses. *Geological Society of America Special Paper*, 317, pp. 127–142.

Liu, T., An, Z.S., Yuan, B. & Han, J. (1985). The loess-paleosol sequence in China and climatic history. *Episodes*, **8**, 21–27.

Machette, M.N. (1985). Calcic soils of the southwestern United States. *Geological Society of America Special Paper*, 203, pp. 1–21.

Mack, G.H., James, W.C. & Monger, H.C. (1993). Classification of paleosols. *Geological Society of America Bulletin*, **105**, 129–136.

Mahaney, W.C. (Ed.) (1978). *Quaternary soils*. GeoAbstracts Ltd., Norwich, England.

Markewich, H.W. (Ed.) (1993). Progress report on chronostratigraphic and paleoclimatic studies, Middle Mississippi River Valley, Eastern Arkansas and Western Tennessee. *U.S. Geological Survey Open-file Report*, 93–273, 61 pp.

Markewich, H.W. (Ed.) (1994). Second progress report on chronostratigraphic and paleoclimatic studies, Middle Mississippi River Valley, Eastern Arkansas, Western Tennessee, and Northwestern Mississipi. *U.S. Geological Survey Open-File Report*, 94–208, 55 pp.

Markewich, H.W. & Markewich, W. (1994). An overview of Pleistocene and Holocene inland dunes in Georgia and the Carolinas-morphology, distributions, age and paleoclimate. *U.S. Geological Survey Bulletin*, **2069**, 32.

Markewich, H.W., Pavich, M.J., Mausbach, M.J., Hall, R.L., Johnson, R.G. & Hearn, P.P. (1987). Age relations between soils and geology in the Coastal Plain of Maryland and Virginia. *U.S. Geological Survey Bulletin*, 1589-A, 34 pp.

Markewich, H.W., Pavich, M.J., Mausbach, M.J., Stuckey, N., Johnson, R.G. & Gonzalez, V. (1986). Soil development and its relation to ages of morphostratigraphic units in Horry County, South Carolina. *U.S. Geological Survey Bulletin*, 1589-B, 61 pp.

Markewich, H.W., Wysocki, D.A., Pavich, M.J., Rutledge, E.M., Millard, H.T., Jr., Rich, F.J., Maat, P.B., Rubin, M. & McGeehin (1998). Paleopedology plus TL, ^{10}Be and ^{14}C dating as tools in stratigraphic and paleoclimatic investigations, Mississippi River Valley, USA. *Quaternary International*, **51/52**, 143–167.

Martinson, D.G., Pisias, N.G., Hays, J.D., Imbrie, J., Moore, T.C. & Shackleton, N.J. (1987). Age dating and the orbital theory of the ice ages: Development of a high- resolution 0–300,000-year chronostratigraphy. *Quaternary Research*, **27**, 1–29.

McAuliffe, J. (1994). Landscape evolution, soil formation, and ecological patterns and processes is Sonoran Desert bajadas. *Ecological Monographs*, **64**, 111–148.

McDonald, E.V., Pierson, F.B., Flerchinger, G.N. & McFadden, L.D. (1996). Application of a process-based soil-water balance model to evaluate the influence of Late Quaternary climate change on soil-water movement. *Geoderma*, **74**, 167–192.

McFadden, L.D. (1988). Climatic influences on rates and processes of soil development in Quaternary deposits of southern California. *Geological Society of America Special Paper*, **216**, 153–177.

McFadden, L.D. & Knuepfer, P.L.K. (1990). Soil geomorphology: The linkage of pedology and surficial processes. *In*: McFadden, L.D. & Knuepfer, P.L.K. (Eds), Soils and Landscape Evolution, *Geomorphology*, Vol. 3, No. 3/4, pp. 197–206.

McFadden, L.D., McDonald, E.V., Wells, S.G., Anderson, K., Quade, J. & Forman, S.L. (1998). The vesicular layer and carbonate collars of desert soils and pavements: Formation, age and relation to climate change. *Geomorphology*, **24**, 101–145.

McFadden, L.D. & Weldon, R.J. (1987). Rates and processes of soil development on Quaternary terraces in Cajon Pass,

California. *Geological Society of America Bulletin*, **98**, 280–293.

McKeague, J.A., Brydon, J.E. & Miles, N.M. (1971). Differentiation of forms of extractable iron and aluminum in soils. *Soil Science Society of America Proceedings*, **35**, 33–38.

Menking, K.M., Bischoff, J.L. & Fitzpatrick, J.A. (1997). Climatic/hydrologic oscillations since 155,000 yr B.P. at Owens lake, California, reflected in abundance and stable isotope composition of sediment carbonate. *Quaternary Research*, **48**, 58–68.

Millard, H.T. & Maat, P.B. (1994). Thermoluminescence dating procedures in use at the U.S. Geological Survey, Denver, Colorado. *U.S. Geological Survey Open-File Report*, 94–249, 112 pp.

Mitchell, J.M. (1976). An overview of climatic variability and its causal mechanisms. *Quaternary Research*, **6**, 481–493.

Morrison, R. (1964). Lake Lahontan: Geology of the southern Carson Desert, Nevada. *U.S. Geological Survey Professional Paper*, 401, 156 pp.

Muhs, D.R. (1984). Intrinsic thresholds in soil systems. *Physical Geography*, **5**, 99–110.

Muhs, D.R. (1985). Age and paleoclimatic significance of Holocene sand dunes in northeastern Colorado. *Annals Association of American Geographers*, **74**(4), 566–582.

Muhs, D.R., Kennedy, G.L. & Rockwell, T.K. (1994). Uranium-series ages of marine terrace corals from the Pacific coast of North America and implications for last-interglacial sea level history. *Quaternary Research*, **42**, 72–87.

Muhs, D.R., Bettis, E.A., Been, J. & McGeehin, J.P. (2001). Imapct of climate on parent material and chemical weathering of soils in the Mississippi River Valley. *Soil Science Society of America Journal*, **65**(6), 1761–1777.

Muhs, D.R. & Holliday, V.T. (1995). Evidence of active dune sand on the great plains in the 19th century from accounts of early explorers. *Quaternary Research*, **43**.

Muhs, D.R., Simmons, K.R. & Steinke, B. (2002). Timing and warmth of the last interglacial period: New U-series evidence from Hawaii and Bermuda and a new fossil compilation for North America. *Quaternary Science Reviews*, **21**, 1355–1383.

Nahon, D.B. (1991). *Introduction to the petrology of soils and chemical weathering*. New York, John Wiley and Sons, 313 pp.

Nettleton, W.D., Gamble, E.E., Allen, B.L., Borst, G. & Peterson, F.F. (1989). Relict soils of Subtropical Regions of the United States. *In*: Bronger, A. & Catt, J. (Eds), *Paleopeology*, Catena Supplement, 16, pp. 59–94.

Olson, C.G. (1989). Soil geomorphic research and the importance of paleosol stratigraphy in Quaternary investigations, Midwestern U.S. *In*: Bronger, A. & Catt, J. (Eds), *Paleopedology*, Catena Supplement, 16, pp. 129–142.

Oppo, D.W., Horowitz, M. & Lehman, S.J. (1997). Marine core evidence for reduced deep water production during Termination II followed by a relatively stable sbstage 5e (Eemian). *Paleoceanography*, **12**, 51–63.

Oppo, D.W. & Lehman, S.J. (1995). Suborbital timescale variability of North Atlantic deep water during the past 200,000 years. *Paleoceanography*, **10**, 901–910.

Pan, Z., Arritt, R.W., Gutowski, W.J., Jr. & Takle, E.S. (2001). Soil moisture in a regional climate model: Simulation and projection. *Geophysical Research Letters*, **28**(15), 2947–2950.

Pavich, M.J. (1986). Processes and rates of saprolite production and erosion on a foliated granitic rock of the Virgina Piedmont *In*: Colman, S.M. & Dethier, D.P. (Eds), *Rates of Chemical Weathering of Rocks and Minerals*. Academic Press, Orlando, pp. 522–590.

Pavich, M.J. & Vidic, N. (1993). Application of paleomagnetic and ^{10}Be analyses to chronostrtigraphy of Alpine glacio-fluvial terraces, Sava River Valley, Slovenia. *In*: Swart, P. (Ed.), Climate change in continental isotopic records, *American Geophysical Union Monograph*, 78, pp. 263–275.

Petit, J.R. *et al.* (1990). Paleoclimatological and chronological implications of the Vostok core dust record. *Nature*, **343**, 56–58.

Phillips, F.M. *et al.* (1996). Chronology for fluctuations in late Pleistocene Sierra Nevada glaciers and lakes. *Science*, **274**, 749–751.

Poore, R.Z., Dowsett, H.J., Verardo, S. & Quinn, T.M. (2003). Millenial to century–scale variability of gulf of Mexico Holocene climate records. *Paleoceanography*, **18**(2), 1048.

Poore, R.Z., Osterman, L., Curry, W.B. & Phillips, R.L. (1999). Late Pleistocene and Holocene meltwater events in the western Arctic Ocean. *Geology*, **27**, 759–762.

Porter, S.C., Hallet, B., Wu, X.H. & An, Z.S. (2001). Dependence of near-surface magnetic susceptibility on dust accumulation rate and precipitation on the Chinese Loess Plateau. *Quaternary Research*, **55**, 271–283.

Quade, J. & Cerling, T.E. (1995). Expansion of C_4 grasses in the Late Miocene of Nothern Pakistan: Evidence from stable isotopes in paleosols. *Paleogeography Paleoclimatology, Paleoecology*, **115**, 91–116.

Reheis, M.C., Sowers, J.M., Taylor, E.M., McFadden, L.D. & Harden, J.W. (1992). Morphology and genesis of carbonate soils on the Kyle Canyon fan, Nevada, USA. *Geoderma*, **52**, 303–342.

Retallack, G.J. (1990). *Soils of the Past: An Introduction to Paleopedology*. Unwin-Hyman, London, 520p.

Reynolds, R.L. & King, J.W. (1995). Magnetic records of climate change. *Reviews of Geophysics, American Geophysical Union Supplement*, pp. 101–110.

Richmond, G.M. (1962). Quaternary stratigraphy of the La Sal Mountains, Utah. *U.S. Geological Survey Professional Paper*, 324, 135 pp.

Robock, A., Schlosser, C.A., Vinnikov, K.Y., Speranskaya, N.A. & Entin, J.K. (1998). Evaluation of AMIP soil moisture simulations. *Global and Planetary Change*, **19**, 181–208.

Rodbell, D.T., Forman, S.L., Pierson, J. & Lynn, W.C. (1997). The stratigraphy and chronology of Mississippi Valley loess in western Tennessee. *Geological Society of America Bulletin*, **109**, 1134–1148.

Rohling, E.J., Mayewski, P.A. & Challenor, P. (2003). On the timing and mechanism of millennial-scale climate variability during the last glacial cycle. *Climate Dynamics*, **20**, 257–267.

Ruhe, R.V. (1965). Quaternary Paleopedology. *In*: Wright, H.E. & Frey, D.G. (Eds), *The Quaternary of the United States*. Princeton, N.J., Princeton University Press, pp. 755–764.

Ruhe, R.V. (1976). Stratigraphy of mid-continent loess. *In*: Mahaney, W.C. (Ed.), *Quaternary Stratigraphy of North America*. Stroudsburg, PA, Dowden, Hutchinson and Ross, pp. 197–211.

Ruhe, R.V., Daniels, R.B. & Cady, J.G. (1967). Landscape evolution and soil formation in southwestern Iowa. *Soil Conservation Service Technical Bulletin*, 1349, 242 pp.

Ruhe, R.V. & Olson, C.G. (1980). Clay mineral indicators of glacial and nonglacial sources of Wisconsin loesses in southern Indiana, USA. *Geoderma*, **24**, 283–297.

Runge, E.C.A. (1973). Soil development sequences and energy models. *Soil Science*, **115**, 183–193.

Schwertman, U. (1993). Relations between iron oxides, soil color and soil formation. *In*: Bigham, J.M. & Ciolkozs, E.J. (Eds), *Soil Color, Soil Science Society of America Special Publication*, Vol. 31, pp. 51–70.

Schwertman, U., Murad, E. & Schulze, D.G. (1982). Is there Holocene reddening (hematite formation) in soils of axeric temperate areas? *Geoderma*, **23**, 191–208.

Sharp, W.D., Ludwig, K.R., Chadwick, O.A., Amundson, R. & Glaser, L.L. (2003). Dating fluvial terraces by ^{230}Th/U on pedogenic carbonate, Wind River Basin, Wyoming. *Quaternary Research*, **59**, 139–150.

Sharpenseel, H.W. (1971). Radiocarbon dating of soils-problems, troubles, hopes. *In*: Yaalon, D.H. (Ed.), *Paleopedology: Origin, Nature and Dating of Paleosols*. Jerusalem, Israel Universities Press, pp. 77–88.

Sibrava, V., *et al.* (Eds) (1986). *Quaternary glaciations in the northern hemisphere*. Oxford, Pergamon Press, 511 pp.

Simonson, R.W. (1978). A multiple process model of soil genesis. *In*: Mahaney, W.C. (Ed.), *Quaternary Soils, GeoAbstracts*. Norwich, England, University of East Anglia, pp. 1–25.

Simonson, R.W. (1995). Airborne dust and its significance to soils. *Geoderma*, **65**, 1–43.

Singer, A. (1980). Paleoclimatic interpretation of clay minerals in soils and weathering profiles. *Earth Science Reviews*, **15**, 303–326.

Sirocko, F., Garbe-Schoenberg, C.D., McIntyre, A. & Molfino, B. (1996). Teleconnections between the subtropical monsoons and high latitude climates during the last deglaciation. *Science*, **272**, 526–529.

Smith, G.I., Bischoff, J.L. & Bradbury, J.P. (1997). Synthesis of paleoclimatic records from Owens Lake Core OL–92. *In*: Smith, G.I. & Bischoff, J.L. (Eds), Paleoclimate record from core OL–92, Owens Lake, Southeast California. *Geological Society of America Special Paper*, 317, pp. 143–160.

Smith, G.I. & Street-Perrott, F.A. (1983). Pluvial lakes of the western United States. *In*: Porter, S.C. & Wright, H.E. (Eds), *Late Quaternary Environments of the United States*. Minneapolis, University of Minnesota Press, pp. 190–212.

Soil Survey Staff (1999). Soil taxonomy (2nd ed.). USDA Natural Resources Conservation Service, *Agricultural Handbook*, No. 436, 871 pp.

Szabo, B.J., Ludwig, K.R., Muhs, D.R. & Simmons, K.R. (1994). Thorium–230 ages of corals and duration of the last interglacial sea-level high stand on Oahu, Hawaii. *Science*, **266**, 93–96.

Trimble, S.W. (1999). Decreased rates of alluvial sediment storage in the Coon Creek Basin, Wisconsin, 1975–93. *Science*, **285**, 1244–1246.

Valentine, K.W.G. & Dalrymple, J.B. (1976). Quaternary buried paleosols: A critical review. *Quaternary Research*, **6**, 209–222.

Velde, B. (1992). *Introduction to clay minerals*. London, Chapman and Hall, 198 pp.

Vidic, N.J. (1998). Soil-age relationships and correlations: Comparison of chronosequences in the Ljubljana Basin, Slovenia and USA. *In*: Catt, J. & Bronger, A. (Eds), *Reconstruction and Climatic Implications of Paleosols*. Catena, Vol. 34, No. 1–2, pp. 113–130.

Vidic, N.J., Verosub, K.L. & Singer, M.J. (2003). The Chinese loess perspective on marine isotope stage 11 as an extreme interglacial. *In*: Droxler, A.W., Poore, R.Z. & Burkle, L.H. (Eds), *Earth's climate and obital history: The maine isotope stage 11 question. Geophysical Monograph 37*, American Geophysical Union, Washinghton, D.C. pp. 231–240.

Wang, Y., Amundson, R. & Trumbore, S. (1996). Radiocarbon dating of soil organic matter. *Quaternary Research*, **45**, 282–288.

Wang, Y., McDonald, E., Amundson, R., McFadden, L. & Chadwick, O. (1994). An isotopic study of soils in chronological sequences of alluvial deposits, Providence Mountains, California. *Geological Society of America Bulletin*, **108**(4), 379–391.

Whipple, K.X. (2001). Fluvial landscape response time: How plausible is steady state denudation? *American Journal of Science*, **301**, 313–325.

Winograd, I.J. (2001). The magnitude and proximate cause of ice-sheet growth since 35,000 yr B.P. *Quaternary Research*, **56**, 299–307.

Winograd, I.J., Landwehr, J.M., Ludwig, K.R., Coplen, T.B. & Riggs, A.C. (1997). Duration and structure of the past four interglaciations. *Quaternary Research*, **48**, 141–154.

Woillard, G.M. (1978). The Grand Pile peat bog: A continuous pollen record for the last 140,000 years. *Quaternary Research*, **9**, 1–21.

Wolman, M.G. & Gerson, R. (1978). Relative scales of time and effectiveness of climate in watershed geomorphology. *Earth Surface Processes*, **3**, 189–208.

Woodhouse, C.A. & Overpeck, J.T. (1998). 2000 years of drought variability in the Central United States. *Bulletin of the American Meteorological Society*, **79**, 2693–2714.

Yaalon, D.H. (1971). Soil forming processes in time and space. *In*: Yaalon, D.H. (Ed.), *Paleopedology*. Jerusalem, Israel Univiversities Press, pp. 29–39.
Yaalon, D.H. (1983). Climate, time and soil development. *In*: Wilding, L.P., Smeck, N.E. & Hall, G.F. (Eds), *Pedogenesis and Soil Taxonomy I. Concepts and interactions*. Amsterdam, Elsevier, pp. 233–251.
Yaalon, D.H. & Berkowicz, S. (Eds) (1997). History of Soil Science. *Advances in Geoecology* (Vol. 29). Verlag, Catena, 438 pp.
Zhang, X.Y., Arimoto, R. & An, Z.S. (1999). Glacial and interglacial patterns for Asian dust transport. *Quaternary Science Reviews*, **18**(6), 811–819.

Earthquake recurrence inferred from paleoseismology

Brian F. Atwater[1], Martitia P. Tuttle[2], Eugene S. Schweig[3], Charles M. Rubin[4], David K. Yamaguchi[5] and Eileen Hemphill-Haley[6]

[1] *U.S. Geological Survey at Department of Earth and Space Sciences, University of Washington, Seattle, WA 98195-1310, USA*
[2] *M. Tuttle & Associates, 128 Tibbetts Lane, Georgetown, ME 04548, USA*
[3] *U.S. Geological Survey, 3876 Central Ave., Ste. 2, Memphis, TN 38152-3050, USA*
[4] *Department of Geological Sciences, Central Washington University, 400 East Eighth Avenue, Ellensburg, WA 98926, USA*
[5] *Department of Dental Public Health Sciences, University of Washington, Seattle, WA 98195-7475, USA*
[6] *Department of Geology, Humboldt State University, Arcata, CA 95521, USA*

Introduction

Earthquakes threaten the United States, as illustrated by hazard maps for the 48 conterminous states (Fig. 1). Much of the threat comes from unusually large earthquakes that recur hundreds or thousands of years apart. Engineering designs, insurance rates, and emergency plans depend on national maps that forecast seismic shaking at various probability levels (Fig. 1b). The study of prehistoric earthquakes – paleoseismology – provides long-term rates of earthquake occurrence to improve confidence in such forecasts.

Paleoseismology emerged in the last decades of the 20th century, after 1965. It draws on many kinds of research, including geomorphology, stratigraphy, structural geology, geochronology, paleoecology, oceanography, civil engineering, archaeology, ethnology, and documentary history. Its literature includes collected papers and workshop proceedings (Crone & Omdahl, 1987; Ettensohn *et al.*, 2002; Hancock & Michetti, 1997; Masana & Santanach, 2001; Ota *et al.*, 1992; Pavlides *et al.*, 1999; Serva & Slemmons, 1995; Shiki *et al.*, 2000; Yeats & Prentice, 1996), national and regional overviews (Camelbeeck, 2001; Clague, 1996; Grant & Lettis, 2002; Ota & Okumura, 1999; Research Group for Active Faults of Japan, 1992; Talwani & Schaeffer, 2001), topical reviews (Jacoby, 1997; Obermeier, 1996), textbooks (McCalpin, 1996; Noller *et al.*, 2000; Yeats *et al.*, 1997), and narratives intended for general audiences (Nance, 1988; Sieh & LeVay, 1998).

This chapter describes three North American examples of earthquake history inferred from Quaternary geology. The examples resemble one another by providing long-term perspectives unavailable from traditional seismological records. Each example includes multiple earthquakes inferred from widespread paleoseismic evidence. These earthquakes suggest rates and patterns of recurrence that help define earthquake hazards. The examples differ in tectonic setting, in the kinds of features that record prehistoric earthquakes, and in overlap with instrumental and written records.

Described first is evidence for infrequent surface rupture on faults in a small part of California's diffuse boundary between the Pacific and North America plates. The faults form a 50-km-wide shear zone east of the San Andreas fault. Collectively termed the eastern California shear zone, these faults accommodate lateral motion not absorbed by the San Andreas. Movement on some of them produced surface ruptures and two large, instrumentally recorded earthquakes in the 1990s. Prehistoric offsets exposed in trenches show that thousands of years probably separated such ruptures in the Holocene. Age ranges of the prehistoric ruptures overlap among the faults. These findings suggest that the shear zone produces large earthquakes in infrequent series.

Next we discuss earthquakes in the interior of the North America plate – in the New Madrid seismic zone of Missouri, Arkansas, and Tennessee. This region's low relief and slow rates of modern deformation belie a late Holocene history of large earthquakes more frequent than those in the eastern California shear zone. A series of three large earthquakes in 1811 and 1812, known from historical accounts, produced thousands of sand blows in an alluvial area at least 200 km by 80 km. Sand blows similarly record earlier series of New Madrid earthquakes in A.D. 800–1000 and 1300–1600.

Our final example comes from the Cascadia subduction zone, where oceanic lithosphere descends beneath the North America plate in California, Oregon, Washington, and British Columbia. Though unknown from this region's written history, great subduction earthquakes repeatedly lowered much of its Pacific coast by at least 0.5 m, most recently in A.D. 1700. The subsidence is marked by buried soils at estuaries. Such soils from the past 3500 years in Washington imply that the earthquakes recur at irregular intervals ranging from a few hundred years to about one thousand years.

Eastern California Shear Zone

The eastern California shear zone, centered about 150 km northeast of Los Angeles (Fig. 2), exhibits geologic evidence for prehistoric surface ruptures during episodes thousands of years apart.

Modern Deformation and Earthquakes

According to geodetic measurements, the shear zone absorbed right-lateral slip at 11–14 mm/yr during the 1990s (McClusky *et al.*, 2001; Miller *et al.*, 2001; Sauber *et al.*, 1994). This slip rate accounts for a quarter of the interplate motion, which averages 50 mm/yr (DeMets *et al.*, 1990, 1994).

DEVELOPMENTS IN QUATERNARY SCIENCE
VOLUME 1 ISSN 1571-0866
DOI:10.1016/S1571-0866(03)01015-7

(a) **Index map**

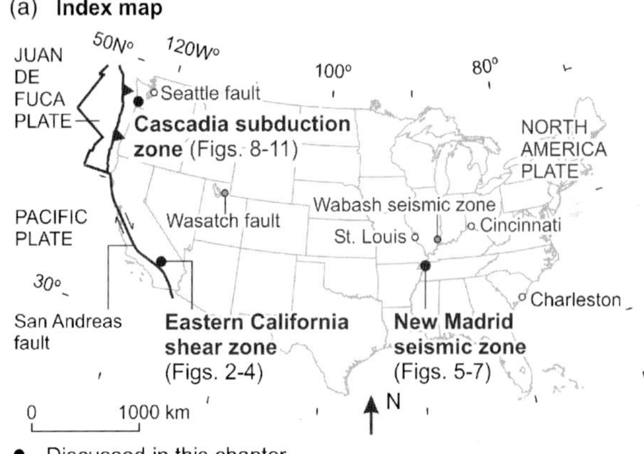

(b) **Earthquake hazard**—Estimated probability of exceeding 0.2 g horizontal acceleration in any 50-year period

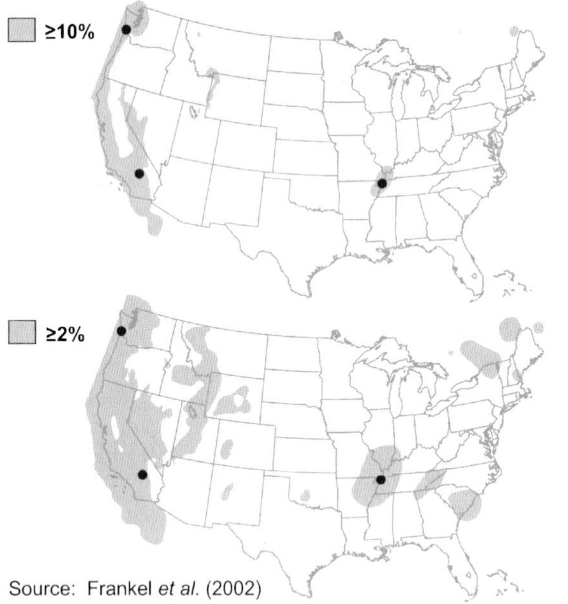

Acceleration of 0.2 g can cause partial collapse of ordinary buildings. Ground motion with a 10% probability of being exceeded in 50 years would be expected to be exceeded once in 500 years on average. For 2% in 50 years, the ground motion would be exceeded once in 2500 years on average.

Fig. 1. Overview of earthquake hazards in the conterminous United States.

The geodetic measurements coincided with a decade in which the eastern California shear zone produced two unusually large earthquakes (Figs 2 and 3). The first and largest, the 1992 Landers earthquake of moment magnitude (M) 7.3, ruptured several north- to northwest-striking right-lateral faults along a total length of 70 km (Sieh *et al.*, 1993). Within a few tens of seconds, rupture started on the southern Johnson Valley fault, progressed northward, slowed at stepovers to and from the Homestead Valley fault, and finally ended along the Camp Rock fault (Cohee & Beroza, 1994; Wald & Heaton, 1994). Coseismic dextral slip at the ground surface commonly exceeded 3 m; it reached a maximum of 6 m along the northern Emerson fault. Seven years later on a parallel trend 30 km to the northeast, the M 7.1 Hector Mine earthquake produced as much as 5 m of surface dextral slip on the Lavic Lake and Bullion faults (Treiman *et al.*, 2002). This 1999 earthquake also triggered small earthquakes over much of southern California (Hauksson *et al.*, 2002; Rymer *et al.*, 2002a).

The 1992 Landers and 1999 Hector Mine earthquakes have few historical precedents in eastern California. Before 1992, no large ($M > 7.0$) earthquakes had ruptured eastern California faults since the 1872 Owens Valley earthquake, centered 200 km north-northwest (Fig. 2a). Instead, the shear zone's largest events were moderate earthquakes of M_L 6.1 (1947 Manix), M_L 5.5 (1975 Galway Lake and 1979 Homestead Valley), and M 6.2 (1992 Joshua Tree; Fig. 2b and c). (M_L, local Richter magnitude, is similar to M in this size range.) All these earthquakes were exceeded in size by the 1992 M 6.5 Big Bear earthquake, an aftershock to the Landers earthquake.

Earthquakes of the 1990s thus define an uncommon episode of seismic activity in the eastern California shear zone. An earlier seismic episode, farther north in east-central California and western Nevada, occurred between 1872 and 1954 in a shear zone 500 km long (Wallace, 1978, 1984). These historical examples raise the question, do M 6–7 earthquakes in the eastern California shear zone typically come in clusters?

Prehistoric Earthquakes

Paleoseismic studies of the eastern California shear zone began a few weeks after the 1992 Landers earthquake and eventually involved more than 17 trenches across eleven faults (Fig. 2c). The studies focused on playas where the vertical component of slip produced stratigraphic offsets in fine-grained, stratified deposits of Holocene age (Fig. 3). Evidence for surface rupture includes faults and fissures that terminate at buried land surfaces, folding and warping of beds, deposits that resulted from ponding against fault scarps, and scarp-derived colluvium. Laminated lacustrine deposits allow detection of vertical separation as small as several centimeters. Prehistoric faulting also produced noticeable offsets in alluvium, colluvium, and buried soils. Detrital charcoal and peat beds have yielded radiocarbon ages that limit inferred times of the prehistoric ruptures and related earthquakes.

The paleoseismic studies confirm that the Landers and Hector Mine earthquakes were rare events (Fig. 4). Few of the faults trenched show evidence for more than two surface ruptures between 10,000 years ago and A.D. 1992. No prehistoric rupture of Holocene age has been found where the Lavic Lake fault ruptured in 1999, with the possible exception of minor slip after than A.D. 260 (Rymer *et al.*, 2002b).

The inferred earthquakes can be grouped into three Holocene episodes on the basis of overlapping radiocarbon ages (episodes A, B, and C in Fig. 4). The episodes are loosely defined because uncertainties in dating the prehistoric ruptures commonly span centuries (Fig. 4a). Episode A, about 8000–9000 cal yr B.P., includes the most recent

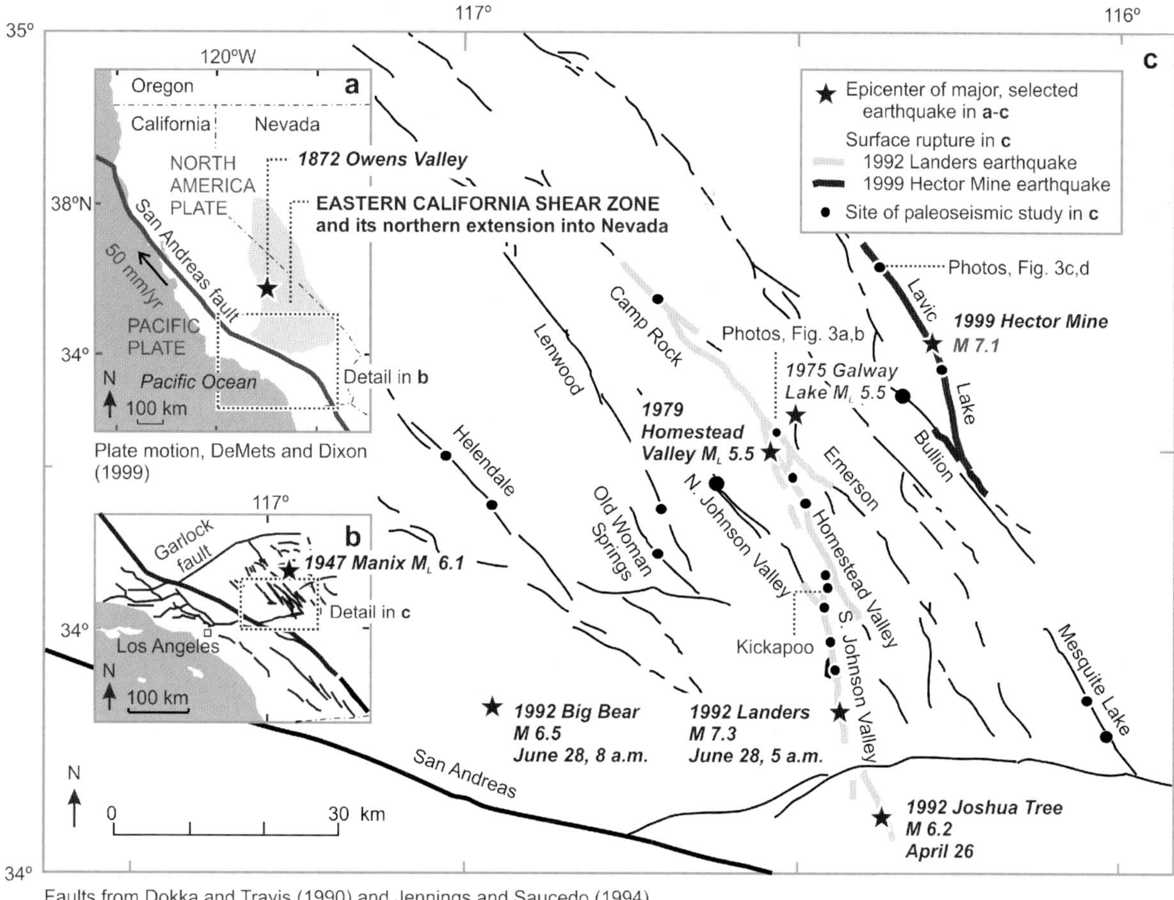

Fig. 2. Faults of the eastern California shear zone and vicinity. Faults in (c) have documented or presumed evidence for surface rupture within the past 10,000 years.

pre-1992 events on the Kickapoo and northern Emerson faults. These events produced fault scarps similar in height to the 1992 scarp. Episode A may also include the penultimate prehistoric surface rupture on the Helendale and Mesquite Lake faults, as well as ruptures on the Lenwood, Camp Rock, and southern Johnson Valley faults. Episode B, about 5000–6000 cal yr B.P., followed several thousand years of apparent quiescence. Surface ruptures occurred on the Lenwood, Johnson Valley, Bullion, and Mesquite Lake faults. The shear zone became active again in the past 1000 years, during episode C. This latest series of earthquakes, which continued into the 1990s, produced surface rupture on many faults in the shear zone. Though represented by a single rupture at most sites, episode C includes both a prehistoric rupture and the 1992 Landers rupture on the Camp Rock fault (Fig. 4, site 4).

Implications and Challenges

The long intervals between episodes imply that the earthquakes of the 1990s represent an unusual peak in seismic activity in the eastern California shear zone. However, episode C differs too much from A and B for any of the episodes to define the likely size and pattern of future surface ruptures (Rockwell *et al.*, 2000). Viewed as part of episode C (Fig. 4), the shear zone's 20th-century earthquakes imply either that additional earthquakes are likely, or that episode C is drawing to a close. The Working Group of California Earthquake Probabilities (1995), without much paleoseismic information about the eastern California shear zone, presumed that in coming decades, the zone would continue producing earthquakes like those since 1970 (Fig. 2c).

Geophysicists have proposed various triggers for swift series of earthquakes in the eastern California shear zone (Freed & Lin, 2001; Harris & Simpson, 2002; Hudnut *et al.*, 2002; Pollitz & Sacks, 2002; Zeng, 2001). The zone's history of episodic Holocene earthquakes suggests that a realistic trigger will permit thousands of years to elapse between earthquake series.

New Madrid Seismic Zone

Paleoseismology can clarify fault location and earthquake recurrence far from plate boundaries, in continental regions where tectonic activity has less geomorphic or seismological

(a) **Ruptures** on the Emerson fault from the 1992 Landers earthquake. A degraded older scarp runs parallel to them.

(b) **Trench** across 1992 ruptures on playa in **a**. Maximum dextral slip, 2.3 m; maximum uplift, 0.8 m. Site *6* in Figure 4a.

(c) **Uplift** at a bend in the Lavic Lake fault, 1999 Hector Mine earthquake. Maximum uplift, 1 m; dextral slip nearby, 2 m.

(d) **Oblique slip** along bend in Lavic Lake fault, 1999. Maximum dextral slip, 2.5 m; maximum uplift, 1.2 m.

Fig. 3. Surface ruptures of the 1992 Landers and 1999 Hector Mine earthquakes (locations, Fig. 2c).

expression than in the eastern California shear zone. One such region is the lower Mississippi River valley (Schweig *et al.*, 2002). This valley contains the New Madrid seismic zone (Fig. 5a), which during the winter of 1811–1812 produced some of the most widely felt earthquakes in the written history of the United States. Studies of prehistoric earthquakes in the New Madrid region have shown that the 1811–1812 earthquakes were not freak, one-time events.

The three largest shocks of the 1811–1812 sequence, of *M* 7.5–8.0 (Atkinson *et al.*, 2000; Hough *et al.*, 2000; Johnston, 1996), rank among Earth's largest intraplate quakes (Johnston & Kanter, 1990). They destroyed settlements along the Mississippi River, damaged buildings as far away as Cincinnati and St. Louis (Fig. 1), and were felt at distances as great as 1,800 km (Nuttli, 1973). They induced severe liquefaction and related ground failure throughout the New Madrid region (Fig. 5b; Fuller, 1912; Obermeier, 1989; Saucier, 1977) and locally as far as 250 km from inferred epicenters (e.g. Johnston & Schweig, 1996; Street & Nuttli, 1984).

Although few faults have geomorphic expression in the New Madrid region, numerous small modern earthquakes illuminate several interseting faults (Fig. 5a; Chiu *et al.*, 1992; Pujol *et al.*, 1997). Most of these earthquakes occur beneath Late Wisconsin and Holocene deposits of the Mississippi River and its tributaries. Many of the fluvial deposits liquefied during the A.D. 1811–1812 earthquakes, venting water and sand that formed sand blow deposits across about 10,000 km^2 (Figs 5 and 6). Prehistoric sand blows in this area provide the main evidence for two earlier episodes of New Madrid earthquakes during the past 1200 years.

Paleoseismic Evidence

According to oral traditions of Native Americans in the Mississippi River valley, a great earthquake devastated the region centuries before 1811 (Lyell, 1849). Geologic evidence for such an earthquake was first reported by Fuller (1912), who noted liquefaction-related ground failures and a history of uplift and erosion predating 1811. He inferred that the region had experienced "early shocks of an intensity equal to if not greater than that of the last."

Detailed study of pre-1811 earthquakes in the region began at the Reelfoot scarp (Fig. 5a). This landform coincides with a northwest-trending zone of microseismicity and may be a monocline above a blind thrust fault (Russ, 1982). As inferred from deformed sediments exposed in trenches across the scarp, prehistoric folding and earthquake-induced liquefaction occurred at least twice in the past 2000 years (Russ, 1979), probably in A.D. 780–1000 and A.D. 1260–1650 (Kelson *et al.*, 1992, 1996).

Archeological studies contributed to the recognition that many sand blows in the New Madrid region predate 1811–1812. For example, 30 km northeast of Reelfoot scarp at Towosahgy State Park (Fig. 5a), sand-filled fissures and two related sand blow deposits underlie a Native American mound

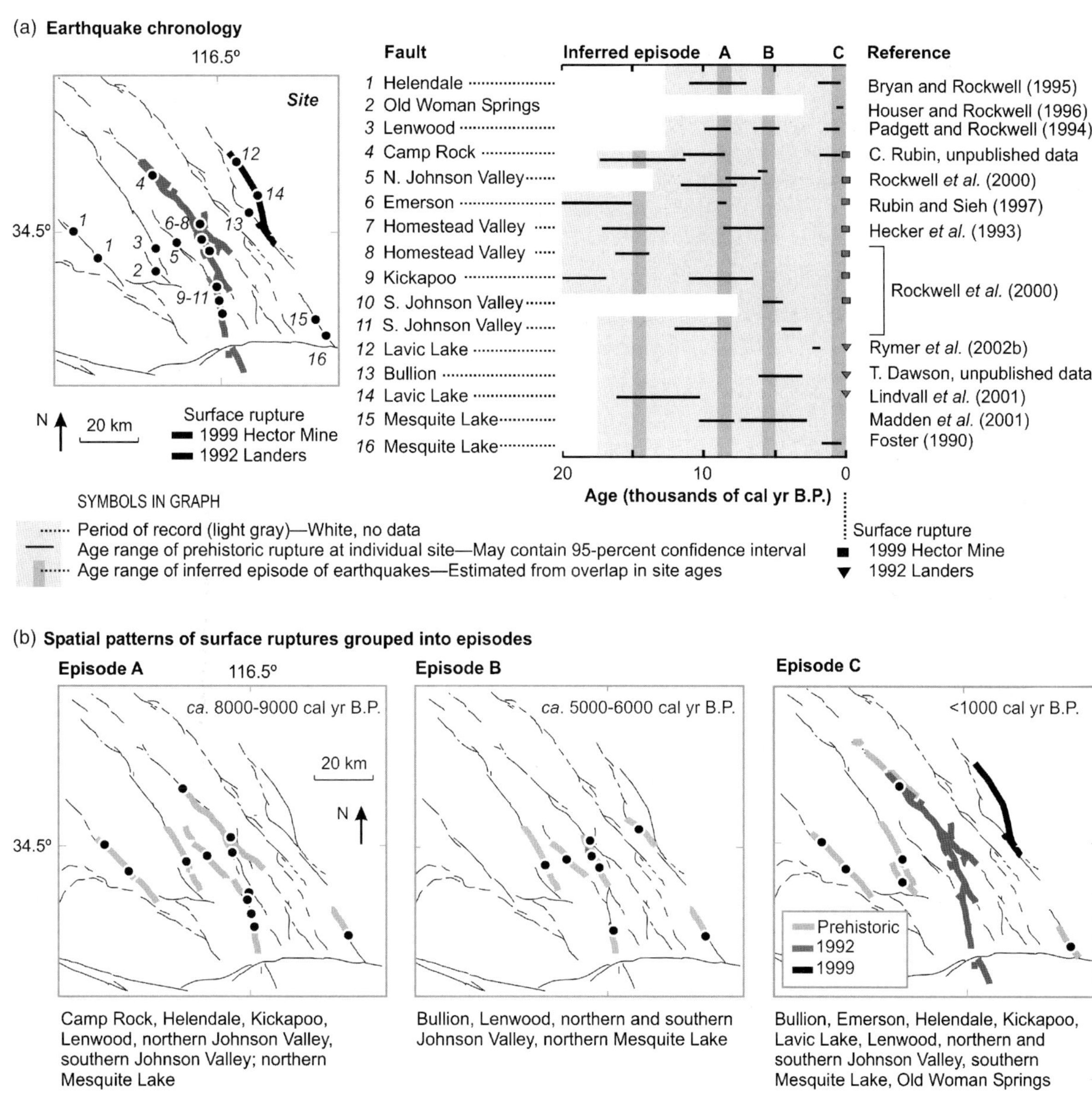

Fig. 4. Chronology and spatial patterns of earthquakes that produced surface ruptures in the eastern California shear zone.

(Saucier, 1991). The liquefaction features were attributed to two large earthquakes between about A.D. 400 and A.D. 1000.

Since the 1980s, hundreds of liquefaction features have been examined in the New Madrid region. These include more than 50 sand blows that have been studied in detail, many at archeological sites (Broughton *et al.*, 2001; Li *et al.*, 1998; Tuttle *et al.*, 1996, 1999, 2002; Vaughn, 1994; Wesnousky & Leffler, 1992). The combination of regional reconnaissance and detailed investigations has advanced the dating of the region's prehistoric earthquakes and the assessment of its earthquake potential (Tuttle *et al.*, 2002).

The challenge has not been finding sand blows, which abound in the region (Figs 5 and 6), but rather finding sand blows that can be dated well. In this agricultural region, plowing and grading have disturbed the upper 15–20 cm of soils at most sites. Soils developed on 1811–1812 sand blows are commonly thin enough to have been completely reworked by plowing. Soils developed on prehistoric sand blows, however, can be thick enough to retain cultural materials below the plow zone (Fig. 7). The New Madrid region contains thousands of Native American sites occupied at various times during the past 2000 years (Morse & Morse, 1983). Remains of these sites – including fire pits, storage pits, post molds, and trench fills – have been found on or beneath sand blows. The cultural horizons contain wood, charcoal, and plant remains that yield minimum and

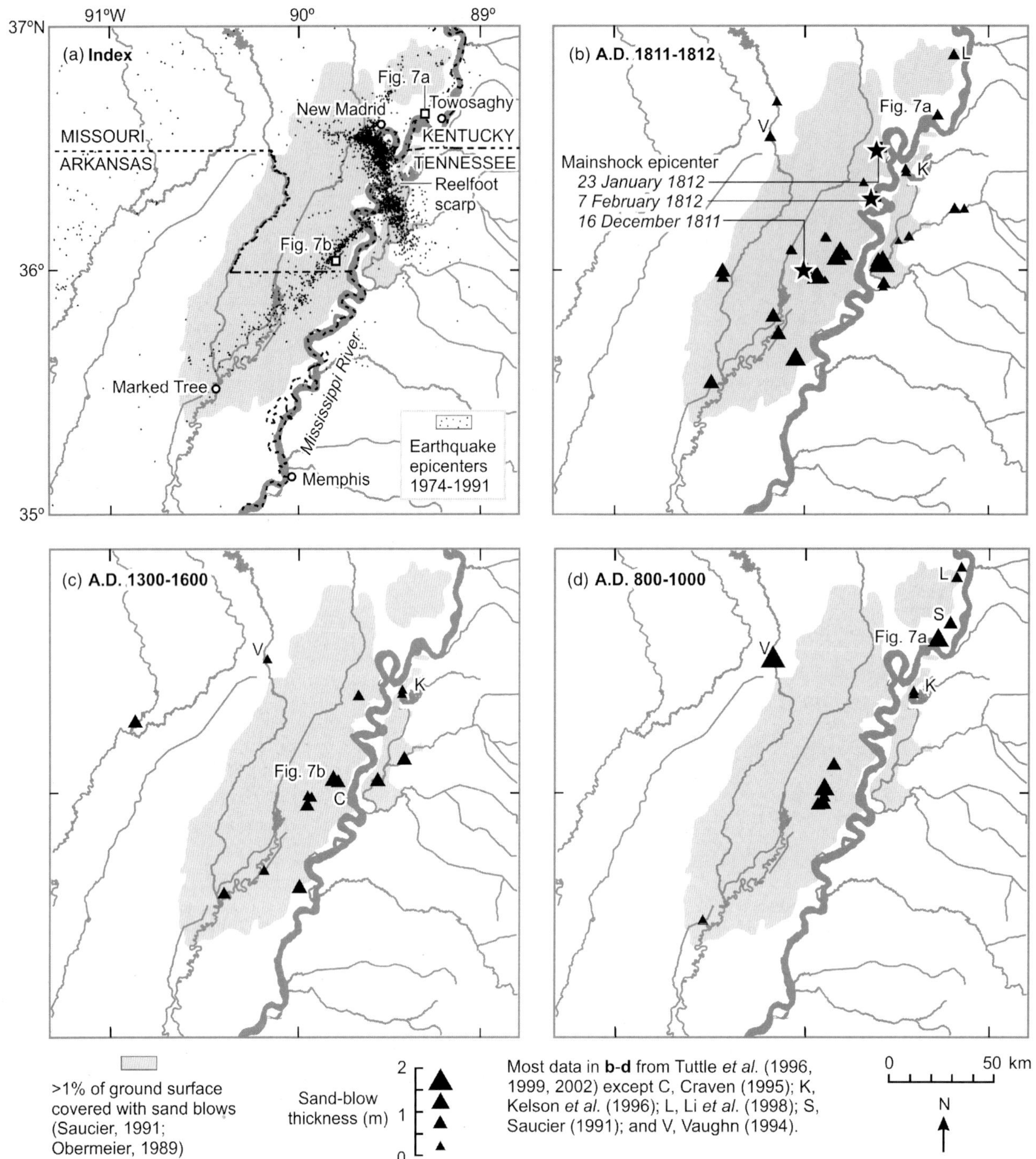

Fig. 5. Index map (a) and distribution and sizes of sand blows (b–d) at the New Madrid seismic zone.

maximum ages for earthquake-induced liquefaction features (Tuttle, 2001).

In addition to archaeological sites, the New Madrid region contains natural and artificial drainages that expose cross sections through historic and prehistoric sand blows. Reconnaissance of river and ditch banks has yielded some of the information about the size and spatial distribution of liquefaction features summarized in Figs 5 and 6.

Prehistoric Earthquakes

In the New Madrid region, prehistoric liquefaction features commonly date to A.D. 800–1000 or 1300–1600. In size and distribution, features in these age ranges resemble the sand blows from the earthquakes of 1811–1812 (Fig. 5b–d). Additional liquefaction features date from at least two earlier time intervals since 3000 B.C., but too few sites have been

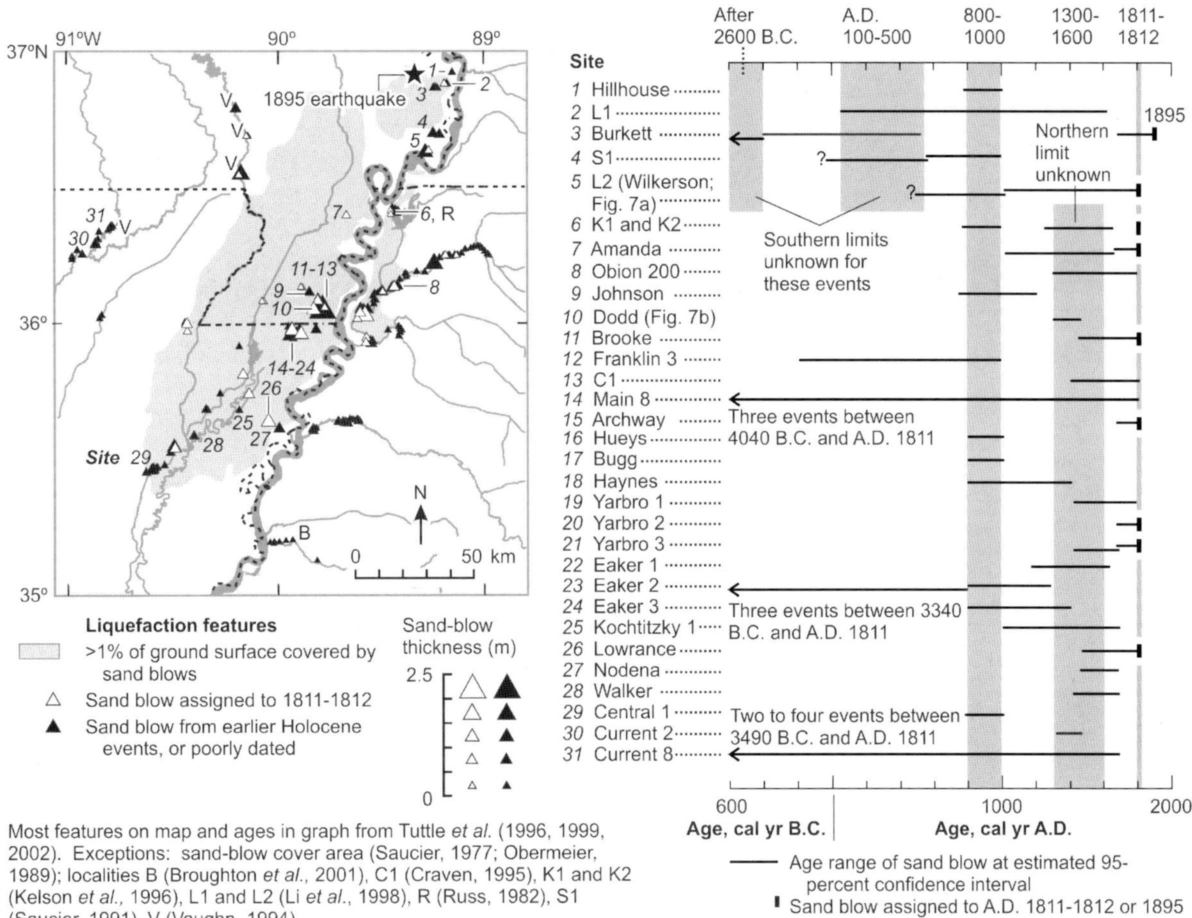

Fig. 6. Chronology of earthquakes at the New Madrid seismic zone.

studied to estimate the locations and sizes of the earthquakes that produced them (Fig. 6).

The episodes of A.D. 800–1000 and 1300–1600 each contained earthquakes in swift series. The serial earthquakes of 1811–1812 produced multiple, upward-fining depositional units, each of which probably represents an individual earthquake (Saucier, 1989). Most prehistoric sand blows also contain such multiple units, both from the years 800–1000 and from 1300–1600 (Tuttle *et al.*, 2002).

Improved estimates of earthquake recurrence may be obtained by further studying the liquefaction features older than A.D. 800. These efforts may also help address other issues at the New Madrid seismic zone, such as long-term fault behavior (Tuttle *et al.*, 2002), causes for large earthquakes in a mid-plate region (Grollimund & Zoback, 2001; Kenner & Segall, 2000; Pollitz *et al.*, 2001; Stuart, 2001), and slowness of present-day deformation (Newman *et al.*, 1999).

Implications and Challenges

The New Madrid events of A.D. 800–1000, 1300–1600, and 1811–1812 together indicate recurrence intervals as short as 200 years or as long as 800 years, with a two-interval average of about 500 years (Fig. 6). This average has been incorporated into the latest national earthquake hazard maps as the recurrence interval for New Madrid earthquakes like those in 1811–1812 (Fig. 1; Frankel *et al.*, 2002, p. 3). In previous mapping of the region's earthquake hazards, the interval used was 1000 years.

Cascadia Subduction Zone

In our eastern California and New Madrid examples, geologic records of prehistoric earthquakes resemble those produced by historical earthquakes known from instrumental records and eyewitness accounts. In some other places, paleoseismic evidence has no local analog in written history. Paleoseismology provides the only detailed knowledge of surface ruptures on Utah's Wasatch fault (Gori & Hayes, 1992, 2000; McCalpin & Nishenko, 1996), as was anticipated by Gilbert (1883). Prehistoric liquefaction features record

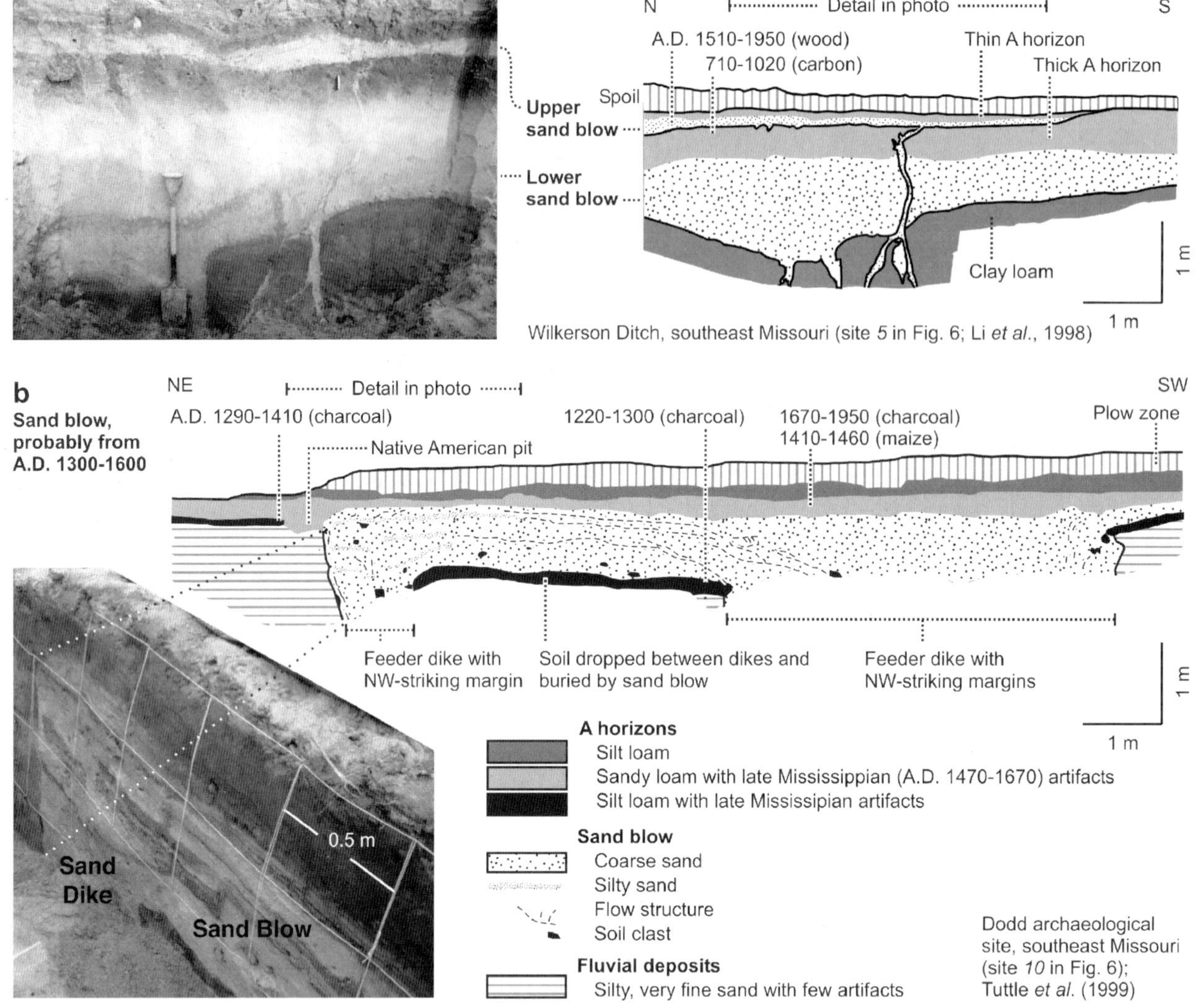

Fig. 7. Examples of sand blows at the New Madrid seismic zone.

the most recent large earthquake at the Wabash seismic zone of Illinois and Indiana, north of the New Madrid seismic zone (Obermeier *et al.*, 1991). Coastal geology shows that Washington's Seattle fault produced its most recent large earthquake about A.D. 900 (Bucknam *et al.*, 1992; fault locations in Fig. 1a).

Likewise at the Cascadia subduction zone (Fig. 8), all earthquakes of *M* 8–9 predate the region's written history. These great earthquakes ruptured the boundary between the subducting Juan de Fuca plate and the overriding North America plate. Although few earthquakes attain *M* 9 – the 20th century had no more than three or four examples (Kanamori, 1977; Ruff, 1989, p. 273) – the Cascadia earthquake in A.D. 1700 probably did. In the 1990s, this and other great earthquakes inferred from paleoseismology elevated the hazard mapped along the Pacific coast from northern California to southern British Columbia (Petersen *et al.*, 2002; Fig. 1b).

Coseismic Subsidence

Cascadia's great-earthquake hazard escaped detection until the last two decades of the 20th century. Geophysicists deduced that Cascadia can produce great earthquakes (Heaton & Kanamori, 1984; Savage *et al.*, 1981). Geologists then began finding evidence that great Cascadia earthquakes have happened (reviewed by Clague, 1997). Much of the geologic evidence consists of the buried soils of former forests and marshes that subsided into estuaries during earthquakes (Fig. 9).

Such subsidence can lower entire regions. During great thrust earthquakes at subduction zones, the upper plate lurches seaward above the rupture. Where this motion elastically stretches and thins the upper plate, the land surface drops. The grandest modern examples of coseismic subsidence come from earthquakes in Chile (1960, *M* 9.5) and Alaska (1964, *M* 9.2). Each of these earthquakes produced a largely coastal

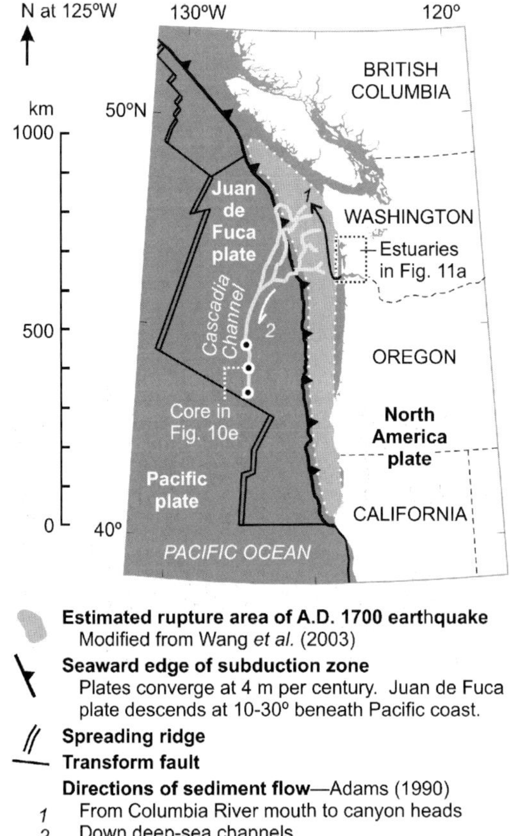

Fig. 8. Cascadia subduction zone.

downwarp more than 800 km long, many tens of kilometers wide, and as much as 2.3 m deep (Plafker, 1972). The Alaskan earthquake quickly entered the stratigraphic record at the head of a macrotidal estuary, where post-earthquake tides killed subsided forests and meadows while burying their soils with silt (Atwater *et al.*, 2001; Ovenshine *et al.*, 1976).

Estuarine stratigraphic records of coseismic subsidence can commonly be distinguished from those of other kinds of coastal change, such as gradual rise in sea level, sudden breaching of sand spits, and anomalous deposition by storms or floods (Nelson *et al.*, 1996b). To be considered evidence for coseismic subsidence, the top of a buried soil must mark a change from a relatively high environment (such as a forest or the upper part of a tidal marsh) to a relatively low one (such as an unvegetated tidal flat). Growth-position fossils of vascular plants can record such a drop (Atwater & Hemphill-Haley, 1997, p. 44), as can assemblages of diatoms, foraminifers, and pollen (Guilbault *et al.*, 1996; Hemphill-Haley, 1995; Hughes *et al.*, 2002; Kelsey *et al.*, 2002; Nelson *et al.*, 1996a; Shennan *et al.*, 1996). The change, moreover, must have happened suddenly. Sediment texture and fossils differ across a sharp contact, wide outer rings show trees healthy until their last year or two, and growth-position stems and leaves of herbaceous plants imply rapid burial (Atwater & Yamaguchi, 1991; Jacoby *et al.*, 1995; Fig. 9c).

If an earthquake produces a tsunami or liquefaction, the earthquake may be further marked by sand that mantles a buried soil in a stratigraphic section otherwise free of sand. At many Cascadia estuaries, a sand sheet suggests that burial of a freshly subsided soil began with a tsunami (e.g. Clague *et al.*, 2000). Marine diatoms within and landward of such sand sheets strengthen the case for tsunami inundation (Fig. 9d; Hemphill-Haley, 1996). In this same stratigraphic position at a few Cascadia estuaries, sand lenses fed by sand dikes show that soil burial began with venting of water and sand in response to earthquake-induced liquefaction that happened about the time the soil subsided (Kelsey *et al.*, 2002, p. 309; Obermeier, 1996, p. 43).

Prehistoric Earthquakes

In the late Holocene, coseismic subsidence in coastal Washington and Oregon has recurred at intervals mostly 300–800 years long (Kelsey *et al.*, 2002). Because of uncertainties in correlations based solely on numerical ages, little is known about the coastwise extent of individual subsidence events before A.D. 1700. However, at least three estuaries of southern Washington probably share a 3000-year history of repeated coseismic subsidence at irregular intervals (Figs 10 and 11).

This earthquake history is based on a widely correlative stack of buried soils exposed in low-tide outcrops (Fig. 10a and b). The stacked soils consistently differ from one another in organic-matter preservation and fossil-forest extent, in ways that imply differing lengths of time between earthquakes (Fig. 10c and d). The better preserved a buried organic horizon and its herbaceous fossils, the shorter the time when this buried organic matter remained subject to degradation in the profile of the next soil. The farther downstream a forested site, the longer the interseismic time when gradual uplift and sedimentation allowed forests to spread seaward along estuarine salinity gradients (Fig. 10d; Atwater & Hemphill-Haley, 1997, pp. 95–99; Benson *et al.*, 2001).

These relative measures of interseismic time agree with numerical estimates from radiocarbon dating (Fig. 10d). Radiocarbon ages from estuaries of southwest Washington anchor an earthquake chronology of uncommon precision – not only because most of the ages have reported errors of just 10–20 ^{14}C yr, but also because many ages were measured on the rings of earthquake-killed trees (Fig. 11; Appendix 1). Such tree-ring samples allow exact correction for the age of dated material relative to the time of an inferred earthquake (Nelson *et al.*, 1995). Other materials set only limiting ages for the earthquakes: maximum ages from detritus in pre-earthquake soils, minimum ages from rhizomes (below-ground stems) of plants that colonized post-earthquake tidal flats.

The individual ages are grouped in Fig. 11 by stratigraphic position defined by soil preservation and paleoecology – by field correlation of seven buried soils named J, L, N, S, U, W and Y (Fig. 10a–c; Atwater & Hemphill-Haley, 1997). The individual ages, many previously unpublished, yield combined age ranges for the field-correlated events (gray columns, Figs 10d and 11c and d). Most of these event age

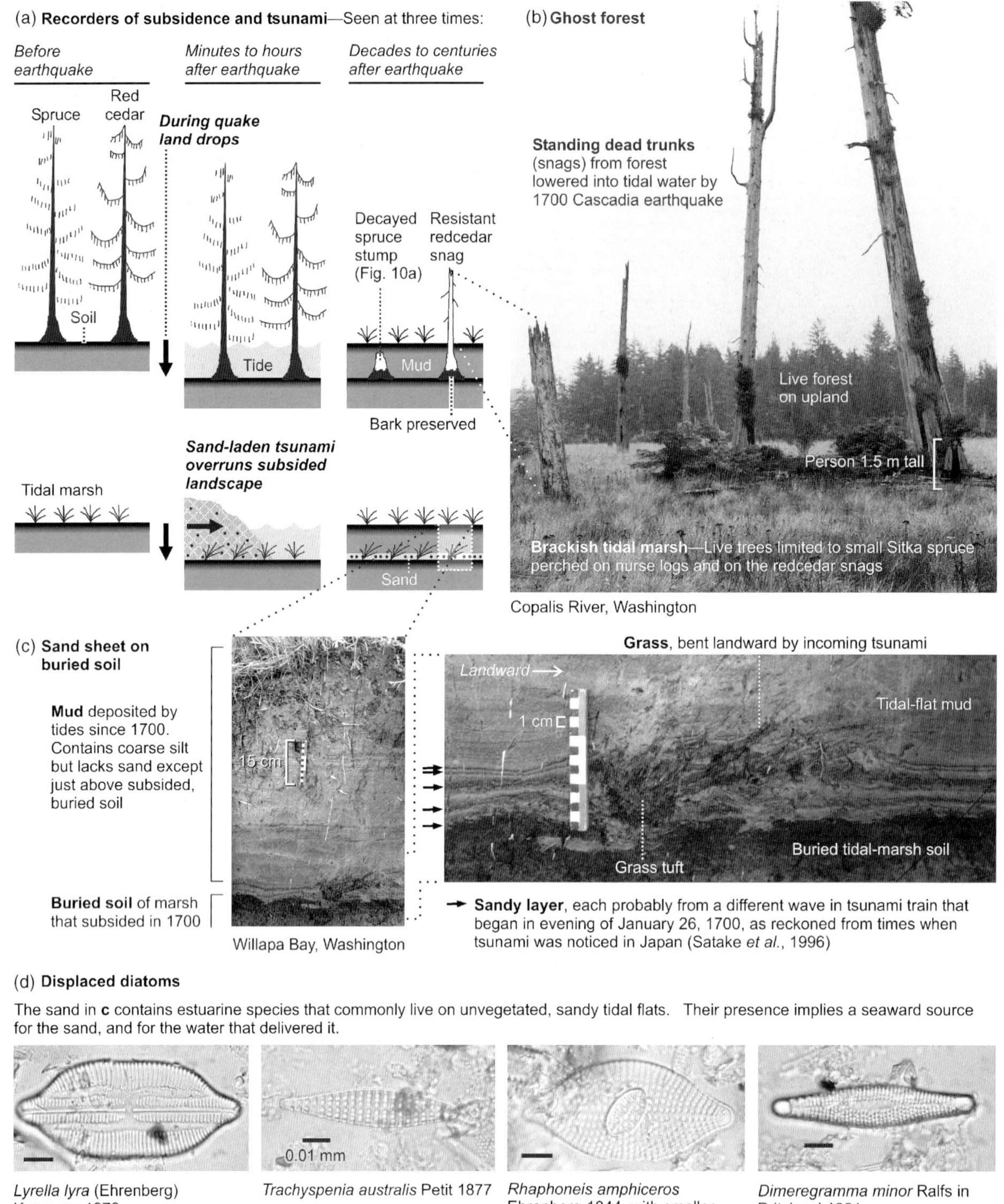

Fig. 9. Evidence for coseismic subsidence and tsunami occurrence at the Cascadia subduction zone (locations, Fig. 11a).

ranges are governed by times of tree death (black bars in Fig. 11d); some are limited also by ages from pre-event detritus or from post-event rhizomes (arrows in Fig. 11d).

To derive the age range for each event, its field-correlated individual ages were combined under the key assumption that they all refer to the same event – either an earthquake or a swift series of earthquakes. The combining is based on Bayesian statistics, which applied to an ordered sequence of ages can yield event ages with narrowed confidence limits (Biasi & Weldon, 1994; Ramsey, 2000). The ranges include generous

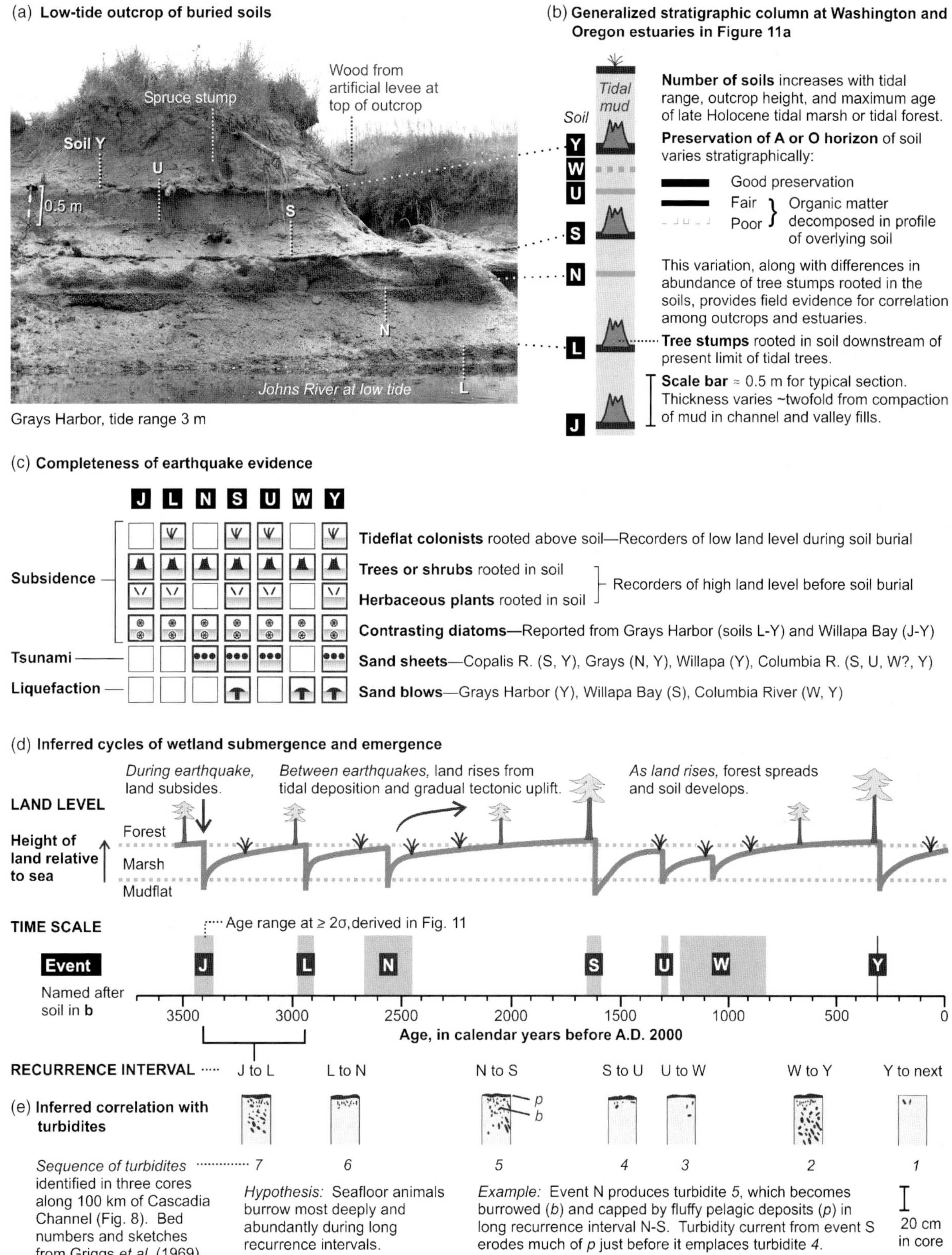
(a) Low-tide outcrop of buried soils
Wood from artificial levee at top of outcrop
Spruce stump
Soil Y
U
0.5 m
S
N
L
Johns River at low tide
Grays Harbor, tide range 3 m
(b) Generalized stratigraphic column at Washington and Oregon estuaries in Figure 11a
Tidal mud
Soil
Y
W
U
S
N
L
J
Number of soils increases with tidal range, outcrop height, and maximum age of late Holocene tidal marsh or tidal forest.
Preservation of A or O horizon of soil varies stratigraphically:
Good preservation
Fair
Poor
Organic matter decomposed in profile of overlying soil
This variation, along with differences in abundance of tree stumps rooted in the soils, provides field evidence for correlation among outcrops and estuaries.
Tree stumps rooted in soil downstream of present limit of tidal trees.
Scale bar ≈ 0.5 m for typical section. Thickness varies ~twofold from compaction of mud in channel and valley fills.
(c) Completeness of earthquake evidence
J L N S U W Y
Subsidence
Tideflat colonists rooted above soil—Recorders of low land level during soil burial
Trees or shrubs rooted in soil
Herbaceous plants rooted in soil
Recorders of high land level before soil burial
Contrasting diatoms—Reported from Grays Harbor (soils L-Y) and Willapa Bay (J-Y)
Tsunami
Sand sheets—Copalis R. (S, Y), Grays (N, Y), Willapa (Y), Columbia R. (S, U, W?, Y)
Liquefaction
Sand blows—Grays Harbor (Y), Willapa Bay (S), Columbia River (W, Y)
(d) Inferred cycles of wetland submergence and emergence
LAND LEVEL
During earthquake, land subsides.
Between earthquakes, land rises from tidal deposition and gradual tectonic uplift.
As land rises, forest spreads and soil develops.
Height of land relative to sea
Forest
Marsh
Mudflat
TIME SCALE
Age range at ≥ 2σ, derived in Fig. 11
Event
Named after soil in b
J L N S U W Y
3500 3000 2500 2000 1500 1000 500 0
Age, in calendar years before A.D. 2000
RECURRENCE INTERVAL
J to L
L to N
N to S
S to U
U to W
W to Y
Y to next
(e) Inferred correlation with turbidites
p
b
Sequence of turbidites identified in three cores along 100 km of Cascadia Channel (Fig. 8). Bed numbers and sketches from Griggs et al. (1969).
7 6 5 4 3 2 1
Hypothesis: Seafloor animals burrow most deeply and abundantly during long recurrence intervals.
Example: Event N produces turbidite 5, which becomes burrowed (b) and capped by fluffy pelagic deposits (p) in long recurrence interval N-S. Turbidity current from event S erodes much of p just before it emplaces turbidite 4.
20 cm in core

Fig. 10. Evidence for recurrent earthquakes in southwest Washington and northwest Oregon.

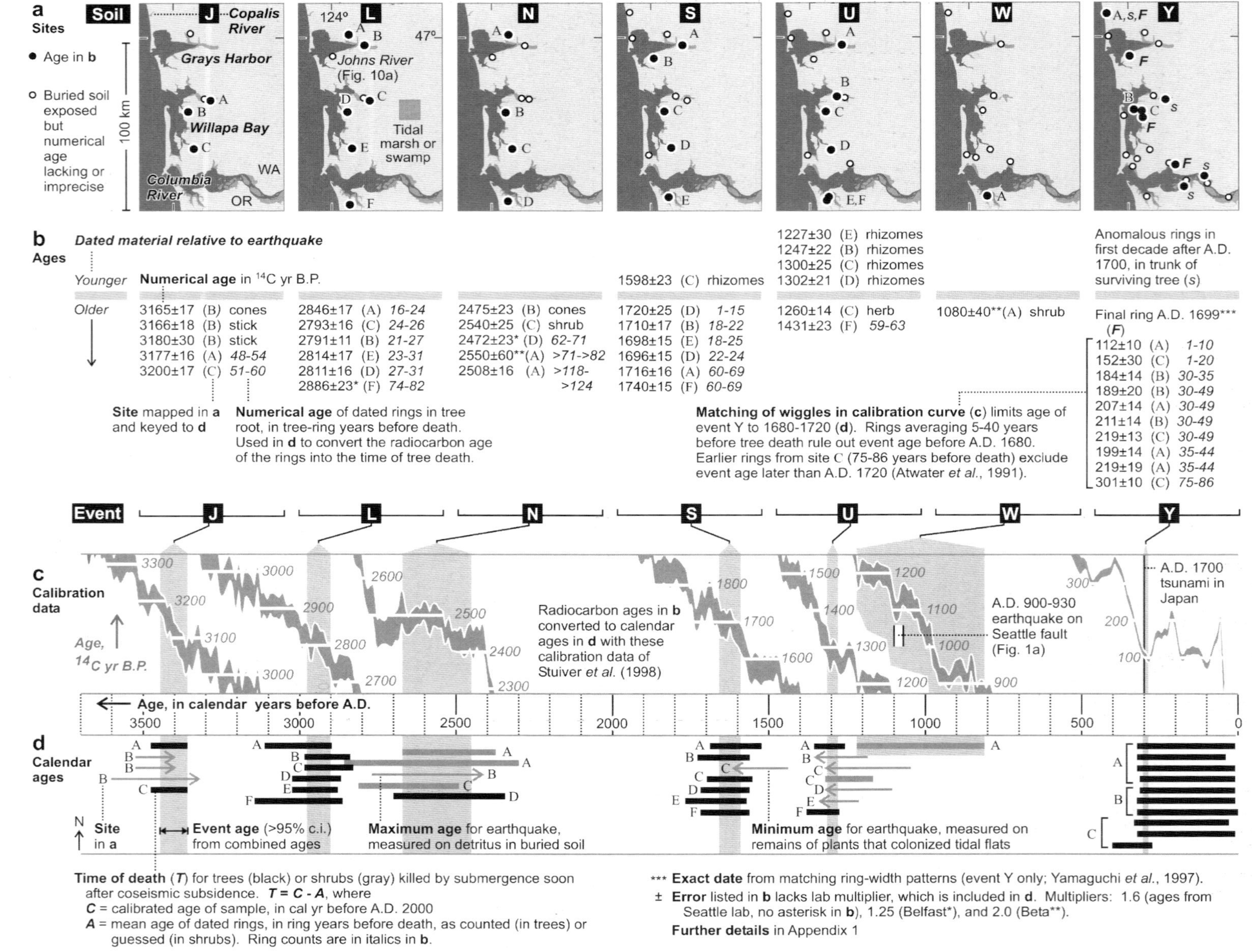

Fig. 11. Chronology of prehistoric great earthquakes in southwest Washington and northwest Oregon.

estimates of uncertainty in radiocarbon analysis (augmented by error multipliers listed at bottom right in Fig. 11). These procedures, in the calibration program Oxcal, yield event age ranges that probably include 95% confidence intervals.

The age ranges for the seven events in Figs 10 and 11 define six recurrence intervals that vary in length from a few centuries (intervals S–U and U–W) to one millennium (N–S). Though the intervals average 500–540 years, only one of the recurrence intervals (J–L) is close to this average. During the longest intervals, which exceeded this average by several centuries (N–S and W–Y), tidal forests advanced seaward as the shallowest buried soils decomposed (Figs 9b and 10a–d).

This history of aperiodic earthquakes probably correlates with turbidity-current deposits off the Oregon coast in Cascadia deep-sea channel (Atwater & Hemphill-Haley, 1997, pp. 102–103). The deposits, derived from Columbia River sediment on the continental shelf and slope, apparently originated at submarine canyon heads above the fault ruptures that caused coseismic subsidence in coastal Washington (Fig. 8). The turbidity currents repeated at intervals that averaged close to 600 years in the past 8000 years. Because eroded pelagic deposits between turbidites are similar in thickness (*p* in Fig. 10e), the repetition was first interpreted as periodic (Adams, 1990; Griggs *et al.*, 1969). However, the successive turbidites vary in their depth and abundance of animal burrows (*b* in Fig. 10e). This variability links the turbidites with the aperiodic earthquakes inferred from estuarine stratigraphy in coastal Washington and adjacent Oregon.

The most recent great Cascadia earthquake was dated by radiocarbon methods to the decades around A.D. 1700 (Nelson *et al.*, 1995; event Y in Figs 10 and 11). This era precedes, by almost a century, the Spanish and English exploration that marks the beginning of written history at Cascadia (Hayes, 1999). Along nearly 1000 km of Japan's Pacific coast, however, government officials and merchants noted a puzzling tsunami in A.D. 1700 that lacked a nearby earthquake. The time of this orphan tsunami suggests that a great Cascadia earthquake occurred on the evening of January 26, 1700. The tsunami's height of several meters further suggests that this earthquake attained *M* 9 (Satake *et al.*, 1996).

Tree-ring studies in southwest Washington and adjacent Oregon support these inferences from Japan. Death and stress in subsided trees date to the first few years after the 1699 growing season (Jacoby *et al.*, 1997; Yamaguchi *et al.*, 1997). Except for a few dozen survivors of the earthquake (Jacoby *et al.*, 1997), all trees in the region's modern tidal forests postdate 1700 (Benson *et al.*, 2001).

Implications and Challenges

The latest version of the national earthquake hazard map gives equal weight to two patterns of great-earthquake recurrence at Cascadia (Frankel *et al.*, 2002, p. 11). In one, *M* 9.0 earthquakes rupture the full 1100-km length of the subduction zone every 500 years. In the other, the subduction zone breaks in segments 250 km long that behave independently of one another. This style or rupture produces one earthquake of *M* 8.3 every 110 years somewhere along the zone (Frankel *et al.*, 1996). The shorter recurrence intervals for the independent *M* 8.3 earthquakes yield higher probabilistic ground motions than does the 500-year interval for *M* 9.0 events.

Such hazard estimates are likely to improve as great-earthquake history becomes better documented along the Cascadia subduction zone. Does the zone contain segments that sometimes rupture independently, decades or centuries out of phase with other segments? Along a single part of the subduction zone, do long recurrence intervals commonly precede short ones, much as long interval N–S preceded short intervals S–U and U–W (Fig. 10d)? Does long interval W–Y thereby justify increasing the probabilistic hazard on the national map (Fig. 1b)? Paleoseismic studies at Cascadia are just beginning to address such questions.

Also needed at Cascadia – and elsewhere – are estimates of the smallest earthquake and shortest recurrence interval that paleoseismic records resolve. Such estimates are likely to affect recurrence intervals and the probabilistic hazard inferred from them.

Summary

Paleoseismology has provided engineers and public officials with long histories of recurrent earthquakes (or histories of recurrent series of earthquakes). Typical intervals between the earthquakes (or series) span hundreds of years in our New Madrid and Cascadia examples and thousands of years in our eastern California example. In addition to enabling such estimates of recurrence intervals, paleoseismology can provide evidence for regional clustering of earthquakes in seismic zones (eastern California, New Madrid) and for aperiodic rupture along the same part of a fault (Cascadia). Such findings have made paleoseismology an essential part of earthquake-hazard assessment in the United States.

Acknowledgments

For critical reviews we thank Tony Crone, Carol Prentice, and Tom Brocher. Authors' responsibilities: southeastern California, Rubin; New Madrid, Tuttle and Schweig; Cascadia, Atwater, Yamaguchi, and Hemphill-Haley; compilation, Atwater and Yamaguchi.

References

Adams, J. (1990). Paleoseismicity of the Cascadia subduction zone – evidence from turbidites off the Oregon-Washington margin. *Tectonics*, **9**, 569–583.

Atkinson, G., Bakun, B., Bodin, P., Boore, D., Cramer, C., Frankel, A., Gasperini, P., Gomberg, J., Hanks, T., Herrmann, B., Hough, S., Johnston, A., Kenner, S., Langston, C., Linker, M., Mayne, P., Petersen, M., Powell, C., Prescott, W., Schweig, E., Segall, P., Stein, S., Stuart, B., Tuttle, M. & VanArsdale, R. (2000). Reassessing the New Madrid seismic zone. *EOS, Transactions of the American Geophysical Union*, **81**(397), 402–403.

Atwater, B.F. (1992). Geologic evidence for earthquakes during the past 2000 years along the Copalis River, southern coastal Washington. *Journal of Geophysical Research*, **97**, 1901–1919.

Atwater, B.F. (1996). Coastal evidence for great earthquakes in western Washington. *In*: Rogers, A.M., Walsh, T.J., Kockelman, W.J. & Priest, G.R. (Eds), Assessing earthquake hazards and reducing risk in the Pacific Northwest. *U.S. Geological Survey Professional Paper*, 1560, pp. 77–90.

Atwater, B.F. & Hemphill-Haley, E. (1997). Recurrence intervals for great earthquakes of the past 3,500 years at northeastern Willapa Bay, Washington. *U.S. Geological Survey Professional Paper*, 1576, 108 pp.

Atwater, B.F., Stuiver, M. & Yamaguchi, D.K. (1991). Radiocarbon test of earthquake magnitude at the Cascadia subduction zone. *Nature*, **353**, 156–158.

Atwater, B.F. & Yamaguchi, D.K. (1991). Sudden, probably coseismic submergence of Holocene trees and grass in coastal Washington State. *Geology*, **19**, 706–709.

Atwater, B.F., Yamaguchi, D.K., Bondevik, S., Barnhardt, W.A., Amidon, L.J., Benson, B.E., Skjerdal, G., Shulene, J.A. & Nanayama, F. (2001). Rapid resetting of an estuarine recorder of the 1964 Alaska earthquake. *Geological Society of America Bulletin*, **113**, 1193–1204.

Benson, B.E., Atwater, B.F., Yamaguchi, D.K., Amidon, L.J., Brown, S.L. & Lewis, R.C. (2001). Renewal of tidal forests in Washington State after a subduction earthquake in A.D. 1700. *Quaternary Research*, **56**, 139–147.

Biasi, G.P. & Weldon, R., III (1994). Quantitative refinement of calibrated ^{14}C distributions. *Quaternary Research*, **41**, 1–18.

Broughton, A.T., Van Arsdale, R.B. & Broughton, J.H. (2001). Liquefaction susceptibility mapping in the City of Memphis and Shelby County, Tennessee. *Engineering Geology*, **62**, 207–222.

Bryan, K.A. & Rockwell, T.K. (1995). Holocene character of the Helendale fault zone. Lucerne Valley, San Bernardino County, California [abstract]. *Geological Society of America Abstracts with Programs*, **27**, 7.

Bucknam, R.C., Hemphill-Haley, E. & Leopold, E.B. (1992). Abrupt uplift within the past 1700 years at southern Puget Sound, Washington. *Science*, **258**, 1611–1614.

Camelbeeck, T. (Ed.) (2001). Proceedings of the workshop, Evaluation of the potential for large earthquakes in regions of present day low seismic activity in Europe, March 12th to 17th, 2000, Han-sur-Lesse (Belgium). Luxembourg, Centre Européen de Géodynamique et de Séismologie, Musée National d'Histoire Naturelle, Section Astrophysique et Géophysique, 174 pp.

Chiu, J.M., Johnston, A.C. & Yang, Y.T. (1992). Imaging the active faults of the central New Madrid seismic zone. *Seismological Research Letters*, **63**, 375–393.

Clague, J.J. (Ed.) (1996). Paleoseismology and seismic hazards, southwestern British Columbia. *Geological Survey of Canada Bulletin*, 494, 88 pp.

Clague, J.J. (1997). Evidence for large earthquakes at the Cascadia subduction zone. *Reviews of Geophysics*, **35**, 439–460.

Clague, J.J., Bobrowsky, P.T. & Hutchinson, I. (2000). A review of geological records of large tsunamis at Vancouver Island, British Columbia, and implications for hazard. *Quaternary Science Reviews*, **19**, 849–863.

Cohee, B.P. & Beroza, G.C. (1994). Slip distribution of the 1992 Landers earthquake and its implications for earthquake source mechanics. *Bulletin of the Seismological Society of America*, **84**, 692–712.

Craven, J.A. (1995). Evidence of paleoseismicity within the New Madrid seismic zone at a Late Mississippian Indian occupation site in the Missouri Bootheel [abstract]. *Geological Society of America Abstracts with Programs*, **27**(6), 394.

Crone, A.J. & Omdahl, E.M. (Eds) (1987). Directions in paleoseismology. *U.S. Geological Survey Open-File Report*, 87–673, 456 pp.

DeMets, C. & Dixon, T.H. (1999). New kinematic models for Pacific-North America motion from 3 Ma to present, I: evidence for steady motion and biases in the NUVEL-1A model. *Geophysical Research Letters*, **26**, 1921–1924.

DeMets, C., Gordon, R.G., Argus, D.F. & Stein, S. (1990). Current plate motions. *International Journal of Geophysics*, **101**, 425–478.

DeMets, C., Gordon, R.G., Argus, D.F. & Stein, S. (1994). Effect of recent revisions to the geomagnetic reversal timescale on estimates of current plate motions. *Geophysical Research Letters*, **21**, 2191–2194.

Dokka, R.K. & Travis, C.J. (1990). Late Cenozoic strike-slip faulting in the Mojave Desert, California. *Tectonics*, **9**, 311–340.

Ettensohn, F.R., Rast, N. & Brett, C.E. (Eds) (2002). Ancient seismites. *Geological Society of America Special Paper*, 359, 200 pp.

Foster, J.H. (1990). Structural relations at the Twentynine Palms Marine Corps Base, Twentynine Palms, CA [abstract]. *Geological Society of America Abstracts with Programs*, **22**(7), 22.

Frankel, A., Mueller, C., Barnhard, T., Perkins, D., Leyendecker, E.V., Dickman, N., Hanson, S. & Hopper, M. (1996). National seismic hazard maps, June 1996 documentation. *U.S. Geological Survey Open-File Report*, 96–532, 69 pp. http://geohazards.cr.usgs.gov/eq/hazmapsdoc/junecover.html

Frankel, A.D., Petersen, M.D., Mueller, C.S., Haller, K.M., Wheeler, R.L., Leyendecker, E.V., Wesson, R.L., Harmsen, S.C., Cramer, C.H., Perkins, D.M. & Rukstales, K.S. (2002). Documentation for the 2002 update of the national seismic hazard maps. *U.S. Geological Survey Open-File Report*, 02–420, 33 pp. (report at http://geohazards.cr.usgs.gov/eq/of02–420/OFR02–420.pdf; maps at http://geohazards.cr.usgs.gov/eq/).

Freed, A.M. & Lin, J. (2001). Delayed triggering of the 1999 Hector Mine earthquake by viscoelastic stress transfer. *Nature*, **411**, 180–183.

Fuller, M.L. (1912). The New Madrid earthquakes. *U.S. Geological Survey Bulletin*, **494**, 119.

Gilbert, G.K. (1883). A theory of the earthquakes of the Great Basin, with a practical application. *American Journal of Science, 3rd series*, **27**, 49–53.

Gori, P.L. & Hayes, W.W. (Eds) (1992, 2000). Assessment of regional earthquake hazards and risk along the Wasatch Front, Utah. *U.S. Geological Survey Professional Paper*, 1500, parts A–J (1992) and K–R (2000), various paginating.

Grant, L.B. & Lettis, W.R. (Eds) (2002). Special issue on paleoseismology of the San Andreas fault system. *Bulletin of the Seismological Society of America*, **92**(7).

Griggs, G.B., Carey, A.G. & Kulm, L.D. (1969). Deep-sea sedimentation and sediment-fauna interaction in Cascadia Channel and on Cascadia Abyssal Plain. *Deep-Sea Research*, **16**, 157–170.

Grollimund, B. & Zoback, M.D. (2001). Did deglaciation trigger intraplate seismicity in the New Madrid seismic zone? *Geology*, **29**, 175–178.

Guilbault, J.-P., Clague, J.J. & Lapointe, M. (1996). Foraminiferal evidence for the amount of coseismic subsidence during a late Holocene earthquake on Vancouver Island, west coast of Canada. *Quaternary Science Reviews*, **15**, 913–937.

Hancock, P.L. & Michetti, A.M. (Eds) (1997). Paleoseismology: understanding past earthquakes using Quaternary geology. *Journal of Geodynamics*, **24**(1–4), 304.

Harris, R.A. & Simpson, R.W. (2002). The 1999 Mw 7.1 Hector Mine earthquake: a test of the stress shadow hypothesis? *Bulletin of the Seismological Society of America*, **92**, 1497–1512.

Hauksson, E., Jones, L. & Hutton, K. (2002). The 1999 Mw 7.1 Hector Mine, California, Earthquake sequence: complex conjugate strike-slip faulting. *Bulletin of the Seismological Society of America*, **92**, 1154–1170.

Hayes, D. (1999). *Historical atlas of the Pacific Northwest; maps of exploration and discovery*. Seattle, Sasquatch Books, 208 pp.

Heaton, T.H. & Kanamori, H. (1984). Seismic potential associated with subduction in the northwestern United States. *Bulletin of the Seismological Society of America*, **74**, 335–344.

Hecker, S., Fumal, T.E., Powers, T.J., Hamilton, J.C., Garvin, C.D. & Schwartz, D.P. (1993). Late Pleistocene-Holocene behavior of the Homestead Valley fault segment-1992 Landers, CA surface rupture [abstract]. *EOS (Transactions of the American Geophysical Union)*, **74**(43), 612.

Hemphill-Haley, E. (1995). Intertidal diatoms from Willapa Bay, Washington: Application to studies of small-scale sea-level changes. *Northwest Science*, **69**(1), 29–45.

Hemphill-Haley, E. (1996). Diatoms as an aid in identifying late-Holocene tsunami deposits. *The Holocene*, **6**, 439–448.

Hendey, N.I. (1964). An introductory account of the small algae of British coastal waters, part V, Bacillariophyceae (Diatoms). *Fishery Investigations Series*, Vol. 41, 317 pp.

Hough, S.E., Armbruster, J.G., Seeber, L. & Hough, J.F. (2000). On the modified Mercalli intensities and magnitudes of the 1811–1812 New Madrid. *Journal of Geophysical Research*, **105**, 23,839–23,864.

Houser, C.E. & Rockwell, T.K. (1996). Tectonic geomorphology and paleoseismicity of the Old Woman Springs fault, San Bernadino County, California [abstract]. *Geological Society of America Abstracts with Programs*, **28**(5), 76.

Hudnut, K.W., King, N.E., Galetzka, J.E., Stark, K.F., Behr, J.A., Aspiotes, A., van Wyk, S., Dockter, S. & Wyatt, F. (2002). Continuous GPS observations of postseismic deformation following the 16 October 1999 Hector Mine, California earthquake (M_W 7.1). *Bulletin of the Seismological Society of America*, **92**, 1403–1422.

Hughes, J.F., Mathewes, R.W. & Clague, J.J. (2002). Use of pollen and vascular plants to estimate coseismic subsidence at a tidal marsh near Tofino, British Columbia. *Paleaeogeography, Palaeoclimatology, and Palaeoecology*, **185**, 145–161.

Jacoby, G.C. (1997). Application of tree ring analysis to paleoseismology. *Reviews of Geophysics*, **35**, 109–124.

Jacoby, G.C., Bunker, D.E. & Benson, B.E. (1997). Tree-ring evidence for an A.D. 1700 Cascadia earthquake in Washington and northern Oregon. *Geology*, **25**, 999–1002.

Jacoby, G.C., Carver, G. & Wagner, W. (1995). Trees and herbs killed by an earthquake ~300 yr ago at Humboldt Bay, California. *Geology*, **23**, 77–80.

Jennings, C.W. & Saucedo, G.J. (compilers) (1994). Fault activity map of California and adjacent areas with locations and ages of recent volcanic eruptions. California Division of Mines and Geology, California geologic data map no. 6, scale, 1:750,000.

Johnston, A.C. (1996). Seismic moment assessment of earthquakes in stable continental regions – I. Instrumental seismicity. *Geophysical Journal International*, **124**, 381–414.

Johnston, A.C. & Kanter, L.R. (1990). Earthquakes in stable continental crust. *Scientific American*, **262**(3), 68–75.

Johnston, A.C. & Schweig, E.S. (1996). The enigma of the New Madrid earthquakes of 1811–1812. *Annual Review of Earth and Planetary Sciences*, **24**, 339–384.

Kanamori, H. (1977). The energy release in great earthquakes. *Journal of Geophysical Research*, **82**, 2981–2987.

Kelsey, H.M., Witter, R.C. & Hemphill-Haley, E. (2002). Plate-boundary earthquakes and tsunamis of the past 5500 yr, Sixes River estuary, southern Oregon. *Geological Society of America Bulletin*, **114**, 298–314.

Kelson, K.I., Simpson, G.D., Van Arsdale, R.B., Harris, J.B., Haraden, C.C. & Lettis, W.R. (1996). Multiple Holocene earthquakes along the Reelfoot fault, central New Madrid seismic zone. *Journal of Geophysical Research*, **101**, 6151–6170.

Kelson, K.I., Van Arsdale, R.B., Simpson, G.D. & Lettis, W.R. (1992). Assessment of the style and timing of late Holocene surficial deformation along the central Reelfoot scarp, Lake County, Tennessee. *Seismological Research Letters*, **63**, 349–356.

Kenner, S.J. & Segall, P. (2000). A mechanical model for intraplate earthquakes; application to the New Madrid seismic zone. *Science*, **289**, 2329–2332.

Li, Y., Schweig, E.S., Tuttle, M.P. & Ellis, M.A. (1998). Evidence for large prehistoric earthquakes in the northern New

Madrid seismic zone, central United States. *Seismological Research Letters*, **69**, 270–276.

Lindvall, S., Rockwell, T., Schwartz, D., Dawson, J., Helms, J., Madden, C., Yule, D., Stenner, H., Ragona, D., Kasman, G., Seim, M., Meltzner, A. & Caffee, M. (2001). Paleoseismic investigations of the 1999 M7.1 Hector Mine earthquake surface rupture and adjacent Bullion fault, Twentynine Palms Marine Corps Base, California. *Geological Society of America Abstracts with Programs*, **33**, 79.

Lyell, C. (1849). *A second visit to the United States of North America*. London, John Murray, 238 pp.

Madden, C., Rubin, C.M. & Streig, A. (2001). Preliminary paleoseismic results from the Mesquite Lake fault, Twentynine Palms, California. *Geological Society of America Abstracts with Programs*, **33**, 79.

Masana, E. & Santanach, P. (2001). Paleoseismology in Spain. *Acta Geological Hispanica*, **36**(3–4), 193–354.

McCalpin, J.P. (Ed.) (1996). *Paleoseismology*. San Diego, Academic Press, 588 pp.

McCalpin, J.P. & Nishenko, S.P. (1996). Holocene paleoseismicity, temporal clustering, and probabilities of future large (M>7) earthquakes on the Wasatch fault zone, Utah. *Journal of Geophysical Research*, **101**, 6233–6253.

McClusky, S., Bjornstad, S., Hager, B., King, R., Meade, B., Miller, M., Monastero, F. & Souter, B. (2001). Present day kinematics of the eastern California shear zone from a geodetically constrained block model. *Geophysical Research Letters*, **28**, 3369–3372.

Miller, M., Johnson, D., Dixon, T. & Dokka, R. (2001). Refined kinematics of the Eastern California shear zone from GPS observations, 1993–1998. *Journal Geophysical Research*, **106**, 2245–2263.

Morse, D.F. & Morse, P.A. (1983). *Archaeology of the central Mississippi Valley*. San Diego, Academic Press, 345 pp.

Nance, J.J. (1988). *On shaky ground*. New York, Morrow, 416 pp.

Nelson, A.R., Atwater, B.F., Bobrowsky, P.T., Bradley, L.-A., Clague, J.J., Carver, G.A., Darienzo, M.E., Grant, W.C., Krueger, H.W., Sparks, R., Stafford, T.W. & Stuiver, M. (1995). Radiocarbon evidence for extensive plate-boundary rupture about 300 years ago at the Cascadia subduction zone. *Nature*, **378**, 371–374.

Nelson, A.R., Jennings, A.E. & Kashima, K. (1996a). Holocene intertidal stratigraphy, microfossils, rapid submergence, and earthquake recurrence at Coos Bay, southern coastal Oregon, USA. *Geological Society of America Bulletin*, **108**, 141–154.

Nelson, A.R., Shennan, I. & Long, A.J. (1996b). Identifying coseismic subsidence in tidal-wetland stratigraphic sequences at the Cascadia subduction zone of western North America. *Journal of Geophysical Research*, **101**, 6115–6135.

Newman, A.V., Stein, S., Weber, J., Engeln, J., Mao, A. & Dixon, T.H. (1999). Slow deformation and low seismic hazard at the New Madrid seismic zone. *Science*, **284**, 619–621.

Noller, J.S., Sowers, J.M. & Lettis, W.R. (2000). *Quaternary geochronology; methods and applications*. Washington, DC, American Geophysical Union, AGU reference shelf, 4, 582 pp.

Nuttli, O.W. (1973). The Mississippi Valley earthquakes of 181 and 1812; intensities, ground motions, and magnitudes. *Bulletin of the Seismological Society of America*, **63**, 227–248.

Obermeier, S.F. (1989). The New Madrid earthquakes: An engineering-geologic interpretation of relict liquefaction features. *U.S. Geologic Survey Professional Paper*, 1336-B, 114 pp.

Obermeier, S.F. (1996). Use of liquefaction-induced features for paleoseismic analysis – an overview of how seismic liquefaction features can be distinguished from other features and how their regional distribution and properties of source sediment can be used to infer the location and strength of Holocene paleo-earthquakes. *Engineering Geology*, **44**, 1–76.

Obermeier, S.F., Bleuer, N.R., Munson, C.A., Munson, P.J., Martin, W.S., McWilliams, K.M., Tabaczynski, D.A., Odum, J.K., Rubin, M. & Eggert, D.L. (1991). Evidence of strong earthquake shaking in the lower Wabash Valley from prehistoric liquefaction features. *Science*, **251**, 1061–1063.

Ota, Y., Nelson, A.R. & Berryman, K.R. (Eds) (1992). Impacts of tectonics on Quaternary coastal evolution. *Quaternary International*, Vols 15/16, 184 pp.

Ota, Y. & Okumura, K. (1999). Progress in paleoseismology in Japan during the 1990s, in Special issue on the XV INQUA Congress; Recent progress of Quaternary studies in Japan. *Daiyonki Kenkyu (The Quaternary Research)*, **38**, 253–261.

Ovenshine, A.T., Lawson, D.E. & Bartsch-Winkler, S.R. (1976). The Placer River Silt – an intertidal deposit caused by the 1964 Alaska earthquake. *Journal of Research of the U.S. Geological Survey*, **4**, 151–162.

Padgett, D.C. & Rockwell, T.K. (1994). Paleoseismology of the Lenwood fault, San Bernardino County, California, in Mojave Desert. *In*: Murbach, D. (Ed.), *South Coast Geological Society Annual Fieldtrip Guidebook*, pp. 222–238.

Pavlides, S.B., Pantosti, D. & Zhang, P. (1999). Earthquakes, paleoseismology and active tectonics. *Tectonophysics*, **308**(1–2), 1–298.

Petersen, M.D., Cramer, C.H. & Frankel, A.D. (2002). Simulations of seismic hazard for the Pacific Northwest of the United States from earthquakes associated with the Cascadia subduction zone. *Pure and Applied Geophysics*, **159**, 2147–2168.

Plafker, G. (1972). Alaskan earthquake of 1964 and Chilean earthquake of 1960: implications for arc tectonics. *Journal of Geophysical Research*, **77**, 901–925.

Pollitz, F.P., Kellogg, L. & Bürgmann, R. (2001). Sinking mafic body in a reactivated lower crust: a mechanism for stress concentration at the New Madrid seismic zone. *Bulletin of the Seismological Society of America*, **92**, 1882–1887.

Pollitz, F.F. & Sacks, I.S. (2002). Stress triggering of the 1999 Hector Mine earthquake by transient deformation following the 1992 Landers earthquake. *Bulletin of the Seismological Society of America*, **92**, 1487–1496.

Pujol, J., Johnston, A., Chiu, J. & Yang, Y. (1997). Refinement of thrust faulting models for the central New Madrid seismic zone. *Engineering Geology*, **46**, 281–298.

Ramsey, C.B. (2000). *OxCal program v3.5*. University of Oxford, Radiocarbon Accelerator Unit, http://www.rlaha.ox.ac.uk/orau/

Research Group for Active Faults of Japan (1992). Maps of active faults in Japan with an explanatory text. Tokyo, University of Tokyo Press, 71 pp. and four oversize maps [in Japanese and English].

Rockwell, T., Lindvall, S., Herzberg, M., Murback, D., Dawson, T. & Berger, G. (2000). Paleoseismology of the Johnson Valley, Kickapoo, and Homestead Valley faults. Clustering of earthquakes in the eastern California shear zone. *Bulletin of the Seismological Society of America*, **90**, 1200–1236.

Rubin, C.M. & Sieh, K. (1997). Long dormancy, low slip rate and similar slip-per-event for the Emerson fault, Eastern California shear zone. *Journal Geophysical Research*, **102**, 15,319–15,333.

Ruff, L.J. (1989). Do trench sediments affect great earthquake occurrence? *Pure and Applied Geophysics*, **129**, 263–282.

Russ, D.P. (1979). Late Holocene faulting and earthquake recurrence in the Reelfoot lake area, northwestern Tennessee. *Geological Society of America Bulletin*, **90**(Pt. I), 1013–1018.

Russ, D.P. (1982). Style and significance of surface deformation in the vicinity of New Madrid, Missouri. *In*: Investigations of the New Madrid, Missouri, earthquake region. *U.S. Geological Survey Professional Paper*, 1236-H, pp. 95–114.

Rymer, M.J., Boatwright, J., Seekins, L.C., Yule, J.D. & Liu, J. (2002a). Triggered surface slips in the Salton Trough associated with the 1999 Hector Mine, California, earthquake. *Bulletin of the Seismological Society of America*, **92**, 1300–1317.

Rymer, M.J., Seitz, G., Weaver, K.D., Orgil, A., Faneros, G., Hamilton, J.C. & Goetz, C. (2002b). Geologic and paleoseismic study of the Lavic Lake fault at Lavic Lake, Mojave Desert. *Bulletin of the Seismological Society of America*, **92**, 1577–1591.

Satake, K., Shimazaki, K., Tsuji, Y. & Ueda, K. (1996). Time and size of a giant earthquake in Cascadia inferred from Japanese tsunami record of January 1700. *Nature*, **379**, 246–249.

Sauber, J., Thatcher, W., Solomon, S.C. & Lisowski, M. (1994). Geodetic slip rate for the eastern California shear zone and the recurrence time of Mojave desert earthquakes. *Nature*, **367**, 264–266.

Saucier, R.T. (1977). Effects of the New Madrid earthquake series in the Mississippi alluvial valley. Vicksburg, Mississippi, *U.S. Army Engineers Waterways Experiment Station Miscellaneous Paper S-77–5*.

Saucier, R.T. (1989). Evidence for episodic sand-blow activity during the 1811–12 New Madrid (Missouri) earthquake series. *Geology*, **17**, 103–106.

Saucier, R.T. (1991). Geoarchaeological evidence of strong prehistoric earthquake in the New Madrid (Missouri) seismic zone. *Geology*, **19**, 198–296.

Savage, J.C., Lisowski, M. & Prescott, W.H. (1981). Geodetic strain measurements in Washington. *Journal of Geophysical Research*, **86**, 4929–4940.

Schweig, E.S., Tuttle, M.P., Crone, A.J., Machette, M.N. & Cramer, C. (2002). Seismic hazards of stable continents: the important role of paleoseismology. *Seismological Research Letters*, **73**, 244.

Serva, L. & Slemmons, D.B. (Eds) (1995). Perspectives in paleoseismology. *Association of Engineering Geologists Special Publication*, 6, 139 pp.

Shennan, I., Long, A.J., Rutherford, M.M., Green, F.M., Innes, J.B., Lloyd, J.M., Zong, Y. & Walker, K.J. (1996). Tidal marsh stratigraphy, sea-level change and large earthquakes, I: a 5000 year record in Washington, USA. *Quaternary Science Reviews*, **15**, 1023–1059.

Shiki, T., Cita, M.B. & Gorsline, D.S. (Eds) (2000). Sedimentary features of seismites, seismo-turbidites and tsunamiites. *Sedimentary Geology*, Vol. 135, pp. 1–320.

Sieh, K. & LeVay, S. (1998). *The Earth in turmoil*. New York, W.H. Freeman, 324 pp.

Sieh, K.E., Jones, L.M., Hauksson, E., Hudnut, K.W., Eberhart-Phillips, D., Heaton, T.H., Hough, S.E., Hutton, L.K., Kanamori, H., Lilje, A., Lindvall, S.C., McGill, S.F., Mori, J.J., Rubin, C.M., Spotila, J.A., Stock, J.M., Thio, H.K., Treiman, J.A., Wernicke, B.P. & Zachariasen, J. (1993). Near-field investigations of the Landers earthquake sequence, April to July 1992. *Science*, **260**, 171–176.

Street, R. & Nuttli, O. (1984). The central Mississippi Valley earthquakes of 1811–1812. *In*: Gori, P.L. & Hays, W.W. (Eds), Proceedings, Symposium on the New Madrid earthquakes. *U.S. Geological Survey Open-File Report*, 84–770, pp. 33–63.

Stuart, W.D. (2001). GPS constraints on M7–8 earthquake recurrence times for the New Madrid seismic zone. *Seismological Research Letters*, **72**, 745–753.

Stuiver, M., Reimer, P.J., Bard, E., Beck, J.W., Burr, G.S., Hughen, K.A., Kromer, B., McCormac, F.G., v.d. Plicht, J. & Spurk, M. (1998). INTCAL98 Radiocarbon age calibration 24,000–0 cal BP. *Radiocarbon*, **40**, 1041–1083.

Talwani, P. & Schaeffer, W.T. (2001). Recurrence rates of large earthquakes in the South Carolina Coastal Plain based on paleoliquefaction data. *Journal of Geophysical Research*, **106**, 6621–6642.

Treiman, J.A., Kendrick, K.J., Bryant, W.A., Rockwell, T.K. & McGill, S.F. (2002). Primary surface rupture associated with the M_W 7.1 16 October 1999 Hector Mine earthquake, San Bernardino County, California. *Bulletin of the Seismological Society of America*, **92**, 1171–1191.

Tuttle, M.P. (2001). The use of liquefaction features in paleoseismology: Lessons learned in the New Madrid seismic zone, central United States. *Journal of Seismology*, **5**, 361–380.

Tuttle, M.P., Collier, J., Wolf, L.W. & Lafferty, R.H. (1999). New evidence for a large earthquake in the New Madrid seismic zone between A.D. 1400 and 1670. *Geology*, **27**(9), 771–774.

Tuttle, M.P., Lafferty, R.H., III, Guccione, M.J., Schweig, E.S., III, Lopinot, N., Cande, R.F., Dyer-Williams, K. &

Haynes, M. (1996). Use of archaeology to date liquefaction features and seismic events in the New Madrid seismic zone, central United States. *Geoarchaeology*, **11**, 451–480.

Tuttle, M.P., Schweig, E.S., Sims, J.D., Lafferty, R.H., Wolf, L.W. & Haynes, M.L. (2002). The earthquake potential of the New Madrid seismic zone. *Bulletin of the Seismological Society of America*, **92**, 2080–2089.

Vaughn, J.D. (1994). Paleoseismology studies in the Western Lowlands of southeast Missouri, Final Report to the U.S. Geological Survey for grant number 14–08–0001-G1931, 27 pp.

Wald, D.J. & Heaton, T.H. (1994). Spatial and temporal distribution of slip for the 1992 Landers, California earthquake. *Bulletin of the Seismological Society of America*, **84**, 668–691.

Wallace, R.E. (1978). Patterns of faulting and seismic gaps in the Great Basin Province: Proceedings of Conference VI: Methodology for identifying seismic gaps and soon-to-break gaps. *U.S. Geological Survey Open-File Report*, 78–943, pp. 857–868.

Wallace, R.E. (1984). Patterns and timing of late Quaternary faulting in the Great Basin Province and relation to some regional tectonic features. *Journal of Geophysical Research*, **89**, 5763–5769.

Wang, K., Wells, R., Mazzotti, S., Hyndman, R.D. & Sagiya, T. (2003). A revised dislocation model of interseismic deformation of the Cascadia subduction zone. *Journal of Geophysical Research*, **108**(B1), 2026.

Wesnousky, S.G. & Leffler, L. (1992). The repeat time of the 1811 and 1812 New Madrid earthquakes: a geological perspective. *Bulletin of the Seismological Society of America*, **84**, 1756–1785.

Working Group on California Earthquake Probabilities (1995). Seismic hazards in southern California: Probable earthquakes, 1994 to 2024. *Bulletin of the Seismological Society of America*, **85**, 379–439.

Yamaguchi, D.K., Atwater, B.F., Bunker, D.E., Benson, B.E. & Reid, M.S. (1997). Tree-ring dating the 1700 Cascadia earthquake. *Nature*, **389**, 922–923, 1017.

Yeats, R.S. & Prentice, C.S. (1996). Introduction to special section: paleoseismology. *Journal of Geophysical Research*, **101**, 5847–5853.

Yeats, R.S., Sieh, K. & Allen, C.R. (1997). *The geology of earthquakes*. New York, Oxford University Press, 568 pp.

Zeng, Y. (2001). Viscoelastic stress triggering of the 1999 Hector Mine earthquake by the 1992 Landers earthquake. *Geophysical Research Letters*, **28**, 3007–3010.

Appendix 1 Radiocarbon ages in Figure 11.

Event	Estuary	Site (Fig. 11a)	Lab no.	Sample age (radiocarbon yr B.P.). One-standard deviation counting error excludes [E] or includes [I] the multiplier at lower right in Fig. 11			Material dated (Fig. 11b; ring number 1 adjoins bark; spruce, *Picea sitkensis*; redcedar, *Thuja plicata*)	Sample age with respect to time of earthquake (–, not applicable)		Earthquake age not shaved in Oxcal (individual ages in Fig. 11d; calibrated with data of Stuiver *et al.*, 1998; curves in Fig. 11c)[a]		Earthquake age shaved in Oxcal (event age ranges plotted in Figs 10d and 11d; italicized, individual ages not plotted)[a]		Location (AHH, Atwater &; Hemphill-Haley, 1997; A96, Atwater, 1996; A92, Atwater, 1992, Table 1)
				Age	E	I		Numerical mean or range (ring years before plant death)	Qualitative (loosely limiting ages)					
Y[b]												1698	1715	Combined age for event
	Copalis River	A	QL-4408	112	11	17.6	Spruce root in soil, rings 1–10	5.5	–	1680	1960	*1698*	*1715*	47°07.08′N, 124°10.01′W
	Willapa Bay	C	QL-4405	152	30	48	Spruce root in soil, rings 1–20	10.5	–	1670	1970	*1697*	*1715*	Niawiakum River near Pool locality (AHH, p. 12)
	Willapa Bay	B	QL-4401	184	14	22.4	Spruce root in soil, rings 30–35	32.5	–	1690	1990	*1697*	*1716*	Bay Center beach (A96 p. 83)
	Willapa Bay	B	QL-4403	189	20	32	Spruce root in soil, rings 30–49	39.5	–	1680	2000	*1697*	*1716*	Bay Center beach (A96 p. 83)
	Copalis River	A	QL-4400	207	14	22.4	Spruce root in soil, rings 30–49	39.5	–	1680	1990	*1697*	*1716*	47°07.08′N, 124°09.96′W
	Willapa Bay	B	QL-4402	211	14	22.4	Spruce root in soil, rings 30–49	39.5	–	1680	1990	*1697*	*1716*	Bay Center beach (A96 p. 83)
	Willapa Bay	C	QL-4404	219	13	20.8	Spruce root in soil, rings 30–49	39.5	–	1680	1990	*1697*	*1716*	Niawiakum River near Pool locality (AHH, p. 12); 46°36.68′N, 123°53.61′W
	Copalis River	A	QL-4410	199	15	24	Spruce root in soil, rings 35–44	39.5	–	1690	1990	*1697*	*1716*	47°07.11′N, 124°09.96′W
	Copalis River	C	QL-4409	219	19	30.4	Spruce root in soil, rings 35–44	39.5	–	1680	1990	*1697*	*1716*	47°07.08′N, 124°10.01′W
	Willapa Bay	C	QL-4406	301	10	16	Spruce root in soil, rings 75–86	80.5	–	1600	1730	*1697*	*1716*	Niawiakum River near Pool locality (AHH, p. 12); 46°36.69′N, 123°53.74′W
W												780	1190	Based on single age
	Columbia River	A	Beta-121421	1080	40	80	Shrub root in soil	10–30	–	780	1190	*780*	*1190*	Lewis and Clark River, 46°07.90′N, 123°52.52′W
U												686	721	Combined age for event
	Columbia River	E	QL-4924	1227	30	48	*Triglochin* rhizomes above soil	–	Younger than earthquake	Before 680	Before 950	*Before 690*	*Before 940*	Lewis and Clark River, 46°09.37′N, 123°51.28′W
	Willapa Bay	B	QL-4822	1247	22	35.2	*Triglochin* rhizomes above soil	–	Younger than earthquake	Before 680	Before 890	*Before 700*	*Before 890*	Willapa River, Airport locality of AHH (p. 12, 78)
	Willapa Bay	D	QL-4827	1300	25	40	*Triglochin* rhizomes above soil	–	Younger than earthquake	Before 650	Before 810	*Before 690*	*Before 860*	Naselle River, 46°24.25′N, 123°50.42′W
	Willapa Bay	C	QL-4798	1302	21	33.6	*Triglochin* rhizomes above soil	–	Younger than earthquake	Before 660	Before 780	*Before 690*	*Before 810*	Niawiakum River, Oyster locality (AHH, p. 12, 44)
	Willapa Bay	C	QL-4795	1260	14	22.4	*Salicornia* stems in growth position within and <5 cm above soil	5–15	–	680	830	*686*	*721*	Niawiakum River, Oyster locality (AHH, p. 12, 44)
	Columbia River	F	UB-4499	1431	28	28	Spruce root in soil, rings 59–63	61	–	620	725	*685*	*721*	Lewis and Clark River, 46°07.05′N, 123°52.35′W
	Grays Harbor	A	QL-4913	1449	14	22.4	Spruce root in soil, rings 85–89	87	–	645	745	*687*	*721*	Chehalis River, 46°58.70′N, 123°46.87′W
S												340	410	Combined age for event
	Willapa Bay	C	QL-4797	1598	23	36.8	*Triglochin* rhizomes above soil	–	Younger than earthquake	Before 380	Before 560	*Before 390*	*Before 550*	Niawiakum River, Oyster locality (AHH, p. 12, 44)
	Willapa Bay	D	QL-4826	1720	25	40	Spruce root in soil, rings 1–15	8	–	240	430	*340*	*410*	Naselle River, 80 m upstream from locality 20 of A92
	Grays Harbor	B	QL-4882	1710	17	27.2	Spruce root in soil, rings 18–22	20	–	270	440	*340*	*410*	Johns River, 1.8 m depth at locality 14 of A92, site JR-1 of Shennan and others (1996)
	Columbia River	E	QL-4922	1698	15	24	Spruce root in soil, rings 18–25	21.5	–	280	440	*340*	*410*	Lewis and Clark River, 46°07.07′N, 123°51.65′W
	Willapa Bay	D	QL-4915	1696	15	24	Spruce root in soil, rings 22–24	23	–	280	440	*340*	*410*	Naselle River, locality 20 of A92, 46°23.23′N, 123°49.71′W

Appendix 1 (Continued)

Event	Estuary	Site (Fig. 11a)	Lab no.	Sample age (radiocarbon yr B.P.). One-standard deviation counting error excludes [E] or includes [I] the multiplier at lower right in Fig. 11			Material dated (Fig. 11b; ring number 1 adjoins bark; spruce, *Picea sitkensis*; redcedar, *Thuja plicata*)	Sample age with respect to time of earthquake (–, not applicable)		Earthquake age not shaved in Oxcal (individual ages in Fig. 11d; calibrated with data of Stuiver *et al.*, 1998; curves in Fig. 11c)[a]		Earthquake age shaved in Oxcal (event age ranges plotted in Figs 10d and 11d; italicized, individual ages not plotted)[a]		Location (AHH, Atwater &; Hemphill-Haley, 1997; A96, Atwater, 1996; A92, Atwater, 1992, Table 1)
				Age	E	I		Numerical mean or range (ring years before plant death)	Qualitative (loosely limiting ages)					
	Grays Harbor	A	QL-4912	1716	16	25.6	Spruce root in soil, rings 60–69	64.5	–	310	480	*340*	*410*	Chehalis River, 46°58.70′N, 123°46.87′W
	Willapa Bay	C	QL-4796	1740	15	24	Redcedar root in soil, rings 60–69	64.5	–	300	450	*340*	*410*	Niawiakum River, Pool locality of AHH (1997, p. 12, 64)
N												−670	−470	Combined age for event
	Willapa Bay	C	QL-4824	2540	25	40	Shrub root in soil	10–30	–	−800	−500	*−670*	*−470*	Naselle River, 46°23.34′N, 123°50.19′W
	Columbia River	D	UB-4497	2472	29	29	Spruce root in soil, rings 62–71	67	–	−700	−340	*−670*	*−470*	Lewis and Clark River, 46°07.05′N, 123°52.35′W
	Grays Harbor	A	Beta-113267	2550	60	120	Spruce root in soil, rings ≥ (71–82)	60–100	–	−850	−300	*−670*	*−470*	East Fork Hoquiam River, 47°01.07′N, 123°52.51′W
	Grays Harbor	A	QL-4930	2508	16	25.6	Spruce root in soil, rings ≥ (118–124)	120–160	–	−670	−390	*−670*	*−470*	East Fork Hoquiam River, 47°01.07′N, 123°52.51′W
	Willapa Bay	B	QL-4715	2475	23	36.8	Spruce cones on soil	–	Older than earthquake	After 770	After 410	*After −790*	*After−530*	Niawiakum River, Redtail locality (AHH, p. 12, 28)
L												−975	−895	Combined age for event
	Grays Harbor	A	QL-4916	2846	17	27.2	Spruce root in soil, rings 16–24	20		−1110	−890	*−975*	*−895*	East Fork Hoquiam River, 47°01.07′N, 123°52.51′W
	Willapa Bay	C	QL-4917	2793	16	25.6	Spruce root in soil, rings 20–26	23		−980	−820	*−975*	*−895*	Willapa River, Jensen locality at horizontal coordinate 52 m (AHH, p. 12, 70)
	Grays Harbor	B	QL-4883	2791	11	17.6	Spruce root in soil, rings 21–27	24		−980	−830	*−975*	*−895*	Blue Slough, 2.9 m depth at locality 10 of A92, 46°56.85′N, 123°43.42′W
	Willapa Bay	E	QL-4914	2814	17	27.2	Spruce root in soil, rings 23–27	25		−1020	−870	*−975*	*−895*	Naselle River, 46°23.34′N, 123°50.19′W
	Willapa Bay	D	QL-4923	2811	16	25.6	Spruce root in soil, rings 27–31	29		−1020	−860	*−975*	*−895*	Niawiakum River, Pool locality at horizontal coordinate 7 m (AHH, p. 12, 64)
	Columbia River	F	UB-4496	2886	29	29	Spruce root in soil, rings 74–82	78		−1140	−860	*−975*	*−895*	Lewis and Clark River, 46°07.05′N, 123°52.35′W
J												−1440	−1355	Combined age for event
	Willapa Bay	A	QL-4919	3177	16	25.6	Spruce root in soil, rings 48–54	51		−1470	−1350	*−1440*	*−1355*	South Fork Willapa River, 3.2 m depth at locality 15 of A92, 46°40.34′N, 123°59.18′W
	Willapa Bay	C	QL-4884	3200	17	27.2	Spruce root in soil, rings 51–60	54.5		−1470	−1350	*−1440*	*−1360*	Naselle River, 46°23.34′N, 123°50.19′W
	Willapa Bay	B	QL-4718	3165	17	27.2	Spruce cones on soil		Older than earthquake	After 1520	After 1390	*After −1520*	*After −1400*	Niawiakum River, Redtail locality (AHH, p. 12, 28)
	Willapa Bay	B	QL-4717	3166	18	28.8	Twigs on soil		Older than earthquake	After 1520	After 1390	*After −1520*	*After −1400*	Niawiakum River, Redtail locality (AHH, p. 12, 28)
	Willapa Bay	B	QL-4716	3180	30	48	Stick on soil		Older than earthquake	After 1600	After 1310	*After −1600*	*After −1380*	Niawiakum River, Redtail locality (AHH, p. 12, 28)

[a] Probably contains 95-percent confidence interval. In cal yr A.D. [+] and cal yr B.C. [−] (converted to cal yr before A.D. 2000 in Figs 10d and 11d).
[b] Ages reported by Atwater *et al.* (1991), recalibrated in this table.

Quaternary volcanism in the United States

William E. Scott

Cascades Volcano Observatory, U.S. Geological Survey, 1300 SE Cardinal Ct., Vancouver, WA 98683, USA

Introduction

The last third of the 20th century witnessed great advances in understanding the Quaternary history of volcanic areas in the United States. Contributions to the 1965 INQUA volume, "The Quaternary of the United States" (Wright & Frey, 1965), highlighted tephrochronology as an emerging method for dating and correlating continental sediments and paleoenvironmental change in the Cordillera and Great Plains. In the same volume, regional summaries about the eastern Sierra Nevada, Cascade Range, Snake River Plain, desert Southwest, and Alaska described growing knowledge about the Quaternary volcanic history of areas in the Cordillera, aided especially by development of potassium–argon (K–Ar) and paleomagnetic dating. Subsequent reviews, such as those written for the Decade of North American Geology and other major syntheses, demonstrated continued improvements in tephrochronology and its widespread applications to Quaternary geology (Baker *et al.*, 1991; Dupre *et al.*, 1991; Gustavson *et al.*, 1991; Izett, 1981; Morrison, 1991; Sarna-Wojcicki, 2000; Sarna-Wojcicki & Davis, 1991; Sarna-Wojcicki *et al.*, 1983, 1991; Wayne *et al.*, 1991).

During these same decades, societal problems related to energy supplies and natural hazards spurred studies of young volcanic systems in the western conterminous United States, Alaska, and Hawai'i. Concerns about the future of fossil fuels led to a vigorous program to evaluate geothermal resources in the western U.S. (e.g. Muffler, 1979). A basic element of this program was detailed geologic mapping and geochronology of Cascade volcanoes, large calderas such as Long Valley, Valles, and Yellowstone, and numerous other volcanic fields in the Cordillera. The 1980–1986 eruption of Mount St. Helens, Washington, the 1983 to present eruption of Kīlauea, Hawai'i, the 1984 eruption of Mauna Loa, Hawai'i, numerous eruptions in Alaska – including the 1976 and 1986 eruptions of Augustine Volcano, the 1989–1990 eruption of Redoubt Volcano, and the 1992 eruption of Mount Spurr – increased public awareness of volcanic hazards on the ground and in the flight paths of commercial aircraft. Eruptions with tragic consequences at El Chichón (Mexico) in 1982 and Ruiz (Colombia) in 1985 mobilized the international volcanologic community to work to reduce the impacts of future eruptions. Notable hazard-assessment, eruption-forecasting, and risk-mitigation successes at Pinatubo (Philippines) in 1991, Unzen (Japan) in 1991–1996, Rabaul (Papua New Guinea) in 1994, Ruapehu (New Zealand) in 1995–1996, Soufrière Hills (Montserrat) in 1995 to present, Usu (Japan) in 2000–2001, and numerous other volcanoes attest to the growing understanding of eruption precursors and hazardous volcanic processes.

Because tephrochronology in the conterminous United States has been the focus of numerous recent reviews (e.g. Sarna-Wojcicki, 2000; Sarna-Wojcicki *et al.*, 1991), this paper concentrates chiefly on advances in understanding of the Quaternary history of volcanic areas in the United States and key processes in the evolution of volcanoes. Tephrochronology in Alaska is discussed because it is a field that has grown rapidly in the past few decades.

This review is organized around the concept of *volcanic loci* – vents or groups of vents that define logical units of volcanism in space and time (Smith & Luedke, 1984). Loci of Quaternary volcanism are found in the western Cordillera of the conterminous United States, in southern and western Alaska, and in the Hawaiian Islands (Fig. 1). In the conterminous United States, linear arrangements of Quaternary loci lie within major tectonic zones of the Cascade arc, western and northern margins of the Basin and Range province, Snake River Plain-Yellowstone, and western and southern margins of the Colorado Plateau (Fig. 2; Luedke & Smith, 1991). Quaternary volcanism in Alaska has occurred in the Aleutian arc, Wrangell Mountains, a few localities in southeastern Alaska, and scattered basalt fields in western Alaska (Fig. 6; Fournelle *et al.*, 1994; Miller & Richter, 1994; Moll-Stalcup, 1994). The Hawaiian Islands are the latest Tertiary and Quaternary products of the Hawaiian Ridge-Emperor Seamounts volcanic chain, a 6000-km-long chain of shield volcanoes erupted on the sea floor beginning about 80 myr ago (Fig. 7; Clague & Dalrymple, 1987, 1989). The chain originates from the northward and later westward movement of the Pacific Plate over the Hawaiian hot spot, a melting anomaly in the asthenosphere. Each of these areas is discussed in greater detail below. Where possible, citations are limited to recent reviews or comprehensive papers wherein references to earlier works are available.

Cascade Volcanic Arc

Oblique convergence of the North American and Juan de Fuca Plates has maintained a volcanic arc in the northwestern United States for the past 40 myr. The Quaternary Cascades, which extend from southern British Columbia to northern California, are restricted to a relatively narrow area of the Cenozoic arc. Variation in tectonic regime along the arc results in several segments that are defined by differences in density and distribution of volcanic vents, volcanic production rate, and character of volcanism (Fig. 3; Guffanti & Weaver, 1988; Sherrod & Smith, 1990). The northern segment, which is apparently the most strongly compressional, is characterized by isolated composite centers, including

DEVELOPMENTS IN QUATERNARY SCIENCE
VOLUME 1 ISSN 1571-0866
DOI:10.1016/S1571-0866(03)01016-9

PUBLISHED BY ELSEVIER B.V.

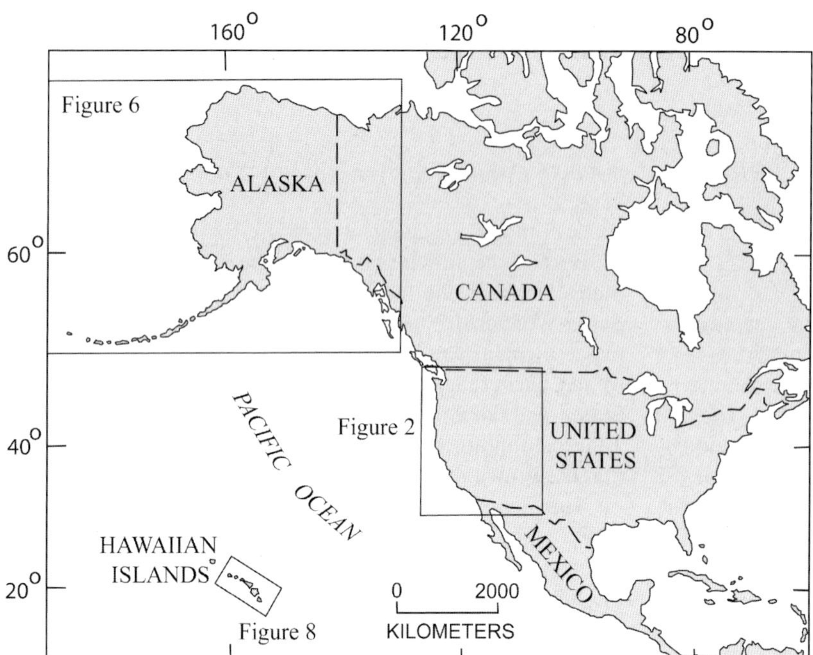

Fig. 1. Index of Quaternary volcanic areas in the United States that are shown in identified figures.

Mount Baker and Glacier Peak. To the south, several major composite centers, Mount Rainier, Mount St. Helens, Mount Adams, and Mount Hood, dominate a broad zone of monogenetic vents of chiefly mafic (basalt and basaltic andesite) composition that occur between these centers and extend locally to the west and east. South of Mount Hood, the major composite centers of central and southern Oregon – Mount Jefferson, Three Sisters, Newberry, and Crater Lake (Mount Mazama) – lie within a dense cluster of andesitic and basaltic monogenetic vents. This relation of composite centers and surrounding monogenetic vents repeats in the Mount Shasta-Medicine Lake area and at the Lassen Peak volcanic center in northern California. The main effect of this segmentation is that, in the northern segments, eruptive activity is focused in a relatively small area resulting in repeated growth of a composite volcano with perhaps a few satellite vents nearby. In southern segments, the denser distribution of vents throughout the arc produces overall growth of the range, largely as an array of overlapping shield volcanoes.

The Quaternary history of the Cascades encompasses a wide range of eruptive styles. These include highly explosive ignimbrite eruptions that created calderas; long-lived, recurrently active composite volcanoes that display a wide range of behaviors from explosive to effusive; and the effusive growth of small monogenetic mafic volcanoes.

Caldera-Forming Events

Large-volume, caldera-forming eruptions have been relatively rare in the Cascades during Quaternary time. The 7700 cal yr B.P. eruptive sequence at Mount Mazama that created Crater Lake caldera, Oregon, is the youngest and best known and represents the culmination of tens to hundreds of thousands of years of petrologic evolution (Bacon, 1983; Druitt & Bacon, 1986).

The climactic eruption produced about 50 km^3 (dense-rock equivalent, or DRE) of roughly equal proportions of rhyodacitic tephra-fall deposits and chemically zoned ignimbrite. The tephra-fall deposit provides an important mid-Holocene tephrostratigraphic horizon over much of the Pacific Northwest, the northern Rocky Mountains, and adjacent southern Canada (e.g. Sarna-Wojcicki *et al.*, 1983). Analysis of proximal fall deposits reveals the initial Plinian phase of the eruption rose from a single vent, and lasted perhaps 20 hr with the eruption column at times exceeding 50 km in altitude (Young, 1990). Collapse of the Plinian column formed a modest volume of welded ignimbrite, which was followed by formation of ring vents as decompression of the magma reservoir by magma withdrawal initiated caldera subsidence (Bacon, 1983; Suzuki-Kamata *et al.*, 1993). Ignimbrites of the ring-vent phase swept outwards more than 50 km and left deposits locally more than 100 m thick of chiefly rhyodacitic composition with a late, subordinate andesitic component. Continued subsidence ultimately created a caldera ~5 by 6 km, which has enlarged through landsliding and erosion to its present ~8 by 10 km. Recent explorations of the lake floor reveal post-caldera eruptive products, submerged shorelines, a variety of deposits related to landslides and other mass-wasting events, and lacustrine sediments (Bacon *et al.*, 2002; Nelson *et al.*, 1994).

The only other Cascade eruptions of Quaternary age tied directly to caldera formation were similar to the Mazama eruption in scale and composition of products. The 1.15-myr eruption of Kulshan caldera at the northeast base of Mount Baker was the source of the Lake Tapps tephra in the western Washington (Hildreth, 1996). Enclosing lacustrine sediments

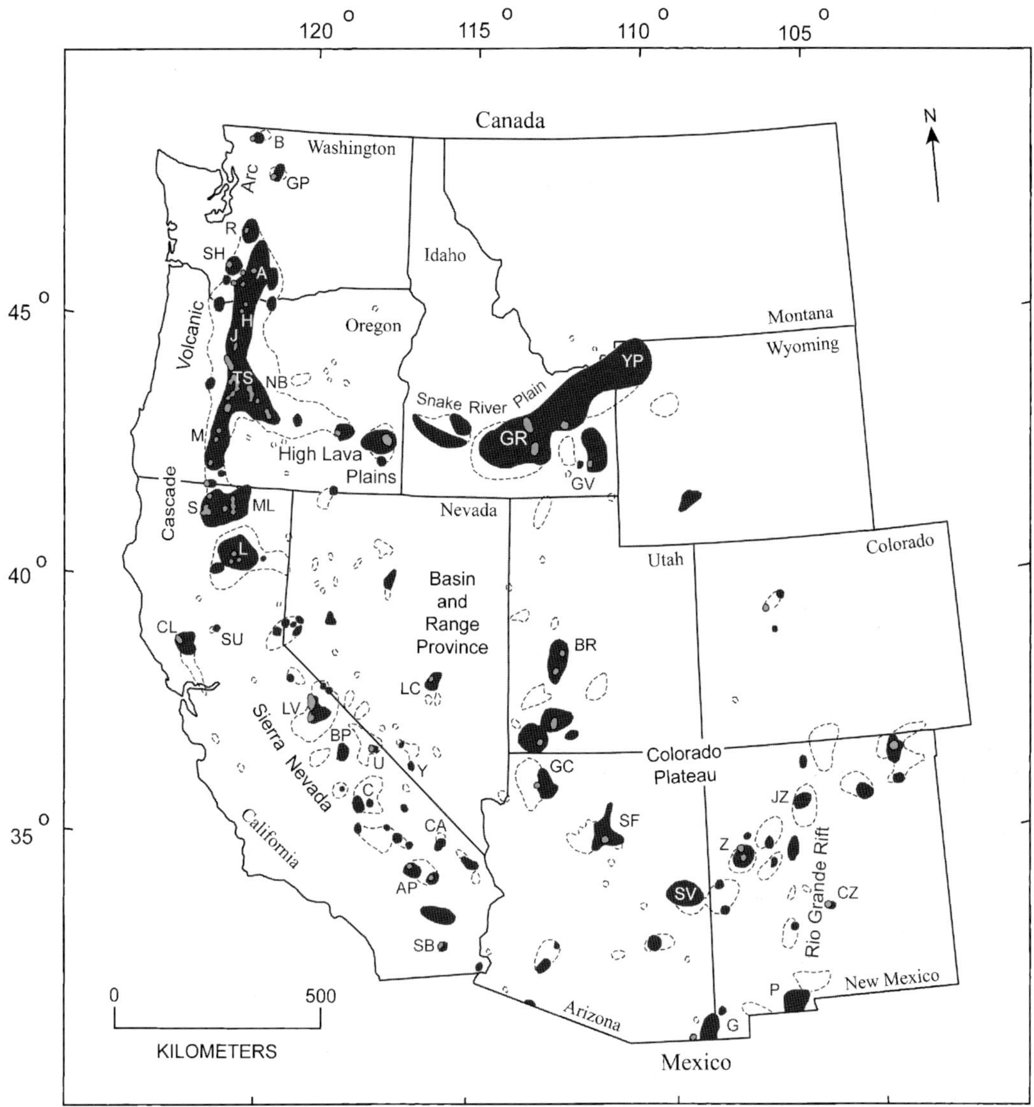

Fig. 2. Volcanic loci in the conterminous United States that have been active during the Quaternary (black); gray areas within loci have been active during Holocene time. Dashed line outlines areas of volcanic loci active during the past 5 myr (modified from Luedke & Smith, 1991, Figs 4 and 6). Abbreviations on map: A, Adams; AP-Amboy-Pisgah; B, Baker; BP, Big Pine; BR, Black Rock; C, Coso; CA, Cima; CL, Clear Lake; CZ, Carrizozo; G, Geronimo; GC, Grand Canyon; GP, Glacier Peak; GR, Great Rift; GV, Gem Valley; H, Hood; J, Jefferson; JZ, Jemez Mts.; L, Lassen; LC, Lunar Crater; LV, Long Valley; M, Mazama (Crater Lake); ML, Medicine Lake; P, Potrillo; R, Rainier; S, Shasta; SB, Salton Buttes; SF, San Francisco field; SU, Sutter Buttes; SV, Springerville; U, Ubehebe; Y, Yucca Mt.; YP, Yellowstone Plateau; Z, Zuni-Bandera.

and pollen evidence suggest the tephra accumulated in a near-glacial environment (Westgate *et al.*, 1987). On the basis of this and other evidence, the caldera eruption occurred while the Cordilleran ice sheet covered the North Cascades. Such an origin, coupled with 1 myr of repeated advances of alpine glaciers and the Cordilleran Ice Sheet, explains the lack of preservation of proximal ignimbrite or fall deposits except for those within the caldera (Hildreth, 1996). The third known caldera-forming eruption occurred at the Lassen volcanic center about 600,000 yr ago and produced the Rockland tephra, an important middle Pleistocene marker bed in the northwestern Basin and Range province and northern California (Clynne, 1990; Lanphere *et al.*, 1999b; Sarna-Wojcicki *et al.*, 1983).

Other ignimbrites of Quaternary age at Three Sisters, Newberry, and Medicine Lake volcanoes may be related to caldera-forming eruptions, but direct geologic evidence is lacking. Voluminous younger eruptive products obscure the source regions of four major ignimbrites of middle Pleistocene age exposed east of Three Sisters (Hill, 1991; Hill & Taylor, 1990). Two of the four have coeval fall deposits that correlate chemically with distal tephra marker beds in the western U.S. (Sarna-Wojcicki *et al.*, 1991). The oldest unit, Desert Springs Tuff, matches the Rye Patch Dam

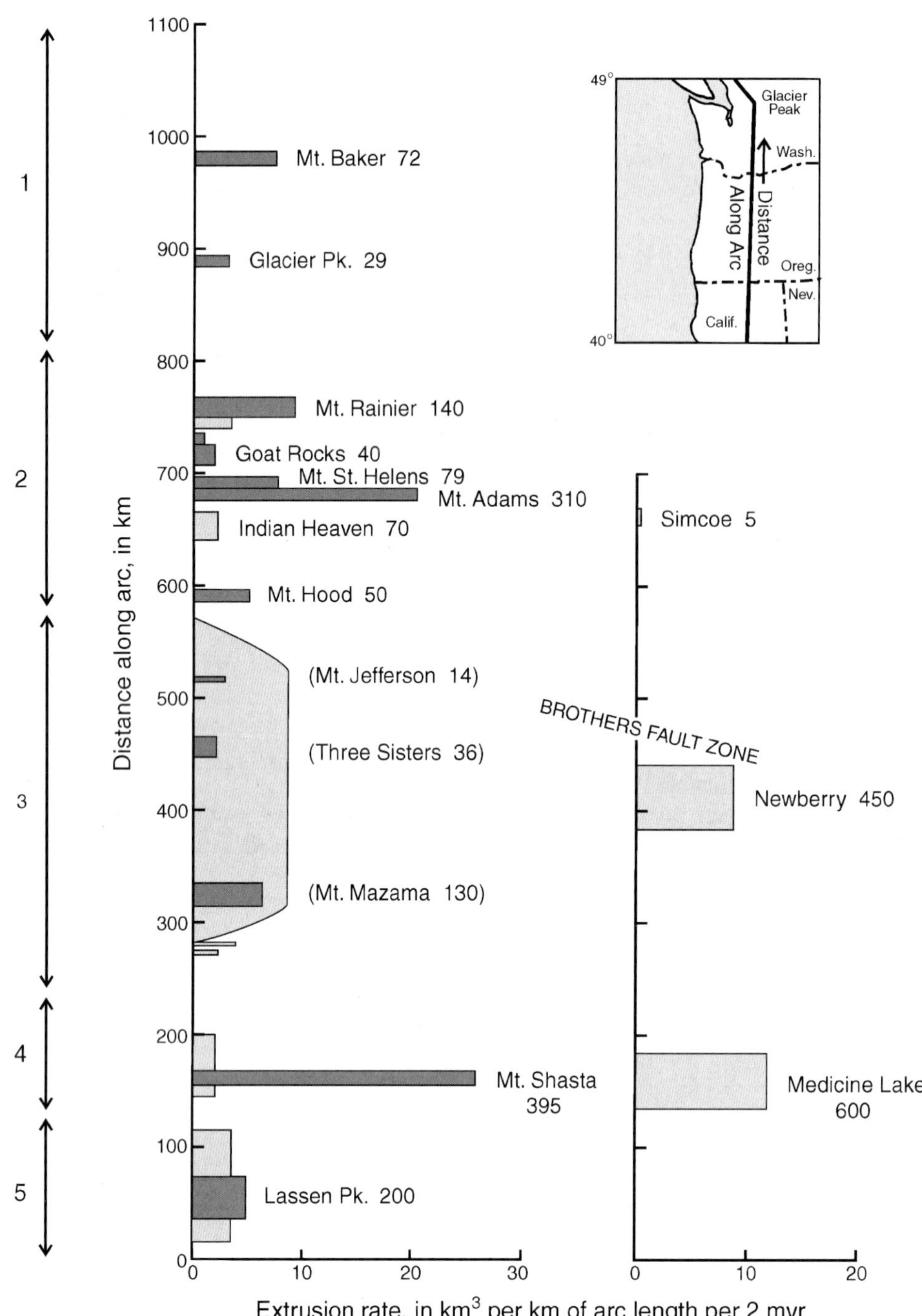

Fig. 3. Quaternary extrusion rate for Cascade volcanic arc (modified from Sherrod & Smith, 1990, Fig. 3). Distance is measured north from latitude 40°N (see inset), and the area under the curve is the total volume of Quaternary volcanic rocks in the arc. Light gray indicates chiefly mafic rocks, dark gray chiefly andesite to rhyolite. Boxes for individual volcanoes drawn such that bases are equal to basal diameters of volcanoes along arc and box areas are equal to Quaternary volume. Volumes for volcanoes in parentheses are included in continuous belt of Quaternary rocks in the High Cascades of central Oregon. Simcoe, Newberry, and Medicine Lake volcanic centers are offset about 50–80 km east of the arc axis. Vertical numbered lines represent segments of arc identified on basis of vent distribution and tectonic considerations (Guffanti & Weaver, 1988).

ash bed, which underlies with little stratigraphic separation Yellowstone's ~640,000-yr-old Lava Creek B ash bed. The next youngest ignimbrite, the Tumalo Tuff, overlies an immediately preceding tephra-fall deposit, the Bend Pumice, which correlates chemically with the widespread Loleta ash bed of northern California and the adjacent Pacific Ocean. The age of these units is about 400,000 yr (Lanphere *et al.*, 1999a). The youngest ignimbrite, Shevlin Park Tuff, is about 260,000 yr old (Lanphere *et al.*, 1999a), but its correlation to distal tephra-fall deposits

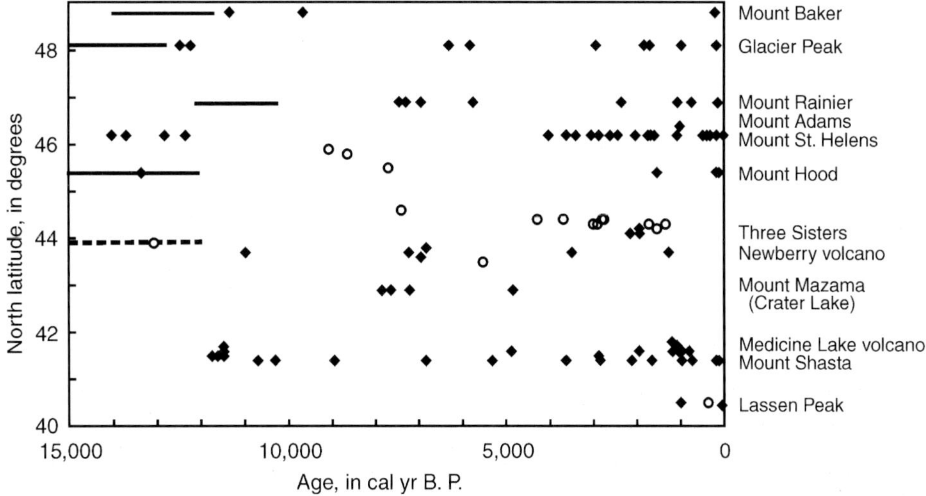

Fig. 4. Dated eruptive periods of major Cascade volcanic centers (diamonds) and mafic volcanoes (circles) during the past 15,000 years. Eruptive periods typically include a number of individual eruptions closely spaced in time. Mount Adams and Medicine Lake volcano are plotted slightly north of their true position to differentiate their symbols from those of Mounts St. Helens and Shasta, respectively. Solid and dashed lines represent time intervals that encompass poorly dated late-glacial eruptions of major centers and mafic volcanoes, respectively (modified from Scott, 1990, Fig. 4).

remains unresolved (Conrey *et al.*, 2001; Sarna-Wojcicki *et al.*, 1991).

Both Newberry and Medicine Lake volcanoes have broad summit calderas and locally exposed ignimbrites, but caldera subsidence was probably complex in both cases. Newberry volcano has been the source of more potentially voluminous ignimbrites and widespread silicic tephras than has Medicine Lake (Kuehn, 2002; MacLeod *et al.*, 1995). Some or all of which may have contributed to caldera formation. Only one ignimbrite is known at Medicine Lake volcano – an andesitic unit erupted while glacier ice covered the upper flanks of the volcano (Donnelly-Nolan & Nolan, 1986), but it appears not to be associated with caldera formation (Donnelly-Nolan, 1988). More likely the caldera formed and has been maintained by eruptions of voluminous mafic lava flows.

Evolution of Major Composite Cones

During the past several decades, detailed geologic mapping and stratigraphic studies, bolstered by wide application of radiometric- and paleomagnetic-dating techniques, have advanced understanding of the ages and eruptive histories of Cascade composite cones. Concurrent efforts to interpret emplacement mechanisms of various types of deposits through field, experimental, and theoretical studies, especially those at erupting volcanoes, have also contributed to these advancements. Current views of the major cones are highlighted by a new appreciation for geologically rapid rates of growth and destruction.

Much of the work in the Cascades in the 1960s and 1970s focused on unraveling the postglacial history of composite cones. Pioneering work by Dwight Crandell and Donal Mullineaux at Mounts Rainier and St. Helens merged Quaternary stratigraphy, geomorphology, and volcanology to reveal the eruptive and mass-wasting history of the cones. Such information was used as the basis for assessing potential hazards from future eruptions (e.g. Crandell, 1987; Crandell & Mullineaux, 1978; Mullineaux, 1996). The results of

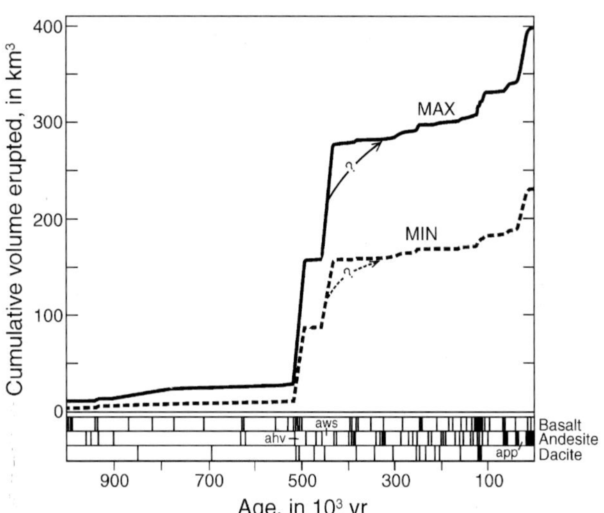

Fig. 5. Cumulative erupted volume vs. time for the Mount Adams volcanic center in southern Washington (modified from Hildreth & Lanphere, 1994, Fig. 6). Cumulative curves labeled MAX and MIN are based on liberal and conservative estimates of volumes of units. Composition of units is shown at base of diagram. Shaded boxes labeled ahv, aws, and app represent the three major cone-building episodes of the middle and late Quaternary, during which eruption rates greatly exceeded long-term average rates. Queried arrows show possible alternatives.

such investigations throughout the U.S. portion of the arc accentuate the episodic behavior of these types of volcanoes (Fig. 4). Frequent eruptions over centuries to millennia are separated by periods of relative inactivity of similar length. For example, Mount St. Helens, which has been the most active Cascade volcano during the past 4000 yr, was apparently dormant for much of early and middle Holocene time. This quiescence followed a period of frequent eruptions during latest Pleistocene time. As a result of a high rate of late Holocene eruptions, the upper flanks of the pre-1980 cone were composed of lava flows and domes emplaced during the past few thousand years. In contrast, Mount Hood, which was also apparently dormant for much of the Holocene, experienced only localized changes in form as a result of two relatively modest eruptive periods since 1500 cal yr B.P.

Longer-term views of eruptive histories of the Cascade volcanoes are emerging from detailed geologic mapping and from geochronology that relies greatly on K–Ar and $^{40}Ar/^{39}Ar$ dating techniques. Such techniques, developed and refined over the past few decades, allow high-precision dating of young volcanic products (e.g. Lanphere, 2000; Renne, 2000). The emerging pattern is one of recurrently active, long-lived (10^5 to 10^6 yr) centers, with many cones having significant construction during the past 100,000 yr (Table 1). Many imposing volcanoes, including Mounts Baker, Rainier, Adams, and Hood, and Lassen Peak, hosted major growth spurts shortly before, during, or shortly after the last glaciation. To date, the most detailed published record is that from the Mount Adams volcanic field (Fig. 5; Hildreth & Lanphere, 1994). Key points about the long-term behavior of the field appear to be generally applicable to all of the Cascade centers, as supported by ongoing studies.

(1) Stratovolcanoes grow in spurts. The construction of an imposing cone need only takes a few percent of the lifetime of a volcano. Such spurts may be separated in time by tens or hundreds of millennia.
(2) These volcanic systems may remain active at greatly reduced eruption rates between the pulses of peak productivity.
(3) Large stratovolcanoes can remain recurrently active for half a million years or more. There is no evidence that such systems need evolve toward a climactic caldera-forming stage. Systems that host

Table 1. Ages of Cascade composite centers based on calibrated ^{14}C, K–Ar, and $^{40}Ar/^{39}Ar$ dating.

Volcano or Volcanic Center	Oldest Products from Proximal[a] Vents (10^3 yr)	Youngest Major Cone Building Prior to End of MIS[b] 2 (10^3 yr)	References
Mount Baker	455	43–10	Hildreth & Lanphere (2000)
Glacier Peak	<780	13[c]	Tabor & Crowder (1969), Begét (1982), Begét (1983), Foit *et al.* (1993)
Mount Rainier	500	40–10	Lanphere & Sisson (1995), Sisson *et al.* (2001)
Mount St. Helens	50[d]	16–12	Crandell (1987), Berger & Busacca (1995), Mullineaux (1996)
Mount Adams	520	40–10	Hildreth & Lanphere (1994), Lanphere (2000)
Mount Hood	600	55–10	Scott *et al.* (1997)
Mount Jefferson	280	70–40	Conrey (1991)
Three Sisters[e]	<260	<90	Hill & Taylor (1990), Hill & Duncan (1990), Lanphere *et al.* (1999a, b), Calvert & Hildreth (2002)
Newberry volcano	500–600		Linneman (1990), MacLeod *et al.* (1995)
Mount Mazama (Crater Lake)	420	35	Bacon (1983), Bacon & Lanphere (2000), Bacon *et al.* (2002)
Mount Shasta	590	<130	Miller (1980), Christiansen & Miller (1989)
Medicine Lake volcano	500–600		Donnelly-Nolan (1988), Lanphere, Turrin & Donnelly-Nolan *in* Donnelly-Nolan (1998)
Lassen Peak	>600	28	Clynne (1990), Turrin *et al.* (1998), Lanphere *et al.* (1999a, b), Christiansen *et al.* (2002a)

[a] Vent known to lie within a few kilometers of youngest vent or inferred to be so on basis of distribution of products.
[b] Marine oxygen isotope stage.
[c] Assumes that substantial growth accompanied episode of explosive eruptions at 13,000 cal yr B.P.
[d] Earlier eruptions are known from the St. Helens region, but vent areas are either unknown or known to be off the current edifice (Berger & Busacca, 1995; Evarts, 2001).
[e] South and Middle Sisters.

such eruptions do so for a variety of tectonic and magma-supply reasons.

(4) Long-lived arc volcanoes display eruptive patterns that encompass a wide range of recurrence times. Centuries or decades of frequent eruptions may be separated by centuries or millennia of relative quiet.

A corollary to the view of episodic rapid growth of the large composite cones is that they can erode rapidly. Layers with contrasting physical properties, steep dips, and high degrees of fracturing combined with steep slope angles, high precipitation rates, and glaciers have long been known to create conditions favorable for high rates of erosion of Cascade composite volcanoes (Mills, 1976, 1992). At Mount Rainier, avalanches of hydrothermally altered material from the upper flanks of the volcano generated clay-bearing lahars, some as large as several cubic kilometers in volume and flowing more than 100 km from source (Crandell, 1971). Subsequent work supports and refines this view, and stresses the importance of large volumes of pore water and mechanically weakened material in facilitating rapid transformation of a debris avalanche into a far-traveled lahar (Frank, 1995; Scott *et al.*, 1995; Vallance & Scott, 1997). Recent incorporation of geologic and geophysical data into a three-dimensional stability analysis of Mount Rainier accentuates the contributions of hydrothermally weakened material and steep slopes to avalanche formation, and identifies areas at greatest risk from future collapse (Finn *et al.*, 2001; Reid *et al.*, 2000, 2001; Sisson *et al.*, 2001). As witnessed at Mount St. Helens in 1980, magma intrusion and related seismicity, ground deformation, and minor eruptions can also play important roles in triggering large-volume debris avalanches (Lipman & Mullineaux, 1981; Voight & Elsworth, 1997).

Such avalanches rapidly remove substantial portions of composite volcanoes; thereby creating large depressions to be filled by later eruptive products. About 300,000 yr ago, ancestral Mount Shasta spawned one of the largest known (about 50 km^3) debris avalanches from a composite cone (Crandell, 1989). Several generations of edifice building have occurred since then (Christiansen & Miller, 1989). Exposures created by the 1980 evisceration of Mount St. Helens by a debris avalanche reveal evidence of similar events having occurred about 2800 cal yr B.P. (Hausback, 2000) that were followed by rebuilding of the edifice during several eruptive periods.

Mafic Fields

More than 1000 mafic (chiefly basalt and basaltic andesite) volcanoes of Quaternary age lie scattered through the Cascade Range from southeast of Mount Rainier to Lassen Peak (Guffanti & Weaver, 1988; Sherrod & Smith, 2000; Smith, 1993). The density of vents decreases in three areas, which are the only areas in which rivers draining the interior have been able to maintain a course through the arc – the Columbia, Klamath, and Pit. The highest density of vents is in the central Oregon Cascade Range from just north of Mount Jefferson to south of Crater Lake, also encompassing the major composite centers of Three Sisters and Newberry. Quaternary extrusion rates measured as lava volume per unit length of arc per unit time show this area of the arc to be the most productive (Fig. 3), suggesting that greater extension across the arc in this area allows a greater proportion of magma to reach the surface (Sherrod & Smith, 1990). Current tectonic models of Cascadia explain this enhanced extension as a consequence of northward migration of the fore arc and its fragmentation into large rotating blocks (e.g. Wells *et al.*, 1998). A large, clockwise-rotating block that forms much of the Oregon Coast Range accounts for rapid extension of the central Oregon portion of the arc. North of this block, the arc is under increasing compression, reflected in lower extrusion rates and eventually an absence of vents between the major composite centers. To the south, the arc is in an intermediate state of stress, resulting from northwest translation of the Sierra Nevada block into the southern end of the arc. Mafic vents are locally abundant around the composite centers of Mount Shasta, Medicine Lake, and Lassen Peak.

The central Oregon segment of the arc is constructed almost entirely of overlapping shield volcanoes, scoria cones, and lava flows. These, along with the major composite centers, variably fill a complex graben formed in arc rocks of latest Miocene to early Pleistocene age (Taylor, 1990). North-south alignments of coeval vents attest to the importance of mafic dikes intruded into extending crust in feeding Quaternary eruptions.

The Mount Bachelor volcanic chain is a latest Pleistocene example of the types of eruptive activity that have formed this arc segment (Gardner, 1994; Scott & Gardner, 1992). Eruptions began there during the waning phase of the last glaciation, when the Cascades of central Oregon lay beneath a mountain ice sheet. The eventual 25-km-long alignment of vents of the chain developed along the east margin of a major outlet glacier that flowed south from the Three Sisters area. Paleomagnetic evidence suggests that the chain grew in four episodes. Individual episodes lasted from years to perhaps as long as a millennium and produced a total of about 40 km^3 of basalt and basaltic andesite, chiefly in three major shield volcanoes and one chain of scoria cones that fed a broad apron of lava flows. The bulk of activity ended by about 13,000 cal yr B.P., the age of a widely recognized glacier still-stand or re-advance in the Pacific Northwest (see Kaufman *et al.*, this volume). By then, Mount Bachelor, the youngest and largest of the shields, had grown high enough to support a small glacier fronted by moraines that were partly overridden by the latest lava flows. The history of the Mount Bachelor chain attests to the rapidity with which the central Oregon Cascade landscape has changed as the result of mafic volcanism.

Western Cordillera Volcanic Loci Beyond the Cascades

As in the Cascade volcanic arc, the locations and styles of Quaternary volcanism elsewhere in the conterminous United States mimicked the late Tertiary patterns (Fig. 2; Luedke &

Smith, 1991). Most are fields of chiefly basaltic lava flows and cones associated with extensional tectonic regimes, but a few are long-lived silicic centers that produced the largest Quaternary eruptions in the U.S. The following discussion proceeds through three broadly defined geographic regions – northern Basin and Range province, western and southern margins of the Colorado Plateau, and western Basin and Range province.

Northern Basin and Range Province

The northern margin of the Basin and Range province marks a diffuse, transform-tectonic boundary between well-defined basin-and-range extension to the south and more stable regions to the north (Christiansen & Yeats, 1992). The region encompasses a broad swath of volcanic vents of late Neogene age that stretches from the Cascade arc near Newberry volcano, across the High Lava Plains of eastern Oregon, Snake River Plain of Idaho, to the Yellowstone volcanic plateau of Wyoming. The relatively less productive volcanic areas in the western part of the zone include a few early Pleistocene rhyolite domes just east of Newberry. The domes are the youngest in a broad belt of rhyolitic volcanism that spread westward from southeastern Oregon-southwestern Idaho beginning about 15 myr ago. In contrast, contemporaneous basaltic volcanism in the same areas lacks such a trend. Rather, basaltic fields of Quaternary age are scattered throughout the northern part of the Tertiary belt with several prominent fields of latest Pleistocene to Holocene age. In the Christmas Lake Valley, directly east of Newberry, they include several prominent maar volcanoes – Big Hole, Hole-in-the-Ground, and Fort Rock. These were formed by hydrovolcanic eruptions driven by magma interacting with shallow ground water or lakes, perhaps during wetter climatic periods (Heiken *et al.*, 1981; Lorenz, 1970). Broad shield volcanoes and lava-dammed and lava-capped sedimentary sequences of early and middle Pleistocene age are scattered across the western Snake River Plain, Idaho, where they record the interplay between volcanism and the Snake River striving to maintain its course along the southern margin of the plain (Malde, 1991).

The eastern Snake River Plain starts at about longitude 115°W and extends for 350 km northeast to Yellowstone as a largely undissected volcanic plain about 100 km wide, composed chiefly of basaltic shield volcanoes surrounded by broad fields of tube-fed lava (Kuntz *et al.*, 1992; Link & Mink, 2002; Malde, 1991). Individual shields cover several hundred square kilometers and have volumes as large as 7 km^3. Non-shield portions of the plain are covered with small lava cones, scoria cones, and spatter ramparts along fissures and their related lava flows. Landforms resulting from hydrovolcanism are relatively rare, probably owing to the lack of shallow ground water except along the plain margins and in playas fed by marginal drainages.

A widely accepted model for the origin of the eastern Snake River Plain and the Yellowstone volcanic system has developed over the past several decades (Armstrong *et al.*, 1975; Christiansen, 2001; Christiansen *et al.*, 2002b; Pierce & Morgan, 1992; Smith & Braile, 1994). While debate continues about many details, researchers agree that silicic volcanism, involving the formation of large rhyolitic calderas, was prevalent in southwestern Idaho 15–12 myr ago and propagated progressively northeastward, reaching the Yellowstone region in late Pliocene and Quaternary time. As activity at each silicic system ended, regional subsidence, basaltic volcanism, and marginal sedimentation ensued, eventually creating a fill as thick as 2 km over the silicic rocks. Essentially the entire surface of the eastern Snake River Plain is younger than 780,000 yr (boundary between Bruhnes and Matuyama Polarity Chrons); volcanic fields of late Pleistocene and Holocene age cover about 13% of the Plain (Kuntz *et al.*, 1992). Most of these fields lie along rift zones transverse to the trend of the plain. The rift zones are parallel to and in places collinear with basin and range faults in the adjoining ranges, and reflect the same regional extension with dike injection accommodating the strain (Kuntz, 1992). The Great Rift, the longest and most continuous rift, has hosted the majority of latest Quaternary activity (Kuntz *et al.*, 1986a, b). During that time, quasi-steady-state, volume-predictable eruptions of basalt and higher-silica contaminated magma have occurred with a recurrence interval of several hundred to ~3000 yr. The last such eruption occurred about 2200 ^{14}C yr B.P.

During the past 2 myr, the northeast-migrating silicic focus produced three volcanic cycles of rhyolitic volcanism from the Yellowstone volcanic system, each lasting between one-half and one million years (Christiansen, 2001). Each cycle began with eruptions of rhyolite from vents in the area that would eventually undergo caldera collapse and eruptions of basalt in marginal areas. Large ignimbrites, ~300–2500 km^3 in volume and covering several thousand square kilometers, and associated widespread tephra-fall deposits were erupted at the climax of each cycle, producing calderas tens of kilometers across. Such caldera-forming eruptions are among the largest-known volcanic events on Earth. The caldera of the most recent cycle, which overlaps part of the first-cycle caldera, measures 85 by 45 km. It formed by the subsidence of two major cauldron blocks, each of which produced, in rapid succession, a recognizable flow unit. Recently revised $^{40}Ar/^{39}Ar$ ages for the climactic eruptions are 2.059 ± 0.004 myr, 1.285 ± 0.004 myr, and 0.639 ± 0.002 myr (Lanphere *et al.*, 2002). Distal tephra-fall deposits from these eruptions are widespread in the west and Great Plains and form important markers beds (Izett & Wilcox, 1982) that aid in stratigraphic correlations (e.g. Williams, 1994) and studies of surficial processes (e.g. Dethier, 2001). Each caldera collapse was followed by hundreds of thousands of years of resurgent doming of the subsided caldron blocks and rhyolite extrusions, some of which were preceded by explosive eruptions that produced pyroclastic-flow and tephra-fall deposits. Some of the post-collapse rhyolite obsidian flows are exceptionally large, covering up to 800 km^2 with thickness as great as 300 m. The margins of a few have large lobate reentrants, indicative of emplacement against a glacier, including one dated at 110,000 yr (Christiansen, 2001; Richmond, 1986). Such an age suggests that a substantial glacier existed in the interior of Yellowstone during Marine

Oxygen-Isotope Stage (MIS) 5d, shortly after the end of the peak of the last interglaciation (Kukla *et al.*, 2002).

Ongoing basaltic intrusions keep the Yellowstone magmatic system hot and restless and drive a large hydrothermal system with its well-known surface geothermal expressions in Yellowstone National Park (Fournier *et al.*, 1994; Smith & Braile, 1994; White *et al.*, 1988). Hydrothermal explosions have formed craters on size scales of a few meters to a few kilometers during postglacial time. Historical crustal movements of centimeters per year over broad areas of the caldera (Dzurisin *et al.*, 1994; Wicks *et al.*, 1998) and millennial-scale cycles of inflation and deflation measured in meters to tens of meters around Yellowstone Lake and its outlet attest to the ongoing dynamic behavior of the Yellowstone magmatic-hydrothermal system (Locke & Meyer, 1994; Pierce *et al.*, 2002).

South of the eastern Snake River Plain, several basins in southeastern Idaho have been sites of recurring basaltic volcanism and minor rhyolitic volcanism during the Quaternary (Luedke & Smith, 1991). Lava flows in the Gem Valley area diverted the Bear River into the Bonneville Basin about 100,000 yr ago (Bouchard *et al.*, 1998), thereby increasing substantially the inflow to Lake Bonneville and sending the lake to its highest Pleistocene level and catastrophic overflow.

Margin of Colorado Plateau-Rio Grande Region

A broad arcuate belt of Quaternary volcanic loci wraps around the west and south sides of the Colorado Plateau and through the Rio Grande region to northeastern New Mexico (Luedke & Smith, 1991). Most are basaltic fields, but a few produced silicic rocks, including the long-lived Jemez field of north-central New Mexico.

Volcanic fields along the Colorado Plateau-Basin and Range boundary lie in areas of crustal extension. The Black Rock Desert contains chiefly basaltic cones and lava flows (Luedke & Smith, 1991), two of which erupted into the Sevier Desert arm of Lake Bonneville (Oviatt & Nash, 1989). Pavant Butte was formed about 15,500 ^{14}C yr B.P., when the lake was approaching its maximum level. Hydrovolcanic eruptions in water depths as great as 85 m created a broad submerged mound of tephra that consists of a complex sequence of lithofacies produced by pyroclastic fall, sediment gravity-flow, syneruptive water waves, and mass wasting (White, 1996). Once the mound breached the lake's surface, a subaerial pyroclastic cone was built and subsequently notched by waves during the stillstand at the Bonneville shoreline. Tephra-fall deposits of the eruption form a key stratigraphic marker as far as 100 km away and permit detailed correlation of shoreline and deeper-water deposits (Oviatt *et al.*, 1994). About 1000 yr later and 25 km farther south, eruptions at Tabernacle Hill built a small tuff cone and lava flow that spread as far as 3 km radially outward into the lake, which stood at or slightly below the recessional Provo shoreline (Oviatt & Nash, 1989). As a result of its emplacement in the lake, the flow's margin is locally pillowed, maintains an even altitude, and is locally buried in lacustrine sediments.

Quaternary basaltic and more mafic lava flows form many cinder cones in fields scattered along the faulted western margin of the Colorado Plateau in southwestern Utah and northwestern Arizona (Luedke & Smith, 1991). The western Grand Canyon region provides a 20-myr record of migration of basaltic volcanism 100 km northeastward, coincident with crustal extension (Wenrich *et al.*, 1995). Lava flows provide important datums for assessing rates of landscape processes including faulting and uplift, river downcutting, and soil formation (Hamblin, 1994; Lucchitta *et al.*, 2000). Cosmogenic ^{3}He, K/Ar, ^{40}Ar/^{39}Ar, and thermoluminescence ages constrain Quaternary downcutting rates of about 400 m/myr for the Colorado River in the Grand Canyon east of the faulted zone (Fenton *et al.*, 2001).

North-central Arizona's San Francisco volcanic field, which ranges in age from $>$6 myr to $<$1000 cal yr B.P., is rare among volcanic fields along the Colorado Plateau margin because of the broad compositional range of its lavas. In addition to hundreds of basaltic scoria cones and scattered monogenetic vents of intermediate and rhyolitic composition, it contains five long-lived composite volcanoes formed of varying proportions of andesite, dacite, and rhyolite (Ulrich *et al.*, 1989). Such assemblages, more akin to those of composite volcanoes in arc settings, are thought to result from a combination of fractional crystallization of basalt and partial melting of thick continental crust by basalt to produce rhyolite, which then mixes with basalt to form the intermediate magmas (Arculus & Gust, 1995). Vents of the volcanic field become younger in a northeastward direction, which led Tanaka *et al.* (1986) to hypothesize that the coincidence of basement lineaments and a thermal anomaly in the asthenosphere or lower lithosphere, coupled with southwestward movement of the North American plate, produced the observed patterns of vents and magma genesis.

Considerable interest has centered on the origin of the topography of San Francisco Mountain, one of the long-lived composite centers in the San Francisco field. Relatively smooth south, west, and north outer slopes rise more than 1 km above the Colorado Plateau, but a large horseshoe-shaped basin and valley incise the east slope. A $\sim$200,000-yr-old rhyolite dome postdates development of at least part of the basin and valley, which contain evidence of multiple glaciations. Besides erosion, proposed origins of the basin include collapse resulting from magma withdrawal or evisceration by debris avalanches, similar to that at Mount St. Helens in 1980 (Duffield, 1997; Holm, 1987). Depositional evidence for the latter has not been found to date.

The wide array of ages of volcanic landforms in the San Francisco volcanic field, combined with geochronologic data (Conway *et al.*, 1997; Tanaka *et al.*, 1986), has attracted studies aimed at using degree of landform change as a method of estimating age, especially of scoria cones by applying diffusion models (Conway *et al.*, 1998; Hooper & Sheridan, 1998). Such studies have shed details on the pattern and recurrence of eruptions in fields of monogenetic basaltic volcanoes. For instance, the SP cluster of vents north of San Francisco Mountain has displayed relatively steady-state volcanism during the Bruhnes Normal-Polarity Chron of about one volcano (typically a scoria cone and lava flow) per 15,000 yr. Information of this type allows probabilistic assessment of future events. For the SP cluster, Conway *et al.* (1998) estimate with 90% confidence that an eruption will occur within a delineated 250-km^2 area within the next 22,000–26,000 yr.

The southeastern margin of the Colorado Plateau is the site of numerous late Cenozoic volcanic fields, most with some Quaternary activity, that define the northeast-trending Jemez zone from south-central Arizona to northeast New Mexico (Luedke & Smith, 1991; Smith & Luedke, 1984). Among them, the Springerville volcanic field in east-central Arizona is similar to the San Francisco field in terms of an eastward age progression of vents through the late Cenozoic (Condit & Connor, 1996). The last major pulse of activity, which lasted from about 2.1 myr to 300,000 yr ago, began with tholeiitic basalt and changed to increasingly more alkalic basalt. The Zuni-Bandera field in west-central New Mexico apparently began erupting ~4 myr ago, although most exposed flows are of Quaternary age and were fed from a dense concentration of basaltic vents that form a northeast-trending alignment (Baldridge *et al.*, 1989a, b). The youngest activity, whose vent lay east of the alignment, produced the 50-km-long, ~1000 cal yr B.P. McCarty's flow, a tholeiitic tube-fed basalt characterized by a fresh glassy pahoehoe surface.

The Jemez zone intersects the Rio Grande Rift at the Jemez Mountains volcanic field, which was the site of two Pleistocene caldera-forming eruptions of high-silica rhyolite. Early studies of the products of these eruptions (Bandelier Tuff) and their calderas, were instrumental in illuminating the concepts of ash-flow tuffs (or large-volume pumiceous ignimbrites) and resurgent calderas (Smith & Bailey, 1968; Smith *et al.*, 1970). Additional work suggests that the caldera outlines of both eruptions were nearly identical, as were the locations of initial vents for the Plinian phases of each eruption (Self *et al.*, 1986). The eruptions occurred about 1.6 and 1.2 myr ago and produced important tephra marker beds in the southwestern United States (Izett, 1981; Izett & Obradovich, 1994). The younger was followed by extrusion of rhyolite domes as late as 500,000 yr ago. More than 400,000 yr of apparent dormancy followed before rhyolitic eruptions resumed about 50,000–60,000 yr ago (Reneau *et al.*, 1996). This long hiatus, coupled with petrologic evidence that the younger magmas were newly generated melts and not related to the older moat rhyolites, suggested to Wolff & Gardner (1995) that the Valles system may be entering a new cycle of eruptive activity.

The northeast end of the Jemez zone was the site of minor early Quaternary basaltic volcanism, as were areas south of the Jemez zone (Luedke & Smith, 1991). The Carrizozo field produced a 60-km-long flow that is probably only about 1000 yr old. Fields of basalt and more mafic lavas scattered south along the Rio Grande rift and adjacent Basin and Range province contain some well-known hydrovolcanic features, including Pleistocene maars up to 5 km in diameter in the Potrillo field on the Mexico-New Mexico border and Geromino field on the Mexico-New Mexico-Arizona border. The maars occur in basin settings and probably formed during times of wetter climate and higher water tables. Several of the maars are significant localities for mantle and lower crustal xenoliths (e.g. Kempton & Dungan, 1989).

Western Basin and Range Province

Numerous late Cenozoic volcanic fields, many with Quaternary eruptions, define a broad belt in the western Basin and Range province along the east side of the Sierra Nevada to southeastern California (Luedke & Smith, 1991). The fields lie in areas of crustal extension, and most of the fields produced only mafic rocks, but several erupted silicic rocks, including Long Valley caldera.

Several basaltic fields in the region have received attention in ongoing evaluations of volcanic and intrusive hazards in the Yucca Mountain area of southern Nevada, which has recently been selected to be the nation's high-level, nuclear-waste repository (Connor *et al.*, 2000; Smith *et al.*, 2002). About eight scoria cones and related lava flows that lie within 50 km of the repository site were formed about 1 myr, 300,000 yr, and ≤100,000 yr ago. Probabilistic hazard assessments based on the age, distribution, and structural control of past eruptions suggest that the annual probability of volcanic disruption of the planned repository is on the order of 10^{-8} to 10^{-7}. Work in these volcanic fields has produced spirited debate on several issues. Estimates of the age of the Lathrop Wells cone, the youngest vent near Yucca Mountain, have ranged from Holocene to more than 100,000 yr. A widely accepted age is about 80,000 yr on the basis of $^{40}Ar/^{39}Ar$ and surface-exposure dating techniques (Heizler *et al.*, 1999). Some workers have also argued that the Lathrop Wells cone produced multiple eruptions separated by significant time intervals (Wells *et al.*, 1990), rather than behaving in the monogenetic mode considered typical for such vents. Secular-variation paleomagnetic evidence and an inferred non-volcanic origin of scoria-rich deposits interpreted initially as primary tephra-fall deposits have cast doubt on the multiple-eruption hypothesis (Turrin *et al.*, 1991; Whitney & Shroba, 1991).

The Cima volcanic field, 180 km south of Yucca Mountain, and the Lunar Craters field, a similar distance north, are larger basaltic fields, each of which contains tens of vents of Quaternary age (Dohrenwend *et al.*, 1987). The availability of numerous dated lava flows and scoria cones ranging in age from >1 myr to ~15,000 cal yr has provided a natural laboratory for studies of landscape and drainage-system development (Dohrenwend *et al.*, 1986; Wells *et al.*, 1985). Rates of geomorphic change are initially slow on the lava flows owing to their high permeability. Mantling of flows with eolian sediment and formation of soils are important processes in reducing permeability and promoting runoff and development of drainage networks.

Two well-known basaltic scoria cones and flows of probable Holocene age, Pisgah and Amboy, lie south of the Cima field (Glazner *et al.*, 1991). Lava flows in these fields display a wide range in surface characteristics and have been used to test and calibrate numerous remote-sensing techniques for Earth and planetary studies (e.g. Arvidson *et al.*, 1993).

Salton Buttes, a cluster of 5 small rhyolite domes of Holocene age, lie in a pull-apart basin formed by right-lateral faulting in a zone of transform faults between the spreading center in the Gulf of California and the San Andreas fault (Herzig & Jacobs, 1994). A high-temperature geothermal system is present at shallow depth below the buttes. There may be other Quaternary volcanoes in the area that have been buried by sedimentary deposits of the Colorado River, which are accumulating at a rapid rate in this area of regional subsidence.

The Coso volcanic field, located near the southern end of the Sierra Nevada, also hosts an active geothermal system. Quaternary eruptions of basalt and rhyolite ranging in age from about 1 myr to ~40,000 yr reflect the familiar magmatic processes in areas of crustal extension: the emplacement of basalt into extending lithosphere; eruption of some of the basalt; and partial melting of crustal rocks by heat from unerupted basalt lodged in the lithosphere to form rhyolite (Duffield *et al.*, 1980). The relation between extension and magmatism is expressed as a time-predictable pattern, wherein the time interval between eruptions during the past 500,000 yr has been proportional to the volume produced during the previous eruption (Bacon, 1982). Extensional strain is thought to accumulate in roof rocks at a constant rate. Fractures relieve the strain and serve as conduits, or dikes, for rising magma, which erupts in a volume proportional to the width of the dikes. This model predicts that the next eruption of rhyolite at Coso is expected in 60,000 ± 35,000 yr, and that the next eruption of rhyolite should occur within the next 55,000 yr.

Several basaltic volcanic fields of Quaternary age lie along the east side of the Sierra Nevada north of Coso, and one contains a minor amount of rhyolite (Luedke & Smith, 1991). Lava flows from these fields provide chronologic control for stratigraphic and process studies, including a ~120,000-yr-old basalt lava flow in the Big Pine field that is instrumental in dating the local Sierran glacial record (Gillespie *et al.*, 1984). Farther east along the California-Nevada border, Ubehebe Crater, a classic tuff ring composed of base-surge deposits (Crowe & Fisher, 1973), may record the youngest known eruption in the southwestern part of the Basin and Range province. Tephra of Ubehebe Crater has a radiocarbon age of 140–300 cal yr (1-σ uncertainty) and provides a maximum-limiting age for the last surface rupture along the northern Death Valley fault (Klinger, 2001).

The Long Valley caldera and Mono-Inyo Craters of east-central California is by far the largest and currently most active volcanic system in the western Basin and Range province (Bailey, 1989; Bailey *et al.*, 1976; Hill *et al.*, 2002). Long Valley caldera, a 15- by 30-km elliptical depression, was formed by the eruption of about 600 km^3 (DRE) of rhyolitic ignimbrite and widespread tephra-fall deposits. $^{40}Ar/^{39}Ar$ ages of multiple samples of pumice from five localities allow precise dating of the caldera-forming eruption at 758,900 ± 1800 yr (Sarna-Wojcicki *et al.*, 2000). The proximity of the tephra-fall deposit and the Matuyama-Brunhes paleomagnetic reversal in many sedimentary sequences yields a calculated age of the reversal of 774,200 ± 2800 yr under assumptions of constant sedimentation rates. Recent investigations in the Long Valley area reveal details of the ~90-hr-long caldera-forming eruption. Vent position migrated around the caldera rim in a complex pattern, and ignimbrite and plinian-fall deposits were emplaced concurrently rather than sequentially as had been interpreted previously (Wilson & Hildreth, 1997). Caldera collapse was followed by formation of a central resurgent dome and episodic emplacement of surrounding rhyolite domes during middle to late Pleistocene time. From about 200,000–50,000 yr ago, eruptions of basaltic and silicic lavas covered much of the western caldera floor. Concurrent eruptions of rhyodacite from a cluster of vents built the composite volcano of Mammoth Mountain on the southwestern caldera margin.

Extending north from Long Valley caldera, the Inyo and Mono Craters define a 40-km-long chain of rhyolite vents that formed during the past 40,000 yr. About 20 vents in the chain produced eruptions during the past 5000 yr, including 10 that erupted between about 600 and 250 yr ago (Miller, 1985; Sieh & Bursik, 1986; Stine, 1987). Rhyolite dikes fed most of these eruptions. The dikes are thought to be accommodating the regional extensional strain that, prior to 40,000 yr ago, was being accommodated by range-front faulting along the nearby Sierra Nevada frontal-fault system (Bursik & Sieh, 1989).

Since 1978, elevated levels of seismicity, including intense earthquake swarms, more than 80 cm uplift of the resurgent dome, and increased efflux of carbon dioxide, have generated concern about future eruptive activity (Hill *et al.*, 2002). These observed signs of volcanic unrest are being caused by magma intrusion under the resurgent dome, the south moat of the caldera, and to depths as shallow as 3–4 km under Mammoth Mountain in the southwestern part of the caldera. The outcome of the ongoing quarter-century of volcanic unrest is uncertain, but the past history of the region suggests that the range of possible future eruption scenarios is wide in terms of vent location, magma type, explosivity, and volume (Hill *et al.*, 2002).

Western California

Two Quaternary eruptive loci in northern California lie west of the Sierra Nevada. Sutter Buttes are a cluster of andesitic to rhyolitic lava domes and shallow intrusive masses surrounded by an apron of fluvially reworked pyroclastic-flow and lahar deposits in the Sacramento River valley (Hausback & Nilsen, 1999). The Clear Lake locus is a long-lived center that has produced lavas from basalt to rhyolite, last erupted about 10,000 yr ago, and hosts an active geothermal system (McLaughlin & Donnelly-Nolan, 1981). It is thought to originate from mantle upwelling as the Mendocino triple junction migrates northward (Levander *et al.*, 1998). Resulting intrusion of basalt into the crust produces partial melting and generates a variety of intermediate and silicic magmas.

Alaska

Quaternary volcanic loci in Alaska lie in four tectonic regions (Fig. 6; Luedke & Smith, 1986). Most are in the

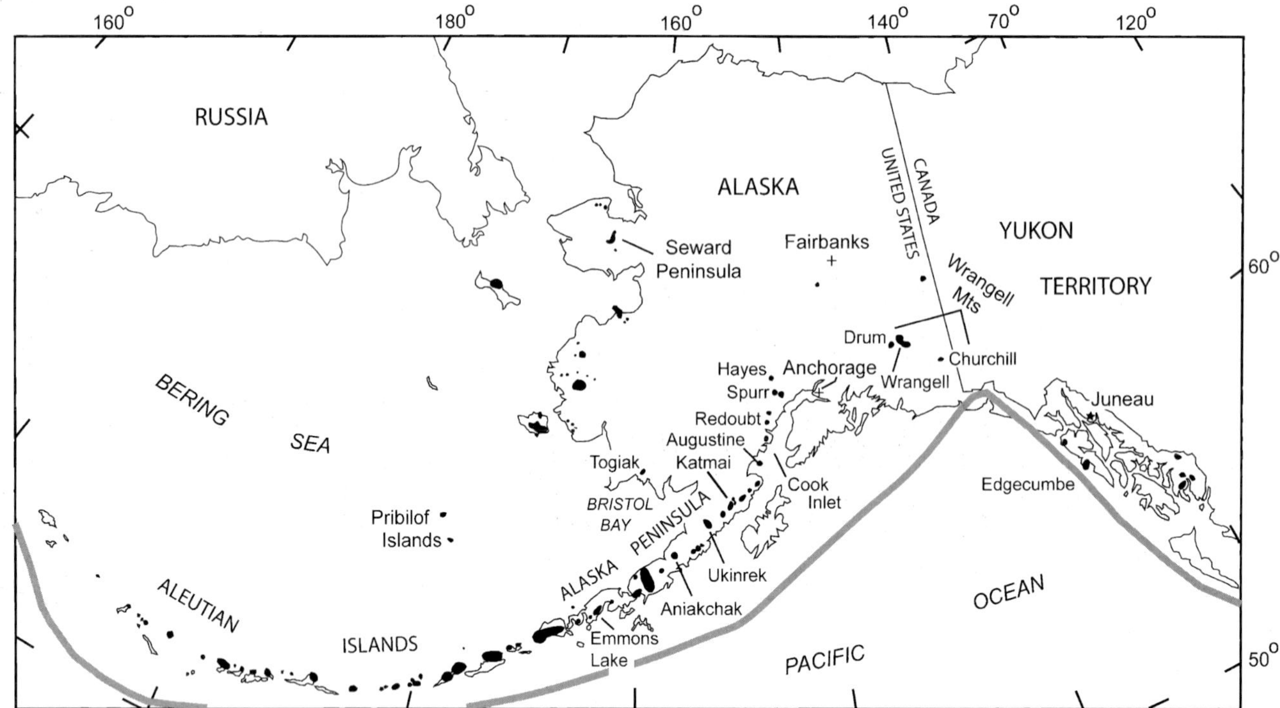

Fig. 6. Loci of volcanic vents of Quaternary age and tectonic-plate boundaries in Alaska (vent data from Luedke & Smith, 1986).

Aleutian volcanic arc, which stretches 2600 km west from central Alaska through the Alaska Peninsula (continental-arc segment; Miller & Richter, 1994) and Aleutian Islands (island-arc segment; Fournelle *et al.*, 1994), and result from subduction of the Pacific Plate beneath North America. The large composite volcanoes of the Wrangell Mountains of eastern Alaska lie in a tectonically complex area at the junction of the Aleutian arc and a system of transform faults that mark the boundary between the Pacific and North American plates (Miller & Richter, 1994). Isolated volcanoes in southeastern Alaska lie a short distance east of the transform boundary. The fourth region is a broad belt of basaltic volcanic fields that trends northeast from the Pribilof Islands in the Bering Sea to the Seward Peninsula in west-central Alaska (Moll-Stalcup, 1994).

Aleutian Volcanic Arc

The Aleutian arc is among the most active volcanic areas on Earth – hardly a year passes without an eruption at one of its volcanoes. Strategic interests following World War II focused volcanic studies on several Aleutian Islands during the 1950s (e.g. Coats, 1950). Opportunities to better understand relations among global tectonics, magmatic systems, and evolution of continents, as well as well-publicized eruptions of Cook Inlet volcanoes in 1989 and 1992, broadened geologic interest in the arc. These recent eruptions also highlighted concerns about hazards of Aleutian volcanic ash clouds to international aviation in the north Pacific region and elsewhere (Casadevall, 1994). About 80 major Aleutian composite volcanic centers have erupted in Holocene time and at least half of those have erupted during the past few centuries of recorded history (Miller *et al.*, 1998). The largest eruption on Earth during the 20th century was the 1912 caldera-forming eruption in the Katmai region of the Alaska Peninsula (Hildreth & Fierstein, 2000), one of about nine caldera-forming eruptions of postglacial age in the Aleutians that produced more than 10 km^3 of ejecta (Miller & Richter, 1994). Five of these events may each have erupted more than 50 km^3 of ejecta as ignimbrite and tephra-fall deposits, and thus have the potential for providing widespread tephrostratigraphic markers for use in geologic, archeological, and paleoecological studies (e.g. Begét, 2001).

The high level of eruptive activity in the Aleutian arc has provided numerous eruptions during recent decades for detailed study, which in turn has led to significant advances in volcanology. Eruptions at Augustine Volcano in 1976 and 1986 (Miller *et al.*, 1998), Ukinrek Maars in 1977 (Self *et al.*, 1980), Redoubt Volcano in 1989–1990 (Miller & Chouet, 1994), and the Crater Peak vent of Mount Spurr in 1992 (Keith, 1995), along with several small eruptions at more remote volcanoes in the central and western Aleutians, have provided opportunities to observe several key eruptive processes and their resulting deposits.

Pacific-North America Plate interactions have focused magmatism in the Aleutian region throughout Cenozoic time; the present-day Aleutian arc is a product of ongoing volcanism since the middle Miocene (Miller & Richter, 1994). Studies of major centers demonstrate that most have erupted recurrently throughout Quaternary time, although some, such as Augustine Volcano (Waitt & Begét, 1996) have a record of only latest Pleistocene and Holocene activity. A brief discussion of several Aleutian centers follows.

Aniakchak Volcano

Aniakchak is a relatively low (1.4 km), broad volcano surmounted by a circular 10-km-diameter caldera of late Holocene age. Neal *et al.* (2000) summarize past and recent studies as part of a volcano-hazards assessment. The caldera walls and radial ridges expose lava flows and volcaniclastic deposits of ancestral stratovolcanoes of Pleistocene age that predate caldera formation. One sequence is about 850,000–550,000 yr old, while the younger is about 440,000–10,000 yr old. About 20 distal tephra beds were produced by explosive eruptions of Aniakchak from 10,000 to 3500 ^{14}C yr B.P., but little record of these events is found near the volcano. A pyroclastic-flow deposit of suspected early Holocene age might document a caldera-collapse event, but the main caldera producer was a massive eruption about 3700 cal yr B.P. that vented 50–100 km^3 of compositionally zoned (dacitic to andesitic) ignimbrite and tephra-fall deposits. Ignimbrites swept out as far as 60 km from the volcano and flowed into both Bristol Bay of the Bering Sea and the Pacific Ocean.

As much as 15 m above sea level on the north shore of Bristol Bay, pumiceous sand beds lie immediately above distal tephra of the eruption. The sand beds are interpreted as deposits of tsunamis generated by the ignimbrite entering the bay 130 km away (Waythomas & Neal, 1998). Currently accepted models of ignimbrite deposition by progressive aggradation (e.g. Branney & Kokelaar, 1992), rather than by *en masse* emplacement, suggest that simple displacement of water by the ignimbrite was not, by itself, capable of generating a tsunami of the recorded height. The tephra-fall deposit of the climactic eruption occupies a relatively narrow area that trends slightly west of north across western Alaska and the Seward Peninsula, and is useful in archeological and paleoecological studies (Table 2; Begét *et al.*, 1992). It has been found more than 1000 km from the caldera as a layer 1–2 cm thick.

Tens of explosive and effusive eruptions occurred in the caldera after its formation. A ~100-m-deep lake existed in the caldera during some earlier eruptive episodes, but it drained rapidly sometime before 900 ^{14}C yr B.P., creating a catastrophic flood (Neal *et al.*, 2000; Waythomas *et al.*, 1996).

Katmai Volcanic Cluster

The remarkable eruption and caldera collapse in the Katmai region of the Alaska Peninsula in 1912 rank as the largest of

Table 2. Key tephra layers in Alaska and Yukon Territory.

Name	Source Vent or Area	Age[a]	Known Distribution	References[b]
1912 Katmai	Novarupta	A.D. 1912	E Aleutian Peninsula, lower Cook Inlet, Kodiak Island	7
White River Ash (E)	Mount Churchill?	1.2	E from Churchill-Bona, Alaska; S Yukon and W Northwest Territories	10, 13
White River Ash (N)	Mount Churchill?	1.8	E-central Alaska, W-central Yukon Territory	10, 13
Aniakchak	Aniakchak caldera	~3.7	N from Aniakchak, W Alaska Seward Peninsula	3
Hayes; multiple beds	Hayes volcano	~4.0	Several lobes, S to lower Cook Inlet, NE to central Alaska Range, north flank of Alaska Range	2, 14
Fisher-Funk	Fisher caldera	~10.3	Unimak Island, SW Alaska Peninsula	6
Mt. Edgecumbe; tephra set ME	Mount Edgecumbe	~13	N; Sitka, Juneau, Glacier Bay, Yakutat Bay	4, 15
Dawson	Emmons Lake volcanic center?	27	Alaska Peninsula, NE to W-central Yukon Territory	8, 16, 18
Old Crow	Aleutian arc?	120–160	Seward Peninsula, southwest Alaska, central Alaska, W Yukon Territory	1, 9, 11, 12, 17
Sheep Creek	Wrangell Mts	~190	E-central Alaska, W-central Yukon Territory	1, 5, 11

Note: See Fig. 7 for location of volcanoes and geographic areas of tephra distribution.

[a] In thousands of calendar years, unless A.D.

[b] References: 1, Begét (2001); 2, Begét *et al.* (1991); 3, Begét *et al.* (1992); 4, Begét & Motyka (1998); 5, Berger *et al.* (1996); 6, Carson *et al.* (2002); 7, Fierstein & Hildreth (1992); 8, Froese *et al.* (2002); 9, Kaufman *et al.* (2001); 10, Lerbekmo & Campbell (1969); 11, Muhs *et al.* (2001); 12, Preece *et al.* (1999); 13, Richter *et al.* (1995); 14, Riehle *et al.* (1990); 15, Riehle *et al.* (1992 a, b); 16, Waythomas *et al.* (2001); 17, Westgate *et al.* (1985); 18, Westgate *et al.* (2000).

the 20th century on the basis of quantity of ejecta (13 km^3 DRE). This 60-hr event occurred within a dense concentration of stratovolcanoes and other vents called the "Katmai volcanic cluster" (Hildreth & Fierstein, 2000). Such volcanic clusters occur elsewhere in the Aleutians, but Katmai is the best known. Within 15 km of the 1912 vent, called Novarupta, are 5 andesite to dacite stratovolcanoes with Holocene eruptions. Four of these volcanoes were active from middle or late Pleistocene to Holocene time, whereas one appears to be entirely of Holocene age. Except for events related to the 1912 eruption, only one historical eruption is known, the construction of New, or Southwest, Trident volcano between 1953 and 1974.

Detailed investigations of the Katmai cluster reveal how rare an eruption like that of 1912 is in the Aleutians (Fierstein & Hildreth, 1992; Hildreth & Fierstein, 2000). Most extraordinarily, the eruption was focused at a new vent off the long-lived stratovolcanoes, whereas caldera collapse occurred on the summit of one of them, Mount Katmai, 10 km away. A high level of seismicity accompanied and immediately followed the eruption, including 50–100 earthquakes $\geq$M 5.0 and 3 earthquakes M 6.8 to 7.0, which together released the energy of an M 7.4 earthquake. Most of this energy release is thought to be tied to caldera collapse. The products of the eruption include andesite and dacite, which together account for about 40% of the erupted volume, and the majority high-silica rhyolite, the only known occurrence of such rhyolite in the Quaternary Aleutian arc. At the Novarupta vent, rhyolite erupted first, followed by an intricate mixture of dacite, andesite, and minor rhyolite, and then chiefly dacite; final explosive eruptions involved an andesite-dacite mixture. Some unknown time later, dacite domes appeared in Katmai caldera, while a rhyolite dome was extruded in the Novarupta vent. The model that best explains the 1912 eruption requires withdrawal of considerable magma from a shallow zoned reservoir beneath Mount Katmai to account for caldera collapse. To drive the eruption at Novarupta 10 km distant, magma must have migrated from the chamber through a sill and then dike to the vent.

Recent Eruptions

Well-monitored and well-studied eruptions in the Cook Inlet region in 1989–1990 at Redoubt Volcano (Miller & Chouet, 1994) and in 1992 at the Crater Peak vent of Mount Spurr volcano (Keith, 1995) provided new insights into several volcanic processes, especially interactions of pyroclastic fallout and flows with snow pack and glaciers (Redoubt) and tephra-cloud dynamics (Spurr).

Redoubt Volcano. Redoubt is a glacier-clad, basaltic to dacitic stratovolcano, which has a 900,000-yr eruptive history. A large debris avalanche, similar in volume to the 1980 debris avalanche at Mount St. Helens, occurred early in postglacial time (Begét & Nye, 1994). Numerous lahar deposits and lithic-tephra layers of Holocene age probably record rebuilding of the edifice following the avalanche.

The 1989–1990 eruptions began with several brief explosions in December that largely destroyed a lava dome formed in 1966, followed by 4 months of repeated lava-dome growth and collapse. The eruptive vent lay near the head of 9-km-long Drift Glacier, which flows through a steep narrow valley for about 5 km before broadening into a piedmont lobe. Initial explosions produced mixed avalanches of snow, glacier ice, pyroclasts, and other rock debris that swept as far as 14 km down the north and south flanks of the volcano (Pierson & Janda, 1994; Trabant *et al.*, 1994; Waitt *et al.*, 1994). The impact and loading of pyroclastic material on steep, thickly snow-covered slopes probably triggered the avalanches. About 65 million m^3 of snow and ice was removed from the breached summit crater at the glacier head and from the steep reaches below. Resulting deposits were ice diamicts composed of gravel-sized clasts of glacier ice, rock, and pumice in a matrix of snow and ice grains, ash, and frozen pore water. Also present were huge tabular snow clasts and ice blocks. Lithic and pumice clasts constituted only about 10–20% of the deposit. The relatively low content of hot pyroclasts could not generate sufficient melt water to produce debris flows, although a trailing watery flood terraced the diamicts.

Eruptions during the next few months consisted of lava-dome growth punctuated by repeated collapse and generation of lithic pyroclastic flows (Gardner *et al.*, 1994). Such flows continued to erode the glacier thereby developing steep-walled ice canyons in the piedmont lobe. High contents of hot material relative to snow and ice produced huge quantities of melt water that mixed with abundant debris to form lahars, or volcanic debris flows, that swept far down valley to Cook Inlet. By the time eruptions ended, more than 100 million m^3 of snow and ice had been removed, including the upper 3 km of Drift Glacier. Thick deposits of pyroclastic flows and lahars locally mantled the piedmont lobe of the glacier.

Crater Peak. Three 4-hr-long subplinian eruptions of modest volume ($\sim$12–15 million m^3 DRE each of basaltic andesite) burst from the Crater Peak vent on Mount Spurr in June, August, and September 1992 (Eichelberger *et al.*, 1995). Detailed field and laboratory studies of the eruptive products (Gardner *et al.*, 1998; McGimsey *et al.*, 2001; Neal *et al.*, 1995) were combined with data from several satellite remote-sensing techniques to yield the following three-stage evolutionary model of the volcanic ash cloud for 5 days after the eruption (Rose *et al.*, 2001). During an eruption and for 1–2 hr thereafter, clouds grew rapidly in area, partly through initial lateral spreading by gravity flow but mostly by wind advection. About 30 min after an eruption stopped, all coarse ash- and lapilli-sized particles had fallen out blanketing areas near the volcano with tephra loads of 10–250 kg/m^2. The second stage lasted for about one day, during which the cloud continued to grow in area, but its fine-particle ($<$25 μm) concentration decreased by an order of magnitude or more. Such rapid fallout of fine ash in at least two of the eruptions can only be explained by particle aggregation, which resulted in deposition of secondary thickness maxima 150–350 km downwind. Ice from magmatic, hydrothermal, and atmospheric sources probably played an important role in formation of aggregates. Sulfur dioxide released by the eruption was initially sequestered in ice and its highest concentration was measured on the second day after the eruption, as ice evaporated during fallout of aggregates. The third stage lasted for several more days, during which the cloud drifted thousands of kilometers. Its ash concentration decreased slowly as the remainder of fine ash, amounting to only a few percent of the total volume of erupted magma, fell out to form very thin deposits.

Wrangell Mountains

The Quaternary volcanoes of the Wrangell Mountains lie chiefly at the northwest end of a long-lived (~25 myr), dominantly andesitic, volcanic field that extends into the western Yukon of Canada (Miller & Richter, 1994). This distribution, coupled with few dated lavas in the field younger than about 200,000 yr, is thought to reflect a decreasing rate of subduction through time as the transform plate boundary shifted westward (Fig. 6; Richter *et al.*, 1990). Through this process, the terrain south and east of the Wrangells that was being subducted and driving volcanism became decoupled from the Pacific Plate. Subduction and related volcanic activity then may have ceased, or nearly so. Seemingly at odds with this model is the occurrence of two large-volume explosive eruptions of late Holocene age, discussed in a later section, from suspected vents in the Bona-Churchill area of the St. Elias Mountains at the eastern end of the Wrangell volcanic field.

The best-known volcanoes in the Wrangell field, Mounts Wrangell and Drum, are large (10^3 to 10^4 km^3), glacier covered, broadly shield shaped, and composed chiefly of lava flows (Richter *et al.*, 1990, 1994). Each center has been recurrently active for hundreds of thousands of years during early to middle Pleistocene time. Evidence is sparse for eruptive activity in the last 100,000 yr, but, owing to extensive glacier cover, the potential for preservation of evidence of minor eruptions is small.

In addition to thick sequences of lava flows that form the massive volcanoes, other aspects of the history of the Wrangell field are recorded in deposits of large debris avalanches and lahars in the lowland south and west of Mounts Drum and Wrangell. The ~300,000-yr-old Chetaslina mass-flow deposit consists of classic debris-avalanche facies, including a proximal block-rich facies with blocks up to tens of meters in size and a medial mixed facies in which blocks are suspended in a lahar-like matrix (Waythomas & Wallace, 2002). About 40 km from source, the 4-km^3 avalanche entered a major river valley; debris flows and hyperconcentrated flows continued farther downstream. It is not known whether the transformation to far-traveled debris flows was the result of mixing with the river water or of temporary damming and outburst flooding.

Southeast Alaska

Several Quaternary eruptive centers lie along the Fairweather transform fault zone in southeast Alaska (Luedke & Smith, 1986). The majority are small-volume basalt fields, but the Mount Edgecumbe volcanic field is a long-lived composite center (Riehle *et al.*, 1992a). The bulk of postglacial activity occurred about 13,000 cal yr B.P. (calibration of radiocarbon ages in Begét & Motyka, 1998) and persisted for centuries to a millennium. Products encompass a wide range in composition from basalt to rhyolite and total about 5–6 km^3 of magma (Riehle *et al.*, 1992a, b). A widespread dacitic tephra of this sequence provides an important stratigraphic marker in postglacial deposits over most of southeastern Alaska (Table 2).

West-Central Alaska

A broad (500-km-wide) belt of basaltic volcanic fields of Quaternary age extends about 1000 km northeast from the Pribilof Islands in the Bering Sea across western Alaska to the Seward Peninsula (Moll-Stalcup, 1994). Several of these fields each cover thousands of square kilometers and include vents and lava flows of a variety of ages. But others consist of a single shield volcano or scoria cone and lava flow. One particular eruption of several recognized in the Togiak River valley occurred in a lake melted into glacier ice at least 300 m thick. Hydrovolcanic eruptions filled the lake with hyaloclastite, which in turn was capped by a subaerial lava flow (Kaufman *et al.*, 2001). The resulting flat-topped form, called a tuya, is about 2.5 by 6 km. The capping basalt lava flow has a ^{40}Ar/^{39}Ar age of about 260,000 yr, indicating that the coeval glacier probably dates from MIS 8. Some of the basaltic fields in western Alaska and on the Seward Peninsula have had Holocene eruptions as shown by radiocarbon ages and by little-weathered cones, lava flows, and maars.

Alaskan Tephrochronology

During the past few decades, an increasing number of regionally important tephra layers of Quaternary age have been recognized (Fig. 7; Table 2). They are now used widely in stratigraphic studies, especially of sequences of eolian and glacial deposits in central and eastern Alaska and Yukon Territory (e.g. Preece *et al.*, 1999) and in southern Alaska (e.g. Carson *et al.*, 2002; Riehle, 1985). Many of the deposits found in interior Alaska and the Yukon are dacitic to rhyolitic in composition and typically centimeters to decimeters thick hundreds of kilometers from possible source vents. Such units appear to be products of voluminous explosive eruptions of Aleutian or Wrangell volcanoes. Perhaps many are related to caldera-forming eruptions for which evidence abounds, especially in the Aleutian arc. Chemical and mineralogical data allow differentiation of two families of tephra, termed type I and II (Preece *et al.*, 2000). Type I tephras are characterized by relatively lower crystal content, with orthopyroxene and clinopyroxene, and well-developed Eu anomaly in rare-earth-element profiles. Type II tephras are more crystal rich, with orthopyroxene and amphibole, and have either a weakly

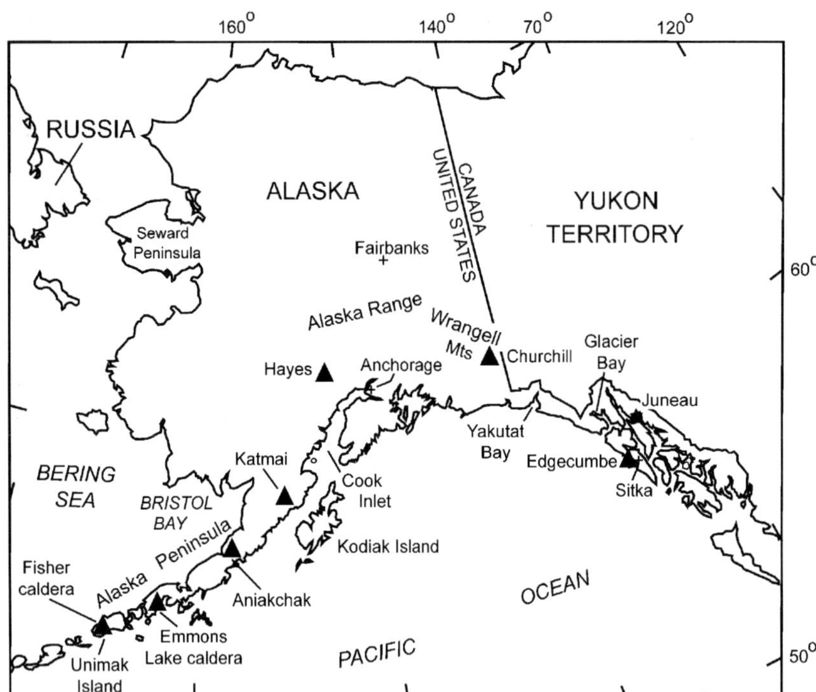

Fig. 7. Alaskan volcanoes that have produced widely recognized Quaternary tephra layers and geographic areas mentioned in text and Table 2.

developed or no Eu anomaly. The source of most tephra beds is not known, but type I tephras are thought to come from Aleutian sources and type II from Wrangell sources. Three widespread and voluminous tephras are discussed below.

Old Crow tephra is a light-colored, fine-grained ash bed found in Pleistocene sediments across central Alaska from the Seward Peninsula to western Yukon Territory (Westgate *et al.*, 1985). It has recently been described from southwest Alaska as well (Kaufman *et al.*, 2001). It is as much as 30 cm thick in the Fairbanks area, 10 cm thick on the Seward Peninsula, and 1–5 cm thick in the western Yukon, and its volume is likely on the order of 50 km^3. A fission-track age of about 140,000 ± 10,000 yr (Preece *et al.*, 1999) and stratigraphic relations between the tephra and glacial (Begét & Keskinen, 2003) and interglacial (Muhs *et al.*, 2001) deposits near Fairbanks suggest emplacement late in MIS 6 or early in MIS 5. The exact source of Old Crow tephra is unknown, but is likely from the Alaska Peninsula or Aleutian Islands on the basis of its chemical composition.

Dawson tephra is found in west-central Yukon Territory and is similar in appearance and composition to the much more widespread Old Crow tephra (Froese *et al.*, 2002; Westgate *et al.*, 2000). But it is substantially thicker, as much as 30 cm vs. 1–5 cm for the Old Crow. The Dawson, dated by radiocarbon at 27,000 cal yr B.P., is found chiefly in loess that was deflated from nearby outwash deposits. It was erupted from an Aleutian source, perhaps Emmons Lake volcanic center in southwest Alaska Peninsula, >700 km distant (Waythomas *et al.*, 2001). Its relatively young age and restricted distribution, although far from source, suggest that the tephra-fall zone may be relatively narrow. Regardless, the 30-cm thickness at such a great distance implies a large-magnitude eruption.

The Holocene White River ash comprises two broad lobes of tephra, one extending northward and the other eastward from the Mount Bona-Mount Churchill area of the eastern Wrangell Mountains, close to the Alaska-Yukon border (Lerbekmo & Campbell, 1969). The northern lobe is dated at about 1800 cal yr B.P., the more voluminous eastern lobe at about 1200 cal yr B.P. (Lerbekmo *et al.*, 1975). Together they constitute a bulk volume of 25–30 km^3. Recent geologic and geochemical evidence support Mount Churchill as the source of the eastern and presumably both lobes of the deposit (Richter *et al.*, 1995).

Hawaiian Islands

More than 100 volcanoes of the Hawaiian Ridge-Emperor Seamounts volcanic chain decrease in age progressively toward the southeast, where the active subaerial shield volcanoes on the Island of Hawai‘i, Kīlauea and Mauna Loa, and the active submarine volcano, Lō‘ihi Seamount, lie above the Hawaiian hot spot (Fig. 8; Clague & Dalrymple, 1987, 1989). As the Pacific Plate moves northwestward away from the hot spot, the volcanoes slowly subside owing to thermal aging of the lithosphere and isostatic depression in response to the crustal load imparted by growth of the volcanoes. As a result, most of the volcanoes in the chain now lie below sea level, many capped by a thick sequence of coral. The latest Tertiary and Quaternary volcanoes of the Hawaiian Islands have received the closest scrutiny, most recently by submarine investigations (e.g. Takahashi *et al.*, 2002).

Growth Stages of a Hawaiian Volcano

During the past half-century, detailed geologic mapping, geochronologic, geomorphic, petrologic, marine geologic, and geophysical studies of the Hawaiian Islands and surrounding sea floor have greatly refined the multi-stage eruptive and erosional evolution of a typical shield volcano (Fig. 9; see summaries by Clague & Dalrymple, 1987, 1989; Langenheim & Clague, 1987; Moore & Clague, 1992; Peterson & Moore, 1987; Takahashi *et al.*, 2002). A distinct association of lavas and eruptive styles characterizes each eruptive stage.

A pre-shield stage is inferred on the basis of current activity at Lō'ihi Seamount, which lies about 30 km southeast of the Island of Hawai'i (Figs 9 and 10; Malahoff, 1987). Lō'ihi is a moderately steep-sided edifice about 3 km high; the rim of its summit caldera is at a depth of about 1 km. Early lavas are alkalic basalt, whereas younger lavas are of tholeiitic and transitional character. Recent deep submarine investigations reveal evidence of a pre-shield stage of Kīlauea that began more than 275,000 yr ago and lasted more than 100,000 yr (Lipman *et al.*, 2002). The evidence consists of a large volume of volcaniclastic debris shed from the early alkalic edifice. At least in Kīlauea's case, the pre-shield stage may account for significantly more than the few percent of the ultimate volume of a Hawai'ian shield thought previously.

A shield-building stage of chiefly tholeiitic basalt volcanism constructs the bulk (>95% by some estimates) of each edifice over a period of roughly 1 million years. Eruptions build a broad submarine volcano of pillow lava several kilometers high that eventually breaches the sea surface. As the volcano nears the sea surface, explosive interaction of erupting lava with water creates large quantities of hyaloclastite that is transported seaward by waves, currents, and landslides. Slopes of 10–20° are typical for the submarine part of the edifice. The subaerial part of the shield, largely composed of fluid lava flows, but including an unknown volume of intruded magma, then grows on the clastic platform and may reach an altitude of more than 4 km above sea level. Although most eruptions are effusive, historical examples and the geologic record offer numerous examples of past explosive eruptions, many tied to hydrovolcanic processes (e.g. Fiske *et al.*, 1999; Mastin, 1997; McPhie *et al.*, 1990). The volumes of individual shields range from $\sim 1 \times 10^4$ km^3 to $>4 \times 10^4$ km^3, with 80 to more than 95% of the total below sea level (Bargar & Jackson, 1974). For some perspective on their great size, each shield is roughly one-tenth the total estimated volume of the Pacific Northwest's Columbia

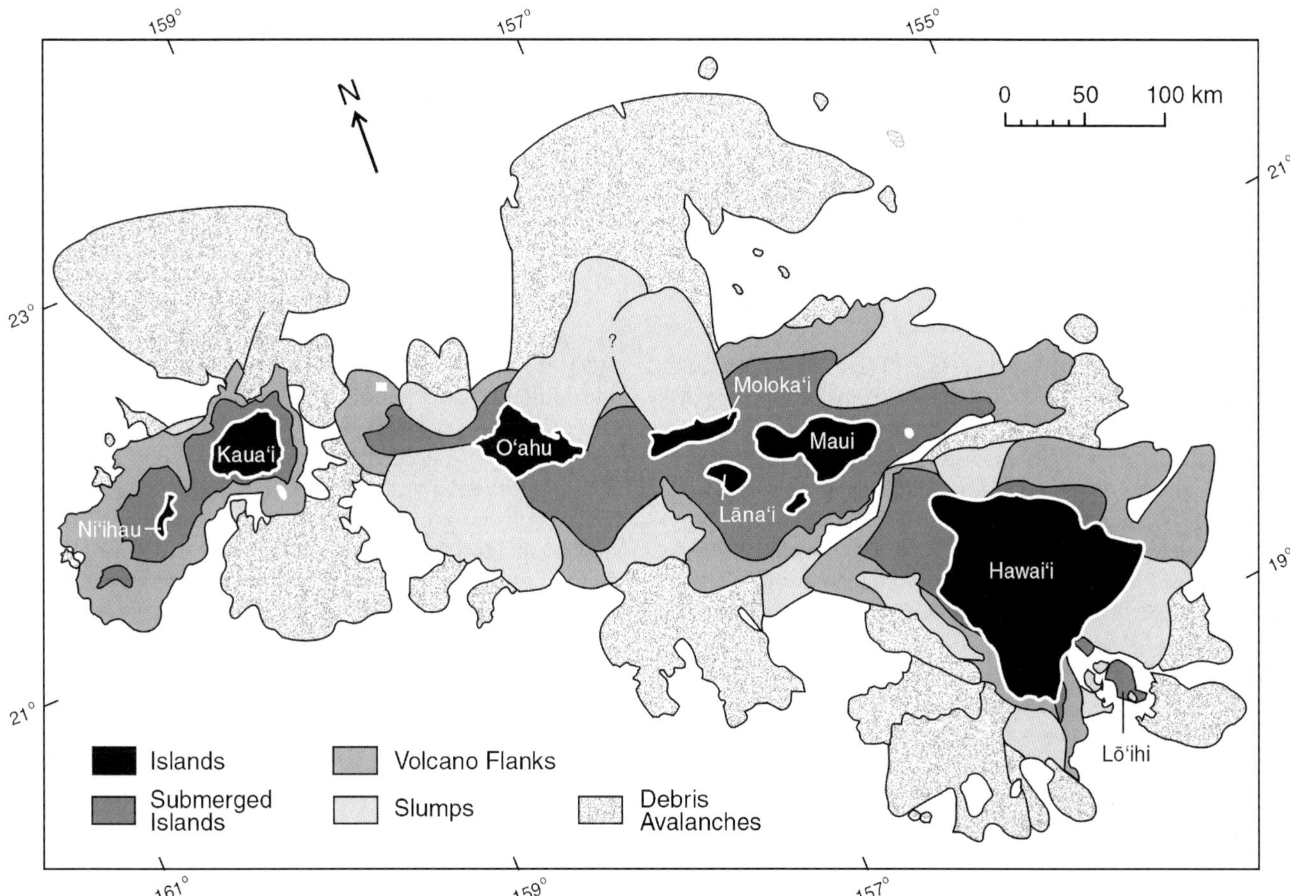

Fig. 8. Southeast end of the Hawaiian Ridge-Emperor Seamounts volcanic chain showing major geologic features (modified from Moore et al., 1994).

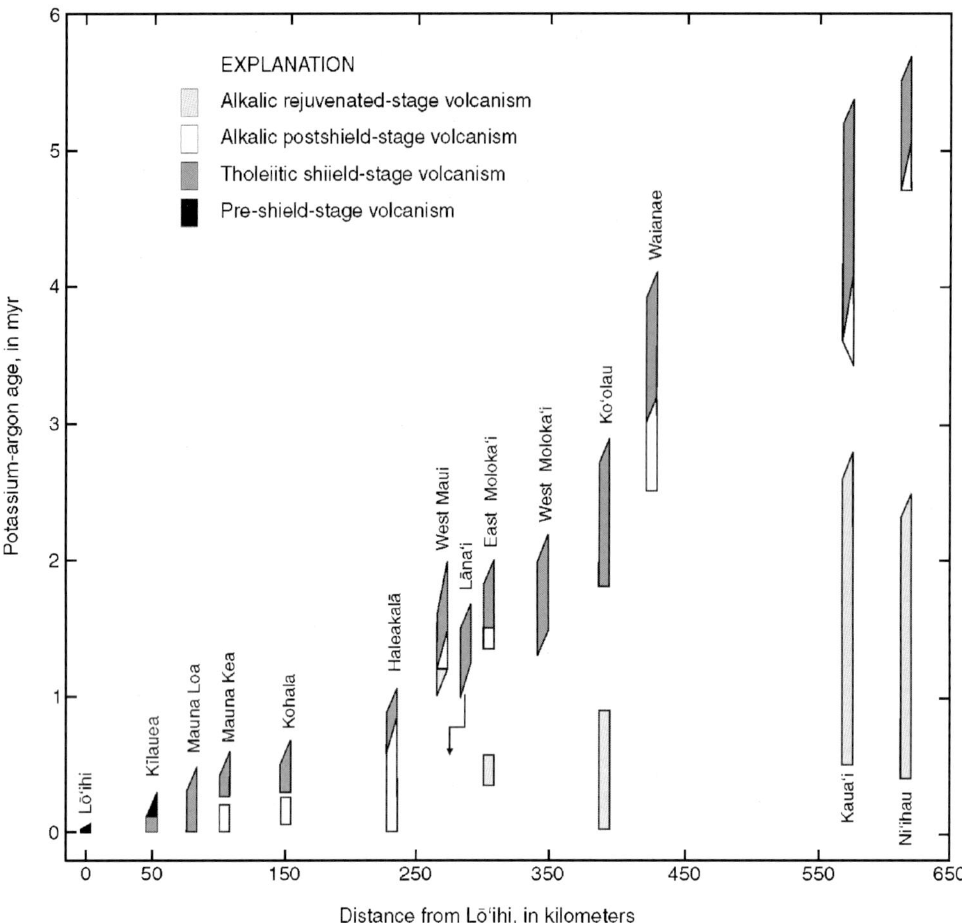

Fig. 9. Late Cenozoic history of the Hawaiian volcanoes (modified from Clague & Dalrymple, 1987, Fig. 1.6). Patterns show duration of stages of development for individual shield volcanoes. Angled lines show age or geologic uncertainty. A pre-shield stage, which is underway at Lō'ihi Seamount about 30 km offshore of the southeastern coast of the Island of Hawai'i, has also been recognized recently at Kīlauea (Lipman et al., 2002). Extended post-shield stage for Haleakalā from Sherrod et al. (2003).

River Basalt Group of Miocene age (Tolan *et al.*, 1989). The great crustal loads imposed by their rapid and massive growth causes isostatic subsidence that continues at a decreasing rate for perhaps as long as one million years after eruptions cease (Moore, 1987). Kīlauea and Mauna Loa are still in the shield-building stage (Rhodes & Lockwood, 1995; Tilling & Dvorak, 1993; Wolfe & Morris, 1996) and are subsiding at rates of about 2–3 mm/yr on the basis of tide-gage records and dating of submerged coral reefs (Moore *et al.*, 1996).

Lavas of the postshield, or capping, stage account for about 1% of the total edifice volume. Eruption rates of alkalic basalt and later higher-silica differentiates are considerably lower than those of tholeiite during the shield-building stage. Large scoria cones and tephra blankets, coupled with relatively more viscous lava flows, construct a summit cap that is somewhat steeper than the original shield, with slopes up to 20°. The frequency of eruptions declines and eruptive activity finally ends. Mauna Kea began its postshield stage about 200,000–250,000 yr ago, began producing more silicic hawaiitic lavas 70,000 yr ago, and last erupted about 4400 ^{14}C yr B.P. (Wolfe *et al.*, 1997). The average eruption rate during the latter period was about one order of magnitude lower than that during the former.

As eruption rate declines substantially during the postshield stage, erosion and subsidence ultimately exceed the rate of volcano growth. Recent investigations at East Maui volcano suggest that the considerable erosion displayed by the creation of Haleakalā Crater probably occurred in less than 10^5 yr, and that postshield volcanism continues (Sherrod *et al.*, 2003). The volcano has erupted tens of times in the Holocene (Bergmanis *et al.*, 2000). During its postshield stage, Mauna Kea was subjected to multiple glaciations (Wolfe *et al.*, 1997). Now-subsided shields that once reached above about 3800 m, the estimated equilibrium-line altitude of glaciers during MIS2 and earlier glacial stages (Porter, 1979), would also have been subjected to episodes of glacial erosion. Windward slopes of the volcanoes typically receive several meters of annual rainfall, locally more than 7 m. Intense weathering, canyon incision, shoreline erosion, eustatic sea level changes, and mass wasting accompanied by ongoing subsidence reduce the smooth shields to highly

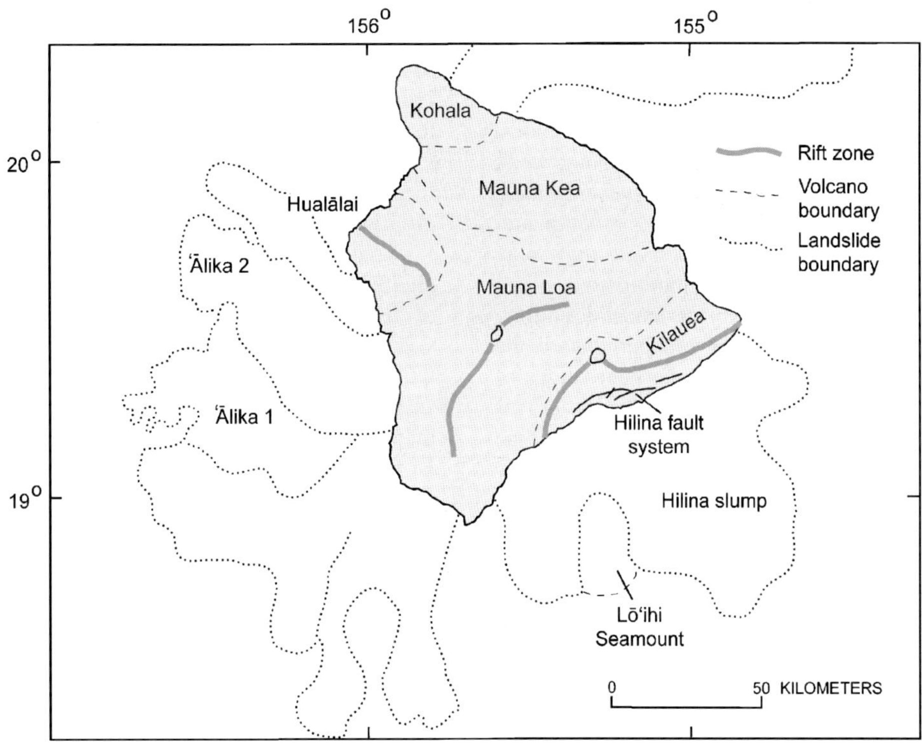

Fig. 10. Island of Hawaiʻi showing five major subaerial shield volcanoes and summit calderas on Mauna Loa and Kīlauea; the submarine volcano, Lōʻihi Seamount; and submarine slumps and landslides (modified from Lipman et al., 2002, Fig. 1).

dissected landforms. Little recognizable original form remains as shown by the condition of volcanoes on islands such as Oʻahu and Kauaʻi whose major shield building occurred about 2–5 myr ago (Clague & Dalrymple, 1989). Continued thermal contraction of the lithosphere that was heated during its time over the hot spot causes ongoing subsidence as volcanoes evolve to coral atolls and ultimately guyots and seamounts.

On some, but not all, of the shield volcanoes in the Hawaiian Islands, a rejuvenated stage of volcanism follows 10^5 to 10^6 yr of erosion and quiescence. This stage, which may last several million years, is characterized by sporadic, small-volume eruptions of alkalic basalt and more-silica-poor lavas from vents not typically associated with central conduits or rift zones of the shield or postshield stages. The scoria cones, tuff cones, and lava flows of the Pleistocene Honolulu Series on O'ahu are well-known examples of rejuvenated-stage volcanism (MacDonald *et al.*, 1983). Rejuvenation followed growth of the Koʻolau shield in late Pliocene time without an intervening postshield stage (Fig. 9; Langenheim & Clague, 1987).

Giant Submarine Landslides

Side-scan sonar imagery of the seafloor around the Hawaiian Islands obtained during the 1980s and 1990s revealed clearly what had been suspected from earlier bathymetric surveys – the presence of numerous (about 70) giant landslides on the submarine flanks of the volcanoes (Figs 8 and 10; Moore *et al.*, 1994; Smith *et al.*, 1999). The areas (up to 10^4 km^2) and volumes (up to 10^3 km^3) of the slides dwarf landslides on subaerial volcanoes. They represent two types of flank failure, slump and debris avalanche.

Slumps are gradual mass movement in which the flanks maintain structural coherence as blocks kilometers thick creep and lurch seaward along relatively shallow (<10 km), landward-dipping decollements (Lipman *et al.*, 2002). Evidence of active slumping of the south flank of Kīlauea (Fig. 10) includes historical earthquakes approaching magnitude 8, high seaward-facing arcuate scarps of the Hilina fault system, ongoing surface deformation (e.g. Cannon & Bürgmann, 2001; Swanson *et al.*, 1976), and numerous submarine benches, folds, and thrust faults (e.g. Hills *et al.*, 2002).

In contrast, debris avalanches are shallower and longer than slumps and move much more rapidly (Moore *et al.*, 1994). They typically have well-defined arcuate headwalls tens of kilometers across and can travel 100 to more than 200 km over very gentle gradients (Figs 8 and 10). Proximal portions contain kilometer-size megablocks, whereas more distal portions are characterized by hummocky terrain similar to that of subaerial volcanic debris avalanches. Whether or not any of the prominent subaerial scarps on Hawaiian shields are headwalls of debris avalanches, as opposed to slumps, remains unresolved. Recent investigations of the precipitous north shore of Molokaʻi suggest that it is not the headwall of the 1.5-myr-old Wailau debris avalanche, but more likely a fault scarp similar to Kīlauea's Hilina fault (Clague &

Moore, 2002). Apparently little, if any, of the headwall for the avalanche was above sea level at the time of failure.

Clague & Moore (2002) review the numerous factors that have been hypothesized to contribute to the instability of the unbuttressed flanks of Hawaiian shield volcanoes. Forceful dike injection into rift zones; sliding along weak zones within fragmental deposits of submarine eruptions, marine sediments, or cumulate dunites; and groundwater pressurization caused by magma intrusion all may play a role, but the exact mechanisms contributing to a given failure remain elusive. The age of giant landslides along the Hawaiian-Emperor chain becomes younger toward the southeast; they are typically coeval with the later stages of shield growth when the volcanoes are active and subject to intrusion (Moore *et al.*, 1994). This relation suggests that large slumps and debris avalanches of Quaternary age lie chiefly on the islands southeast of O‘ahu, from which the great Nuuanu landslide was spawned at about the Plio-Pleistocene boundary (Fig. 8; Herrero-Bervera *et al.*, 2002). The youngest giant debris avalanches in the Hawai‘ian chain are probably the ~100,000-yr-old ‘Ālika landslides on the west flank of Mauna Loa (Fig. 10; Moore & Chadwick, 1995).

An expected consequence of large submarine landslides is a disturbance of the sea surface and generation of tsunami. The M7.2 earthquake in November 1975 caused by sudden seaward movement on the Hilina slump (Fig. 10) generated a devastating tsunami that reached almost 15 m above sea level locally and about 1 m on more distant islands (Tilling *et al.*, 1976). The giant debris avalanches may produce substantially larger tsunami. Moore & Moore (1988) interpret deposits containing coral and other marine fossils at altitudes more than 300 m above present sea level on several islands as evidence of such giant waves. Their interpretation of the origin of some of the deposits has been challenged (Keating & Helsley, 1999). In addition, a model proposed recently suggests that localized uplift of the central Hawai‘ian Islands may account for the high-altitude coral (Grigg & Jones, 1997). But the landslide-tsunami link appears to be strengthening as investigations proceed. In one case, detailed analysis of cores of sediment that bury an avalanche deposit, as well as of turbidites in more distant cores, support the case that the ~100,000-yr-old ‘Ālika 2 debris avalanche (Fig. 10) off the west coast of Hawai‘i was responsible for generating a giant wave that struck L̄ana‘i (McMurty *et al.*, 1999) and perhaps other places in the Pacific basin (Young & Bryant, 1992). Mathematical models of tsunami that use input parameters from giant Hawaiian avalanches generate waves of the order inferred from local geological evidence, and forecast waves meters to tens of meters high striking the west coast of North America about 5 hr after an avalanche (Satake *et al.*, 2002).

Summary

Quaternary volcanism in the United States followed patterns well established by late Cenozoic tectonic elements (Luedke & Smith, 1991). In the conterminous U.S., eruptions occurred along the length of the Cascade arc and widely around the Basin and Range province and southern margin of the Colorado Plateau to northeastern New Mexico. Composite volcanic centers in the Cascades generated relatively high eruption rates and rapid volcano growth over periods of 10^3 to 10^4 yr alternating with periods of much reduced activity. Large flank failures and caldera-forming eruptions catastrophically changed the form of some edifices; subsequent eruptions rebuilt many of them.

Volcanism elsewhere in the Cordillera was restricted chiefly to areas of crustal extension and consisted dominantly of basaltic eruptions in long-lived fields of monogenetic volcanoes. Volcanic loci of Quaternary age lay scattered within broader zones that hosted eruptions over the past several million years. Some of these long-lived fields show evidence of migration of active vent areas through time as the North American Plate moved relative to a deeper fixed zone of magma genesis. Tectonic conditions at a few locations in the interior allowed development of long-lived, large-volume silicic magmatic systems (Yellowstone, Long Valley, Valles) that produced by far the largest Quaternary eruptions in North America. Widely dispersed tephra of these events provide stratigraphic markers over much of the western half of the United States.

In Alaska, the Aleutian arc and Queen Charlotte-Fairweather strike-slip plate boundaries localized most volcanism. Owing to westward migration of the transform boundary, the Wrangell volcanic field perhaps became progressively more isolated from magma-genesis processes during the Quaternary and eruption rates declined. In contrast, the Aleutian arc has sustained a high eruption rate resulting in about eight Holocene caldera-forming eruptions and numerous historical eruptions. Additional Quaternary volcanism occurred in west-central Alaska, in widely scattered basalt fields behind the Aleutian arc.

Quaternary eruptions in the Hawaiian Islands reflected the long-lived volcanism related to movement of the Pacific Plate across the Hawaiian hot spot. Each major Hawaiian volcano follows a general evolution from submarine volcano to rapidly growing emergent shield volcano. Decreasing eruption rate through postshield and rejuvenated stages is accompanied by reductions in island area through erosion as well as isostatic and cooling-induced subsidence. Deposits of giant slumps and debris avalanches, formed largely during the shield-building stage, cover much of the submarine flanks of the islands. Such failures were accompanied by tsunamis, some of which swept several hundred meters above sea level locally.

Advances in understanding volcanoes and volcanic processes accelerated in the U.S. and elsewhere during the last few decades of the 20th century as a result of several major eruptions worldwide, some with profound human impact and societal consequences. Many were closely monitored by a wide variety of geophysical, geochemical, and remote-sensing techniques, and subjected to intense studies of eruptive products both during and following eruptive activity. In addition, detailed mapping, geochronologic, geophysical, and petrologic investigations of Quaternary volcanic centers in the U.S. revealed new insights about long-term eruptive history; magma-genesis, -transport, and -eruption processes; and influence of tectonics on volcanic systems. Besides the satisfaction of pursuing interesting and challenging scientific problems, much of this advance has been driven by societal needs to mitigate the risk posed by volcanoes and to assess geothermal-energy resources associated with geologically young volcanic systems. Such needs will continue to drive volcano research in the 21st century.

Acknowledgments

I thank Robert Tilling, Edward Wolfe, and David Yamaguchi for timely and constructive reviews of all or part of the manuscript, and Bobbie Myers for preparing most of the illustrations.

References

Arculus, R.J. & Gust, D.A. (1995). Regional petrology of the San Francisco volcanic field, Arizona, USA. *Journal of Petrology*, **36**, 827–861.

Armstrong, R.L., Leeman, W.P. & Malde, H.E. (1975). K-Ar dating, Quaternary and Neogene volcanic rocks of the Snake River Plain, Idaho. *American Journal of Science*, **275**, 225–251.

Arvidson, R.E., Shepard, M.K., Guiness, E.A., Petry, S.B., Plaut, J.J., Evans, D.L., Farr, T.G., Greeley, R., Lancaster, N. & Gaddis, L.R. (1993). Characterization of lava-flow degradation in the Pisgah and Cima volcanic fields, California, using Landsat Thematic Mapper and AIRSAR data. *Geological Society of America Bulletin*, **105**, 175–188.

Bacon, C.R. (1982). Time-predictable bimodal volcanism in the Coso Range, California. *Geology*, **10**, 65–69.

Bacon, C.R. (1983). Eruptive history of Mount Mazama and Crater Lake caldera, Cascade Range, U.S.A. *Journal of Volcanology and Geothermal Research*, **18**, 57–115.

Bacon, C.R. & Lanphere, M.A. (2000). Quaternary eruption, intrusion, and alteration at Crater Lake, Oregon. *Geological Society of America Abstracts with Programs*, **32**(7), 50–51.

Bacon, C.R., Gardner, J.V., Mayer, L.A., Buktenica, M.W., Dartnell, P., Ramsey, D.W. & Robinson, J.E. (2002). Morphology, volcanism, and mass wasting in Crater Lake, Oregon. *Geological Society of America Bulletin*, **114**, 692–765.

Bailey, R.A. (1989). Geologic map of Long Valley caldera, Mono Inyo Craters volcanic chain, and vicinity, eastern California. U.S. Geological Survey Miscellaneous Investigations Map I-1933, scale 1:62,500, 2 sheets.

Bailey, R.A., Dalrymple, G.B. & Lanphere, M.A. (1976). Volcanism, structure, and geochronology of Long Valley caldera, Mono County, California. *Journal of Geophysical Research*, **81**, 725–744.

Baker, V.R., Bjornstad, B.N., Busacca, A.J., Fecht, K.R., Kiver, E.P., Moody, U.L., Rigby, J.G., Stradling, D.F. & Tallman, A.M. (1991). Quaternary geology of the Columbia Plateau. *In*: Morrison, R.B. (Ed.), *Quaternary Nonglacial Geology: Conterminous U.S.* Boulder, CO, Geological Society of America, The Geology of North America, K-2, pp. 215–250.

Baldridge, W.S., Perry, F.V., Vaniman, D.T., Nealey, L.D., Laughlin, A.W. & Wohletz, K.H. (1989a). Field guide to excursion 8A: Oligocene to Holocene magmatism and extensional tectonics, central Rio Grande rift and southeastern Colorado Plateau, New Mexico and Arizona. *In*: Chapin, C.E. & Zidek, J. (Eds), Field excursions to volcanic terranes in the western United States, Volume I: Southern Rocky Mountain region. *New Mexico Bureau of Mines and Mineral Resources Memoir*, Vol. 46, pp. 202–230.

Baldridge, W.S., Perry, F.V., Vaniman, D.T., Nealey, L.D., Leavy, B.D., Laughlin, A.W., Kyle, P., Bartov, Y., Stenitz, G. & Gladney, E.S. (1989b). Excursion 8A: Magmatism associated with lithospheric extension: Middle to late Cenozoic magmatism of the southeastern Colorado Plateau and central Rio Grande rift, New Mexico and Arizona. *In*: Chapin, C.E. & Zidek, J. (Eds), Field excursions to volcanic terranes in the western United States, Volume I: Southern Rocky Mountain region: New Mexico. *Bureau of Mines and Mineral Resources Memoir*, Vol. 46, pp. 187–202.

Bargar, K.E. & Jackson, E.D. (1974). Calculated volumes of individual shield volcanoes along the Hawaiian-Emperor chain. *U.S. Geological Survey Journal of Research*, **2**, 545–550.

Begét, J.E. (1982). Postglacial volcanic deposits at Glacier Peak, Washington, and potential hazards from future eruptions. *U.S. Geological Survey Open-File Report*, 82–830, 77 pp.

Begét, J.E. (1983). Glacier Peak, Washington: A potentially hazardous Cascade volcano. *Environmental Geology*, **5**, 83–92.

Begét, J.E. (2001). Continuous late Quaternary proxy climate records from loess in Beringia. *Quaternary Science Reviews*, **20**, 499–507.

Begét, J.E. & Keskinen, M.J. (2003). Trace-element geochemistry of individual glass shards of the Old Crow tephra and the age of the Delta glaciation, central Alaska. *Quaternary Research*, **60**, 63–69.

Begét, J.E. & Motyka, R.J. (1998). New dates on late Pleistocene dacitic tephra from the Mount Edgecumbe volcanic field, southeastern Alaska. *Quaternary Research*, **49**, 123–125.

Begét, J.E. & Nye, C.J. (1994). Postglacial eruption history of Redoubt Volcano, Alaska. *Journal of Volcanology and Geothermal Research*, **62**, 31–54.

Begét, J.E., Mason, O. & Anderson, P. (1992). Age, extent and climatic significance of the c. 3400 BP Aniakchak tephra, western Alaska, USA. *The Holocene*, **2**, 51–56.

Begét, J.E., Reger, R.D., Pinney, D.-A., Gillespie, T. & Campbell, K. (1991). Correlation of the Holocene Jarvis Creek, Tangle Lakes, Cantwell, and Hayes tephras in south-central and central Alaska. *Quaternary Research*, **35**, 174–189.

Berger, G.W. & Busacca, A.J. (1995). Thermoluminescence dating of late Pleistocene loess and tephra from eastern Washington and southern Oregon and implications for the eruptive history of Mount St. Helens. *Journal of Geophysical Research*, **100**, 22,361–22,374.

Berger, G.W., Péwé, T.L., Westgate, J.A. & Preece, S.L. (1996). Age of Sheep Creek tephra (Pleistocene) in central Alaska from thermoluminescence dating of bracketing loess. *Quaternary Research*, **45**, 263–270.

Bergmanis, E.C., Sinton, J.M. & Trusdell, F.A. (2000). Rejuvenated volcanism along the southwest rift zone, East Maui, Hawaii. *Bulletin of Volcanology*, **62**, 239–255.

Bouchard, D.P., Kaufman, D.S., Hochberg, A. & Quade, J. (1998). Quaternary history of the Thatcher Basin, Idaho, reconstructed from the $^{87}Sr/^{86}Sr$ and amino acid composition of lacustrine fossils; implications for the diversion of the Bear River into the Bonneville basin. *Palaeogeography, Palaeolimnology, and Palaeoclimatology*, **141**, 95–114.

Branney & Kokelaar (1992). A reappraisal of ignimbrite emplacement; progressive aggradation and changes from particulate to non-particulate flow during emplacement of high-grade ignimbrite. *Bulletin of Volcanology*, **54**, 504–520.

Bursik, M. & Sieh, K. (1989). Range front faulting and volcanism in the Mono Basin, eastern California. *Journal of Geophysical Research*, **94**, 15,587–15,609.

Calvert, A. & Hildreth, W. (2002). Radiometric dating of Three Sisters volcanic field. *Geological Society of America Abstracts with Programs*, **34**(5), A-90.

Cannon, E.C. & Bürgmann, R. (2001). Prehistoric fault offsets of the Hilina fault system, south flank of Kilauea Volcano, Hawaii. *Journal of Geophysical Research*, **106**, 4207–4219.

Carson, E.C., Fournelle, J.H., Miller, T.P. & Mickelson, D.M. (2002). Holocene tephrochronology of the Cold Bay area, southwest Alaska Peninsula. *Quaternary Science Reviews*, **21**, 2213–2228.

Casadevall, T.J. (Ed.) (1994). Volcanic ash and aviation safety: Proceedings of the First International Symposium on Volcanic Ash and Aviation Safety. *U.S. Geological Survey Bulletin*, 2047, 450 pp.

Christiansen, R.L. (2001). The Quaternary and Pliocene Yellowstone Plateau volcanic field of Wyoming, Idaho, and Montana. *U.S. Geological Survey Professional Paper*, 729-G, 145 pp.

Christiansen, R.L. & Miller, C.D. (1989). Mount Shasta and vicinity. *In*: Chapin, C.E. & Zidek, J. (Eds), Field excursions to volcanic terranes in the western United States, Volume II: Cascades and intermountain west. *New Mexico Bureau of Mines and Mineral Resources Memoir*, Vol. 47, pp. 216–225.

Christiansen, R.L. & Yeats, R.S. (1992). Post-Laramide geology of the U.S. Cordilleran region. *In*: Burchfiel, B.C. *et al.* (Eds), *The Cordilleran orogen: Conterminous U.S.* Boulder, CO, Geological Society of America, The Geology of North America, G-3, pp. 261–406.

Christiansen, R.L., Clynne, M.A. & Muffler, L.J.P. (2002a). Geologic map of the Lassen Peak, Chaos Crags, and upper Hat Creek area, California. U.S. Geological Survey Geologic Investigations Series Map I-2723, scale 1:24,000 and 1:2500, 1 sheet.

Christiansen, R.L., Foulger, G.R. & Evans, J.R. (2002b). Upper-mantle origin of the Yellowstone hotspot. *Geological Society of America Bulletin*, **114**, 1245–1256.

Clague, D.A. & Dalrymple, G.B. (1987). The Hawaiian-Emperor volcanic chain, Part I, Geologic evolution. *In*: Decker, R.W. *et al.* (Eds), Volcanism in Hawaii. *U.S. Geological Survey Professional Paper*, 1350, pp. 5–54.

Clague, D.A. & Dalrymple, G.B. (1989). Tectonics, geochronology, and origin of the Hawaiian-Emperor volcanic chain. *In*: Winterer, E.L. *et al.* (Eds), *The Eastern Pacific Ocean and Hawaii.* Boulder, CO, Geological Society of America, The Geology of North America, N, pp. 188–217.

Clague, D.A. & Moore, J.G. (2002). The proximal part of the giant submarine Wailau landslide, Molokai, Hawaii. *Journal of Volcanology and Geothermal Research*, **113**, 259–287.

Clynne, M.A. (1990). Stratigraphic, lithologic, and major element geochemical constraints on magmatic evolution at Lassen volcanic center. *Journal of Geophysical Research*, **95**, 19,651–19,669.

Coats, R.R. (1950). Volcanic activity in the Aleutian arc. *U.S. Geological Survey Bulletin*, 974-B, pp. 35–49.

Condit, C.D. & Connor, C.B. (1996). Recurrence rates of volcanism in basaltic volcanic fields: An example from the Springerville volcanic field, Arizona. *Bulletin of the Geological Society of America*, **108**, 1225–1241.

Connor, C.B., Stamatakos, J.A., Ferrill, D.A., Hill, B.E., Ofoegbu, G.I., Conway, F.M., Sagar, B. & Trapp, J. (2000). Geologic factors controlling patterns of small-volume basaltic volcanism: Application to a volcanic hazards assessment at Yucca Mountain, Nevada. *Journal of Geophysical Research*, **105**, 417–432.

Conrey, R.M. (1991). Geology and petrology of the Mt. Jefferson area, High Cascade Range, Oregon. Ph.D. dissertation. Pullman, Washington State University, 357 pp.

Conrey, R.M., Donnelly-Nolan, J.D., Taylor, E.M., Champion, D. & Bullen, T. (2001). The Shevlin Park Tuff, central Oregon Cascade Range: Magmatic processes recorded in an arc-related ash-flow tuff. *Eos, Transactions of the American Geophysical Union*, **82**(47), F1345–F1356.

Conway, F.M., Connor, C.B., Hill, B.E., Condit, C.D., Mullaney, K. & Hall, C.M. (1998). Recurrence rates of basaltic volcanism in SP cluster, San Francisco volcanic field, Arizona. *Geology*, **26**, 655–658.

Conway, F.M., Ferrill, D.A., Hall, C.M., Morris, A.P., Stamatakos, J.A., Connor, C.B., Halliday, A.N. & Condit, C. (1997). Timing of basaltic volcanism along the Mesa Butte fault in the San Francisco volcanic field, Arizona, from $^{40}Ar/^{39}Ar$ dates: Implications for longevity of cinder cone alignments. *Journal of Geophysical Research*, **102**, 815–824.

Crandell, D.R. (1971). Postglacial lahars from Mount Rainier volcano, Washington. *U.S. Geological Survey Professional paper*, 677, 75 pp.

Crandell, D.R. (1987). Deposits of pre-1980 pyroclastic flows and lahars from Mount St. Helens volcano, Washington. *U.S. Geological Survey Professional Paper*, 1444, 93 pp.

Crandell, D.R. & Mullineaux, D.R. (1978). Potential hazards from future eruptions of Mount St. Helens volcano, Washington. *U.S. Geological Survey Bulletin*, 1383-C, 26 pp.

Crandell, D.R. (1989). Gigantic debris avalanche of Pleistocene age from ancestral Mount Shasta volcano, California, and debris-avalanche hazard zonation. *U.S. Geological Survey Bulletin*, 1861, 32 pp.

Crowe, B.M. & Fisher, R.V. (1973). Sedimentary structures in base-surge deposits with special reference to cross-bedding, Ubehebe Craters, Death Valley, California. *Geological Society of America Bulletin*, **84**, 663–682.

Dethier, D.P. (2001). Pleistocene incision rates in the western United States calibrated using Lava Creek B tephra. *Geology*, **29**, 783–786.

Dohrenwend, J.C., Abrahams, A.D. & Turrin, B.D. (1987). Drainage development on basaltic lava flows, Cima volcanic field, southeast California, and Lunar Crater volcanic field, south-central Nevada. *Geological Society of America Bulletin*, **99**, 405–413.

Dohrenwend, J.C., Wells, S.G. & Turrin, B.D. (1986). Degradation of Quaternary cinder cones in the Cima volcanic field, Mojave Desert, California. *Geological Society of America Bulletin*, **97**, 421–427.

Donnelly-Nolan, J.D. (1998). Abrupt shift in $\delta^{18}O$ values at Medicine Lake volcano (California, USA). *Bulletin of Volcanology*, **59**, 529–536.

Donnelly-Nolan, J.M. (1988). A magmatic model of Medicine Lake volcano, California. *Journal of Geophysical Research*, **93**, 4412–4420.

Donnelly-Nolan, J.M. & Nolan, K.M. (1986). Catastrophic flooding and eruption of ash-flow tuff at Medicine Lake volcano, California. *Geology*, **14**, 875–878.

Druitt, T.H. & Bacon, C.R. (1986). Lithic breccia and ignimbrite erupted during the collapse of Crater Lake caldera, Oregon. *Journal of Volcanology and Geothermal Research*, **29**, 1–32.

Duffield, W.A. (1997). *Volcanoes of Northern Arizona*. Grand Canyon Association, 68 pp.

Duffield, W.A., Bacon, C.R. & Dalrymple, G.B. (1980). Late Cenozoic volcanism, geochronology, and structure of the Coso Range, Inyo County, California. *Journal of Geophysical Research*, **85**, 2381–2404.

Dupre, W.R., Morrison, R.B., Clifton, H.E., Lajoie, K.R., Ponti, D.J., Powell, C.L., II, Mathieson, S.A., Sarna-Wojcicki, A.M., Leithold, E.L., Lettis, W.R., McDowell, P.F., Rockwell, T.K., Unruh, J.R. & Yeats, R.S. (1991). Quaternary geology of the Pacific margin. *In*: Morrison, R.B. (Ed.), *Quaternary Nonglacial Geology: Conterminous U.S.* Boulder, CO, Geological Society of America, The Geology of North America, K-2, pp. 141–214.

Dzurisin, D., Yamashita, K.M. & Kleinman, J.W. (1994). Mechanisms of crustal uplift and subsidence at the Yellowstone caldera, Wyoming. *Bulletin of Volcanology*, **56**, 261–270.

Eichelberger, J.C., Keith, T.E.D., Miller, T.P. & Nye, C.J. (1995). The 1992 eruptions of Crater Peak vent, Mount Spurr volcano, Alaska: Chronology and summary. *In*: Keith, T.E.C. (Ed.), The 1992 eruptions of Crater Peak vent, Mount Spurr volcano, Alaska. *U.S. Geological Survey Bulletin*, 2139, pp. 1–18.

Evarts, R.C. (2001). Geologic map of the Silver Lake Quadrangle, Cowlitz County, Washington. U.S. Geological Survey Miscellaneous Field Studies Map MF-2371, scale, 1:24,000, 1 sheet.

Fenton, C.R., Webb, R.H., Pearthree, P.A., Cerling, T.E. & Poreda, R.J. (2001). Displacement rates on the Toroweap and Hurricane faults: Implications for Quaternary downcutting in the Grand Canyon, Arizona. *Geology*, **29**, 1035–1038.

Fierstein, J. & Hildreth, W. (1992). The Plinian eruptions of 1912 at Novarupta, Katmai National Park, Alaska. *Bulletin of Volcanology*, **54**, 646–684.

Finn, C.A., Sisson, T.W. & Desszcz-Pan (2001). Aerogeophysical measurements of collapse-prone hydrothermally altered zones at Mount Rainier volcano. *Nature*, **409**, 600–603.

Fiske, R.A., Rose, T.R., Swanson, D.A. & McGeehin, J.C. (1999). The Kulanaokuaiki tephra: Product of newly recognized pyroclastic eruptions at Kilauea volcano, Hawaii. *Eos, Transactions of the American Geophysical Union*, **80**(46), 1196–1197.

Foit, F.F., Jr., Mehringer, P.J., Jr. & Sheppard, J.C. (1993). Age, distribution, and stratigraphy of Glacier Peak tephra in eastern Washington and western Montana, United States. *Canadian Journal of Earth Sciences*, **30**, 535–552.

Fournelle, J.H., Marsh, B.D. & Myers, J.D. (1994). Age, character, and significance of Aleutian arc volcanism. *In*: Plafker, G. & Berg, H.C. (Eds), *The Geology of Alaska*. Boulder, CO, Geological Society of America, The Geology of North America, G-1, pp. 723–758.

Fournier, R.O., Christiansen, R.L., Hutchinson, R.A. & Pierce, K.L. (1994). A field-trip guide to Yellowstone National Park, Wyoming, Montana, and Idaho: Volcanic, hydrothermal, and glacial activity in the region. *U.S. Geological Survey Bulletin*, 2099, 46 pp.

Frank, D. (1995). Surficial extent and conceptual model of hydrothermal system at Mount Rainier, Washington. *Journal of Volcanology and Geothermal Research*, **65**, 51–80.

Froese, D., Westgate, J., Preece, S. & Storer, J. (2002). Age and significance of the late Pleistocene Dawson tephra in eastern Beringia. *Quaternary Science Reviews*, **21**, 2137–2142.

Gardner, C.A. (1994). Temporal, spatial, and petrologic variations of lava flows from the Mount Bachelor volcanic chain, central Oregon High Cascades. *U.S. Geological Survey Open-File Report*, 94–261, 99 pp.

Gardner, C.A., Cashman, K.V. & Neal, C.A. (1998). Tephra-fall deposits from the 1992 eruption of Crater Peak, Alaska: Implications for clast textures fro eruptive products. *Bulletin of Volcanology*, **59**, 537–555.

Gardner, C.A., Neal, C.A., Waitt, R.B. & Janda, R.J. (1994). Proximal pyroclastic deposits from the 1989–1990 eruption of Redoubt Volcano, Alaska – stratigraphy, distribution, and physical characteristics. *Journal of Volcanology and Geothermal Research*, **62**, 213–250.

Gillespie, A.R., Huneke, J.C. & Wasserburg, G.J. (1984). Eruption age of a approximately 100,000-year-old basalt from $^{40}Ar/^{39}Ar$ analysis of partially degassed xenoliths. *Journal of Geophysical Research*, **89**, 1033–1048.

Glazner, A.F., Farmer, G.L., Hughes, W.T., Wooden, J.L. & Pickthorn, W. (1991). Contamination of basaltic magma by mafic crust at Amboy and Pisgah Craters, Mojave Desert, California. *Journal of Geophysical Research*, **96**, 13,673–13,691.

Grigg, R.W. & Jones, A.T. (1997). Uplift caused by lithospheric flexure in the Hawaiian Archipelago as revealed by elevated coral deposits. *Marine Geology*, **141**, 11–25.

Guffanti, M.A. & Weaver, C.S. (1988). Distribution of late Cenozoic volcanic vents in the Cascade Range: Volcanic arc segmentation and regional tectonic considerations. *Journal of Geophysical Research*, **93**, 6513–6529.

Gustavson, T.C., Baumgartner, R.W., Jr., Caran, S.C., Holliday, V.T., Mehnert, H.H., O'Neill & Reeves, C.C., Jr. (1991). Quaternary geology of the southern Great Plains and an adjacent segment of the Rolling Plains. *In*: Morrison, R.B. (Ed.), *Quaternary Nonglacial Geology: Conterminous U.S.* Boulder, CO, Geological Society of America, The Geology of North America, K-2, pp. 477–501.

Hamblin, W.K. (1994). Late Cenozoic lava dams in the western Grand Canyon. *Geological Society of America Memoir*, 183, 139 pp.

Hausback, B.P. (2000). Geologic map of the Sasquatch Steps area, north flank of Mount St. Helens, Washington. U.S. Geological Survey Geologic Investigations Series Map I-2463, scale 1:4,000, 1 sheet.

Hausback, B.P. & Nilsen, T.H. (1999). Sutter Buttes. *In*: Wagner, D.L. & Graham, S.A. (Eds), *Geologic Field Trips in Northern California*; centennial meeting of the Cordilleran Section of the Geological Society of America. Special Publication, California Department of Conservation, Division of Mines and Geology, 119, pp. 246–254.

Heiken, G.H., Fisher, R.V. & Peterson, N.V. (1981). A field trip to the maar volcanoes of the Fort Rock-Christmas Lake Valley basin, Oregon. *In*: Johnston, D.A. & Donnelly-Nolan, J.D. (Eds), Guides to some volcanic terranes in Washington, Idaho, Oregon, and northern California. *U.S. Geological Survey Circular*, Vol. 838, pp. 119–140.

Heizler, M.T., Perry, F.V., Crowe, B.M., Peters, L. & Appelt, R. (1999). The age of Lathrop Wells volcanic center: An $^{40}Ar/^{39}Ar$ dating investigation. *Journal of Geophysical Research*, **104**, 767–804.

Herrero-Bervera, E., Canon-Tapia, E., Walker, J.P.L. & Guerrero-Garcia, J.C. (2002). The Nuuanu and Wailau giant landslides; insights from paleomagnetic and anisotropy of magnetic susceptibility (AMS) studies. *Physics of the Earth and Planetary Interiors*, **129**, 83–98.

Herzig, C.T. & Jacobs, D.C. (1994). Cenozoic volcanism and two-stage extension in the Salton trough, southern California and northern Baja California. *Geology*, **22**, 991–994.

Hildreth, E.W. & Lanphere, M.A. (2000). Eruptive history and geochronology of the Mount Baker volcanic district, Washington. *Geological Society of America Abstracts with Programs*, **32**(6), 19.

Hildreth, W. (1996). Kulshan caldera: A Quaternary subglacial caldera in the North Cascades, Washington. *Geological Society of America Bulletin*, **108**, 786–793.

Hildreth, W. & Fierstein, J. (2000). Katmai volcanic cluster and the great eruption of 1912. *Geological Society of America Bulletin*, **112**, 1594–1620.

Hildreth, W. & Lanphere, M.A. (1994). Potassium-argon geochronology of a basalt-andesite-dacite arc system: The Mount Adams volcanic field, Cascade Range of southern Washington. *Geological Society of America Bulletin*, **106**, 1413–1429.

Hill, B.E. (1991). Petrogenesis of compositionally distinct silicic volcanoes in the Three Sisters region of the Oregon Cascade Range; the effects of crustal extension on the development of continental arc silicic magmatism. Ph.D. dissertation. Corvallis, Oregon State University, 247 pp.

Hill, B.E. & Duncan, R.A. (1990). The timing and significance of silicic magmatism in the Three Sisters region of the Oregon High Cascades. *Eos, Transactions of the American Geophysical Union*, **71**(43), 1614.

Hill, B.E. & Taylor, E.M. (1990). Oregon central High Cascade pyroclastic units in the vicinity of Bend, Oregon. *Oregon Geology*, **52**, 125–139.

Hill, D.P., Dzurisin, D., Ellsworth, W.L., Endo, E.T., Galloway, D.L., Gerlach, T.M., Johnston, M.J.S., Langbein, J., McGee, K.A., Miller, C.D., Oppenheimer, D. & Sorey, M.L. (2002). Response plan for volcano hazards in the Long Valley caldera and Mono Craters region, California. *U.S. Geological Survey Bulletin*, 2185, 57 pp.

Hills, D.J., Morgan, J.K., Moore, G.F. & Leslie, S.C. (2002). Structural variability along the submarine south flank of Kilauea Volcano, Hawaii, from a multichannel seismic reflection survey. *In*: Takahashi, E. *et al.* (Eds), Hawaiian volcanoes: Deep underwater perspectives. *American Geophysical Union Monograph*, Vol. 128, pp. 105–124.

Holm, R.F. (1987). San Francisco Mountain: A late Cenozoic composite volcano in northern Arizona. *Geological Society of America Centennial Field Guide – Rocky Mountain Section*, 389–392.

Hooper, D.M. & Sheridan, M.F. (1998). Computer-simulation models of scoria cone degradation. *Journal of Volcanology and Geothermal Research*, **83**, 241–267.

Izett, G.A. (1981). Volcanic ash beds: Recorders of upper Cenozoic silicic pyroclastic volcanism in the western United States. *Journal of Geophysical Research*, **86**, 10,200–10,222.

Izett, G.A. & Obradovich, J.D. (1994). $^{40}Ar/^{39}Ar$ age constraints for the Jaramillo Normal Subchron and the Matuyama-Brunhes geomagnetic boundary. *Journal of Geophysical Research*, **99**, 2925–2934.

Izett, G.A. & Wilcox, R.E. (1982). Map showing localities of the Huckleberry Ridge, Mesa Falls, and Lava Creek ash beds (Pearlette family ash beds) of Pliocene and Pleistocene age in the western United States and southern Canada. U.S. Geological Survey Miscellaneous Field Investigations Map I-1325, scale 1:4,000,000, 1 sheet.

Kaufman, D.S., Manley, W.F., Forman, S.L. & Layer, P.W. (2001). Pre-late-Wisconsin glacial history, coastal Ahklun Mountains, southwestern Alaska – new amino acid, thermoluminescence, and $^{40}Ar/^{39}Ar$ results. *Quaternary Science Reviews*, **20**, 337–352.

Keating, B.H. & Helsley, C.E. (1999). The ancient shorelines of Lanai, Hawaii, revisited. *Sedimentary Geology*, **150**, 3–15.

Keith, T.E.C. (Ed.) (1995). The 1992 eruptions of Crater Peak vent, Mount Spurr volcano, Alaska. *U.S. Geological Survey Bulletin*, 2139, 220 pp.

Kempton, P.D. & Dungan, M.A. (1989). Geology and petrology of basalts and included mafic, ultramafic, and granulitic xenoliths of the Geronimo volcanic field, southeastern Arizona. *In*: Chapin, C.E. & Zidek, J. (Eds), Field excursions to volcanic terranes in the western United States, Volume I: Southern Rocky Mountain region. *New Mexico Bureau of Mines and Mineral Resources Memoir*, Vol. 46, pp. 161–174.

Klinger, R.E. (2001). Late Quaternary volcanism of Ubehebe Crater. *In*: Machette, M.N. *et al.* (Eds), Quaternary and late Pliocene geology of the Death Valley region: Recent observations on tectonics, stratigraphy, and lake cycles (Guidebook for the 2001 Pacific Cell – Friends of the Pleistocene fieldtrip). *U.S. Geological Survey Open-File Report*, 01–51, pp. A21–A24.

Kuehn, S.C. (2002). Stratigraphy, distribution, and geochemistry of the Newberry volcano tephra. Ph.D. dissertation. Pullman, Washington State University, 701 pp.

Kukla, G.J., Bender, M.L., de Beaulieu, J.-L., Bond, G., Broecker, W.S., Cleveringa, P., Gavin, J.E., Herbert, T.D., Imbrie, J., Jouzel, J., Keigwin, L.D., Knudsen, K.-L., McManus, J.F., Merkt, J., Muhs, D.R., Müller, H., Poore, R.Z., Porter, S.C., Seret, G., Shackleton, N.J., Turner, C., Tzedakis, P.C. & Winograd, I.J. (2002). Last interglacial climates. *Quaternary Research*, **58**, 2–13.

Kuntz, M.A. (1992). A model-based perspective of basatic volcanism, eastern Snake River Plain, Idaho. *In*: Link, P.K. *et al.* (Eds), Regional geology of eastern Idaho and western Wyoming. *Geological Society of America Memoir*, Vol. 179, pp. 289–304.

Kuntz, M.A., Champion, D.E., Spiker, E.C. & Lefebvre, R.H. (1986a). Contrasting magma types and steady-state, volume-predictable, basaltic volcanism along the Great Rift, Idaho. *Geological Society of America Bulletin*, **97**, 579–594.

Kuntz, M.A., Covington, H.D. & Schorr, L.J. (1992). An Overview of basaltic volcanism of the eastern Snake River Plain, Idaho. *In*: Link, P.K. *et al.* (Eds), Regional geology of eastern Idaho and western Wyoming. *Geological Society of America Memoir*, Vol. 179, pp. 227–267.

Kuntz, M.A., Spiker, E.C., Rubin, M., Champion, D.E. & Lefebvre, R.H. (1986b). Radiocarbon studies of latest Pleistocene and Holocene lava flows on the Snake River Plain, Idaho: Data, lessons, and interpretations. *Quaternary Research*, **25**, 163–176.

Langenheim, V.A.M. & Clague, D.A. (1987). The Hawaiian-Emperor volcanic chain, Part II, Stratigraphic framework of volcanic rocks of the Hawaiian Islands. *In*: Decker, R.W. *et al.* (Eds), Volcanism in Hawaii. *U.S. Geological Survey Professional Paper*, 1350, pp. 55–84.

Lanphere, M.A. (2000). Comparison of conventional K-Ar and $^{40}Ar/^{39}Ar$ dating of young mafic volcanic rocks. *Quaternary Research*, **53**, 294–301.

Lanphere, M.A. & Sisson, T.W. (1995). K-Ar ages of Mount Rainier volcanics. *Eos, Transactions of the American Geophysical Union*, **76**(45), 651.

Lanphere, M.A., Champion, D.E., Christiansen, R.L., Donnelly-Nolan, J.D., Fleck, R.J., Sarna-Wojcicki, A.M., Obradovich, J.D. & Izett, G.A. (1999a). Evolution of tephra dating in the western United States. *Geological Society of America Abstracts with Programs*, **31**(6), 73.

Lanphere, M.A., Champion, D.E., Christiansen, R.L., Izett, G.A. & Obradovich, J.D. (2002). Revised ages for tuffs of the Yellowstone Plateau volcanic field: Assignment of the Huckleberry Ridge Tuff to a new geomagnetic polarity event. *Geological Society of America Bulletin*, **114**, 559–568.

Lanphere, M.A., Champion, D.E., Clynne, M.A. & Muffler, L.J.P. (1999b). Revised age of the Rockland tephra, northern California: Implications for climate and stratigraphic reconstructions in the western United States. *Geology*, **27**, 135–138.

Lerbekmo, J.F. & Campbell, F.A. (1969). Distribution, composition, and source of the White River Ash, Yukon Territory. *Canadian Journal of Earth Sciences*, **6**, 109–116.

Lerbekmo, J.F., Westgate, J.A., Smith, D.G.W. & Denton, G.H. (1975). New data on the character and history of the White River volcanic eruption, Alaska. *In*: Suggate, R.P. & Cresswell (Eds), *Quaternary Studies*. Wellington, Royal Society of New Zealand, pp. 203–209.

Levander, A., Henstock, T.J., Meltzer, A.S., Beaudoin, B.C., Trehu, A.M. & Klemperer, S.L. (1998). Fluids in the lower crust following Mendocino triple junction migration: Active basaltic intrusion? *Geology*, **26**, 171–174.

Link, P.K. & Mink, L.L. (Eds) (2002). Geology, Hydrogeology, and environmental remediation: Idaho National Engineering and Environmental Laboratory, eastern Snake River Plain, Idaho. *Geological Society of America Special Paper*, 353, 311 pp.

Linneman, S.R. (1990). The petrologic evolution of the Holocene magmatic system of Newberry volcano, central Oregon. Ph.D. dissertation. Laramie, University of Wyoming, 312 pp.

Lipman, P.W. & Mullineaux, D.R. (1981). The 1980 eruptions of Mount St. Helens, Washington. *U.S. Geological Survey Professional Paper*, 1250, 844 pp.

Lipman, P.W., Sisson, T.W., Ui, T., Naka, J. & Smith, J.R. (2002). Ancestral submarine growth of Kilauea volcano and instability of its south flank. *In*: Takahashi, E. *et al.* (Eds), Hawaiian volcanoes: Deep underwater perspectives. *American Geophysical Union Monograph*, Vol. 128, pp. 161–191.

Locke, W.W. & Meyer, G.A. (1994). A 12,000-year record of vertical deformation across the Yellowstone caldera margin: The shorelines of Yellowstone Lake. *Journal of Geophysical Research*, **99**, 20,079–20,094.

Lorenz, V. (1970). Some aspects of the eruption mechanism of the Big Hole maar, central Oregon. *Geological Society of America Bulletin*, **81**, 1823–1830.

Lucchitta, I., Curtis, G.H., Davis, M.E., Davis, S.W. & Turrin, B. (2000). Cyclic aggradation and downcutting, fluvial response to volcanic activity, and calibration of soil-carbonate stages in the western Grand Canyon, Arizona. *Quaternary Research*, **53**, 23–33.

Luedke, R.G. & Smith, R.L. (1986). Map showing distribution, composition, and age of late Cenozoic volcanic centers in Alaska. U.S. Geological Survey Miscellaneous Investigations Series Map I-1091-F, scale, 1:1,000,000, 3 sheets.

Luedke, R.G. & Smith, R.L. (1991). Quaternary volcanism in the western conterminous United States. *In*: Morrison, R.B. (Ed.), *Quaternary Nonglacial Geology: Conterminous U.S.* Boulder, CO, Geological Society of America, The Geology of North America, K-2, pp. 75–92.

MacDonald, G.A., Abbott, A.T. & Peterson, F.L. (1983). *Volcanoes in the sea – The geology of Hawaii*. Honolulu, Hawaii, The University Press of Hawaii, 517 pp.

MacLeod, N.S., Sherrod, D.R., Chitwood, L.A. & Jensen, R.A. (1995). Geologic map of Newberry volcano, Deschutes, Klamath, and Lake counties, Oregon. U.S. Geological Survey Miscellaneous Investigations Series Map I-2455, scale 1:62,500 and 1:24,000 1 sheet.

Malahoff, A. (1987). Geology of the summit of Loihi submarine volcano. *In*: Decker, R.W. *et al.* (Eds), Volcanism in Hawaii. *U.S. Geological Survey Professional Paper*, 1350, pp. 133–144.

Malde, H.E. (1991). Quaternary geology and structural history of the Snake River Plain, Idaho and Oregon. *In*: Morrison, R.B. (Ed.), *Quaternary Nonglacial Geology: Conterminous U.S.* Boulder, CO, Geological Society of America, The Geology of North America, K-2, pp. 251–281.

Mastin, L.G. (1997). Evidence for water influx from a crater lake during the explosive hydromagmatic eruption of 1790, Kilauea volcano, Hawaii. *Journal of Geophysical Research*, **102**, 20,093–20,109.

McGimsey, R.G., Neal, C.A. & Riley, C.M. (2001). Areal distribution, thickness, mass, volume, and grain size of tephra-fall deposits from the 1992 eruptions of Crater Peak vent, Mt. Spurr volcano, Alaska. *U.S. Geological Survey Open-File Report*, 01–370, 32 pp.

McLaughlin, R.J. & Donnelly-Nolan, J.M. (Eds) (1981). Research in the Geysers-Clear Lake geothermal area, northern California. *U.S. Geological Survey Professional Paper*, 1141, 259 pp.

McMurty, G.M., Herrero-Bervera, E., Cremer, M.D., Smith, J.R., Resig, J., Sherman, C.E. & Torresan, M.E. (1999). Stratigraphic constraints on the timing and emplacement of the Alika 2 giant submarine landslide. *Journal of Volcanology and Geothermal Research*, **94**, 35–58.

McPhie, J., Walker, G.P.L. & Christiansen, R.L. (1990). Phreatomagmatic and phreatic fall and surge deposits from explosions at Kilauea volcano, Hawaii, 1790 A.D.: Keanakakoi Ash Member. *Bulletin of Volcanology*, **52**, 334–354.

Miller, C.D. (1980). Potential hazards from future eruptions in the vicinity of Mount Shasta volcano, northern California. *U.S. Geological Survey Bulletin*, 1503, 43 pp.

Miller, C.D. (1985). Holocene eruptions at the Inyo volcanic chain, California: Implications for possible eruptions in Long Valley caldera. *Geology*, **13**, 14–17.

Miller, T.P. & Chouet, B.A. (Eds) (1994). The 1989–1990 eruptions of Redoubt Volcano, Alaska. *Journal of Volcanology and Geothermal Research*, **62**, 1–530.

Miller, T.P. & Richter, D.H. (1994). Quaternary volcanism in the Alaska Peninsula and Wrangell Mountains, Alaska. *In*: Plafker, G. & Berg, H.C. (Eds), *The Geology of Alaska*. Boulder, CO, Geological Society of America, The Geology of North America, G-1, pp. 759–779.

Miller, T.P., McGimsey, R.G., Richter, D.H., Riehle, J.R., Nye, C.J., Yount, M.E. & Dumoulin, J.A. (1998). Catalog of the historically active volcanoes of Alaska. *U.S. Geological Survey Open-File Report*, 98–582, 104 pp.

Mills, H.H. (1976). Estimated erosion rates on Mount Rainier, Washington. *Geology*, **4**, 401–406.

Mills, H.H. (1992). Post-eruption erosion and deposition in the 1980 crater of Mount St. Helens, Washington, determined from digital maps. *Earth Surface Processes and Landforms*, **17**, 739–754.

Moll-Stalcup, E.J. (1994). Latest Cretaceous and Cenozoic magmatism in mainland Alaska. *In*: Plafker, G. & Berg, H.C. (Eds), *The Geology of Alaska*. Boulder, CO, Geological Society of America, The Geology of North America, G-1, pp. 589–619.

Moore, G.W. & Moore, J.G. (1988). Large-scale bedforms in boulder gravel produced by giant waves in Hawaii. *In*: Clifton, H.E. (Ed.), Sedimentologic consequences of convulsive geologic events. *Geological Society of America Special Paper*, 229, pp. 101–110.

Moore, J.G. (1987). Subsidence of the Hawaiian Ridge. *In*: Decker, R.W. *et al.* (Eds), Volcanism in Hawaii. *U.S. Geological Survey Professional Paper*, 1350, pp. 85–100.

Moore, J.G. & Chadwick, W.W., Jr. (1995). Offshore geology of Mauna Loa and Adjacent areas, Hawaii. *In*: Rhodes, J.M. & Lockwood, J.P. (Eds), Mauna Loa revealed: Structure, composition, history, and hazards. *American Geophysical Union Monograph*, Vol. 92, pp. 21–44.

Moore, J.G. & Clague, D.A. (1992). Volcano growth and evolution of the island of Hawaii. *Geological Society of America Bulletin*, **104**, 1471–1484.

Moore, J.G., Ingram, B.L., Ludwig, K.R. & Clague, D.A. (1996). Coral ages and island subsidence, Hilo drill hole. *Journal of Geophysical Research*, **101**, 11,599–11,605.

Moore, J.G., Normark, W.R. & Holcomb, R.T. (1994). Giant Hawaiian landslides. *Annual Reviews of Earth and Planetary Sciences*, **22**, 119–144.

Morrison, R.B. (1991). Quaternary stratigraphic, hydrologic, and climatic history of the Great Basin, with emphasis on Lakes Lahontan, Bonneville, and Tecopa. *In*: Morrison, R.B. (Ed.), *Quaternary Nonglacial Geology: Conterminous U.S.* Boulder, CO, Geological Society of America, The Geology of North America, K-2, pp. 283–320.

Muffler, L.J.P. (Ed.) (1979). Assessment of geothermal resources of the United States, 1978. *U.S. Geological Survey Circular*, 790, 163 pp.

Muhs, D.R., Ager, T.A. & Begét, J.E. (2001). Vegetation and paleoclimate of the last interglacial period, central Alaska. *Quaternary Science Reviews*, **20**, 41–61.

Mullineaux, D.R. (1996). Pre-1980 tephra-fall deposits erupted from Mount St. Helens, Washington. *U.S. Geological Professional Paper*, 1563, 99 pp.

Neal, C.A., McGimsey, R.G., Gardner, C.A., Harbin, M.L. & Nye, C.J. (1995). Tephra-fall deposits from the 1992 eruptions of Crater Peak, Mount Spurr volcano, Alaska. *In*: Keith, T.E.C. (Ed.), The 1992 eruptions of Crater Peak vent, Mount Spurr volcano, Alaska. *U.S. Geological Survey Bulletin*, 2139, pp. 65–80.

Neal, C.A., McGimsey, R.G., Miller, T.P., Riehle, J.R. & Waythomas, C.F. (2000). Preliminary volcano-hazard assessment for Aniakchak Volcano, Alaska. *U.S. Geological Survey Open-File Report*, 00–519, 35 pp.

Nelson, C.H., Bacon, C.R., Robinson, S.W., Adam, D.P., Bradbury, J.P., Barber, J.H., Jr., Schwartz, D. & Vagenas, G. (1994). The volcanic, sedimentologic, and paleolimnologic history of the Crater Lake caldera floor, Oregon: Evidence for small caldera evolution. *Geological Society of America Bulletin*, **106**, 684–704.

Oviatt, C.G. & Nash, W.P. (1989). Late Pleistocene basaltic ash and volcanic eruptions in the Bonneville basin, Utah. *Geological Society of America Bulletin*, **101**, 292–303.

Oviatt, C.G., McCoy, W.D. & Nash, W.P. (1994). Sequence stratigraphy of lacustrine deposits: A Quaternary example from the Bonneville basin, Utah. *Geological Society of America Bulletin*, **106**, 133–144.

Peterson, D.W. & Moore, R.B. (1987). Geologic history and evolution of geologic concepts, island of Hawaii. *In*: Decker *et al.* (Eds), Volcanism in Hawaii. *U.S. Geological Survey Professional Paper*, 1350, pp. 149–189.

Pierce, K.L. & Morgan, L.A. (1992). The track of the Yellowstone hot spot: Volcanism, faulting, and uplift. *In*: Link, P.K. *et al.* (Eds), Regional geology of eastern Idaho and western Wyoming. *Geological Society of America Memoir*, Vol. 179, pp. 1–53.

Pierce, K.L., Cannon, K.P., Meyer, G.A., Trebesch, M.J. & Watts, R.D. (2002). Post-glacial inflation-deflation cycles, tilting, and faulting in the Yellowstone caldera based on Yellowstone Lake shorelines. *U.S. Geological Survey Open-File Report*, 01–0142, 55 pp.

Pierson, T.C. & Janda, R.J. (1994). Volcanic mixed avalanches: A distinct eruption-triggered mass-flow process at snow-clad volcanoes. *Geological Society of America Bulletin*, **106**, 1351–1358.

Porter, S.C. (1979). Hawaiian glacial ages. *Quaternary Research*, **12**, 161–187.

Preece, S.J., Westgate, J.A., Alloway, B.V. & Milner, M.W. (2000). Characterization, identity, distribution, and source of late Cenozoic tephra beds in the Klondike district of the Yukon, Canada. *Canadian Journal of Earth Sciences*, **37**, 983–996.

Preece, S.J., Westgate, J.A., Stemper, B.A. & Péwé, T.L. (1999). Tephrochronology of late Cenozoic loess at Fairbanks, central Alaska. *Geological Society of America Bulletin*, **111**, 71–90.

Reid, M.E., Christian, S.B. & Brien, D.L. (2000). Gravitational stability of three-dimensional stratovolcano edifices. *Journal of Geophysical Research*, **105**(B3), 6043–6056.

Reid, M.E., Sisson, T.W. & Brien, D.L. (2001). Volcano collapse promoted by hydrothermal alteration and edifice shape, Mount Rainier, Washington. *Geology*, **29**, 779–782.

Reneau, S.L., Gardner, J.N. & Forman, S.L. (1996). New evidence for the age of the youngest eruptions in the Valles caldera, New Mexico. *Geology*, **24**, 7–10.

Renne, P.R. (2000). K-Ar and $^{40}Ar/^{39}Ar$ dating. *In*: Noller, J.S., Sowers, J.M. & Lettis, W.R. (Eds), *Quaternary Geochronology: Methods and Applications*. American Geophysical Union Reference Shelf 4, pp. 77–100.

Rhodes, J.M. & Lockwood, J.P. (Eds) (1995). Mauna Loa revealed: Structure, composition, history, and hazards. *American Geophysical Union Monograph*, Vol. 92, 348 pp.

Richmond, G.M. (1986). Stratigraphy and chronology of glaciations in Yellowstone National Park. *In*: Šibrava, V., Brown, D.Q. & Richmonds, G.M. (Eds), Quaternary glaciations in the Northern Hemisphere. *Quaternary Science Reviews*, Vol. 5, pp. 83–98.

Richter, D.H., Moll-Stalcup, E.J., Miller, T.P., Lanphere, M.A., Dalrymple, G.B. & Smith, R.L. (1994). Eruptive history and petrology of Mount Drum volcano, Wrangell Mountains, Alaska. *Bulletin of Volcanology*, **56**, 29–46.

Richter, D.H., Preece, S.J., McGimsey, R.G. & Westgate, J.A. (1995). Mount Churchill, Alaska: Source of the late Holocene White River Ash. *Canadian Journal of Earth Sciences*, **32**, 741–748.

Richter, D.H., Smith, J.G., Lanphere, M.A., Dalrymple, G.B., Reed, B.L. & Shew, N. (1990). Age and progression of volcanism, Wrangell volcanic field, Alaska. *Bulletin of Volcanology*, **53**, 29–44.

Riehle, J.R. (1985). A reconnaissance of the major Holocene tephra deposits in the upper Cook Inlet region, Alaska. *Journal of Volcanology and Geothermal Research*, **26**, 37–74.

Riehle, J.R., Bowers, P.M. & Ager, T.A. (1990). The Hayes tephra deposits, an upper Holocene marker horizon in south-central Alaska. *Quaternary Research*, **33**, 276–290.

Riehle, J.R., Champion, D.E., Brew, D.A. & Lanphere, M.A. (1992a). Pyroclastic deposits of the Mount Edgecumbe volcanic field, southeast Alaska: Eruptions of a stratified magma chamber. *Journal of Volcanology and Geothermal Research*, **53**, 117–143.

Riehle, J.R., Mann, D.J., Peteet, D.M., Engstrom, D.R., Brew, D.A. & Meyer, C.E. (1992b). The Mount Edgecumbe tephra deposits, a marker horizon in southeastern Alaska near the Pleistocene-Holocene boundary. *Quaternary Research*, **37**, 183–202.

Rose, W.I., Bluth, G.J.S., Schneider, D.J., Ernst, G.G.J., Riley, C.M., Henderson, L.J. & McGimsey, R.G. (2001). Observations of volcanic clouds in their first few days of atmospheric residence: The 1992 eruptions of Crater peak, Mount Spurr volcano, Alaska. *Journal of Geology*, **109**, 677–694.

Sarna-Wojcicki, A.M. (2000). Tephrochronology. *In*: Noller, J.S., Sowers, J.M. & Lettis, W.R. (Eds), *Quaternary Geochronology: Methods and Applications*. American Geophysical Union Reference Shelf 4, pp. 357–377.

Sarna-Wojcicki, A.M. & Davis, J.O. (1991). Quaternary tephrochronology. *In*: Morrison, R.B. (Ed.), *Quaternary Nonglacial Geology: Conterminous U.S.* Boulder, CO, Geological Society of America, The Geology of North America, K-2, pp. 93–116.

Sarna-Wojcicki, A.M., Champion, D.E. & Davis, J.O. (1983). Holocene volcanism in the conterminous United States and the role of silicic volcanic ash layers in correlation of latest-Pleistocene and Holocene deposits. *In*: Wright, H.E., Jr. (Ed.), *The Holocene, Late-Quaternary Environments of the United States*. Minneapolis, University of Minnesota Press, Vol. 2, pp. 52–77.

Sarna-Wojcicki, A.M., Lajoie, K.R., Meyer, C.E., Adam, D.P. & Rieck, H.J. (1991). Tephrochronologic correlation of upper Neogene sediments along the Pacific margin, conterminous United States. *In*: Morrison, R.B. (Ed.), *Quaternary Nonglacial Geology: Conterminous U.S.* Boulder, CO, Geological Society of America, The Geology of North America, K-2, pp. 117–140.

Sarna-Wojcicki, A.M., Pringle, M.S. & Wijbrans, J. (2000). New ^{40}Ar/^{39}Ar age of the Bishop Tuff from multiple sites and sediment calibration for the Matuyama-Brunhes boundary. *Journal of Geophysical Research*, **105**, 21,431–21, 443.

Satake, K., Smith, J.R. & Shinozaki, K. (2002). Three-dimensional reconstruction and tsunami model of the Nuuanu and Wailau giant landslides, Hawaii. *In*: Takahashi, E. *et al.* (Eds), Hawaiian volcanoes: Deep underwater perspectives. *American Geophysical Union Monograph*, Vol. 128, pp. 333–346.

Scott, K.M., Vallance, J.W. & Pringle, P.T. (1995). Sedimentology, behavior, and hazards of debris flows at Mount Rainier, Washington. *U.S. Geological Survey Professional Paper*, 1547, 56 pp.

Scott, W.E. (1990). Patterns of volcanism in the Cascade arc during the past 15,000 years. *Geoscience Canada*, **17**, 179–183.

Scott, W.E. & Gardner, C.A. (1992). Geologic map of the Mount Bachelor volcanic chain and surrounding area, Cascade Range, Oregon. U.S. Geological Survey Miscellaneous Investigations Series Map I-1967, scale 1:50,000, 1 sheet.

Scott, W.E., Gardner, C.A., Sherrod, D.R., Tilling, R.I., Lanphere, M.A. & Conrey, R.M. (1997). Geologic history of Mount Hood volcano, Oregon – A field-trip guidebook. *U.S. Geological Survey Open-File Report*, 97–263, 38 pp.

Self, S., Kienle, J. & Huot, J.P. (1980). Ukinrek Maars, Alaska, II. Deposits and formation of the 1977 craters. *Journal of Volcanology and Geothermal Research*, **7**, 39–65.

Self, S., Goff, F., Gardner, J.N., Wright, J.V. & Kite, W.M. (1986). Explosive rhyolitic volcanism in the Jemez Mountains: Vent locations, caldera development and relation to regional structure. *Journal of Geophysical Research*, **91**, 1779–1798.

Sherrod, D.R. & Smith, J.G. (1990). Quaternary extrusion rates from the Cascade Range, northwestern United States and British Columbia. *Journal of Geophysical Research*, **95**, 19,465–19,474.

Sherrod, D.R. & Smith, J.G. (2000). Geologic map of upper Eocene to Holocene volcanic and related rocks of the Oregon Cascade Range. U.S. Geological Survey Geologic Investigations Series I-2569, scale 1:500,000, 2 sheets.

Sherrod, D.R., Nishimitsu, Y. & Tagami, T. (2003). New K-Ar ages and the geologic evidence against rejuvenated-stage volcanism at Haleakala, East Maui, a postshield-stage volcano of the Hawaiian island chain. *Geological Society of America Bulletin*, **116**, 683–694.

Sieh, K. & Bursik, M. (1986). Most recent eruption of the Mono Craters, eastern central California. *Journal of Geophysical Research*, **91**, 12,539–12,571.

Sisson, T.W., Vallance, J.W. & Pringle, P.T. (2001). Progress made in understanding Mount Rainier's hazards. *Eos, Transactions of the American Geophysical Union*, **82**, 113–120.

Smith, E.I., Keenan, D.L. & Plank, T. (2002). Episodic volcanism and hot mantle: Implications for volcanic hazard studies at the proposed nuclear waste repository at Yucca Mountain, Nevada. *GSA Today*, **12**, 4–10.

Smith, J.G. (1993). Geologic map of upper Eocene to Holocene volcanic rocks and related rocks in the Cascade Range, Washington. U.S. Geological Survey Miscellaneous Investigations Map I-2005, scale 1:500,000, 2 sheets.

Smith, J.R., Malahoff, A. & Shor, A.N. (1999). Submarine geology of the Hilina Slump and morpho-structural evolution of Kilauea Volcano, Hawaii. *Journal of Volcanology and Geothermal Record*, **94**, 59–88.

Smith, R.B. & Braile, L.W. (1994). Crustal structure and evolution of an explosive silicic system at Yellowstone National Park. *In*: Boyd, F.R. (Ed.), *Explosive Volcanism: Inception, Evolution, and Hazards*. Washington, DC, National Academy Press, pp. 96–109.

Smith, R.L. & Bailey, R.A. (1968). Stratigraphy, structure, and volcanic evolution of the Jemez Mountains, New Mexico. *In*: Epis, R.C. (Ed.), Cenozoic volcanism in the southern Rocky Mountains. *Colorado School of Mines Quarterly*, Vol. 63(3), pp. 259–260.

Smith, R.L. & Luedke, R.G. (1984). Potentially active volcanic lineaments and loci in the western conterminous United States. *In*: Boyd, F.R. (Ed.), *Explosive Volcanism: Inception, Evolution, and Hazards*. Washington, DC, National Academy Press, pp. 47–66.

Smith, R.L., Bailey, R.A. & Ross, C.S. (1970). Geologic map of the Jemez Mountains, New Mexico. U.S. Geological Survey Miscellaneous Geologic Investigations Map I-571, scale 1:125,000, 1 sheet.

Stine, S.W. (1987). Mono Lake: The past 4,000 years [Ph.D. thesis]. Berkeley, University of California, 732 pp.

Suzuki-Kamata, K., Kamata, H. & Bacon, C.R. (1993). Evolution of the caldera-forming eruption at Crater Lake, Oregon, indicated by component analysis of lithic fragments. *Journal of Geophysical Research*, **98**, 14,059–14, 074.

Swanson, D.A., Duffield, W.A. & Fiske, R.S. (1976). Displacement of the south flank of Kilauea Volcano: The result of forceful intrusion of magma into the rift zones. *U.S. Geological Survey Professional paper*, 963, 39 pp.

Tabor, R.W. & Crowder, D.F. (1969). On batholiths and volcanoes – Intrusion and eruption of late Cenozoic magmas in the Glacier Peak area, North Cascades, Washington. *U.S. Geological Survey Professional Paper*, 604, 67 pp.

Takahashi, E., Lipman, P.W., Garcia, M.O., Naka, J. & Aramaki, S. (Eds) (2002). Hawaiian volcanoes: Deep underwater perspectives. *American Geophysical Union Monograph*, 128, 418 pp.

Tanaka, K.L., Shoemaker, E.M., Ulrich, G.E. & Wolfe, E.W. (1986). Migration of volcanism in the San Francisco volcanic field, Arizona. *Geological Society of America Bulletin*, **97**, 129–141.

Taylor, E.M. (1990). Volcanic history and tectonic development of the central High Cascade Range, Oregon. *Journal of Geophysical Research*, **95**, 19,611–19,622.

Tilling, R.I. & Dvorak, J.J. (1993). Anatomy of a basaltic volcano. *Nature*, **363**, 125–133.

Tilling, R.I., Koyanagi, R.Y., Lipman, P.W., Lockwood, J.P., Moore, J.G. & Swanson, D.A. (1976). Earthquake and related catastrophic events, Island of Hawaii, November 29, 1975: A preliminary report. *U.S. Geological Survey Circular*, 740, 33 pp.

Tolan, T.L., Reidel, S.P., Beeson, M.H., Anderson, J.L., Fecht, K.R. & Swanson, D.A. (1989). Revisions to the estimates of areal extent and volume of the Columbia River Basalt Group. *In*: Reidel, S.P. & Hooper, P.R. (Eds), Volcanism and tectonism in the Columbia River flood-basalt province. *Geological Society of America Special Paper*, 239, pp. 1–20.

Trabant, D.C., Waitt, R.B. & Major, J.J. (1994). Disruption of Drift glacier and origin of floods during the 1989–1990 eruptions of Redoubt Volcano, Alaska. *Journal of Volcanology and Geothermal Research*, **62**, 369–385.

Turrin, B.D., Champion, D.E. & Fleck, R.J. (1991). $^{40}Ar/^{39}Ar$ age of the Lathrop Wells volcanic center, Yucca Mountain, Nevada. *Science*, **253**, 654–657.

Turrin, B.D., Christiansen, R.L., Clynne, M.A., Champion, D.E., Gerstel, W.J., Muffler, L.J.P. & Trimble, D. (1998). Age of Lassen Peak, California, and implications for the ages of late Pleistocene glaciations in the southern Cascade Range. *Geological Society of America Bulletin*, **110**, 931–945.

Ulrich, G.E., Condit, C.D., Wenrich, K.J., Wolfe, E.W., Holm, R.F., Nealey, L.D., Conway, F.M., Aubele, J.C. & Crumpler, L.S. (1989). Miocene to Holocene volcanism of the southern Colorado Plateau, Arizona. *In*: Chapin, C.E. & Zidek, J. (Eds), Field excursions to volcanic terranes in the western United States, Volume I: Southern Rocky Mountain region. *New Mexico Bureau of Mines and Mineral Resources Memoir*, Vol. 46, pp. 1–41.

Vallance, J.W. & Scott, K.M. (1997). The Osceola Mudflow from Mount Rainier: Sedimentology and hazard implications of a huge clay-rich debris flow. *Geological Society of America Bulletin*, **109**, 143–163.

Voight, B. & Elsworth, D. (1997). Failure of volcano slopes. *Géotechnique*, **47**, 1–31.

Waitt, R.B. & Begét, J.E. (1996). Provisional geologic map of Augustine Volcano, Alaska. *U.S. Geological Survey Open-File Report*, 96–516, 44 pp., scale 1:25,000, 1 sheet.

Waitt, R.B., Gardner, C.A., Pierson, T.C., Major, J.J. & Neal, C.A. (1994). Unusual ice diamicts emplaced during the December 15, 1989 eruption of Redoubt Volcano, Alaska. *Journal of Volcanology and Geothermal Research*, **62**, 409–428.

Wayne, W.J., Aber, J.S., Agard, S.S., Bergantino, R.N., Bluemle, J.P., Coates, D.A., Cooley, M.E., Madole, R.F., Martin, J.E., Mears, B., Jr., Morrsion, R.B. & Sutherland, W.M. (1991). Quaternary geology of the northern Great Plains. *In*: Morrison, R.B. (Ed.), *Quaternary Nonglacial Geology: Conterminous U.S.* Boulder, CO, Geological Society of America, The Geology of North America, K-2, pp. 441–476.

Waythomas, C.F. & Neal, C.A. (1998). Tsunami generation by pyroclastic flow during the 3500-B.P. caldera-forming eruption of Aniakchak Volcano, Alaska. *Bulletin of Volcanology*, **60**, 110–124.

Waythomas, C.F. & Wallace, K.L. (2002). Flank collapse at Mount Wrangell, Alaska, recorded by volcanic mass-flow deposits in the Copper River lowland. *Canadian Journal of Earth Sciences*, **39**, 1257–1279.

Waythomas, C.F., Mangan, M.T., Miller, T.P., Layer, P.L. & Trusdell, F.A. (2001). Caldera-forming eruptions of the Emmons lake volcanic center, Alaska Peninsula, Alaska: Probable source of the Dawson Creek tephra in Yukon Territory, Canada. *Eos Transactions of the American Geophysical Union*, **82**(47), F1405.

Waythomas, C.F., Walder, J., McGimsey, R.G. & Neal, C.A. (1996). A catastrophic flood caused by drainage of a caldera lake at Aniakchak volcano, Alaska, and implications for volcanic-hazards assessment. *Geological Society of America Bulletin*, **108**, 861–871.

Wells, R.E., Weaver, C.S. & Blakely, R.J. (1998). Fore-arc migration in Cascadia and its neotectonic significance. *Geology*, **26**, 759–762.

Wells, S.G., Dohrenwend, J.C., McFadden, L.D., Turrin, B.D. & Mahrer, K.D. (1985). Late Cenozoic landscape evolution on lava flow surfaces of the Coma volcanic field, Mojave Desert, California. *Geological Society of America Bulletin*, **96**, 1518–1529.

Wells, S.G., McFadden, L.D., Renault, C.E. & Crowe, B.M. (1990). Geomorphic assessment of late Quaternary volcanism in the Yucca Mountain area, southern Nevada: Implications for the proposed high-level radioactive waste repository. *Geology*, **18**, 549–553.

Wenrich, K.J., Billingsley, G.H. & Blackerby, B.A. (1995). Spatial migration and compositional changes of Miocene-Quaternary volcanism in the western Grand Canyon. *Journal of Geophysical Research*, **100**, 10,417–10,440.

Westgate, J.A., Easterbrook, D.J., Naeser, N.D. & Carson, R.J. (1987). Lake Tapps tephra: An early Pleistocene stratigraphic marker in the Puget lowland, Washington. *Quaternary Research*, **28**, 340–355.

Westgate, J.A., Preece, S.J., Kotler, E. & Hall, S. (2000). Dawson tephra: A prominent stratigraphic marker of late Wisconsinan age in west-central Yukon, Canada. *Canadian Journal of Earth Sciences*, **37**, 621–627.

Westgate, J.A., Walter, R.C., Pearce, G.W. & Gorton, M.P. (1985). Distribution, stratigraphy, petrochemistry, and palaeomagnetism of the late Pleistocene Old Crow tephra in Alaska and the Yukon. *Canadian Journal of Earth Sciences*, **22**, 893–906.

White, D.E., Hutchinson, R.A. & Keith, T.E.C. (1988). The geology and remarkable thermal activity of Norris Geyser Basin, Yellowstone National Park, Wyoming. *U.S. Geological Survey Professional Paper*, 1456, 84 pp.

White, J.D.L. (1996). Pre-emergent construction of a lacustrine basaltic volcano, Pahvant Butte, Utah (USA). *Bulletin of Volcanology*, **58**, 249–262.

Whitney, J.W. & Shroba, R.R. (1991). Comment on "Geomorphic assessment of late Quaternary volcanism in the Yucca Mountain area, southern Nevada: Implications for the proposed high-level radioactive waste repository." *Geology*, **19**, 661.

Wicks, C., Jr., Thatcher, W. & Dzurisin, D. (1998). Migration of fluids beneath Yellowstone Caldera inferred from satellite radar interferometry. *Science*, **282**, 458–462.

Williams, S.K. (1994). Late Cenozoic tephrostratigraphy of deep sediment cores from the Bonneville basin, northwest Utah. *Geological Society of America Bulletin*, **105**, 1517–1530.

Wilson, C.J.N. & Hildreth, W. (1997). The Bishop Tuff: New insights from eruptive stratigraphy. *Journal of Geology*, **105**, 407–439.

Wolfe, E.W. & Morris, J. (compilers) (1996). Geologic map of the island of Hawaii. U.S. Geological Survey Miscellaneous Investigations Series Map I-2524-A, scale 1:100,000, 3 sheets.

Wolfe, E.W., Wise, W.S. & Dalrymple, G.B. (1997). The geology and Petrology of Mauna Kea volcano, Hawaii – A study of postshield volcanism. *U.S. Geological Survey Professional Paper*, 1557, 129 pp.

Wolff, J.A. & Gardner, J.N. (1995). Is the Valles caldera entering a new cycle of activity? *Geology*, **23**, 411–414.

Wright, H.E., Jr. & Frey, D.G. (Eds) (1965). *The Quaternary of the United States*. Princeton, NJ, Princeton University Press, 922 pp.

Young, R.W. & Bryant, E.A. (1992). Catastrophic wave erosion on the southeastern coast of Australia: Impact of the Lanai tsunami ca. 105 ka? *Geology*, **20**, 199–202.

Young, S.R. (1990). Physical volcanology of Holocene airfall deposits from Mt. Mazama, Crater Lake, Oregon. Ph.D. dissertation. Lancaster, England, University of Lancaster, 298 pp.

Late-Quaternary vegetation history of the eastern United States

Eric C. Grimm[1] and George L. Jacobson Jr.[2]

[1] *Illinois State Museum, Research and Collections Center, 1011 East Ash Street, Springfield, IL 62703, USA; grimm@museum.state.il.us*
[2] *Institute for Quaternary and Climate Studies, Department of Biological Sciences, University of Maine, Orono, ME 04469, USA; jacobson@maine.edu*

Introduction

In 1965, M.B. Davis, E.J. Cushing, and D.R. Whitehead published seminal papers in the review volume for the 7th INQUA Congress (Wright & Frey, 1965) summarizing late-Quaternary vegetational and climatic change in the eastern United States (Cushing, 1965; Davis, 1965; Whitehead, 1965). These papers summarized existing data, presented current controversies, discussed problems and limitations of pollen analysis, and suggested future directions of research. Prior to that time, many inferences about late-Quaternary vegetation history had been derived from biogeography, and these three authors stressed the importance of paleobotanical evidence for reconstructing vegetation history. Cushing and Davis emphasized the dangers in drawing inferences from modern distributions, especially disjunctions. Biogeographers had drawn many dubious conclusions about plant migrations from disjunct populations, but, as Cushing and Davis argued, disjuncts could either be relics, as often assumed, or, just as likely, outposts of advancing populations. Disjunct populations may occur in unusual habitats, such as in the Wisconsin driftless area and the St. Lawrence Gulf region, precisely because these habitats are unusual, not because they are stranded relics. All three authors disputed the popular ideas of some biogeographers, notably Lucy Braun (1950, 1951, 1955), who maintained that the vegetation south of the glacial border had changed little during the glacial periods. Although the fossil evidence clearly indicated that boreal species had moved far to the south, these authors cautioned that vegetation zones may not have moved intact, and, in fact, they noted examples of taxa that are allopatric today but were sympatric in the past. Cushing in particular noted that the paleobotanical evidence supported the individualistic hypothesis of Gleason (1926, 1939) and Whittaker's (1957) community concept of "loosely ordered complexity," in contrast to Clementsian ideas of communities as distinct units in time and space.

Whitehead stated that the southward displacement of boreal species was indisputable, but he also recognized that unusual combinations of species seemed to have occurred together in the Southeast during the Pleistocene. Moreover, although boreal species had certainly moved south, there were problems with identifications, and the cones from *Picea* in Louisiana, although similar to *P. glauca*, were unusually large. The "non-analog" communities of the Pleistocene were to become an important research topic and challenge for paleoclimatic reconstruction. If boreal species had moved south, where had temperate and southern species gone? Florida seemed important, but at the time of Whitehead's review, only W.A. Watts' unpublished pollen diagram from Mud Lake (Fig. 1) existed from Florida, and it was missing the full-glacial interval. Watts published this diagram in 1969, and he was to add greatly to the vegetation history of Florida over the next three decades. Much has been learned about Florida vegetation, but questions still remain.

When Davis, Cushing, and Whitehead reviewed the late-Quaternary vegetation history of eastern North America, data were available from only a small number of sites, and many of these had few or no radiocarbon dates. In addition, the standards of palynology were quite low for the early studies. Thus, these authors realized that correlations with climatic events in Europe or even across eastern North America, although tempting, were quite speculative. Although the number of radiocarbon-dated pollen diagrams has grown enormously since 1965, temporal correlation remains a fundamental challenge of Quaternary paleoecology. The development of AMS radiocarbon dating has greatly increased opportunities for precise correlations, but even today only a small fraction of studied sites are so dated.

Although investigators had long made climatic inferences from paleobotanical data, Davis noted that few had actually tried to establish quantitative relationships between species distributions and climatic variables. She suggested that vegetation might be omitted from the equation and that direct relationships be established between fossil pollen spectra and surface samples. During the ensuing decades, the relationships between modern and fossil pollen spectra, along with those between modern pollen spectra and modern climatic variables, proved to be major research foci. The analysis of statistical relationships between pollen and climate necessitated the development of pollen databases, which were initially project-based but are now in the public domain. Perhaps unforeseen in 1965 was the development of computer climate models (General Circulation Models or GCMs) and the need for validation of model climatic predictions. The quantification of climatic variables from pollen data has been of paramount importance for GCM validation and was a major research objective of the COHMAP (Cooperative Holocene Mapping Project) group, which involved dozens of palynologists and climate modelers (COHMAP Members, 1988).

Davis was especially concerned with possible human influences on prehistoric vegetation. She noted the increase

DEVELOPMENTS IN QUATERNARY SCIENCE
VOLUME 1 ISSN 1571-0866
DOI:10.1016/S1571-0866(03)01017-0

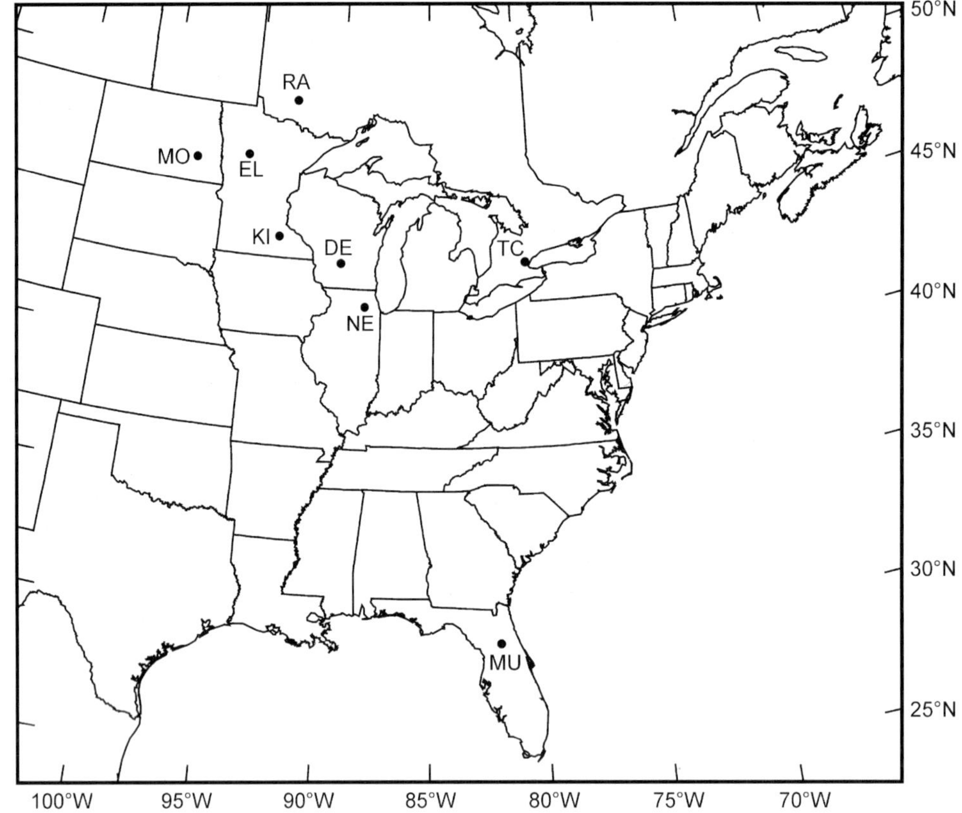

Fig. 1. Map of the eastern United States and southeastern Canada. Sites specifically discussed in the text are: DE – Devils Lake, EL – Elk Lake, KI – Kirchner Marsh, MO – Moon Lake, MU – Mud Lake, NE – Nelson Lake, RA – Rattle Lake, TC – Twiss Marl Pond and Crawford Lake, TU – Lake Tulane.

in pollen from *Ambrosia*, introduced weeds, and crops in the upper portions of many sediment cores but worried about potential impacts of Europeans and Native Americans below these horizons. She noted the *Tsuga* decline, which is dated to about 5000 years ago and is evident in many pollen diagrams from the Northeast, and suggested that human activity or fire rather than climate might have been the cause. Cushing discussed the Prairie Peninsula and noted Gleason's (1922) hypothesis that burning by Native Americans was the cause, a view widely held by ecologists today. Davis touched upon the issue of migrational lag, which she developed further in later papers. The extent to which equilibrium exists between vegetation and climate was to become a major topic related to development of quantitative reconstructions of paleoclimate from fossil pollen data. Her later papers stimulated considerable debate, which has led to a clearer understanding of the temporal and spatial scales of vegetation-climate equilibration.

A number of recent papers have summarized the late-Quaternary vegetational and climatic history of eastern North America (Davis, 1983a; Grimm *et al.*, 2001; Jackson *et al.*, 1997; Jacobson *et al.*, 1987; Webb, 1988; Webb *et al.*, 1983, 1987, 1993b). We do not intend to duplicate these efforts, but instead will focus here on a number of outstanding issues and problems.

Chronology

Chronology is paramount for reconstructing past environments. The development of radiocarbon dating was a great advance for late-Quaternary studies and has permitted temporal correlation across space. Spatially extensive time-stratigraphic markers, such as tephras, are scarce in the eastern U.S. The *Tsuga* decline (Bennett & Fuller, 2002; Davis, 1981a) is a useful time-stratigraphic marker within the range of *Tsuga*. This marked decline is thought to have resulted from a pathogen outbreak (Allison *et al.*, 1986; Davis, 1981a), although an ultimate cause of summer drought has been suggested (Haas & McAndrews, 2000). In any case, it has an age of about 4750 ± 50 ^{14}C yr B.P. or about 5490 ± 60 cal yr B.P. (Bennett & Fuller, 2002).

Most radiocarbon dates from cores analyzed for pollen from the eastern U.S. are from bulk sediment. Many of these dates are highly problematic (Fowler *et al.*, 1986). The "hard-water" error from dissolved carbonate rocks is well known (Broecker & Walton, 1959; Deevey *et al.*, 1954; Ogden, 1967a, b; Saarnisto, 1988). A common practice has been to radiocarbon date the European settlement horizon, which is usually known to within a few decades, then apply the correction to all dates from the core. Although this correction is probably better than none, it likely is not applicable to the entire core because of changing limnological conditions (Ogden, 1967b); because of large variations in atmospheric $\Delta^{14}C$, dates from the last few centuries typically have a large calendar age range (Stuiver, 1993; Stuiver & Quay, 1980). Even sediment from lakes in noncarbonate surroundings can have significant errors, perhaps owing to old groundwater or slow CO_2 exchange between the lake and atmosphere (Håkansson, 1979). In the Midwest, a much under-appreciated source of error is ancient organic carbon derived from coal, shale, and other sedimentary rocks. The late-glacial sections of many cores contain numerous pre-Quaternary pollen and spores (Cushing, 1964), which probably indicate significant pre-Quaternary carbon content (Nambudiri *et al.*, 1980). The source of some of this carbon may be wind-blown loess. This problem is particularly severe in sediments with low organic matter, which typify late-glacial sediments from many sites. Thus, all bulk-sediment radiocarbon dates must be regarded with caution.

Annually laminated (varved) sediments have the potential to provide absolute dating, but lakes with varved sediments do not exist in much of the region. In addition, counting varves can be problematic; false or missing varves are common, and accumulative error is a problem. Elk Lake, Minnesota, is perhaps the most studied varved lake in the eastern U.S. (Bradbury & Dean, 1993). The *Picea* decline, which is well-dated to about 10,000 ^{14}C yr B.P. in the Elk Lake region, corresponds to about 10,000 varve years. However, 10,000 ^{14}C yr B.P. is about 11,500 cal yr B.P., which implies a 1500-yr offset between the varve years and "calendar" years in that case. Varves are certainly useful for establishing the time-span of events, but for chronology they should be confirmed with alternative dating methods.

Radiocarbon dating by accelerator mass spectrometry (AMS) has revolutionized the capability for deriving reliable chronologies. This technique, which requires only a few mg of carbon, can date macrofossils and charcoal from terrestrial plants, thereby avoiding many of the problems with decay-count dates. The opportunity, for example, to date a single *Picea* needle now allows reliable chronologies to be developed for late-glacial sediments, which have been a particular problem. Even now, however, AMS dates are available for only a small fraction of the hundreds of sites in the eastern U.S. The potential now exists for AMS dating of pollen grains themselves, so perhaps the plant taxa of interest can be dated directly and independently in future studies (Brown *et al.*, 1989, 1992; Long *et al.*, 1992; Regnéll, 1992; Regnéll & Everitt, 1996; Richardson & Hall, 1994).

Databases

The study of vegetation change over large geographic areas has required the compilation of pollen and macrofossil databases. The COHMAP project assembled a research database for its purposes (COHMAP Members, 1988; Wright *et al.*, 1993). This "database" was not public, contained minimal metadata, and simply comprised a large number of text files. Because it was not relational, queries and searches were difficult. To rectify this situation, the North American Pollen Database and the North American Plant Macrofossil Database were created with support from the NOAA Paleoclimate Program. These databases, which are relational and public, greatly facilitate synoptic paleoenvironmental studies (Jackson *et al.*, 2000a).

Quantitative Techniques

Vegetation Reconstruction

Since the papers of Davis, Cushing and Whitehead in 1965, major advances have been made in the quantification of pollen data. Early efforts involved zonation of individual pollen diagrams and comparison of pollen diagrams with ordination and other techniques (Birks, 1986; Birks & Gordon, 1985; Gordon & Birks, 1972, 1974; Grimm, 1987, 1988). The application of quantitative techniques to climate and vegetation reconstruction has been another avenue of research, and eastern North America has been a major testing ground. Early quantitative techniques attempted to calibrate the relationship between pollen percentages and some measure of actual abundances in the vegetation. Building on the work of Fagerlind (1952), Davis (1963) developed the R-value model, in which R is the ratio of the pollen percentage of a taxon to its vegetational percentage. Other approaches are linear regression (Webb *et al.*, 1981) and an extended R-value method that estimates a background component (Parsons & Prentice, 1981; Prentice, 1986; Prentice & Parsons, 1983; Prentice & Webb, 1986). Although research into these methods has provided a clearer understanding of the relationship between pollen percentages and tree abundances, the calibrations have not been much applied beyond these methodological papers. Lack of confidence in the calibrations arises from changing pollen source areas and changing patterns of vegetation, which are unknown and which affect the calibrations. Moreover, the calibrations are not generally applicable to savanna and non-forested vegetation types. In general, palynologists have found modern analog techniques to be more useful.

Analog techniques quantitatively compare fossil pollen assemblages with modern assemblages, and the modern vegetation is then an analog for the past vegetation. These techniques quantify the reconstructions of vegetation that palynologists typically conceptualize. Quantification of biomass or plant cover is perhaps most useful at the stand level. However, because pollen data from most sites represent larger areas with significant diversity of habitats and community types that are combined in the pollen rain reaching a site, vegetation reconstruction tends to be at the formation or biome level. The interest is not so much in the exact quantification of cover, which will vary with the landscape and successional status, but with the biome.

Analog techniques fall into two categories, one implicit the other explicit. The first is to plot fossil and modern samples together in an ordination (Jacobson & Grimm, 1986; Lamb, 1984, 1985). The distances between samples represent dissimilarities. The second approach is to analyze dissimilarities (DC's) directly (Jackson *et al.*, 2000b; Overpeck *et al.*, 1985, 1992). In this method, modern pollen surface-samples are assigned to biomes; the DC's (usually squared chord distances) are calculated between fossil and modern samples; and modern samples that have DC's with a fossil sample less than a critical value are considered to be analogs for the fossil sample. If a fossil sample has several modern analogs within the critical value, the biome most represented in the analog modern samples is considered to be the analog biome for a fossil sample. If no modern samples are less than the critical value, then the fossil sample has no analog in the modern vegetation.

Overpeck *et al.* (1992) used the modern analog technique to map biomes in eastern North America for the past 18,000 ^{14}C yr at intervals of 3000 ^{14}C yr. They mapped eight biomes that have modern analogs plus an additional "no analog" biome. These maps show shifting positions of biomes through time and space and not simple northward movement of biomes after deglaciation. By 9000 ^{14}C yr B.P., the modern biomes were well established, but the modern configuration of biomes, although not the exact distribution, does not appear until about 6000 ^{14}C yr B.P., which the overlapping isopoll maps of Jacobson *et al.* (1987) also illustrate. Even by 3000 years ago, the modern distribution of biomes was not fully realized. The biome maps also show a large area of non-analog vegetation at 12,000 ^{14}C yr B.P. and smaller areas of non-analog vegetation persisting until 3000 years ago.

Important feedbacks occur between vegetation and climate, and sophisticated climate models require vegetation cover as a parameter. Ideally, rather than prescribing vegetation, the climate output would be input into a biome model, which would then generate the vegetation. This need plus a general paleoecological emphasis on biomes has led to biome modeling. Rather than using actual species, these biome models use plant functional types (PFT's), which are based on life form, physiology, phenology, leaf form, and climatic tolerances. The biomization method entails assigning PFT's to biomes, assigning pollen taxa to PFT's, and using these assignments to assign pollen taxa to biomes. For fossil samples, an "affinity score" is calculated between pollen samples and biomes. The biome with the highest affinity score is assigned to the fossil pollen sample, subject to a tie-breaking rule (Prentice *et al.*, 1992, 1996; Williams *et al.*, 1998, 2000, 2001). Advantages of biomization are that non-analog biomes are definable and that no reliance is placed on comparison with modern samples, which may be biased by anthropogenic activities.

Climate Reconstruction

Climate and vegetation are both multivariate phenomena. In eastern North America, two general classes of methods for quantitatively relating climate and pollen data have been popular: transfer functions and response surfaces. Transfer or calibration functions generally involve various forms of multiple regression. Functions are empirically derived from modern pollen and climate data and are used to generate estimates for past climate variables with fossil pollen data (Bartlein & Webb, 1985; Bartlein *et al.*, 1984; Bernabo, 1981; Gajewski, 1988; Howe & Webb, 1983; Sachs *et al.*, 1977; Webb, 1985; Webb & Bryson, 1972; Webb & Clark, 1977). A major problem with transfer functions is that they are linear; whereas pollen variables often vary nonlinearly with climate. As a result, the functions vary regionally, and for past pollen assemblages the correct set of functions must be used (Bartlein *et al.*, 1986). Regional variation compelled Bartlein & Webb (1985) to partition eastern North America into 13 subregions with different sets of equations. To avoid this problem, Bartlein *et al.* (1986) introduced a response-surface method, which utilizes nonlinear functions or multidimensional surfaces that describe the pollen-climate relationship for all of eastern North America. In essence, these surfaces map pollen variables in climate space rather than geographic space. Bartlein *et al.* (1986) used polynomial regression to fit pollen surfaces to climate variables, but later studies used locally weighted regression. Response surfaces have proven very useful for climate reconstruction and for data-model comparisons (Huntley *et al.*, 1989; Prentice *et al.*, 1991; Webb *et al.*, 1987, 1993a, b, 1998). Response surfaces can be used both for inferring climate from fossil pollen (reverse modeling) and for simulating pollen distributions from climate-model results (forward modeling) (Prentice *et al.*, 1991; Webb *et al.*, 1987, 1993b, 1998). Both of these techniques are useful for comparing climate-model results to data.

Rates of Change

Palynologists have long described rates of change in pollen diagrams in qualitative terms, especially the rapid changes at the Pleistocene-Holocene transition. Jacobson & Grimm (1986) introduced a method for quantitatively measuring the rate of change in a pollen diagram, which they later modified and applied to an array of sites in eastern North America (Grimm & Jacobson, 1992; Jacobson *et al.*, 1987). These methods involve measuring dissimilarities between pollen spectra evenly spaced in time and averaging at each time step. Other approaches involve event correlation (Overpeck, 1987) or calculation of dissimilarities between gridded data points at evenly spaced time intervals (Overpeck *et al.*, 1991, 1992; Williams *et al.*, 2001). These methods simply identify times of rapid change across large geographic areas regardless of the local nature of qualitative changes. Despite dating uncertainties during late-glacial time, the results of Jacobson *et al.* (1987) and Grimm & Jacobson (1992) show peaks of change corresponding with the Younger Dryas interval, the beginning of the Bølling, and coeval with Heinrich event 1, suggesting synchrony of climate change across eastern North America and the North Atlantic.

Plant Migration and Range Expansion

If climate changes or ice sheets retreat and expose new ground, plant species respond by moving into new regions and perhaps abandoning others. Often called "migrations," these shifts in plant populations are really colonizations of new regions by seed dispersal and establishment (Davis & Shaw, 2001). Geographic shifts in populations can be mapped with pollen data from a geographic array of sites. Two styles of maps have been popular: isochrone or "migration" maps (Davis, 1976, 1981b, 1983a, b; Davis *et al.*, 1986; Davis & Jacobson, 1985; Jacobson, 1979) and "isopoll" maps, either contoured or gridded (Bernabo & Webb, 1977; Delcourt *et al.*, 1984; Dexter *et al.*, 1987; Huntley & Webb, 1989; Jacobson *et al.*, 1987; Webb, 1987, 1988; Webb *et al.*, 1983, 1987, 1993b).

Isochrone maps attempt to show the range limits of a taxa at different time periods. Establishing absolute range limits can be difficult, however. In eastern North America, influx and percentage data have been used to detect range limits (Davis *et al.*, 1991). With influx data, a sharp increase in pollen influx from low values is considered to mark local arrival (Davis, 1976, 1981b, 1983a). With percentage data, pollen percentages in surface samples are compared with modern range maps, and a minimum pollen percentage that indicates local presence is determined (Davis & Jacobson, 1985; Jacobson, 1979). Both percentage and influx curves tend to have initial tails of low values before sharply increasing. Opinions differ on when taxa actually arrived. Woods & Davis (1989) argued that the initial values indicate long-distance transport (>20 km) from approaching populations, whereas Bennett (1985, 1988a, b) believed that the earliest occurrence of a continuous pollen curve indicates local presence and that a later sharp increase represents exponential population growth. In truth, either scenario could be true, and neither one probably applies universally. The above authors studied common or dominant species that are good pollen producers (*Fagus* and *Tsuga*). Another possibility is that certain species, including those now typically abundant, existed at low population densities for long periods, perhaps in favorable but restricted habitats, and then expanded when climate conditions became more favorable (Dexter *et al.*, 1987). In this case, a species may have been present in the vegetation for a long period but below the pollen-detection level.

The northern Great Plains provide an example. Although their aerial coverage is small, a number of tree species are widespread in the Great Plains, mainly along streams and in ravines (e.g. *Ulmus americana*, *Fraxinus pennsylvanica*, *Acer negundo*, *Quercus macrocarpa*). The occurrence of these taxa in the pollen rain at sites in the Great Plains such as Moon Lake in North Dakota (Laird *et al.*, 1996, 1998) and in surface samples (McAndrews & Wright, 1969) is sporadic and below any reasonable detection level, lower than taxa such as *Alnus* and *Pinus*, which do not occur in the Great Plains. These trees have undoubtedly occurred in the Great Plains throughout the Holocene, but they never did expand abruptly or exponentially. Nevertheless, they are widespread and are poised to expand from innumerable foci if climate should suddenly become more favorable for them. *Ulmus* and *Quercus* did expand rapidly and did become dominant in the forests eastward in Minnesota. Thus, species that have the capacity to become forest dominants do not necessarily become dominants if climate is not sufficiently favorable, although they may be widespread on the landscape in restricted habitats.

Detailed studies of *Fagus* and *Tsuga* in Wisconsin and the Upper Peninsula of Michigan (Davis *et al.*, 1986, 1991, 1998; Woods & Davis, 1989) and of *Pinus strobus* in Minnesota (Jacobson, 1979), in which the trees are big pollen producers and where numerous lakes and hollows exist in various landscape positions for pollen analysis, have shown that the ecological controls and the actual range expansion of species can be determined with reasonable confidence. However, in most other cases the isochrone maps are really population expansion maps, which may bear little relationship to actual plant migration.

If pollen isochrone maps are assumed to be migration maps (i.e. show range limits), then migration rates can be calculated. The rates for major forest trees in North America typically lie within the limits of 100–500 m/yr (Davis, 1976, 1981b; Huntley & Webb, 1989; King & Herstrom, 1997). Perhaps surprisingly, large seeded trees such as *Quercus* and *Fagus* are no slower than trees with wind dispersed seeds such as *Tsuga* and *Ulmus*. Birds, especially blue jays (*Cyanocitta cristata*), effectively disperse and bury the large seeds of Fagaceous trees and were probably instrumental in the range expansion of these trees (Darley-Hill & Johnson, 1981; Johnson & Adkisson, 1985; Johnson & Webb, 1989). Regardless of whether isochrone maps show actual range or population expansions, they vividly illustrate the individualistic behavior of different taxa and the recency of modern plant associations. Following deglaciation, different taxa did not simply move northwards but expanded in different directions, at different rates, and at different times. Taxa such as *Fagus* and *Tsuga*, which today have similar ranges, have had much different postglacial histories.

In contrast to isochrone maps, which purport to show taxon ranges at different times, isopoll maps make no assumptions about presence or absence of the species but instead simply map the abundance of pollen in space, either by contouring the data (Bernabo & Webb, 1977; Delcourt *et al.*, 1984; Dexter *et al.*, 1987; Huntley *et al.*, 1989; Huntley & Webb, 1989; Webb, 1987, 1988; Webb *et al.*, 1983) or by gridding (Prentice *et al.*, 1991; Webb *et al.*, 1987, 1993b). Another application of isopolls has been to show major ecotones, such as the prairie-forest border with a minimum sum of prairie forbs or the boreal forest with a minimum *Picea* sum (Bernabo & Webb, 1977; Webb *et al.*, 1983). A related approach has been with different colors for different taxa to map areas that are enclosed by isopolls that indicate significant presence in the vegetation. Overlapping areas combine into secondary colors, which indicate different plant associations (Jacobson *et al.*, 1987; Williams *et al.*, 2001). All of these maps show plant populations shifting through time and space as they track climate changes. The colored maps vividly illustrate shifting plant associations, including past associations that have no analogs in the modern vegetation.

Migrational Lag Versus Equilibrium

After a major climate change, some time is required for vegetation to establish quasi-equilibrium with climate. The length of this time has been a major controversy, and its importance is paramount because the utility of pollen data to serve as a climate proxy depends on a fairly short time. Early investigators invoked migrational lag as an explanation for the existence of pollen assemblages in the late-glacial period that have no modern analog. For example, in a number of early papers, H.E. Wright Jr. suggested that the absence of *Pinus* from the late-glacial spruce forest, in contrast to its ubiquity in the modern boreal forest, was due to migrational lag. He hypothesized that *Pinus* was stranded in the Appalachian highlands during the Pleistocene and migrated northward more slowly than *Picea* after deglaciation, with its immigration into Minnesota blocked by the Great Lakes (Wright, 1964, 1968b; Wright *et al.*, 1963). Cushing (1965) noted this hypothesis but also suggested the alternative hypothesis that climate during the time of the *Picea* zone may have been "radically different" from today and unsuitable for *Pinus* but still within the climatic tolerances of *Picea*.

The assumption that a taxon did not exist outside the isochrones of its "migration map" has underlain considerable controversy over whether significant migrational lags (on the order of millennia) followed major climate changes or whether vegetation remained in equilibrium with climate, with only short lags (centuries). In a number of papers, Davis (1976, 1978, 1981b, 1983a; Davis *et al.*, 1986) argued that following abrupt warming at the end of the Pleistocene trees migrated northward for several thousand years, and during this time vegetation was in disequilibrium with climate. Davis (1976) suggested that even the time span of an interglaciation was too short for equilibrium to be established. Davis *et al.* (1986) suggested that *Tsuga* was limited by dispersal from reaching its climatic limits in Michigan and Wisconsin until about 5000 years ago. Davis (1981b, 1983a) maintained that because trees did not simply move northwards en masse as climate warmed after deglaciation but moved at different rates and in different directions, their absence throughout much of the Holocene south of their modern limits was due to migrational lag.

For New England, Davis (1976, 1981b, 1983a) argued that seed-dispersal limitations delayed the immigrations of the major forest dominants *Carya*, *Castanea*, *Fagus*, and *Tsuga*. The fact that the isochrone maps indicated that *Carya* had reached its northern limits after the warmest part of the Holocene was particularly critical to her argument (Davis, 1981b). This argument assumes that the isochrone maps are, in fact, migration maps, but it is false if the trees existed in the vegetation in small populations below the pollen detection limit. In an elegant study, Gaudreau (1986) showed that migrational lag was unlikely and that these trees probably did exist in small populations before their arrival as defined by a sudden increase in pollen influx. These taxa all had the same maximum rates of "migration" during the Holocene, although in different regions at different times. For example, *Carya* expanded northwards rapidly west of the Appalachians (Bettis *et al.*, 1990) but apparently slowly east of the Appalachians. Thus, the slow expansion of *Carya* in the Northeast was not an inherent limitation on dispersal. Gaudreau also showed that even at low and sporadic frequencies, these taxa showed topographic contrasts, which must result from local populations, not long-distance dispersal. For example, *Carya* was consistently more abundant at lower elevation. She concluded that population expansion was often considerably decoupled in time from population establishment.

Direct evidence for minor, outlying populations and for stand-scale patterns of vegetation has emerged from studies that take advantage of the collecting properties of small basins (Bradshaw, 1988; Jacobson & Bradshaw, 1981). The most important sites for revealing the mosaic patterns of vegetation have been small forest hollows – typically tiny ponds or permanently wet depressions in which pollen grains avoid oxidation. Because these sites are so small (generally overhung by the tree canopy), they receive a large fraction of their pollen from plants within a few hundred meters (Calcote, 1995; Jackson & Wong, 1994; Sugita, 1995). In Wisconsin, studies of small hollows show that the modern local mosaic of *Tsuga* and hardwoods has persisted *in situ* for as much as several thousand years (Davis *et al.*, 1998). Farther east, Schauffler & Jacobson (2002) used data from small hollows to demonstrate that *Picea* trees existed in a narrow belt along the cool, moist coast of Maine during most of the Holocene, when otherwise *Picea* was almost absent from warmer, drier inland landscapes. These stand-scale studies have illustrated that small numbers of individuals were indeed present as minor parts of the vegetation considerable distances outside the major populations of those taxa. These were the small, established populations from which rapid expansions could develop following appropriate climate change.

Webb (1986) contended that the response time of vegetation to climate change was short – centuries, not millennia – and that vegetation was never out of equilibrium with climate for extended time spans. Webb and colleagues carried the individualistic hypothesis further by maintaining that differences among taxa in their migration routes and rates resulted from individualistic responses of species to different aspects of climate, not just temperature, but also precipitation and seasonality. In addition, although a large and abrupt climate change did occur at the end of the Pleistocene, climate continued to change throughout the Holocene as perihelion moved from summer to winter, and the ranges of plants continued to shift in response to these complex Holocene climate changes (Huntley *et al.*, 1989; Huntley & Webb, 1989; Prentice, 1983; Prentice *et al.*, 1991; Webb, 1986, 1987).

Prentice *et al.* (1991) tested the hypothesis that multiple climate variables could explain the Holocene history of *Carya*. Using response surfaces for six pollen taxa other than *Carya*, they generated past climate, which, in turn, was used to simulate isopoll maps for *Carya*. This test showed that climate alone could explain the postglacial pattern of *Carya*, and that migrational lag need not be a factor. The early papers by Davis emphasized warming during the Holocene but neglected the importance of precipitation. *Carya* migrated northwards rapidly in the Midwest (Bettis *et al.*, 1990), where the early Holocene was wet, but its expansion (but perhaps not migration) was delayed in the Northeast, which was dry in the early Holocene (Grimm *et al.*, 2001; Prentice *et al.*, 1991; Webb *et al.*, 1993a). Finally, Jackson *et al.* (1997) assembled a macrofossil database to complement the pollen database and mapped both macrofossils and pollen. The macrofossils provide a better resolution of range limits, although not abundance; and they show rapid northward expansion of tree species following deglaciation. Jackson *et al.* (1997) concluded that significant migrational lag did not occur at regional to subcontinental scales.

Much of the discussion regarding disequilibrium has focused on Holocene vegetation following rapid climate change in the late-glacial period. Another issue has been the possibility of disequilibrium during the time of rapid late-glacial climate change, in particular as an explanation for non-analog vegetation during this time. However, the pollen assemblages most dissimilar from modern preceded the greatest rates of climate and vegetation change (Williams *et al.*, 2001), as did the greatest extent of non-analog vegetation (Overpeck *et al.*, 1992). These observations severely weaken the disequilibrium hypothesis for late-glacial non-analog vegetation. In addition, the timing of the greatest rate of vegetation change corresponded with the greatest rate of climate change. Non-analog vegetation was associated with non-analog climate. Thus, plant associations remained in equilibrium with climate at the 1000-yr time scale during the late-glacial period, and non-analog vegetation was the result of non-analog climate (Overpeck *et al.*, 1992; Williams *et al.*, 2001), the alternative hypothesis of Cushing (1965).

Physiological Controls on Paleovegetation

Quantitative reconstruction of climate from paleovegetation data has relied chiefly on correlation between taxon ranges and climate variables. However, paleoecologists have long realized that autecological and physiological characteristics of plants should be useful for interpreting climate history. Frustratingly little is known, however, about the physiology of even common, commercially important forest trees. Consequently, the precise explanations for climate-plant distributional relationships are still sketchy, and range limits of individual taxa frequently are fundamentally understood only when they involve major climate boundaries such as the 0 °C isotherm (freezing) or the –40 °C isotherm, the approximate limit of super-cooling for many species (Woodward, 1987). Many future advances in terrestrial paleoecology and paleoclimatology will benefit from basic information about autecological phenomena such as limits to germination, survival of seedlings and early growth-stages of plants, day-length requirements, and even the autecology of endemic plant pathogens. In addition, because plant groups can have considerable ecological and physiological variability, improvements in palynology and in the use of macrofossils that allow identification of individual species can lead to more highly resolved paleoenvironmental insights. We discuss below examples of two important pollen taxa – *Picea* and *Ambrosia* – for which increased taxonomic resolution and understanding of physiology has contributed to a greater paleoclimate understanding.

Picea

Picea is one of the most important pollen taxa for paleoclimatic interpretation. It dominated vast, but different, geographic regions during the Pleistocene and Holocene. *Picea glauca* and *P. mariana* have closely congruent ranges today but favor different soils (Nienstaedt & Zasada, 1990; Viereck & Johnston, 1990). In southeastern Labrador, discrimination of these two species revealed that *Picea glauca* became established first and that widespread paludification after 5700 cal yr B.P. favored *P. mariana*, which almost completely replaced *P. glauca* by 4400 cal yr B.P. (Engstrom & Hansen, 1985; Lamb, 1980, 1984). The change in *Picea* species implies a change from drier to wetter climate, which is missed by quantitative studies that do not differentiate the species (because most palynologists have not).

In New England, recent improvements in pollen taxonomy and application of species-climate insights have illuminated the paleoecology of the three *Picea* species that occur there. Lindbladh *et al.* (2002) established a reliable method for differentiating among *Picea glauca*, *P. mariana*, and *P. rubens*. Because the ecological adaptations of these three species are quite different, their distributions since the last glacial maximum are of considerable interest. *Picea rubens* has a much more restricted distribution spatially and climatically, occurring only in northern New England and the Maritime provinces of eastern Canada, where it is an important component of the upland vegetation. Initial application of the new method to existing fossil pollen samples from New England indicates that *P. rubens* became abundant in the region only during the late-Holocene expansion of *Picea*; apparently it was not important in the late-glacial *Picea* forests of the Northeast (Lindbladh *et al.*, 2003). *P. rubens* is more sensitive to low winter temperatures than the other *Picea* species. Today, the Maine-Maritimes region is less subject to extreme outbreaks of winter cold than is the mid-continent, and heavy winter snowfall protects the young trees. Precession-driven differences in seasonal insolation would have made late-glacial and early-Holocene winters more extreme and thus less suitable for *P. rubens*. Clearly the late-Holocene emergence of *P. rubens* in the Northeast carries a somewhat different and more restricted climate signal from that of the trans-continental distributions of *P. glauca* and *P. mariana*. Where *P. rubens* lived during the late-glacial interval and the early Holocene remains to be discovered, although the central Appalachians seem a likely possibility.

During the Pleistocene, *Picea* pollen occurred as far south as Louisiana (Jackson & Givens, 1994) and the Florida panhandle (Watts *et al.*, 1992), not with its usual boreal associates of today such as *Larix* and *Abies*, but rather with temperate deciduous trees. As noted by Whitehead (1965), an unusually long-coned Pleistocene form of *Picea* has been known for some time from the Southeast, and Jackson & Weng (1999) have recently demonstrated that this form is an extinct species, which they named *Picea critchfieldii*. During the late Pleistocene, *P. critchfieldii* ranged from Louisiana to Georgia and as far north as Tennessee (Jackson & Weng, 1999). In the Tunica Hills of Louisiana, macrofossil and pollen studies showed that it occurred with temperate deciduous trees such as *Quercus*, *Fagus*, *Fraxinus*, *Carya*, *Juglans nigra*, *Acer*, and *Ulums* (Jackson & Givens, 1994). *Picea critchfieldii* had a more southerly distribution than the extant boreal *Picea* species, and the inclusion of fossil pollen from this species in quantitative paleoclimate reconstructions is a potential source of error.

Ambrosia

Plant-physiological insights from *Ambrosia* have shed light on several mysteries about late-Quaternary climate at both the northern and southern limits of its range. This native taxon is a major pollen producer and is a characteristic indicator of prairies and grasslands. However, because *Ambrosia* is a weed that expanded greatly with European settlement and land disturbance, many investigators have ignored it for paleoclimatic reconstruction. Indeed, modern relationships between *Ambrosia* abundance and climate would not be appropriate for reconstructing past climate via response surfaces or transfer functions. However, because *Ambrosia* is a serious agricultural weed, its physiology has been much studied. *Ambrosia* seeds require cold stratification for germination, which will occur only in spring; they re-enter secondary dormancy, requiring restratification if external requirements for germination are not met by late spring (Baskin & Baskin, 1977a, b; Bazzaz, 1970; Stoller & Wax, 1973; Willemsen, 1975a, b). Germination requires light, moist soils, and a daily cycle of cool nights and warm days (Dickerson, 1968; Raynal & Bazzaz, 1973; Shrestha *et al.*, 1999). After emergence, phenological development of *Ambrosia* is dependent on temperature and photoperiod (Deen *et al.*, 1998a, b; Shrestha *et al.*, 1999). Although *Ambrosia* must germinate in the spring, it is a quantitative short-day plant, which grows slowly during the long days of mid-summer. Ideal day length for *Ambrosia artemisiifolia* is 14 hr or less (Baskin & Baskin, 1977b; Deen *et al.*, 1998b). As day length shortens in late summer, *A. artemisiifolia* grows more rapidly, flowering and fruiting in late summer and autumn (Deen *et al.*, 1998a, b).

During the mid-Holocene on the northern Great Plains, percentages of *Ambrosia* were very high, but fluctuating. The high values were greater than in any protohistoric surface samples. Some modern samples have equally high *Ambrosia* percentages, but they are due to human disturbance. Thus, the mid-Holocene values do not have a protohistoric analog. Today, *Ambrosia* is not an important weed in the northern Great Plains. It increased slightly with European settlement in the eastern part of the region and not at all in the western part. The primary marker of Eurosettlement in the northern plains is a rise in Chenopodiineae. The 14-hr day length necessary for *Ambrosia* maturation is not reached in the northern plains until late August, and the time between the attainment of short day lengths and first frost limits its northern distribution (Grimm, 2001).

Lake levels in the northern Great Plains were low during the mid-Holocene, and a variety of evidence indicates substantially drier climate during this time. However, *Ambrosia* is not very drought tolerant and requires ample summer precipitation to survive. Today, its optimum range is in the Corn Belt, southeast of the northern Great Plains (Grimm, 2001). Of the major prairie pollen-types (*Ambrosia*, *Artemisia*, other Asteraceae, Chenopodiineae, Poaceae), *Ambrosia* is associated with the highest precipitation values (Webb & Bernabo, 1977; Webb *et al.*, 1983). Warmer temperatures during the mid-Holocene may partly explain the higher values of *Ambrosia* in the northern plains, but the very high values are not consistent with summer drought. A suggestion has been that *Ambrosia* was growing on drawdown lake muds, but *Ambrosia* does not occur in these habitats today, whereas Chenopodiineae do. The spring germination requirement for *Ambrosia* also suggests that it was not growing on drawdown mud flats, which typically appear later in summer. The very high percentages of *Ambrosia* imply significant summer precipitation and bare ground on the uplands. The explanation for this seeming paradox is winter drought. Hydrologic and isotopic studies show that winter (late autumn to early spring) precipitation, including snowmelt, controls groundwater and lake levels in the northern plains; summer precipitation, although greater, is mostly evapotranspired (Winter & Rosenberry, 1995; Yu *et al.*, 2002). Thus, low lake levels during the mid-Holocene imply winter drought. Winter precipitation recharges groundwater and soil moisture, thereby buffering the effects of summer drought. Summer precipitation is highly variable on the plains today, and no doubt was in the past. With the elimination of the buffering effect of winter precipitation, an ordinary dry summer would have been more devastating to the vegetation, creating bare ground ideal for *Ambrosia* colonization. Ordinary wet summers would have provided sufficient moisture for *Ambrosia* germination and survival. Thus, summer drought in the mid-Holocene may or may not have been any more severe than today; however, because of a persistent deficit in winter precipitation, the normal variations in summer rainfall produced greater response in the vegetation.

Winter drought also explains another seeming mystery in northern Great Plains paleovegetation. Today, as climate becomes drier westward, *Artemisia* becomes more abundant. However, *Artemisia* did not become more abundant in the mid-Holocene when climate was "drier," as might be expected. The explanation is that the deep-rooted *Artemisia* relies on winter precipitation that soaks deep into the soil, and it cannot compete with shallow-rooted grasses for summer moisture. Thus, winter drought in the mid-Holocene did not favor an increase in *Artemisia* (Grimm, 2001).

The physiological adaptations of *Ambrosia* also clarify interpretation of climate at the southern margin of its range. In the Florida peninsula, the last 60,000 years are characterized by periods of pine dominance alternating with open grassland-scrub oak vegetation (Grimm *et al.*, 1993).

Ambrosia is abundant during grassland phases in the Pleistocene. The Heinrich events occurred during the Pleistocene pine phases, which were very similar palynologically to the late Holocene pine forest. The most distinctive zone of the last 60,000 years was the early Holocene, which was a grassland phase, but without *Ambrosia*. Today, *Ambrosia* is relatively uncommon this far south, probably because of the cold stratification and cool evening temperatures required for successful germination. The abundance of *Ambrosia* during the Pleistocene grassland phases implies cooler temperatures.

The mid-Holocene grasslands of the northern Great Plains and the Pleistocene grasslands of Florida perhaps have modern analogs if prairie pollen types are lumped together as "prairie forbs." Notably, however, *Ambrosia* is often left out of the forb category for paleoclimate reconstruction because of its modern abundance that is related to human disturbance (Webb *et al.*, 1993b, 1987). In fact, the major prairie types have different distributions within the immense North American grasslands, and the lumping of types conceals considerable variability within the grasslands. The high-*Ambrosia* prairie of the mid-Holocene in the northern Great Plains probably does not have a protohistoric analog, nor may the high-*Ambrosia* Pleistocene grasslands of Florida. Nevertheless, knowledge of the physiology of *Ambrosia* contributes greatly to climatic interpretation.

Late-Glacial Biomes in the Midwest

Williams *et al.* (2001) defined two late-glacial non-analog biomes, spruce parkland and mixed parkland. The mixed parkland was widespread in the Midwest, especially at the 14,000 and 13,000 cal yr B.P. time slices, which fell within the Bølling-Allerød chronozone. This well-known non-analog vegetation type consisted of a mixture of conifers and deciduous trees. Important conifers were *Picea*, *Larix*, and *Abies*, but notably not *Pinus*. Important deciduous trees were *Fraxinus nigra*-type, *Ostrya/Carpinus*, and *Populus*. Other deciduous trees were present, including *Ulmus* and *Quercus*. *Fraxinus nigra*-type was particularly abundant. Non-analog vegetation types indicate different climate in the past, but precisely because these vegetation types do not exist anywhere at present, they are a challenge for quantitative reconstruction of past climate. Interpretation can be based on the ecological requirements of the dominant tree species, although non-analog climate may have produced responses not observable today.

Bryson & Wendland (1967) proposed that during the late-glacial period the Laurentide Ice Sheet blocked arctic air from entering the Midwest and that air draining from the ice sheet warmed adiabatically so that winters south of the ice sheet were milder and drier than today. They also proposed that the presence of the ice sheet and tropical air to 40°N would have strengthened the westerlies between 40°N and 45°N, forcing dry Pacific air into the Midwest during summer, resulting in a climate drier than today. The occurrence of "thermophilous" deciduous trees with *Picea* is consistent with the idea of winters mild enough for deciduous trees, which cannot endure extremely cold winter temperatures, but with summers cool enough for *Picea*, which cannot endure very warm temperatures (Woodward, 1987). Many authors have focused on the significant presence of *Artemisia* during the late-glacial period and have suggested that it signifies a drier climate, e.g. Amundson & Wright (1979).

Climatic interpretation from response surfaces indicates an increase in mean January temperatures beginning 15,000 ^{14}C yr B.P. and a substantial increase in mean July temperature beginning 12,000 ^{14}C yr B.P. (Webb *et al.*, 1987). Lake-level data indicate higher-than-present effective moisture in the Midwest at 12,000 ^{14}C yr B.P. (Webb *et al.*, 1987). Thus, these data are consistent with a warm, wet climate in the Midwest during the late-glacial period. However, simulations with Community Climate Model 1 (CCM1) for 16,000 and 14,000 cal yr B.P. indicate a "hypercontinental" climate with winters colder than present, summers warmer than today, and less precipitation (Williams *et al.*, 2001). The data do not support the model simulation. Williams *et al.* (2001) suggest that the mixed parkland biome was most similar to the modern spruce-aspen-oak parklands north of the Canadian prairies in Alberta and Saskatchewan, which have a cold, dry, continental climate. However, they do point out that, significantly, these modern parklands lack *Fraxinus*, a major component of the late-glacial mixed parkland biome, and that *Fraxinus* and other taxa such as *Salix* suggest moist conditions.

The abundance of *Fraxinus nigra*-type pollen in the Midwest during late-glacial time is especially problematic. This pollen type includes *Fraxinus nigra* and *F. quadrangulata*. *Fraxinus quadrangulata* occurs in the southern Midwest and mid-South; *Fraxinus nigra* is a more northern species, and its range overlaps with the boreal *Picea* species. Almost certainly, *Fraxinus nigra* was the source of the late-glacial and early Holocene *Fraxinus nigra*-type pollen. *Fraxinus nigra* grows in swamps and on poorly drained soils (Wright & Rauscher, 1990). Therefore, the great abundance of *Fraxinus nigra* in the late-glacial and early Holocene intervals implies high soil moisture and does not support the simulation of hypercontinental climate.

The late-glacial occurrence of *Fraxinus nigra* and *Artemisia* is puzzling. Perhaps the activities of probocideans created openings in the forest, which *Artemisia* colonized. Timing, however, is an issue. Dating of the late-glacial period is notoriously poor. The well-dated late-glacial section from Nelson Lake in northeastern Illinois shows that prior to the Bølling-Allerød chronozone ($\sim$14,700 cal yr B.P.), *Picea* dominated, with about 10% *Fraxinus nigra*-type pollen and $\sim$5% *Artemisia*. During the Bølling-Allerød (14,700–12,900 cal yr B.P.), *Picea* fell, *Artemisia* fell to insignificant levels, and *Fraxinus nigra* became especially abundant (Grimm & Maher, 2002). Isopoll maps also show that the late-glacial maximum of *Artemisia* occurred before the major expansion of *Fraxinus* (Jacobson *et al.*, 1987). *Abies* increased during the Bølling-Allerød, and it also is an indicator of abundant precipitation (Prentice *et al.*, 1991; Thompson *et al.*, 2000; Webb *et al.*, 1983). *Picea* can flourish at warmer temperatures if precipitation and soil moisture are high (Webb *et al.*, 1993a). Thus, the temperatures suitable for *Picea* in the Midwest may not have been as cool during the wet late-glacial as they are today. The lack of *Pinus* and *Betula* is a striking feature of the late-glacial *Picea* forest compared to the modern boreal forest (Wright, 1964, 1968b; Wright *et al.*, 1963), where these taxa are typically postfire successional species. Thus, moist climate and lack of fire may explain the absence of *Pinus* and *Betula* from the late-glacial *Picea* forest.

In conclusion, we suggest that climate was relatively dry during early late-glacial time prior to 14,700 cal yr B.P., but that wet conditions prevailed afterwards. The high values of *Fraxinus nigra* and the absence or low values of *Pinus* and *Betula* during this time support this interpretation. A reasonable climatic scenario is that during the late-glacial summer, warm tropical air collided with cold air draining from the retreating Laurentide Ice Sheet, producing abundant precipitation in the Midwest. Florida was dry during this time, suggesting placement of the Bermuda High over the Peninsula, and anticyclonic winds spun moisture from the Gulf of Mexico into the continental interior.

Late-Glacial Climatic Oscillations

The now-classic Bølling-Allerød/Younger Dryas climate sequence had become well established in Scandinavia before modern paleoecological research was in North America. Naturally, some of the earliest studies in the Northeast (Deevey, 1951) included comparisons with the northern European patterns. Although Deevey conducted his pioneering studies in Maine before the advent of radiocarbon dating, he nevertheless concluded that changes similar to Younger Dryas cooling were evident in cores from several of the sites. He noted especially that the lithologic sequence of increasing organic content of late-glacial sediments was interrupted by a reversion to low organics and was followed by an abrupt rise to Holocene organic-rich gyttja, closely resembling the Scandinavian pattern.

Subsequently, however, as radiocarbon dating of lake sediments provided the first glimpses of spatial and temporal shifts in vegetation, the patterns of particular biological changes in northeastern North America were recognized to be time-transgressive, with no specific events emerging with broad synchronous change. In fact, closely spaced analyses of the transition from *Picea* to *Pinus* in northern Minnesota revealed an ecologically logical south-to-north change through time during the late-glacial period, with no apparent semblance of the late-glacial climate reversals elsewhere (Amundson & Wright, 1979). Farther east, detailed examination of radiocarbon-dated pollen records from nearly 50 sites in northern New England and adjacent Canada led Davis & Jacobson (1985) to conclude that no return to cooler conditions had occurred after the onset of postglacial warming in that region, and specifically that Younger Dryas cooling was not evident in those plen records.

Recognition that Deevey may well have been correct in his pre-radiocarbon-dating insights came with the discovery by Mott *et al.* (1986) of peaty deposits buried below inorganic diamictons at several sites in New Brunswick and Nova Scotia; these dated 11,000–10,000 ^{14}C yr B.P., suggesting that either a local ice advance or some significant disturbance in the development of stable postglacial landscapes had interrupted the late-glacial onset of plant growth. Studies with high temporal resolution, coupled with the advent of AMS dating of lake sediments in the region, then led to the discovery that the lithologic sequence noted by Deevey (1951) was repeated in virtually every lake in the Maine-New Brunswick region and that the segment of low organics was essentially contemporaneous with the Younger Dryas cooling in Scandinavia (Borns *et al.*, 2003; Cwynar & Levesque, 1995; Doner, 1996; Dorion, 1997a, b; Levesque *et al.*, 1994, 1996; Mayle & Cwynar, 1995a, b; Mayle *et al.*, 1993; Stea & Mott, 1998). In fact, the inorganic-organic-inorganic-organic sequence is present in all sites except those that were still ice covered during the Younger Dryas interval (Dorion *et al.*, 2001). Pollen data indicate that cold-adapted plants returned to prominence during that period (Borns *et al.*, 2003; Doner, 1996; Mayle & Cwynar, 1995a, b; Mayle *et al.*, 1993), so the occurrence of the Younger Dryas cold period in northeastern North America is well supported. The failure of earlier studies to recognize a cold interval coeval with the Younger Dryas was probably owing to non-analog pollen assemblages and to complicated biotic response. In northern Maine and the Maritimes, *Picea* declined while non-arboreal taxa such as Cyperaceae and *Artemisia* increased during the Younger Dryas chronozone (Doner, 1996; Mayle & Cwynar, 1995a, b; Mayle *et al.*, 1993). In addition to the Younger Dryas, high-resolution pollen, midge, loss-on-ignition, and AMS radiocarbon analyses of sites in the Maritimes reveal an earlier cold period – the Killarney Oscillation – that is coeval with the intra-Allerød cold period (Levesque *et al.*, 1993a, b, 1994, 1996, 1997).

In southern Ontario, stable isotope (δ^{18}O and δ^{13}C) studies of carbonates from Crawford Lake and Twiss Marl Pond show a striking detailed correlation of late-glacial climatic events with Europe and the North Atlantic. Not only are the Bølling-Allerød and Younger Dryas evident, but also minor cold events during the Bølling-Allerød, including the intra-Bølling cold period, the Older Dryas, and the intra-Allerød cold period (Yu, 2000; Yu & Eicher, 1998, 2001; Yu & Wright, 2001).

Early studies in the Midwest (Wright *et al.*, 1963) attempted to correlate vegetation changes with glacier advances and retreats, although uncertain radiocarbon dating hampered precise correlation. Investigators in the Midwest found little evidence of a climatic reversal coeval with the European Younger Dryas until Shane (1987) demonstrated a cooling south of the Great Lakes in Ohio and Indiana contemporaneous with the Younger Dryas. A *Picea* recurrence near the top of the *Picea* zone, although often not well dated, occurs at many sites in the upper Midwest (Rind *et al.*, 1986; Shane & Anderson, 1993). High-resolution AMS dating of sites in northern Illinois and southern Wisconsin shows a *Picea* recurrence exactly coeval with the Younger Dryas, as defined in the GISP 2 ice core; in addition, it shows lower *Picea* percentages and high *Fraxinus nigra*-type coincident with the Bølling-Allerød (Grimm & Maher, 2002). Investigators have recently correlated the Marquette readvance of the Laurentide Ice Sheet into Wisconsin and the Upper Peninsula of Michigan with the Younger Dryas (Drexler *et al.*, 1983; Farrand & Drexler, 1985; Larson & Schaetzl, 2001; Lowell *et al.*, 1999). This readvance ended at the end of the Younger Dryas interval. Björck (1985) studied Rattle Lake, which is located on a highland that during and immediately prior to the Younger Dryas was located south of Lake Agassiz and north of the Superior Lobe, which was advancing southward. The late-Allerød pollen assemblage is strange, consisting of the trees *Picea*, *Fraxinus nigra*, and *Ulmus* together with ~30% *Artemisia* and ~4–8% *Ambrosia*. The deciduous trees and *Ambrosia* imply warm summer temperatures – consistent with higher summer insolation at the time – even though Rattle Lake lay between Lake Agassiz and the Superior ice lobe. Alternatively, pollen from thermophilous plants is due to long-distance transport from the south, and the assemblages are quantitatively uninterpretable. Because concentration of pollen is low in the late-Allerød sediments, pollen derived from long distances may be proportionally over-represented. The thermophilous taxa decreased or disappeared during the Younger Dryas, and the vegetation was dwarf shrub tundra. If the pollen from thermophilous taxa were derived from long-distance transport, then their demise during the Younger Dryas at Rattle Lake suggests a major decline in these populations to the south and a significant Younger Dryas climatic event.

Both vegetation and the ice sheet responded to the Younger Dryas cooling event in the Midwest. In addition, stable-isotope analysis of a speleothem from a Missouri cave shows approximately a 4 °C temperature depression associated with the Younger Dryas (Denniston *et al.*, 2001). Thus, evidence for the Younger Dryas in the Midwest extends from Ontario and the Upper Peninsula of Michigan to Missouri. Lowell *et al.* (1999) noted that the massive Laurentide Ice Sheet responded more slowly to Younger Dryas cooling than did smaller glaciers. Conceivably, the response times of vegetation and speleothem isotopic composition could be more rapid, and differences in response times could explain apparent offsets between these rapidly responding proxies and glacier movements.

Yu & Wright (2001) recently reviewed the mounting evidence for the Younger Dryas and other amphi-Atlantic climate events in the interior of North America. Much of this evidence is derived from AMS-dated, high-resolution, multi-proxy studies. Shuman *et al.* (2002) argued that colder sea-surface temperatures during the Younger Dryas altered atmospheric circulation, which modified climate globally. However, the response was spatially varied, not necessarily warm-cold-warm everywhere. Nevertheless, synchronous change was widespread, a conclusion also supported by rates-of-change analyses (Grimm & Jacobson, 1992; Jacobson *et al.*, 1987). However, several of the sites featured in the Shuman *et al.* (2002) analysis have suspect bulk-sediment radiocarbon dates. For one site, Devils Lake, Wisconsin, conventional decay-count dates of bulk sediment had placed the *Picea*/*Pinus* transition at the beginning of the Younger Dryas, suggesting warming, whereas new AMS dates of macrofossils place the transition at the end of the Younger Dryas (Grimm & Maher, 2002), implying cold Younger Dryas climate rather than warm.

In many cases, the inability to correlate late-glacial pollen events with the European-North Atlantic sequence may be the result of poor dating, highly variable sedimentation rates, low-resolution pollen counting, difficulties interpreting non-analog pollen assemblages, and gaps in basal core segments. Interpretation of non-analog vegetation types may be a particular problem. In the southern Great Lakes region, the recurrence of *Picea* and the decline in deciduous trees is an unambiguous indicator of colder climate, even though the vegetation has no modern analogs. In other areas, however, the seemingly bizarre assemblages of prairie forbs (*Artemisia*, *Ambrosia*), wet-soil deciduous trees (*Ulmus*, *Fraxinus nigra*), and boreal taxa (*Picea*) challenge and obfuscate climatic interpretation. Puzzling still is the lack of correlation of vegetation changes with the marked $\delta^{18}O$ signal of climatic change in sites west of Lake Ontario (Yu, 2000; Yu & Eicher, 1998; Yu & Wright, 2001). At Nelson Lake in northern Illinois, the Younger Dryas sediments are thin, and the *Picea* recurrence appears fairly insignificant on a depth scale. However, detailed AMS dating shows that sedimentation was very slow during the Younger Dryas, and that the time span of the *Picea* recurrence was more than a millennium (Grimm & Maher, 2002). Wide-interval counting and dating, typical of many late-glacial pollen profiles, could easily have missed the entire Younger Dryas event. These same problems contribute to the inability, in many cases, to correlate vegetation events south of the ice sheet with ice advances and retreats, which often are also imprecisely dated. Sites such as Kirchner Marsh (Wright *et al.*, 1963) show significant fluctuations during the *Picea* zone, and it may well be that high-resolution AMS dating and pollen counting will reveal a close correspondence with movements of the Laurentide Ice Sheet and with the European-North Atlantic sequence, not only with the larger Bølling-Allerød/Younger Dryas signal, but also with finer-scale events such as the Older Dryas and intra-Allerød cold period.

Convincing evidence now exists from New England to the Midwest that glaciers readvanced during the Younger Dryas interval and that climate was cold. The Younger Dryas event is also evident in cores from Florida, and the timing and duration of this event closely matches the GISP 2 ice-core record. However, in Florida, this event was warm (E.C. Grimm *et al.*, unpublished data). Apparently, the shutdown in thermohaline circulation that caused cooling in northern regions caused warming in the subtropics as heat normally transported from the subtropics by the Gulf Stream remained there. Thus, the Florida data support the general contention by Shuman *et al.* (2002) that the climate change during the Younger Dryas was synchronous but spatially variable.

The Early Holocene Pine Zone in the Northeast

The migration maps of Davis (1976, 1981b, 1983a) focus attention on the absence of deciduous trees such as *Carya* and *Castanea* from the Northeast in the early Holocene, which Davis attributed to migrational lag. *Pinus*, mainly *Pinus strobus*, dominated the Northeast during the early Holocene, and the climate signal of the *Pinus* zone was not fully appreciated. It is now clear that the early Holocene in the Northeast was warm and dry, favorable for *Pinus* and unfavorable for deciduous trees. Pollen and macrofossil studies by Davis *et al.* (1980) show that *Pinus strobus* grew as much as 350 m above its present limit in the White Mountains of New Hampshire soon after arriving ~10,200 cal yr B.P., indicating that climate was already warmer than present by that time. *Tsuga* grew 300–400 m above its modern limit soon after arriving ~7800 cal yr B.P. *Pinus strobus* not only reached is maximal altitudinal limits in the early Holocene, it also reached its maximal latitudinal limits, attaining its modern range by 10,200 cal yr B.P. (Jacobson & Dieffenbacher-Krall, 1995). Webb *et al.* (1993a) used pollen response surfaces to reconstruct soil moisture and mean annual temperature for the Northeast and concluded that climate was drier than today during the *Pinus* zone, ~10,200–6800 cal yr B.P. A number of studies indicate that lake levels were low at the time of the *Pinus* zone (Almquist *et al.*, 2001; Almquist-Jacobson & Sanger, 1995; Dwyer *et al.*, 1996; Newby *et al.*, 2000; Webb, 1990), and fire frequency was higher than in the hardwood forest that succeeded it (Anderson *et al.*, 1992; Clark *et al.*, 1996).

In summary, the climate of the Northeast was warmest and driest during the early-Holocene *Pinus* zone from about 10,200 to 6000 cal yr B.P. – a time of enhanced summer insolation. The climate became moister and cooler after the *Pinus strobus* phase, with the establishment of hemlock-northern hardwoods forest and ultimately the recent rise of the *Picea-Abies* forest (Schauffler & Jacobson, 2002). Thus, the scarcity or absence of deciduous trees from the Northeast in the early Holocene may have been due to dry climate, and not migrational lag.

Early and Mid-Holocene Vegetation and Climate in the Midwest

Although a postglacial "xerothermic" or "hypsithermal" period of more arid climate had been hypothesized for some time (Cooper, 1958; Deevey & Flint, 1957; Gleason, 1922; Transeau, 1935; Wright, 1976), the classic studies of Wright *et al.* (1963) and McAndrews (1966) unequivocally demonstrated the existence and timing of mid-Holocene prairie expansion in Minnesota between approximately 9000 and 4500 cal yr B.P. Lacking sites from the heart of the Prairie Peninsula, Wright (1968a) extrapolated the Minnesota chronology to the Prairie Peninsula as a whole and postulated a postglacial climatic curve that was "one of steady change toward a time of maximum warmth and dryness about [8000 cal yr B.P.], followed by a reversal that was very slow until about [4500 cal yr B.P.] and then somewhat faster." King (1981) essentially followed Wright's climatic reconstruction in his interpretation of his pollen profiles from Volo Bog and Chatsworth Bog, the first radiocarbon-dated pollen diagrams from Illinois. Especially critical is Chatsworth Bog, from which comes the only complete Holocene pollen diagram from the Grand Prairie of Illinois, the heart of the tall-grass prairie. Webb *et al.* (1983) suggested a climatic history of the Prairie Peninsula somewhat more complicated than Wright's unimodal model. With data from an array of sites from across the Midwest, they mapped the 20% isopoll for "prairie forbs," which approximately tracks the prairie-forest border. The unimodal model still held for Minnesota, with maximal prairie development early, about 8000 cal yr B.P., then declining; however, in Illinois prairie development was bimodal, first appearing about 9000 cal yr B.P., contracting somewhat, then reaching maximal development later, 6800–3200 cal yr B.P. This reconstruction was pinned on the Chatsworth Bog profile.

Pollen data from Roberts Creek in northeastern Iowa (Baker *et al.*, 1996; Chumbley *et al.*, 1990) and Money Creek and Pine Creek in southeastern Minnesota (Baker *et al.*, 2002) unambiguously show the later development of prairie in the eastern Prairie Peninsula; *Ulmus* forest predominated throughout the early Holocene from the demise of the late-glacial *Picea* forest until ~6200 cal yr B.P., with maximal development of prairie ~6200–3800 cal yr B.P. Thus, mesic forest prevailed in the driftless region of southeastern Minnesota and northeastern Iowa during the driest part of the prairie period in central Minnesota. Additional sites along the northern fringe of the Prairie Peninsula show this later maximum of aridity extending to northeastern Illinois (Baker *et al.*, 1992). Stable-isotope studies and high-precision uranium-series dating from Cold Water Cave in northeastern Iowa indicate maximal Holocene aridity 5900–3600 yr B.P., consistent with the plant-fossil evidence (Baker *et al.*, 1998; Dorale *et al.*, 1992). However, stable-isotope studies from Spring Valley Cave and Mystery Cave in southeastern Minnesota, which are only about 50 km northwest of Cold Water Cave, indicate an early Holocene development of prairie about 8000 cal yr B.P. Thus, these studies fix the position of prairie-forest border between the caves in southeastern Minnesota and northeastern Iowa during the period 8000–6000 cal yr B.P. The plant-fossil and speleothem $\delta^{13}C$ data show that during these two millennia a fairly sharp gradient existed between arid climate to the west and humid climate to the east. Spring Valley Cave lies very close to the modern prairie-forest border in southeastern Minnesota, which is topographically controlled essentially along the margin of the driftless area. Immediately prior to European settlement, the prairie-forest border lay roughly parallel to a continentally steep moisture gradient but was locally fixed along firebreaks and was much sharper than the moisture gradient itself. In many places firebreaks sharply demarcated prairie and forest, with no climatic difference on opposite sides of the break (Grimm, 1984). Locally the interactions among fire, vegetation, and firebreaks sharpened the boundary between prairie and forest biomes, which are climatically controlled on a continental scale but are physiographically controlled on a more local scale. Undoubtedly these interactions existed in the past. Thus, the climatic gradient 8000–6000 years ago may not have been as sharp as implied by the vegetation gradient.

About 8000 years ago, the prairie-forest border became established near its modern position; about 6000 years ago, prairie rapidly advanced eastward, then returned to its modern position about 3000 years ago. However, this position was a climatic fulcrum throughout the middle Holocene. Prior to 6000 years ago maximal prairie development and aridity ocurred to the west, while *Ulmus* forest and humid climate persisted to the east. Then, after 6000 years ago, prairie retreated and climate became less arid to the west, while maximal prairie and aridity developed to the east. After 3000 years ago, climate became more humid across the entire Prairie Peninsula region.

Originally, Dorale *et al.* (1992) interpreted the $\delta^{18}O$ data from Cold Water Cave to imply a 3–4 °C temperature increase associated with maximal development of prairie. However, subsequent studies of additional speleothems from Cold Water Cave (Denniston *et al.*, 1999a) and from Spring Valley and Mystery Caves (Denniston *et al.*, 1999b) showed that the temperature signal is over-estimated and unreliable. In addition to temperature, $\delta^{18}O$ depends on evaporative enrichment, moisture source, and seasonality of precipitation (Denniston *et al.*, 1999a, b; Yu *et al.*, 1997).

Denniston *et al.* (1999a) found that in the stalagmite from Cold Water Cave with the least evaporative enrichment of ^{18}O was depleted during the time of maximal prairie development, implying cooler temperatures. Yu *et al.* (1997) found depleted ^{18}O values in authigenic marl from Crawford Lake in southern Ontario during the period of lowest lake levels, which would imply less evaporative enrichment (low lake levels imply increased evaporation). Because these implications are unreasonable, both groups of investigators suggested that the moisture source or seasonality of precipitation must have changed. Cool-weather precipitation is depleted in ^{18}O, and a relative increase in winter precipitation could account for the isotopic observations. This interpretation is opposite that for reduced winter precipitation in the Great Plains during the time of maximum drought (Grimm, 2001). The Great Plains are west of the fulcrum, and climate may have been different. If both scenarios are correct, then winter precipitation was relatively reduced across the entire region prior to 6000 years ago. To the west, increased winter precipitation in the mixed prairie after 6000 years ago caused lake levels to rise, recharged soil moisture, buffered summer droughts, and promoted greater biomass production. The question in the east is whether winter precipitation increased after 6000 years ago or summer precipitation decreased; either case would increase the proportion of winter precipitation, which would lighten annual meteoric $\delta^{18}O$. In any event, given the multiplicity of factors that can influence carbonate $\delta^{18}O$, the hypothesis for increased relative winter precipitation in the east must be regarded as tentative.

The late mid-Holocene period of maximal aridity extended at least as far east as Crawford Lake, just east of Lake Ontario, where it occurred between ~5500 and 2000 cal yr B.P. (Yu *et al.*, 1997). This period is somewhat later even than in Iowa and Illinois, suggesting that the spatially time-transgressive maximum of Holocene aridity continued eastward to southern Ontario and possibly even into New England, where the time of maximal aridity (or at least lowest lake levels) was between 6000 and 3000 cal yr B.P. (Almquist *et al.*, 2001; Newby *et al.*, 2000; Webb *et al.*, 1993a).

Following the influential papers by Borchert (1950) and Bryson & Wendland (1967), many authors have recognized the relative dominance of Gulf, Pacific, and Arctic air masses as strongly influencing the amount of precipitation in the Midwest. However, the position of the jet stream and storm tracks is also of paramount importance (Rodionov, 1994), both in summer and winter. Zonal flow may bring dry Pacific air into the region, whereas a more developed trough-ridge system will bring humid air from the Gulf. Humid Gulf air may predominate in summer but still produce little precipitation if the jet stream and colder air masses necessary to cause lift remain too far north. A ridge-trough system in winter will bring in very dry Alberta air, whereas more meridional flow will bring more precipitation. Working out the detailed climate history for the eastern United States will help understand the climate mechanism.

Time-Transgressive Versus Synchronous Change

The isochrone maps pioneered by Davis illuminate the time-transgressive nature of plant migrations, and isopoll maps reinforce this time-transgressiveness as ranges of taxa move across the continent over the millennia. In particular, H.E. Wright emphasized the time-transgressive nature of climate and vegetation change (Watson & Wright, 1980; Wright, 1976, 1981) and specifically rejected attempts of Bryson & Wendland (1967) and Bryson *et al.* (1970) to apply the typological Blytt-Sernander sequence to North America. Yet the temptation to correlate North American vegetation changes with the strong North European signal remained alluring. The temptation became even greater with the publication of the very strong climate signals evident in the Greenland ice cores and with the worldwide record of events in marine cores. Ice-core and marine-core data suggest associated changes in oceanic thermohaline circulation and major reorganizations of atmospheric circulation (Broecker & Denton, 1989). Of particular attraction was the Younger Dryas event, for which evidence was emerging worldwide and which climate models suggested should have a worldwide signal, albeit mixed, colder in some areas, warmer in others (Rind *et al.*, 1986). Rates-of-change analyses of pollen data showed synchronous changes across eastern North America, even though the qualitative and quantitative nature of change varied regionally (Grimm & Jacobson, 1992; Jacobson *et al.*, 1987). These analyses show that although taxa ranges moved time-transgressively, abrupt climatic changes involving major reorganizations of atmospheric circulation affected vegetation everywhere. High-resolution studies with increased dating precision are indicating an increasingly close correspondence with North Atlantic-North European climatic events (Hu *et al.*, 1999; Levesque *et al.*, 1993a; Mayle & Cwynar, 1995b; Wright, 1989; Yu & Eicher, 1998; Yu & Wright, 2001). Major reorganizations of atmospheric circulation inevitably cause synchronous climatic change hemispherically or even globally (Broecker & Denton, 1989). Although synchronous, these changes can be different qualitatively and quantitatively regionally and can be overlain on time-transgressive movements of taxa ranges. The geographic extent of a climatic event is an important scientific question. The large late-glacial fluctuations evident in the ice cores have attracted much attention, but Holocene events, although more subdued, are receiving increased interest, including the 8200-yr event (Alley *et al.*, 1997; Hu *et al.*, 1999) and the 1500-yr climatic oscillation (Bond *et al.*, 1997, 2001; Viau *et al.*, 2002). The danger in event correlation, of course, is that virtually every proxy data source exhibits fluctuations, and probably almost any time-line pulled out of a hat would have coeval changes in proxies over a wide region within the errors of radiocarbon dating. Thus, caution should be exercised.

Conclusions

Much progress has been made since the classic papers of Davis (1965), Cushing (1965), and Whitehead (1965). The individualistic behavior of species is undeniable. Vegetation is an opportunistic assemblage of species with similar tolerances and adaptations to climatic and other environmental variables. Because every species has a different range of tolerance, assemblages of species change as climate changes.

The long debate over the importance of migrational lag has been largely settled; lags may have lasted a few hundred years, but not thousands, and lag is probably not a serious problem for climatic reconstruction on the continental and millennial space and time scales.

Great strides have been made in methods of quantitative analysis, made possible by powerful computers and database software. Still, the most elegant numerical method is no better than the data.

Dating remains a significant, sometimes severe problem, although AMS radiocarbon dating now provides the possibility to obtain much more reliable chronologies.

Climate reconstructions based on modern pollen distributions are limited to some extent by the tremendous human influence on modern vegetation. Because of modern human influence some dominant native types diagnostic of certain biomes, e.g. *Ambrosia*, are left out of the analyses. Certainly, an array of pre-European settlement samples would enable better formulation of pollen-climate relationships.

Increased taxonomic resolution has and probably will continue to contribute to refinements in climate and vegetation reconstruction. We still have a rudimentary understanding of the physiology of many, if not most, taxa. Greater understanding would place more limits on climate-plant distributions and would aid the interpretation of non-analog vegetation types.

We still do not know where many of the eastern deciduous trees were during the Pleistocene, although macrofossil studies (e.g. Jackson & Givens, 1994), have shown that a number of deciduous trees coexisted with *Picea* in the southeast. They were not in Florida. It is likely that many of these species were restricted to specialized habitats in the Appalachian Piedmont, an area with relatively few paleobotanical sites, especially ones strategically located in appropriate habitats.

The *Picea* that occurred in the southeast is an extinct species, one that favored milder climates than extant species. The effect of including this species in climatic reconstructions has not been fully evaluated. *Picea critchfieldii* is exceptional; in contrast to mammals, we know of very few plant extinctions since the Pliocene. Even the Pliocene flora is very similar to that of the Pleistocene. Just a few taxa (e.g. *Pterocarya*, *Sciadopitys*) distinguish Pliocene from Pleistocene floras, and these were relatively minor constituents (Hansen *et al.*, 2001; Rich, 1995). However, the genera *Pterocarya* and *Sciadopitys* were probably monospecific, and their absence is visible. Many species of *Quercus* and *Carya*, for example, occur in eastern North America, and the extinction of some of these species could easily have gone undetected.

The application of polymerase chain reaction (PCR) techniques to fossils offers an exciting possibility to understand the genetic shifts associated with climatic change and migration, and these studies are in their infancy. In her inimitable way, Margaret Davis has recently challenged us to consider evolutionary and genetic shifts as plant populations unceasingly expand, contract, and move over the landscape in response to climatic change (Davis & Shaw, 2001). What characters are mutable? What are the limits to adaptation? What role does hybridization play?

An abrupt change in atmospheric circulation may cause synchronous vegetation change over broad regions. Such abrupt changes overlain on time-transgressive migrations may be evidenced as jumps rather than smooth migrations if sufficient chronologic and spatial detail is available, as Grimm (1983) and Almendinger (1992) reconstructed the advance of the prairie-forest border in Minnesota after the Holocene aridity maximum. However, if genetic adaptation is occurring at the migrational fringe, then migration may continue to occur slowly between climatic changes. Such a scenario would be more likely at a true climatic boundary than at the prairie-forest border, where firebreaks sharply delineate vegetation types with no climatic differences on opposite sides (even though the boundary occurs within a steepened climatic gradient).

Even without anthropogenic forcing, climate will change in the future. Because the world is now already near an interglacial peak in temperatures, greenhouse warming may extend beyond the Quaternary climate space, so that all current forcings are non-analog and responses must be considered hypothetical. That greenhouse gases trap heat is indisputable; that climate will change is indisputable. Paleoecological and paleoclimatic insights may help clarify the future.

Acknowledgments

We thank our Ph.D. advisor Edward J. Cushing, from Missouri, the "Show Me" state, who continually challenged us to question that which is generally believed and to consider alternatives and who taught us everything we know about pollen. We thank Herbert E. Wright, Jr., who taught us to think of the big picture and that the only person who worries about having an idea rejected is the person with only one idea. We thank Margaret B. Davis whose recognition of the important issues has continually challenged paleoecologists to find answers wherever they lie. We also thank all the scientists who have contributed their data to public databases and have thereby advanced the study of past vegetation and climate.

References

Alley, R.B., Mayewski, P.A., Sowers, T., Stuiver, M., Taylor, K.C. & Clark, P.U. (1997). Holocene climatic instability: a prominent, widespread event 8200 yr ago. *Geology*, **25**, 483–486.

Allison, T.D., Moeller, R.E. & Davis, M.B. (1986). Pollen in laminated sediments provides evidence for a mid-Holocene forest pathogen outbreak. *Ecology*, **67**, 1101–1105.

Almendinger, J.C. (1992). The late Holocene history of prairie, brush-prairie, and jack pine (*Pinus banksiana*) forest on outwash plains, north-central Minnesota, USA. *The Holocene*, **2**, 37–50.

Almquist, H., Dieffenbacher-Krall, A.C., Flanagan-Brown, R. & Sanger, D. (2001). The Holocene record of lake levels of Mansell Pond, central Maine, USA. *The Holocene*, **11**, 189–201.

Almquist-Jacobson, H. & Sanger, D. (1995). Holocene climate and vegetation in the Milford drainage basin, Maine, U.S.A. *Vegetation History and Archaeobotany*, **4**, 211–222.

Amundson, D.C. & Wright, H.E., Jr. (1979). Forest changes in Minnesota at the end of the Pleistocene. *Ecological Monographs*, **49**, 1–16.

Anderson, R.S., Jacobson, G.L., Jr., Davis, R.B. & Stuckenrath, R. (1992). Gould Pond, Maine: late-glacial transition from marine to upland environments. *Boreas*, **21**, 359–371.

Baker, R.G., Bettis, E.A., III, Denniston, R.F., Gonzalez, L.A., Strickland, L.E. & Krieg, J.R. (2002). Holocene paleoenvironments in southeastern Minnesota – chasing the prairie-forest ecotone. *Palaeogeography, Palaeoclimatology, Palaeoecology*, **177**, 103–122.

Baker, R.G., Bettis, E.A., III, Schwert, D.P., Horton, D.G., Chumbley, C.A., Gonzalez, L.A. & Reagan, M.K. (1996). Holocene paleoenvironments of northeast Iowa. *Ecological Monographs*, **66**, 203–234.

Baker, R.G., Gonzalez, L.A., Raymo, M., Bettis, E.A., III, Reagan, M.K. & Dorale, J.A. (1998). Comparison of multiple proxy records of Holocene environments in the midwestern United States. *Geology*, **26**, 1131–1134.

Baker, R.G., Maher, L.J., Chumbley, C.A. & Van Zant, K.L. (1992). Patterns of Holocene environmental change in the midwestern United States. *Quaternary Research*, **37**, 379–389.

Bartlein, P.J., Prentice, I.C. & Webb, T., III (1986). Climatic response surfaces from pollen data for some eastern North American taxa. *Journal of Biogeography*, **13**, 35–57.

Bartlein, P.J. & Webb, T., III (1985). Mean July temperature for eastern North America at 6000 yr B.P.: regression equations for estimates based on fossil-pollen data. *Sillogeus*, **55**, 301–342.

Bartlein, P.J., Webb, T., III & Fleri, E. (1984). Holocene climatic change in the northern Midwest: pollen-derived estimates. *Quaternary Research*, **22**, 361–374.

Baskin, J.M. & Baskin, C.C. (1977a). Dormancy and germination in seeds of common ragweed with reference to Beal's buried seed experiment. *American Journal of Botany*, **64**, 1174–1176.

Baskin, J.M. & Baskin, C.C. (1977b). Role of temperature in the germination ecology of three summer annual weeds. *Oecologia*, **30**, 377–382.

Bazzaz, F.A. (1970). Secondary dormancy in the seeds of the common ragweed *Ambrosia artemisiifolia*. *Bulletin of the Torrey Botanical Club*, **97**, 302–305.

Bennett, K.D. (1985). The spread of *Fagus grandifolia* across eastern North America during the last 18,000 years. *Journal of Biogeography*, **12**, 147–164.

Bennett, K.D. (1988a). Holocene geographic spread and population expansion of *Fagus grandifolia* in Ontario, Canada. *Journal of Ecology*, **76**, 547–557.

Bennett, K.D. (1988b). Modelling Holocene changes in beech populations: a reply to Dexter *et al.* (1987). *Review of Palaeobotany and Palynology*, **56**, 361–362.

Bennett, K.D. & Fuller, J.L. (2002). Determining the age of the mid-Holocene *Tsuga canadensis* (hemlock) decline, eastern North America. *The Holocene*, **12**, 421–429.

Bernabo, J.C. (1981). Quantitative estimates of temperature changes over the last 2700 years in Michigan based on pollen data. *Quaternary Research*, **15**, 143–159.

Bernabo, J.C. & Webb, T., III (1977). Changing patterns in the Holocene pollen record of northeastern North America: a mapped summary. *Quaternary Research*, **8**, 64–96.

Bettis, E.A., III, Baker, R.G., Nations, B.K. & Benn, D.W. (1990). Early Holocene pecan, *Carya illinoensis*, in the Mississippi River Valley near Muscatine, Iowa. *Quaternary Research*, **33**, 102–107.

Birks, H.J.B. (1986). Numerical zonation, comparison and correlation of Quaternary pollen-stratigraphic data. *In*: Berglund, B.E. (Ed.), *Handbook of Holocene Palaeoecology and Palaeohydrology*. John Wiley & Sons Ltd., Chichester, pp. 743–774.

Birks, H.J.B. & Gordon, A.D. (1985). *Numerical Methods in Quaternary Pollen Analysis*. Academic Press, London, 317 pp.

Björck, S. (1985). Deglaciation chronology and revegetation in northwestern Ontario. *Canadian Journal of Earth Sciences*, **22**, 850–871.

Bond, G., Kromer, B., Beer, J., Muscheler, R., Evans, M.N., Showers, W., Hoffmann, S., Lotti-Bond, R., Hajdas, I. & Bonani, G. (2001). Persistent solar influence on North Atlantic climate during the Holocene. *Science*, **294**, 2130–2136.

Bond, G., Showers, W., Cheseby, M., Lotti, R., Almasi, P., deMenocal, P., Priore, P., Cullen, H., Hajdas, I. & Bonani, G. (1997). A pervasive millennial-scale cycle in North Atlantic Holocene and glacial climates. *Science*, **278**, 1257–1266.

Borchert, J.R. (1950). Climate of the central North American grassland. *Annals of the Association of American Geographers*, **40**, 1–39.

Borns, H.W., Jr., Dorion, C.C., Jacobson, G.L., Jr., Kreutz, K.J., Thompson, W.B., Weddle, T.K., Doner, L.A., Kaplan, M.R. & Lowell, T.V. (2003). The deglaciation of Maine. *In*: Ehlers, J. & Gibbard, P.L. (Eds), *Quaternary Glaciations – Extent and Chronology. Part II: North America*. Elsevier, Amsterdam.

Bradbury, J.P. & Dean, W.E. (Eds) (1993). Elk Lake, Minnesota: evidence for rapid climate change in the north-central United States. *Geological Society of America*, Special Paper 276, Boulder, Colorado, 336 pp.

Bradshaw, R.H.W. (1988). Spatially-precise studies of forest dynamcis. *In*: Huntley, B. & Webb, T., III (Eds), *Handbook of Vegetation Science, Volume 7: Vegetation History*. Kluwer Academic Publishers, Dordrecht, pp. 725–751.

Braun, E.L. (1950). *Deciduous Forests of Eastern North America*. Blakiston Company, Philadelphia, 596 pp.

Braun, E.L. (1951). Plant distribution in relation to the glacial boundary. *Ohio Journal of Science*, **51**, 139–146.

Braun, E.L. (1955). The phytogeography of unglaciated eastern United States and its interpretation. *Botanical Review*, **21**, 297–375.

Broecker, W.S. & Denton, G.H. (1989). The role of ocean-atmosphere reorganizations in glacial cycles. *Geochimica et Cosmochimica Acta*, **53**, 2465–2501.

Broecker, W.S. & Walton, A. (1959). The geochemistry of ^{14}C in the freshwater systems. *Geochimica et Cosmochimica Acta*, **16**, 15–38.

Brown, T.A., Farwell, G.W., Grootes, P.M. & Schmidt, F.H. (1992). Radiocarbon AMS dating of pollen extracted from peat samples. *Radiocarbon*, **34**, 550–556.

Brown, T.A., Nelson, D.E., Mathewes, R.W., Vogel, J.S. & Southon, J.R. (1989). Radiocarbon dating of pollen by accelerator mass spectrometry. *Quaternary Research*, **32**, 205–212.

Bryson, R.A., Baerreis, D.A. & Wendland, W.M. (1970). The character of late-glacial and post-glacial climatic changes. *In*: Dort, W., Jr. & Jones, J.K., Jr. (Eds), *Pleistocene and Recent Environments of the Central Great Plains*. University of Kansas, Department of Geology, Special Publication 3. The University Press of Kansas, Lawrence, pp. 53–74.

Bryson, R.A. & Wendland, W.M. (1967). Tentative climatic patterns for some late glacial and post-glacial episodes in central North America. *In*: Mayer-Oakes, W.J. (Ed.), *Life, Land and Water: Proceedings of the 1966 Conference on Environmental Studies of the Glacial Lake Agassiz Region*. University of Manitoba Press, Winnipeg, pp. 271–298.

Calcote, R.R. (1995). Pollen source area and pollen productivity: evidence from forest hollows. *Journal of Ecology*, **83**, 591–602.

Chumbley, C.A., Baker, R.G. & Bettis, E.A., III (1990). Midwestern Holocene paleoenvironments revealed by floodplain deposits in northeastern Iowa. *Science*, **249**, 272–274.

Clark, J.S., Royall, P.D. & Chumbley, C. (1996). The role of fire during climate change in an eastern decidous forest at Devil's Bathtub, New York. *Ecology*, **77**, 2148–2166.

COHMAP Members (1988). Climatic changes of the last 18,000 years: observations and model simulations. *Science*, **241**, 1043–1052.

Cooper, W.S. (1958). Terminology of post-Valders time. *Geological Society of America Bulletin*, **69**, 941–945.

Cushing, E.J. (1964). Redeposited pollen in late-Wisconsin pollen spectra from east-central Minnesota. *American Journal of Science*, **262**, 1075–1088.

Cushing, E.J. (1965). Problems in the Quaternary phytogeography of the Great Lakes region. *In*: Wright, H.E., Jr. & Frey, D.G. (Eds), *The Quaternary of the United States*. Princeton University Press, Princeton, NJ, pp. 403–416.

Cwynar, L.C. & Levesque, A.J. (1995). Chironomid evidence for late-glacial climatic reversals in Maine. *Quaternary Research*, **43**, 405–413.

Darley-Hill, S. & Johnson, W.C. (1981). Acorn dispersal by the blue jay (*Crynocitta cristata*). *Oecologia*, **50**, 231–232.

Davis, M.B. (1963). On the theory of pollen analysis. *American Journal of Science*, **261**, 897–912.

Davis, M.B. (1965). Phytogeography and palynology of northeastern United States. *In*: Wright, H.E., Jr. & Frey, D.G. (Eds), *The Quaternary of the Unites States*. Princeton University Press, Princeton, NJ, pp. 377–401.

Davis, M.B. (1976). Pleistocene biogeography of temperate deciduous forests. *Geoscience and Man*, **13**, 13–26.

Davis, M.B. (1978). Climatic interpretation of pollen in Quaternary sediments. *In*: Walker, D. & Guppy, J.C. (Eds), *Biology and Quaternary Environments*. Australian Academy of Science, Canberra, pp. 35–51.

Davis, M.B. (1981a). Outbreaks of forest pathogens in Quaternary history. *In*: *Proceedings of the IV International Palynological Conference (1976–1977), Volume 3*. Birbal Sahni Institute of Paleobotany, Lucknow, pp. 216–227.

Davis, M.B. (1981b). Quaternary history and the stability of forest communities. *In*: West, D.C., Shugart, H.H. & Botkin, D.B. (Eds), *Forest Succession: Concepts and Application*. Springer-Verlag, New York, pp. 132–153.

Davis, M.B. (1983a). Holocene vegetational history of the eastern United States. *In*: Wright, H.E., Jr. (Ed.), *Late-Quaternary environments of the United States, Volume 2. The Holocene*. University of Minnesota Press, Minneapolis, pp. 166–181.

Davis, M.B. (1983b). Quaternary history of deciduous forests of eastern North America and Europe. *Annals of the Missouri Botanical Garden*, **70**, 550–563.

Davis, M.B., Calcote, R.R., Sugita, S. & Takahara, H. (1998). Patchy invasion and the origin of a hemlock-hardwoods forest mosaic. *Ecology*, **79**, 2641–2659.

Davis, M.B., Schwartz, M.W. & Woods, K. (1991). Detecting a species limit from pollen in sediments. *Journal of Biogeography*, **18**, 653–668.

Davis, M.B. & Shaw, R.G. (2001). Range shifts and adaptive responses to Quaternary climate change. *Science*, **292**, 673–679.

Davis, M.B., Spear, R.W. & Shane, L.C.K. (1980). Holocene climate of New England. *Quaternary Research*, **14**, 240–250.

Davis, M.B., Woods, K.D., Webb, S.L. & Futyma, R.P. (1986). Dispersal versus climate: expansion of *Fagus* and *Tsuga* into the Upper Great Lakes region. *Vegetatio*, **67**, 93–103.

Davis, R.B. & Jacobson, G.L., Jr. (1985). Late glacial and early Holocene landscapes in northern New England and adjacent areas of Canada. *Quaternary Research*, **23**, 341–368.

Deen, W., Hunt, L.A. & Swanton, C.J. (1998a). Photothermal time describes common ragweed (*Ambrosia artemisiifolia* L.) phenological development and growth. *Weed Science*, **46**, 561–568.

Deen, W., Hunt, T. & Swanton, C.J. (1998b). Influence of temperature, photoperiod, and irradiance on the phenological development of common ragweed (*Ambrosia artemisiifolia*). *Weed Science*, **46**, 555–560.

Deevey, E.S. & Flint, R.F. (1957). Postglacial hypsithermal interval. *Science*, **125**, 182–184.

Deevey, E.S., Jr. (1951). Late-glacial and post-glacial pollen diagrams from Maine. *American Journal of Science*, **249**, 177–207.

Deevey, E.S., Jr., Gross, M.S., Hutchinson, G.E. & Kraybill, H.L. (1954). The natural C14 contents of materials from hard-water lakes. *Proceedings of the National Academy of Sciences of the United States of America*, **40**, 285–288.

Delcourt, P.A., Delcourt, H.R. & Webb, T., III (1984). *Atlas of mapped distributions of dominance and modern pollen percentages for important tree taxa of eastern North America*. American Association of Stratigraphic Palynologists Contribution Series, **14**, 131 pp.

Denniston, R.F., González, L.A., Asmerom, Y., Baker, R.G., Reagan, M.K. & Bettis, E.A., III (1999a). Evidence for increased cool season moisture during the middle Holocene. *Geology*, **27**, 815–818.

Denniston, R.F., González, L.A., Baker, R.G., Asmerom, Y., Reagan, M.K., Edwards, R.L. & Alexander, E.C. (1999b). Speleothem evidence for Holocene fluctuations of the prairie-forest ecotone, north-central USA. *The Holocene*, **9**, 671–676.

Denniston, R.F., González, L.A., Asmerom, Y., Polyak, V., Reagan, M.K. & Saltzman, M.R. (2001). A high-resolution speleothem record of climatic variability at the Allerød-Younger Dryas transition in Missouri, central United States. *Palaeogeography, Palaeoclimatology, Palaeoecology*, **176**, 147–155.

Dexter, F., Banks, H.T. & Webb, T., III (1987). Modeling Holocene changes in the location and abundance of beech populations in eastern North America. *Review of Palaeobotany and Palynology*, **50**, 273–292.

Dickerson, C.T., Jr. (1968). Studies on the germination, growth, development and control of common ragweed (*Ambrosia artemisiifolia* L.). Unpublished Ph.D. dissestation, Cornell University.

Doner, L.A. (1996). Late-Pleistocene environments in Maine and the Younger Dryas dilemma. Unpublished M.S. thesis, University of Maine.

Dorale, J.A., González, L.A., Reagan, M.K., Pickett, D.A., Murrell, M.T. & Baker, R.G. (1992). A high-resolution record of Holocene climate change in speleothem calcite from Cold Water Cave, northeast Iowa. *Science*, **258**, 1626–1630.

Dorion, C.C. (1997a). Middle to late Wisconsinan glacial chronology and paleoenvironments along a transect from eastern coastal Maine north to New Brunswick and Quebec. *In*: *Canadian Quaternary Association, 8th Biennial Meeting, Program Abstracts*, pp. 19–20.

Dorion, C.C. (1997b). An updated high resolution chronology of deglaciation and accompanying marine transgression in Maine. Unpublished M.S. thesis, University of Maine.

Dorion, C.C., Balco, G.A., Kaplan, M.R., Kreutz, K.J., Wright, J.D. & Borns, H.W., Jr. (2001). Statigraphy, paleoceanography, chronology, and environment during deglaciation of eastern Maine. *In*: Retelle, M.J. & Weddle, T.K. (Eds), *Deglacial History and Relative Sea-Level Changes, Northern New England and Adjacent Canada*. Geological Society of America, Special Paper 351, Boulder, CO, pp. 215–242.

Drexler, C.W., Farrand, W.R. & Hughes, J.D. (1983). Correlation of glacial lakes in the Superior Basin with eastward discharge events from Lake Agassiz. *In*: Teller, J.T. & Clayton, L. (Eds), *Glacial Lake Agassiz*. Geological Association of Canada, Special Paper 26, pp. 309–329.

Dwyer, T.R., Mullins, H.T. & Good, S.C. (1996). Paleoclimatic implications of Holocene lake-level fluctuations, Owasco Lake, New York. *Geology*, **24**, 519–522.

Engstrom, D.R. & Hansen, B.C.S. (1985). Postglacial vegetational change and soil development in southeastern Labrador as inferred from pollen and chemical stratigraphy. *Canadian Journal of Botany*, **63**, 543–561.

Fagerlind, F. (1952). The real signification of pollen diagrams. *Botaniska Notiser*, **105**, 185–224.

Farrand, W.R. & Drexler, C.W. (1985). Late Wisconsinan and Holocene history of the Lake Superior Basin. *In*: Karrow, P.F. & Calkin, P.E. (Eds), *Quaternary Evolution of the Great Lakes*. Geological Association of Canada, Special Paper 30, pp. 17–32.

Fowler, A.J., Gillespie, R. & Hedges, R.E.M. (1986). Radiocarbon dating of sediments. *Radiocarbon*, **28**, 441–450.

Gajewski, K. (1988). Late Holocene climate changes in eastern North America estimated from pollen data. *Quaternary Research*, **29**, 255–262.

Gaudreau, D.C. (1986). Late-Quaternary vegetational history of the Northeast: paleoecological implications of topographic patterns in pollen distributions. Unpublished Ph.D. dissertation, Yale University.

Gleason, H.A. (1922). The vegetational history of the Middle West. *Annals of the Association of American Geographers*, **12**, 39–85.

Gleason, H.A. (1926). The individualistic concept of the plant association. *Bulletin of the Torrey Botanical Club*, **53**, 7–26.

Gleason, H.A. (1939). The individualistic concept of the plant association. *American Midland Naturalist*, **21**, 92–110.

Gordon, A.D. & Birks, H.J.B. (1972). Numerical methods in Quaternary palaeoecology. I. Zonation of pollen diagrams. *New Phytologist*, **71**, 961–979.

Gordon, A.D. & Birks, H.J.B. (1974). Numerical methods in Quaternary palaeoecology. II. Comparison of pollen diagrams. *New Phytologist*, **73**, 221–249.

Grimm, E.C. (1983). Chronology and dynamics of vegetation change in the prairie-woodland region of southern Minnesota, U.S.A. *New Phytologist*, **93**, 311–350.

Grimm, E.C. (1984). Fire and other factors controlling the Big Woods vegetation of Minnesota in the mid-nineteenth century. *Ecological Monographs*, **54**, 291–311.

Grimm, E.C. (1987). CONISS: a FORTRAN 77 program for stratigraphically constrained cluster analysis by the method of incremental sum of squares. *Computers & Geosciences*, **13**, 13–35.

Grimm, E.C. (1988). Data analysis and display. *In*: Huntley, B. & Webb, T., III (Eds), *Handbook of Vegetation Science, Volume 7: Vegetation History*. Kluwer Academic Publishers, Dordrecht, pp. 43–76.

Grimm, E.C. (2001). Trends and palaeoecological problems in the vegetation and climate history of the northern Great Plains, U.S.A. *Biology and Environment: Proceedings of the Royal Irish Academy*, **101B**, 47–64.

Grimm, E.C. & Jacobson, G.L., Jr. (1992). Fossil-pollen evidence for abrupt climate changes during the past 18 000 years in eastern North America. *Climate Dynamics*, **6**, 179–184.

Grimm, E.C., Jacobson, G.L., Jr., Watts, W.A., Hansen, B.C.S. & Maasch, K.A. (1993). A 50,000-year record of climate oscillations from Florida and its temporal correlation with the Heinrich events. *Science*, **261**, 198–200.

Grimm, E.C., Lozano-García, S., Behling, H. & Markgraf, V. (2001). Holocene vegetation and climate variability in the Americas. *In*: Markgraf, V. (Ed.), *Interhemispheric Climate Linkages*. Academic Press, San Diego, pp. 325–370.

Grimm, E.C. & Maher, L.J. (2002). AMS radiocarbon dating documents climate events in the upper Midwest coeval with the Bølling/Allerød and Younger Dryas episodes. *Geological Society of America Abstracts with Programs*, **34**(6), 352.

Haas, J.N. & McAndrews, J.H. (2000). The summer drought related hemlock (*Tsuga canadensis*) decline in eastern North America 5,700 to 5,100 years ago. *In*: McManus, K.A., Shields, K.S. & Souto, D.R. (Eds), *Proceedings: Symposium on Sustainable Management of Hemlock Ecosystems in Eastern North America*. U.S. Department of Agriculture, Forest Service, Northeastern Research Station, General Technical Report NE-267, Newtown Square, PA, pp. 81–88.

Håkansson, S. (1979). Radiocarbon activity in submerged plants from various south Swedish lakes. *In*: Berger, R. & Suess, H.E. (Eds), *Radiocarbon Dating: Proceedings of the Ninth International Conference, Los Angeles and La Jolla, 1976*. University of California Press, Berkeley, pp. 433–443.

Hansen, B.C.S., Grimm, E.C. & Watts, W.A. (2001). Palynology of the Peace Creek site, Polk County, Florida. *Geological Society of American Bulletin*, **113**, 682–692.

Howe, S. & Webb, T., III (1983). Calibrating pollen data in climatic terms: improving the methods. *Quaternary Science Reviews*, **2**, 17–51.

Hu, F.S., Slawinski, D., Wright, H.E., Jr., Ito, E., Johnson, R.G., Kelts, K.R., McEwan, R.F. & Boedigheimer, A. (1999). Abrupt changes in North American climate during early Holocene times. *Nature*, **400**, 437–440.

Huntley, B., Bartlein, P.J. & Prentice, I.C. (1989). Climatic control of the distribution and abundance of beech (*Fagus* L.) in Europe and North America. *Journal of Biogeography*, **16**, 551–560.

Huntley, B. & Webb, T., III (1989). Migration: species' response to climatic variations caused by changes in the earth's orbit. *Journal of Biogeography*, **16**, 5–19.

Jackson, S.T. & Givens, C.R. (1994). Late Wisconsinan vegetation and environment of the Tunica Hills Region, Louisiana/Mississippi. *Quaternary Research*, **41**, 316–325.

Jackson, S.T., Grimm, E.C. & Thompson, R.S. (2000a). Database resources in Quaternary paleobotany. *Sida, Botanical Miscellany*, **18**, 113–120.

Jackson, S.T., Webb, R.S., Anderson, K.H., Overpeck, J.T., Webb, T., III, Williams, J.W. & Hansen, B.C.S. (2000b). Vegetation and environment in eastern North America during the last glacial maximum. *Quaternary Science Reviews*, **19**, 489–508.

Jackson, S.T., Overpeck, J.T., Webb, T., III, Keattch, S.E. & Anderson, K.H. (1997). Mapped plant-macrofossil and pollen records of late Quaternary vegetation change in eastern North America. *Quaternary Science Reviews*, **16**, 1–70.

Jackson, S.T. & Weng, C. (1999). Late Quaternary extinction of a tree species in eastern North America. *Proceedings of the National Academy of Sciences of the United States of America*, **96**, 13,847–13,852.

Jackson, S.T. & Wong, A. (1994). Using forest patchiness to determine pollen source areas of closed-canopy pollen assemblages. *Journal of Ecology*, **82**, 89–99.

Jacobson, G.L., Jr. (1979). The palaeoecology of white pine (*Pinus strobus*) in Minnesota. *Journal of Ecology*, **67**, 697–726.

Jacobson, G.L., Jr. & Bradshaw, R.H.W. (1981). The selection of sites for paleovegetational studies. *Quaternary Research*, **16**, 80–96.

Jacobson, G.L., Jr. & Dieffenbacher-Krall, A. (1995). White pine and climate change: insights from the past. *Journal of Forestry*, **93**(7), 39–42.

Jacobson, G.L., Jr. & Grimm, E.C. (1986). A numerical analysis of Holocene forest and prairie vegetation in central Minnesota. *Ecology*, **67**, 958–966.

Jacobson, G.L., Jr., Webb, T., III & Grimm, E.C. (1987). Patterns and rates of vegetation change during the deglaciation of eastern North America. *In*: Ruddiman, W.F. & Wright, H.E., Jr. (Eds), *The Geology of North America, Volume K-3. North America and Adjacent Oceans During the Last Deglaciation*. Geological Society of America, Boulder, CO, pp. 277–288.

Johnson, W.C. & Adkisson, C.S. (1985). Dispersal of beech nuts by blue jays in fragmented landscapes. *American Midland Naturalist*, **113**, 319–324.

Johnson, W.C. & Webb, T., III (1989). The role of blue jays (*Cyanocitta cristata* L.) in the postglacial dispersal of fagaceous trees in eastern North America. *Journal of Biogeography*, **16**, 561–571.

King, G.A. & Herstrom, A.A. (1997). Holocene tree migration rates objectively determined from fossil pollen data. *In*: Huntley, B., Cramer, W., Morgan, A.V., Prentice, H.C. & Allen, J.R.M. (Eds), *Past and Future Rapid Environmental Changes: The Spatial and Evolutionary Responses of Terrestrial Biota*. NATO ASI Series. Springer-Verlag, Berlin, pp. 91–101.

King, J.E. (1981). Late Quaternary vegetational history of Illinois. *Ecological Monographs*, **51**, 43–62.

Laird, K.R., Fritz, S.C., Cumming, B.F. & Grimm, E.C. (1998). Early-Holocene limnological and climatic variability in the northern Great Plains. *The Holocene*, **8**, 275–285.

Laird, K.R., Fritz, S.C., Grimm, E.C. & Mueller, P.G. (1996). Century-scale paleoclimatic reconstruction from Moon Lake, a closed-basin lake in the northern Great Plains. *Limnology and Oceanography*, **41**, 890–902.

Lamb, H.F. (1980). Late Quaternary vegetational history of southeastern Labrador. *Arctic and Alpine Research*, **12**, 117–135.

Lamb, H.F. (1984). Modern pollen spectra from Labrador and their use in reconstructing Holocene vegetational history. *Journal of Ecology*, **72**, 37–59.

Lamb, H.F. (1985). Palynological evidence for postglacial change in the position of tree limit in Labrador. *Ecological Monographs*, **55**, 241–258.

Larson, G. & Schaetzl, R. (2001). Origin and evolution of the Great Lakes. *Journal of Great Lakes Research*, **27**, 518–546.

Levesque, A.J., Cwynar, L.C. & Walker, I.R. (1994). A multiproxy investigation of late-glacial climate and vegetation change at Pine Ridge Pond, southwest New Brunswick, Canada. *Quaternary Research*, **42**, 316–327.

Levesque, A.J., Cwynar, L.C. & Walker, I.R. (1996). Richness, diversity and succession of late-glacial chironomid assemblages in New Brunswick, Canada. *Journal of Paleolimnology*, **16**, 257–274.

Levesque, A.J., Cwynar, L.C. & Walker, I.R. (1997). Exceptionally steep north-south gradients in lake temperatures during the last deglaciation. *Nature*, **385**, 423–426.

Levesque, A.J., Mayle, F.E., Walker, I.R. & Cwynar, L.C. (1993a). The amphi-Atlantic Oscillation: a proposed late-glacial climatic event. *Quaternary Science Reviews*, **12**, 629–643.

Levesque, A.J., Mayle, F.E., Walker, I.R. & Cwynar, L.C. (1993b). A previously unrecognized late-glacial cold event in eastern North America. *Nature*, **361**, 623–626.

Lindbladh, M., Jacobson, G.L., Jr. & Schauffler, M. (2003). The postglacial history of three Picea species in New England, USA. *Quaternary Research*, **59**, 61–69.

Lindbladh, M., O'Connor, R. & Jacobson, G.L., Jr. (2002). Morphometric analysis of pollen grains for paleoecological studies: classification of *Picea* from eastern North America. *American Journal of Botany*, **89**, 1459–1467.

Long, A., Davis, O.K. & de Lanois, J. (1992). Separation and ^{14}C dating of pure pollen from lake sediments: nanofossil AMS dating. *Radiocarbon*, **34**, 557–560.

Lowell, T.V., Larson, G.J., Hughes, J.D. & Denton, G.H. (1999). Age verification of the Lake Gribben forest bed and the Younger Dryas advance of the Laurentide Ice Sheet. *Canadian Journal of Earth Sciences*, **36**, 383–393.

Mayle, F.E. & Cwynar, L.C. (1995a). Impact of the Younger Dryas cooling event upon lowland vegetation of Maritime Canada. *Ecological Monographs*, **65**, 129–154.

Mayle, F.E. & Cwynar, L.C. (1995b). A review of multi-proxy data for the Younger Dryas in Atlantic Canada. *Quaternary Science Reviews*, **14**, 813–821.

Mayle, F.E., Levesque, A.J. & Cwynar, L.C. (1993). Accelerator-mass-spectrometer ages for the Younger Dryas event in Atlantic Canada. *Quaternary Research*, **39**, 355–360.

McAndrews, J.H. (1966). Postglacial history of prairie, savanna, and forest in northwestern Minnesota. *Torrey Botanical Club Memoir*, **22**(2), 1–72.

McAndrews, J.H. & Wright, H.E., Jr. (1969). Modern pollen rain across the Wyoming basins and the northern Great Plains (U.S.A). *Review of Palaeobotany and Palynology*, **9**, 17–43.

Mott, R.J., Grant, D.R.G., Stea, R.R. & Occhietti, S. (1986). Late-glacial climate oscillation in Atlantic Canada equivalent to the Allerød-Younger Dryas event. *Nature*, **323**, 247–250.

Nambudiri, E.M.V., Teller, J.T. & Last, W.M. (1980). Pre-Quaternary microfossils – a guide to errors in radiocarbon dating. *Geology*, **8**, 123–126.

Newby, P.E., Killoran, P., Waldorf, M.R., Shuman, B.N., Webb, R.S. & Webb, T., III (2000). 14,000 years of sediment, vegetation, and water-level changes at the Makepeace Cedar Swamp, southeastern Massachusetts. *Quaternary Research*, **53**, 352–368.

Nienstaedt, H. & Zasada, J.C. (1990). *Picea glauca* (Moench) Voss white spruce. *In*: Burns, R.M. & Honkala, B.H. (Eds), *Silvics of North America. Volume 1. Conifers*. U.S. Department of Agriculture, Forest Service, Agriculture Handbook 654, Washington, DC, pp. 204–226.

Ogden, J.G., III (1967a). Radiocarbon and pollen evidence for a sudden change in climate in the Great Lakes region approximately 10,000 years ago. *In*: Cushing, E.J. & Wright, H.E., Jr. (Eds), *Quaternary Paleoecology*. Yale University Press, New Haven, CT, pp. 117–127.

Ogden, J.G., III (1967b). Radiocarbon determinations of sedimentation rates from hard and soft-water lakes in northeastern North America. *In*: Cushing, E.J. & Wright, H.E., Jr. (Eds), *Quaternary Paleoecology*. Yale University Press, New Haven, CT, pp. 175–183.

Overpeck, J.T. (1987). Pollen time series and Holocene climate variability of the Midwest United States. *In*: Berger, W.H. & Labeyrie, L.D. (Eds), *Abrupt Climate Change: Evidence and Implications*. D. Reidel Publishing Company, Dordrecht, pp. 137–143.

Overpeck, J.T., Bartlein, P.J. & Webb, T., III (1991). Potential magnitude of future vegetation change in eastern North America: comparisons with the past. *Science*, **254**, 692–695.

Overpeck, J.T., Webb, R.S. & Webb, T., III (1992). Mapping eastern North American vegetation change of the past 18 ka: no-analogs and the future. *Geology*, **20**, 1071–1074.

Overpeck, J.T., Webb, T., III & Prentice, I.C. (1985). Quantitative interpretation of fossil pollen spectra: dissimilarity coefficients and the method of modern analogs. *Quaternary Research*, **23**, 87–108.

Parsons, R.W. & Prentice, I.C. (1981). Statistical approaches to *R*-values and the pollen-vegetation relationship. *Review of Palaeobotany and Palynology*, **32**, 127–152.

Prentice, I.C. (1983). Postglacial climatic change: vegetation dynamics and the pollen record. *Progress in Physical Geography*, **7**, 273–286.

Prentice, I.C. (1986). Forest-composition calibration of pollen data. *In*: Berglund, B.E. (Ed.), *Handbook of Holocene Palaeoecology and Palaeohydrology*. John Wiley & Sons Ltd., Chichester, pp. 799–816.

Prentice, I.C., Bartlein, P.J. & Webb, T., III (1991). Vegetation and climate change in eastern North America since the last glacial maximum. *Ecology*, **72**, 2038–2056.

Prentice, I.C., Cramer, W., Harrison, S.P., Leemans, R., Monserud, R.A. & Solomon, A.M. (1992). A global biome model based on plant physiology and dominance, soil properties and climate. *Journal of Biogeography*, **19**, 117–134.

Prentice, I.C., Guiot, J., Huntley, B., Jolly, D. & Cheddadi, R. (1996). Reconstructing biomes from palaeoecological data: a general method and its application to European pollen data at 0 and 6 ka. *Climate Dynamics*, **12**, 185–194.

Prentice, I.C. & Parsons, R.W. (1983). Maximum likelihood linear calibration of pollen spectra in terms of forest composition. *Biometrics*, **39**, 1051–1059.

Prentice, I.C. & Webb, T., III (1986). Pollen percentages, tree abundances and the Fagerlind effect. *Journal of Quaternary Science*, **1**, 35–43.

Raynal, D.J. & Bazzaz, F.A. (1973). Establishment of early successional plant populations on forest and prairie soil. *Ecology*, **54**, 1335–1341.

Regnéll, J. (1992). Preparing pollen concentrates for AMS dating – a methodological study from a hard-water lake in southern Sweden. *Boreas*, **21**, 373–377.

Regnéll, J. & Everitt, E. (1996). Preparative centrifucation – a new method for preparing pollen concentrates suitable for radiocarbon dating by AMS. *Vegetation History and Archaeobotany*, **5**, 201–205.

Rich, F.J. (1995). Palynological characteristics of near-shore shell-bearing Pliocene through Holocene sediments of Florida, Georgia and South Carolina. *Tulane Studies in Geology and Paleontology*, **28**, 97–111.

Richardson, F. & Hall, V.A. (1994). Pollen concentrate preparation from highly organic Holocene peat and lake deposits for AMS dating. *Radiocarbon*, **36**, 407–412.

Rind, D., Peteet, D., Broecker, W., McIntyre, A. & Ruddiman, W. (1986). The impact of cold North Atlantic sea surface temperatures on climate: implications for the Younger Dryas cooling (11–10 k). *Climate Dynamics*, **1**, 3–33.

Rodionov, S.N. (1994). Association between winter precipitation and water level fluctuations in the Great Lakes and atmospheric circulation patterns. *Journal of Climate*, **7**, 1693–1706.

Saarnisto, M. (1988). Time-scales and dating. *In*: Huntley, B. & Webb, T., III (Eds), *Handbook of Vegetation Science, Volume 7: Vegetation History*. Kluwer Academic Publishers, Dordrecht, pp. 77–112.

Sachs, J.P., Webb, T., III & Clark, D.R. (1977). Paleoecological transfer functions. *Annual Review of Earth and Planetary Sciences*, **5**, 159–178.

Schauffler, M. & Jacobson, G.L., Jr. (2002). Persistence of coastal spruce refugia during the Holocene in northern New England, USA, detected by stand-scale pollen stratigraphies. *Journal of Ecology*, **90**, 235–250.

Shane, L.C.K. (1987). Late-glacial vegetational and climatic history of the Allegheny Plateau and the till plains of Ohio and Indiana, USA. *Boreas*, **16**, 1–20.

Shane, L.C.K. & Anderson, K.H. (1993). Intensity, gradients and reversals in late glacial environmental change in east-central North America. *Quaternary Science Reviews*, **12**, 307–320.

Shrestha, A., Roman, E.S., Thomas, A.G. & Swanton, C.J. (1999). Modeling germination and shoot-radicle elongation of *Ambrosia artemisiifolia*. *Weed Science*, **47**, 557–562.

Shuman, B., Webb, T., III, Bartlein, P. & Williams, J.W. (2002). The anatomy of a climatic oscillation: vegetation change in eastern North America during the Younger Dryas chronozone. *Quaternary Science Reviews*, **21**, 1777–1791.

Stea, R.R. & Mott, R.J. (1998). Deglaciation of Nova Scotia: stratigraphy and chronology of lake sediment cores and buried organic sections. *Géographie physique et Quaternaire*, **52**, 3–21.

Stoller, E.W. & Wax, L.M. (1973). Periodicity of germination and emergence of some annual weeds. *Weed Science*, **21**, 574–580.

Stuiver, M. (1993). A note on single-year calibration of the radiocarbon time scale, AD 1510–1954. *Radiocarbon*, **35**, 67–72.

Stuiver, M. & Quay, P.D. (1980). Changes in atmospheric carbon-14 attributed to a variable sun. *Science*, **207**, 11–19.

Sugita, S. (1995). Pollen representation of vegetation in Quaternary sediments: theory and method in patchy vegetaion. *Journal of Ecology*, **82**, 879–898.

Thompson, R.S., Anderson, K.H. & Bartlein, P.J. (2000). *Atlas of relations between climatic parameters and distributions of important trees and shrubs in North America*. U.S. Geological Survey Professional Paper 1650 A-B.

Transeau, E.N. (1935). The Prairie Peninsula. *Ecology*, **16**, 423–437.

Viau, A.E., Gajewski, K., Fines, P., Atkinson, D.E. & Sawada, M.C. (2002). Widespread evidence of 1500 yr climate variability in North America during the past 14 000 yr. *Geology*, **30**, 455–458.

Viereck, L.A. & Johnston, W.F. (1990). *Picea mariana* (Mill.) B.S.P black spruce. *In*: Burns, R.M. & Honkala, B.H. (Eds), *Silvics of North America. Volume 1. Conifers*. U.S. Department of Agriculture, Forest Service, Agriculture Handbook 654, Washington, DC, pp. 227–237.

Watson, R.A. & Wright, H.E., Jr. (1980). The end of the Pleistocene: a general critique of chronostratigraphic classification. *Boreas*, **9**, 153–163.

Watts, W.A. (1969). A pollen diagram from Mud Lake, Marion County, north-central Florida. *Geological Society of American Bulletin*, **80**, 631–642.

Watts, W.A., Hansen, B.C.S. & Grimm, E.C. (1992). Camel Lake: a 40,000-yr record of vegetational and forest history from northwest Florida. *Ecology*, **73**, 1056–1066.

Webb, R.S. (1990). Late Quaternary water-level fluctuations in the northeastern United States. Unpublished Ph.D. dissertation, Brown University.

Webb, R.S., Anderson, K.H. & Webb, T., III (1993a). Pollen response-surface estimates of late-Quaternary changes in the moisture balance of the northeastern United States. *Quaternary Research*, **40**, 213–227.

Webb, T., III, Bartlein, P.J., Harrison, S.P. & Anderson, K.H. (1993b). Vegetation, lake levels, and climate in eastern North America for the past 18,000 years. *In*: Wright, H.E., Jr., Kutzbach, J.E., Webb, T., III, Ruddiman, W.F., Street-Perrott, F.A. & Bartlein, P.J. (Eds), *Global Climates Since the Last Glacial Maximum*. University of Minnesota Press, Minneapolis, pp. 415–467.

Webb, T., III (1985). Holocene palynology and climate. *In*: Hecht, A.D. (Ed.), *Paleoclimate Analysis and Modeling*. John Wiley & Sons, Inc., New York, pp. 163–195.

Webb, T., III (1986). Is vegetation in equilibrium with climate? How to interpret late-Quaternary pollen data. *Vegetatio*, **67**, 75–91.

Webb, T., III (1987). The appearance and disappearance of major vegetational assemblages: long-term vegetational dynamics in eastern North America. *Vegetatio*, **69**, 177–187.

Webb, T., III (1988). Eastern North America. *In*: Huntley, B. & Webb, T., III (Eds), *Handbook of Vegetation Science, Volume 7: Vegetation History*. Kluwer Academic Publishers, Dordrecht, pp. 385–414.

Webb, T., III, Anderson, K.H., Bartlein, P.J. & Webb, R.S. (1998). Late Quaternary climate change in eastern North America: a comparison of pollen-derived estimates with climate model results. *Quaternary Science Reviews*, **17**, 587–606.

Webb, T., III, Bartlein, P.J. & Kutzbach, J.E. (1987). Climatic change in eastern North America during the past 18,000 years; comparisons of pollen data with model results. *In*: Ruddiman, W.F. & Wright, H.E., Jr. (Eds), *The Geology of North America, Volume K-3. North American and Adjacent Oceans During the Last Deglaciation*. The Geological Society of America, Boulder, CO, pp. 447–462.

Webb, T., III & Bernabo, J.C. (1977). The contemporary distribution and Holocene stratigraphy of pollen types in eastern North America. *In*: Elsik, W.C. (Ed.), *Contributions of Stratigraphic Palynology (With Emphasis on North America). Part 1. Cenozoic Palynology*. American Association of Stratigraphic Palynologists Contributions Series 5a, pp. 130–146.

Webb, T., III & Bryson, R.A. (1972). Late- and post-glacial climatic change in the northern Midwest, U.S.A: quantitative estimates derived from fossil pollen spectra by multivariate statistical analysis. *Quaternary Research*, **2**, 70–115.

Webb, T., III & Clark, D.R. (1977). Calibrating micropaleontological data in climatic terms: a critical review. *Annals of the New York Academy of Sciences*, **288**, 93–118.

Webb, T., III, Cushing, E.J. & Wright, H.E., Jr. (1983). Holocene changes in the vegetation of the Midwest. *In*: Wright, H.E., Jr. (Ed.), *Late-Quaternary environments of the United States, Volume 2. The Holocene*. University of Minnesota Press, Minneapolis, pp. 142–165.

Webb, T., III, Howe, S.E., Bradshaw, R.H.W. & Heide, K.M. (1981). Estimating plant abundances from pollen percentages: the use of regression analysis. *Review of Palaeobotany and Palynology*, **34**, 269–300.

Whitehead, D.R. (1965). Palynology and Pleistocene phytogeography of unglaciated eastern North America. *In*: Wright, H.E., Jr. & Frey, D.G. (Eds), *The Quaternary of the United States*. Princeton University Press, Princeton, NJ, pp. 417–432.

Whittaker, R.H. (1957). Recent evolution of ecological concepts in relation to the eastern forests of North America. *American Journal of Botany*, **44**, 197–206.

Willemsen, R.W. (1975a). Dormancy and germination of common ragweed seeds in the field. *American Journal of Botany*, **62**, 639–643.

Willemsen, R.W. (1975b). Effect of stratification temperature and germination temperature on germination and the induction of secondary dormancy in common ragweed seeds. *American Journal of Botany*, **62**, 1–5.

Williams, J.W., Shuman, B.N. & Webb, T., III (2001). Dissimilarity analyses of late-Quaternary vegetation and climate in eastern North America. *Ecology*, **82**, 3346–3362.

Williams, J.W., Summers, R.L. & Webb, T., III (1998). Applying plant functional types to construct biome maps from eastern North American pollen data: comparisons with model results. *Quaternary Science Reviews*, **17**, 607–627.

Williams, J.W., Webb, T., III, Richard, P.H. & Newby, P. (2000). Late Quaternary biomes of Canada and the eastern United States. *Journal of Biogeography*, **27**, 585–607.

Winter, T.C. & Rosenberry, D.O. (1995). The interaction of ground water with prairie pothole wetlands in the Cottonwood Lake area, east-central North Dakota, 1979–1990. *Wetlands*, **15**, 193–211.

Woods, K.D. & Davis, M.B. (1989). Paleoecology of range limits: beech in the Upper Peninsula of Michigan. *Ecology*, **70**, 681–696.

Woodward, F.I. (1987). *Climate and Plant Distribution*. Cambridge University Press, Cambridge, 174 pp.

Wright, H.E., Jr. (1964). Aspects of the early postglacial forest succession in the Great Lakes region. *Ecology*, **45**, 439–448.

Wright, H.E., Jr. (1968a). History of the Prairie Peninsula. *In*: Bergstrom, R.E. (Ed.), *The Quaternary of Illinois: A Symposium in Observance of the Centennial of the University of Illinois*. Univerisity of Illinois College of Agriculture, Special Publication 14, Urbana, pp. 78–88.

Wright, H.E., Jr. (1968b). The roles of pine and spruce in the forest history of Minnesota and adjacent areas. *Ecology*, **49**, 937–955.

Wright, H.E., Jr. (1976). The dynamic nature of Holocene vegetation: a problem in paleoclimatology, biogeography, and stratigraphic nomenclature. *Quaternary Research*, **6**, 581–596.

Wright, H.E., Jr. (1981). Holocene chronostratigraphy for United States and southern Canada. *In*: Mangerud, J., Birks, H.J.B. & Jäger, K.-D. (Eds), *Chronostratigraphic Subdivision of the Holocene*. Striae. Societas Upsaliensis pro Geologia Quaternaria, Uppsala, pp. 53–55.

Wright, H.E., Jr. (1989). The amphi-Atlantic distribution of the Younger Dryas paleoclimatic oscillation. *Quaternary Science Reviews*, **8**, 295–306.

Wright, H.E., Jr. & Frey, D.G. (Eds) (1965). *The Quaternary of the Unites States: A review volume for the VII Congress of the International Association for Quaternary Research*. Princeton University Press, Princeton, NJ, 922 pp.

Wright, H.E., Jr., Kutzbach, J.E., Webb, T., III, Ruddiman, W.F., Street-Perrott, F.A. & Bartlein, P.J. (1993). *Global Climates Since the Last Glacial Maximum*. University of Minnesota Press, Minneapolis, 569 pp.

Wright, H.E., Jr., Winter, T.C. & Patten, H.L. (1963). Two pollen diagrams from southeastern Minnesota: problems in the regional late-glacial and postglacial vegetational history. *Geological Society of American Bulletin*, **74**, 1371–1396.

Wright, J.W. & Rauscher, H.M. (1990). *Fraxinus nigra* Marsh. black ash. *In*: Burns, R.M. & Honkala, B.H. (Eds), *Silvics of North America. Volume 2. Hardwoods*. U.S. Department of Agriculture, Forest Service, Agriculture Handbook 654, Washington, DC, pp. 344–347.

Yu, Z. (2000). Ecosystem response to lateglacial and early Holocene climate oscillations in the Great Lakes region of North America. *Quaternary Science Reviews*, **19**, 1723–1747.

Yu, Z. & Eicher, U. (1998). Abrupt climate oscillations during the last deglaciation in central North America. *Science*, **282**, 2235–2238.

Yu, Z. & Eicher, U. (2001). Three amphi-Atlantic century-scale cold events during the Bølling-Allerød warm period. *Géographie physique et Quaternaire*, **55**, 175–183.

Yu, Z., Ito, E. & Engstrom, D.R. (2002). Water isotopic and hydrochemical evolution of a lake chain in the northern Great Plains and its paleoclimatic implications. *Journal of Paleolimnology*, **28**, 207–217.

Yu, Z., McAndrews, J.H. & Eicher, U. (1997). Middle Holocene dry climate caused by change in atmospheric circulation patterns: evidence from lake levels and stable isotopes. *Geology*, **25**, 251–254.

Yu, Z. & Wright, H.E., Jr. (2001). Response of interior North America to abrupt climate oscillations in the North Atlantic region during the last deglaciation. *Earth-Science Reviews*, **52**, 333–369.

Quaternary vegetation and climate change in the western United States: Developments, perspectives, and prospects

Robert S. Thompson[1], Sarah L. Shafer[2], Laura E. Strickland[1], Peter K. Van de Water[3] and Katherine H. Anderson[3]

[1] *U.S. Geological Survey, MS 980, Denver, CO 80225-0046, USA; rthompson@usgs.gov, lstrickland@usgs.gov*
[2] *U.S. Geological Survey, 200 SW 35th Street, Corvallis, OR 97333, USA; sshafer@usgs.gov*
[3] *University of Colorado, Institute of Arctic and Alpine Research, 1560 30th Street, Boulder, CO 80309, USA; p.vandewater@comcast.net, katherine.anderson@colorado.edu*

Introduction

At the time of the 1965 INQUA Congress a relatively small group of energetic and insightful scientists was rapidly expanding knowledge of how vegetation and climate had changed in North America during the late Quaternary. Continued work by these scientists, their students, and others has improved understanding of the patterns and causes of past changes in vegetation and climate. Overall, understanding of the strengths and potential limitations of paleobotanical data has come far in the last four decades. In addition, there is now a much greater understanding of the complex interactions among changes in climate, vegetation, atmospheric chemistry, earth surface processes, and human activities. The ongoing development of numerical models that portray these interacting systems provides a new avenue for exploration of the past and estimation of the future.

In this chapter we review some of this progress since 1965, examine the strengths and shortcomings of the major sources of data on Quaternary vegetation and climate change, and discuss the use of models as a means to explore past and potential future environmental changes. We focus on the western United States without attempting to repeat recent reviews that summarize the wealth of paleoecological data from this region. Table 1 provides a list of some of the major reviews of vegetation and climate change, both for all of the West and for subregions therein.

The Western United States

By the "western United States" (or simply "the West") we refer to the conterminous United States west of the 100th meridian (Fig. 1). It is a region of high physiographic relief (see Bartlein & Hostetler, this volume) and of strong local and regional climatic gradients. In this paper we divide the West into seven provinces, based on climate and physiography (Fig. 1):

- The *Pacific Northwest* comprises most of the states of Oregon and Washington. The western portion of this province, adjacent to the Pacific Ocean, has major North-South trending mountain ranges (the Coast and Cascade Ranges), mild winter temperatures, heavy winter precipitation, and cool, dry summers. The eastern portion lies in the rain shadow of the Cascade Range, and consequently receives less precipitation. It also has colder winters and hotter summers than the western portion.
- The *California* Province, as used here, refers to the portion of the state of California and adjacent Oregon that lies west of the Sierra Nevada and Transverse Ranges. It is a region of mild winter temperatures, with hot summers occurring in the Central Valley of California and near the Mexican border. The coast and higher elevations have cooler summers, and maritime fog is a significant factor in maintaining cool and moist summers in coastal northern California. The California Province receives most of its precipitation in the winter and much of the region experiences severe summer drought.
- The *Great Basin/Intermountain Region* (or Province) includes that part of the Basin and Range region that occurs north of the Mojave Desert, thus including southeastern Oregon, the Snake River Plain of southern Idaho, portions of easternmost California (east of the Sierra Nevada), most of Nevada, and most of western Utah. This is an arid to semiarid region with cold winters and hot summers (at least in the intermountain basins).
- The *Rocky Mountains* include the major mountain ranges and intervening high valley basins from northern Idaho and Montana, western Wyoming, northeastern Utah, central and western Colorado, and northern New Mexico. The northern portion of this Province receives abundant winter precipitation, and especially toward the south, a secondary summer precipitation maximum. The high mountains throughout the region harbor some of the few moist environments in the West outside of the Pacific Northwest. Winter temperatures can be cold, especially in some of the northern intermountain basins, whereas summers are relatively mild.
- The *Colorado Plateau* Province includes the buttes, mesas, canyonlands, and mountain ranges of southeastern Utah, southwestern Colorado, northwestern New Mexico, and northeastern Arizona. This physiographically complex region includes the Grand

DEVELOPMENTS IN QUATERNARY SCIENCE
VOLUME 1 ISSN 1571-0866
DOI:10.1016/S1571-0866(03)01018-2

PUBLISHED BY ELSEVIER B.V.

Table 1. Reviews of late Neogene and late Quaternary vegetation and climate changes in the Western United States since the publication of "The Quaternary of the United States" in 1965.

Late Neogene-Quaternary vegetation change in Western United States	Leopold & Denton (1987), Thompson (1991), Adam (1995), Thompson & Fleming (1996)
Late Quaternary regional summaries of palynological data	
California	Adam (1985), Anderson (1990a), Adam (1995)
Great Basin and interior Pacific Northwest	Mehringer (1985)
Northern Rocky Mountains	Whitlock (1993), Whitlock & Bartlein (1993)
Pacific Northwest	Heusser (1983), Heusser *et al.* (1985), Whitlock (1992)
Pacific Northwest and northern Rocky Mountains	Barnosky *et al.* (1987)
Southwest	Hall (1985)
Western United States	Baker (1983)
Western hemisphere (including Western United States)	Grimm *et al.* (2001), Whitlock *et al.* (2001)
Late Quaternary regional summaries of packrat midden data	
Chihuahuan Desert	Van Devender (1990a)
Colorado Plateau	Betancourt (1984, 1990)
Grand Canyon and lower Colorado River drainage	Cole (1982, 1990a, b)
Great Basin	Thompson & Mead (1982), Wells (1983), Thompson (1990)
Mojave Desert	Spaulding (1990a, 1995)
Sonoran Desert	Van Devender (1990b)
Southwest	Van Devender (1977), Van Devender & Spaulding (1979), Van Devender *et al.* (1987), Betancourt *et al.* (1990a, b)
Late Quaternary regional summaries of palynological and packrat midden data	
Colorado Plateau	Anderson *et al.* (2000)
Great Basin	Wigand & Rhode (2002)
North America (including Western United States)	Bartlein *et al.* (1998)
Southwestern United States	Spaulding *et al.* (1983), Van Devender *et al.* (1987)
Western United States	Thompson *et al.* (1993), Mock & Brunelle-Daines (1999), Thompson & Anderson (2000)

Canyon, as well as innumerable smaller strongly incised drainages. This province experiences freezing winters and hot summers, and is arid to semiarid, receiving modest precipitation in both the cold and warm seasons.

- The *Southwest*, as used here, refers to the hot deserts of the southern part of the West, as well as interspersed desert grasslands and mountain ranges. The Mojave Desert occupies the northwest portion of the Southwest in southern Nevada, northwest Arizona, and the Death Valley region of eastern California. This desert experiences freezing winters and very hot summers and receives its small allotment of precipitation generally in the winter season. The Sonoran Desert straddles the Mexican border from southeastern California across southwestern and southcentral Arizona. It has a unique climate with winter temperatures rarely dipping below freezing, very hot summers, and a biseasonal (winter-summer) precipitation regime. As seen in Fig. 1, the parts of the Sonoran Desert that occur in northern Mexico and adjacent southernmost Arizona receive significant amounts of summer rainfall. The Chihuahuan Desert and associated desert grasslands occur at somewhat higher elevations than the Sonoran Desert and occupies portions of southern New Mexico, southeasternmost Arizona, west Texas, and adjacent Mexico. This desert has colder winters than the Sonoran Desert, combined with hot summers. Summer rainfall is dominant over winter precipitation in this subprovince.
- The *Great Plains* lie along the eastern margin of our area of interest. As the name implies, this is a region of generally low relief (although elevations may reach 1500–2000 m on the Plains in Colorado and Wyoming). The northern Great Plains have an extremely continental climate with cold winters and hot summers, while the southern Great Plains have mild winters and hot summers. Relatively little winter precipitation occurs across much of the province, which receives the majority of its precipitation in the spring and summer months.

The climatic descriptions in the above paragraphs refer to commonly used climatic variables such as temperature and precipitation. Increasingly ecologists and paleoecologists also utilize "bioclimatic" variables that may better represent the physiological requirements and tolerances of plant species (and hence of the ecosystems and vegetation types discussed

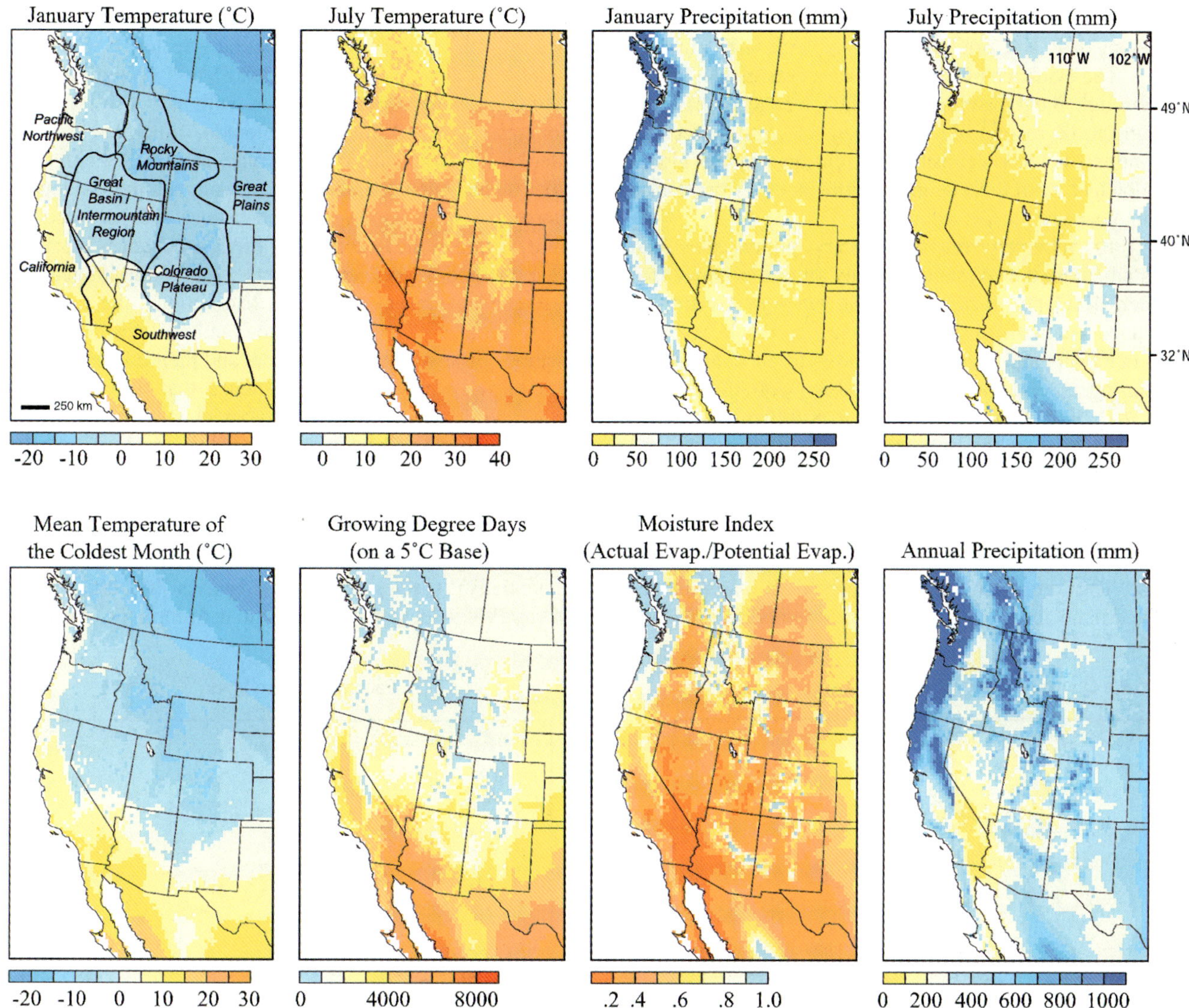

Fig. 1. Climate of western North America, with geographic regions as used in this paper. These data are shown on a 25-km equal-area grid for the period 1951–1980 (modified from Thompson et al., 1999a; Thompson & Anderson, 2000).

below; see Fig. 1 for maps of these variables in the West). These variables include "mean temperature of the coldest month" (MTCO), which in the West closely resembles January temperature. The "growing degree days" variable (on a 5 °C base: GDD5) is a measure of the total amount of energy received over the growing season (Newman, 1980), and provides a way to compare regions with (for example) long growing seasons with moderate temperatures to regions with shorter but hotter growing seasons. As seen in Fig. 1, this variable illustrates the long-hot growing seasons of the Sonoran and Chihuahuan Deserts of the Southwest, in contrast to the short-cool growing seasons of the high elevation Rocky Mountains, Sierra Nevada, and Cascade Range. The "moisture index" (actual evaporation/potential evaporation) variable provides an integrated measure of the annual moisture budget (see Thompson *et al.*, 1999a for further discussion). By comparing the portrayal of this variable with annual precipitation in Fig. 1, the reader can see how the influences of temperature, solar radiation, and other factors interact with annual precipitation to produce regional patterns of moisture and aridity. The moisture index highlights the few areas of the West (primarily in the Pacific Northwest, northern Rocky Mountains, and other high mountain regions) that harbor relatively moist habitats.

State of Knowledge in 1965

The occasion of the 1965 INQUA Congress, coupled with the publication of "The Quaternary of the United States" (Wright & Frey, 1965), set the stage for much of what was to come in the following decades in the rapidly expanding fields of paleoecology, Quaternary geology, and paleoclimatology. The number of North American specialists in past vegetation change was still relatively small in 1965, analytical and dating methods were still under development, and the scientific world had still to accept now-widely-held concepts such as continental drift and orbital forcing of long-term climatic changes. The evidence of multiple glacial-interglacial cycles back through the Quaternary and late Neogene was just beginning to be recognized from deep-sea isotopic records and continental loess deposits. Past changes in the global carbon budget, with their attendant influences on vegetation and climate, were not yet recognized.

In 1965 relatively few Quaternary pollen sites had been studied in the West, and many of the records that had been developed were from what are now considered to be unconventional depositional settings (such as playa lakes, archaeological sites, arroyo walls, fossil dung, forest soils, spring mounds, and caves). Subsequent studies (e.g. Fall, 1987) have demonstrated that it is difficult to assess the influences of taphonomic processes and changing (and unknown) source areas on these unconventional deposits and, as indicated in Fig. 2, many of them are now ignored by present-day investigators. Radiocarbon dates were relatively few in 1965, and tephrochronology was not yet widely applied. The remarkable plant macrofossil assemblages preserved in ancient *Neotoma* (packrat or woodrat) middens had only recently been reported, and the full potential of these new data sources had not yet been realized.

The "Quaternary of the United States" (Wright & Frey, 1965), published in conjunction with the VIIth INQUA Congress held in Boulder, Colorado, presented a broad interdisciplinary vision of Quaternary studies. As with many other fields encompassed in this volume, the authors of chapters dealing with vegetation history and paleoecology were writing at the leading edge of great advances in knowledge, methodologies, and insights. Despite the limited data available on present-day plant distributions, the relatively small number of well-dated sites, and the still-developing methods of analysis, these authors and their contemporaries established a new branch of science that has profoundly influenced paleoecological, ecological, and paleoclimatic thought. These authors recognized: the complex interplay among climatic, geologic, and biotic factors that influenced the present and past distributions and abundances of plant species (Davis, 1965; Heusser, 1965); the importance of current biogeographic patterns in reflecting climatic and (potentially) species dispersal histories (Cushing, 1965; Davis, 1965; Heusser, 1965; Martin & Mehringer, 1965); the

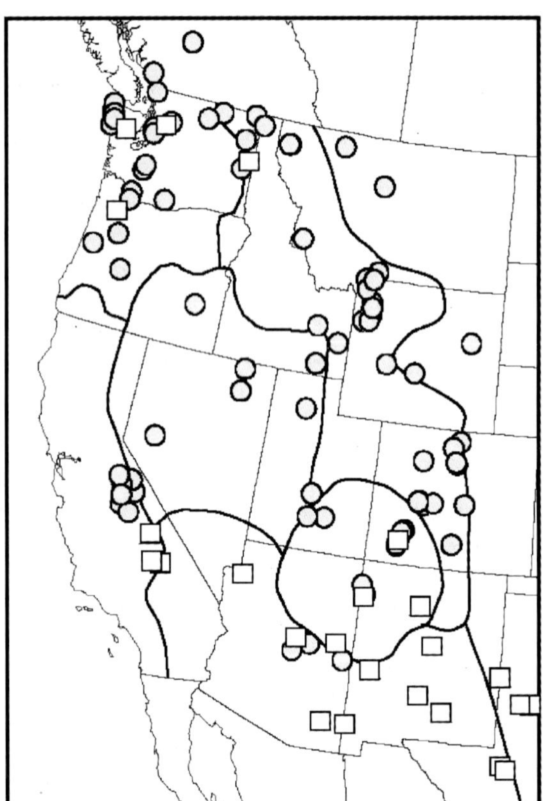

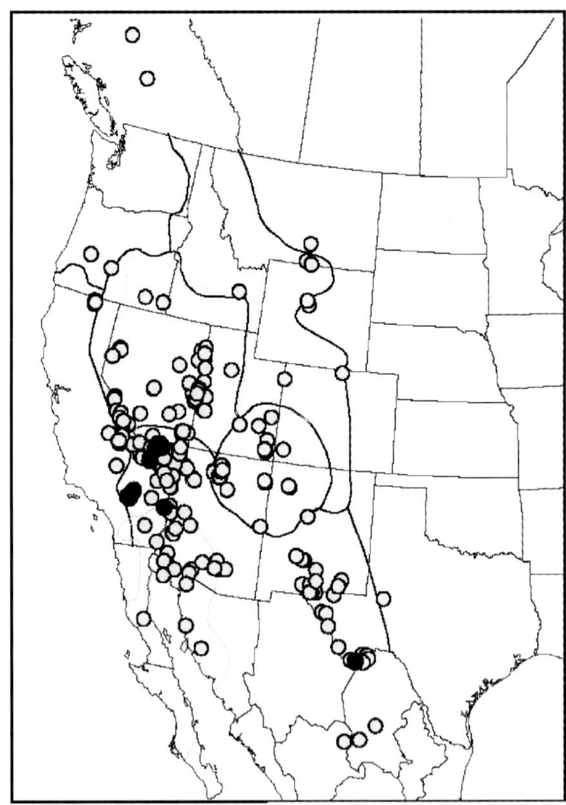

Fig. 2. Late Quaternary paleobotanical study sites in western North America. In the panel on the left, gray circles represent the distribution of palynological sites from lakes and wetlands (modified from Thompson & Anderson, 2000), whereas squares represent the palynological sites available to Martin & Mehringer (1965) and Heusser (1965). As discussed in the text, many of these latter sites are from playas, caves, archaeological sites, and other settings that are now usually viewed as presenting interpretive problems. In the panel on the right, black circles represent packrat midden sites available to Martin & Mehringer (1965), whereas gray circles represent sites studied since 1965 (data from the USGS-NOAA Packrat Midden Database).

potential significance of fire on ecosystems (Davis, 1965; Heusser, 1965); and the role of prehistoric and historic land- and resource- use on vegetation (Davis, 1965; Martin & Mehringer, 1965). These authors examined concepts of individualistic species vs. community responses to climatic change (e.g. Cushing, 1965), and the relative importance of climatic controls and dispersal rates in determining plant distributions (especially at their northern limits following the retreat of glaciers). They recognized the potential importance of mixtures of boreal and temperate elements in late Pleistocene assemblages, presaging coming decades of studies of 'non-analogue' late Pleistocene vegetation.

In the following sections we discuss aspects of the advances in Western paleobotanical studies since the 1965 INQUA meeting. Our objectives here are to explain the methods used in the reconstructions of past vegetation and climate, to explore the strengths and weaknesses of sensors, and address some of the problems of spatial, temporal, and taxonomic resolution that occur within Western data sets and their interpretations.

Advances in Reconstructing Paleoclimate

The science of reconstructing past changes in vegetation and climate involves the use of present-day climate and vegetation data, coupled with a diverse array of palynological, macrofossil, and isotopic data. Below we describe advances in data acquisition, analysis, and interpretation that have occurred in this field in the western United States over the past several decades.

Present-Day Plant Distributions and Climate Data

The authors of the 1965 INQUA vegetation history chapters had little access to organized information on the present-day distributions of major plant species and their climatic tolerances. In the decades that followed, E.L. Little, Jr. and his co-authors at the U.S. Forest Service published maps of the distributions of more than 600 tree and shrub species (Little, 1971, 1976, 1981). Additionally, detailed regional examinations were made of the ecology and climatic tolerances of many species (see Turner *et al.*, 1995 for an excellent example). Unfortunately, distribution maps still do not exist for many of the plant species that are represented in the Quaternary paleoecological record, largely because they are of little economic importance.

Reconstructions of past climates from paleobotanical data rely on the present-day relationships between the distribution (or abundance) of plant species (or traits) and present-day climatic parameters. The climate of the period of instrumental weather records (approximately the last century) has not been stationary, and unfortunately there is no generally agreed-upon time interval to use as the standard for modern calibration exercises. In addition, the high physiographic relief of most of the West, coupled with the relatively small number of long-term weather stations (most of which are in valley bottoms) has made it difficult to assess the climatic conditions across the elevational and geographic gradients of the Western mountains. New climate data sets have become available over the past decade (e.g. Daly *et al.*, 1994; New *et al.*, 1999, 2000), and new estimation techniques continue to be developed. An organized effort has been made to understand and illustrate the relations between the distributions of plant species in North America and climatic and bioclimatic parameters (Thompson *et al.*, 1999a, b, 2000), primarily using the range maps provided by the work of Little and colleagues coupled with climatic data on a 25-km equal area grid (this gridded data set, shown in part in Fig. 1, was constructed by Patrick J. Bartlein and colleagues at the University of Oregon – see Thompson *et al.*, 1999a for further discussion). As discussed below, the present-day vegetation-climate relations determined by this effort provide the basis for reconstructing past climates from vegetation data resolved at the species level from packrat middens and other macrofossil assemblages.

Dating Techniques

The primary sources of information on past vegetation changes in western North America are studies of stratigraphic changes in palynological assemblages preserved in lake sediments or other permanently wet sites and studies of the plant macrofossil assemblages preserved in the permanently arid interiors of packrat middens. Radiocarbon dating is the principal method of dating both kinds of paleoecological records (e.g. Taylor *et al.*, 1992), with tephrochronology providing important chronological data for sedimentary deposits in regions that had active volcanoes during the Quaternary (primarily in and near the Pacific Northwest; Sarna-Wojcicki & Davis, 1991). Tephrochronology and paleomagnetic dating (Verosub, 2000) provide independent corroboration of late Quaternary isotope-derived chronologies and provide dating of Quaternary deposits outside the limits of existing isotopic techniques.

Since 1965, radiocarbon dating has become increasingly precise and is now widely available as a commercial laboratory service (see Taylor, 1987; Trumbore, 2000 for a description of radiocarbon dating techniques). The introduction of accelerator mass spectrometry (AMS) dating in the late 1970s significantly increased geological precision and reduced the amount of material needed to as little as 200 μg to 2 mg of material, compared with the several grams minimum for conventional radiocarbon dating. The small sample size required for AMS dating has allowed investigators to apply radiocarbon dating to increasingly specific targets, such as the dating of individual plant macrofossils (e.g. Van Devender *et al.*, 1985) or pollen grains (Brown *et al.*, 1989; Mensing & Southon, 1999). Additionally, the calibration of radiocarbon dates in terms of calendric ages has been significantly enhanced, both in terms of precision and in extending farther back in time (Stuiver *et al.*, 1998a, b).

Pollen Records

Many methodological advances have occurred in palynology in the past several decades and we confine our discussion to aspects specific to the West. Over the past several decades the number of western Quaternary pollen records has increased greatly, especially in the Pacific Northwest, Rocky Mountains, and other montane settings (Fig. 2). As mentioned above, many of the pollen records developed by 1965 (particularly from the arid interior and Southwest) were from unconventional deposits (e.g. Martin, 1963; Martin & Mehringer, 1965). Although these authors recognized the potential shortcomings of many of these data sources, the lakes, bogs, and mires widely studied in northern Europe and the eastern United States were virtually unavailable in the West (outside of the Pacific Northwest and high mountains throughout the region). Martin & Mehringer (1965) documented many shortcomings of the western records, including contamination from older deposits and the widespread occurrence of long-distance pollen transport, especially in the relatively low pollen productivity of desert environments. Their pollen records were dominated by a relatively small number of wind-dispersed pollen types that could usually be resolved only to genus or family. They struggled with large hard-water effects on radiocarbon ages from some of the few permanently wet sites that they could study, and they foresaw the potential that plant macrofossils from packrat middens could overcome many of their difficulties. Packrat midden studies developed rapidly in the subsequent decades, and supplanted pollen studies in much of the arid and semi-arid West (e.g. Betancourt *et al.*, 1990a, b, c). As a result of the advent of packrat midden analysis, coupled with the problems discussed above, many of the early pollen records from unconventional sites of the Southwest and adjacent areas are now viewed with suspicion.

As seen in Fig. 3, most of the lake and wetland palynological study sites in the West are set in cool-moist environments. These sites provide the great majority of paleoecological information for the moist regions of the Pacific Northwest, Rocky Mountains, and other high mountain regions. Conversely, packrat middens, discussed in the next section, are best preserved in warm and dry environments and provide the most reliable data source for arid and semiarid environments. Pollen data from stratigraphic deposits (usually lakes) provide quasi-continuous records that are spatially integrated over varying scales (and thus the settings and taphonomies of the collecting basin are important). Many investigators have been attracted to the potential for very long (hundreds of thousands to millions of years) pollen records from the subsiding deep structural basins of the West (e.g. Adam *et al.*, 1989; Davis, 1998). Unfortunately, these records are commonly difficult to interpret, due both to the dominance of a few wide-ranging pollen types and to the large regions censused by the pollen being deposited in these basins. Additionally, the records commonly suffer from imperfect preservation of pollen and recycling of materials from older deposits.

In contrast, the lake and wetland environments of portions of the Pacific Northwest, Rocky Mountains, and other

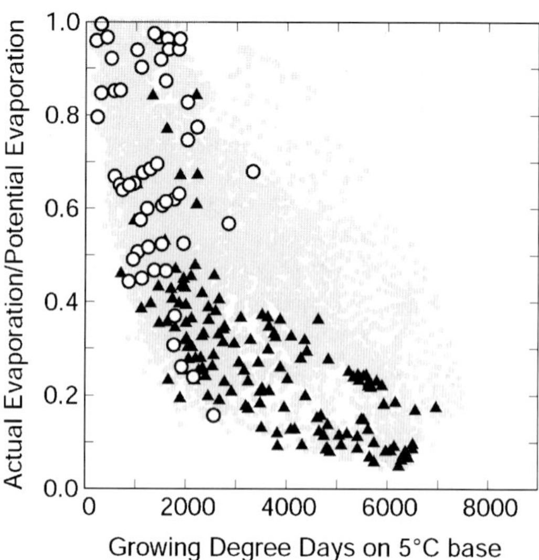

Fig. 3. The occurrence of late Quaternary paleobotanical sites plotted against growing degree days and the moisture index. The small gray circles represent points on the 25-km grid of western North American climate where neither pollen sites nor packrat middens have been studied. The open circles represent pollen study sites (data from Thompson & Anderson, 2000), whereas the black triangles represent packrat midden localities (data from the USGS-NOAA Packrat Midden Database). As seen in this illustration, palynological sites generally occur in relatively cool and moist environments, whereas packrat middens are generally preserved in warmer and drier environments (see Betancourt et al., 1990b, c; Spaulding, 1990b; Thompson et al., 1993 for comparisons of pollen- and midden-based reconstructions).

montane parts of the West provide excellent materials for conventional Quaternary pollen analysis. In many of these regions the density of pollen sites is great enough to support elevational and geographic examinations of changes in vegetation (and inferred climates) through time. Such studies have been carried out in the Pacific Northwest (Barnosky *et al.*, 1987), the Sierra Nevada Range of California (Anderson, 1990a), and the Yellowstone area (Whitlock, 1993; Whitlock & Bartlein, 1993) and central Colorado (Fall, 1985) areas of the Rocky Mountains. These studies provide unique four-dimensional views of vegetation and climate change, with plants changing their geographic and elevational ranges in response to climatic changes through the late Quaternary.

The pollen flora of much of the western United States is dominated by a few pollen types (notably *Pinus*, Taxaceae-Cupressaceae-Taxodiaceae (TCT – usually *Juniperus* in the arid and semiarid West), *Quercus*, *Artemisia*, other Asteraceae, Chenopodiaceae-*Amaranthus*, and Poaceae) that each encompass many species representing a wide variety of environmental and climatic conditions. In addition to the coarse taxonomic resolution of the palynological data, it is commonly difficult (especially in mountain regions) to assess the "region" represented by the pollen data, or how it has changed through time. Investigators have explored these

Table 2. Studies of present-day pollen distributions.

Pacific Northwest
- Washington state – Pacific Coast east to crest of Cascade Range: Heusser (1969, 1973, 1978a, b, c), Heusser, L.J. (1983)
- Washington state east of the Cascade Range: Mack & Bryant (1974)
- Oregon and southern Washington state: Minckley & Whitlock (2000)

California
- Adam & West (1983), Anderson & Davis (1988), Anderson (1990b)

Great Basin/intermountain region
- Southern Idaho: Davis (1984)
- Eastern Nevada: Thompson (1992)
- Eastern California: Solomon & Silkworth (1986)

Rocky Mountains
- Wyoming basins: McAndrews & Wright (1969)
- Western Wyoming: Fall (1994)
- Colorado mountains: Fall (1992)

Colorado plateau
- Northern Arizona: Hevley (1968), Fall (1987)
- Northern New Mexico: Bent & Wright (1963)
- Southwestern Colorado: Maher (1963)

Southwest
- Mojave Desert: Mehringer (1967)
- Sonoran Desert: Hevley *et al.* (1965), Schoenwetter & Doerschlag (1971), O'Rourke (1986)
- Desert grassland: Potter & Rowley (1960), Martin (1963)
- Chihuahuan Desert: Meyer (1977)

Great Plains
- McAndrews & Wright (1969), Hoyt (2000)

Western United States
- Davis (1995)

potential difficulties by examining the pollen contents of present-day surface samples and comparing these data with the surrounding vegetation and the current climate.

Palynological studies of such "modern" samples were begun in the early 1960s, and have expanded to cover major portions of the western United States (see Table 2 for selected surface sample studies). Permanently wet lakes and wetlands are uncommon across much of the arid and semiarid West, and unfortunately many of the surface sample studies of the present-day distributions of pollen types in the West have had to utilize surface soil ("pinch") samples, moss polsters, sediments from cattle watering tanks, or other less-than-ideal data sources. Figure 4 provides a coarse-scale illustration of the relations between relative abundances of pollen taxa and bioclimatic parameters in the western United States (based on the core-top samples from the data set used by Thompson & Anderson, 2000). As seen in this illustration, the pollen percentages of some major pollen taxa (such as *Pinus*) show little relationship to bioclimatic parameters. As seen in the left-hand column, the highest pollen percentages of *Picea*, *Artemisia*, and Chenopodiaceae-*Amaranthus* generally occur in regions with relatively cold winters, whereas the highest percentages of *Tsuga*, and TCT, occur in regions with relatively warm winters. In the central column, *Picea* has its highest pollen percentages in regions with relatively low GDD5 values (GDD5 < 1000), whereas many of the other taxa shown here have their highest pollen percentages when GDD5 is near 2000. In the right-hand column, the highest percentages of *Picea* and *Tsuga* occur in moist environments, whereas the highest percentages of *Artemisia* and Chenopodiaceae-*Amaranthus* occur in dry conditions. Collectively, these data demonstrate that there are measurable relations between the proportions of major pollen types and climatic conditions in the western United States, although the coarse taxonomic resolution of the pollen data may limit paleoclimatic interpretations to a fairly general level.

Packrat Middens

Prior to the development of packrat (*Neotoma*) midden analysis, paleoecologists working in the Southwest and other dry regions of the West were the "poor cousins" of their colleagues working in humid eastern North America and Europe, where pollen records from permanent lakes provided excellent records of past vegetation changes. With the advent and spread of midden analysis, Western investigators could, in many cases, obtain a level of taxonomic detail and chronological precision that was unattainable in other settings. As shown in Figs 2 and 3, packrat middens provide the major source of paleovegetation information for the warm arid and semiarid regions of the West. The methods of midden analysis and regional summaries are described in detail in the chapters in Betancourt *et al.* (1990a), and we review only major methodological concepts here.

Packrat middens are dry, rock-hard, urine-cemented masses of fecal pellets, plant remains, and a variety of other materials collected by the animals that made them. Middens are usually preserved in permanently dry caves and rock shelters, and paleoecological information from these deposits thus may be biased toward the environments surrounding these rocky substrates. Plant macrofossil assemblages from middens provide detailed inventories of the species living within a few tens of meters of the midden sites at the time that the middens were constructed (Dial & Czaplewski, 1990; Finley, 1990). The plant remains in these assemblages are generally extremely well preserved, can usually be identified to the species level, and middens provide excellent material for radiocarbon dating, isotopic analyses, pollen analysis, studies of vertebrates and insects, plant-morphological examinations (such as changes in stomatal density), and a variety of other still-developing studies. Packrat midden assemblages thus can provide detailed information on small areas at discrete points in time (in contrast to most pollen records, which provide quasi-continuous records of

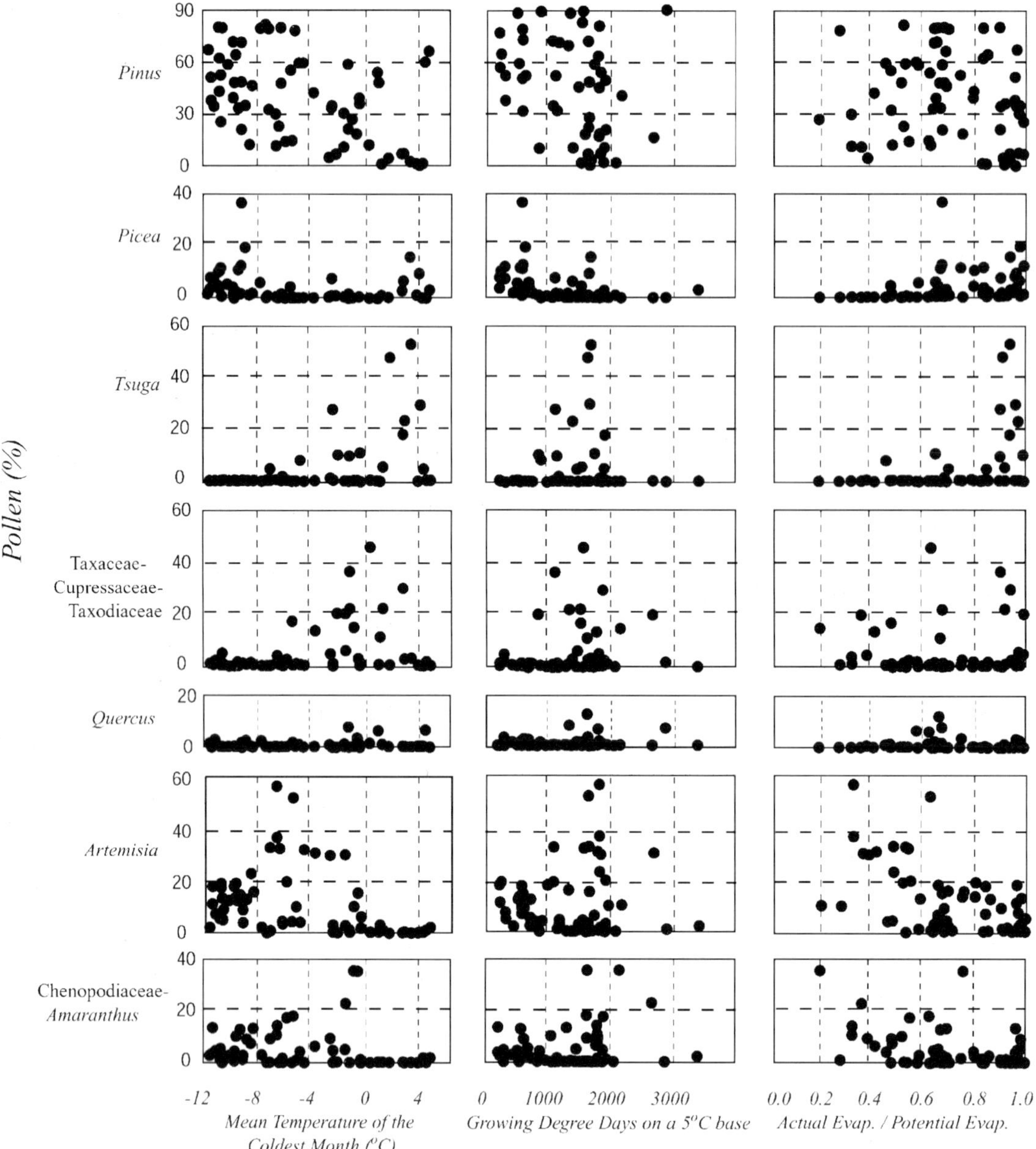

Fig. 4. Pollen percentage data from core-tops of lake and wetland sites in western North America (data from Thompson & Anderson, 2000) plotted against mean temperature of the coldest month (MTCO), growing degree days (on a 5 °C base – GDD5), and a moisture index (actual evaporation/potential evaporation).

vegetation change at a more generalized spatial scale and at relatively-coarse taxonomic resolution). Unfortunately, the geographic and climatic overlap between pollen sites and midden sites is small (Figs 2 and 3), with the result that the western Quaternary paleovegetation data set largely consists of two essentially non-overlapping data sources of differing temporal, spatial, and taxonomic resolution. Intensive regional and local collecting of packrat middens can provide time-series of vegetation change, despite temporal gaps.

Since packrat midden analysis emerged in the early 1960s as a tool for understanding late Pleistocene and Holocene paleoenvironments, a variety of methods have developed to sample, date, and analyze the contents of middens. Also, studies of modern packrat behavior and the botanical contents of recent middens in relation to local plant communities have illuminated potential biases in the midden record which can affect the accuracy of middens as indicators of past vegetation.

Sampling

Midden samples have been collected for a range of specific projects, and this may have resulted in temporal bias in the

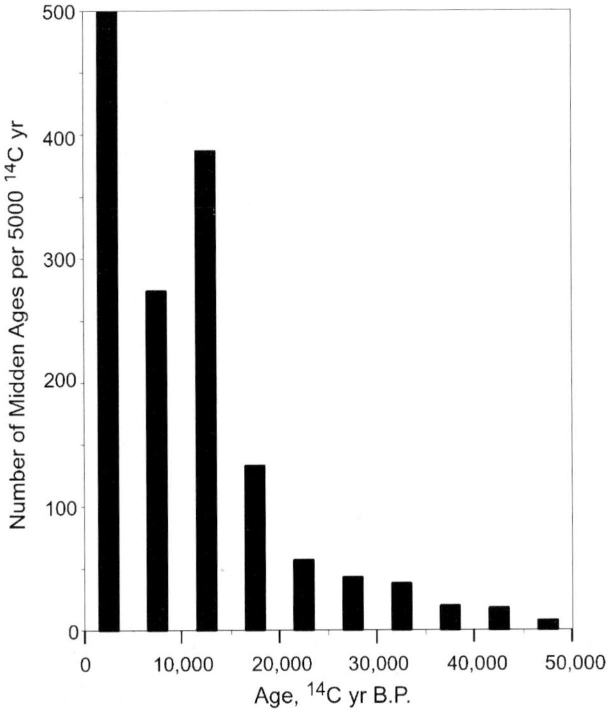

Fig. 5. Histograms of radiocarbon dates on packrat middens from western North America (data from the USGS-NOAA Packrat Midden Database).

collection and analysis of packrat middens: as seen in Fig. 5 there are a large number of middens from the late Holocene and late Pleistocene, with a smaller number of early to middle Holocene middens. This difference may reflect regional aridity during the latter interval, or may reflect the interaction between the long-term trend of decreasing number of middens preserved with increasing age, and selective collecting (or analysis) of middens that are dominated by extralocal vegetation. In other words, although in some cases there may be fewer latest Pleistocene middens than Holocene middens, investigators interested in late Pleistocene vegetation would preferentially sample middens that contained plant remains that differed from the plants living at the site today.

Analysis

When packrat midden plant macrofossil assemblages were first reported by P.V. Wells and colleagues (Wells, 1966; Wells & Berger, 1967; Wells & Jorgensen, 1964) the plant remains were tallied on semi-quantitative scales based on actual counts. Since that time there have been more methods of reporting these data than there are investigators, with widely used measures including: presence-absence, raw counts of numbers of specimens, raw weights of each taxon, weighted percentages of conifers, number of macrofossils per kilogram of either washed or unwashed midden matrix, and percentages of the total number of identified specimens. It is not clear that any of these methods provides either a superior measure of abundance or is more replicable than any other. Fossil pollen is usually well preserved in the matrices of packrat middens and can provide information to augment that available from macrofossil assemblages (Anderson & Van Devender, 1991, 1995; Jennings, 1996; Mehringer & Wigand, 1990; Thompson, 1985).

Radiocarbon Dating

Most packrat middens sampled before 1985 were dated using conventional gas proportional or liquid scintillation techniques (Webb & Betancourt, 1990). These methods are still being used today. However, many investigators prefer to use AMS dating (discussed above), which permits them to date very small samples. This is desirable both to limit the destruction of important paleobotanical materials and to identify contaminants and mixed-age assemblages (see Van Devender *et al.*, 1985 for further discussion). Materials used for dating include unwashed and washed midden matrix, twigs, fecal pellets, fragments of a single species, remains of juniper, or total conifer remains.

The most suitable part of a midden for radiocarbon dating has been debated (Webb & Betancourt, 1990). Wells (1976) advocated dating unwashed midden debris collected from a single, horizontal layer, whereas Van Devender (1977) suggested dating either a single plant fragment or a collection of fragments belonging to a single taxon removed from the matrix by washing, screening, and sorting, which requires a large sample size (Webb & Betancourt, 1990). Prior to 1970 most middens were dated using unwashed midden matrix. In the 1970s monospecific dates, mainly on juniper, provided the majority of the dates. During the 1980s there was a significant rise in the dating of fecal pellets, which provide a sizable amount of contemporaneous material for dating (Webb & Betancourt, 1990).

Calibration

Paleoecological reconstructions depend upon the accuracy with which plant remains from middens represent species richness and relative abundance in the local flora. Techniques, such as similarity coefficients, have been used to estimate how well midden assemblages represent species richness – Sorensen's index is the most widely used, although Jaccard's index, and Simpson's index have also been used (Spaulding *et al.*, 1990). Species richness is highly dependent on sample size and the ability of the investigator to identify plant specimens. Attempts have been made to relate abundance of plant macrofossils in middens to abundance of individual species on the landscape, but this is not easily measured. Abundance data have been analyzed using the Bray-Curtis index comparing percentage of number of identified specimens (%NISP) and percentage of total plant cover and density (Cole & Webb, 1985). Frequencies of some species in middens seem to correlate with their importance in the community while others do not. Relating the number or weight of macrofossils to cover and density does not provide a reliably accurate measure of modern plant community attributes (Spaulding *et al.*, 1990). Cover and density may not be the most appropriate measures for this kind of comparison.

Recent studies using paired samples of modern vegetation communities and paired samples of paleocommunities investigate the probability that the absence of a taxon from a woodrat midden implies absence from the paleo-landscape. The probability of a false inference occurring was calculated as between 7 and 11%, with higher or lower probabilities for certain taxa (Nowak *et al.*, 2000). Study of modern middens and woodrat species (Dial & Czaplewski, 1990) show that dietary specialization affects the contents of modern middens, and biases the representation of relative abundance of the vegetation. However, if several contemporaneous middens are sampled, species diversity may be fairly represented. Fortunately for paleoecologists, interpretations of species richness are fairly robust while interpretations of species abundance are uncertain (Dial & Czaplewski, 1990) because differences in midden composition resulting from woodrat selectivity are mainly in the relative proportions of material collected and not in the species richness.

Paleoclimatic Reconstructions

The identification of plant remains to the species level in ancient packrat midden assemblages, in conjunction with ongoing efforts to understand the present-day relations between plant distributions and climatic parameters (discussed above), potentially provides the basis for detailed reconstructions of past climates. Fig. 6 illustrates some examples of this potential: (1) Several species that today live either exclusively in the mountains of the Great Basin (*Pinus longaeva*) or there and farther north (*Pinus flexilis, Juniperus communis, Artemisia tridentata*) were present in the Mojave Desert of southern Nevada and adjacent California (where *Yucca brevifolia* now lives). Comparison of the histograms of the bioclimatic parameters associated with these species suggests that the late Pleistocene climate was characterized by colder winters, cooler summers, and wetter conditions than those of today; (2) Plants that today live on the middle to lower mountain slopes or valleys of the Great Basin, California, the Mojave Desert, or Colorado Plateau (and/or farther north) such as *Juniperus scopulorum, Pinus monophylla, Juniperus osteosperma*, and *Juniperus californica* during the late Pleistocene lived in the now extremely hot and dry region of the Sonoran Desert occupied today by *Cereus giganteus*. Again, the Pleistocene climate appears to have had colder winters, cooler summers, and higher moisture levels than that of today.

Databases

Quaternary pollen and plant macrofossil data for North America have been compiled and curated in cooperative databases (Jackson *et al.*, 2000). Three separate databases are available in electronic format on the World Wide Web; The North American Pollen Database (NAPD, http://www.ngdc.noaa.gov/paleo/pollen.html) for pollen assemblages from sites throughout the United States and Canada; The North American Plant Macrofossil Database (NAPMD, http://uwadmnweb.uwyo.edu/Botany/NAPMD/index.htm) for plant macrofossils derived from wet sediments in the United States, Canada, and Greenland; and The USGS/NOAA Western North American Packrat Midden Database (WNAPMD, http://climchange.cr.usgs.gov/data/midden) for plant macrofossils from packrat middens in arid regions of the western United States, Canada, and Mexico (Strickland *et al.*, 2001). The data will be permanently archived in a publicly accessible digital format at the World Data Center-A for Paleoclimatology, located at the NOAA National Geophysical Data Center (NGDC) in Boulder, Colorado (Anderson, 1995; Webb *et al.*, 1994).

Isotopes from Plant Tissues

Since 1965, there has been a rapid expansion in the use of stable isotope analysis of plant material (^{13}C, ^{18}O, D [Deuterium]) to determine biochemical, photosynthetic, and ecological processes and influences. Initial surveys used stable isotopes to identify three different types of photosynthetic physiology (C_3, C_4, and CAM), each showing unique isotopic values. For example, C_3 and C_4 plant values show distinct $\delta^{13}C$ values (Ehleringer, 1991) whereas C_4 and CAM plants can be differentiated comparing $\delta^{13}C$ and δD (Sternberg & DeNiro, 1983; Sternberg *et al.*, 1985). Within each photosynthetic group isotopic variability reflects environmental effects resulting from biochemical differences in carbon acquisition, plant-water use and photosynthetic carbon fixation (Farquhar, 1983; Farquhar *et al.*, 1980). In western North America, the importance of water to plant communities has focused much of the isotopic work on communities residing in arid environments (Ehleringer, 1993a, b). More recently, this theoretical framework constructed for modern plant-isotope relations has been applied to fossil plant material to track past plant response to changing atmospheric and climatic conditions (Beerling, 1996a, b; Jahren *et al.*, 2001; Jennings & Elliott-Fisk, 1993; Long *et al.*, 1994; Marino *et al.*, 1992; Pendall *et al.*, 1999; Siegel, 1983; Van de Water *et al.*, 1994). These paleoenvironmental reconstructions provide new insights into historic and prehistoric ecosystem adjustments brought about by local, regional, and global environmental shifts especially during, but not limited to, the transition from the Pleistocene to Holocene.

Paleoenvironmental research using stable isotopes has centered on western North America primarily because of the abundant supply of packrat midden macrofossils. Excellent preservation coupled with species-specific identifications make these macrofossil collections unequaled in their use for stable isotopic reconstructions. The studies to date have used stable isotope values to determine past conditions by establishing modern relationships between the growth environment of current western plant communities (Ehleringer, 1993a, b;

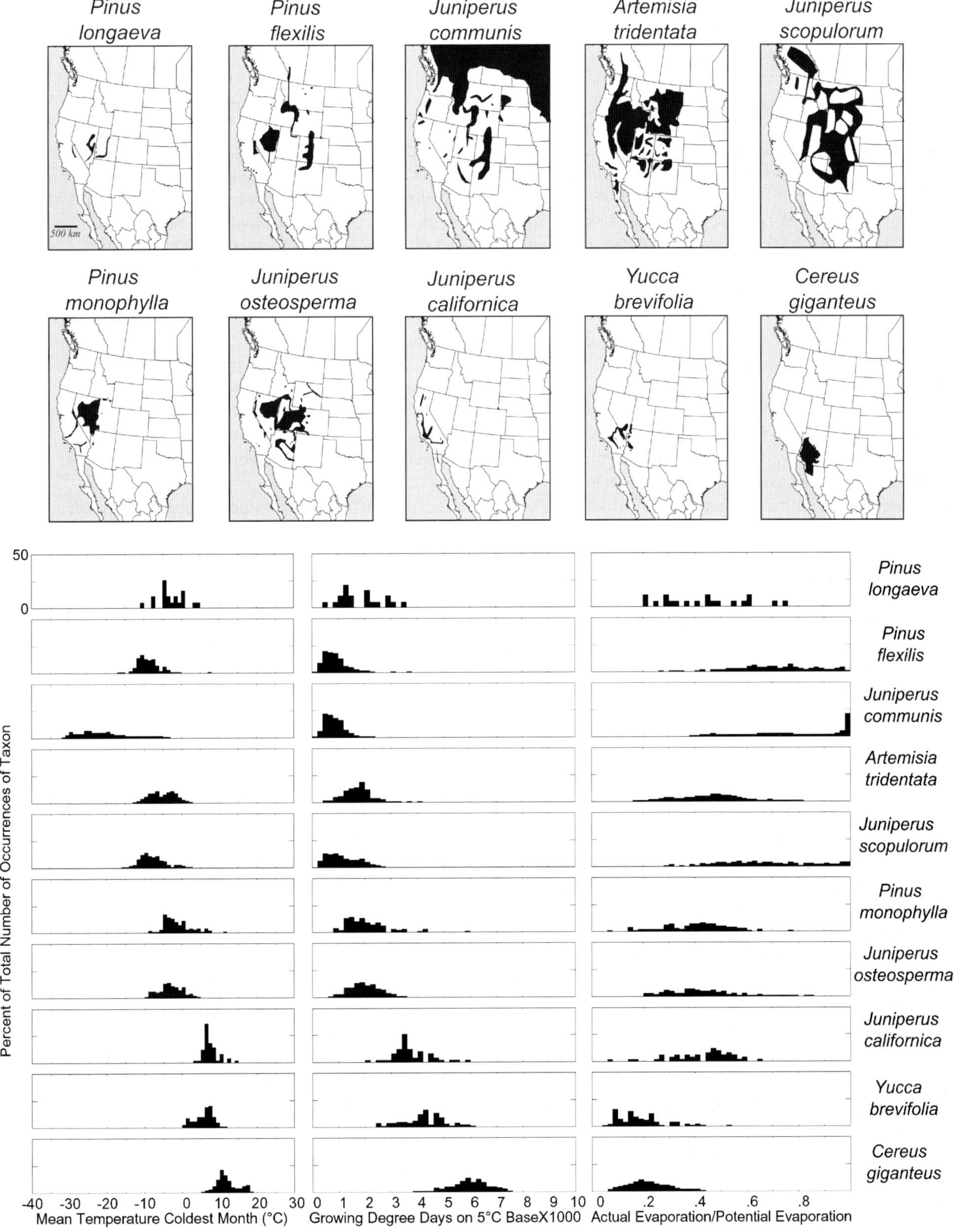

Fig. 6. Distribution maps (generalized and simplified from Little, 1971, 1976, 1981) and associated present-day bioclimatic data (from Thompson et al., 1999a, b) for plant species recovered from late Pleistocene packrat middens from the Sonoran Desert, Mojave Desert, and Great Basin. The bioclimatic data are presented as percentages of the total number of occurrences for a given species that occur within the specified range of the bioclimatic variable.

Pendall, 2000; Terwilliger *et al.*, 2002; Van de Water *et al.*, 2002; Williams & Ehleringer, 2000) and those during the historic past (Pedicino *et al.*, 2002). With these relationships established, analysis of fossil material has used single stable isotopes or combinations to reconstruct past environmental conditions. For example, carbon acquisition physiology and the trade offs with plant water-use are reported for C_3 plants over the past 30,000 years using $\delta^{13}C$ (Van de Water, 1999; Van de Water *et al.*, 1994). Theoretical equations for carbon acquisition in C_4 plants (Farquhar, 1983) suggest that past atmospheric $\delta^{13}C$ values are captured as a portion of the $\delta^{13}C$ plant tissue values. Initial reconstructions of past atmospheric $\delta^{13}C$ values using midden macrofossils showed promising results (Marino *et al.*, 1992) but additional analysis of midden and modern material has revealed significant levels of natural variability (Toolin & Eastoe, 1993; Van de Water *et al.*, 2002) greater than atmospheric $\delta^{13}C$ changes found since the last glacial maximum (LGM: approximately 21,000 cal yr B.P. or 18,000 ^{14}C yr B.P.) from ice core records (Friedli *et al.*, 1984, 1986; Indermühle *et al.*, 1999; Leuenberger *et al.*, 1992; Smith *et al.*, 1999).

The potential of both $\delta^{18}O$ and δD to provide climatic information has long been recognized (Epstein & Yapp, 1976; Epstein *et al.*, 1976, 1977; Yapp & Epstein, 1977). The relevance of $\delta^{18}O$ and δD in plant tissue relies upon an understanding of systematic fractionation of water sources at each site prior to plant uptake (Rozanski *et al.*, 1993) along with fractionation factors occurring during synthesis of plant tissue (Sternberg, 1989). Isotopic variability in plant material is related to temperatures and relative humidity through environmental interactions during evapotranspiration (Pendall, 2000). This is especially true in photosynthesizing tissues, such as leaves, because of the direct linkage between atmospheric gas exchange and evapotranspiration during carbon acquisition. The complex relationships between source water fractionation and plant environment interactions have limited the application of $\delta^{18}O$ and δD to the fossil record. In addition, the use of packrat midden material has been limited, however, early studies using both $\delta^{18}O$ and δD show promise as a means to add temperature, relative humidity, and precipitation sourcing to past environmental reconstructions (Pendall *et al.*, 1999; Siegel, 1983).

Phytoliths

Many plant species produce phytoliths ("plant stones"), structures of silica or (less commonly) calcium oxalte. These structures are resistent to decay and are commonly preserved as microfossils in soils. Sediment analysis for opal phytoliths has been shown to add important information previously unavailable from grasslands and other areas poorly represented in pollen and midden records (e.g. Blinnikov *et al.*, 2002; Fredlund & Tieszen, 1994). Phytolith analysis also may allow differentiation within the Poaceae unavailable from pollen analysis. Studies show that phytoliths contained within a stratigraphic section represent the vegetation at the local stand level (Fredlund & Tieszen, 1994), however, most phytoliths occur within the size range of silts (10 to $>50\,\mu m$), and may become airborne for downwind deposition.

Other Evidence of Vegetation Change

Wood fragments and standing snags recovered from above present treeline indicate that early to middle Holocene treelines were as much as 100 m higher than today during portions of the Holocene across much of the West (Carrara *et al.*, 1991; Epstein *et al.*, 1999; Friedman *et al.*, 1988; LaMarche, 1973; LaMarche & Mooney, 1972; Lloyd & Graumlich, 1997) suggesting mean summer temperatures nearly 2 °C above present. Plant macrofossils from dry cave sediments, lake sediments, and the dung of extinct animals have also been reported from the western United States, although they remain minor sources of data.

Paleoclimatic Reconstructions

The pollen, plant macrofossil, and isotopic data discussed above have been analyzed by a variety of qualitative and quantitative approaches to estimate past climatic conditions. These methods seek to interpret aspects of past climates from changes in geographic range boundaries, changes in elevational limits, shifts from dry to moist sites, changes in relative abundance, shifts in dominance of plants with different carbon assimilation pathways, and/or changes in isotopic ratios within plant tissues. As discussed below, there is increasing awareness that these phenomena may also be influenced (or controlled) by either climatic parameters, non-climatic factors (such as fire, human activities, atmospheric chemistry, etc.), or combinations of the two. Attempts at quantitative reconstructions have involved examinations of regional lapse rates (e.g. Betancourt, 1984), regression-based analyses of fossil and modern pollen data (Adam & West, 1983; Heusser *et al.*, 1980, 1985), identification of the nearest living populations of extirpated taxa (and assigning the climate of these sites to the fossil locality: Spaulding, 1985); use of "climatic envelopes" that describe the present range of climatic conditions associated with plant taxa (Rhode & Madsen, 1995; Sharpe, 2002); GIS-based analysis of the climatic factors that exclude the occurrences of certain taxa (Arundel, 2002); and, identification of the most similar extant vegetation to the fossil assemblage (and assigning climatic values based on the calibration data set [modern analogue analysis], e.g. Thompson *et al.*, 1999c).

Non-Climatic Factors That Influence Vegetation

Among the most important advances of the past few decades is the recognition and elucidation of the role(s) that non-climatic factors may play in determining the composition and distribution of plant communities. It will take many decades more to understand fully the complex interactions within the large array of potentially interacting climatic, environmental, and human processes involved in vegetation change. Below are brief discussions of some of the currently recognized factors.

Atmospheric CO_2 Concentrations

Past changes in atmospheric CO_2 concentrations have played a significant role in vegetation changes observed in the paleoenvironmental record. CO_2 concentrations have indirectly affected vegetation by altering global climate. In addition to this indirect effect, changes in atmospheric CO_2 concentrations also have a direct physiological effect on plants via the role of CO_2 in photosynthesis. Empirical studies have shown that increases in atmospheric CO_2 concentrations lead to increases in net primary production in some plant species, potentially improving their success in competing with other species (Amthor, 1995). Additionally, increases in CO_2 concentrations can increase the water-use efficiency of some taxa, allowing them to better tolerate low moisture conditions (Amthor, 1995; Polley, 1997). Increased water-use efficiency has been implicated in the expansion of woody vegetation (i.e. trees and shrubs) into arid regions of the interior western United States (Amthor, 1995; Polley, 1997). Changing levels of atmospheric carbon dioxide may also influence the frost sensitivity of plants (Royer *et al.*, 2002).

One prediction that comes from reduced atmospheric CO_2 concentrations is that plants have to adjust their morphology to compensate for lower carbon assimilation rates (Beerling *et al.*, 1995; Beerling & Chaloner, 1992; Woodward, 1987a, b; Woodward & Bazzaz, 1988). The packrat midden record allows for detailed analysis of individual plant parts to test these predictions. Van de Water *et al.* (1994) found that *Pinus flexilis* needles were significantly thinner during the full glacial conditions compared to Holocene samples. Similar results occur in *Pinus edulis* and *Pinus remota* packrat midden macrofossils along a transect running from southeastern Utah to the Big Bend region of Texas (Van de Water, 1999). Reduced late Pleistocene atmospheric CO_2 is predicted to also have resulted in increased stomata on leaf surfaces to aid the diffusion of CO_2 into the leaf interior. Increased stomatal density (stomata per leaf area) is documented in *Pinus flexilis* (Van de Water *et al.*, 1994), *Pinus edulis* and *Pinus remota* (Van de Water, 1999). Changes in stomatal density and index (stomata per mesophyll cells) occur in other plant species from packrat middens, but the characteristics show differential responses at the species level and need further analysis of additional collections (Van de Water, 1999).

Species respond individualistically to changes in CO_2, and their individual responses are affected by a number of factors, including nutrient availability and interactions with other species (e.g. Smith *et al.*, 2000). The species-specific responses of plants to past changes in atmospheric CO_2 would have altered competitive interactions among species, contributing to the vegetation changes observed in the paleoecological record. Various studies have shown that both climate change and changes in atmospheric CO_2 need to be considered to explain past vegetation changes (e.g. Jolly & Haxeltine, 1997).

Influence of Disturbance Regimes

A variety of disturbance processes have influenced vegetation distributions in the western United States. Past changes in climate variability and the occurrence of extreme climate events, such as droughts, have affected vegetation on seasonal to decadal time scales (Allen & Breshears, 1998; Woodhouse & Overpeck, 1998). Records from multiple climate proxies, including tree-ring data, ENSO records, lake-level data, etc., have been used to determine the magnitude and frequency of past droughts and the resulting response of vegetation. Disease and pest outbreaks in the western United States have also affected vegetation on a variety of temporal and spatial scales. Pest outbreaks in the region have been documented in both tree-ring and lake-sediment records, and are associated with periods of drought as well as periods of increased precipitation (e.g. Swetnam & Betancourt, 1998).

Fire regimes have had large impacts on vegetation patterns in the western United States. Information on the magnitude and frequency of fire regimes over the last millennium is available from fire-scarred tree-ring records and stand-age analysis. On longer time-scales, charcoal samples from lake-sediment cores have recorded changes in fire frequencies on centennial to millennial time scales. These lake-sediment records indicate that climate-vegetation-fire interactions in the western United States have been complex throughout the Holocene. For example, in some areas changes in fire frequency have been accompanied by vegetation changes (e.g. Veblen & Lorenz, 1991), while in other areas vegetation has been relatively unresponsive to changes in fire frequency (Whitlock, 1993; Whitlock & Bartlein, this volume).

Prehistoric Humans

There is still a great deal of debate over the magnitude of the impact of prehistoric people on ecosystems in western North America. Evidence exists for local vegetation changes around areas of large settlements in the southwestern United States (e.g. Jones *et al.*, 1999). In some areas, burning by Native Americans is cited as having maintained pre-Euro-American settlement vegetation patterns, although the extent to which anthropogenic fires simply augmented non-anthropogenic fire regimes is still debated (Vale, 2002; Whitlock & Bartlein, this volume). Humans have also been implicated in the Pleistocene extinction of North American megafauna, such as mammoths (Martin, 1984). The relatively abrupt extinction of these species would have altered grazing and browsing pressure on vegetation over large areas of North America, particularly in areas of grasslands and savannas (Martin & Burney, 1999).

Historic Humans

Euro-American settlement, beginning in the late-1500s and then rapidly expanding in the 1800s and 1900s, initiated wide-scale land use and land cover changes in western North America (Marsh, 1871; Thomas, 1956). Land use activities associated with Euro-American settlement resulted in large-scale disturbance and fragmentation of native vegetation, changes in species distributions, and alteration of disturbance regimes. Many of these land use activities continue to affect vegetation on the landscape today.

Forest and woodland areas throughout western North America were cleared during Euro-American settlement to provide land for farming; timber for buildings, railroads, and mines; and fuel for cooking, heating, and industrial processes, such as ore smelting (Greeley, 1925). Large areas of cropland were created in the western United States, particularly after 1850 (Ramankutty & Foley, 1999). Western Canada experienced a similar expansion of cropland at the beginning of the 1900s (Ramankutty & Foley, 1999). Farming, settlement, and other land use activities often resulted in the draining of wetlands.

Grazing, primarily by cattle and sheep, was historically significant in western North America and continues in the region today, affecting plant communities ranging from low-elevation grasslands to high-elevation meadows and forests (Bogan *et al.*, 1998; McPherson, 1997). Historic over-grazing, particularly pronounced prior to the Taylor Grazing Act of 1934 (43 U.S.C. sec. 315), has been attributed with a number of vegetation changes, including the decline of native bunchgrasses in the American Southwest, alteration of fire regimes, and channel incision affecting riparian vegetation. Grazing also continues to play an important role in the spread of non-native species, such as *Bromus tectorum* (cheatgrass) (Knapp, 1996).

Settlement of western landscapes is continuing as the population of western North America continues to grow. Urban areas are expanding and rural populations are growing as well, destroying and fragmenting remaining areas of native vegetation. The transportation networks that connect cities and rural areas have been major contributors to habitat fragmentation, as have the networks of logging roads on private, state, and federal lands. Roads and railroads create corridors that not only fragment habitat, but also provide pathways for the introduction of invasive species (Forman, 1995).

Non-Native Invasive Species

The introduction of many non-native species to western North America has been facilitated by past and on-going changes in land use and land cover. Non-native species introduced by Euro-American settlement have been in western North America for centuries. Adobe bricks produced in California prior to 1800 contain non-native grasses, indicating the long history of invasive plant species in the region (Clark, 1956). In general, as one moves northward in North America, the number of non-native species tends to decrease. Rejmánek & Randall (1994) estimated that California has approximately 1025 non-native vascular plant species, British Columbia has approximately 646, and Alaska has approximately 144 non-native vascular plant species. Withers *et al.* (1998) attribute this decline in non-native species with increasing latitude to a combination of cold temperatures, low human population densities, and the fact that many high-latitude species are already pan-Arctic in their distribution.

Many non-native plant species have been deliberately introduced to western North America for a variety of purposes, ranging from use as ornamentals to species introduced to provide increased forage and soil erosion control (McPherson, 1997; OTA, 1993). OTA (1993) estimates that at least 36 of approximately 300 non-indigenous weed species in the western United States escaped from agriculture or horticulture. In addition to affecting the species composition of plant communities, some introduced species, such as cheatgrass, have altered fire regimes (OTA, 1993). Introduced plant pests and pathogens have also led to the decline of native plant species.

Human Alteration of Fire Regimes

Fire suppression in western North America continues to alter the vegetation density and species composition of many plant communities. Fire suppression was always prevalent around human settlements, but expanded when U.S. government agencies initiated policies of fire suppression in the forests of the western U.S. in the early 1900s. Fire suppression has led to the filling in of many forests with younger trees and fire-intolerant species. These increased fuel loads can create more intense fires when fires occur, resulting in changes in species composition as species that are intolerant to severe fires decline (e.g. Barton, 2002). In grasslands and savannas, grazing may have reduced fire frequency by reducing the amount and connectivity of fine fuels, such as grasses. This reduction in fire frequency is contributing to the invasion of woody vegetation into grasslands, and the expansion of trees into grasslands and shrublands in certain areas (McPherson, 1997).

The combinations of human activities have made it increasingly difficult to determine the causes of vegetation changes occurring on the landscape. For example, many woody species in western North America, including pinyon pines and junipers, have expanded into grasslands and savannas over the last century (Polley, 1997). Are these expansions primarily due to local-scale land use changes, or regional-scale forcings, such as climate change (e.g. Flannigan *et al.*, 2000)? Juniper expansion has been attributed to changes in grazing pressures, fire suppression, and range adjustments in response to climate changes. Increased atmospheric CO_2 concentrations may be increasing juniper water-use efficiency and enhancing the re-growth of juvenile trees after injury by fire or herbivory, potentially providing junipers with a competitive advantage over other species (Bond & Midgley, 2000).

Development and Application of Models

Numerical models increasingly provide the means to explore the potential impacts of (and interactions between) climatic and nonclimatic factors in vegetation change. Advancements in computing technology are providing researchers with a larger range of analytical tools, including the ability to run increasingly complex numerical models. A variety of general circulation models (GCMs) are now used to simulate atmosphere and ocean dynamics, providing insights into the dynamics of paleoclimates (see Bartlein & Hostetler, this volume). Complementing GCMs are a wide variety of physically-based process models for simulating a variety of ecosystem processes, including vegetation dynamics. Among the vegetation models are dynamic global vegetation models (DGVMs; e.g. Sitch *et al.*, 2003). DGVMs can characterize a variety of ecosystem processes, including disturbance regimes and the physiological response of plants to changes in atmospheric CO_2 concentrations. DGVMs have been used to investigate both Quaternary vegetation change and potential future vegetation change.

Models can be run at a variety of temporal and spatial scales. They are also used to simulate both past and future climate and environments. Model simulations can be compared with paleoenvironmental proxy data sets to develop and test hypotheses of paleoenvironmental change. Figure 7 provides a simple example of such a data-model comparison. In this example, the BIOME4 equilibrium vegetation model (Kaplan, 2001) was used to simulate the occurrence of forest vegetation in the West during the LGM, the middle Holocene

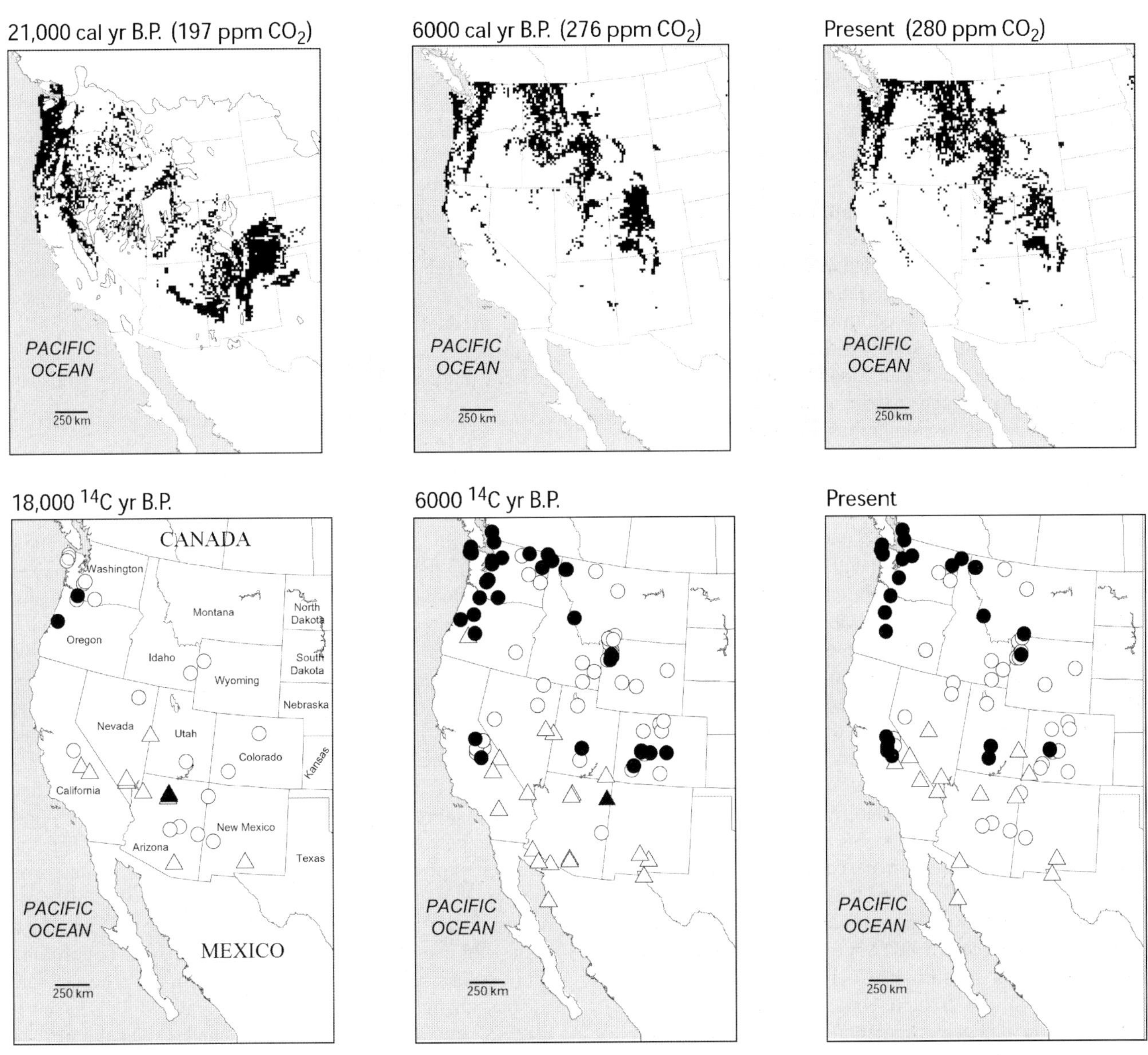

Fig. 7. Comparison of simulated (top row) and observed (bottom row) changes in forest cover through time in the western United States. The top row illustrates BIOME4 vegetation model simulations of the occurrence of forest vegetation for 21,000 cal yr B.P. (approximately 18,000 ^{14}C yr B.P.), 6000 cal yr B.P. ($\sim$6000 ^{14}C yr B.P.), and the present day. In the bottom row, circles represent pollen data and triangles represent packrat midden data (modified from Thompson & Anderson, 2000). Black symbols indicate the occurrence of forest, whereas white symbols indicate the occurrence of other vegetation types. Model: BIOME4 (Kaplan, 2001). Climate data: 6000 and 21,000 cal yr B.P. (CCM3, Bonan, 1996; CCM3 data interpolation, Patrick J. Bartlein, pers. comm.), present (Thompson et al., 1999a, b, c). Soil data: CONUS-Soil (Miller & White, 1998).

(approximately 6000 cal yr B.P. or 6000 ^{14}C yr B.P.), and the present day. The simulated results include soil variability, changes in atmospheric carbon dioxide concentrations, and GCM simulations of past climate. The simulated changes in forest cover can be compared to observed occurrences of forest vegetation from fossil pollen and packrat midden data (modified from Thompson & Anderson, 2000). In this example, forest cover is adequately simulated for the present and for 6000 years ago when compared with observed data, while forest cover is overestimated for the LGM.

The increase in the amount of both data generated by models and observed data sets has led to the development of new data visualization techniques. Of particular importance are the visualization techniques that allow the depiction of data through space and time. These range from small multiple maps (e.g. Bartlein *et al.*, 1998) to animations of paleoenvironmental changes through time.

Potential Future Vegetation Change

There is increasing evidence that human activities are affecting the Earth's climate and that anthropogenic-induced climate changes will likely continue for centuries to come (IPCC, 2001). Coupled atmosphere-ocean general circulation models (AOGCMs) have been used to simulate these potential future climate changes to the end of the century (IPCC, 2001). Recent AOGCM simulations indicate potential future increased warming for much of western North America in both winter and summer, and increases in winter precipitation by 2071–2100 (Giorgi *et al.*, 2001). Just as past climate changes have affected the distribution of plant species, these potential future climate changes will likely significantly affect the distribution of species throughout western North America.

Much of what is understood about how plant taxa may respond to future climate change is based on the paleoenvironmental record of how vegetation responded to climate changes in the past. One of the most important lessons from the paleoenvironmental record is that species respond individualistically to variations in climate (Davis, 1981; Huntley, 1995; Webb, 1995). As climate changes in the future, species populations will respond by trying to track suitable climate conditions. Some species may be able to expand their range under future climate conditions, while suitable climate conditions for other species, particularly those with restricted ranges, may disappear entirely from western North America (Bartlein *et al.*, 1997; Shafer *et al.*, 2001; Thompson *et al.*, 1998). As species distributions change, new plant communities will be created, some of which may have no analogue in the paleoenvironmental record.

The future rate of climate change is estimated to be more rapid than can be accommodated by the natural migration rates of many plant species. Paleoenvironmental records indicate that past migration rates for trees were only on the order of 1000 m/year (Pitelka & the Plant Migration Workshop, 1997). The ability of particular species to disperse to new habitat may also depend on the effects of climate change on other species. For example, whitebark pine (*Pinus albicaulis*) is dependent on Clark's nutcracker (*Nucifraga columbiana*) for seed dispersal and thus negative climate change impacts on Clark's nutcrackers could significantly affect whitebark pine as well (Tomback, 2001). Increases in atmospheric CO_2 concentrations may also affect the migration rates of certain species. Empirical studies indicate that changes in CO_2 concentrations contribute to increased fecundity, including increased seed production for some species (e.g. LaDeau & Clark, 2001). Changes to reproductive success would alter species competitive interactions and population dynamics.

Future climate change will also affect the magnitude and frequency of a variety of disturbance regimes that affect vegetation. Warmer and drier summer conditions in many parts of western North America could increase the frequency of drought stress on plants, creating fuel conditions suitable for fires. Changes in species distributions will also affect fire regimes. Dukes (2000) points out that the potential future changes to the fire regime will be dependent on both the timing of precipitation and the timing of plant senescence. For example, if increased CO_2 concentrations lead to increased water-use efficiency in perennials, thereby delaying senescence longer into the fire season, then the potential for fires may be reduced (Dukes, 2000). In contrast, increased plant productivity, due to increased CO_2 and water-use efficiency, could increase the amount of fuels available, thereby potentially increasing fire potential on the landscape.

Summary

The large number of scientific investigations since the 1965 INQUA provide a complex and detailed view of vegetation change, its relation to climate, atmospheric chemistry, and human influences. The flora and major vegetation types of the western United States have been present for several million years (Leopold & Denton, 1987; Thompson, 1991, 1996). Glacial periods (with low levels of atmospheric carbon dioxide) favored expansions of woodlands and shrublands, whereas full interglacials (which represent a much smaller portion of the past million years) favored the expansion of forests and deserts. Ongoing changes in atmospheric chemistry, climate, and human activities may lead to major vegetation changes over the coming decades to centuries. The combination of observations from the paleoenvironmental record, modern ecological studies, and modeling now permit assessments of the magnitude of potential future changes in the context of natural variability. They also provide opportunities for hypothesis testing and identification of the processes driving past changes in vegetation and climate. Models have been used to simulate potential future changes over the next few decades to centuries. Although models cannot predict future climate changes with accuracy, they do improve understanding of the potential magnitude and direction of future change.

Understanding the dynamics of paleoenvironmental change can contribute to current conservation and natural resource management efforts. The paleoenvironmental record clearly indicates that species respond individualistically to climate change (Davis, 1981; Huntley, 1995; Thompson,

1988; Webb, 1995). In response to future climate change, taxa will attempt to track climate and disperse to areas with suitable climate habitat. Appropriate climate conditions for some plant species may disappear entirely from western North America, while other species may be able to expand their ranges. In addition to responding to changes in temperature and precipitation, vegetation will also respond to changes in the concentrations of atmospheric CO_2. The combination of warmer temperatures and increased concentrations of atmospheric CO_2 may, for example, allow woodland species to replace steppe and grassland species over much of the region and result in more rapid vegetation changes than have occurred in the last several million years. Understanding the dynamics of paleovegetation change will help conservation and natural resource managers anticipate the potential rate, magnitude, and complexity of future vegetation change.

References

Adam, D.P. (1985). Quaternary pollen records from California. *In*: Bryant, V.M., Jr. *et al.* (Eds), *Pollen Records of Late Quaternary North American Sediments: Dallas, Texas, USA*. American Association of Stratigraphic Palynologists Foundation, pp. 125–140.

Adam, D.P. (1995). Reflections on the development of the California pollen record. *In*: Steadman, D.W. *et al.* (Eds), *Late Quaternary Environments and Deep History: The Mammoth Site of Hot Springs, South Dakota*, Inc. Scientific Papers, Vol. 3, pp. 117–130.

Adam, D.P., Sarna-Wojcicki, A.M., Rieck, H.J., Bradbury, J.P., Dean, W.E. & Forester, R.M. (1989). Tule Lake, California: The last 3 million years. *Palaeogeography, Palaeoclimatology, Palaeoecology*, **72**, 89–103.

Adam, D.P. & West, G.J. (1983). Temperature and precipitation estimates through the last glacial cycle from Clear Lake, California, pollen data. *Science*, **219**, 168–170.

Allen, C.D. & Breshears, D.D. (1998). Drought-induced shift of a forest-woodland ecotone: Rapid landscape response to climate variation. *Proceedings of the National Academy of Science USA*, **99**, 14839–14842.

Amthor, J.S. (1995). Terrestrial higher-plant response to increasing atmospheric [CO_2] in relation to the global carbon cycle. *Global Change Biology*, **8**, 243–274.

Anderson, D.M. (1995). Global paleoenvironmental data. Workshop Report Series 95-2.

Anderson, R.S. (1990a). Holocene forest development and paleoclimates within the central Sierra Nevada, California. *Journal of Ecology*, **78**, 470–489.

Anderson, R.S. (1990b). Modern pollen rain within and adjacent to two giant sequoia (Sequoiadendron giganteum) groves, Yosemite and Sequoia national parks, California. *Canadian Journal of Forest Research*, **20**, 1289–1305.

Anderson, R.S., Betancourt, J.L., Mead, J.I., Hevly, R.H. & Adam, D.P. (2000). Middle-and late-Wisconsin paleobotanic and paleoclimatic records from the southern Colorado Plateau, USA. *Palaeogeography, Palaeoclimatology, Palaeoecology*, **155**, 31–57.

Anderson, R.S. & Davis, O.K. (1988). Contemporary pollen rain across the central Sierra Nevada, California USA: Relationship to modern vegetation types. *Arctic and Alpine Research*, **20**, 448–460.

Anderson, R.S. & Van Devender, T.R. (1991). Comparison of pollen and macrofossils in packrat (Neotoma) middens: A chronological sequence from the Waterman Mountains of southern Arizona, USA. *Review of Palaeobotany and Palynology*, **68**, 1–28.

Anderson, R.S. & Van Devender, T.R. (1995). Vegetation history and paleoclimates of the coastal lowlands of Sonora, Mexico – pollen records from packrat middens. *Journal of Arid Environments*, **30**, 295–306.

Arundel, S.T. (2002). Modeling climate limits of plants found in Sonoran Desert packrat middens. *Quaternary Research*, **58**, 112–121.

Baker, R.G. (1983). Holocene vegetational history of the western United States. *In*: Wright, H.E., Jr. (Ed.), *Late Quaternary Environments of the United States, The Holocene*. Vol. 1, pp. 109–127.

Barnosky, C.W., Anderson, P.M. & Bartlein, P.J. (1987). The northwestern U.S. during deglaciation; Vegetational history and paleoclimatic implications. *In*: Ruddiman, W.F. *et al.* (Eds), *North America and Adjacent Oceans During the Last Deglaciation, The Geology of North America*. Geological Society of America, Vol. K-3, pp. 289–322.

Bartlein, P.J., Anderson, K.H., Anderson, P.M., Edwards, P.E., Mock, C.J., Thompson, R.S., Webb, R.S., Webb, T., III & Whitlock, C. (1998). Paleoclimate simulations for North America over the past 21,000 years: Features of the simulated climate and comparisons with paleoenvironmental data. *Quaternary Science Reviews*, **17**, 549–585.

Bartlein, P.J. & Hostetler, S.W. (this volume). Modeling paleoclimates.

Bartlein, P.J., Whitlock, C. & Shafer, S.L. (1997). Future climate in Yellowstone National Park region and its potential impact on vegetation. *Conservation Biology*, **11**(3), 782–792.

Barton, A.M. (2002). Intense wildfire in southeastern Arizona: Transformation of a Madrean oak-pine forest to oak woodland. *Forest Ecology and Management*, **165**, 205–212.

Beerling, D.J. (1996a). ^{13}C discrimination by fossil leaves during the late-glacial climate oscillation 12–10 ka BP: Measurements and physiological controls. *Oecologia*, **108**, 29–37.

Beerling, D.J. (1996b). Ecophysiological responses of woody plants to past CO_2 concentrations. *Tree Physiology*, **16**, 389–396.

Beerling, D.J., Birks, H.H. & Woodward, F.I. (1995). Rapid late-glacial atmospheric CO_2 changes reconstructed from the stomatal density record of fossil leaves. *Journal of Quaternary Science*, **10**, 379–384.

Beerling, D.J. & Chaloner, W.G. (1992). Stomatal density as an indicator of atmospheric CO_2 concentration. *The Holocene*, **2**, 71–78.

Bent, A.M. & Wright, H.E. (1963). Pollen analysis of surface materials and lake sediments from the Chuska Mountains,

New Mexico. *Geological Society of America Bulletin*, **74**, 491–500.

Betancourt, J.L. (1984). Late quaternary plant zonation and climate in southeastern Utah. *Great Basin Naturalist*, **44**, 1–35.

Betancourt, J.L. (1990). Late quaternary biogeography of the Colorado Plateau. *In:* Betancourt, J.L. *et al.* (Eds), *Packrat Middens: The Last 40,000 Years of Biotic Change*. Tucson, University of Arizona Press, pp. 259–292.

Betancourt, J.L., Van Devender, T.R. & Martin, P.S. (Eds) (1990a). *Packrat Middens: The Last 40,000 Years of Biotic Change*. Tucson, University of Arizona Press, 467 pp.

Betancourt, J.L., Van Devender, T.R. & Martin, P.S. (1990b). Introduction. *In*: Betancourt, J.L. *et al.* (Eds), *Packrat Middens: The Last 40,000 Years of Biotic Change*. Tucson, University of Arizona Press, pp. 2–11.

Betancourt, J.L., Van Devender, T.R. & Martin, P.S. (1990c). Synthesis and prospectus. *In*: Betancourt, J.L. *et al.* (Eds), *Packrat Middens: The Last 40,000 Years of Biotic Change*. Tucson, University of Arizona Press, pp. 435–447.

Blinnikov, M., Busacca, A. & Whitlock, C. (2002). Reconstruction of the late Pleistocene grassland of the Columbia basin, Washington, USA, based on phytolith records in loess. *Palaeogeography, Palaeoclimatology, Palaeoecology*, **177**, 1–25.

Bogan, M.A., Allen, C.D., Muldavin, E.H., Platania, S.P., Stuart, J.N., Farley, G.H., Mehlhop, P. & Belnap, J. (1998). Southwest. *In*: Mac, M.J., Opler, P.A., Puckett Haecker, C.E. & Doran, P.D. (Eds), *Status and Trends of the Nation's Biological Resources*. Reston, VA, U.S. Geological Survey, 2 vols.

Bonan, G.B. (1996). A Land Surface Model (LSM Version 1.0) for Ecological, Hydrological, and Atmospheric Studies: Technical description and user's guide. NCAR Technical Note NCAR/TN-417+STR: Boulder, CO, NCAR, 150 pp.

Bond, W.J. & Midgley, G.F. (2000). A proposed CO_2-controlled mechanism of woody plant invasion in grasslands and savannas. *Global Change Biology*, **6**, 865–869.

Brown, T.A., Nelson, D.E., Mathewes, R.W., Vogel, J.S. & Southon, J.R. (1989). Radiocarbon dating of pollen by accelerator mass spectrometry. *Quaternary Research*, **32**, 205–212.

Carrara, P.E., Trimble, D.A. & Meyer, R. (1991). Holocene treeline fluctuations in the northern San Juan Mountains, Colorado, U.S.A., as indicated by radiocarbon-dated conifer wood. *Arctic and Alpine Research*, **23**(3), 233–246.

Clark, A.H. (1956). The impact of exotic invasion on the remaining New World mid-latitude grasslands. *In*: Thomas, W.L., Jr. (Ed.), *Man's Role in Changing the Face of the Earth*. Chicago, IL, University of Chicago Press.

Cole, K. (1982). Late Quaternary zonation of vegetation in the eastern Grand Canyon. *Science*, **217**, 1142–1145.

Cole, K.L. (1990a). Late Quaternary vegetation gradients through the Grand Canyon. *In*: Betancourt, J.L. *et al.* (Eds), *Packrat Middens: The last 40,000 Years of Biotic Change*. Tucson, University of Arizona Press, pp. 240–258.

Cole, K.L. (1990b). Reconstruction of past desert vegetation along the Colorado River using packrat middens. *Palaeogeography, Palaeoclimatology, Palaeoecology*, **76**, 349–366.

Cole, K.L. & Webb, R.H. (1985). Late Holocene vegetation changes in Greenwater Valley, Mojave desert, California. *Quaternary Research*, **23**, 227–235.

Cushing, E.J. (1965). Problems in the Quaternary phytogeography of the Great Lakes Region. *In*: Wright, H.E., Jr. *et al.* (Eds), *The Quaternary of the United States*. Princeton, New Jersey, Princeton University Press, pp. 403–416.

Daly, C., Neilson, R.P. & Phillips, D.L. (1994). A statistical-topographic model for mapping climatological precipitation over mountainous terrain. *Journal of Applied Meteorology*, **33**, 140–158.

Davis, M.B. (1965). Phytogeography and palynology of northeastern United States. *In*: Wright, H.E., Jr. *et al.* (Eds), *The Quaternary of the United States*. Princeton, New Jersey, Princeton University Press, pp. 377–401.

Davis, M.B. (1981). Quaternary history and the stability of forest communities. *In*: West, D.C. *et al.* (Eds), *Forest Succession: Concepts and Application*. New York, NY, Springer-Verlag, pp. 132–153.

Davis, O.K. (1984). Pollen frequencies reflect vegetation patterns in a Great Basin (USA) mountain range. *Review of Palaeobotany and Palynology*, **40**, 295–315.

Davis, O.K. (1995). Climate and vegetation patterns in surface samples from arid western USA: Application to Holocene climate reconstructions. *Palynology*, **19**, 95–117.

Davis, O.K. (1998). Palynological evidence for vegetation cycles in a 1.5 million year pollen record from the Great Salt Lake, Utah, USA. *Palaeogeography, Palaeoclimatology, Palaeoecology*, **138**, 175–185.

Dial, K.P. & Czaplewski, N.J. (1990). Do woodrat middens accurately represent the animals' environments and diets. *In*: Betancourt, J.L. *et al.* (Eds), *Packrat Middens: The Last 40,000 Years of Biotic Change*. Tucson, University of Arizona Press, pp. 43–58.

Dukes, J.S. (2000). Will the increasing atmospheric CO_2 concentration affect the success of invasive species? *In*: Mooney, H.A. *et al.* (Eds), *Invasive Species in a Changing World*. Washington, DC, Island Press, pp. 95–113.

Ehleringer, J.R. (1991). $^{13}C/^{12}C$ fractionation and its utility in terrestrial plant studies. *In*: Cole, D.C. & Fry, B. (Eds), *Carbon Isotope Techniques*. New York, Academic Press, pp. 187–200.

Ehleringer, J.R. (1993a). Carbon and water relations in desert plants: An isotopic perspective. *In*: Ehleringer, J.R. *et al.* (Eds), *Stable Isotopes and Plant Carbon/Water Relations*. San Diego, Academic Press, pp. 155–172.

Ehleringer, J.R. (1993b). Gas-exchange implication of isotopic variation in arid land plants. *In*: Griffiths *et al.* (Eds), *Plant Responses to Water Deficit*. Environmental Plant Biology Series: London, BIOS Scientific Publications, pp. 265–284.

Epstein, S., Thompson, P. & Yapp, C.J. (1977). Oxygen and hydrogen isotopic ratios in plant cellulose. *Science*, **198**, 1209–1215.

Epstein, S., Xu, X. & Carrara, P. (1999). A climatic record from ^{14}C-dated wood fragments from southwestern Colorado. *Global Biogeochemical Cycles*, **13**, 781–784.

Epstein, S. & Yapp, C.J. (1976). Climatic implications of the D/H of hydrogen in C-H groups in tree cellulose. *Earth and Planetary Science Letters*, **30**, 252–261.

Epstein, S., Yapp, C.J. & Hall, J.H. (1976). The determination of the D/H ratio of non-exchangeable hydrogen in cellulose extracted from aquatic and land plants. *Earth and Planetary Science Letters*, **30**, 241–251.

Fall, P.L. (1985). Holocene dynamics of the subalpine forest in central Colorado. *In*: Jacobs, B.F. *et al.* (Eds), *Late Quaternary Vegetation and Climate of the American Southwest*. American Association of Stratigraphic Palynologists, Contributions Series No. 16, pp. 31–46.

Fall, P.L. (1987). Pollen taphonomy in a canyon stream. *Quaternary Research*, **28**, 393–406.

Fall, P.L. (1992). Pollen accumulation in a montane region of Colorado, USA: A comparison of moss polsters, atmospheric traps, and natural basins. *Review of Palaeobotany and Palynology*, **72**, 169–197.

Fall, P.L. (1994). Modern pollen spectra and vegetation in the Wind River Range, Wyoming. *Arctic and Alpine Research*, **26**, 383–392.

Farquhar, G.D. (1983). On the nature of carbon isotope discrimination in C4 species. *Australian Journal of Plant Physiology*, **10**, 205–226.

Farquhar, G.D., von Caemmerer, S. & Berry, J.A. (1980). A biochemical model of photosynthetic CO_2 assimilation in leaves of C3 species. *Planta*, **149**, 78–90.

Finley, R.B., Jr. (1990). Woodrat ecology and behavior and the interpretation of paleomiddens. *In*: Betancourt, J.L. *et al.* (Eds), *Packrat Middens: The Last 40,000 Years of Biotic Change*. Tucson, University of Arizona Press, pp. 28–42.

Flannigan, M.D., Stocks, B.J. & Wotton, B.M. (2000). Climate change and forest fires. *The Science of the Total Environment*, **262**, 221–229.

Forman, R.T.T. (1995). *Land mosaics: The ecology of landscapes and regions*. Cambridge, UK, Cambridge University Press.

Fredlund, G.G. & Tieszen, L.L. (1994). Modern phytolith assemblages from the North American Great Plains. *Journal of Biogeography*, **21**, 321–335.

Friedli, H., Lötscher, H., Oeschger, H., Siegenthaler, U. & Stauffer, B. (1986). Ice core record of the $^{13}C/^{12}C$ ratio of atmospheric CO_2 in the past two centuries. *Nature*, **324**, 237–238.

Friedli, H., Moor, E., Oeschger, H., Siegenthaler, U. & Stauffer, B. (1984). $^{13}C/^{12}C$ ratios in CO_2 extracted from Antarctic ice. *Geophysical Research Letters*, **11**, 1145–1148.

Friedman, I., Carrara, P.E. & Gleason, J. (1988). Isotopic evidence of Holocene climatic change in the San Juan Mountains, Colorado. *Quaternary Research*, **30**, 350–353.

Giorgi, F., Whetton, P.H., Jones, R.G., Christensen, J.H., Mearns, L.O., Hewitson, B., von Storch, H., Francisco, R. & Jack, C. (2001). Emerging patterns of simulated regional climatic changes for the 21st century due to anthropogenic forcings. *Geophysical Research Letters*, **28**, 3317–3320.

Greeley, W.B. (1925). The relation of geography to timber supply. *Economic Geography*, **1**, 1–11.

Grimm, E.C., Lozano-Garcia, S., Behling, H. & Markgraf, V. (2001). Holocene vegetation and climate in the Americas. *In*: Markgraf, V. (Ed.), *Interhemispheric Climate Linkages*. Academic Press, pp. 325–370.

Hall, S.A. (1985). Quaternary pollen analysis and vegetational history of the southwest. *In*: Bryant, V.M., Jr. *et al.* (Eds), *Pollen Records of Late Quaternary North American Sediments*. Dallas, American Association of Stratigraphic Palynologists Foundation, pp. 95–123.

Heusser, C.J. (1965). A Pleistocene phytogeographical sketch of the Pacific Northwest and Alaska. *In*: Wright, H.E., Jr. *et al.* (Eds), *The Quaternary of the United States*. Princeton, NJ, Princeton University Press, pp. 469–483.

Heusser, C.J. (1969). Modern pollen spectra from the Olympic Peninsula, Washington. *Bulletin of the Torrey Botanical Club*, **96**, 407–417.

Heusser, C.J. (1973). Modern pollen spectra from Mount Ranier, Washington. *Northwest Science*, **47**, 1–8.

Heusser, C.J. (1978a). Modern pollen rain in the Puget Lowland of Washington. *Bulletin of the Torrey Botanical Club*, **105**, 296–305.

Heusser, C.J. (1978b). Modern pollen rain of Washington. *Canadian Journal of Botany*, **56**, 1510–1517.

Heusser, C.J. (1978c). Modern pollen spectra from western Oregon. *Bulletin of the Torrey Botanical Club*, **105**, 14–17.

Heusser, C.J. (1983). Vegetational history of the northwestern United States including Alaska. *In*: Wright, H.E., Jr. (Ed.), *Late Quaternary Environments of the United States*. Porter, S.C. (Ed.), *The Late Pleistocene*, Vol. 1, pp. 239–258.

Heusser, C.J., Heusser, L.E. & Peteet, D.M. (1985). Late-Quaternary climatic change on the American north Pacific coast. *Nature*, **315**, 485–487.

Heusser, C.J., Heusser, L.E. & Streeter, S.S. (1980). Quaternary temperatures and precipitation for the northwest coast of North America. *Nature*, **286**, 702–704.

Heusser, L.E. (1983). Contemporary pollen distribution in coastal California and Oregon. *Palynology*, **7**, 19–42.

Hevley, R.H. (1968). Modern pollen rain in northern Arizona. *Journal of the Arizona Academy of Science*, **5**, 116–126.

Hevley, R.H., Mehringer, P.J. & Yokum, H. (1965). Modern pollen rain in the Sonoran desert. *Journal of the Arizona Academy of Science*, **3**, 125–135.

Hoyt, C.A. (2000). Pollen signatures of the arid to humid grasslands of North America. *Journal of Biogeography*, **27**, 687–696.

Huntley, B. (1995). How vegetation responds to climate change: Evidence from palaeovegetation studies. *In*: Pernetta, J.C. *et al.* (Eds), *Impacts of Climate Change on Ecosystems and Species: Environmental Context*. IUCN, Gland, Switzerland, pp. 43–63.

Indermühle, A., Stocker, T.F., Joos, F., Fischer, H., Smith, H.J., Wahlen, M., Deck, B., Mastroianni, D., Tschumi, J., Blunier, T., Meyer, R. & Stauffer, B. (1999). Holocene carbon-cycle dynamics based on CO_2 trapped in ice at Taylor Dome, Antarctica. *Nature*, **398**, 121–126.

IPCC (2001). *Climate Change 2001: The Scientific Basis*. Houghton, J.T., Ding, Y., Griggs, D.J., Noguer, M., van der Linden, P.J., Dai, X., Maskell, K. & Johnson, C.A. (Eds). Cambridge, UK, Cambridge University Press, 881 pp.

Jackson, S.T., Grimm, E.C. & Thompson, R.S. (2000). Database resources in Quaternary paleobotany. *SIDA Botanical Miscellany*, **18**, 113–120.

Jahren, A.H., Amundson, R., Kendall, C. & Wigand, P. (2001). Paleoclimatic reconstruction using the correlation in $\delta^{18}O$ of hackberry carbonate and environmental water, North America. *Quaternary Research*, **56**, 252–263.

Jennings, S.A. (1996). Analysis of pollen contained in middens from the White Mountains and Volcanic Tableland of eastern California. *Palynology*, **20**, 5–13.

Jennings, S.A. & Elliott-Fisk, D.L. (1993). Packrat midden evidence of late Quaternary vegetation change in the White Mountains, California – Nevada. *Quaternary Research*, **39**, 214–221.

Jolly, D. & Haxeltine, A. (1997). Effect of low atmospheric CO_2 on tropical African montane vegetation. *Science*, **276**, 786–788.

Jones, T.L., Brown, G.M., Raab, L.M., McVickar, J.L., Spaulding, W.G., Kennett, D.J., York, A. & Walker, P.L. (1999). Environmental imperatives reconsidered: Demographic crises in western North America during the Medieval climatic anomaly. *Current Anthropology*, **40**, 137–170.

Kaplan, J.O. (2001). Geophysical Applications of Vegetation Modeling. Ph.D. dissertation, Lund, University of Lund, 128 pp.

Knapp, P.A. (1996). Cheatgrass (Bromus tectorum L) dominance in the Great Basin Desert: History, persistence, and influences to human activities. *Global Environmental Change*, **6**, 37–52.

LaDeau, S. & Clark, J.S. (2001). Rising CO_2 levels and the fecundity of forest trees. *Science*, **292**, 95–98.

LaMarche, V.C., Jr (1973). Holocene climatic variations inferred from treeline fluctuations in the White Mountains, California. *Quaternary Research*, **3**, 632–660.

LaMarche, V.C., Jr & Mooney, H.A. (1972). Recent climatic change and development of the bristlecone pine (P. longaeva Bailey) krummholz zone, Mt. Washington, Nevada. *Arctic and Alpine Research*, **4**, 61–72.

Leopold, E.B. & Denton, M.F. (1987). Comparative age of grassland and steppe east and west of the northern Rocky Mountains. *Annals of the Missouri Botanical Garden*, **74**, 841–867.

Leuenberger, M., Siegenthaler, U. & Langway, C.C. (1992). Carbon isotope composition of atmospheric CO_2 during the last ice age from and Antarctic ice core. *Nature*, **357**, 488–490.

Little, E.L., Jr. (1971). *Atlas of United States Trees, Volume 1, Conifers and Important Hardwoods*. U.S. Department of Agriculture Miscellaneous Publication, Vol. 1146, 9 pp., 200 maps.

Little, E.L., Jr. (1976). *Atlas of United States Trees, Volume 3, Minor Western Hardwoods*. U.S. Department of Agriculture miscellaneous publication, Vol. 1314, 13 pp., 290 maps.

Little, E.L., Jr. (1981). *Atlas of United States Trees, Volume 6, Supplement*. U.S. Department of Agriculture miscellaneous publication, Vol. 1410, 31 pp., 39 maps.

Lloyd, A.H. & Graumlich, L.J. (1997). Holocene dynamics of treeline forests in the Sierra Nevada. *Ecology*, **78**, 1199–1210.

Long, A., Warneke, L.A., Betancourt, J.L. & Thompson, R.S. (1994). Deuterium variations in plant cellulose from fossil packrat middens. *In*: Betancourt *et al.* (Eds), *Packrat Middens: The Last 40,000 years of Biotic Change*. Tucson, The University of Arizona Press, pp. 380–396.

Mack, R.N. & Bryant, V.M., Jr. (1974). Modern pollen spectra from the Columbia Basin, Washington. *Northwest Science*, **48**, 183–194.

Maher, L.J. (1963). Pollen analysis of surface materials from the southern San Juan Mountains, Colorado. *Geological Society of America Bulletin*, **74**, 1485–1504.

Marino, B.D., McElroy, M.B., Salawitch, R.J. & Spaulding, W.G. (1992). Glacial-to-interglacial variations in the carbon isotopic composition of atmospheric CO_2. *Nature*, **357**, 461–466.

Marsh, G.P. (1871). *Man and Nature; Or, Physical Geography as Modified by Human Action*. New York, NY, Charles Scribner and Company.

Martin, P.S. (1963). *The Last 10,000 Years: a Fossil Pollen Record of the American Southwest*. Tuscon, University of Arizona Press, 87 pp.

Martin, P.S. (1984). Prehistoric overkill: The global model. *In*: Martin, P.S. *et al.* (Eds), *Quaternary Extinctions: A Prehistoric Revolution*. Tucson, University of Arizona Press, pp. 354–403.

Martin, P.S. & Burney, D.A. (1999). Bring back the elephants! *Wild Earth* (Spring issue), pp. 57–64.

Martin, P.S. & Mehringer, P.J., Jr. (1965). Pleistocene pollen analysis and biogeography of the southwest. *In*: Wright, H.E., Jr. *et al.* (Eds), *The Quaternary of the United States*. Princeton, NJ, Princeton University Press, pp. 433–451.

McAndrews, J.H. & Wright, H.E., Jr. (1969). Modern pollen rain across the Wyoming basins and the northern Great Plains (USA). *Review of Palaeobotany and Palynology*, **9**, 17–43.

McPherson, G.R. (1997). *Ecology and Management of North American Savannas*. Tucson, AZ, University of Arizona Press, 208 pp.

Mehringer, P.J., Jr. (1967). Pollen analysis of the Tule Springs Site, Nevada. *In*: Wormington, H.M. *et al.* (Eds), *Pleistocene Studies in Southern Nevada*. Carson City, Nevada State Museum Anthropological Papers 13, pp. 129–200.

Mehringer, P.J., Jr. (1985). Late-Quaternary pollen records from the interior Pacific Northwest and northern Great Basin of the United States. *In*: Bryant, V.M., Jr. & Holloway, R.G. (Eds), *Pollen Records of Late Quaternary North American Sediments*. Dallas, American Association of Stratigraphic Palynologists Foundation, pp. 167–189.

Mehringer, P.J. & Wigand, P.E. (1990). Comparison of Late Holocene environments from woodrat middens and pollen: Diamond Craters, Oregon. *In*: Betancourt, J.L. *et al.* (Eds),

Packrat Middens: The Last 40,000 Years of Biotic Change: Tucson, University of Arizona Press, pp. 294–325.

Mensing, S.A. & Southon, J.R. (1999). A simple method to separate pollen for AMS radiocarbon dating and its application to lacustrine and marine sediments. *Radiocarbon*, **41**(1), 1–8.

Meyer, E.R. (1977). A reconnaissance survey of pollen rain in Big Bend National Park, Texas: modern control for a paleoenvironmental study. *In*: Wauer, R.H. *et al.* (Eds), *Transactions of the Symposium on the Biological Resources of the Chihuahuan Desert Region. United States and Mexico*. Sul Ross State University; Alpine, Texas (October 17–18, 1974). Washington, DC, U.S. Department of the Interior, National Park Service Transactions and Proceedings, Series 3, pp. 115–123.

Miller, D.A. & White, R.A. (1998). A conterminous United States multilayer soil characteristics dataset for regional climate and hydrology modeling. *Earth Interactions*, **2**, 1–26.

Minckley, T. & Whitlock, C. (2000). Spatial variation of modern pollen in Oregon and Southern Washington, USA. *Review of Palaeobotany and Palynology*, **112**, 97–123.

Mock, C.J. & Brunelle-Daines, A.R. (1999). A modern analogue of western United States summer paleoclimate at 6000 years before present. *The Holocene*, **9**, 541–545.

New, M.G., Hulme, M. & Jones, P.D. (1999). Representing twentieth-century space-time climate variability. Part I: Development of a 1961–1990 mean monthly terrestrial climatology. *Journal of Climate*, **12**, 829–856.

New, M.G., Hulme, M. & Jones, P.D. (2000). Representing twentieth-century space-time climate variability. Part II: Development of 1901–1996 monthly grids of terrestrial surface climate. *Journal of Climate*, **13**, 2217–2238.

Newman, J.E. (1980). Climate change impacts on the growing season of the North American "corn belt". *Biometeorology*, **7**(2), 128–142.

Nowak, R.S., Nowak, C.L. & Tausch, R.J. (2000). Probability that a fossil absent from a sample is also absent from the paleolandscape. *Quaternary Research*, **54**, 144–154.

O'Rourke, M.K. (1986). The Implications of Atmospheric Pollen Rain for Fossil Pollen Profiles in the Arid Southwest. Ph.D. dissertation, Tucson, University of Arizona, 186 pp.

OTA [Office of Technology Assessment] (1993). Harmful Non-Indigenous Species in the United States, OTA-F-565: Washington, DC, U.S. Government Printing Office.

Pedicino, L.C., Leavitt, S.W., Betancourt, J.L. & Van de Water, P.K. (2002). Historical variations in $\delta^{13}C_{leaf}$ of herbarium specimens in the southwestern U.S. *Western North American Naturalist*, **62**, 348–359.

Pendall, E. (2000). Influence of precipitation seasonality on piñon pine cellulose δD values. *Global Change Biology*, **6**, 287–301.

Pendall, E., Betancourt, J.L. & Leavitt, S.W. (1999). Paleoclimatic significance of δD and $\delta^{13}C$ values in piñon pine needles from packrat middens spanning the last 40,000 years. *Palaeogeography, Palaeoclimatology, Palaeoecology*, **147**, 53–72.

Pitelka, L.F. & the Plant Migration Workshop (1997). Plant migration and climate change. *American Scientist*, **85**, 464–473.

Polley, H.W. (1997). Implications of rising atmospheric carbon dioxide concentration for rangelands. *Journal of Range Management*, **50**, 562–577.

Potter, L.D. & Rowley, J.S. (1960). Pollen rain and vegetation, San Augustin Plains, New Mexico. *Botanical Gazette*, **122**, 1–25.

Ramankutty, N. & Foley, J.A. (1999). Estimating historical changes in global land cover: Croplands from 1700 to 1992. *Global Biogeochemical Cycles*, **13**, 997–1027.

Rejmánek, M. & Randall, J.M. (1994). Invasive alien plants in California: 1993 summary and comparison with other areas in North America. *Madroño*, **41**, 161–177.

Rhode, D. & Madsen, D.B. (1995). Late Wisconsin/Early Holocene vegetation in the Bonneville Basin. *Quaternary Research*, **44**, 246–256.

Royer, D.L., Osborne, C.P. & Beerling, D.J. (2002). High CO_2 increases the freezing sensitivity of plants: Implications for paleoclimatic reconstructions from fossil floras. *Geology*, **30**, 963–966.

Rozanski, K. Araguas-Araguas, L. & Gonfiantini, R. (1993). Isotopic patterns in modern global precipitation. *In*: Swart, P.K. *et al.* (Eds), *Climate Change in Continental Isotopic Records*: Washington DC. American Geophysical Union Monograph, Vol. 78, pp. 1–36.

Sarna-Wojcicki, A.M. & Davis, J.O. (1991). Quaternary tephrochronology. *In*: Morrison, R.B. (Ed.), *Quaternary Nonglacial Geology, Conterminous U.S.* Boulder, CO, Geological Society of America, pp. 93–116.

Schoenwetter, J. & Doerschlag, L.A. (1971). Surficial pollen records from central Arizona I. Sonoran desert scrub. *Journal of the Arizona Academy of Science*, **6**, 216–221.

Shafer, S.L., Bartlein, P.J. & Thompson, R.S. (2001). Potential changes in the distributions of western North America tree and shrub taxa under future climate scenarios. *Ecosystems*, **4**, 200–215.

Sharpe, S.E. (2002). Constructing seasonal climograph overlap envelopes from Holocene packrat midden contents, Dinosaur National Monument, Colorado. *Quaternary Research*, **57**, 306–313.

Siegel, R.D. (1983). Paleoclimatic Significance of D/H and $^{13}C/^{12}C$ Ratios in Pleistocene and Holocene Wood. M.S. Thesis, Tucson, University of Arizona, 105 pp.

Sitch, S., Smith, B., Prentice, I.C., Arneth, A., Bondeau, A., Cramer, W., Kaplan, J.O., Levis, S., Lucht, W., Sykes, M.T., Thonicke, K. & Venevsky, S. (2003). Evaluation of ecosystem dynamics, plant geography and terrestrial carbon cycling in the LPJ dynamic global vegetation model. *Global Change Biology*, **9**, 161–185.

Smith, H.J., Fischer, H., Wahlen, M., Mastroianni, D. & Deck, B. (1999). Dual modes of the carbon cycle since the last glacial maximum. *Nature*, **400**, 248–250.

Smith, S.D., Huxman, T.E., Zitzer, S.F., Charlet, T.N., Housman, D.C., Coleman, J.S., Fenstermaker, L.K., Seeman, N.R. & Nowak, R.S. (2000). Elevated CO_2

increases productivity and invasive species success in an avid ecosystem. *Nature*, **408**, 79–82.

Solomon, A.M. & Silkworth, A.B. (1986). Spatial patterns of atmospheric pollen transport in a montane region. *Quaternary Research*, **25**, 150–162.

Spaulding, W.G. (1985). Vegetation and Climates of the Last 45,000 Years in the Vicinity of the Nevada Test Site, South-Central Nevada. *U.S. Geological Survey Professional Paper 1329*, 83 pp.

Spaulding, W.G. (1990a). Vegetational and climatic development of the Mojave Desert: The last glacial maximum to the present. *In*: Betancourt, J.L. *et al.* (Eds), *Packrat Middens: The Last 40,000 Years of Biotic Change*. Tucson, University of Arizona Press, pp. 166–199.

Spaulding, W.G. (1990b). Comparison of pollen and macrofossil based reconstructions of late Quaternary vegetation in western North America. *Review of Palaeobotany and Palynology*, **64**, 359–366.

Spaulding, W.G. (1995). Environmental change, ecosystem responses, and the late Quaternary development of the Mojave Desert. *In*: Steadman, D.W. *et al.* (Eds), *Late Quaternary Environments and Deep History: A tribute to Paul S. Martin. The Mammoth Site of Hot Springs, South Dakota, Inc*. Scientific Papers, Vol. 3, pp. 139–164.

Spaulding, W.G., Betancourt, J.L., Croft, L.K. & Cole, K.L. (1990). Packrat middens: Their composition and methods of analysis. *In*: Betancourt, J.L. *et al.* (Eds), *Packrat Middens: The Last 40,000 Years of Biotic Change*: Tucson, University of Arizona Press, pp. 59–84.

Spaulding, W.G., Leopold, E.B. & Van Devender, T.R. (1983). Late Wisconsin paleoecology of the American southwest. *In*: Wright, H.E., Jr. (Ed.), *Late Quaternary Environments of the United States*. Porter, S.C. (Ed.), *The Late Pleistocene*, Vol. 1, pp. 259–293.

Sternberg, L.S.L. (1989). Oxygen and hydrogen isotope ratios in plant cellulose: Mechanisms and applications. *In*: Rundel *et al.* (Eds), *Stable Isotopes in Ecological Research*. Berlin, Springer-Verlag, pp. 124–141.

Sternberg, L.S.L. & DeNiro, M.J. (1983). Isotopic composition of cellulose from C_3, C_4 and CAM plants growing near one another. *Science*, **220**, 947–949.

Sternberg, L.S.L., DeNiro, M.J. & Johnson, H.B. (1985). Oxygen and hydrogen isotope ratios of water from photosynthetic tissues of CAM and C3 plants. *Plant Physiology*, **82**, 428–431.

Strickland, L.E., Thompson, R.S. & Anderson, K.H. (2001). NOAA/USGS North American Packrat Midden Database Data Dictionary, *U.S. Geological Survey Open File Report* 01-022, 28 pp.

Stuiver, M., Reimer, P.J., Bard, E., Beck, J.W., Burr, G.S., Hughen, K.A., Kromer, B., McCormac, G., van der Plicht, J. & Spurk, M. (1998a). INTCAL98 radiocarbon age calibration, 24000-0 cal BP. *Radiocarbon*, **40**, 1041–1083.

Stuiver, M., Reimer, P.J. & Braziunas, T.F. (1998b). High-precision radiocarbon age calibration for terrestrial and marine samples. *Radiocarbon*, **40**, 1127–1151.

Swetnam, T.W. & Betancourt, J.L. (1998). Mesoscale disturbance and ecological response to decadal climatic variability in the American Southwest. *Journal of Climate*, **11**, 3128–3147.

Taylor, R.E. (1987). *Radiocarbon dating, an archaeological perspective*. London, Academic Press, 212 pp.

Taylor, R.E., Long, A. & Kra, R. (Eds) (1992). *Radiocarbon After Four Decades: An Interdisciplinary Perspective*. New York, Springer-Verlag, 596 pp.

Terwilliger, V.J., Betancourt, J.L., Leavitte, S.W. & Van de Water, P.K. (2002). Leaf cellulose δD and δ^{18}O trends with elevation differ in direction among co-occurring, semiarid plant species. *Geochimica et Cosmochimica Acta*, **66**, 3887–3900.

Thomas, W.L. (Ed.) (1956). *Man's Role in Changing the Face of the Earth*. Chicago, IL, University of Chicago Press, pp. 11–56.

Thompson, R.S. (1985). Palynology and *Neotoma* middens. *In*: Jacobs, B.F. *et al.* (Eds), *Late Quaternary Vegetation and Climates of the American Southwest*, American Association of Stratigraphic Palynologists, Contributions Series 16, pp. 89–112.

Thompson, R.S. (1988). Western North America – vegetation dynamics in the western United States: modes of response to climatic fluctuations. *In*: Huntley, B. *et al.* (Eds), *Vegetation History, Volume 7 of the Handbook of Vegetation Science*. Dordrecht/Boston/London, Kluwer Academic Publishers, pp. 415–458.

Thompson, R.S. (1990). Late Quaternary vegetation and climate in the Great Basin. *In:* Betancourt, J.L. *et al.* (Eds), *Packrat Middens: The Last 40,000 Years of Biotic Change*. Tucson, University of Arizona Press, pp. 200–239.

Thompson, R.S. (1991). Pliocene environments and climates in North America. *Quaternary Science Reviews*, **10**, 115–132.

Thompson, R.S. (1992). Late Quaternary environments in Ruby Valley, Nevada. *Quaternary Research*, **37**, 1–15.

Thompson, R.S. (1996). Pliocene and early Pleistocene environments and climates of the western Snake River Plain, Idaho. *Marine Micropaleontology*, **27**, 141–156.

Thompson, R.S. & Anderson, K.H. (2000). Biomes of western North America at 18,000, 6,000, and 0 yr B.P. reconstructed from pollen and packrat midden data. *Journal of Biogeography*, **27**, 555–584.

Thompson, R.S., Anderson, K.H. & Bartlein, P.J. (1999a). Atlas of Relations Between Climatic Parameters and Distributions of Important Trees and Shrubs in North America – Introduction and Conifers. *U.S. Geological Survey Professional Paper 1650-A*, 269 pp.

Thompson, R.S., Anderson, K.H. & Bartlein, P.J. (1999b). Atlas of Relations Between Climatic Parameters and Distributions of Important Trees and Shrubs in North America – Hardwoods. *U.S. Geological Survey Professional Paper 1650-B*, 423 pp.

Thompson, R.S., Anderson, K.H. & Bartlein, P.J. (1999c). Quantitative Paleoclimatic Reconstructions from Late Pleistocene Plant Macrofossils of the Yucca Mountain region. *U.S. Geological Survey Open-File Report 99–338*, 39 pp.

Thompson, R.S., Anderson, K.H., Bartlein, P.J. & Smith, S.A. (2000). Atlas of Relations Between Climatic Parameters and Distributions of Important Trees and Shrubs in North America – *Conifers, Hardwoods, and Monocots. U.S. Geological Survey Professional Paper 1650-C*, 386 pp.

Thompson, R.S. & Fleming, R.F. (1996). Middle Pliocene vegetation: Reconstructions, paleoclimatic inferences, and boundary conditions for climate modeling. *Marine Micropaleontology*, **27**(1/4), 27–50.

Thompson, R.S., Hostetler, S.W., Bartlein, P.J. & Anderson, K.H. (1998). A Strategy for Assessing Potential Future Changes in Climate, Hydrology, and Vegetation in the Western United States. *U.S. Geological Survey Circular 1153*, 20 pp.

Thompson, R.S. & Mead, J.I. (1982). Late Quaternary environments and biogeography in the Great Basin. *Quaternary Research*, **17**, 39–55.

Thompson, R.S., Whitlock, C., Bartlein, P.J., Harrison, S.P. & Spaulding, W.G. (1993). Climatic changes in the western United States since 18,000 yr B.P. *In*: Wright, H.E., Jr. *et al.* (Eds), *Global Climates Since the Last Glacial Maximum*. University of Minnesota Press, pp. 468–513.

Tomback, D.F. (2001). Clark's Nutcracker: Agent of regeneration. *In*: Tomback, D.F. *et al.* (Eds), *Whitebark Pine Communities: Ecology and Restoration*. Washington, DC, Island Press, pp. 89–104.

Toolin, L.J. & Eastoe, C.J. (1993). Late Pleistocene-Recent atmospheric $\delta^{13}C$ record in C_4 grasses. *Radiocarbon*, **35**, 263–269.

Trumbore, S.E. (2000). Radiocarbon geochronology. *In*: Noller, J. (Ed.), *Quaternary Geochronology: Methods and Applications*. Washington DC, American Geophysical Union, Reference Shelf 4, pp. 71–102.

Turner, R.M., Bowers, J.E. & Burgess, T.L. (1995). *Sonoran desert plants: An ecological atlas*. University of Arizona Press.

Vale (2002). *Fire, Native Peoples and the Natural Landscape*. Washington, DC, Island Press, 315 pp.

Van de Water, P.K. (1999). $\delta^{13}C$ and Stomatal Density Variability in Modern and Fossil Leaves of Key Plants in the Western United States. Ph.D. Dissertation, Tucson, University of Arizona, 326 pp.

Van de Water, P.K., Leavitt, S.W. & Betancourt, J.L. (1994). Trends in stomatal density and $^{13}C/^{12}C$ ratios of Pinus flexilis needles during last glacial-interglacial cycle. *Science*, **264**, 239–243.

Van de Water, P.K., Leavitt, S.W. & Betancourt, J.L. (2002). Leaf $\delta^{13}C$ variability with elevation, slope aspect, and precipitation in the southwest United States. *Oecologia*, **132**, 332–343.

Van Devender, T.R. (1977). Holocene woodlands in the southwestern deserts. *Science*, **198**, 189–192.

Van Devender, T.R. (1990a). Late quaternary vegetation and climate of the Chihuahuan Desert, United States and Mexico. *In*: Betancourt, J.L. *et al.* (Eds), *Packrat Middens: The Last 40,000 Years of Biotic Change*. Tucson, University of Arizona Press, pp. 104–133.

Van Devender, T.R. (1990b). Late quaternary vegetation and climate of the Sonoran Desert, United States and Mexico. *In*: Betancourt, J.L. *et al.* (Eds), *Packrat Middens: The Last 40,000 Years of Biotic Change*. Tucson, University of Arizona Press, pp. 134–165.

Van Devender, T.R., Martin, P.S., Thompson, R.S., Cole, K.L., Jull, A.J.T., Long, A., Toolin, L.J. & Donahue, D.J. (1985). Fossil packrat middens and the tandem accelerator mass spectrometer. *Nature*, **317**, 610–613.

Van Devender, T.R. & Spaulding, W.G. (1979). Development of vegetation and climate in the southwestern United States. *Science*, **204**, 701–710.

Van Devender, T.R., Thompson, R.S. & Betancourt, J.L. (1987). Vegetation history of the deserts of southwestern North America; The nature and timing of the late Wisconsin – Holocene transition. *In*: Ruddiman, W.F. *et al.* (Eds), *North America and Adjacent Oceans During the Last Deglaciation*. The Geology of North America. Geological Society of America, Vol. K-3, pp. 323–352.

Veblen, T.T. & Lorenz, D.C. (1991). *The Colorado Front Range: A Century of Ecological Change*. University of Utah Press, Salt Lake City, 186 pp.

Verosub, K.L. (2000). Paleomagnetic dating. *In:* Noller, J. (Ed.), *Quaternary Geochronology; Methods and Applications*. Washington DC, American Geophysical Union, Reference Shelf 4, pp. 339–356.

Webb, R.H. & Betancourt, J.L. (1990). The spatial and temporal distributions of radiocarbon ages from packrat middens. *In*: Betancourt, J.L. *et al.* (Eds), *Packrat Middens: The Last 40,000 Years of Biotic Change*. Tucson, University of Arizona Press, pp. 85–102.

Webb, R.S., Anderson, D.M. & Overpeck, J.T. (1994). Editorial: Archiving data at the World Data Centre – for paleoclimatology. *Paleoceanography*, **9**, 391–393.

Webb, T., III (1995). Pollen records of late Quaternary vegetation change: Plant community rearrangements and evolutionary implications. *In*: National Research Council Commission on Geosciences, Environment, and Resources (Eds), *Effects of Past Global Change on Life*. Washington, DC, National Academy Press, pp. 221–232.

Wells, P.V. (1966). Late Pleistocene vegetation and degree of pluvial climatic change in the Chihuahuan Desert. *Science*, **153**, 970–975.

Wells, P.V. (1976). Macrofossil analysis of wood rat (Neotoma) middens as a key to the Quaternary vegetational history of arid America. *Quaternary Research*, **6**, 223–248.

Wells, P.V. (1983). Paleogeography of montane islands in the Great Basin since the last glaciopluvial. *Ecological Monographs*, **53**, 341–382.

Wells, P.V. & Berger, R. (1967). Late Pleistocene history of coniferous woodlands in the Mohave Desert. *Science*, **155**, 1640–1647.

Wells, P.V. & Jorgensen, C.D. (1964). Pleistocene woodrat middens and climatic change in Mohave Desert – a record of juniper woodlands. *Science*, **143**, 1171–1174.

Whitlock, C. (1992). Vegetational and climatic history of the Pacific Northwest during the last 20,000 years:

Implications for understanding present-day biodiversity. *The Northwest Environmental Journal*, **8**, 5–28.

Whitlock, C. (1993). Postglacial vegetation and climate of Grand Teton and southern Yellowstone National Parks. *Ecological Monographs*, **63**(2), 173–198.

Whitlock, C. & Bartlein, P.J. (1993). Spatial variations in Holocene climatic change in the Yellowstone region. *Quaternary Research*, **43**, 231–238.

Whitlock, C. & Bartlein, P.J. (this volume). Variations in Holocene fire activity as a record of past environmental change.

Whitlock, C., Bartlein, P.J., Markgraf, V. & Ashworth, A.C. (2001). The mid-latitudes of North and South America during the last glacial maximum and early Holocene: Similar paleoclimatic sequences despite differing large-scale controls. *In*: Markgraf, V. (Ed.), *Interhemispheric Climate Linkages*. Academic Press, pp. 391–416.

Wigand, P.E. & Rhode, D. (2002). Great Basin vegetation history and aquatic systems: The last 150,000 years. *In*: Hershler, R., Madsen, D.B. & Currey, D.R. (Eds), *Great Basin Aquatic Systems History*. Smithsonian Contributions to the Earth Sciences, No. 33, pp. 309–367.

Williams, D.G. & Ehleringer, J.R. (2000). Intra- and interspecific variation for summer precipitation use in pinyon – juniper woodlands. *Ecological Monographs*, **70**, 517–537.

Withers, M.A., Palmer, M.W., Wade, G.L., White, P.S. & Neal P.R. (1998). Changing patterns in the number of species in North American floras. *In*: Sisk, T.D. (Ed.), *Perspectives on the Land Use History of North America: A Context for Understanding our Changing Environment*. U.S. Geological Survey, Biological Resources Division, Biological Science Report USGS/BRD/BSR-1998-0003.

Woodhouse, C.A. & Overpeck, J.T. (1998). 2000 years of drought variability in the central United States. *Bulletin of the American Meteorological Society*, **79**, 2693–2714.

Woodward, F.I. (1987a). *Climate and Plant Distribution*. Cambridge University Press, 174 pp.

Woodward, F.I. (1987b). Stomatal numbers are sensitive to increases in CO_2 from pre-industrial levels. *Nature*, **327**, 617–618.

Woodward, F.I. & Bazzaz, F.A. (1988). The responses of stomatal density to CO_2 partial pressure. *Journal of Experimental Botany*, **39**, 1771–1781.

Wright, H.E. & Frey, D.G. (1965). The Quaternary of the United States: A Review Volume for the VII Congress of the International Association for Quaternary Research. Princeton, Princeton University Press, **9**, 22 pp.

Yapp, C.J. & Epstein, S. (1977). Climatic implications of D/H ratios of meteoric water over North America (9500–22000) as inferred from ancient wood cellulose C-H hydrogen. *Earth and Planetary Science Letters*, **34**, 333–350.

Results and paleoclimate implications of 35 years of paleoecological research in Alaska

Patricia M. Anderson[1], Mary E. Edwards[2] and Linda B. Brubaker[3]

[1] *University of Washington, Quaternary Research Center, Box 35-1360, Seattle, WA 98195-1360, USA*
[2] *Norwegian University of Science and Technology, Department of Geography N-7491 Trondheim Norway and University of Alaska, Institute of Arctic Biology, Fairbanks, AK 99775-7000, USA*
[3] *University of Washington, College of Forest Resources, Box 35-2100, Seattle, WA 98195-2100, USA*

Introduction

Since the first comprehensive summary (Heusser, 1965), tremendous strides have been made in defining the late Quaternary paleoenvironmental history of Alaska. The number of paleoecological records has increased greatly (Fig. 1, Table 1), and major shifts have occurred in research questions and interpretive frameworks. The earliest stages of Alaskan research focused on documenting past changes by defining the basic glacial, marine, and bio-stratigraphies (e.g. Ager & Brubaker, 1985; Heusser, 1965; Hopkins, 1967). By the early 1980s, the paleo-database was sufficiently large to allow more question-driven investigations, such as those that examined the mammoth-steppe biome and its biological paradox (Hopkins *et al.*, 1982). The growing database also clearly demonstrated that major plant taxa responded differently to climate fluctuations, and the previously held idea of intact plant communities simply migrating latitudinally in response to the waxing and waning of glacial conditions was no longer tenable (e.g. Anderson & Brubaker, 1994). Current Alaskan research is concerned with the processes and mechanisms responsible for observed changes and often involves comparisons of the paleo-records to numerical or conceptual models (e.g. Bartlein *et al.*, 1991, 1998). These and related studies highlight the complexity of arctic climate change, specifically showing that northern regions responded individualistically to global-scale climate forcings (e.g. CAPE Project Members, 2001), physical and biological feedbacks mediated the effects of those forcings (e.g. Bartlein *et al.*, 1998; Gallimore & Kutzbach, 1996; Kutzbach & Gallimore, 1988; Peteet *et al.*, 1997), and vegetation change was a function of multiple climate factors (e.g. Edwards & Barker, 1994).

Shifts in Alaskan research paradigms are directly linked to advances in techniques, particularly regarding paleovegetation studies. For example, little was known in the 1960s about the extent to which the modern vegetation types in Alaska could be distinguished by their pollen spectra (Colinvaux, 1964; Livingstone, 1955) or about the climate factors controlling modern vegetation patterns (Hopkins, 1959). A series of papers published in the 1980s demonstrated that major pollen taxa accurately delimited modern vegetation zones and variations within these zones (e.g. Anderson & Brubaker, 1986; Anderson *et al.*, 1989; Heusser, 1985; Peteet, 1986; see also Anderson & Brubaker, 1993; Oswald, 2002). Response surface analyses, which explore climate-vegetation-pollen relationships, were a direct outgrowth of the availability of an extensive modern data set. This approach clearly illustrated the heterogeneity of vegetation response possible under specified climate conditions. For example, response surfaces suggest that *Picea* treeline in northwestern Canada is sensitive to changes in growing season (Anderson *et al.*, 1991), whereas seasonality is likely of more importance in western Alaska (Edwards & Barker, 1994). Despite the application of numerical approaches, pollen-based, quantitative estimates of past climate remain rare (e.g. Heusser *et al.*, 1985). However, quantitative reconstructions from fossil insect assemblages (see Elias, 2001), which use mutual climatic range (MCR) limits, have provided mean temperatures of the warmest and coldest months and mean annual precipitation for many areas of Alaska.

The compilation of an extensive modern pollen data set was also used to improve paleovegetation interpretations based on similarity of modern and fossil spectra. Prior to the 1990s, analog-based inferences often rested on personal knowledge of a landscape and/or presumptions about the representation of that landscape by the pollen. The improved modern data set permitted statistical appraisals (primarily using square chord distances), thereby eliminating the qualitative aspects of the previously defined vegetation histories. For example, sites from Alaska and adjacent northwestern Canada show that, in contrast to previous thinking, full-glacial spectra have reasonable analogs with some modern arctic sites, whereas late-glacial records are the least analogous to present (e.g. Anderson *et al.*, 1989). Other studies (Anderson *et al.*, 1994a; Oswald, 2002) traced the development of modern tundra communities, demonstrating a richness and complexity in their histories that early researchers had not believed discernible (e.g. Colinvaux, 1964).

The advent of the data-model comparison approach (COHMAP, 1988) encouraged all Quaternary investigators to systematically explore in an iterative fashion the mechanisms (provided by the models) likely responsible for the observed changes in the paleo-data. As the Alaskan database continued to grow in the late 1980s and 1990s, the model simulations became more central to paleoclimate interpretations (e.g. Anderson & Brubaker, 1993; Edwards & Barker, 1994). Initially, the paleo-data were used to "test" the reasonableness of model simulations (e.g. Barnosky *et al.*, 1987). However, further development of climate and vegetation models led to more sophisticated applications, such as exploring

DEVELOPMENTS IN QUATERNARY SCIENCE
VOLUME 1 ISSN 1571-0866
DOI:10.1016/S1571-0866(03)1019-4

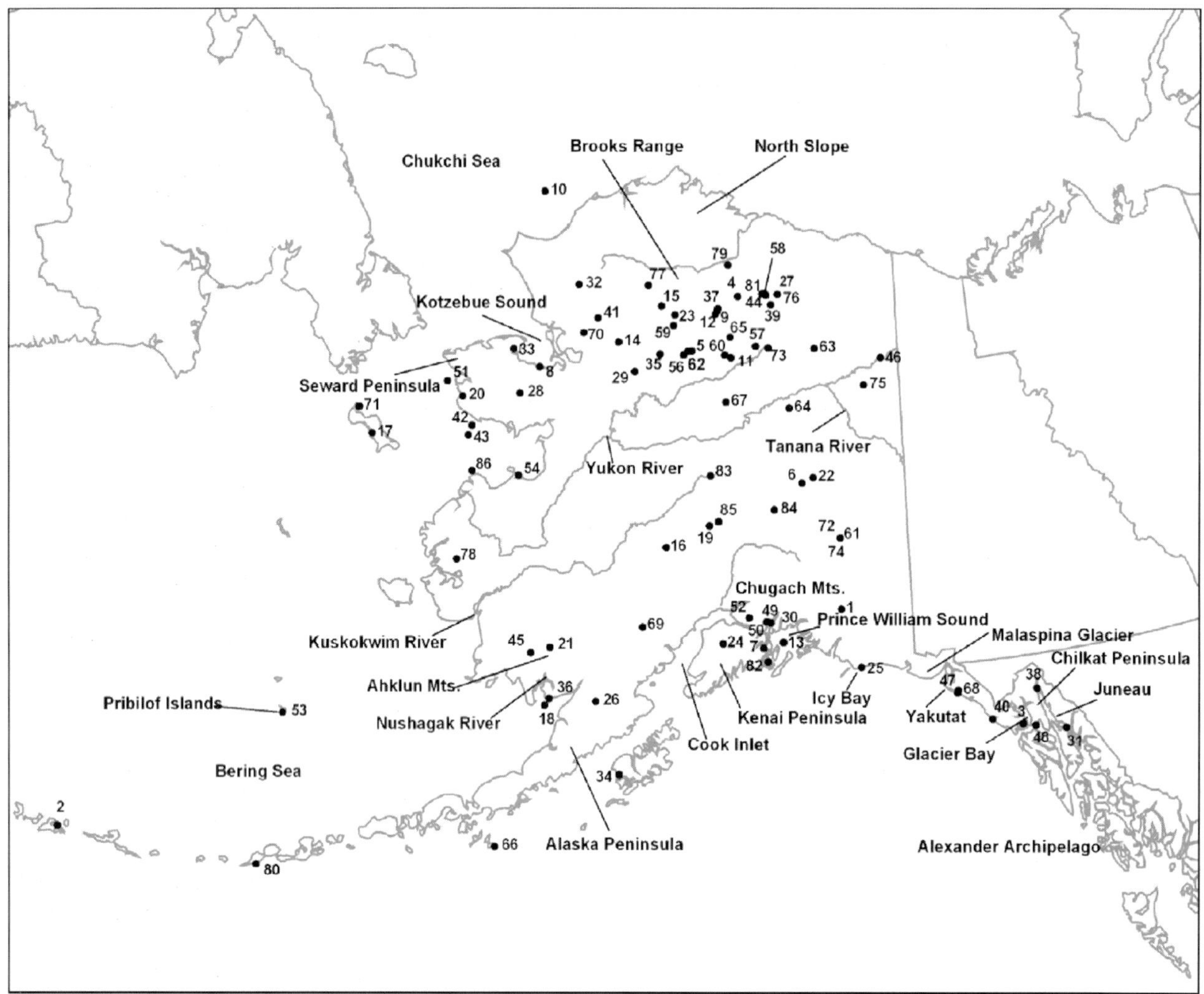

Fig. 1. Map of Alaska showing select site locations and place names. See Table 1 for key to sites.

high-latitude feedbacks between the biosphere and atmosphere (e.g. Foley *et al.*, 1994; Gallimore & Kutzbach, 1996; TEMPO, 1996; see Kutzbach & Guetter, 1996, and Peteet *et al.*, 1997, for influences of other feedbacks) and using plant functional groups to enhance understanding of vegetation response to past climate change (e.g. Edwards *et al.*, 2000a).

The above improvements in uses of biological data in paleoenvironmental research have been paralleled by equal advances in the physical sciences. Because the breadth of the latter studies are too great to summarize here, we limit our review to the vegetation and climate histories of Alaska over the last 18,000 ^{14}C yr B.P. Excellent, up-to-date summaries of all aspects of Beringian research are available in Elias & Brigham-Grette (2001), including several papers about the last glaciation and the mid-Wisconsinan interstade. Other regional to statewide overviews are provided by Ager & Brubaker (1985), Ager (1983), Heusser (1985), Lamb & Edwards (1988), Anderson & Brubaker (1993, 1994), Brubaker *et al.* (1995), Mann & Hamilton (1995), and Mann *et al.* (1998). Readers unfamiliar with the modern vegetation, physiography, and climate of Alaska (Fig. 2) are referred to reviews by Barnosky *et al.* (1987), Anderson & Brubaker (1993), and Mock *et al.* (1998). The following discussion of the last glacial maximum and the post-glaciation differ, because research emphases have varied between the two periods, with the former focused on conflicting interpretations of faunal and paleobotanical data and the latter concerned with regional patterns of vegetation and climate change. To facilitate describing post-glacial vegetation change, we have summarized geographic patterns for 14,000–9000 ^{14}C yr B.P. but focused on individual taxa changes for the remainder of the Holocene. Only select sites are cited as examples in the text (see Fig. 1 for more complete site coverage). Timing of vegetational changes are based on multiple sites with chronologies drawn from both conventional and AMS dates.

The Last Glacial Maximum

Virtually all palynological records dating to the last glacial maximum (LGM; ~25,000–14,000 ^{14}C yr B.P.) are dominated by Poaceae, Cyperaceae, and *Artemisia*, with lesser

Table 1. The key to sites.

	Site Locations	References		Site Locations	References
1.	70 Mile Lake	Ager & Brubaker (1985)	41.	Niliq Lake	Anderson (1988)
2.	Adak Island	Heusser (1978)	42.	Nome section	Hopkins *et al.* (1960)
3.	Glacier Bay/Adams Inlet	Mckenzie (1970)	43.	Norton Sound marine cores	Elias *et al.* (1996a), Elias *et al.* (1997)
4.	Ahaliorak Lake	Eisner & Colinvaux (1990)	44.	Oil Lake	Eisner & Colinvaux (1992)
			45.	Ongivinuk Lake	Hu *et al.* (1996)
5.	Angal Lake	Brubaker *et al.* (1983)	46.	Ped Pond	Edwards & Brubaker (1986)
6.	Birch Lake	Ager (1975)	47.	Pike Lakes muskeg	Peteet (1991)
7.	Blackstone Bay	Heusser (1983a)	48.	Pleasant Island/Glacial Bay	Hansen & Engstrom (1996)
8.	Cape Deceit	Matthews (1974)			
9.	Chandler Lake	Livingstone (1955)	49.	Point Quick, Harriman	Heusser (1983a)
10.	Chukchi Sea marine cores	Elias *et al.* (1996a), Elias *et al.* (1997)	50.	Point Woronzof	Miller & Dobrovolny (1959)
11.	Death Valley	Livingstone (1955)	51.	Point Clarence cores	Elias *et al.* (1996a)
12.	Eight Lake	Livingstone (1955)	52.	Port Wells sites	Heusser (1983a)
13.	Emerald Cove	Heusser (1983a)	53.	Pribilof Island sites	Colinvaux (1967b, 1981)
14.	Epiguruk	Schweger (1982)	54.	Puyuk Lake	Ager (1982)
15.	Etivlik Lake	Oswald *et al.* (1999)	55.	Ranger Lake	Brubaker *et al.* (1983)
	Lake of the Pleistocene	Mann *et al.* (2002)	56.	Rebel Lake	Edwards *et al.* (1985)
16.	Farewell Lake	Hu *et al.* (1996)	57.	Red Green Lake	Oswald (2002)
17.	Flora Lake	Colinvaux (1967a)	58.	Redondo Lake	Brubaker *et al.* (1983)
18.	Flounder Flat	Short *et al.* (1992)	59.	Redstone Lake	Brubaker *et al.* (1983)
19.	Foraker Slump	Elias *et al.* (1996b)	60.	Rock Creek	Schweger (1981)
20.	Glacial Lake	Lozhkin *et al.* (1996)	61.	Ruppert Lake	Brubaker *et al.* (1983)
21.	Grandfather	Hu *et al.* (1996)	62.	Sakana Lake	Anderson & Brubaker (1994)
	Nimgun Lakes	Hu *et al.* (2002)			
22.	Harding Lake	Ager & Brubaker (1985)	63.	Sands of Time Lake	Edwards & Barker (1994)
23.	Headwaters Lake	Brubaker *et al.* (1983)	64.	Screaming Yellowlegs Pond	Edwards *et al.* (1985)
24.	Kenai Peninsula sites: Hidden Lake	Ager & Brubaker (1985)	65.	Shumagin Island and adjacent Alaska Peninsula sites	Heusser (1983b)
	Homer Spit peat and Circle Lake	Ager (2000)			
	Tern Lake peat	Ager (2001)	66.	Sithylemenkat Lake	Anderson *et al.* (1990)
25.	Icy Cape/Icy Bay	Heusser (1960), Peteet (1986)	67.	Slate Mesa	Peteet (1991)
			68.	Snipe Lake	Brubaker *et al.* (2001)
26.	Idavain Lake	Brubaker *et al.* (2001)	69.	Squirrel Lake	Anderson (1985)
27.	Imnaviat Creek	Eisner (1991)	70.	St. Lawrence Island peat sections	Lozhkin *et al.* (1998)
28.	Imuruk Lake	Colinvaux (1964)			
29.	Joe Lake	Anderson (1988), Anderson *et al.* (1994a)	71.	Tangle Lake	Schweger (1981)
			72.	Tangled Up Lake	L. Anderson *et al.* (2001)
30.	Jonah Bay	Heusser (1983a)	73.	Tenmile Lake	Anderson *et al.* (1994b)
31.	Juneau sites	Heusser (1952)	74.	Tiinkdhul Lake	Anderson *et al.* (1988)
32.	Kaiyak Lake	Anderson (1985)	75.	Toolik Lake	Bergstrom (1984)
33.	Kitluk	Höfle *et al.* (2001)	76.	Tukuto Lake	Oswald *et al.* (1999)
34.	Kodiak Island sites	Peteet & Mann (1994)	77.	Tungak Lake	Ager (1982)
35.	Kollioksak Lake	Anderson & Brubaker (1994)	78.	Umiat	Livingstone (1957)
			79.	Umnak Island	Heusser (1973)
36.	Kvichak Peninsula sites	Ager (1982), Short *et al.* (1992)	80.	Upper Capsule Lake	Oswald (2002)
			81.	Verdant Island	Heusser (1983a)
37.	Lake A	Livingstone (1955)	82.	Wien Lake	Hu *et al.* (1993)
38.	Lily Lake	Cwynar (1990)	83.	Windmill Lake	Bigelow & Edwards (2001)
	Meli Lake	L. Anderson *et al.* (2001)	84.	Wonder Lake	Anderson *et al.* (1994b)
40.	Muskeg Cirque	Mann (1983)	85.	Yukon delta cores	Elias *et al.* (1996a)

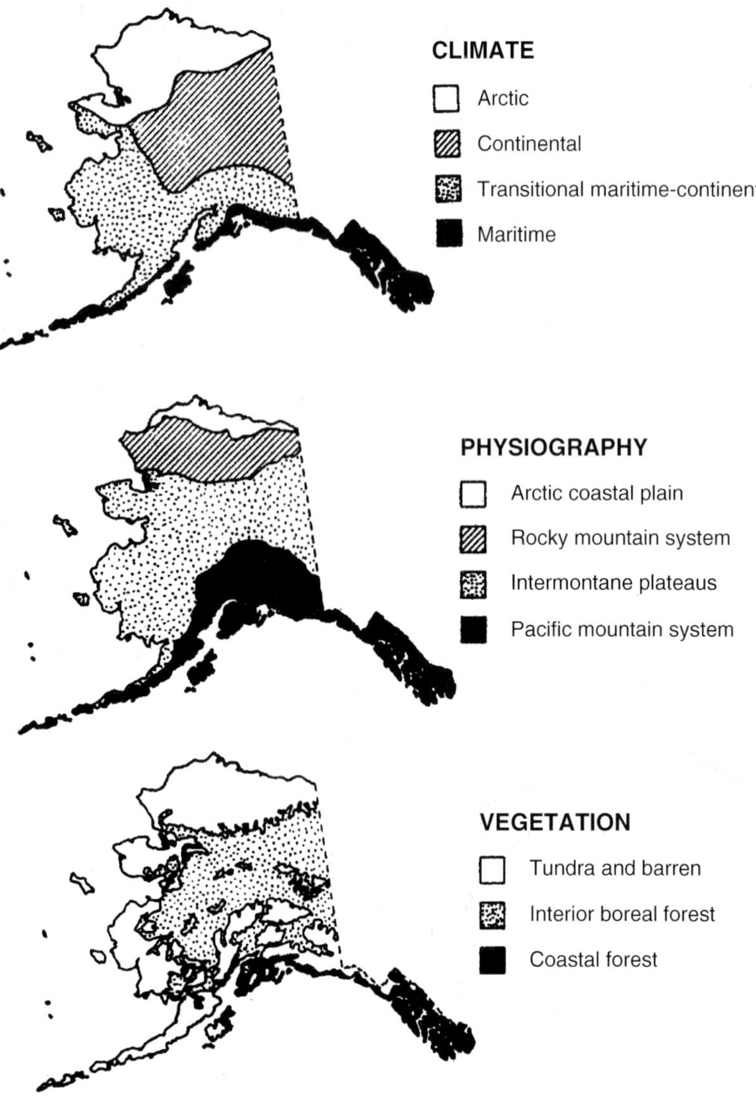

Fig. 2. Maps of climate (Watson, 1959), physiography (Wahrhaftig, 1965), and vegetation (Viereck & Little, 1975) of Alaska. The northern and western tundras support graminoid meadows in wet areas (especially near the coasts) and shrub Betula, Alnus and/or Salix tundra in better drained sites at low to mid-elevations. Sparse alpine tundra is restricted to higher elevations in the mountains. The interior boreal forest is dominated by Picea marina and Picea glauca, with Betula papyrifera, Populus balsamifera, Populus tremuloides, and Larix laricina being of local importance. Tsuga heterophylla and Picea sitchensis are the most common trees of the southern coastal forests, with scattered occurrences of Tsuga mertensiana, Pinus contorta, Populus trichocarpia, and Alnus rubra.

but ecologically significant amounts of forbs and *Salix*. Qualitative interpretations of these unusual pollen spectra invoked the presence of dry tundra (Birch Lake), polar desert (Imuruk Lake), and/or a tundra mosaic (Epiguruk; Squirrel and Kaiyak lakes). These reconstructions were questioned based on the functional ecology of the LGM megafauna (e.g. *Mammuthus, Equus, Bison, Rangifer*; Guthrie, 1968, 1982). High nutritional requirements, particularly of caecal grazers such as *Equus*, argued for a productive vegetation more akin to steppe than tundra. Thus began the "steppe-tundra" debate, which at times became polarized into camps favoring extremes of high and low plant productivity (Hopkins *et al.*, 1982; see also Guthrie, 2001). This controversy reflected the ambiguity in the LGM pollen flora (i.e. percentages of the dominant pollen taxa could be derived reasonably from steppe or tundra) and the absence of analytical techniques that conclusively proved one viewpoint over another.

This long-standing impasse received little attention by palynologists until the recent application of a plant functional group approach (Bigelow *et al.*, in press). LGM pollen taxa were assigned systematically to one or more plant functional types (PFTs, defined by growth form, phenology, morphology, and bioclimatic traits) that, in turn, were assigned to biomes using a rule-based algorithm (Prentice *et al.*, 1996). Most Alaskan spectra were classified as either prostrate shrub tundra or Poaceae-forb tundra. When a vegetation model (Haxeltine & Prentice, 1996) using the same approach to PFTs and biomes was coupled with the output of an atmospheric general circulation model, similar biomes were predicted to those reconstructed solely from the LGM pollen data (Kaplan, 2001; Kaplan *et al.*, in press). Productivity of Poaceae-forb tundra, as estimated by the vegetation model, was high relative to modern tundra. The results of these analyses strongly suggest that LGM vegetation in Alaska was tundra, not steppe. Furthermore, seeking LGM analogs in present-day tundra (see Anderson *et al.*, 1989) is not warranted because of climatically induced differences in productivity.

A LGM landscape preserved by a volcanic ash-fall on the rolling interfluves of northern Seward Peninsula provides a unique opportunity to evaluate the above conclusions

(Goetcheus & Birks, 2001). Pollen samples collected from this surface were classified as Poaceae-forb or prostrate shrub tundra biomes (Bigelow *et al.*, in press), in agreement with the regional vegetation previously reconstructed from PFTs and a climate-vegetation model. Plant macrofossils indicate a species-rich vegetation dominated by Poaceae, Cyperaceae, and tundra forbs with a dense ground layer of acrocarpous mosses. Soil gleying, active layers with depths similar to modern (~50 cm), and intermittent, hummock-like features further support a tundra interpretation.

Despite the regional consistency of biome results (Bigelow *et al.*, in press), detailed studies of local sites, such as represented by the Kitluk surface and alluvial sections, confirm the mosaic hypothesis of a heterogeneous LGM landscape (e.g. Elias *et al.*, 1996a; Schweger, 1982; see also Guthrie, 1982). For example, vegetation mats of prostrate *Salix* characterized the lee sides of interfluves, indicating the herb-dominated tundra of northern Seward Peninsula supported pockets of woody plant growth (Goetcheus & Birks, 2001). Given the paleobotanical data from the Kitluk site, it is easy to imagine slight topographic variations causing scattered areas of snow accumulation that provided the necessary moisture and protection for shrubs to survive the harsh LGM conditions.

The widespread occurrence of productive, dry, Poaceae-forb tundra implies LGM conditions that were significantly drier and cooler than present. Estimates derived from lake-level reconstructions and hydrological budgets suggest that precipitation was 40–75% lower than present (Barber & Finney, 2000). Loess deposition, active sand dunes and sand seas, and the development of sand wedges are additional evidence of arid climates (e.g. Hopkins, 1982). MCR analyses of fossil insects indicate mean summer temperatures that were 1–4 °C cooler than modern, although minimum temperature of the coldest months may have been higher than present (Elias, 2001). The presence of mountain glaciers and widespread development of ice wedges in the Alaskan lowlands further argues for significant decreases in temperature (e.g. Hopkins, 1982).

Spatial and Temporal Patterns of Change Since the LGM

Although many late Quaternary changes in Alaska are time transgressive, we have divided the late-glacial and Holocene vegetation history into four broad intervals: late glaciation (replacement of full-glacial herb-dominated vegetation by shrub tundra); the late glacial-early Holocene transition (establishment of the first post-glacial forested landscapes); the early to mid-Holocene (first appearances of modern plant communities); and the mid- to late Holocene (present-day distributions of modern plant communities). Geographic regions are northern (north of the Brooks Range), central (between the Brooks and Alaska ranges), and southern Alaska (south of the Brooks Range, including the Aleutian, Pribilof, Shumagin, and Kodiak islands, lower Kuskokwim and lower Yukon drainages, Ahklun Mountains, Kenai Peninsula, lands bordering Prince William Sound, and the Alaskan panhandle).

Late Glaciation (14,000–11,000 ^{14}C yr B.P./16,800–13,000 cal yr B.P.)

This interval marks the onset of post-glacial climate amelioration in Alaska and is characterized by the increased presence of woody taxa on the landscape. Although the replacement of full-glacial Poaceae-forb or prostrate-shrub communities by shrub *Betula* tundra spanned several thousand ^{14}C yr, the earliest vegetation changes (between ~14,000 and 13,000 ^{14}C yr B.P.) occurred in discrete "pockets," such as in the then-exposed Chukchi and Bering shelves (Chukchi Sea and Norton Sound marine cores), Kotzebue Sound basin (Squirrel and Kaiyak lakes), modern Yukon delta (Tungak Lake), and Kenai Peninsula (Hidden Lake). This pattern suggests that shrub *Betula* survived in small, discrete populations during the LGM.

Full-glacial tundra predominated through the late glaciation in the west-central Brooks Range (Tukuto, Etivlik, Red Green, Upper Capsule lakes), although shrub *Betula* tundra perhaps established locally between ~13,000 and 12,500 ^{14}C yr B.P. (Lake of the Pleistocene and Oil Lake). Isolated populations of *Populus* may have been present on the North Slope by this time (Lake of the Pleistocene). In contrast to the far north, *Betula* pollen percentages increase in most central Alaskan records beginning ~13,000 ^{14}C yr B.P., but pollen accumulation rates indicate that shrub *Betula* probably was not common across the intermontane region until 12,000 ^{14}C yr B.P. (Ped, Birch, Wien, Joe lakes; Screaming Yellowlegs Pond).

In southwestern Alaska, the first evidence for the establishment of shrub *Betula* tundra is in the Yukon Delta (Tungak Lake) at ~14,000 ^{14}C yr B.P., with the latest arrival on the St. Paul Islands by ~11,000 ^{14}C yr B.P. Shrub *Betula* tundra occupied the Kuskokwim and Nushagak lowlands (Idavain, Farewell, Snipe lakes; Flounder Flat 4, Kvichak Peninsula 5) and was at least present locally in the foothills of the Ahklun Mountains (Nimgun Lake) by ~12,000 ^{14}C yr B.P. Dense shrub tundra in the northernmost part of this area (also including Wien Lake) graded southward into predominantly graminoid tundra near the Ahklun Mountains (Grandfather and Ongivinuk lakes). *Populus* was present in southwestern Alaska by ~12,000 ^{14}C yr B.P., but significant populations apparently were confined to areas near Puyuk and Farewell lakes.

Moist meadows were common on Kodiak Island until ~12,500 ^{14}C yr B.P., with a shift to *Empetrum-Salix* heath, which continued until ~10,000 ^{14}C yr B.P. (Phalarope Pond). Shrub *Betula* tundra was present on Kenai Peninsula (Hidden and Circle lakes; Homer Spit) by late-glacial times. In southeastern Alaska, the recently deglaciated terrain supported herb-dominated tundra, with *Pinus contorta* and *Alnus* establishing in the northern Alexander Archipelago by 12,500 ^{14}C yr B.P (Lily Lake; see also Mann & Hamilton, 1995) and in Glacier Bay (Pleasant Island Lake) and the

Juneau area (Juneau sites) by ~11,000 ^{14}C yr B.P. Ages of *Pinus* macrofossils in sites from Yakutat (Pike Lakes muskeg), Lituya (Muskeg Cirque) and Glacier (Pleasant Island Lake) bays, Chilkat Peninsula (Lily Lake), and Juneau (Juneau sites) argue for an early and widespread late-glacial occurrence of the trees, suggestive of either a full-glacial refugium in southeastern Alaska or rapid northwestward migration following ice retreat (Peteet, 1991).

Climate interpretations of the widespread shift from herb-dominated to shrub *Betula* tundra indicate an increase in summer warmth, summer precipitation, and/or snow depth (e.g. Ager, 1983; Anderson & Brubaker, 1994). The importance of moisture changes during this interval is particularly clear in east-central Alaska, where the regional increase in shrub *Betula* at ~12,000 ^{14}C yr B.P. correlates with times of rapid rises in lake levels (Bigelow & Edwards, 2001; Edwards *et al.*, 2001). This shift may have been gradual as *Salix* and Cyperaceae, indicators of more mesic conditions, increased just prior to the *Betula* rise. The early spread (~14,000–13,000 ^{14}C yr B.P.) of shrub *Betula* in west-central Alaska indicates that an east-west moisture gradient postulated for the LGM (Barnosky *et al.*, 1987) perhaps continued into the late glaciation. The persistence of herb-dominated tundra in areas of southwestern and northern Alaska and the islands of the Bering strait suggests other regional to sub-regional temperature and/or effective moisture gradients. Late-glacial climates of south-central and southeastern Alaska are difficult to interpret because of the paucity of sites and the local influence of glacial retreat on the vegetation records. However, Peteet (1991) suggested that the generally widespread occurrence of *Pinus contorta* during the latter part of the late glaciation may reflect greater seasonality.

Although pollen studies do not typically provide quantitative paleoclimate estimates, such information is available from MCR reconstructions of fossil beetles. Trends in these data indicate increasing temperature and moisture for the late glaciation (Elias *et al.*, 1996a). Data from Flounder Flat, dated to ~12,500 ^{14}C yr B.P., indicate mean summer temperature of 13.5 °C and mean winter temperature of −30 °C (Elias, 2001). For the period 11,500–11,000 ^{14}C yr B.P., mean summer temperatures were 11.5–13 °C and 9.25–10.25 °C to the south and north of Bering Strait, respectively. Mean winter temperatures ranged from −28.5° to −23.5 °C and −32.25° to −27.25 °C for each region.

The Late Glacial-Early Holocene Transition (11,000–9000 ^{14}C yr B.P./~13,000–10,000 cal yr B.P.)

These 2000 ^{14}C years are some of the most intriguing in the late Quaternary history of Alaska. Forests make their first post-glacial appearances in northern, central, and southern Alaska. Unlike today, the interior forest was dominated by deciduous (*Populus*) and not evergreen (*Picea*) species. Key Holocene taxa, such as *Pinus contorta*, *Picea sitchensis*, *Picea glauca*, and *Alnus* make their initial appearances, although the latter two only near the end of this interval. Many of the pollen assemblages throughout the state lack modern analogs, making interpretations of past vegetation and climate particularly challenging. However, some records contain the first biological evidence for temperature reversals (i.e. a Younger Dryas-type climate oscillation) within what had been described previously as either a steady post-glacial warming or an interval of warmer than present climates.

In northern Alaska, the extent of shrub tundra likely was restricted until ~10,000 ^{14}C yr B.P., when *Betula* spread through much of the northwestern Brooks Range (Tukuto and Etivlik lakes). Moist graminoid-prostrate shrub tundra established in the north-central Brooks Range minimally by the early Holocene (Red Green, Upper Capsule, Oil lakes; Imnavait Creek). Scattered evidence for *Populus* on the North Slope indicates the tree probably was restricted to floodplains and low-elevation, south-facing slopes (Hopkins *et al.*, 1981; Mann *et al.*, 2002; Nelson & Carter, 1987; Oswald *et al.*, 1999). Based on a ~20 radiocarbon-dated macrofossil remains, Mann *et al.* (2002) suggest that *Populus* twice expanded on to the North Slope (between 11,500 and 10,900 ^{14}C yr B.P. and 10,000 and 7200 ^{14}C yr B.P.), perhaps the result of changes in fluvial geomorphic regimes.

Populus forest or woodlands and shrub *Betula-Salix* tundra were common in central Alaska, although interpretations of forest extent vary from relatively restricted populations along slopes or floodplains to rather extensive, dense forests covering lower elevations (e.g. Ager, 1983; Anderson *et al.*, 1994a; Brubaker *et al.*, 1983; Hu *et al.*, 1993). This forest possibly included thickets of high shrub *Betula* and *Salix* as seen today in areas of southwestern Alaska. Shrub *Alnus* and *Picea* make their first significant albeit spatially limited appearances at the end of the transitional period. The earliest ages for increased *Alnus* percentages are from scattered sites near Kotzebue Sound (e.g. Kaiyak Lake, ~9000 ^{14}C yr B.P.) and the south-central Brooks Range (e.g. Toolik Lake, ~9500 ^{14}C yr B.P.). Based on the presence of 10% pollen (Anderson & Brubaker, 1986; see Hu *et al.*, 1993 for alternative interpretation), *Picea* established ~9500 ^{14}C yr B.P. at Birch and Wien lakes (Note: the Birch Lake date may be erroneously old; Bigelow & Edwards, 2001). The presence of at least small populations of *Picea* in central Alaska is confirmed by radiocarbon-dated macrofossils (e.g. Wien Lake ~9500 ^{14}C yr B.P.; Tangle Lakes ~9100 ^{14}C yr B.P.).

The most spatially complex and diverse vegetation occurred in southern Alaska. In far southwestern Alaska, Aleutian records (Adak, Umnak), which begin between ~11,000 and 10,000 ^{14}C yr B.P., indicate the presence of graminoid-*Salix-Empetrum* communities. Shrub *Betula* established on the Alaskan Peninsula by 10,000 ^{14}C yr B.P. (Peteet & Mann, 1994). While *Alnus* occurred in the Yukon delta as early as ~12,000–11,000 ^{14}C yr B.P., the shrub was not a significant element on the landscape until 7500 ^{14}C yr B.P. (Puyuk Lake). *Populus* was present in the southwestern lowlands by ~11,000 ^{14}C yr B.P., although tree density was likely greatest towards the interior (Farewell Lake) with only scattered populations at sites closer to the present-day coast (Ongivinuk and Grandfather lakes). Shrub *Betula* tundra was replaced briefly in southwestern Alaska by more herb-dominated vegetation, with the decline in shrub *Betula* being most pronounced in near-coastal sites (Ongivinuk and Grandfather lakes). A shift to colder vegetation was also noted for Kodiak Island ~10,800–10,000 ^{14}C yr B.P. (Phalarope Pond).

Populus-Alnus-Salix scrub forest occupied lower elevations of Kenai Peninsula (Hidden and Circle lakes; Homer Spit peat) beginning ~10,300–9500 ^{14}C yr B.P. However, *Betula-Alnus* shrub tundra, which included a significant component of *Shepherdia canadensis*, was present in the mountains (Tern Lake peat). In Prince William Sound (Port Wells sites), Cyperaceae tundra with thickets of *Salix* and *Alnus* dominated mid- to late portions of this transitional period. Vegetation of Icy Cape was mesic Cyperaceae-Ericales tundra, with the addition of *Alnus* by ~9000 ^{14}C yr B.P. Near Yakutat, these treeless landscapes gave way to coniferous forest or woodland, with *Pinus contorta* establishing ~11,000–10,000 ^{14}C yr B.P. (Pike Lakes muskeg) and *Picea sitchensis* ~10,000 ^{14}C yr B.P. (Slate Mesa; Pike Lakes muskeg), although the latter did not become common until the mid-Holocene. Tundra remained at high elevations (Slate Mesa), with invasion by *Alnus* ~10,000 ^{14}C yr B.P. *Pinus contorta* established near Lituya Bay (Muskeg Cirque) ~10,500 ^{14}C yr B.P., followed quickly by an increase in *Alnus* and *Picea sitchensis*. Plant macrofossils confirm the presence of *Picea sitchensis* in Glacial Bay (Adams Inlet) ~11,200 ^{14}C yr B.P. and of *Pinus contorta* in the Alexander Archipelago (Pike Lakes muskeg) ~10,000 ^{14}C yr B.P.

The increased abundance of *Populus* during the late glacial-early Holocene transition persuaded many investigators that summer temperatures were likely warmer than present in central and northern Alaska (e.g. Anderson & Brubaker, 1994; Brubaker *et al.*, 1983; Edwards & Barker, 1994). High *Juniperus* pollen percentages sometimes associated with the *Populus* period are also indicative of warm, dry conditions. Range extensions of *Populus* and other thermophilous plant species, extended limits of *Castor*, increased eolian activity, and melting of ice wedges lent further support for climate that was warmer than present (Anderson, 1988; Carter & Hopkins, 1982; Edwards & Barker, 1994; Edwards *et al.*, 1985; Hopkins *et al.*, 1981; McCulloch & Hopkins, 1966; Nelson & Carter, 1987). Insect and plant macrofossil studies suggest at least a 3.5 °C increase in summer temperature (Carter *et al.*, 1984; Elias, 2001). Deeper snows were postulated for parts of northern and west-central Alaska, based on stream incisions and apparently stronger spring floods, resulting in greater disturbance and potential habitats for *Populus* (Brubaker *et al.*, 1983; Carter & Hopkins, 1982; Carter *et al.*, 1984). However, lake-level evidence from east-central Alaska suggests that climate remained drier than modern (Edwards *et al.*, 2001). The subsequent *Populus* decline that occurred ~9000 ^{14}C yr B.P. has been attributed to decreased summer temperatures and/or reduction in spring flooding (Brubaker *et al.*, 1983; Edwards & Dunwiddie, 1985).

Other studies (e.g. Bartlein *et al.*, 1995; Brubaker *et al.*, 1983; Brubaker *et al.*, 1995; Hu *et al.*, 1993) noted the presence or absence of *Populus* can be strongly determined by edaphic factors, and thus the paleobotanical data should be used with some caution when making paleoclimatic inferences. For example, the invasion of *Populus-Salix* communities into shrub *Betula* tundra would result eventually in production of organic-rich and acid soils that ultimately would no longer support *Populus* forest. Furthermore, differences among records for the time of high *Populus* pollen percentages suggest non-climate factors may more correctly explain the presence of these deciduous forests, although issues related to chronologies could account for some of this temporal variation (Bartlein *et al.*, 1995).

Although much research has focused on the warm climates of the late glacial-Holocene transition, there is growing paleoecological evidence that southern and possibly portions of south-central and northern Alaska experienced a Younger-Dryas type oscillation (Bigelow & Edwards, 2001; Brubaker *et al.*, 2001; Engstrom *et al.*, 1990; Hansen & Engstrom, 1996; Hu *et al.*, 1995; Hu *et al.*, 2002; Mann *et al.*, 2002; Peteet & Mann, 1994). In some of these records, the vegetational changes are subtle (e.g. Windmill Lake), whereas in others the shifts are clearer (e.g. Phalarope Pond and Nimgun Lake). With the exception of a single North Slope site (Lake of the Pleistocene), the clearest vegetational responses are restricted primarily to the North Pacific rim, suggesting a link to cooling sea-surface temperatures (Bigelow & Edwards, 2001; Peteet & Mann, 1994; see also Ager, 2000, and Mann *et al.*, 2002). The influence of a cool North Pacific may have been moderated in the interior by increasing summer insolation, as paleo-records to the north generally show no or weak evidence of cooling.

The Early to Mid-Holocene (9000–6000 ^{14}C yr B.P./~10,000–6800 cal yr B.P.)

During this interval, modern forest (traced here with the history of *Picea*) and shrub tundra communities (described using the history of *Alnus*) made their first appearances as climatic patterns continued to adjust to post-glacial conditions. The time-transgressive nature of the observed vegetation shifts is more apparent than in previous periods (e.g. Anderson & Brubaker, 1994), which is partially a reflection of improved site-distributions and chronological control.

Between ~9000 and 8000 ^{14}C yr B.P., mapped pollen percentages of *Picea*, supported in some cases by macrofossils, indicate an east to west spread of trees from the Tanana valley to the south-central Brooks Range and upper Kuskokwim valley (Anderson & Brubaker, 1993, 1994; Brubaker *et al.*, 2001). At least in the Tanana valley, both *Picea glauca* and *Picea mariana* were present during the early Holocene (Windmill and Birch lakes), but *Picea glauca* probably was the only or dominant conifer in areas farther west (Screaming Yellowlegs Pond and Wien Lake). During this early migration, *Picea* likely was restricted to valley bottoms and warm, well-drained slopes (Brubaker *et al.*, 1995). Between 8000 and 7000 ^{14}C yr B.P., *Picea* range-limits retreated eastward or *in situ* populations declined (Ruppert and Sands of Time lakes, Screaming Yellowlegs Pond). A second migration of *Picea*, predominantly *Picea mariana*, occurred between 7000 and 6000 ^{14}C yr B.P (Screaming Yellowlegs Pond; Ruppert and Joe lakes). By the mid-Holocene, *Picea* was well established throughout most of its current range in east-central Alaska (Ped, Jan, Tiinkdhul, Windmill, Sands of Time lakes), had extended westward into the upper reaches of Kotzebue Sound (Ruppert and Joe lakes), and had colonized areas of the Kuskokwim lowlands (Farewell Lake). Evidence for the northward or westward extension of *Picea* beyond modern limits is absent, with tundra persisting in northern and far western Alaska throughout the Holocene. However, the arrival dates of ~9100 ^{14}C yr B.P. near Tangle Lake, ~8700 ^{14}C yr B.P. in the northern Chugach Mountains, ~8000 ^{14}C yr B.P. in upper Cook Inlet and central Kenai Peninsula, and ~4000 ^{14}C yr B.P. in southern Kenai Peninsula indicate that *Picea* (probably *Picea glauca*) spread from the interior to south-central Alaska (Ager, 1983, 2000, 2001). A third species, *Picea sitchensis,* was present in southeastern Alaska (Slate Mesa) by 10,000 ^{14}C yr B.P. However, these populations never expanded appreciatively during the early to mid-Holocene (Peteet, 1991).

By the mid-Holocene, *Alnus* was found across the Brooks Range, throughout the central interior, and in southern Alaska (Anderson & Brubaker, 1993; Brubaker *et al.*, 2001; Mann & Hamilton, 1995). Between 9000 and 8000 ^{14}C yr B.P., small populations of the shrub occurred in the northwestern (Squirrel and Kaiyak lakes) and south-central Brooks Range (Ruppert Lake and Screaming Yellowlegs Pond), at isolated sites within the Yukon basin (Sands of Time Lake), and in the Kuskokwim-Ahklun region (Ongivinuk, Snipe, Farewell lakes). By 7000 ^{14}C yr B.P., *Alnus* was common in much of the intermontane region and by 6500 ^{14}C yr B.P. (Windmill and Eightmile lakes), had spread to the North Slope (e.g. Tukuto and Etvilik lakes; Imnaviat Creek) and the north-central Alaska Range (e.g. Tenmile, Wonder, Windmill lakes). High pollen percentages (>30%) suggest *Alnus* was particularly abundant in the western and central Brooks Range by 6000 ^{14}C yr B.P. (Anderson & Brubaker, 1994).

Alnus was present in the northern Alexander Archipelago by ~12,500 ^{14}C yr B.P. (Lily Lake; see also Mann & Hamilton, 1995), established at Icy Cape between ~10,800 and 10,000 ^{14}C yr B.P. (see also Peteet, 1986), Prince William Sound (Port Wells sites) after ~8300 ^{14}C yr B.P., and on Kenai Peninsula by ~9500 ^{14}C yr B.P. (Circle Lake; Homer Spit; Tern Lake peat). Ager (2001) suggested that refugial populations of *Alnus tenuifolia* possibly survived the LGM in south-central Alaska based on consistent albeit minor amounts of *Alnus* in the pollen records. In some areas (e.g. Slate Mesa; Circle Lake; Tern Lake peat; Homer Spit), *Alnus* communities clearly dominated the early to mid-Holocene vegetation.

The early to mid-Holocene climate is characterized by significant temperature and/or precipitation variations within any given region. In northern and central Alaska, the decline in *Populus* pollen (Ruppert, Wien, Tiinkdhul lakes), in some cases the reversion to shrub tundra (Joe, Pleistocene, Tukuto lakes), and reduced evidence for the extra-limital presence of warm plant and animal taxa have been cited as evidence for summers that were cooler than previously (i.e. ~11,000–9000 ^{14}C yr B.P., the regional post-glacial thermal maximum). In contrast, warmer than present conditions probably occurred throughout this interval in east-central Alaska (Abbott *et al.*, 2000; Bigelow, 1997; Edwards *et al.*, 2001). Evidence from southwestern Alaska does not indicate warmer than present summer temperatures. However, the north-south gradient observed in the transitional period continued during the early to mid-Holocene, indicating that cooler growing-season temperatures persisted nearer the coast (Brubaker *et al.*, 2001).

Summer temperature initially was proposed as the main factor that limited the spread of *Picea* to west-central Alaska until the mid- to late Holocene (e.g. Anderson & Brubaker, 1993). However, more recent studies suggest this westward migration was strongly influenced by water stress (e.g. Bigelow & Edwards, 2001; Edwards & Barker, 1994; Hu *et al.*, 1998). Lake-level data from east-central Alaska indicate that climates were dry during the earliest Holocene, with precipitation reaching near-modern levels by the mid-Holocene (Abbott *et al.*, 2000; Bigelow, 1997; Edwards *et al.*, 2001). Furthermore, the expansion of *Alnus* throughout much of central Alaska during this interval is viewed as evidence for increased moisture (e.g. Anderson & Brubaker, 1993). This conclusion is not supported by paleohydrological studies in the central Brooks Range (Meli, Tangled Up lakes), which indicate that conditions were relatively arid prior to ~5000–6000 ^{14}C yr B.P. The Brooks Range trend, however, correlates well with that described for east-central Alaska.

In southeastern Alaska, transfer function reconstructions for Icy Cape indicate July temperatures were warmer than present beginning ~10,000 ^{14}C yr B.P., with maximum warmth ~9000–7000 ^{14}C yr B.P., and maximum aridity ~8000 ^{14}C yr B.P. (Heusser *et al.*, 1985; see also Mann *et al.*, 1998). This paleoclimate scenario differs from that proposed earlier by Heusser (1960), where southeastern Alaska was thought to be cool and moist during the early to mid-Holocene, experiencing maximum warmth and minimum precipitation between 5000 and 2000 ^{14}C yr B.P. Peteet (1991) noted that under early Holocene warming conditions, *Picea sitchensis* should have been common in far southeastern Alaska. However, the dominance of *Alnus* in the Yakutat region suggests an increased seasonality that possibly resulted in an abundance of favorable habitats caused by dry summers and severe winters.

The Mid- to Late Holocene (6000 ^{14}C yr B.P. to Present/~6800 cal yr B.P. to Present)

Unlike previous intervals, vegetation and climate changes are more marked in southeastern Alaska than in areas north of the Alaska Range. Although the modern composition of boreal and tundra communities of central and northern Alaska was achieved by ~6000 ^{14}C yr B.P., their present-day distributions were not in place in western Alaska and portions of the Alaska Range until ~4000–3000 ^{14}C yr B.P. In contrast, several important tree species first appeared in southeastern Alaska only at the beginning of this period, with modern landscapes occurring throughout the region by ~3000 ^{14}C yr B.P.

The mid-Holocene begins with a second expansion of *Picea* (likely dominated by *Picea mariana*) across north-central Alaska (Anderson & Brubaker, 1994). By

~4000 ^{14}C yr B.P., populations advanced into the southwestern lowlands (e.g. Grandfather and Snipe lakes) and across the southwestern flanks of the Brooks Range, reaching modern limits in the Kotzebue Sound region (Squirrel and Kaiyak lakes). *Picea* populations probably also increased within previously forested portions of the interior (Edwards *et al.*, 2000a). Mapped pollen percentages of *Alnus* suggest a greater presence of these shrubs in the western and central Brooks Range from ~6000–5000 ^{14}C yr B.P., with a subsequent decline to modern pollen values by 2000 ^{14}C yr B.P. (Anderson & Brubaker, 1994). Additional data suggest changes in boreal and tundra vegetation during the late Holocene (e.g. Bigelow & Edwards, 2001; Hu *et al.*, 2001b), but they are too inconclusive and/or sparsely distributed to discuss broad trends.

The mid- to late Holocene of southeastern Alaska witnessed significant expansions of several key conifer species. *Picea sitchensis*, which established at Lituya Bay ~10,500 ^{14}C yr B.P. (Muskeg Cirque) and at Icy Cape by 7600 ^{14}C yr B.P., arrived in Prince William Sound between 4000 and 3000 ^{14}C yr B.P. (Port Wells sites and Jonah Bay), and southern Kenai Peninsula ~1650 ^{14}C yr B.P. (Circle Lake, Homer Spit). *Tsuga heterophylla* is dated to ~7100 ^{14}C yr B.P. at Lituya Bay (Muskeg Cirque), ~5000 ^{14}C yr B.P. in the Yakutat area (Pikes Lake muskeg), and ~3800 ^{14}C yr B.P. at Icy Cape. Arrival dates for *Tsuga mertensiana* are ~5000 ^{14}C yr B.P. at Lituya Bay and Yakutat, ~3500 ^{14}C yr B.P. at Icy Cape, ~2700 ^{14}C yr B.P. near Prince William Sound, and ~2900 ^{14}C yr B.P. in the northern Kenai Mountains. These patterns suggest a north-to-south migration of both *Tsuga* species.

Pollen records for this interval mostly address climate changes within forested regions of Alaska. For example, the westward expansion of *Picea mariana* is attributed to an increase in moisture and/or decrease in summer temperatures (Anderson & Brubaker, 1993; Edwards *et al.*, 2001; Hopkins *et al.*, 1981). Alternative explanations related to paludification, disturbance regime, and bedrock geology caution that *Picea mariana* distribution may not be solely climate-driven (e.g. Cwynar, 1982; Mann *et al.*, 1995). Yet the spread of *Picea mariana* is not the only evidence for climate changes in central and northern Alaska. Oxygen isotope studies of lake sediments in the central Brooks Range suggest mid- to late Holocene conditions that were cooler and wetter than previously (L. Anderson *et al.*, 2001). Maximum abundances of *Alnus* occurred ~5000 ^{14}C yr B.P. in the southwestern and south-central Brooks Range, suggestive of high precipitation (possibly increased snow cover; Hu *et al.*, 1995) and/or lower summer temperatures (Anderson, 1985; Anderson & Brubaker, 1994; Anderson *et al.*, 1991). Increased peat accumulation in north-central Alaska may also indicate moister conditions (Hamilton & Robinson, 1977). Modern lake-levels were common in east-central Alaska ~6000 ^{14}C yr B.P., a further indication of increased effective moisture during the mid-Holocene (Edwards *et al.*, 2001).

The movement of most conifer species in southeastern Alaskan has been linked to storm frequencies and precipitation (Heusser, 1983a; Peteet, 1986, 1991). Transfer function reconstructions suggest a Holocene precipitation maximum ~4000 ^{14}C yr B.P. and a gradual decline in July temperature beginning ~8000 ^{14}C yr B.P., reaching near modern values by ~4000 ^{14}C yr B.P. (Heusser *et al.*, 1985). Evidence of Neoglacial cooling is better documented in glacial than paleobotanical records (Mann & Hamilton, 1995; Mann *et al.*, 1998; see also Barclay *et al.*, 2001; Calkin *et al.*, 2001; Wiles *et al.*, 2002). Glaciers expanded after 6000–5000 ^{14}C yr B.P., suggesting a progressive rather than abrupt cooling. Transfer function results indicate a difference of ~3 °C between the post-glacial thermal maximum and the coldest intervals of the Neoglaciation (Heusser *et al.*, 1985). The Neoglaciation itself was characterized by warm and cool periods of a few 100–1000-yr duration, with changes of several °C (Mann *et al.*, 1998). Precipitation evidently also varied, altering the location of snow line by 100s of meters (Mann, 1986; Wiles *et al.*, 1995). Modern or near-modern conditions probably established by ~3000 ^{14}C yr B.P. (Mann *et al.*, 1998).

Key Issues for Further Study

High-quality radiocarbon chronologies are fundamental for describing environmental histories and for comparing these histories with paleoclimate models. Radiocarbon dating has always been problematic in the Arctic, where organic content of sediment is often low and carbon has a long residence time on the landscape, potentially delaying the input of organic material to sedimentary basins and thus altering sediment ages (Abbott & Stafford, 1996). AMS radiocarbon dating permits the use of extremely low-organic content samples (e.g. pollen and small plant macrofossils), thereby reducing the reliance on bulk sediment dates. AMS ages are now available for many Alaskan sites, but in some records the AMS ages are considerably younger than those from bulk sediments (e.g. Bigelow & Edwards, 2001), implying that many dates for Alaskan records may be incorrect. While the previously defined patterns of change will likely remain the same (e.g. the east-to-west migration of *Picea* across central Alaska), they will undoubtedly shift in time, necessitating a re-examination of the climatic causes behind the paleo-observations. Additionally, paleoenvironmental histories based on calibrated rather than ^{14}C yr will further alter the current understanding of mechanisms responsible for shifts in past environments (Bartlein *et al.*, 1995; Edwards & Barker, 1994).

Palynological data, though the key source of paleoenvironmental information in Alaska, are limited in scope, and multi-proxy approaches are quickly becoming the scientific standard in Alaskan research. The use of multiple indicators provides opportunities for addressing a broader array of climate and paleoecological questions than is possible using a single proxy. For example, geochemical studies in southwestern Alaska reconstruct growing-season temperatures over the last 2000 ^{14}C yr B.P. (Hu *et al.*, 2001a) and within-watershed ecosystem changes during the last 12,000 ^{14}C yr B.P. (Hu *et al.*, 1993, 1996, 2001b). In east-central Alaska, paleohydrological investigations have integrated information from sedimentology, geochemistry, sediment magnetic susceptibility, paleobotany, and lake-level models to document water-level changes and to obtain quantitative estimates of precipitation over the last 12,500 ^{14}C yr B.P. (Abbott *et al.*, 2000; Barber & Finney, 2000; Edwards *et al.*, 2000a, b; Edwards *et al.*, 2001). Although multi-proxy records are still relatively rare, these interdisciplinary investigations will likely form the basis of future syntheses of late Quaternary environments of Alaska.

Modeling provides an important means to evaluate the current knowledge of processes driving the paleoenvironmental changes documented in the Alaskan records. However, previous examination of paleoclimate simulations and paleodata has clearly demonstrated the need for more sophisticated earth-system models (e.g. Bartlein *et al.*, 1998), particularly for understanding the complexity of land-atmosphere feedbacks. For example, vegetation and other terrestrial surface attributes can significantly alter energy, water, and carbon exchanges between the land and the atmosphere (e.g. Chapin & Starfield, 1997; Levis *et al.*, 1999; Smith & Shugart, 1993; TEMPO, 1996). The distribution, physiognomic characteristics, and physiological status of arctoboreal vegetation and the complexity of northern landscapes (e.g. tundra vs. forest; barrens vs. vegetation; continuous vs. discontinuous permafrost) play important roles in determining energy, water, and carbon budgets at northern high latitudes; changes in any of these budgets can have significant impacts on the seasonal and annual climatology of much of the Northern Hemisphere (e.g. Bonan *et al.*, 1992, 1995; Foley *et al.*, 1994; Harvey, 1988). Thus, considerable potential exists for northern land-cover status to enhance or mitigate climate depending on the nature of the feedback mechanisms, and investigations of these mechanisms, past and present, likely will be of high priority in Alaskan research over the next decade. The current suite of computer models applied in Alaska has the ability to simulate past plant taxon (e.g. Bartlein *et al.*, 1998) or biome equilibrium distributions based on particular climate scenarios and improved model input of edaphic and plant physiological factors (e.g. Bigelow *et al.*, in press; Edwards *et al.*, 2000a). The parallel development of ecological models that address climate effects on processes directly modifying plant competition, dispersal, and disturbance regime provide opportunities to investigate transient, as opposed to equilibrium, responses of vegetation to climate change (e.g. Rupp *et al.*, 2000a, b). Overall, the emerging array of models affords a rich and exciting potential for unraveling causes of past environmental change, including complex feedback interactions, and for providing critical tools for predicting system responses to a warming, global climate.

Acknowledgments

We thank two unnamed reviewers for their comments on an earlier version of this manuscript. We also thank Matt Duvall and Lyn Gualitieri for their help with the figures. This work was supported by the National Science Foundation (OPP-001874).

References

Abbott, M.B. & Stafford, T.W., Jr. (1996). Radiocarbon geochemistry of modern and ancient lake systems, Baffin Island, Canada. *Quaternary Research*, **45**, 300–311.

Abbott, M.B., Finney, B.F., Edwards, M.E. & Kelts, K.R. (2000). Lake-level reconstructions and paleohydrology of Birch Lake, central Alaska, based on seismic reflection profiles and core transects. *Quaternary Research*, **53**, 154–166.

Ager, T.A. (1975). *Late Quaternary environmental history of the Tanana valley, Alaska*. Ohio State University Institute of Polar Studies, Report 54, Columbus, 117 pp.

Ager, T.A. (1982). Vegetational history of western Alaska during the Wisconsin glacial interval and the Holocene. *In*: Hopkins, D.M., Matthews, J.V., Jr., Schweger, C.E. & Young, S.B. (Eds), *Paleoecology of Beringia*. New York, Academic Press, pp. 75–93.

Ager, T.A. (1983). Holocene vegetational history of Alaska. *In*: Wright, H.E., Jr. (Ed.), *Late Quaternary environments of the United States*. The Holocene. Minneapolis, University of Minnesota Press, Vol. 2, pp. 128–141.

Ager, T.A. (2000). Postglacial vegetation history of the Kachemak Bay area, Cook Inlet, south-central Alaska. *U.S. Geological Survey Professional Paper*, 1615, pp. 147–165.

Ager, T.A. (2001). Holocene vegetation history of the northern Kenai Mountains, south-central Alaska. *U.S. Geological Survey Professional Paper*, 1633, pp. 91–107.

Ager, T.A. & Brubaker, L.B. (1985). Quaternary palynology and vegetational history of Alaska. *In*: V.M. Bryant, V.M. & Holloway, R.G. (Eds), *Pollen Records of Late Quaternary North American Sediments*. Dallas, American Association of Stratigraphic Palynologists, pp. 353–384.

Anderson, L., Abbott, M.B. & Finney, B.P. (2001). Holocene climate inferred from oxygen isotope ratios in lake sediments, central Brooks Range, Alaska. *Quaternary Research*, **55**, 313–321.

Anderson, P.M. (1985). Late Quaternary vegetational change in the Kotzebue Sound area, northwestern Alaska. *Quaternary Research*, **24**, 307–321.

Anderson, P.M. (1988). Late Quaternary pollen records from the Kobuk and Noatak River drainages, northwestern Alaska. *Quaternary Research*, **29**, 263–276.

Anderson, P.M. & Brubaker, L.B. (1986). Modern pollen assemblages from northern Alaska. *Review of Palaeobotany and Palynology*, **46**, 273–291.

Anderson, P.M. & Brubaker, L.B. (1993). Holocene vegetation and climate histories of Alaska. *In*: Wright, H.E., Jr., Kutzbach, J.E., Webb, T., III, Ruddiman, W.F., Street-Perrott, F.A. & Bartlein, P.J. (Eds), *Global Climates Since the Last Glacial Maximum*. Minneapolis, University of Minnesota Press, pp. 386–400.

Anderson, P.M. & Brubaker, L.B. (1994). Vegetation history of northcentral Alaska: A mapped summary of late-Quaternary pollen data. *Quaternary Science Reviews*, **13**, 71–92.

Anderson, P.M., Reanier, R.E. & Brubaker, L.B. (1988). Late Quaternary vegetational history of the Black River region in northeastern Alaska. *Canadian Journal of Earth Science*, **25**, 84–94.

Anderson, P.M., Bartlein, P.J., Brubaker, L.B., Gajewski, K. & Ritchie, J.C. (1989). Modern analogues of late-Quaternary pollen spectra from the western interior of North America. *Journal of Biogeography*, **16**, 573–596.

Anderson, P.M., Reanier, R.E. & Brubaker, L.B. (1990). A 14,000-year pollen record from Sithylemenkat Lake, north-central Alaska. *Quaternary Research*, **33**, 400–404.

Anderson, P.M., Bartlein, P.J., Brubaker, L.B., Gajewski, K. & Ritchie, J.C. (1991). Vegetation-pollen-climate relationships for the arcto-boreal region of North America and Greenland. *Journal of Biogeography*, **18**, 565–582.

Anderson, P.M., Bartlein, P.J. & Brubaker, L.B. (1994a). Late Quaternary history of tundra vegetation in northwestern Alaska. *Quaternary Research*, **41**, 306–315.

Anderson, P.M., Lozhkin, A.V., Eisner, W.R., Hopkins, D.M., Brubaker, L.B. & Colinvaux, P.A. (1994b). Pollen records from Ten Mile and Wonder Lake, Alaska. *Geographie Physique et Quaternaire*, **48**, 131–141.

Barber, V.A. & Finney, B.P. (2000). Late Quaternary paleoclimatic reconstructions for interior Alaska based on paleolake-level data and hydrologic models. *Journal of Paleolimnology*, **24**, 29–41.

Barclay, D.J., Calkin, P.E. & Wiles, G.C. (2001). Holocene history of Hubbard glacier in Yakutat Bay and Russell Fiord, southern Alaska. *Geological Society of America Bulletin*, **113**, 388–402.

Bartlein, P.J., Anderson, P.M., Edwards, M.E. & McDowell, P.F. (1991). A framework for interpreting paleoclimatic variations in eastern Beringia. *Quaternary International*, **10–12**, 73–83.

Bartlein, P.J., Edwards, M.E., Shafer, S.L. & Barker, E.D., Jr. (1995). Calibration of radiocarbon ages and the interpretation of paleoenvironmental records. *Quaternary Research*, **44**, 417–424.

Bartlein, P.J., Anderson, K.H., Anderson, P.M., Edwards, M.E., Mock, C.J., Thompson, R.S., Webb, R.S., Webb, T., III & Whitlock, C. (1998). Paleoclimate simulations for North America over the past 21,000 years: Features of the simulated climate and comparisons with paleoenvironmental data. *Quaternary Science Reviews*, **17**, 549–585.

Barnosky, C.W., Anderson, P.M. & Bartlein, P.J. (1987). The northwestern U.S. during deglaciation; vegetational history and paleoclimatic implications. *In*: Ruddiman, W.F. & Wright, H.E., Jr. (Eds), *North America and Adjacent Oceans During the Last Deglaciation*. Boulder, Geological Society of America, The Geology of North America, K-3, Geological Society of America, pp. 289–321.

Bergstrom, M.F. (1984). Late Wisconsin and Holocene history of a deep arctic lake, north-central Brooks Range, Alaska. M.S. Thesis, Columbus, Ohio, Ohio State University. 112 pp.

Bigelow, N.H. (1997). Late Quaternary vegetation and lake level changes in central Alaska. Ph.D. dissertation, Fairbanks, University of Alaska, 212 pp.

Bigelow, N.H. & Edwards, M.E. (2001). A 14,000 yr paleoenvironmental record from Windmill Lake, central Alaska: Late glacial and Holocene vegetation in the Alaska Range. *Quaternary Science Reviews*, **20**, 203–215.

Bigelow, N.H., Brubaker, L.B. Edwards, M.E., Harrison, S.P., Prentice, I.C., Anderson, P.M., Andreev, A.A., Bartlein, P.J., Christiansen, T.R., Cramer, W., Kaplan, J.O., Lozhkin, A.V., Matveyeva, N.V., Murray, D.F., McGuire, A.D., Razzhivin, V.Y., Ritchie, J.C., Smith, B., Walker, D.A., Clayden, S.L., Ebel, T., Gajewski, K., Hahne, J., Holmqvist, B.H., Igarashi, Y., Jordan, J.W., Kremenetski, K.V., Melles, M., Oswald, W.W., Paus, A., Pisaric, M.F.J., Shilova, G.N., Seigert, C., Volkova, V.S. & Wolf, V.G. (in press). Climate change and arctic ecosystems I: Biome reconstructions of tundra vegetation types at 0, 6, and 18 radiocarbon kyr in the Arctic. *Journal of Geophysical Research*.

Bonan, G.B., Pollard, D. & Thompson, S.L. (1992). Effects of boreal forest vegetation on global climate. *Nature*, **359**, 716–718.

Bonan, G.B., Chapin, F.S., III & Thompson, S.L. (1995). Boreal forest and tundra ecosystems as components of the climate system. *Climatic Change*, **29**, 145–167.

Brubaker, L.B., Garfinkel, H.L. & Edwards, M.E. (1983). A late-Wisconsin and Holocene vegetation history from the central Brooks Range: Implications for Alaskan paleoecology. *Quaternary Research*, **20**, 194–214.

Brubaker, L.B., Anderson, P.M. & Hu, F.S. (1995). Arctic tundra biodiversity: A temporal perspective from late Quaternary pollen records. *In*: Chapin, F.S., III & Körner, E. (Eds), Arctic and alpine biodiversity. *Ecological Studies*. Berlin, Springer-Verlag, Vol. 113, pp. 111–125.

Brubaker, L.B., Anderson, P.M. & Hu, F.S. (2001). Vegetation ecotone dynamics in southwest Alaska during the late Quaternary. *Quaternary Science Reviews*, **20**, 175–188.

Calkin, P.E., Wiles, G.C. & Barclay, D.J. (2001). Holocene coastal glaciation of Alaska. *Quaternary Science Reviews*, **20**, 449–461.

CAPE Project Members (2001). Holocene paleoclimate data from the Arctic: Testing models of global climate change. *Quaternary Science Reviews*, **20**, 1275–1287.

Carter, L.D. & Hopkins, D.M. (1982). Late Wisconsinan winter snow cover and sand-moving winds on the Arctic coastal plains of Alaska [abs.]. *Eleventh Annual Arctic Workshop Abstracts*, 8 pp.

Carter, L.D., Nelson, R.E. & Galloway, J.P. (1984). Evidence for early Holocene increased precipitation and summer warmth in arctic Alaska [abs.]. *Abstracts of the Geological Society of America Cordilleran Section Meeting*, 80, 274.

Chapin, F.S., III & Starfield, A.M. (1997). Time lags and novel ecosystems in response to transient climatic change in arctic Alaska. *Climatic Change*, **35**, 449–461.

COHMAP Members (1988). Climatic changes of the last 18,000 Years: Observations and model simulations. *Science*, **241**, 1043–1052.

Colinvaux, P.A. (1964). The environment of the Bering land bridge. *Ecological Monographs*, **34**, 297–329.

Colinvaux, P.A. (1967a). A long pollen record from St. Lawrence Island, Bering Sea, Alaska. *Palaeogeography, Palaeoclimatology, and Palaeoecology*, **3**, 29–43.

Colinvaux, P.A. (1967b). Bering land bridge: Evidence of spruce in late Wisconsin times. *Science*, **156**, 380–383.

Colinvaux, P.A. (1981). Historical ecology in Beringia: The south land bridge coast at St. Paul Island. *Quaternary Research*, **16**, 18–36.

Cwynar, L.C. (1982). A late-Quaternary vegetation history from Hanging Lake, northern Yukon. *Ecological Monographs*, **52**, 1–24.

Cwynar, L.C. (1990). A late Quaternary vegetation history from Lily Lake, Chilkat Peninsula, southeast Alaska. *Canadian Journal of Botany*, **8**, 1106–1112.

Edwards, M.E. & Barker, E.D., Jr. (1994). Climate and vegetation in northeastern Alaska 18,000 yr B.P.-present. *Palaeoecology, Palaeogeography, Palaeoclimatology*, **109**, 127–135.

Edwards, M.E. & Brubaker, L.B. (1986). Late Quaternary vegetation history of the Fishhook Bend area, Porcupine River, Alaska. *Canadian Journal of Earth Sciences*, **23**, 1765–1773.

Edwards, M.E. & Dunwiddie, P.W. (1985). Dendrochronological and palynological observations on Populus balsamifera in northern Alaska, U.S.A. *Arctic and Alpine Research*, **17**, 271–278.

Edwards, M.E., Anderson, P.M., Garfinkel, H.L. & Brubaker, L.B. (1985). Late Wisconsin and Holocene vegetation history of the upper Koyukuk region, Brooks Range, Alaska. *Canadian Journal of Botany*, **63**, 616–646.

Edwards, M.E., Anderson, P.M., Brubaker, L., Ager, T., Andreev, A., Bigelow, N., Cwynar, L., Eisner, W., Harrison, S., Hu, F.-S., Jolly, D., Lozhkin, A., MacDonald, G., Mock, C., Ritchie, J., Sher, A., Spear, R., Williams, J. & Yu, G. (2000a). Pollen-based biomes for Beringia 18,000, 6000 and 0 ^{14}C yr BP. *Journal of Biogeography*, **27**, 521–554.

Edwards, M.E., Bigelow, N.H., Finney, B.P. & Eisner, W.R. (2000b). Records of aquatic pollen and sediment properties as indicators of late-Quaternary Alaskan lake levels. *Journal of Paleolimnology*, **24**, 55–68.

Edwards, M.E., Mock, C.J., Finney, B.P., Barber, V.A. & Bartlein, P.J. (2001). Potential analogues for paleoclimatic variations in eastern interior Alaska during the past 14,000 yr: Atmospheric-circulation controls of regional temperature and moisture responses. *Quaternary Science Reviews*, **20**, 189–202.

Eisner, W.R. (1991). Palynological analysis of a peat core from Imnavait Creek, the North Slope, Alaska. *Arctic*, **44**, 279–282.

Eisner, W.R. & Colinvaux, P.A. (1990). A long pollen record from Ahaliorak Lake, arctic Alaska. *Review of Palaeobotany and Palynology*, **63**, 35–52.

Eisner, W.R. & Colinvaux, P.A. (1992). Late Quaternary pollen records from Oil Lake and Feniak Lake, Alaska.U.S.A. *Arctic and Alpine Research*, **24**, 56–63.

Elias, S.A. (2001). Mutual climatic range reconstructions of seasonal temperatures based on late Pleistocene fossil beetle assemblages in Eastern Beringia. *Quaternary Science Reviews*, **20**, 77–92.

Elias, S.A. & Brigham-Grette, J. (Eds) (2001). *Beringian paleoenvironments festschrift in honour of D.M. Hopkins*. Oxford, Elsevier Science Ltd., 574 pp.

Elias, S.A., Short, S.K., Nelson, C.H. & Birks, H.H. (1996a). Life and times of the Bering land bridge. *Nature*, **382**, 60–63.

Elias, S.A., Short, S.K. & Waythomas, C.F. (1996b). Late Quaternary environments, Denali National Park and Preserve, Alaska. *Arctic*, **49**, 292–305.

Elias, S.A., Short, S.K. & Birks, H.H. (1997). Late Wisconsin environments of the Bering Lake Bridge. *Palaeogeography, Palaeoclimatology, Palaeoecology*, **136**, 293–308.

Engstrom, D.R., Hansen, B.C.S. & Wright, H.E., Jr. (1990). A possible Younger Dryas record in southeastern Alaska. *Science*, **250**, 1383–1385.

Foley, J.A., Kutzbach, J.E., Coe, M.T. & Levis, S. (1994). Climate and vegetation feedbacks during the mid-Holocene. *Nature*, **371**, 52–54.

Gallimore, R.G. & Kutzbach, J.E. (1996). Role of orbitally induced changes in tundra area in the onset of glaciation. *Nature*, **381**, 503–505.

Goetcheus, V.G. & Birks, H.H. (2001). Full-glacial upland tundra vegetation preserved under tephra in the Beringia National Park, Seward Peninsula, Alaska. *Quaternary Science Reviews*, **20**, 135–147.

Guthrie, R.D. (1968). Paleoecology of the large-mammal community in interior Alaska during the late Pleistocene. *American Midland Naturalist*, **79**, 346–363.

Guthrie, R.D. (1982). Mammals of the mammoth steppe as paleoenvironmental indicators. *In*: Hopkins, D.M., Matthews, J.V., Jr., Schweger, C.E. & Young, S.B. (Eds), *Paleoecology of Beringia*. New York, Academic Press, pp. 307–329.

Guthrie, R.D. (2001). Origin and causes of the mammoth steppe: A story of cloud cover, woolly mammoth tooth pits, buckles, and inside-out Beringia. *Quaternary Science Reviews*, **20**, 549–574.

Hamilton, T.D. & Robinson, S. (1977). Late Holocene (Neoglacial) environmental changes in central Alaska [abs.]. *Geological Society of America Abstracts with Programs*, 9, 1003.

Hansen, B.S.C. & Engstrom, D.R. (1996). Vegetation history of Pleasant Island, southeastern Alaska, since 13,000 yr B.P. *Quaternary Research*, **46**, 161–175.

Harvey, L.D.D. (1988). On the role of high latitude ice, snow and vegetation feedbacks in the climatic response to external forcing changes. *Climatic Change*, **13**, 191–224.

Haxeltine, A. & Prentice, I.C. (1996). BIOME3: An equilibrium terrestrial model based on ecophysiological constraints, resource availability, and competition among plant functional types. *Global Biogeochemical Cycles*, **10**, 693–709.

Heusser, C.J. (1952). Pollen profiles from southeastern Alaska. *Ecological Monographs*, **22**, 331–352.

Heusser, C.J. (1960). Late Pleistocene environments of north Pacific North America. New York, *American Geographical Society Special Publication*, 35 pp.

Heusser, C.J. (1965). A Pleistocene phytogeographical sketch of the Pacific Northwest and Alaska. *In*: Wright,

H.E., Jr. & Frey, D.G. (Eds), *The Quaternary of the United States*. Princeton, Princeton University Press, pp. 469–483.

Heusser, C.J. (1973). Postglacial palynology of Umnak Island, Aleutian Islands, Alaska. *Review of Palaeobotany and Palynology*, **15**, 277–285.

Heusser, C.J. (1978). Postglacial vegetation of Adak Island, Aleutian Islands, Alaska. *Bulletin of the Torrey Botanical Club*, **15**, 18–23.

Heusser, C.J. (1983a). Holocene vegetation history of the Prince William Sound region, south-central Alaska. *Quaternary Research*, **19**, 337–355.

Heusser, C.J. (1983b). Pollen diagrams from the Shumagin Islands and adjacent Alaska Peninsula, southwestern Alaska. *Boreas*, **12**, 279–295.

Heusser, C.J. (1985). Quaternary pollen records from the Pacific Northwest coast: Aleutians to the Oregon-California boundary. *In*: Bryant, V.M., Jr. & Holloway, R.G. (Eds), *Pollen Records of Late Quaternary North American Sediments*. Dallas, American Association of Stratigraphic Palynologists, pp. 141–166.

Heusser, C.J., Heusser, L.E. & Peteet, D.M. (1985). Late Quaternary climatic change on the American North Pacific coast. *Nature*, **315**, 485–487.

Höfle, C., Edwards, M.E., Hopkins, D.M., Mann, D.M. & Ping, C.-L. (2001). The full-glacial environment of the northern Seward Peninsula, Alaska, reconstructed from the 21,500-year-old Kitluk paleosol. *Quaternary Research*, **53**, 143–153.

Hopkins, D.M. (1959). Some characteristics of the climate in forest and tundra regions in Alaska. *Arctic*, **12**, 215–220.

Hopkins, D.M. (Ed.) (1967). *The Bering Land Bridge*. Stanford, Stanford University Press.

Hopkins, D.M. (1982). Aspects of paleogeography of Beringia during the late Pleistocene. *In*: Hopkins, D.M., Matthews, J.V., Jr., Schweger, C.E. & Young, S.B. (Eds), *Paleoecology of Beringia*. New York, Academic Press, pp. 3–28.

Hopkins, D.M., MacNeil, F.S. & Leopold, E.B. (1960). The coastal plain at Nome, Alaska: A late Cenozoic type section for the Bering Strait region. *Report of the 21st International Geological Congress*, Pt. 4, pp. 44–57.

Hopkins, D.M., Smith, P.A. & Matthews, J.V., Jr. (1981). Dated wood from Alaska and the Yukon: Implications for forest refugia in Beringia. *Quaternary Research*, **15**, 217–249.

Hopkins, D.M., Matthews, J.V., Jr., Schweger, C.E. & Young, S.B. (Eds) (1982). *Paleoecology of Beringia*. New York, Academic Press, 489 pp.

Hu, F.S., Brubaker, L.B. & Anderson, P.M. (1993). A 12,000-yr record of vegetation change and soil development from Wien Lake, central Alaska. *Canadian Journal of Botany*, **71**, 1133–1142.

Hu, F.S., Brubaker, L.B. & Anderson, P.M. (1995). Postglacial vegetation and climate change in the northern Bristol Bay region, southwestern Alaska. *Quaternary Research*, **43**, 382–392.

Hu, F.S., Brubaker, L.B. & Anderson, P.M. (1996). Boreal ecosystem development in the northwestern Alaska Range since 11,000 yr B.P. *Quaternary Research*, **45**, 188–201.

Hu, F.S., Ito, E., Brubaker, L.B. & Anderson, P.M. (1998). Ostracode geochemical record of Holocene climatic change and implications for vegetational response in the northwestern Alaska Range. *Quaternary Research*, **49**, 86–95.

Hu, F.S., Ito, E., Brown, T.A., Curry, B.B. & Engstrom, D.R. (2001a). Pronounced climatic variations in Alaska during the last two millennia. *Proceedings of the National Academy of Science*, **98**, 10552–10556.

Hu, F.S., Finney, B.P. & Brubaker, L.B. (2001b). Effects of Holocene Alnus expansion on aquatic productivity, nitrogen cycling, and soil development in southwestern Alaska. *Ecosystems*, **4**, 358–368.

Hu, F.S., Lee, B.Y., Kaufman, D.S., Yoneji, S., Nelson, D.M. & Henne, P.D. (2002). Response of tundra ecosystem in southwestern Alaska to Younger-Dryas climatic oscillation. *Global Change Biology*, **8**, 1156–1163.

Kaplan, J.O. (2001). Geophysical applications of vegetation modeling. Ph.D. dissertation, Lund, Sweden, Lund University.

Kaplan, J.O., Bigelow, N.H., Bartlein, P.J., Christiansen, T.R., Cramer, W., Harrison, S.P., Matveyeva, N.V., McGuire, A.D., Murray, D.F., Prentice, I.C., Razzhivin, V.Y., Smith, B., Walker, D.A., Anderson, P.M., Andreev, A.A., Brubaker, L.B., Edwards, M.E., Lozhkin, A.V. & Ritchie, J.C. (in press). Climate change and arctic ecosystems II: Modeling paleodata-model comparisons, and future projections. *Journal of Geophysical Research*.

Kutzbach, J.E. & Gallimore, R.G. (1988). Sensitivity of a coupled atmosphere/mixed-layer ocean model to changes in orbital forcing at 9000 years B.P. *Journal of Geophysical Research*, **93**(D1), 803–821.

Kutzbach, J.C. & Guetter, P.J. (1996). The influence of changing orbital parameters and surface boundary conditions on climate simulations for the past 18000 years. *Journal of Atmospheric Sciences*, **43**, 1726–1759.

Lamb, H.F. & Edwards, M.E. (1988). The Arctic. *In*: Huntley, B. & Webb, T., III (Eds), *Vegetation History*. Boston, Kluwer Academic Press, pp. 519–555.

Levis, S., Foley, J.A. & Pollard, D. (1999). Potential high-latitude vegetation feedbacks on CO_2-induced climate change. *Geophysical Research Letters*, **26**, 747–750.

Livingstone, D.A. (1957). Pollen analysis of a valley fill near Umiat, Alaska. *American Journal of Sciences*, **255**, 254–260.

Livingstone, D.A. (1955). Some pollen profiles from arctic Alaska. *Ecology*, **36**, 587–600.

Lozhkin, A.V., Anderson, P.M., Eisner, W.R., Hopkins, D.M. & Brubaker, L.B. (1996). The change of vegetation cover of western Alaska during the 18000 years. *In*: Bichkov, Y.M. (Ed.), *Quaternary Climates and Vegetation of Beringia*. Magadan, Russia, North East Interdisciplinary Research Institute, Far East Branch, Russian Academy of Sciences, pp. 31–42. In Russian.

Lozhkin, A.V., Hopkins, D.M., Solomatkina, T.B., Eisner, W.R. & Brigham-Grette, J. (1998). Radiocarbon dates and palynological characteristics of peat from St. Lawrence

Island, Alaska. *In*: Simakov, K.V. (Ed.), *Environmental Changes in Beringia During the Quaternary*. Magadan, Russia, North East Interdisciplinary Research Institute, Far East Branch, Russian Academy of Sciences, pp. 9–27. In Russian.

Mann, D.H. (1983). The Quaternary history of the Lituya glacial refugium, Alaska. Ph.D. dissertation, Seattle, University of Washington.

Mann, D.H. (1986). Reliability of a fjord glacier's fluctuations for paleoclimatic reconstructions. *Quaternary Research*, **25**, 10–24.

Mann, D.H. & Hamilton, T.D. (1995). Late Pleistocene and Holocene paleoenvironments of the north Pacific coast. *Quaternary Science Reviews*, **14**, 449–471.

Mann, D.H., Fastie, C.L., Rowland, E.L. & Bigelow, N.H. (1995). Spruce succession, disturbance, and geomorphology on the Tanana River floodplain, Alaska. *Ecoscience*, **2**, 184–199.

Mann, D.H., Crowell, A.L., Hamilton, T.D. & Finney, B.F. (1998). Holocene geologic and climatic history around the Gulf of Alaska. *Arctic Anthropology*, **35**, 112–131.

Mann, D.H., Peteet, D.M., Reanier, R.E. & Kunz, M.L. (2002). Responses of an arctic landscape to Lateglacial and early Holocene climatic changes: The importance of moisture. *Quaternary Science Reviews*, **21**, 997–1021.

Matthews, J.V., Jr. (1974). Quaternary environments at Cape Deceit (Seward Peninsula Alaska) evolution of a tundra ecosystem. *Geological Society of America Bulletin*, **85**, 1353–1384.

McCulloch, D.S. & Hopkins, D.M. (1966). Evidence for an early Recent warm interval in northwestern Alaska. *Geological Society of America Bulletin*, **77**, 1089–1108.

Mckenzie, G.D. (1970). Glacial geology of Adams Inlet, southeastern Alaska. *Columbus, Institute of Polar Studies Report*, 25. Ohio State University.

Miller, R.D. & Dobrovolny, E. (1959). Surficial geology of Anchorage and vicinity, Alaska. *U.S. Geological Survey Bulletin*, 1093, 128 pp.

Mock, C.J., Bartlein, P.J. & Anderson, P.M. (1998). Atmospheric circulation patterns and spatial climatic variations in Beringia. *International Journal of Climatology*, **18**, 1085–1104.

Nelson, R.E. & Carter, L.D. (1987). Paleoenvironmental analysis of insects and extralimital Populus from an early Holocene site on the Arctic Slope of Alaska. *Arctic and Alpine Research*, **19**, 230–241.

Oswald, W.W., Brubaker, L.B. & Anderson, P.M. (1999). Late Quaternary vegetational history of the Howard Pass area, northwestern Alaska. *Canadian Journal of Botany*, **77**, 570–581.

Oswald, W.W. (2002). Holocene vegetational history of the central Arctic Foothills, northern Alaska: Pollen representation of tundra and edaphic controls on the response of tundra to climate change. Ph.D. dissertation, Seattle, University of Washington, 152 pp.

Peteet, D.M. (1986). Vegetational history of the Malaspina Glacier district. *Quaternary Research*, **25**, 100–120.

Peteet, D.M. (1991). Postglacial migration history of lodgepole pine near Yakutat, Alaska. *Canadian Journal of Botany*, **69**, 786–796.

Peteet, D.M. & Mann, D.H. (1994). Late-glacial vegetational, tephra, and climatic history of southwestern Kodiak Island, Alaska. *Ecosciences*, **1**, 255–267.

Peteet, D.M., Del Genio, A. & Lo, K.K.W. (1997). Sensitivity of northern hemisphere air temperatures and snow expansion to north Pacific sea surface temperatures in the Goddard Institute for Space Studies general circulation model. *Journal of Geophysical Research*, **102**, 23781–23791.

Prentice, I.C., Guiot, J., Huntley, B., Jolly, D. & Cheddadi, R. (1996). Reconstructing biomes from paleoecological data: A general method and its application to European pollen data at 0 and 6 ka. *Climate Dynamics*, **12**, 185–194.

Rupp, T.S., Chapin, F.S., III & Starfield, A.M. (2000a). Response of subarctic vegetation to transient climatic change on the Seward Peninsula in north-west Alaska. *Global Change Biology*, **6**, 541–555.

Rupp, T.S., Starfield, A.M. & Chapin, F.S., III (2000b). A frame-based spatially explicit model of subarctic vegetation response to climatic change: Comparison with a point model. *Landscape Ecology*, **15**, 383–400.

Schweger, C.E. (1981). Chronology of late glacial events from the Tangle Lakes, Alaska Range. *Arctic Anthropology*, **18**, 97–101.

Schweger, C.E. (1982). Late Pleistocene vegetation of eastern Beringia: Pollen analysis of dated alluvium. *In*: Hopkins, D.M., Matthews, J.V., Jr., Schweger, C.E. & Young, S.B. (Eds), *Paleoecology of Beringia*. New York, Academic Press, pp. 95–112.

Short, S.K., Elias, S.A., Waythomas, C.F. & Williams, N.E. (1992). Fossil pollen and insect evidence for postglacial environmental conditions, Nushagak and Holitna lowland regions, southwest Alaska. *Arctic*, **45**, 381–392.

Smith, T.M. & Shugart, H.H. (1993). The transient response of terrestrial carbon storage to a perturbed climate. *Nature*, **361**, 523–526.

TEMPO (1996). Potential role of vegetation feedback in the climate sensitivity of high-latitude regions, a case study at 6000 years B.C. *Global Biogeochemical Cycles*, **10**, 727–736.

Viereck, L.A. & Little, E.L., Jr. (1975). *Atlas of United States trees: Alaska trees and common shrubs*. Washington, D.C., U.S. Department of Agriculture, Vol. 2.

Wahrhaftig, C. (1965). Physiographic divisions of Alaska. *U.S. Geological Survey*, Paper 482, 52 pp.

Watson, C.E. (1959). *Climates of the states, Alaska*. U.S. Weather Bureau, Climatography of the United States, No. 60–49.

Wiles, G.C., Calkin, P.E. & Post, A. (1995). Glacier fluctuations in the Kenai fjords, Alaska, U.S.A.: An evaluation of controls on iceberg-calving glaciers. *Arctic and Alpine Research*, **27**, 234–245.

Wiles, G.C., Jacoby, G.C., Davi, N.K. & McAllister, R.P. (2002). Late Holocene glacier fluctuations in the Wrangell Mountains, Alaska. *Geological Society of America Bulletin*, **114**, 896–908.

Quaternary history from the U.S. tropics

Sara Hotchkiss

Department of Botany and Center for Climatic Research, University of Wisconsin, 430 Lincoln Drive, Madison, WI 53706, USA

Introduction

Because tropical climate exerts a controlling influence on general circulation, understanding the response of the tropics to climate forcing is critical for predicting global climatic change. The tropical latitudes also house a large proportion of the Earth's biological and cultural diversity and face some of the greatest rates of human population increase, economic development, and land use change in the world today. The interactions of human cultures with ecosystem processes, sea level, and climate are complex – we influence our environment, while changes in the environment influence our cultures. These interactions make the future of the tropics difficult to predict, and the influence of the tropical latitudes on global climate and biological and cultural diversity makes their history particularly important to understand.

Thirty years ago, the prevailing understanding was that the tropics had been relatively stable for millions of years, allowing a gradual accumulation of new species to generate extremely high species diversity. New data and new strategies for dating, environmental reconstruction, and climate modeling developed over the last 30 years have greatly improved understanding of tropical climate and vegetation history, human migrations, and other environmental changes during the Quaternary. This chapter summarizes a few major issues in tropical climate history, sea level, and biogeography, focusing on data from islands in the Caribbean and the Pacific Ocean associated with the United States (Table 1). These scattered bits of land are not a geographically coherent entity, but they have contributed valuable information and methodological advances to studies of Quaternary history of the tropics. Much of the discussion focuses on the Hawai'ian Islands, which have been studied in the greatest detail, but islands in the central and southern Pacific and the Caribbean are considered where data are available.

Climate History

Tropical Temperatures During the Last Glacial Maximum

The magnitude of tropical temperature change during glacial periods has been one of the most controversial facets of Quaternary climate history. The CLIMAP (1981) project's reconstruction of last glacial maximum sea surface temperature (SST) has been a benchmark against which accumulating evidence and model simulations are measured. CLIMAP found little change in tropical SST during the last glacial maximum (LGM), with temperatures even slightly warmer than present in the Pacific Ocean gyres.

Several lines of evidence from tropical locations have challenged the CLIMAP (1981) reconstruction, collectively suggesting that LGM tropical temperatures were considerably cooler than present. High tropical mountains supported LGM glaciers with equilibrium line altitudes about 900 ± 135 m below present, which would require about 4–5 °C cooling under constant precipitation (Broecker, 1997; Porter, 2001; Rind & Peteet, 1985; Thompson, 2000). Temperature reconstructions from noble gases in groundwater (Stute *et al.*, 1992, 1995) and geochemistry of corals (Beck *et al.*, 1992, 1997; Guilderson *et al.*, 1994) suggest that low-elevation tropical temperatures were about 5 °C cooler than present during the LGM. Terrestrial pollen data from many sites show an elevation shift in vegetation near sea level of about 500–700 m, or about 2.9–3.6 °C, during the LGM (Farrera *et al.*, 1999).

While the *magnitude* of cooling in terrestrial data is generally greater than the CLIMAP SST estimates, the *global pattern* of LGM tropical temperatures from terrestrial regions appears consistent with the CLIMAP SST patterns. The largest cooling in terrestrial pollen records appears in the neotropics (5.6–5.7 °C), with moderate cooling in the Indian Ocean region (2.2–2.6 °C), and only slight cooling in the southern Pacific Ocean (1.0–1.1 °C), where slight warming was predicted by CLIMAP (Farrera *et al.*, 1999). A resolution of the differences in magnitude of tropical temperature change from various data sources is important for estimating the sensitivity of global climate to future climate forcing.

The Hawaiian Islands have long been cited as a model of the LGM tropical climate problem (Rind & Peteet, 1985). Located in the central north Pacific, an area CLIMAP reconstructed to be warmer than present during the LGM, Mauna Kea nonetheless supported an ice cap during the last glacial period (Porter, 1979a, b). Two ^{36}Cl dates on boulders from the full-glacial moraine suggest the glacier was at its greatest extent 18,900 and 20,300 ^{36}Cl yr B.P.; deglaciation was probably complete by about 15,000 ^{36}Cl yr B.P. (Dorn *et al.*, 1991; Kaufman *et al.*, 2003, this volume). Full-glacial equilibrium line depression for the Mauna Kea ice cap is estimated at 935 m; if precipitation is held constant, this estimate yields a 4–6 °C drop in high-elevation temperature during the LGM (Kaufman *et al.*, 2003, this volume; Porter, 1979b). In a run of the GENESIS model (Hostetler & Mix, 1999; Mix *et al.*, 1999) in which the SST near Hawai'i was set to the CLIMAP estimates, but temperatures in the eastern tropical Pacific and equatorial Atlantic were lower than CLIMAP values, it was necessary to reduce LGM air

DEVELOPMENTS IN QUATERNARY SCIENCE
VOLUME 1 ISSN 1571-0866
DOI:10.1016/S1571-0866(03)01020-0

Table 1. Tropical land masses associated with the United States, either as states, commonwealths, territories, or under compacts of free association. Land area, approximate number of islands, latitude, and longitude are listed for U.S.-associated lands in Polynesia, Micronesia, and the Caribbean.

	Area (km^2)	Number of Islands	Latitude	Longitude
Polynesia				
Hawai'ian Islands	16,558	132	21 18 N	157 51 W
Midway Islands	6.2	3	28 13 N	177 22 W
American Samoa	199	7	14 20 S	170 00 W
Baker Island	2	1	0 13 N	176 31 W
Howland Island	1.6	1	0 48 N	176 38 W
Jarvis Island	4.5	1	0 22 S	160 03 W
Johnston Atoll	2.8	1	16 45 N	169 31 W
Kingman Reef	1	1	6 24 N	162 24 W
Palmyra Atoll	10	1	5 52 N	162 06 W
Micronesia				
Commonwealth of the Northern Mariana Islands	759	9	15 12 N	145 45 E
Territory of Guam	541	1	13 28 N	144 47 E
Republic of Palau	458	200	7 30 N	134 30 E
Federated States of Micronesia	180	31	6 55 N	158 15 E
Wake Island	6.5	3	19 17 N	166 36 E
Caribbean				
Puerto Rico	8959	7+	18 15 N	66 30 W
United States Virgin Islands	349	6	18 20 N	64 50 W
Navassa Island	5.2	1	18 25 N	75 02 W

temperatures by 2.0–3.4 °C to obtain a Mauna Kea ice cap of the reconstructed size (Hostetler & Clark, 2000).

A high-resolution analysis of ocean sediments near Hawai'i suggests that the Pacific temperature problem may have been a result of the suite of sediments available to the CLIMAP project. All of the CLIMAP (1981) cores were deep-ocean sediments affected by the corrosive Pacific Deep Water, and calcite dissolution, bioturbation of slowly accumulating sediments, and low dating resolution may have affected the faunal assemblages used to reconstruct temperature history (Lee & Slowey, 1999). Modern analog technique interpretations of planktonic foraminifera from a shallower, calcite-saturated site on the submarine flank of O'ahu show cooling of ~1.5 °C in August and nearly 3 °C in February during the LGM. The $\delta^{18}O$ record of the planktonic foraminifer *Globigerinoides ruber* from the same core has been interpreted to reflect LGM cooling of 1–2 °C (Lee & Slowey, 1999). A pollen record from 463 m elevation (~580 m during the LGM low sea level stand) on O'ahu, interpreted using indices based on elevational range limits of plants and pollen response surfaces, implies cooling of about 2–4 °C during the LGM (Figs 1 and 2; Hotchkiss & Juvik, 1999).

A survey of mean temperature of the coldest month from global tropical terrestrial data shows that LGM lapse rates were steeper than present where data are sufficient to test for a change (Farrera *et al.*, 1999). The midpoints of the three available Hawai'ian temperature estimates – temperature depression of 1–3 °C at the ocean surface (Lee & Slowey, 1999), 2–4 °C at ca. 580 m LGM elevation (Hotchkiss & Juvik, 1999), and 4–6 °C at about 3780 m LGM elevation on Mauna Kea (Porter, 1979b) – yield a lapse rate of about 5.4 °C/km, only slightly steeper than the modern overall lapse rate of about 5.3 °C/km. Actual modern lapse rates are affected by the persistent trade wind temperature inversion, and are about 6.9 °C/km below about 1500 m and 3.3 °C/km above (Hotchkiss *et al.*, 2000), but the existing temperature estimates from the Hawai'ian Islands are too sparse to test for a change in lapse rate with elevation during the LGM.

Late Quaternary Regional Circulation in the North Pacific

Several lines of evidence suggest that the northeast trade winds persisted during the late Quaternary in the Hawaiian Islands. Orientations of lithified sand dunes formed during glacial low sea level stands indicate northeasterly wind directions, as do cinder cones deposited during low sea stands on O'ahu (Winchell, 1947). Cinder cones on Mauna Kea indicate persistent trade winds during the final eruptive phase. Many cones are asymmetrical towards the west, suggesting somewhat more easterly winds than the present northeasterly pattern (Porter, 1972). Loess and dune sands on the northwest flank of Mauna Kea show two phases of deposition, with the older phase beginning after 103,000 ± 10,000 K-Ar yr ago. A paleosol separating the two phases of loess deposition dates at about 48,000 ^{14}C yr B.P., while the youngest ages from the loess are about 17,000–18,000 ^{14}C yr B.P (ca.

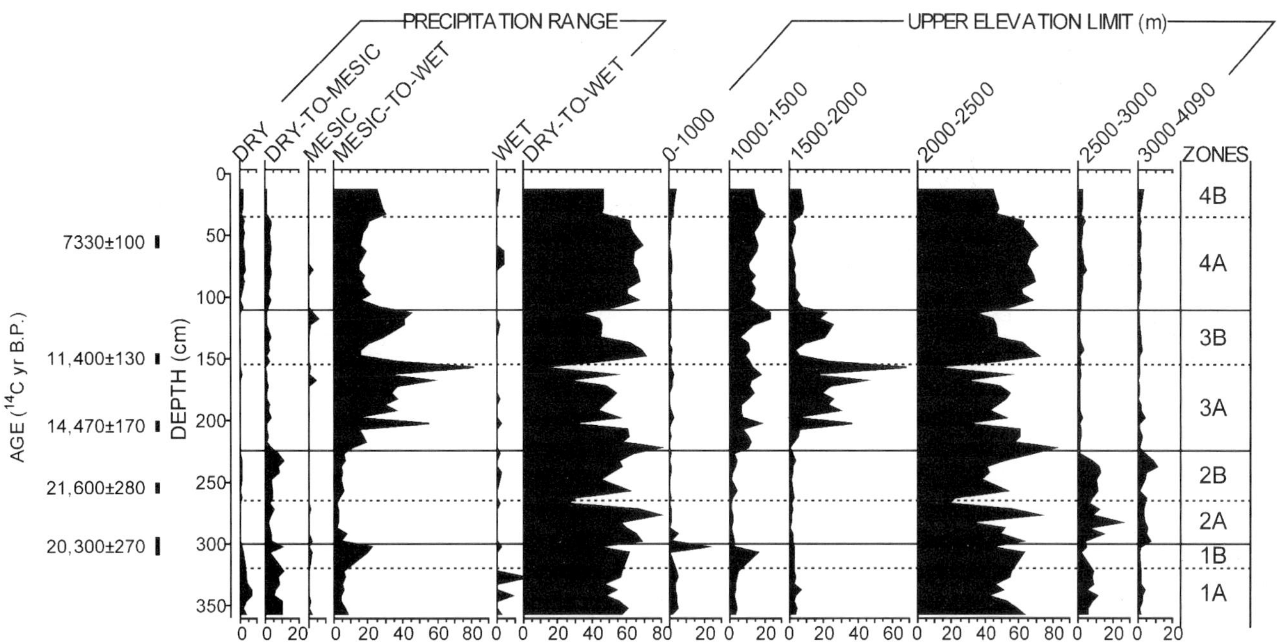

Fig. 1. Summary pollen percentage diagram from Ka'au Crater, O'ahu. Flowering plant pollen percentages are summed according to the modern precipitation and elevation ranges of the taxa they include. Zones are derived by CONISS stratigraphically constrained sum-of-squares cluster analysis of pollen assemblages; the last glacial maximum falls within zone 2 (Hotchkiss & Juvik, 1999). The uppermost three samples follow a sedimentary hiatus and contain pollen of European introductions.

20,000–22,000 cal yr B.P.). Average paleowind directions from the dunes and loess sequences agree with cinder cone asymmetry, suggesting dominance by easterly winds from about 48,000–20,000 years ago, and a depositional environment that was more arid than it is today (Porter, 1973, 1997).

Soils on leeward Kohala Mountain, Hawai'i, also suggest consistently arid conditions over the past 350,000 years (Chadwick *et al.*, in press). The oxygen isotopic composition of halloysite and soil carbonates formed during the entire period of pedogenesis on leeward Kohala corresponds well with the composition of modern soil water in these presently arid sites (Hsieh, 1997; Hsieh *et al.*, 1998; Ziegler *et al.*, in press). Whereas leeward Kohala soils appear to have been under arid conditions throughout their formation, soils on the wetter slopes near the summit of Kohala may have spent much of their pedogenic history under conditions drier than present (Hotchkiss *et al.*, 2000). If so, the past 10,000 years of wetter conditions have been sufficient to reset soil properties by erasing the mineralogical and chemical traces

Fig. 2. Pollen abundance index estimates of median annual precipitation and mean annual temperature for fossil pollen assemblages. Stars mark samples containing taxa introduced within the last 200 years. Age is estimated using linear interpolation of uncalibrated radiocarbon dates. The line is a 3-point running average of the sample values. From Hotchkiss & Juvik (1999).

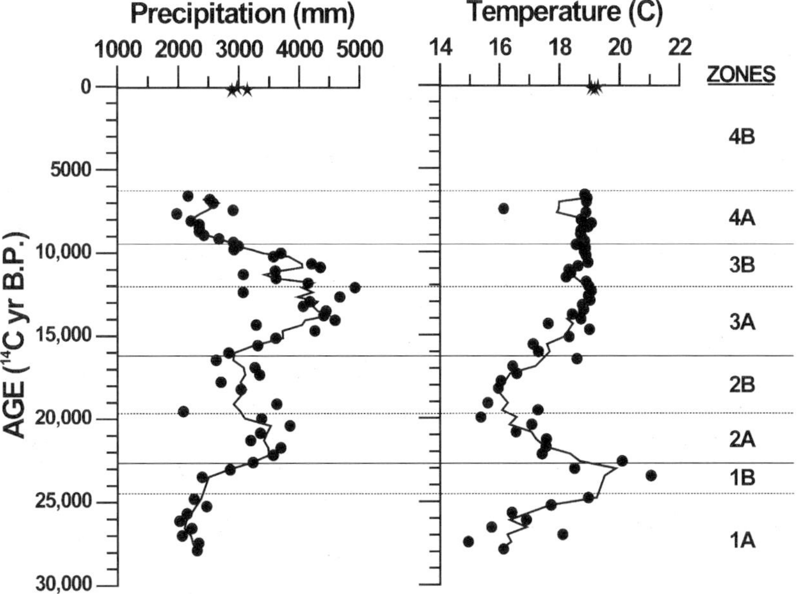

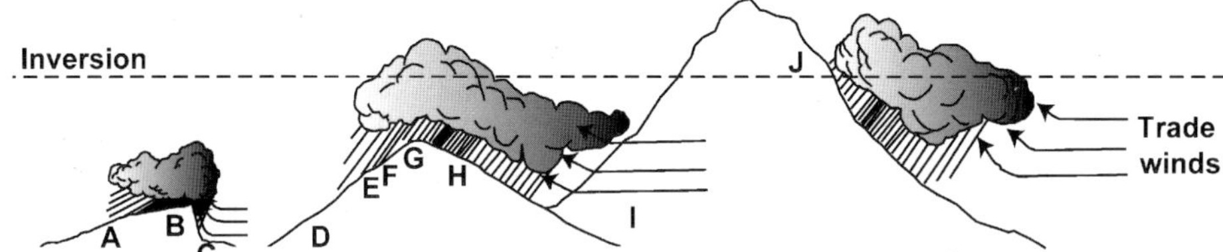

Fig. 3. Positions of Hawai'ian sites mentioned in the text, relative to schematic representations of mountains, trade winds, and the trade wind inversion. Site A represents central O'ahu (Gavenda, 1992; Ruhe, 1964), site B represents Ka'au Crater (Hotchkiss & Juvik, 1999), site C represents the north shore of O'ahu (Athens, 1997; Athens & Ward, 1995; Athens et al., 1995), site D represents lowland leeward Kohala (Chadwick et al., in press; Hsieh, 1997; Hsieh et al., 1998; Ziegler et al., in press), sites E–H represent upland Kohala pollen records (Hotchkiss, 1998; S. Hotchkiss, unpublished data), site I represents lowland leeward Maui sites (Athens, 1997; Kolb & Murakami, 1994; Nagahara, 1995), and site J represents Flat Top Bog (Burney et al., 1995).

of past dry environments (Chadwick *et al.*, in press). This conclusion agrees with observations of a chronosequence of Hawaiian soils under similar rainfall conditions, in which the primary minerals are largely exhausted on all substrates older than 20,000 years (Vitousek *et al.*, 1997).

Other soils and paleoecological records paint a more complex picture of past Hawaiian climates. Oxisols in the Wahiawa Basin, central O'ahu, record greater weathering than occurs under the current semi-arid climate (Gavenda, 1992). Remains of trees and palms in the Salt Lake Tuff, deposited during a low sea stand 350,000 years ago (Lyon, 1930), and tree fern molds in full- and late-glacial ash deposits on Kilauea Volcano (Easton, 1987), imply wetter conditions at low-elevation sites during glacial periods. A sample from likely late-glacial sediments from lowland leeward Maui contains pollen of mesic to wet vegetation types, while early Holocene samples from the same core show drier vegetation (Athens, 1997).

Four middle-elevation sites on Kohala Mountain (sites E–H, Fig. 3; Hotchkiss, 1998 and unpublished data) and one on Maui (site J, Fig. 3; Burney *et al.*, 1995) preserved little or no peat during the last glacial maximum. The parsimonious explanation for this lack of peaty sediment is that conditions were dry. Drier conditions are also implied by open grassland vegetation during middle- and late-glacial time at the Kohala summit and two leeward Kohala sites that were surrounded by wet forest during the Holocene (sites E–G, Fig. 3; Hotchkiss, 1998, S. Hotchkiss, unpublished data). Early Holocene pollen assemblages that may contain older pollen support the hypothesis of a drier last glacial maximum for high-elevation Flat Top Bog, Maui (site J, Fig. 3; Burney *et al.*, 1995), lowland leeward Maui (site I, Fig. 3; Athens, 1997) and windward Kohala (site H, Fig. 3; S. Hotchkiss, unpublished data). Ka'au Crater, at ~580 m elevation on O'ahu, accumulated sediment during the last glacial period. The reconstructed vegetation suggests somewhat drier conditions during the last glacial maximum (Figs 1 and 2 and site B, Fig. 3; Hotchkiss & Juvik, 1999).

Drier full-glacial conditions at middle- and high-elevation windward sites are not necessarily inconsistent with wetter conditions in semi-arid low-elevation leeward sites. Presently, the trade winds encounter the islands from the northeast, and orographic uplift produces increasing rainfall with elevation on the windward side below about 2000 m elevation (ranging from 1500–3000 m; Giambelluca & Schroeder, 1998), where orographic precipitation is blocked by the trade wind inversion (Fig. 3). Leeward areas remain relatively dry and receive most of their annual precipitation during storms, when the trade winds and inversion are weak or absent. A 120-m drop in sea level during the LGM would have effectively shifted present rainfall patterns downward, and could produce apparent changes in climate that are actually the result of changes in the relationship of a particular place on an island to the predominant orographic precipitation patterns. A 100-m lowering of sea level could produce a more than 100% increase in precipitation in parts of central O'ahu due to increased elevation relative to sea level and greater exposure to trade wind precipitation (Gavenda, 1992; Ruhe, 1964). Similarly, a drop in the elevation of the inversion could decrease precipitation at presently wet sites just below the present modal elevation of the inversion, or leeward of summits near that elevation (sites E–H, Fig. 3; Hotchkiss, 1998; S. Hotchkiss, unpublished data). Low-elevation leeward sites would have remained dry regardless of changes in the inversion or sea level, unless there was a major shift in average wind direction (sites D and I, Fig. 3).

Holocene records from Hawai'i also produce a mixed but coherent picture. Flat Top Bog, on Haleakala, Maui, records changes in vegetation at 2270 m elevation since about 10,600 cal yr B.P. (9380 ± 70 ^{14}C yr B.P.; site J, Fig. 3; Burney *et al.*, 1995). The site is now surrounded by grasslands in the relatively dry climate above the trade wind inversion. Conditions were similar to present between 10,600 and 6600 cal yr B.P., but between about 6600 and 2300 cal yr B.P. (5800–2200 ^{14}C yr B.P.) pollen of forest taxa predominated, interrupted by a thick tephra layer. After about 2200 cal yr B.P., pollen of grass and sedges again dominates the assemblages, suggesting a return to the drier conditions above the trade wind inversion. The change from grassland/shrubland to fern-dominated vegetation with very little grass implies an increase in the elevation of the trade wind inversion between about 6600 and 2200 cal yr B.P. (Burney *et al.*, 1995). Two low-elevation cores from the north shore of O'ahu and one from urban Honolulu show drier conditions 5500–2200 and 3500–2200 cal yr B.P. (sites C and A, Fig. 3;

Athens, 1997; Athens & Ward, 1993; Athens *et al.*, 1995). These results are consistent with an upward shift in both sea level and orographic precipitation patterns. They may also be consistent with long-duration El Niño-Southern Oscillation events (Diaz & Markgraf, 1993), which would produce dry winters in Hawai'i, disproportionately affecting leeward areas where most of the annual precipitation occurs in the winter.

South Pacific and Caribbean Climate History

High-resolution records of decadal and interannual climate variability in the late Holocene have been obtained recently to assess the frequency and intensity of changes in tropical climate, especially quasi-periodic phenomena with global teleconnections such as the Pacific Decadal Oscillation (PDO) and El Niño-Southern Oscillation (ENSO). Most of these records have come from tropical islands not associated with the U.S., but a few have been documented from U.S.-associated islands of the Pacific and Caribbean.

Context for high-resolution records of Hobcene climate change in the Caribbean is provided by a low-resolution record from sand on the southern shelf of St. Thomas, United States Virgin Islands. The sand is derived mostly from calcareous algae and mollusks, and the faunal composition did not change by much over the past 5000 years – despite dramatic changes throughout the Caribbean since the 1980s (e.g. Aronson & Precht, 2001) – but patterns of water movement shifted and barnacles disappeared, implying that water depth has increased, perhaps as a result of subsidence (Kindinger *et al.*, 1983). $\delta^{18}O$ variations in *Halimeda* pieces in the sand suggest that water temperature may also have increased by about 4 °C during the late Holocene (Holmes, 1983).

A nearby record with greater temporal resolution, in sediment cores from the insular shelf off southwestern Puerto Rico (Nyberg *et al.*, 2001), consists of planktonic foraminifera, magnetic properties, carbonate, and organic carbon. The sediments accumulated during the last 2000 years, and the magnetic properties of minerals have a periodicity of about 200 years over that time. Before about A.D. 850–1000, higher hematite content in the sediments suggests drier conditions and increased atmospheric export of dust from North Africa. Major changes in the geochemical and magnetic records about A.D. 850–1000 suggest an increase in precipitation, in agreement with terrestrial records from the Caribbean and Mesoamerica (Hodell *et al.*, 1991; Hodell *et al.*, 1995; Horn & Sandford, 1992; Metcalfe, 1995; Metcalfe *et al.*, 1994). About A.D. 1300, just before and during the beginning of the Little Ice Age, ratios of organic carbon to total nitrogen were high, suggesting high input from terrestrial sources, probably caused by high precipitation and runoff. Trade wind strength, inferred from the abundance of the planktonic foraminifer *Globigerina bulloides*, was low from about A.D. 1325–1400 (Nyberg *et al.*, 2001). Sea-surface salinity, inferred from $\delta^{18}O$ of the planktonic foraminifer *Globigierinoides ruber* and neural-network estimates of temperature from planktonic foraminiferal assemblages, decreased around A.D. 1400 (Nyberg *et al.*, 2002). The reduced salinity correlates with a rapid increase in inferred trade-wind strength, and total organic carbon increased at the same time, suggesting increased coastal upwelling driven by stronger meridional circulation in the north Atlantic (Nyberg *et al.*, 2001).

A high-resolution analysis of Mg/Ca ratios and $\delta^{18}O$ in coral skeletons from Puerto Rico provides evidence for seasonal changes in sea-surface temperature and salinity over the past several hundred years (Watanabe *et al.*, 2001). The inferred sea-surface temperature was about 2 °C lower between A.D. 1699 and 1703, and sea-surface salinity was more seasonal during those years.

Two other U.S.-affiliated Caribbean sites have contributed to the development of methods for reconstructing high-resolution climate records from corals. High-resolution samples from *Montastrea annularis* near La Parguera, Puerto Rico, show that intra-annual changes in $\delta^{18}O$ correlate well with water temperature records over the period A.D. 1964–1982. Changes in mean annual values of $\delta^{13}C$ and $\delta^{18}O$ show similar trends and may reflect interannual variability in North Atlantic circulation (Winter *et al.*, 1991). Plutonium from nuclear testing peaked in coral years A.D. 1959 and 1964, compared with natural radionuclides in 30 annual growth bands created by a coral from the U.S. Virgin Islands between A.D. 1951 and 1980 (Benninger & Dodge, 1986), demonstrating that corals can preserve a high-resolution record of atmospheric plutonium fallout.

High-resolution coral records from the tropical Pacific are contributing new perspective on interannual and decadal climate variability associated with ENSO and PDO. A 112-year record of $\delta^{18}O$ in a *Porites* coral from Palmyra Island, in the central tropical Pacific, demonstrates decadal climate variability with amplitude of about 0.3 °C in sea-surface temperature, and a strong warming trend since the middle 1970s (Cobb *et al.*, 2001). The decadal pattern is coherent with equatorial climate records from the Atlantic and Indian oceans and suggests the influence of the central tropical Pacific on global decadal climate variability over the last century (Cobb *et al.*, 2001).

The Palmyra record complements several high-resolution coral records from Pacific islands not affiliated with the U.S. A 96-year $\delta^{18}O$ *Porites* record from Tarawa Atoll in the Gilbert Islands (Cole *et al.*, 1993) suggests that ENSO weakened between A.D. 1930 and 1965. The Tarawa $\delta^{18}O$ record is coherent with an index of central Pacific rainfall (Wright, 1989) and a Galápagos coral $\delta^{18}O$ record (Shen *et al.*, 1992), but records fewer ENSO cool extremes in the central equatorial Pacific than in the Galápagos. A 271-year SST record derived from Sr/Ca variability in a *Porites* coral core from Rarotonga, in the South Pacific gyre, shows a pattern of decadal and interdecadal SST change with amplitude >0.75 °C. Several of the largest decadal SST variations at Rarotonga are coherent with the PDO index in the North Pacific over the period A.D. 1900–1997 (Linsley *et al.*, 2000). *Porites* can harbor algal symbionts, which can alter the incorporation of Sr into aragonite lattice, overprinting the SST signal with a signal more closely related to photosynthetic activity of the symbiont (Cohen *et al.*, 2002). Linsley *et al.* (2000) demonstrated coherence between SST and Sr/Ca ratios over the period A.D. 1960–1997, while $\delta^{18}O$ explained significantly less of the SST variance in a linear least-squares regression; however, methods for detecting the presence of algal symbionts over time have yet to be developed.

A comparison of 15–45 year *Pavona* and *Porites* coral $\delta^{18}O$ records from coastal Panama, Costa Rica, and the Galápagos shows a strong correlation with ENSO in the Galápagos where salinity varies little, and a significant positive correlation between $\delta^{18}O$ and $\delta^{13}C$, except where water column clarity varies greatly (Wellington & Dunbar, 1995). $\delta^{18}O$ in two *Porites* corals from Nauru records interannual variability reflecting the Pacific warm pool, the Indonesian Low, and ENSO variation, but only one of the two $\delta^{13}C$ records correlates with cloud cover records which should affect photosynthesis of algal symbionts (Guilderson & Schrag, 1999).

High-resolution coral records have the potential to provide floating chronologies of tropical climate variability throughout the Quaternary. Sr/Ca records from uplifted coral terraces on the Huon Peninsula, Papua New Guinea, record SST 2–3 °C cooler than present during the early Holocene, with annual fluctuations of ± 1 °C and occasional excursions of ± 2 °C, suggesting more frequent strong ENSO events (McCulloch *et al.*, 1996). These high-resolution coral "windows" on past climates show promise for improving understanding of interannual, decadal, and even seasonal climate variability in the context of global climatic change.

Sea Level

Many of the tropical islands associated with the U.S. have contributed to the record of relative sea-level changes in the Caribbean and the Pacific. Data from the Hawai'ian Islands are most abundant. Harold T. Stearns and his coworkers mapped the distribution of dozens of past shorelines in the Hawaiian Islands (Stearns, 1935, 1939, 1966, 1973, 1974, 1975, 1978; many other authors are cited in Stearns, 1978). This monumental work identified dozens of emergent or submerged reef limestones, beach deposits, wave-cut notches, and platforms, and other evidence for Pleistocene shorelines. Figure 4 is Stearns' summary of the positions of 24 mapped Pleistocene shorelines from the Hawaiian Islands in relation to sea level fluctuations (Stearns, 1978).

Interglacial Sea-level Maxima

Limestones deposited during the last interglacial sea-level maximum now stand about 7.5 m above mean sea level in Hawai'i (Lum & Stearns, 1970; Muhs & Szabo, 1994; Stearns, 1939, 1974, 1978). Stearns (1978) described the Waimanalo and Leahi limestones on O'ahu as two distinct shorelines above sea level, separated by a regression (Fig. 4). Uranium-series ages range from 137,000 ± 8000 yr to 112,000 ± 6000 yr (Ku *et al.*, 1974; Stearns, 1974). Several workers have interpreted the U-series ages in two groups, with cemented corals averaging about 133,000 yr, and coral conglomerates averaging about 119,000 yr (Chappell, 1983; Chappell & Shackleton, 1986; Chappell & Veeh, 1978; Moore, 1982; Smart & Richards, 1992; Stearns, 1976). To test the evidence for a two-stage interglacial high stand, Muhs & Szabo (1994) remapped the Waimanalo formation east of Kaena point. They collected multiple well-preserved specimens of *Pocillopora* and *Porites* and dated them with uranium-series methods. The results support a long last interglacial high stand (Muhs & Szabo, 1994), in agreement with work by Kauffman (1986), Chen *et al.* (1991), and Muhs *et al.* (1994), but they do not conclusively rule out the presence of multiple higher-resolution sea-level oscillations during the high stand.

A carbonate section at Barbers Point, O'ahu, preserves a continuous last-interglacial sequence including two transgressions conformably separated by a regression within marine oxygen isotope stage (MIS) 5e (Sherman *et al.*, 1993). Elsewhere, the Waimanolo formation is truncated by an erosional unconformity, which Sherman *et al.* (1993) ascribe to the mid-MIS 5e regression. The two-stage last interglacial sea-level high recorded at Barbers Point agrees with interpretations of emergent coral reefs from the Huon Peninsula in New Guinea (Bloom *et al.*, 1974; Chappell, 1974; Chappell & Veeh, 1978; Moore, 1982; Stein *et al.*, 1993), Baja California (Sirkin *et al.*, 1990), and submerged speleothems in the Bahamas (Richards *et al.*, 1994). Last interglacial sea-level records from western

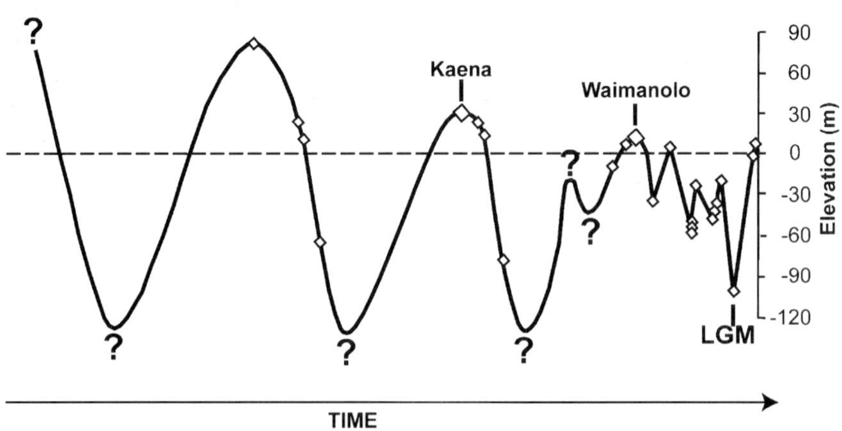

Fig. 4. Stearns' (1978) summary of past sea levels in the Hawaiian Islands. Last glacial maximum, last-interglacial Waimanolo limestones, and the earlier interglacial Kaena high stand are labeled and discussed in the text.

Australia, the Bahamas, Barbados, and the Huon Peninsula show rapid sea-level rise at the beginning of the last interglacial, with the last 80 m of sea-level rise occurring in only a few thousand years (McCulloch & Esat, 2000).

The Waimanolo limestones stand about 1.5 m higher than most other last interglacial sea-level data. If the last interglacial mean sea level was +6 m (e.g. Bloom *et al.*, 1974; Broecker *et al.*, 1968; Chen *et al.*, 1991; Dodge *et al.*, 1983; Harmon *et al.*, 1983; Lindberg *et al.*, 1980; Mattews, 1973; Ota & Omura, 1992), O‘ahu has been since uplifted at an average rate of 0.03–0.05 mm/yr, probably due to lithospheric flexure created by volcanic loading on the island of Hawai‘i (Muhs & Szabo, 1994). Lithospheric modeling of the Hawaiian Islands is in agreement, placing O‘ahu near the zone of lithospheric uplift due to loading on Hawai‘i (Watts & ten Brink, 1989).

The +30 m Kaena high stand reported by Stearns (1973, 1978; Fig. 4) on O‘ahu may correlate with MIS 11 (Hearty, 2002), 13, 15, or 17 (Muhs & Szabo, 1994). Dates for the Kaena high stand based on amino acid racemization, uranium/thorium, and electron spin resonance fall in the range 300,000–550,000 yr (Brückner & Radke, 1989; Hearty, 2002; Hearty *et al.*, 2000; Jones, 1993). One section on western O‘ahu suggests a progressive rise in sea level, with three stages (+5–6 m, +13.5 m, and +28 m) correlating with estimates from Bermuda and the Bahamas (Hearty, 2002). The maximum MIS-11 sea level in the Bahamas and Bermuda is +20 m; attributing 20 m of the Hawaiian data to eustatic sea-level rise leaves 8 m to be explained by lithospheric uplift. This estimate gives O‘ahu a long-term uplift rate of about 0.02 mm/yr over the past ca. 400,000 yr, slightly less than the estimates of about 0.03–0.05 mm/yr over the past 125,000 yr from MIS 5e Waimanolo limestones (Hearty, 2002; Muhs & Szabo, 1994).

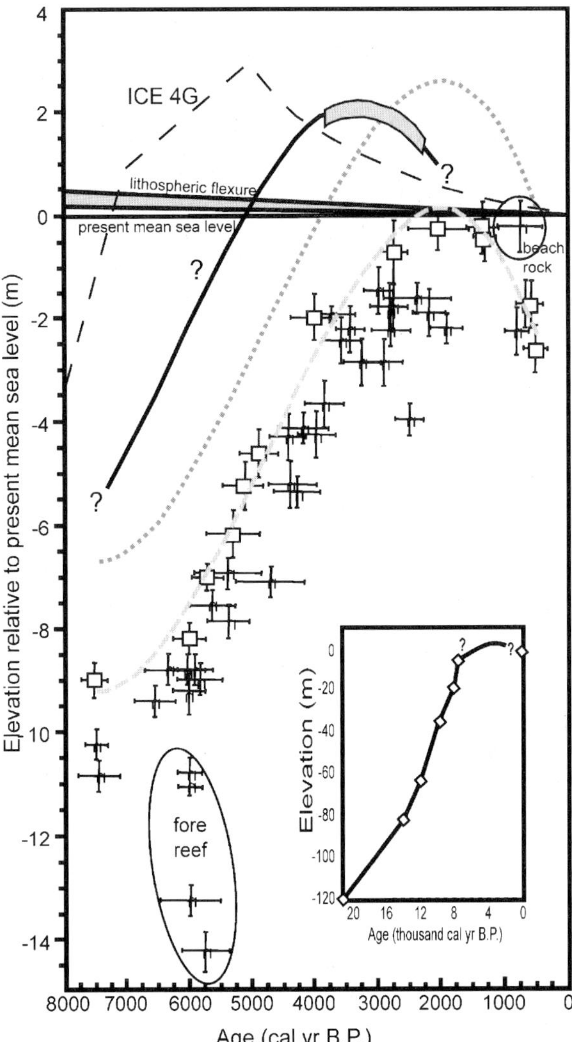

Fig. 5. Hawaiian sea-level history since the last glacial maximum (inset) and during the Holocene (main panel), after Grossman & Fletcher (1998) and Grigg et al. (2002). Dark symbols are dates on reef cores and open symbols are dates on reef surface samples, both from Haunama Bay (Easton & Olson, 1976). Grey dashed line is a best fit third-degree polynomial estimate of the Haunama reef growth curve (Grossman & Fletcher, 1998). Grey dotted line is a correction for coral habitat position (Montaggioni, 1988), black solid line is proposed sea level, incorporating results of a study of Kapapa Island (Grossman & Fletcher, 1998), and in agreement with studies from windward O‘ahu (Fletcher & Jones, 1996) and Kaua‘i (Calhoun & Fletcher, 1996). The thin dashed line is a reconstruction of sea level using the ICE-4G data set. Figure redrawn from Grossman & Fletcher (1998). Inset includes proposed sea level from the larger panel and sea-level estimates from submerged reefs and limestones in the Au‘au Channel (Grigg et al., 2002).

Late-Glacial and Holocene Sea Level History

Submerged reefs and karst below the Au‘au channel, between Maui and Lana‘i, suggest about 120 m of sea-level rise since the LGM (inset, Fig. 5; Grigg *et al.*, 2002). About 14,000 cal yr B.P., sea level stood at −82 m, rising 15 mm/yr between 14,000 and 10,000 cal yr B.P. Sea level was −63 m 12,000 cal yr B.P., −37 m 10,000 cal yr B.P., and −18 m 8000 cal yr B.P. Most of the remaining Holocene sea-level rise occurred 8000–6000 cal yr B.P. at about 9 mm/yr, a rate the authors estimate to be 2–4 times too great for reefs in the channel to keep up (Grigg *et al.*, 2002). Little reef accumulated on O‘ahu during the Holocene outside areas sheltered from large open-ocean long-period swell (Grigg, 1998).

Trends in middle and late Holocene sea levels have been controversial. Stearns (1935, 1974, 1977) argued that Hawaiian shorelines record a middle Holocene relative sea level higher than present, while Easton & Olson (1976), Ku *et al.* (1974), and Bryan & Stevens (1993) found no evidence for Holocene sea levels higher than present in Hawai‘i. In a review of Hawaiian Holocene marine terraces, Jones (1993) used a compilation of 85 ^{14}C dates to infer that sea level rose rapidly from a low of about −18 m about 8370 ^{14}C yr B.P. (~9350 cal yr B.P.) to a high of about +0.8 m to +1.8 m 4200–1200 ^{14}C yr B.P. (~4800–1500) cal yr B.P.). Geomorphic evidence from O'ahu (Athens & Ward, 1991, 1993; Fletcher & Jones, 1996) and Kaua'i (Calhoun & Fletcher, 1996; Jones, 1992; Matsumoto *et al.*, 1988) also supports a Hawaiian middle Holocene high stand. Grossman & Fletcher (1998) made a detailed stratigraphic analysis of emerged Holocene shorelines on Kapapa Island, off the coast of O'ahu. Using radiocarbon dating of fossil beach samples, they reconstructed sea level based on the surveyed elevation of the beach samples, the mid-points of fossil beach faces, and the surface of the emerged intertidal beach. All three methods imply relative sea level higher than present 5000–2000 cal yr B.P., with a maximum of +2 m about 3500 cal yr B.P. (Fig. 5; Grossman & Fletcher, 1998).

Other U.S.-affiliated islands in the Pacific have contributed to the Holocene sea-level research. Several studies suggest sea levels were 0.5–2.4 m above present within the last few thousand years. Relative sea-level rise apparently peaked ~6800–3100 cal yr B.P. in the Northern Mariana Islands (Curray *et al.*, 1970; Kayanne *et al.*, 1993; Tracey *et al.*, 1964), ~5600–930 cal yr B.P. in the central equatorial zone (Falkland & Woodroffe, 1997; Schofield, 1977; Tracey, 1972; Woodroffe & MacLean, 1998); and ~3760–1850 cal yr B.P. in the Marshall Islands (Buddemeier *et al.*, 1975; Curray *et al.*, 1970; Menard, 1964; Tracey & Ladd, 1974). The peak came somewhat later in the western equatorial zone: ~2900–700 cal yr B.P. on Kosrae (Athens, 1995), ~1370 cal yr B.P. on Kure Atoll (Gross *et al.*, 1969), and ~2400–1200 cal yr B.P. on Midway Atoll (Ladd *et al.*, 1970). In the Caroline Islands, reefs and tidal swamps contain undated platforms that have been interpreted as lithified storm deposits (Newell & Bloom, 1970; Shepard *et al.*, 1967), but Curray *et al.* (1970) and Matsumoto *et al.* (1986) reported possible evidence of higher Holocene sea-stands. American Samoa also has evidence of an undated highstand ~2 m above present sea level (Stice & McCoy, 1968).

Data from these and other islands in the tropical Pacific have contributed to regional analyses of Holocene sea level history. A summary of data meeting strict criteria shows that relative sea level in the equatorial Pacific peaked 1–2 m above present sea level between about 5000 and 1500 cal yr B.P., with the greatest elevation and rates of change toward the west and near the equator (Grossman *et al.*, 1998). Dickinson (2001) discussed local and regional hydro-isostatic and tectonic adjustments that influence local sea-level records in the Pacific, summarizing a regional rise in eustatic sea level in the early Holocene that was followed by more variable late Holocene rates of sea-level fall, influenced by regional hydro-isostacy.

Biogeography

Sea Level and Biogeography

Pleistocene sea-level changes influenced migrations and extinctions in island chains and can affect the assumptions of equilibrium used to explain modern distributions of species with island biogeographic theory (MacArthur & Wilson, 1963). For example, many tropical islands had extensive exposed lowland plains during the late Pleistocene that would have provided a higher proportion of lowland, dry habitat than is currently available. High diversity of modern xeric scrubland communities in southwestern Puerto Rico, which have twice as many species of birds and insect individuals per square kilometer as montane wet forests, may be a relict of the greater land area available to species of warm xeric habitats during glacial periods (Kepler, 1977; Pregill & Olson, 1981).

Habitat loss and fragmentation due to rising sea level is probably also a primary cause of the absence or restricted modern distributions of many species in the Caribbean. During the last glacial maximum, Puerto Rico and the Virgin Islands (except St. Croix) formed a single connected land mass about twice the size of modern Puerto Rico (Fig. 6A; Heatwole & MacKenzie, 1967). Puerto Rico has a high degree of endemism, perhaps due to a size greater than that of the other remaining islands. Many Puerto Rican endemics, including the Puerto Rican woodpecker, may have been more widespread during the last glacial period (Cruz, 1974; Wetmore, 1922). Excavations in Puerto Rico have revealed several fossil vertebrate species that appear to have been more widespread during the early Holocene, including curly-tailed lizards, rock iguanas, a burrowing owl, the Bahaman mockingbird, a caracara, and a palm swift (Olson, 1976; Olson & Hilgartner, 1982; Olson & McKitrick, 1981; Pregill, 1981; Pregill & Olson, 1981; Wetmore, 1920). An analysis of geographic distributions of reptiles and amphibians endemic to the Puerto Rican shelf showed that islands separated for a long time have fewer species in common and greater endemism, relative to islands that have been recently connected (Heatwole & MacKenzie, 1967).

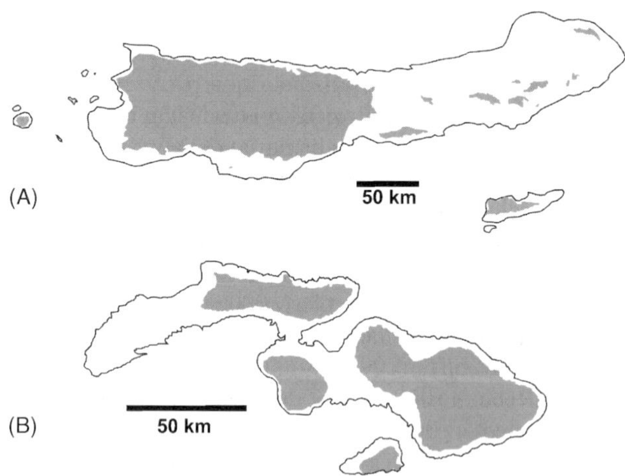

Fig. 6. Sea level change and connections between islands during the last glacial maximum. Solid line marks last glacial maximum sea level, and modern islands are shaded grey. (A) Puerto Rico and adjacent islands, after Heatwole & MacKenzie (1967). (B) Islands of Maui Nui, after Price (2002).

A similar pattern can be seen on the Hawai'ian islands of Maui Nui (Maui, Lana'i, Moloka'i, and Kaho'olawe), which were connected during several glacial low sea-level stands (Fig. 6B; Kaho'olawe was not connected during the LGM). Hawaiian dry forests in general are anomalously species-poor, compared to mesic and wet forests (Price, 2002), presumably because human land use has disproportionately destroyed the relict Hawaiian dry forests that remained after sea-level rise in the Holocene. The islands of Maui Nui have more dry forest taxa per area than other Hawaiian islands, however, and that pattern may be a result of the extensive lowland dry habitat that was exposed during low sea levels of the past few interglacial periods. Former land connections within Maui Nui and between Moloka'i and O'ahu may also explain the distributions of the flightless moa-nalos (*Thambetochen* and *Ptaiochen*; Olson & James, 1991) and some poorly dispersed plant species (Price, 2002), whereas closely related species of *Drosophila* and plants on the individual volcanoes of Maui Nui may have diverged during sea level high stands (Bonacum *et al.*, 2002; Givnish, 1995; Price, 2002).

Quaternary Extirpations, Extinctions, and Introductions

Much research on the history of tropical lands associated with the United States has to do with the interactions of people with their environments. Tropical islands colonized by humans during the Holocene follow a general pattern of loss of lowland forest cover, disappearance of endemic species, erosion, and introduction of widespread species. Diverse native floras and faunas in the tropics often underwent three distinct phases of change: increased disturbance during the initial period of human colonization; increasing rates of vegetation change, extinctions, extirpations, and introduction of competing species as land use was intensified; and a further wave of extinctions and introductions following contact with the global culture and economy (Burney, 1997; Burney *et al.*, 1994; Burney *et al.*, 2001; Dodson, 1992; Kirch, 1982, 1983, 2000). Although most of the evidence discussed here has to do with biological impacts of indigenous and European peoples, physical impacts of humans have also been important at both local and landscape scales (e.g. Athens & Ward, 1993; Burney, 2002; Hammatt, 1978; Hommon, 1980; Kirch & Kelly, 1975; Kraft, 1980; Rosendahl, 1972; Yen *et al.*, 1972).

Initial Settlement

Before human settlement, tropical islands harbored many species found nowhere else. The pre-human flora and fauna of the Hawaiian Islands are the best known among U.S.-affiliated tropical islands, although extinctions of endemic species are still being discovered. Fossil assemblages from the Māhā'ulepu sinkhole on Kaua'i include a diverse fauna of molluscs and vertebrates, with many diatom and vascular plant species, including the native *Pritchardia* palms (Burney *et al.*, 2001). Several cores from lowland O'ahu also show abundant pollen of *Pritchardia* palms shortly before Polynesian settlement (Athens, 1997; Athens *et al.*, 1992). Several lines of evidence record strong vegetation gradients pre-dating human arrival.

Charcoal from two low-elevation sites on windward Maui includes dryland or coastal shrubs, dryland trees, and mesic forest trees, implying a gradient of dry to mesic forest near the sites just before Polynesian colonization (Murakami, 1996). Evidence from charcoal, pollen, and phytoliths suggests that dry to mesic montane forest also existed above about 660 m elevation on leeward Maui before human arrival, with lowland dry forest or shrubland below 660 m (Kolb & Murakami, 1994; Nagahara, 1995). On the leeward side of Kohala Volcano, on the island of Hawai'i, analysis of phytoliths and charcoal in agricultural soils shows a similar gradient of pre-agricultural vegetation assemblages. Inland, woody plants were abundant in the wettest area, with more grass part-way down the precipitation and elevation gradient, while woody vegetation was more abundant in the xeric lowland area. Near the coast, grass-type phytoliths were more abundant again (Murakami, 1994a, b; Pearsall & Trimble, 1984). Early historic landscape descriptions and the distribution of relict native plant species on Kohala are consistent with the vegetation gradient reconstructed from fossil records (McEldowney, 1983). All the lowland areas just described are now dominated by introduced grasses or shrubs; very few native species remain. The pattern of change over time reveals both direct and indirect effects of two waves of human colonization.

Increasing Disturbance and Development of Agriculture

Often the first suggestion of human arrival on tropical Pacific islands and conclusive evidence for forest clearance and agriculture are separated by a thousand years or more. An increase in the abundance of charcoal and other disturbance indicators in six cores from Palau suggests human settlement 5360–5760 cal yr B.P. on Ngerekebesang Island, and 4160–4300 cal yr B.P. on nearby Babeldaob Island (Athens & Ward, 2001). The differences in dates from the two neighboring islands strengthens the interpretation of human settlement rather than regional drought as the cause of the increase in disturbance indicators. Pollen analysis from Palau shows that after the initial increase in disturbance indicators, mangroves decreased and ferns, sedges, grasses, *Pandanus*, coconut palm, and charcoal increased about 2650 cal yr B.P. The pattern of change suggests forest clearance and transformation to savanna, perhaps established and maintained by fire (Athens & Ward, 2001). An increase in charcoal at several sites on Kaua'i supports inference of human presence in leeward coastal areas ca. 830 $\pm$ 50 ^{14}C yr B.P. (1050–1095, 1140–1280 cal yr B.P.; Burney, 2002; Burney & Burney, 2003), followed by a decline in native species in sediments of the Māhā'ulepu sinkhole (Burney *et al.*, 2001).

A core from Laguna Tortuguero, on the northern coast of Puerto Rico, shows a sudden large increase in charcoal abundance ca. 5300 cal yr B.P (Burney *et al.*, 1994). While the earliest archaeological documentation of human presence on Puerto Rico dates considerably later than the increase in charcoal (ca. 3200 cal yr B.P.; Rouse, 1992), people were present in Cuba and Hispaniola at the time (Kozlowski, 1974; Moore, 1991). Little is yet known about changes in vegetation at the time of the increase in charcoal (but see Ogle, 1970).

Many Pacific islands show a similar pattern of increasing rates of disturbance, with loss of lowland forest and indigenous species following the invasion by Polynesians and their accompanying species. Kosrae, in the Caroline Islands, may have been colonized as early as 2300–2500 cal yr B.P. (Ward, 1988) and appears to have been transformed quickly from lowland forest to agroforest by 1550–1350 cal yr B.P. (Athens *et al.*, 1996). On the Orote Peninsula, on the west side of Guam, charcoal first appears in sediment dated to about 3600 cal yr B.P. (Athens & Ward, 1995), roughly coincident with the evidence for arrival of humans in the Mariana Islands (on Saipan; Hunter-Anderson & Butler, 1991). *Pandanus* and other forest trees declined rapidly after about 2600 cal yr B.P. and were virtually absent by about 2300 cal yr B.P. (Athens & Ward, 1995). About 1400 cal yr B.P. grass pollen suddenly increased, coincident with increasing charcoal and decreasing tree pollen. Coconut palm was present 4460 cal yr B.P., before human arrival, but it increased as other tree taxa decreased (Athens & Ward, 1995). Early archaeological soils on presently grass-dominated Guam and nearby Rota contain few grass phytoliths relative to phytoliths of woody plants or forbs (Pearsall, 1987, 1990). Apparently humans destroyed the forests using fire and agricultural practices that encouraged grass and coconut palms.

Lowland Hawai'i also lost much of its native vegetation during the early development and intensification of agriculture. Lowland forests on O'ahu began to be converted to human-dominated landscapes about A.D. 800–1200 (Athens, 1997; Athens & Ward, 1993), and by A.D. 500 little forest vegetation remained (Athens, 1997). On Kaho'olawe and Lāna'i, dryland forest taxa declined with Polynesian settlement, and those that survived must have been extirpated by exotic herbivores after European contact (Allen, 1983, 1984; Allen & Murakami, 1999; Hommon, 1983; Murakami, 1983). The decline of palm phytoliths and increase of grass pollen and phytoliths during the Polynesian period on Maui attest to the gradual clearing of land for agriculture (Cummings, 1993, 1995), and pollen and phytoliths record the clearing of woody vegetation and alteration of the native vegetation gradient on Kohala (Murakami, 1994a, b; Pearsall & Trimble, 1984). After abandonment of agriculture, woody vegetation increased again in northern Kohala (Murakami, 1994a, b). Nearly all remaining forest and shrubland was cleared from lowland areas in Hawai'i after European contact and has now been replaced by exotic species.

Changes in animal communities also followed colonization by humans. During the period before European contact, Pacific rats, chickens, pigs, dogs, candlenut, coconut palm, skinks, snails, geckos, and various weeds were introduced on many Pacific islands, and many species of native birds were locally extirpated or driven extinct (James, 1995; Kirch, 2000; Steadman, 1995). More than 50% of endemic Hawai'ian land birds became extinct during prehistoric human occupation (Burney *et al.*, 2001; James & Olson, 1991; James *et al.*, 1987; Olson & James, 1982, 1984, 1991). Clearing of woody vegetation for agriculture on Kohala resulted in burn layers containing many native snail species; sediments above the burn layers contain very few snails (Christensen, 1983). Many native snails and one native bat species also ceased to appear in the fossil record during the Polynesian period on lowland Kaua'i (Burney *et al.*, 2001). The extirpation of endemic Hawai'ian land snails on Kaua'i appears to follow the entire three-phase pattern, with extinctions during the period of initial settlement, the late prehistoric period, and the period following European contact, when three species of introduced carnivorous snails first appear in sediments (Burney *et al.*, 2001). Additional endemic species disappeared after contact with the global economy, and introductions such as European rats, boars, cane toads, bullfrogs, and several exotic plant species dominate the fossil assemblages of the last 200 years (Burney *et al.*, 2001).

Secondary Effects of Animal Extinctions and Introductions

These animal introductions and extinctions also affected other aspects of island ecosystems. For example, pollen and spores extracted from bird coprolites in Holocene cave sediments on Maui strongly suggest that the extinct flightless moa-nalos ate ferns, at least when trapped in sinkholes (James & Burney, 1997). If the moa-nalos were major consumers of ferns, their extinction must have changed plant community dynamics in places where the birds had formerly been abundant. Similar indirect effects of extinctions of pollinators, predators, or obligate host plants must have cascaded through island ecosystems during periods of rapid environmental change or human occupation.

Athens *et al.* (2002) proposed an interesting reinterpretation of the patterns of anthropogenic introductions and extinctions in Hawai'i. They presented evidence from a stratigraphic pollen record from Ordy Pond, O'ahu, that indicates decline of native forests before any evidence of increased charcoal or extensive human settlement of the Ewa plain. They also cited the lack of certain native forest species or large quantities of bird bone in early archaeological sites immediately after Polynesian colonization. The implication is that the forest birds disappeared as a result of habitat destruction, rather than hunting, and that the main cause of forest decline was the introduction of the seed-eating Polynesian rat, rather than direct effects of humans. Further tests of these ideas require integrated analysis of paleoecology, archaeology, geomorphology, and climate history to reconstruct the temporal and spatial correlations among several interacting components of an ecosystem. Few landscapes are yet known well enough for such an analysis, but island ecosystems are ideally suited. Because islands feel the ecological consequences of extinctions, introductions, environmental change, and resource limitations earlier and more intensely than continents, the integrated Quaternary history of island ecosystems and cultures may prove a valuable model for the problems facing global ecosystems and cultures in the coming centuries.

Summary

Over the past 30 years, the tropical latitudes have yielded increasing understanding of changes in sea level, climate, biogeography, and human migrations during the Quaternary. A persistent paradox regarding full glacial temperatures in the tropics prompted the development of several new techniques for reconstructing past temperature. With new data accumulating, a global reevaluation of the sensitivity of tropical latitudes to global climatic forcing is possible. A new generation of coupled atmosphere-ocean climate models is using the new paleoclimatic data to explore the global implications of tropical climate change and teleconnections arising from sources of finer-scale climate variability such as ENSO and PDO.

Some of the most valuable information to be found in the Quaternary history of the tropics is perspective on ourselves as parts of integrated ecosystems. Paleoecological and archaeological records show clearly that humans have dramatic direct effects on the ecosystems they inhabit. Where we have invaded previously human-free islands in the tropics, many of the changes we have produced are unlike any changes over the previous thousands of years. Our indirect effects are likely to be even more pervasive than our direct effects on our environment. The challenge of the next thirty years may be to understand not only how our actions affect the climate system and other aspects of the physical and biological environment, but how these changes feed back to alter the culture, economics and politics that influence how we affect our environment. Tropical islands have intertwined biological, physical, and cultural histories that foreshadow global issues.

Acknowledgments

Constructive comments by D.A. Burney, R. Calcote, E. Grossman, and S.C. Porter greatly improved earlier drafts of this manuscript. Thanks also to D.K. Alexander, G.S. Brush, O.A. Chadwick, J. Coil, M. Finkelstein, S.P. Horn, M. Jeraj, S.W. Kaplan, P.V. Kirch, M. Twieten, P.M. Vitousek, and many others for conversations that contributed to the material presented here.

References

Allen, M.S. (1983). *Analysis of Archaeobotanical Materials*. Honolulu, Hawaii, Department of Anthropology, Bernice Pauahi Bishop Museum.

Allen, M.S. (1984). A review of archaeobotany and palaeoethnobotany in Hawaii. *Hawaiian Archæology*, **1**, 19–30.

Allen, M.S. & Murakami, G.M. (1999). Lana'i Island's arid lowland vegetation in late prehistory. *Pacific Science*, **53**(1), 88–112.

Aronson, R.B. & Precht, W.F. (2001). Applied paleoecology and the crisis on Caribbean coral reefs. *PALAIOS*, **16**, 195–196.

Athens, J.S. (1995). *Landscape Archaeology: Prehistoric Settlement, Subsistence, and Environment of Kosrae, Eastern Caroline Islands, Micronesia*. Honolulu, Hawaii, International Archaeological Research Institute, 474 pp.

Athens, J.S. (1997). Hawaiian native lowland vegetation in prehistory. *In*: Kirch, P.V. & Hunt, T.L. (Eds), *Historical Ecology in the Pacific Islands. Prehistoric Environmental and Landscape Change*. New Haven and London, Yale University Press, pp. 248–270.

Athens, J.S. & Ward, J.V. (1991). Paleoenvironmental and archaeological investigations, Kawainui March flood control project, Oahu, U.S. Army Engineering Division, Pacific Ocean Environmental Section, Ft. Shaafter, Hawaii. Honolulu, Hawaii, International Archaeological Research Institute, 129 pp.

Athens, J.S. & Ward, J.V. (1993). Environmental change and prehistoric Polynesian settlement in Hawai'i. *Asian Perspectives*, **32**(2), 205–223.

Athens, J.S. & Ward, J.V. (1995). Paleoenvironment of the Orote Peninsula, Guam. *Micronesica*, **28**, 51–76.

Athens, J.S. & Ward, J.V. (2001). Paleoenvironmental evidence for early human settlement in Palau: The Ngerchau core. *Pacific 2000: Proceedings of the Fifth International Conference on Easter Island and the Pacific*. Los Osos, California, Bearsville Press.

Athens, J.S., Ward, J.V. & Wickler, S. (1992). Late Holocene lowland vegetation, O'ahu, Hawai'i. *New Zealand Journal of Archaeology*, **14**, 9–34.

Athens, J.S., Ward, J.V. & Blinn, D.W. (1995). Paleoenvironmental investigations at Uko'a pond, Kawailoa ahupua'a, O'ahu, Hawaii. Report prepared for Engineers Surveyors Hawaii, Honolulu. Honolulu, International Archaeological Research Institute.

Athens, J.S., Ward, J.V. & Murakami, G.M. (1996). Development of an agroforest on a Micronesian high island: Prehistoric Kosraean agriculture. *Antiquity*, **70**, 834–846.

Athens, J.S., Tuggle, H.D., Ward, J.V. & Welch, D.J. (2002). Avifaunal extinctions, vegetation change, and Polynesian impacts in prehistoric Hawai'i. *Archaeologia Oceania*, **37**, 57–78.

Beck, J.W., Edwards, R.L., Ito, E., Taylor, F.W., Recy, J., Rougerie, F., Joannot, P. & Henin, C. (1992). Sea surface temperature from skeletal Sr/Ca ratios. *Science*, **257**, 644–647.

Beck, J.W., Recy, J., Taylor, F., Edwards, R.L. & Gabioch, G. (1997). Abrupt changes in early Holocene tropical sea surface temperature derived from coral records. *Nature*, **385**, 705–707.

Benninger, L.K. & Dodge, R.E. (1986). Fallout plutonium and natural radionuclides in annual bands of the coral Montastrea annularis, St. Croix, U.S. Virgin Islands. *Geochimica et Cosmochimica Acta*, **50**, 2785–2797.

Bloom, A.L., Broecker, W.S., Chappell, J.M.A., Mattews, R.K. & Mesolella, K.J. (1974). Quaternary sea level fluctuations on a tectonic coast: New 230Th/234U dates from the Huon Peninsula, New Guinea. *Quaternary Research*, **4**, 185–205.

Bonacum, J., O'Grady, P.M. & DeSalle, R. (2002). Phylogenetic relationships of the endemic Hawaiian Drosophilidae. *Pacific Science*, in review.

Broecker, W.S., Thurber, D.L., Goddard, J., Ku, T.-L., Mattews, R.K. & Mesolella, K.J. (1968). Milankovitch hypothesis supported by precise dating of coral reefs and deep-sea sediments. *Science*, **159**, 297–300.

Broecker, W.S. (1997). Mountain glaciers: Recorders of atmospheric water vapor content? *Global Biogeochemical Cycles*, **11**, 589–597.

Brückner, H. & Radke, U. (1989). Fossile Strande und Korallenbanke auf Oahu, Hawaii. *Essener Geographische Arbeiten*, **17**, 291–308.

Bryan, W.B. & Stevens, R.S. (1993). Coastal bench formation at Hanauma Bay, Oahu, Hawaii. *Geological Society of America Bulletin*, **87**, 711–719.

Buddemeier, R.W., Smith, S.V. & Kinzie, R.A. (1975). Holocene windward reef-flat history, Enewetok Atoll. *Geological Society of America Bulletin*, **86**, 1581–1584.

Burney, D.A. (1997). Tropical islands as paleoecological laboratories: Gauging the consequences of human arrival. *Human Ecology*, **25**, 437–457.

Burney, D.A. (2002). Late Quaternary chronology and stratigraphy of twelve sites on Kaua'i. *Radiocarbon*, **44**, 13–44.

Burney, D.A., Burney, L.P. & MacPhee, R.D.E. (1994). Holocene charcoal stratigraphy from Laguna Tortuguero, Puerto Rico, and the timing of human arrival on the island. *Journal of Archaeological Science*, **21**, 273–281.

Burney, D.A., DeCandido, R.V., Burney, L.P., Kostel-Hughes, F.N., Stafford, T.W., Jr. & James, H.F. (1995). A Holocene record of climate change, fire ecology, and human activity from montane Flat Top Bog, Maui. *Journal of Paleolimnology*, **13**, 209–217.

Burney, D.A., James, H.F., Burney, L.P., Olson, S.L., Kikuchi, W., Wagner, W.L., Burney, M., McCloskey, D., Kikuchi, D., Grady, F.V., Gage, R., III & Nishek, R. (2001). Fossil evidence for a diverse biota from Kauai and its transformation since human arrival. *Ecological Monographs*, **71**(4), 615–641.

Burney, L.P. & Burney, D.A. (2003). Charcoal stratigraphies for Kaua'i and the timing of human arrival. *Pacific Science*, **57**, 211–226.

Calhoun, R.S. & Fletcher, C.H. (1996). Late Holocene coastal plain stratigraphy and sea-level history at Hanalei, Kauai, Hawaiian Islands. *Quaternary Research*, **45**, 47–58.

Chadwick, O.A., Gavenda, R.T., Kelly, E.F., Ziegler, K., Olson, C.G., Elliott, W.C. & Hendricks, D.M. (2003). The impact of climate on the biogeochemical functioning of volcanic soils. *Chemical Geology*, in press.

Chappell, J. (1974). Geology of coral terraces, Huon Peninsula, New Guinea: A study of Quaternary tectonic movements and sea level changes. *Geological Society of America Bulletin*, **85**, 553–570.

Chappell, J. (1983). A revised sea level record for the last 300,000 years from Papua New Guinea. *Search*, **4**, 99–101.

Chappell, J. & Veeh, H.H. (1978). Late Quaternary tectonic movements and sea-level changes at Timor and Atauro Island. *Geological Society of America Bulletin*, **89**, 356–367.

Chappell, J. & Shackleton, N.J. (1986). Oxygen isotopes and sea level. *Nature*, **324**, 137–140.

Chen, J.H., Curran, H.A., White, B. & Wasserburg, G.J. (1991). Precise chronology of the last interglacial period: ^{234}U-^{230}Th data from fossil coral reefs in the Bahamas. *Geological Society of America Bulletin*, **103**, 82–97.

Christensen, C.C. (1983). Analysis of land snails. *In*: Clark, J.T. & Kirch, P.V. (Eds), *Archaeological Investigations of the Mudlane-Waimea-Kawaihae Road Corridor, Island of Hawai'i: An Interdisciplinary Study of an Environmental Transect*. Report 83–1, Honolulu, Department of Anthropology, Bernice P. Bishop Museum, pp. 449–471.

CLIMAP (1981). Seasonal reconstruction of the earth's surface at the last glacial maximum. *Geological Society of America Map and Chart Series*, MC-36.

Cobb, K.M., Charles, C.D. & Hunter, D.E. (2001). A central tropical Pacific coral demonstrates Pacific, Indian, and Atlantic decadal climate connections. *Geophysical Research Letters*, **28**(11), 2209–2212.

Cohen, A.L., Owens, K.E., Layne, G.D. & Shimizu, N. (2002). The effect of algal symbionts on the accuracy of Sr/Ca paleotemperatures from coral. *Science*, **296**, 331–333.

Cole, J.E., Fairbanks, R.G. & Shen, G.T. (1993). Recent variability in the Southern Oscillation: Isotopic results from a Tarawa Atoll coral. *Science*, **260**, 1790–1793.

Cruz, A. (1974). Distribution, probable evolution, and fossil record of West Indian woodpeckers (family Picidae). *Caribbean Journal of Science*, **14**, 183–188.

Cummings, L.S. (1993). Pollen analysis at habitation and agricultural sites near Waiohuli ahupua'a on the island of Maui, Hawai'i. Denver, Colorado, PaleoResearch Laboratories.

Cummings, L.S. (1995). Phytolith analysis at Keokea ahupua'a, Kula district, Maui. Denver, Colorado, Paleo Research Laboratories.

Curray, J.R., Shepard, F.P. & Veeh, H.H. (1970). Late Quaternary sea-level studies in Micronesia: CARMARSEL expedition. *Geological Society of America Bulletin*, **81**, 1865–1880.

Diaz, H.F. & Markgraf, V. (Eds) (1993). *El Niño: Historical and Paleoclimatic Aspects of the Southern Oscillation*. Cambridge, Cambridge University Press.

Dickinson, W.R. (2001). Paleoshoreline record of relative Holocene sea levels on Pacific islands. *Earth-Science Reviews*, **55**, 191–234.

Dodge, R.E., Fairbanks, R.G., Benninger, L.K. & Maurrasse, F. (1983). Pleistocene sea levels from raised coral reefs of Haiti. *Science*, **219**, 1423–1425.

Dodson, J. (Ed.) (1992). *The Naive Lands: Prehistory and Environmental Change in Australia and the Southwest Pacific*. Melbourne, Longman Cheshire.

Dorn, R.I., Phillips, F.M., Zreda, M.G., Wolfe, E.W., Jull, A.J.T., Donahue, D.J., Kubik, P.W. & Sharma, P. (1991). Glacial chronology. *National Geographic Research and Exploration*, **7**(4), 456–471.

Easton, W.H. (1987). Volcanism in Hawaii. Stratigraphy of Kilauea Volcano. *U.S. Geological Survey Professional Paper*, **1350**, 243–260.

Easton, W.H. & Olson, E.A. (1976). Radiocarbon profile of Hanauma Reef, Oahu, Hawaii. *Geological Society of America Bulletin*, **87**, 711–719.

Falkland, A.C. & Woodroffe, C.D. (1997). Geology and hydrogeology of Tarawa and Christmas Island, Kiribati. *Developments in Sedimentology*, **54**, 577–610.

Farrera, I., Harrison, S.P., Prentice, I.C., Ramstein, G., Guiot, J., Bartlein, P.J., Bonnefille, R., Bush, M., Cramer, W., von Grafenstein, U., Holmgren, K., Hooghiemstra, H., Hope, G., Jolly, D., Lauritzen, S.-E., Ono, Y., Pinot, S., Stute, M. & Yu, G. (1999). Tropical climates at the last glacial maximum: A new synthesis of terrestrial palaeoclimate data. I. Vegetation, lake-levels and geochemistry. *Climate Dynamics*, **15**, 823–856.

Fletcher, C.H. & Jones, A.T. (1996). Sea-level high-stand recorded in Holocene shoreline deposits on Oahu, Hawaii. *Journal of Sedimentary Research*, **66**, 632–641.

Gavenda, R.T. (1992). Hawaiian Quaternary paleoenvironments: A review of geological, pedological, and botanical evidence. *Pacific Science*, **46**(3), 295–307.

Giambelluca, T.W. & Schroeder, T.A. (1998). Climate. *In*: Juvik, S.P. & Juvik, J.O. (Eds), *Atlas of Hawai'i*, (3rd ed.). Honolulu, University of Hawai'i Press.

Givnish, T.J. (1995). Molecular evolution, adaptive radiation, and geographic speciation in *Cyanea* (Campanulaceae, Lobelioidae). *In*: Wagner, W.L. & Funk, V.A. (Eds), *Hawaiian Biogeography: Evolution on a Hot Spot Archipelago*. Washington, DC, Smithsonian Institution Press, pp. 288–337.

Grigg, R.W. (1998). Holocene coral reef accretion in Hawaii: A function of wave exposure and sea level history. *Coral Reefs*, **17**, 263–272.

Grigg, R.W., Grossman, E.E., Earle, S.A., Gittings, S.R., Lott, D. & McDonough, J. (2002). Drowned reefs and antecedent karst topography, Au'au Channel, S.E. Hawaiian Islands. *Coral Reefs*, **21**, 73–82.

Gross, M.G., Milliman, J.D., Tracey, J.I. & Ladd, H.S. (1969). Marine geology of Kure and Midway atolls, Hawaii: A preliminary report. *Pacific Science*, **23**, 1725.

Grossman, E.E., Fletcher, C.H., III & Richmond, B.M. (1998). The Holocene sea-level highstand in the equatorial Pacific: Analysis of the insular paleosea-level database. *Coral Reefs*, **17**, 309–327.

Grossman, E.E. & Fletcher, C.H., III (1998). Sea level higher than present 3500 years ago on the northern main Hawaiian Islands. *Geology*, **26**(4), 363–366.

Guilderson, T.P., Fairbanks, R.G. & Rubenstone, J.L. (1994). Tropical temperature variations since 20,000 years ago: Modulating interhemispheric climate change. *Science*, **263**, 663–665.

Guilderson, T.P. & Schrag, D.P. (1999). Reliability of coral isotope records from the western Pacific warm pool: A comparison using age-optimized records. *Paleoceanography*, **14**, 457–464.

Hammatt, H.H. (1978). Geoarchaeological stratigraphy in the Hawaiian Islands. Paper, 43d Annual Meeting, Society of American Archaeologists, Tucson. Cited in Kirch, P.V. (1982). The impact of prehistoric Polynesians on the Hawaiian ecosystem. *Pacific Science*, **36**, 1–14.

Harmon, R.S., Mitterer, R.M., Kriausakul, N., Lands, L.S., Schwarcz, H.P., Garrett, P., Larson, G.J., Vacher, H.L. & Rowe, M. (1983). U-series and amino-acid racemization geochronology of Bermuda: Implications for eustatic sea-level fluctuation over the past 250,000 years. *Palaeogeography, Palaeoclimatology, Palaeoecology*, **44**, 41–70.

Hearty, P.J. (2002). The Ka'ena highstand of O'ahu, Hawai'i, further evidence of Antarctic ice collapse during the middle Pleistocene. *Pacific Science*, **56**(1), 65–81.

Hearty, P.J., Kaufman, D.S., Olson, S.L. & James, H.F. (2000). Stratigraphy and whole-rock amino acid geochronology of key Holocene and Last Interglacial carbonate deposits in the Hawaiian Islands. *Pacific Science*, **54**, 423–442.

Heatwole, H. & MacKenzie, F. (1967). Herpetogeography of Puerto Rico. IV. Paleogeography, faunal similarity, and endemism. *Evolution*, **21**, 429–438.

Hodell, D.A., Curtis, J.H., Jones, G.A., Higuera-Gundy, A., Brenner, M., Binford, M.W. & Dorsey, K.T. (1991). Reconstruction of Caribbean climate change over the past 10,500 years. *Nature*, **352**, 790–793.

Hodell, D.A., Curtis, J.H. & Brenner, M. (1995). Possible role of climate in the collapse of Classic Maya civilization. *Nature*, **375**, 391–394.

Holmes, C.W. (1983). $\delta^{18}O$ variations in the *Halimeda* of Virgin Islands sands: Evidence of cool water in the northeast Caribbean, Late Holocene. *Journal of Sedimentary Petrology*, **53**(2), 429–438.

Hommon, R.J. (1980). Multiple resources nomination, Kaho'olawe Archaeological Sites, National Register of Historic Places, Washington, DC.

Hommon, R.J. (1983). Kaho'olawe archaeological excavations, 1981. Honolulu, Hawaii, Science Management.

Horn, S.P. & Sandford, R.L., Jr (1992). Holocene fires in Costa Rica. *Biotropica*, **24**, 354–361.

Hostetler, S.W. & Mix, A.C. (1999). Reassessment of ice-age cooling of the tropical ocean and atmosphere. *Nature*, **399**, 673–676.

Hostetler, S.W. & Clark, P.U. (2000). Tropical climate at the last glacial maximum inferred from glacier mass-balance modeling. *Science*, **290**, 1747–1750.

Hotchkiss, S.C. (1998). Quaternary vegetation and climate of Hawai'i. Unpublished Ph.D. dissertation, University of Minnesota, Saint Paul.

Hotchkiss, S.C. & Juvik, J.O. (1999). A Late-Quaternary pollen record from Ka'au Crater, O'ahu, Hawai'i. *Quaternary Research*, **52**, 115–128.

Hotchkiss, S., Vitousek, P.M., Chadwick, O.A. & Price, J. (2000). Climate cycles, geomorphological change, and the interpretation of soil and ecosystem development. *Ecosystems*, **3**, 522–533.

Hsieh, J.C.C. (1997). An oxygen isotopic study of soil water and pedogenic clays in Hawaii. Unpublished Ph.D. thesis, Pasadena, California Institute of Technology.

Hsieh, J.C.C., Chadwick, O.A., Kelly, E.F. & Savin, S.M. (1998). Oxygen isotopic composition of soil water: Quantifying evaporation and transpiration. *Geoderma*, **82**, 269–293.

Hunter-Anderson, R.L. & Butler, B.M. (1991). An overview of Northern Marianas prehistory. Report prepared for the Division of Historic Preservation, Commonwealth of the Northern Mariana Islands, Saipan. Guam, Micronesian Archaeological Research Services.

James, H.F. (1995). Prehistoric extinctions and ecological changes on oceanic islands. *Ecological Studies*, **115**, 87–102.

James, H.F. & Olson, S.L. (1991). Descriptions of thirty-two new species of Hawaiian birds. Part II. Passeriformes. *Ornithological Monographs*, **46**, 1–88.

James, H.F. & Burney, D.A. (1997). The diet and ecology of Hawaii's extinct flightless waterfowl: Evidence from coprolites. *Biological Journal of the Linnean Society*, **62**, 279–297.

James, H.F., Stafford, T., Steadman, W.D., Olson, S., Martin, P., Jull, A. & McCoy, P. (1987). Radiocarbon dates on bones of extinct birds from Hawaii. *Proceedings of the National Academy of Sciences, USA*, **84**, 2350–2354.

Jones, A.T. (1992). Holocene coral reef on Kauai Hawaii: Evidence for a sea-level highstand in the central Pacific. Quaternary coasts of the United States: Marine and lacustrine systems. *SEPM Special Publication*, **48**, 267–271.

Jones, A.T. (1993). Review of the chronology of marine terraces in the Hawaiian archipelago. *Quaternary Science Reviews*, **12**, 811–823.

Kauffman, A. (1986). The distribution of 230Th/234U ages in corals and the number of last interglacial high-sea stands. *Quaternary Research*, **25**, 55–62.

Kaufman, D.S., Porter, S.C. & Gillespie, A.R. (2003). Quaternary alpine glaciation in Alaska, the Pacific Northwest, Sierra Nevada, and Hawaii, this volume.

Kayanne, H., Ishii, T., Matsumoto, E. & Yonekura, N. (1993). Late Holocene sea-level change on Rota and Guam, Mariana Islands, and its constraint on geophysical predictions. *Quaternary Research*, **40**, 189–200.

Kepler, A.K., (1977). Comparative study of todies (Todidae) with emphasis on the Puerto Rican Tody, *Todus mexicanus*. *Publications of the Nuttall Ornithological Club*, **16**, 1–190.

Kindinger, J.L., Miller, R.J. & Holmes, C.W. (1983). Sedimentology of southwestern roads region, U.S. Virgin Islands – origin and rate of sediment accumulation. *Journal of Sedimentary Petrology*, **53**, 0439–0447.

Kirch, P.V. (1982). The impact of the prehistoric Polynesians on the Hawaiian ecosystem. *Pacific Science*, **36**(1), 1–13.

Kirch, P.V. (1983). Man's role in modifying tropical and subtropical Polynesian ecosystems. *Archaeology in Oceania*, **18**, 26–31.

Kirch, P.V. (2000). *On the road of the winds. An archaeological history of the Pacific islands before European contact*. Berkeley, Los Angeles, and London, University of California Press.

Kirch, P.V. & Kelly, M. (Eds) (1975). Prehistory and ecology in a windward Hawaiian valley: Halawa Valley, Molokai. *Pacific Anthropological Records*. Honolulu, Bernice P. Bishop Museum, Vol. 24.

Kolb, M.J. & Murakami, G.M. (1994). Cultural dynamics and the ritual role of woods in pre-contact Hawai'i. *Asian Perspectives*, **33**(1), 58–78.

Kozlowski, J.K. (1974). Preceramic cultures in the Caribbean. Zeszyty Naukowe, Uniwerstytetu Jagiellonskiego 386, Prace Archeologiczne, Zezyt 20. Kraków, Poland.

Kraft, J.C. (1980). Summary of results of the Kawainui Marsh study. Letter to Coastal Zone Management Program, State of Hawaii, 18 December 1980.

Ku, T.L., Kimmel, M.A., Easton, W.H. & O'Neil, T.J. (1974). Eustatic sea level 120,000 years ago on Oahu, Hawaii. *Science*, **183**, 959–962.

Ladd, H.S., Tracey, J.I. & Gross, A.G. (1970). Deep drilling on Midway Atoll. *USGS Professional Paper*, **680A**, 1–22.

Lee, K.E. & Slowey, N.C. (1999). Cool surface waters of the subtropical North Pacific Ocean during the last glacial. *Nature*, **397**, 512–514.

Lindberg, D.R., Roth, B., Kellogg, M.G. & Hubbs, C.L. (1980). Invertebrate megafossils of Pleistocene (Sangamon interglacial) age from Isla de Guadalupe, Baja California, Mexico. *In*: Power, D.M. (Ed.), *The California Islands: Proceedings of a Multidisciplinary Symposium*. Santa Barbara, California, Santa Barbara Museum of Natural History, pp. 41–62.

Linsley, B.K., Ren, L., Dunbar, R.B. & Howe, S.S. (2000). El Niño Southern Oscillation (ENSO) and decadal-scale climate variability at 10°N in the eastern Pacific from 1893 to 1994: A coral-based reconstruction from Clipperton Atoll. *Paleoceanography*, **15**, 322–335.

Lum, D. & Stearns, H.T. (1970). Pleistocene stratigraphy and eustatic history based on cores at Waimanalo, Oahu, Hawaii. *Geological Society of America Bulletin*, **81**, 1–16.

Lyon, H.L. (1930). The flora of Moanalua 100,000 years ago. *Proceedings of the Hawaiian Academy of Sciences, B.P. Bishop Museum Special Publication*, **16**, 6–7.

MacArthur, R.H. & Wilson, E.O. (1963). An equilibrium theory of insular zoogeography. *Evolution*, **17**(4), 373–387.

Matsumoto, E., Matsushima, Y. & Miyata, T. (1986). Holocene sea-level studies of swampy coastal plains in Truk and Ponape, Micronesia. *In*: Sugimura, Y. (Ed.), *Sea-Level Changes and Tectonics in the Middle Pacific. Report of the HIPAC Project in 1984 and 1985*. Kobe, Japan, Kobe University Press, pp. 95–110.

Matsumoto, E., Matsushima, Y. & Miyata, T. (1988). Holocene sea-level studies of swampy coastal plains in Truk and Ponape, Micronesia. *In*: Sugimura, Y. (Ed.), *Sea-Level Changes and Tectonics in the Middle Pacific. Report 6 of the HIPAC Project in 1986 and 1987*. University of Tokyo, pp. 91–99.

Mattews, R.K. (1973). Relative elevation of late Pleistocene high sea level stands: Barbados uplift rates and their implications. *Quaternary Research*, **3**, 147–153.

McCulloch, M.T. & Esat, T. (2000). The coral record of last interglacial sea levels and sea surface temperatures. *Chemical Geology*, **169**, 107–129.

McCulloch, M., Mortimer, G., Esat, T., Xianhua, L., Pillans, B. & Chappell, J. (1996). High resolution windows into early Holocene climate: Sr/Ca coral records from the

Huon Peninsula. *Earth and Planetary Science Letters*, **138**, 169–178.

McEldowney, H. (1983). A description of major vegetation patterns in the Waimea-Kawaihae region during the early historic period. *In*: Clark, J.T. & Kirch, P.V. (Eds), *Archaeological Investigations of the Mudlane-Waimea-Kawaihae Road Corridor, Island of Hawai'i: An Interdisciplinary Study of an Environmental Transect*. Report 83–1, Honolulu, Department of Anthropology, Bernice P. Bishop Museum, pp. 407–448.

Menard, H.W. (1964). *Marine geology of the Pacific*. McGraw Hill, New York, 271 pp.

Metcalfe, S.E. (1995). Holocene environmental change in the Zacapu Basin, Mexico: A diatom-based record. *The Holocene*, **5**, 196–208.

Metcalfe, S.E., Street-Perrott, F.A., O'Hara, S.L., Hales, P.E. & Perrott, R.A. (1994). The paleolimnological record of environmental change: Examples from the arid frontier of Mesoamerica. *In*: Millingon, A.C. & Pye, K. (Eds), *Environmental Change in Drylands: Biogeographical and Geomorphological Perspectives*. Chichester, Wiley, pp. 131–145.

Mix, A.C., Morey, A.E. & Pisias, N.G. (1999). Foraminiferal faunal estimates of paleotemperature: Circumventing the no-analog problem yields cool ice age tropics. *Paleoceanography*, **14**(3), 350–359.

Montaggioni, L.F. (1988). Holocene reef growth history in mid-plate high volcanic islands. *Proceedings of the 6th International Coral Reef Symposium, Australia*, **3**, 455–460.

Moore, C. (1991). Cabaret: Lithic workshop sites in Haiti. *In*: Ayubi, E.N. & Haviser, J.B. (Eds), Proceedings of the 13th International Congress of Caribbean Archaeology. *Reports of the Archaeological-Anthropological Institute of the Netherlands Antilles*, **9**, 92–104.

Moore, W.S. (1982). Late Pleistocene sea-level history. *In*: Ivanovitch, M. & Harmon, R.S. (Eds), *Uranium-Series Disequilibrium: Applications to Environmental Problems*. Oxford, Clarendon Press, pp. 481–496.

Muhs, D.R., Kennedy, G.L. & Rockwell, T.K. (1994). Uranium-series ages of marine terrace corals from the Pacific coast of North America and implications for last-interglacial sea level history. *Quaternary Research*, **42**, 72–87.

Muhs, D.R. & Szabo, B.J. (1994). New uranium-series ages of the Waimanalo limestone, Oahu, Hawaii: Implications for sea level during the last interglacial period. *Marine Geology*, **118**, 315–326.

Murakami, G.M. (1983). *Identification of Charcoal from Kaho'olawe Archaeological Sites*. Honolulu, Hawaii, Science Management.

Murakami, G.M. (1996). Identification of Charcoal from the Luluku Banana Farmers' Relocation Area, Maunawili, O'ahu. *In*: Williams (Ed.), *Report from the Luluku Banana Farmers' Relocation Area*. Maunawili, O'ahu.

Murakami, G.M. (1994a). Identification of charcoal from Kuluipahu, North Kohala, Hawai'i. Appendix 5. *In*: Adams, J. & Athens, J.S. (Eds), *Archaeological Inventory Survey, Upland Portions of Kuluipahu and Awalua, North Kohala, Island of Hawai'i*. Honolulu, International Archaeological Research Institute.

Murakami, G.M. (1994b). Wood charcoal identification. Appendix E. *In*: Erkelens, C. & Athens, J.S. (Eds), *Archaeological Inventory Survey, Kohala Plantation Village, North Kohala, Hawai'i*. Honolulu, International Archaeological Research Institute.

Nagahara, V. (1995). Material culture and social change in Kula. *In*: Kolb, M. (Ed.), *Kula: The Archaeology of Upcountry Maui in Waiohuli and Keokea*. Honolulu, State Historic Preservation Division, pp. 202–210.

Newell, N.D. & Bloom, A.L. (1970). The reef flat and 'two-meter eustatic terrace' of some Pacific atolls. *Geological Society of America Bulletin*, **81**, 1881–1894.

Nyberg, J., Kuijpers, A., Malmgren, B.A. & Kunzendorf, H. (2001). Late Holocene changes in precipitation and hydrography recorded in marine sediments from the northeastern Caribbean Sea. *Quaternary Research*, **56**, 87–102.

Nyberg, J., Malmgren, B.A., Kuijpers, A. & Winter, A. (2002). A centennial-scale variability of North Atlantic surface hydrography during the late Holocene. *Palaeogeography, Palaeoclimatology, Palaeoecology*, **183**, 25–41.

Ogle, C.J. (1970). Pollen analysis of selected *Sphagnum*-bog sites in Puerto Rico. *In*: Odum, H.T. & Pigeon, R.F. (Eds), *A Tropical Rain Forest: A Study of Irradiation and Ecology at El Verde, Puerto Rico*. Book 1. Office of Information Services, U.S. Atomic Energy Commission.

Olson, S.L. (1976). A new species of Milvago from Hispaniola, with notes on other fossil caracaras from the West Indies (Aves: Falconidae). *Proceedings of the Biological Society of Washington*, **88**, 355–366.

Olson, S.L. & James, H.F. (1982). Fossil birds from the Hawaiian Islands: Evidence for wholesale extinction by man before Western contact. *Science*, **217**, 633–635.

Olson, S.L. & James, H.F. (1984). The role of Polynesians in the extinction of the avifauna of the Hawaiian Islands. *In*: Martin, P.S. & Klein, R.L. (Eds), *Quaternary Extinctions: A Prehistoric Revolution*. Tucson, University of Arizona Press, pp. 768–780.

Olson, S.L. & James, H.F. (1991). Descriptions of 32 new species of birds from the Hawaiian Islands: Part I, non-Passeriformes. *Ornithological Monographs*, **45**, 1–88.

Olson, S.L. & McKitrick, M.C. (1981). A new genus and species of emberizine finch from Pleistocene cave deposits in Puerto Rico (Aves: Passeriformes). *Journal of Vertebrate Paleontology*, **1**, 279–283.

Olson, S.L. & Hilgartner, W. (1982). Fossil and subfossil birds from the Bahamas. *Smithsonian Contributions in Paleobiology*, **48**, 22–60.

Ota, Y. & Omura, A. (1992). Contrasting styles and rates of tectonic uplift of coral reef terraces in the Ryukyu and Daito Islands, southwestern Japan. *Quaternary International*, **15/16**, 17–29.

Pearsall, D.M. (1987). Analysis of phytolith samples. *In*: Butler, B.M. (Ed.), *Archaeological Investigations on the North Coast of Rota, Mariana Islans: The Airport Road Project*. Center for Archaeological Investigations, Southern Illinois University at Carbondale, pp. 348–354.

Pearsall, D.M. (1990). Application of phytolith analysis to reconstruction of past environments and subsistence: Recent research in the Pacific. *Micronesica Suppl.*, **2**, 65–74.

Pearsall, D.M. & Trimble, M.K. (1984). Identifying past agricultural activity through soil phytolith analysis: A case study from the Hawaiian Islands. *Journal of Archaeological Science*, **11**, 119–133.

Porter, S.C. (1972). Distribution, morphology, and size frequency of cinder cones on Mauna Kea volcano, Hawaii. *Geological Society of America Bulletin*, **83**, 3607–3612.

Porter, S.C. (1973). Stratigraphy and chronology of late Quaternary tephra along the south rift zone of Mauna Kea volcano, Hawaii. *Geological Society of America Bulletin*, **84**, 1923–1940.

Porter, S.C. (1979a). Hawaiian glacial ages. *Quaternary Research*, **12**, 161–187.

Porter, S.C. (1979b). Quaternary stratigraphy and chronology of Mauna Kea, Hawaii: A 380,000-yr record of mid-Pacific volcanism and ice-cap glaciation. *Geological Society of America Bull. Pt. II*, **90**, 908–1093.

Porter, S.C. (1997). Late Pleistocene eolian sediments related to pyroclastic eruptions of Mauna Kea Volcano, Hawaii. *Quaternary Research*, **47**, 261–276.

Porter, S.C. (2001). Snowline depression in the tropics during the last glaciation. *Quaternary Science Reviews*, **20**, 1067–1091.

Pregill, G. (1981). Late Pleistocene herpetofaunas from Puerto Rico. *Miscellaneous Publications of the University of Kansas Museum of Natural History*, **71**, 1–72.

Pregill, G. & Olson, S.L. (1981). Zoogeography of West Indian vertebrates in relation to Pleistocene climatic cycles. *Annual Reviews of Ecology and Systematics*, **12**, 75–98.

Price, J. (2002). Paleogeography and floristic biogeography of the Hawaiian Islands. Unpublished Ph.D. thesis, Graduate Program in Geography, University of California-Davis.

Richards, D.A., Smart, P.L. & Edwards, R.L. (1994). Maxiumum sea levels for the last glacial period from U-series ages of submerged speleothems. *Nature*, **367**, 357–360.

Rind, D. & Peteet, D. (1985). Terrestrial conditions at the last glacial maximum and CLIMAP sea-surface temperature estimates: Are they consistent? *Quaternary Research*, **24**, 1–22.

Rosendahl, P.H. (1972). Aboriginal agriculture and residence patterns in upland Lapakahi, Island of Hawaii. Unpublished Ph.D. dissertation, University of Hawaii.

Rouse, I. (1992). *The Tainos: Rise and decline of the people who greeted Columbus*. New Haven, Yale University Press.

Ruhe, R.V. (1964). An estimate of paleoclimate in Oahu, Hawaii. *American Journal of Science*, **262**, 1098–1115.

Schofield, J.C. (1977). Late Holocene sea level, Gilbert and Ellice islands, west central Pacific Ocean. *New Zealand Journal of Geology and Geophysics*, **20**, 503–529.

Shen, G.T., Linn, L.J., Campbell, T.M., Cole, J.E. & Fairbanks, R.G. (1992). *Journal of Geophysical Research – Oceans*, **97**, 12689–12697.

Shepard, F.P., Curray, J.R., Newman, W.A., Bloom, A.L., Newell, N.D., Tracey, J.I. & Veeh, H.H. (1967). Holocene changes in sea level: Evidence in Micronesia. *Science*, **157**, 542–544.

Sherman, A.D., Glenn, C.R., Jones, A.T., Burnett, W.C. & Schwarcz, H.P. (1993). New evidence for two highstands of the sea during the last interglacial, oxygen isotope substage 5e. *Geology*, **21**, 1079–1082.

Sirkin, L., Szabo, B., Padill, A., Pedrin, A. & Diaz, R. (1990). Uranium-series ages of marine terraces, La Paz Peninsula, Baja California Sur, Mexico. *Coral Reefs*, **9**, 25–30.

Smart, P.L. & Richards, D.A. (1992). Age estimates for the late Quaternary high sea-stands. *Quaternary Science Reviews*, **11**, 687–696.

Steadman, D.W. (1995). Prehistoric extinctions of Pacific island birds: Biodiversity meets zooarchaeology. *Science*, **267**, 1123–1130.

Stearns, H.T. (1935). Pleistocene shore lines on the islands of Oahu and Maui, Hawaii. *Bulletin of the Geological Society of America*, **46**, 1927–1956.

Stearns, H.T. (1939). Geologic map and guide to the island of Oahu, Hawaii. *Hawaii Division of Hydrography Bulletin*, **49**, 615–628.

Stearns, H.T. (1966). *Geology of the State of Hawaii*. Palo Alto, California, Pacific Books, 266 pp.

Stearns, H.T. (1973). Potassium-argon ages of lavas from the Hawi and Pololu volcanic series, Kohala volcano, Hawaii: Discussion. *Geological Society of America Bulletin*, **85**, 795–804.

Stearns, H.T. (1974). Submerged shorelines and shelves in the Hawaiian Islands and a revision of some of the eustatic emerged shorelines. *Geological Society of America Bulletin*, **85**, 795–804.

Stearns, H.T. (1975). PCA 25-ft stand of the sea on Oahu, Hawaii. *Geological Society of America Bulletin*, **86**, 1279–1280.

Stearns, H.T. (1976). Estimates of the position of sea level between 140,000 and 75,000 years ago. *Quaternary Research*, **6**, 445–449.

Stearns, H.T. (1977). Radiocarbon profile of Hanauma Reef, Oahu, Hawaii: A discussion. *Geological Society of America Bulletin*, **88**, 1535.

Stearns, H.T. (1978). Quaternary Shorelines in the Hawaiian Islands. *Bernice P. Bishop Museum Bulletin*, **237**, 1–57.

Stein, M., Wasserburg, G.J., Aharon, P., Chen, J.H., Zhu, Z.R., Bloom, A. & Chappell, J. (1993). TIMS U-series dating and stable isotopes of the last interglacial event in Papua New Guinea. *Geochimica et Cosmochimica Acta*, **57**, 2541–2554.

Stice, G.D. & McCoy, F.W. (1968). The geology of the Manu'a Islands, Samoa. *Pacific Science*, **22**, 427–457.

Stute, M., Forster, M., Frischkorn, H., Serejo, A., Clark, J.F., Schlosser, P., Broecker, W.S. & Bonani, G. (1995). Cooling of tropical Brazil (5°C) during the last glacial maximum. *Science*, **269**, 379–383.

Stute, M., Schlosser, P., Clark, J.F. & Broecker, W.S. (1992). Paleotemperatures in the southwestern United States derived from noble gases in ground water. *Science*, **256**, 1000–1003.

Thompson, L.G. (2000). Ice core evidence for climate change in the tropics: Implications for our future. *Quaternary Science Reviews*, **19**, 18–35.

Tracey, J.I. (1972). Holocene emergent reefs in the central Pacific. 2nd Conference of the American Qutaternary Association, Abstracts 51–5.

Tracey, J.I. & Ladd, H.S. (1974). Quaternary history of Enewetok and Bikini atolls, Marshall Islands. *Proceedings of the 2nd International Coral Reef Symposium*, **2**, 537–550.

Tracey, J.I., Schlanger, S.O., Stark, J.T., Doan, D.B. & May, H.G. (1964). General Geology of Guam. *USGS Professional Paper*, 403-A, 104 pp.

Vitousek, P.M., Chadwick, O.A., Crews, T.E., Fownes, J.H., Hendricks, D.M. & Herbert, D. (1997). Soil and ecosystem development across the Hawaiian Islands. *Geological Society America Today*, **7**(9), 1–8.

Ward, J.V. (1988). Palynology of Kosrae, Eastern Caroline Islands: Recoveries from pollen rain and Holocene deposits. *Review of Palaeobotany and Palynology*, **55**, 247–271.

Watanabe, T., Winter, A. & Oba, T. (2001). Seasonal changes in sea surface temperature and salinity during the Little Ice Age in the Caribbean Sea deduced from Mg/Ca and $^{18}O/^{16}O$ ratios in corals. *Marine Geology*, **173**, 21–35.

Watts, A.B. & ten Brink, U.S. (1989). Crustal structure, flexure, and subsidence history of the Hawaiian Islands. *Journal of Geophysical Research*, **94**, 10473–10500.

Wellington, G.M. & Dunbar, R.B. (1995). Stable isotopic signature of El Niño-Southern Oscillation events in eastern tropical Pacific reef corals. *Coral Reefs*, **14**, 5–25.

Wetmore, A. (1920). Five new species of birds from cave deposits in Porto Rico. *Proceedings of the Biological Society of Watshington*, **33**, 77–81.

Wetmore, A. (1922). Remains of birds from caves of Porto Rico. *Bulletin of the American Museum of Natural History*, **46**, 297–333.

Winchell, H. (1947). Honolulu Series, Oahu, Hawaii. *Geological Society of America Bulletin*, **58**, 1–48.

Winter, A., Goenaga, C. & Maul, G.A. (1991). Carbon and oxygen isotope time series from an 18-year Caribbean reef coral. *Journal of Geophysical Research*, **96**(C9), 16673–16678.

Woodroffe, C. & MacLean, R. (1998). Pleistocene morphology and Holocene emergence of Christmas (Kiritimati) Island, Pacific Ocean. *Coral Reefs*, **17**, 235–238.

Wright, P.B. (1989). Homogenized long-period Southern Oscillation indexes. *International Journal of Climatology*, **9**, 33–54.

Yen, D.E., Kirch, P.V., Rosendahl, P. & Riley, T. (1972). Prehistoric agriculture in the upper Makaha Valley, Oahu. *In*: Yen, D.E. & Ladd, E. (Eds), Makaha Valley historical project: Interim report number 3. *Pacific Anthropological Records*. Honolulu, Bernice P. Bishop Museum, Vol. 18, pp. 59–94.

Ziegler, K., Hsieh, J.C.C., Chadwick, O.A., Kelly, E.F., Hendricks, D.M. & Savin, S.M. (2003). Halloysite as a kinetically controlled end product of arid-zone basalt weathering. *Chemical Geology*, in press.

Climatically forced vegetation dynamics in eastern North America during the late Quaternary Period

Thompson Webb III, Bryan Shuman and John W. Williams

Department of Geological Sciences, Brown University, Providence, RI 02912-1846, USA; thompson_webb_iii@brown.edu

Introduction

Vegetation dynamics span multiple spatial and temporal scales, and the changes involved manifest themselves in a variety of ways depending upon the ecological unit (from individuals to biomes) and/or taxonomic level (e.g. species, genera, families and orders) of description. Many biotic phenomena contribute to vegetation change including: (1) the establishment, growth, and death of individual plants within stands; (2) changes in the frequency, size, and genetic make-up of populations within landscapes; (3) changes in the distribution of species, genera, and plant functional types across regions and continents; and (4) the evolution and extinction of species. These biotic phenomena cause the vegetation to change in structure, density, extent, and composition, and they lead to and result from a variety of biospheric dynamics (such as variations in net primary production and carbon sequestration). Depending on scale, vegetation changes are caused by some combination of external (i.e. environmental) forcing and the biotic phenomena themselves. The multiple competing forcings (at work at different scales) and many nonlinear linkages (including feedbacks) can make the cause-and-effect explanations difficult to sort out at certain temporal and spatial scales. Across long-time spans, however, such as the late Quaternary, environmental variations are large and well known and their effect on vegetation history is relatively easy to recognize.

In this chapter, we consider vegetation dynamics at regional to continental scales and across millennia, scales at which vegetation change is primarily forced by centennial to orbital scales of climate change. The vegetation changes show up as the changing abundance, geographic extent, location, and association of plant taxon populations, which we record as changing pollen percentages. Only by linking the forces and induced responses can we convert the study of vegetation change and history into an analysis of vegetation dynamics, because to do so we must relate the apparent "motion" in these taxon populations to underlying forces, which is the very definition of dynamics. Motion by definition is temporal change in location, which requires temporal sequences of maps, difference maps, and/or isochrone maps to illustrate. Mapping temporal change in the vegetation is therefore central to studies of climatically forced vegetation dynamics. Here, we map fossil pollen data, as a proxy for vegetation data, from eastern and northern North America and compare both continental-scale and local records of the pollen-recorded vegetation change to maps and time series of independently observed or estimated paleoclimate data. These comparisons are key to our empirical understanding of late-Quaternary vegetation dynamics. We admit that the "motion" of taxon populations shown on our maps is an epiphenomenon of the differential carbon sequestration in the different taxa in different locations, but we focus here on the motion apparent in the time series of pollen maps and use it and other pollen-recorded changes to represent how the vegetation changed. Many studies show how well pollen data from surficial sediments represent plant taxon abundances today and thus underpin our interpretative step here (Bradshaw & Webb, 1985; Jackson, 1994; Webb, 1995, 1974).

Datasets of lake-level variations, chironomid-inferred temperatures, and stable isotope ratios, as well as climate model output, help us to show the "forces" behind vegetation changes and to identify dynamics. We therefore take advantage of advances in paleoclimate data, analysis, and modeling that are providing an increasingly detailed picture of late-Quaternary climate changes. Just as radiocarbon dating freed pollen data from a correlation-based time frame, newly developed paleoclimate datasets now allow pollen data to be interpreted within an independently derived climate framework. We can therefore describe how the vegetation responded to multivariate changes in climate involving temperature, moisture, and seasonality.

We use both time series and maps of pollen data and climate estimates: (1) to illustrate a strong connection between climate and vegetation change; (2) to document continental- and regional-scale vegetation dynamics that result from millennial- and orbital-scale climate forcing; and (3) to demonstrate that the conditions held for dynamic equilibrium between vegetation and climate at orbital time scales and possibly at millennial scales. By mapping both individual taxa and assemblages of taxa, we describe vegetation responses to independently documented climatic forcing at several levels of ecological organization from taxon movements to shifts in biome position, extent, and composition. Our chapter focuses on examples from North American vegetation history that illustrate key climatically forced vegetation dynamics. In doing so, we aim to complement the discussion of vegetation history by Grimm & Jacobson (this volume), Thompson *et al.* (this volume), and Anderson *et al.* (this volume), and build on the critical reviews written by Cushing (1965), Davis (1965), and Whitehead (1965) that Grimm & Jacobson (this volume) so ably review in their introduction. Too few pollen diagrams with radiocarbon dates existed for mapping the data on an independent time frame in 1965. Since then palynologists

DEVELOPMENTS IN QUATERNARY SCIENCE
VOLUME 1 ISSN 1571-0866
DOI:10.1016/S1571-0866(03)01021-2

have published over 500 pollen diagrams with radiocarbon dates in eastern and northern North America. Other researchers have generated data independent of pollen data for estimating past changes in climate, and climate modeling has yielded valuable simulations of late Quaternary climates and climate change (Webb, 1998; Wright *et al.*, 1993). These developments allow a fresh understanding of vegetation dynamics and testing of many of the hypotheses posed by Cushing (1965), Davis (1965), and Whitehead (1965).

Data and Methods

Records of Pollen, Lake Levels, and Vegetation Change

We apply the zoom-lens concept (i.e. linking maps and time-series data in order to span spatial and temporal scales, Figs 1–11, Webb, 1995, 2001) to describe vegetation dynamics across several levels of ecological organization. We zoom from individual sites up to continental patterns and from centuries up to several millennia in order to incorporate local changes for taxon populations up to continental scale rearrangements of taxa, tree-cover groups, and biomes. We use time sequences of continental-scale maps to show long-term spatial trends (e.g. range shifts and changing abundances) that most likely result from macroclimatic patterns (Figs 1–3) and subsume all of the short-term, small-scale vegetation dynamics most often studied by ecologists, such as succession. Contouring the pollen data (via locally weighted spatial smoothing) emphasizes the broad-scale patterns. Uncertainties in: (a) radiocarbon dates; (b) their calibration into calendar years (Stuiver *et al.*, 1998); and (c) the interpolation of time between dates limit the temporal resolution of maps to ~500–1000 yr, but much higher temporal resolution can be obtained in individual pollen diagrams. By combining maps and time series, we can zoom in spatially to individual sites and gain temporal resolution that is not possible when correlating data between sites at the broader map scale (Webb, 2000, 2001). Pairing time series of pollen data with independent climate records from the same site allows focused

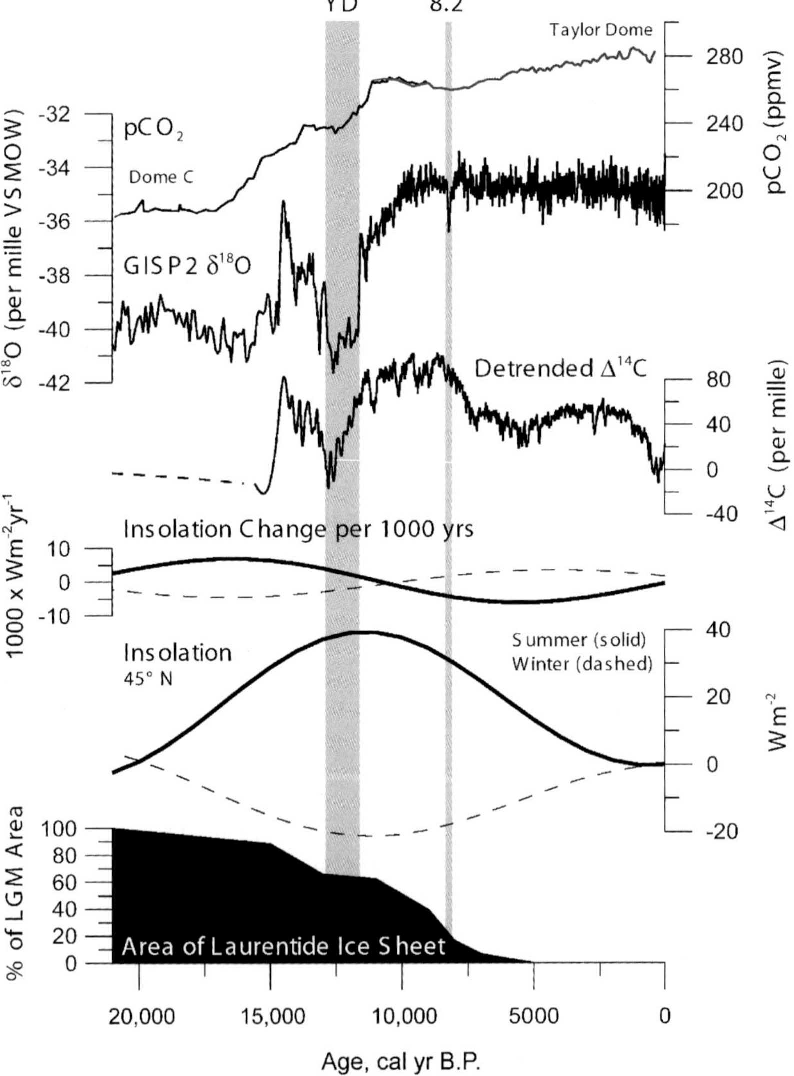

Fig. 1. Controls on North American climates and climate changes over the past 21,000 years. Atmospheric carbon dioxide concentration (pCO_2) is shown from Antarctic ice core records (Taylor Dome, Indermühle et al., 1999; Dome C, Monnin et al., 2001). Oxygen isotope ratios from a Greenland ice core (GISP2) record abrupt temperature changes in the North Atlantic region (Stuiver et al., 1995). The linearly detrended record of change in atmospheric radiocarbon concentration (Stuiver et al., 1998) records long-term changes in radiocarbon production, but also rapid changes in oceanic ventilation and North Atlantic thermohaline circulation (Clark et al., 2001; Stuiver et al., 1995). Long-term changes in summer and winter insolation at 45°N latitude are shown as calculated by Berger (1978). The area of the Laurentide ice sheet was estimated from maps by Dyke & Prest (1987) and shifted in calendar years in accordance with Barber et al. (1999).

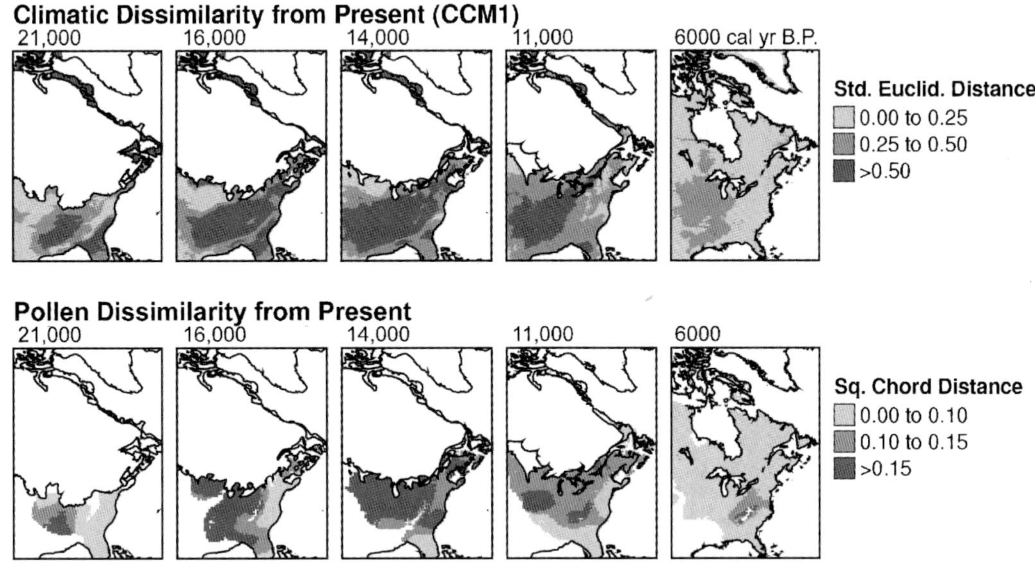

Fig. 2. Dissimilarity values for the fossil pollen data and CCM1 climate simulations for different times. The rows of maps indicate the dissimilarity of the simulated climates (top) and fossil pollen assemblages (bottom) from their best modern counterpart. (See Williams et al., 2001, for explanation of the dissimilarity measures used.)

study of the nature and rates of vegetation change associated with millennial- and finer-scale changes in climate. These temporal changes are the local and regional manifestations of continental-scale phenomena shown on the maps.

The contoured maps of pollen percentages show changes in range and abundance for individual taxa and associations of taxa (Figs 3 and 4). Biome maps (Fig. 4) derived from a pollen-to-biome classification method (Prentice *et al.*, 1996; Prentice & Webb, 1998; Williams *et al.*, 1998) and tree cover maps (Fig. 5) characterize vegetation physiognomy and structure (Williams, 2003; Williams & Jackson, 2003). We also assess the changes in vegetation between time intervals by mapping: (1) the difference in the percent abundance of individual pollen types (Fig. 6); (2) biome distribution changes (Fig. 6); and (3) dissimilarity measures for assemblage changes at each grid point (Figs 2 and 7). To highlight long-term changes, we focus on the changes that took place between both 21,000 and 6000 cal yr B.P. and presettlement time (500 cal yr B.P.) (Fig. 6). Squared-chord distances (Overpeck *et al.*, 1985) provide a multivariate measure of the difference between pollen samples, either between fossil and modern samples (Fig. 2) or between consecutive time intervals spaced apart by 1000, 3000, and 5000 years (Fig. 7). These latter maps provide an update to the histograms of chord distances between samples at 100-yr intervals from selected sites in Jacobson *et al.* (1987) and between 500-, 1000-, and 3000-year intervals at all sites in Overpeck *et al.* (1991).

Our maps of the pollen and lake-level data update those of Bernabo & Webb (1977), Webb *et al.* (1983), Jacobson *et al.* (1987), Webb (1988), Harrison (1989), R.S. Webb (1990), R.S. Webb *et al.* (1993), T. Webb *et al.* (1993), and Jackson *et al.* (1997) by showing the data after their radiocarbon dates are calibrated to calendar years. We also map paleogeographic features like ice-sheet extent, shorelines, and glacial lakes. Shuman (2001) and Shuman *et al.* (2002b) describe the lake-level data and their mapping, and T. Webb *et al.* (1993), Shuman (2001), and Williams *et al.* (2000, 2001) list the pollen sites and dating choices. The pollen data are available from (http://www.ngdc.noaa.gov/paleo/napd.html).

Independent Climate Estimates

A multi-proxy paleoclimate framework for the past 21,000 calendar years is emerging from recent studies that is suitable for comparison with the maps of fossil pollen data at millennial and longer time-scales and with selected individual pollen records at shorter time-scales. Each type of paleoclimate data is sensitive to different aspects of climate, and together they provide a rich source of information about past climatic variations. Many aspects of past climates can be deduced, independently of the fossil pollen record, using paleolimnological data to infer past lake levels (e.g. Digerfeldt, 1986), to obtain stable isotopes ratios (e.g. Fritz *et al.*, 2000; Stuiver, 1968; Yu *et al.*, 1997), and to yield assemblages of aquatic biota such as diatoms, ostracodes, chironomids, and testate amoebae (e.g. Booth & Jackson, 2003; Fritz *et al.*, 2000; Smith, 1993; Walker *et al.*, 1991). Pairing such records with fossil pollen assemblages enables the study of ecological responses to environmental change, particularly when all records are collected from the same site (Ammann *et al.*, 2000; Bradbury & Dean, 1993; Shuman *et al.*, 2003; Williams *et al.*, 2002; Yu *et al.*, 1997).

We compare the fossil pollen data to local and regional temperature trends inferred from hydrogen isotope ratios (Fig. 8) (Huang *et al.*, 2002) and from chironomid

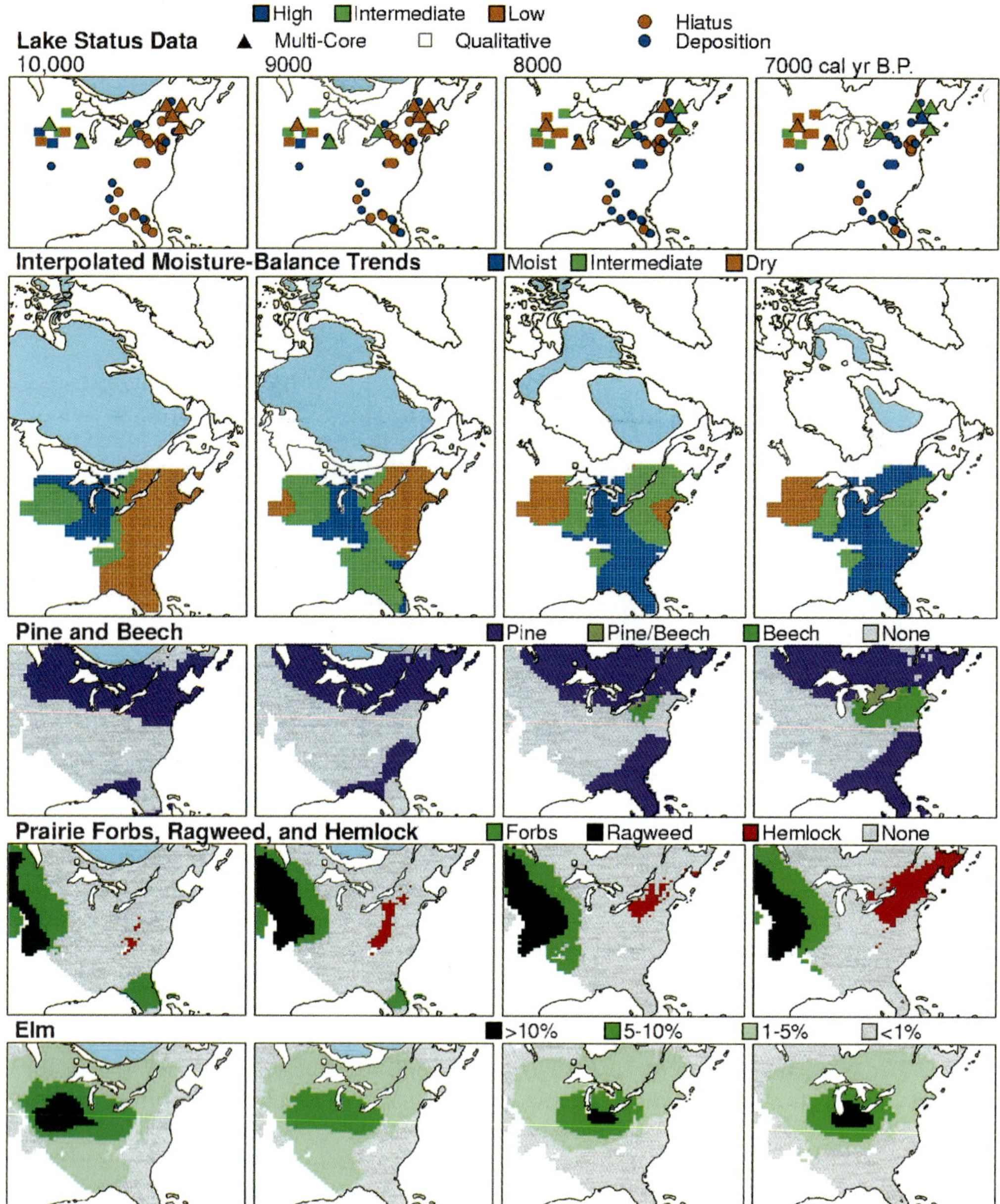

Fig. 3. Maps from 10,000 to 7000 cal yr B.P. illustrate changing moisture-balance patterns and vegetation distributions as the Laurentide ice sheet collapsed. The uppermost panel shows three types of lake-level data: multi-core, multi-proxy studies like Digerfeldt (1986), qualitative assessments of lake-level indicators following Harrison (1989), and hiatuses in published pollen stratigraphies. The second panel shows the general trends in moisture balance according to a locally weighted interpolation of the lake-level data. Two lower panels show parallel changes in the extent of regions with >25% pine (Pinus), 5% beech (Fagus), 15% prairie forb (Asteraceae, Chenopodiaceae/Amaranthaceae, and Artemisia), 10% ragweed (Ambrosia), and 10% hemlock (Tsuga) pollen (from Shuman et al., 2002b).

assemblages (Fig. 9) (Levesque *et al.*, 1993; Walker *et al.*, 1991; Williams *et al.*, 2002). Both types of data can be controlled by factors other than temperature, but in certain settings, such as those considered here, each can yield useful paleotemperature estimates. We also map past moisture balance trends derived from lake-level data (Fig. 3) (e.g. Harrison, 1989; Shuman, 2001; Shuman *et al.*, 2002b; R.S. Webb *et al.*, 1993). Striking regional similarities among the lake-level histories of multiple lakes studied by the Digerfeldt (1986) method likely reflect long-term changes in regional moisture-balance (e.g. among New England and Quebec lakes studied by Almquist *et al.*, 2001; Lavoie & Richard, 2000; Newby *et al.*, 2000; Shuman, 2001; Shuman *et al.*, 2001). The maps also include: (1) evidence for drier-than-modern

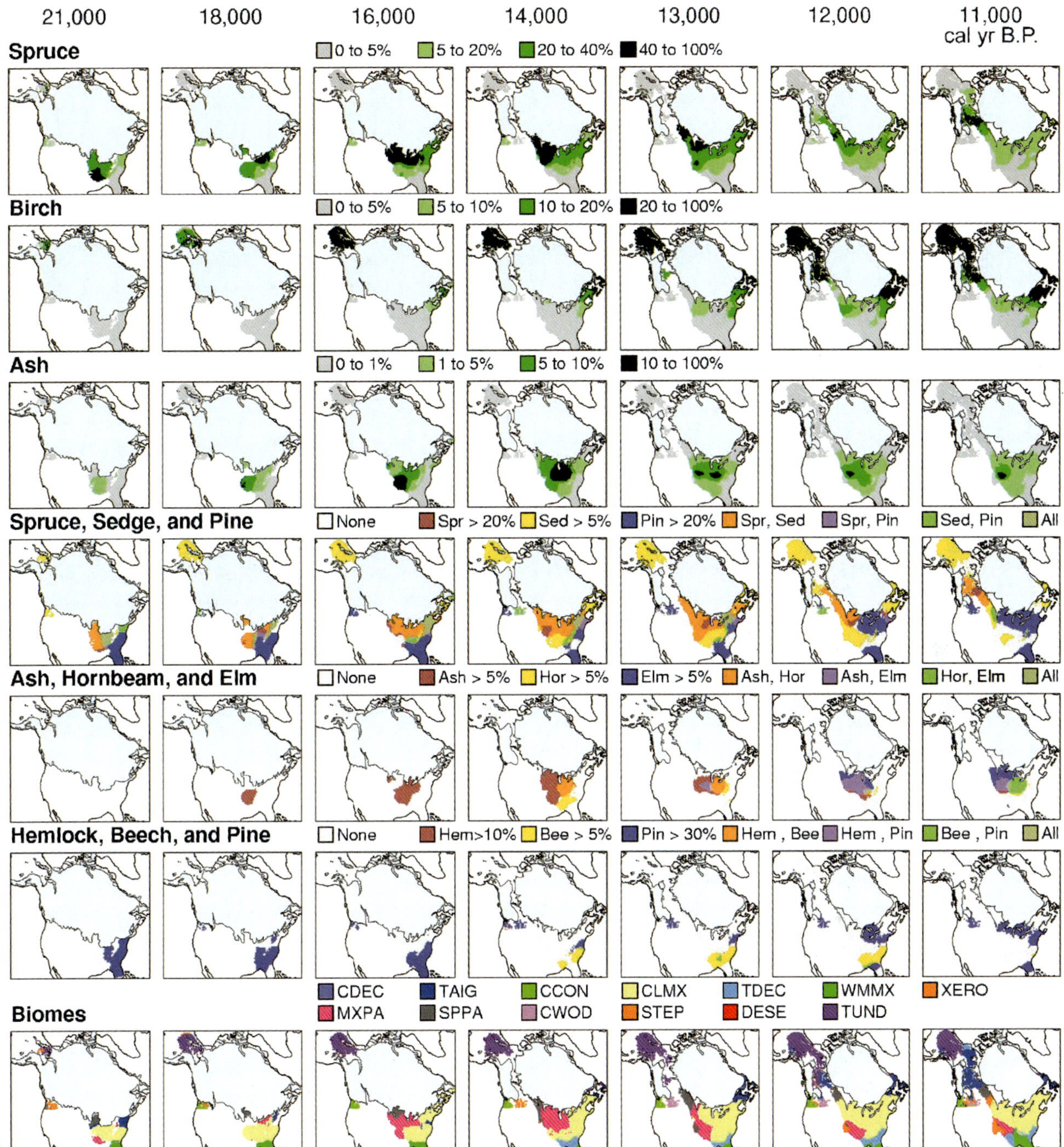

Fig. 4. Maps of individual taxa, combinations of three taxa, and biomes for selected dates during the past 21,000 years. Upper three rows show the distributions of individual taxa: spruce (Picea), birch (Betula), and ash (Fraxinus). Three progressively darker shades of green represent progressively higher relative abundance. Grey indicates the absence of the taxa at the mapped abundance level, and white indicates regions with no data. The fourth through sixth rows contain maps of three combinations of three taxa: spruce, sedge (Cyperaceae), and pine (Pinus); ash, hornbeam (Ostrya-type), and elm (Ulmus); pine, hemlock (Tsuga) and beech (Fagus). Individual taxa are plotted, where they grow alone, as either red, yellow, or blue, and combinations of the taxa are plotted as combinations of these primary colors. The bottom row shows maps of biome distribution. Biomes include cool deciduous forest (CDEC, light blue), taiga (TAIG, dark blue), cool conifer forest (CCON, light green), cool mixed forest (CLMX, yellow), temperate deciduous forest (TDEC, pale blue), warm mixed forest (WMMX, bright green), xerophytic woodland (XERO, orange), mixed parkland (MXPA, red), spruce parkland (SPPA, gray), conifer woodland (CEOD, pink), steppe (STEP, orange), desert (DESE, bright orange), and tundra (TUND, purple). Maps are accurate within an envelope of 500 years about the time assigned to them.

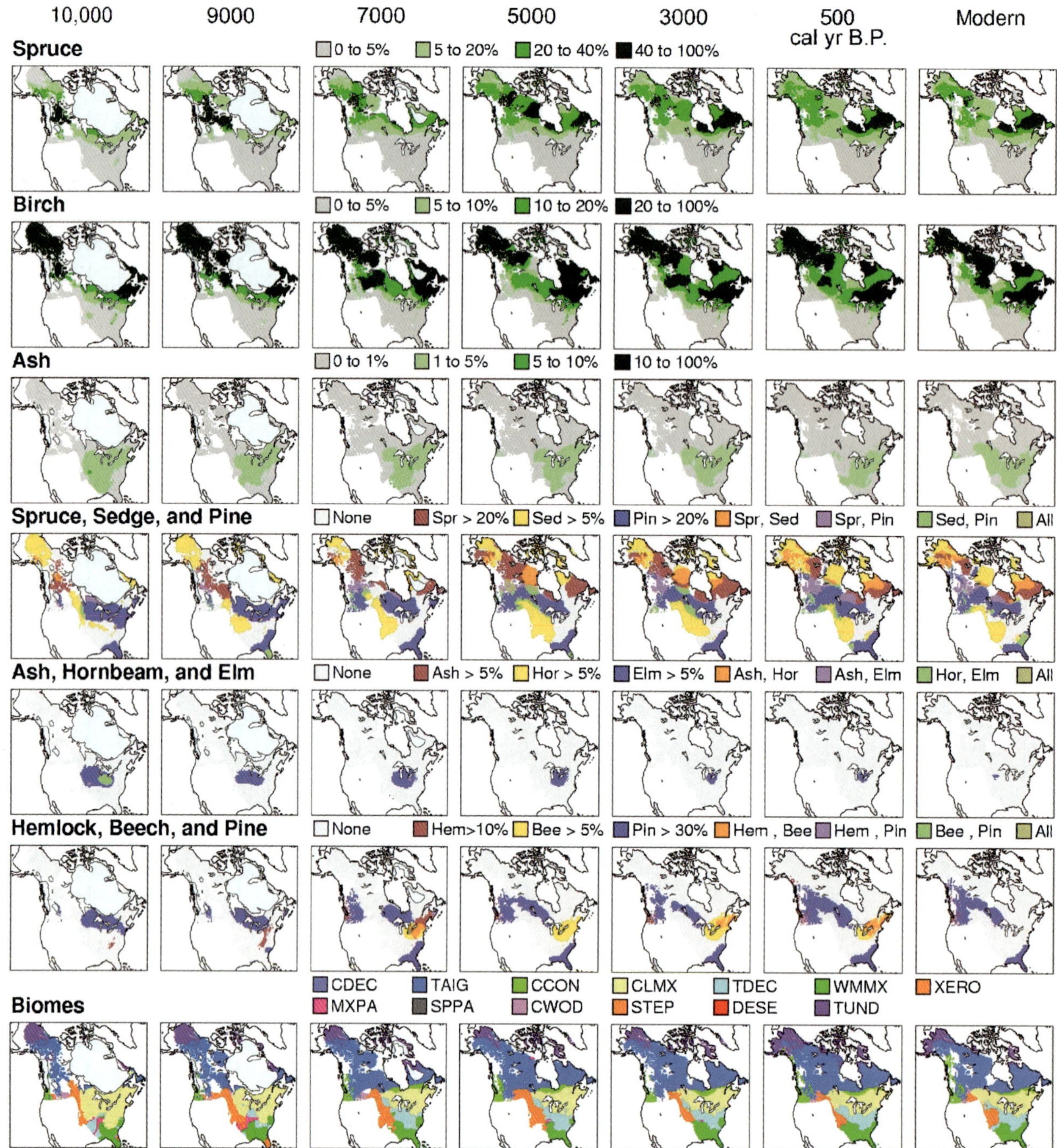

Fig. 4. (Continued)

conditions from sedimentary hiatuses in lake cores (Webb & Webb, 1988); and (2) qualitative estimates of lake-level change based upon sediment type, aquatic macrofossils, and other indicators (e.g. Harrison, 1989; Webb, 1990).

General circulation model experiments have provided another source of independently derived information about past climates at broad scales. The experiments show possible climate responses to known changes in climatic boundary conditions (Fig. 1), such as the long-term changes in insolation (Berger, 1978), glacial extent (Dyke & Prest, 1987), and atmospheric carbon dioxide concentrations (Indermühle *et al.*, 1999; Monnin *et al.*, 2001). Model simulations may not be accurate at all scales or for each climatic variable, but previous data-model comparisons support the general simulation of glacial-to-interglacial climate change (Webb, 1998; Whitlock *et al.*, 2001; Wright *et al.*, 1993) and of certain millennial-scale climate changes (Rind *et al.*, 1986; Rutter *et al.*, 2000).

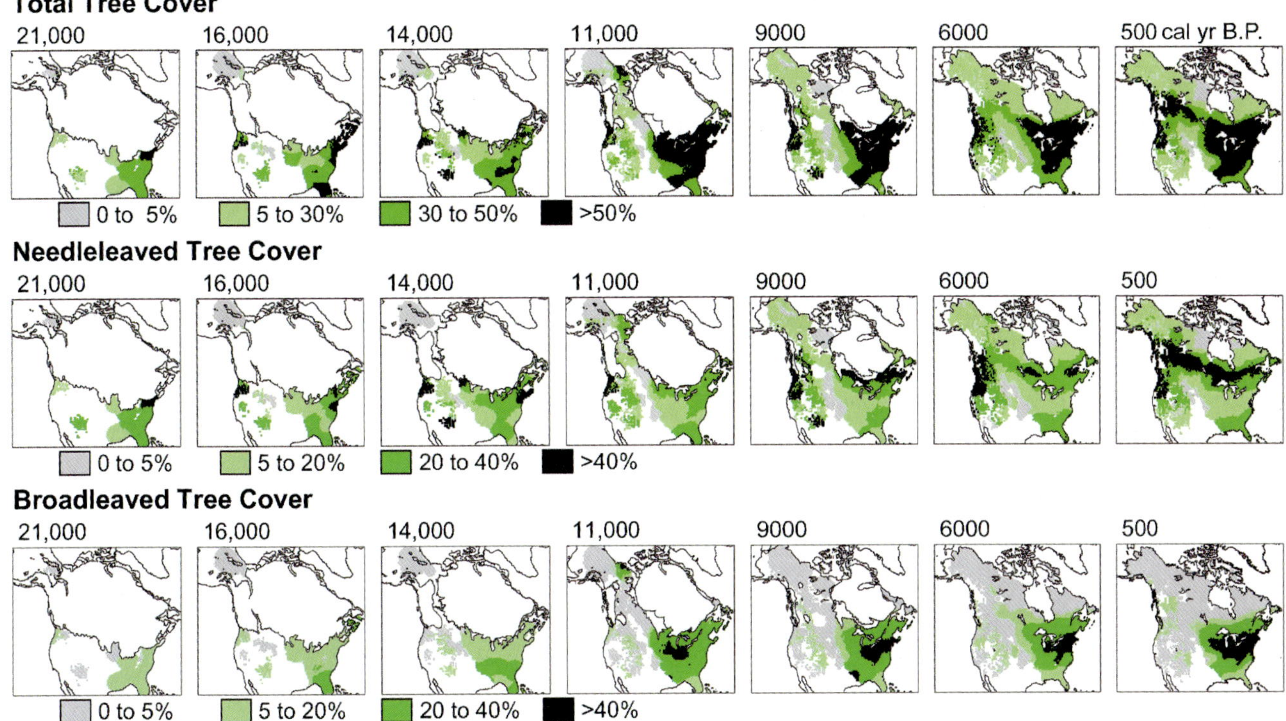

Fig. 5. Tree cover maps for all woody taxa, needle-leaved taxa, and broad-leaved taxa. Tree coverages are expressed as percentages of total area within each 50 × 50 km grid cell (from Williams, 2003).

Results

North American Climates Over the Past 21,000 Years

In the time since the last glacial maximum (LGM), 21,000 years ago, North American climates have changed significantly at both orbital and millennial time scales. Between 21,000 and 6000 years ago, climates warmed and full-glacial conditions gave way to interglacial conditions (Bartlein *et al.*, 1998; COHMAP, 1988). Changes in seasonal insolation, ice-sheet extent, and atmospheric carbon dioxide and dust concentration are the main controls for these orbital-scale changes (Kutzbach *et al.*, 1993, 1998) (Fig. 1). The glacial-interglacial transition was punctuated, however, by fast changes in global climate controls (Fig. 1), such as: (1) the reorganization of oceanic heat transport ca. 14,600, 12,900, and 11,600 cal yr B.P. at the beginning and end of the Bølling/Allerød and Younger Dryas chronozones (Broecker *et al.*, 1985; Clark *et al.*, 2002; Ruddiman & McIntyre, 1981; Rühlemann *et al.*, 1999), which is tracked by changes in atmospheric ^{14}C concentration (Clark *et al.*, 2001; Stuiver *et al.*, 1995); (2) the rapid increases in atmospheric carbon dioxide concentration ca. 15,000 and 11,000 cal yr B.P. (Monnin *et al.*, 2001); and (3) the final collapse of the Laurentide ice sheet ca. 8200 cal yr B.P. (Barber *et al.*, 1999).

At orbital time scales, data and model syntheses by COHMAP (1988; Webb, 1998; Wright *et al.*, 1993) show how: (1) ice-sheet retreat; and (2) the shift from low to high to low seasonality in insolation (Fig. 1) created spatially variable patterns of climate change across North America and the globe. In North America, the southern branch of the jet and winter storm track moved north, and the glacial anticyclone lessened and then disappeared as continental ice sheets retreated. From 16,000 to 9000 cal yr B.P., the combination of ice sheet coverage, CO_2 levels, and increased insolation seasonality (Fig. 1) was different enough from today to induce the NCAR CCM1 to simulate climates unlike any today in the North American mid-continent (Fig. 2). These simulated climates without modern analogs were characterized by warmer-than-present summers, colder-than-present winters, and low precipitation levels relative to present (Kutzbach *et al.*, 1998; Webb *et al.*, 1998). The timing of maximum warmth, cooling, dryness, or wetness varied geographically, however, in response to the spatial variations in: (a) forcing; and (b) atmospheric circulation changes, which differentially altered the advection of heat and moisture (Bartlein *et al.*, 1998). For example, during the early Holocene, higher-than-present seasonality of insolation and the retreating ice sheet altered radiation and temperature gradients. As a result, circulation patterns changed. The maps of lake-level variations (Fig. 3) in eastern North America illustrate the resulting spatial patterns of moisture-balance change by showing how the Midwest dried out while the Southeast and then the Northeast became wetter. Part of this gradual change was punctuated by an abrupt change in atmospheric circulation at 8200 cal yr B.P. induced by the final collapse of the Laurentide ice sheet (Shuman *et al.*, 2002b).

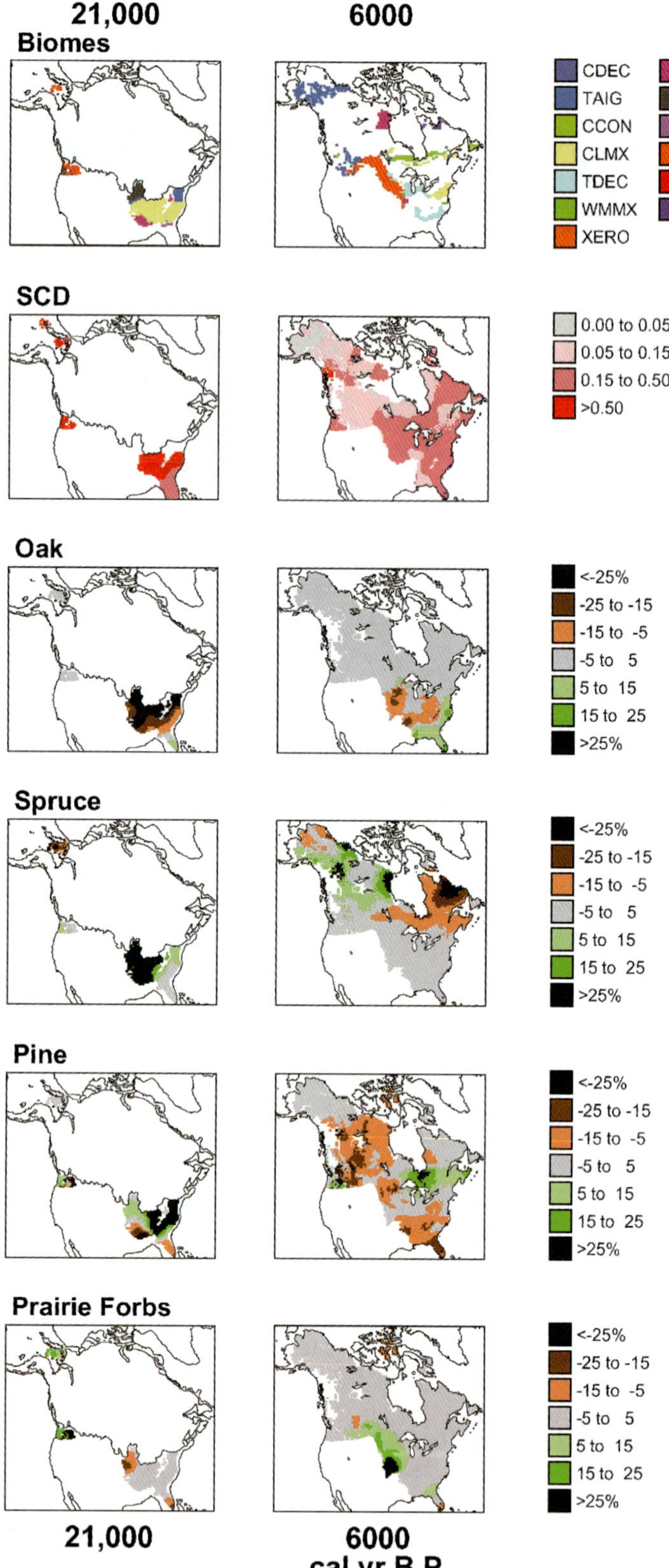

Fig. 6. Vegetation anomaly maps for the last glacial maximum and mid-Holocene, expressed in terms of biomes, oak, spruce, pine, and prairie forbs. All anomalies are expressed as differences between the past time period and pre-settlement vegetation (21,000–500 cal yr B.P.; 6000–500 cal yr B.P.). Blank areas in the biome maps indicate no data or no change in biome type. In the top panels, colored areas show the biome assignments for the past interval in grid cells that have changed between the present and past. Biome abbreviations same as in Fig. 4. Green colors in the individual taxon anomaly maps indicate that a taxon was locally more abundant in the past; browns indicate areas where a taxon is more abundant at present.

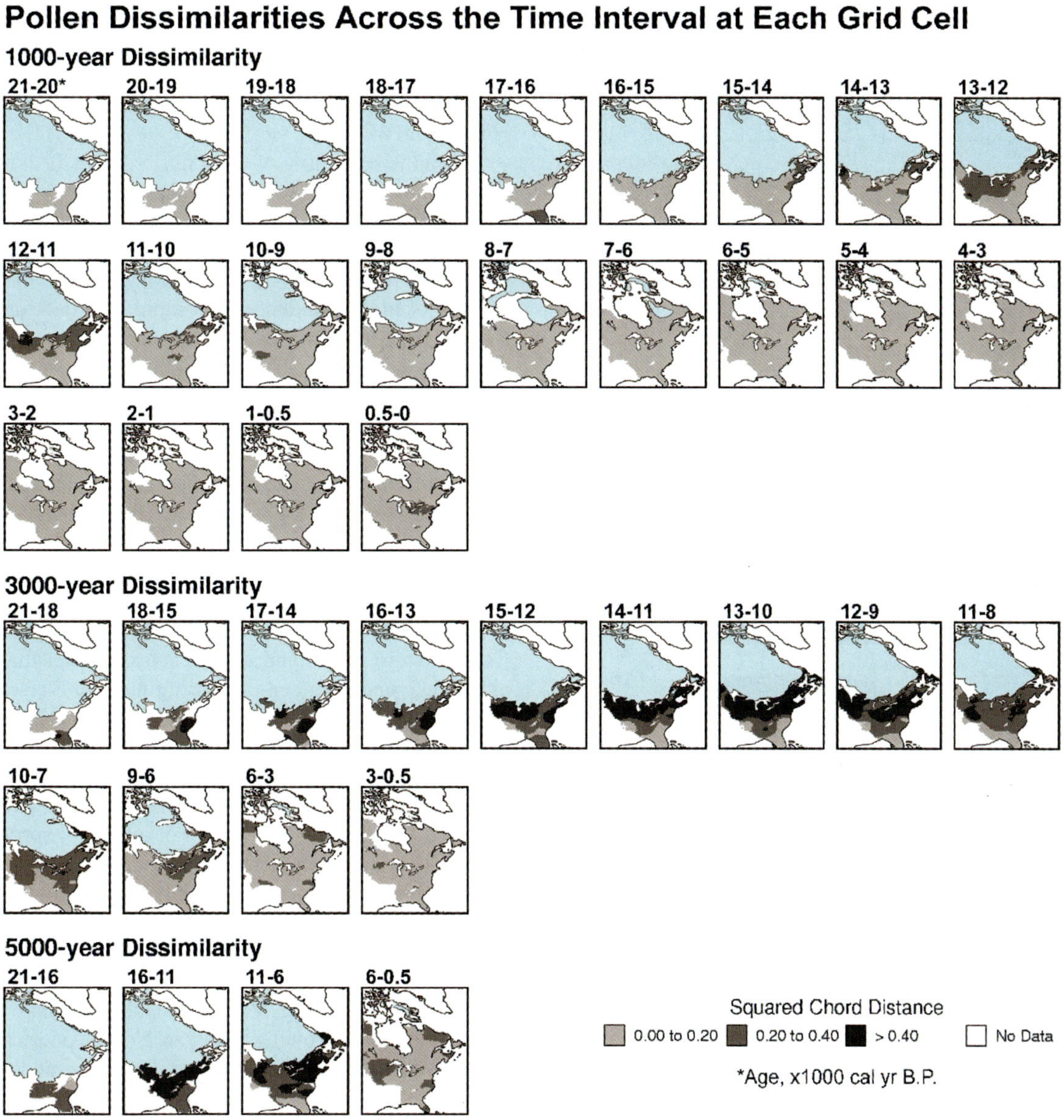

Fig. 7. Maps of square-chord distance between fossil pollen spectra at each mapped 50 × 50 km grid point and pollen spectra at the same grid point 1000, 3000, and 5000 years later. The difference between modern (0 yr ago) and just before European settlement (500 yr ago) is also shown. Large differences between intervals are estimated by large square-chord distances, and plotted as dark gray or black. Distances greater than 0.20 are greater than the modern difference among biomes (see Overpeck et al., 1985). The contour intervals were chosen because they allow the maps to illustrate the contrast in the spatial patterns between the 1000-yr difference maps and the 3000-yr and 5000-yr difference maps.

Millennial-scale climate variability also resulted in distinct spatial patterns of change. Changes during the Younger Dryas chronozone (12,900–11,600 cal yr B.P.) provide an illustrative example. Sea-surface temperatures (SST's) in the North Atlantic were colder then than during the previous two millennia (Ruddiman & McIntyre, 1981) and directly cooled proximate regions of North America (Mott *et al.*, 1986; Peteet *et al.*, 1990; Walker *et al.*, 1991). These patterns of climate change during the Younger Dryas chronozone resulted from reorganized ocean and atmospheric heat transport, which also led to seasonally warmer climates elsewhere in North America (Kneller & Peteet, 1999; Shuman *et al.*, 2002a) and in other regions of the world (Bluiner *et al.*, 1998; Clark *et al.*, 2002; Rühlemann *et al.*, 1999). Globally, Younger Dryas climate patterns were unique within the past 21,000 years because the changes in ocean circulation were embedded within a context of radiation forcing, ice-sheet extent, and carbon dioxide concentrations that was strikingly different from other times (Figs 1 and 4) (see also figures in Whitlock *et al.*, 2001).

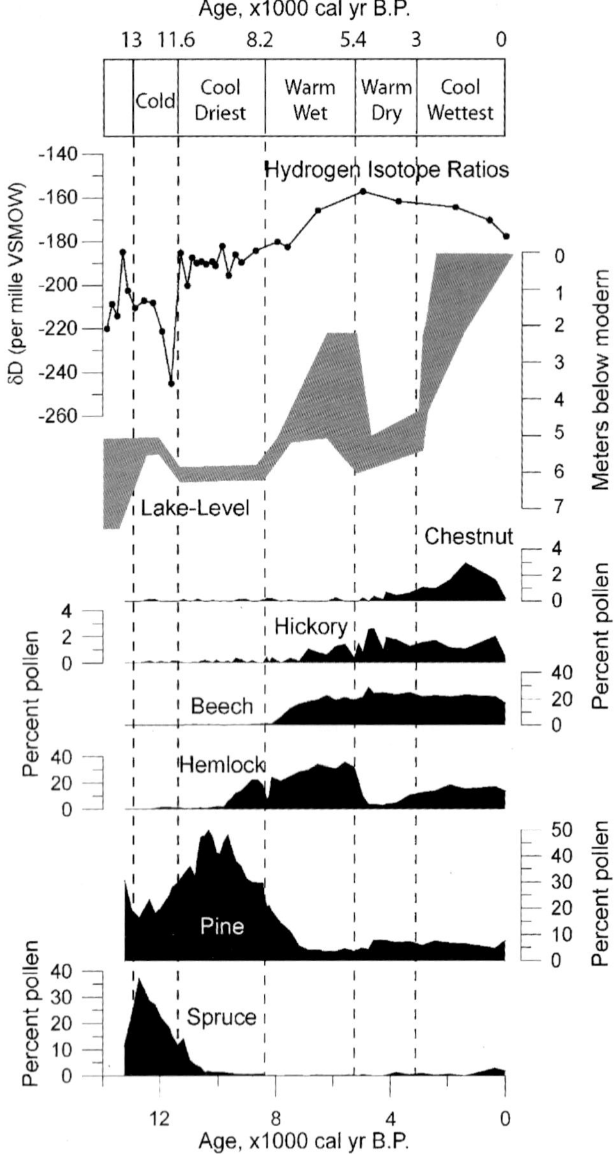

Fig. 8. Summary of New England climate and vegetation history. Hydrogen isotope ratios and lake-level estimates from Crooked Pond (Huang et al., 2002; Shuman et al., 2001) in southeastern MA are compared with pollen percentages from North Pond (Whitehead & Crisman, 1978) in western MA 206 km to the northwest of Crooked Pond. Inferred climate phases are given at top (from Shuman et al., 2003).

Individualistic Responses by Taxa to Orbital-Scale Climate Change

Our maps match those of Jackson *et al.* (2000) in showing how much the vegetation patterns at the LGM differed from those today (Figs 4 and 6). Conifer parklands grew south of the ice sheet grading into pine woodlands along the southeast coast into Florida (Fig. 4). Deciduous trees and shrubs were not numerous, but grasses and sedges were relatively abundant implying an openness to the vegetation (Figs 4 and 5). The full-glacial climatic conditions produced a distribution of biomes that differed markedly from today. Several modern biomes (e.g. taiga, cool conifer forest, and temperate deciduous forest) were not evident or were much restricted in range (Figs 4 and 6).

With the beginning of the long-term ice-sheet retreat and temperature increase (Fig. 1), both the range boundaries and the regions of peak abundance for many taxa shifted northward as well as east or west (Fig. 4). For example, spruce was abundant from 21,000 to 13,000 cal yr B.P. in the east and its area of high abundance then rapidly shifted into western Canada by 11,000 cal yr B.P. By the early Holocene, the range of spruce covered much of boreal Canada as it does today, but was restricted by the ice sheet. Following 7000 years of low abundance in the east, however, the peak of spruce abundance shifted eastward from western to eastern Canada after 6000 cal yr B.P. The maximum abundance of many taxa also increased as ranges shifted. Birch abundance began increasing in Beringia after 18,000 cal yr B.P. and south of the ice sheet only after 16,000 cal yr B.P. In eastern North America, hemlock only began increasing as it spread north through the Appalachians after 10,000 cal yr B.P. (Figs 3 and 4). Other taxa (e.g. ash and elm) reached greater-than-modern abundance and range extent during the late Pleistocene and early Holocene from 16,000 to 10,000 cal yr B.P. (Fig. 4). As a consequence of these varying independent movements among the taxa in terms of their geographic range and the location of their abundance maxima, new assemblages emerged and then disappeared. For example, spruce, sedge, and ash overlapped in their area of abundance from 16,000 to 12,000 cal yr B.P. to create a mixed parkland biome that does not exist today (Fig. 4). The mixed forest that contains both hemlock and beech today only formed after the ice sheet collapsed ca. 8000 cal yr B.P. (Figs 3 and 4).

The associations of plant taxa and functional types that were unlike any growing today in North America (i.e. no-analog vegetation) appear to have developed under climates that were also dissimilar from today (Fig. 2). From 16,000 to 11,000 cal yr B.P., the NCAR CCM1 simulated climates without modern analogs just where the pollen data indicate that plant assemblages without modern equivalents grew (Fig. 2) (Williams *et al.*, 2001). The observed match between the pollen data and model simulations supports earlier hypotheses that the no-analog plant associations formed as a consequence of plant taxa responding individualistically to multivariate changes in climate (Cushing, 1965; Overpeck *et al.*, 1992).

Besides changes in temperature and the seasonality of insolation, variations in moisture balance also played a large role in altering vegetation patterns. For example, within the early Holocene, lake-level maps show the Midwest becoming dry between 10,000 and 7000 cal yr B.P. and the vegetation responding with the eastward expansion of the prairie (Fig. 3). In the Northeast, conditions were dry at 10,000 cal yr B.P., and then became wetter as mesic taxa like birch, hemlock, maple, and beech replaced the white pine populations that had been dominant there (Fig. 3; Davis & Jacobson, 1985; Jackson *et al.*, 1997; Newby *et al.*, 2000; Shuman *et al.*, 2001, 2002b). The Southeast was

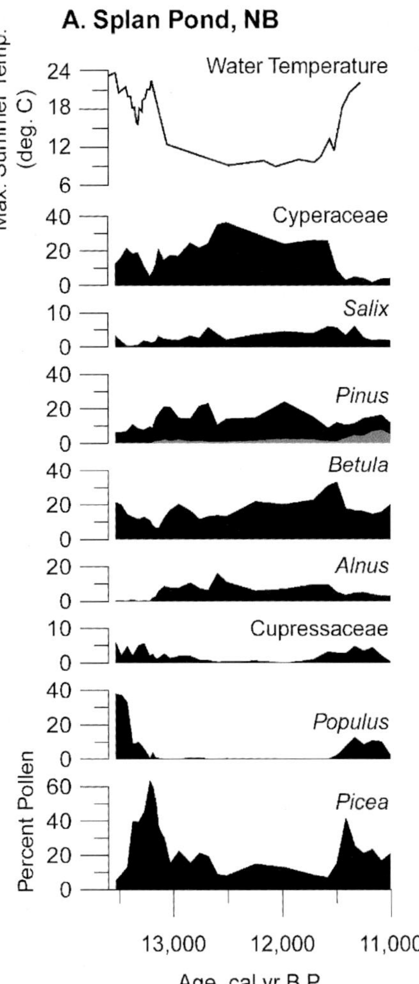

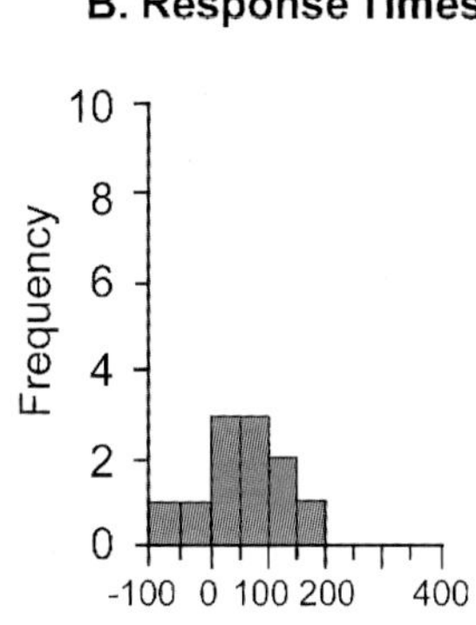

Fig. 9. (a) Summer temperatures at Splan Pond, New Brunswick, spanning the Younger Dryas chronozone, inferred from fossil chironomid evidence (Levesque et al., 1993) and compared to pollen data from the same site (Mayle & Cwynar, 1995). Arboreal pollen types shown are spruce (Picea), aspen (Populus), cedars (Cupressaceae), alder (Alnus), birch (Betula), and total pines (Pinus) with white pine (P. strobus-type) shown in gray; nonarboreal pollen types are willow (Salix) and sedges (Cyperaceae). (b) A histogram of the time lags (positive if pollen data are lagging) associated with the most significant (i.e. lowest p-value) cross-correlations between the climate proxy record and pollen principal components done down-core for each of eleven sites in Maritime Canada and Europe (Williams et al., 2002).

also drier than today before 10,000 cal yr B.P. and then its moisture balance became more positive as southern pines replaced oaks there after 8000 cal yr B.P. These maps show how closely the vegetation changes match those in moisture balance and how well taxa favoring wet or dry conditions respond to the appropriate changes in moisture (Fig. 3). The climate response surfaces for these pollen taxa support these interpretations (T. Webb *et al.*, 1993, 1998).

The adjustments of individual plant taxa to climate change also led to new arrangements of vegetation cover, density, and functional composition. Biome maps (Figs 4 and 6) show large changes in the position and area of biomes between the last glacial maximum and present, with a major reorganization between 14,000 and 9000 cal yr B.P. Compared to the individual taxa, the reconstructed biomes did not move much, however. Instead of migrating large distances as climate changed, biomes grew and shrank in places where novel associations of taxa and functional types appeared and disappeared (Fig. 4). Biomes characteristic of the Holocene (e.g. temperate deciduous forest, taiga) were absent or occurred only in limited areas during the late Pleistocene, whereas the mixed parkland biome largely disappeared after 11,000 cal yr B.P. (Fig. 4). In some cases, shifts in biome position and type can be traced to the distribution of individual plant taxa – for example, the northward spread of spruce populations caused the end of its association with sedge, ash, and hornbeam (the end of the mixed parkland) and the beginning of an association with alder and birch (the rise of taiga) (Fig. 4). Here, not only range shifts but the increases in maximum abundances of alder and birch after 16,000 cal yr B.P. also played a key role in the development of the new biome. In other cases, variations in the distribution of individual taxa are subsumed as internal variations in biome composition (Williams *et al.*, 2003). As categorical representations of the vegetation, the biome maps limit apparent vegetational change to ecotonal regions between biomes (see the 6000 cal yr B.P. to present differences in Fig. 6). Continuous measures of the vegetation variation such as shown on isopoll maps (Fig. 4) and difference maps (Fig. 6) demonstrate that this emphasis on ecotones is an artifact of the biome categories; actual change was far more widespread (Figs 4–7).

Describing the relative density of various plant life forms avoids limitations inherent to categorical classifications of the vegetation while still providing an estimate of vegetation structure (DeFries *et al.*, 2000). Reconstructions of % tree cover based on calibrating pollen records to AVHRR

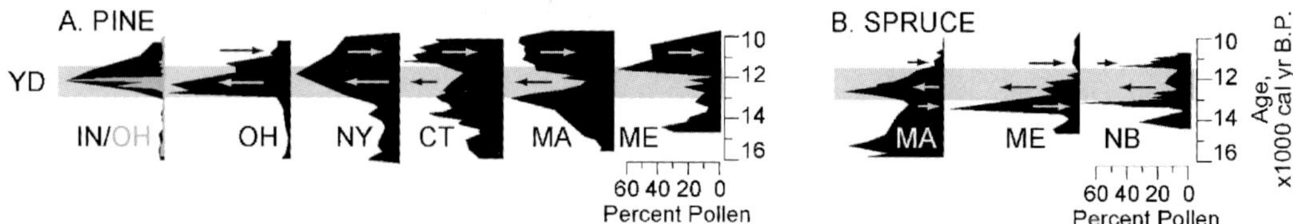

Fig. 10. Pine and spruce pollen percentages track abrupt range shifts during the Younger Dryas chronozone (12,900–11,600 cal yr B.P.) Records of pine (Pinus) (A) and spruce (Picea) (B) pollen percentages are plotted with time on a vertical axis to show changes in their geographic distributions. Arrows indicate the east-west range shifts of pine (A) and the north-south range shifts of spruce (B). Each site is labeled by state as in Shuman et al. (2002a). The Stotzel-Leis site in Ohio (Shane, 1987) is shown in gray in inset A, and superimposed upon the stratigraphy from Pretty Lake in Indiana (Williams, 1974) (from Shuman et al., 2002a).

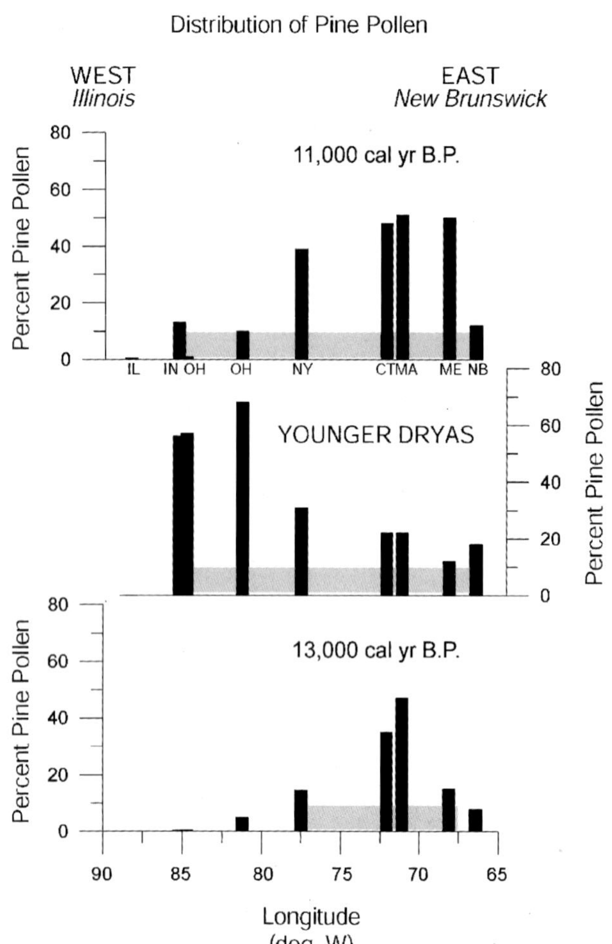

Fig. 11. The abundance of pine pollen across eastern North America before, during, and after the Younger Dryas chronozone. Histograms represent the % pine (Pinus) pollen at each of nine sites (Shuman et al., 2002a) at 13,000, 12,000 (middle of the Younger Dryas chronozone), and 11,000 cal yr B.P. The percentages are plotted with respect to longitude to show changes in abundance across the range of pine. Each site is labeled by state. Horizontal gray bars mark the range of sites where the presence of pine populations is indicated by pollen percentages of >10% (from Shuman et al., 2002a).

(Advanced Very High Resolution Radiometer) observations of the modern vegetation indicate that total tree cover has increased since the last glacial maximum (Fig. 5) (Williams, 2003; Williams & Jackson, 2003). Total tree cover in eastern North America at the last glacial maximum was generally <50%, and forests and parklands were dominated by needle-leaved tree taxa. Much of the subsequent increase in total tree cover is due to the rise in the abundances of broadleaved deciduous tree taxa as temperatures increased. Needle-leaved tree coverage, initially high across most of eastern North America, separated into northern and southern bands by 11,000 cal yr B.P. The tree cover maps are able to show subtler gradients and features in the vegetation than shown by biome maps (Figs 4 and 5). For example, the prevalence of cool mixed forest across much of eastern North America during the late Pleistocene does not hint at the increase in tree-cover densities or at the general shift from needle-leaved to broad-leaved taxa during this period (Figs 4 and 5).

Zoomed-in View of Climate and Vegetation History in New England

When we zoom in to specific time series in the Northeast (Shuman *et al.*, 2003), stable isotope ratios and lake-level estimates show six phases of climate history in southern New England during the past 15,000 years, corresponding with six regional pollen zones (Fig. 8). The different arrival times of the dominant taxa match well with the development of favorable conditions for their growth and do not appear to be controlled by dispersal limitations (Johnson & Webb, 1989). The new climate records support Deevey's (1939) original climatic interpretation of the sequence of pollen zones and add to the climate history in Davis *et al.* (1980).

Highly negative hydrogen isotope ratios and lower-than-modern lake levels (Huang *et al.*, 2002; Newby *et al.*, 2000; Shuman *et al.*, 2001) indicate that New England climates were colder and drier than present before 13,000 cal yr B.P., when cold-tolerant spruce and pine populations were common. An expansion of spruce populations and a decline of pine populations during the Younger Dryas chronozone followed a cooling indicated by lowered stable isotope ratios (Huang *et al.*, 2002; Shemesh & Peteet, 1998). After 11,600 cal yr B.P., a rapid positive shift in hydrogen isotope ratios and a decrease in lake levels indicate a rapid shift to warmer- and drier-than-earlier conditions coincident with the decimation of spruce populations and expansion of white pines (Huang *et al.*, 2002; Lavoie & Richard, 2000; Newby *et al.*, 2000; Shuman *et al.*, 2001; R.S. Webb *et al.*, 1993). The pines remained abundant until the populations of mesic types like hemlock and beech expanded. This change coincided with a rise of lake levels ca. 8000 cal yr B.P. (Shuman *et al.*, 2002b). Hickory populations did not expand in New England until more than a thousand years later when the hydrogen isotope ratios indicate that conditions became at least as warm as modern. Chestnut populations, which prefer higher soil moisture levels, expanded after 3000 cal yr B.P. when lake-levels indicate that conditions became as wet as modern (Shuman *et al.*, 2003). Warm and dry conditions from 5400 to 3000 cal yr B.P. likely contributed to the decline of hemlock populations at 5400 cal yr B.P. by contributing to moisture stress in the trees and making them more susceptible to disease (Bhiry & Filion, 1996; Davis, 1981a, b; Haas & McAndrews, 2000; Newby *et al.*, 2000; Shuman *et al.*, 2001, 2003; Yu *et al.*, 1997). The independent climate records clarify many of the issues that Davis (1965) critically evaluated.

Individualistic Responses by Taxa to Millennial-Scale Climate Change

Paired records of fossil pollen data and independent climate estimates at individual sites indicate rapid vegetation responses to the abrupt climate changes associated with the Younger Dryas chronozone (Fig. 9). Many North American sites show a shift to cold-tolerant vegetation between 12,900 and 11,600 cal yr B.P. (e.g. Mayle & Cwynar, 1995; Mott *et al.*, 1986; Peteet *et al.*, 1990; Shane, 1987; Shuman *et al.*, 2002a). High-resolution records reveal that such vegetation changes were nearly as fast as the climate changes that caused them. For example, cross-correlation analysis between pollen and chironomid data from Splan Pond, New Brunswick, shows that vegetation lagged climate by less than 100 years (Williams *et al.*, 2002) (Fig. 9). Similarly rapid responses (< 100 years to ca. 200 years) to century-scale climate changes are observed for European lakes (Ammann *et al.*, 2000; Birks & Ammann, 2000; Tinner & Lotter, 2001), implying that abrupt climate change led to rapid widespread vegetation change in areas adjacent to the North Atlantic.

Rapid changes in taxon abundance associated with the Younger Dryas interval are equally evident across eastern North America (Shuman *et al.*, 2002a) (Figs 10 and 11), and changes within the ranges of most taxa (and within biomes) are qualitatively no slower than responses quantified near range limits and ecotones, such as in New Brunswick. The rapid decimation of spruce populations at Splan Pond, New Brunswick (Mayle & Cwynar, 1995; Williams *et al.*, 2002), occurred as conditions cooled and tree line shifted southward in the first century of the Younger Dryas chronozone (Figs 9 and 10). Simultaneously, spruce populations increased again to the south in southern New England (Peteet *et al.*, 1990; Shuman *et al.*, 2002a) (Fig. 10). Because peak spruce abundance occurs only under a narrow, optimal range of climate conditions, intermediate levels of spruce abundance were maintained in Maine where conditions were neither too harsh nor optimal for spruce growth from 12,900 to 11,600 cal yr B.P. (Shuman *et al.*, 2002a) (Fig. 10).

A similar shift in the range and peak abundance of pine populations also occurred during the Younger Dryas chronozone and extended well inland (Fig. 10). Pine populations declined at the center of their former range (Fig. 11), but expanded their range into new sites in Ohio and Indiana (Figs 10 and 11). A unimodal distribution of pine abundance was maintained because, as the range of pine expanded westward, pine populations decreased in New England (where climates had become less favorable) and increased in western New York (Webb *et al.*, 2003) and the Midwest (where climates had become warm-enough for pine growth). As the climatic gradients shifted, the narrow range of optimal conditions for pine growth shifted westward in parallel with the range of tolerable conditions. Significant vegetational change, therefore, occurred not only at range boundaries or ecotones, but within the ranges of taxa (and within biomes) as well (Figs 10 and 11). Maps of plant assemblages show that the westward expansion of pine was part of a broad-scale reorganization of North American vegetation patterns between 13,000 and 12,000 cal yr B.P. (Shuman *et al.*, 2002a) (Figs 4 and 7). By classifying the vegetation into biomes, however, some of these abrupt vegetation changes can be hidden as within-biome changes in composition (compare maps of spruce, sedge, and pine assemblages with biome maps in Fig. 4), and mapped biomes changes are emphasized at ecotones, even though dramatic ecosystem changes were much more widespread (Fig. 6) (Williams *et al.*, 2003).

The responses by spruce and pine were similar in that both taxa rapidly shifted their distribution during the Younger Dryas chronozone, but the responses represent distinct and individualistic plant responses (Figs 4 and 10). The change in the distribution of spruce reflects a north-to-south shift in abundance (Fig. 10b), largely within a region previously colonized by spruce, whereas the changes in pine distribution reflect an east-to-west shift in abundance (Fig. 10a) that included the colonization of many new sites, although rapid expansion from *in situ* midwestern populations cannot be ruled out. The difference between spruce and pine response appears to arise because pines require warmer summer temperatures than spruce, and mid-continent summers may have warmed at the onset of the Younger Dryas chronozone when summer conditions cooled in New England (see GISS GCM results of Rind *et al.*, 1986; Shuman *et al.*, 2002a).

Linking Orbital and Millennial Scales of Vegetation Change

In the last 21,000 years, the magnitude of vegetation changes has not been uniform in space or time (Fig. 7). Maps of the differences between subsequent millennia (i.e. 1000-year dissimilarity maps, Fig. 7) show the most widespread, large magnitude changes going into (13,000–12,000 cal yr B.P.)

and out of (12,000–11,000 cal yr B.P.) the Younger Dryas chronozone. This pattern of temporal changes agrees with the histograms of Overpeck *et al.* (1991). Mapping the same dissimilarity data with an alternative scale that highlights smaller magnitude changes agrees with the histograms of Jacobson *et al.* (1987) and shows that change was widespread near the ice sheet from 15,000 to 9000 cal yr B.P. (not shown). The differences from one millennium to the next are dwarfed, however, by those over 3000- and 5000-year intervals (Fig. 7). These latter maps show the cumulative impact of orbitally forced changes in climate. They also show that the 1000-year dissimilarity maps slice up the vegetation changes too finely for the slow, large-magnitude changes to become manifest.

Over 3000-year intervals, the largest areas were affected from 18,000 to 6000 cal yr B.P. with the largest magnitude changes from 15,000 to 9000 cal yr B.P. focused in the 500 km south of the retreating ice sheet (Fig. 7). Changes across 3000-year timesteps before (17,000–14,000; 16,000–13,000 cal yr B.P.) and after the Younger Dryas (11,000–8000; 10,000–7000 cal yr B.P.) exceed the millennial differences at the beginning and end of the Younger Dryas (Fig. 7). And even the 3000-year changes across the Younger Dryas chronozone (14,000–11,000; 13,000–10,000 cal yr B.P.), which entirely ignore any change into and out of the Younger Dryas chronozone, far exceed the magnitude of the more rapid changes. Here we see the prominence of the orbital-scale changes over the millennial-scale climate variability in overall impact. This dominance of the orbital-scale is also evident in the maps of individual taxa and biomes (Fig. 4), which show long-term migrations of taxa and reorganization of biomes more clearly than the rapid migrations of some taxa across hundreds of kilometers at the beginning and end of the Younger Dryas (Figs 10 and 11). But on the millennial scale maps, we also see how the changes going into and out of the Younger Dryas stand out as the largest differences between millennia.

Summary of Results

The maps and diagrams have shown some of the patterns and processes active across time and space scales, as well as across different levels of ecological organization. Vegetation appears: (a) to track multivariate changes in climate (i.e. the combination of temperature, moisture, seasonality) (Figs 2, 3 and 8); and (b) to respond widely and rapidly to climate change (Figs 6, 7 and 9–11). Analyses of time-series of pollen data show that vegetation changed rapidly at times of century-scale climate change (i.e. beginning and end of the Younger Dryas chronozone) with a non-linear but continuous response across the region where climate changed (Figs 9–11). On a longer-time scale, the combination of temperature and moisture changes in New England is sufficient to explain the different arrival and expansion times of the key taxa that distinguish the classic sequence of pollen zones (Fig. 8). Seasonality is also a factor driving vegetation change given the parallel distributions of vegetation types and simulated climates without modern equivalent (Fig. 2).

Vegetation changed in structure and composition as dictated by the climate and by the differential response of individual taxa to climate. The changes yielded plant associations unlike those today when the combination of climate variables differed from those of today (Figs 2–4). Biomes are one way of summarizing vegetation structure but mapped time-series of biomes hide much of the variation evident in isopoll and plant-cover maps. Biomes, unlike individual taxa, do not shift large distances, but form and expand or contract in area as the result of changes in vegetation composition that result from individualistic changes in plant distributions (Fig. 4). Map patterns of pollen data are dominated by progressive vegetation responses to long-term orbitally forced climate changes (Fig. 7), but vegetation also rapidly reorganized in response to century- and millennial-scale climate variability (Figs 9–11). Such rapid responses show that rates of orbital-scale climate change were far slower than potential rates of vegetation change and that the impact of abrupt events is more pronounced at the local scale than at the continental scale (compare Figs 9 and 10 with Figs 4 and 7).

Discussion

The growing set of climate estimates constructed from sources other than pollen data has enabled new assessment of the degree to which climate controls vegetation dynamics. Changes in temperature, moisture-balance, and seasonality have all affected the vegetation of North America. As Williams *et al.* (2002) – building on the work of Tinner & Lotter (2001), and Ammann *et al.* (2000) – have shown, vegetation responds rapidly (within ~100 years) when climate changes rapidly on centennial time scales. The rapid responses are not just restricted to mountainous terrain but also occur in lowlands where dispersal distances can be large. These response times support the hypothesis that vegetation has been and is in a dynamic equilibrium with climate (Webb, 1986), particularly at the temporal and spatial scales of our 1000-year interval continental maps. For the conditions of dynamic equilibrium to be met, the vegetation response time must be short relative to the time scale of forcing. Orbitally forced climate variations set a beat throughout Earth history as certain as the diurnal and annual cycles (though perhaps more complex in their generation and hence more broad band). Therefore, plant taxa not only have had to evolve in the face of orbitally forced climate changes but have had to develop coping mechanisms that facilitate rapid-enough responses to changes in the position of their optimal climates (Davis & Shaw, 2001; Huntley & Webb, 1989). Taxa maintain distinct climate preferences over long time periods (Huntley *et al.*, 1989), and to do so, they must have the ability to disperse long-distances while being flexible enough to adapt to other environmental changes (e.g. changes in day length as taxa migrate northward – see Davis, 1965, and Davis & Shaw, 2001, for a discussion of this requirement).

Plant taxa also had to develop mechanisms to cope with frequent millennial-scale climate variations. That so many pollen diagrams record big changes during the Younger Dryas chronozone illustrates century-scale response times for most taxa (Shuman *et al.*, 2002a). We recognize, however, that range shifts for some taxa may significantly lag century-scale and even some millennial-scale climate oscillations. If so, our claim for dynamic equilibrium between the vegetation and climate at millennial scales involves just the Type A response (sensu, Webb, 1986) but not the full Type B response involving equilibrium range shifts and complete soil development. The vegetation changes significantly within centuries but its full response in remote areas can take longer.

The results of Shuman *et al.* (2002a) have further shown that the rapid responsiveness to climate change is widespread and not just limited to ecotones. In fact the perception of action being focused along ecotones presupposes the existence of biomes or communities, which the data show to be ephemeral because of individualistic taxon changes well beyond any ecotones (Figs 6, 7, 10 and 11). Here we can see how our human choice to classify the vegetation and pollen data affects what we conclude and perceive. Paleoecology is fraught with such traps. Because choices of how to analyze, summarize, and display the data can influence interpretation, multiple and sometimes conflicting conclusions can be and often have been drawn from the same data. We are keen for the focus on ecotones (as areas for high sensitivity of the vegetation to climate change) to disappear. The sensitivity of vegetation to climate is widespread, and the processes that enable rapid vegetation change at sites well away from ecotones (and range limits) need to be better understood. Greater climatic sensitivity at ecotones is not a feature of the data but rather a feature of how the data are summarized (Fig. 6). If the climate changes at a site, the vegetation will respond regardless of where the site is located within a biome or taxon's range (Figs 6–8, 10 and 11).

The equilibrium responses involve individualistic-type changes in range and abundance among taxa (Davis, 1981a, b; Gleason, 1926; Jacobson *et al.*, 1987; Jackson & Overpeck, 2000) that in turn lead to changes in association among taxa (Figs 2–4). Plant assemblages are never resilient to big changes in climate and either persist while undergoing large changes in the relative abundances of dominant taxa or emerge or disappear (Fig. 4). Some of the past climates were so different from those today that the contemporary plant-assemblages have no modern equivalents (Overpeck *et al.*, 1992), and the Williams *et al.* (2001) analysis of the climate model results helps to confirm this conclusion (Fig. 2). Williams *et al.* (2000) also showed that one advantage of the biome-classification method of Prentice *et al.* (1996) is that it allows biomes that do not exist today to be defined and identified in the past, and thus provides one way of classifying the no-analog vegetation in terms of biomes (Figs 4 and 6).

A major feature of our continental and regional maps is that they show vegetation at spatial and temporal scales where climate forcing is strong enough to drive the major vegetational dynamics. These maps contrast with pollen diagrams that have the time span of 10,000 to 15,000 years for major changes in climate but record vegetation changes over several spatial scales some of which are too small for climate to be controlling. Soil development, fires, disease, and disturbance can locally dominate over climate in forcing changes in vegetation. When such is the case, changes in local pollen assemblages do not record climate. A zoom lens view that allows for both the map and diagram perspectives helps paleoecologists sort out the different factors controlling change. When local changes fit together as regional trends, then climate is likely the dominant control (as noted even by von Post, 1916). When local trends are anomalous or unique, then other factors prevail (Graumlich & Davis, 1993; Webb, 1974); and difference diagrams (Gaudreau *et al.*, 1989; Jacobson, 1979), charcoal records (Green, 1981; Patterson & Backman, 1988; Swain, 1973), and knowledge of human disturbance (Davis, 1965) can help identify the cause for the vegetation change. Furthermore, comparing vegetation and climate changes across temporal scales also requires spanning spatial scales. At the local scale, rapid climate changes alter the vegetation patterns as greatly as orbital-scale climate changes; whereas on continental scales, orbital-scale changes dwarf millennial-scale differences (Figs 4 and 7).

The warming at the end of the Younger Dryas Chronozone is large and results from both orbital and millennial-scale forcing. To argue that the 4–8 °C temperature increase is all millennial scale variation is to ignore the long-term impacts of orbital scale changes and ice-sheet retreat. There is no question that large millennial-scale differences going into and out of the Younger Dryas chronozone reflect the rapid climatic reorganization associated with changes in North Atlantic circulation (Fig. 1). But these abrupt changes are imbedded within the rapid part of glacial-interglacial transition, and occur on the scale of decades to a millennium, which are too short for the orbital-scale controls to cause significant differences. Even the large long-term differences measured between 11,000 and 6000 cal yr B.P. (3000- and 5000-yr distance maps in Fig. 7), which can be attributed to rapidly decreasing summer insolation and the rapid melting of the Laurentide ice sheet (Fig. 1), are not distinguishable on the square-chord distance maps when divided into 1000-year segments (Fig. 7). Progressive changes in insolation, ice volume, and atmospheric carbon dioxide concentration were also faster before the Younger Dryas than during it (Fig. 1), and yet, the large long-term differences between 16,000 and 11,000 cal yr B.P., evident across >3000 years, did not register as millennial-scale differences (Fig. 7). Likewise, slow orbitally forced climate changes probably contributed little to the large changes calculated for the beginning of the Younger Dryas chronozone (1000-yr distance maps in Fig. 7). Much of the difference from 13,000 to 12,000 and from even 12,000 to 11,000 cal yr B.P. resulted from century-scale changes in vegetation composition that were synchronous at many sites ca. 12,900 and 11,600 cal yr B.P. (Fig. 10) (Shuman *et al.*, 2002a).

The vegetation changes across 3000- and 5000-year intervals (Fig. 7) and across the past 21,000 years (Fig. 6), however, dwarfed even the large millennial-scale differences associated with the Younger Dryas climate changes. The glacial-to-interglacial climate warming was far larger than the reorganization of climate patterns during rapid climate changes caused by ocean and atmospheric circulation changes. Maps of the long-term changes in insolation (Berger,

1978; Whitlock *et al.*, 2001) and net radiation (Whitlock *et al.*, 2001) show that the mid-latitudes underwent a large long-term increase in summer radiative forcing between 21,000 and 11,000 cal yr B.P. (as well as between LGM and modern). Consequently, even a large reorganization of the thermohaline circulation and oceanic heat transport could not widely alter climatic patterns of ca. 12,000 cal yr B.P. to be like those of 21,000 cal yr B.P. In some regions, cooling was significant at the onset of the Younger Dryas (e.g. Greenland, Stuiver *et al.*, 1995) (Fig. 1) and local vegetation patterns changed significantly (Figs 9–11), but the continental-scale climate gradients (and vegetation patterns) immediately before, during, and after the Younger Dryas remained more like each other than those even 3000 years later (Figs 4–7). The long-term northward migration of taxa, such as spruce (Fig. 4), was altered only subtly because the latitudinal band where climate was appropriate for them remained north of their LGM positions and well south of their Holocene positions.

Comparison between pollen diagrams and maps shows that large local changes are not always indicative of large continental-scale changes. On local scales, orbital- to centennial-scale climate changes had a large impact on ecosystem composition (Figs 8–10), but only the larger magnitude, long-term changes appear important in continental-scale maps of individual taxa and assemblages (Fig. 4). The geographic displacement of vegetation types associated with century-scale climate changes (even when those that involved range shifts of hundreds of kilometers, Figs 10 and 11) resulted in large local changes in vegetation composition, but was small compared to the range shifts of thousands of kilometers from 17,000 to 6000 cal yr B.P. (Fig. 4). On the continental scale, rapid change is only evident (e.g. 13,000–12,000 and 12,000–11,000 cal yr B.P.) when many individual sites show rapid change simultaneously (Fig. 10). Many pollen diagrams show several rapid (century-scale) changes during the past 21,000 years, but most of these changes resulted from long-term time-transgressive changes that rapidly impacted individual sites while gradually shifting species' ranges (see spruce decline discussion in Bernabo & Webb, 1977). The inference that large local changes reflect global climate changes is "iffy" at best and can only be sustained when maps show how the local changes link to other changes over a large area.

Conclusions

The emerging view from independently inferred paleoclimate estimates and patterns is allowing new assessment of climate-induced vegetation dynamics. Forcing factors at both orbital and millennial scales cause climate to change which in turn affects the pattern of major vegetation changes across the continent and at individual sites. The data show that the vegetation has changed when climate changed both at continental scales and at individual sites. The rates of migration, abundance change, and assemblage change have all varied at the recorded rate of climate change. Different recorders of the climate along with climate modeling results are allowing: (a) the influence of temperature changes to be separated from that of moisture change (Huang *et al.*, 2002; Shuman *et al.*, 2002b, 2003); and (b) the impact of the changing seasonality to be assessed (Williams *et al.*, 2001). The multivariate nature of climate change combined with multivariate response surfaces that are unique for each plant taxon allow for widely varying responses among taxa to similar changes in climate (Bartlein *et al.*, 1986; T. Webb *et al.*, 1993). Differential migration paths and rates among taxa are expected as the result of such multivariate climate forcing and responsiveness (Prentice *et al.*, 1991), and taxa today that grow together should have in general arrived or at least expanded in abundance at different times (Shuman *et al.*, 2002a, b, 2003; Williams *et al.*, 2003).

Individualistic plant behavior scales up to cause the emergence and disappearance of plant assemblages, of combinations of plant functional types, and hence of biomes. Changes at these higher levels of ecological organization ultimately result in feedbacks to the climate system (Kutzbach *et al.*, 1996; Street-Perrott *et al.*, 1990). The individual taxa move farther and more dramatically, however, than the biomes that they comprise. Changes by individual taxa combine and scale up to become differences in vegetation structure and composition not captured by biome classification (Williams, 2003; Williams *et al.*, 2003). Multiple depictions and visualizations of vegetation change are therefore needed to portray its full diversity of changes. As new data sets and increasingly sophisticated vegetation and climate models come on-line to improve past climate estimates, we foresee many new insights into vegetation dynamics that will include how: (1) vegetation-climate feedbacks (Kutzbach *et al.*, 1996); (2) changing atmospheric concentrations of carbon dioxide (Cowling & Sykes, 1999; Davis, 1991); and (3) evolution and extinction of taxa (Davis & Shaw, 2001; Jackson & Weng, 1999) enhance or inhibit the shifting of taxon abundances and distributions as climate changes. A current fascination with understanding the impact of global warming on ecosystems, landscapes, and human livelihood has placed a premium on studying decade-to-millennial scale climate changes. We are hopeful that the reductionism of this approach will in time give way to a more holistic approach to understanding vegetation and climate dynamics. Such will be necessary if we are to understand the roles of millennial and orbital scale climate and vegetation dynamics in evolution and speciation. Studies of the genetic make-up of fossil plants are just beginning and should open many exciting lines of inquiry (Davis & Shaw, 2001). Such studies will allow us to understand how Hutchinson's (1965) evolutionary play unfolds in the continuously changing ecological theater (Webb, 1995).

Acknowledgments

Grants from the Earth System History Program at NSF supported this research. Earlier support came from the Office of Energy Research at DOE. B. Shuman was supported during the writing by a NOAA Climate and Global Change Postdoctoral Fellowship and the University of Oregon. J. Williams was supported by the National Center for

Ecological Analysis and Synthesis, a Center funded by NSF (Grant #DEB-0072909), the University of California, and the Santa Barbara campus. We thank P. Leduc and P. Newby for technical support and P.J. Bartlein for cartographic advice and paleogeographic information. Our mapping relied on excellent work by E. Grimm and J. Keltner in developing the North American Pollen Database (NAPD). We thank J.T. Overpeck and R.S. Webb for their work at NGDC at NOAA to support the development of this database. We also thank numerous palynologists for making their data available to the NAPD and Pierre Richard for giving access to data from Base de Donnees polliniques et Macrofossiles du Québec.

References

Almquist, H., Dieffenbacher-Krall, A.C., Brown, R. & Sanger, D. (2001). An 8000-yr Holocene Record of Lake-levels at Mansell Pond, Central Maine, USA. *The Holocene*, **11**, 189–201.

Ammann, B., Birks, H.J.B., Brooks, S.J., Eicher, U., von Grafenstein, U., Hofmann, W., Lemdahl, G., Schwander, J., Tobolski, K. & Wick, L. (2000). Quantification of biotic responses to rapid climatic changes around the Younger Dryas – a synthesis. *Palaeogeography, Palaeoclimatology, Palaeoecology*, **159**, 313–347.

Anderson, P., Edwards, M.E. & Brubaker, L.B. (2003). Results and paleoclimate implications of 35 years of paleoecological research in Alaska (this volume).

Barber, D.C., Dyke, A., Hillaire-Marcel, C., Jennings, A.E., Andrews, J.T., Kerwin, M.W., Bilodeau, G., McNeely, R., Southon, J., Morehead, M.D. & Gagnon, J.-M. (1999). Forcing of the cold event 8,200 years ago by catastrophic drainage of Laurentide lakes. *Nature*, **400**, 344–348.

Bartlein, P.J., Prentice, I.C. & Webb, T., III (1986). Climatic response surfaces based on pollen from some eastern North America taxa. *Journal of Biogeography*, **13**, 35–57.

Bartlein, P.J., Anderson, P.M., Anderson, K.H., Edwards, M.E., Thompson, R.S., Webb, R.S., Webb, T., III & Whitlock, C. (1998). Paleoclimate simulations for North America for the past 21,000 years: features of the simulated climate and comparisons with paleoenvironmental data. *Quaternary Science Reviews*, **17**, 549–585.

Bernabo, J.C. & Webb, T., III (1977). Changing patterns in the Holocene pollen record from northeastern North America: a mapped summary. *Quaternary Research*, **8**, 64–96.

Berger, A. (1978). Long-term variations of caloric insolation resulting from the Earth's orbital elements. *Quaternary Research*, **9**, 139–167.

Birks, H.H. & Ammann, B. (2000). Two terrestrial records of rapid climatic change during the glacial-Holocene transition (14,000–9,000 calendar years B.P.) from Europe. *Proceedings of the National Academy of Sciences*, **97**, 1390–1394.

Booth, R.K. & Jackson, S.T. (2003). A high resolution record of late Holocene moisture variability from a Michigan raised bog. *The Holocene*.

Bradbury, J.P. & Dean, W.E. (1993). Elk Lake, Minnesota: evidence for rapid climate change in the north-central United States. *Geological Society of America Special Paper 276*, 336 pp.

Broecker, W.S., Peteet, D. & Rind, D. (1985). Does the ocean-atmosphere system have more than one stable mode of operation? *Nature*, **315**, 21–25.

Bluiner, T., Chappellax, J., Schwander, J., Dallenbach, A., Stauffer, B., Stocker, T.F., Raynaud, D., Jouzel, J., Clausen, H.B., Hammer, C.U. & Johnson, S.J. (1998). Asynchrony of Antarctic and Greenland climate change during the last glacial period. *Nature*, **394**, 739–743.

Bhiry, N. & Filion, L. (1996). Mid-Holocene hemlock decline in eastern North America linked with phytophagous insect activity. *Quaternary Research*, **45**, 312–320.

Bradshaw, R.H.W. & Webb, T., III (1985). Relationships between contemporary pollen and vegetation data from Wisconsin and Michigan, USA. *Ecology*, **66**, 721–737.

Clark, P.U., Marshall, S.J., Clarke, G.J.C., Hostetler, S.W., Licciardi, J.M. & Teller, J.M. (2001). Freshwater forcing of abrupt climate change during the last deglaciation. *Science*, **293**, 283–287.

Clark, P.U., Pisias, N.G., Stocker, T.F. & Weaver, A.J. (2002). The role of the thermohaline circulation in abrupt climate change. *Nature*, **415**, 863–869.

Cowling, S.A. & Sykes, M.T. (1999). Physiological significance of low atmospheric CO_2 for plant-climate interactions. *Quaternary Research*, **52**, 237–242.

COHMAP Members (1988). Climatic changes of the last 18,000 years: observations and model simulations. *Science*, **241**, 1043–1052.

Cushing, E.J. (1965). Problems in the Quaternary phytogeography of the Great Lakes region. *In*: Wright, H.E., Jr. & Frey, D.G. (Eds), *The Quaternary of the United States*. Princeton, NJ, USA, Princeton University Press, pp. 403–416.

Davis, M.B. (1965). Phytogeography and palynology of northeastern United States. *In*: Wright, H.E., Jr. & Frey, D.G. (Eds), *The Quaternary of the Unites States*. Princeton, NJ, USA, Princeton University Press, pp. 377–401.

Davis, M.B. (1981a). Outbreaks of forest pathogens in Quaternary history. *In*: Bharadwaj, D., Vishnu-Mittre & Maheshwari, H. (Eds), *Proceedings of the Fourth International Palynological Conference*. Lucknow, India, Birbal Sahni Institute of Paleobotany, Vol. 3, pp. 261–227.

Davis, M.B. (1981b). Quaternary history and the stability of forest communities. *In*: West, D.C., Shugart, H.H. & Botkin, D.B. (Eds), *Forest Succession: Concepts and Application*. New York, Springer Verlag, pp. 132–177.

Davis, M.B. (1991). Research questions posed by the paleoecological record of global change. *In*: Bradley, R. (Ed.), *Global Changes of the Past*. Boulder, CO, UCAR/Office of Interdisciplinary Earth Sciences, pp. 385–395.

Davis, M.B., Spear, R. & Shane, L. (1980). Holocene climate of New England. *Quaternary Research*, **14**, 240–250.

Davis, M.B. & Shaw, R.G. (2001). Range shifts and adaptive responses to Quaternary climate change. *Science*, **292**, 673–679.

Davis, R.B. & Jacobson, G.L., Jr. (1985). Late-glacial and early Holocene landscapes in northern New England and adjacent areas. *Quaternary Research*, **23**, 341–368.

Deevey, E.S., Jr. (1939). Studies on Connecticut lake sediments. I. A postglacial climatic chronology for southern New England. *American Journal of Science*, **237**, 691–724.

DeFries, R.S., Hansen, M.C. & Townshend, J.R.G. (2000). Global continuous fields of vegetation characteristics: a linear mixture model applied to multi-year 8 km AVHRR data. *International Journal of Remote Sensing*, **21**, 1389–1414.

Digerfeldt, G. (1986). Studies on past lake-level fluctuations. *In*: Berglund, B.E. (Ed.), *Handbook of Holocene Palaeoecology and Palaeohydrology*. Chichester, John Wiley and Sons, pp. 127–142.

Dyke, A.S. & Prest, V.K. (1987). Late-Wisconsinan and Holocene history of the Laurentide Ice Sheet. *Géographie physique et Quaternaire*, **41**, 237–263.

Fritz, S.C., Ito, E., Yu, Z., Laird, K.R. & Engstrom, D. (2000). Hydrologic variation in the northern Great Plains during the last two millennia. *Quaternary Research*, **53**, 175–184.

Gaudreau, D., Jackson, S.J. & Webb, T., III (1989). Spatial scale and sampling strategy in paleoecological studies of vegetation patterns in mountainous terrain. *Acta Bot. Neerl.*, **38**, 369–390.

Gleason, H.A. (1926). The individualistic concept of the plant association. *Bulletin of the Torrey Botanical Club*, **53**, 7–26.

Graumlich, L.J. & Davis, M.B. (1993). Holocene variation in spatial scales of vegetation pattern in the upper Great Lakes. *Ecology*, **74**, 826–839.

Green, D.G. (1981). Time series and postglacial forest ecology. *Quaternary Research*, **15**, 265–277.

Grimm, E. & Jacobson, G. (this volume). Late Quaternary Vegetation history of the eastern United States.

Haas, J.N. & McAndrews, J.H. (2000). The summer drought related hemlock (*Tsuga canadensis*) decline in Eastern North America 5700 to 5100 years ago. *In*: McManus, K. (Ed.), *Proceedings: Symposium on Sustainable Management of Hemlock Ecosystems in Eastern North America* (June 22–24, 1999). Durham, NH: United States Department of Agriculture, Forest Service, Northeastern Research Station, General Technical Report NE-267, pp. 81–88.

Harrison, S.P. (1989). Lake level and climatic change in eastern North America. *Climate Dynamics*, **3**, 157–167.

Huang, Y., Shuman, B., Yang, Y. & Webb, T., III (2002). Hydrogen isotope ratios of palmitic acid in lacustrine sediments record late Quaternary climate variations. *Geology*, **30**, 1103–1106.

Huntley, B. & Webb, T., III (1989). Migration: species' response to climatic variations caused by changes in the earth's orbit. *Journal of Biogeography*, **16**, 5–19.

Huntley, B., Bartlein, P.J. & Prentice, I.C. (1989). Climatic control of the distribution and abundance of beech (*Fagus* L.) in Europe and North America. *Journal of Biogeography*, **16**, 551–560.

Hutchinson, G.E. (1965). *The ecological theater and the evolutionary play*. New Haven, CT, Yale University Press, 139 pp.

Indermühle, A., Stocker, T.F., Joos, F., Fischer, H., Smith, H.J., Wahlen, M., Deck, B., Mastroianni, D., Tschumi, J., Blunier, T., Meyer, R. & Stauffer, B. (1999). Holocene carbon-cycle dynamics based on CO_2 trapped in ice at Taylor Dome, Antarctica. *Nature*, **398**, 121–126.

Jackson, S.T. (1994). Pollen and spores in Quaternary lake sediments as sensors of vegetation composition: theoretical models and empirical evidence. *In*: Traverse, A. (Ed.), *Sedimentation of Organic Particles*. Cambridge, UK, Cambridge University Press, pp. 253–286.

Jackson, S.T. & Overpeck, J.T. (2000). Responses of plant populations and communities to environmental changes of the Late Quaternary. *Paleobiology*, **26**(Suppl.), 194–220.

Jackson, S.T. & Weng, C. (1999). Late Quaternary extinction of a tree species in eastern North America. *Proceedings of the National Academy of Sciences*, **96**, 13,847–13,852.

Jackson, S.T., Overpeck, J.T., Webb, T., III, Keattch, S. & Anderson, K.H. (1997). Mapped plant macrofossil and pollen records of late-Quaternary vegetation change in eastern North America. *Quaternary Science Reviews*, **16**, 1–70.

Jackson, S.T., Webb, R.S., Anderson, K.H., Overpeck, J.T., Webb, T., III, Williams, J.W. & Hansen, B.C.S. (2000). Vegetation and environment in eastern North America during the last glacial maximum. *Quaternary Science Reviews*, **19**, 489–508.

Jacobson, G.L., Jr. (1979). The palaeoecology of white pine (*Pinus strobus*) in Minnesota. *Journal of Ecology*, **67**, 697–726.

Jacobson, G.L., Jr., Webb, T., III & Grimm, E.C. (1987). Patterns and rates of vegetation change during the deglaciation of eastern North America. *In*: Ruddiman, W.F. & Wright, H.E., Jr. (Eds), *North American and Adjacent Oceans During the Last Deglaciation: The geology of North America*. Boulder, CO, USA, The Geological Society of America, pp. 277–288.

Johnson, W.C. & Webb, T., III (1989). The role of blue jays in the postglacial dispersal of Fagaceous trees in eastern North America. *Journal of Biogeography*, **16**, 561–571.

Kneller, M. & Peteet, D. (1999). Late-glacial to early Holocene climate changes from a central Appalachian pollen and macrofossil record. *Quaternary Research*, **51**, 133–147.

Kutzbach, J.E., Guetter, P.J., Behling, P.J. & Selin, R. (1993). Simulated climatic changes: results of the COHMAP climate-model experiments. *In*: Wright, H.E., Jr., Kutzbach, J.E., Webb, T., III, Ruddiman, W.F., Street-Perrott, F.A. & Bartlein, P.J. (Eds), *Global Climates since the Last Glacial Maximum*. Minneapolis, University of Minnesota Press, pp. 24–93.

Kutzbach, J.E., Bartlein, P.J., Foley, J., Harrison, S., Hostetler, S., Liu, Z., Prentice, I.C. & Webb, T., III (1996). The potential role of vegetation feedback in the climate sensitivity of high-latitude regions: a case study at 6000 years before present. *Global Biogeochemical Cycles*, **10**, 727–736.

Kutzbach, J.E., Gallimore, R., Harrison, S.P., Behling, P., Selin, R. & Laarif, F. (1998). Climate and biome simulations for the past 21,000 years. *Quaternary Science Reviews*, **17**, 473–506.

Lavoie, M. & Richard, P.J.H. (2000). Postglacial water-level changes of a small lake in southern Quebec, Canada. *The Holocene*, **10**, 621–634.

Levesque, A.J., Mayle, F.E., Walker, I.R. & Cwynar, L.C. (1993). A previously unrecognized late-glacial cold event in eastern North America. *Nature*, **361**, 623–626.

Mayle, F.E. & Cwynar, L.C. (1995). Impact of the younger Dryas cooling event upon lowland vegetation of Maritime Canada. *Ecological Monographs*, **65**, 129–154.

Monnin, E., Indermühle, A., Dällenbach, A., Flückiger, J., Stauffer, B., Stocker, T.F., Raynaud, D. & Barnola, J.-M. (2001). Atmospheric CO_2 concentrations over the last glacial termination. *Science*, **291**, 112–114.

Mott, R.J., Grant, D.R., Stea, R. & Occhietti, S. (1986). Late-glacial climatic oscillation in Atlantic Canada equivalent to the Allerød/younger *Dryas* event. *Nature*, **323**, 247–250.

Newby, P., Killoran, P., Waldorf, M.R., Webb, T., III & Webb, R.S. (2000). 11,500 years of sediment, vegetation, and water level changes at Makepeace Cedar Swamp, southeastern Massachusetts. *Quaternary Research*, **53**, 352–368.

Overpeck, J.T., Webb, T., III & Prentice, I.C. (1985). Quantitative interpretation of fossil pollen spectra: dissimilarity coefficients and the method of modern analogs. *Quaternary Research*, **23**, 87–108.

Overpeck, J.T., Bartlein, P.J. & Webb, T., III (1991). Potential magnitude of future vegetation change in eastern North America: Comparisons with the past. *Science*, **254**, 692–695.

Overpeck, J.T., Webb, R.S. & Webb, T., III (1992). Mapping eastern North American vegetation change over the past 18,000 years: no-analogs and the future. *Geology*, **20**, 1071–1074.

Patterson, W.A., III & Backman, A.E. (1988). Fire and disease history of forests. *In*: Huntley, B. & Webb, T., III (Eds), *Vegetation History*. Dordrecht, Kluwer Academic Publishers, pp. 603–632.

Peteet, D.M., Vogel, J.S., Nelson, D.E., Southron, J.R., Nickmann, R.J. & Heusser, L.E. (1990). Younger Dryas climatic reversal in northeastern USA? AMS ages for an old problem. *Quaternary Research*, **33**, 219–230.

Prentice, I.C. & Webb, T., III (1998). BIOME 6000: Global paleovegetation maps and testing global biome models. *Journal of Biogeography*, **25**, 997–1005.

Prentice, I.C., Bartlein, P.J. & Webb, T., III (1991). Vegetational climate change in eastern North America since the last glacial maximum. *Ecology*, **72**, 2038–2056.

Prentice, I.C., Guiot, J., Huntley, B., Jolly, D. & Cheddadi, R. (1996). Reconstructing biomes from palaeoecological data: a general method and its application to European pollen data at 0 and 6 ka. *Climate Dynamics*, **12**, 185–194.

Rind, D., Peteet, D., Broecker, W., McIntyre, A. & Ruddiman, W. (1986). The impact of cold North Atlantic sea-surface temperatures on climate: implications for the Younger Dryas cooling (11–10k). *Climate Dynamics*, **1**, 3–33.

Ruddiman, W.F. & McIntyre, A. (1981). The North Atlantic Ocean during the last deglaciation. *Paleogeography, Paleoclimatology, Paleoecology*, **35**, 145–214.

Rühlemann, C., Mulitza, S., Müller, P.J., Wefer, G. & Zahn, R. (1999). Warming of the tropical Atlantic Ocean and slowdown of thermohaline circulation during the last deglaciation. *Nature*, **402**, 511–514.

Rutter, N.W., Weaver, A.J., Rokosh, D., Fanning, A.F. & Wright, D.G. (2000). Data-model comparison of the Younger Dryas event. *Canadian Journal of Earth Sciences*, **37**, 811–830.

Shane, L.C.K. (1987). Late glacial vegetational and climatic history of the Allegheny Plateau and the Till Plains of Ohio and Indiana, USA. *Boreas*, **16**, 1–20.

Shemesh, A. & Peteet, D. (1998). Oxygen isotopes in fresh water biogenic opal: Northeastern U.S. Alleröd-Younger Dryas temperature shift. *Geophysical Research Letters*, **25**, 1935–1938.

Shuman, B.N. (2001). Vegetation responses to moisture balance and abrupt climate change in North America during the Late Quaternary. Ph.D. dissertation. Providence, RI, Brown University, 312 pp.

Shuman, B.N., Bravo, J., Lynch, J., Kaye, J., Newby, P. & Webb, T., III (2001). Late quaternary water level variations and vegetation history at Crooked Pond, southeastern Massachusetts. *Quaternary Research*, **56**, 401–410.

Shuman, B., Webb, T., III, Bartlein, P.J. & Williams, J.W. (2002a). The anatomy of a climatic oscillation: vegetation change in eastern North America during the Younger Dryas Chronozone. *Quaternary Science Reviews*, **21**, 1777–1791.

Shuman, B., Bartlein, P.J., Logar, N., Newby, P. & Webb, T., III (2002b). Parallel vegetation and climate responses to the early-Holocene collapse of the Laurentide Ice Sheet. *Quaternary Science Reviews*, **21**, 1793–1805.

Shuman, B., Newby, P., Huang, Y. & Webb, T., III (2003). Evidence for the close climatic control of New England vegetation history. *Ecology*.

Smith, A.J. (1993). Lacustrine ostracodes as hydrochemical indicators in lakes of the north-central United States. *Journal of Paleolimnology*, **8**, 121–134.

Street-Perrott, F.A., Mitchell, J.F.B., Marchand, D.S. & Brunner, J.S. (1990). Milankovitch and albedo forcing of the tropical monsoons: a comparison of geologic evidence and numerical simulations for 9,000 yr B.P. *Transactions of the Royal Society of Edinburgh, Earth Sciences*, **81**, 407–427.

Stuiver, M. (1968). Oxygen-18 content of atmospheric precipitation during last 11,000 years in the Great Lakes region. *Science*, **162**, 994–997.

Stuiver, M., Grootes, P.M. & Braziunas, T.F. (1995). The GISP2 $\delta^{18}O$ climate record of the past 16,500 years and the role of sun, ocean, and volcanoes. *Quaternary Research*, **44**, 341–354.

Stuiver, M., Reimer, P.J., Bard, E., Beck, J.W., Burr, G.S., Hughen, K.A., Komar, B., McCormac, F.G., Plicht, J.v.d. & Spurk, M. (1998). INTCAL98 Radiocarbon age calibration 24,000-0 cal BP. *Radiocarbon*, **40**, 1041–1083.

Swain, A.M. (1973). A history of fire and vegetation in northeastern Minnesota as recorded in lake sediments. *Quaternary Research*, **3**, 383–396.

Thompson, R.S., Shafer, S.L., Strickland, L.E., Van de Water, P.K. & Anderson, K.H. (this volume). Quaternary vegetation and climate change in the western United States: developments, perspectives, and prospects.

Tinner, W. & Lotter, A.F. (2001). Central European vegetation response to abrupt climate change at 8.2 ka. *Geology*, **29**, 551–554.

von Post, L. (1967) [1916]. Forest tree pollen in south Swedish peat bog deposits. *Pollen et Spores*, **9**, 378–401.

Walker, I.R., Mott, R.J. & Smol, J.P. (1991). Allerød-Younger Dryas lake temperature from midge fossils in Atlantic Canada. *Science*, **253**, 1010–1012.

Webb, R.S. (1990). Late Quaternary water fluctuations in the northeastern United States. Ph.D. dissertation. Providence, RI, Brown University, 351 pp.

Webb, R.S., Anderson, K.H. & Webb, T., III (1993). Pollen response-surface estimates of late-Quaternary changes in the moisture balance of the northeastern United States. *Quaternary Research*, **40**, 213–227.

Webb, R.S. & Webb, T., III (1988). Rates of sediment accumulation in pollen cores from small lakes and mires of eastern North America. *Quaternary Research*, **30**, 284–297.

Webb, T., III (1974). Corresponding patterns of pollen and vegetation in lower Michigan: a comparison of quantitative data. *Ecology*, **55**, 17–28.

Webb, T., III (1986). Is vegetation in equilibrium with climate? How to interpret Late-Quaternary pollen data. *Vegetatio*, **67**, 75–91.

Webb, T., III (1988). Eastern North America. *In*: Huntley, B. & Webb, T., III (Eds), *Vegetation History*, Vol. VII *in Handbook of Vegetation Science*. Dordrecht, The Netherlands, Kluwer Academic Publishers, pp. 385–414.

Webb, T., III (1995). Pollen records of Late Quaternary vegetation change: plant community rearrangements and evolutionary implications. *In*: Board of Earth Sciences and Resources, Commission on Geosciences, Environment, and Resources, National Research Council (Eds), *Effects of Past Global Change on Life*. Washington, DC, National Academy Press, pp. 221–232.

Webb, T., III (Ed.) (1998). Late quaternary climates: Data syntheses and model experiments. *Quaternary Science Reviews*, **17**(6–7), 465–688.

Webb, T., III (2000). Exploration of biogeographic databases: zoom lenses, space travel, and scientific imagination. *Journal of Biogeography*, **27**, 7–9.

Webb, T., III (2001). Paleoecology. *In*: Levine, S. (Ed.), *Encyclopedia of Biodiversity*. New York, Academic Press, Vol. 4, pp. 451–462.

Webb, T., III, Cushing, E.J. & Wright, H.E., Jr. (1983). Holocene changes in the vegetation of the Midwest. *In*: Wright, H.E., Jr. (Ed.), *Late-Quaternary Environments of the United States. The Holocene*. Minneapolis, University of Minnesota Press, Vol. 2, pp. 142–165.

Webb, T., III, Bartlein, P.J., Harrison, S.P. & Anderson, K.H. (1993). Vegetation, lake level, and climate change in eastern North America. *In*: Wright, H.E., Jr., Kutzbach, J.E., Webb, T., III, Ruddiman, W.F., Street-Perrott, F.A. & Bartlein, P.J. (Eds), *Global Climates Since the Last Glacial Maximum*. Minneapolis, University of Minnesota Press, pp. 415–467.

Webb, T., III, Anderson, K.H., Webb, R.S. & Bartlein, P.J. (1998). Late Quaternary climate changes in eastern North America: a comparison of pollen-derived estimates with climate model results. *Quaternary Science Reviews*, **17**, 587–606.

Webb, T., III, Shuman, B.N., Leduc, P., Newby, P. & Miller, N. (2003). Late Quaternary climate history of western New York State. *Bulletin of the Buffalo Society of Natural Sciences* (in press).

Whitehead, D.R. (1965). Palynology and Pleistocene phytogeography of unglaciated eastern North America. *In*: Wright, H.E., Jr. & Frey, D.G. (Eds), *The Quaternary of the United States*. Princeton, NJ, USA, Princeton University Press, pp. 417–432.

Whitehead, D.R. & Crisman, T. (1978). Paleolimnological studies of small New England (USA) ponds. Part 1. Late glacial and postglacial trophic oscillations. *Polskie Archiwum Hydrobiologii*, **25**, 471–481.

Whitlock, C., Bartlein, P.J., Markgraf, V. & Ashworth, A.C. (2001). Late glacial vegetation records in the Americas and climatic implications. *In*: Markgraf, V. (Ed.), *Interhemispheric Climate Linkages*. San Diego, Academic Press, pp. 391–416.

Williams, A.S. (1974). Late-glacial – postglacial vegetational history of the Pretty Lake region, northeastern Indiana: Washington, DC, U.S. Government Printing Office, *U.S. Geological Survey Professional Paper 686-B*.

Williams, J.W. (2003). Needleleaved and broadleaved tree cover distributions in North America since the last glacial maximum. *Global and Planetary Change*, **35**, 1–23.

Williams, J.W. & Jackson, S.T. (2003). Palynological and AVHRR observations of modern vegetational gradients in eastern North America. *The Holocene*, **13**, 485–497.

Williams, J.W., Summers, R.L. & Webb, T., III (1998). Applying plant functional types to construct biome maps from eastern North American pollen data: comparisons with model results. *Quaternary Science Reviews*, **17**, 607–627.

Williams, J.W., Webb, T., III, Richard, P.J.H. & Newby, P. (2000). Late Quaternary biomes of Canada and the eastern United States. *Journal of Biogeography*, **27**, 585–607.

Williams, J.W., Shuman, B.N. & Webb, T., III (2001). Dissimilarity analyses of late-Quaternary vegetation and climate in eastern North America. *Ecology*, **82**, 3346–3362.

Williams, J.W., Shuman, B.N., Webb, T., III, Bartlein, P.J. & Leduc, P.L. (2003). Late Quaternary vegetation dynamics in North America: scaling from taxa to biomes. *Ecological Monographs*.

Williams, J.W., Post, D.M., Cwynar, L.C., Lotter, A.F. & Levesque, A.J. (2002). Rapid and widespread vegetation responses to past climate change in the North Atlantic. *Geology*, **30**, 971–974.

Wright, H.E., Jr., Kutzbach, J.E., Webb, T., III, Ruddiman, W.F., Street-Perrott, F.A. & Bartlein, P.J. (Eds) (1993). *Global climates since the last glacial maximum*. Minneapolis, University of Minnesota Press, 569 pp.

Yu, Z., Andrews, J.H. & Eicher, U. (1997). Middle Holocene dry climate caused by change in atmospheric circulation patterns: Evidence from lake levels and stable isotopes. *Geology*, **25**, 251–254.

Holocene fire activity as a record of past environmental change

Cathy Whitlock and Patrick J. Bartlein

Department of Geography, University of Oregon, Eugene, OR 97403, USA; whitlock@oregon.uoregon.edu; bartlein@uoregon.edu

Introduction

Fire is the dominant form of natural disturbance in temperate forests, and, as such, it serves as a process that modulates forest susceptibility to climate change, disease, and other forms of disturbance. Fire has been identified as an important catalyst of vegetation change during rapid climate shifts in the past (e.g. T. Clark *et al.*, 1996; Swetnam & Betancourt, 1998), and it has been implicated as the primary agent of ecosystem change in the future (e.g. Overpeck *et al.*, 1990, 2003; Watson *et al.*, 2000). At the global scale, biomass burning is considered an important but poorly understood process in the global carbon cycle, one that releases greenhouse gases, aerosols, and particulates to the atmosphere but also sequesters carbon as inert charred matter and ash (Cofer *et al.*, 1997; Watson *et al.*, 2000). At the regional scale, fire plays an essential role in maintaining the integrity of forest ecosystems (MacNeil, 2000; Mills & Lugo, 2001; Nature, 2000). Because of fire's importance as an ecosystem process at large and small scales, it is necessary to understand: (1) the response of fires to past, present, and future climate change for global change assessments; and (2) the role of fire in maintaining forest health and promoting ecosystem change for better forest management.

Like many types of paleoenvironmental data, information on past fires can be interpreted in climatic terms as well as used as an indicator of how particular ecosystems respond to known climate changes. The ultimate objective of paleoenvironmental research is to do both – understand the cause and ecological consequences of climate change. Two sources of paleoecological data provide information on fire-climate interactions. One source, dendrochronological data, includes records of fire-scarred tree-rings and maps showing the distribution of forest-stand ages following fire (see Agee, 1993; Johnson & Gutsell, 1994, for information on methods). Dendrochronological methods provide highly resolved spatial reconstructions of past fire activity, but they are limited by the age of living trees, which spans only the last few centuries in most places. This relatively short duration makes it difficult to examine the role of fire during periods of major climate change. Moreover, tree-ring records are best suited to reconstruct low-intensity ground fires that do not kill trees and often offer little information on the frequency of stand-replacing crown fires, which have become more widespread in western forests in the last two decades.

The second data source is the record of particulate charcoal deposited in lakes and wetlands during and shortly after a fire (see Whitlock & Anderson, 2003; Whitlock & Larsen, 2002, for information on methods). Fire occurrence is identified by sedimentary layers with abundant or above-background levels of charcoal particles. The size and exact location of fires cannot be resolved with the specificity of dendrochronological studies, but charcoal records have the advantage of providing a fire reconstruction that spans thousands of years and encompasses periods of major climate change and vegetation reorganization. Annually resolved fire reconstructions are possible, but in most cases fire history is described in terms of fire episodes (one or more fires) during a time span of years to a few decades.

Fire was recognized as a past and present link between climate change and vegetation response in one chapter (Davis, 1965) of the review volume for the VII Congress of the International Association for Quaternary Research (Wright & Frey, 1965). Since 1965, research in fire history has undergone a renaissance that has improved the use of fire data as both a paleoclimatic and paleoecologic tool. Recent studies consider fire as a proximal cause of vegetation changes and also recognize that vegetation patterns (both spatially and temporally) help shape fire regimes. The role of climate as the ultimate control of both vegetation composition and fire regimes is also widely recognized. In this chapter, we discuss some of the recent advances, including efforts to: (1) better understand the processes that introduce charcoal into lakes and wetlands; (2) refine the methods to interpret these deposits; and (3) evaluate the response of fire to climate and vegetation controls operating on different time scales based on paleoecological evidence, paleoclimate simulations, and modern assessments. We focus this review on research in North America.

Refinements in Charcoal Analysis

The use of charcoal data to reconstruct fire history in North America began in the late 1960s and early 1970s when microscopic charcoal particles (generally $<100\,\mu m$ in diameter) were tallied as part of routine pollen analysis. Early on, Swain (1973, 1978) developed a fire reconstruction based on peaks in the ratio of charcoal-to-pollen accumulation in varved-sediment lakes in Wisconsin. Since then, numerous studies have tallied microscopic charcoal particles, sometimes referred to as pollen-slide charcoal, and included the time series on pollen diagrams. Because small particles are carried aloft during a fire and may travel long distances, the source of microscopic charcoal is generally ascribed to regional fires, i.e. fires occurring in the region but not necessarily in the local watershed. The initial research, nicely summarized by Tolonen (1986) and Patterson *et al.* (1987), demonstrates the value of microscopic charcoal for regional fire reconstructions, and Smith & Anderson (1992), Fall (1997), Delcourt

DEVELOPMENTS IN QUATERNARY SCIENCE
VOLUME 1 ISSN 1571-0866
DOI:10.1016/S1571-0866(03)01022-4

et al. (1998), Reasoner & Huber (1999) and Carcaillet *et al.* (2001a) provide recent examples of this approach.

Macroscopic Charcoal Analysis

Interest in fire history has shifted from general descriptions of regional-scale incidence or frequency of regional fires to more spatially specific reconstructions of local fires. Local fire records are based on the interpretation of macroscopic particles (generally $>100\,\mu m$ in diameter) recovered in sieved residues (e.g. Carcaillet *et al.*, 2001b; Long *et al.*, 1998) and on petrographic thin-sections (e.g. Anderson & Smith, 1997; Clark, 1988b, 1990) from contiguous core samples. Studies that use macroscopic charcoal emphasize the fire history of the local watershed, and regional reconstructions are based on networks of local records. The attention on local fire reconstructions has necessitated a better understanding of charcoal taphonomy, i.e. the processes that introduce and deposit charcoal to a lake or wetland. The principles of particle motion physics have been used to describe the transport of charcoal particles from a point source, suggesting that particles $>1000\,\mu m$ in diameter are deposited near a fire, particles $<100\,\mu m$ travel well beyond 100 m, and very small particles are carried great distances before settling (Clark, 1988a; Patterson *et al.*, 1987). Studies of recent fires indicate a sharp decline in macroscopic charcoal beyond the fire margin (Anderson *et al.*, 1986; Clark *et al.*, 1998; Gardner & Whitlock, 2001; Ohlson & Tryterud, 2000; Whitlock & Millspaugh, 1996), and charcoal abundance displays a negative logarithmic distribution away from the source area (Fig. 1A). Within the burned region, charcoal continues to accumulate in lakes in the years after a fire as a result of secondary material transported and deposited from burned slopes and lake margins (Bradbury, 1996; Whitlock & Millspaugh, 1996) (Fig. 1B). These processes of sediment focusing as well as bioturbation may account for the fact that charcoal peaks in sediment cores often span several centimeters.

Calibration of charcoal records comes from comparing the age of charcoal peaks with the timing of known fire events. In the case of varved-sediment records, the correspondence between peaks and known fires may be accurate to the year (Clark, 1990). Macroscopic charcoal records in non-laminated sediments are dated with ^{210}Pb and AMS ^{14}C methods. Charcoal peaks in these sediments match the age of known fire events but with less precision; the best results come from regions with infrequent, high-severity fires that produce abundant charcoal (Millspaugh & Whitlock, 1995; Mohr *et al.*, 2000).

Identification of past fire episodes from charcoal records generally involves decomposing the influx data into a background component and a "peaks" component (Clark & Patterson, 1997). The background component is considered to represent secondary material introduced to the lake during intervals between fires, or to measure variations in fuel availability and characteristics, or changes in charcoal delivery processes to the lake. The peaks component represents the charcoal introduced into the lake sediments during and immediately after a particular fire in the "charcoal catchment" of the lake. The sequence of charcoal peaks is the inferred record of fire episodes, from which fire frequency and the inverse, fire return intervals (or time between fires), are calculated. Two approaches have been proposed for the decomposition. Some researchers use a Fourier-series filter (Press *et al.*, 1992) to define the background component, and identify the peaks as positive deviations of the influx series from the background component (e.g. Carcaillet *et al.*, 2001b; Clark & Royall, 1996). The background series is assumed to be composed of many slowly varying sinusoidal components, and a filter used in a particular application can be designed to remove specific long-term variations in the charcoal influx. Another strategy uses a locally weighted moving average to define the background and a threshold ratio to identify peaks (e.g. Brunelle & Anderson, 2003; Hallett & Walker, 2000; Long *et al.*, 1998) (Fig. 2) The first approach is a "global" one in the sense that the background component is determined using information from the entire record, whereas the second is obviously a "local" one that allows for changes in the structure (the variability of the time series of charcoal influx and the shape of its spectrum) over time. Other approaches for performing the decomposition can be envisioned and in practice no single approach may be optimal for all cases.

Other Proxy Indicators of Fire

Other fire proxies complement charcoal-based fire reconstructions and have been used to infer the location and sometimes the size or severity of past fires. Gavin *et al.* (2003) examined charcoal in soil profiles in coastal British Columbia to better identify the location of past fire events recorded by lake-sediment charcoal. The location and the time-since-fire, determined by the age of charcoal particles in the soil sites, varied in wet and dry settings within the watershed. Some wet north-facing exposures, for example, had not burned since the warm, relatively dry conditions of the early Holocene.

Close correspondence between charcoal peaks and changes in key pollen taxa (Green, 1981; Patterson & Backman, 1988; Swain, 1973) or assemblages of pollen taxa that represent different stages of forest succession (Larsen & MacDonald, 1998; Rhodes & Davis, 1995) has been used to infer fire severity. Cross-correlations between pollen taxa and charcoal have helped identify vegetation responses to past fire events. For example, cross-correlation results from a site in northern Alberta revealed increases in the abundance of particular pollen taxa after peaks in pollen-slide charcoal. The succession of pollen types was used as a guide for identifying local fire events and post-fire recovery (Larsen & MacDonald, 1998). On longer time scales, changes in the abundance of fire-adapted and fire-sensitive taxa in the pollen record have also matched general trends in fire frequency through the Holocene (J. Clark *et al.*, 1996; Hallett & Walker, 2000; Long *et al.*, 1998).

Sugita *et al.* (1997) tried to evaluate the source area of the charcoal signal through the use of computer models that

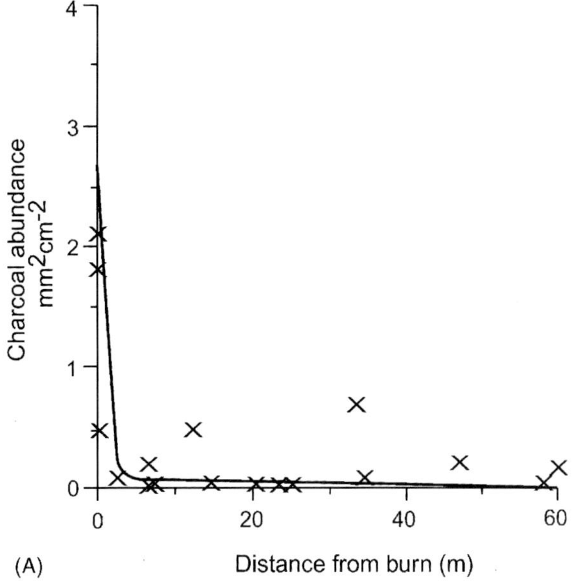

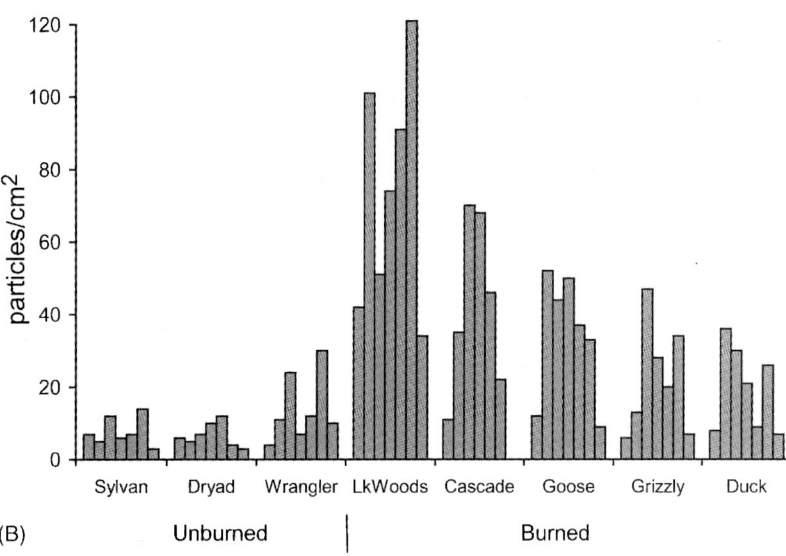

Fig. 1. Relationship between distance from a fire and the deposition of charcoal. A. Empirical evidence from Siberia, which shows that large charcoal particles display a negative exponential distribution away from their source and thus provide information on local watershed fires (after Clark et al., 1998). B. Charcoal accumulation (macroscopic particle abundance in the upper 2 cm of deep-water sediments) into lakes in Yellowstone National Park following the 1988 fires (data from 1989–1993, 1995 and 1997). Lakes in burned and unburned watersheds received charcoal during and immediately after the fire, but burned sites continued to accrue charcoal in the few years after the fire, and levels declined by 1997 (Whitlock & Millspaugh, 1996; unpublished data).

simulate pollen source area. The magnitude of stratigraphic changes in particular fire-sensitive pollen taxa was used to infer the size and location of the fire. The assumption was that local fires would register a change in pollen abundance, but the magnitude of the response would depend on the fire size and distance from the lake and the lake size. These modeling approaches hold great promise for improving the spatial specificity of the fire reconstruction based on charcoal data.

Lithologic variations have also been used to infer the intensity of the fires and confirm their location within a watershed. Peaks of high-magnetic susceptibility in lake-sediment and wetland records have been used to detect pulses of erosion associated with fires and the formation of ferromagnetic minerals (Gedye *et al.*, 2000; Millspaugh & Whitlock, 1995). Fire-induced erosion has also been inferred from increases in the content of aluminum, vanadium, and silt in sediment associated with charcoal peaks (Cwynar, 1978) and increases in varve thickness (Clark, 1990; Larsen & MacDonald, 1998). The usefulness of sediment magnetism and geochemistry as fire proxies depends on the fire location, fire type and intensity, and soils and substrate type.

In summary, the use of charcoal data to reconstruct past fires has become grounded by studies of modern charcoal transport and deposition. High-resolution charcoal analysis of contiguous samples in varved- and non-laminated sediment records provide detailed time series that can be used to reconstruct fire frequency. Microscopic charcoal records offer information on fire activity at the regional scale, whereas macroscopic charcoal data reveal the frequency of local fires. Both approaches have been used to study vegetation

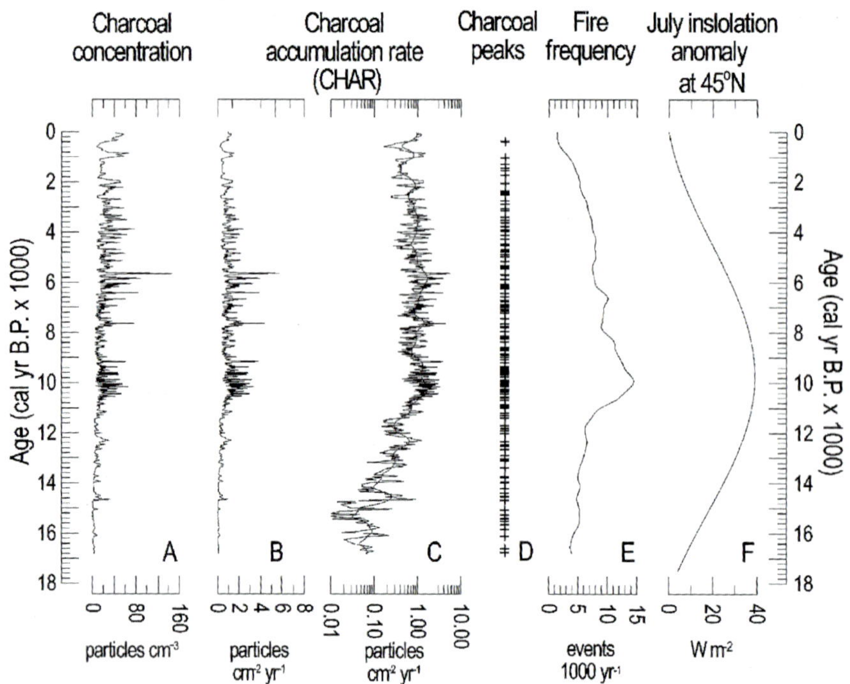
Cygnet Lake, Yellowstone National Park
Charcoal concentration
Charcoal accumulation rate (CHAR)
Charcoal peaks
Fire frequency
July insolation anomaly at 45°N
Age (cal yr B.P. x 1000)
0 2 4 6 8 10 12 14 16 18
A B C D E F
0 80 160
particles cm-3
0 2 4 6 8
particles cm2 yr-1
0.01 0.10 1.00 10.00
particles cm2 yr-1
0 5 10 15
events 1000 yr-1
0 20 40
W m-2

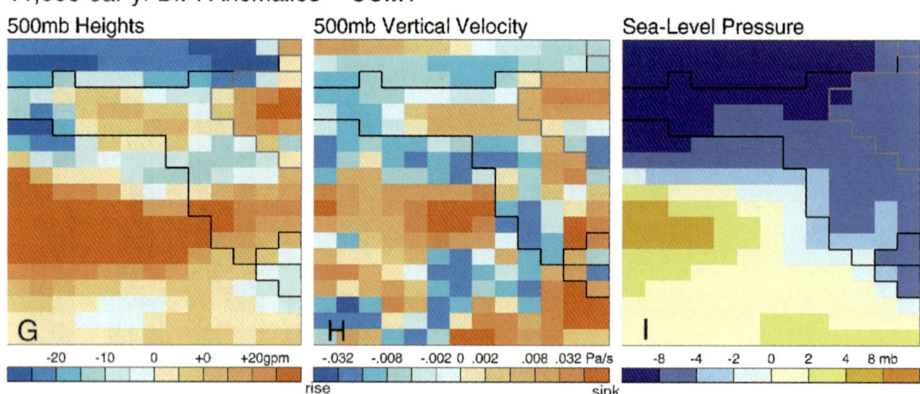
11,000 cal yr B.P. Anomalies -- CCM1
500mb Heights
500mb Vertical Velocity
Sea-Level Pressure
G
H
I
-20 -10 0 +0 +20gpm
-.032 -.008 -.002 0 .002 .008 .032 Pa/s
rise
sink
-8 -4 -2 0 2 4 8 mb

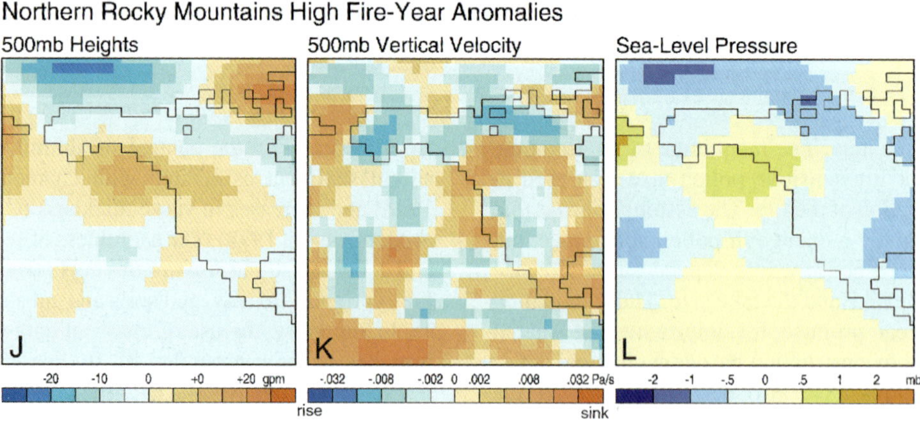
Northern Rocky Mountains High Fire-Year Anomalies
500mb Heights
500mb Vertical Velocity
Sea-Level Pressure
J
K
L
-20 -10 0 +0 +20 gpm
-.032 -.008 -.002 0 .002 .008 .032 Pa/s
rise
sink
-2 -1 -.5 0 .5 1 2 mb

and climate change. In addition, explicit consideration of the components that comprise charcoal records has improved fire reconstructions. The trends in macroscopic charcoal time-series provide information on changes in biomass and fuel availability through time, while the high-frequency signal denotes variations in fire frequency. As a result of these advances, charcoal and related fire proxy data are effective tools for studying both fire-vegetation and fire-climate relations in the past.

Modern and Past Fire-Climate Linkages

Fire Weather and Fire Climate

A better understanding of fire weather and fire climatology offers opportunities to refine the use of fire history data to discern fire-climate relationships in the past. Like many other environmental events (e.g. floods, avalanches), fires reflect both concurrent and antecedent conditions. *Fire weather* is the assemblage of specific meteorological conditions that vary on hourly and daily time scales and control the ignition, spread, and suppression of individual fires; *fire climate* consists of atmospheric circulation configurations and the attendant surface-climate anomalies that persist over time spans of weeks to seasons to years. The likelihood of fire in a given year is determined by particular weather conditions and their influence on fuel moisture, precipitation, relative humidity, lightning, air temperature, and wind (Pyne *et al.*, 1996; Weber & Flannigan, 1997). These weather conditions are embedded in large-scale climate anomalies, involving the strength of subtropical high-pressure systems, the location of upper-level ridges and troughs (and the moisture fluxes and vertical motions they govern), and the specific location of convectional storms during the fire season (Johnson & Wowchuk, 1993; Nash & Johnson, 1995; Skinner *et al.*, 1999, 2002).

Specific mechanisms link both large-scale and secondary circulation features with what happens at the surface in a particular region. This link has been described for specific regions in North America (e.g. Skinner *et al.*, 1999, 2002; Weber & Flannigan, 1997) to infer: (1) the anomalous components of mid-tropospheric flow, which in turn enhance or deflect the flow of moisture into a region; (2) the location of persistent ridges and troughs, which enhance or suppress precipitation through their influence on vertical motions; and (3) the effects of these components on cloudiness and surface energy balances, which influence soil and fuel moisture. Some studies have attempted to bridge these scales by considering statistical relationships between synoptic-scale atmospheric circulation patterns and specific fire-related meteorological variables (Flannigan & Harrington, 1988; Klein *et al.*, 1996; Roads *et al.*, 2000). Others have considered the remote controls of regional climate anomalies (e.g. North Pacific sea-surface temperatures) and regional area burned (Flannigan *et al.*, 2000).

Modern studies disclose the nature of fire-climate linkages on short time scales. On inter-annual time scales, fire activity inferred from dendrochronological data in many areas of North America is correlated with climate variations arising from atmosphere/ocean interactions, like the El Niño-Southern Oscillation (ENSO) (Swetnam & Betancourt, 1990, 1992; Veblen *et al.*, 2000). Likewise, decades of higher-than-normal precipitation (and fuel build-up) are often associated with significant fires when they are interrupted by drought episodes that reduce fuel moisture (Swetnam & Betancourt, 1998). Shifts between El Niño and La Niña phases or in the decadal-scale climate variability determine drought severity during a particular fire season or years, as well as the accumulation of fuels in previous years. This alternation of wet and dry episodes has been shown to be important in shaping the fire regime of the last few centuries, especially in low-elevation conifer forests (e.g. Clark, 1988c; Grissino-Mayer & Swetnam, 2000) and grasslands (Clark *et al.*, 2002).

Fire-Climate Linkages on Holocene Time Scales

On century and longer time scales ($>10^2$ years), large-scale changes in the climate system caused by variations in the seasonal cycle of insolation, atmospheric composition, and atmosphere-ocean interactions emerge as important controls of the fire regime. Variations in the seasonal cycle of insolation on orbital time scales (10^3–10^5 years) govern the slowly varying components of the climate system, and these apparently shape the long-term fire regime and vegetation. High-resolution charcoal records provide an opportunity to examine changes in fire activity in response to changes in the large-scale controls. For example, a record from Cygnet Lake in Yellowstone National Park was examined to reconstruct the fire history of the last 17,000 years, and compare it with changes in summer insolation that varied the intensity of

Fig. 2. A–F. Cygnet Lake, Yellowstone National Park (Millspaugh et al., 2000) provides an example of the charcoal decomposition approach of Long et al. (1998). A. Charcoal concentration calculated in contiguous samples; B. Data re-expressed as charcoal accumulation rates; C–D. Data decomposed into background trend (smooth curve) and charcoal peaks above background (Peaks are inferred to be fire episodes); E. Locally weighted frequency of charcoal peaks per 1000 years; F. Summer insolation anomaly. (Comparison of E and F suggests that time of maximum fires and the broad rise and fall of fire frequency is governed by the influences of the insolation anomaly.) G–L. Climate patterns shown in global climate model (GCM) simulations and modern NCEP data provide an explanation for high-fire activity noted at Cygnet Lake. G–I. 11,000 cal yr B.P.-minus-present anomalies of July 500 mb heights, vertical velocity, sea-level pressure, as simulated by a GCM (Kutzbach et al., 1998; see also Bartlein et al., 1998). J–L. Composite anomalies for same variables in NCEP data (Kalnay et al., 1996; Kistler et al., 2001) for recent years of large-area burned in Montana.

drought in the region (Millspaugh *et al.*, 2000) (Fig. 2). The record suggested that fire frequency in the early Holocene (between 10,000 and 11,000 cal yr B.P.) reached a maximum of 10–15 episodes/1000 years, in contrast to frequencies of <5 episodes/1000 years over the last few thousand years (Millspaugh *et al.*, 2000).

A comparison of modern climate during high-area burned years with paleoclimate model simulations produced by general circulation models elucidates the large-scale climate features that give rise to fires in the region, both at present and in the early Holocene (Fig. 2). Despite the radical difference in the boundary conditions between the present and 11,000 cal yr B.P. (most notably the still-large ice sheet and greater-than-present summer insolation at 11,000 cal yr B.P.), the anomaly patterns look broadly similar. They are also similar in scale and patchiness, suggesting that the large-scale settings of years (or millennia) with frequent fires are similar (see figure legend for further discussion). In this example, the large-scale control – variations in summer insolation – affected past fire activity through changes in the same features of atmospheric circulation that influence the intensity of summer aridity and fire occurrence at present. The similarity suggests the intriguing hypothesis that circulation patterns that characterized the mean climatic state during the early Holocene resemble anomalous circulation patterns that occur as components of inter-annual circulation variations at present.

In the western U.S., Holocene fire reconstructions have been used to study the fire and climate histories of two precipitation regimes that are evident in the climate at present. One regime is centered in the Pacific Northwest and receives little precipitation in summer and hence the summer-to-annual precipitation ratio is low. In this summer-dry regime, summer conditions are influenced by the northerly position of the northeast Pacific subtropical high-pressure system, which brings dry stable conditions. Cygnet Lake lies in this precipitation regime. In contrast, the summer-wet regime receives relatively high summer precipitation at present as a result of the influence of monsoonal precipitation. This regime is well developed in the American Southwest, but summer precipitation maxima are also evident in the Great Plains and parts of the northern Rockies. During the early Holocene, these two precipitation regimes intensified in the western U.S. as a result of greater-than-present insolation in summer (Bartlein *et al.*, 1998; Fall *et al.*, 1995; Mock & Brunelle-Daines, 1999; Whitlock & Bartlein, 1993). Areas under the influence of the subtropical high became drier as summer insolation directly increased temperatures and decreased effective moisture and indirectly strengthened the subtropical high. In the summer-wet regime, temperatures were higher as a result of the direct effects of greater insolation, and summer precipitation increased as a result of stronger onshore flow.

Shifts in long-term fire frequency in summer-dry and summer-wet regions probably reflect variations in fuel moisture as a result of shifts in the relative importance of winter and summer precipitation. The influence of a stronger-than-present subtropical high in the early Holocene is evident in fire reconstructions from the summer-dry regions of the Pacific Northwest and northern Rockies. Fire frequency was high before and at ca. 7000 cal yr B.P., coincident with the timing of maximum summer drought, and declined to present day as the climate became wetter (Fig. 3A). Fire activity declined in the late Holocene as the climate became cooler and wetter and present vegetation was established. This response is also registered at sites in the mixed conifer forests of the Klamath Mountains of northern California, although less clearly because fires are generally more frequent there and the sedimentation rates of the lakes are very slow. Especially high periods of fire activity took place at ca. 8000, 4000, and 1000 cal yr B.P. (Fig. 3A). The number of fire episodes (charcoal peaks) varied among regions depending on the fuel conditions and severity of drought. For example, fires were less frequent in the relatively wet Coast Range, more common in the summer-dry regions of the Rocky Mountains, and very common in the dry forests of the Klamath region. Despite differences in fire frequency among summer-dry regions, the temporal trends in fire occurrence, governed by large-scale circulation features, show similarities.

The summer-wet sites of the northern Rockies (Pintlar and Baker lakes) and in Yellowstone National Park (Slough Creek Lake) featured low fire occurrence in the early Holocene and highest fire activity in the late Holocene. This shift in fire frequency is consistent with the effects of the amplification of the seasonal cycle of insolation in areas strongly influenced by the summer monsoon. Higher-than-present summer precipitation suppressed fires in the early Holocene as a result of stronger onshore flow; as summer insolation decreased and monsoons weakened in the middle and late Holocene, fire activity increased. The summer-wet sites also record high fire frequency in the late-glacial period, perhaps in response to higher-than-present summer insolation but weaker monsoonal circulation than in the early Holocene.

Charcoal records from 30 sites in Quebec also disclose variations in summer precipitation through the Holocene (Carcaillet & Richard, 2000; Fig. 4). Examination of recent years of high area burned in eastern Canada suggests that the fire seasons are associated with the development of an anomalous upper-level ridge immediately over and upstream of east-central Canada (Skinner *et al.*, 1999, 2002). During these years, more meridional circulation deflects moisture-bearing systems away from southern Quebec. The records show higher fire incidence in the early Holocene, followed by a shift to fewer fires after 8000 cal yr B.P. (Carcaillet & Richard, 2000). Fire activity increased after 3000 cal yr B.P. and remained high to the present at several sites. Early-Holocene drought is attributed to the effects of higher-than-present summer insolation and locally dry conditions at the margin of the ice sheet. Inferred greater dominance of humid Atlantic air masses over Canada in middle Holocene summers led to fewer fires than before. More frequent incursion of dry Pacific air or cool Arctic air in summers resulted in more fires after 3000 cal yr B.P. The fire regime shifts occur independently of the vegetation changes, suggesting the control of climate for fire conditions is different from that for vegetation change (Carcaillet *et al.*, 2001a).

Fire variations on centennial scales are evident in dendrochronological and lake-sediment records across

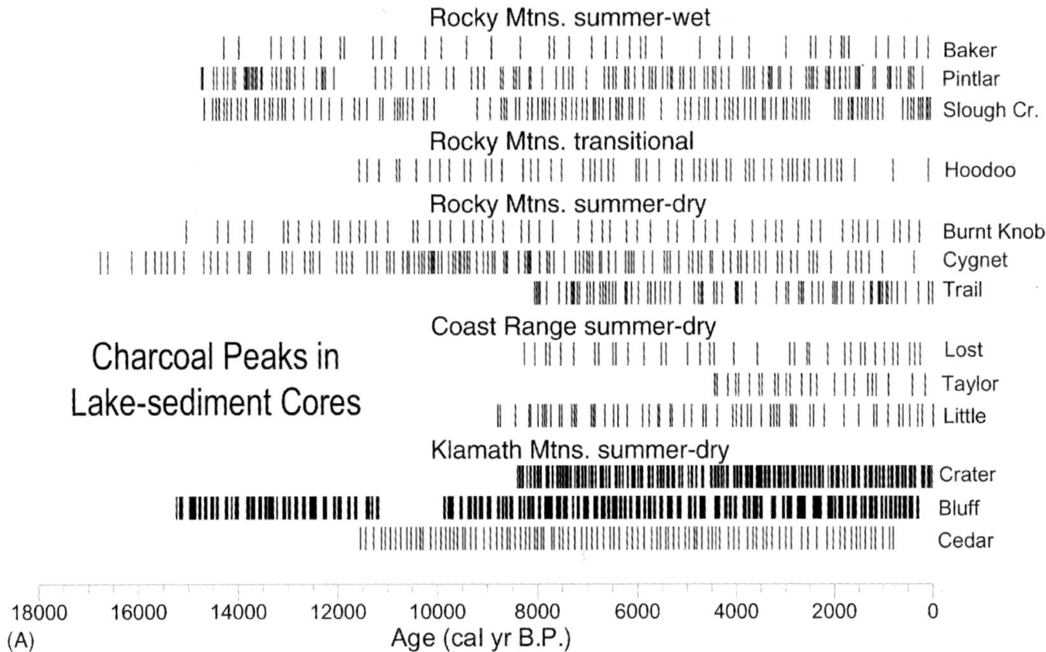

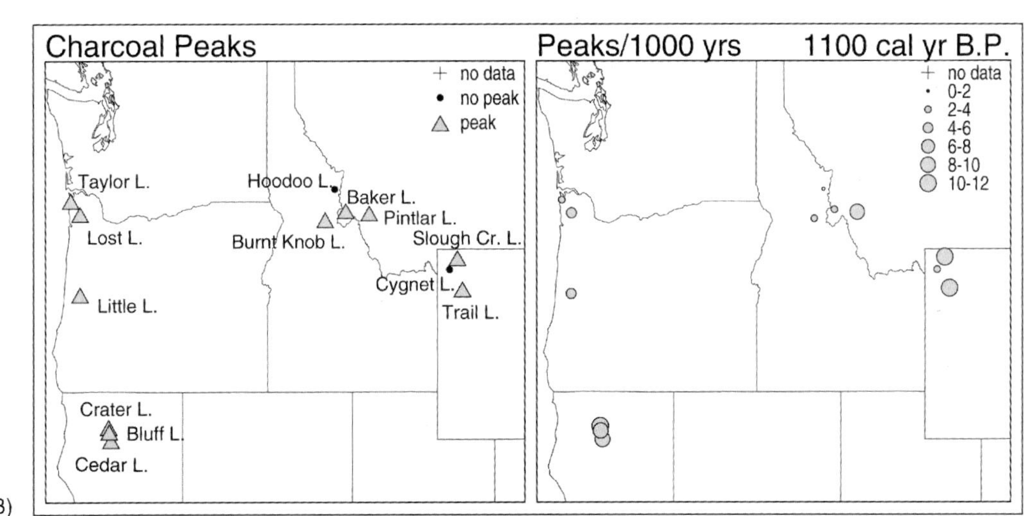

Fig. 3. A. Macroscopic charcoal peaks in lake-sediment records from summer-wet and summer-dry climate regions in the northwestern United States. Charcoal peaks are the occurrence of charcoal accumulation values above background values and interpreted as "fire episodes" within the watersheds of the lakes. B. Sites with fire activity at ca. 1100–1200 cal yr B.P. as shown by the presence of charcoal peaks. This period falls within the so-called Medieval Warm Period when fire frequency was high at many sites, as evidenced by the number of charcoal peaks/1000 years. Reference to lake sites: Baker, Pintlar, and Hoodoo lakes (Brunelle, 2002); Burnt Knob Lake (Brunelle & Whitlock, 2003); Slough Creek Lake (Millspaugh & Whitlock, 2003); Cygnet Lake (Millspaugh et al., 2000); Trail Lake and Cedar Lake (Whitlock, R. Sherriff, and T. Minckley, unpublished data); Taylor Lake (Long & Whitlock, 2002); Lost Lake (Long, 2003); Little Lake (Long et al., 1998); Crater and Bluff lakes (Mohr et al., 2000).

North America. For example, increased fire activity is noted between 800 and 1200 cal yr B.P. – the so-called Medieval Warm Period – in Yellowstone National Park (Meyer *et al.*, 1995; Millspaugh & Whitlock, 2003), the Sierra Nevada (Swetnam, 1993; Brunelle & Anderson, 2003), the Klamath Mountains (Mohr *et al.*, 2000), and the northern Rocky Mountains (Brunelle, 2002; Hallett & Walker, 2000) (Fig. 3B). Synchronization of fire activity across regions that display different climate responses on millennial time scales suggests that short-term variations in climate can override the long-term controls. The situation is not unlike recent years (e.g. 1988, 1996, and 2000) when large areas burned in the West as a result of anomalous drought and fuel buildup.

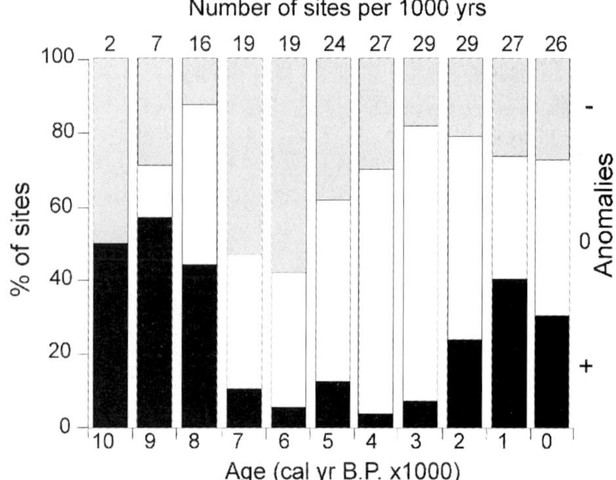

Fig. 4. Variations in fire activity during the Holocene based on microscopic charcoal records from Quebec. The data are presented as charcoal abundance anomalies calculated at 1000-year intervals, in which charcoal influx at a particular interval is compared with the mean charcoal influx over entire Holocene at each site. Below average fire activity is gray; above average is black (based on Carcaillet & Richard, 2000).

Discussion

The growing network of high-resolution fire reconstructions, based on charcoal analysis of lake-sediment records, offers a new proxy by which to study climate change and climate variability in the Holocene. The strength of charcoal data as a paleoclimatic tool rests on understanding the controls of fire regimes and the climate and weather conditions that accumulate and dessicate fuels and create ignitions. In grassland regions, wet years give rise to frequent fires because they allow build-up of fine fuels necessary to spread fires, whereas in forested regions, fire activity increases in dry seasons when fuels dry out. In xeric forests, previous years or seasons are as critical in determining the fire regime as is the season of fire occurrence; in mesic forests, fire events correlate best with climate anomalies of the fire season, with little antecedent influence (Westerling *et al.*, 2003). In other words, one rule does not apply to all fire-climate systems, and this complexity makes the reconstruction of past climate conditions from fire history information particularly challenging.

Charcoal analysis has undergone considerable refinement in the recent decades with improvements in the understanding of charcoal taphonomy and the interpretation of charcoal data. Local fire histories of relatively high precision are obtained from an examination of macroscopic charcoal records, and these records are used to infer variations in fire episodes through time. Microscopic charcoal data also provide fire-history information, but at a more generalized regional scale. Additional fire proxies, including high-resolution pollen data, soil charcoal, and lithologic and geochemical information, enhance fire-history interpretations. The growing network of charcoal records from North America shows variations in fire history that are fairly coherent within climate regions and reasonable in light of our understanding of changes in climate and vegetation during the Holocene.

Better appreciation of the climate conditions that give rise to fires on daily-to-annual time scales has furthered fire-history research from the standpoints of description (Where did fires occur and when?) and hypothesis testing (How does the fire history record provide a series of natural experiments to test fire-climate relationships under different climatic controls?). Modern climate data help reveal the large-scale atmospheric circulation features that promoted large severe fires in recent years. With the use of paleoclimate model simulations, it is possible to examine the frequency of such conditions through the Holocene and ascertain how past variations in fire frequency are related to droughts on annual-to-decadal times scales (e.g. ENSO and PDO [Pacific Interdecadal Oscillation]) and on centennial time scales (e.g. Medieval Warm Period). On longer time scales, the role of more slowly varying components of the climate system, such as changes in the seasonal cycle of insolation, atmospheric composition, sea-surface temperatures, and ice-sheet size, on fire regimes comes into play.

Current collaborations among fire historians, climatologists, and paleoclimatologists promise to lead to the development of new conceptual models that explain the causes of variations in fire incidence on different spatial and temporal scales. Insights gained from understanding fire-climate-vegetation relations in the past are an essential part of this effort. For example, information on the climate conditions conducive to past fires helps assess the role of fire in the face of projected changes in climate resulting from global warming. Moreover, close comparison of fire and other paleoecological records discloses the ecological consequences of changing fire regimes and plant communities, and this information can help guide management decisions that consider the role of natural disturbance in current and future forests. Examination of a network of fire-history records that span a range of past conditions will help discern the relative importance of fuel build-up vs. climate change during current fire years. Thus, the paleoecological perspective provides an understanding of the ecological and environmental conditions that promote individual fires, create particular fire regimes, and determine the natural range of variability in this important ecosystem process.

Summary

Recent advances in fire-history research have enabled the use of charcoal records both as a proxy of climate change and as a tool for examining the response of ecosystems to disturbance. High-resolution charcoal records, obtained from the sediments of lakes and wetlands, are available from several sites in North America and provide information on changes in fire regime occurring on annual-to-millennial time scales during the Holocene. Such reconstructions are supported by studies of recent fires that identify the charcoal source area and processes of charcoal deposition, as well as modern fire-climate-vegetation interactions. When charcoal time series are decomposed, it is possible to distinguish between a slowly varying component that represents variation in fuel composition and hydrology, and charcoal peaks that signify fire episodes. Associated pollen records complement the reconstruction by disclosing past changes in vegetation following individual fires, the role of vegetation in shaping fire regimes, and the long-term consequences of changes in the fire regime on vegetation. The influence of large-scale controls of climate on fire patterns, including variations in summer insolation during the Holocene, is evident in the temporal and spatial variations in fire reconstructions in western and northeastern North America and from an analysis of atmospheric circulation features that give rise to fires.

Acknowledgments

The research described in this paper was supported by grants from the National Science Foundation (ATM-0117160, EAR-9906100). We appreciate the use of unpublished data provided by Tom Minckley, and Rosemary Sherriff, and the help of Andrea Brunelle and Jennifer Marlon in assembling the charcoal database. We thank Bryan Shuman and an anonymous reviewer for helpful comments on the manuscript.

References

Agee, J.K. (1993). *Fire Ecology of Pacific Northwest Forests*. Washington, DC, Island Press, 493 pp.

Anderson, R.S., Davis, R.B., Miller, N.G. & Stuckenrath, R. (1986). History of the late- and post-glacial vegetation and disturbance around Upper South Branch Pond, northern Maine. *Canadian Journal of Botany*, **64**, 1977–1986.

Anderson, R.S. & Smith, S.J. (1997). The sedimentary record of fire in montane meadows, Sierra Nevada, California, USA: A preliminary assessment. *In*: Clark, J.S., Cachier, H., Goldammer, J.G. & Stocks, B. (Eds), *Sediment Records of Biomass Burning and Global Change*. Berlin, Springer Verlag, NATO ASI Series 1: Global Environmental Change, **51**, 313–328.

Bartlein, P.J., Anderson, K.H., Anderson, P.M., Edwards, M.E., Mock, C.J., Thompson, R.S., Webb, R.S., Webb, T., III & Whitlock, C. (1998). Paleoclimate simulations for North America over the past 21,000 years: Features of the simulation climate and comparisons with paleoenvironmental data. *Quaternary Science Reviews*, **17**, 549–585.

Bradbury, J.P. (1996). Charcoal deposition and redeposition in Elk Lake, Minnesota, USA. *The Holocene*, **6**, 339–344.

Brunelle, A. (2002). Holocene changes in fire, climate and vegetation in the northern Rocky Mountains of Idaho and western Montana. Ph.D. dissertation. Eugene, University of Oregon.

Brunelle, A.R. & Anderson, R.S. (2003). Sedimentary charcoal as an indicator of late-Holocene drought in the Sierra Nevada, California, and its relevance to the future. *The Holocene*, **13**, 21–28.

Brunelle, A. & Whitlock, C. (2003). Postglacial fire, vegetation, and climate history in the Clearwater Range, Northern Idaho, USA. *Quaternary Research* (in press).

Carcaillet, C. & Richard, P.J.H. (2000). Holocene changes in seasonal precipitation highlighted by fire incidence in eastern Canada. *Climate Dynamics*, **16**, 549–559.

Carcaillet, C., Bergeron, Y., Richard, P.J.H., Fréchette, B., Gauthier, S. & Prairie, Y.T. (2001a). Change of fire frequency in the eastern Canadian boreal forests during the Holocene: Does vegetation composition or climate trigger the fire regime? *Journal of Ecology*, **89**, 930–946.

Carcaillet, C., Bouvier, M., Larouche, A.C. & Richard, P.J.H. (2001b). Comparison of pollen-slide and sieving methods in lacustrine charcoal analyses for local and regional fire history. *The Holocene*, **11**, 476–547.

Clark, J.S. (1988a). Particle motion and the theory of stratigraphic charcoal analysis: Source area, transport, deposition, and sampling. *Quaternary Research*, **30**, 67–80.

Clark, J.S. (1988b). Stratigraphic charcoal analysis on petrographic thin sections: Applications to fire history in northwestern Minnesota. *Quaternary Research*, **30**, 81–91.

Clark, J.S. (1988c). Drought cycles, the 'Little Ice Age' and fuels: A 750 yr record of fire in northwestern Minnesota. *Nature*, **334**, 233–235.

Clark, J.S. (1990). Fire and climate change during the last 750 years in northern Minnesota. *Ecological Monographs*, **60**, 135–159.

Clark, J.S. & Patterson, W.A. (1997). Background and local charcoal in sediments: Scales of fire evidence in the paleorecord. *In*: Clark, J.S., Cachier, H., Goldammer, J.G. & Stocks, B. (Eds), *Sediment Records of Biomass Burning and Global Change*. Berlin, Springer Verlag, NATO ASI Series 1: Global Environmental Change, **51**, 23–48.

Clark, J.S. & Royall, P.D. (1996). Local and regional sediment charcoal evidence for fire regimes in presettlement north-eastern North America. *Journal of Ecology*, **84**, 365–382.

Clark, J.S., Royall, P.D. & Chumbley, C. (1996). The role of fire during climate change in an eastern deciduous forest at Devil's Bathtub, New York. *Ecology*, **77**, 2148–2166.

Clark, J.S., Lynch, J., Stocks, B. & Goldammer, J. (1998). Relationships between charcoal particles in air and sediments in West-central Siberia. *The Holocene*, **8**, 19–29.

Clark, J.S., Grimm, E.C., Donovan, J.J., Fritz, S.C., Engstrom, D.R. & Almendinger, J.E. (2002). Drought cycles and landscape responses to past aridity on prairies of the northern Great Plains, USA. *Ecology*, **83**, 595–601.

Clark, T.L., Jenkins, M.A., Coen, J. & Packham, D. (1996). A coupled atmosphere-fire model: Convective feedback on fire-line dynamics. *Journal of Applied Meteorology*, **35**, 875–901.

Cofer, W.R., Koutzenogii, K.P., Kokorin, A. & Excurra, A. (1997). Biomass burning emissions and the atmosphere. *In*: Clark, J.S., Cachier, H., Goldammer, J.G. & Stocks, B.

(Eds), *Sediment Records of Biomass Burning and Global Change*. Berlin, Springer Verlag. NATO ASI Series 1: Global Environmental Change, **51**, 169–188.

Cwynar, L.C. (1978). Recent history of fire and vegetation from annually laminated sediment of Greenleaf Lake, Algonquin Park, Ontario. *Canadian Journal of Botany*, **56**, 10–12.

Davis, M.B. (1965). Phytogeography and palynology of northeastern United States. *In*: Wright, H.E., Jr. & Frey, D.G. (Eds), *The Quaternary of the United States*. Princeton New Jersey, Princeton University Press, 377–401.

Delcourt, P.A., Delcourt, H.R., Ison, C.R., Sharp, W.E. & Gremillion, K.J. (1998). Prehistoric human use of fire, the eastern agricultural complex, and Appalachian oak-chestnut forests: Paleoecology of Cliff Palace Pond, Kentucky. *American Antiquity*, **63**, 263–278.

Fall, P.L. (1997). Fire history and composition of the subalpine forest of western Colorado during the Holocene. *Journal of Biogeography*, **24**, 309–325.

Fall, P.L., Davis, P.T. & Zielinski, G.A. (1995). Late Quaternary vegetation and climate of the Wind River Range, Wyoming. *Quaternary Research*, **43**, 393–404.

Flannigan, M.D. & Harrington, J.B. (1988). A study of the relation of meteorological variables to monthly provincial area burned by wildfire in Canada 1953–1980. *Journal of Applied Meteorology*, **27**, 66–72.

Flannigan, M., Todd, B., Wotton, M., Stocks, B., Skinner, W. & Martell, D. (2000). Pacific sea surface temperatures and their relation to area burned in Canada. *Third Symposium on Fire and Forest Meteorology*. Boston, Massachusetts, American Meteorological Society, 151–157.

Gardner, J.J. & Whitlock, C. (2001). Charcoal accumulation following a recent fire in the Cascade Range, northwestern USA, and its relevance for fire-history studies. *The Holocene*, **11**, 541–549.

Gavin, D.G., Brubaker, L.B. & Lertzman, K.P. (2003). Holocene fire history of a coastal temperate rainforest based on soil charcoal radiocarbon dates. *Ecology*, **84**, 186–201.

Gedye, S.J., Jones, R.T., Tinner, W., Ammann, B. & Oldfield, F. (2000). The use of mineral magnetism in the reconstruction of fire history: A case study from Lago di Origlio, Swiss Alps. *Palaeogeography, Palaeoclimatology, and Palaeoecology*, **164**, 101–110.

Green, D.G. (1981). Time series and postglacial forest ecology. *Quaternary Research*, **15**, 265–277.

Grissino-Mayer, H.D. & Swetnam, T.W. (2000). Century-scale climate forcing of fire regimes in the American Southwest. *The Holocene*, **10**, 213–220.

Hallett, D.J. & Walker, R.C. (2000). Paleoecology and its application to fire and vegetation management in Kootenay National Park, British Columbia. *Journal of Paleolimnology*, **24**, 401–414.

Johnson, E.A. & Gutsell, S.L. (1994). Fire frequency models, methods and interpretations. *Advances in Ecological Research*, **25**, 239–287.

Johnson, E.A. & Wowchuk, D.R. (1993). Wildfires in the southern Canadian Rocky Mountains and their relationship to mid-tropospheric anomalies. *Canadian Journal of Forest Research*, **23**, 1213–1222.

Kalnay, E., Kanamitsu, M., Kistler, R., Collins, W., Deaven, D., Gandin, L., Iredell, M., Saha, S., White, G., Woollen, J., Zhu, Y., Leetmaa, A., Reynolds, B., Chelliah, M., Ebisuzaki, W., Higgins, W., Janowiak, J., Mo, K.C., Ropelewski, C., Wang, J., Jenne, R. & Joseph, D. (1996). The NCEP/NCAR 40-year reanalysis project. *Bulletin of the American Meteorological Society*, **77**, 437–472.

Kistler, R., Kalnay, E., Collins, W., Saha, S., White, G., Woollen, J., Chelliah, M., Ebisuzaki, W., Kanamitsu, M., Kousky, V., van den Dool, H., Jenne, R. & Fiorino, M. (2001). The NCEP-NCAR 50-year reanalysis: Monthly means CD-ROM and documentation. *Bulletin of the American Meteorological Society*, **82**, 247–267.

Klein, W.H., Charney, J.J., McCutchan, M.H. & Benoit, J.W. (1996). Verification of monthly mean forecasts for fire weather elements in the contiguous United States. *Journal of Climate*, **9**, 3317–3327.

Kutzbach, J.E., Gallimore, R., Harrison, S., Behling, P., Selin, R. & Laarif, R. (1998). Climate and biome simulations for the past 21,000 years. *Quaternary Science Reviews*, **17**, 473–506.

Larsen, C.P.S. & MacDonald, G.M. (1998). An 840-year record of fire and vegetation in a boreal white spruce forest. *Ecology*, **79**, 106–118.

Long, C.J. (2003). Holocene fire and vegetation history of the Oregon Coast Range. Ph.D. dissertation. Eugene, University of Oregon.

Long, C.J. & Whitlock, C. (2002). Fire and vegetation history from the coastal rain forest of the western Oregon Coast Range. *Quaternary Research*, **58**, 215–225.

Long, C.J., Whitlock, C., Bartlein, P.J. & Millspaugh, S.H. (1998). A 9000-year fire history from the Oregon Coast Range, based on a high-resolution charcoal study. *Canadian Journal of Forest Research*, **28**, 774–787.

MacNeil, J.S. (2000). Forest fire plan kindles debate. *Science*, **289**, 1448–1449.

Meyer, G.A., Wells, S.G. & Jull, A.J.T. (1995). Fire and alluvial chronology in Yellowstone National Park: Climatic and intrinsic controls on Holocene geomorphic processes. *Geological Society of America Bulletin*, **107**, 1211–1230.

Mills, T.J. & Lugo, A.E. (2001). The long-term answer: Fight fire with research. *Nature*, **409**, 452.

Millspaugh, S.H. & Whitlock, C. (1995). A 750-year fire history based on lake sediment records in central Yellowstone National Park. *The Holocene*, **5**, 283–292.

Millspaugh, S.H. & Whitlock, C. (2003). Postglacial fire, vegetation, and climate history of the Yellowstone-Lamar and Central Plateau provinces, Yellowstone National Park. *In*: Wallace, L. (Ed.), *After the Fires: The Ecology of Change in Yellowstone National Park*. New Haven, Connecticut, Yale University Press (in press).

Millspaugh, S.H., Whitlock, C. & Bartlein, P.J. (2000). Variations in fire frequency and climate over the last 17,000 years in central Yellowstone National Park. *Geology*, **28**, 211–214.

Mock, C.J. & Brunelle-Daines, A.R. (1999). A modern analogue of western United States summer palaeoclimate at 6000 years before present. *The Holocene*, **9**, 541–545.

Mohr, J.A., Whitlock, C. & Skinner, C.J. (2000). Postglacial vegetation and fire history, eastern Klamath Mountains, California. *The Holocene*, **10**, 587–601.

Nash, C. & Johnson, E.A. (1995). Synoptic climatology of lightning-caused forest fires in subalpine and boreal forests. *Canadian Journal of Forest Research*, **26**, 1859–1874.

Nature (2000). Seeing the wood for the trees, **406**, 661.

Ohlson, M. & Tryterud, E. (2000). Interpretation of the charcoal record in forest soils: Forest fires and their production and deposition of macroscopic charcoal. *The Holocene*, **10**, 519–525.

Overpeck, J.T., Rind, D. & Goldberg, R. (1990). Climate-induced changes in forest disturbance and vegetation. *Nature*, **343**, 51–53.

Overpeck, J.T., Whitlock, C. & Huntley, B. (2003). Terrestrial biosphere dynamics in the climate system: Past and future. *In*: Bradley, R.S. & Pedersen, T. (Eds), *Causes and Consequences of Past Climate Variations*. International Geosphere Biosphere Project Book Series: New York: Springer, 81–103.

Patterson, W.A., III, Edwards, K.J. & MacGuire, D.J. (1987). Microscopic charcoal as a fossil indicator of fire. *Quaternary Science Reviews*, **6**, 3–23.

Patterson, W.A. III & Backman, A.E. (1988). Fire and disease history of forests. *In*: Huntley, B. & Webb, T. III (Eds), *Vegetation History*. Dordrecht, Kluwer Academic Publishers, 603–632.

Press, W.H., Flannery, B.P., Teukolsky, S.A. & Vetterling, W.T. (1992). *Numerical recipes*. Cambridge, Cambridge University Press, 963 pp.

Pyne, S.J., Andrews, P.L. & Laven, R.D. (1996). *Introduction to Wildland Fire*. New York, John Wiley and Sons, 769 pp.

Reasoner, M.A. & Huber, U.M. (1999). Postglacial palaeoenvironments of the upper Bow Valley, Banff National Park, Alberta, Canada. *Quaternary Science Reviews*, **18**, 475–492.

Rhodes, T.E. & Davis, R.B. (1995). Effects of late Holocene forest disturbance and vegetation change on acidic Mud Pond, Maine, USA. *Ecology*, **76**, 734–746.

Roads, J.O., Chen, S.-C., Fujioka, F.M. & Burgan, R.E. (2000). Development of a seasonal fire weather forecast for the contiguous United States. *Third Symposium on Fire and Forest Meteorology*. Boston Massachusetts, American Meteorological Society, 99–102.

Skinner, W.R., Stocks, B.J., Martell, D.L., Bonsal, B. & Shabbar, R. (1999). The association between circulation anomalies in the mid-troposphere and area burned by wildland fire in Canada. *Theoretical and Applied Climatology*, **63**, 89–105.

Skinner, W.R., Flannigan, M.D., Stock, B.J., Martell, D.L., Wooton, B.M., Todd, J.B., Mason, J.A., Logan, K.A. & Bosch, E.M. (2002). A 500hPa synoptic wildland fire climatology for large Canadian forest fires, 1959–1996. *Theoretical and Applied Climatology*, **71**, 157–169.

Smith, S.J. & Anderson, R.S. (1992). Late Wisconsin paleoecologic record from Swamp Lake, Yosemite National Park, California. *Quaternary Research*, **38**, 91–102.

Sugita, S., MacDonald, G.M. & Larsen, C.P.S. (1997). Reconstruction of fire disturbance and forest succession from fossil pollen in lake sediments: Potential and limitations. *In*: Clark, J.S., Cachier, H., Goldammer, J.G. & Stocks, B. (Eds), *Sediment Records of Biomass Burning and Global Change*. Berlin, Springer Verlag, NATO ASI Series 1: Global Environmental Change, **51**, 387–412.

Swain, A.M. (1973). A history of fire and vegetation in northeastern Minnesota as recorded in lake sediments. *Quaternary Research*, **3**, 383–396.

Swain, A.M. (1978). Environmental changes during the past 2000 yr in north-central Wisconsin: Analysis of pollen, charcoal and seeds from varved lake sediments. *Quaternary Research*, **10**, 55–68.

Swetnam, T.W. (1993). Fire history and climate change in Giant Sequoia groves. *Science*, **262**, 885–889.

Swetnam, T.W. & Betancourt, J.L. (1990). Fire-Southern Oscillation relations in the Southwestern United States. *Science*, **249**, 1017–1020.

Swetnam, T.W. & Betancourt, J.L. (1992). Temporal patterns of El Nino/Southern Oscillation – wildfire teleconnections in the southwestern United States. *In*: Diaz, H.F. & Markgraf, V. (Eds), *El Nino: Historical and Paleoclimatic Aspects of the Southern Oscillation*. Cambridge, U.K., Cambridge University Press, 258–270.

Swetnam, T.W. & Betancourt, J.L. (1998). Mesoscale disturbance and ecological response to decadal climatic variability in the American Southwest. *Journal of Climate*, **11**, 3128–3147.

Tolonen, K. (1986). Charred particle analysis. In: Berglund, B.E. (Ed.), *Handbook of Holocene Palaeoecology and Palaeohydrology*. New York, John Wiley and Sons, 485–496.

Veblen, T.T., Kitzberger, T. & Donnegan, J. (2000). Climatic and human influences on fire regimes in ponderosa pine forests in the Colorado Front Range. *Ecological Applications*, **10**, 1178–1195.

Watson, R.T., Noble, I.R., Bolin, B., Ravindranath, N.H., Verardo, D.J. & Dokken, D.J. (2000). *Land Use, Land-Use Change, and Forestry*. Intergovernmental Panel on Climate Change: Cambridge, UK, Cambridge University Press, 377 pp.

Weber, M.G. & Flannigan, M.D. (1997). Canadian boreal forest ecosystem structure and function in a changing climate: Impact on fire regimes. *Environmental Review*, **5**, 145–166.

Westerling, A.L., Gershunov, A., Brown, T.T., Cayan, D.R. & Dettinger, M.D. (2003). Climate and wildfire in the western United States. *Bulletin of the American Meteorological Society*, May issue, 595–604.

Whitlock, C. & Anderson, R.S. (2003). Methods and interpretation of fire history reconstructions based on sediment records from lakes and wetlands. *In*: Swetnam, T.W., Montenegro, G. & Veblen T.T. (Eds), *Fire and Climate Change in the Americas*. New York, Springer, 3–31.

Whitlock, C. & Bartlein, P.J. (1993). Spatial variations of Holocene climatic change in the Yellowstone region. *Quaternary Research*, **39**, 231–238.

Whitlock, C. & Larsen, C. (2002). Charcoal as a fire proxy. *In*: Smol, J.P., Birks, H.J.B. & Last, W.M. (Eds), *Tracking Environmental Change Using Lake Sediments: Terrestrial, Algal, and Siliceous Indicators*. Dordrecht: Kluwer Academic Publishers, **3**, 75–97.

Whitlock, C. & Millspaugh, S.H. (1996). Testing assumptions of fire history studies: An examination of modern charcoal accumulation in Yellowstone National Park. *The Holocene*, **6**, 7–15.

Wright, H.E., Jr. & Frey, D.G. (1965). *The Quaternary of the United States*. Princeton, New Jersey, Princeton University Press, 377 pp.

Interannual to decadal climate and streamflow variability estimated from tree rings

David W. Stahle, Falko K. Fye and Matthew D. Therrell

Tree-Ring Laboratory, Ozark Hall 113, University of Arkansas, Fayetteville, AR 72701, USA

Introduction

Tree-ring dating (dendrochronology) is the most accurate and precise dating method in geochronology. Tree-ring chronologies are dated to the exact calendar year, and can be directly calibrated with instrumental climatic data with seasonal to annual resolution. Hundreds of moisture-sensitive tree-ring chronologies have now been developed for North America, and most are available from the International Tree-Ring Data Bank at the National Geophysical Data Center. These annually resolved time series present many unique opportunities for socially relevant research into the dynamics and impacts of regional to global climate systems. Tree-ring data can be used to document natural hydroclimatic variability prior to the period of anthropogenic climate modification, including severe drought extremes not witnessed in the instrumental record. Climate reconstructions from tree rings can be linked with social, economic, and ecological data to explore the human and environmental impact of past climate extremes. This chapter highlights selected tree-ring research in the continental United States, with a particular emphasis on hydroclimatic variability, environmental and human impacts, and recent developments with new multi-century tree-ring chronologies in the eastern and western United States.

Hydroclimatic Applications of the Continental Network of Tree-Ring Chronologies

The network of centuries-long tree-ring chronologies now available for North America extends from the tropics of southern Mexico to the boreal forest, and from the Pacific to Atlantic coasts (Fig. 1). This may be the finest large-scale network of high-resolution paleoclimatic proxies yet produced worldwide, and its development has been funded primarily by the National Science Foundation, Climate Dynamics and Paleoclimatology Programs. This abundance and wide distribution of centuries-long tree-ring chronologies (Fig. 1) testifies to the fact that ancient forests still survive across North America, particularly on remote rugged terrain and on xeric sites where tree growth is slow and the woodlands unsuited for lumber production or agriculture (Stahle, 2002).

All of the tree-ring chronologies located in Fig. 1 span at least the last 200 years, and the vast majority date back to A.D. 1700 or earlier. The chronologies are coded by species, and several moisture-sensitive species in the white oak group dominate the network in the eastern United States (especially *Quercus alba, Q. stellata*, and *Q. prinus*). Eastern hemlock (*Tsuga canadensis*) and baldcypress (*Taxodium distichum*) are heavily represented in the eastern network. Ponderosa pine (*Pinus ponderosae*) and Douglas-fir (*Pseudotsuga menziesii*) are the most important species in the western network. Recent sampling of Douglas-fir in small isolated higher elevation microenvironments of central and southern Mexico (Therrell *et al.*, 2002), and near the northern limit of the species in British Columbia (Watson & Luckman, 2002) provide a single-species array that extends over nearly 40 degrees of latitude. Few tree species have such an impressive latitudinal range, and even fewer are as valuable for climatic reconstruction as Douglas-fir.

Many of the chronologies illustrated in Fig. 1 were used by E.R. Cook *et al.* (1999) to reconstruct the summer Palmer Drought Severity Index (PDSI) on a $2 \times 3^\circ$ latitude/longitude grid covering the entire coterminous United States from A.D. 1700–1979. These reconstructions have been validated in most regions against independent PDSI data not used in the calibration process, and some of the reconstructions extend much earlier over sub-regions of the country (e.g. back to A.D. 1500 over most of the West and Southeast). These gridded tree-ring reconstructions of summer PDSI faithfully capture the main regional modes of drought and wetness variability across the United States, as identified with instrumental PDSI data from A.D. 1895–1981 by Karl & Koscielny (1982).

Fye *et al.* (2003) have used composite analysis of the Cook *et al.* (1999) summer PDSI reconstructions to map the largest decadal moisture excursions of the 20th century and place them into the historical perspective of the past 500 years. The tree-ring reconstructions generally reproduce the geography, timing, and relative intensity of the major moisture regimes actually observed over the United States during the 20th century, but not the absolute intensity. Thus, the largest dry and wet regimes seen in the instrumental record during the 20th century were also the largest in the reconstructions during the same interval. These decadal regimes included the early 20th-century pluvial, the Dust Bowl drought, and the 1950s Southwestern-Southern Plains drought (also note Karl, 1988; Trenberth, 1991).

The drought of 1934 was the worst that was recorded by both the instruments and tree rings from A.D. 1895 to present, and over the past 500 years may have been surpassed only once by the extreme and widespread drought of A.D. 1580 (Fye *et al.*, 2003). The prolonged Dust Bowl drought of the 1930s is one of the few decadal droughts to impact most of the United States. The tree-ring reconstructions suggest that the Dust Bowl drought may have been the most severe, sustained, and widespread drought to impact the continental

DEVELOPMENTS IN QUATERNARY SCIENCE
VOLUME 1 ISSN 1571-0866
DOI:10.1016/S1571-0866(03)01023-6

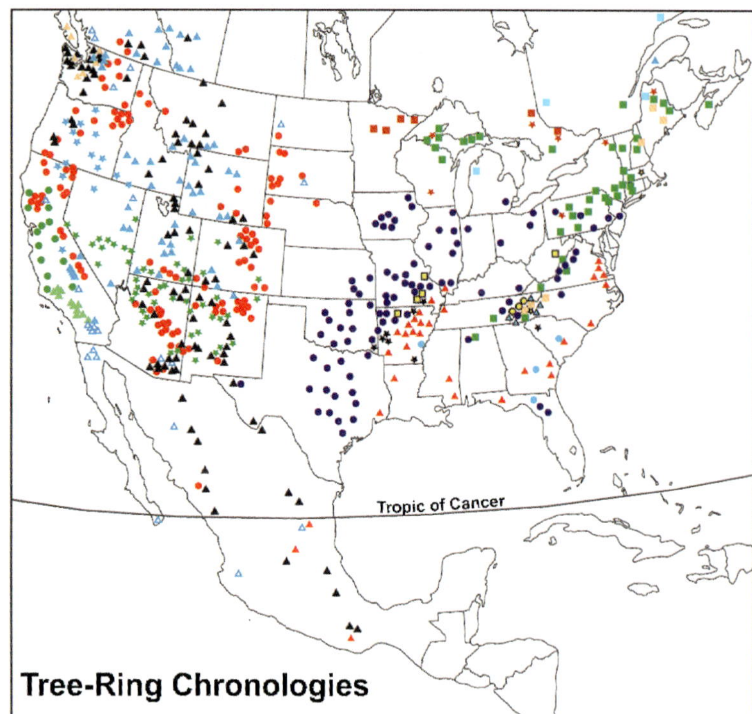

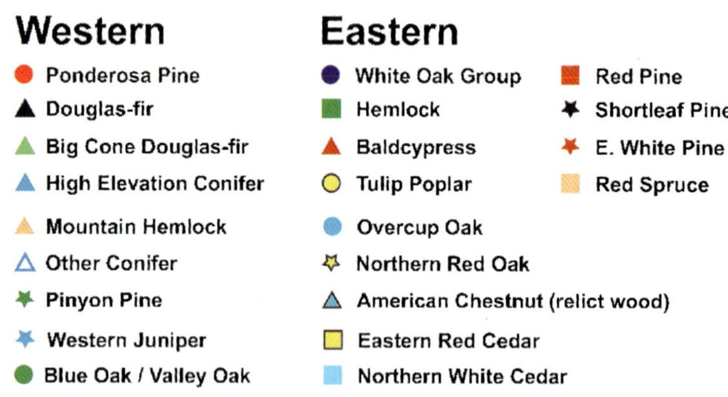

Fig. 1. Locations of tree-ring chronologies at least 200 years long that are now available for the United States, southern Canada, and Mexico. These chronologies have been developed by the combined efforts of the tree-ring community, following the initial work of A.E. Douglass and E. Schulman. The chronologies are coded by species and most are available from the International Tree-Ring Data Bank at the National Geophysical Data Center in Boulder, Colorado.

United States since the mega-drought of the 16th century (Fye *et al.*, 2003). The 1950s drought was the second most severe and sustained drought of the 20th century, and was concentrated over the Southwest-Southern Plains (Fig. 2). The geography of the 1950s drought is reasonably well reproduced by the tree-ring reconstructions, and appears to have been replicated by ten similar decadal droughts over the past 500 years (Fig. 2). These 1950s-like droughts varied from 5- to 21-years in duration, and include the 16th-century mega-drought from A.D. 1567–1587.

Because drought is limiting to tree growth, tree-ring chronologies are generally more accurate proxies of drought than wet conditions. Nevertheless, the Cook *et al.* (1999) PDSI estimates provide a good representation of the early 20th-century pluvial actually recorded in the instrumental PDSI data from approximately 1905–1923 (Fig. 3). The long tree-ring reconstructions suggest that the early 20th-century pluvial may have been the most intense and prolonged wet anomaly across the western two thirds of the country in 500 years, although two near analogs occurred in the early 19th and early 17th centuries (Fye *et al.*, 2003; Fig. 3). Thus the 20th century period of instrumental climate observation appears broadly representative of the annual and decadal PDSI variability of the past 400 years, but not before (e.g. Cook & Evans, 2000).

The extended wet episode during the early 20th century certainly biased estimates of mean runoff during the negotiations leading up to the Colorado River Compact. Compact negotiators assumed a mean flow at Lee Ferry of 17.5 million acre feet/year (MAF/yr) and allocated 7.5 MAF/yr to the upper basin, 7.5 MAF/yr to the lower basin, 1.5 MAF/yr to Mexico, and the remaining 1.0 MAF/yr as a bonus to the lower basin (Reisner, 1986). California alone was entitled to 4.4 MAF/yr under the compact. However, a tree-ring reconstruction of annual stream flow at Lee Ferry by Stockton (1975) indicated that the period from 1906–1930 included the highest runoff totals estimated for the Colorado in 400 years, and from 1930–1952 "the discharge of the Colorado

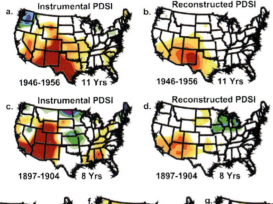

Fig. 2. The summer PDSI reconstructions of Cook et al. (1999) have been used to average and map consecutive episodes of drought that resembled the 1950s drought over the past 500 years (Fye et al., 2003). The instrumental measurements of the 1950s drought are illustrated (top left), along with the tree-ring reconstructions for 1946–1956 (top right). These 1950s-like droughts varied from 6- to 18-years in duration, and were concentrated over the Southwest.

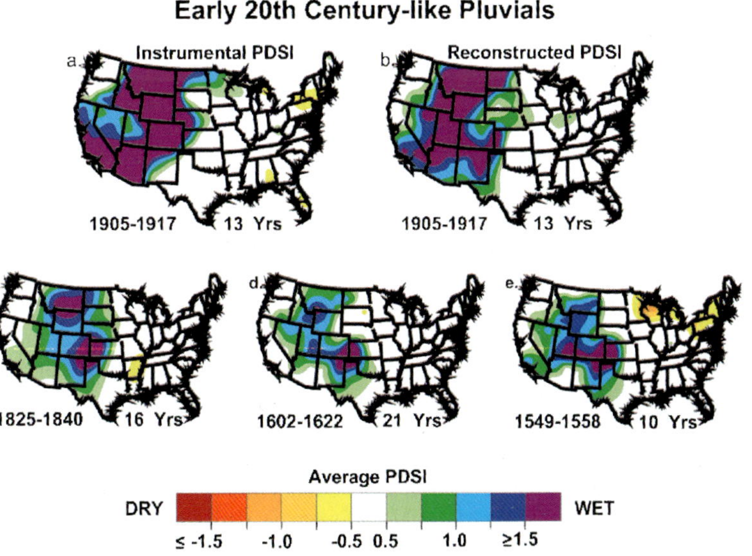

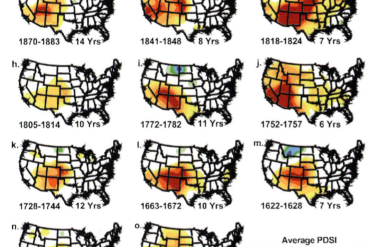

Fig. 3. The summer PDSI reconstructions of Cook et al. (1999) have been used to average and map prolonged wet episodes over the past 500 years, including the early 20th century pluvial (Fye et al., 2003). The instrumental measurements for the early 20th century pluvial are illustrated (top left), along with the tree-ring reconstructions for 1905–1917 (top right). Prolonged and widespread wet episodes occurred over the western United States in the 16th, 17th and 19th centuries, but the early 20th century pluvial as estimated from tree rings appears to have been unmatched in magnitude over the past 500 years (Fye et al., 2003).

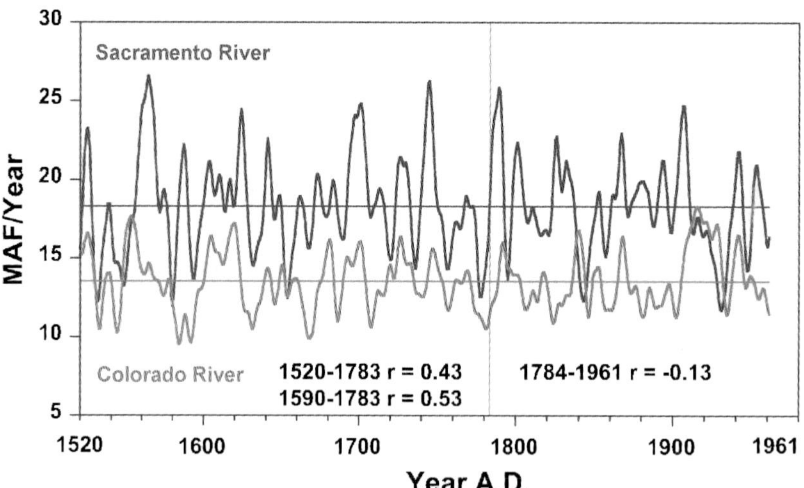

Fig. 4. Tree-ring reconstructed annual streamflow for the Sacramento and Colorado Rivers in millions of acre feet per year (from Meko et al., 2001; Stockton, 1975). To highlight the decadal variability in the data, both time series have been smoothed with a cubic smoothing spline reducing 50% of the variance in a sine wave with a period of 10 years (Cook & Peters, 1981). The full Sacramento River record extends from 869–1977, with an estimated mean annual flow of 17.9 MAF/yr (Meko et al., 2001). Only the period from 1520 to 1961 is illustrated here for the Sacramento, when the estimated mean flow was 18.3 MAF/yr. The long term mean streamflow for the Colorado is estimated at 13.5 MAF/yr for 1520–1961 (see Meko et al., 1991, 2001, for discussion of the estimated error associated with both reconstructions). Note the apparent break in decadal covariance between these two series in the 1780s (the correlation between segments of the smoothed time series is also reported).

River at Lee Ferry (near the Arizona-Utah border) averaged only 11.7 MAF" (Raymond Hill quoted in Reisner, 1986). Stockton's (1975) tree-ring-based estimate of long-term mean runoff for the Colorado River at Lee Ferry of only 13.5 MAF/yr remains today as perhaps the most famous single number ever calculated with tree-ring data (Fig. 4, bottom).

Meko *et al.* (2001) report a new high-quality reconstruction of Sacramento River flow using an expanded network of precipitation-sensitive chronologies that includes blue oak (*Q. douglasii*) and western juniper (*Juniperus occidentalis*). This reconstruction explicitly accounts for the uncertainty associated with the changing suite of tree-ring predictors back in time, and indicates that certain single-year droughts and extended drought episodes before A.D. 1600 may have exceeded any witnessed on Sacramento during the period of instrumental observation.

As noted by Meko *et al.* (1991) the co-occurrence of drought over the major watersheds of the Sierra Nevada and Rocky Mountains could seriously impact water resource allocations in Southern California and the Southwest. The overall correlation between decadally smoothed versions of the long Colorado and Sacramento River reconstructions is only $r = 0.31$, but the two estimated flow series were more coherent at decadal scales during the 17th and 18th centuries than afterward (the correlation is $r = 0.53$ for A.D. 1585–1783, Fig. 4). The 16th century mega-drought, for example, does appear to have impacted runoff in both the Rockies and Sierra Nevada (Figs 2 and 4). These results suggest that there may be natural large-scale climate conditions under which decadal drought and wetness regimes can impact both watersheds simultaneously. The water storage facilities constructed for both the Sacramento and Colorado have enormous capacity to deal with extended drought and wetness regimes (Dracup & Kendall, 1991), however, a repeat of the severe sustained droughts of the late 16th, mid-17th, and late 18th centuries could seriously constrain water supply and allocation across the Southwest, especially if they were to occur over both the Sacramento and Colorado watersheds.

Tree-Ring Estimates of Climate Forcing Factors

H.C. Fritts first demonstrated the usefulness of the North American tree-ring chronology network for estimation of large-scale ocean-atmospheric forcing of inter-annual and decadal climate variability (e.g. Fritts, 1976). In fact, tree-ring data from subtropical North America record one of the strongest ENSO signals in climate proxies worldwide, and are highly correlated with oxygen isotope chronologies developed from annually banded corals in the equatorial Pacific (Cleaveland *et al.*, 2003). Lough & Fritts (1985) developed the first tree-ring reconstruction of the El Nino/Southern Oscillation (ENSO), estimating the Southern Oscillation Index (SOI) from North American tree-ring chronologies from A.D. 1600–1965. Stahle *et al.* (1998a) used tree-ring data from North America and Indonesia to reconstruct the winter SOI from A.D. 1706–1979, and estimated a significant shift in the envelope of reconstructed winter SOI variance during the 19th century. However, both SOI reconstructions rely heavily on the ENSO teleconnection to subtropical North America, which Cole & Cook (1998) show may not have been stationary during the last 150 years.

Lough & Fritts (1985) point out that the potential extra-tropical proxies of ENSO have large uncertainties associated with the teleconnection filter through which the tropical ENSO signal must propagate into the extra-tropics. This is a problem that underlies many attempts to use remote proxies to estimate climate-forcing mechanisms that reside in the coupled ocean-atmospheric system. One solution for the paleoclimatic reconstruction of ENSO variability might be the development of annually or seasonally resolved marine and terrestrial proxies from the centers of action of ENSO in the equatorial Pacific. Certain annually banded coral and tree species found in the Indonesian archipelago offer some hope of defining equatorial ENSO variability for several centuries (e.g. Allan & D'Arrigo, 1999). These extended ENSO estimates would then be valuable in conjunction with climate-sensitive tree-ring chronologies in the extra-tropics for mapping past teleconnection patterns. For example, Cole *et al.* (2002) use coral data from Miana atoll to identify extended cold conditions in the central equatorial Pacific during the mid-19th century, that they link to the tree-ring reconstructed drought that occurred simultaneously over North America (see Fig. 9, bottom right, in Cole & Cook, 2002).

The dramatic shift in North Pacific/North American climate at 1976–1977 has been linked to the Pacific Decadal Oscillation (Mantua *et al.*, 1997). Gedalof & Smith (2001) used tree-ring chronologies of mountain hemlock (*Tsuga mertensiana*) near the North Pacific and Gulf of Alaska to reconstruct a PDO index. They used an intervention detection algorithm to identify the three suspected regime shifts in instrumental PDO during the 20th century, and nine earlier shifts in the reconstructed record that extends from A.D. 1600 to the present. Their results suggest that the North Pacific has experienced regimes of alternately warm and cold sea-surface temperatures (SST) over the last 400 years, and that transitions between regimes tend to be abrupt, similar to the regime shifts witnessed during the 20th century in instrumental data. Other tree-ring reconstructions of the PDO have been reported (e.g. Biondi *et al.*, 2001; D'Arrigo *et al.*, 2001). These reconstructions to some degree share large-scale paleoclimate variance that is influenced by the PDO and other large-scale climate-forcing mechanisms, and Biondi *et al.* (2001) specifically attempt to remove ENSO variability from their estimate of inter-decadal variability. It is interesting that the PDO reconstructions of Biondi *et al.* (2001) and Gedalof & Smith (2001), using completely different tree-ring data sets from southern California-Baja California and the Pacific Northwest-Canada-Alaska, respectively, both located immediately downstream of the North Pacific Ocean, identify the same large decade-scale reversals of North Pacific climate near 1750, 1946 and 1977, and perhaps lesser events as well.

The Southwest or Mexican Monsoon (Douglas *et al.*, 1993; Higgins *et al.*, 1999) is an important component of summer precipitation over the United States and northern Mexico, and seems to exhibit a weak out-of-phase relationship with summer precipitation amounts over the Midwest and Florida (Douglas & Englehart, 1996; Higgins *et al.*, 1999). Unfortunately, many tree-ring chronologies from western North American conifers do not have a strong response to mid- to late-summer precipitation totals, and may be more highly correlated with winter-spring precipitation totals. This response is believed to relate to soil-moisture recharge at the onset of the spring growing season and to physiological factors controlling stored photosynthate during the summer, fall, and winter preceding growth (e.g. Fritts, 1966). However, separate chronologies of earlywood (EW) and latewood (LW) width have been developed in attempts to isolate a useful record of summer precipitation in the monsoon region. EW chronologies in the Tex-Mex sector (the southwestern USA and northern Mexico) have been shown to have stronger winter precipitation and ENSO signal than the total-ring-width chronologies derived from the same trees (Cleaveland *et al.*, 2003). LW chronologies from the Tex-Mex sector are more highly correlated with summer precipitation, and some appear to be correlated with the onset of the Mexican Monsoon (Therrell *et al.*, 2002). Meko & Baisan (2001) have shown considerable promise for the use of latewood width and other latewood properties for the estimation of summer monsoon precipitation over Arizona.

The potential role of solar-lunar forcing on climate variability over the United States occupied much of A.E. Douglass' research during the early development of dendrochronology (Douglass, 1920). Cook *et al.* (1997) recently revisited the analyses of drought area across the United States conducted by Mitchell *et al.* (1979). Using a greatly expanded set of tree-ring chronologies, Cook *et al.* (1997) identified the statistically significant bi-decadal drought rhythm reported in the earlier work. This bi-decadal drought rhythm over western North America is perhaps the most stable and statistically robust solar-lunar band signal yet detected in annually resolved paleo data. But it does not appear to explain more than 10% of the variance in drought area over western North America and the physical mechanisms involved in this hypothesized solar-lunar forcing of surface climate remain unclear (Cook *et al.*, 1997).

Multiproxy Paleoenvironmental Research and Tree Rings

Tree-ring chronologies have many virtues for climate reconstruction, including exact calendar dating, a well-specified climatic signal, massive replication of tree-ring time series per collection site, and repetition of chronologies across the landscape. But these time series are derived from living organisms with certain characteristics that limit their usefulness for reconstructing all dimensions of climate variability, particularly lower-frequency secular changes in climate and long-duration dry or wet regimes. These biological characteristics are discussed below and include a degree of tree-growth adaptability to changing environmental conditions, a survivor effect, and the nonstationarity of radial growth attending the increasing size and age of the plant.

For an example of growth adaptability, field observations suggest that baldcypress root systems may act as a natural high-pass filter on the derived time series of radial growth. The fine feeder root systems of baldcypress tend to stratify in the well oxygenated zone of near surface waters, and appear capable of tracking persistent multi-decadal changes in mean water level with the adventitious growth of new root mass (e.g. Stahle & Cleaveland, 1992). This growth tracking tends to smooth out the tree-ring registration of low-frequency shifts between dry and wet regimes, and the resulting baldcypress chronologies are dominated by high frequency inter-annual to decadal changes in precipitation.

A survivor effect may contribute to an underestimation of drought severity from tree-ring data. Old living trees in a forest do record past droughts. But the tree-ring chronologies derived from still-living trees may underestimate the true severity of a past drought because some trees experience such severe growth reductions that they do not survive the drought to contribute to the derived chronology.

The systematic change in mean growth removed with statistical detrending of individual ring-width time series prior to computation of the mean index chronology can also reduce the tree-ring registration of low-frequency climatic trends. The magnitude of this potential detrending bias varies inversely with the length of the individual ring-width time series used to compile the mean chronology (the so-called "segment length curse" Cook *et al.*, 1995). Recent analyses have attempted to deal more explicitly with the detrending issue, basically exploiting the coherence or "cross-dating" of low-frequency growth excursions sometimes seen among multiple trees (the so-called RCS or regional curve standardization method, Briffa *et al.*, 1992a). These studies have reconstructed greater low-frequency temperature variability that appears to agree better with independent indications of large temperature excursions of the past millennium (e.g. Briffa *et al.*, 1992a; Esper *et al.*, 2002).

Regardless of the statistical methodologies employed, the biological realities of tree growth may always to some degree limit the tree-ring registration of long climate regimes and trends. When feasible, the multiproxy approach can improve the registration of the full range of paleoenvironmental variability. For example, Woodhouse & Overpeck (1998) employed several paleoenvironmental indicators including early instrumental records, historical documents, tree-ring reconstructions, lake and alluvial sedimentary records, eolian deposits, and archaeological remains to construct a drought chronology for the central United States. They concluded that several droughts more severe than those recorded during the 20th century occurred over the past 2000 years, and that droughts before A.D. 1600 may have been more persistent and widespread. Large decadal drought variability before A.D. 1600 is also indicated by regional tree-ring data (Meko *et al.*, 1995), and by an analysis of the gridded PDSI reconstructions for the southwestern United States extended back to A.D. 1200 (Cook & Evans, 2000).

The colonial and historic periods offer many opportunities for multiproxy reconstruction of past climate and environmental variability. Guyette (2002) has integrated tree-ring-dated fire scars with rural population levels and cultural practices to construct a long historical sequence of natural and human-caused fire and landscape change for the Ozark Plateau and forest transition zone of Missouri. Brunstein (1996) developed a detailed event chronology of latewood frost rings in bristlecone pine of Colorado, and linked them to historic late-summer snowstorms and subfreezing temperatures. These frost-ring records were integrated with independent tree-ring reconstructions and historical accounts of winter temperature, snowstorms, and glacier advance to identify the terminal Little Ice Age period of early 19th century as one of the most anomalous in the past 2000 years (also see Briffa *et al.*, 1992b; Mann *et al.*, 1998). The large-magnitude volcanic eruptions during the early 19th century, including Tambora, Indonesia, and Cosequina, Nicaragua, have been implicated in this large-scale cold interval, and the bristlecone pine frost-ring record has itself been linked to the chronology of cataclysmic eruptions (LaMarche & Hirschboeck, 1984).

The potential for detailed meteorological and climatological reconstructions using ring-width data, frost rings, documentary accounts, and early instrumental measurements has been illustrated for A.D. 1828. Frost rings in oaks of the eastern United States (Stahle, 1990) indicate that 1828 was one of the major "false spring" episodes of the past 350 years. False spring includes both a climatologically warm winter, followed by a meteorologically significant spring freeze event. The available instrumental and documentary records for 1828 confirm and greatly elaborate on the anomalous climate and weather events of 1828 (Fig. 5). Mean winter temperatures (Dec–Mar) were anomalously warm across eastern North America in 1828, peach trees blossomed at Christmas 1827 in Little Rock, Arkansas, and temperatures exceeded 22 °C (70 °F) on several occasions in January and February at Nashville, Tennessee. David Ludlum (1968) described the winter of 1827–1828 as "the warmest winter in the American experience." At the same time, fur trapper reports and a few thermometer records from trading posts in Canada indicate that the winter of 1828 was colder than average in the Northern Rockies. These winter surface temperature anomalies suggest an upper-air flow pattern marked by a persistent ridge over the East and trough over the West. Conditions changed abruptly in early April of 1828, when a severe outbreak of cold air swept across the United States, and caused frost damage to oak trees, extensive fruit tree and crop damage, widespread reports of frost and freeze, and sub-freezing temperatures readings across the South (C.J. Mock, pers. comm.).

Human Impacts of Tree-Ring Reconstructed Climate Extremes

Henri Grissino-Mayer's (1996) Douglas-fir reconstruction of annual precipitation from El Malpais, New Mexico, is particularly long and outstanding (Fig. 6). The El Malpais reconstruction has been used to document past drought extremes that exceed anything witnessed during the 20th century, and provides a long-term climatological framework for the major social and environmental changes that have occurred in this region over the past two millennia. For example, the 8th-century mega-drought at El Malpais extended from approximately A.D. 735 to 765, and was one of the two or three most severe and prolonged

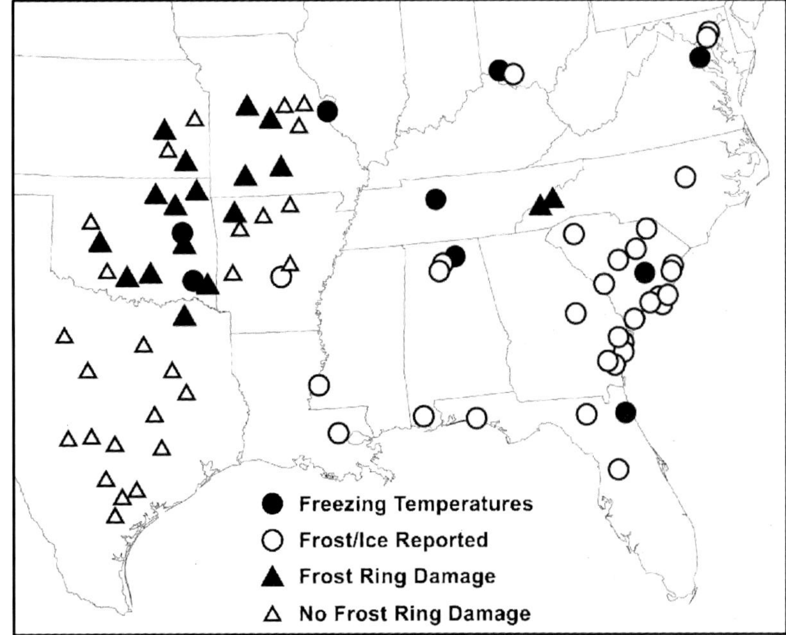

Fig. 5. The false-spring episode of 1828 was recorded by tree ring, documentary, and thermometer data across the southeastern United States. Temperatures below −5 °C are required to cause frost ring damage in oak trees of the eastern United States, and the absence of frost damage to oaks south of the Red River indicates that the −5 °C isotherm did not penetrate into Texas during the early April event (C.J. Mock et al., 2003, pers. comm.).

droughts to impact the Southwest region in the past 2000 yr (Fig. 6).

The "Great Pueblo Drought," during the late 13th century, was registered at El Malpais (Fig. 6), and has been implicated in Anasazi migrations. Burns (1983) used tree-ring data and dryland crop yields from southwest Colorado to estimate reduced crop yields over the Four Corners region during the Great Drought. Salzer (2000) developed a long record of bristlecone pine growth for the San Francisco Peaks in northern Arizona and identified a period of extended warm-dry conditions during the late 13th century, contemporaneous with the Anasazi depopulation of the Colorado Plateau.

The 16th-century mega-drought at El Malpais may have been the most severe sustained drought to impact New Mexico in the last 2000 yr (Figs 2 and 6; Cook & Evans, 2000; Grissino-Mayer, 1996; Meko *et al.*, 1995; Stahle *et al.*, 2000). The summer PDSI reconstructions of Cook *et al.* (1999) cover most of the conterminous USA back to A.D. 1500 (114 out of 154 reconstructed grid points), with good chronology coverage over the Southeast and West. The onset and end of a prolonged drought can be difficult to specify, but a composite of 18 consecutive years during the heart of the mega-drought (A.D. 1570–1587) illustrates the magnitude of this drought that was centered over the southwestern USA (Fig. 2). However, the 16th-century mega-drought certainly did not end at the Mexican border (Fig. 2n). In fact, existing tree-ring data from Durango, Mexico (Stahle *et al.*, 2000), and a new chronology from Puebla, indicate that the mega-drought may have begun over the highlands of Mexico in the 1540's, and then may have expanded north and eastward to impact most of the continent before it played out in the 1590s.

The 16th-century mega-drought occurred at the dawn of European colonization in North America, but the available evidence indicates that there were significant socioeconomic and environmental impacts of this record drought. Some Tewa and Keres Pueblo villages were abandoned during this time, and may have been particularly vulnerable to prolonged drought because they depended on dryland farming (Schroeder, 1968). Taos Pueblo is believed to have been continuously occupied since the 15th century, and may have survived the drought with irrigation agriculture.

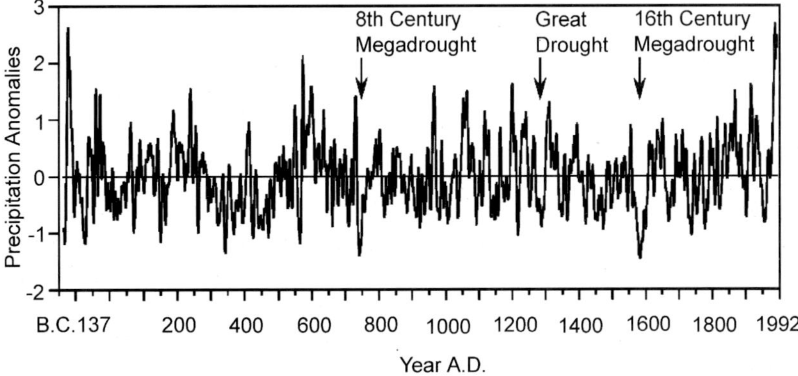

Fig. 6. The tree-ring reconstruction of annual precipitation at El Malpais, New Mexico (normalized units), based on ancient Douglas-fir trees and relic wood (Grissino-Mayer, 1996, Grissino-Mayer et al., 1997). The reconstruction has been smoothed to highlight decadal variability (as in Fig. 4).

The Spanish sphere of influence extended from Florida north into the Carolinas during the 16th century, and included the colony of Santa Elena on Parris Island, South Carolina. Santa Elena was occupied from A.D. 1565 to 1587, and documents survive which describe severe drought during the 1560s (Parr, 1999).

The year 1587 was momentous over the southeastern USA. Santa Elena was abandoned by the Spanish, and Sir Walter Raleigh's Lost Colony on Roanoke Island (North Carolina) disappeared from history. The tree-ring data indicate that A.D. 1587 was the driest year in 800 years in the Tidewater region, and the period A.D. 1587–1589 was the driest three-year episode in 700 years (Stahle *et al.*, 1998b). The Spanish and English colonist alike depended heavily on trade and tribute from the Native Americans, but the documentary record indicates that these tribesmen suffered heavily during drought. The fate of the Lost Colonists may be partly attributed to hardship arising from extreme drought, which may have been part of the continental-scale mega-drought of the 16th century extending into the eastern United States.

The most extreme consequences of the mega-drought during the tumultuous 16th century may have occurred in Mexico, where extreme drought interacted with conquest, colonization, enslavement of the native population under the encomienda system of New Spain, and with terrible outbreaks of epidemic disease to result in one of the great demographic catastrophes in world history (Acuna-Soto *et al.*, 2002). Precipitation estimates from tree-ring data for Durango, Mexico, indicates that the 16th-century mega-drought was the worst drought over Mexico in the past 600 years (Cleaveland *et al.*, 2003).

A wealth of historical, early instrumental, and proxy data enrich the study of the environmental and human impacts of climate extremes during the 19th century. West (1995) has described the 19th-century Cheyenne and Arapaho of the central High Plains as a people in crisis. West (1992) advanced the hypothesis that environmental cycles of wetness and drought interacted with emigration by Native Americans and Europeans, over-exploitation of bison, and the destruction of critical riparian habitat and to result in the disappearance of bison from the central High Plains by A.D. 1860. West (1995) cited tree-ring evidence for an extended wet period during the early 19th century, followed by intense and prolonged drought over the Central Plains as a key element in the bison population decline.

Fye *et al.* (2003) use the gridded PDSI reconstructions of Cook *et al.* (1999) to map the spatial structure of the early 19th century wet period over the central United States (Fig. 7, top). This pluvial period lasted from approximately 1825 to 1840, and was one of the longest and most widespread wet episodes over the United States in the last 500 years (perhaps second only to the early 20th-century pluvial period).

The environmental impact of the early 19th-century pluvial is only beginning to emerge. The longest fire-free interval in the composite fire-scar chronology for two Kipuka fire records from El Malpais National Monument occurred during the early 19th-century pluvial (Grissino-Mayer & Swetnam, 1997). The Kipuka fire record represents a very

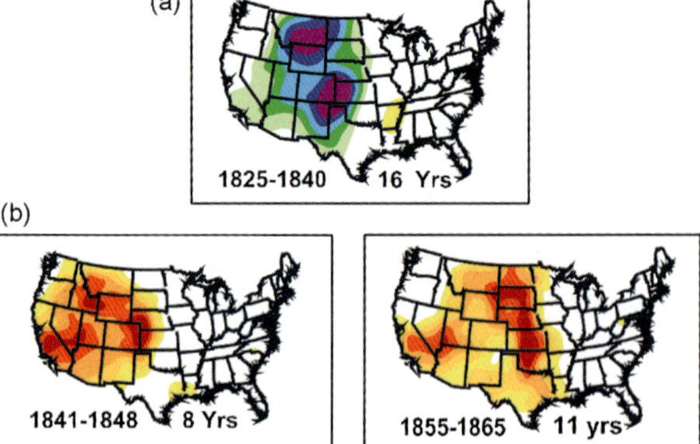

Fig. 7. The Cook et al. (1999) reconstructions of summer PDSI have been used to map the extended wet and dry regimes over the United States during the 19th century. The early 19th-century pluvial period appears to have had a large effect on ecosystem dynamics over much of the West (a), and the subsequent droughts (b) may have interacted with human activities to result in habitat degradation over the central High Plains (West, 1995).

isolated habitat type and may be one of the only fire-scar chronologies in western North America without major anthropogenic effects on fire frequency. The wet conditions during the early 19th century were also reflected by abundant forage and large animal populations across the Central Plains (West, 1995). However, severe drought developed in the mid-1840s over Colorado and continued into the 1860s over portions of the Central Plains (Fig. 7; Woodhouse *et al.*, 2002). These dry conditions occurred while human and livestock utilization of the critical riparian corridors was increasing, and West (1995) argues that this convergence of events led to extensive deterioration of habitat and may have contributed to bison decline in the region.

Multi-Century Long Tree-Ring Chronologies

The remarkable sub-fossil tree-ring records from Western Europe now span the past 10,000 years, and include some of the longest tree-ring chronologies in the world (e.g. Briffa & Matthews, 2002; Leuschner *et al.*, 2002). But a well-replicated network of millennia-long tree-ring chronologies has also been developed for the western United States, including several exceptional lower-elevation bristlecone pine chronologies developed by Don Graybill (e.g. Graybill & Funkhouser, 2000). Other millennium-long tree-ring chronologies have been developed in the western United States for western juniper (Meko *et al.*, 2001), Foxtail pine (*P. balforiana*; Graumlich, 1993), Limber pine (*P. flexilis*; Schulman, 1956), and Douglas-fir (Grissino-Mayer, 1996).

Chronologies at least 700 yr long are located in Fig. 8, but Hughes & Graumlich (1996) report at least 80 tree-ring chronologies over 1000 yr long, and 22 chronologies over 2000 yr long from the greater Southwest. They also used the Methuselah Walk bristlecone chronology to develop an

Fig. 8. Same as Fig. 1, except only those chronologies dating before A.D. 1300 are included.

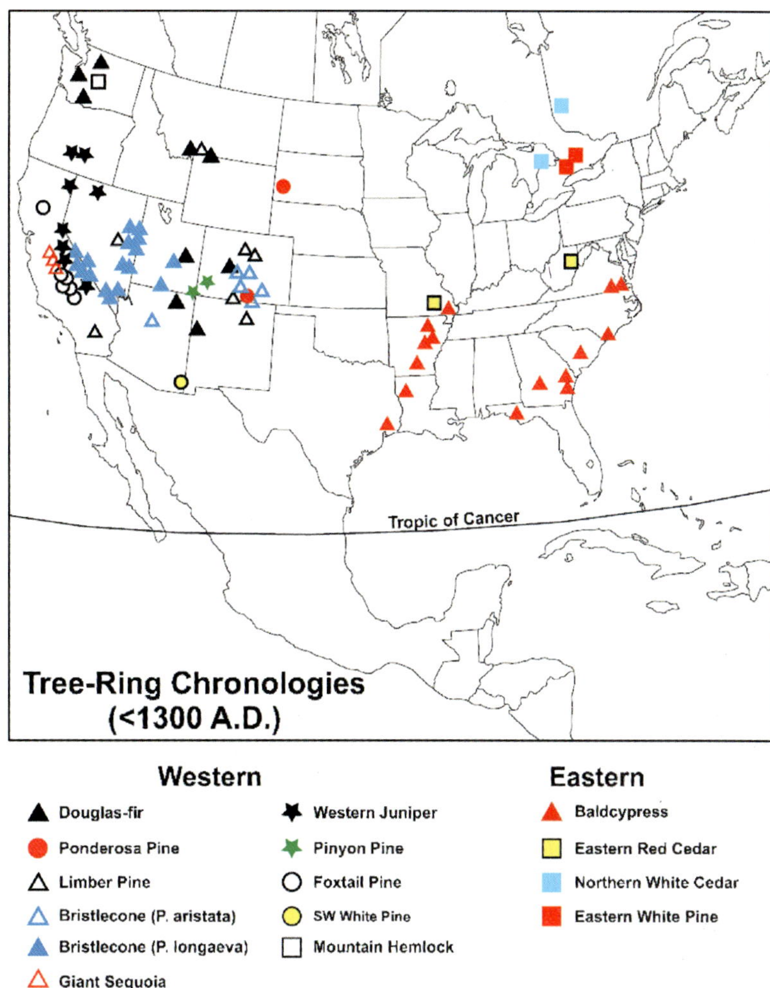

8000-yr-long annual precipitation reconstruction for southern Nevada, the longest calibrated tree-ring reconstruction yet produced (Hughes *et al.*, 2002). Hughes & Funkhouser (1998) selected a subset of six well replicated, climate-sensitive bristlecone pine chronologies at least 1700 yr long to repeat the precipitation reconstruction for southern Nevada, and identify extended dry periods from the 10th to 14th centuries that roughly correspond with geomorphologic evidence for low stands at Mono Lake reported by Stine (1994).

Brown *et al.* (1992) confirmed and expanded upon A.E. Douglass' classic tree-ring work with giant sequoia (*Sequoiadendron giganteum*) (Douglass, 1920). Three sequoia chronologies at least 2300 yr long, and a regional composite chronology 3200 yr long are now available for sequoia in central California. Due in part to strong fire-related impacts on growth, these remarkable sequoia chronologies are not linearly related to precipitation amounts. But Hughes & Brown (1992) used a threshold of low growth to construct an

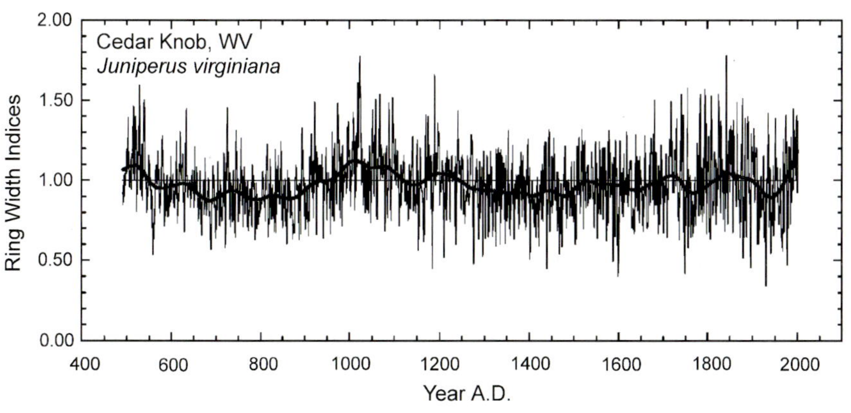

Fig. 9. The long tree-ring chronology of eastern red cedar from Cedar Knob, West Virginia, is plotted, along with a smooth curve highlighting century-scale variability in the time series (E.R. Cook & B. Buckley, pers. comm.).

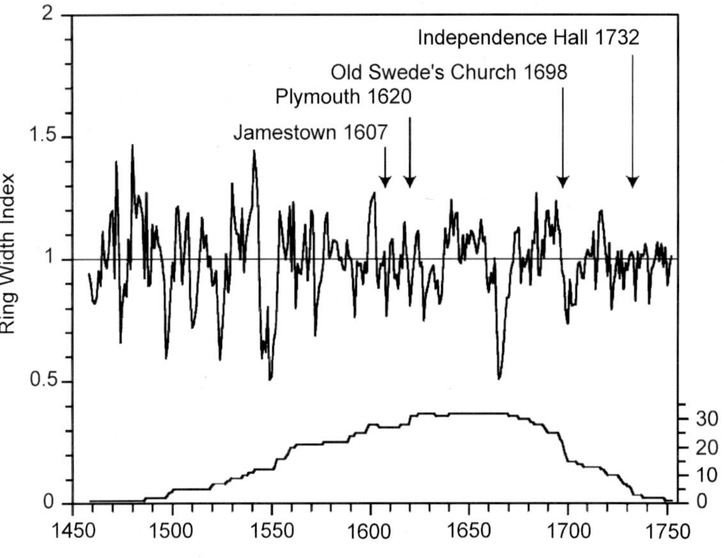

Fig. 10. The exactly dated composite tree-ring chronology developed from white oak timbers preserved in historic buildings at Philadelphia, Pennsylvania (E.R. Cook, personal communication). The changing sample size in the chronology, tree-ring dated construction episodes for selected structures, and other significant historical events are also noted.

event chronology of extreme drought over the Sierra Nevada for the past 2100 yr.

The longevity records of the western conifers are unmatched by species native to the eastern woodlands, but eastern species over 1000 yr old have been discovered, and chronologies 1500 to over 2000 yr long have been developed (Stahle *et al.*, 1988; Larson & Kelly, 1991). Currently there are 22 tree-ring chronologies that are at least 700 yr long, and ten that are over 1000 yr long in the eastern woodlands (Fig. 8).

The longest tree-ring chronologies now available for the eastern woodlands have been developed from northern white cedar (*Thuja occidentalis*) found on limestone-dolomite cliffs and talus slopes of the Niagara Escarpment (Larson & Kelly, 1991). A 2787-yr-long white-cedar chronology has been developed from living trees and relic white-cedar wood on islands off the Bruce Peninsula in Lake Huron, and the outer well-replicated portion of this chronology has been calibrated with the Palmer drought index (B. Buckley pers. comm.). Ancient white-cedar sites extend from New England across the Great Lake region and promises to provide a new high-resolution climate record for the late Holocene.

Guyette *et al.* (1989) have developed an 850-yr-long red-cedar (*Juniperus virginiana*) chronology for the Missouri Ozarks, and another red-cedar chronology has recently been developed for Cedar Knob, West Virginia, that dates back to the 5th century A.D., using living trees and relic wood (Fig. 9). The Cedar Knob chronology has been processed to retain low-frequency variability (the RCS method), and appears to reflect the Medieval Warm Epoch, Little Ice Age, and the 20th century warming trend. The climate response of this chronology is still under investigation, but growing season length may partly explain the long-term trends (E.R. Cook, 2002, pers. comm.).

Blackgum (*Nyssa sylvatica*) may also contribute to the development of a network of multi-century tree-ring chronologies in the Northeast. Individual blackgum trees over 600 yr old have been found, making it one of the longest-lived hardwoods known (N. Pederson, pers. comm.).

Historic buildings across the eastern United States represent an important potential resource for long chronology development. The early historic buildings of the eastern United States were often constructed with native timbers cut

Fig. 11. The radiocarbon dates derived for 24 sub-fossil hardwood logs recovered from buried alluvial deposits in northern Missouri (after Guyette & Dey, 2000). Note the clusters of sub-fossil wood recovered from ca. 8000 ^{14}C yr B.P. and between 3500 and 2000 ^{14}C yr B.P.

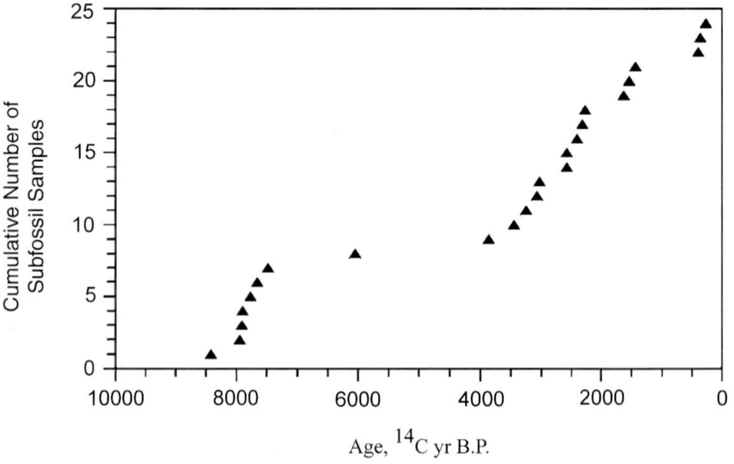

from virgin forests, and some structures preserve valuable tree-ring records of climate variability for colonial and pre-colonial America (e.g. Therrell, 2000). The Tree-Ring Lab at Lamont-Doherty Earth Observatory is developing the early historic dendrochronology for the vicinity of Boston and Philadelphia. The master dating chronology for Philadelphia white oak timbers now dates back to A.D. 1460 (Fig. 10), and includes tree-ring samples extracted from the trusses in Independence Hall and other significant 17th and 18th century structures (E.R. Cook, pers. comm.).

Old logging debris and sub-fossil wood represent another important resource for developing long chronologies in the East. In fact, most long baldcypress, red-cedar, and white-cedar chronologies have been augmented and extended with relic or sub-fossil wood sometimes still found on the ground surface. Baldcypress sinker logs (cut logs lost during historic logging operations) and sub-fossil logs (natural deadfall) are often found submerged in river channels or swamps (Stahle & Cleaveland, 1992). Sinker logs and sub-fossil timbers are also occasionally found in the Great Lakes and other natural streams and lakes throughout the East. Guyette & Cole (1999) recovered well-preserved eastern white pine (*P. strobus*) logs from natural lakes in Ontario. These logs represent old-growth lake-margin pine that fell into the lake and were preserved. Using these timbers, Guyette & Cole (1999) have developed two long white-pine chronologies dating back to A.D. 982 and 1187.

Buried wood of great age has also been reported for several eastern species (e.g. Lyell, 1849), including sub-fossil hardwoods from late glacial and post-glacial deposits in Mississippi (Grissinger *et al.*, 1982; Jackson & Givens, 1994). Recently, Guyette & Dey (2000) have recovered quantities of course woody debris from small streams in the glaciated terrain of northern Missouri. The wood debris includes large logs to nearly whole trees, and is dominated by the genus *Quercus*. Radiocarbon dating indicates that the available samples span the past 8000 ^{14}C yr, with a cluster of samples dating near 8000 ^{14}C yr B.P. (Fig. 11). The prospects for developing an exactly dated, continuous chronology for 8000 ^{14}C yr from these buried timbers are daunting indeed, because most specimens contain only 100–200 annual rings, and because significant temporal gaps may have occurred in the deposition and/or preservation of this buried wood resource (R.P. Guyette, 2002, pers. comm.). However, the outstanding long sub-fossil oak and pine chronologies in Germany and Ireland were constructed from similar relatively short ring sequences, so a major chronology development effort with the buried wood resources in Missouri and Mississippi may well be justified.

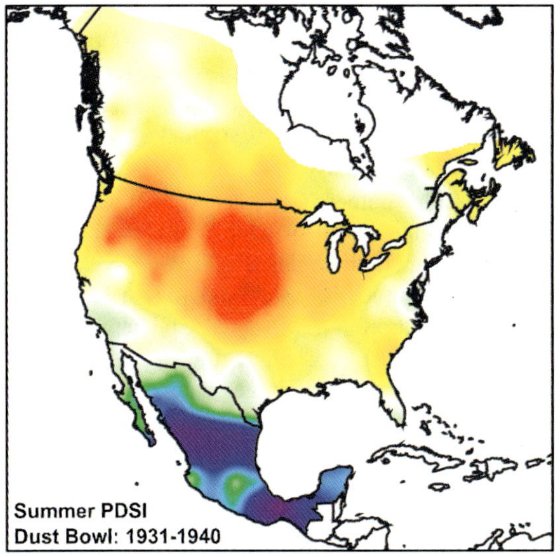

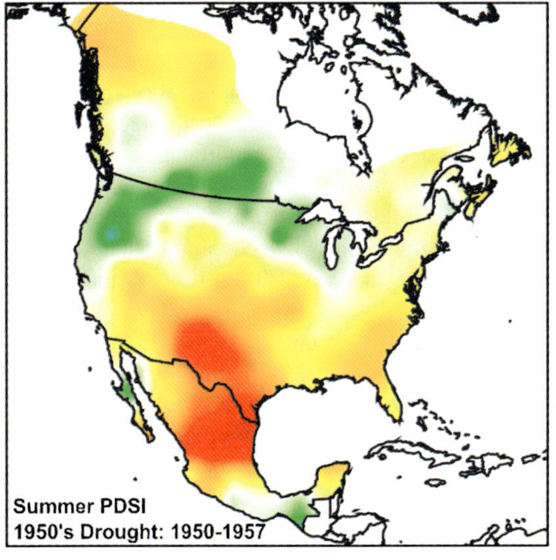

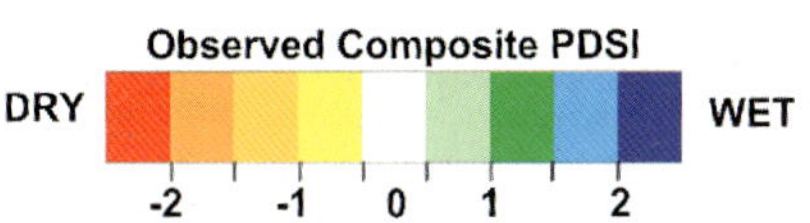

Fig. 12. The 2.5 × 2.5° latitude/longitude grid of instrumental summer PDSI data compiled for the United States, southern Canada, and Mexico by E.R. Cook (personal communication) have been used to map the 10-year average moisture anomalies across the continent during the Dust Bowl (1931–1940) and during eight years of the 1950's drought (1950–1957). Note the prolonged wetness in Mexico during the Dust Bowl drought, and over the northern United States and southern Canada during the 1950s drought.

Conclusions

The tree-ring community has been actively pursuing the development of long, climate-sensitive tree-ring chronologies in the United States, including novel species and habitat types. However, a comparison of the chronology network illustrated in Fig. 1 with the continent-scale moisture anomalies illustrated in Fig. 12 makes it clear that additional tree-ring chronology development across North America will be needed to capture the full spatial detail of the two most intense drought regimes of the 20th century, and presumably for many decadal extremes during the pre-instrumental era (e.g. Figs 2 and 3). For example, the composite maps of the instrumental summer PDSI network recently compiled by E.R. Cook (2002, personal communication) illustrate that the Dust Bowl drought of the 1930s was attended by wetness well above average in Mexico, while the decadal drought over the Tex-Mex sector in the 1950s included a pronounced band of persistent wetness across portions of the northern United States and southern Canada (Fig. 12).

The emerging continent-wide network of moisture-sensitive tree-ring chronologies (Fig. 1) will be exceptionally valuable for documenting the geographical impact and recurrence of great drought extremes. The hemispheric footprint of these decadal moisture regimes might also be used to constrain the concurrent large-scale ocean-atmospheric circulation (e.g. Fritts, 1976). But new chronologies are still needed in the United States (including Alaska), Canada, and Mexico to complete a representative hemispheric array. Much greater effort will also be needed to extend selected tree-ring records across the continent before A.D. 1600 when the variability of past climate appears to have been fundamentally greater at inter-annual and decadal timescales than that estimated for the past 400 years.

One of the most important applications of the North American tree-ring network has been to place the industrial era of anthropogenic climate modification into long-term perspective. Unfortunately, many of the chronologies located in Fig. 1 where collected in the 1980's and therefore do not represent the climate and environmental changes that have occurred in the last 20 years. The dendroclimatic community has focused primarily on expanding the geographic scope and time depth of this outstanding network. But to maximize the social and scientific value of this unparalleled array of natural environmental proxies we need to institute procedures and secure funding for the timely updating of selected high quality chronologies across North America to monitor climatic variability and change as they occur.

Acknowledgments

Tree-ring research at the University of Arkansas has been supported by the National Science Foundation, Paleoclimatology Program (Grant number ATM-9986074). We thank Dave Meko, Ed Cook, Brendan Buckley, and Rich Guyette for data and advice. Two anonymous reviewers offered editorial assistance that have helped improve this paper.

References

Acuna-Soto, R., Stahle, D.W., Cleaveland, M.K. & Therrell, M.D. (2002). Mega-drought and megadeath in 16th century Mexico. *Emerging Infectious Diseases*, **8**, 360–362.

Allan, R.J. & D'Arrigo, R.D. (1999). "Persistent" ENSO sequences: How unusual was the 1990–1995 El Nino? *Holocene*, **9**, 101–118.

Biondi, F., Gershunov, A. & Cayan, D.R. (2001). North Pacific decadal climate variability since 1661. *Journal of Climate*, **14**, 5–10.

Briffa, K.R. & Matthews, J.A. (2002). ADVANCE-10K: A European contribution towards a hemispheric dendroclimatology for the Holocene. *The Holocene*, **12**, 639–642.

Briffa, K.R., Jones, P.D. & Schweingruber, F.H. (1992b). Tree-ring density reconstructions of summer temperature patterns across western North America since 1600. *Journal of Climate*, **5**, 735–764.

Briffa, K.R., Jones, P.D., Bartholin, T.S., Eckstein, D., Schweingruber, F.H., Karlen, W., Zetterberg, P. & Eronen, M. (1992a). Fennoscandian summers from A.D. 500: Temperature changes on short and long timescales. *Climate Dynamics*, **7**, 111–119.

Brown, P.M., Hughes, M.K., Baisan, C.H., Swetnam, T.W. & Caprio, A.C. (1992). Giant sequoia ring-width chronologies from the central Sierra Nevada, California. *Tree-Ring Bulletin*, **52**, 1–14.

Brunstein, F.C. (1996). Climatic significance of the bristlecone pine latewood frost-ring record at Almagre Mountain, Colorado, USA. *Arctic and Alpine Research*, **28**, 65–76.

Burns, B.T. (1983). Simulated Anasazi storage behavior using crop yields reconstructed from tree rings: A.D. 652–1968. Ph.D. dissertation. Tucson, University of Arizona, 739 pp.

Cleaveland, M.K., Stahle, D.W., Therrell, M.D., Villanueva-Diaz, J. & Burns, B.T. (2003). Tree-ring reconstructed winter precipitation and tropical teleconnections in Durango, Mexico. *Climate Change*, **59**, 369–388.

Cole, J.E. & Cook, E.R. (1998). The changing relationship between ENSO variability and moisture balance in the continental United States. *Geophysical Research Letters*, **25**, 4529–4532.

Cole, J.E., Overpeck, J.T. & Cook, E.R. (2002). Multiyear La Nina events and persistent drought in the contiguous United States. *Geophysical Research Letters. 29*, No. 13, 10.1029/2001 GRL 013561, 2002.

Cook, E.R. & Evans, M. (2000). Improving estimates of drought variability and extremes from centuries-long tree-ring chronologies: A PAGES/CLIVAR example. *PAGES Newsletter*, **8**, 10–12.

Cook, E.R. & Peters, K. (1981). The smoothing spline: A new approach to standardizing forest interior ring-width series for dendroclimatic studies. *Tree-Ring Bulletin*, **41**, 45–53.

Cook, E.R., Briffa, K.R., Meko, D.M., Graybill, D.A. & Funkhouser, G. (1995). The segment length curse in long tree-ring chronology development for paleoclimatic studies. *The Holocene*, **5**, 229–237.

Cook, E.R., Meko, D.M. & Stockton, C.W. (1997). A new assessment of possible solar and lunar forcing of the bidecadal drought rhythm in the western United States. *Journal of Climate*, **10**, 1343–1356.

Cook, E.R., Meko, D.W., Stahle, D.W. & Cleaveland, M.K. (1999). Drought reconstructions for the continental United States. *Journal of Climate*, **12**, 1145–1162.

D'Arrigo, R.D., Villalba, R. & Wiles, G.C. (2001). Tree-ring estimates of Pacific decadal climate variability. *Climate Dynamics*, **18**, 219–224.

Douglas, A.V. & Englehart, P.J. (1996). Variability of the summer monsoon in Mexico and relationships with drought in the United States. *Proc. 21st Annual Climate Diagnostics Workshop*. Huntsville, AL, Climate Prediction Center, pp. 296–299.

Douglas, M.W., Maddox, R.A., Howard, K.W. & Reyes, S. (1993). The Mexican monsoon. *Journal of Climate*, **6**, 1665–1677.

Douglass, A.E. (1920). Evidence of climatic effects in the annual rings of trees. *Ecology*, **1**, 24–32.

Dracup, J.A. & Kendall, D.R. (1991). Climate uncertainty: Implications for operation of water control systems. *In*: Burges S.J. (Ed.), *Managing Water Resources in the West Under Conditions of Climate Uncertainty*. Washington, DC, National Academy Press, pp. 158–176.

Esper, J., Cook, E.R. & Schweingruber, F. (2002). Low-frequency signals in long tree-ring chronologies for reconstructing past temperature variability. *Science*, **295**, 2250–2253.

Fritts, H.C. (1966). Growth Rings of Trees: Their correlation with climate. *Science*, **154**, 873–979.

Fritts, H.C. (1976). *Tree rings and climate*. London, Academic Press, 567 pp.

Fye, F.K., Stahle, D.W. & Cook, E.R. (2003). Paleoclimatic analogs to 20th century moisture regimes across the USA. *Bulletin of the American Meteorological Society*, **84**, 901–909.

Gedalof, Z. & Smith, D.J. (2001). Interdecadal climate variability and regime-scale shifts in Pacific North America. *Geophysical Research Letters*, **28**, 1515–1518.

Graumlich, L.J. (1993). A 1000-year record of temperature and precipitation in the Sierra Nevada. *Quaternary Research*, **39**, 249–255.

Graybill, D.A. & Funkhouser, G. (2000). Dendroclimatic reconstructions during the past millennium in the southern Sierra Nevada and Owens Valley, California. *In*: Lavenberg, R. (Ed.), *Southern California Climate: Trends and extremes of the past 2000 years*. Los Angeles, CA, Natural History Museum of Los Angeles County.

Grissinger, E.H., Murphey, J.B. & Little, W.C. (1982). Late-Quaternary valley-fill deposits in north-central Mississippi. *Southeastern Geology*, **23**, 147–162.

Grissino-Mayer, H.D. & Swetnam, T.W. (1997). Multi-century history of wildfire in the ponderosa pine forests of El Malpais National Monument. *New Mexico Bureau of Mines and Minerals Bulletin*, **156**, 163–172.

Grissino-Mayer, H.D. (1996). A 2129-year reconstruction of precipitation for northwestern New Mexico, USA. *In*: Dean, J.S., Meko, D.M. & Swetnam, T.W. (Eds), *Tree Rings, Environment, and Humanity*. Radiocarbon and University of Arizona Press, pp. 191–204.

Grissino-Mayer, H.D., Swetnam, T.W. & Adams, R.K. (1997). The rare, old-aged conifers of El Malpais-their role in understanding climatic change in the American Southwest. *New Mexico Bureau of Mines and Minerals Bulletin*, **156**, 155–162.

Guyette, R.P. & Cole, W.G. (1999). Age characteristics of coarse woody debris (Pinus strobus) in a lake littoral zone. *Canadian Journal of Fisheries and Aquatic Sciences*, **56**, 496–505.

Guyette, R.P. (2002). A successional perspective of anthropogenic fire regimes. *Ecosystems*, **5**, 427–486.

Guyette, R.P. & Dey, D. (2000). Ancient woods uncovered. Unpublished research report, Center for Agroforestry, Columbia, University of Missouri, 15 pp.

Guyette, R.P., Cutter, B.E. & Henderson, G.S. (1989). Long-term relationships between molybdenum and sulfur concentrations in redcedar tree rings. *Journal of Environmental Quality*, **18**, 385–389.

Higgins, R.W., Chen, Y. & Douglas, A.V. (1999). Interannual variability of the North American warm season precipitation regime. *Journal of Climate*, **12**, 653–680.

Hughes, M.K. & Brown, P.M. (1992). Drought frequency in central California since 101 B.C. recorded in giant sequoia tree rings. *Climate Dynamics*, **6**, 161–167.

Hughes, M.K. & Funkhouser, G. (1998). Extremes of moisture availability reconstructed from tree rings for recent millennia in the Great Basin of western North America. *In*: Beniston, M. & Innes, J.I. (Eds), *The Impacts of Climatic Variability on Forests*. Berlin, Springer, pp. 99–107.

Hughes, M.K. & Graumlich, L.J. (1996). Mulitmillennial dendroclimatic records from western North America. *In*: Bradley, R.S., Jones, P.D. & Jouzel, J. (Eds), *Climatic Variations and Forcing Mechanisms of the Last 2000 Years*. Berlin, Springer Verlag, pp. 109–124.

Hughes, M.K., Funkhouser, G. & Ni, F. (2002). The ancient bristlecone pines of Methuselah Walk, California, as a natural archive of past environment. *PAGES News*, **10**, 16–17.

Jackson, S.T. & Givens, C.R. (1994). Late Wisconsin vegetation and environment of the Tunica Hills region, Louisiana/Mississippi. *Quaternary Research*, **41**, 316–325.

Karl, T.R. (1988). Multiyear fluctuations of temperature and precipitation: The gray area of climate change. *Climatic Change*, **12**, 179–197.

Karl, T.R. & Koscielny, A.J. (1982). Drought in the United States. *Journal of Climatology*, **2**, 313–329.

LaMarche, V.C. & Hirschboeck, K.K. (1984). Frost rings in trees as records of major volcanic eruptions. *Nature*, **307**, 121–128.

Larson, D.W. & Kelly, P.E. (1991). The extent of old-growth Thuja occidentalis on cliffs of the Niagara Escarpment. *Canadian Journal of Botany*, **69**, 1628–1636.

Leuschner, H.H., Sass-Klaassen, U., Jansma, E., Baillie, M.G.L. & Spurk, M. (2002). Subfossil European bog oaks: Population dynamics and long-term growth depressions as indicators of changes in the Holocene hydro-regime and climate. *The Holocene*, **12**, 695–706.

Lough, J.M. & Fritts, H.C. (1985). The southern oscillation and tree rings: 1600–1961. *Journal of Climate and Applied Meteorology*, **24**, 952–966.

Ludlum, D. (1968). *Early American Winters II, 1821–1870*. Boston, American Meteorological Society.

Lyell, C. (1849). *A Second Visit to the United States of North America*. London, John Murray.

Mann, M.E., Bradley, R.S. & Hughes, M.K. (1998). Global-scale temperature patterns and climate forcing over the past six centuries. *Nature*, **392**, 779–787.

Mantua, N.J., Hare, S.R., Zhang, Y., Wallace, J.M. & Francis, R.C. (1997). A Pacific interdecadal climate oscillation with impacts on salmon production. *Bulletin of the American Meteorological Society*, **78**, 1069–1079.

Meko, D.M. & Baisan, C.H. (2001). Pilot study of latewood-width of conifers as an indicator of variability of summer rainfall in the North American monsoon region. *International Journal of Climatology*, **21**, 697–708.

Meko, D.M., Hughes, M.K. & Stockton, C.W. (1991). Climate change and climate variability: The paleo record. *In*: Burges, S.J. (Ed.), *Managing Water Resources in the West Under Conditions of Climate Uncertainty*. National Academy Press, Washington, DC, pp. 71–100.

Meko, D.M., Stockton, C.W. & Boggess, W.R. (1995). The tree-ring record of severe sustained drought. *Water Resources Bulletin*, **31**, 789–801.

Meko, D.W., Therrell, M.D., Baisan, C.H. & Hughes, M.K. (2001). Sacramento River flow reconstructed to A.D. 869 from tree rings. *Journal of the American Water Resources Association*, **37**, 1029–1039.

Mitchell, J.M., Jr., Stockton, C.W. & Meko, D.M. (1979). Evidence of a 22-year rhythm of drought in the western United States related to the Hale solar cycle since the 17th century. *In*: McCormac, B.M. & Seliga, T.A. (Eds), *Solar-Terrestrial Influences on Weather and Climate*. D. Reidal. pp. 125–144.

Parr, K.L. (1999). *"To settel is to conquer": Spaniards, native Americans, and the colonization of Santa Elena in sixteenth-century Florida*. Ph.D. dissertation. Chapel Hill, University of North Carolina, 321 pp.

Reisner, M. (1986). *Cadillac desert*. New York, Penguin Books, 582 pp.

Salzer, M.W. (2000). Dendroclimatology in the San Francisco Peaks region of northern Arizona, USA. Dissertation, Tucson, University of Arizona.

Schroeder, A.H. (1968). Shifting for survival in the Spanish Southwest. *New Mexico Historical Review*, **4**, 291–310.

Schulman, E. (1956). *Dendroclimatic Changes in Semiarid America*. Tucson, University of Arizona Press, 142 pp.

Stahle, D.W. (1990). The tree-ring record of false spring in the southcentral United States. Ph.D. dissertation, Tempe, Arizona State University.

Stahle, D.W. (2002). The unsung ancients. *Natural History*, **111**, 48–53.

Stahle, D.W. & Cleaveland, M.K. (1992). Reconstruction and analysis of spring rainfall over the Southeastern U.S. for the past 1000 years. *Bulletin of the American Meteorological Society*, **73**, 1947–1961.

Stahle, D.W., Cleaveland, M.K. & Hehr, J.G. (1988). North Corolina climate changes reconstructed from tree rings: A.D. 372 to 1985. *Science*, **240**, 1517–1519.

Stahle, D.W., Cleaveland, M.K., Blanton, D.B., Therrell, M.D. & Gay, D.A. (1998b). The lost colony and Jamestown droughts. *Science*, **280**, 564–567.

Stahle, D.W., Cook, E.R., Cleaveland, M.K., Therrell, M.D., Meko, D.M., Grissino-Mayer, H.D., Watson, E. & Luckman, B.H. (2000). Tree-ring data document 16th century mega-drought over North America. *Eos*, **81**, 12, 121, 125.

Stahle, D.W., D'Arrigo, R.D., Krusic, P.J., Cleaveland, M.K., Cook, E.R., Allan, R.J., Cole, J.E., Dunbar, R.B., Therrell, M.D., Gay, D.A., Moore, M.D., Stokes, M.A., Burns, B.T., Villanueva-Diaz, J. & Thompson, L.G. (1998a). Experimental dendroclimatic reconstruction of the Southern Oscillation. *Bulletin of the American Meteorological Society*, **79**, 2137–2152.

Stine, S. (1994). Extreme and persistent drought in California and Patagonia during mediaeval time. *Nature*, **269**, 546–549.

Stockton, C.W. (1975). *Long-Term Streamflow Records Reconstructed from Tree Rings*. Tucson, University of Arizona Press, 111 pp.

Therrell, M.D. (2000). The historic and paleoclimatic significance of log buildings in southcentral Texas. *Historical Archaeology*, **34**, 25–37.

Therrell, M.D., Stahle, D.W., Cleaveland, M.K. & Villanueva-Diaz, J. (2002). Warm season tree growth and precipitation over Mexico. *Journal of Geophysical Research* 107, No. D14, 10.1029/2001 DJ000851, 2002.

Trenberth, K.E., 1991. Climate change and climate variability: The climate record. *In*: Burges S.J. (Ed.), *Managing Water Resources in the West Under Conditions of Climate Uncertainty*. Washington, DC, National Academy Press, pp. 47–70.

Watson, E. & Luckman, B.H. (2002). The development of a moisture-stressed tree-ring chronology network for the southern Canadian cordillera. *Tree-Ring Research*.

West, E. (1995). *The way to the West*. Albuquerque, University of New Mexico Press.

Woodhouse, C.A. & Overpeck, J.T. (1998). 2000 years of drought variability in the central United States. *Bulletin of the American Meteorological Society*, **79**, 2693–2714.

Woodhouse, C.A., Lukas, J.J. & Brown, P.M. (2002). Drought in the Western Great Plains, 1845–1846. *Bulletin of the American Meteorological Society*, **83**, 1485–1493.

Quaternary Coleoptera of the United States and Canada

Allan C. Ashworth

Department of Geosciences, North Dakota State University, Fargo, ND 58105-5517, USA

These noble knights of the insect realm
Pronota, elytra, their armour laid down.
In bits and pieces, it's their finest hour
They are the key, they have the power
They tell of forests, or glacial climes,
Of lakes or tundra or ancient mires
Take time, look, they are there to find,
And listen, they tell of another time.

Maureen Marra

Introduction

When Herbert Ross wrote the chapter "Pleistocene Events and Insects" for the 1965 volume of "The Quaternary of the United States," he did so without referring to a single fossil record. He had available to him reports of Pleistocene fossil beetles from glacial deposits in Ontario, Massachusetts, and Illinois (Scudder, 1890, 1898; Wickham, 1917), cave deposits in Pennsylvania (Horn, 1876), and the La Brea asphalt deposits in California (e.g. Pierce, 1946). It is impossible to know why he didn't refer to any of the paleontological literature but he may have considered it irrelevant to his purpose. He was interested in documenting the effects of climate change and glaciation on the evolution of caddisflies. In his chapter, Ross provided several examples of how the disjunct distributions in many species were best explained by climate change. His purpose was to demonstrate that the isolation of populations caused by Quaternary climate changes had resulted in evolution. Ross would be very pleased to learn that fossils of mandibles, pronota and frontoclypeal schlerites of caddisflies (Trichoptera) from Quaternary deposits are providing indisputable evidence of the types of changes he could only theorize about (Williams, 1989; Williams & Eyles, 1995).

Quaternary entomology in North America has come a long way since Ross wrote his article. In addition to the caddisflies already mentioned, and to the beetles, the subject of this report, fossil head capsules of midges (Diptera: Chironomidae) have been shown to be extremely valuable indicators of paleosalinity and paleotemperature (Walker, 1995). The fossils, however, that have received most of the attention are the disarticulated skeletons of beetles (Coleoptera). As a graduate student, I was regailed by Russell Coope's and Fred Shotton's stories about their 1965 INQUA field trip to study the glacial deposits of the Midwest. During the trip, Coope collected sediments from the cliff sections on Lake Erie at Garfield Heights, Cleveland, from which he later described a few ground-beetle species (Coope, 1968). In the same year, John Matthews' (1968), described a late Pleistocene assemblage from Fairbanks, Alaska. Since the pioneering studies, numerous late Tertiary and fossil beetle assemblages have been studied from various parts of the continent. Those contributions have been invaluable in the information they have provided about the response of beetles to climate change (Ashworth, 2001).

The purpose of this report is to provide a region-by-region account of the status of paleoentomological studies. Undoubtedly, it covers some of the same ground as earlier reviews (Ashworth, 1979; Elias, 1994; Morgan & Morgan, 1980; Morgan *et al.*, 1984). The beetle species and their ages and locations referred to in this report are a small fraction of the data that are available. Every record in the literature has been compiled in a database "The Fossil Coleoptera of North and Central America, including Greenland." A fully searchable web version of the database (www.ndsu.edu\fossilbeetles\) will be launched in July 2003, in conjunction with the INQUA Congress.

Early Pleistocene and Late Tertiary Assemblages

Several fossil assemblages from northernmost Greenland, Ellesmere, Meighen and Banks islands in the Canadian High Arctic, and the Seward Peninsula, Alaska, are comprised of well-preserved fossils of late Miocene, Pliocene and Early Pleistocene age (Böcher, 1995; Hopkins *et al.*, 1971; Matthews, 1976a, b, 1977, 1979a, b; Matthews *et al.*, 1986; Matthews & Telka, 1997). Two of the assemblages stand out because of their excellent preservation of chitinous remains and their species richness. The oldest of these is the Pliocene (~3 myr) Beaufort Formation, on Meighen Island, from which 149 species have been reported (Matthews & Telka, 1997), and the Plio-Pleistocene Kap København Formation of northern Peary Land, Greenland, from which 142 species have been identified (Böcher, 1995). Both fossil assemblages are dominated by extant species that currently inhabit tundra and tundra-forest habitats. Several of the assemblages indicate that treeline was several hundred kilometers north of its present position.

A recent Mutual Climatic Range (MCR) analysis of the Tertiary assemblages indicates mean July temperatures consistently higher than present values for those locations. In contrast, the oldest Pleistocene assemblage has a mean July temperature about 5 °C lower than present (Elias & Matthews, 2002). The Plio-Pleistocene Kap Købnhaven assemblage represents a mean July temperature about 9 °C higher than today, and the Pliocene Meighen Island assemblage, a mean July temperature about 15 °C higher.

Only a few species from the Tertiary assemblages are considered to be extinct; Elias & Matthews (2002) list the ground-beetles *Diacheila matthewsi* Böcher and

DEVELOPMENTS IN QUATERNARY SCIENCE
VOLUME 1 ISSN 1571-0866
DOI:10.1016/S1571-0866(03)01024-8

Asaphidion yukonense sp. nov., the hydrophilid *Helophorus meighenensis* Matthews, and the micropeplid *Micropeplus hopkinsi* Matthews (Elias & Matthews, 2002). The list should also include the hydrophilid *Helophorus coopei* Matthews, from the late Miocene (~5.7 myr) Lava Camp site in Alaska (Matthews, 1976b). The micropeplid, *Micropeplus hoogendorni* Matthews, had to be removed from the list because living descendants were discovered in Siberia, where they had been identified as *M. dokuchaevi* (Elias, 1994). Based on the Arctic fossil assemblages, the number of extant taxa increases with time from about 85% in the late Miocene to 100% in the Pleistocene (Ashworth, 2001). To emphasize this, the oldest known early Pleistocene (~1.8 myr) assemblage from Cape Deceit, Alaska, contains no extinct species (Matthews & Telka, 1997).

Morgan *et al.* (1993) reported a diverse assemblage of beetles from north-central Baffin Island that was formerly considered to be of Sangamonian age. The species currently inhabit tundra patches near treeline, which is located about 1325 km south of the fossil site. Efforts to date the assemblage have not been successful, but the absence of any extinct species in the assemblage may indicate that they are of Pleistocene rather than Pliocene or older age.

The evidence for evolution in the fossil record has been even more elusive than for extinction. Matthews (1974) reported that the non-functional flight wings in the staphylinid *Tachinus apterus* Mäcklin, from the Cape Deceit assemblage, were slightly longer than those of their modern counterparts. No doubt many of the existing tundra species evolved during the Miocene, but not necessarily all of them. Matthews & Telka (1997) speculated that the weevil *Lepidophorus lineaticollis* Kirby may have evolved at the beginning of the Pleistocene. The species is relatively abundant in later Pleistocene deposits but is unknown in any of the Tertiary assemblages.

Late Tertiary and early Pleistocene fossil assemblages are important because they provide evidence for stasis in some species of about five million years. Paleoentomologists tend to view long-term stasis as the norm. In contrast, many entomologists believe that the isolation of populations caused by climate changes must have resulted in elevated rates of allopatric speciation and extinction during the Pleistocene. They find it difficult to accept stasis and low extinction rates as the norm, even though the fossil record is difficult to interpret in any other way. Paleoentomologists are also probably guilty of over-interpreting their data. The only observations for constancy in morphology over millions of years are for species of the northern temperate fauna. Whether or not stasis is the norm for the tropics, where diversity is orders of magnitude greater, is unknown.

The cause of stasis is poorly understood. Coope (1978) proposed the attractive idea that it was the inconstancy of the Pleistocene that ultimately resulted in the constancy of species. He suggested that the forced dispersal of beetles as they responded to climate changes kept their gene pools mixed, and thereby prevented the accumulation of mutations that would result in speciation. This idea is still being propagated (Elias & Matthews, 2002), although there is genetic evidence indicating that it doesn't work (Ashworth, 2001; Reiss *et al.*, 1999). A study of the mitochondrial DNA of *Amara alpina* (Paykull), a ground-beetle identified as a fossil from a Pliocene deposit on the North Slope of Alaska (Elias & Matthews, 2002), indicates that Beringian, Rocky Mountain, and Appalachian populations of the species have been genetically isolated from one another for hundreds of thousands of years and have still maintained their morphological integrity (Reiss *et al.*, 1999).

Beringian Assemblages

During the Wisconsinan and earlier glaciations, large expanses of Alaska, the Yukon and Northwest territories, and northern British Columbia were unglaciated. The region is faunistically important for two reasons. First, Beringia was a refugium for arctic species during glaciations (Schwert & Ashworth, 1988). This is reflected today by the region having a much greater species and genetic diversity than any other arctic or alpine area in North America. In a study of the molecular genetics of *Amara alpina*, an obligate tundra species, eight haplotypes were defined from Beringian populations compared to two and five, respectively, for Appalachian and Rocky Mountain populations, and only one for Hudson Bay populations (Ashworth, 1996; Reiss *et al.*, 1999).

Second, the Bering Land Bridge has periodically facilitated faunal exchanges between Asia and North America. Flightless species of beetles that have amphiberingian distributions (populations on both sides of the Bering Straits) would seem to be indisputable evidence that populations have dispersed across the land bridge during glacial episodes. Surprisingly, however, there is not much similarity between the faunas of western (Siberia) and eastern Beringia (northwestern North America) at the species level. Species in the genus *Carabus* are medium- to large-size, brachypterous (flightless) ground-beetles. Species catalogues for easternmost Siberia, Alaska, and the Yukon and Northwest territories list 12 species, of which only two (17%) are shared (Bousquet & Larochelle, 1993; Kryzhanovskij *et al.*, 1995). Similarly, only three (18%) species of the 12 listed for the macropterous (winged) ground-beetle species *Elaphrus* are shared. Elias *et al.* (2000) list 216 species of predatory and scavenging beetles that have been identified as fossils in Quaternary deposits from sites in eastern and western Beringia; of these only 26 (12%) are shared.

Flightless ground-beetles of the pterostichine subgenus *Cryobius* are among the most characteristic beetles of North American tundra and alpine habitats. Ball & Currie (1997) discuss how the subgenus probably evolved in the high mountains of eastern Asia. After speciation, the beetles dispersed eastward across the Bering Land Bridge into North America, where they underwent further speciation in Beringia and in the tundra east of the Mackenzie River. While the land bridge has undoubtedly resulted in faunal exchanges between Asia and North America, e.g. *Cryobius*, the relatively low percentages of shared species between eastern and western Beringia also suggest that the Bering and Chukchi seas have been formidable barriers to beetle dispersal.

Numerous fossil beetle assemblages have been examined from eastern Beringia. Elias (2001) lists 46 fossil sites ranging in age from about 150,000 yr (marine oxygen isotope stage (MIS) 6) to the Holocene. Lists of species identified from fossils are provided by Matthews & Telka (1997) and by Elias *et al.* (2000). Fossil assemblages from the older glacial intervals have similar species compositions to those of the last glacial maximum, and it is impossible to tell if minor variations are the result of taphonomy or subtle climatic differences. The species of mid-Wisconsinan assemblages are more diverse than those of the glacial intervals, reflecting a warmer climate, but not as warm as the last interglaciation (Matthews & Telka, 1997). Spatial differences have been noted in the fossil record, with the coastal regions of southwestern Alaska serving as refugia for mesic taxa during the last glaciation (Elias, 1992a).

For the most part, Wisconsinan assemblages are dominated by taxa of treeless habitats. Fossils of scolytid beetles that burrow beneath the bark of coniferous trees, are only represented in deposits of the last interglaciation and the Holocene. Assemblages from the last interglaciation and the early Holocene are also the only ones to contain fossils of species that migrated northward in response to climates that must have been warmer than those of today (Matthews & Telka, 1997; Nelson & Carter, 1987). MCR analyses for Beringia have been reported by Elias (2000, 2001) who employed a calibration developed by Elias *et al.* (1999). Mean July temperatures for the glacial intervals were estimated to be 5–6 °C colder, and for the last interglaciation, about 3.5 °C warmer than today. The mean July temperature for the mid-Wisconsinan interval was reported to be similar to that of today.

One of the most controversial topics in Beringian paleoecology involves the nature of the vegetation cover during the last glacial maximum. Conflicting hypotheses for a steppe- or mammoth tundra vs. a tundra mosaic, similar to that of today, have been widely discussed (Guthrie, 1990; Ritchie, 1984). The question that has driven much of the debate is whether or not a tundra vegetation, similar to that of today, could have supplied the nutritional needs to support herds of large grazing mammals. Fossil beetle assemblages have been one of the lines of evidence used to support the mammoth steppe hypothesis. Matthews (1982) observed that assemblages consisting of the weevil *Lepidophorus lineaticollis* and byrrhids of the genus *Morychus* characterized glacial fossil assemblages. He interpreted the "*Lepidophorus-Morychus* complex" to be representive of better-drained, xeric soils, and therefore more probably associated with grassland than tundra. Elias *et al.* (2000), on the basis of autecological data for *L. lineaticollis* and species of *Morychus*, do not believe that the complex can be used as an unequivocal indicator of steppe-tundra. There may be some hair splitting involved in their argument, however, as they concede that the complex is more representative of less-mesic tundra.

Elias *et al.* (1992, 1996) argued that mesic tundra, not steppe-tundra, was the vegetation of the Bering Land Bridge during the last glaciation until it was submerged by rising sea level at about 11,000 ^{14}C yr B.P. They based their argument on assemblages of fossil beetles and macroscopic plant remains obtained from cores that sampled submerged peat deposits. The assemblages, not surprisingly, are dominated by aquatic and hygrophilous beetle and plant species; no species of grasslands were found.

The related questions of whether the *Lepidophorus – Morychus* complex of the interior of eastern Beringia is an indicator of steppe-tundra, or the absence of the complex on the land bridge denies its existence, should remind paleoentomologists that the nature of fossil beetle assemblages is such that they can only be used to interpret the immediate environment surrounding the depositional basin. The absence of grassland species in the land bridge peat deposits does not mean that grasslands teeming with species were not part of the landscape. Extrapolation of biomes from point source fossil beetle assemblages, especially the absence of data, is inherently difficult.

Pacific Northwest Assemblages

Relatively few fossil assemblages have been studied from British Columbia, Washington, and Oregon. A.V. Morgan (in Karrow *et al.*, 1995) identified a few genera from some fragments isolated from the interglacial Whidbey Formation at Point Wilson, Washington. The identifications, however, provided no information about the interglacial climate. Hebda *et al.* (1983) described a few species from a mid-Wisconsinan site north of Vancouver. The species occur within the region today indicating a mid-Wisconsinan climate that was similar or slightly cooler than today. A much larger mid-Wisconsinan assemblage, consisting of 56 taxa, was described from Kalaloch on the Olympic Peninsula, Washington (Cong & Ashworth, 1996). A majority of the species are those associated with lowland and montane forests of the region today. A few of the species, however, are members of the northern boreal forest fauna and have not been recorded as extant in the Pacific Northwest. Based on these taxa, the mean July temperature was estimated to be about 1 °C cooler than at present.

Two of the more intriguing fossils from this assemblage consist of a head and left elytron of an undescribed species of a blind trechine (Cong, 1997). The head lacks eyes and the elytron also possesses features that are characteristic of subterranean beetles. Blind, cave-dwelling trechines are common in the southeastern United States but are unknown in the Pacific Northwest. The fossil is closest to species of the *Trechiama* series, a group of Japanese cave-dwelling beetles. It is unknown whether the species is extinct or extant, but possibly it could still be living in cracks in rocks buried deep beneath leaf litter. A biogeographic relationship to Asia is represented in other beetle taxa from the Pacific Northwest and possibly indicates a shared history in the ancient Tertiary forests of Beringia.

Only two fossil beetle assemblages dating to the last glacial maximum have been examined from the Pacific Northwest. At Kalaloch, the assemblage is that of a treeless environment, probably tundra, that is very different than the mid-Wisconsinan forested assemblages (Cong & Ashworth,

1996). The most distinctive beetle in the assemblage is an alpine chrysomelid beetle, *Asiorestia pallida* (Fall), that inhabits meadows above 1600 m in the Olympic Mountains; its occurrence near sea level indicates a mean July temperature that was about 3 °C cooler than present between about 20,000 and 17,000 ^{14}C yr B.P. At Port Moody and Mary Hill in British Columbia, a small fossil beetle assemblage with ages of between 18,000 to 19,000 ^{14}C yr B.P. has a rich scolytid fauna associated with a subalpine forest (Miller *et al.*, 1985).

The only late Wisconsinan assemblage that has been studied from the Pacific Northwest is from near the top of the Kitsap Formation (Fig. 1). The rich assemblage of fossils from Discovery Park, Seattle, with an age of 14,760 ^{14}C yr B.P., accumulated in an abandoned channel on the margins of a braided channel when the Puget Lobe of the Cordilleran Ice Sheet was advancing southward during the Vashon Stade (Ashworth *et al.*, 2000; Nelson & Coope, 1982). Most of the species have existing geographic ranges within the Puget Sound region, including species such as the micropeplid *Kalissus nitidus* LeConte, that is restricted to the Pacific Northwest. A few of the species, including the ground-beetles *Pelophila borealis* (Paykull) and *Notiophilus aquaticus* (Linné), are from northern boreal and western montane forested regions. The micropeplid *Micropeplus cribratus* LeConte is from east of the Rocky Mountains.

The mixture of species from open-ground and forested habitats indicates a mosaic vegetation of grassland and spruce woodland. No obligate tundra species have been identified in the fossil assemblage. Mean July temperature is estimated to have been as warm, or only slightly cooler, than that of Seattle today. Based on the low representation of obligate forest species, the climate is also inferred to have been drier than that of today.

Rocky Mountain Assemblages

Several late Wisconsinan and Holocene fossil assemblages have been examined from the Rocky Mountains (Elias, 1991). During the last glaciation, ice caps developed on the summits of many of the ranges and alpine species were displaced to lower elevations; at Marias Pass, Montana, the displacement was about 450 m (Elias, 1987a). In a MCR analysis involving 20 assemblages from Utah, Colorado and Montana, the mean July temperature at 14,500 ^{14}C yr B.P. was reported to have been about 10–11 °C lower than today (Elias, 1996). By 10,000 ^{14}C yr B.P., however, mean July temperature was reported to be between 2 and 6 °C warmer than today. The amount of late-glacial temperature increase is much greater than has reported for any other region of the continent. How much of the change is real, and how much might be the result of normalizing the data from assemblages that have a large latitudinal and elevational range (1548–3325 m), is uncertain.

California Assemblages

The only Quaternary fossil assemblages to be examined from California are from the asphalt deposits in the southern part of the state in and around Los Angeles and Santa Barbara

Fig. 1. Microscope view of a dish of Quaternary beetle fossils from the Kitsap Formation, Discovery Park site, Seattle. The fossils consist of three pronota and 13 elytra (wingcases). The large right elytron on the right belongs to the ground-beetle Opithius richardsoni Kirby, a species that inhabits sand and gravel bars on fast flowing rivers where it feeds on water-borne insect materials. The AMS age of 14,760 ^{14}C yr B.P. for the assemblage is based on dating three specimens of the larger elytra. The scale bar at the bottom of the image is 5 mm.

(Miller, 1983, 1997). In a series of 14 papers, published between 1945 and 1954, W.D. Pierce described large numbers of new species from the asphalt-impregnated sediments at Rancho La Brea and McKittrick. He considered the majority of the species to be extinct. The stratigraphy and age of most of Pierce's specimens is unknown (Miller, 1983). Pierce was a "splitter" and many of the species and subspecies he erected have been synonymized; e.g. nine of Pierce's new species and subspecies of the tenebrionid *Coniontis* have been reassigned to the extant species *Coniontis abdominalis* LeConte. Miller (1983) lists the revisions that had been made of the staphylinids (Moore & Miller, 1978), silphids (Miller & Peck, 1979), tenebrionids (Doyen & Miller, 1980), and scarabs (Miller *et al.*, 1981). Two of the extinct species that Pierce described, however, are noteworthy in that they are the only known cases of species extinction in North America. The species are *Copris pristus* Pierce and *Onthophagus everestae* Pierce, both dung-feeding scarab beetles (Miller *et al.*, 1981). The age of these specimens is unknown but they are believed to date from the end of the Pleistocene when several of the La Brea mammals and birds became extinct. Miller (1983) suggested that a reduction of dung of large mammals may have played a role in the extinction of the scarabs.

Fig. 2. A late Wisconsinan fossil beetle from a 14,400 ^{14}C yr B.P. wood rat (Neotoma) midden from the Sonoran Desert in the Kofa Mountains, Arizona. The fossil is an exceptionally well-preserved specimen of the darkling beetle Stibia tuckeri Casey (Tenebrionidae). The main exoskeletal parts, the head, pronotum, elytra, and basal leg segments are still articulated. The vertical scale bar is 5 mm.

Southwestern Desert Assemblages

The first report of fossil beetles from a woodrat (*Neotoma*) midden was from the Chihuahan Desert of west Texas (Ashworth, 1973). Since then several assemblages, ranging in age from mid-Wisconsinan to Holocene, have been studied from middens in Arizona, New Mexico, Texas, Utah, and the adjoining provinces of Mexico (Elias, 1987b, 1990, 1992; Elias & Van Devender, 1990, 1992; Elias *et al.*, 1992; Hall *et al.*, 1988, 1989, 1990; Spilman, 1976). The fossils are unusually well-preserved and with a few exceptions the ages of the assemblages are well-known (Fig. 2). Midden beetle assemblages have proved to be disappointing for paleoclimatic analyses, certainly when compared to studies of macroscopic plant remains (Betancourt *et al.*, 1990). The reason is not the skills of the researchers but the restricted amount of information that is available on the ecology and geographic distribution of modern desert beetles.

Holocene assemblages appear to be comprised of species that are mostly within their existing geographical and elevational ranges. There is some evidence that during the Wisconsinan, beetle species, like plant species, moved to lower elevations in both the Sonoran Desert (Hall *et al.*, 1989) and the Colorado Plateau (Elias *et al.*, 1992). In the Chihuahan Desert, grassland species of ground- and scarab beetles migrated into the region from the north in response to the cooler and wetter conditions. They were replaced about 12,000 years ago by desert species (Elias & Van Devender, 1990). In the lowlands of the Sonoran Desert, however, the Wisconsinan beetle fauna is very similar to that of the present suggesting only a minimal response to climate change (Hall *et al.*, 1988).

Mid-Continental Assemblages

The fossil assemblages of this vast region, which extends from Manitoba and North Dakota in the west, to New York State and Quebec in the east, and to Missouri in the south, have been reviewed by Morgan *et al.* (1984), Morgan (1987), Schwert & Ashworth (1988), Schwert (1992) and Elias (1994). MCR analyses are also available for the mid and late Wisconsinan assemblages (Elias, 1999; Elias *et al.*, 1996).

The oldest assemblage is from the County Line site, west-central Illinois. The age of the fragmentary and poorly preserved fossils that lie within a bedrock depression is estimated to be between 830,000 and 730,000 yr (Miller *et al.*, 1994). The species identified include the large ground-beetle *Carabus sylvosus* Say, a forest floor species of mesic eastern and southern deciduous woodland, and the very rare scarab beetle, *Micraegialia pusilla* Horn, a species of mesic prairie sites. Late Wisconsinan fossils of *M. pusilla* are known from wetlands surrounded by open spruce forest (Ashworth *et al.*, 1981; Ashworth & Schwert, 1992). The strange mix of boreal and southern faunistic elements was also detected in plant, mollusk and vertebrate fossils (Miller *et al.*, 1994). The four species identified in the assemblage are all extant.

One other pre-Sangamonian assemblage has been reported from the mid-continent from the Gastropod Silts in northeastern South Dakota (Garry *et al.*, 1997). An Illinoian age is assigned mostly on the basis of an amino-acid analysis and stratigraphic position. The few species that were identified, namely the ground-beetles *Diacheila polita* Faldermann, *Bembidion morulum* LeConte, and *Agonum quinquepunctatum* Motschulsky, and the hydrophilid *Helophorus arcticus* Brown, all are well-represented in later Sangamonian and Wisconsinan assemblages in the mid-continent. All of the species are extant and occur in tundra-forest habitats today, although the geographic ranges of *D. polita* and *H. arcticus* no longer overlap.

Several assemblages from the region are assigned a Sangamonian (last interglaciation) age. The ages are tentative and some of the assemblages may be of early or even mid-Wisconinan age. Pilny & Morgan (1987) and Churcher *et al.* (1990) reported a diverse assemblage of about 200 species from a warm interval at Innerkip, southern Ontario. There are several species in the assemblage, e.g. the ground-beetle *Carabus vinctus* Weber for which the fossil site is at the northern limits of their existing range. The climate is estimated to have been as warm as that of today. Anderson *et al.* (1990) considered the Pointe-Fortune assemblage, from a site on the southern Ontario-Quebec border, also to be of Sangamonian age. Unlike the Innerkip assemblage, the Pointe-Fortune assemblage contains boreal forest species, such as the ground-beetle *Bembidion transparens* (Gebler) and the rove beetle *Olophrum boreale* (Paykull), that indicate a colder climate than at present. The Henday assemblage from deposits on the Nelson River, near Gillam, Manitoba, is also considered to be Sangamonian based on its stratigraphic position (Nielsen *et al.*, 1986). The fossils, however, are mostly tundra species, e.g. *Pterostichus costatus* (Ménétriés), from habitats far north of the fossil site. Similarly, the tundra ground-beetle *Diachela polita*, from a borehole in the Sangamonian Owl Creek beds, near Timmins, northern Ontario, indicates a climate that was colder than today (Mott & DiLabio, 1990).

Assemblages from early and mid-Wisconsinan interstades (MIS 4–3) have been described from several localities in the Great Lakes region (Ashworth, 1980; Cong *et al.*, 1996; Karrow *et al.*, 2001; Matthews *et al.*, 1987; Morgan, 1972; Morgan & Morgan, 1980; Warner *et al.*, 1988). In the absence of any real chronologic control, it is difficult to determine how these assemblages relate to one another. The early Wisconsinan assemblage, from Woodbridge, northeast of Toronto, contains the tundra and treeline ground-beetles *Diacheila polita* and *Elaphrus lapponicus* Gyllenhal, and the hydrophilid *Helophorus lapponicus* (Karrow *et al.*, 2001). Another tundra species, *Amara alpina* (Paykull), was reported from an assemblage also believed to be of early Wisconsinan age in southern Quebec (Matthews *et al.*, 1987). These species also occur in late-Wisconsinan assemblages in the region making it a reasonable assumption that tundra and treeline species inhabited the margin of the Laurentide Ice Sheet continually during Wisconsinan time, responding to glacier advances and retreats by relatively short southward and northward movements.

Changes in the species compositions of assemblages at considerable distances from the ice sheet suggest that the margin was responding to regional and perhaps even global climate changes. Cong *et al.* (1996) reported a series of changes in a mid-Wisconsinan beetle assemblage from Titusville, Pennsylvania, in which a northern boreal fauna was replaced by a mixed conifer-hardwood fauna. In turn, the mixed forest fauna was replaced by a boreal forest fauna. All of the changes occurred within the interval from 43,500 to 39,000 ^{14}C yr B.P. The short warm interval was estimated to have occurred at about 40,000 ^{14}C yr B.P., coeval with the warm Upper Warren Interstade of the British Isles, and the warm sea-surface temperature interval between Heinrich events 4 and 5.

In the late mid-Wisconsinan Farmdalian interval, from 28,000 to 24,000 ^{14}C yr B.P., fossil assemblages from Athens, Illinois, and Biggsville, Iowa, are typical of closed spruce forests (Carter, 1985). An older assemblage with an age of 33,000 ^{14}C yr B.P. from St. Charles in south-central Iowa, is that of prairie or savanna with patches of a conifer-hardwood forest (Baker *et al.*, 1991). The sketchy picture that emerges for the early and mid-Wisconsinan is one of spatial and temporal heterogeneity in which populations of plants and insects were responding more dynamically than at any time during the Holocene.

In the continental interior, arctic beetle species replaced boreal forest species along the southern margin of the Laurentide Ice Sheet at about 21,500 ^{14}C yr B.P. Colonization was probably from populations that dispersed southward in front of the growing ice sheet and from populations that dispersed westward and eastward from montane refugia in the Appalachian and Rocky mountains, respectively (Schwert & Ashworth, 1988). During the interval from 21,500 until 14,500 ^{14}C yr B.P., arctic and subarctic species, representing tundra and treeline environments, inhabited a discontinuous zone along the margin of the ice sheet. Fossil assemblages typical of this time have been reported from Iowa to New York (Ashworth & Willenbring, 1998; Baker *et al.*, 1986; Garry *et al.*, 1990). The Conklin Quarry assemblage, Iowa, dating between 18,090 and 16,710 ^{14}C yr B.P., contains several species of tundra ground-beetles, including *Amara alpina* and *Diacheila polita*. A similar community of modern ground beetles was collected in a series of pitfall traps set on a well-drained slope of shrub tundra located on the east side of Atigun Pass, Brooks Range, Alaska (A.C. Ashworth & D.P. Schwert, unpublished data).

Tundra species continued to be well-represented in fossil assemblages until about 14,500 ^{14}C yr B.P. (Morgan, 1987; Schwert, 1992; Schwert & Ashworth, 1988). After that time, there are only occasional records. Barnosky *et al.* (1988) reported *Diacheila polita* from a 12,080 ^{14}C yr B.P. assemblage from Bradford County, Pennsylvania, and Mott *et al.* (1981) reported *Amara alpina* from a 11,050 ^{14}C yr B.P. assemblage from St. Eugene, Quebec. The St. Eugene assemblage was from an isolated patch of tundra on the shores of the Champlain Sea.

The most southerly fossil beetle assemblages that have been studied are from near St. Louis, Missouri, with ages ranging from 22,300 to 17,250 ^{14}C yr B.P. (Schwert *et al.*, 1997). Several species associated with spruce forest in those assemblages indicate that during the late Wisconsinan cold air masses covered a large area of the continent, far south of the ice sheet.

Climatic warming, in combination with the ice sheet acting as a barrier to northward dispersal, is posited as the cause of a widespread regional extinction of tundra species (Schwert & Ashworth, 1988). Fossil assemblages from the ice margin between about 14,500 and 12,500 ^{14}C yr B.P. are composed of a combination of subarctic and boreal species of a treeless vegetation for which there is no modern analog. *Helophorus arcticus* Brown, a subarctic hydrophilid of tundra and treeline habitats from Hudson Bay to Labrador, is one of the most distinctive beetles of this fauna. What physiological adaptations enabled the species to survive extinction in the ice-marginal zone at 14,500 ^{14}C yr B.P. is uncertain (Ashworth, 2000); by 12,500 ^{14}C yr B.P., however, spruce forest and beetle species associated with it had replaced the open ground biota.

In addition to faunal changes that were primarily driven by climate change, others resulted from landscape and hydrologic changes. Rivers flowing in the prairies and glaciated terrains of midcontinental North America tend to have low velocities. During the last glaciation, huge volumes of meltwater coalesced to form large braided river systems. Species that inhabit various riparian habitats on fast-flowing rivers were able to expand their ranges eastward from the Rocky Mountains to the mid-continent (Ashworth & Schwert, 1991). Two of these species, *Opisthius richardsoni* Kirby, which inhabits barren gravel bars in foothills rivers in the western mountains from New Mexico to Alaska, and *Asaphidion yukonense* Wickham, which inhabits willow-vegetated sand bars in the northwestern Rocky Mountains, were still surviving in the mid-continent at 10,100 ^{14}C yr B.P. in the Rainy River area of western Ontario (Bajc *et al.*, 2000). These species probably continued to disperse northward into Manitoba and Ontario following the retreating ice margin.

Schwert & Ashworth (1988) proposed that the most likely region for tundra species to have survived glaciation, following their extirpation from ice-marginal habitats along the southern margin of the Laurentide ice sheet, would have been Beringia. They also postulated that the postglacial colonization of the Arctic would have been from species that survived in Beringia. To test this hypothesis, a molecular genetic study of the tundra ground-beetle *Alpina alpina* was undertaken. DNA was examined from specimens from the North Slope of Alaska, from the eastern and western shores of Hudson Bay, and from alpine zones in the Rocky Mountains and the Appalachians in New England and Quebec. The study confirmed that the founder population for the postglacial Hudson Bay populations was most probably from the southern part of the Beringian refugium in British Columbia (Ashworth, 1996; Reiss *et al.*, 1999).

Acadian Assemblages

Late-glacial fossil assemblages have been studied from several localities in Nova Scotia and New Brunswick (Miller, 1995, 1996, 1997a, b, 2000; Miller & Morgan, 1991). This is the only region of the continent where changes in beetle assemblages are reported to represent climatic cooling during the Younger Dryas Stade. As the ice sheet retreated from maritime Canada about 13,000 ^{14}C yr B.P., the deglaciated areas were colonized by herb tundra (Mott, 1994). The beetles that inhabited this open terrain were species which currently inhabit the treeline and tundra of arctic Canada. They included the ground beetles *Diacheila arctica*, *Elaphrus lapponicus*, and *Stereocerus haematopus* (Dejean), and the non-aquatic hydrophilid beetle *Helophorus arcticus*, which contrary to several literature references, is not a halophile (Ashworth, 2000).

Miller (1996) reported that starting at about 11,800 ^{14}C yr B.P., birch and spruce invaded the tundra zone and the more cold-adapted beetle taxa were extirpated from the lowlands. The most characteristic feature of fossil assemblages during this time is the occurrence of several species of scolytid beetles associated with conifers. About 10,800 ^{14}C yr B.P., individuals of the omaliine staphylinid beetle, *Olophrum boreale*, appeared in fossil assemblages and continued until about 10,400 ^{14}C yr B.P. They appeared at a time when lithological and pollen evidence indicate a return to a tundra-like vegetation, a change that is attributed to the cooler and wetter conditions of the Younger Dryas Stade (Mott, 1994). Changes in the species composition of Chironomidae, represented by fossils of their head capsules, indicate a 2–3 °C cooling of surface lake waters during the Younger Dryas Stade (Walker *et al.*, 1991).

Miller (1996) attributed *O. boreale* with the importance of being the name-bearer of the "*Olophrum boreale* fauna" which he considered to be the biozone for the Younger Dryas Stade. He noted that the response of beetles to Younger Dryas cooling appeared to be less than that represented by either pollen or lake sediments and suggested that an extirpation of the majority of cold-adapted taxa during the initial warming might have precluded their presence during theYounger Dryas. An MCR analysis of the assemblages indicated a mean July temperature during Younger Dryas time about 5 °C lower than today (Miller & Elias, 2000).

Holocene Assemblages

Holocene assemblages from different regions of the continent share the characteristic of being composed mostly of species for which the fossil locations are within the existing geographic ranges (e.g. Hall & Nelson, 1990; Hebda *et al.*, 1990; Lavoie, 2001; Lavoie *et al.*, 1997; Morgan *et al.*, 1985; Schwert & Ashworth, 1985). This is simply a measure of the small amount of temperature variation during the last 10,000 years. There are some notable exceptions. Nelson & Carter (1987) reported a early Holocene assemblage from the North Slope of Alaska in which seven species were north of their existing distributions in the boreal forest zone. The beetles inhabit well-drained grassy areas with mean summer temperatures 2–3 °C warmer than is represented at the fossil site today. Lavoie *et al.* (1997) reported a mid- to late Holocene treeline assemblage from northern Quebec in which there were several species that also were north of their existing ranges. Both of these assemblages are from ecotones and it is possible that small changes in geographic range go undetected because the majority of fossil sites are not located on sensitive ecological boundaries.

The most dramatic changes detected in Holocene fossil assemblages are not caused by climate change but by humans. Baker *et al.* (1993) and Schwert (1996) showed that massive changes occurred in the beetle fauna of Iowa between 1840 and 1880 at the time the area was being settled. On land, a diverse ground-beetle fauna was disturbed and replaced by beetles associated with humans: dung-feeding scarabs associated with cattle and stored product pests associated with granaries. Plowing the native prairie led to large amounts of sediments being washed into rivers, increasing suspended sediments, and turning the waters turbid. Some of the most abundant pre-settlement fossils in the assemblages are those of elmids (riffle beetles). Elmid populations, associated with clear water, unpolluted rivers, and lakes, collapsed as a result of the agricultural activities. Today, it is almost impossible to conceive that the sluggish rivers of the mid-continent, with large suspended sediment loads, were once as clear as mountain streams.

Conclusions

(1) Studies of Quaternary fossil assemblages have made invaluable contributions to biogeographical and paleoclimatological studies. The late Tertiary and early Pleistocene assemblages provide proof that species can exist for millions of years although a convincing explanation of how genetically isolated populations maintain morphological constancy is as elusive as ever.

(2) The primary response of beetles to climate change is survival by dispersal to suitable habitats (Ashworth, 2001). The effect of this is that species "migrate" across vast distances or "track" climate changes.

(3) Many species of beetles are adapted to narrow ranges of climatic factors making them excellent indicators of past climates. Numerous paleoclimatic studies have been completed and semi-quantitative Mutual Climatic Range (MCR) analyses summarized for many regions of the continent (Elias, 1996, 1999, 2000, 2001; Elias *et al.*, 1996; Elias & Matthews, 2002; Miller & Elias, 2000).

Future Studies

(1) The response of beetles to climate change is much better known in some regions than others. We need to find out how the fauna of eastern and southern North America, south of the ice margin, responded to climate change. No fossil assemblages have been studied along the coast from New Jersey to Florida. No fossil assemblages have been studied from the central and southern Appalachian states even though it is likely that these areas were refugia during previous glaciations.

(2) The Pacific Northwest, from northern British Columbia to Oregon, is an area in which there is an abundance of peat bogs and buried peat horizons. The little we know of the beetle response to climate change suggests that it was different than in the central and eastern parts of the continent. Studies of fossil assemblages from the unglaciated Queen Charlotte Island and sites south of the ice margin in Washington and Oregon are needed to interpret the climates associated with the last glacial cycle. The hypothesis that the northwest coast was a migration route for early Americans provides an additional reason for paleoenvironmental analyses.

(3) Rancho La Brea is an incredible archive and records the only two extinctions of Quaternary beetles in North America. There is a need for a study with good stratigraphic and age control of the specimens. No studies have been made of the asphalt deposits since ^{14}C AMS dating was developed and making it possible to date the fossil beetles directly. Exactly when did those two dung beetle species go extinct? Was it really coeval with the extinction of the large mammals? These are two questions for which we should be able to obtain better answers.

(4) Beetles are one of the few groups of living organisms for which there is an extensive and well-dated Quaternary fossil record. Molecular DNA studies are a tool for examining the phylogeographic relationships of modern species. Only one study has so far attempted to link molecular and paleontological studies (Reiss *et al.*, 1999). The combination of studies provides a powerful tool for examining various questions in biogeography.

(5) Large amounts of climate data used in MCR analyses are available from the NOAA paleoclimatological archives. These data, however, are not in a user-friendly form. There is a need to develop a searchable web database that would link species range data to climatic data for the purpose of MCR and other paleoclimatic analyses.

(6) Quaternary paleoentomology is a speciality that requires a lot of time in training to develop identification skills. Skills are concentrated in a few individuals and they may or may not be passed on to the next generation. Paleoentomologists need to develop other ways to provide a legacy of their knowledge. In a world of 3-D scanners and supercomputers, we need to incorporate emerging technologies to aid in identification training.

References

Anderson, T.W., Matthews, J.V., Jr., Mott, R.J. & Richard, S.H. (1990). The Sangamonian Pointe-Fortune site, Ontario-Quebec border. *Géographie Physique et Quaternaire*, **44**, 271–287.

Ashworth, A.C. (1973). Fossil beetles from a fossil wood rat midden in western Texas. *Coleopterists Bulletin*, **27**, 139–140.

Ashworth, A.C. (1979). Quaternary Coleoptera studies in North America: Past and present. *In*: Erwin, T.L., Ball, G.E. & Whitehead, D.R. (Eds), *Carabid Beetles, Their Evolution, Natural History and Classification*. Dr.W. Junk bv Publishers, The Hague, pp. 395–406.

Ashworth, A.C. (1980). Environmental implications of a beetle assemblage from the Gervais Formation (early Wisconsinan?) Minnesota. *Quaternary Research*, **13**, 200–212.

Ashworth, A.C. (1996). The response of arctic Carabidae (Coleoptera) to climate change based on the fossil record of the Quaternary Period. *Annales Zoologici Fennici*, **33**, 125–131.

Ashworth, A.C. (2000). The ecology of Helophorus arcticus Brown (Coleoptera: Hydrophilidae) reconsidered. *The Coleopterists Bulletin*, **54**, 370–378.

Ashworth, A.C. (2001). Perspectives on Quaternary beetles and climate change. *In*: Gerhard, L.C., Harrison, W.E. & Hanson, B.E. (Eds), *Geological Perspectives of Global Climate Change*. American Association of Petroleum Geologists Studies in Geology 47, Tulsa, Oklahoma, pp. 153–168.

Ashworth, A.C. & Schwert, D.P. (1991). On the occurrences of *Opisthius richardsoni* Kirby and *Asaphidion yukonense* Wickham (Coleoptera, Carabidae) as late Pleistocene fossils. *Proceedings of the Entomological Society of Washington*, **93**, 511–514.

Ashworth, A.C. & Schwert, D.P. (1992). The Johns Lake site: A late Quaternary fossil beetle (Coleoptera) assemblage from the Missouri Coteau of North Dakota. *In*: Erickson, J.M. & Hoganson, J.W. (Eds), Proceedings of the F.D. Holland Jr., Geological Symposium. North Dakota Geological Survey Miscellaneous Series, Vol. 76, pp. 257–265.

Ashworth, A.C. & Willenbring, J. (1998). Fossil beetles and climate change at Sixmile Creek, Ithaca, New York. *American Paleontologist*, **6**, 2–3.

Ashworth, A.C., Gutenkunst, M. & Nelson, R.E. (2000). The climate during the last advance of the Puget Sound glacier (Vashon Stade, Fraser Glaciation) based on fossil beetles from Seattle. *Geological Society of America Abstracts With Programs*, **32**, A-404.

Ashworth, A.C., Schwert, D.P., Watts, W.A. & Wright, H.E., Jr. (1981). Plant and insect fossils at Norwood in south-central Minnesota: A record of late-glacial succession. *Quaternary Research*, **16**, 66–79.

Bajc, A.F., Schwert, D.P., Warner, B.G. & Williams, N.E. (2000). A reconstruction of Moorhead and Emerson Phase environments along the eastern margin of glacial Lake Agassiz, Rainy River basin, northwestern Ontario. *Canadian Journal of Earth Sciences*, **37**, 1335–1353.

Baker, R.G., Rhodes, R.S., II, Schwert, D.P., Ashworth, A.C., Frest, T.J., Hallberg, G.R. & Janssens, J.A. (1986). A full-glacial biota from southeastern Iowa, USA. *Journal of Quaternary Science*, **1**, 91–108.

Baker, R.G., Schwert, D.P., Bettis, E.A., III, Kemmis, T.J., Horton, D.G. & Semken, H.A. (1991). Mid-Wisconsinan stratigraphy and paleoenvironments at the St. Charles site in south-central Iowa. *Geological Society of America Bulletin*, **103**, 210–220.

Baker, R.G., Schwert, D.P., Bettis, E.A., III & Chumbley, C.A. (1993). Impact of Euro-American settlement on a riparian landscape in northeast Iowa, midwestern USA: An integrated approach based on historical evidence, floodplain sediments, fossil pollen, plant macrofossils and insects. *The Holocene*, **3**, 314–323.

Ball, G.E. & Currie, D.C. (1997). Ground beetles (Coleoptera: Trachypachidae and Carabidae) of the Yukon: Geographical distribution, ecological aspects, and origin of the extant fauna. *In*: Danks, H.V. & Downes, J.A. (Eds), *Insects of the Yukon*. Biological Survey of Canada (Terrestrial Arthropods), Ottawa, pp. 445–489.

Barnosky, A.D., Barnosky, C.W., Nickmann, R.J., Ashworth, A.C., Schwert, D.P. & Lantz, S.W. (1988). Late Quaternary paleoecology at the Newton site, Bradford Co., northeastern Pennsylvania: Mammuthus columbi, palynology, and fossil insects. *In*: Laub, R.S., Miller, N.G. & Steadman, D.W. (Eds), *Late Pleistocene and Early Holocene Paleoecology and Archeology of the Eastern Great Lakes Region*. Bulletin of the Buffalo Society of Natural Sciences 33, Buffalo, New York, pp. 173–184.

Betancourt, J.L., Van Devender, T.R. & Martin, P.S. (1990). *Packrat Middens the last 40,000 years of Biotic Change*. University of Arizona Press, Tucson.

Böcher, J. (1995). Palaeoentomology of the Kap København Formation, a Plio-Pleistocene sequence in Peary Land, North Greenland, Meddelelser Om Grønland. *Geoscience*, **33**, 1–82.

Bousquet, Y. & Larochelle, A. (1993). Catalogue of the Geadephaga (Coleoptera: Trachypachidae, Rhysodidae, Carabidae including Cicindelina) of America, north of Mexico. *Memoirs of the Entomological Society of America*, **167**, 1–397.

Carter, K.D. (1985). Middle and late Wisconsinan (Pleistocene) insect assemblages from Illinois. Unpublished M.Sc. Thesis, University of North Dakota, Grand Forks, North Dakota.

Churcher, C.S., Pilny, J.J. & Morgan, A.V. (1990). Late Pleistocene vertebrate, plant and insect remains from the Innerkip site, southwestern Ontario. *Géographie physique et Quaternaire*, **44**, 299–308.

Cong, S. (1997). Fossils of an undescribed blind trechine (Coleoptera:Carabidae) from near Kalaloch, Olympic Peninsula, Washington. *The Coleopterists Bulletin*, **51**, 208–211.

Cong, S. & Ashworth, A.C. (1996). Palaeoenvironmental interpretation of Middle and Late Wisconsinan fossil coleopteran assemblages from western Olympic Peninsula, Washington. *Journal of Quaternary Science*, **11**, 345–356.

Cong, S., Ashworth, A.C., Schwert, D.P. & Totten, S.M. (1996). Fossil beetle evidence for a short warm interval near 40,000 yr B.P. at Titusville, Pennsylvania. *Quaternary Research*, **45**, 216–225.

Coope, G.R. (1968). Insect remains from silts below till at Garfield Heights, Ohio. *Bulletin of the Geological Society of America*, **79**, 753–756.

Coope, G.R. (1978). Constancy of insect species vs. inconstancy of Quaternary environments. *In*: Mound, L.A. & Waloff, N. (Eds), *Diversity of Insect Faunas*. Blackwell, Oxford, pp. 176–187.

Doyen, J.T. & Miller, S.E. (1980). Review of Pleistocene darkling ground beetles of the California asphalt deposits (Coleoptera: Tenebrionidae, Zopheridae). *Pan-Pacific Entomologist*, **56**, 1–10.

Elias, S.A. (1987a). Climatic significance of late Pleistocene insect fossils from Marias Pass, Montana. *Canadian Journal of Earth Sciences*, **25**, 922–926.

Elias, S.A. (1987b). Paleoenvironmental significance of late Quaternary insect fossils from packrat middens in south-central Mexico. *Southwestern Naturalist*, **32**, 383–390.

Elias, S.A. (1990). Observations on the taphonomy of late Quaternary insect fossil remains in pack rat middens of the Chihuahuan Desert. *Palaios*, **5**, 356–363.

Elias, S.A. (1991). Insects and climatic change. Fossil evidence from the Rocky Mountains. *Bioscience*, **41**, 552–559.

Elias, S.A. (1992). Late Quaternary zoogeography of the Chihauhuan Desert insect fauna, based on fossil records from packrat middens. *Journal of Biogeography*, **19**, 285–297.

Elias, S.A. (1992a). Late Quaternary beetle fauna of southwestern Alaska: Evidence of a refugium for mesic and hygrophilous species. *Arctic and Alpine Research*, **24**, 133–144.

Elias, S.A. (1994). *Quaternary insects and their environments*. Smithsonian Institution Press, Washington.

Elias, S.A. (1996). Late Pleistocene and Holocene seasonal temperatures reconstructed from fossil beetle assemblages in the Rocky Mountains. *Quaternary Research*, **46**, 311–318.

Elias, S.A. (1999). Mid-Wisconsin seasonal temperatures reconstructed from fossil beetle assemblages in eastern North America: Comparisons with other proxy records from the Northern Hemisphere. *Journal of Quaternary Science*, **14**, 255–262.

Elias, S.A. (2000). Late Pleistocene climates of Beringia, based on analysis of fossil beetles. *Quaternary Research*, **53**, 229–235.

Elias, S.A. (2001). Mutual climatic range reconstructions of seasonal temperatures based on Late Pleistocene fossil beetle assemblages in eastern Beringia. *Quaternary Science Reviews*, **20**, 77–91.

Elias, S.A. & Matthews, J.V., Jr. (2002). Arctic North American seasonal temperatures from the latest Miocene to the Early Pleistocene, based on mutual climatic range analysis of fossil beetle assemblages. *Canadian Journal of Earth Sciences*, **39**, 911–920.

Elias, S.A. & Van Devender, T.R. (1990). Fossil insect evidence for late Quaternary climatic change in the Big Bend region, Chihuahuan Desert, Texas. *Quaternary Research*, **34**, 249–261.

Elias, S.A. & Van Devender, T.R. (1992). Insect fossil evidence of late Quaternary environments in the northern Chihuahuan Desert of Texas and New Mexico: Comparisons with the paleobotanical record. *Southwestern Naturalist*, **37**, 101–116.

Elias, S.A., Anderson, K.H. & Andrews, J.T. (1996). Late Wisconsin climate in northeastern USA and southeastern Canada, reconstructed from fossil beetle assemblages. *Journal of Quaternary Science*, **11**, 417–421.

Elias, S.A., Andrews, J.T. & Anderson, K.H. (1999). Insights on the climatic constraints on the beetle fauna of coastal Alaska, USA, derived from the mutual climatic range method of paleoclimate reconstruction, Arctic, Antarctic. *and Alpine Research*, **31**, 94–98.

Elias, S.A., Berman, D. & Alfimov, A. (2000). Late Pleistocene beetle faunas of Beringia: Where east meets west. *Journal of Biogeography*, **27**, 1349–1363.

Elias, S.A., Mead, J.I. & Agenbroad, L.D. (1992). Late Quaternary arthropods from the Colorado Plateau, Arizona and Utah. *Great Basin Naturalist*, **52**, 59–67.

Elias, S.A., Short, S.K. & Phillips, R.L. (1992). Paleoecology of late-glacial peats from the Bering Land Bridge, Chukchi Sea shelf region, northwestern Alaska. *Quaternary Research*, **38**, 371–378.

Elias, S.A., Short, S.K., Nelson, C.H. & Birks, H.H. (1996). Life and times of the Bering Land Bridge. *Nature*, **382**, 60–63.

Garry, C.E., Baker, R.W., Gilbertson, J.P. & Huber, J.K. (1997). Fossil insects and pollen from late Illinoian sediments in midcontinental North America. *In*: Ashworth, A.C., Buckland, P.C. & Sadler, J.P. (Eds), *Studies in Quaternary Entomology – an Inordinate Fondness for Insects*. Quaternary Proceedings 5, Wiley, Chichester, pp. 113–124.

Garry, C.E., Schwert, D.P., Baker, R.G., Kemmis, T.J., Horton, D.G. & Sullivan, A.E. (1990). Plant and insect remains from the Wisconsinan interstadial/stadial transition at Wedron, north-central Illinois. *Quaternary Research*, **33**, 387–399.

Guthrie, R.D. (1990). *Frozen fauna of the mammoth steppe. The story of blue babe*.

Hall, H.A. & Nelson, R.E. (1990). *The Maine Geologist*, **16**, 5.

Hall, W.E., Olson, C.A. & Van Devender, T.R. (1989). Late Quaternary and modern arthropods from the Ajo Mountains of southwestern Arizona. *Pan-Pacific Entomologist*, **65**, 322–347.

Hall, W.E., Van Devender, T.R. & Olson, C.A. (1988). Late Quaternary arthropod remains from Sonoran Desert middens, southwestern Arizona and northwestern Sonora. *Quaternary Research*, **29**, 277–293.

Hall, W.E., Van Devender, T.R. & Olson, C.A. (1990). Arthropod history of the Puerto Blanco Mountains, Organ Pipe Cactus National Monument, southwestern Arizona. *In*: Betancourt, J.L., Van Devender, T.R. & Martin, P.S. (Eds), *Packrat Middens: The Last 40,000 Years of Biotic Change*. University of Arizona Press, Tucson, pp. 363–379.

Hebda, R.J., Warner, B.G. & Cannings, R.A. (1990). Pollen, plant macrofossils, and insects from fossil woodrat (*Neotoma cinerea*) middens in British Columbia. *Géographie Physique et Quaternaire*, **44**, 227–234.

Hebda, R., Hicock, S.R., Miller, R.F. & Armstrong, J.E. (1983). Paleoecology of Mid-Wisconsin sediments from Lynn Canyon, Fraser Lowland, British Columbia. *Geologist's Association of Canada Annual Meeting, Program With Abstracts*, **8**, 31.

Hopkins, D.M., Matthews, J.V., Wolfe, J.A. & Silberman, M.L. (1971). A Pliocene flora and insect fauna from the Bering Strait region, Palaeogeography, Palaeoclimatology. *Palaeoecology*, **9**, 211–231.

Horn, G.H. (1876). Notes on some coleopterous remains from the bone cave at Port Kennedy, Pennsylvania. *Transactions of the American Entomological Society*, **5**, 241–245.

Karrow, P.F., Ceska, A., Hebda, R.J., Miller, B.B., Seymour, K.L. & Smith, A.J. (1995). Diverse nonmarine biota from the Whidbey Formation (Sangamonian) at Point Wilson, Washington. *Quaternary Research*, **44**, 434–437.

Karrow, P.F., McAndrews, J.H., Miller, B.B., Morgan, A.V., Seymour, K.L. & White, O.L. (2001). Illinoian to Late Wisconsinan stratigraphy at Woodbridge, Ontario. *Canadian Journal of Earth Sciences*, **38**, 921–942.

Kryzhanovskij, O.L., Belousov, I.A., Kabak, I.I., Kataev, B.M., Makarov, K.V. & Shilenkov, V.G. (1995). *A checklist of the ground-beetles of Russia and adjacent lands (Insecta, Coleoptera, Carabidae)*. Pennsoft Series Faunistica 3. Pennsoft Publishers, Sofia.

Lavoie, C. (2001). Reconstructing the late-Holocene history of a subalpine environment (Charlevoix, Québec) using fossil insects. *The Holocene*, **11**, 89–99.

Lavoie, C., Elias, S.A. & Filion, L. (1997). A 7000-year record of insect communities from a peatland environment, southern Quebec. *Ecoscience*, **4**, 394–403.

Lavoie, C., Elias, S.A. & Payette, S. (1997). Holocene fossil beetles from a treeline peatland in subarctic Quebec. *Canadian Journal of Zoology*, **75**, 227–236.

Matthews, J.V., Jr. (1968). A paleoenvironmental analysis of three late Pleistocene coleopterous assemblages from Fairbanks, Alaska. *Quaestiones Entomologicae*, **4**, 202–224.

Matthews, J.V., Jr. (1974). Quaternary environments at Cape Deceit (Seward Peninsula, Alaska): Evolution of a tundra ecosystem. *Bulletin of the Geological Society America*, **85**, 1353–1384.

Matthews, J.V., Jr. (1976a). Insect fossils from the Beaufort Formation: Geological and biological significance. *Geological Survey of Canada Papers*, **76–1b**, 217–227.

Matthews, J.V., Jr. (1976b). Evolution of the subgenus *Cyphelophorus* (Genus *Helophorus*, Hydrophilidae, Coleoptera): Description of two new fossil species and discussion of *Helophorus tuberculatus* Gyll. *Canadian Journal of Zoology*, **54**, 652–673.

Matthews, J.V., Jr. (1977). Tertiary Coleoptera fossils from the North American Arctic. *Coleopterists Bulletin*, **31**, 297–308.

Matthews, J.V., Jr. (1979a). Late Tertiary carabid fossils from Alaska and the Canadian Archipelago. *In*: Erwin, T.L., Ball, G.E. & Whitehead, D.R. (Eds), *Carabid Beetles, Their Evolution, Natural History and Classification*. Dr. W. Junk bv Publishers, The Hague, pp. 425–445.

Matthews, J.V., Jr. (1979b). Fossil beetles and the Late Cenozoic history of the tundra environment. *In*: Gray, J. & Boucot, A.J. (Eds), *Historical Biogeography, Plate Tectonics and the Changing Environment*. Oregon State University Press, pp. 371–378.

Matthews, J.V., Jr., Mott, R.J. & Vincent, J.-S. (1986). Preglacial and Interglacial environments of Banks Island: Pollen and macrofossils from Duck Hawk Bluffs and related sites. *Géographie Physique et Quaternaire*, **40**, 279–298.

Matthews, J.V., Jr., Smith, S.L. & Mott, R.J. (1987). Plant macrofossils, pollen, and insects of arctic affinity from Wisconsinan sediments in Chaudiere Valley, southern Quebec. *Current Research Part A, Geological Survey of Canada Papers*, **87–1a**, 165–175.

Matthews, J.V., Jr. (1982). East Beringia during late Wisconsin time. A review of the biotic evidence. *In*: Hopkins, D.M., Matthews, J.V., Jr., Schweger, C.E. & Young, S.B. (Eds), *Paleoecology of Beringia*. Academic Press, New York, pp. 127–150.

Matthews, J.V., Jr. & Telka, A. (1997). Insect fossils from the Yukon. *In*: Danks, H.V. & Downes, J.A. (Eds), *Insects of the Yukon*. Biological Survey of Canada (Terrestrial Arthropods), Ottawa, pp. 911–962.

Miller, B.B., Graham, R.W., Morgan, A.V., Miller, N.G., McCoy, W.D., Palmer, D.F., Smith, A.J. & Pilny, J.J. (1994). A biota associated with Matuyama-age sediments in west-central Illinois. *Quaternary Research*, **41**, 350–365.

Miller, R.F. (1995). Late-Glacial Coleoptera and the paleoclimate at Hirtles, Nova Scotia. *Atlantic Geology*, **31**, 95–101.

Miller, R.F. (1996). Allerød – Younger Dryas Coleoptera from western Cape Breton Island, Nova Scotia, Canada. *Canadian Journal of Earth Sciences*, **33**, 33–41.

Miller, R.F. (1997a). Late-glacial (Allerød) Coleoptera from Joggins, Lantz and Blomiden, central Nova Scotia, Canada. *Atlantic Geology*, **33**, 223–229.

Miller, R.F. (1997b). Late-Glacial Coleoptera in maritime Canada and the transition to the Younger Dryas chronozone. *In*: Ashworth, A.C., Buckland, P.C. & Sadler, J.P. (Eds), *Studies in Quaternary Entomology – An Inordinate Fondness for Insects*. Quaternary Proceedings 5, Wiley, Chichester, pp. 177–184.

Miller, R.F. (2000). Late-glacial (Younger Dryas) Coleoptera from Saint John, New Brunswick (NTS 21 G/1). *In*: Carroll, B.M.W. (ed.), *Current Research 1999*. New Brunswick Department of Natural Resources, Mineral resource Report 2000-4, pp. 31–38.

Miller, R.F. & Elias, S.A. (2000). Late-glacial climate in the Maritimes region, Canada, reconstructed from mutual climatic range analysis of fossil Coleoptera. *Boreas*, **29**, 79–88.

Miller, R.F. & Morgan, A.V. (1991). Late-glacial Coleoptera fauna from Lismore, Nova Scotia. *Atlantic Geology*, **27**, 193–197.

Miller, R.F., Morgan, A.V. & Hicock, S.R. (1985). Pre-Vashon fossil Coleoptera of Fraser age from the Fraser lowland,

British Columbia. *Canadian Journal of Earth Sciences*, **22**, 498–505.

Miller, S.E. (1983). Late Quaternary insects of Rancho La Brea and McKittrick, California. *Quaternary Research*, **20**, 90–104.

Miller, S.E. (1997). Late Quaternary insects of Rancho La Brea, California, USA. *In*: Ashworth, A.C., Buckland, P.C. & Sadler, J.P. (Eds), *Studies in Quaternary Entomology – An Inordinate Fondness For Insects*. Quaternary Proceedings 5, Wiley, Chichester, pp. 185–193.

Miller, S.E. & Peck, S.B. (1979). Fossil carrion beetles of Pleistocene California asphalt deposits, with a synopsis of Holocene California Silphidae (Insecta: Coleoptera: Silphidae). *Transactions of the San Diego Society of Natural History*, **19**, 85–106.

Miller, S.E., Gordon, R.D. & Howden, H.F. (1981). Reevaluation of Pleistocene scarab beetles from Rancho La Brea, California (Coleoptera, Scarabaeidae). *Proceedings of the Entomological Society of Washington*, **83**, 625–630.

Moore, I. & Miller, S.E. (1978). Fossil rove beetles from Pleistocene California asphalt deposits (Coleoptera Staphylinidae). *Coleopterists Bulletin*, **32**, 37–39.

Morgan, A.V., Kuc, M. & Andrews, J.T. (1993). Paleoecology and age of the Flitaway and Isortoq interglacial deposits, north-central Baffin Island, Northwest Territories, Canada. *Canadian Journal of Earth Sciences*, **30**, 954–974.

Morgan, A.V. (1987). Late Wisconsin and early Holocene paleoenvironments of east-central North America based on assemblages of fossil Coleoptera. *In*: Ruddiman, W.F. & Wright, H.E., Jr. (Eds), *North America and Adjacent Oceans during the Last Deglaciation*. The Geology of North America K-3. Geological Society of America, Boulder, Colorado, pp. 353–370.

Morgan, A.V. & Morgan, A. (1980). Faunal assemblages and distributional shifts of Coleoptera during the Late Pleistocene in Canada and the northern United States. *Canadian Entomologist*, **112**, 1105–1128.

Morgan, A.V., Morgan, A., Ashworth, A.C. & Matthews, J.V., Jr. (1984). Late Wisconsin fossil beetles in North America. *In*: Wright, H.E., Jr. (ed.), *Late Quaternary Environments of the United States*. Volume 1: The Late Pleistocene (Porter, S.C. (ed.). Longman, London, pp. 354–363.

Morgan, A., Morgan, A.V. & Elias, S.A. (1985). Holocene insects and paleoecology of the Au Sable River, Michigan. *Ecology*, **66**, 1817–1828.

Morgan, A. (1972). The fossil occurrence of *Helophorus arcticus* Brown (Coleoptera Hydrophilidae) in Pleistocene deposits of the Scarborough Bluffs, Ontario. *Canadian Journal of Zoology*, **50**, 555–558.

Mott, R.J. (1994). Wisconsinan late-glacial environmental change in Nova Scotia a regional synthesis. *Journal of Quaternary Science*, **9**, 155–160.

Mott, R.J. & DiLabio, N.W. (1990). Paleoecology of organic deposits of probable last interglacial age in northern Ontario. *Géographie physique et Quaternaire*, **44**, 309–318.

Mott, R.J., Anderson, T.W. & Matthews, J.V., Jr. (1981). Late-glacial paleoenvironments of sites bordering the Champlain Sea based on pollen and macrofossil evidence. *In*: Mahaney, W.C. (ed.), *Quaternary Paleoclimate*. GeoAbstracts, Norwich, England, pp. 129–172.

Nelson, R.E. & Carter, L.D. (1987). Paleoenvironmental analysis of insects and extralimital *Populus* from an early Holocene site on the arctic slope of Alaska, USA. *Arctic and Alpine Research*, **19**, 230–241.

Nelson, R.E. & Coope, G.R. (1982). A late-Pleistocene insect fauna from Seattle, Washington. *Abstracts, 7th Biennial Meeting, American Quaternary Association Conference, Seattle, Washington*, 146 pp.

Nielsen, E., Morgan, A.V., Morgan, A., Mott, R.J. & Rutter, N.W. (1986). Stratigraphy, palaeoecology, and glacial history of the Gillam area, Manitoba. *Canadian Journal of Earth Sciences*, **23**, 1641–1661.

Pierce, W.D. (1946). Fossil arthropods of California. 11. Descriptions of the dung beetles (Scarabaeidae) of the tar pits. *Bulletin of the Southern California Academy of Science*, **45**, 119–131.

Pilny, J.J. & Morgan, A.V. (1987). Paleoentomology and paleoecology of a possible Sangamonian site near Innerkip, Ontario. *Quaternary Research*, **28**, 157–174.

Reiss, R.A., Ashworth, A.C. & Schwert, D.P. (1999). Molecular genetic evidence for the post-Pleistocene divergence of populations of the arctic-alpine ground beetle *Amara alpina* (Paykull) (Coleoptera:Carabidae). *Journal of Biogeography*, **26**, 785–794.

Ritchie, J.C. (1984). *Past and Present Vegetation of the far Northwest of Canada*. University of Toronto Press, Toronto.

Ross, H.H. (1965). Pleistocene events and insects. *In*: Wright, H.E. & Frey, D.G. (Eds), *The Quaternary of the United States*. Princeton University Press, New Jersey, pp. 583–595.

Schwert, D.P. (1992). Faunal transitions in response to an ice age: The late Wisconsinan record of Coleoptera in the north-central United States. *Coleopterists Bulletin*, **46**, 68–94.

Schwert, D.P. (1996). Effect of Euro-American settlement on an insect fauna – paleontological analysis of the recent chitin record of beetles (Coleoptera) from northeastern Iowa. *Annals of the Entomological Society of America*, **89**, 53–63.

Schwert, D.P. & Ashworth, A.C. (1988). Late Quaternary history of the northern beetle fauna of North America: A synthesis of fossil and distributional evidence. *Memoirs of the Entomological Society of Canada*, **144**, 93–107.

Schwert, D.P. & Ashworth, A.C. (1985). Fossil evidence of late Holocene faunal stability in southern Minnesota (Coleoptera). *Coleopterists Bulletin*, **39**, 67–79.

Schwert, D.P., Torpen-Kreft, H.J. & Hajic, E.R. (1997). Characterisation of the late Wisconsinan tundra/forest transition in midcontinental North America using assemblages of fossil beetles. *In*: Ashworth, A.C., Buckland, P.C. & Sadler, J.P. (Eds), *Studies in Quaternary Entomology – an Inordinate Fondness for Insects*. Quaternary Proceedings 5, Wiley, Chichester, pp. 237–243.

Scudder, S.H. (1890). Tertiary insects of North America. *Bulletin of the United States Geological Survey of the Territories (Hayden)*, **13**, 1–663.

Scudder, S.H. (1898). The Pleistocene beetles of Fort River, Massachusetts. *Monographs of The United States Geological Survey*, **24**, 740–746.

Spilman, T.J. (1976). A new species of fossil Ptinus from fossil wood rat nests in California and Arizona (Coleoptera, Ptinidae), with a postscript on the definition of a fossil. *Coleopterists Bulletin*, **30**, 239–244.

Walker, I.R. (1995). Chironomids as indicators of past environmental change. *In*: Armitage, P.D., Cranston, P.S. & Pinder, L.C.V. (Eds), *The Chironomidae: Biology and Ecology of Non-Biting Midges*. London: Chapman and Hall, pp. 405–422.

Walker, I.R., Mott, R.J. & Smol, J.P. (1991). Allerød-Younger Dryas lake temperatures from midge fossils in Atlantic Canada. *Science*, **253**, 1010–1012.

Warner, B.G., Morgan, A.V. & Karrow, P.F. (1988). A Wisconsinan interstadial arctic flora and insect fauna from Clarksburg, southwestern Ontario, Canada. *Palaeoegeography, Palaeoclimatology, Palaeoecology*, **68**, 27–47.

Wickham, H.F. (1917). Some fossil beetles from the Sangamon peat, Champaign County, Illinois. *American Journal of Science*, **44**, 137–145.

Williams, N.E. (1989). Factors affecting the interpretation of caddisfly assemblages from Quaternary sediments. *Journal of Paleolimnology*, **1**, 241–248.

Williams, N.E. & Eyles, N. (1995). Sedimentary and paleoclimatic controls on caddisfly (Insecta: Trichoptera) assemblages during the last interglacial-to-glacial transition in southern Ontario. *Quaternary Research*, **43**, 90–105.

Vertebrate paleontology

S. David Webb[1], Russell W. Graham[2], Anthony D. Barnosky[3], Christopher J. Bell[4], Richard Franz[1], Elizabeth A. Hadly[5], Ernest L. Lundelius Jr.[6], H. Gregory McDonald[7], Robert A. Martin[8], Holmes A. Semken Jr.[9] and David W. Steadman[1]

[1] *Florida Museum of Natural History, University of Florida, Gainesville, FL 32611-7800, USA; Webb@flmnh.ufl.edu*
[2] *Department of Earth Sciences, Denver Museum of Natural History, Denver, CO 80205, USA*
[3] *Museum of Paleontology, University of California, Berkeley, CA 94720, USA*
[4] *Department of Geological Sciences, University of Texas at Austin, Austin, TX 78712, USA*
[5] *Department of Biological Sciences, Stanford University, Stanford, CA 94303, USA*
[6] *Vertebrate Paleo Lab, J.J. Pickle Research Campus, University of Texas, Austin, TX 78712, USA*
[7] *Geological Resources Division, National Park Service, Denver, CO 80225, USA*
[8] *Department of Biological Science, Murray State University, Murray, KY 42071, USA*
[9] *Department of Geology, University of Iowa, Iowa City, IA 52242, USA*

Introduction

When the United States last hosted a congress of the International Quaternary Union (INQUA), Wright & Frye (1965) edited the well-known volume The Quaternary of the United States. There were five chapters on vertebrates including "Quaternary Mammals" by Hibbard *et al.* (1965) and "Quaternary Reptiles" by Auffenberg & Milstead (1965). In the subsequent four decades new developments altered the practice of vertebrate paleontology. A fundamental obvious achievement, yet easily overlooked, consisted of doubling the number of Quaternary fossil localities and vastly augmenting the fossil samples. A quantum increase in microvertebrate samples stemmed from Hibbard's (1949) revolutionary introduction of efficient screening techniques. The resulting progress can be measured by the fact that virtually all species of mammals now living in North America are recorded as fossils, and that approximate geographic distributions of many can be credibly mapped back into the past. Other advances came with the deep-sea definition of glacial and interglacial cycles (marine oxygen isotope stages, MIS), improved methods of dating bone and fossil deposits, new electronic databases, access to ancient DNA, improved methods of phylogenetic reconstruction, and use of isotopic signatures in studies of diet and taphonomy. We briefly comment on the impact of these new facets.

North America is extraordinarily diverse eco-geographically. There is therefore a strong tendency to develop separate regional traditions in studying the succession of vertebrates from the Quaternary. This tendency is reinforced by the simple fact that nature provides different resources in each region. For this reason we devote the last part of this review article to some paleontological highlights from individual states and regions.

Biostratigraphy and Chronostratigraphy

Half a century ago Savage (1951) extended North American Land Mammal Ages (NALMA) into the Quaternary by characterizing the Irvingtonian and Rancholabrean. He led the way in providing NALMA with stage definitions and formal stratotypes. Distinguishing these NALMA from one another, and also the Irvingtonian from the preceding Blancan (largely Pliocene), still receives considerable emphasis. It remains traditional among this continent's Quaternary paleontologists to consider the Pliocene as well as the Pleistocene and Holocene. Thus, their purview generally spans nearly five million years.

Biostratigraphers studying the terrestrial faunal succession in North America embrace two somewhat different philosophies. Some emphasize an Oppelian approach in which overlapping range zones of multiple taxa characterize an assemblage. Others rely more heavily on a single widespread taxon, most often a "geologically instantaneous" intercontinental immigrant, to define a biochron.

Arvicoline rodents characterize the Oppelian approach in Eurasia and North America. Since these small grazers are diverse within the three Holarctic continents, evolved rapidly, and are easily identified by their molar teeth, they have excellent potential for biostratigraphic correlation and stage definition. As noted, widespread use of Hibbard's (1949) screening techniques amplified the availability of rodents in the North American record. One arvicoline biochronology for North America divides the Blancan & Irvingtonian into eight subdivisions (Repenning, 1987).

One serious limitation is that arvicolines often have narrower ecological tolerances and more limited distributions than many large mammals. For example, several proposed arvicoline zones turn out to be restricted to the High Plains. Similarly, Porcupine Cave in south-central Colorado (Fig. 1) has produced a rich assemblage of arvicolines at an elevation of 2900 m, yet that fauna does not readily correlate with other assemblages of similar age from farther east and at lower elevations (Bell & Barnosky, 2000).

North American Quaternary paleontologists generally support the quest for stage-defining immigrants because of that system's simplicity (Woodburne, 1987). *Mammuthus* (mammoths) define the Irvingtonian and *Bison* the Rancholabrean in North America (Savage, 1951). The first mammoths appear in North America fairly early in the

DEVELOPMENTS IN QUATERNARY SCIENCE
VOLUME 1 ISSN 1571-0866
DOI:10.1016/S1571-0866(03)01025-X

Fig. 1. Entrance to Porcupine Cave, south central Colorado, 2900 m, 1988.

Pleistocene. The strongest association of North American mammoths with radiometric dates occurs in the Rock Creek local fauna, Texas, where the fossiliferous sediments are overlain by the 1.2-myr Cerro Toledo ash. Other early mammoths occur interbedded with 1.4-myr Guaje Pumice in the Santa Fe Formation in New Mexico, and just above the 1.36-myr Bruneau Basalt in Idaho. In Nebraska, early mammoths occur in the Rushville and Gordon local faunas of early Irvingtonian age. The largest sample of early *Mammuthus* comes from the Leisey Shell Pit in Florida (Fig. 2), with an estimated age of 1.5 myr (Hulbert *et al.*, 1995). The appearance of *Mammuthus* in North America was essentially contemporaneous over the whole continent, with differences in first appearance dates in various regions attributable to discontinuous deposition and the vagaries of associating productive fossil sites with good dates.

Fig. 2. Leisey Bone Bed overlying marine mollusks, Hillsborough County, Florida, 1983.

A problem similar to the restriction of certain arvicoline biochrons to the High Plains may also occur for the first appearance of *Bison* as the definition of the Rancholabrean. The genus was present in Alaska very early in the Pleistocene, but appeared much later south of the Laurentide Ice Sheet (Bell *et al.*, in press). A second difficulty stems from the late Blancan appearance of *Bison* recently reported from horn cores in two Florida sites (McDonald & Morgan, 1999). Possibly these two horn cores represent other bovids; their identification must be more fully discussed.

The oldest well-determined and well-dated *Bison* records from conterminous United States represent long-horned species. These are from American Falls, Idaho, the Mt. Scott local fauna in Meade County, Kansas, and the Ten Mile Hill beds near Charleston, South Carolina. The Idaho occurrence is estimated to be 200,000 yr old. It is found in lake sediments that accumulated behind a lava dam with a much younger K/Ar date of 72,000 yr (Pinsof, 1991). The Mt. Scott local fauna is estimated to be about 160,000 yr old. The South Carolina *Bison* is dated at about 240,000 yr by a U/Th date of interbedded corals (Szabo, 1985).

The end of the Rancholabrean NALMA is readily defined by the great extinction of most megafauna at 13,000 cal yr B.P. (ca. 11,000 ^{14}C yr B.P.). Specialists who prefer the Oppelian approach characterize the Rancholabrean by the association of modern insectivores, rodents, and carnivores with the now-extinct megafauna. The importance and impact of that extinction episode are discussed below.

A major concern of Hibbard *et al.* (1965) was to correlate fossil mammals with the four glacial and three interglacial cycles then recognized, and to reconstruct environmental conditions for them. It is now clear from deep-sea cores and their oxygen isotopic stages that there were about 20 major climatic fluctuations in the last 2 myr (Ruddiman & Wright, 1987). Consequently, efforts to correlate vertebrate faunal history with continental glacial episodes were abandoned. Now the great divide lies at about 50,000 years ago. Younger faunas generally can be dated accurately by ^{14}C; older faunas, representing 90% of Quaternary time, must be dated by other chronostratigraphic techniques.

The best approaches to dating older faunas are multidisciplinary, for example, integrating radiometric dates of tephra layers with paleomagnetic profiles and faunal evolution. The best example of such work, in a remarkably long section, comes from the Anza Borrego Desert in southern California (Fig. 3). There terrestrial sediments from the upper part of the Imperial Formation and the entire Palm Spring Formation attain a thickness of more than 4000 m, as a consequence of tectonic subsidence of the Salton Trough. Based on 150 magnetic samples from the area, Johnson *et al.* (1983) identified 12 paleomagnetic intervals spanning the upper Gilbert, the Gauss, and the lower Matuyama magnetic chrons. These were cross-correlated with one tephra that yielded a K/Ar date of 2.3 myr and with the rich biostratigraphic framework spanning Blancan and Irvingtonian NALMA (Cassiliano, 1999).

In Florida and South Carolina the primary biostratigraphic determinations of Blancan and Irvingtonian terrestrial faunas are supplemented by interfingering relationships with marine deposits. For example, at Macasphalt Shell Pit near Sarasota, Unit 4 yields about 100 species of terrestrial and freshwater vertebrates of late Blancan age. Because this unit is bracketed by many other units bearing a rich marine fauna, it was possible to conduct an integrative geochronological study combining magnetostratigraphy and strontium isotope history, along with ostracod, mollusk, and mammal biochronologies (Jones *et al.*, 1991).

Vertebrate history in the latest Quaternary is greatly illuminated by new methods of applying ^{14}C dating to amino acids in bone. This contrasts with earlier methods that provided dates associated with bone or bracketing the age of the bone. In 1965 it was recognized that dates on bone were unreliable. The organic fraction was often contaminated by foreign proteins from rootlets or humic acid, and the inorganic fraction was susceptible to postdepositional isotopic exchange. In the late 1980s, two major breakthroughs radically improved bone dating, one technological and the other methodological. First, accelerators in tandem with mass spectrometers (TAMS) could directly measure the ^{14}C in a sample, thus requiring only small samples (micrograms vs. kilograms). An individual rodent tooth can now be dated, whereas previously an ungulate limb bone might have been required (Stafford *et al.*, 1991). Also, a direct measurement of ^{14}C, rather than the decay rate of this radioactive isotope by scintillation counters, yields a more precise date. In the initial phases of TAMS dating, standard errors for samples between 15,000 and 20,000 yr old were several hundred years. More recently, strengthening of the AMS magnets improved precision in that age range to give standard errors of less than 50 years.

The second, methodological improvement involved selecting individual amino acids for dating (Stafford *et al.*, 1982). Dates from organic molecules, found in a wide variety of compounds like humates, were rejected in favor of

Fig. 3. The long succession of fossiliferous terrestrial sediments in the Anza Borrego Desert, California, 1991.

those from proline and hydroxyproline, which are generally intrinsic to bone. Examination of other isotope ratios, especially nitrogen, can help determine if isotopic exchange is a problem. Finally, filtering collage samples though an inert resin known as XAD-2 allows sample purification without requiring separation of specific amino acids (Stafford *et al.*, 1988, 1999). Thus, technological and methodological advancements in bone ^{14}C-dating provide a far more precise and more accurate chronology.

Biogeography

As the chronological framework improved and vertebrate samples increased in number, paleontologists gave greater emphasis to distribution patterns at several different scales. On the broadest scale, intercontinental dispersals of vertebrate species are of great interest, especially because their magnitude and frequency seem to have increased during the Pliocene and Pleistocene. Shifts in intracontinental vertebrate species ranges have major implications for climatic and ecological history.

Intercontinental dispersals had major impacts on the history of North American vertebrates, especially during the Quaternary Period when these effects seem to have accelerated. Trans-Beringian interchange greatly influenced the character of the North American vertebrate fauna, with increasingly larger percentages being derived from Eurasia from the Pliocene into the Holocene. By Irvingtonian time, the North American mammal fauna had incorporated the influence of treeless tundra-like environments as indicated by the immigration of *Soergelia* (a muskox) and diverse arvicolines, including red-backed voles (*Clethrionomys*), brown lemmings (*Lemmus*), and collared lemmings (*Dicrostonyx*), as well as Caribou (*Rangifer*). Other taxa such as elk (*Cervus*), meadow and prairie voles (*Microtus*), and mammoths (*Mammuthus*) may not be critical indicators of tundra but they do reflect open environments that support ample supplies of grass. For this reason, some paleoecologists have referred to this environment as arctic steppe (Guthrie, 1982). Some Eurasian immigrants, including *Bison*, *Saiga*, *Praeovibos*, *Alces*, and *Xenocyon*, were prevented, for reasons not fully understood, from migrating south of the Laurentide Ice Sheet. Many paleontologists consider the Alaskan-Yukon fauna effectively non-North American, and affiliate it more closely with the Olyorian of Siberia, at least until after the Brunhes magnetochron (Bell *et al.*, in press).

The Great Interamerican Interchange is noted for dispersing more terrestrial vertebrates southward from North America than northward from South America. Nonetheless, during the late Blancan, about 3 myr ago, a substantial group of South American immigrants reached this continent (Webb, 1991). Perhaps the most exotic among them were three families of shelled edentates including a small armadillo (*Dasypus*), a giant armadillo (*Pampatherium*), and the tank-like glyptodont (*Glyptotherium*). Ground sloths of three sizes, representing three different families, also dispersed northward. The ground sloth, *Megalonyx*, notable historically for having been studied by Thomas Jefferson in the White House, spread very widely, even into Alaska. Large rodents including capybaras (*Neochoerus* and *Hydrochoerus*) and porcupines (*Erethizon*) also moved north. The only large ungulate to come against the tide of northern forms was a toxodont genus (*Mixotoxodon*), unless one wishes to count the aquatic manatees (*Trichechus*) as a kind of ungulate. The vampire bats (*Desmodus*) presumably followed some large Neotropical mammals that they regularly parasitized. Finally, the most astonishing immigrant, previously known from the Tertiary of southern South America, was a flightless predatory bird (*Titanis*) that stood more than 3 m tall. Four additional genera from the Neotropical Realm appeared in the early Irvingtonian, notably opossums (*Didelphis*), a small ground sloth (*Nothrotheriops*), a small glyptodont (*Pachyarmatherium*), and a giant anteater (*Myrmecophaga*). Still other important tropical groups, including many kinds of birds, rodents and monkeys, reached Central America without leaving a fossil record (Webb & Rancy, 1996).

Avian migrations on both intercontinental and intracontinental scales constitute another major feature of vertebrate biogeography. Although these migrations are often attributed to Quaternary climatic events, new evidence suggests that they began by the middle Miocene (Steadman & Martin, 1984). Strong global cooling with increased seasonality at mid and high latitudes during the Miocene (Zachos *et al.*, 2001) intensified the differentiation of temperate and boreal vegetation zones in North America (Graham, 1999, p. 267). These effects triggered the diversification of passerines, including migratory forms (Becker, 1987; Olson, 1989, 2001). Late Miocene and early Pliocene evidence consists of genera whose living species are migratory passerines, for example the sparrow (*Ammodramus*) from the late Clarendonian to early Hemphillian of Kansas (Steadman, 1981) and a bunting (*Passerina*) from the late Hemphillian of Chihuahua, Mexico (Steadman & Martin, 1984). This latter record gains chronological credence because the estimated divergence time among living *Passerina* species based on mtDNA is 2.4–3.7 myr (Klicka *et al.*, 2001).

Much of the modern long-distance migration from the Nearctic to the Neotropics involves nine-primaried oscines (warblers, tanagers, buntings, sparrows, and orioles; Emberizidae *sensu lato*). Unfortunately, this group of songbirds is very difficult to characterize at the species level with postcranial osteology, and therefore is poorly represented in the published fossil record. This problem might be partially offset in the future by analyses of ancient DNA (selected late Quaternary fossils only) and a larger investment of time in passerine comparative osteology. The ~20 Quaternary glacial-interglacial cycles (the past 1.8 myr) represent the coolest interval in the history of living genera and species of birds. Long-distance migration may have involved more species of birds, traveling over longer distances, during warm (interglacial) intervals than during times of expanded continental ice sheets. When most of North America north of 40°N was covered with ice during the last glacial maximum, the latitudinal extent of long-distance migration must

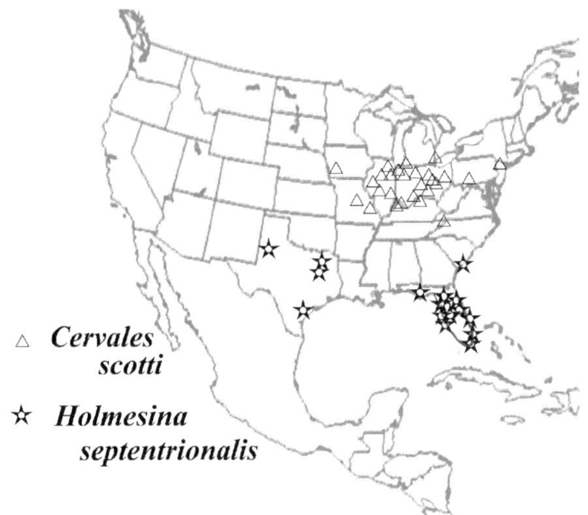

Fig. 4. FAUNMAP I shows the distribution in the U.S. of two late Pleistocene mammalian species, Cervalces scotti and Hokmesina septentronalis, during the Wisconsin Glaciation. Database as of June 25, 1993. Map redrawn from original provided courtesy of Illinois State Museum.

have been substantially shorter than today for the various migratory waterfowl, raptors, shorebirds, and passerines that currently nest north of 40°N. Likewise when currently arid parts of the western United States were filled with major lakes during the glacial maxima (Bachhuber, 1992), there may have been major alternatives to the four principle flyways followed today by migratory birds.

It is increasingly evident that mammals also exhibit interesting intracontinental dispersal patterns, presumably in response to climate change. Such range shifts helped Hibbard *et al.* (1965) distinguish their presumed glacial and interglacial faunas. Today digital databases like FAUNMAP (1994) greatly facilitate detailed biogeographic analyses (Figs 4 and 5). Replacing earlier efforts such as Martin & Hoffmann (1987) is a relational attribute database that utilizes the Geographic Information System (GIS) at the Illinois State Museum to map changes through time. The records can track individual species, species groups, or any other combination of parameters in the database. FAUNMAP I compiles information on location, geological and numerical ages, species composition, archaeological associations, and taphonomic data on faunas in the 48 contiguous United States for the last 40,000 years (FAUNMAP, 1994). More recently, FAUNMAP II has extended geographic coverage to include Alaska and Canada, and has extended the chronologic scope back through the Blancan NALMA.

The analyses resulting from FAUNMAP I demonstrated that each species responded to environmental change individually in accordance with its own tolerance limits. This reconstruction differs fundamentally from the wholesale shift of life zones and assemblages proposed in the older literature (e.g. Blair, 1965). Faunal provinces defined statistically during the latest Pleistocene are different from those that exist today. Finally, the analyses by the FAUNMAP Working

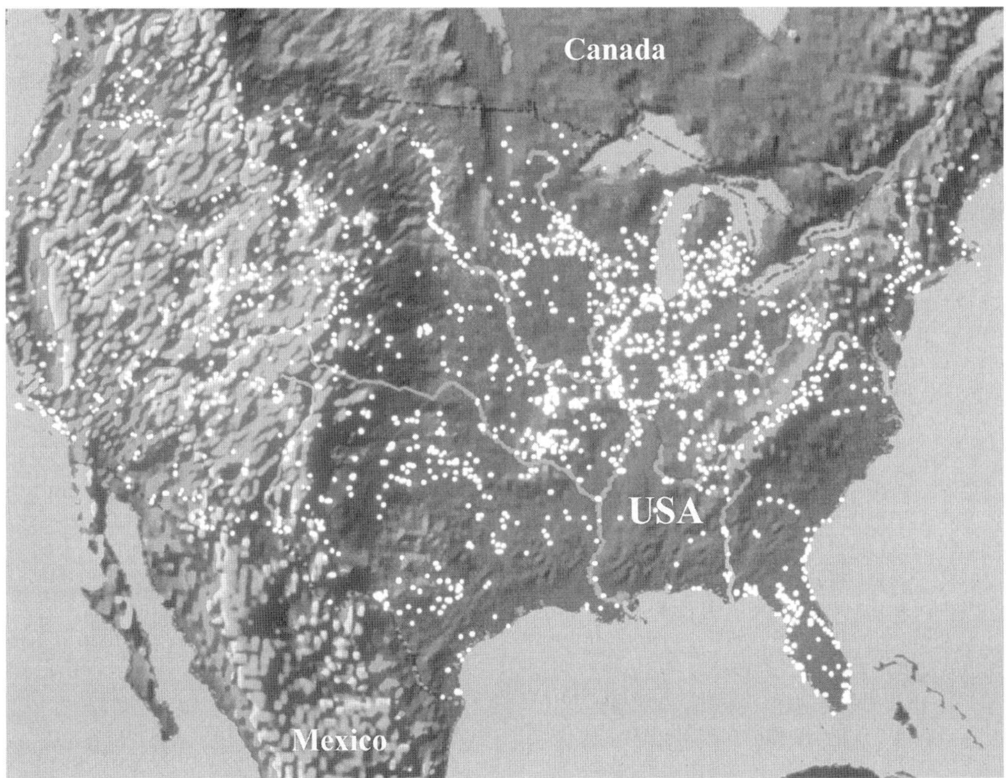

Fig. 5. FAUNMAP I sites in database as of June 25, 1993. Map photo courtesy of Illinois State Museum.

Group (1996) showed that late Pleistocene faunas and environments were more heterogeneous that those of the Holocene. This richness extends beyond the loss attributable to large mammal extinctions, discussed below, and is evident among the smaller vertebrates in every region where latest Pleistocene faunas have been adequately sampled.

Non-Analog Faunas and Taphonomy

Glacial-interglacial shifts in climate rearranged the composition of plant communities and consequently that of vertebrate faunas as well. It is not surprising that one of the great floristic conclusions from palynological studies, that plant species shifted individualistically, is also reflected in vertebrate faunas (FAUNMAP Working Group, 1996). In these studies, communities did not appear to move in lock-step, but appeared to pull apart in very complex patterns (Owen-Smith, 1989).

In each region of North America, one can track in detail the latest Pleistocene shifts of plants. Throughout the west, high-elevation conifers lived at much lower elevations during glacial times than today. On the other hand, by Holocene time the late Pleistocene pinyon-juniper woodlands of the southwest were replaced by desert scrub (Betancourt *et al.*, 1990). In most of the northeastern United States, a decline in spruce (*Picea*) and an increase in oak (*Quercus*) are two of the most striking vegetational changes during the transition from the late Pleistocene to the Holocene (Davis & Shaw, 2001; Jackson *et al.*, 1997). This represents the replacement of a largely coniferous woodland with a mixture of hardwoods and conifers, the latter consisting largely of pine (*Pinus*) and/or hemlock (*Tsuga*) rather than spruce. Some of the warmer floral elements, such as hickories (*Carya*) and black gum (*Nyssa*), did not reach New England until several millennia after the oaks. The late Pleistocene fossil plant data from eastern North America also suggest that open wooded habitats, variously called woodlands, parklands, or savannas, were more widespread than the closed-canopy forests that, in the absence of logging, characterize much of the region today. Quite possibly, the open character of these forests reflects the impact of North America's large mammals, especially mammoths and mastodons. Conversely, the present density of pristine forests may represent an abnormal aspect attributable to the missing megafauna (Graham & Grimm, 1990; Owen-Smith, 1989).

The individualistic response resulted in late Pleistocene vertebrate assemblages that, as already pointed out by Hibbard (1960, 1970), frequently consist of extant taxa that do not overlap geographically today. For instance, one finds arctic and boreal voles (*Clethrionomys*, *Phenacomys*, and various species of *Microtus*), lemmings (*Dicrostonyx*, *Lemmus* and *Synaptomys*), and shrews (*Sorex*) associated with temperate forest-dwelling woodrats (*Neotoma*), squirrels (*Sciurus* and *Tamias*), and shrews (*Cryptotis* and *Blarina*), also accompanied by grasslands taxa such as ground squirrels (*Spermophilus*), prairie dogs (*Cynomys*), and pocket gophers (*Thomomys* and/or *Geomys*). Such associations have been called variously "disharmonious," "out-of-step," or "non-analog" faunas. Recent studies have confirmed this by directly ^{14}C-dating individual bones of non-analog taxa from the same stratigraphic horizon (Stafford & Semken, 1990; Stafford *et al.*, 1999). In other words, these faunas represent true paleocommunities reflecting environments that no longer exist (Graham, 1985, 1986; Graham *et al.*, 1987; Semken, 1988; Stafford *et al.*, 1999).

In other instances, ranges of mammalian species move together through time, suggesting that some communities do remain coherent (Lyons, 2003). A good example consists of late Pleistocene sites in the northern Great Plains that contain modern tundra analogs. Of particular interest is the fauna from Prairieburg, Cedar County, Iowa where the most common mammal, *Lemmus sibiricus*, is associated with other tundra taxa that include collared lemming (*Dicrostonyx torquatus*), singing vole (*Microtus miurus*), and arctic ground squirrel (*Spermophilus undulatus*), but no grassland or forest taxa (Foley & Raue, 1987). As expected, such true tundra faunas lived immediately south of the Wisconsinan ice sheet during full-glacial time.

Another small but intriguing set of full-glacial sites in the Great Plains sample the modern prairie fauna. Present analogs for these sites are located well to the north of their recovery location, and support the conclusion of Stewart (1987) that Wisconsinan faunas of the Central and Northern Great Plains are composed of steppe and boreal taxa. Typical examples include the Domebo fauna (Slaughter, 1965) in Oklahoma; the Jones (Davis, 1987), Robert (Schultz, 1967), and Trapshoot (Stewart, 1987) in Kansas; the Lang/Ferguson (Martin, 1987) in South Dakota; and the Agate Basin (Walker, 1987) in Wyoming.

A very pressing question concerns whether the well-studied faunal patterns of the last glacial termination might approximate patterns from earlier glacial cycles. At least one analog of a modern prairie fauna with much greater antiquity has been recognized in the Sandahl local fauna of Irvingtonian age (Semken, 1966). Two well-studied examples from the late Irvingtonian with long stratigraphic records occur in south-central Colorado. These are the sites in the San Luis Valley studied by Rogers *et al.* (1992) and the work of Bell & Barnosky (2000) at Porcupine Cave (Fig. 1), discussed further below. Ongoing analyses reveal non-analog faunas that persisted through glacial/interglacial transitions, even while they experienced substantial extinction of small species (Barnosky *et al.*, 1996). Nonetheless, the Porcupine Cave samples show that numbers of species within trophic levels have not changed, indicating a modicum of stability, at least in these montane settings.

Another rich microvertebrate fauna samples an even earlier interval in the midcontinent. This is the study of Martin *et al.* (2002) detailing the Deer Park fauna in Meade County, Kansas, during its transition from wetter to drier conditions within a glacial interval in the late Blancan. As discussed below, the numbers of rodents, and even the role of arvicolines, were roughly comparable then to the present rodent fauna of southwestern Kansas.

The hypothesized stability of most amphibian and reptile faunas in the North American Quaternary record was recently called into question. Traditionally, the stability of herpetofaunas was seen as their major distinction from later Cenozoic mammalian faunas. "If there is any one thing that characterizes the reptiles of the entire last half of the Cenozoic it is their stability" (Auffenberg & Milstead, 1965, p. 557). Recently, however, Bell & Gauthier (2002) cast serious doubt on this generalization by noting that present geography played a key role in the identification of most Quaternary herp specimens. Now that the circularity of this approach is clear, paleoherpetologists have adopted a more rigorous approach to their identifications, using the principles of phylogenetic systematics.

This more rigorous approach, however, leads to a practical problem in herpetofaunal taxonomic resolution. Using current understanding of morphological synapomorphies, the majority of herpetofaunal elements cannot be identified to species. This was first demonstrated for snakes from Papago Springs Cave in Arizona by Czaplewski *et al.* (1999). More recently, it also became evident in an exhaustive study of morphology and taxonomic utility of the ilium in North American toads of the genus *Bufo* (Bever, 2002). A sustained effort must be made to seek out morphological synapomorphies in the skeletal elements typically recovered from the fossil record. In many cases, this will simply mean re-evaluating traditionally utilized characters in the context of broader phylogenetic comparisons. New synapomorphies have incisively delineated the fossil distribution of various species of horned lizards, genus *Phrynosoma* (Norell, 1989).

These preliminary results are leading some Quaternary paleontologists to reconsider the herpetofaunal stability hypothesis and to undertake fundamental new character analyses. Meanwhile many herps can be paleoecologically informative even without species-level resolution. For example, fossil tortoises in areas north of their modern distribution definitely carry climatic significance (King & Saunders, 1986). The patterns of many such changing Quaternary distributions are indicated by Holman (1995) and by Bell & Gauthier (2002).

Taphonomy, meaning the processes of site formation and modification after burial, is another factor that must be considered in non-analog faunas. As shown by an extensive literature developed during the last two decades (e.g. Behrensmeyer & Hill, 1980; Bonnichsen & Sorg, 1989; Haynes, 1991), mixing processes before and after deposition are major concerns. Before a fossil assemblage can be considered non-analog, the contemporaneity of its members should be demonstrated by ^{14}C dating. In the study by Stafford *et al.* (1999) a considerable number of supposed non-analog faunas failed this test. Other important taphonomic factors are sample size and distribution of elements. Poorly sampled faunas may incorrectly appear to have modern analogs.

Extinctions and Conservation

Abrupt loss of nearly all large late Pleistocene land mammals in the Americas constitutes one of the unresolved mysteries in the history of life. The first concerted effort to search for the cause of this event was an INQUA symposium (Martin & Wright, 1967). The symposium documented the scope and speed of the great megafaunal extinction in North America, with loss of at least 35 genera of large mammals ca. 13,000 cal yr B.P. The roster of losses includes armadillos, glyptodonts, four families of ground sloths, giant beavers, capybaras, lions, pumas, sabertooths, peccaries, llamas, camels, pronghorns, musk-oxen, goats, horses, mammoths, mastodons, and gomphotheres (Alroy, 1999; Graham & Lundelius, 1984; Martin & Steadman, 1999). That symposium found no clear resolution between the two major causal hypotheses, climate change and human hunting. A central position on megafaunal extinctions was that of Kurtén & Anderson (1980, p. 363): "No one cause can account for it; rather a mosaic of adverse conditions prevailed . . . making the [megafauna] vulnerable to environmental pressures, including man, the hunter, who probably delivered the *coup de grace* to some of the megafauna between 12,000 and 9000 years ago." Another major effort to comprehend late Quaternary extinctions (Martin & Klein, 1984) still found this issue unresolved in North America, although on balance the efficacy of human environmental modification had gained many adherents in a global perspective.

While most accounts of late Quaternary extinctions feature the three dozen species of large mammals, it is also important to consider other groups. Although the North American herpetofaunal record does not disclose the loss of many species during the Pleistocene, southward restriction and ultimate loss of giant tortoises, genus *Hesperotestudo*, constitutes one striking herpetofaunal example. It is not generally appreciated that the number of avian generic extinctions in the latest Quaternary is about half of that affecting mammals. Most affected were carrion-feeders such as condors, vultures, storks, eagles, and caracaras, as well as other commensals that depended on large mammals (Steadman & Martin, 1984). Thus, extinction of large mammalian herbivores set off a trophic cascade that also wiped out large carnivores and scavengers, including many kinds of large birds.

Brown (1995) noted that large body size may be associated with extinction episodes via two major causal connections. First, large organisms have fewer individuals on any given landscape, and thus any environmental perturbations are more likely to degrade their numbers below the threshold for survival. Secondly large animals have more severe life history constraints, including low litter size and long intervals between reproduction. Even so, not every large mammal species became extinct. It will be useful to determine if some special life history features permitted bison, black bears, pronghorns, and several kinds of deer, goats, and sheep to survive? Kiltie (1984) suggested that study of breeding schedules and reproductive rates may explain the survival of the bovids and cervids noted above. Fisher (1987, 1996) reads from individual mammoth and mastodont tusks a wealth of information about life histories, the season of death, and possible distinction between human and climatic population pressures. Frison's work (e.g. 1992) with such prey species as mammoth and bison in Paleoindian contexts points an essential direction in which megafaunal population studies must evolve. Reconstruction of life histories in critical fossil populations holds the secret of how each species slides into extinction.

Two primary schools continue to vie for a complete explanation of the late Pleistocene vertebrate extinctions. One of these, as noted above, features the vast and various environmental shifts that coincided with the last deglaciation in North America. The other, emphasizing the role of human hunters, was formally presented by Paul S. Martin in a 1965 INQUA symposium as the "Overkill Hypothesis" (Martin, 1967). This concept was later amended to the "Blitzkrieg Hypothesis" (Martin & Klein, 1984). It makes the testable prediction that the appearance of Paleoindians in North America should be essentially concurrent with the megafaunal extinction. It does not follow, however, that a close correlation in time, now somewhat dubious, would demonstrate a causal relationship.

Computer simulations of North American megafaunal extinction by Alroy (2001) present a perfect challenge for more population studies of extinct megafauna. His analyses go well beyond previous attempts to model Paleoindian populations as they colonized the New World, adding interactions among declining herbivore species and population-level responses of large herbivores as they were presumably hunted. Alroy's model seems quite robust in most parameters, and correctly predicts the timing of the mass extinction, without resorting to local or *ad hoc* environmental scenarios. It is nonetheless a model that needs to be iteratively tested with revised records from FAUNMAP (FAUNMAP Working Group, 1996).

Another extinction hypothesis, articulated by MacPhee & Marx (1997), connects indirectly with the arrival of humans. When Paleoindians entered North America their dogs brought a series of diseases new to the resident mammal fauna. The major difficulty with this hypothesis is that it will be very difficult to test. The search for evidence has concentrated in the frozen ground of Beringia, and it is difficult to see how it can be extended into lower latitudes.

Still another hypothesis that begins with the human impact on mammoths (*Mammuthus*) and mastodonts (*Mammut*) is the "Keystone Species Hypothesis" of Norman Owen-Smith (1989), developing from his extensive experience with African ungulate communities. If human hunters rapidly and repeatedly decimated proboscidean populations, the reduction and ultimate demise of these great (keystone) herbivores would produce a cascade of detrimental changes depriving other herbivores in that ecosystem of their primary resources. This model might be refuted if it were shown that the proboscideans were among the last of the megafauna to go extinct.

Whatever the cause, the terminal Pleistocene extinction of large mammals and birds removed from the Americas vast faunas resembling African game parks. With increasing precision the megafaunal extinction episode is documented at ca. 13,000 cal yr B.P., that is during the last 1% of Quaternary time. Preliminary studies by Stafford *et al.* (1999) precisely dating of the youngest megafauna suggest that the extinctions may have occurred within a few hundred years. That perspective explains why Quaternary vertebrate extinction studies are beginning to merge with conservation studies. Graham & Grimm (1990) point out that studies of past habitat and climate change often illuminate conservationists' studies of present and future change. For example, a re-examination of the paleoecology of the endangered black-footed ferrets (*Mustela nigripes*) shows that they often lived without prairie dogs, contrary to the view (based only on historic observations) that *Cynomys* was their "obligate" prey (Owen *et al.*, 2000). Quaternarists must participate in modeling past and present species ranges, habitats, and genetic makeup to represent the full potential range of niche parameters. Such a combined approach of past and present perspectives gives the best hope for understanding and managing future faunal change.

Isotopes, Ecology, and Diet

Recognition that stable isotopes in bone retain a reliable record of an animal's diet opened a major new field of study illuminating both evolutionary and ecological aspects of ancient vertebrates. Ratios between ^{12}C and ^{13}C in tooth enamel establish the alternative photosynthetic pathways (C_3 or C_4) for different types of plants herbivores have eaten. The following example helps to explain the sympatry between three proboscideans that were the most abundant taxa at the West Palm Beach Site in the late Wisconsinan of south Florida. There Koch *et al.* (1998) showed that 92% of the diet of *Mammuthus columbi* consisted of C_4 grasses, while 90% of that of *Mammut americanum* was C_3 (presumably browse). The third proboscidean, the Neotropical gomphothere (*Cuvieronius tropicus*), consumed 49% C_4 grasses, exactly the intermediate strategy predicted by resource partitioning theory. Similarly, MacFadden *et al.* (1999) employed carbon isotopes to demonstrate a substantial (and partly unexpected) spread among feeding strategies of hypsodont horses in the late Hemphillian Palmetto Fauna of central Florida.

Several isotopes other than C also contribute insights into vertebrate paleoecology. Stable isotopes of N document the degree of carnivory of an animal and whether it has fed on marine organisms. $^{87}Sr/^{86}Sr$ ratios have varied through Earth history, and that ratio in most instances faithfully records the geological substrate inhabited by an organism. For example, the enamel of *Mammut americanum* from Florida registered levels too high to represent the Gulf Coastal Plain sedimentary environments. The only likely explanation is that they migrated northward across the Fall Line to granitic terrain (Hoppe *et al.*, 1999).

Finally, the oxygen isotopic composition of bone is determined by the source and temperature of the water utilized by an animal. Such data can provide a significant key to microclimatic patterns. Rubenstein *et al.* (2002) used oxygen isotopes to link the breeding and wintering ranges of a migratory songbird. Fisher (1987) applied such analysis to successive lamellae in proboscidean tusks to determine their seasonal growth patterns and to determine the season of death. In a broader application, Smith & Patterson (1994) utilized oxygen isotope ratios to assess climatic history of successive fish faunas from lake beds in the Glenns Ferry Formation of Idaho. In another long-term stratigraphic study, Rogers & Wang (2002) traced a late Matuyama shift in climate by proxy evidence extracted from dental enamel of pocket gophers (*Thomomys*).

Phylogenetic Studies and DNA

Phylogenetic studies have advanced substantially in two different modalities. The first involves theoretical advances, such as showing how molecular and morphological data can be integrated and studying patterns of faunal dynamics (Webb & Barnosky, 1985). The second advances have come via methodological innovations, such as cladistic analysis using new software such as PAUP (Phylogenetic Analysis Using Parsimony), and Norell's (1989) work on fossil and recent iguanid lizards. As a result, many of the evolutionary patterns that were interpreted as orthogenetic evolution several decades ago are now seen as bushes with many branches (e.g. Guthrie, 1970, for *Bison*; Steadman, 1980, for turkeys; and Martin & Barnosky, 1993, for a sampler of other groups). Such tree-like patterns are more concordant with our current understanding of the complexity of evolutionary systems in geological time.

The increasing possibility of obtaining and amplifying DNA from late Pleistocene and Holocene fossil bones offers an exciting supplement to more traditional phylogenetic methods used by vertebrate paleontologists. Although such studies are still in their infancy, ancient DNA evidence is profoundly reshaping perceptions of phylogenetic patterns. Quaternary vertebrate populations often offer an ideal halfway house between modern samples and ancient fossil samples. For instance, in her study of pocket gophers (*Thomomys talpoides*), Hadly (1997) shows no change in their genetics in spite of major spatial shifts. Evidently, their body size changes are due entirely to phenotypic plasticity and not to migration. On the other hand voles (*Microtus*) in the same settings show substantial shifts in their DNA.

Genetic studies of modern vertebrates have also helped unravel evolutionary and biogeographic shifts in populations during the rapid environmental changes of the Quaternary. *Puma concolor* extends the entire length of both American continents, thus occupying the largest range of any New World land mammal. It surely lived in North America first, extending its range into South America during the Pliocene-Pleistocene Great American Interchange. Yet the North American *Puma* populations share a remarkably homogeneous genome, in contrast with the several genetically diverse subspecies in Central and South America (Culver *et al.*, 2000). These authors conclude that North America lost its puma population and later was recolonized from a reservoir in eastern South America. Genetics thus demonstrates that the mountain lion is the fifth large cat swept away in North America's late Pleistocene extinction episode, the others being two sabercats (*Smilodon fatalis* and *Dinobastis serus*), the North American cheetah (*Miracinonyx truman*), and the American lion (*Panthera atrox*). It is also possible to include the jaguar (*Panthera onca*) as the sixth large cat species so eclipsed, for it experienced at least a substantial southward range reduction at the end of the Wisconsinan.

DNA from permafrost-preserved bones of brown bears in northwestern North America (eastern Beringia), integrated with modern samples, reveals a very complex history of local extinctions, reinvasions, and possible competition with the extinct short-faced bear (*Arctodus*). "Late Pleistocene histories of mammalian taxa may be more complex than those that might be inferred from the fossil record or contemporary DNA sequences alone" (Barnes *et al.*, 2002, p. 2267).

Morphological studies can be effectively combined with ancient DNA comparisons to better distinguish simple phenotypic shifts from fundamental evolutionary changes. Speciation events may be frequently produced by climatic shifts during the Quaternary, but these require a more direct demonstration of genetic discontinuity than traditional speculative models often present. Klicka *et al.* (2001) clearly demonstrate rapid genomic evolution among buntings (genus *Passerina*) during the Quaternary. Similarly, Martin's (1979) morphological work on populations of cotton rats (genus *Sigmodon*) documents highly punctuated evolution. Many more careful studies combining genetic and morphometric evidence are needed before generalizations about evolutionary rates and patterns are warranted.

Some Regional Highlights

Long stratigraphic sequences with superpositional control provide an ideal basis for working out the continental succession of Pliocene and Pleistocene vertebrate faunas. Fortunately there are about ten such sections, mainly in western North America. These include the Anza-Borrego Desert (Fig. 2) and the San Timoteo Badlands in California; the San Pedro Valley and the Gila River Valley in Arizona and western New Mexico; the Snake River Plain of Idaho; the Ringold Formation in Washington; Meade County, Kansas; and several sections in Nebraska. These sections provide a framework for biostratigraphic and biochronologic correlation of the many other unique fossil localities that have been studied across the United States. Some highlights of current research are presented below by states and regions.

Alaska

Increasingly, Quaternary paleontologists have recognized that the far north, beyond the continental glaciers, must be considered separately from temperate North America. Its denomination as eastern Beringia emphasizes its ecological continuity with western Beringia, especially during glacial intervals when sea level was low. The large mammals of Beringia, dominated by *Equus*, *Bison*, *Rangifer*, and *Mammuthus*, extended all the way into southern France and northern Spain, where they appear in lifelike color on the walls of Lascaux and Altamira (Guthrie, 2001). During the past 30 years, the mammoth steppe has been characterized by Guthrie (1990) and others as a fossil ecosystem quite different from the present tundra, taiga, and muskeg, which cannot support such a vast grazing fauna. An abundance of grass and forbs, along with a climate of drastically limited snowfall, supported immense multispecies herds, of which woolly mammoth was the keystone species. In eastern

Beringia the drainage routes trended westward, directing migratory corridors from Canada into Asia and back.

Only after an ice-free corridor opened southward, at the very end of the Pleistocene, did surviving elements of the Beringian fauna enter temperate North America (including the lower 48 states). Among the wave of new species that appeared in the latest Pleistocene and early Holocene, filtering in from Alaska, were humans, wolves, grizzly bear, elk, moose, and bison. This is a bit confusing in cases where an older sibling species was already present: for example, *Canis dirus* was supplanted by *Canis lupus*, and native *Cervalces* was followed by immigrant *Alces*. It is intriguing to note that a few of the mammoth steppe species had already edged through to ranges south of the great ice sheets, including such wide-ranging herbivores as caribou (*Rangifer*) and muskoxen (*Ovibos*) and also the giant lion (*Felis leo atrox*).

The Prince of Wales Island off southeastern Alaska has an extensive karstland in the Tongass National Forest. Large collections of vertebrates from On Your Knees Cave represent the mid-Wisconsin interstade and the last glacial maximum (Heaton *et al.*, 1996). Sediments of the earlier period yield evidence of denning carnivores, including black bear (*Ursus americanus*), brown bear (*Ursus arctos*), Arctic fox (*Alopex lagopus*), and river otter (*Lontra canadensis*), as well as caribou (*Rangifer tarandus*), a bovid (*Saiga* or *Oreamnos*), and abundant rodents, including hoary marmot (*Marmota caligata*), heather vole (*Phenacomys intermedius*), long-tailed vole (*Microtus longicaudus*), and brown lemming (*Lemmus trimucronatus*). This fauna suggests a variety of local landscapes with the last species indicating expanding tundra conditions at the approach of the last glacial maximum. The best indicator of the latter time is the ringed seal (*Phoca hispida*), which represents the presence of land-fast sea ice. Apparently, foxes scavenged remains of these and also harbor seal (*Phoca vitulina*) and Steller's sea lion (*Eumatopias jubatus*).

Great Basin

The unique basin-and-range physiography of the Great Basin has a clear influence on the geographic distribution of the extant flora and fauna. Furthermore, this geographic setting interacted in a dramatic manner with climatic change at the end of the Pleistocene. Brown (1978) proposed that boreal mammal and bird species extended widely across lowland areas during the late Pleistocene when boreal vegetation extended to lower elevations. As temperatures warmed during the transition to the Holocene, boreal biota retreated to higher elevations where they ultimately became isolated and in some cases suffered local extirpation. There was no possibility of subsequent colonization of these "montane islands" by boreal land mammals from "continental sources" such as the Rocky Mountains or Sierra Nevada. A similar pattern was suggested for birds (Johnson, 1978).

This model sets up predictions that can be tested by examining the fossil record of boreal mammals in the Great Basin. Four predictions are the following: (1) boreal mammals now isolated on Great Basin mountains must have occupied lowland valleys in the past; (2) boreal mammals now found only on certain ranges were present on others in the past; (3) boreal mammals no longer occuring in the Great Basin were there in the past; and (4) no Holocene colonizations by boreal mammals. These predictions have been supported elegantly by analyses of various late Pleistocene and early Holocene mammal faunas from the Great Basin (e.g. Grayson, 1987, 1993; Heaton, 1990, 1999; Mead & Mead, 1989; Mead *et al.*, 1992).

Abundant data derived from such diverse sources as ice cores, speleothems, and lake sediments indicate that similar climatic cycles extend into the Great Basin's past throughout the Quaternary (e.g. Winograd *et al.*, 1992). Thus, the boreal faunal dynamics outlined by Brown (1978) and supported by paleontological studies spanning the Pleistocene/Holocene boundary ought to be extended into earlier ages.

Recent recognition of a Middle Pleistocene (Irvingtonian mammal age) fauna from Cathedral Cave in the eastern Nevada is thus of particular significance (Bell, 1995). The cave entrance has an elevation of 1950 m and opens near the mouth of a steep canyon overlooking the Snake Valley, assuring that fossils within its sediments were derived from a wide range of elevations. Taphonomic vectors probably included raptorial carnivoran mammals, raptorial birds, and woodrats. The latter are rodents of the genus *Neotoma* with a proclivity to incorporate natural objects from their surroundings into their large well-cemented nests. Woodrat nests have provided an extraordinarily rich record of Great Basin biota during the past 40,000 years (Betancourt *et al.*, 1990). Preliminary studies of the extraordinarily rich Cathedral Cave vertebrates indicate that its arvicoline rodent assemblage is essentially identical to that of the Porcupine Cave in Colorado, discussed further below. Evidently Cathedral Cave extends the record of Great Basin faunal dynamics, with cycles of lowland expansion and montane isolation, back to an interval between 750,000 and 850,000 years ago (Bell & Barnosky, 2000).

In the canyon country of Utah dry alcoves provide the first intact boli of *Mammuthus columbi* along with dung blankets representing several other extinct ungulates. These digesta, dating to as recently as 11,670 ^{14}C yr B.P., include plant macrofossils, hair, and insect fauna (Davis *et al.*, 1985).

Idaho

A great wealth of late Pliocene material comes from the Glenns Ferry Formation in the western Snake River Plain. The focus of classic and current research is the local fauna especially the famous "Horse Quarry" where more than 200 individuals of the early horse, *Equus simplicidens*, have been recovered (Fig. 6). Additionally, 112 other species of vertebrates are known, making Hagerman the richest known Blancan fauna. A series of younger faunas, including Froman Ferry and Tyson Ranch, in the Glenns Ferry Formation document the transition to the Irvingtonian land mammal age. In addition to terrestrial faunas, this formation includes extensive fish faunas associated with ancient Lake Idaho (Smith &

Fig. 6. Excavations at Hagerman Horse Quarry, Hagerman, Idaho, 1997.

Patterson, 1994). Subsequently, the modern Snake River drainage was created by capture of that lake by the Columbia River system.

A wealth of later Quaternary faunas are associated with the eastern Snake River Plain, most notably the American Falls fauna of last-interglacial age (Pinsof, 1991). Recently described from a higher elevation near the Wyoming border is the Booth Canyon fauna. Among distinctive features of this fauna is the higher proportion of the extinct muskox, *Bootherium bombifrons*. Finally, from the northern part of the state, at Tolo Lake, is a recent excavation of a large death assemblage (representing more than ten individuals) of Columbian mammoth, *M. columbi*, in association with *Bison antiquus*.

Colorado

A critical new site for middle Pleistocene paleontology in the Rocky Mountain region and adjacent parts of the Great Basin is Porcupine Cave in the central Rocky Mountains of Colorado (Fig. 1). The cave is situated at 2900 m in a large intermontane basin and presumably sampled a wide range of the ambient faunas in that area through the time duration of the accumulated sediments. Age assessments were based on paleomagnetic data and the known temporal ranges of recovered fauna, especially the arvicoline rodents (Bell & Barnosky, 2000). The arvicoline rodent fauna consists of at least seven extinct taxa including *Phenacomys* cf. *gryci*, *Mimomys* cf. *virginianus*, *Mictomys* cf. *meltoni*, *Allophaiomys pliocaenicus*, *Microtus* (=*Terricola*) *meadensis*, *Microtus paroperarius*, and an extinct and extant morphotypes of sagebrush vole, *Lemmiscus*. The estimated age of the middle portion of the Pit Locality is between 750,000 and 850,000 yr (Barnosky, 2003; Bell & Barnosky, 2000). This same suite of arvicoline rodents, presumably representing a unique biochronologic signature, appears in Cathedral Cave, Nevada, discussed above.

Wyoming

Paleontological work in the Yellowstone ecosystem, one of the largest virtually intact temperate ecosystems on Earth, helps bridge the gap between the fossil record and the present. With the time dimension thus expanded, one can develop new insights into how future climates will influence the ecology and evolution of the native western biota. Holocene cave sites, especially Lamar Cave and Waterfall Locality, sample over 80% of the present Yellowstone vertebrate fauna (Hadly, 1999). Lamar Cave contains over 10,000 identified mammal specimens representing 41 species. There are about as many avian specimens, but identifications are still in progress. Besides skeletal remains, isotopic and genetic samples are studied. Some of the findings are as follows:

(1) The late Holocene fauna still exists within 5–7 km of the fossil sites.
(2) Small mammal abundances fluctuate in the same manner as trapping studies today.

(3) Habitat generalists retain more constant population abundances than do habitat specialists during periods of cliimatic change.
(4) Relative abundances in these sites correlate well with estimates of effective populations size derived from genetic variation studies.
(5) Species persist through long periods of time with similar abundances, suggesting stable community assembly rules (within this time frame).
(6) Body sizes of mammals respond to climatic change by increasing during cold periods, while amphibian body sizes get smaller.
(7) Migration rates estimated from genetics are lower than expected in pocket gophers and higher than expected in voles.
(8) Elks and wolves are long-term natives of the Yellowstone fauna.
(9) Well-sampled fossil sites can more comprehensively document the local fauna than many trapping studies today.

Walker (1987) presented the first broad review of late-glacial faunas in the Wyoming Basin based on ~70 localities. The vertebrate samples and their stratigraphic contexts range from the long chronological coverage of Natural Trap Cave in the northern Bighorn Mountains to small faunas associated with four cultural levels in the Agate Basin of northeastern Wyoming. The full-glacial samples are marked by such tundra species as *Ovibos moschatus*, *Rangifer tarandus*, *Ovis canadensis*, and abundant *Dicrostonyx torquatus*, this last species exteding some 2000 km south of its present range. The common large grazers are *Mammuthus columbi*, *Camelops hesternus*, *Equus conversidens*, and *Bison antiquus*. The associated large carnivores, *Arctodus simus* (short-faced bear), *Ursus arctos* (grizzly bear), *Miracinonyx trumani* (American cheetah), and *Felis atrox* (American lion), exhibit adaptations for open habitats and were presumably predators on the large grazers.

South Dakota

In the town of Hot Springs, South Dakota a sediment-filled sinkhole occurs within carbonate deposits of the Spearfish Formation. During the last glaciation, thermal artesian water formed a pond which trapped remains of more than 50 juvenile and young adult males of *Mammuthus columbi*. In addition, 37 other vertebrate species have been recovered including *Arctodus simus*, *Camelops* sp. and *Hemiauchenia macrocephala* (Agenbroad *et al.*, 1990).

Kansas

The Meade Basin in Kansas is once more a critical center of research on the succession of Pliocene and Pleistocene vertebrates, recalling the heyday of Hibbard and his hard-working crews. Critical new evidence of a rich late Blancan microfauna adds a slightly younger facies to the classic Deer Park assemblage (Martin *et al.*, 2002). An abundance of *Baiomys* in Kansas during the Blancan, and its subsequent retreat to Mexico and the southwestern United States, suggests a slightly drier regime during that earlier interval. Minus aquatic species, the number of rodent species (18) from southwestern Kansas some 3 myr ago closely resembled that living there today.

The next-younger sample of the Borchers Badlands fauna boasts critical relationships between superposed localities and dated tephra. The Borchers site lies directly on the Huckleberry Ridge Ash, which has a radiometric date of 2.10 myr and records the last appearances in the region of the dwarf cotton mouse (*Sigmodon minor*) and the giant tortoise (*Geochelone*). An early Irvingtonian fauna from Nash 72 Quarry (Fig. 6), which lies 2.5 m above Huckleberry Ridge ash, records the first-known appearance of *Microtus* in the Meade Basin (Eshelman & Hibbard, 1981). Many fundamental patterns of rodent evolution and systematics seem to be centered in the rich faunal sequence from Meade County, Kansas.

Missouri Valley

In Missouri and Iowa, intensive waterscreening of matrix from 95 Holocene sites, most with cultural associations, has yielded 78 vertebrate species. Archaeological chronology permits subdivision of some sites into 50-yr intervals with a resultant 147 interpretable faunules. Paleoenvironmental analysis of these data suggests that: (1) Holocene conditions started on the western plains and transgressed eastward over a 1000-yr period; (2) the altithermal intensified in western Iowa between 8400 and 6350 yr B.P., but did not affect eastern Iowa forests during this interval; and (3) the drought predicted for the western United States approximately A.D. 1200 also impacted the plains. The faunas associated with sites occupied during this time and into the beginning of the Little Ice Age indicate dominance of Pacific air, a cool dry air mass, over the region. Although rainfall was reduced, decreased evaporation/transpiration partially balanced this effect so that human settlements persisted on the North Central Plains (Semken & Falk, 1987). A number of sites within the Northern Plains contained one or two mammalian species that presently are allopatric by up to 160 km, a configuration that is typical for Holocene local faunas.

Texas

In the southern Great Plains one crosses the Llano Estacado southeasterward onto the Edwards Plateau which is bounded on the east and south by the Balcones Escarpment. The plateau is underlain by flat-lying, mid-Cretaceous limestones in which numerous solution caves record the vertebrate biota of that area over a considerable portion of the Pleistocene. Careful study of more than 40 cave faunas has illuminated patterns of faunal change during the last 25,000 years (Lundelius, 1967). In addition, Fyllan and Kitchen Door caves produce faunas of Irvingtonian age in sediments with reversed remnant magnetism (Holman & Winkler, 1987; Taylor, 1982). The age of the Fyllan Cave Local Fauna is estimated to be about 830,000 yr. The fauna contains a number of extinct taxa including *Dasypus bellus*, *Aztanolagus agilis*, *Atopomys texensis*, *Allophaiomys* sp., and *Ondatra annectans*. It also has at least six extralimital taxa including *Synaptomys cooperi*, *Glaucomys cf. G. volans*, *Notiosorex*, and *Tapirus* sp. The fauna indicates a warm climate more humid that present but without the present climatic extremes.

One of the earliest studied faunas is from Friesenhahn Cave. It provides unusually well-preserved specimens, including articulated skeletons of *Homotherium serum*, *Mylohyus nasutus*, and *Sylvilagus*. In addition, there are skeletons of juvenile (probably neonates) of *Homotherium* as well as complete ontogenetic series of the dentition (Rawn-Schatzinger, 1992). An older unit also contains large numbers of teeth and bones of young *Mammuthus* exhibiting putative tooth marks that may represent the work of *Homotherium*.

Most of the Pleistocene faunas date from the last glacial maximum and contain three groups of species: extinct taxa, extant extralimital taxa, and extant taxa that still occur in the area. Level 3 from Laubach Cave, with a date of $23{,}230 \pm 490\ ^{14}C$ yr B.P. thus precedes the last glacial maximum and contains a few taxa such as *Tremarctos* sp., *Didelphis virginiana*, and *Glyptotherium* sp. not represented in most of the younger faunas (Lundelius, 1985). It also contains the only record of the Mexican free tailed bat, *Tadarida brasiliensis*, older than 2000 yr B.P. (Toomey, 1993).

The Pleistocene faunas of the Edwards Plateau share a great many extinct taxa with faunas of other areas of North America. The common widespread forms include *Canis dirus*, *Mammuthus*, *Paramylodon*, *Megalonyx*, *Smilodon*, antilocaprids, *Bison antiquus*, *Camelops*, *Hemiauchenia*, and *Equus* sp. They lack eastern forms such as *Bootherium*, *Castoroides*, and *Cervalces* and Rocky Mountain taxa such as *Nothrotheriops* and *Euceratherium*. A few taxa such as *Mammut*, *Glyptotherium* and *Chlamytherium*, which are widespread in faunas from the Gulf Coastal Plain and eastern North America, are found only along the eastern edge of the Edwards Plateau. Few examples of western taxa are known from the western edge of the plateau, an exception being *Navahoceras* from Cueva Quebrada in Val Verde County (Lundelius, 1984).

The extant taxa from the Pleistocene faunas show much the same pattern as the extinct ones. Carnivores, many rodents, and insectivores are shared with many faunas from other areas. Extralimital species include *Sorex cinereus*, *Synaptomys cooperi*, *Tamias striatus*, *Microtus pennsylvanicus*, and *Microtus ochrogaster*. The extralimital forms are now found in areas to north and east of the Edwards Plateau that have more mesic climates. To date no species from the Rocky Mountains are known from deposits on the Edwards Plateau.

The deposits in many of the Texas caves contain Holocene deposits with abundant fossils. The longest continuous sequence, from about 17,000 years ago to the present, is in Hall's Cave, Kerr County, Texas (Toomey, 1993). This shows the expected change from mesic to more xeric conditions about 10,000 years ago, with the disappearance of the extinct and many extralimital taxa. It also shows fluctuation in the percentages of several environmentally sensitive species. For example, changing proportions of *Notiosorex* (desert shrew) and *Cryptotis* (little short tailed shrew) indicate a relatively dry period between 11,600 and 10,125 ^{14}C yr B.P. From 10,125 to 9200 ^{14}C yr B.P. conditions were more mesic. This was followed by long period of gradual drying until about 2000 ^{14}C yr B.P. when there was return to conditions similar to present.

Florida

The richest Quaternary faunas in eastern North America come from diverse sedimentary deposits along the eastern Gulf of Mexico and the southern Atlantic Coastal Plain, spanning a broad range of Pliocene and Pleistocene formations. The richest example is an early Irvingtonian fauna from Leisey Shell Pits on the southeastern edge of Tampa Bay. There an extensive bonebed lying between marine shell beds produced some 50,000 catalogued specimens, including the richest early Pleistocene samples of sabercats, mammoths, tapirs, llamas, and many others from more than 100 vertebrate species. Its estimated age of about 1.5 myr is based on integrating mammalian biostratigraphy, marine mollusk biostratigraphy, magnetostratigraphy and Sr-isotope stratigraphy (Hulbert *et al.*, 1995).

Possibly the most diverse late Blancan vertebrate site is Inglis 1A at the mouth of the Withlacoochee River (Figs 7 and 8). Although only a small sinkhole, this site provides the most complete census of the mingled interamerican fauna thus far available. The site evidently sampled a subtropical savanna that extended around the Gulf of Mexico and expanded markedly during a low sea level of the latest Pliocene. The herpetofauna consists of 47 species which indicate "a mixed habitat of mature longleaf pine with xeric hammock interspersed" (Meylan, 1982, p. 67). Many of the species at Inglis 1A still live in similar associations along the belt of relatively xerophytic habitats in the central uplands of Florida. Study of these stable associations over a span of nearly 3 myr can illuminate present conservation efforts in these threatened habitats today.

Florida and the southern Appalachians produce rich samples of fossil birds, several of which exemplify "disharmonious" faunas, like those of small mammals described above. Major changes in distribution during the Quaternary are evident for many extant, non-resident species, including certain hawks, grouse, columbids, and corvids (Emslie, 1998; Lundelius *et al.*, 1983; Parmalee, 1992; Steadman, 2001). A common feature of these displaced avifaunas from the southeast is that they now live well to the north or west, often in grassland or northern coniferous habitats.

From the Plio-Pleistocene of Florida (Blancan through Rancholabrean land mammal ages), Emslie (1998) listed 93 species of non-passerine birds, 24 species of probable resident passerines, and only five species of probable Nearctic/Neotropical migrant passerines, the latter consisting of two thrushes, a catbird, warbler, and bunting.

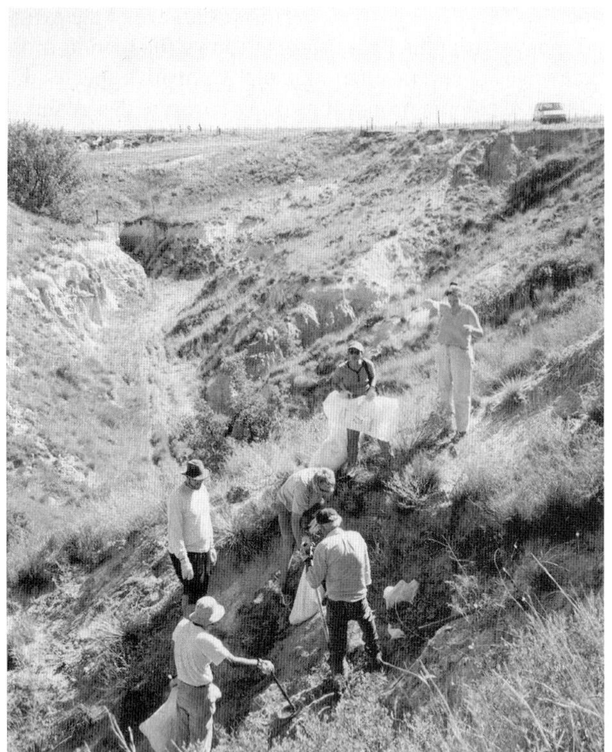

Fig. 7. Nash 72 Quarry, 3.5 m above the Huckleberry Ridge ash, seen in background. Meade County, Kansas, 2000.

Of the 122 species of birds from the Blancan through Rancholabrean of Florida, 21 (17%) are extinct, namely a cormorant, anhinga, stork, teratorn, three condors, duck, three eagles, accipitrid vulture, cracid, turkey, two rails, phorusracid, woodcock, pigeon, and two owls (Emslie, 1998). The six extinct genera from Florida (the teratorn *Teratornis*, condor *Aizenogyps*, duck *Anabernicula*, eagle *Amplibuteo*, accipitrid vulture *Neophronotops*, and phorusrhacid *Titanis*) have all been found elsewhere in North America. Just as is the case with living species today, many of the largest extinct taxa of Pleistocene birds were very widespread.

Appalachian Mountains and Allegheny Plateau

The Appalachian Mountains and Allegheny Plateau have been famous for their caves and associated Quaternary fossil-bearing deposits since the classic studies of Edward Cope and Joseph Leidy. A century passed before a Carnegie Museum team carried out the first detailed excavation of one of these sites, New Paris No. 4 (Guilday *et al.*, 1964). Their stratigraphic transect ranged from latest Wisconsinan through Holocene time. The late Wisconsinan fauna included arctic species such as caribou, yellow-cheeked vole, and collared lemmings, well south of their modern distributions, along with more southern taxa such as peccary, spectacled bear and beautiful armadillo, north of their implied distribution. This and other Appalachian sites also exhibited plains species, including thirteen-lined ground squirrel and plains pocket gopher well east of their present ranges and Rocky Mountain alpine taxa, notably the pika, in the Appalachians. Level-by-level maps, possibly the first paleoecological centerfolds, show the constriction in biogeographical distribution of species up section. The changing vertebrate faunules imply a succession of landscapes from boreal forest parkland to boreal forest, and then, after a quantum reduction in the early Holocene, to mixed forest and finally to the present deciduous forest conditions. Guilday *et al.* (1964) also integrated a level-by-level palynological analysis with paleoenvironmental interpretations based on fossil vertebrates.

Fig. 8. Screenwashing at Inglis 1A, Mouth of Withlacoochee River, Florida, 1974. Photo by Howard C. Converse, Jr.

Outlook Prospectus

There are still many unresolved issues, raised by Hibbard *et al.* (1965), that continue to challenge Quaternary vertebrate paleontologists. First, ^{14}C dating is generally limited to the last 50,000 yr. For deposits and faunas between this age and approximately 200,000 yr ago, there are no techniques that have been accepted as completely reliable for determining age. Electron spin resonance dating of enamel, optically stimulated resonance dating of sand grains, U-series dating of bone, and young age ranges of $^{40}Ar–^{39}Ar$ dating, each may become a more reliable clock for the great range of environments studied by Quaternary vertebrate paleontologists. Perhaps one of these methods will provide the key to accurately dating the boundary between the Irvingtonian and Rancholabrean NALMA.

There are also unresolved questions about the dispersal of species between and within continents. Perhaps the most intriguing concerns the ecogeographic barrier that frequently prohibited faunal continuity between Beringia and temperate latitudes of North America. Paleontologists have not yet identified clearly which mechanisms led to its occasional breakdown, chronicled by the sudden influx of Eurasian boreal beasts into the conterminous United States. Looking southward, reciprocal faunal mingling between North America and South America, and the changing ecological signatures that these migrations represent, require much refinement, especially in the American tropics (Webb & Rancy, 1996).

Finally, the cause of Pleistocene extinctions remains hotly debated (Alroy, 2001; Martin & Klein, 1984). The popular press attempts to capture the debate with three buzzwords, "Chill, kill, and ill." The prospect of fuller resolution depends on developing refined ecological and population studies of individual species and their communities. Only at that level can future extinction models be rigorously tested. The same work projected from the Quaternary into the future will also provide important insights into vertebrate conservation strategies.

Many of these issues point to the need for more complete faunal data from a broader spectrum of time and space. North American vertebrate paleontologists have undertaken FAUNMAP as a partial answer to this need, but there is still a loud call for tighter integration of floristic with faunistic data.

Dedication

This review reminds us that the work of paleontologists, as in many branches of historical science, is cumulative. It prompts us to honor the generations of dedicated scholars and scientists who have laid the foundations upon which we continue to build. Here we dedicate our work to four outstanding Quaternary vertebrate paleontologists who illuminated our paths.

Claude W. Hibbard (Fig. 9) stands out as the most productive Quaternary vertebrate paleontologist in North America, both in the field and in the museum. Björn Kurtén and Elaine Anderson undertook their magnificent synthesis in the 1970s, when the discipline had become perhaps too kaleidoscopic. The resulting volume on "North American Quaternary Mammals" inspired the entire profession and gave it new impetus. John E. Guilday, predecessor in the eastern region, led an extraordinarily productive team to new levels of multidisciplinary environmental reconstruction based on meticulously excavated spelean systems.

Fig. 9. Claude W. Hibbard preparing burlap strips for a plaster jacket, Blancan deposits at Hagerman, Idaho, 1973. Photo by Mick Hagger.

Even as we dedicate this chapter to these four predecessors, we note that there were many other key participants. Kurtén & Anderson (1980) cited 45 institutional collections and 124 professional colleagues as contributors to the success of their enterprise. More than ever, the best practices in Quaternary vertebrate paleontology continue to require collaboration of many scholars with diverse expertise.

References

Agenbroad, L.D., Mead, J.I. & Nelsom, L.W. (Eds) (1990). *Megafauna and man: Discovery of America's heartland.* Mammoth Site of Hot Springs, South Dakota, Inc., Scientific Papers 1, 150 pp.

Alroy, J. (1999). Putting North America's End-Pleistocene megafaunal extinction in context: large-scale analyses of spatial patterns, extinction rates, and size distribution. *In*: MacPhee, R.D. (Ed.), *Extinctions in Near Time: Causes, Contexts and Consequences*. Kluwer Academic/Plenum Publishers, New York, pp. 105–143.

Alroy, J. (2001). A multispecies overkill simulation of the end-Pleistocene megafaunal mass extinction. *Science*, **292**, 1893–1896.

Auffenberg, W. & Milstead, W.W. (1965). Reptiles in the Quaternary of North America. *In*: Wright, H.E., Jr. & Frey, D.G. (Eds), *The Quaternary of the United States: A Review Volume for the VII Congress of the International Association for Quaternary Research*. Princeton University Press, Princeton, NJ, pp. 557–568.

Bachhuber, F.W. (1992). A pre-late Wisconsin paleolimnologic record from the Estancia Valley, central New Mexico. *Geological Society of America Special Paper 270*, pp. 289–307.

Barnes, I., Matheus, P., Shapiro, B., Jensen, D. & Cooper, A. (2002). Dynamics of Pleistocene population extinctions in Beringian brown bears. *Science*, **295**, 2267–2270.

Barnosky, A.D. (Ed.) (2003). *Biodiversity response to environmental change in the middle Pleistocene: The Porcupine cave fauna from Colorado*. University of California Press, Berkeley, CA (in press).

Barnosky, A.D., Rouse, T.I., Hadly, E.A., Wood, D.L., Keesing, F.L. & Schmidt, V.A. (1996). Comparison of mammalian response to glacial-interglacial transitions in the middle and late Pleistocene. *In*: Stewart, K.M. & Seymour, K.L. (Eds), *Palaeoecology and Palaeoenvironments of Late Cenozoic Mammals: Tributes to the Career of C.S. (Rufus) Churcher*. University of Toronto Press, pp. 16–33.

Becker, J.J. (1987). *Neogene avian localities of North America*. Smithsonian Institution Press, Washington, DC.

Bell, C.J. (1995). A middle Pleistocene (Irvingtonian) microtine rodent fauna from White Pine County, Nevada, and its implication for microtine rodent biochronology. *Journal of Vertebrate Paleontology*, **15**, 18A.

Bell, C.J. & Barnosky, A.D. (2000). The microtine rodents from the Pit Locality in Porcupine Cave, Park County, Colorado. *Annals of Carnegie Museum*, **69**, 93–134.

Bell, C.J. & Gauthier, J.A. (2002). North American Quaternary Squamata: Re-evaluation of the stability hypothesis. *Journal of Vertebrate Paleontology*, **22**(Suppl. 3), 35A.

Bell, C.J., Lundelius, E.L., Jr., Barnosky, A.D., Graham, R.W., Lindsay, E.H., Ruez, D.R., Jr., Semken, H.A., Jr., Webb, S.D. & Zakrzewski, R.J. (in press). The Blancan, Irvingtonian,and Rancholabrean land mammal ages. *In*: Woodburne, M.O. (Ed.), *Late Cretaceous and Cenozoic Mammals of North America: Geochronology and Biostratigraphy*. Columbia University Press.

Betancourt, J.L., Van Devender, T.R. & Martin, P.S. (Eds) (1990). *Packrat middens: The last 40,000 years of biotic change*. University of Arizona Press, Tucson, 469 pp.

Bever, G.S. (2002). Variation in the anuranilium and its implications for species-level identification of fragmentary fossil specimens. *Journal of Vertebrate Paleontology*, **22**(Suppl. 3), 36A.

Behrensmeyer, A.K. & Hill, A.P. (Eds) (1980). *Fossils in the making*. University of Chicago Press, Chicago, pp. 1–338.

Blair, W.F. (1965). Amphibian speciation. *In*: Wright, H.E., Jr. & Frey, D.G. (Eds), *The Quaternary of the United States: A Review Volume for the VII Congress of the International Association for Quaternary*. Princeton University Press, Princeton, NJ, pp. 543–556.

Bonnichsen, R. & Sorg, M.H. (1989). *Bone modification*. Center for the Study of First Americans Institute for Quaternary Studies, University of Maine, Orono.

Brown, J.H. (1978). The theory of insular biogeography and the distribution of boreal birds and mammals. *Great Basin Naturalist Memoirs*, **2**, 209–227.

Brown, J.H. (1995). *Macroecology*. University of Chicago Press, Chicago, 269 pp.

Cassiliano, M.L. (1999). Biostratigraphy of Blancan and Irvingtonian mammals in the Fish Creek-Vallecito Creek section, Southern California, and a review of the Blancan-Irvingtonian boundary. *Journal of Vertebate Paleontology*, **19**, 169–186.

Culver, M., Johnson, W.E., Pecon-Slattery, J. & O'Brien, S.J. (2000). Genomic ancestry of the American Puma (*Puma concolor*). *Journal of Heredity*, **91**, 186–197.

Czaplewski, N.J., Mead, J.I., Bell, C.J., Peachey, W.D. & Ku, T.-L. (1999). Papago Springs Cave revisited, Part II: Vertebrate paleofauna. *Occasional Papers of the Oklahoma Museum of Natural History*, **5**, 1–41.

Davis, L.C. (1987). Late Pleistocene/Holocene environmental changes in the Central Great Plains of the United States: The mammalian record. *In*: Graham, R.W., Semken, H.A., Jr. & Graham, M.A. (Eds), *Late Quaternary Mammalian Biogeography and Environments of the Great Plains and Prairies*. Illinois State Museum Scientific Papers, Springfield, IL, Vol. 22, pp. 88–143.

Davis, M.B. & Shaw, R.G. (2001). Range shifts and adaptive responses to Quaternary climate change. *Science*, **292**, 673–679.

Davis, O.K., Mead, J.I., Martin, P.S. & Agenbroad, L.D. (1985). Riparian plants were a major component of the diet of mammoths of southern Utah. *Current Research in the Pleistocene*, **2**, 81–82.

Emslie, S.D. (1998). Avian community, climate, and sea-level changes in the Plio-Pleistocene of the Florida peninsula. Ornithological Monographs No. 50. American Ornithologists' Union, Washington, DC.

Eshelman, R.E. & Hibbard, C.W. (1981). Nash Local Fauna (Pleistocene: Aftonian) of Meade County, Kansas: Contributions of Museum of Paleontology, University of Michigan, Vol. 25, pp. 317–326.

FAUNMAP Working Group (1994). FAUNMAP: A database documenting late Quaternary distributions of mammal species in the United States: Illinois State Museum Scientific Papers XXV, Vol. 1.

FAUNMAP Working Group (1996). Spatial response of mammals to late Quaternary environmental fluctuations. *Science*, **272**, 1601–1606.

Fisher, D.C. (1987). Mastodont procurement by Paleoindians of the great Lakes Region: Hunting or scavenging? *In*: Nitecki, M.H. & Nitecki, D.V. (Eds), *Evolution of Human Hunting*. Plenum Press, New York, pp. 309–421.

Fisher, D.C. (1996). Proboscidean extinctions in North America. *In*: Shoshani, J. & Tassy, P. (Eds), *The Proboscidea: Evolution and Paleoecology of Elephants and their Relatives*. Oxford Universisty Press, Oxford, pp. 296–315.

Foley, R.L. & Raue, L.E. (1987). *Lemmus sibiricus* from the late Quaternary of the midwestern United States. *Current Research in the Pleistocene*, **4**, 105–106.

Frison, G.C. (1992). The Foothills-Mountains and the Open Plains: The dichotomy in Paleoindian subsistence strategies between two Ecosystems. *In*: Stanford, D.J. & Day, J.S. (Eds), *Ice Age Hunters of the Rockies*. University of Colorado Press, Boulder.

Graham, A. (1999). *Late cretaceous and cenozoic history of North American vegetation: North of Mexico*. Oxford University Press, Oxford, UK.

Graham, R.W. (1985). Response of mammalian communities to environmental Changes during the late Quaternary. *In*: Diamond, J. & Case, T.J. (Eds), *Community Ecology*. Harper and Row, New York, pp. 300–313.

Graham, R.W. (1986). Plant-animal interactions and Pleistocene extinctions. *In*: Elliot, D.K. (Ed.), *The Dynamics of Extinction*. John Wiley and Sons, New York, pp. 131–154.

Graham, R.W. & Grimm, E.C. (1990). Effects of global climatic change on the patterns of terresterial biological communities. *Trends in Ecology and Evolution*, **5**, 289–292.

Graham, R.W. & Lundelius, E.L., Jr. (1984). Coevolutionary disequilibrium and Pleistocene extinctions. *In*: Martin, P.S. & Klein, R.G. (Eds), *Quaternary Extinctions: A Prehistoric Revolution*. University of Arizona Press, Tucson, 892 pp., 223–249.

Graham, R.W., Semken, H.A., Jr. & Graham, M.A. (Eds) (1987). *Late quaternary mammalian biogeography and environments of the great plains and prairies*. Illinois State Museum Scientific Papers, Vol. 22, Springfield, IL, 491 pp.

Grayson, D.K. (1987). The biogeographic history of small mammals in the Great Basin: Observations on the last 20,000 years. *Journal of Mammalogy*, **68**, 359–375.

Grayson, D.K. (1993). The desert's past: A natural prehistory of the Great Basin. Smithsonian Institution Press, Washington, 356 pp.

Guilday, J.E., Martin, P.S. & McCrady, A.D. (1964). New Paris No. 4: A Late Pleistocene Cave Deposit in Bedford County, Pennsylvania. *National Speleological Society Bulletin*, **26**, 121–194.

Guthrie, R.D. (1970). Bison evolution and zoogeography during the Pleistocene. *Quarterly Review of Biology*, **45**, 1–15.

Guthrie, R.D. (1982). Mammals of the Mammoth Steppe as paleoenvironmental indicators. *In*: Hopkins, D.M., Matthewes, J.V., Jr., Schweger, C.E. & Young, S.B. (Eds), *Paleoecology of Beringia*. Academic Press, Inc., NY, pp. 307–329.

Guthrie, R.D. (1990). *Frozen fauna of the mammoth steppe*. University of Chicago Press, Chicago, 298 pp.

Guthrie, R.D. (2001). Origin and causes of the mammoth steppe: a cloud cover, woolly mammal tooth pits, buckles and inside-out Beringia. *Quaternary Science Reviews*, **20**, 549–574.

Hadly, E.A. (1997). Evolutionary and ecological response of pocket gophers (Thomomys talpoides) to late Holocene climate change. *Biological Journal of the Linnean Society*, **60**, 277–296.

Hadly, E.A. (1999). Fidelity of terrestrial vertebrate fossils to a modern ecosystem. *Palaeogeography, Palaeoclimatology, Palaeoecology*, **149**, 389–409.

Haynes, G. (1991). *Mammoths, mastodonts, and elephants – biology, behavior, and the fossil record*. Cambridge University Press, Cambridge.

Heaton, T.H. (1990). Quaternary mammals of the Great Basin: Extinct giants, Pleistocene relicts, and recent immigrants. *In*: Ross, R.M. & Allmon, W.D. (Eds), *Causes of Evolution: A Paleontological Perspective*. The University of Chicago Press, Chicago, pp. 422–465.

Heaton, T.H. (1999). Late Quaternary vertebrate history of the Great Basin. *In*: Gillette, D.D. (Ed.), *Vertebrate Paleontology in Utah*. Utah Geological Survey Miscellaneous Publication 99–1, pp. 501–507.

Heaton, T.H., Talbot, S.L. & Shields, G.F. (1996). An ice age refugium for large mammals in the Alexander Archipelago, southwestern Alaska. *Quaternary Research*, **46**, 186–192.

Hibbard, C.W. (1949). Pleistocene vertebrate paleontology in North America. *Geological Society of America Bulletin*, **60**, 1417–1428.

Hibbard, C.W. (1960). An interpretation of Pliocene and Pleistocene climates in North America. *Annual Report, Michigan Academy of Science, Arts and Letters*, **62**, 5–30.

Hibbard, C.W. (1970). Pleistocene mammalian local faunas from the Great Plains and central lowland provinces of the United States. *In*: Dort, W., Jr. & Jones, J.K., Jr. (Eds), *Pleistocene and Recent Environments of the Central Great Plains*. Department of Geology, University of Kansas Special Publication 3. Lawrence, 433 pp., 395–433.

Hibbard, C.W., Ray, C.E., Savage, D.E. Taylor, D.W. & Guilday, J.E. (1965). *Quaternary Mammals of North America*. *In*: Wright, H.E., Jr. & Frye, D.G. (Eds), *The Quaternary of the United States*. Princeton Univ. Press, Princeton, NJ, 922 pp., 509–525.

Holman, J.A. (1995). *Pleistocene amphibians and reptiles in North America*. Oxford Monographs on Geology and Geophysics 32. Oxford University Press, New York, 243 pp.

Holman, J.A. & Winkler, A.J. (1987). *A mid-Pleistocene (Irvingtonian) herpetofauna from a cave in southcentral Texas*. Pearce-Sellards Series, Texas Memorial Museum, the University of Texas at Austin, Vol. 44, pp. 1–17.

Hoppe, K.A., Koch, P.L., Carlson, R.W. & Webb, S.D. (1999). Tracking mammoths and mastodons: Reconstruction of migratory behavior using strontium istotope ratios. *Geology*, **27**, 439–442.

Hulbert, R.C., Jr., Morgan, G.S. & Webb, S.D. (Eds) (1995). Paleontology and Geology of the Leisey Shell Pits, Early Pleistocene of Florida. *Florida Museum of Natural History Bulletin*, **37**, 1–660.

Jackson, S.T., Overpeck, J.T., Webb, T., III, Keattch, S.E. & Anderson, K.H. (1997). Mapped plant-microfossil and pollen records of late Quaternary vegetation change in eastern North America. *Quaternary Science Reviews*, **16**, 1–70.

Johnson, N.K. (1978). Patterns of avian geography and speciation in the intermountain region. *Great Basin Naturalist Memoirs*, **2**, 137–215.

Johnson, N.M., Officer, C.B., Opdyke, N.D., Woodard, G.D., Zeither, P.K. & Lindsay, E.H. (1983). Rates of late Cenozoic tectonism in the Vallecito-Fish Creek Basin, western Imperial Valley, California. *Geology*, **11**, 664–667.

Jones, D.S., MacFadden, B.J., Webb, S.D., Mueller, P.A., Hodell, D.A. & Cronin, T.M. (1991). Integrated Geochronology of a classic Pliocene fossil site in Florida: Linking marine and terrestrial biochronologies. *Journal of Geology*, **99**, 637–648.

Kiltie, R.A. (1984). Seasonality, gestation time and large mammal extinctions. *In*: Martin, P.S. & Klein, R.G. (Eds), *Quaternary Extinctions: A Prehistoric Revolution*. University of Arizona Press, Tucson, 892 pp., 250–258.

King, J.E. & Saunders, J.J. (1986). *Geochelone* in Illinois and the Illinoian-Sangamonian vegetation of the type region. *Quaternary Research*, **25**, 89–99.

Klicka, J., Fry, A.J., Zink, R.M. & Thompson, C.W. (2001). A cytochrome-*b* perspective on *Passerina* bunting relationships. *Auk*, **118**, 611–623.

Koch, P.L., Hoppe, K.A. & Webb, S.D. (1998). The Isotopic ecology of late Pleistocene mammals in North America. Part 1: Florida. *Chemical Geology*, **152**, 119–138.

Kurtén, B. & Anderson, E. (1980). *Pleistocene mammals of North America*. Columbia University Press, New York, 442 pp.

Lundelius, E.L., Jr. (1984). A late Pleistocene mammalian fauna from Cueva Quebrada, Val Verde County, Texas. *Special Publication of the Carnegie Museum of Natural History*, **8**, 456–481.

Lundelius, E.L., Jr. (1985). Pleistocene vertebrates from Laubach Cave. *In*: Woodruff, C.M. *et al.* (Eds), *Edwards Aquifer-northern segment, Travis, Williamson and Bell Counties*. Austin Geological Society Guidebook, Texas, Vol. 8, pp. 41–45.

Lundelius, E.L., Jr. (1967). Late-Pleistocene and Holocene faunal history of central Texas. *In*: Martin, P.S. & Wright, H.E., Jr. (Eds), *Pleistocene Extinctions – The Search for a Cause*. Proceedings of the VII congress of the International Association for Quaternary Research, Yale University Press, New Haven, Vol. 6, pp. 287–319.

Lundelius, E.L., Jr., Graham, R.W., Anderson, E., Guilday, J., Holman, J.A., Steadman, D.W. & Webb, S.D. (1983). Terrestrial vertebrate faunas. *In*: Porter, S.C. (Ed.), *Late-Quaternary Environments of the United States. Volume 1, The Late Pleistocene*. University of Minnesota Press, Minneapolis, pp. 311–353.

Lyons, S.K. (2003). A quantitative Assessment of the range shifts of Pleistocene Mammals. *Journal of Mammalogy*, **84**.

MacFadden, B.J., Solounias, N. & Cerling, T.E. (1999). Ancient diets, ecology, and extinction of 5-million-year-old horses from Florida. *Science*, **283**, 824–827.

MacPhee, R.D.E. & Marx, P.A. (1997). The 40,000-year plague: humans, hyperdisease, and first contact extinctions. *In*: Goodman, S. & Patterson, B. (Eds), *Natural Change and Human Impact in Madagascar*. Smithsonian Institution Press, Washington, DC, pp. 169–217.

McDonald, J.N. & Morgan, G.S. (1999). The appearance of Bison in North America. *Current Research in the Pleistocene*, **16**, 127–129.

Martin, J.E. (1987). Paleoenvironment of the Lange-Ferguson Clovis kill site in the badlands of South Dakota. *In*: Graham, R.W., Semken, H.A., Jr. & Graham, M.A. (Eds), *Late Quaternary Mammalian Biogeography and Environments of the Great Plains and Prairies*. Illinois State Museum Scientific Papers, Springfield, Vol. 22, 491 pp. 314–333.

Martin, L.D. & Hoffmann, R.S. (1987). Pleistocene faunal Provinces and Holocene biomes of the central Great Plains. *In*: Johnson, W.C. (Ed.), *Quaternary Environments of Kansas*. Kansas Geological Survey Guidebook Series 5, Lawrence, Kansas, pp. 159–165.

Martin, P.S. (1967). Prehistoric overkill. *In*: Martin, P.S. & Wright, H.E., Jr. (Eds), *Pleistocene Extinctions: The Search for a Cause*. Yale University Press, New Haven, pp. 75–120.

Martin, P.S. & Wright, H.E., Jr. (Eds) (1967). *Pleistocene extinctions – The search for a cause*. Vol. 6 of the Proceedings of the VII Congress of the International Association for Quaternary Research. Yale University Press, New Haven.

Martin, P.S. & Klein, R.G. (Eds) (1984). *Quaternary extinctions: A prehistoric revolution*. University of Arizona Press, Tucson, 892 pp.

Martin, P.S. & Steadman, D.W. (1999). Prehistoric extinctions on islands and continents. *In*: MacPhee, R.D. (Ed.), *Extinctions in Near Time: Causes, Contexts and Consequences*. Kluwer Academic/Plenum Publishers, New York, pp. 17–55.

Martin, R.A. (1979). Fossil history of the rodent genus Sigmodon. *Evolutionary Monographs*, **2**, 1–36.

Martin, R.A. & Barnosky, A.D. (Eds) (1993). *Morphological change in quaternary mammals of North America*. University of Cambridge Press Syndicate, Cambridge, 413 pp.

Martin, R.A., Honey, J.G., Pelaez-Campomanes, P., Goodwin, H.T., Baskin, J.A. & Zakrewski, R.J. (2002). Blanacan lagomorphs and rodents of the Deer Park assemblages, Meade County, Kansas. *Journal of Paleontology*, **76**, 1072–1090.

Mead, E.M. & Mead, J.I. (1989). Snake Creek Burial Cave and a review of the quaternary mustelids of the Great Basin. *The Great Basin Naturalist*, **49**, 143–154.

Mead, J.I., Bell, C.J. & Murray, L.K. (1992). *Mictomys borealis* (northern bog lemming) and the Wisconsin paleoecology of the east-central Great Basin. *Quaternary Research*, **37**, 229–238.

Meylan, P.A. (1982). The squamate Reptiles of Inglis IA fauna, Citrus County, Florida. *Florida State Museum Bulletin, Biological Sciences*, **27**, 1–85.

Norell, M.A. (1989). Late Cenozoic lizards of the Anza Borrego Desert, California. *Natural History Museum of Los Angeles County Contributions in Science*, **414**, 1–31.

Olson, S.L. (1989). Aspects of global avifaunal dynamics during the Cenozoic. *Acta XIX Congressus Internationalis Ornithologici*, **2**, 2023–2029.

Olson, S.L. (2001). Why so many kinds of passerine birds? *Bioscienc*, **51**, 268–269.

Owen, P.R., Bell, C.J. & Mead, E.M. (2000). Fossils, diet and conservation of black-footed Ferrets (*Mustela nigripes*). *Journal Mammalogy*, **81**, 422–433.

Owen-Smith, N. (1989). Pleistocene extinctions: the pivotal role of megaherbivores. *Paleobiology*, **13**, 351–362.

Parmalee, P.W. (1992). A late Pleistocene avifauna from northwestern Alabama. *Natural History Museum Los Angeles County Science Serial*, **36**, 307–318.

Pinsof, J.D. (1991). A cranium of *Bison alaskensis* (Mammalia: Artiodactyla) and comments on fossil Bison diversity within the American Falls area, southeastern Idaho. *Journal Vertebrate Paleontology*, **11**, 509–514.

Rawn-Schatzinger, V. (1992). The scimitar cat *Homotherium serum* Cope: Osteology, functional morphology, and predatory behavior. *Illinois State Museum Reports of Investigations*, **47**, 1–80.

Repenning, C.A. (1987). Biochronology of the microtine rodents of the United States. *In*: Woodburne, M.O. (Ed.), *Cenozoic Mammals of North America: Geochronology and Biostratigraphy*. University of Califoria Press, Berkeley, pp. 236–268.

Rogers, K.L., Larson, E.E., Smith, G., Katzman, D., Smith, G.R., Cerling, T., Wang, Y., Baker, R.G., Lohmann, K.C., Repenning, C.A., Patterson, P. & Mackie, G. (1992). Pliocene and Pleistocene geologic and climatic evolution in the San Luis Valley of south-central Colorado. *Palaeogeography, Palaeoclimatology, Palaeoecology*, **94**, 55–86.

Rogers, K.L. & Wang, Y. (2002). Stable isotopes in pocket gopher teeth as evidence of a late Matuyama climate shift in the southern Rocky Mountains. *Quaternary Research*, **57**, 200–207.

Rubenstein, D.R., Chamberlain, C.P., Holmes, R.T., Ayers, M.P., Waldbauer, J.R., Graves, G.R. & Tuross, N.C. (2002). Linking breeding and wintering ranges of a migratory songbird using stable isotopes. *Science*, **295**, 1062–1065.

Ruddiman, W.F. & Wright, H.E., Jr. (Eds) (1987). *North America and adjacent oceans during the last deglaciation*. The Geology of North America, Geological Society of America, Vol. K-3.

Savage, D.E. (1951). Late Cenozoic vertebrates of the San Francisco Bay region: University of California Publications. *Bulletin of the Department of Geological Sciences*, **28**, 215–314.

Schultz, G.E. (1967). Four superimposed Late-Pleistocene vertebrate faunas from southwest Kansas. *In*: Martin, P.S. & Wright, H.E., Jr. (Eds), *Pleistocene Extinctions: The Search for a Cause*. Yale University Press, New Haven, pp. 321–348.

Semken, H.A., Jr. (1966). *Stratigraphy and paleontology of the McPherson Equus beds (Sandahl Local Fauna), McPherson County, Kansas*. Contributions Museum Paleontology, University of Michigan, Vol. 20, pp. 121–178.

Semken, H.A., Jr. (1988). Environmental interpretation of the "disharmonious" late Wisconsinan Biome of Southeastern North America. *In*: Laub, R.S., Miller, N.G. & Steadman, D.W. (Eds), *Late Pleistocene and Early Holocene Paleoecology and Archaeology of the Eastern Great Lakes Region*. Bulletin Buffalo Society of Natural Sciences, Vol. 33, 316 pp. 185–194.

Semken, H.A., Jr. & Falk, C.R. (1987). Late Pleistocene/Holocene mammalian faunas and environmental changes on the Northern Plains of the United States. *In*: Graham, R.W., Semken, H.A., Jr. & Graham, M.A. (Eds), *Late Quaternary Mammalian Biogeography and Environments of the Great Plains and Prairies*. Illinois State Museum Scientific Papers 22, Springfield, IL, pp. 176–313.

Slaughter, R.E. (1965). The vertebrates of the Domebo local fauna, Pleistocene of Oklahoma. *In*: Leonhardy, F.C. (Ed.), *A Paleo-Indian Mammoth Kill in the Prairie-Plains*. Contributions of the Museum of the Great Plains, Lawton, Vol. 1, pp. 31–35.

Smith, G.R. & Patterson, W.P. (1994). Mio-Pliocene seasonality on the Snake River Plain: comparison of faunal and oxygen isotopic evidence. *Palaeogeography, Palaeoclimatology, Palaeoecology*, **107**, 291–302.

Stafford, T.W., Jr. & Semken, H.A., Jr. (1990). Accelerator ^{14}C dating of two micromammal species representative of the late Pleistocene disharmonious fauna from Peccary Cave, Newton County, Arkansas. *Current Research in the Pleistocene*, **7**, 129–132.

Stafford, T.W., Jr., Duhamel, R.C., Haynes, C.V., Jr. & Brendel, K. (1982). Isolation of proline and hydroxyproline from fossil bone. *Life Sciences*, **31**, 931–938.

Stafford, T.W., Jr., Brendel, K. & Duhamel, R.C. (1988). Radiocarbon, ^{13}C and ^{15}N analysis of fossil bone: Removal of humates with XAD-2 resin. *Geochemica et Cosmochemica Acta*, **52**, 2257–22667.

Stafford, T.W., Jr., Hare, P.E., Currie, L., Jull, A.J.T. & Donahue, D.J. (1991). Accelerator radiocarbon dating at the molecular level. *Journal of Archaeological Science*, **18**, 35–72.

Stafford, T.W., Jr., Semken, H.A., Jr., Graham, R.W., Klippel, W.F., Markova, A., Smirnov, N.G. & Southon, J. (1999). First accelerator mass spectrometry ^{14}C dates documenting contemporaneity of nonanalog species in late Pleistocene mammal communities. *Geology*, **27**, 903–906.

Steadman, D.W. (1980). A review of the osteology and paleontology of Turkeys (Aves: Meleagridinae). *Natural History Museum Los Angeles County Contribution Science*, **330**, 131–207.

Steadman, D.W. (1981). A re-examination of *Palaeostruthis hatcheri* (Shufeldt), a late Miocene sparrow from Kansas. *Journal of Vertebrate Paleontology*, **1**, 171–173.

Steadman, D.W. (2001). A long-term history of terrestrial birds and mammals in the Chesapeake/Susquehanna drainage. *In*: Curtin, P.D., Brush, G.S. & Fisher, G.W. (Eds), *Discovering the Chesapeake: The History of an*

Ecosystem. Johns Hopkins University Press, Baltimore, MD, pp. 83–108.

Steadman, D.W. & Martin, P.S. (1984). Extinction of birds in the late Pleistocene of North America. *In*: Martin, P.S. & Klein, R.G. (Eds), *Quaternary Extinctions: A Prehistoric Revolution*. University of Arizona Press, Tucson, pp. 466–477.

Stewart, J.D. (1987). *Latitudinal effects in Wisconsinan mammalian faunas of the plains*. Kansas Geological Survey, Guidebook Series 5, pp. 153–158.

Szabo, B.J. (1985). Uranium-series dating of fossil corals from marine sediments of southeastern United States Atlantic Coastal Plain. *Geological Society America Bulletin*, **96**, 398–406.

Taylor, A.J. (1982). The mammalian fauna from the Mid-Irvingtonian Fyllan Cave local fauna, Travis County, Texas. Masters Thesis, University of Texas at Austin, 106 pp.

Toomey, R. (1993). Late Pleistocene and Holocene faunal and environmental changes at Hall's Cave, Kerr County, Texas. Ph.D. Dissertation, University of Texas at Austin. 2 vols.

Walker, D.N. (1987). Late Pleistocene/Holocene environmental changes in Wyoming: The mammalian record. *In*: Graham, R.W., Semken, H.A., Jr. & Graham, M.A. (Eds), *Late Quaternary Mammalian Biogeography and Environments of the Great Plains and Prairies*. Illinois State Museum Scientific Papers 22, Springfield, IL, pp. 334–393.

Webb, S.D. (1991). Ecogeography and the Great American Interchange. *Paleobiology*, **17**, 266–280.

Webb, S.D. & Barnosky, A.D. (1985). Faunal dynamics of quaternary mammals. *Annual Review Earth and Planteary Sciences*, **17**, 413–439.

Webb, S.D. & Rancy, A. (1996). Late Cenozoic evolution of the Neotropical mammal fauna. *In*: Jackson, J.B.C., Budd, A.F. & Coates, A.G. (Eds), *Evolution and Environment in Tropical America*. University of Chicago Press, Chicago, pp. 335–358.

Winograd, I.J., Coplen, T.B., Landwehr, J.M., Riggs, A.C., Ludwig, K.R., Szabo, B.J., Kolesar, P.T. & Revesz, K.M. (1992). Continuous 500,000-year climate record from vein calcite in Devils Hole, Nevada. *Science*, **258**, 255–260.

Woodburne, M.O. (1987). *Cenozoic mammals of North America – Geochronology and biostratigraphy*. University of California Press, Berkeley.

Wright, H.E., Jr. & Frye, D.G. (Eds) (1965). *The quaternary of the United States*. Princeton University Press, Princeton, NJ, 922 pp.

Zachos, J., Pagani, M., Sloan, L., Thomas, E. & Billups, K. (2001). Trends, rhythms, and aberrations in global climate 65 Ma to Present. *Science*, **292**, 686–693.

Peopling of North America

David J. Meltzer

Department of Anthropology, Southern Methodist University, Dallas, TX 75275-0336, USA

Introduction

In the nearly forty years since INQUA last convened in the United States, our knowledge of the peopling of North America has expanded dramatically in some areas – less so in others. Indeed, some of the questions unanswered then remain unanswered now, despite an increase in the number of sites, a battery of new and sophisticated methodological tools and analytical techniques brought to bear on the problem, and the contribution of disciplines like molecular biology which, in 1965, might have seemed irrelevant to prehistory. Nonetheless, much has been learned, and though the peopling process is now better understood, it is also proving to have been far more complicated than once thought. In order to assess where matters stand today regarding the peopling of North America, it is instructive to first summarize where they were when INQUA was here in 1965.

Peopling of North America – ca. 1965

In 1965, the chronology of the Clovis occupation – then the earliest secure archaeological presence in North America – was just coming into focus (Haynes, 1964). Half a dozen Clovis sites from the Great Plains and Southwest had been radiocarbon dated (Lehner, Dent, Clovis, Naco, Domebo, Ventana Cave), and with one exception all fell in the brief period between 11,500 and 11,000 ^{14}C yr B.P. – here, as elsewhere, all ages are in uncalibrated radiocarbon years (Haynes, 1964; Stephenson, 1965). Clovis and Clovis-like fluted points and tools were found in other parts of North America, especially in eastern North America. But these were mostly surface sites lacking suitable or sufficient material for radiocarbon dating (Griffin, 1965). Still, their typological similarity to Clovis assemblages in the west (Williams & Stoltman, 1965), along with their position relative to Pleistocene landscape features (such as the beaches of proglacial lakes [e.g. Mason, 1958; Quimby, 1958]) suggested these were the same age as their western counterparts – if not slightly older (Griffin, 1965, p. 657).

This apparently sudden appearance of Clovis across the continent ~11,500 ^{14}C yr B.P. coincided neatly with geological evidence then emerging that indicated that just a few centuries earlier an "Ice free" corridor opened along the eastern flank of the Canadian Rockies between the Laurentide and Cordilleran ice sheets. That corridor connected Alaska with the rest of continental United States, perhaps for the first time in 15,000 years, suggesting the first Americans arrived in the conterminous United States fast on the heels of glacial retreat (Haynes, 1964; Martin, 1973; cf. Wendorf, 1966).

Precisely where these groups originated was unknown. There had long been consensus that they must have come out of Asia (e.g. Boas, 1912), and as early as the 1930s a dry land route into America had been identified – the Bering Land Bridge, itself the subject of an important INQUA session in 1965 (Hopkins [Ed.], 1967). But, as of 1965, no Clovis points had been found in western Beringia (northeast Asia west to the Lena River). Fluted points had been recovered in Alaska, but their ages were proving notoriously difficult to pin down, and already there was a dispute over whether these pre-dated or post-dated ones from the conterminous United States. Kreiger (1954) had already suggested the Alaskan points might result from a migratory "backwash." Yet, few if any had appeared along the possible migration route linking Alaska through Canada (Stephenson, 1965; Wendorf, 1966). The technology of fluting was therefore thought to have been invented soon after arrival south of the ice sheets (Haynes, 1964; Stephenson, 1965). The greater variety and abundance of fluted points in the southeastern United States – indicating a longer period of occupation, a larger population, or both – pointed to this as a likely area of origin (Griffin, 1965; Mason, 1962; Williams & Stoltman, 1965). But, of course, there were no radiocarbon ages to back up that assertion.

Regardless of where Clovis originated and when, in 1965 it appeared that it marked the initial human presence in North America. The absence of any sites in the lower 48 states dating to a time before passage south from Alaska was blocked by late Wisconsin ice sheets cast doubt on the "illusive" claims (made most explicitly by Krieger, 1964) that there had been a pre-Clovis presence in the Americas. Nonetheless, the possibility of a pre-Clovis occupation remained open (Haynes, 1964; Stephenson, 1965). But having watched several highly touted pre-Clovis claims (e.g. Carter, 1957) fail to withstand critical scrutiny, archaeologists in 1965 were starting to show signs of what would become a deep-rooted skepticism about accepting any such claims at face value (e.g. Meighan, 1965).

Once south of the ice sheets, Clovis groups apparently spread rapidly, moving across the continent in less than 1000 years (Griffin, 1965; Haynes, 1964). The relative rapidity of their dispersal was attributed to their wide-ranging pursuit of Pleistocene big game (Williams & Stoltman, 1965). Yet, in 1965 the only sites with artifacts in direct association with the remains of now-extinct fauna were – again – in the Great Plains and Southwest, where several sites produced fluted points and butchering tools with mammoth skeletons (Stephenson, 1965). Outside those regions, a few localities had yielded artifacts and extinct fauna (especially mastodon in the eastern woodlands), but in no case were these demonstrably the result of a predator-prey relationship – nor was it certain the association was anything more than a coincidental

DEVELOPMENTS IN QUATERNARY SCIENCE
VOLUME 1 ISSN 1571-0866
DOI:10.1016/S1571-0866(03)01026-1

occurrence on the same geological surface (Baumhoff & Heizer, 1965; Griffin, 1965; Meighan, 1965; Williams & Stoltman, 1965). In fact, evidence for the subsistence activities of these early groups was singularly lacking (Griffin, 1965). Even so, the geographic distribution of fluted points and mastodons in places like the southeastern states seemed to coincide (Williams & Stoltman, 1965; also Martin, 1967), a "paradox" that to some became "intelligible under the hypothesis that . . . fluted-point makers roved the countryside in pursuit of big game" (Williams & Stoltman, 1965, p. 677). Finding supporting evidence for that seemed just "a matter of time and more concerted effort" (Williams & Stoltman, 1965, p. 674).

That hypothesis, of course, fueled the long-held suspicion (Grayson, 1984) that human hunters may have been responsible for the extinction of the Pleistocene megafauna, a notion that at the 1965 INQUA conference was actively being revived by Martin (1967). He faced considerable obstacles in doing so: namely, the absence of widespread archaeological evidence of human hunting; the complete lack of any association of Clovis artifacts with animals other than mammoth and mastodon; and the uncertainty about when the extinctions occurred. Resolving the chronology was critical, since Martin's overkill argument then (as now), hinged on the temporal correlation between the arrival of hunters and the demise of the megafauna. Martin (1967) reasoned that if all the megafauna disappeared from the landscape at the same time at the end of the Pleistocene it implied a cause that could strike down animals of very different physiology and adaptation across many distinct habitants and do so essentially instantaneously: like voracious, fast-moving hunters. If different genera disappeared over a long span of time, that opened the possibility their demise resulted from a more complicated cause: like late glacial climate change, which played out across North America over thousands of years in different ways with different consequences in different environments.

Yet, quite a number of mastodon localities in the eastern United States had radiocarbon ages indicating survival of the fauna well into the mid-Holocene (Griffin, 1965; Hester, 1960; Meighan, 1965; Williams & Stoltman, 1965), indicating this species, at least, was apparently little affected by the intrusion of Clovis "hunters." Considering that, and in light of how little evidence there was of megafaunal hunting, Griffin concluded (with tongue firmly in cheek):

> If man was responsible for the disappearance of some of the Late Pleistocene fauna in the Northeast, he must have used magic rather than implements. This magic was not very effective for it took some 6000 years to eliminate the animals (Griffin, 1965, p. 658).

Martin, recognizing the threat of those late survivals to his model of Pleistocene overkill, began vetting (and rejecting) all cases of postglacial megafaunal survival in his own INQUA contribution (Martin, 1967).

Fast-moving big-game hunters or not, the Clovis occupation did not last very long. Across the continent, new fluted point forms appeared in the latest Pleistocene and Early Holocene – the timing varying by area. In the 1960s, changes in point types were considered to mark culture change (at best, an ill-defined concept), but it was unclear if the disappearance of Clovis and its replacement by a variety of regional forms signaled the emergence of distinctive adaptations, population movements, responses to climate change, changes in artifact style, or some combination thereof (Griffin, 1965; Williams & Stoltman, 1965). In some areas, notably the Great Plains, there appeared to be a corresponding shift in subsistence to bison hunting – a change that, implicitly if not explicitly, was attributed to the demise of the Pleistocene megafauna and the requisite shift in diet (Meighan, 1965; Stephenson, 1965).

Research Directions Since 1965

In the decades after 1965, research into the peopling of North America expanded on multiple fronts. The literature over this period is substantial, but the major themes and direction of the discussion can be tracked via a series of synthetic articles and edited volumes. These include, in chronological order, Haynes (1969), Martin (1973), MacNeish (1976), Bryan (Ed.) (1978), Humphrey & Stanford (Eds) (1979), Rutter & Schweger (Eds) (1980), Ericson *et al.* (Eds) (1982), Stanford (1982), Shutler (Ed.) (1983), Dincauze (1984), Irving (1985), Kirk & Szathmary (Eds) (1985), Mead & Meltzer (Eds) (1985), Bryan (Ed.) (1986), Greenberg *et al.* (1986), Carlisle (Ed.) (1988), Meltzer (1989), Bonnichsen & Turnmire (Eds) (1991), Dillehay & Meltzer (Eds) (1991), Meltzer (1993), Szathmary (1993), Whitley & Dorn (1993), Bonnichsen & Steele (Eds) (1994), Meltzer (1995), Akazawa & Szathmary (Eds) (1996), West (Ed.) (1996), Bonnichsen & Turnmire (Eds) (1999), Goebel (1999), Powell & Neves (1999), Fiedel (2000), Renfrew (Ed.) (2000), Jablonski (Ed.) (2002). These citations should provide entry into the relevant literature, save for that in the fast-moving arena of molecular biology, where few broad syntheses have been published (cf. Schurr, 2004).

Extensive field investigations greatly enlarged the late Pleistocene archaeological record, and not just from the conterminous United States, but also from Canada, Alaska, and Siberia. This included many Clovis-age artifacts and sites, along with a handful of Late Pleistocene/Early Holocene human skeletal remains (Fig. 1). Several of the latter, including some that do not resemble contemporary Native Americans, became lightning rods of political and legal controversy, controversy fueled by ambiguities in the Native American Graves Protection and Repatriation Act (NAGPRA), which had been signed into law in 1990 (NAGPRA, in brief, stipulated that human skeletal remains are to be returned on request to the modern Native American tribes who were their "lineal descendants." Identifying lineal descendants can be a relatively straightforward process where the remains are only centuries old. It is no easy task where the skeleton and the modern populations are hundreds of generations apart, and separated by 10,000 years of gene flow, drift, selection and migration,

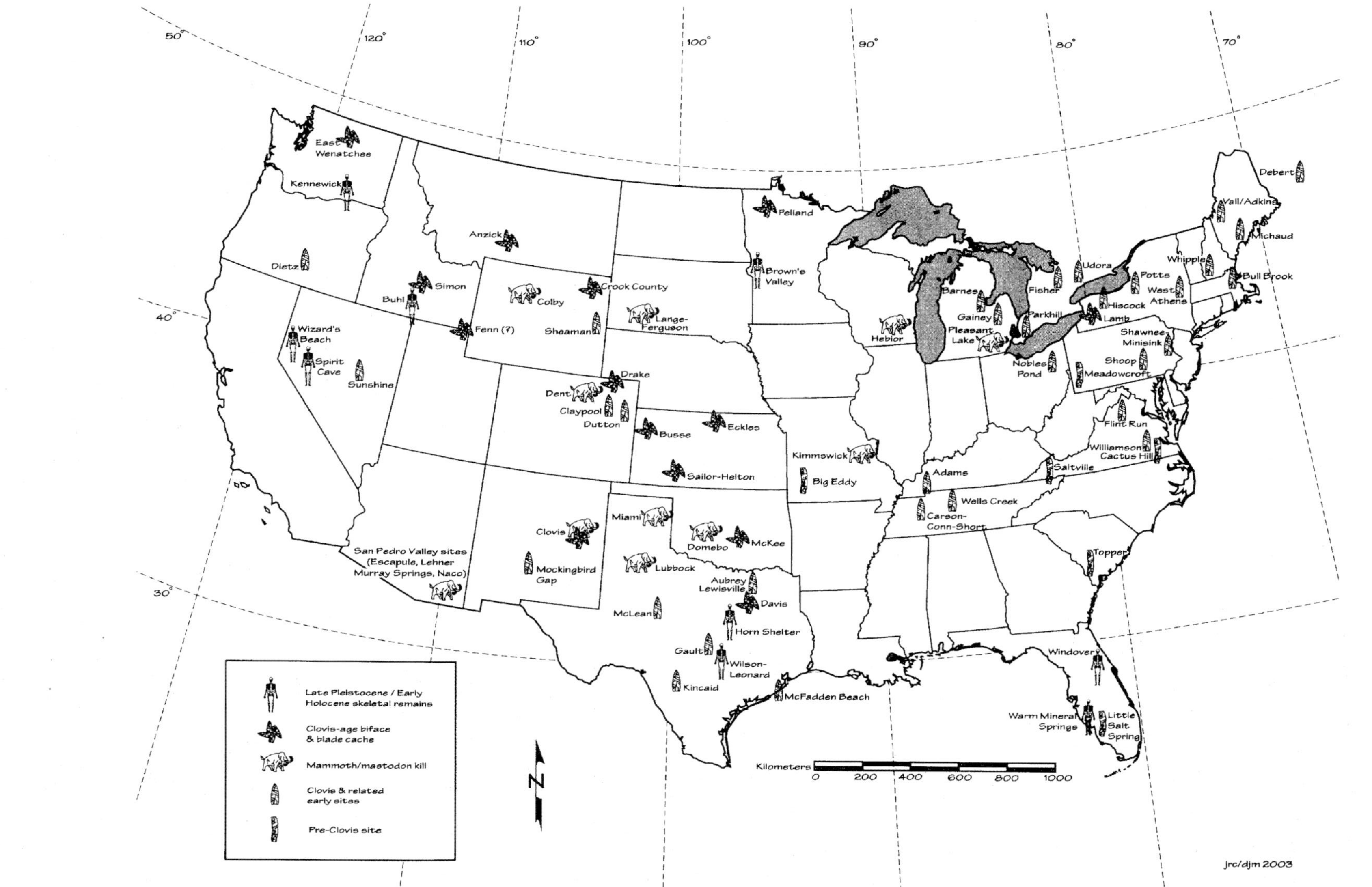

Fig. 1. Map of United States showing the location of selected late Pleistocene archaeological sites, including (a) possible pre-Clovis localities, not all of which have been accepted by the archaeological community (see text); (b) western Clovis and related fluted point sites in eastern North America; (c) sites for which there is secure evidence of human hunting of mammoth or mastodon (as identified by Grayson & Meltzer, 2002); (d) Clovis and Clovis-like caches; and (e) sites with late Pleistocene and early Holocene human skeletal remains. See key for symbols. Fig. by J. Cooper and D. Meltzer.

as well as five centuries of post European contact warfare, famine, dislocation, admixture, and demographic collapse).[1]

Those investigations also included a battery of ostensibly pre-Clovis age sites (Fig. 1), and these also proved controversial, at least within academic circles. By the 1980s, dozens had been reported, some estimated to be as much as 200,000 years old (e.g. Bryan [Ed.], 1986; Irving, 1985). Each of these sites was carefully evaluated to determine whether it had (1) genuine artifacts or human skeletal remains in (2) unmixed geological deposits accompanied by (3) reliable pre-Clovis age radiometric ages (criteria which, in one form or another, have been used for the last century to evaluate ostensibly-ancient sites [e.g. Chamberlin, 1903; Haynes, 1969]). Virtually all proved flawed, and fatally so (Dincauze, 1984).

Nonetheless, the possibility of a pre-Clovis presence was given new life in the last two decades from an unexpected quarter. Geneticists reconstructing phylogenetic histories of contemporary Native Americans from uni-parentally inherited genetic markers in mitochondrial DNA (mtDNA, inherited mother to child) and (more recently) on the non-recombining portion of the Y chromosome (NRY, inherited father to son), inferred these groups must have split from their Asian ancestors and departed for America sometime prior to the Last Glacial Maximum (LGM) (e.g. Karafet *et al.*, 1999, Merriwether, 2002; Schurr, 2004; Torroni *et al.*, 1993, 1994).[2] Studies of Native American languages indicated a similar antiquity (e.g. Nichols, 1990, 2002; cf. Greenberg, 1987). These are not, of course, secure absolute ages.

Developments in radiometric dating, especially the advent of AMS dating in the early 1980s and the subsequent extension of the calibration curve via U-series and varved sediments into late Pleistocene times, enhanced the details of the chronology of Clovis and later Paleoindian occupations. But it also showed these occupations occurred during a period of complex changes in atmospheric ^{14}C (Kitagawa & van der Plicht, 1998), complicating efforts at developing a fine-scale cultural chronology (Fiedel, 1999, 2000; Holliday, 2000; Mann *et al.*, 2001; Taylor *et al.*, 1996). The effort also helped correct many erroneous ages on alleged pre-Clovis sites and human remains derived from then-experimental or unreliable dating techniques (amino acid racemization and cation-ratio dating) which flourished in archaeology in the 1970s and 1980s (reviewed in Taylor, 1991; cf. Whitley & Dorn, 1993). More recently, luminescence dating (OSL and TL) has been applied with some success to several early sites (Feathers, 1997).

Attention also focused on the ecological stage across which colonizing groups dispersed. There was particular emphasis on the timing, character, and viability of the late Pleistocene landscapes of Beringia and northern North America, the likely routes by which colonists traveled from Asia into the Americas (Hopkins *et al.* [Eds], 1982; also Elias, 2002; Elias & Brigham-Grette [Eds], 2001; Fedje & Josenhans, 2000; Mandryk, 1996, 2001; Rutter & Schweger [Eds], 1980).

Over these decades the Pleistocene overkill model was elaborated by Martin (e.g. Martin, 1967, 1984, 1990; Mosimann & Martin, 1975), and put to the test on several points, not least by close scrutiny of the archaeological record for telltale "smoking Clovis points." That scrutiny was enhanced by developments in zooarchaeology and taphonomy (spurred by Binford, 1981), which helped establish criteria useful in evaluating the role, if any, humans played in the accumulation of a fauna at a site (e.g. Grayson & Meltzer, 2002, 2003; Haynes & Stanford, 1984), and by increasing refinement of the chronology of faunal extinction. By the 1980s, it was clear that no megafaunal genera had survived into Holocene times; that same vetting of the radiocarbon record also showed that more than half the genera in question did not have ^{14}C ages indicating survival into Clovis times (FAUNMAP, 1994; Grayson, 1991; Meltzer & Mead, 1983, 1985). Until their terminal ages could be demonstrated, one could not accept at face value Martin's fundamental assumptions that extinctions occurred simultaneously across all three dozen genera, let alone were coincident with Clovis.

Much of the discussion of the peopling of North America over the last four decades focused on finding things, such as the oldest archaeological site or the youngest occurrence of a now-extinct fauna. Recently, however, attention shifted to finding things out – notably, the processes of range expansion of hunter-gatherers across a rich, empty, and dynamic late Pleistocene landscape; the nature of human foraging strategies and the role of big-game in human hunting; and, on modeling the signature that colonization leaves on archaeological landscapes, and in the genes and languages and skeletal morphology of contemporary and descendant populations (e.g. Anderson, 1995; Anderson & Gillam, 2000; Jablonski [Ed.], 2002; Kelly, 1999; Meltzer, 1995; Steele *et al.*, 1998; Surovell, 2000).

Yet, by any measure the most significant development these last four decades was the breaking of the Clovis barrier

[1] To complicate matters further, we lack sufficiently precise genetic markers to be able to link DNA from ancient human skeletons with specific modern tribes (Merriwether, 2000). However, for many Native American groups and, for that matter by the provisions of NAGPRA, skeletal or genetic affinity are just one set of criteria for defining lineal descent; one can also use tribal traditions and geographic proximity. This has had the unfortunate result in some instances of pitting science against tribal tradition, and archaeologists against Native Americans (Thomas, 2000). Little good can come of that save, perhaps, clarification of the ambiguities and procedures of NAGPRA.

[2] Because mtDNA and NRY are uni-parentally inherited, change over time as a function of mutation, rather than mutation and recombination as is characteristic of autosomal exchange that occurs across the vast majority of the human genome where inheritance is from both parents; and, because those mutations arise at a known and (relatively) rapid rate, differences on these loci (mtDNA and NRY) between two or more populations becomes a molecular clock marking the time elapsed since they were part of the same original parent population. Although straightforward in principle, the method is not uncomplicated and there are caveats attached to its use (Merriwether, 2002; Schurr, 2004): not least, the recent realization that mtDNA variation may not be selectively neutral, and thus rates of mtDNA change on which the molecular clock is based may be driven by factors other than simply genetic mutation (Wallace *et al.*, 1999).

in the late 1990s. Excavations by Dillehay (1989, 1997) at Monte Verde, Chile, have convinced much of the archaeological community there was a human presence in the Americas earlier than Clovis (Adovasio & Pedler, 1997; Meltzer *et al.*, 1997). Although located in South America, Monte Verde's age (~12,500 ^{14}C yr B.P.) and distance from the Beringian gateway have profound implications for the peopling of North America. Monte Verde has wrought a sea change in American archaeology, and in its wake there has been a flurry of discussion, debate, new ideas and new interpretations.

The implications of Monte Verde will doubtless reverberate for years to come. In the meantime, I offer a brief summary of where matters stand today on the peopling of North America (see also Meltzer, 2002, 2004). I do so with the caveat that any summary in this volatile and often contentious arena is unavoidably idiosyncratic and inevitably ephemeral.

Peopling of North America – A Current Perspective

The Geography of Colonization

By about 35–25,000 years ago, humans had reached northeast Asia west of the Lena River and Verhoyansk Mountains, but within these regions had scarcely ventured above 55° N latitude. Only after ~25,000 ^{14}C yr B.P., and over the next 10,000 years, did they expand north and east of that region (Slobodin, 2001; Vasil'ev, 2001). That expansion slowed, and perhaps stalled during the LGM, when it appears – based on a scarcity of sites – that northeast Asia had only a sparse human presence. Humans did not reach far western Beringia, the jumping off point for migration to Americas, in significant numbers until ~15–14,000 ^{14}C yr B.P. (Derev'anko, 1998; Goebel, 1999; Goebel & Slobodin, 1999; Slobodin, 2001; Vasil'ev, 2001; papers in West [Ed.], 1996). By then, of course, lowered global sea levels had exposed the large expanse of continental shelf beneath what is now the Bering Sea, creating a ~1500-km-wide land link connecting Asia and America (Schweger *et al.*, 1982).

The Bering Land Bridge existed from ~25,000 to nearly 10,000 years ago (Elias, 2002; Elias *et al.*, 1996; Mann & Hamilton, 1995; Schweger *et al.*, 1982). The terrain was flat, largely unglaciated, cold, dry, and covered in grassy steppe-tundra (Clague *et al.*, 2004; Guthrie, 1990, 2001; Schweger *et al.*, 1982), across which people and animals could (and did) walk from Siberia to Alaska – and back – with relative ease. The Pleistocene faunas of western and eastern Beringia are virtually identical, testifying to the fullness of the Holarctic biotic exchange (Matheus, 2001; although for still unknown reasons, the woolly rhinoceros and giant short-faced bear are among the major Siberian and North American mammals [respectively] that failed to make the Pleistocene passage across Beringia [Kurten & Anderson, 1980]).

But if migrating from Siberia to Alaska was relatively easy, travelling south from Alaska to the conterminous United States may not have been – depending on the timing (Wendorf, 1966). For thousands of years before and after the LGM, glaciers buried the intervening terrain, thus blocking passage southward (Jackson & Duk-Rodin, 1996; Mann & Peteet, 1994). Following deglaciation, two routes south opened, though not simultaneously: one was an interior route along the eastern edge of the Rocky Mountains, the other along the Pacific coast.

The interior route along the Rocky Mountain flanks (which had several approaches [Mandryk, 2001]), was effectively blocked by the North American ice sheet perhaps as early as 30,000 ^{14}C yr B.P., and remained closed until ~11,500 ^{14}C yr B.P. (Catto, 1996; Clague *et al.*, 2004; Jackson & Duk-Rodin, 1996). But even after the Laurentide and Cordilleran glaciers retreated and an ice-free corridor began to open between them (which it did, zipper-like, from its northern and southern ends), the deglaciated terrain was initially inhospitable. The lingering influence of still-extant ice sheets and katabatic winds kept temperatures depressed, while recent deglaciation left behind a desolate region covered in meltwater lakes and unvegetated glacial deposits. This biologically unproductive landscape probably was not a viable passageway for human colonization (Clague *et al.*, 2004; Mandryk, 1993). Palynological and paleontological evidence suggests it took several thousand years following glacial retreat before conditions ameliorated and the land was re-colonized by the plants and animals on which hunter-gatherers could subsist (Mandryk *et al.*, 2001; Wilson & Burns, 1999).

A coastal route was also impassable during the LGM, as unbroken glacial ice extended to the outer edge of the continental shelf in the northern edge of the Gulf of Alaska (Clague *et al.*, 2004; Mann & Peteet, 1994). Travel down the coast could have been impeded by ice bergs and possible sea ice, and calving glacier fronts – the latter occasionally hundreds of kilometers across, with deep crevasses, and unstable leading edges that calved into the sea. Subsistence resources were scarce (Mann & Hamilton, 1995; Wright, 1991). Post LGM coastal deglaciation was complex and occurred at different rates, owing to out-of-phase ice advance/retreat at different points along the coastline between southwestern Alaska and western Washington (Mann & Hamilton, 1995; Stright, 1990). However, by approximately 13,000 ^{14}C yr B.P., most of the outer coast from Alaska to the continental United States was ice-free, and plants and animals necessary to support hunter-gatherers were beginning to re-colonize much of the landscape (Fedje & Christensen, 1999; Mandryk *et al.*, 2001). Early in that process, the productivity of the tidal environment may have been dampened as sediment-laden rivers draining melting ice sheets poured into littoral zones (T. Hamilton, pers. comm., 2001).

Opinion is divided over whether in-migrating groups coming down the coast would have traveled on foot or by boat (Anderson & Gillam, 2000; Dixon, 1999; Erlandson, 2002; Hamilton & Goebel, 1999; Workman, 2001). So far those are but opinions, supported only by circumstantial evidence: thus, the fact that groups traveled over water to Australia tens of thousands of years earlier is taken as testimony that watercraft could have been used to ply to coast of the Americas as well. Of course, the southern Pacific waters are far more hospitable than those of the northern Pacific and Gulf of Alaska, where death by hypothermia would be an almost inevitable outcome of falling out of a boat (Workman, 2001). For that matter, it is presumed by some that coastal colonizers would have relied partly on marine resources, and thus the antiquity of the use of marine resources here and elsewhere can be taken as indirect evidence of a coastal colonization by boat (Dixon, 1999; Erlandson, 2002). However, whether these groups had boats is a separate empirical issue from whether they also practiced a maritime economy, so the presence or absence of a maritime economy is not, per force, evidence for/against the use of boats, or otherwise is proof of their mode of travel. Forestalling the empirical resolution of this issue, and indeed the larger question of whether colonization proceeded via the coast, is the fact that much of the late Pleistocene Alaskan and Canadian coast is under water, drowned by rising Holocene seas (Dixon, 1999; Erlandson, 2002; Fedje & Christensen, 1999; Fedje & Josenhans, 2000; Josenhans *et al.*, 1997). However, not all of that coast was drowned: rapid isostatic rebound and regional post-glacial tectonic uplift have resulted in continual subaerial exposure of segments of the coast (Clague, 1989; Clague *et al.*, 2004). They await systematic survey and testing for possible archaeological remains.

A Matter of Timing

When people first trekked south from Alaska is uncertain, as the oldest archaeological sites along both the coastal and interior routes are younger than ~10,500 ^{14}C yr B.P. (Erlandson, 2002; Fedje *et al.*, 1995; Mandryk *et al.*, 2001; Wilson & Burns, 1999). Yet, based on evidence from south of the ice sheets, the initial colonists must have passed through this part of North America at least 2000 years earlier.

That evidence comes from archaeological and non-archaeological sources. In terms of the latter, divergence estimates for the five major mtDNA lineages (haplogroups A-D, X) found among Native Americans range from 20,000 to 36,000 ^{14}C yr B.P. A similar antiquity is estimated for haplotypes on the NRY, although there the temporal range is even wider (Schurr, 2004). Leaving aside specific questions about mutation rates and divergence times, especially in light of the possibility that selection – and not just mutation – may drive some genetic changes in mtDNA (Wallace *et al.*, 1999); and how well or whether molecular divergence and population splits coincided, that is, whether the molecular clock started ticking the moment groups departed Asia for America or whether groups split in Asia prior to migrating separately to North America (Meltzer, 1995; Merriwether, 2002; Schurr, 2004); the genetic evidence clearly supports an early (pre-Clovis) arrival of ancestral Americans. Coincidentally, some linguists have estimated an arrival of ancestral Americans at ~35,000 ^{14}C yr B.P., based on the number (~1000) and diversity of languages spoken in the Americas at European Contact (if it is assumed that languages diverged over time at set rates from the ancestral language(s) spoken on arrival) (Nichols, 1990).

Yet, non-archaeological sources provide only circumstantial evidence of great antiquity: neither the genes nor languages of contemporary peoples can be directly dated; only archaeological remains can (Meltzer, 1989; Nichols, 2002; Schurr, 2004). Thus, though the non-archaeological data are suggestive, the determination of whether people were here in pre-Clovis times is strictly an archaeological matter. For that reason, if genetic evidence suggests an arrival earlier than the archaeological record currently supports, then either the molecular clock needs to be re-calibrated or there are gaps in the archaeological record. Both, of course, are likely to be true.

At the moment, archaeological evidence indicates people were here in pre-Clovis times. The evidence comes from the Monte Verde site in Chile (Dillehay, 1989, 1997), one component of which (MVII) has radiocarbon ages that average ~12,500 ^{14}C yr B.P., and which yielded an extraordinary array of inorganic and organic artifacts and features. These include wooden foundation timbers, planks, and pegs used in the construction of a series of rectangular huts; wooden mortars containing charred and uncharred skins and seeds of edible plants; finely woven string; a wide range of plants, some exotic, some chewed, some occurring in presumed human coprolites; hearths with burned and unburned plant and animal remains; and the burned and/or broken and split bones of mastodon, along with pieces of its meat and hide (some of the hide adhering to wooden timbers, the apparent remnants of coverings that once draped over the huts). Also found were hundreds of stone, bone, tusk, and wooden artifacts – some with plant residues and tar (obtained from the distant coast) still adhering to their surfaces (Dillehay, 1997). Indeed, owing to its spectacular preservation, the bulk of the artifacts and material at Monte Verde are organic remains, not stone tools. Even so, the absolute size of the Monte Verde stone tool assemblage is not unusual by any measure, and its relative proportion may only demonstrate just how much of the non-stone tool component we are missing from other, more poorly preserved sites.

Monte Verde is the oldest known site in the hemisphere, and its distance from Siberia (~16,000 km) naturally raises questions about when the ancestors of this group crossed Beringia, and by what route they traveled south from Alaska. If the Beringian crossing took place after the LGM, ca. 14,000 ^{14}C yr B.P. (as suggested by the current estimate of the arrival of hunter-gatherers in far northeastern Siberia), then presumably the colonizers came south along the coast, since the ice-free corridor was then impassable, and would remain so until well after Monte Verde was abandoned (e.g. Dillehay, 2000; Erlandson, 2002; Mandryk *et al.*, 2001). Alternatively, if one accepts the suggestive but unconfirmed evidence from the earlier (MVI) component at Monte Verde dated to ~33,000 ^{14}C yr B.P. (Dillehay, 1997, 2000), groups might have come south well before the LGM – in which case they could have come via the interior or coast, both passageways being open at that time.

Having people at Monte Verde in southern Chile at 12,500 ^{14}C yr B.P. implies they ought to be in North America at a comparable antiquity (which is not to say that sites of this age ought to be in North America in comparable numbers to their contemporary abundance in Europe or Africa, as has been argued [e.g. Klein, 1999; Martin, 1987]. After all, the Old World was occupied earlier than the New no matter how

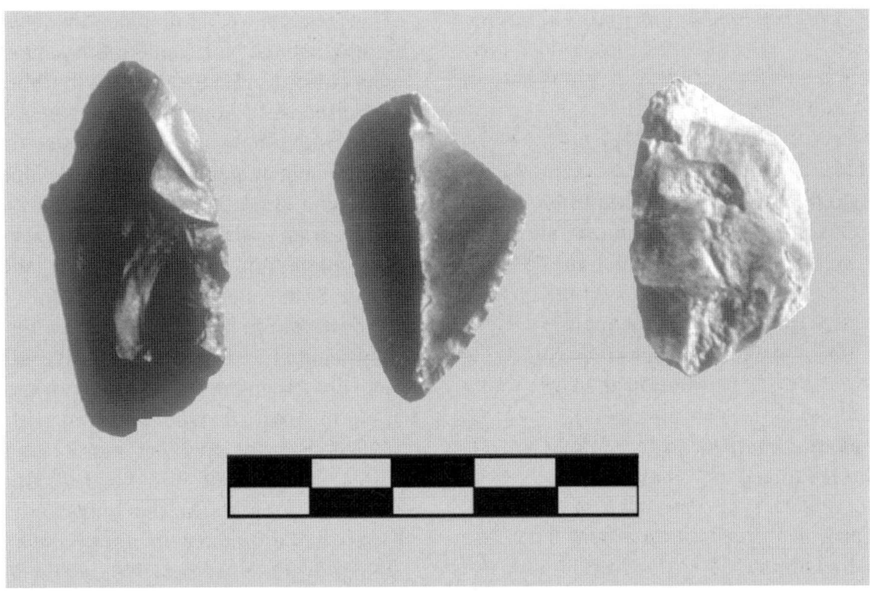

Fig. 2. Bifacial and unifacial flake tools from lower Stratum IIa, the pre-Clovis age levels at Meadowcroft Rockshelter, Pennsylvania. The specimen in the center – called a Mungai knife after a local resident – is the oldest stone tool found at the site, and was recovered from a ~16,000 ^{14}C B.P. surface (Adovasio and Page 2002:159). Adapted by J. Cooper from a photograph courtesy of J. Adovasio.

old the New, and by late Pleistocene times had a relatively larger population occupying a much smaller area, their density on the ground in turn enhancing their archaeological visibility [Meltzer, 1995]).There are a number of sites in North America said to be as old or older (Fig. 1), including Big Eddy, Missouri (Lopinot *et al.*, 1998, 2000); Cactus Hill, Virginia (MacAvoy & McAvoy, 1997; MacAvoy *et al.*, 2000); Meadowcroft Rockshelter, Pennsylvania (Adovasio *et al.*, 1990, 1999; a full listing of the Meadowcroft Rockshelter publications and presentations is in Adovasio & Page, 2002); Saltville, Virginia (McDonald, 2000); and Topper, South Carolina (Goodyear, 2001). With the exception of Meadowcroft, all of these have only recently been discovered and/or reported – some are still undergoing excavation and analysis. As a result, their evidence has not been fully published or evaluated.

The long-running debate over Meadowcroft, a site which has produced unmistakable artifacts (Fig. 2) in deposits dated to perhaps as much as 14,250 ^{14}C yr B.P., remains unresolved (although one matter is certain: by weathering nearly three decades of debate, Meadowcroft has cheated the pre-Clovis actuarial tables). Lingering doubts about possible contamination of the radiocarbon dates at the site have now been effectively rebutted (Goldberg & Arpin, 1999), but questions remain about the position of the artifacts and organic remains relative to the radiocarbon dated charcoal, which should be answered when the final report on this site is published.

None of these sites is fully accepted by an archaeological community which maintains a residue of skepticism toward pre-Clovis claims, even after Monte Verde. For now, we face the curious situation that the oldest acceptable site in the Americas (MVII at Monte Verde) is almost as far from the Beringian gateway as one can reach, with the vast territory in between lacking sites old enough or similar enough to represent traces of colonizers left along the way. Of course, candor compels the admission that it is no easy task deciding what stone tool assemblages belong together, especially ones separated by large gaps in space and time. For that matter, we are even harder pressed to link archaeologically-detectable technological patterns or changes with those identified among contemporary languages or genes.

The earliest widely-accepted archaeological evidence in North America therefore is still Clovis, which occurs across North America and into Central America, but extends no further south than Panama (Anderson & Faught, 2000; Ranere & Cooke, 1991). Contemporaneous South American sites and artifacts are unlike Clovis in many respects (Dillehay, 2000; Politis, 1991; cf. Morrow & Morrow, 1999).

Questions of Origin

What, then, to make of the population histories of North and South America, in so far as those can be read from current archaeological evidence? Sorting the matter along two dimensions – timing and number of migrations – four hypotheses suggest themselves, and these are shown in Table 1. (The situation is almost certainly more complex than this schematic suggests, but it is a useful format for discussion).

Thus sorted, the first pair of hypotheses (**H1** and **H3**) propose there was but a single migration to the Americas in pre- or post-LGM times, and all occupations throughout the hemisphere were derived from it. A single migration is a position to which some geneticists subscribe, and they tend to favor that migration occurring in pre-LGM times (e.g.

Table 1. Four hypotheses for the number and timing of migrations to America.

Timing	Number of Migrations	
	Single	Multiple
pre-LGM	**H1:** Single migration in pre-LGM times	**H2:** Multiple migrations, earliest of which in pre-LGM times. Subsequent migrations could occur then or post-LGM
post-LGM	**H3:** Single migration in post-LGM times	**H4:** Multiple migrations in post-LGM times

Kolman *et al.*, 1996; Merriwether, 2002; Merriwether *et al.*, 1996; Silva *et al.*, 2002) – in which case it could have utilized either a coastal or interior route. If that single migration occurred in post-LGM times, it presumably occurred prior to 12,500 ^{14}C yr B.P. – Monte Verde providing a minimum age – and therefore likely utilized a coastal route, given what is presently known of post-LGM glacial history. In either case, the dissimilarity of artifact assemblages hemisphere-wide during Late Glacial times (ca. 14,000–10,000 ^{14}C yr B.P.) would thus be the result of temporal and spatial gaps in the archaeological record. With the discovery of additional sites, these gaps should eventually disappear.

The alternative pair of hypotheses (**H2** and **H4**) is that there were two (or more) migrations to the Americas. If that is the case, then the present dissimilarity in assemblages could be real, and thus the traces found at Monte Verde and perhaps in terminal Pleistocene South America represent a separate and earlier migratory pulse (pre- or post-LGM) than that which created the Clovis archaeological record. Pre-LGM migratory groups could have traveled via coastal or interior routes into the conterminous United States, while the routes utilized by post-LGM groups would depend on the timing of their arrival (again, given what is presently known of post-LGM glacial history).

It is not possible at present to determine the number and timing of migratory pulses, largely because we cannot yet ascertain the historical relationship between Clovis and the groups who occupied Monte Verde. Still, hypotheses proposing two (or more) migrations appears to have a higher probability on current skeletal and genetic evidence (Brace *et al.*, 2001; Karafet *et al.*, 1999; Lell *et al.*, 2002; Powell & Neves, 1999; Schurr, 2004), as well as on archaeological grounds. In regard to the latter, if Clovis was derived from an as-yet invisible pre-Clovis presence, that would imply its rapid spread represents diffusion of a technology through an existing population, and evidence of that new technology being grafted on to other kinds of extant assemblages. But we have no evidence of either (Meltzer, 1995; Storck, 1991). If Clovis represents a separate migratory pulse, not only would that explain the lack of an obvious technological link between late Pleistocene assemblages of North and South America (Dillehay, 2000), it would also suggest that by then the North American continent was effectively empty. Clovis spread far and fast across a wide area, much farther and faster than one would expect had they encountered other people along the way (Anderson & Gillam, 2000; Meltzer, 1995, 2002). Naturally, this raises the question – yet unanswered – of what might have become of a colonizing group that passed through North America earlier. Some have suggested that the colonists who ended up at Monte Verde had traveled down the Pacific coast without significant inland movement until reaching, say, Panama, effectively leaving neither people nor archaeological footprints in North America (Dillehay, 2000; Erlandson, 2002).

Whether one, two, or more late Pleistocene migratory pulses, there continues to be compelling skeletal and linguistic support for the longstanding view that the first Americans originated in Asia (Brace *et al.*, 2001; Nichols, 2002; Powell & Neves, 1999; Steele & Powell, 2002; Turner, 2002). That view was taken a step further in recent years by comparative studies of Native American and Asian DNA, which pinpointed Mongolia (using mtDNA) or Siberia (using NRY) as likely ancestral homelands (e.g. Merriwether, 2002; Merriwether *et al.*, 1996; Santos *et al.*, 1999; cf. Brace *et al.*, 2001). That Native American mtDNA and Y chromosome DNA show affinities to populations in different areas (e.g. Lell *et al.*, 2002) is not altogether surprising for several reasons, not least that these loci track the molecular histories and lineages of females vs. males (respectively). There is genetic as well as considerable anthropological evidence to suggest that marriage and residence patterns can differ significantly between females and males (e.g. Seielstad *et al.*, 1998).

Regardless, those claims are based on where Asian populations genetically closest to contemporary Native Americans presently live. There is neither archaeological nor genetic evidence to indicate they have inhabited these same regions since the Pleistocene, or that the genetic composition of the ancient population(s) was the same as the modern inhabitants (this hypothesis can be put to the test if ancient DNA is recovered from fossil human remains in those regions). On its face, that possibility seems unlikely, given the challenge of adapting to this uncompromising climate and landscape and the population movements and genetic drift that would result. If human groups en route to North America were travelling through northeast Asia in small numbers and rapidly, and did not leave populations along the way, there would be no descendants among the modern inhabitants of the region to preserve that ancestral genetic signature.

For much the same reasons, there might be few archaeological traces as well (in northeast Asia or America). The number of archaeological sites produced in a given period/region is a product of many things, not the least of which, all other things being equal (and they rarely are), is the size of the human population (Butzer, 1991). In turn, the number of sites recovered archaeologically is determined, among other factors, by the antiquity of the period, erosional and depositional processes over that lapse of time, search techniques, and especially the population of sites relative to the size of the area being searched by archaeologists (Butzer, 1991; Dillehay & Meltzer, 1991). We do not know how large the colonizing group of the Americas was, but it was almost certainly not large. We assume – in the absence of any evidence whatsoever but in keeping with results of studies of modern hunter-gatherers – that minimally viable numbers for the initial population ranged from scores to hundreds (Anderson & Gillam, 2000; Dillehay, 2000; Fiedel, 2000; Steele *et al.*, 1996, 1998). This implies, of course, that in the earliest centuries and millennia of colonization, large areas were simply unoccupied, from which we draw the corollary that people were likely present in Siberia, Beringia, and the Americas before we see them archaeologically.

Viewed another way, that suggests the oldest archaeological sites found will not necessarily represent the first people in the region. This inference perhaps explains why the oldest late Pleistocene sites in Siberia and western Beringia are dominated by microcores and microblades (Goebel, 1999; Slobodin, 2001). These are tools that have little in common with the initial assemblages of the Americas south of the ice sheets, and that includes Clovis. A fluted biface was found at the Magadan site in Russia (King & Slobodin, 1996), but whether this lone specimen has historical significance (ancestral Clovis?), as opposed to being merely an accident of flaking on this particular specimen, is quite another matter.

On the other side of Beringia, there are Alaskan and Canadian sites with fluted points, but these are relatively few, and appear to be younger than Clovis (Reanier, 1995; Hamilton & Goebel, 1999; Wilson & Burns, 1999), as they ought not to be if they were left behind by Clovis colonizers moving south. It still appears, as it did in 1965 (Griffin, 1965; Williams & Stoltman, 1965), that the technology of fluting was invented south of the ice sheets, and moved north (Hamilton & Goebel, 1999; Meltzer, 1995).

Goebel *et al.* (1991) suggested Clovis had its roots in the Nenana Complex of central Alaska, as evinced by similarities between the respective tool assemblages. Yet, Nenana assemblages lack artifacts diagnostic of Clovis (most obviously, fluted points), and include tool classes absent from Clovis (e.g. microblades and microcores) (Bonnichsen, 1991; Holmes, 2001; Meltzer, 1995). The artifact classes that are common to both are tools types in use for long periods of prehistory in many different settings, and thus are an unreliable basis for establishing historical affinities or movements (Hoffecker, 2001; Holmes, 2001). Finally, with the recent re-assessment of the Nenana chronology (Hamilton & Goebel, 1999) this northern complex may be no older than 11,300 ^{14}C yr B.P., making it a poor candidate for a historical antecedent of a complex that dates back to at least 11,550 ^{14}C yr B.P.

That specific archaeological "footprints" of the first Americans cannot be tracked back to Alaska and across the Bering Sea to northeast Asia does not, of course, exclude the possibility the Americas were colonized by groups coming out of Asia: again, the preponderance of evidence points precisely in that direction (Merriwether, 2002; Turner, 2002). Rather, it only means we cannot yet specify the place(s) in Asia where these populations originated, or where certain technologies first developed. It could also imply that the route through Alaska was not via the Interior but along the southern coastline where sites of Late Glacial antiquity may remain undiscovered.

Although we do not need to seek the origins of the first Americans elsewhere, recent years have seen the revival of a very old idea (e.g. Abbott, 1878): that Pleistocene groups colonized America from Europe, across North Atlantic pack ice (Stanford & Bradley, 2002). The modern version of this notion is based on similarities between Clovis artifacts and those of the Solutrean period of Upper Paleolithic Spain and France. Yet, the few attributes held in common (e.g. the use of red ocher and the presence of outre passé flaking in biface manufacture) are ones that can result from convergence rather than common ancestry (Fiedel, 2000; Straus, 2000). Moreover, there are many and pronounced differences between the two in artifact forms, technologies, and materials. Archaeologically demonstrating contact between peoples and continents requires the list of similarities be long and the list of differences short; that situation does not obtain here. And despite a flurry of recent claims (mostly in the media, especially surrounding the Kennewick, Washington, find[3]), there is neither skeletal nor genetic evidence to support the idea the Americas were colonized by Pleistocene Europeans (Brown *et al.*, 1998; Merriwether, 2002; Smith *et al.*, 1999; Turner, 2002). Finally, five thousand years and several thousand miles of ocean separate the Solutrean and Clovis – formidable gaps for any scenario linking the two.

The Clovis Archaeological Record

As North America currently lacks a widely accepted or extensive pre-Clovis archaeological record, one can do little more than speculate about the timing, entry route(s), or adaptations of pre-Clovis colonizers. Assuming, for the

[3] An early Holocene skeleton was found near Kennewick, Washington, in 1996, which differed morphologically from modern Native Americans (Chatters, 2000). Although this prompted some questionable speculations about its affinities, as well as a long-running lawsuit in U.S. federal court to allow study of the remains (*Bonnichsen et al. v. United States*, Civil No. 96-1481JE, District of Oregon), it should come as no surprise that Kennewick – or any other skeletal remains of this age – might appear morphologically dissimilar from modern Native Americans. After all, skeletal remains from this period are rare (thus we lack a representative sample of the morphological variation within the population), effective gene pools were smaller and more isolated (perhaps leading to greater variability owing to local founder effects), and some 10,000 years of population history and evolutionary change (drift, etc.) have occurred in the interim (Powell & Neves, 1999). Still, more data are required to resolve the scientific issues surrounding the biology of the earliest Americans. More litigation is in the offing as well: the August 30, 2002 *Opinion and Order* issued by John Jelderks, the U.S. Magistrate in the case, allowed study of the remains by the plaintiffs, but motions to appeal the order were subsequently filed with the 9th Circuit Court of Appeals by both the U.S. Department of Justice and four Northwest tribes. Such are the scientific and legal consequences of ambiguities in NAGPRA. A history of the Kennewick controversy and links to the relevant documents can be found at: http://www.kennewick-man.com

sake of discussion, that Clovis itself represents a separate migratory pulse, its more substantial archaeological record sheds light on the adaptations and evolutionary history of dispersal across what may have been a new (or at least people-free) North American landscape in the centuries after 11,500 ^{14}C yr B.P. It is to that record I now turn.

The hallmark of Clovis is its distinctive projectile point, lanceolate in shape, fluted on each face, with the flute generally extending about one-third to one-half the length of the face (Fig. 3). Finished points are ground smooth on their base and partway up the sides, presumably with the intent to dull the edges where the point was attached (hafted) to the spear, to insure the sinew or other material binding the point was not inadvertently cut during use. Although sharing these attributes, Clovis points continent-wide vary in their morphology (Anderson & Faught, 2000; Morrow & Morrow, 1999; Tankersley, 1994), a variation that likely bespeaks divergence of populations and knapping styles and techniques over time and space (a form of cultural "drift").

Other elements of the Clovis tool kit include end and side scrapers, gravers, knives, and occasional bone tools (Haynes, 1982; Stanford, 1991). Distinctive artifact classes (e.g. ivory objects, limaces, adze-like forms) occur in some geographical regions but not others.

Decorative art is exceedingly rare among Clovis sites, the entire corpus consisting of only a few dozen specimens of limestone incised with parallel or intersecting lines, all but two specimens of which come from a single site (Gault site, Texas [Collins, 1998]). Art like that routinely found in contemporary Paleolithic Europe – cave paintings and sculpted figurines of extinct animals and humans appearing in both natural and abstract forms – does not occur in late Pleistocene North America. Although examples of such have been innocently reported (e.g. Kraft & Thomas, 1976), closer examination revealed these to be 19th century frauds (Griffin *et al.*, 1988; Meltzer & Sturtevant, 1983).

Clovis points and tools were manufactured primarily on bifaces, but a recently recognized blade-based technology is proving to be common in the southern midcontinent, though absent from other places (e.g. Collins, 1999; Stanford, 1999). More differences in the Clovis tool kit and technology will likely emerge, as additional details of the archaeological record emerge. At present, we lack representative assemblages from many geographic areas, and have found only a limited number of large habitation sites, where – by virtue of longer periods of occupation and greater numbers of activities – a wider range of tool classes would be expected.

In making stone tools, Clovis knappers relied almost exclusively on high-quality crypto- and non-crystalline stone, including chert, jasper, chalcedony and obsidian. The stone was usually acquired at bedrock outcrops rather than from secondary sources (e.g. fluvial or glacial gravels). That preference for outcrops rather than more widely scattered secondary sources reveals the extent of their knowledge of the geology of the continent. Clovis knappers found chert sources we have yet to locate which, as Moeller (2002) observes, should serve as a caution when making claims about the distances these groups traveled across the landscape. That preference for outcrop sources also helped ensure stone of adequate size and quality could be obtained. Size and quality were critical because Clovis bifaces readied for use were upwards of 20 cm in length and width and >600 gms in mass (e.g. Frison & Bradley, 1999), requiring quarried masses of stone of even greater size. Such high-quality stone was less failure-prone, and more easily re-worked as supplies dwindled, both of which would have been important to mobile hunter-gatherers who could not predict when they would next be near a stone source (Goodyear, 1979). Because the stone can often be identified to the outcrop from which it was obtained, it is apparent Clovis groups were routinely moving hundreds of kilometers across the landscape (Tankersley, 1991). The scale of their mobility varied by region, depending on the nature and density of resources being exploited, but in general Clovis groups had far more extensive ranges than groups in later prehistoric times. There have been several efforts to determine more precisely the size of those ranges

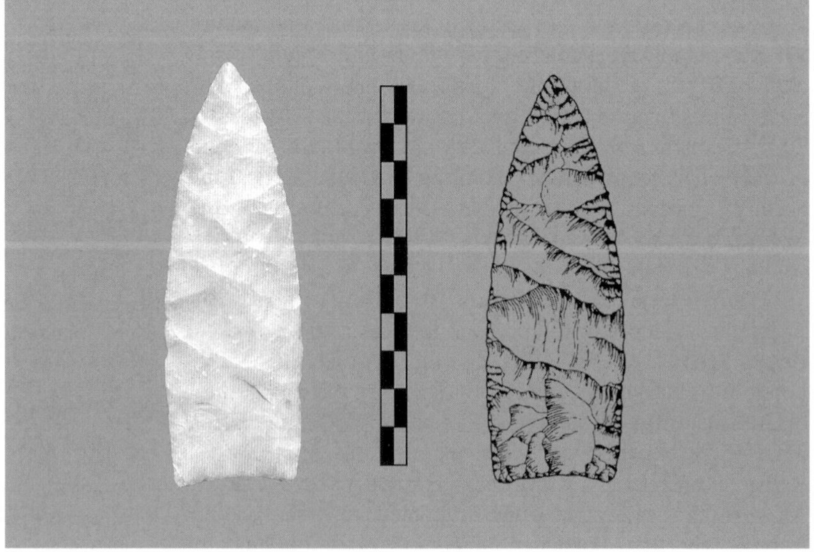

Fig. 3. Clovis point from the type site near Portales, New Mexico (Blackwater Draw Site Specimen No. A186), displaying the characteristic features, including basal fluting and outre passe flaking. The specimen is made of (ed.)wards Formation chert, likely from an outcrop source near Big Springs, Texas, ~300 km southeast of the site. This point was recovered during the El Llano Archaeological Society excavations in July 1963, associated with the vertebra and ribs of Mammoth No. 4 (although is not shown on the published plan map of those excavations [Warnica, 1966, p. 346]). It is 11.10 cm in length. Line drawing by F. Sellet, photograph by D. Meltzer; montage by J. Cooper.

and the populations occupying them (e.g. Anderson, 1995; Fiedel, 2000), and although interesting, these are necessarily speculative equations with many uncontrolled variables.

The high mobility of Clovis groups is reflected in their sites (Kelly & Todd, 1988; Surovell, 2000). For the most part these are small, lack site furniture (items that go with a place and not with the persons occupying the place, Binford, 1979), and rarely include storage pits or evidence of habitation structures. Those structures that do occur are insubstantial. All of which indicates these groups were not staying for long periods of time at particular places, nor returning to specific places repeatedly (the exception here are their stone sources, which were often returned to repeatedly [e.g. Gardner, 2002]). This, in turn, suggests they were exploiting resources that were widely available and not place-restricted: such is the advantage of colonists on a landscape without other people, a landscape without social boundaries.

Clovis artifacts and sites have been found across the continent in a variety of environments, from the rich grasslands of the western Plains to the complex boreal/deciduous forests of the American southeast (papers in Bonnichsen & Turnmire [Eds], 1991). No subsequent North American occupation was so widespread or occupied such diverse habitats. It is important to add, however, many areas appear only sparsely occupied, including the Great Basin, the Columbia and Colorado Plateaus, northern Great Plains, northern Rockies, and the uppermost and lowermost reaches of the Mississippi Valley (e.g. Anderson & Faught, 2000). This spotty distribution is undoubtedly biased by differences in the ages of exposed geomorphic surfaces, contemporary land use patterns, and the amount of archaeological and/or collector activity (Shott, 2002). Still, a more representative sample would likely only change the details, and not the essential fact that the Clovis presence on the landscape was broad, not deep – the manifestation of highly mobile people at low population densities.

The geographic spread of Clovis across the continent was often envisioned as a more-or-less uniform diffusion across space, perhaps in an expanding wave front (e.g. Martin, 1973; Young & Bettinger, 1995). More recent GIS-based studies using the details of North American topography and terrain, the inferred viability of different Pleistocene habitats, and the continent-wide distribution of Clovis sites and isolated fluted points, predictably paint a more complicated picture (e.g. Anderson & Gillam, 2000; Steele *et al.*, 1996, 1998; cf. Fiedel, 2000). Anderson & Gillam (2000), for example, propose that expansion threaded across the continent through areas of relatively low topography, along major river valleys, in proximity to glacial and pluvial lakes and, in eastern North America, along the coastal margin. Expansion could have proceeded in either a "string of pearls" or a "leap-frog" mode, the former a spatially continuous expansion, the latter rapid jumps across large stretches of landscape, possibly with pauses in "staging areas" along the way (Anderson & Gillam, 2000; also Anderson, 1995). A "leap-frog" model seems to better fit the dense but widely separated clusters of Clovis and Clovis-age materials across the continent, but must have entailed strategies to maintain a critical mass of population and/or interaction with distant kin, so as to offset the demographic danger of living in small numbers over large areas devoid of other people (Anderson, 1995; Anderson & Gillam, 2001; Moore, 2001; Moore & Moseley, 2001).

The chronology of the Clovis occupation varies across the continent and, as was the case in 1965, remains better known for some areas than others. The oldest sites are still those on the Great Plains and in the Southwest, which range in age from 11,570 to 10,900 ^{14}C yr B.P. (Holliday, 2000; Stanford, 1999; Taylor, 2000). Significantly, the earliest appearance of Clovis continues to coincide with the opening of a viable ice-free corridor – granting that the timing of the opening is made fuzzy by uncertainty about the timing of deglaciation, draining, revegetation, etc. of the emerging corridor (Mandryk, 2001; Mandryk *et al.*, 2001).

There are now reliable radiocarbon dates on a dozen or so Clovis and Clovis-like fluted point sites in the eastern United States, but despite the continued suspicion that Clovis technology emerged in this area (Stanford, 1991), no eastern fluted point site yet pre-dates 11,500 ^{14}C yr B.P. (Anderson *et al.*, 2002). A very few approach that antiquity (e.g. Shawnee-Minisink, Pennsylvania; Dent, 2002]), but the majority fall between 10,600 and 10,200 ^{14}C B.P. – a period contemporaneous with the Folsom (post-Clovis) occupation on the Great Plains (Bonnichsen & Will, 1999; Haynes *et al.*, 1984; Lepper, 1999; Meltzer, 1988). Clovis-like materials occur in the Great Basin and far west, perhaps as early as 11,500 ^{14}C yr B.P., although their ages remain uncertain, and the cultural chronology is confused by the possibly contemporaneous occurrence of large unfluted stemmed points (Beck & Jones, 1997; Bryan & Tuohy, 1999; Grayson, 1993).

The radiocarbon record supports the long-held suspicion these groups radiated rapidly across the continent, the process taking perhaps no more than 500 radiocarbon years. The latter part of this episode, of course, overlaps the Younger Dryas (YD), which includes several radiocarbon plateaus that distort radiocarbon ages (Hughen *et al.*, 2000; Kitagawa & van der Plicht, 1998; Peteet, 2000; Taylor *et al.*, 1996). Although calibration may ultimately change the apparent speed of dispersal (perhaps "slowing" it to, say, approaching 1000 calendar years – see Dincauze, 2002; cf. Fiedel, 2000), it will nonetheless remain one of the fastest expansions of an archaeological complex known in prehistory (among the few cases faster were the expansion of prehistoric Thule across northern Canada in the centuries after 900 A.D. [Meltzer, 2002, discusses similarities and differences between the Thule and Clovis dispersals], and the dispersal of Lapita groups throughout Near and Remote Oceania beginning some 3000 years ago [Kirch, 1997]).

It seems reasonable to suppose, given human population density was lower at this time than at any subsequent period in American prehistory, and that these groups were occupying a relatively rich landscape, that demographic pressure was not fueling that dispersal. These groups traveled much farther and faster than they had to if they were just looking for new land to siphon off burgeoning populations that would have otherwise put a strain on local resources (Kelly, 1996; Mandryk, 1993; Meltzer, 1995). But why (or how) did they move so far, so fast?

Human Hunting and Pleistocene Mammalian Extinctions

The traditional explanation is that Clovis people were specialized hunters in pursuit of wide-ranging big-game, notably the now-extinct Pleistocene megafauna, which were themselves able to override ecological boundaries (Kelly & Todd, 1988; Mason, 1962; Martin, 1973). Clovis colonizers could compensate for their lack of knowledge about the landscape by exploiting the same prey-species niche through all the habitats they traversed (Kelly, 1996). Harvesting the same food resources in new locations using traditional weaponry, hunting skills, and tactics would allow efficient and rapid dispersal (Keegan & Diamond, 1987; Kelly & Todd, 1988). Such a subsistence strategy and the rapid dispersal it could permit would buttress the claim humans were responsible for the extinction of the megafauna at the end of the Pleistocene in North and South America (Martin, 1973). By this scenario the herbivores met their demise directly as a result of intensive hunting, while carnivore population temporarily boomed with the sudden increase in scavengeable carcasses on the landscape, then went bust when the supply ran out. If hunters were killing off the megafauna at rates proposed by the overkill model (Martin, 1973; Mosimann & Martin, 1975), that would be incentive enough for rapid expansion, for their prey were always in front of them, not behind them.

However, there are many reasons to doubt that scenario. First, recent studies of faunal remains from Clovis sites, as well as of the isotope geochemistry of rare human bone from this period, show that these late Pleistocene groups exploited a greater variety of animal and plant resources than traditionally supposed (Green *et al.*, 1998; Johnson, 1991; Meltzer, 1993; Spiess *et al.*, 1985; Stanford, 1991, 1999). Second, it is unlikely hunters could convert search and processing strategies and tactics successful against one prey species for use on another (Frison, 1991, 1999). Third, ethnographic and archaeological evidence demonstrates that specialized big-game hunting – let alone the hunting of a continent of animals to extinction – was rare among hunter-gatherers, and linked to particular habitat types (Binford, 2001; Hofman & Todd, 2001; Meltzer, 1993). This is not surprising: although models of foraging theory as well as empirical evidence suggest animals of large body size are high-ranked prey, that is not necessarily true of animals of the largest body sizes (risk comes into play); and, on productive, game-rich landscapes essentially devoid of other people – late Pleistocene North America – foragers would likely abandon a patch before extinction of the local fauna (Broughton, 1994; Grayson, 2001; Kaplan & Hill, 1992; Kelly, 1995). Winterhalder & Lu (1997) model circumstances under which depletion and extinctions are quite plausible, but it is doubtful those circumstances characterized this place and time, given the richness of the late Pleistocene landscape.

Finally, and most telling, in spite of four decades of "time and more concerted effort" (Williams & Stoltman, 1965, p. 674), few additional Clovis big-game kill sites have been found, and most of those are still in western North America. Clovis kills continue to be conspicuously scarce in eastern North America, despite the rich record there of terminal Pleistocene fossil localities. In fact, it is only a slight exaggeration to say that the overall tally of kill sites may even be less today than it was in 1965. For we now have greater knowledge of the various natural processes that can fracture or disarticulate skeletal remains in ways that mimic human activity. We can employ more stringent criteria to differentiate claims of association between artifacts and extinct faunal remains that are compelling, from claims that are not; and we can better differentiate evidence of hunting from evidence for other behaviors – such as scavenging (Binford, 1981; Grayson & Meltzer, 2002; G. Haynes, 1991; Haynes & Stanford, 1984; Lyman, 1994). Even recently, for example, Laub re-evaluated the Hiscock (NY) mastodon "kill" and concluded there is no evidence to support that interpretation. Instead, he now argues the site was a quarry where Clovis groups obtained mastodon bone and ivory from geological deposits (Laub, 2002).

Applying such criteria to all purported Clovis or Clovis-age megafaunal "kill" sites – of which there are more than 75 – leaves only 14 sites in North America (Fig. 1) for which there is secure and unambiguous evidence of human hunting (Grayson & Meltzer, 2003). When measured against the archaeological record of the hunting of extinct mammals of, for example, Upper Paleolithic France, that is a remarkably thin record (Grayson & Meltzer, 2002). Of those 14 sites, twelve contained mammoth, and the other two mastodon. There are no unequivocal kill sites for any of the other 33 genera of North American large mammals that went extinct at the end of the Pleistocene (Table 2a), again despite their abundance in the late Pleistocene paleontological record (Grayson, 2001). That remains of a few of those other genera are occasionally found in small numbers in archaeological sites is intriguing, but proves little more than their contemporaneity with Clovis people on the late Pleistocene landscape (Grayson & Meltzer, 2002).

Turning the matter around, bison and other large mammals (Table 2b and c) were hunted in North America beginning as early as $\sim$10,900 ^{14}C yr B.P. In the case of bison, there is abundant archaeological evidence of planned hunts, bone beds containing hundreds of slaughtered animals, impact-fractured projectile points and skinning and butchering tools (e.g. Frison, 1991). Such intensive predation was often highly wasteful: of the 200 bison stampeded by hunters into an arroyo at the Early Holocene-age Olsen-Chubbuck site (Colorado), some 25% of the animals at the bottom of the carcass pile were left to rot untouched (Wheat, 1972). Nearly 11,000 years of human predation culminated in widespread slaughter by buffalo hide hunters in the late 19th century. Yet, despite being hunted for millennia bison (and, for that matter, the taxa in Table 2c) failed to go extinct; while 35 genera of animals that were not hunted at all or very little (mammoths and mastodons) did go extinct.

Of course, as noted above, it has not been demonstrated that all 35 of those now extinct genera were contemporaries of humans, or even lasted until the terminal Pleistocene – let alone, that all went extinct simultaneously. Although we often assume as much (Martin, 1984), only 14 of the 35

Table 2. North American late Pleistocene mammals. Taxa that are in bold are ones for which there is secure archaeological evidence of human predation.

Order	Family	Genus & Species	Common Name
2a. North American Genera that went extinct in the Late Pleistocene			
Xenartha	Dasypodidae	*Pampatherium* sp.	Southern pampathere
		Holmesina septentrionalis	Northern pampathere
	Glyptodontidae	*Glyptotherium floridanus*	Simpson's glyptodont
	Megalonyhiae	*Megalonyx jeffersonii*	Jefferson's ground sloth
	Megatheriidae	*Eremotherium rusconii*	Ruscon's ground sloth
		Nothrotheriops shastensis	Shasta ground sloth
	Mylontidae	*Glossotherium harlani*	Harlan's ground sloth
Carnivora	Mustelidae	*Brachyprotoma obtusata*	Short-faced skunk
	Canidae	*Cuon alpinus*[a]	Dhole
	Ursidae	*Tremarctos floridanus*[a]	Spectacled bear
		Arctodus simus	Giant short-faced bear
	Felidae	*Smilodon fatalis*	Sabertooth cat
		Homotherium serum	Scimitar cat
		Miracinonyx trumani	American cheetah
Rodentia	Castoridae	*Castoroides ohioensis*	Giant beaver
	Hydrochoeridae	*Hydrochoerus holmesi*[a]	Holmes's capybara
		Neochoerus pinckneyi	Pinckney's capybara
Lagomorpha	Leporidae	*Aztlanolagus*	Aztlan rabbit
Perissodactyla	Equidae	*Equus* spp.[a]	horses
	Tapiridae	*Tapirus* spp.[a]	tapirs
Artiodactyla	Tayussuidae	*Mylohyus nasutus*	Long-nosed peccary
		Platygonus compressus	Flat-headed peccary
	Camelidae	*Camelops hesternus*	Western camel
		Hemiauchenia macrocephala	Large-headed llama
		Paleolama mirifica	Stout-legged llama
	Cervidae	*Navahoceros fricki*	Mountain deer
		Cervalces scotti	Stag-moose
	Antilocapridae	*Capromeryx minor*	Diminutive pronghorn
		Tetrameryx shuleri	Shuler's pronghorn
		Stockoceros spp.	Pronghorns
	Bovidae	*Saiga tatarica*[a]	Saiga
		Euceratherium collinum	Shrub ox
		Bootherium bombifrons	Harlan's muskox
Proboscidea	Mammutidae	***Mammut americanum***	**American mastodon**
	Elephantidae	***Mammuthus* spp.**	**Mammoth**
2b. North American species that went extinct, while other members of the same genus survived in North America			
Xenartha	Dasypodidae	*Dasypus bellus*	Beautiful armadillo
Carnivora	Canidae	*Canis dirus*	Dire wolf
	Ursidae	*Temarctos floridanus*	Spectacled bear
	Felidae	*Panthera leo atrox*	American lion
Artiodactyla	Bovidae	*Oreamnos harringtonii*	Harrington's mountain goat
		Bison antiquus	**Bison**
2c. Select North American large mammal genera/species that survived			
Artiodactyla	Cervidae	***Alces alces***	**Moose**
		Cervus elaphus	**Elk**
		***Odocoileus* spp.**	**Deer**
		Rangifer tarandus	**Caribou**
	Antilocapridae	***Antilocapra americana***	**Pronghorn**
	Bovidae	***Ovibos moschatus***	**Musk ox**
		***Ovis* spp.**	**Mountain sheep**

Sources: Taxonomic data from Anderson (1984), Grayson (1991), Kurten & Anderson (1980). Information on archaeological occurrences from Frison (1991), Grayson & Meltzer (2002), and papers in Damas [Ed.] (1984), Gerlach & Murray [Eds] (2001), Helm [Ed.] (1981).
[a] Others members of the same genus survived outside of North America.

genera have reliable ^{14}C ages indicating survival past 12,000 years ago, the other 21 do not (FAUNMAP, 1994; Grayson, 2001, using criteria developed in Meltzer & Mead, 1983). In fact, the youngest reliable ages we have on some of those other 21 genera predate the LGM (Grayson, 2001). Even though that opens the possibility that some extinctions took place long prior to the appearance of Clovis in an area – and possibly on the continent (Goodyear, 1999; Grayson, 1991, 2001) – confirming that possibility will require additional evidence. The reason, simply, is that many of those same genera lacking terminal Pleistocene ages are also relatively rare in the fossil record, and the number of radiocarbon ages we have for a particular genus is strongly determined by how many fossils of that genus have been found (plotting the number of fossil occurrences against the number of radiocarbon ages [by genera] yields a highly significant correlation, $r^2 = 0.903$ [data from FAUNMAP, 1994; Meltzer & Mead, 1985]). Until we get more radiocarbon dates, and can better discern the timing of their disappearance, we cannot conclude all genera disappeared simultaneously, or gradually, or let assumptions about the timing of extinctions be marshaled in support of arguments about its cause.

Recent years have seen renewed efforts to bolster the case for human overkill: Alroy (2001) provides an elegant simulation model which he believes proves extinctions were an "unavoidable" consequence of the arrival of human hunters, while G. Haynes (2002) argues overkill was an inevitable outcome of late Pleistocene environmental change. However, the test of any simulation model, Alroy's included, is not whether it can show a process could have occurred, but how it fares against the empirical evidence it purports to explain. In the absence of kill sites, the model fails. Haynes' argument depends on there having been a "near continental drought" in late Clovis times (C. Haynes, 1991) which caused the megafauna to crowd together at water-holes where they were easy-picking for human hunters (Haynes, 2002). Unfortunately, there is no evidence of a drought of this magnitude or extent in Clovis times (Holliday, 2000), of crowding at the supposed megafaunal oases, or of slaughter thereat.

The matter returns, as it must, to the empirical record, and there the facts are clear: very few kill sites have been found – and then only of mammoth and mastodon – this in spite of decades of intensive searching, and a rich paleontological record of many of these animals (we do not lack for fossils of this age). All of which makes it hard to avoid hearing the echo of Griffin's (1965) conclusion: if human hunters had a role in killing off the megafauna they must have used magic, and then carefully hid the evidence.

Living and Learning on a New Landscape

Since specialized big-game hunting was not a significant component in Clovis subsistence, we still face the question of why (or how) Clovis groups moved so far, so fast. Haynes (1987) proposed the engaging scenario that curiosity, a charismatic leader(s) with the urge to see what was over the next hill or around the bend, and a landscape teeming with megafauna, lured Clovis groups across the continent. Although curiosity and charisma assuredly played a role in individual cases, it does not provide a robust model for expansion across an entire continent over many centuries.

Others attribute the fast dispersal to the changing climates and environments of Late Glacial North America, including the shift from patchy to zonal environments, a decline in faunal biomass, extinction of megafauna, and a change from equable to more continental regimes (Fiedel, 2000; Kelly, 1996, 1999; Kelly & Todd, 1988). Clovis groups had to move long distances and quickly, as local game populations declined precipitously in response those changes. Yet, the scale of those changes was on the order of centuries. Hunter-gatherers respond to the local weather – primarily on a daily and weekly basis – but also as it varied seasonally, annually, or over the course of their lifetimes. How or whether they respond depends on whether those changes triggered prey population fluctuations, reduced surface water, or otherwise restructured resource availability in ways that would have been detectable to and directly impacted their foraging activities. Long-term patterns of low frequency climatic variation over centuries (and many human generations) of the colonization process may not have been detectable on a human scale, and thus not directly relevant.

Efforts have been made to link more rapid late Pleistocene climatic excursions – the Younger Dryas most prominently (Severinghaus & Brook, 1999; Steig, 2001) – to patterns in the contemporary archaeological record (e.g. Fiedel, 1999, 2000; C. Haynes, 1991). The YD has even been invoked by geneticists to explain the high frequency and reduced diversity of mtDNA haplogroup A2 among Na-Dene and Eskimo groups, on the assumption that harsh YD climates forced an occupational hiatus and thus a population bottleneck (Forster *et al.*, 1996). However, evidence of a YD impact on humans in the high Arctic, where there is little doubt YD climate change was rapid and potentially significant, is equivocal (but see Mann *et al.*, 2001). It is not apparent the YD triggered an occupational hiatus, or that changes in the distribution and abundance of sites during this period are real (as opposed to a vagary of sampling in this still little known region) (Bigelow & Powers, 2001). Yet, the Arctic may be the most likely region to see a YD impact on humans, if one is to be seen. So far, there is little evidence climate and landscape changes were as dramatic in mid-latitude, temperate North America (Grimm *et al.*, 1993; Peteet, 2000), or had any impact on Clovis and later Pleistocene groups (Holliday, 2000; cf. C. Haynes, 1991).

For my part, I have attributed the rapid and widespread movement of Clovis groups in part to their unfamiliarity with the landscape (Meltzer, 2002, 2003, 2004). Anthony (1990:901) observes that immigrants are not likely to move into areas about which they have no secure prior knowledge. Although the point is well taken, at some time in the North American past there was little choice in the matter. If, in fact, Clovis groups were colonizing a diverse and unfamiliar new continent, there was likely strong selective pressure to learn their landscapes (Meltzer, 2002). Landscape learning has at least three elements (entailments of which are discussed in Meltzer, 2003): wayfinding, tracking weather and climate, and mapping resources (of all kinds: food, water, stone, etc.).

Landscape learning would be especially important early in the colonization process, when environmental uncertainty was high, environments were patchy and varied temporally and at large scales (as they would relative to colonizers on a new landscape), and when human population numbers were low and groups were most vulnerable to extinction (Kaplan & Hill, 1992; also Kelly, 1995; Moore, 2001; Stephens & Krebs, 1986). Under these circumstances, selection would favor rapid and extensive exploration in order to reduce environmental uncertainty and forager risk, and provide foragers with the knowledge that would enable rapid niche shifts.

There are demographic costs to moving that far that fast, for on a continent the size of North America populations would have been stretched thinly across the landscape (evidence of which may appear in the genetics of their descendants [Malhi *et al.*, 2002]). To avoid inbreeding or, worse, extinction, groups would have to maintain a "critical mass" of population and an accessible source of potential mates, by participating in a larger effective gene pool (cf. Surovell, 2000). This would have been more or less difficult depending on the local group's size, population growth rates, kin structure, age and sex composition, as well as how rapidly it was moving away from its geographic homeland and/or from other groups, and on environmental constraints on group size and population densities (Moore, 2001).

Demography and landscape learning are tightly linked, as the decision to stay in a patch or move onto the next is in part based on the suitability of a new patch relative to the current one, after factoring in the costs of moving (Kelly, 1995). Those foragers who can better calculate those costs increase their chances of success and survival. By gaining information about a landscape one potentially reduces risk and mortality, and thus can increase population growth and recruitment rates.

Arguably, then, the colonization process on a new landscape involved trade-offs between multiple competing demands (Meltzer, 2002): *maintaining resource returns*, or keeping food on the table, particularly as preferred or high-ranked resources declined, and in the face of limited knowledge of the landscape; *maximizing mobility*, to learn as much as possible, as quickly as possible about the landscape and its resources (in order to reduce environmental uncertainty in space and time), while *maximizing residence time in resource-rich habitats* to enhance knowledge of specific changes in resource abundance and distribution; *minimizing group size*, to buffer environmental uncertainty or risk on an unknown landscape; and, finally and most critically, *maximizing the effective gene pool by maintaining contact between dispersed groups*, in order to sustain information flow, social relations and, most especially, demographic viability. Colonizers had to balance the equation of moving to learn and explore, and staying to observe.

Under this model we expect to see among colonizers large scale exploration to map the landscape (which, arguably, might be marked by stone tool caches); periodic aggregations of widely dispersed groups, to exchange mates, resources, and information; and extensive mating networks, in which spouses can be drawn from distant groups. Central to making all this work would be high settlement mobility to maintain contacts with distant groups, map the landscape, and monitor resources and environmental conditions beyond the social and geographic boundaries of the local group; and open social networks, to enable individuals to move easily between and be readily integrated within distant groups. Highly territorial behavior would be decidedly disadvantageous. Although certain of those expectations are met (tool caches indeed occur in Clovis [Fig. 1], but not later Paleoindian times), the model has not been fully put to the test, largely for lack of sufficient data with the requisite temporal resolution (Meltzer, 2004).

The End of the Era

One element common to many models of colonization (in North America and elsewhere), is that colonizers on a landscape with few other people not only had to be able to track great distances to find mates and exchange information and resources, they also had to be able to get along with near and distantly related groups they encountered (Kirch, 1997; Lourandos, 1997). Having large and open social networks based on flexible and fluid social and kin relations, fewer languages, the easy integration of individuals and groups, and sometimes long-distance exchange and alliance networks – all combine to diminish differences among peoples who need to be able to readily renew ties under geographic circumstances that might keep them apart for years at a time.

One way these open social systems are manifest in the archaeological realm is by the widespread distribution, use, and exchange of instantly recognizable, and sometimes highly symbolic artifacts – such as unique styles of projectile points or, in the case of prehistoric Oceania, ceramic vessels (Kirch, 1997; Whallon, 1989). These forms served as a "currency" (a term not to be taken too literally) for social and ritual functions, and over long spans and large areas served to maintain recognition and alliances. Early in the Paleoindian period Clovis points are broadly similar stylistically, technologically, and typologically across a vast area of North America. The extent of Clovis distribution is likely a by-product of the size of the dispersal, but their similarity across that range may well reflect common symbols of an extensive social and mating network, which helped to check the attenuating effects of distance.

But those effects were inevitable, cultural drift becoming more pronounced. Although the timing varies by area, new stylistic variants begin appearing sometime after 10,900 ^{14}C yr B.P. in the central and western portions of the continent, and after 10,600 ^{14}C yr B.P. in eastern North America. By 10,500 ^{14}C yr B.P. the once pan-North American form is replaced by a variety regionally-distinctive point forms (Anderson, 1995; Anderson & Faught, 2000; Meltzer, 2002).

Archaeologists have learned in recent years not to place undue weight on style and stylistic change in projectile points, for these may not be telling us about on-the-ground groups, dispersals, or adaptations (Dillehay, 2000;

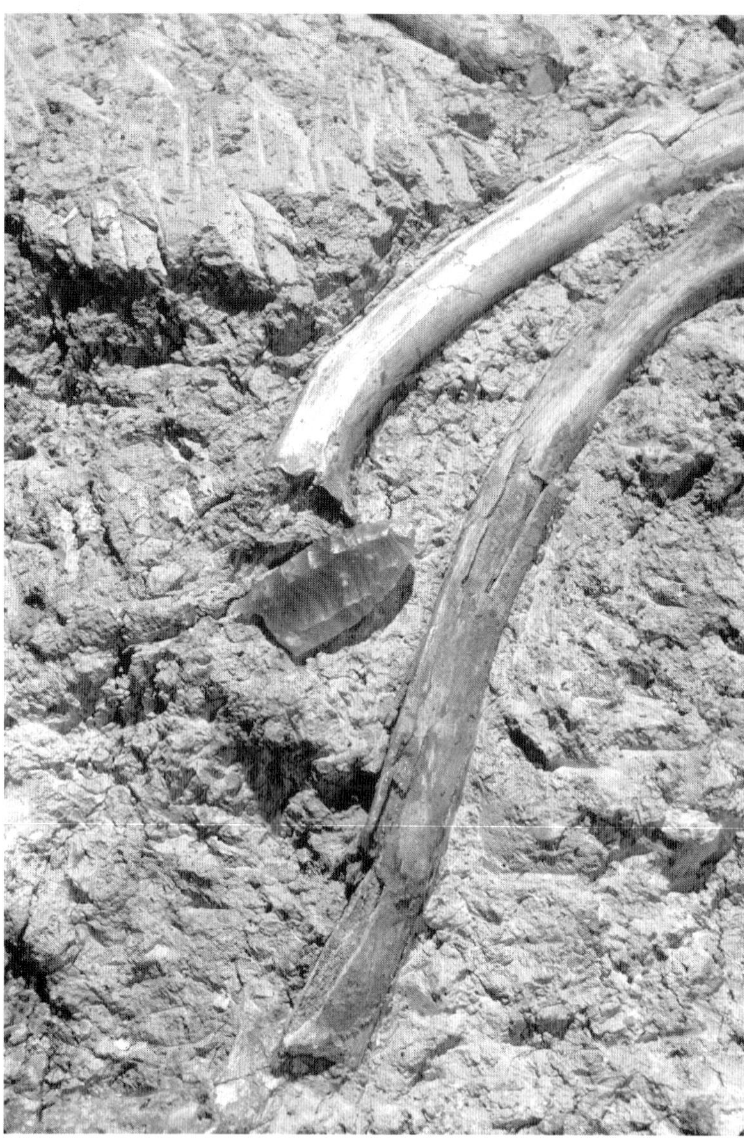

Fig. 4. Folsom projectile point in situ between the ribs of a Bison antiquus at the type site near Folsom, New Mexico. Like Clovis points, Folsom points are also fluted, but these points are routinely smaller, thinner, and more finely made, sometimes showing fine pressure flaking along the margins. This specimen, which is 4.54 cm in length, and is made of Flattop chert, which outcrops ~450 km north of the site. Photograph taken ca. September, 1927, and it marks the first occasion in which artifacts were found in unequivocal association with an extinct species – although in this instance, the genus survived. See the text. Photograph courtesy of Denver Museum of Nature & Science, Denver, Colorado.

Pluciennik, 1996). That said, these more-regionally specific styles that appear in mid-latitude North America starting after 11,000 ^{14}C yr B.P. do seem to correspond with distinct adaptive strategies. Unlike Clovis, these later forms are more restricted geographically, have new and sometime prey- or region-specific foraging strategies which occasionally involved new technologies (the two often co-occur [Binford, 2001]), and in places relied more on locally available stone indicating more restricted mobility. Folsom occupations, for example, appear on the Plains and Rocky Mountains of western North America, with a subsistence strategy tied to exploitation of *Bison antiquus* (Fig. 4) which, owing to a combination of competitive release following megafaunal extinction and the postglacial expansion of C4 grasslands, exploded in numbers. In other areas, late Paleoindian adaptations are different or sometimes less well known (Anderson & Sassaman [Eds], 1996; Beck & Jones, 1997; Frison, 1991).

Assume, for the moment, these stylistic forms mark cultural groups (however defined), and that different forms mark different groups in time and space (leaving aside how or whether these may have been isomorphic with languages or demes). Viewed that way, the shift from a single broad and relatively homogeneous form to multiple regional forms in the span from 11,500 to 10,500 ^{14}C yr B.P., can be seen as the settling of colonizers in specific areas; a relaxation in the pressure to maintain contact with distant kin, and thus a reduction in the spatial scale and openness of the social systems (Meltzer, 2002).

Stepping further out onto this speculative limb, one could attribute these patterns to an overall, continent-wide increase in population which reduced the need to maintain large and open social systems critical to insuring access to resources, information, and mates. Thus, once descendant populations were no longer demographically vulnerable and had little incentive to sustain long distance mating networks, the very vastness of North America and its topographic and geographic barriers would have conspired to impede interaction. The isolation of populations that resulted might help explain the apparently anomalous human skeletons we see in early Holocene times at places like Kennewick, and the variability evident in the genetics of their descendants (Malhi *et al.*, 2002; Powell & Neves, 1999). In effect, these early remains may not resemble contemporary Native Americans, but are ancestral nonetheless (Powell & Neves, 1999).

The challenge, of course, is to devise ways of testing such hypotheses.

Where Do We Go Next? – Unresolved Issues in the Peopling of North America

The last forty years, and especially the last half dozen, have seen great changes in our understanding of the first Americans. Much of what we knew or thought we knew about when the process of colonization began, how often migrations occurred, and how they played out on the late Pleistocene landscape, has been turned on its head. The conventional view of fast moving big-game hunters exploding on to the continent as the ice-free corridor zipped opened in terminal Pleistocene times is no longer a tenable scenario for the initial peopling of North America. Yet, though we now believe the peopling process to have been far more complex – beginning much earlier, possibly involving multiple migratory pulses, and involving a different entry route – fundamental questions about the antiquity, number, timing, and adaptive strategies of the Pleistocene peoples of North America remain. In the scramble to address those questions many ideas – some controversial – are being tried on for size. Time will tell which ones fit.

In the meantime, there is much to do (cf. Fiedel, 2000; Meltzer, 2002). At the most basic level, we need more archaeological data. Although we may never detect the very first archaeological "footprints" of people on the North American continent, for reasons already noted, the lesson of Monte Verde is there must be unequivocal pre-Clovis sites here. The search for these will force the methodological question of whether we are (or have been) looking in the right places and in the right ways for early sites (Butzer, 1991; Collins, 1991; Dillehay & Meltzer, 1991), the answer to which will likely demand a greater contribution of geological and geophysical tools and techniques than now employed in archaeological field programs. Doing so should help fill in the temporal and spatial gaps in the archaeological record between the currently oldest evidence from Monte Verde, and the later and better known Clovis record.

To resolve the question of entry routes, whether via the coast or interior, and if along the coast whether on foot or by boat or some combination therein, we will need direct archaeological evidence: sites, of course, but one can always hope for the discovery of a late-Pleistocene boat. Useful as well would be a firmer handle on the Quaternary geology and environmental history along the possible routes. There has been an increased effort in recent years to find sites marking a coastal entry using coring, remote sensing, bathymetry, and sampling of the submerged paleo-terrain (Fedje & Christensen, 1999; Fedje & Josenhans, 2000). None older than 10,500 ^{14}C yr B.P. have so far been found, but this is an effort with great potential. So too are efforts to model landscape evolution on the mainland, to help find uplifted Pleistocene coast. Although heavy vegetation here and on off shore islands (where remnant coastline may also be preserved) limits surface visibility, these are obvious areas to search.

Ironically, we also still need more data on Clovis. That Clovis groups were not big game hunters begs the question of what they were eating. There is much to be gained by applying intensive recovery methods to Clovis age sites to capture organic remains of their diet (Dent, 2002; Ferring, 2001), by isotopic analysis of recovered human bone, and by searching for habitation and camp sites that can round out the picture of their adaptations.

A critical part of that effort will be the derivation of finer-grained data on the climates and environments of terminal Pleistocene times. What were the available resources, and how they were being exploited by these groups? As a subset of this point, we must address the question of whether late Pleistocene climatic and environmental change was sufficiently rapid and severe that it would have been detectable to and had an impact on human foragers. Ultimately, we need to close the gap between the archaeological and paleoecological temporal scales (recording change on a scale of decades or centuries), and our real-time (daily, weekly, yearly) models of human colonization processes.

With finer-grained climatic and ecological data in hand, we are then poised to apply models of foraging theory to understand the adaptive strategies of hunter-gatherer populations as they may have responded to changes in resource abundance and distribution while moving across space and through time. Although these models were built for real-time activities of hunter-gatherer groups, properly scaled and employing archaeologically measurable variables (Grayson & Cannon, 1999), they could have great value here. For these models can give us insight – testable insight – into Clovis subsistence patterns, as well as the closely linked issue of how foragers on a people-free and unfamiliar landscape grapple with incomplete information and uncertainty. Developing and testing models of subsistence and landscape learning should ultimately answer the still-looming larger question of how and why Clovis groups expanded so far and so fast across late Pleistocene North America.

Finally, that expansion and the potential processes of drift – cultural and genetic – have clear implications for both the archaeological record and the genetics of the descendant populations. It would be fruitful to model how population processes and social mechanisms of initial colonizers (wide-ranging mobility, the maintenance of long-distance mating networks, etc.) may have played out on the late Pleistocene North American landscape, and how these and possible later migrations might be manifest archaeologically, and what might be reflected in the genetic diversity of their descendants. Once the gaps in the archaeological record are filled in, and we possess a more extensive and detailed record of mtDNA and NRY diversity, it would then be especially interesting to determine whether (or at what points) the archaeological and genetic results converge.

One obvious place to seek convergence would be in ancient DNA from late Pleistocene North American human skeletal remains. So far, however, none has been recovered, and the effort to do so will face problems of sample size, the difficulty of amplifying damaged DNA, the potential for contamination with modern DNA – not to mention concerns over scientific access to ancient human skeletal remains (Kolman & Tuross, 2000; cf. Merriwether, 2000). Even so, if (when) ancient DNA is recovered, it will provide the opportunity to test hypotheses derived from DNA of living populations, and reveal whether there were genetic lineages that entered the New World but subsequently went extinct (Meltzer, 1989; Merriwether, 2002). And that might go a long way toward explaining why patterns in the archaeological record and the genetic evidence from modern populations at the moment give different conclusions about the antiquity, number, and timing of migrations to the Americas.

Acknowledgments

I am most grateful to Michael Cannon, Donald Grayson, Vance Holliday, Richard Klein, and Daniel Mann, for their valuable and constructive comments on a draft of this paper. Susan Bruning and Joanne Dickenson supplied important information; Judy Cooper provided much needed expertise in creating the figures. I thank Stephen Porter and Alan Gillespie for the opportunity to write this chapter, and especially their editorial patience and pointers. If all this collective expertise could not save me from error, the fault must surely be mine.

References

Abbott, C.C. (1878). Second report on the Paleolithic implements from the glacial drift in the valley of the Delaware River, near Trenton, New Jersey. *Peabody Museum Annual Report*, **11**, 225–257.

Adovasio, J. & Page, J. (2002). *The first Americans: In pursuit of archaeology's greatest mystery*. New York, Random House, 328 pp.

Adovasio, J. & Pedler, D. (1997). Monte Verde and the antiquity of humankind in the Americas. *Antiquity*, **71**, 573–580.

Adovasio, J., Donahue, J. & Stuckenrath, R. (1990). The Meadowcroft Rockshelter radiocarbon chronology 1975–1990. *American Antiquity*, **55**, 348–354.

Adovasio, J., Pedler, D., Donahue, J. & Stuckenrath, R. (1999). No vestige of a beginning nor prospect for an end: Two decades of debate on Meadowcroft Rockshelter. *In*: Bonnichsen, R. & Turnmire, K. (Eds), *Ice-Age People of North America: Environments, Origins, and Adaptations*. Corvallis, Oregon, Center for the Study of the First Americans, pp. 415–431.

Akazawa, T. & Szathmary, E. (Eds) (1996). *Prehistoric Mongoloid dispersals*. Oxford, Oxford University Press, 389 pp.

Alroy, J. (2001). A multispecies overkill simulation of the end-Pleistocene megafaunal mass extinction. *Science*, **292**, 1893–1896.

Anderson, D. (1995). Paleoindian interaction networks in the eastern woodlands. *In*: Nassaney, M. & Sassaman, K. (Eds), *Native American Interaction: Multiscalar Analyses and Interpretations in the Eastern Woodlands*. Knoxville, University of Tennessee Press, pp. 1–26.

Anderson, D. & Faught, M. (2000). Palaeoindian artifact distribution: Evidence and implications. *Antiquity*, **74**, 507–513.

Anderson, D. & Gillam, J.C. (2000). Paleoindian colonization of the Americas: Implications from an examination of physiography, demography, and artifact distribution. *American Antiquity*, **65**, 43–66.

Anderson, D. & Gillam, J.C. (2001). Paleoindian interaction and mating networks: Reply to Moore and Moseley. *American Antiquity*, **66**, 530–535.

Anderson, D. & Sassaman, K. (Eds) (1996). *The Paleoindian and early Archaic southeast*. Tuscaloosa, University of Alabama Press, 526 pp.

Anderson, D., Brose, D., Dincauze, D., Grumet, R. Shott, M. & Waldbauer, R. (2002). *The earliest Americans theme study for the eastern United States*. National Park Service, National Historic Landmark Series, 422 pp.

Anderson, E. (1984). Who's who in the Pleistocene: A mammalian bestiary. *In:* Martin, P. & Klein, R. (Eds), *Quaternary Extinctions: A Prehistoric Revolution*. Tucson: University of Arizona Press, pp. 40–89.

Anthony, D. (1990). Migration in archaeology: The baby and the bathwater. *American Anthropologist*, **92**, 895–914.

Baumhoff, M. & Heizer, R. (1965). Postglacial climate and archaeology in the desert west. *In*: Wright, H. & Frey. D. (Eds), *The Quaternary of the United States*. Princeton, Princeton University Press, pp. 697–707.

Beck, C. & Jones, G. (1997). The terminal Pleistocene/Early Holocene archaeology of the Great Basin. *Journal of World Prehistory*, **11**, 161–236.

Bigelow, N. & Powers, R. (2001). Climate, vegetation, and archaeology 14,000–9,000 cal yr B.P. *Arctic Anthropology*, **38**, 171–195.

Binford, L. (1979). Organization and formation processes: Looking at curated technologies. *Journal of Anthropological Research*, **35**, 172–197.

Binford, L. (1981). *Bones: Ancient men and modern myths*. New York, Academic Press, 320 pp.

Binford, L. (2001). *Constructing frames of reference*. Berkeley, University of California Press, 563 pp.

Boas, F. (1912). Migrations of Asiatic races and cultures to North America. *Science Monthly*, **28**, 110–117.

Bonnichsen, R. (1991). Clovis origins. *In*: Bonnichsen, R. & Turnmire, K. (Eds), *Clovis Origins and Adaptations*. Corvallis, Oregon, Center for the Study of the First Americans, pp. 309–329.

Bonnichsen, R. & Steele, D. (Eds) (1994). *Method and theory for investigating the peopling of the Americas*. Corvallis, Oregon, Center for the Study of the First Americans, 264 pp.

Bonnichsen, R. & Turnmire, K.L. (Eds) (1991). *Clovis: Origins and adaptations*. Corvallis, Oregon, Center for the Study of the First Americans, 344 pp.

Bonnichsen, R. & Turnmire, K.L. (Eds) (1999). *Ice-Age People of North America: Environments, Origins, and Adaptations*. Corvallis, Oregon, Center for the Study of the First Americans, 536 pp.

Bonnichsen, R. & Will, R. (1999). Radiocarbon chronology of northeastern Paleoamerican sites: Discriminating natural and human burn features. *In*: Bonnichsen, R. & Turnmire, K. (Eds), *Ice-Age People of North America: Environments, Origins, and Adaptations*. Corvallis, Oregon, Center for the Study of the First Americans, pp. 395–415.

Brace, C.L., Nelson, A., Seguchi, N., Oe, H., Sering, L., Qifeng, P., Yongyi, L. & Tumen, D. (2001). Old World sources of the first New World human inhabitants: A comparative craniofacial view. *Proceedings of the National Academy of Sciences*, **98**, 10017–10022.

Broughton, J. (1994). Declines in mammalian foraging efficiency during the Late Holocene, San Francisco Bay, California. *Journal of Anthropological Archaeology*, **13**, 371–401.

Brown, M., Hoseini, S., Torroni, A., Bandeldt, H., Allen, J., Schurr, T. & Wallace, D. (1998). mtDNA haplogroup X: An ancient link between Europe/western Asia and North America? *American Journal of Human Genetics*, **63**, 1852–1870.

Bryan, A. & Tuohy, D. (1999). Prehistory of the Great Basin/Snake River Plain to about 8,500 years ago. *In*: Bonnichsen, R. & Turnmire, K. (Eds), *Ice-Age People of North America: Environments, Origins, and Adaptations*. Corvallis, Oregon, Center for the Study of the First Americans, pp. 249–263.

Bryan, A. (Ed.) (1978). *Early Man in America from a Circum-pacific Perspective*. Edmonton, Archaeological Researches International, 327 pp.

Bryan, A. (Ed.) (1986). *New Evidence for the Pleistocene Peopling of the Americas*. Orono, Maine, Center for the Study of Early Man, 368 pp.

Butzer, K.W. (1991). An Old World perspective on potential mid-Wisconsinan settlement of the Americas. *In*: Dillehay, T. & Meltzer, D. (Eds), *The First Americans: Search and Research*. Boca Raton, CRC Press, pp. 137–156.

Carlisle, R. (Ed.) (1988). Americans before Columbus: Ice Age origins. *Ethnology Monographs*, **12**, 123 pp.

Carter, G. (1957). *Pleistocene Man at San Diego*. Baltimore, Johns Hopkins Press, 400 pp.

Catto, N. (1996). Richardson Mountains, Yukon-Northwest Territories: The northern portal of the postulated "ice-free corridor." *Quaternary International*, **32**, 3–19.

Chamberlin, T. (1903). The criteria requisite for the reference of relics to a glacial age. *Journal of Geology*, **11**, 64–85.

Chatters, J. (2000). The recovery and first analysis of an Early Holocene human skeleton from Kennewick, Washington. *American Antiquity*, **65**, 291–316.

Clague, J. (1989). Quaternary sea levels. *In*: Fulton, R. (Ed.), Quaternary geology of Canada and Greenland: Geological Survey of Canada, *Geology of Canada*, **1**, 43–47.

Clague, J., Mathewes, R. & Ager, T. (2004). Environments of northwest North America before the Last Glacial Maximum *In*: Madsen, D. (Ed.), *Entering America: Northeast Asia and Beringia Before the Last Glacial Maximum*. Salt Lake City, University of Utah Press, in press.

Collins, M. (1991). Rockshelters and the early archaeological record in the Americas. *In*: Dillehay, T. & Meltzer, D. (Eds), *The First Americans: Search and Research*. Boca Raton, CRC Press, pp. 157–182.

Collins, M. (1998). Interpreting the Clovis artifacts from the Gault site. *TARL Research Notes*, **6**, 5–12.

Collins, M. (1999). *Clovis Blade Technology*. Austin, University of Texas Press, 234 pp.

Damas, D. (Ed.) (1984). *Handbook of North American Indians* (Vol. 5). Arctic: Washington, DC, Smithsonian Institution Press, 829.

Dent, R. (2002). Paleoindian occupation of the Upper Delaware Valley: Revisiting Shawnee-Minisink and nearby sites. *In*: Carr, K. & Adovasio, J. (Eds), *Ice Age Peoples of Pennsylvania*. Harrisburg, Pennsylvania Historical and Museum Commission, pp. 51–78.

Derev'anko, A. (1998). *The Paleolithic of Siberia*. Urbana, University of Illinois Press, 406 pp.

Dillehay, T. (1989). *Monte Verde: A Late Pleistocene Settlement in Chile, Volume 1: Palaeoenvironment and Site Context*. Washington, DC, Smithsonian Institution Press, 306 pp.

Dillehay, T. (1997). *Monte Verde: A Late Pleistocene Settlement in Chile, Volume 2: The Archaeological Context and Interpretation*. Washington, D.C., Smithsonian Institution Press, 1071 pp.

Dillehay, T. (2000). *The Settlement of the Americas: A New Prehistory*. New York, Basic Books, 371 pp.

Dillehay, T. & Meltzer, D. (1991). Finale: Processes and prospects *In*: Dillehay, T. & Meltzer, D. (Eds), *The First Americans: Search and Research*. Boca Raton, CRC Press, pp. 287–294.

Dillehay, T. & Meltzer, D. (Eds) (1991). *The first Americans: Search and research*. Boca Raton, CRC Press, 310 pp.

Dincauze, D. (1984). An archaeo-logical evaluation of the case for pre-Clovis occupations. *Advances in World Archaeology*, **3**, 275–323.

Dincauze, D. (2002). Northeast context. *In*: Anderson, D., Brose, D., Dincauze, D., Grumet, R., Shott, M. & Waldbauer, R. (Eds), *The Earliest Americans Theme Study for the Eastern United States*. National Park Service, National Historic Landmark Series, pp. 26–41.

Dixon, E.J. (1999). *Bones, Boats & Bison. Archaeology and the First Colonization of Western North America*. Albuquerque, University of New Mexico Press, 322 pp.

Elias, S. (2002). Setting the stage: Environmental conditions in Beringia as people entered the New World. *In*: Jablonski, N. (Ed.), *The First Americans, the Pleistocene Colonization of the New World*. Memoirs of the California Academy of Sciences, Number 27, pp. 9–25.

Elias, S. & Brigham-Grette, J. (Eds) (2001). Beringian paleoenvironments. *Quaternary Science Reviews*, **20**, 1–574.

Elias, S., Short, S., Nelson, C. & Birks, H. (1996). Life and times of the Bering Land Bridge. *Nature*, **382**, 60–63.

Ericson, J., Taylor, R. & Berger, R. (Eds) (1982). *Peopling of the New World*. San Diego, Ballena Press, 364 pp.

Erlandson, J. (2002). Anatomically modern humans, maritime voyaging, and the Pleistocene colonization of the Americas. *In*: Jablonski, N. (Ed.), *The First Americans, the Pleistocene Colonization of the New World*. Memoirs of the California Academy of Sciences, Number 27, pp. 59–92.

FAUNMAP Working Group (1994). FAUNMAP: A database documenting late Quaternary distributions of mammal species in the United States. *Illinois State Museum Scientific Papers*, 25.

Feathers, J. (1997). The application of luminescence dating in American archaeology. *Journal of Archaeological Method and Theory*, **4**, 1–66.

Fedje, D. & Christensen, T. (1999). Modeling paleoshorelines and locating Early Holocene coastal sites in Haida Gwaii. *American Antiquity*, **64**, 635–652.

Fedje, D. & Josenhans, H. (2000). Drowned forests and archaeology on the continental shelf of British Columbia, Canada. *Geology*, **28**, 99–102.

Fedje, D., White, J., Wilson, M., Nelson, D., Vogel, J. & Southon, J. (1995). Vermilion Lakes Site: Adaptations and Environments in the Canadian Rockies During the Latest Pleistocene and Early Holocene. *American Antiquity*, **60**, 81–108.

Ferring, C.R. (2001). *The Archaeology and Paleoecology of the Aurbrey Clovis Site (41DN479) Denton County, Texas*. Center for Environmental Archaeology, Department of Geography, University of North Texas, Denton.

Fiedel, S. (1999). Older than we thought: Implications of corrected dates for Paleoindians. *American Antiquity*, **64**, 95–116.

Fiedel, S. (2000). The peopling of the New World: Present evidence, new theories, and future directions. *Journal of Archaeological Research*, **8**, 39–103.

Forster, P., Harding, R., Torroni, A. & Bandelt, H.-J. (1996). Origin and evolution of Native American mtDNA variation: A reappraisal. *American Journal of Human Genetics*, **59**, 935–945.

Frison, G. (1991). *Prehistoric Hunters of the High Plains*. New York, Academic Press, 532 pp.

Frison, G. (1999). The late Pleistocene prehistory of the northwestern Plains, the adjacent mountains, and intermontane basins *In*: Bonnichsen, R. & Turnmire, K. (Eds), *Ice-Age People of North America: Environments, Origins, and Adaptations*. Corvallis, Oregon, Center for the Study of the First Americans, pp. 264–280.

Frison, G. & Bradley, B. (1999). *The Fenn Cache: Clovis weapons and tools*. Santa Fe, One Horse Land and Cattle Company.

Gardner, W. (2002). The Paleoindian problem revisited: Observations on Paleoindian in Pennsylvania (a slightly southern slant) *In*: Carr, K. & Adovasio, J. (Eds), *Ice Age Peoples of Pennsylvania*. Harrisburg, Pennsylvania Historical and Museum Commission, pp. 97–103.

Gerlach, C. & Murray, M. (Eds) (2001). *People and Wildlife in Northern North America: Essays in Honor of R. Dale Guthrie*. BAR International Series 944.

Goebel, T. (1999). Pleistocene human colonization of Siberia and peopling of the Americas: An ecological approach. *Evolutionary Anthropology*, **8**, 208–227.

Goebel, T., Powers, R. & Bigelow, N. (1991). The Nenana Complex of Alaska and Clovis origins. *In*: Bonnichsen, R. & Turnmire, K. (Eds), *Clovis Origins and Adaptations*. Corvallis, Oregon, Center for the Study of the First Americans, pp. 49–79.

Goebel, T. & Slobodin, S. (1999). The colonization of western Beringia: Technology, ecology, and adaptations. *In*: Bonnichsen, R. & Turnmire, K. (Eds), *Ice-Age People of North America: Environments, Origins, and Adaptations*. Corvallis, Oregon, Center for the Study of the First Americans, pp. 104–155.

Goldberg, P. & Arpin, T. (1999). Micromorphological analysis of sediments from Meadowcroft Rock shelter, Pennsylvania: Implications for radiocarbon dating. *Journal of Field Archaeology*, **26**, 325–342.

Goodyear, A. (1979). A hypothesis for the use of cryptocrystalline raw materials among Paleo-indian groups of North America. Research Manuscript Series No. 156. Institute of Archaeology and Anthropology, University of South Carolina, Columbia.

Goodyear, A. (1999). The Early Holocene occupation of the southeastern United States. *In*: Bonnichsen, R. & Turnmire, K. (Eds), *Ice-Age People of North America: Environments, Origins, and Adaptations*. Corvallis, Oregon, Center for the Study of the First Americans, pp. 432–481.

Goodyear, A. (2001). The 2001 Allendale Paleoindian expedition and beyond. *Legacy*, **6**, 18–21.

Grayson, D. (1984). Nineteenth-century explanations of Pleistocene extinctions: A review and analysis. *In*: Martin, P. & Klein, R. (Eds), *Quaternary Extinctions: A Prehistoric Revolution*. Tucson, University of Arizona Press. pp. 5–39.

Grayson, D. (1991). Late Pleistocene mammalian extinctions in North America: Taxonomy, chronology, and explanations. *Journal of World Prehistory*, **5**, 193–231.

Grayson, D. (1993). *The Desert's Past: A Natural Prehistory of the Great Basin*. Washington, Smithsonian Institution Press, 356 pp.

Grayson, D. (2001). The archaeological record of human impacts on animal populations. *Journal of World Prehistory*, **15**, 1–68.

Grayson, D. & Cannon, M. (1999). Human paleoecology and foraging theory in the Great Basin. *In*: Beck, C. (Ed.), *Models for the Millennium: Great Basin Anthropology Today*. Salt Lake, University of Utah Press, pp. 141–151.

Grayson, D. & Meltzer, D. (2002). Clovis hunting and large mammal extinctions. *Journal of World Prehistory*, **16**, 313–359.

Grayson, D. & Meltzer, D. (2003). Requiem for North American overkill. *Journal of Archaeological Science*, **30**, 585–593.

Green, T., Cochran, B., Fenton, T., Woods, J., Titmus, G., Tieszen, L., Davis, M. & Miller, S. (1998). The Buhl burial: A Paleoindian woman from southern Idaho. *American Antiquity*, **63**, 437–456.

Greenberg, J. (1987). *Language in the Americas*. Stanford, Stanford University Press, 438 pp.

Greenberg, J., Turner, C. & Zegura, S. (1986). The settlement of the Americas: A comparison of the linguistic, dental, and genetic evidence. *Current Anthropology*, **27**, 477–497.

Griffin, J. (1965). Late Quaternary prehistory in the northeastern woodlands. *In*: Wright, H. & Frey, D. (Eds), *The Quaternary of the United States*. Princeton, Princeton University Press, pp. 655–667.

Griffin, J., Meltzer, D., Smith, B. & Sturtevant, W. (1988). A mammoth fraud in science. *American Antiquity*, **53**, 578–582.

Grimm, E., Jacobson, G., Watts, W., Hansen, B., Maasch, K. *et al.* (1993). A 50,000 year record of climate oscillations from Florida and its temporal correlation with the Heinrich events. *Science*, **261** 198–200.

Guthrie, R.D. (1990). *Frozen Fauna of the Mammoth Steppe*. Chicago, University of Chicago Press, 323 pp.

Guthrie, R.D. (2001). Origin and causes of the mammoth steppe: A story of cloud cover, wooly mammoth tooth pits, buckles, and inside-out Beringia. *Quaternary Science Reviews*, **20**, 549–574.

Hamilton, T. & Goebel, T. (1999). Late Pleistocene peopling of Alaska. *In*: Bonnichsen, R. & Turnmire, K. (Eds), *Ice-Age People of North America: Environments, Origins, and Adaptations*. Corvallis, Oregon, Center for the Study of the First Americans, pp. 156–199.

Haynes, C. (1964). Fluted projectile points: Their age and dispersion. *Science*, **145**, 1408–1413.

Haynes, C. (1969). The earliest Americans. *Science*, **166**, 709–715.

Haynes, C. (1982). Where Clovis progenitors in Beringia? *In*: Hopkins, D., Mathews, J., Schweger, C. & Young, S. (Eds), *Paleoecology of Beringia*. New York, Academic Press, pp. 383–398.

Haynes, C. (1987). Clovis origins update. *The Kiva*, **52**, 83–93.

Haynes, C. (1991). Geoarchaeological and paleohydrological evidence for a Clovis-age drought in North America. *Quaternary Research*, **35**, 438–450.

Haynes, C., Donahue, D., Jull, A. & Zabel, T. (1984). Application of accelerator dating to fluted point Paleoindian sites. *Archaeology of Eastern North America*, **12**, 184–191.

Haynes, G. (1991). *Mammoths, mastodonts, and elephants*. Cambridge, Cambridge University Press, 413 pp.

Haynes, G. (2002). The catastrophic extinction of North American mammoths and mastodonts. *World Archaeology*, **33**, 391–416.

Haynes, G. & Stanford, D. (1984). On the possible utilization of Camelops by early man in North America. *Quaternary Research*, **22**, 216–230.

Helm, J. (Ed.) (1981). *Handbook of North American Indians* (Vol. 6, Subarctic 837 pp). Washington, DC, Smithsonian Institution Press.

Hester, J. (1960). Late Pleistocene extinction and radiocarbon dating. *American Antiquity*, **26**, 58–76.

Hoffecker, J. (2001). Late Pleistocene and Early Holocene sites in the Nenana Valley, Alaska. *Arctic Anthropology*, **38**, 139–153.

Hofman, J. & Todd, L. (2001). Tyranny in the archaeological record of specialized hunters. *In*: Gerlach, C. & Murray, M. (Eds), *People and Wildlife in Northern North America: Essays in Honor of R. Dale Guthrie*. BAR International Series 944, pp. 200–215.

Holliday, V. (2000). The evolution of Paleoindian geochronology and typology on the Great Plains. *Geoarchaeology*, **15**, 227–290.

Holmes, C. (2001). Tanana Valley archaeology circa 12,000 to 8500 yrs B.P. *Arctic Anthropology*, **38**, 154–170.

Hopkins, D. (Ed.) (1967). *The Bering Land Bridge*. Stanford, Stanford University Press, 495 pp.

Hopkins, D., Matthews, J., Schweger, C. & Young, S. (Eds) (1982). *Paleoecology of Beringia*. New York, Academic Press, 489 pp.

Hughen, K., Southon, J., Lehman, S. & Overpeck, J. (2000). Synchronous radiocarbon and climate shifts during the last deglaciation. *Science*, **290**, 1951–1954.

Humphrey, R. & Stanford, D. (Eds) (1979). *Pre-Llano cultures of the Americas: Paradoxes and Possibilities*. Washington, DC, Anthropological Society of Washington, 150 pp.

Irving, W. (1985). Context and chronology of early man in the Americas. *Annual Review of Anthropology*, **14**, 529–555.

Jablonski, N. (Ed.) (2002). The first Americans, the Pleistocene colonization of the New World. *Memoirs of the California Academy of Sciences*, Number 27, 331 pp.

Jackson, L. & Duk-Rodin, A. (1996). Quaternary geology of the ice-free corridor: Glacial controls on the peopling of the New World. *In*: Akazawa, T. & Szathmary, E. (eds), *Prehistoric Mongoloid Dispersals*. Oxford, Oxford University Press, pp. 214–227.

Johnson, E. (1991). Late Pleistocene cultural occupation on the southern Plains. *In*: Bonnichsen, R. & Turnmire, K. (Eds), *Clovis Origins and Adaptations*. Corvallis, Oregon, Center for the Study of the First Americans, pp. 133–152.

Josenhans, H., Fedje, D., Pienitz, R. & Southon, J. (1997). Early humans and rapidly changing Holocene sea levels in the Queen Charlotte Islands – Hecate Strait, British Columbia. *Science*, **277**, 71–74.

Kaplan, H. & Hill, K. (1992). The evolutionary ecology of food acquisition. *In*: Smith, E. & Winterhalder, E. (Eds), *Evolutionary Ecology and Human Behavior*. New York, Aldine de Gruyter, pp. 167–201.

Karafet, T., Zegura, S., Posukh, O., Ospiva, L., Templeton, A. & Hammer, M. (1999). Ancestral Asian source(s) of New World Y-chromosome founder haplotypes. *American Journal of Human Genetics*, **64**, 817–831.

Keegan, W. & Diamond, J. (1987). Colonization of islands by humans: A biogeographical perspective. *Advances in Archaeological Method and Theory*, **10**, 49–92.

Kelly, R. (1995). *The Foraging Spectrum: Diversity in Hunter-Gatherer Lifeways*. Washington, DC, Smithsonian Institution Press, 446 pp.

Kelly, R. (1996). Ethnographic analogy and migration to the western hemisphere. *In*: Akazawa, T. & Szathmary, E. (Eds), *Prehistoric Mongoloid Dispersals*. Oxford, Oxford University Press, pp. 228–240.

Kelly, R. (1999). Hunter-gatherer foraging and colonization of the western hemisphere. *Anthropologie*, **37**, 143–153.

Kelly, R. & Todd, L. (1988). Coming into the country: Early Paleoindian hunting and mobility. *American Antiquity*, **53**, 231–244.

King, M. & Slobodin, S. (1996). A fluted point from the Uptar site, northeastern Siberia. *Science*, **273**, 634–636.

Kirch, P. (1997). *The Lapita Peoples: Ancestors of the Ocean World*. Cambridge, Blackwell Publishers, 353 pp.

Kirk, R. & Szathmary, E. (Eds) (1985). Out of Asia: Peopling the Americas and the Pacific. Canberra, Australia. *The Journal of Pacific History*, 226 pp.

Kitagawa, H. & van der Plicht, J. (1998). Atmospheric radiocarbon calibration to 45,000 yr B.P. Late Glacial fluctuations and cosmogenic isotope production. *Science*, **279**, 1187–1190.

Klein, R. (1999). *The Human Career: Human Biological and Cultural Origins* (2nd ed.). Chicago, University of Chicago Press, 810 pp.

Kolman, C., Sambuughin, N. & Bermingham, E. (1996). Mitochondrial DNA analysis of Mongolian populations and implications for the origin of New World founders. *Genetics*, **142**, 1321–1334.

Kolman, C. & Tuross, N. (2000). Ancient DNA analysis of human populations. *American Journal of Physical Anthropology*, **111**, 5–23.

Kraft, J. & Thomas, R. (1976). Early man at Holly Oak, Delaware. *Science*, **192**, 756–761.

Kreiger, A. (1954). A comment on "Fluted point relationships" by John Witthoft. *American Antiquity*, **19**, 273–275.

Krieger, A. (1964). Early man in the New World. *In*: Jennings, J. & Norbeck, E. (Eds), *Prehistoric Man in the New World*. Chicago, University of Chicago Press, pp. 23–81.

Kurten, B. & Anderson, E. (1980). *Pleistocene Mammals of North America*. New York, Columbia University Press, 442 pp.

Laub, R. (2002). The Paleoindian presence in the northeast: A view from the Hiscock site. *In*: Carr, K. & Adovasio, J. (Eds), *Ice Age Peoples of Pennsylvania*. Harrisburg, Pennsylvania Historical and Museum Commission, pp. 105–121.

Lell, J., Sukernik, R., Starikovskaya, Y., Su, B., Jin, L., Schurr, T., Underhill, P. & Wallace, D. (2002). The dual origin and Siberian affinities of Native American Y chromosomes. *American Journal of Human Genetics*, **70**, 192–206.

Lepper, B. (1999). Pleistocene peoples of midcontinental North America. *In*: Bonnichsen, R. & Turnmire, K. (Eds), *Ice-Age People of North America: Environments, Origins, and Adaptations*. Corvallis, Oregon, Center for the Study of the First Americans, pp. 362–394.

Lopinot, N., Ray, J. & Conner, M. (1998). The 1997 excavations at the Big Eddy site (23CE426) in southwest Missouri. Southwest Missouri State University, *Special Publication*, No. 2.

Lopinot, N., Ray, J. & Conner, M. (2000). The 1999 excavations at the Big Eddy site (23CE426). Southwest Missouri State University, *Special Publication*, No. 2.

Lourandos, H. (1997). *Continent of Hunter-Gatherers: New Perspectives in Australian Prehistory*. Cambridge, Cambridge University Press, 390 pp.

Lyman, R. (1994). *Vertebrate Taphonomy*. Cambridge, Cambridge University Press, 524 pp.

MacAvoy, J. & McAvoy, L. (1997). Archaeological investigations of site 44SX202, Cactus Hill, Sussex County, Virginia. Research Report Series No. 8, Department of Historic Resources, Commonwealth of Virginia.

MacAvoy, J., Baker, J., Feathers, J., Hodges, R., McWeeney, L. & Whyte, T. (2000). Summary of research at the Cactus Hill archaeological site, 44SX202, Sussex County, Virginia. Report to the National Geographic Society in compliance with stipulations of Grant #6345–98.

MacNeish, R. (1976). Early man in the New World. *American Scientist*, **63**, 316–327.

Malhi, R., Eshelman, J., Greenberg, J., Weiss, D., Shook, B., Kaestle, F., Lorenz, J., Kemp, B., Johnson, J. & Smith, D. (2002). The structure and diversity within New World mitochondrial DNA haplogroups: Implications for the prehistory of North America. *American Journal of Human Genetics*, **70**, 905–919.

Mandryk, C. (1993). Hunter-gatherer social costs and the nonviability of submarginal environments. *Journal of Anthropological Research*, **49**, 39–71.

Mandryk, C. (1996). Late Wisconsinan deglaciation of Alberta: Processes and paleogeography. *Quaternary International*, **32**, 79–85.

Mandryk, C. (2001). The ice-free corridor (or not?): An inland route by any other name is not so sweet nor adequately considered. *In*: Gillespie, J., Tupakka, S. & de Mille, C. (Eds), *On Being First: Cultural Innovation and Environmental Consequences of First Peoplings*. Calgary, The Archaeological Association of the University of Calgary, pp. 575–588.

Mandryk, C., Josenhans, H., Fedje, D. & Mathewes, R. (2001). Late Quaternary paleoenvironments of northwestern North America: Implications for inland vs. coastal migration routes. *Quaternary Science Reviews*, **20**, 301–314.

Mann, D. & Hamilton, T. (1995). Late Pleistocene and Holocene environments of the North Pacific Coast. *Quaternary Science Reviews*, **14**, 449–471.

Mann, D. & Peteet, D. (1994). Extent and timing of the Last Glacial maximum in Southwestern Alaska. *Quaternary Research*, **42**, 136–148.

Mann, D., Reanier, R., Peteet, D., Kunz, M. & Johnson, M. (2001). Environmental change and arctic Paleoindians. *Arctic Anthropology*, **38**, 119–138.

Martin, P. (1967). Prehistoric overkill. *In*: Martin, P. & Wright, H. (Eds), *Pleistocene Extinctions: The Search for a Cause*. New Haven, Yale University Press, pp. 75–120.

Martin, P. (1973). The discovery of America. *Science*, **179**, 969–974.

Martin, P. (1984). Prehistoric overkill: The global model. *In*: Martin, P. & Klein, R. (Eds), *Quaternary Extinctions: A Prehistoric Revolution*. Tucson, University of Arizona Press, pp. 354–403.

Martin, P. (1987). Clovisia the beautiful. *Natural History*, **96**, 10–13.

Martin, P. (1990). Who or what destroyed our mammoths? *In*: Agenbroad, L., Mead, J. & Nelson, L. (Eds), *Megafauna and Man: Discovery of America's Heartland*. Hot Springs, The Mammoth Site of Hot Springs, South Dakota. pp. 109–117.

Mason, R. (1958). Late Pleistocene geochronology and the Paleo-indian penetration into the lower Michigan peninsula. University of Michigan Museum of Anthropology, *Anthropological Papers*, 11, 48 pp.

Mason, R. (1962). The Paleo-Indian tradition in eastern North America. *Current Anthropology*, **3**, 227–278.

Matheus, P. (2001). Pleistocene predators and people in eastern Beringia: Did short-faced bears really keep humans out of North America? *In*: Gerlach, C. & Murray, M. (Eds), *People and Wildlife in Northern North America: Essays in Honor of R. Dale Guthrie*. BAR International Series 944, pp. 79–101.

McDonald, J. (2000). An outline of the pre-Clovis archaeology of SV-2, Saltville, Virginia, with special attention to a bone tool dated 14,510 yr B.P. *Contributions from the Virginia Museum of Natural History*, **9**.

Mead, J. & Meltzer, D. (Eds) (1985). *Environments and Extinctions: Man in Late Glacial North America*. Orono, Maine, Center for the Study of Early Man, 209 pp.

Meighan, C. (1965). Pacific coast archaeology. *In*: Wright, H. & Frey, D. (Eds), *The Quaternary of the United States*. Princeton, Princeton University Press, pp. 709–719.

Meltzer, D. (1988). Late Pleistocene human adaptations in eastern North America. *Journal of World Prehistory*, **2**, 1–52.

Meltzer, D. (1989). Why don't we know when the first people came to North America? *American Antiquity*, **54**, 471–490.

Meltzer, D. (1993). Is there a Clovis adaptation? *In*: Soffer, O. & Praslov, N. (Eds), *From Kostenki to Clovis: Upper Paleolithic – Paleo-Indian Adaptations*. New York, Plenum Press, pp. 293–310.

Meltzer, D. (1995). Clocking the first Americans. *Annual Review of Anthropology*, **24**, 21–45.

Meltzer, D. (2002). What do you do when no one's been there before? Thoughts on the exploration and colonization of new lands. *In*: Jablonski, N. (Ed.), The first Americans, The Pleistocene colonization of the New World. *Memoirs of the California Academy of Sciences*, Number 27, pp. 27–58.

Meltzer, D. (2003). Lessons in landscape learning. *In*: Rockman, M. & Steele, J. (Eds), *Colonization of Unfamiliar Landscapes: The Archaeology of Adaptation*. London, Routledge, pp. 224–241.

Meltzer, D. (2004). Modeling the initial colonization of the Americas: Issues of scale, demography, and landscape learning. *In*: Clark, G. & Barton, M. (Eds), *Pioneers on the Land: The Initial Human Colonization of the Americas*. Tucson, University of Arizona Press.

Meltzer, D., Grayson, D., Ardila, G., Barker, A., Dincauze, D., Haynes, C., Mena, F., Núñez, L. & Stanford, D. (1997). On the Pleistocene antiquity of Monte Verde, southern Chile. *American Antiquity*, **62**, 659–663.

Meltzer, D. & Mead, J. (1983). The timing of Late Pleistocene mammalian extinctions in North America. *Quaternary Research*, **19**, 130–135.

Meltzer, D. & Mead, J. (1985). Dating late Pleistocene extinctions: Theoretical issues, analytical bias and substantive results. *In*: Mead, J. & Meltzer, D. (Eds), *Environments and Extinctions: Man in Late Glacial North America*. Orono, Maine, Center for the Study of Early Man, pp. 145–174.

Meltzer, D. & Sturtevant, W. (1983). The Holly Oak shell game: An historic archaeological fraud. *In:* Dunnell, R. & Grayson D. (Eds), *Lulu Linear Punctated: Essays in Honor of George Irving Quimby*. University of Michigan Museum of Anthropology, Anthropological Papers No. 72, pp. 325–352.

Merriwether, D. (2000). Ancient DNA and Kennewick man: A review of Tuross and Kolman's Kennewick man ancient DNA report. *Current Research in the Pleistocene*, **17**, 97–100.

Merriwether, D. (2002). A mitochondrial perspective on the peopling of the New World. *In*: Jablonski, N. (Ed.), The first Americans, the Pleistocene colonization of the New World. *Memoirs of the California Academy of Sciences*, Number 27, pp. 295–310.

Merriwether, D., Hall, W., Vahlne, A. & Ferrell, R. (1996). mtDNA variation indicates Mongolia may have been the source for the founding population for the New World. *American Journal of Human Genetics*, **59**, 204–212.

Moeller, R. (2002). Paleoindian settlement pattern: Just a stone's throw from the lithic source. *In*: Carr, K. & Adovasio, J. (Eds), *Ice Age Peoples of Pennsylvania*. Harrisburg, Pennsylvania Historical and Museum Commission, pp. 91–95.

Moore, J. (2001). Evaluating five models of human colonization. *American Anthropologist*, **103**, 395–408.

Moore, J. & Moseley, M. (2001). How many frogs does it take to leap around the Americas? Comments on Anderson and Gillam. *American Antiquity*, **66**, 526–529.

Morrow, J. & Morrow, T. (1999). Geographic variation in fluted projectile points: A hemispheric perspective. *American Antiquity*, **64**, 215–231.

Mosimann, J. & Martin, P. (1975). Simulating overkill by Paleoindians. *American Scientist*, **63**, 304–313.

Nichols, J. (1990). Linguistic diversity and the first settlement of the New World. *Language*, **66**, 475–521.

Nichols, J. (2002). The first American languages. *In*: Jablonski, N. (Ed.), The first Americans, the Pleistocene colonization of the New World. *Memoirs of the California Academy of Sciences*, Number 27, pp. 273–293.

Peteet, D. (2000). Sensitivity and rapidity of vegetational response to abrupt climate change. *Proceedings of the National Academy of Sciences*, **97**, 1359–1361.

Pluciennik, M. (1996). Genetics, archaeology, and the wider world. *Antiquity*, **70**, 13–14.

Politis, G. (1991). Fishtail projectile points in the southern cone of South America: An overview. *In*: Bonnichsen, R. & Turnmire, K. (Eds), *Clovis Origins and Adaptations*. Corvallis, Oregon, Center for the Study of the First Americans, pp. 287–301.

Powell, J. & Neves, W. (1999). Craniofacial morphology of the first Americans: Pattern and process in the peopling of the New World. *Yearbook of Physical Anthropology*, **42**, 153–188.

Quimby, G. (1958). Fluted points and geochronology of the Lake Michigan basin. *American Antiquity*, **23**, 247–254.

Ranere, A. & Cooke, R. (1991). Paleoindian occupation in the Central American tropics. *In*: Bonnichsen, R. &

Turnmire, K. (Eds), *Clovis Origins and Adaptations*. Corvallis, Oregon, Center for the Study of the First Americans, pp. 237–253.

Reanier, R. (1995). The Antiquity of Paleoindian Materials in Northern Alaska. *Arctic Anthropology*, **32**, 31–50.

Renfrew, C. (Ed.) (2000). *America past, America present: Genes and languages in the Americas and beyond*. Cambridge, McDonald Institute for Archaeological Research, 175 pp.

Rutter, N. & Schweger, C. (Eds) (1980). The Ice-Free corridor and peopling of the New World. *Canadian Journal of Anthropology*, **1**.

Santos, F., Pandya, A., Tyler-Smith, C., Crawford, M. & Mitchell, R. (1999). The central Siberian origin for Native American Y chromosomes. *American Journal of Human Genetics*, **64**, 619–628.

Schurr, T. (2004). Molecular genetic diversity in Siberians and native Americans suggests an early colonization of the New World. *In*: Madsen, D. (Ed.), *Entering America: Northeast Asia and Beringia Before the Last Glacial Maximum*. Salt Lake City, University of Utah Press.

Schweger, C., Matthews, J., Hopkins, D. & Young, S. (1982). Paleoecology of Beringia – a synthesis. *In*: Hopkins, D., Matthews, J., Schweger, C. & Young, S. (Eds), *Paleoecology of Beringia*. New York, Academic Press, pp. 425–444.

Seielstad, M., Minch, E. & Cavalli-Sforza, L. (1998). Genetic evidence for a higher female migration rate in humans. *Nature Genetics*, **20**, 278–280.

Severinghaus, J. & Brook, E. (1999). Abrupt climate change at the end of the last glacial period inferred from trapped air in polar ice. *Science*, **286**, 930–934.

Shott, M. (2002). Sample bias in the distribution and abundance of Midwestern fluted bifaces. *Midcontinental Journal of Archaeology*, **27**, 89–123.

Shutler, R. (Ed.) (1983). *Early Man in the New World*. Beverly Hills, Sage Publications, 223 pp.

Silva, W., Bonatto, S., Holanda, A., Paixao, A., Goldman, G., Abe-Sandes, K., Rodriguez-Delfin, L. *et al.* (2002). Mitochondrial genome diversity of Native Americans supports a single early entry of founder populations into America. *American Journal of Human Genetics*, **71**, 187–192.

Slobodin, S. (2001). Western Beringia at the end of the Ice Age. *Arctic Anthropology*, **38**, 31–47.

Smith, D., Malhi, R., Eshelman, J., Lorenz, J. & Kaestle, F. (1999). Distribution of mtDNA Halogroup X among native North Americans. *American Journal of Physical Anthropology*, **110**, 271–284.

Spiess, A., Curran, M. & Grimes, J. (1985). Caribou (Rangifer tarandus L.) bones from New England Paleo-indian sites. *North American Archaeologist*, **6**, 145–159.

Stanford, D. (1982). A critical review of archaeological evidence relating to the antiquity of human occupation in the New World. *Smithsonian Contributions to Anthropology*, **30**, 202–218.

Stanford, D. (1991). Clovis origins and adaptations: An introductory perspective. *In*: Bonnichsen, R. & Turnmire, K. (Eds), *Clovis Origins and Adaptations*. Corvallis, Oregon, Center for the Study of the First Americans, pp. 1–13.

Stanford, D. (1999). Paleoindian archaeology and late Pleistocene environments in the Plains and southwestern United States. *In*: Bonnichsen, R. & Turnmire, K. (Eds), *Ice-Age People of North America: Environments, Origins, and Adaptations*. Corvallis, Oregon, Center for the Study of the First Americans, pp. 281–339.

Stanford, D. & Bradley, B. (2002). Ocean trails and prairie paths? Thoughts about Clovis origins. *In*: Jablonski, N. (Ed.), The first Americans, the Pleistocene colonization of the New World. *Memoirs of the California Academy of Sciences*, Number 27, pp. 255–271.

Steele, G. & Powell, J. (2002). Facing the past: A view of the North American human fossil record. *In*: Jablonski, N. (Ed.), *The First Americans, the Pleistocene Colonization of the New World*. Memoirs of the California Academy of Sciences, Number 27, pp. 93–122.

Steele, J., Sluckin, T., Denholm, D. & Gamble, C. (1996). Simulating hunter-gatherer colonization of the Americas. *In*: Kamermans H. & Fennema, K. (Eds), Interfacing the past: Computer applications and quantitative methods in archaeology. *Analecta Praehistorica Leidensia*, **28**, 223–227.

Steele, J., Sluckin, T., Denholm, D. & Gamble, C. (1998). Modelling Paleoindian dispersals. *World Archaeology*, **30**, 286–305.

Steig, E. (2001). No two latitudes alike. *Science*, **293**, 2015–2016.

Stephens, D. & Krebs, J. (1986). *Foraging theory*. Princeton, Princeton University Press.

Stephenson, R. (1965). Quaternary occupation of the Plains. *In*: Wright, H. & Frey, D. (Eds), *The Quaternary of the United States*. Princeton, Princeton University Press, pp. 685–696.

Storck, P. (1991). Imperialists without a state: The cultural dynamics of early Paleoindian colonization as seen from the Great Lakes region. *In*: Bonnichsen, R. & Turnmire, K. (Eds), Clovis Origins and Adaptations. Corvallis, Oregon, Center for the Study of the First Americans, pp. 153–162.

Straus, L. (2000). Solutrean settlement of North America? A review of reality. *American Antiquity*, **65**, 219–226.

Stright, M. (1990). Archaeological sites on the North American continental shelf. *In*: Lasca, N. & J. Donahue (Eds), *Archaeological geology of North America*. Boulder, Colorado, Decade of North American Geology, Special Volume 4, pp. 439–465.

Surovell, T. (2000). Early Paleoindian women, children, mobility, and fertility. *American Antiquity*, **65**, 493–508.

Szathmary, E. (1993). mtDNA and the peopling of the Americas. *American Journal of Human Genetics*, **53**, 793–799.

Tankersley, K. (1991). A geoarchaeological investigation of distribution and exchange in the raw material economies of Clovis groups in eastern North America. *In*: Montet-White, A. & Holen, S. (Eds), *Raw Material Economies Among Prehistoric Hunter-Gatherers*. University of Kansas Publications in Anthropology, Vol. 19, pp. 285–303.

Tankersley, K. (1994). The effects of stone and technology on fluted-point morphometry. *American Antiquity*, **59**, 498–510.

Taylor, R. (1991). Framework for dating the late Pleistocene peopling of the Americas. *In*: Dillehay, T. & Meltzer, D. (Eds), *The First Americans: Search and Research*. Boca Raton, CRC Press, pp. 77–111.

Taylor, R. (2000). The contribution of radiocarbon dating to New World archaeology. *Radiocarbon*, **42**, 1–21.

Taylor, R., Haynes, C. & Stuiver, M. (1996). Clovis and Folsom age estimates: Stratigraphic context and radiocarbon calibration. *Antiquity*, **70**, 515–525.

Thomas, D. (2000). *Skull Wars: Kennewick Man, Archeology, and the Battle for Native American Identity*. New York, Basic Books, 270 pp.

Torroni, A., Neel, J., Barrantes, R., Schurr, T. & Wallace, D. (1994). Mitochondrial DNA "clock" for the Amerinds and its implications for timing their entry into North America. *Proceedings of the National Academy of Sciences*, **91**, 1158–1162.

Torroni, A., Schurr, T., Cabell, M., Brown, M., Neel, J., Larson, M., Smith, D., Vullo, C. & Wallace, D. (1993). Asian affinities and continental radiation of the four founding Native American mtDNA. *American Journal of Human Genetics*, **53**, 563–590.

Turner, C. (2002). Teeth, needles, dogs, and Siberia: Bioarchaeological evidence for the colonization of the New World. *In*: Jablonski, N. (Ed.), The first Americans, the Pleistocene colonization of the New World. *Memoirs of the California Academy of Sciences*, Number 27, pp. 123–158.

Vasil'ev, S. (2001). Final Pleistocene northern Asia: Lithic assemblage diversity and explanatory models. *Arctic Anthropology*, **38**, 3–30.

Wallace, D., Brown, M. & Lott, M.T. (1999). Mitochondrial DNA variation in human evolution and disease. *Gene*, **238**, 211–230.

Warnica, J. (1966). New discoveries at the Clovis site. *American Antiquity*, **31**, 345–357.

Wendorf, F. (1966). Early man in the new world: Problems of migration. *The American Naturalist*, **100**, 253–270.

West, F. (Ed.) (1996). *American Beginnings: The Prehistory and Palaeoecology of Beringia*. Chicago, University of Chicago Press, 576 pp.

Whallon, R. (1989). Elements of culture change in the later Palaeolithic. *In*: Mellars, P. & Stringer, C. (Eds), *The human revolution: Behavioral and biological perspectives on the origins of modern humans*. Princeton, Princeton University Press, pp. 433–454.

Wheat, J. (1972). *The Olsen-Chubbuck Site: A Paleo-Indian Bison Kill*. Society for American Archaeology Memoir 26.

Whitley, D. & Dorn, R. (1993). New Perspectives on the Clovis vs. Pre-Clovis Controversy. *American Antiquity*, **58**, 626–647.

Williams, S. & Stoltman, J. (1965). An outline of southeastern United States prehistory with particular emphasis on the Paleo-indian era. *In*: Wright, H. & Frey, D. (Eds), *The Quaternary of the United States*. Princeton, Princeton University Press, pp. 669–683.

Wilson, M. & Burns, J. (1999). Searching for the earliest Canadians: Wide corridors, narrow doorways, small windows. *In*: Bonnichsen, R. & Turnmire, K. (Eds), *Ice-Age People of North America: Environments, Origins, and Adaptations*. Corvallis, Oregon, Center for the Study of the First Americans, pp. 213–248.

Winterhalder, B. & Lu, F. (1997). A forager-resource population ecology model and implications for indigenous conservation. *Conservation Biology*, **11**, 1354–1364.

Workman, W. (2001). Reflections on the utility of the coastal migration hypothesis in understanding the peopling of the New World. Paper presented at the Annual meeting of the Alaska Anthropological Association, Fairbanks.

Wright, H. (1991). Environmental conditions for Paleoindian immigration. *In*: Dillehay, T. & Meltzer, D. (Eds), *The First Americans: Search and Research*. Boca Raton, CRC Press, pp. 113–136.

Young, D. & Bettinger, R. (1995). Simulating the Global Human Expansion in the Late Pleistocene. *Journal of Archaeological Science*, **22**, 89–92.

Modeling paleoclimates

Patrick J. Bartlein[1] and Steven W. Hostetler[2]

[1] *Department of Geography, University of Oregon, Eugene, OR 97403-1251, USA*
[2] *Geological Survey at Geosciences Department, Oregon State University, Corvallis, OR 97331, USA*

Introduction

Data describe, models explain. Both are required to document and understand the past variations of Earth's climate, and to help address the present problem of assessing climate change that may result from human activities. Models (for the most part conceptual as opposed to numerical) have long been applied for understanding climate variations during the Quaternary. Indeed, over a century ago, in a set of papers that contributed to the foundation of scientific method in the geosciences (multiple working hypotheses), T.C. Chamberlin (1897, 1899) provided a comprehensive conceptual model for explaining long-term climatic changes that is remarkably modern in some of its elements. What is regarded as a "climate model" today is generally a computerized numerical representation of the physical processes involved in the climate system, but conceptual models still play an important role in paleoclimate research. Whenever any kind of paleoclimatic data is interpreted, either quantitatively or qualitatively, some kind of model is invoked.

Paleoclimatic data (of the kinds reviewed in this volume) and climate models play a complimentary role in understanding climate change. The data record how climate has changed, but data alone cannot provide an unambiguous explanation of why a particular climate state occurred or changed. This situation arises because most climatic variations recorded geologically have multiple, hierarchial causes (e.g. there is more than one way to create drought in a region) and because environmental subsystems display generally nonlinear responses to climate. Consequently, multiple cause-and-effect pathways can produce the same response in a paleoclimatic indicator. This indeterminacy of the "climate signal" is mitigated somewhat by considering networks of paleoclimatic data and by examining multiple indicators at individual sites, but such "multi-proxy mapping" cannot in itself eliminate the indeterminacy.

Models based on physical principles (or widely accepted empirical representations of those physical principles) do have the potential to provide mechanistic explanations of past climatic variations, provided they are known to work, are applied in an appropriately designed experiment, and (perhaps most importantly) explicitly account for all of the components of the climate system that are involved in a particular climate change. Although comprehensive models of the climate system and its individual components (the atmosphere, oceans, biosphere, hydrosphere, and cryosphere) are evolving rapidly, the development of a comprehensive model that can simulate the temporal and spatial variations of climate on both global and local scales, using as input only the records of the external controls of climate (i.e. an "Earth-system model"), is perhaps a decade away. The indeterminacy of the data and the present limitations of the models thus dictate a synergistic approach for understanding climate variations that relies on integrating paleodata with paleoclimate model simulations.

In this chapter we review the process of climate-system modeling and present a taxonomy of the models recently applied in the study of Quaternary climate change and variation. We also briefly trace the development of climate modeling since the 1965 INQUA volume and its companions were published. A synopsis of climate-modeling results for North America is provided, and we conclude with a discussion of some the emerging issues in the application of models for understanding climatic variations.

Climate-System Modeling

Many conceptual and numerical models that describe the workings of the climate system and its components have been developed, and there probably are as many taxonomies of those models as there are reviews of them. Primary reviews of climate modeling in general include Trenberth (1992), McGuffie & Henderson-Sellers (1997), and Randall (2000). Substantial information for paleoclimate modeling in particular can be found in chapters by Crowley & North (1991, Chaps 1 and 2), Kutzbach (1992), and Peteet (2001). Saltzman (2002) provides a coherent framework for understanding past, present and future climatic variations. Earlier modeling reviews include those by Schneider & Dickinson (1974), NRC (1974), Hecht (1985) and Kutzbach (1985). Because no two sources classify climate models or modeling studies in the same way, the task of providing an overview of the field is complicated. One way to organize a discussion of climate models and their application (climate modeling) is to consider first the nature of the climate system and what controls its variations through time, and then to describe a few large classes of climate models and their applications.

Traditional definitions of climate are typically couched in statistics. For example, climate can be thought of as "... a set of averaged quantities completed with higher moment statistics (such as variances, covariances, correlations, etc.) that characterize the structure and behavior of the atmosphere, hydrosphere, and cryosphere over a period of time" (Piexoto & Oort, 1992), or, less explicitly, as "... the synthesis of

DEVELOPMENTS IN QUATERNARY SCIENCE
VOLUME 1 ISSN 1571-0866
DOI:10.1016/S1571-0866(03)01027-3

weather in a particular region." (Hartmann, 1994). Such statistics-based definitions are being replaced in practice (e.g. IPCC, 2001) by one in which climate is regarded as the collection of individual environmental components (jointly the *climate system*), and the record of their interactions and variations through time.

Although the number of major components of the climate system is relatively small, the number of variables that describe these components is quite large, making a full cataloging of the climate variables that might be represented by models tedious and not very informative. However, the many variables involved in fully describing the climate system generally fall into one of three categories (Saltzman, 2002): those that describe the external forcing of the system (*boundary conditions*), those that describe the slowly varying aspects of the system (e.g. the size of ice sheets) that have traditionally been the focus of Quaternary paleoclimatology (*slow-response variables*), and those that describe the internal variables that are ordinarily thought of as weather (*fast-response variables*). A fourth category of variables, which we call *subsystem variables*, describes the state and function of the many environmental subsystems that are governed by climate and which provide paleoclimatic evidence or "paleodata" (Fig. 1).

Climate-System Variables

Boundary Conditions

In theory, the external controls of climate (or boundary conditions, a term borrowed from numerical analysis) include variables beyond the influence of climate. Such boundary conditions include: (1) the latitudinal and seasonal distribution of insolation (incoming solar radiation), as determined by variations in solar output and the elements of Earth's orbit; (2) the configuration of continents and ocean basins including their topography and bathymetry, and the location of mountain chains and gateways between ocean basins; (3) the abiotic component of atmospheric composition, as determined by volcanic emissions; (4) a small flux of geothermal heat; and (5) human activities not controlled by climate. In practice, what is regarded as an external control as opposed to an internal response depends on both the experimental design of a modeling study and the timescale that it focuses on. Ice sheets, the terrestrial biosphere, and ocean temperature and salinity, for example, are most appropriately regarded as variables internal to the climate system. The areal extent and volume of the major ice sheets are ultimately controlled by

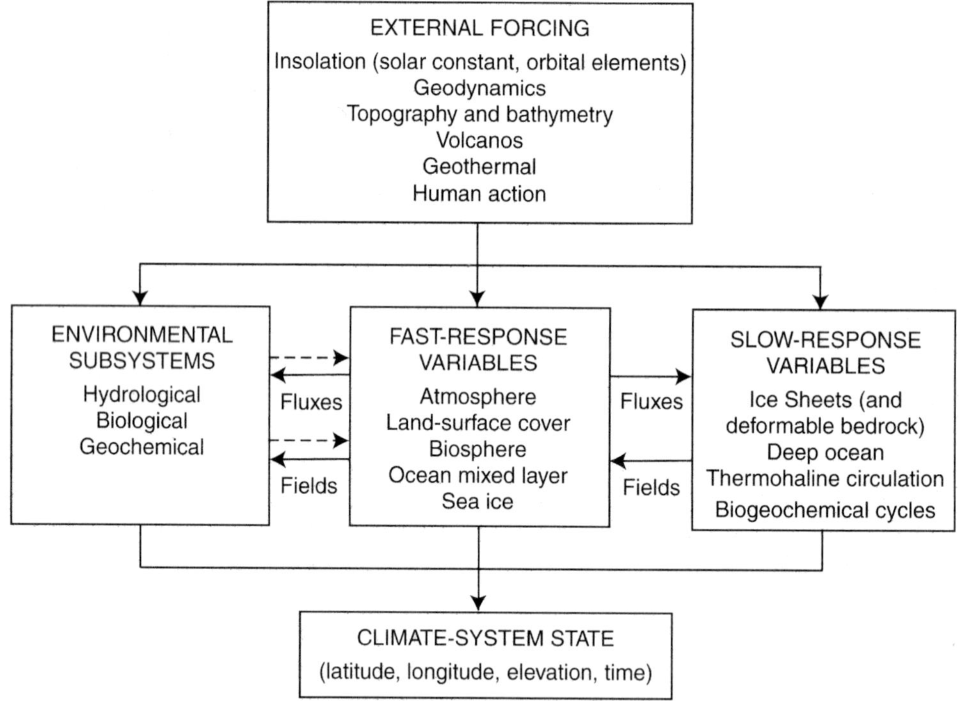

Fig. 1. The climate system (after Fig. 5–3 in Saltzman, 2002). The variables that describe the external forcing of the system (or boundary conditions) directly or indirectly influence the slow-response, fast-response, and environmental-subsystem variables, which in turn influence each other and determine the state of the climate system as a function of longitude, latitude, elevation and time. The arrows labeled "fields" indicate that one set of variables influences the other through patterns of atmospheric circulation, moisture and heat, while those labeled "fluxes" indicate that one set of variables influences another through the transfer of mass and energy. The dashed arrows indicate that the influence of the fast-response variable on environmental-subsystem variables is currently unidirectional in climate models, but that eventually environmental-subsystem variables will interact with the fast-response variables as climate models develop.

external forcing and (over long time spans) can be thought of as an index of the internal state of the climate system. On century-to-annual time scales, however, the sizes of the ice sheets are comparatively constant. In simulations of particular times (e.g. the Last Glacial Maximum) or of sequences of times that are century in length or shorter, the ice sheets can therefore be regarded as one of the external controls that must be "prescribed" (or specified ahead of time) in those simulations. The same is also true for the temperature of the deep ocean and the components of atmospheric composition.

Slow-Response Variables

The slow-response variables of the climate system include ice sheets, large ice shelves, crust and mantle deformable by ice sheets, sea level, temperature and salinity of the deep ocean and the long-term state of its thermohaline and horizontal circulation, and slowly varying reservoirs involved in biogeochemical cycling that determine atmospheric composition. These slowly varying components are most often visualized as time series (as opposed to sequences of maps), and together describe the state of the climate system on century and longer time scales. As is the case for the boundary conditions, the particular role of these variables in climate models can be ambiguous and depends on the specific model and experimental design.

Fast-Response Variables

The three-dimensional states of the atmosphere, land and ocean surfaces are represented by the fast-response variables. For the atmosphere, the key properties that are either observed or simulated include the distribution of temperature, pressure, wind and moisture (including clouds), and its trace-gas and mineral-aerosol composition. At the surface, the fast-response variables include seasonally varying sea-ice extent, soil-moisture content, vegetation cover (including evapotranspiration rate, rooting depth, and albedo), and the temperature, depth and other physical and biological characteristics of the mixed layer of the ocean. On the shortest of time scales (hours-to-days) the fast-response variables can be thought of as describing the weather. In most instances, the fast-response variables are visualized using maps or sequences of maps such as those commonly used in weather forecasts.

Subsystem Variables

A large number of environmental systems and processes respond to variations of climate, and many of these, like those included in the terrestrial or marine biospheres, or surficial hydrologic systems, also provide the principal lines of paleoclimatic evidence that are used to reconstruct past climatic changes (Bradley, 1999). These systems have many variables, including some that play a role in the interaction and feedback between the atmosphere and the surface (and might therefore be thought of as fast-response variables), and some that are dependent on climate but do not feed back to the climate system except in limited ways. There is no common mode for the visualization of these subsystem variables.

Some variables are not easily categorized. Vegetation, for example, plays a key role in the instantaneous coupling of the atmospheric boundary layer and land surface by controlling the exchanges of energy and moisture. The rates of these exchanges depend on the structure of the vegetation and on the states of the atmosphere and underlying soil (including atmospheric humidity, wind, net radiation at the surface and soil-moisture availability) that together influence plant physiology (e.g. stomatal conductance). It was formerly thought that vegetation structure responds slowly to climate changes (on the order of hundreds to thousands of years), placing it in the category of slow-response variables. It is now clear that vegetation structure responds rapidly to climate changes over time spans of years to decades (Tinner & Lotter, 2001; Webb *et al.*, this volume). Soils are dependent on climate and vegetation, but also have strongly expressed geological and geomorphic controls. Key attributes of the soil such as water-holding capacity (WHC) may be dominated by parent material (as in arenaceous soils), and so WHC might be regarded as a boundary condition; in other situations WHC is dominantly controlled by soil morphology, and hence acts like a slow-response variable. The particular category a variable falls into is thus largely dependent on context, location and scale.

The Climate-Modeling Problem

The ultimate goal of climate modeling is to consider simultaneously the first three groups of variables listed above, and, as necessary, also to treat the other environmental systems that depend on climate (e.g. those described by the fourth group, the subsystem variables) – all in order to provide both a description and an explanation of the variations of climate through time. The result of modeling may be a single map or a series of maps or one or more time series. One can decompose the basic problem of climate modeling into a sequence of tasks: (1) use the record of boundary conditions to simulate the time history of the slow-response variables, (2) use the boundary conditions and state of the slow-response variables to simulate the fast-response variables, and (3) use the fast-response variables to understand and simulate the environmental subsystem variables.

Climate Models

Climate models can be classified by describing the particular *applications* to which they are put (e.g. simulating the variations of the second-through-fourth set of variables described above), or by describing the *comprehensiveness* of different models – the number of processes and major components of the climate system they include and the (temporal and spatial) resolution at which those processes and components are represented. These classification schemes provide different ways of organizing the various models and are applicable to both conceptual and numerical models. Here we emphasize numerical models, but we do not underestimate the necessity, utility, and power of conceptual models.

Model Applications

Most applications of numerical climate models can be categorized as having one of three general goals: (1) simulating the evolution over time of the climate system or one of its major components; (2) simulating the spatial patterns of climate-system components; and (3) simulating the detailed function of a single component or process. However, the present trends in climate-model development, which are leading to more comprehensive models of the climate system (see below), are in fact aimed at blurring the distinction among these applications.

Time-Evolution Applications

Numerical climate models have been developed to simulate the evolution of climate over a range of time scales from geological (i.e. those that treat Cenozoic cooling or the onset of Quaternary glaciation) to inter- (and intra-) annual variations, as well as the trends in climate over the past several millennia (see Crowley & North, 1991, Chap. 1). The main goal of time-evolution modeling experiments is to simulate some macro-scale feature of the climate system, such as global ice volume or hemispheric-average temperature, using the record of external controls of climate as input (e.g. Birchfield *et al.*, 1994). Models used in simulating time evolution commonly represent components of the climate-system at low resolution, as in box models or energy-balance models that include representations of continental and marine reservoirs or active layers without being spatially explicit (e.g. Harvey & Huang, 2001).

Spatial-Pattern Applications

Models that focus on simulating the spatial patterns of climate include general circulation models (GCMs) and spatially resolved energy-balance models (EBMs) that include realistic geography (Crowley & North, 1991, Chap. 1). General circulation models, which simulate the three-dimensional structure of the atmosphere, ocean and land surface on time steps of minutes to hours, were originally developed for operational weather forecasting. The models continue to evolve and the operational models being run, for example, by the National Oceanic and Atmospheric Administration and the European Center for Medium-Range Weather Forecasts are now also being used for "reanalysis" projects (Kistler *et al.*, 2001) an approach in which observed parameters of global weather (e.g. atmospheric soundings), sea-surface temperatures (SSTs), and ice cover are incorporated into GCMs, which then provide simulations of the 3-D structure of the atmosphere and state of the land surface.

The acronym "GCM" is often being taken to mean "global climate model." A true global model, however, would be one in which all "internal" climate system components (i.e. variable groups 2 through 4 above) are explicitly represented (as opposed to being prescribed). Global climate models now exist in preliminary form as "EMICs" (Earth-system Models of Intermediate Complexity; Claussen *et al.*, 2002). In practice, the extent to which other components of the climate system are included in a GCM (in addition to the atmosphere and land and ocean surfaces) is represented by additions to the GCM acronym. For example, AOGCMs include explicit representation of a three-dimensional ocean, while AVGCMs not only represent the regular physiological variations that must be included in any GCM, but also allow for variations in vegetation structure (Sellers *et al.*, 1997). As more interactive components are added to GCMs, they will gradually evolve toward full Earth-system Models (ESMs).

An important aspect of both GCMs and EBMs is their potential for simulating climate variables that may be crucial for understanding the response of particular paleoclimatic indicators to climatic variations. For example, many terrestrial subsystems (vegetation, lakes) are governed directly by the surface water and energy balances, or by "bioclimatic variables" (e.g. Prentice *et al.*, 1992), as opposed to standard climate variables like temperature and precipitation. The former, mechanistic, variables are not commonly observed, or may indeed be unobservable, especially over regional scales. The potential for GCMs and EBMs, along with process models, to simulate presently unobservable variables also contributes to understanding the mechanistic controls of the variations in paleoclimatic indicators.

Simulation of the time-evolution of the climate system can also be approached by conducting a series of spatial-pattern simulations, or "snap-shots," wherein the boundary conditions at a number of key times are established from the geologic record and are used to initialize models that produce "equilibrium" simulations for those times (e.g. COHMAP Members, 1988; Valdes, 2000; see also Peteet, 2001, for a discussion of the distinction between "snap-shot" and "sensitivity-test" experimental designs).

Subsystem (Process-Model) Applications

Models that attempt to represent the variables that describe individual environmental subsystems are often called "process models" inasmuch as they are designed to incorporate, either explicitly or implicitly, the actual climatic mechanisms that govern a subsystem such as a watershed, lake, or plant. We exclude from this category statistical relationships that are developed by screening a large number of potential predictors of the distribution or variation of some paleoclimatic indicator, but which do include statistical relationships between predictors and mechanistically related responses (i.e. using mean July temperature or growing-degree days as a proxy for the heat and energy requirements of plants). Many process models exist and we do not attempt an exhaustive review here. Process models range in scope (and scale of application) from those that represent the dynamics of ice sheets (e.g. Marshall, 2002) or global vegetation patterns to those that simulate the responses to climate variations within individual watersheds, lakes, or forest stands.

Many subsystem process models are run in a "stand alone" or "off line" mode in which the there is no feedback with the climate model and input for the models is derived from the output of climate simulations. Process models are generally applied to individual points, such as a particular lake or forest stand, but they can also be run over a global network of sites or grid points, thereby producing simulations that are comparable with the spatial-pattern applications of global or regional climate models. Examples include equilibrium vegetation models (e.g. Harrison *et al*., 1998; Prentice *et al*., 1992) and ice-sheet mass-balance models (Pollard *et al*., 2000), applied over a grid, that use the results of a paleoclimatic simulation as input. The individual model grid points do not communicate with one another as they do in an AGCM or AOGCM.

Model Comprehensiveness

A second way to classify climate models is by their comprehensiveness or scope, as was done by Claussen (2001) and Claussen *et al.* (2002) (Fig. 2). Claussen (2001) considers three attributes or axes: (1) the degree of model *integration*, or the number of interacting components of the climate system that are coupled within the model; (2) the number of processes explicitly simulated in the model, which can also be thought of as the cumulative *dimension* of the model; and (3) the detail of description in both time and space, commonly thought of as model *resolution* (Fig. 2). Highly integrated models account for the interactions and feedbacks among multiple components of the climate system, like the atmosphere, ocean, and terrestrial biosphere. In contrast, less integrated models explicitly represent one component (e.g. the atmosphere in an AGCM), and "parameterize" (or represent by very simple, sometimes empirically based relationships) the behavior of others (like the ocean and land surfaces in an AGCM). Models that represent many processes in a spatially explicit way (e.g. multiple-layer soil-moisture storage simulation) have higher cumulative dimensionality than others that include fewer processes in a more generalized way (e.g. the single-layer or "bucket" approach to soil-moisture storage). The third dimension of Claussen's scheme, resolution, which has always been limited by computing resources, is easiest to envision.

When individual models are placed in Claussen's three-dimensional framework, several clusters emerge: (1) conceptual models that are of low spatial resolution (treating, for example, the whole Earth or greatly generalized continents and oceans) and in process dimensionality; (2) elemental (or low-dimensional) models of an individual or small number of the major components of the climate system; (3) EMICs, which are spatially explicit, and often represent multiple components, but generally at low spatial resolution (Claussen *et al*., 2002); and (4) comprehensive models,

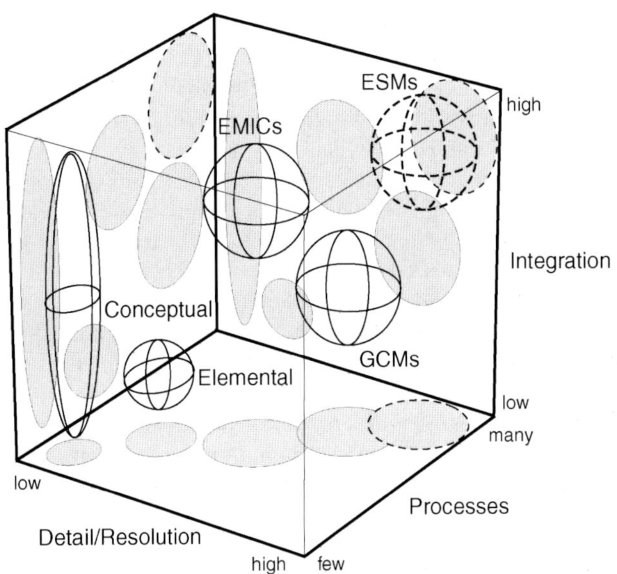

Fig. 2. Classes of climate models plotted in a three-dimensional space that describes model comprehensiveness (after Claussen, 2001; Claussen et al., 2002). The box is defined by the degree of model integration, or the number of interacting components of the climate system that are coupled together in the model, the number of processes explicitly simulated in the model, and the detail of description in time and space, commonly thought of as the resolution of the model. The gray shadows on the wall of the box represent the projections of each cluster of models. ESMs are shown using dashed lines because they currently do not exist.

such as high-resolution "coupled" GCMs (Grassl, 2000). (Claussen *et al.* (2002) recognized clusters 1, 3, and 4, but we found enough differentiation among models to add cluster 2.) We also show a fifth cluster in Fig. 2 (represented by dashed lines) to illustrate the position of a full Earth-system model (ESM).

In a progression of models across the three-dimensional space that defines comprehensiveness, the complexity of the models is measured to a large extent by their resolution and by the number of individual climate-system components described. This progression, however, is not a measure of the worthiness or value of the models. For example, conceptual models, which could be viewed as simple pencil-and-paper "thought experiments," are actually among the more sophisticated of models in use in Quaternary science, whereas GCMs, which appear to be exotic and computationally demanding, are actually familiar to us as the source of day-to-day weather forecasts.

Conceptual Models

Conceptual models include ideas and facts that we know are true (or that we are pretty sure are true, e.g. that Earth has experienced repeated glacial/interglacial variations during the Quaternary, related in some way to orbital variations), but they also include basic statements about how things work, often phrased in terms of hypotheses. Conceptual models are the oldest and most frequently applied kind of model used for understanding past climatic variations. The models are routinely applied by Quaternary scientists when looking at "raw" paleoclimatic data, no matter what kind (geologic, geomorphic, sedimentologic, paleoecologic, isotopic, etc.). Inasmuch as they often consider the Earth system as a whole, conceptual models cover the entire range of integration in the climate-model comprehensiveness space.

As will be illustrated in the next section, conceptual models can take the straightforward form of an annotated list of potential controls of climatic variations (e.g. Mitchell, 1965), but most consist of a description of processes and interactions usually applied to a common sequence of events (e.g. Ruddiman, 2003). The common medium for reporting these models is a commentary in *Science* or *Nature* (e.g. Clark *et al.*, 2002) or the summary chapter of a proceedings volume (e.g. Alley *et al.*, 1999; Stocker *et al.*, 2001). These latter discussions are most always supported by a number of less conceptual, more explicitly numerical modeling studies. Applications of conceptual models span the full range of temporal and spatial scales in Quaternary paleoclimatology and address issues that range from the question of why there are ice ages at all, to what controls individual wiggles in a single time series.

Some conceptual models are expressed numerically. Examples include the Imbrie & Imbrie (1980) or Paillard (1998) models of glacial/interglacial variations in which global ice volume is related to insolation through simple differential equations, with the specific form of the relation determined by the state of the climate system (as represented by ice volume) in a geologically reasonable fashion.

Elemental (or Low-Dimensional) Models

The second cluster includes models that numerically represent one or more components of the climate system in some kind of elemental or single-component fashion, and have lower spatial resolution, integration, or process dimensionality compared with models that are more comprehensive. This simplification results in lower computational demands (again relative to more comprehensive models). The number of models that potentially can be assigned to this cluster is quite large. The models range from highly spatially aggregated energy-balance models – like those of Budyko & Sellers (Budyko, 1982), which attempt to simulate global-average temperature from first principles and can be implemented with spreadsheet software – to spatially explicit energy-balance models like that of Crowley & North (1991). This cluster also includes simulating the slowly varying components of the climate system that must be viewed over relatively long time spans (Saltzman, 2002).

So-called "box models" (e.g. Toggweiler, 1999) in which a few very large-scale geochemical reservoirs (and the flows among them) are simulated are also included in this cluster, as are models that feature coupling between components of the climate system such as the atmosphere and ocean, but for only part of the globe (e.g. the tropics, as in Clement *et al.*, 2001). Elemental or low-dimensional models have also been used extensively to assess potential future climate changes (e.g. Houghton *et al.*, 1997), and to estimate of the sensitivity of global climate to carbon dioxide variations (e.g. Berger *et al.*, 1998; Harvey & Huang, 2001).

EMICs (Earth-System Models of Intermediate Complexity)

The cluster of models described as EMICs occupies an important position in the continuum of climate-model comprehensiveness because it offers a bridge between simple, low-resolution models and more comprehensive spatially explicit models (Claussen *et al.*, 2002). In some ways, EMICs are very low-resolution AOVGCMs; in other ways, they are simple, low-dimensional models (like EBMs) to which some kind of simplified depiction of atmospheric circulation dynamics has been added. Fundamentally, EMICs simulate the interactions among more components of interest than can be done with the current generation of AGCMs or AOGCMs. The low spatial resolution of the EMICs, and the "parameterization" of many processes explicitly represented by GCMs, permits very long simulations of the temporal evolution of climate – an advantage over "snap-shot" simulations. Similarly, when used in "sensitivity-test" mode, EMICs allow a large number of combinations of inputs or parameters to be explored.

Comprehensive Models – GCMs

The fourth cluster of climate models includes those that attempt to simulate the three-dimensional structure of the climate system and its variation over time. This cluster is exemplified by general circulation models (Randall, 2000), which in the current generation, include coupling among one or more climate-system components (although not as in as many combinations as the EMICs). In the most common form of application to paleoclimate, GCM experiments are designed by providing a set of boundary conditions of interest (such as insolation, atmospheric composition or the distribution of ice sheets), and the resulting simulations are analyzed as a sequence of gridded observations of a set of climate variables.

In practice, GCMs are less integrated than the EMICs, owing to the smaller number of climate-system components that are directly coupled. They do, however, define the cluster of models that currently have the highest resolution and greatest number of explicitly represented physical processes among the several that we have described. GCMs play a major role in the assessment of potential future climate changes because they are able to provide spatially explicit simulations of climate under different scenarios of atmospheric trace gas and aerosol composition.

GCMs have been applied in sequences of "time-slice" or snapshot simulations (i.e. Charbit *et al.*, 2002; COHMAP Members, 1988; Valdes, 2000), with the goal of revealing the mechanisms responsible for the regional patterns of climate change. Suites of models have been run under the same set of boundary conditions (i.e. Joussaume *et al.*, 1999) to understand the role (if any) that the structures of individual models play in adequately simulating past climates. Both kinds of studies have featured comparisons with synthesis of paleoclimatic data (Kohlfeld & Harrison, 2000). Such comparisons have implicated the absence of feedbacks among components of the climate system in the mismatches between simulations and observations (Harrison *et al.*, 2002).

Regional climate models (RCMs) can be viewed as a subset of the GCMs, and have a relationship to regional or "fine-mesh" weather forecasting models that parallels that of GCMs to global forecasting models. Regional models require "lateral boundary conditions" – a temporal sequence of three-dimensional fields of variables describing the atmosphere and surface (i.e. SSTs) that are usually provided by a GCM simulation. Regional models can be considered to be "nested" within the lower-resolution global model in that the driving GCM fields are ingested by the RCM along the model's boundary. Compared with GCMs, RCMs allow more spatially explicit simulation of the sensitivity of regional climate and subsystems to large-scale atmospheric controls (Hostetler & Bartlein, 1999). GCMs further allow evaluation of surface-atmosphere feedbacks, such as those associated with large lakes (Hostetler *et al.*, 2000), that are not resolved in coarser-resolution GCMs.

Comprehensive Models – ESMs

GCMs are evolving toward greater comprehensiveness through the development of coupled models that include, for example, explicit representation circulation of the ocean (i.e. AOGCMs), and the terrestrial biosphere (AVGCMs). Coupled models will eventually fill in the region in model continuum now occupied mainly by the EMICs, and will extend into a presently unoccupied region of highly integrated models that incorporate many processes on a high-resolution grid. However, because of computational limits on comprehensive-model simulations, models that feature coupling between a small number of climate-system components will be the rule for the near term. Ultimately, however, "super-GCMs" (Saltzman, 2002), coupled with models of slowly varying components of the climate system, will form a true climate-system model (CSM), or Earth-system model (ESM).

VII INQUA Congress (1965)

The seventh congress of the International Association for Quaternary Research marked a stage in understanding of climate variations in general, and paleoclimate modeling in particular. The congress took place at an interesting time, because both plate-tectonic theory and the astronomical (Milankovitch) theory of climate change emerged in their present forms and evolved over the following decade, as did the fuller depiction of the climate system and its variations and controls that by 1974 resulted in the National Research Council report *Understanding Climatic Change*, which is essentially modern in its scope and outlook.

The principal materials related to climate modeling from the seventh congress include reviews by Broecker (1965) and Mitchell (1965) in the Wright & Frey (1965) "INQUA volume," and a 1968 volume in the *Meteorological Monographs* series (Causes of Climatic Change; Vol. 8, Number 3), which was edited by Mitchell. Several other proceedings volumes were also published, but the majority of papers or chapters with climate modeling content are found in the volumes edited by Wright, Frey, and Mitchell.

The Wright & Frey (1965) volume, *The Quaternary of the United States*, focused on the paleoclimate of the United States, and as a consequence it does not offer a comprehensive review of the fields of paleoclimatology or climate modeling of the time. It can be supplemented, however, by nearly contemporaneous books by Lamb (1966) and Budyko (1982, a summarization of his earlier work). With the organization of the Intergovernmental Panel on Climate Change in 1990, the publication of proceedings from NATO Advanced Study Institutes (e.g. Berger, 1980), and joint U.S./USSR syntheses (e.g. Porter, 1983; Velichko *et al.*, 1984; Wright, 1983) and reports (MacCracken *et al.*, 1990), the study of global climate change was shown to be truly global in perspective and participation.

Climate Modeling in the VII INQUA Congress Materials

With the exception of EMICs, it is possible to see the same classes of models we described above both in the INQUA publications, and in journal articles not part of the formal proceedings but related to the scientific threads of the meeting.

Conceptual Models

Mitchell's review of the causes of climate change in the chapter "Theoretical Paleoclimatology" in the Wright & Frey (1965) volume represents a comprehensive listing of the conceptual models of Quaternary climate variations that were current at the time (see Mitchell's Tables 1 and 2). The particular "causative factors" and the mechanisms through which they control climate reviewed include:

(1) Autovariation, or internal variations of the climate system stemming from its highly nonlinear nature.
(2) Air-sea interaction, including the role of the thermohaline circulation in transporting heat throughout the climate system.
(3) Continental drift, which Mitchell viewed as an indirect cause of Quaternary glaciation.
(4) Orogeny and continental uplift, and their potential effects on large-scale atmospheric circulation patterns.
(5) Carbon-cycle variations, in which the potential for human action to have a significant impact on climate was discussed.

(6) Volcanism, and the effect of dust and aerosols on incoming solar radiation.
(7) Solar variability, including long-term and periodic variations of the solar constant.
(8) Orbital variations, including the potential role of land-surface feedback in amplifying insolation changes.
(9) Feedbacks, in which Mitchell reviewed a number of hypotheses that attempt to explain glacial/interglacial variations from combinations of internal variations and external forcing, including the one advanced by Chamberlin in 1899.

Chamberlin's hypothesis, which we would regard today as one expressed in terms of biogeochemical cycles, is particularly interesting in its consideration of the above controls (although not necessarily in modern terms), as well as the interactions among them. Interactions were referred to by Chamberlin as "intercurrent agencies," an idea now usually described as "coupling" among systems.

The only large gap in Mitchell's list, though filled implicitly, is the potential role of land cover in controlling climate. Changes in land cover, on both the Quaternary and historical time scales, have the potential to influence significantly the emission of dust and mineral aerosols to the atmosphere (Harrison *et al.*, 2001; Mahowald *et al.*, 1999) and to change surface energy balances (DeFries *et al.*, 2002) and consequently other components of the climate system (Chase *et al.*, 2001).

Elemental Models

Several elemental or low-resolution models were discussed in the INQUA proceedings. In the Wright & Frey (1965) volume, Broecker (1965) used a number of elemental models while reviewing the isotopic record of paleoclimatic variations. The chapter by Schumm (1965) included discussion of several elemental geomorphic and hydrologic models to examine the impact of climate changes while Meier (1965) presented an analysis of the response of glacier mass-balance and flow to variations in climatic controls.

In the *Meteorological Monographs* volume, elemental models were used in the discussion of the thermohaline circulation (Weyl, 1968), and the surface energy balance (Ericksson, 1968). Saltzman (1968) considered the surface forcing of atmospheric circulation, and Kutzbach *et al.* (1968) examined the effects of changes in the latitudinal temperature gradient on atmospheric circulation. Both these latter studies examined with simple models the sensitivity of one component of global climate (atmospheric circulation) to changes in forcing in a manner that anticipates those authors' later work.

Comprehensive Models

GCMs appeared in a chapter in the *Meteorological Monographs* volume by Mintz (1968), which suggested how AGCMs might be used to investigate paleoclimatic questions. Although not explicitly part of the INQUA materials, the importance of contemporaneous work by Smagorinsky (1963), and Smagorinsky *et al.* (1965) to later work with GCMs is evident in subsequent publications. In the decade following the VII INQUA Congress, routine application of GCMs in paleoclimatic studies emerged (e.g. Gates, 1976; Williams *et al.*, 1974).

Subsequent Developments

It is apparent that many of the questions and issues that were raised in 1965 the INQUA Congress are still relevant today. Moreover, the specific contributions of the congress and its proceedings contributed to the foundation of the U.S. National Research Council (1974) report *Understanding Climatic Change* and its successors. These reports initiated research agendas for the study of global change (e.g. Malone *et al.*, 1985) that remain relevant today (National Research Council, 1999).

Synopsis of Results from Modeling Quaternary Paleoclimates of North America

Climate models have been applied to advance understanding of many of the aspects of Quaternary climate changes in North America. The presence of the Laurentide Ice Sheet (LIS) makes climatic variations over North America a key component of the general description of long-term climatic variations. As a spatially heterogeneous region subject to the influence of the major northern hemisphere atmospheric circulation mechanisms, the patterns of regional climate changes across the continent have also been of interest. Paleoclimate modeling studies that have focused on North America fall into two general groups: those that focus on the slow-response components of the climate system like the LIS, and those that focus on the spatial patterns of the fast-response components at key times.

Temporal Variations of Climate

Studies of the temporal variations of the climate system have addressed the onset of glaciation over the Cenozoic, as well as the nature of individual glacial cycles, and have also been used to examine the potential controls of glacial/interglacial variations and the genesis of millennial-scale variability. The long-term cooling during the Cenozoic, which ultimately led to Quaternary glacial/interglacial variations, and the higher-frequency variations superimposed on them, present several features or "targets" for which explanations have been attempted using various classes of models. These targets include:

(1) the cooling itself, and the reorganization of the ocean, atmosphere, and cryosphere that is implied.

(2) the non-reversing steps toward more extensive glaciation, such as those around 35 myr ago, 12 myr ago; and within the past 5 myr.
(3) the onset of extensive northern hemisphere glaciation around 2.8 myr ago.
(4) the changes in periodicity and amplitude of global ice volume variations during the last 3 myr.
(5) the sequence of global climate changes during a single glacial cycle.
(6) "sub-millennial"-scale variations in climate.

Cenozoic Cooling and the Quaternary Ice Age

The long-term cooling of the Cenozoic, leading ultimately to the onset of extensive glaciation in the northern hemisphere around 2.8 myr ago, has been examined using a variety of approaches, which have generally featured conceptual models supported by syntheses of data, elemental models of particular components of the climate system, or more comprehensive models used to simulate key times or to explore particular combinations of controls (using "snap-shot" simulations). Examples of the first application include the examinations of isotopic records by Miller *et al.* (1987) and Zachos *et al.* (2001) for the entire Cenozoic, or by Driscoll & Haug (1998) and Haug & Tiedemann (1998) for the past 5 myr. The transition to a more glacial state described by the latter two studies was also examined in simulations with Saltzman's model of paleoclimatic dynamics (Saltzman, 2002; Saltzman & Verbitsky, 1993) and with the LLN 2-D model (Li *et al.*, 1998), both of which are sophisticated elemental models. In another example of a general conceptual model supported by simulations with a more comprehensive model, sensitivity tests by Kutzbach *et al.* (1997) contributed to the evaluation of the tectonic hypotheses of Cenozoic climate change (Ruddiman, 1997). In a similar fashion, the role of the changes in paleogeography from Cretaceous times to present have been explored with coupled AOGCMs (Huber & Sloan, 2001; Otto-Bliesner *et al.*, 2002), which account for the effects of changes in ocean basins and gateways on global climates. Further applications of GCMs and related models to pre-Quaternary climates are described by Parrish (1998).

A number of studies have focused on the onset and maintenance of glacial/interglacial variations, again using a combination of modeling approaches. These include the conceptual (but mechanistic in character) model of Imbrie *et al.* (1992, 1993), and the aforementioned models of Imbrie & Imbrie (1980) and Paillard (1998). In these latter two applications, data analyses or relatively simple numerical models are used to illustrate the features of "thought experiments" that attempt to explain, for example, the features of the oxygen isotopic record.

The inception of a single glaciation, as occurred around 115,000 yr ago, has been examined in several GCM-focused studies. Rind *et al.* (1989) found that the insolation changes between the time of the northern hemisphere summer maximum around 126,000 yr ago, and the relative minimum around 115,000 yr ago were insufficient to initiate permanent snow cover in northeastern North America in their model. In contrast, subsequent simulations by Dong & Valdes (1995), Gallimore & Kutzbach (1996), and deNoblet *et al.* (1996) were able to simulate the accumulation of permanent snowfields, particularly if the models included feedback from climate-induced changes in land cover.

Millennial-Scale Variations

Millennial-scale climate variations have also been examined with combinations of conceptual models, data analyses, and comprehensive models, in particular time-evolving elementary models and EMICs. The conceptual models, which include those described by Alley *et al.* (1999), Stocker *et al.* (2001) and Clark *et al.* (2002), have developed or attempted to test hypotheses for millennial-scale variations that generally involve reorganization of the circulation of the atmosphere and ocean, including the thermohaline circulation, and the global transmission or propagation of climate variations in the North Atlantic.

The kinds of modeling studies used in the development and testing of those hypotheses span the entire continuum of model comprehensiveness. Saltzman (2002) (see also Saltzman & Verbitsky, 1995) showed how millennial-scale variability, like that associated with the Heinrich events, emerged from a dynamical model of the slowly varying components of the climate system. Similar variability emerges in simulations using EMICs (e.g. Crucifix *et al.*, 2002; Ganopolski & Rahmstorf, 2001), which add some spatial specificity to the simulated climate variations. Simulations with GCMs and RCMs that examine the sensitivity of the climate and subsystems at the LGM to imposed changes in North Atlantic sea-surface temperatures reveal further details of the spatial patterns of millennial-scale variations (Hostetler & Bartlein, 1999; Hostetler *et al.*, 1999).

The integrated modeling studies and data analyses of the temporal variations of climate across the different timescales described above support make several generalizations about the temporal variations of the climate of North America leading up to and during the Quaternary:

(1) changes in paleogeography, including changes in mountain belts and oceanic gateways, explain much of the pattern of climate change over the Cenozoic, if the synergistic effects of changes in atmospheric composition and ocean heat transport are considered.
(2) the ice sheets are active components of the climate system, and no realistic account of the temporal and spatial patterns of Quaternary climate change can be made without considering them.
(3) the thermohaline circulation of the ocean seems involved in climate variations across all time scales.
(4) feedback from changes in the land surface and in ocean circulation appear to be involved in amplifying or attenuating the climatic effects of changing boundary conditions such as insolation and the arrangement of continents.

Spatial Patterns of Fast-Response Variables – LGM to Present

The LGM-to-Present "Natural Experiment"

Various modeling studies have focused on the interval between the LGM (Last Glacial Maximum, 21,000 yr ago) and present. During this interval, nature performed experiments with the climate system (Webb & Kutzbach, 1998), and recorded the results in paleoclimatic data sets like those reviewed in this volume and elsewhere (Kohlfeld & Harrison, 2000). Both the nature of the boundary-condition changes over the interval and our knowledge of them has facilitated numerous application of elemental models, EMICs, and, particularly, GCMs.

Comparing the climates of the LGM and 6000 yr ago with present provides an optimal experimental design in which only a few controls are changed from their present settings. At the LGM, there were extensive continental ice sheets, low concentrations of greenhouse gasses in the atmosphere, high aerosol loadings, relatively cold sea-surface temperatures, and land-cover characteristics that featured reduced areas of forest, but the latitudinal and seasonal distribution of insolation was similar to that at present. After the LGM, the amplitude of the seasonal cycle of insolation increased, reaching a maximum around 11,000 yr ago, so that at 6000 yr ago, insolation during the northern summer was greater than at present, while the remainder of the boundary conditions were close to their present (or pre-industrial) values. Because of the elegance of this natural experiment, many simulations have been done for 21,000 and 6000 yr ago, including those in PMIP (Harrison *et al.*, 2002; Joussaume *et al.*, 1999; Palaeoclimate Modeling Intercomparison Project). These two times have also been the focus for simulations with coupled AOGCMs (Braconnot *et al.*, 2000; Harrison *et al.*, 2002; Hewitt & Mitchell, 1998; Hewitt *et al.*, 2001; Shin *et al.*, 2002). Relatively few sequences of simulations with GCMs over this interval have been able to exploit the full natural experiment. The published sequences of experiments include simulations conducted with an early version of the NCAR CCM (National Center for Atmospheric Research, Community Climate Model) (COHMAP Members, 1988; Wright *et al.*, 1993), a subsequent version of the CCM (CCM 1, Webb & Kutzbach, 1998), and with the UGAMP GCM (U.K. Universities Global Atmospheric Modelling Programme General Circulation Model; Valdes, 2000), and the LMD5.3 model (Laboratoire del Météorologie Dynamique; Charbit *et al.*, 2002), models similar to CCM 1 in the degree of coupling among systems.

These sequences of simulations, along with the suites of simulations for 6000 and 21,000 yr ago, jointly show that much of the variation in global and regional climate over this interval can broadly be explained by the influence of the ice sheets on atmospheric circulation and the influence of insolation on circulation and surface water- and energy-balances. Comparisons of model simulations with paleodata demonstrate that, as is the practice of the Intergovernmental Panel on Climate Change (Houghton, 2001), it is indeed feasible to simulate climates different from that at present using the kinds of models reviewed here. The detailed data-model comparisons organized by PMIP do show, however, that the present generation of global models may underestimate the magnitude of the responses of the climate system to changes in its controls (Harrison *et al.*, 2002).

LGM-to-Present Simulations for North America

The sequence of simulations conducted with CCM 1 (Fig. 3), although obsolete by today's standards in terms of model resolution and interactivity of the ocean, still provide the only complete sequence of simulations performed using a coherent experimental design that have been extensively analyzed for North America (see Bartlein *et al.*, 1998; Webb *et al.*, 1998). Figure 3 shows the simulated sequence for several variables for January and July, expressed both as the actual values for each time in the sequence and as anomalies, or differences between each "paleo" simulation and the present-day or "control" simulation. Table 1 summarizes some of the main features in Fig. 3 that were discussed by Bartlein *et al.* (1998) and Webb *et al.* (1998).

The variables in Fig. 3 appear roughly in a sequence that represents cause-and-effect. The top row of maps shows insolation (and anomalies) over the sequence of simulations, along with the outlines of the Laurentide ice sheet in the model – the two major controls of the sequence of regional climatic changes. Net radiation and surface air temperature illustrate the direct effects of insolation and ice-sheet size on the surface energy balance and temperature. The latitudinal and continental-marine contrasts in temperature, in concert with the topographic effects of the ice sheet, influence atmospheric circulation as represented by 500 mb horizontal wind speed and vertical velocity, and sea-level pressure. In turn, atmospheric circulation, in particular the large-scale patterns of vertical motions, and moisture availability (determined mainly by temperature) jointly influence the patterns of precipitation and thus precipitation minus evaporation.

The principal features of the simulated climate over North America include:

(1) displacement by the Laurentide Ice Sheet of the band of fast upper-level winds to the south of its present location in both winter and summer during the LGM and afterward.
(2) development of a "glacial anticyclone" over the ice sheet in eastern North America, and consequent generation of large-scale sinking motions in the eastern and southern quadrants of the ice sheet.
(3) existence of generally drier-than-present conditions during glacial times (when it was colder than present), giving way to wetter-than-present conditions as the continent warmed.
(4) changes in the strength of surface atmospheric circulation features that follow the trends in the boundary conditions: weakening of the Aleutian low

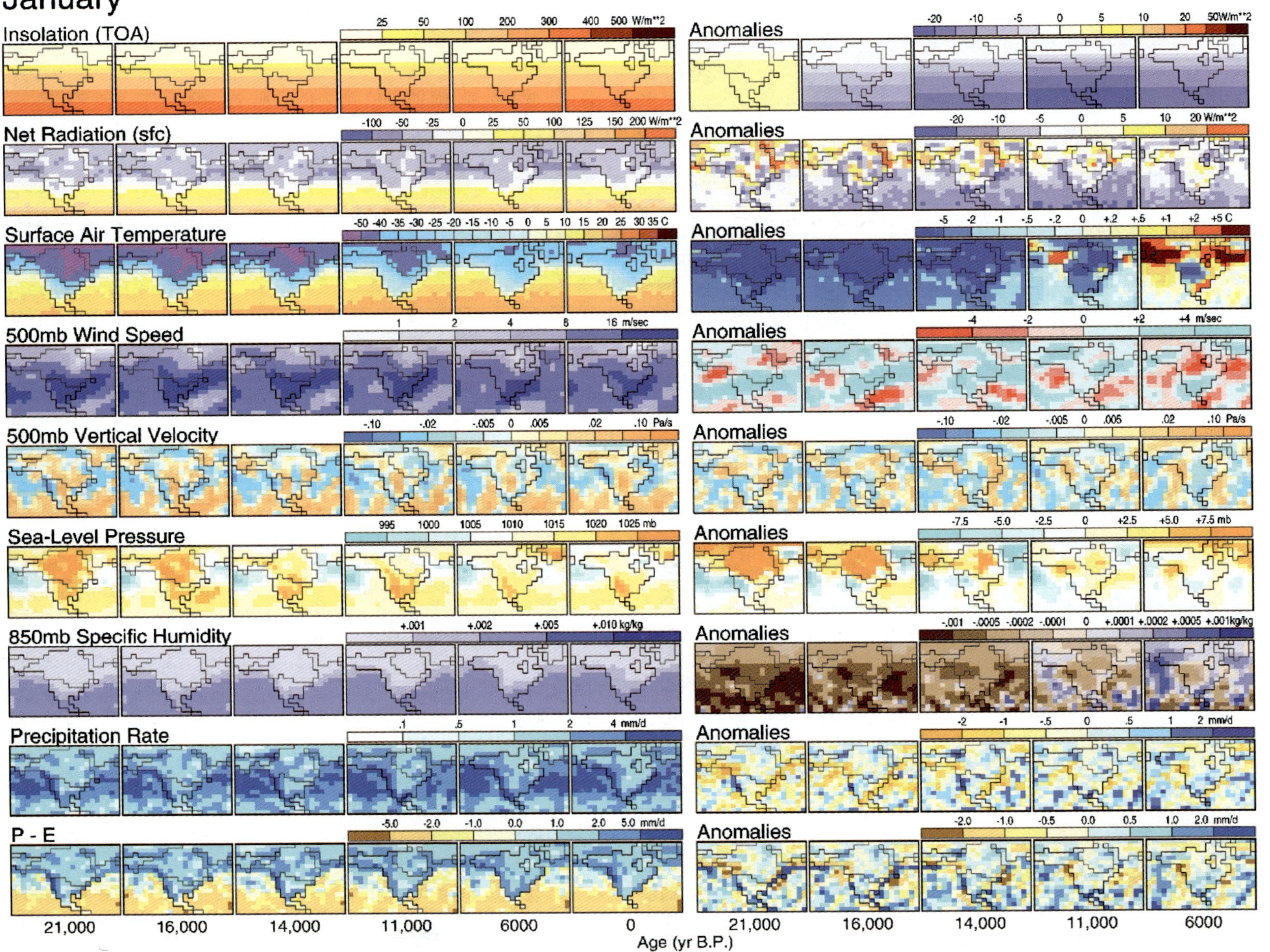

Fig. 3. Sequence of CCM 1 simulations for North America from the LGM to present, for January (top) and July (bottom) (Bartlein et al., 1998; Kutzbach et al., 1998). The simulated values for each experiment are shown on the left-hand side of the figure, while the anomalies ("paleo" experiment minus present "control") are shown on the right. For 500 mb vertical velocity, positive values (orange) on the left-hand panels indicate large-scale sinking motions, while negative values (blue) indicate large-scale rising motions. For 500 mb vertical velocity anomalies, negative values (blue) indicate more rising or less sinking than present, while positive values (orange) indicate more sinking or less rising than present.

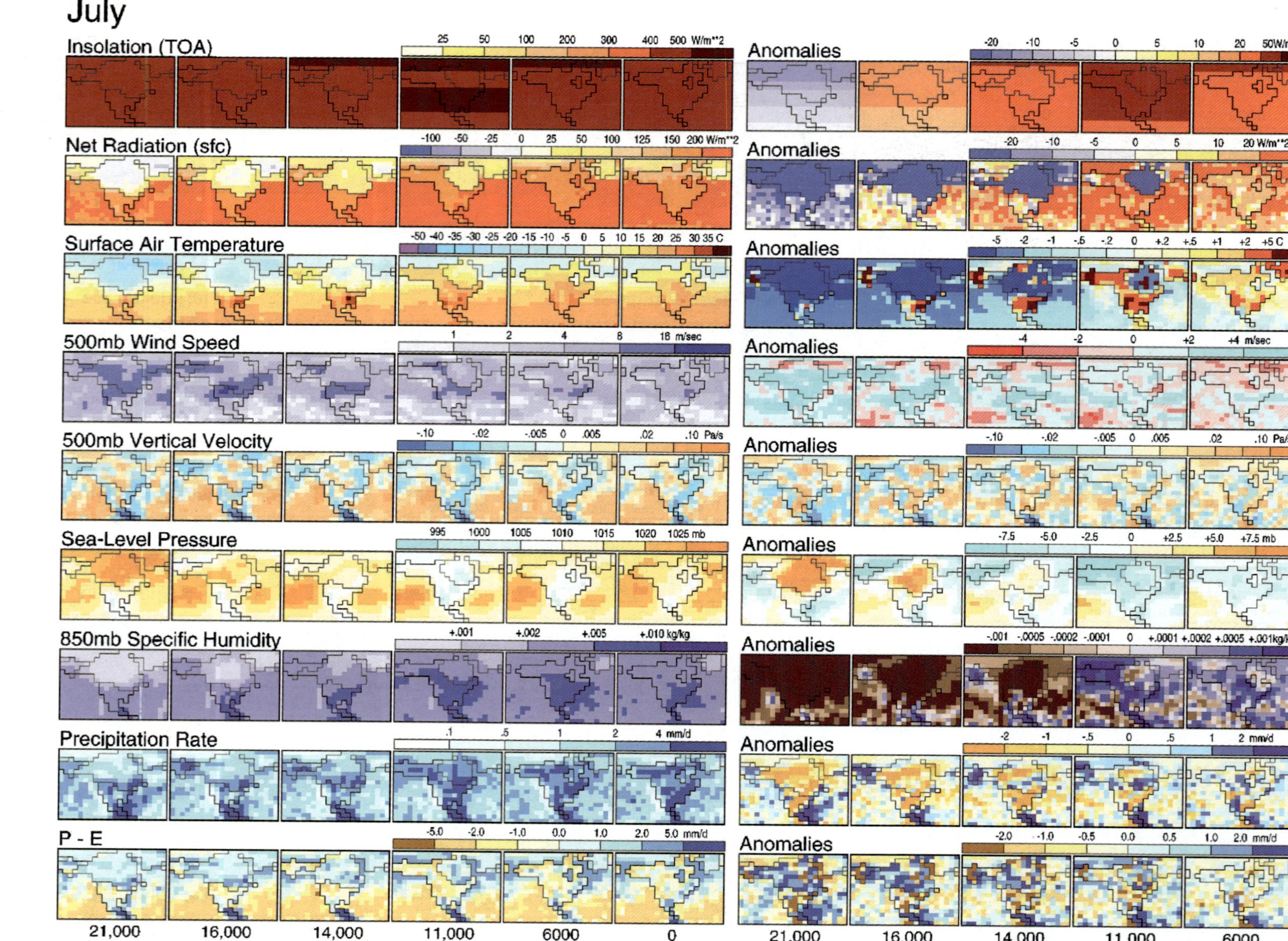

Fig. 3. (Continued)

Table 1. Features of the simulated climate of North America, LGM to present.

Feature	CCM 1 Simulation[a]
Last Glacial Maximum 21,000 cal yr B.P. experiment	
Upper-level winds	Southward displacement and strengthening of the jet stream in both January and July
Sea-level pressure and surface winds	Glacial anticyclone over ice sheet; prevailing easterlies in PNW, strong onshore flow in SW; strong Aleutian low in January, southerly flow into Alaska in January
Large-scale vertical velocity	Strong rising motions in SE in July, and in SW in January
Net radiation	Strong negative anomaly over ice sheet in July
Atmospheric moisture	Much drier than present throughout
Surface climates	Greatest cooling near ice sheet ($\Delta T \sim -10\,^{\circ}$C); less cooling farther away; generally dry to south of ice sheet; wet in SW
Late-Glacial 14,000 cal yr B.P. experiment	
Upper-level winds	Jan. and July jet stream at present latitude, and stronger than present
Sea-level pressure and surface winds	STHs in July as strong or stronger than present
Large-scale vertical velocity	Stronger-than-present rising motions in continental interior in January
Net radiation	Continued strong negative anomaly over ice sheet in July, with positive anomaly along southern edge of continent
Atmospheric moisture	Continued dryness
Surface climates	Continued cold near ice sheet; July temp. near or higher than present in SW U.S., SE U.S., and Alaska; Jan. temp. below or near present throughout
Early Holocene 11,000 cal yr B.P. experiment	
Upper-level winds	Upper-level circulation generally near present configuration; ST ridge over SW U.S. in July
Sea-level pressure and surface winds	Strong STHs in July; strong onshore flow into SW US
Large-scale vertical velocity	Rising motions in the SW and sinking in PNW and eastern North America in July
Net radiation	Continued strong negative anomaly over ice sheet in July, with strong positive anomaly over continental interior in July
Atmospheric moisture	Continued drier than present in Jan; wetter than present over much of continent in July
Surface climates	July temp. higher than present everywhere except along edge of ice sheet; Jan. temp. near present; dry in PNW and Alaska; wet in SW US
Mid-Holocene 6000 cal yr B.P. experiment	
Upper-level winds	Upper-level circulation generally near present configuration; ST ridge over SW U.S. in July
Sea-level pressure and surface winds	Strong STHs in July; strong onshore flow into SW U.S. and S US
Large-scale vertical velocity	Rising motions in continental interior and sinking in PNW in Jan., rising motions in the SW and sinking in PNW and eastern North America in July
Net radiation	Strong positive anomaly at high latitudes and negative in interior in Jan., strong positive anomaly throughout in July
Atmospheric moisture	Generally moister than present
Surface climates	Wetter than present in SW U.S. in July; warmer than present in July throughout

[a] Abbreviations: cal (calendar years before present); PNW (Pacific Northwest); SW (Southwestern U.S.); SE (Southeastern U.S.); STH (subtropical high-pressure system); ST (subtropical).

and the glacial anticyclone as the ice sheet retreated, and strengthening of the East Pacific and Bermuda subtropical high pressure systems in summer as the (positive) insolation anomaly increased, followed by weakening as the insolation anomaly decreased.

(5) increases in summer temperature earlier in the sequence in regions distant from the ice sheet.

(6) development of a thermally induced low surface pressure over the continent in summer when the insolation anomaly was at its maximum, and consequent enhancement of the summer monsoon in the southwestern U.S.

(7) concurrent increases in effective moisture in the southwestern U.S. and decreases in the Pacific Northwest and continental interior when the monsoonal circulation was amplified during the time of summer insolation maximum (Harrison, 2003).

(8) generally lower-than-present winter temperatures over the continent throughout the sequence of simulations.

These responses are quite robust, appearing in most simulations or partial sequences of simulations, and suggest that a substantial part of the regional-scale patterns in paleoclimatic data that have been reviewed in this volume are explainable in terms of the direct and indirect effects of insolation and the direct effect of the ice sheet (see also Bartlein *et al.*, 1998; Thompson *et al.*, 1993; Webb *et al.*, 1993, 1998). More recent simulations and their comparisons with paleoclimatic data have attempted to show how secondary climatic variations across different regions may be mechanistically linked. For example, simulations for 6000 yr ago show a decrease of surface low pressure and an increase in the height upper-level ridge over the western U.S. These changes induce large-scale subsidence (or sinking of air) in the interior of North America (Harrison *et al.*, 2003), thereby linking climate anomalies that are opposite in sign across different regions.

Existing and Emerging Issues in Paleoclimatic Modeling

Comparison of paleoclimatic observations with the conjectures of conceptual models or simulations from numerical models have long been part of the practice of Quaternary science. The rapid development of the different classes of models and syntheses of paleoclimatic data, presented either as maps for key times (e.g. Kohlfeld & Harrison, 2000), or time series at key locations (e.g. Alley & Clark, 1999), ensure that formal comparisons between simulations and observations in data-model comparisons will continue to increase in frequency. There are a number of issues that arise in such comparisons in specific and in paleoclimatic modeling in general that when addressed will enhance the effectiveness of those activities.

Model and Data Resolution

One issue that frequently arises in data-model comparisons is disparity in the spatial and temporal resolutions of paleoclimatic data and model output. Spatial-resolution mismatches are probably most evident in the comparisons of EMIC and GCM output with networks or syntheses of paleoclimatic data. Most present-day GCMs (Fig. 4) have grid-cell resolutions coarser than 2 degrees of latitude or longitude (about 200 km at the equator) while EMICs are still coarser (10 degrees or more). Topography in EMICs and GCMs is highly generalized, and much smoother than the real terrain – topographically complex regions like the Cordillera may be represented as broad featureless domes in a GCM. RCMs mitigate this issue somewhat, and can represent features such as the Sierra Nevada, Cascade Range, or Columbia Basin, but resolutions in RCMs are still fairly coarse (grid cells greater than 25 km on a side).

Paleoclimatic data, in contrast, is often site specific, and a given indicator may represent the environment of a watershed or of a smaller area. Consequently, some kind of "downscaling" of the simulations is necessary, even for RCMs, if the object is the direct comparison of observations and simulations at particular locations. The current approach (Harrison *et al.*, 1998) is to apply the model's "anomalies"

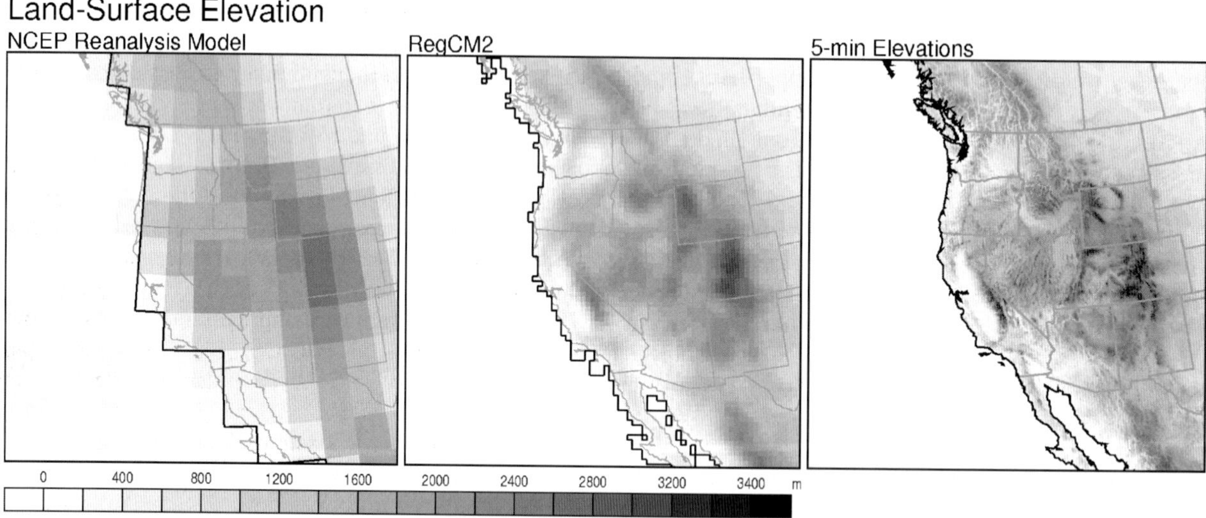

Fig. 4. Land-surface elevations for two climate models and for the western U.S. on a 5-min latitude by longitude grid for comparison. The NCEP Reanalysis Model (Kistler et al., 2001) is a GCM with a rectangular resolution of 2.5 degrees, and represents the topography of the western U.S. as a broad dome centered over western Colorado. Varying elevations over the Pacific Ocean are an artifact of the spectral representation of the atmosphere in the model. RegCM2, a regional climate model (e.g. Hostetler & Bartlein, 1999), as depicted here for the western U.S. has a resolution of 36 km. Viewed at arm's length, the RegCM grid captures much of the physiographic detail for the region, such as the Sierra Nevada and Cascade Range, and the Snake River Plain and Central Valley, that is evident in the actual 5-min elevations.

(differences between a paleoclimatic and modern "control" simulation) to either a gridded higher-resolution observed data set for the present day, or to data for a specific location, both derived by spatial interpolation within a network of modern climate stations. This approach necessarily assumes that the models have sufficient resolution to describe adequately those anomalies, and that mediation of large-scale anomaly patterns does not change over time. Increases in computational capacity, which in turn will allow finer model resolutions, should ultimately minimize this issue.

Simple spatial averaging of the paleoclimatic data to approximate the resolution of a model does not suffice to reduce the mismatch. A topographically complex region, for example, may vary from place to place in its paleoclimatic response to localized physiographic effects on atmospheric circulation (Whitlock & Bartlein, 1993). Paleoclimatic data may also be quite sparse in some regions, and simple interpolation among sites may not be appropriate (Broccoli & Marciniak, 1996).

Mismatches in resolution also arise between time series of climate simulations and observations. Chronological and sampling issues that limit the "downcore" resolution of paleoclimatic data can make comparisons with model output of annual- (or shorter-) timescale resolution difficult. Conversely, there may be limitations in the temporal resolution of a model simulation related to an inability to specify changes in boundary conditions frequently enough.

In both the spatial or temporal cases, some thought must therefore be given to placing the simulations and observations on a similar framework or timescale by appropriate filtering or aggregation. An example of such an approach involving comparisons among time series of data with differing resolutions is given by Tinner & Lotter (2001).

Variables

Another potential mismatch in comparing simulations and observations arises when considering the variables that can be simulated or reconstructed. Climate models such as EMICs and GCMs can be quite specific in what they simulate, including variables (like atmospheric vertical motions) that cannot be directly observed. In contrast, a specific paleoclimatic record may require some kind of transformation in order to be interpreted in quantitative terms, and by themselves, the interpretations cannot discriminate among multiple controls. For example, the observation of a positive glacier mass balance can signify either increased winter precipitation or decreased summer temperature. This ambiguity can be removed, however, through the application of process models (e.g. Hostetler & Clark, 2000) that explicitly quantify the dependence of mass balance on a number of controls.

Experimental Design

The design of a modeling experiment is an important issue that must be considered in comparing simulations and observations. From the paleoclimatic data perspective, this includes the specific protocols that are used to synthesize data, and the scheme for describing chronological control (Kohlfeld & Harrison, 2000). It should be possible for the reader to trace the development of a particular interpretation or reconstruction (see Farrera *et al.*, 1999, for an example; Harrison, 2003).

From the perspective of a model, experimental design is for the most part synonymous with the choice of which boundary conditions are changed and by how much, though it also includes selection of the length of a simulation in the case of spatial-pattern applications of models. Recall that there are two approaches to the design of an experiment – the sensitivity test in which one or more boundary conditions are changed, and the full simulation in which all of the boundary conditions are changed (Peteet, 2001). In the sensitivity-test approach, leaving a boundary condition (say, land-surface cover, or the size and location of ice sheets) unchanged from its present value is the same as assuming that it does not vary over time, or if it does vary, that it does not have any influence on climate. The comparison of a sensitivity-test simulation with observations must therefore consider the extent to which the simulations should be expected to resemble the observations in the first place.

Paleoclimatic Diagnostics

Although it is satisfying when a simulation agrees favorably with some observations, that situation may not be the most informative one – it is the mismatches that indicate something is wrong or needs improvement. There are three sources of apparent mismatches between simulations and observations: inadequacy of the climate model, misinterpretation of the data, and shortcomings in the experimental design, as discussed above. (Note that incorrect "false positive" comparisons could also arise from the same sources.) From a distance, it might be perceived that the goal in data-model comparisons is simply to discriminate among the three sources, but a better way of thinking of the general exercise is as paleoclimatic diagnostics. Analogous to its shorter-timescale cousin climate diagnostics (apart from its longer temporal focus), paleoclimate diagnostics could use almost the same description of its objectives, which are: "... to identify the nature and causes of climate variations on time scales ranging from a month to centuries ... [and] to develop the ability to predict important climate variations on these time scales (NOAA Climate Diagnostics Center Web Page)." If we extend this definition to longer time scales, data syntheses and model simulations can be viewed as complementary tools that can be used to understand past climatic variations.

The motivation for understanding past climatic variations is now stronger than ever, in light of the realization that humans may be producing Quaternary-size changes in climate. The full range of climate models is being used in making projections of future climate, and those models need testing, something that can be done using syntheses of paleoclimatic data and paleoclimatic "natural experiments."

Paleoenvironmental observations also demonstrate that no environmental system is completely insensitive to climatic variations, which raises questions about the magnitude of that sensitivity and how specific systems respond, questions that modeling approaches are well suited to answer. Together, climate models and paleoclimatic data are bringing scientists closer to the goal of understanding climate well enough to predict its future course.

Summary

The synthesis of paleoclimatic data sets and the simulation of past climates using climate models are a complimentary set of activities that lead to better understanding of the climate system. The objective of paleoclimate modeling is to quantify the behavior and variations of the components that describe the climate system. These components include:

(1) boundary conditions or external controls, such as solar irradiance.
(2) slow-response variables that characterize the general state of the climate system, such as continental ice sheets.
(3) fast-response variables that comprise what is ordinarily thought of as weather, such as the configuration of the jet stream and location and strength of surface high- and low-pressure centers.
(4) variables that provide the forcing for the many environmental subsystems that depend on climate, such as lakes.

Climate models can be classified according to the applications to which they are put, which include simulating the temporal evolution and spatial patterns of the climate system, and the attendant responses of environmental subsystems. They may also be classified by their comprehensiveness into several clusters, which include conceptual models, elemental models, Earth-system models of intermediate complexity, comprehensive models represented by coupled general circulation models, and ultimately, full Earth-system models. With one exception, the classes of climate models and the manner in which they are applied were evident in the publications of the INQUA Congress in 1965.

Example simulations of the Quaternary climates of North America illustrate regional patterns of climate. These respond directly to continental ice sheets, and both directly and indirectly to changes in insolation.

Acknowledgments

We thank Bryan Shuman, J.J. Shinker, and Debra Zahnle, and two anonymous reviewers for comments on the manuscript, and Sandy Harrison for discussion. Research was supported by the National Science Foundation (ATM-9910638) and by the U.S. Geological Survey.

References

Alley, R.B. & Clark, P.U. (1999). The deglaciation of the northern hemisphere: a global perspective. *Annual Review of Earth Planet Sciences*, **27**, 149–182.

Alley, R.B., Clark, P.U., Keigwin, L.D. & Webb, R.S. (1999). Making sense of millennial-scale climate change. *In*: Alley, R.B., Clark, P.U., Keigwin, L.D. & Webb, R.S. (Eds), *Mechanisms of Global Climate Change*. Washington, DC, American Geophysical Union, pp. 385–394.

Bartlein, P.J., Edwards, M.E., Mock, C.J., Thompson, R.S., Webb, R.S., Webb, I.T., Whitlock, C., Anderson, K.H. & Anderson, P.M. (1998). Paleoclimate simulations for North America over the past 21,000 years: Features of the simulated climate and comparisons with paleoenvironmental data. *Quaternary Science Reviews*, **17**(6–7), 549–585.

Berger, A. (1980). *Climatic Variations and Variability: Facts and Theories*. Dordrecht, D. Reidel Co., 795 pp.

Berger, A., Loutre, M.F. & Gallee, H. (1998). Sensitivity of the LLN climate model to the astronomical and CO_2 forcings over the last 200 ky. *Climate Dynamics*, **14**(9), 615–629.

Birchfield, E.G., Huaxiao, W. & Rich, J.J. (1994). Century/millennium internal climate oscillations in an ocean-atmosphere-continental ice sheet model. *Journal of Geophysical Research*, **99**(C6), 12,459–12,470.

Braconnot, P., Marti, O., Joussaume, S. & Leclainche, Y. (2000). Ocean feedback in response to 6 kyr BP insolation. *Journal of Climate*, **13**(9), 1537–1553.

Bradley, R.S. (1999). *Quaternary Paleoclimatology*. San Diego, Academic Press, 613 pp.

Broccoli, A.J. & Marciniak, E.P. (1996). Comparing simulated glacial climate and paleodata: A reexamination. *Paleoceanography*, **11**(1), 3–14.

Broecker, W.S. (1965). Isotope geochemistry and the pleistocene climatic record. *In*: Wright, H.E., Jr. & Frey, D.G. (Eds), *The Quaternary of the United States*. Princeton, Princeton University Press, pp. 737–754.

Budyko, M.I. (1982). *The Earth's climate: past and future*. Orlando, Academic Press, 307 pp.

Chamberlin, T.C. (1897). A group of hypotheses bearing on climatic changes. *Journal of Geology*, **5**, 653–683.

Chamberlin, T.C. (1899). An attempt to frame a working hypothesis of the cause of glacial periods on an atmospheric basis. *Journal of Geology*, **7**, 545–584, 667–685, 751–787.

Charbit, S., Ritz, C. & Ramstein, G. (2002). Simulations of Northern Hemisphere ice-sheet retreat: Sensitivity to physical mechanisms involved during the last deglaciation. *Quaternary Science Reviews*, **21**(1–3), 243–265.

Chase, T.N., Pitman, A.J., Running, S.W., Nemani, R.R., Pielke, R.A., Kittel, T.G.F. & Zhao, M. (2001). Relative climatic effects of landcover change and elevated carbon dioxide combined with aerosols: A comparison of model results and observations. *Journal of Geophysical Research D: Atmospheres*, **106**(23), 31,685–31,691.

Clark, P.U., Pisias, N.G., Stocker, T.F. & Weaver, A.J. (2002). The role of the thermohaline circulation in abrupt climate change. *Nature*, **415**(6874), 863–869.

Claussen, M. (2001). Earth system models. *In*: Ehlers, E. & Krafft, T. (Eds), *Understanding the Earth System*. Berlin, Springer, pp. 147–162.

Claussen, M., Mysak, L.A., Weaver, A.J., Crucifix, M., Fichefet, T., Loutre, M.F., Weber, S.L., Alcamo, J., Alexeev, V.A., Berger, A., Calov, R., Ganopolski, A., Goosse, H., Lohmann, G., Lunkeit, F. & Mokhov, II (2002). Earth system models of intermediate complexity: closing the gap in the spectrum of climate system models. *Climate Dynamics*, **18**(7), 579–586.

Clement, A.C., Cane, M.A. & Seager, R. (2001). An orbitally driven tropical source for abrupt climate change. *Journal of Climate*, **14**(11), 2369–2375.

COHMAP Members (1988). Climatic changes of the last 18,000 years: observations and model simulations. *Science*, **241**, 1043–1052.

Crowley, T.J. & North, G.R. (1991). *Paleoclimatology*. New York, Oxford University Press, 339 pp.

Crucifix, M., Berger, A., Loutre, M.F., Tulkens, P. & Fichefet, T. (2002). Climate evolution during the Holocene: A study with an Earth system model of intermediate complexity. *Climate Dynamics*, **19**(1), 43–60.

DeFries, R.S., Bounoua, L. & Collatz, G.J. (2002). Human modification of the landscape and surface climate in the next fifty years. *Global Change Biology*, **8**(5), 438–458.

deNoblet, N.I., Prentice, I.C., Joussaume, S., Texier, D., Botta, A. & Haxeltine, A. (1996). Possible role of atmosphere-biosphere interactions in triggering the last glaciation. *Geophysical Research Letters*, **23**(22), 3191–3194.

Dong, B. & Valdes, P.J. (1995). Sensitivity studies of Northern Hemisphere glaciation using an atmospheric general circulation model. *Journal of Climate*, **8**, 2471–2496.

Driscoll, N.W. & Haug, G.H. (1998). A short circuit in thermohaline circulation: a cause for Northern Hemisphere glaciation? *Science*, **282**(16), October, 436–438.

Ericksson, E. (1968). Air-ocean-icecap interactions in relation to climatic fluctuations and glaciation cycles. *In*: Mitchell, J.M., Jr. (Ed.), *Causes of Climate Change: Meteorological Monographs*. Boston, American Meterological Society, pp. 68–94.

Farrera, I., Guiot, J., Bartlein, P.J., Bonnefille, R., Bush, M., Cramer, W., von Grafenstein, U., Holmgren, K., Hooghiemstra, H., Hope, G., Jolly, D., Lauritzen, S.E., Ono, Y., Pinot, S., Stute, M., Yu, G., Harrison, S.P., Prentice, I.C. & Ramstein, G. (1999). Tropical climates at the Last Glacial Maximum: A new synthesis of terrestrial palaeoclimate data. I. Vegetation, lake-levels and geochemistry. *Climate Dynamics*, **15**(11), 823–856.

Gallimore, R.G. & Kutzbach, J.E. (1996). Role of orbitally induced changes in tundra area in the onset of glaciation. *Nature*, **381**(6), 503–505.

Ganopolski, A. & Rahmstorf, S. (2001). Rapid changes of glacial climate simulated in a coupled climate model. *Nature*, **409**(6817), 153–158.

Gates, W.L. (1976). The numerical simulation of ice-age climate with a global general circulation model. *Journal of the Atmospheric Sciences*, **33**, 1844–1873.

Grassl, H. (2000). Status and improvements of coupled general circulation models. *Science*, **288**(16), June, 1991–1997.

Harrison, S.P. (2003). Contributing to global change science: the ethics, obligations and opportunities of working with paleoenvironmental data bases. *Norsk Geografisk Tidsskrift*.

Harrison, S.P., Braconnot, P., Joussaume, S., Hewitt, C. & Stouffer, R.J. (2002). Comparison of palaeoclimate simulations enhances confidence in models. *Eos*, **83**(40), 447.

Harrison, S.P., Jolly, D., Laarif, F., Abe-Ouchi, A., Dong, B., Herterich, K., Hewitt, C., Joussaume, S., Kutzbach, J.E., Mitchell, J., De Noblet, N. & Valdes, P. (1998). Intercomparison of simulated global vegetation distributions in response to 6 kyr BP orbital forcing. *Journal of Climate*, **11**(11), 2721–2742.

Harrison, S.P., Kohfeld, K.E., Roelandt, C. & Claquin, T. (2001). The role of dust in climate changes today, at the last glacial maximum and in the future. *Earth-Science Reviews*, **54**(1–3), 43–80.

Harrison, S.P., Kutzbach, J.E., Liu, Z., Bartlein, P.J., Otto-Bliesner, B.L., Muhs, D.R., Prentice, I.C. & Thompson, R.S. (2003). Mid-Holocene climates of the Americas: a dynamical response to changed seasonality. *Climate Dymamics*. DOI: 10.1007/s00382–002–0300–6.

Hartmann, D.L. (1994). *Global Physical Climatology*. San Diego, Academic Press, 408 pp.

Harvey, L.D.D. & Huang, Z. (2001). A quasi-one-dimensional coupled climate-change cycle model 1. Description and behavior of the climate component. *Journal of Geophysical Research-Oceans*, **106**(C10), 22,339–22,353.

Haug, G.H. & Tiedemann, R. (1998). Effect of the formation of the Isthmus of Panama on Atlantic Ocean thermohaline circulation. *Nature*, **393**(18), June, 673–676.

Hecht, A.D. (1985). *Paleoclimate Analysis and Modeling*. New York, Wiley, 445 pp.

Hewitt, C.D., Broccoli, A.J., Mitchell, J.F.B. & Stouffer, R.J. (2001). A coupled model study of the last glacial maximum: Was part of the North Atlantic relatively warm? *Geophysical Research Letters*, **28**(8), 1571–1574.

Hewitt, C.D. & Mitchell, J.F.B. (1998). A fully coupled GCM simulation of the climate of the mid-Holocene. *Geophysical Research Letters*, **25**(3), 361–364.

Hostetler, S.W. & Bartlein, P.J. (1999). Simulation of the potential responses of regional climate and surface processes in western North America to a canonical Heinrich event. *American Geophysical Union, Monography*, **112**, 313–327.

Hostetler, S.W. & Clark, P.U. (2000). Tropical climate at the last glacial maximum inferred from glacier mass-balance modeling. *Science*, **290**(5497), 1747–1750.

Hostetler, S.W., Clark, P.U., Bartlein, P.J., Mix, A.C. & Pisias, N.J. (1999). Atmospheric transmission of North Atlantic Heinrich events. *Journal of Geophysical Research*, **104**(D4), 3947–3952.

Hostetler, S.W., Solomon, A.M., Bartlein, P.J., Clark, P.U. & Small, E.E. (2000). Simulated influences of Lake Agassiz on the climate of central North America 11,000 years ago. *Nature*, **405**(6784), 334–337.

Houghton, J.T., Gylvan Meira Filho, L., Griggs, D.J. & Maskell, K. (1997). An Introduction to Simple Climate Models used in the IPCC Second Assessment Report, Intergovernmental Panel on Climate Change, 59 pp.

Houghton, J.T. & Intergovernmental Panel on Climate Change. Working Group I. (2001). Climate change 2001: the scientific basis: contribution of Working Group I to the third assessment report of the Intergovernmental Panel on Climate Change. Cambridge, UK, New York, Cambridge University Press, x, 881 pp.

Huber, M. & Sloan, L.C. (2001). Heat transport, deep waters, and thermal gradients: Coupled simulation of an Eocene Greenhouse Climate. *Geophysical Research Letters*, **28**(18), 3481–3484.

Imbrie, J., Berger, A., Boyle, E.A., Clemens, S.C., Duffy, A., Howard, W.R., Kukla, G., Kutzbach, J.E., Martinson, D.G., McIntyre, A., Mix, A.C., Molfino, B., Morley, J.J., Peterson, L.C., Pisias, N.G., Prell, W.L., Raymo, M.E., Shackleton, N.J. & Toggweiler, J.R. (1993). On the structure and origin of major glaciation cycles, 2. the 100,00-year cycle. *Paleoceanography*, **8**, 699–735.

Imbrie, J., Boyle, E.A., Clemens, S.C., Duffy, A., Howard, W.R., Kukla, G., Kutzbach, J.E., Martinson, D.G., McIntyre, A., Mix, A.C., Molfino, B., Morley, J.J., Peterson, L.C., Pisias, N.G., Prell, W.L., Raymo, M.E., Shackleton, N.J. & Toggweiler, J.R. (1992). On the structure and origin of major glaciation cycles, 1. linear responses to Milankovitch forcing. *Paleoceanography*, **7**, 701–738.

Imbrie, J. & Imbrie, J.Z. (1980). Modeling the climatic response to orbital variations. *Science*, **207**, 943–953.

Joussaume, S., Taylor, K.E., Braconnot, P., Mitchell, J.F.B., Kutzbach, J.E., Harrison, S.P., Prentice, I.C., Broccoli, A.J., Abe-Ouchi, A., Bartlein, P.J., Bonfils, C., Dong, B., Guiot, J., Herterich, K., Hewitt, C.D., Jolly, D., Kim, J.W., Kislov, A., Kitoh, A., Loutre, M.F., Masson, V., McAvaney, B., McFarlane, N., de Noblet, N., Peltier, W.R., Peterschmitt, J.Y., Pollard, D., Rind, D., Royer, J.F., Schlesinger, M.E., Syktus, J., Thompson, S., Valdes, P., Vettoretti, G., Webb, R.S. & Wyputta, U. (1999). Monsoon changes for 6000 years ago: results of 18 simulations from the Paleoclimate Modeling Intercomparison Project (PMIP). *Geophysical Research Letters*, **26**(7), 859–862.

Kistler, R., Kalnay, E., Collins, W., Saha, S., White, G., Woollen, J., Chelliah, M., Ebisuzaki, W., Kanamitsu, M., Kousky, V., van den Dool, H., Jenne, R. & Fiorino, M. (2001). The NCEP-NCAR 50-year reanalysis: Monthly means CD-ROM and documentation. *Bulletin of the American Meteorological Society*, **82**(2), 247–267.

Kohlfeld, K.E. & Harrison, S.P. (2000). How well can we simulate past climates? Evaluating the models using global palaeoenvironmental datasets. *Quaternary Science Reviews*, **19**, 321–346.

Kutzbach, J.E. (1985). Modeling of paleoclimates. *Advances in Geophysics*, **28A**, 159–196.

Kutzbach, J.E. (1992). Modeling large climatic changes of the past. *In*: Trenberth, K.E. (Ed.), *Climate System Modeling*. Cambridge, Cambridge University Press, pp. 669–688.

Kutzbach, J.E., Bryson, R.A. & Shen, W.C. (1968). An evalutation of the thermal Rossby number in the Pleistocene. *In*: Mitchell, J.M., Jr. (Ed.), *Causes of Climate Change: Meteorological Monographs*. Boston, American Meterological Society, pp. 123–138.

Kutzbach, J.E., Ruddiman, W.F. & Prell, W.L. (1997). Possible Effects of Cenozoic Uplift and CO_2 Lowering on Global and Regional Hydrology. *In*: Ruddiman, W.F. (Ed.), *Tectonic Uplift and Climate Change*. New York, Plenum Press, pp. 149–170.

Kutzbach, J., Gallimore, R., Harrison, S., Behling, P., Selin, R. & Laarif, F. (1998). Climate and biome simulations for the past 21,000 years. *Quaternary Science*, **17**, 473–506.

Lamb, H.H. (1966). *The Changing Climate: Selected Papers*. London, Methuen, 236 pp.

Li, X.S., Berger, A. & Loutre, M.F. (1998). CO_2 and northern hemisphere ice volume variations over the middle and late quaternary. *Climate Dynamics*, **14**(7–8), 537–544.

MacCracken, M.C., Budyko, M.I., Hecht, A.D. & Izrael, Y.A. (1990). *Prospects for future climate*. Chelsea, MI, Lewis Publ., 270 pp.

Mahowald, N., Kohfeld, K., Hansson, M., Balkanski, Y., Harrison, S.P., Prentice, I.C., Schulz, M. & Rodhe, H. (1999). Dust sources and deposition during the last glacial maximum and current climate: A comparison of model results with paleodata from ice cores and marine sediments. *Journal of Geophysical Research-Atmospheres*, **104**(D13), 15,895–15,916.

Malone, T.F., Roederer, J.G. & International Council of Scientific Unions and International Council of Scientific Unions. General Assembly (1985). *Global Change: the Proceedings of a Symposium*. New York, Cambridge University Press, 512 pp.

Marshall, S.J. (2002). Modelled nucleation centres of the Pleistocene ice sheets from an ice sheet model with subgrid topographic and glaciologic parameterizations. *Quaternary International*, 95–96, pp. 125–137.

McGuffie, K. & Henderson-Sellers, A. (1997). *A climate modelling primer*. Chichester, Wiley, 253 pp.

Meier, M.F. (1965). Glaciers and climate. *In*: Wright, H.E., Jr. & Frey, D.G. (Eds), *The Quaternary of the United States*. Princeton, Princeton University Press, pp. 795–806.

Miller, K.G., Fairbanks, R.G. & Mountain, G.S. (1987). Tertiary oxygen isotope synthesis, sea level history, and continental margin erosion. *Paleoceanography*, **2**, 1–19.

Mintz, Y. (1968). Very long-term global integration of the primitive equations of atmospheric motion: an experiment in climate simulation. *In*: Mitchell, J.M., Jr. (Ed.), *Causes of Climate Change: Meteorological Monographs*. Boston, American Meterological Society, pp. 20–36.

Mitchell, J.M., Jr. (1965). Theoretical paleoclimatology. *In*: Wright, H.E., Jr. & Frey, D.G. (Eds), *The Quaternary of the United States*. Princeton, Princeton University Press, pp. 881–901.

National Research Council (U.S.). Committee for the Global Atmospheric Research Program (1974). *Understanding Climatic Change*. Washington, DC, National Academy of Sciences, 239 pp.

National Research Council (U.S.). Committee on Global Change Research (1999). *Global Environmental Change: Research Pathways for the Next Decade*. Washington, DC, National Academy Press, 595 pp.

Otto-Bliesner, B.L., Brady, E.C. & Sheilds, C. (2002). Late Cretaceous ocean: Coupled simulations with the National Center for Atmospheric Research Climate System Model. *Journal of Geophysical Research*, 102, D2, 10.1029/2001JD000821.

Paillard, D. (1998). The timing of Pleistocene glaciations from a simple multiple-state climate model. *Nature*, **391**(22), January, 378–381.

Parrish, J.T. (1998). *Interpreting pre-Quaternary Climate from the Geologic Record, The Perspectives in Paleobiology and Earth History Series*. New York, Columbia University Press, xiv, 338 pp.

Peteet, D.M. (2001). Late glacial climate variability and general circulation model (GCM) experiments: an overview. *In*: Markgraf, V. (Ed.), *Interhemispheric Climate Linkages*. San Diego, Academic Press.

Piexoto, J.P. & Oort, A.H. (1992). *Physics of Climate*. New York, American Institute of Physics, 520 pp.

Pollard, D., Krinner, G., Hostetler, S., Oglesby, R., Tarasov, L., Letreguilly, A., Ritz, C., Joussaume, S. & Taylor, K. (2000). Comparisons of ice-sheet surface mass budgets from Paleoclimate Modeling Intercomparison Project (PMIP) simulations. *Global and Planetary Change*, **24**(2), 79–106.

Porter, S.C. (1983). *Late-Quaternary Environments of the United States, Vol. 1: The late Pleistocene*. Minneapolis, University of Minnesota Press, 407 pp.

Prentice, I.C., Cramer, W., Harrison, S.P., Leemans, R., Monserud, R.A. & Solomon, A.M. (1992). A global biome model based on plant physiology and dominance, soil properties and climate. *Journal of Biogeography*, **19**, 117–134.

Randall, D.A. (2000). *General Circulation Model Development*. San Diego, Academic Press, 803 pp.

Rind, D., Peteet, D. & Kukla, G. (1989). Can Milankovitch orbital variations initiate the growth of ice sheets in a general circulation model? *Journal of Geophysical Research*, **94**, 12,851–12,871.

Ruddiman, W.F. (1997). *Tectonic Uplift and Climate Change*. New York, Plenum Press.

Ruddiman, W.F. (2003). Orbital insolation, ice volume and greenhouse gases. *Quaternary Science Reviews*, **22**, 1597–1629.

Saltzman, B. (1968). Surface boundary effects on the general circulation and macroclimate: a review of the theory of the quasi-stationary perturbations in the atmosphere. *In*: Mitchell, J.M., Jr. (Ed.), *Causes of Climate Change. Meteorological Monographs*. Boston, American Meterological Society, pp. 4–19.

Saltzman, B. (2002). *Dynamical Paleoclimatology: Generalized Theory of Global Climate Change*. San Diego, Academic Press, 350 pp.

Saltzman, B. & Verbitsky, M. (1993). Multiple instabilities and models of glacial rhythmicity in the Plio-Pleistocene: a general theory of late Cenozoic climatic change. *Climate Dymamics*, **9**, 1–15.

Saltzman, B. & Verbitsky, M.Y. (1995). Heinrich-scale surge oscillations as an internal property of ice sheets. *Annals of Glaciology*, **23**, 348–351.

Schneider, S.H. & Dickinson, R.E. (1974). Climate modeling. *Rev. Geophysics and Space Physics*, **12**, 447–493.

Schumm, S.A. (1965). Quaternary paleohydrology. *In*: Wright, H.E., Jr. & Frey, D.G. (Eds), *The Quaternary of the United States*. Princeton, Princeton University Press, pp. 783–794.

Sellers, P.J., Dickinson, R.E., Randall, D.A., Betts, A.K., Hall, F.G., Berry, J.A., Collatz, G.J., Denning, A.S., Mooney, H.A., Nobre, C.A., Sato, N., Field, C.B. & Henderson-Sellers, A. (1997). Modeling the exchanges of energy, water, and carbon between continents and the atmosphere. *Science*, **275**(24), January, 502–509.

Shin, S.-I., Liu, Z., Otto-Bliesner, B.L., Brady, E.C., Kutzbach, J.E. & Harrison, S.P. (2002). A Simulation of the Last Glacial Maximum climate using the NCAR-CCSM. *Climate Dymamics*, DOI 10.1007/s00382–002–0260-x.

Smagorinsky, J. (1963). General circulation experiments with the primitive equations. *Monthly Weather Review*, **91**, 99–164.

Smagorinsky, J., Manabe, S. & Holloway, J.L. (1965). Numerical results from a nine-level general circulation model of the atmosphere. *Monthly Weather Review*, **93**, 727–768.

Stocker, T.F., Knutti, R. & Plattner, G.K. (2001). The future of the thermohaline circulation. *In*: Seidov, D., Haupt, B.J. & Maslin, M. (Eds), *The Oceans and Rapid Climate Change: Past, Present and Future*. Washington, DC, American Geophysical Union, pp. 277–293.

Thompson, R.S., Whitlock, C., Bartlein, P.J., Harrison, S.P. & Spaulding, W.G. (1993). Climatic changes in western United States since 18,000 yr B.P. *In*: Wright, H.E., Jr., Kutzbach, J.E., Webb, T., III, Ruddiman, W.F., Street-Perrott, F.A. & Bartlein, P.J. (Eds), *Global Climates Since the Last Glacial Maximum*. Minneapolis, MN, University of Minnesota Press, pp. 468–513.

Tinner, W. & Lotter, A.F. (2001). Central European vegetation response to abrupt climate change at 8.2 ka. *Geology*, **29**(6), 551–554.

Toggweiler, J.R. (1999). Variation of atmospheric CO_2 by ventilation of the ocean's deepest water. *Paleoceanography*, **14**(5), 571–588.

Trenberth, K.E. (1992). *Climate System Modeling*. Cambridge, Cambridge University Press, 788 pp.

Valdes, P. (2000). South American palaeoclimate model simulations: how reliable are the models? *Journal of Quaternary Science*, **13**, 357–368.

Velichko, A.A., Wright, H.E. & Barnosky, C.W. (1984). *Late Quaternary Environments of the Soviet Union*. Minneapolis, University of Minnesota, xxvii, 327 pp.

Webb, I.T., Webb, R.S., Anderson, K.H. & Bartlein, P.J. (1998). Late Quaternary climate change in eastern North America: A comparison of pollen-derived estimates with climate model results. *Quaternary Science Reviews*, **17**(6–7), 587–606.

Webb, T., III, Bartlein, P.J., Harrison, S.P. & Anderson, K.H. (1993). Vegetation, lake levels, and climate in eastern North America for the past 18,000 years. *In*: Wright, H.E., Jr., Kutzbach, J.E., Webb, T., III, Ruddiman, W.F., Street-Perrott, F.A. & Bartlein, P.J. (Eds), *Global Climates Since the Last Glacial Maximum*. Minneapolis, University of Minnesota Press, pp. 415–467.

Webb, T., III, Shuman, B. & Williams, J.W. (this volume). Climatically forced vegetation dynamics in eastern North America during the late Quaternary.

Webb, T. & Kutzbach, J.E. (1998). An introduction to 'Late quaternary climates: Data syntheses and model experiments.' *Quaternary Science Reviews*, **17**(6–7), 465–471.

Weyl, P.K. (1968). The role of the oceans in climatic change: a theory of the ice ages. *In*: Mitchell, J.M., Jr. (Ed.), *Causes of Climate Change. Meteorological Monographs*. Boston, American Meterological Society, pp. 37–64.

Whitlock, C. & Bartlein, P.J. (1993). Spatial variations of Holocene climatic change in the Yellowstone region. *Quaternary Research*, **39**, 231–238.

Williams, J., Barry, R.G. & Washington, W.M. (1974). Simulation of the atmospheric circulation using the NCAR global circulation model with ice age boundery conditions. *Journal of Applied Meteorology*, **13**(3), 305–317.

Wright, H.E., Jr., Kutzbach, J.E., Webb, T., III, Ruddiman, W.F., Street-Perrott, F.A. & Bartlein, P.J. (1993). *Global Climates Since the Last Glacial Maximum*. Minneapolis, Univ. Minnesota Press, 569 pp.

Wright, H.E., Jr. (1983). *Late-Quaternary Environments of the United States, Vol. 2, The Holocene*. Minneapolis, University of Minnesota Press, 277 pp.

Wright, H.E., Jr. & Frey, D.G. (1965). *The Quaternary of the United States*. Princeton, Princeton University Press, 922 pp.

Zachos, J., Billups, K., Pagani, H., Sloan, L. & Thomas, E. (2001). Trends, rhythms, and berrations in global climate 65 Ma to present. *Science*, **292**(5517), 686–693.